Glencoe

Carpentry &
Building Construction

Mark D. Feirer B.S.

John L. Feirer Ph.D

McGraw Hill Education

SAFETY NOTICE

The reader is expressly advised to consider and use all safety precautions described in this book or that might also be indicated by undertaking the activities described herein. In addition, common sense should be exercised to help avoid all potential hazards.

Publisher and Authors assume no responsibility for the activities of the reader or for the subject matter experts who prepared this book. Publisher and Authors make no representation or warranties of any kind, including but not limited to, the warranties of fitness for particular purpose or merchantability, nor for any implied warranties related thereto, or otherwise. Publisher and Authors will not be liable for damages of any type, including any consequential, special, or exemplary damages resulting, in whole or in part, from reader's use or reliance upon the information, instructions, warnings, or other matter contained in this book.

Notice: Information on featured companies, organizations, and their products and services is included for educational purposes only and does not present or imply endorsement of the *Carpentry & Building Construction* program.

Meet the Authors

Mark D. Feirer

Mark D. Feirer is the author of various books and articles on remodeling and new construction. He also writes for Web sites and developed instructional materials for construction videos. Feirer has a B.S. in Industrial Technology, with a specialty in building construction, and was a graduate research assistant in vocational education. After years as a shop woodworker and a licensed general contractor, he became the editor of *Fine Homebuilding* magazine. He is now a Senior Contributor and technical consultant to *This Old House* magazine. Feirer is a cooperating member of the International Code Council.

John L. Feirer

John L. Feirer was chairman of the Industrial Technology and Education Department at Western Michigan University. His textbooks on woodworking, metalworking, building construction, and other technical topics have been widely adopted, and his published work is extensively referenced in educational journals. Dr. Feirer was a Distinguished Faculty Scholar, a Fellow of the International Technology Education Association, and served widely as a consultant to industry and government.

Contributing Writers, Reviewers, and Industry Advisory Board

Contributing Writer

Tom Vessella
Los Angeles Trade – Technical College

Technical Reviewers

Dr. John W. Adcox Jr. CPC
Florida Community College at
Jacksonville
Jacksonville, FL

David Allen
New Castle Chrysler High School
New Castle, IN

Bob Anderson
Wake Forest-Rolesville High School
Wake Forest, NC

Scott Battenfield
El Camino High School
Sacramento, CA

Robert Bolus
The American College of the
Building Arts
North Charleston, SC

Mike Dorey
Jim Hamilton Skills Center
Hodgen, OK

Frank Genello
The American College of the
Building Arts
North Charleston, SC

Tim Lang
Alexandria Technical College
Alexandria, MN

Brian Mate
Central Georgia Technical College
Macon, GA

Andre Pease
Virginia Beach Technical and
Career Education Center
Virginia Beach, VA

Keith Powell
York Technical College
Rock Hill, SC

Robert W. Swegle
International Standards and
Training Alliance
Springfield, IL

Ricky Vickery
Rogers Construction Co. (AGC)
Commerce, GA

Timothy J. Waite, P.E.
Simpson Strong-Tie
Honolulu, HI

Cherri L. Watson
AGC Safety and Education Services
Atlanta, GA

Larry Williams
Steel Framing Alliance
Washing, DC

Leslie Zimmerman
Alexandria Technical College
Alexandria, MN

Math Reviewer

Ken Deforest-Davis
Beloit Memorial High School

Trade and Technical Education Industry Advisory Board

Johan Gallo
Bridgestone/Firestone, Inc.

Cheryl Horn
Northrop Grumman Corporation

Erin Kuhlman
Parsons Corporation

Donna Laquidara-Carr
McGraw-Hill Construction

Robert Ratliff
The Manufacturing Institute

Floyd McWilliams
American Design Drafting
Association

Olen Parker
American Design Drafting
Association

Candice Rogers
McGraw-Hill Construction

Joe Vela
Aedis Architecture & Design

Table of Contents

Ariel Skelley/Blend Images/Corbis

Table of Contents

Table of Contents

Table of Contents

Table of Contents

Table of Contents

Table of Contents

Table of Contents

Find Your Tools for Success...

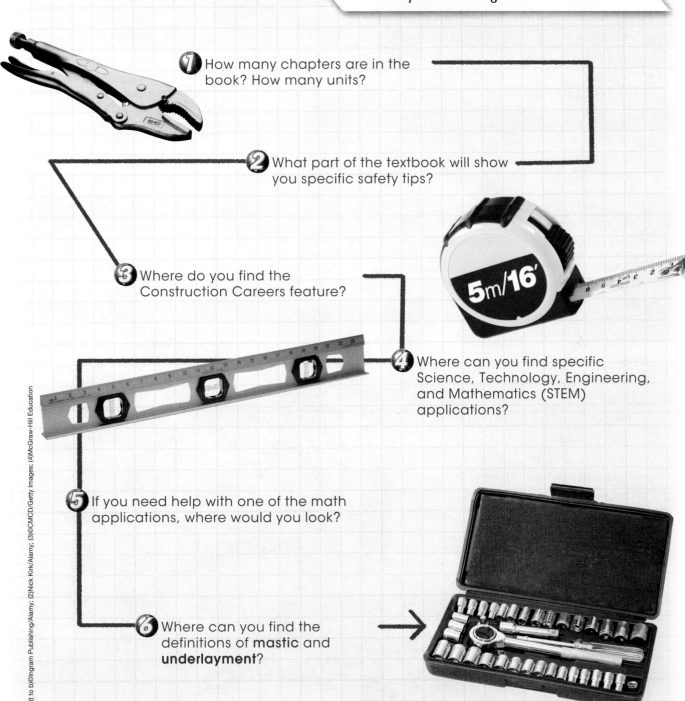

Treasure Hunt

Carpentry & Building Construction contains a wealth of information. The trick is to know where to look to access all the information in the book. If you spend time reviewing this textbook, you get the most out of your reading and study time. Let's begin!

1 How many chapters are in the book? How many units?

2 What part of the textbook will show you specific safety tips?

3 Where do you find the Construction Careers feature?

4 Where can you find specific Science, Technology, Engineering, and Mathematics (STEM) applications?

5 If you need help with one of the math applications, where would you look?

6 Where can you find the definitions of **mastic** and **underlayment**?

UNIT 1
Preparing to Build

In this Unit:

Hands-On Math Project Preview

Green Construction in the Community

After completing this unit, you will research an organization offering community development your area. You will also prepare cost estimates and schedules for a project designed for the organization.

Project Checklist

As you read the chapters in this unit, use this checklist to prepare for the unit project:

✓ Describe the scheduling responsibilities of a general contractor.

✓ Identify the different methods for estimating materials and cost.

✓ Think about the communication skills an entrepreneur will need to be successful.

➤ Go to **connectED.mcgraw-hill.com** for the Unit 1 Web Quest activity.

Huntstock/Getty Images

Construction Careers General Contractor

Profile A general contractor coordinates and supervises the construction process of a structure from the original idea to the completion of construction. The contractor is responsible keeping the project on time and within budget.

Academic Skills and Abilities ....

- mathematics
- interpersonal skills
- presentation skills
- general business management skills
- verbal and written communication skills
- organizing and planning skills

Career Path ....

- on-the-job training
- apprenticeship programs
- certification
- community college courses
- bachelor's degree in construction science
- bachelor's degree in construction management

Explore the Photo

Preparing to Build Construction is a process. *What are the first steps of the construction process?*

The Construction Industry

Section 1.1
Careers in Construction

Section 1.2
Finding a Job

Chapter Objectives

After completing this chapter, you will be able to:

- **List** career specialties in construction.
- **Identify** the education and training that can prepare you for a construction career.
- **Explain** the purpose and function of a business plan for entrepreneurs.
- **Define** employability skills.
- **Describe** how to apply for and obtain a job.
- **Identify** some responsibilities of employers and employees.

Discuss the Photo

The Big Picture Building a house involves many different tasks. *What type of task might the people in this picture be performing?*

Writing Activity: Quick Write

All successful businesses fulfill a need. With a teammate, brainstorm construction needs in your area. If you were to start a construction-related business in your area, what kind of business would it be? Write a paragraph describing your idea for this business.

Huntstock/Getty Images

Chapter 1 Reading Guide

Before You Read Preview

The construction industry has many specialized careers. Choose a content vocabulary or academic vocabulary word that is new to you. When you find it in the text, write down the definition.

Content Vocabulary

- career clusters
- career pathway
- apprentice
- trend
- certification
- entrepreneur
- business plan
- free enterprise
- work ethic
- networking
- résumé
- job application
- interview
- ethics

Academic Vocabulary

You will find these words in your reading and on your tests. Use the academic vocabulary glossary to look up their definitions if necessary.

- visualization
- features
- evaluate
- specific

Graphic Organizer

As you read, use a diagram like the one shown to organize the information, adding ovals as needed.

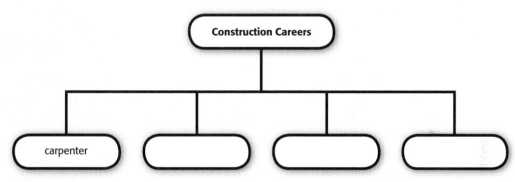

Construction Careers — carpenter

Go to **connectED.mcgraw-hill.com** to download this graphic organizer.

Careers in Construction

Construction & Building

What are the building trades?

The construction industry employs over six million people in the United States. Construction is the process of building, and the construction industry builds both residential and commercial buildings. *Residential construction* involves building houses, remodeling houses, and building additions. *Commercial construction* includes the building of offices, stores, businesses, and government buildings, such as schools and hospitals.

Employment in construction is cyclical, which means that it rises and falls with the economy. When the economy is good, there are many new construction and remodeling projects. When the economy slows, however, the construction industry is one of the first to be affected. When people have less money to spend, there is less demand for new housing, and fewer houses and apartments are built. When interest rates for loans go up, contractors and housing developers may cancel plans for new construction. That means fewer jobs for construction workers. A good economy combines a need for construction with enough financial resources to fund it.

The number of jobs available in the industry fluctuates with the economy, but also depends on the number of qualified workers available. High employee turnover (the number of employees being hired compared to the average number of employees) and shortages of skilled workers and qualified training programs would lead to an excellent job outlook for the construction trades.

Construction has a high rate of self-employment. Self-employed workers may own their own businesses or work as consultants or independent contractors. This allows you to be your own boss and set your own hours. However, you must work hard to build a reputation and customers to ensure that you have enough work to support yourself.

Career Clusters and Pathways

Career clusters are groups of related occupations. Once you have chosen a field that interests you, you must look ahead and consider your career pathway. A **career pathway** is an area of concentration within a career cluster. Two career pathways are shown on page 7. Each pathway contains a group of careers that require similar academic and technical skills as well as similar industry certification or training and education. For example, the construction pathway is one pathway in the Architecture & Construction career cluster. Workers in the construction pathway need good communication, math, and problem-solving skills.

Your career path is your route to a particular career, made up of all of the career moves and job experience that you gain as you work toward your career goal. A career ladder is a **visualization** of your chosen career path. It helps you recognize where you currently are in reaching your goals and what your next career step may be.

Some construction jobs require little or no formal pre-employment training. These workers are trained on the job by more experienced coworkers. Other jobs require several years' experience and sometimes a license or certification to work in the field.

Some construction trade workers begin their careers in residential construction and later move on to other types of construction. Most employers prefer candidates with high school diplomas and who have some formal technical training or education. Apprenticeship programs are common for occupations that require a high level of skill, such as carpenters. An **apprentice** is an inexperienced

Company Owner		Company Owner
Project Manager		Project Manager
Construction Supervisor		Senior Estimator
Master Craftsperson		Estimating/Cost Scheduling Engineer
Journey-level Craftsperson		Field Engineer
Registered Apprentice or On-the-Job Trainee		2-year/4-year Degree
HS Diploma/GED, Laborer/Helper		

 Career Pathways
Step by Step A career ladder shows the steps to a career goal.

worker who learns a trade by working under the guidance of an expert worker. The U.S. Military also offers training for careers in this pathway.

To advance in construction, you need to obtain training, education, and work experience. Employees who take pride in the quality of their work will find many rewarding opportunities. Someone with a carpenter's license can advance to become a construction superintendent. A graduate of a technical school may start as a drafter, creating architectural plans, and become a project manager. An assistant architect may work up to full professional status.

Construction Specialties

Much of the work in construction is done by trained specialists. The three specialized categories are craft, technical, and professional.

Craft Workers Workers in the building crafts or building trades represent the largest group of skilled workers in the United States. There are more than two dozen skilled building crafts, and carpenters make up the largest group. Carpenters are skilled craft workers who work with mostly wooden structures. They erect the wood framework of a building. They install molding, paneling, cabinets, windows, and doors. They build stairs and lay floors.

Members of the building trades often have a high level of skill earned through experience or training. In unions, for example, a worker in the crafts may begin as an apprentice. A journey-level worker has an intermediate level of skill. The highest level is that of *master*. You will learn more about apprentices later in this chapter.

Construction craft work is sometimes grouped into three classifications: structural, finish, and mechanical. Occupations concerned mainly with structural work include carpenter, bricklayer, and cement mason. Finish work is done by painters, glaziers, and roofers. Occupations involving mechanical work include plumber and construction electrician. Many of these workers work outdoors. In some regions, the work tends to be seasonal because of climate or other factors.

Technical Workers Some careers in building construction require additional training and education at a technical institute or community college. Examples include architectural drafter, estimator, and purchasing agent. A drafter works on building plans. An estimator figures the cost of a project. A purchasing agent buys materials according to current needs.

Professional Workers College can prepare you for many professional careers related to construction. These include architect, engineer, and teacher at a trade or vocational school.

Management opportunities in the construction industry are also open to individuals with appropriate work experience, training, and education. For example, the management of a large construction project is the job of a construction supervisor, or construction administrator. This person schedules workers and inspections and arranges for the delivery of materials.

Related Careers There are also opportunities for employment in businesses that serve the construction industry. Many people sell or service tools and equipment. Others supply building materials. Some check for building code enforcement. Still others design, repair, or evaluate new products.

Remodeling and Repair Many jobs in the construction industry do not involve building new houses. This is because many people purchase a house that already exists instead of building a new one. There are some advantages to doing this. The supply of existing housing is huge so there are many choices in many price ranges. In addition,

buying an older house allows a family to move in quickly. An older house does not always fit the needs of its new owners. Sometimes rooms or **features** must be added, enlarged, or updated. Making these changes is called *remodeling*.

There are two general types of remodeling. One type of remodeling involves changing the exterior or interior of a house without adding to the size of the house. The other type of remodeling requires new construction that increases the size of the house. Many builders specialize in one or both types of remodeling and never build a new house. Some builders do both. When the demand for new construction slows, remodeling work often increases. This is because families who cannot afford to buy a new house often decide to improve the house they already live in.

Even a house that has never been remodeled will eventually require repair. On the exterior, weather causes paint to fade, roof shingles to crack, and wood to rot. Inside, years of use will cause surfaces and fixtures to wear out. Many building companies specialize in doing repair work. For example, some companies only replace roof shingles. Others specialize in replacing windows or siding, or installing extra insulation.

Industry Trends

To plan for customers' needs, industry experts track trends. A **trend** is a general development or movement in a certain direction.

Trends affect job opportunities. For example, as the aging population increases, more people will be living in retirement centers and nursing homes. Therefore, more companies will be building these facilities. Three factors influencing residential building trends include family structure, work patterns, and personal preferences.

Family Structure Family structure is changing as the number of single and single-parent households increases. This has increased the demand for apartments and smaller, more energy-efficient houses.

Work Patterns The number of people who work and the increased number of hours they work reflect another trend. People want houses that require less maintenance because their time is limited.

Personal Preferences U.S. residents are increasingly moving to warmer climates. This has led to an increase in construction in those areas.

Education & Training

What is certification?

A high school education is a solid foundation on which to build a career in construction. Construction workers use math skills every day to measure and calculate. Science helps you understand construction methods and materials. Excellent communication skills, such as reading, writing, listening, and speaking, will help you communicate with coworkers and customers.

In both high school and college, you can also take technical education courses and join organizations such as SkillsUSA. SkillsUSA is a national organization that serves teachers and high school and college students enrolled in training programs for technical, skilled, and service occupations. SkillsUSA partners students with industry professionals through the SkillsUSA Championships. In these Championships, carpentry students can participate in contests that test their carpentry skills. Students are judged on their technical abilities. There are also other contests that involve customer service skills, leadership skills, and other work-related skills.

Informal and Formal Training

Another way to learn about the industry is through a part-time, entry-level job in a construction operation. An entry-level position, such as carpenter's helper, requires little or no training or experience. You learn on the job. However, many carpenters learn their trade through formal programs. These include certification programs, apprenticeships, degree programs, and military training programs.

Certification Many community colleges and technical schools offer certificate or certification programs. **Certification** is a formal process that shows that an individual is qualified in a particular job or task. Certification assures potential employers that you have the skills and knowledge they require. Most certification programs for craft workers involve work experience, coursework, and a certification test. Industry or trade organizations often set standards for training and supervise certification programs.

Obtaining certification in any area of construction makes you more employable. Certification is available in specialty areas, such as building code enforcement, framing carpentry, and construction supervision. Certifications may need to be renewed after a certain period of time.

Before enrolling in a school or certification program, **evaluate** the program and the reputation of the school or organization. Find out what jobs are available for people with that particular certificate. Remember, certification programs usually focus on particular skills. Advancement opportunities may require that you obtain more formal education.

 Apprenticeships
Learn the Trade Apprenticeships are a common way of learning the carpentry trade.

Apprenticeships An apprenticeship is an effective way to gain real-life work experience. The apprentice may earn very little or no pay during the learning period. However, workers who have completed apprenticeships are well respected and well paid. In the construction industry, an apprentice carpenter learns under an experienced carpenter. An apprenticeship involves a combination of hands-on training and classroom instruction. Programs vary in length. For example, one carpentry apprenticeship requires an 8,000-hour program that combines on-the-job training and related classroom instruction.

Apprenticeships allow new workers to have plenty of hands-on experience. Professional organizations and industry associations often operate apprenticeship programs. These include the United Brotherhood of Carpenters and Joiners of America and the National Association of Home Builders. The U.S. Bureau of Apprenticeship and Training (BAT) is one place to find information about apprenticeship opportunities.

Associate Degrees Many colleges and universities offer two-year degrees, also called associate degrees, in the construction field. They provide hands-on experience so you can apply the techniques you have learned in the classroom. Select a program that meets your needs. Evaluate the program, the school's credentials, and the employment rate for graduates before making a decision.

Bachelor's Degrees Four-year programs, also called bachelor's degree programs, offer a well-rounded education. They often begin with general education courses in science and the humanities. They also can provide in-depth training in one or more areas that prepares students for supervisory and management positions. General degrees in subjects such as marketing, business, and management provide the basis for learning a wide range of skills and information.

Students may be able to take part in a cooperative education or work experience program at the same time that they are pursuing their degree. Students are matched with a company whose business is related to their interests.

Military Training The military offers training in many construction specialties. They include carpentry specialist, electrician, and plumber.

On-the-Job Training On-the-job training is another option. Some construction managers use a method called job rotation. Entry-level workers are rotated through a series of jobs, which allows them to learn a variety of skills. This is similar to cross-training, in which workers are trained in different skills.

Job shadowing and internships are other forms of on-the-job training. *Job shadowing* is spending time with a worker on the job. It can last anywhere from a few hours to a few days. An *internship* is a short-term job or work experience. Internships can also combine classroom instruction and work experience. Many individuals benefit from finding a mentor, an individual who is willing to guide them and answer their questions.

Remodeling and repair work require many of the same skills needed for building a new home. For this reason, many young people learn the skills needed for new construction by working on older houses.

Reading Check

Recall *What are some forms of on-the-job training?*

Entrepreneurship

What are some advantages of entrepreneurship?

An **entrepreneur** is a person who creates and runs his or her own business. Entrepreneurs take personal and financial risks. However, entrepreneurship can lead to great rewards. Entrepreneurs must be responsible to be successful.

An entrepreneur in the construction industry usually begins by opening a small business, such as a remodeling business. Small businesses are those with fewer than 100 employees. In the United States, more than 53 percent of the workforce works for small businesses. Do the rewards of entrepreneurship outweigh the costs? **Table 1-1** compares the advantages and disadvantages of entrepreneurship.

Types of Business Ownership

After you decide to open a business, you must choose the form of ownership. There are three common types of business ownership: sole proprietorship, partnership, and corporation.

A sole proprietorship is a business with only one owner. The owner, or proprietor, owns all the business's assets and is responsible for all its debts. A partnership is two or more people who share the ownership of the business. Control and profits of the business are divided among partners according to a partnership agreement.

A corporation is a form of business ownership in which a state grants an individual or a group of owners a charter with legal rights. The owners buy shares, or parts of the company. These owners, called shareholders, earn a profit based on the number of shares they own. If the business fails, the owners lose the money they have invested in the business.

◢ Entrepreneurship

Responsibility An entrepreneur is responsible for making sure that all aspects of the job are on schedule. *What are some advantages of entrepreneurship?*

Table 1-1: Advantages and Disadvantages for Entrepreneurs	
Advantages	**Disadvantages**
Ownership: You are the boss and are responsible for making all the decisions.	**No guarantees:** New businesses have a high rate of failure.
Job satisfaction: You can build your reputation and take pride in your accomplishments.	**Competition:** Your competition may have greater skills and experience.
Independence: You decide what jobs to take and set your own hours and schedules.	**Long hours:** It is common for entrepreneurs to work evenings and weekends.
Earning potential: You might make more working for yourself than you could make working for someone else.	**Financial risk:** If you invest your own money, you could lose it all and even go into debt.

Developing a Business Plan

One of the main reasons start-up businesses fail is that a business plan was not made or followed. A **business plan** gives specific information about the business and includes a vision, goals, strategies, and a plan of action.

The entrepreneur's vision should include the goods and services the business will offer. It should indicate how much it will cost to start and run the business and the business location. It must also include a description of the potential customers and an estimate of the profits.

After the vision is described, goals must be stated. These goals must be specific, concrete, and measurable. The plan must also give a timetable for meeting these goals.

A business plan should include strategies for meeting goals. These strategies may include the type of marketing the business will use to attract customers. Marketing is the process of promoting and supplying goods and services to customers. It includes packaging, advertising, selling, and shipping. A business also needs a plan of action. A plan of action helps a business reach its goals by identifying a specific course of action.

Rules and Regulations

The economic system in the United States is known as the free enterprise system. **Free enterprise** is an economic system in which businesses or individuals may buy, sell, and set prices for goods and services. However, businesses are subject to some government controls. For example, the government is responsible for passing and enforcing laws that set safety standards as well as laws that affect prices and wages.

Government also has a voice in building codes and zoning requirements. Zoning divides land into areas used for different purposes. Only certain types of buildings may be built within these defined zones. If you are in the construction business, it is important to understand the zoning process.

Before starting a construction business, you must obtain a license that grants you permission to open a business. Special liability insurance may also be necessary. This protects you from loss in case of damage.

Payment of taxes is another requirement. You must maintain accurate financial records of all income and spending. Many people use record-keeping software to set up and store this information electronically.

Section 1.1 Assessment

After You Read: Self-Check

1. Name the three categories of specialized workers.
2. What three factors influence residential building trends?
3. What is the difference between an internship and an apprenticeship?
4. What are the components of a business plan?

Academic Integration: English Language Arts

5. **Career Information** Work with one or more classmates to find two sources of career information on the Web for your local area or your state. Prepare a three-minute oral presentation that describes the kind of information available in each source. In your presentation, describe how easy the source was to use, what questions the source helped you to answer, and whether you would recommend the source to others.

Go to **connectED.mcgraw-hill.com** to check your answers.

Finding a Job

Employability Skills

What are transferable skills?

Skills are things you know how to do. They can include things you are good at or ways of behaving. Some skills are specific to a certain task. For example, operating a forklift and installing a roof are skills for certain tasks. Skills that you can use in many different situations are called *transferable skills*. Writing, reading, and communicating are examples of transferable skills.

The basic skills you would need for a job in construction are the same transferable skills you would need to find and keep a job. These include academic skills such as mathematics, language arts, and science. Social communication skills, thinking skills, a work ethic, leadership skills, and personal qualities are also important.

Basic Academic Skills

Every employer expects a worker to have certain basic academic skills. These skills provide you with a strong foundation for finding and keeping employment and advancing on the job.

Mathematics The ability to calculate, or work with numbers, is a basic part of every construction job. You will need to be able to add, subtract, multiply, and divide to solve on-the-job problems. You will also need to be able to measure and determine (figure out) lengths and angles. For example:

- Carpenters use math skills to measure and cut lumber, plan stairways, and frame roofs. They often refer to existing tables, such as **Table 1-2**, to help them make calculations.

- Construction managers use math skills to estimate lumber quantities, order lumber and other supplies, schedule deliveries, set up employee work rosters, complete payroll and tax forms, prepare bids, and estimate profits.

- Employees use math skills to keep track of their work hours and pay rates.

Reading Much of the information you receive comes through reading. In construction, you will use reading skills to:

- Interpret building plans by reading schedules and measurements.

- Operate equipment by reading instruction manuals and safety precautions.

- Carry out general job duties by reading workplace policies and communications.

Practice the following steps to develop your reading skills:

1. **Preview**. Read headings and subheads to get an overview.
2. **Look for key points**. This is called *skimming*.
3. **Focus**. Pay full attention to what you are reading.
4. **Visualize**. If the text is not illustrated, try to picture what is being described.
5. **Check**. Ask yourself how well you understand what you have read. If there are words you do not understand, look them up in a dictionary. You can use a special construction dictionary, which contains construction terms.

Table 1-2: Material Needed for Ceiling Plank		
Ceiling Plank	Board Ft Needed for 100 ft² of Surface Area	
Size of Plank	16" OC	24" OC
2×4	59	42
2×6	88	63
2×8	117	84
2×10	3.147	104

Writing Your ability to communicate in writing will help you find a job and perform well on the job.

Before you write anything, picture the audience, or the person or group who will be reading it. Then you can try to write to the audience's needs. Choose language that suits the purpose of your writing. Read what you have written and decide if you have achieved your purpose. The two most common forms of business writing are memos and business letters. Most business communications are intended for one of the following:

- To inform or give instructions.
- To request or ask for information, seek a decision, or call for action.
- To persuade the reader to agree or to pursue a course of action.
- To complain or to protest.

Your writing *style* involves your choice of language and tone. Business communications are written in a direct style with a professional tone.

Proofread documents before sending them. Be sure your ideas are easy to understand and flow in logical order. If you are writing on a computer, use the spell check and grammar check features to check your writing.

Science Construction is really applied science. For example, all the tools and structures you would use as a carpenter operate according to basic scientific principles. Concrete hardens because of a chemical reaction. The frame of a house bears heavy loads based on principles of physics such as measurement, weight, and pressure. Knowledge of scientific principles can help you on the job. It will help you understand why and how things work.

Communication Skills

Communication skills other than writing include speaking and listening. These skills are often called *interpersonal skills* and are used to communicate ideas and to interact with others. They affect how you relate to coworkers and to customers.

Speaking On a construction site, your safety often depends on making yourself understood. How well you are understood depends on how effectively you speak. While speaking, pay attention to the following:

- Pronounce words clearly and correctly.
- Pronounce each syllable of a word.
- Speak slowly enough to be understood.
- Use appropriate volume.
- Avoid using *non-technical slang* (slang that is not related to the workplace) on the job.

When using the telephone, speak clearly and at a moderate volume. How you sound on the telephone may be a customer's first or only impression of your business.

Body Language *Body language* means the messages that your movements send. Body language includes gestures, posture, and expressions, such as the way you sit, stand, move your hands, and smiling or frowning. Examples of positive body language include

 Body Language
Speaking and Listening Active listening involves paying attention and responding.

smiling, looking others directly in the eye, and leaning forward slightly or nodding to show that you are paying attention to a speaker.

Avoid negative body language. Biting your nails and playing with objects such as pens or tools can show that you are nervous. Slouching, fidgeting, and avoiding eye contact can show that you are not paying attention or that you are bored.

Listening *Active listening* is the skill of paying attention to and responding to what someone says. An active listener pays attention and tries to see things from a speaker's point of view. If you are practicing active listening, you listen even if you disagree with the speaker and wait until the speaker has finished before replying or responding. Here are key steps in the listening process:

1. Think about the purpose of the message. Why is the person speaking?

2. Signal your level of understanding with body language, such as nodding your head.

3. Ask questions to help clarify, or make clear, points you do not understand.

4. Listen for the speaker's inflections. *Inflections* are tones that reveal feelings.

5. Look at the speaker's body language.

6. Take notes as needed about important points.

7. Listen for the conclusion of the message.

Thinking Skills

On the job, you also need to think critically, make decisions, and solve problems.

Thinking Critically Thinking critically is the ability to analyze and evaluate. It enables you to respond to a variety of situations.

Making Decisions Making good decisions means carefully weighing all the evidence. It also means considering possible outcomes of your decisions.

Problem Solving You can use the problem-solving process in many different situations.

1. *Clarify the problem*, or state the problem clearly. This helps define what needs to be done.

2. *Gather information*. What is causing the problem? What resources are available?

3. *Identify* possible solutions.

4. *Select* the best solution. Look at the advantages and disadvantages of each.

5. *Test* your solution. This will reveal its strengths and weaknesses.

6. *Evaluate* the solution. Is it effective? If not, select another possible solution. Test each solution until you find one that works.

Work Ethic

Your **work ethic** is the belief that work has value. If you have a good work ethic or a strong work ethic, you take pride in what you do. You know that your skills and efforts have value. A strong work ethic shows your commitment to doing your best on the job. The qualities that are part of a strong work ethic can be developed with practice. They include responsibility, flexibility, commitment, honesty, and cooperation.

Think of responsibility as responding to what a particular situation demands of you. Being responsible means showing up for work on time. It means becoming familiar with the tasks that make up your job and carrying them out correctly. A responsible person is someone who others can rely on. When you are responsible, you also accept the consequences of your choices and actions instead of blaming others.

Flexibility is the ability to adapt willingly to changing circumstances. It can also be a part of your work ethic. Being flexible on the job means adjusting to changes without complaining. The more confident you are in your skills, the easier you will find it to be flexible when circumstances demand it.

You practice honesty on the job when you are truthful and loyal in your words and actions. For example, if you make a mistake on the job, admit your mistake. Then find out how to prevent the same error in the future. Honesty also means not stealing. Stealing might be taking materials or not working while you are on the job.

 Teamwork
Work Together Teamwork builds efficiency and encourages new ideas.

Teamwork As a construction worker, you will often find yourself part of a team. Teamwork means cooperation and trying to get along with everyone. You will practice teamwork on the job by supporting the efforts of your coworkers.

When you have a strong work ethic, you have a commitment to quality and excellence. In construction, a commitment to quality involves using quality materials and methods. When you are committed to quality, you strive to meet the highest standards.

Leadership

Employers look for employees with leadership skills, too. Leadership is the ability to inspire others to accomplish a common task. It also plays a part in making you a good citizen.

You do not need to wait until you are employed to develop leadership skills. Organizations such as SkillsUSA can help.

Managing Resources

Resources are the raw materials with which you do your work. Making the best use of resources is also a skill. The key resources are:

Time You use time effectively when you complete tasks quickly and accurately. You can also learn to prioritize, or put tasks in order of importance. Simplifying tasks is another way to use time well.

Energy Use personal energy resources effectively by getting the right amount of rest.

Money If you are responsible for making purchases, look for good value for the money. If you are receiving money in payment, be honest.

Materials and Equipment The materials, equipment, and tools associated with your job are resources. Use them properly and with safety in mind. Immediately report any problems with or damage to equipment and supplies. Do not waste materials.

People You are a resource. Your employer depends on your labor to accomplish tasks.

Information Information comes in many forms. For example, information is in the building plans you follow, the safety warnings on tools, and the instructions you are given on how to complete a particular task.

Information Resources On the job, you will acquire, use, and share information. Information comes from many sources. Your boss, your coworkers, a drawing, manual, and the Internet all provide information.

Look for clues to tell the difference between useful information and false or useless information. Is it from a reliable source? What evidence is given? Does it seem to make sense? Be careful when using information from the Internet. Some Web sites contain misleading information. Reliable information comes from known sources, such as government agencies.

Use information wisely. If you have been asked to frame a wall, use the information on the plans to do it properly. Ask for advice from experienced coworkers to help make the job go faster or more easily. Follow the instructions your boss has given you.

Share information. Do not keep important information to yourself. If you see a problem, tell your supervisor. If you have a suggestion for doing a better job, share it.

Technology Resources Technology has brought many changes to construction. Computers, improved tools and equipment, and engineered lumber products continue to make construction processes more efficient. These changes affect how jobs are done and who will do them. For example, surveyors now use lasers to lay out a building site. Carpenters use nailers to speed up framing. Supervisors access building plans on their computers.

Stay informed about new technologies and keep your skills up-to-date. Ask your supervisor if you can be trained to use those forms of technology that might be helpful.

Reading Check

List List academic skills used in carpentry.

Finding Employment
What is networking?

Finding a job includes gathering information, applying for a job, having an interview, and responding to an offer.

Gathering Information

The first step in the job-hunting process is gathering information. Many first-time job seekers think that classified ads are the only place to search for a job. Construction jobs are frequently listed in the newspaper. However, other helpful resources are available.

Networking If you have ever followed up on a job tip from someone you know, you have practiced networking. **Networking** means making use of all your personal connections to achieve your career goals. When you receive job information from people you know, you can be informed and confident. You can network with friends and classmates, teachers and mentors, family members, employers and coworkers, and school and professional organizations.

 Gather Information
Be Prepared Before calling to find out about a potential job, prepare any questions ahead of time. *What else could you do to prepare for a phone call?*

Arnold & Brown

When you network, be courteous and respectful. Do not pressure people for information. Follow up on job leads right away. Be on time for interviews. Return phone calls. Always present yourself professionally. Your appearance, communication skills, and behavior reflect upon both you and the person who recommended you. Remember to return the favor. When you become aware of job information, share it with your network.

Trade Publications Extend your job search resources by reading online and print construction trade publications. These professional magazines and newsletters are available by subscription. Most of them list job opportunities. Some of these publications can be found in libraries or on the Internet.

Employment Agencies Employment agencies put employers in touch with potential employees. Many employment agencies charge either the employer or the employee for their services. *Temporary agencies* (or *temp agencies*) are one type of employment agency.

The Internet Thousands of employment resources are available on the Internet. Search engines often list jobs by category. In addition, many companies include job opportunities on their own Web sites. You can network, contact professional organizations, read online versions of trade publications, and register with online employment agencies.

Applying for the Job

Some employers may ask you to begin with a telephone call. Some will ask you to contact them by mail, sending a letter of application and a résumé. A **résumé** is a summary of your career objectives, work experience, job qualifications, education, and training. Other employers will ask you to come in and fill out an application.

Before you are hired, you will be invited to a job interview. Job interviews are formal meetings between you and your potential employer. It is important to perform each step of the job application process in a professional manner.

Responding by Telephone Your job leads may come from listings that give phone numbers and ask you to call for more information. A *hot call* is a call to a specific person or to get specific job information. A *cold call* is a phone call to a possible employer to ask for information about possible jobs.

When making a phone call, follow these guidelines to get the information you need:

- Tell the person who answers the phone that you are calling in response to a job opening. He or she will direct your call to the contact person.

- When you are connected to the contact person, give your name and the name of the job that interests you. If you were referred by someone, mention that person's name.

- The contact person will identify the next steps in the application process. These may include asking you to send a letter of application and a résumé. The contact person may offer to send you a job application or set up an appointment for an interview.

- Write down everything you are told to do. Repeat it to the contact person to make sure you have understood the steps.

- Ask any questions you may have about the application process. Answer any questions the contact person asks you.

- Thank the contact person for his or her time.

Filling Out an Application You may be asked to complete a job application form at some point during the job application process. A **job application** is a form that asks questions about a job applicant's skills, work experience, education, and interests. You can request job applications in person, over the phone, and over the Internet. Job application forms vary, but they all ask for the same basic information. A sample job application form is shown on page 19.

If you are asking for a job application in person, remember to make a good impression from the beginning. Do not enter a

APPLICATION FOR EMPLOYMENT

PERSONAL INFORMATION　　　　　DATE OF APPLICATION: _____

　　　　　　　　　　　　　　　　Social Security Number: _____-_____-_____

Name: _____

　　　　　　Last　　　　　　　　　　First　　　　　　　　　　Middle

Address: _____

　　　　Street　　　　(Apt.)　　　　City　　　　　State　　　　　Zip

Contact Information: _____

　　　Home Telephone　　　　　　　　Cell　　　　　　　　　　Email

POSITION: _____　　Available Start Date: _____

Desired Pay Range: _____　Are you currently employed? _____
　　　　　　　By Hour or Salary

EDUCATION

	Name and Location	Degree/Certification	Major/Subjects of Study
High School			
College or University			
Specialized Training, Trade School, etc.			
Other Education			

Please list your areas of highest proficiency or special skills that may contribute to your abilities in performing the abovementioned position. _____

_____

PREVIOUS EXPERIENCE (Please list beginning from most recent)

Dates Employed	Company Name	Location	Role/Title

Job notes, tasks performed, and reason for leaving:

_____

_____

Job Application

Fill Out the Form You can also fill out job applications online. Instead of using a pen, type your answers into each field.

Local Career Info Requirements for trade workers may vary by region, state, or they may even be specific to a particular construction project. For example, some construction projects may require that workers follow certain safety requirements related to the climate. Another job might require that workers should be able to speak both English and Spanish.

potential workplace unless your clothing is neat and appropriate and you are clean and well groomed.

You can make sure that your application is accurate by creating a list of important information you will need in advance. This list includes your Social Security number (SSN), your driver's license number, and your contact information. You should also bring the contact information for any previous employers, your title, the tasks you did, when you worked there and for how long, and your pay rate.

Always fill out a job application completely, neatly, and accurately. On every job application, be sure to:

- Read and follow directions exactly.
- Use standard English and check your spelling with a dictionary if possible.
- Print neatly, using blue or black ink. Use cursive handwriting for your signature only.
- Read the instructions for completing each blank before responding. Try not to make errors. If you need to correct what you have written, draw a line through it.
- Answer every question. Do not leave any part of the application form blank unless you are asked to do so. If a question does not apply to you, draw a short line or write N/A, which stands for "not applicable," in the space provided.
- Always tell the truth on an application. Submitting false information is illegal.

Applications often request references. *References* are individuals who will recommend you to an employer. Choose references carefully and ask their permission before listing them on an application. Teachers, counselors, and former employers are good references.

Testing Some employers may require you to take one or more tests.

- A skills test or a performance test evaluates how well you can do a particular task. An example might be a basic math skills test.
- A drug test is a blood, hair, or urine test for illegal drugs.

Responding in Writing You may need to write a cover letter when asking for an application form or requesting an interview. A *cover letter* is a brief letter that introduces you to the employer and explains why you are applying for a job. A cover letter should include a brief summary of your education, your experience, and your other qualifications.

If you are sending an e-mail or a paper cover letter, make sure that you have used correct spelling and grammar and that you have the correct contact information. If you are mailing or faxing a paper letter, sign your letter in black ink and make sure that you have the correct contact name, address, and postage on the envelope.

Preparing Your Résumé A résumé is a brief summary of a job applicant's contact information, education, skills, work experience, activities, and interests. The five parts of a résumé are labeled on page 21. Your résumé gives an employer information about your background and gives you a chance to show that you are a good candidate for the job. A chronological résumé organizes information in reverse time order, beginning with your current work experience. Some guidelines for résumés follow on page 22.

Experience A chronological résumé shows your record of work experience. *What else is included in this résumé?*

Contact Information Place yolur name, full address, telephone number (with area code) and e-mail address at the top of your résumé

Job Objective State the job you are applying for. You can change this item if you are using the same résumé to apply for different jobs.

Skills Summary Identify any business or other skills and abilities that your have gained in school, on a job, or in other situations.

Work Experience List your work experience, beginning with your most recent job. Include volunteer or unpaid work (such as internships) if they relate to the job you are applying for.

Education List the schools you have attended and diplomas or degrees you have received, beginning with your most recent education and training. You may also include any special subjects, certification programs, or things you achieved while in these programs.

Alyssa Rodriguez
6400 Old Guilford Road High Point, NC 27260
(336) 555-0135 • a.rodriguez@emails.com

Job Objective   Apprentice-level carpenter for a residential construction firm.

Skills Summary
- Three years' experience in residential construction
- SkillsUSA CareerSafe Card and OSHA 10-hour Safety Course certified
- Strong interpersonal and communication skills; fluent in English and Spanish

Work Experience

| Nov. 2011–present | Venuti Contractors | Greensboro, NC |

Apprentice Trainee
- Cut and shape sidings, moldings, and lumber on various new home and remodeling projects
- Construct and erect frames for multiple new home projects
- Perform administrative tasks: interpreting blueprints, scheduling, creating estimates
- Use hand and power tools safely and efficiently for building tasks

| Feb. 2010–Oct. 2011 | Carolina Builders | High Point, NC |

Carpenter Assistant/Office Assistant
- Performed window, door, and minor roof repairs under supervision
- Installed prefabricated doors, windows, and flooring
- Assisted in translation between carpentry supervisor and laborers
- Performed tool maintenance, door fitting, measurements, and stud location

Education

| 2012–Present | Waverly Community College | Greensboro, NC |

AA, Building Trades Technology (expected graduation date 8/2014)
- Completed NCCER Level I craft completion level certification
- Coursework in Applied Mathematics and Construction Technology
- NCACP Commercial Carpentry certification courses

| 2008–2012 | Liberty High School | High Point, NC |

High School Diploma
- Coursework: Construction Technology I/II, Advanced Algebra, Geometry
- 1st place, 2011 North Carolina SkillsUSA Championship in Carpentry
- Youth Apprenticeship Program, Carolinas Assoc. of General Contractors (CAGC)

- Keep your résumé brief.
- Include accurate contact information.
- Include your career objective.
- Stress relevant work experience, skills, education, and training.
- If you are submitting a résumé online, use keywords that describe your work experience. A *keyword* is a significant word that makes it easier for employers to search for relevant information. If your résumé contains a keyword such as carpentry or framing, employers with construction opportunities will be more likely to call up your résumé in an electronic search.
- Use correct spelling and grammar.
- Use white paper and dark type.
- Avoid using decorative graphics and styles.

Reading Check

Summarize *What is the purpose of a résumé?*

The Interview

You may be asked to come in for a job interview. An **interview** is a meeting between an employer and a job applicant. You may not get an interview every time you apply for a job. During an interview, you will have a chance to convince an employer that you are the right person for the job. Employers will evaluate you based on your appearance, attitude, and your answers to questions.

Before the Interview

The interview process begins when an employer arranges an appointment. Write down the date, time, and place of the interview. Ask for directions if necessary.

The more you know about the employer and the job, the better you will do in the interview. Check community business publications, local newspapers, Internet directories, and professional organizations. Try to find out the size of the business, its

profitability, and its plans for growth. Make notes about what you learn and think about your answers to possible questions.

Focus on Appearance Your appearance will affect the prospective employer's first impression of you. Make sure that you are clean and well-dressed. Make sure your hair and nails are clean and neatly trimmed. Wear only simple jewelry. If you are a man and do not wear a beard, shave before the interview.

Choose appropriate clothing that fits properly and is clean, pressed, and in good condition. If you are not sure what clothing is appropriate for your interview,

Job Interview
Good Impression An interview is your chance to make a good impression. *How can you prepare for an interview?*

call and ask. The person who interviews you will probably appreciate your attention to detail.

Allow plenty of time to locate your destination. Arrive a few minutes early. As you introduce yourself to anyone before meeting with the interviewer, be polite and respectful.

Some common questions asked by interviewers include:

- Why would you like to work here?
- What do you want to be doing in five years?
- What are your qualifications for this job?
- What are your strengths and weaknesses?
- Why did you leave your last job?
- Tell me about a challenge you met or a problem you solved in school or on the job.
- Have you ever been a member of a team or club? What did you like best and least about that experience?
- What questions do you have about the job or this company?
- Why should we hire you?

During the Interview

You will do well in the interview if you are prepared, positive, and relaxed. The interviewer will introduce himself or herself. Smile and introduce yourself in return, and offer your hand for a firm, confident handshake. Remain standing until the interviewer asks you to be seated. He or she will probably ask a few simple questions to help you feel more at ease.

Throughout the interview, maintain eye contact with the interviewer. Eye contact helps show that you are interested in what the interviewer is saying. When you reply, use correct grammar and speak clearly. The interviewer will ask you questions designed to determine if you are the person needed for the job. Do not interrupt the interviewer. If you do not understand a question or do

not know the answer, say so politely. Do not be afraid to ask the interviewer about the nature of the job, your responsibilities, and the work environment. Save questions about pay and benefits, such as vacation time, for the end of the interview.

When the interview ends, thank the interviewer for his or her time. Shake hands as you leave. The interviewer will signal the end of the interview in one of the following ways:

- The interviewer may tell you that you will be contacted later. If the interviewer does not specify a time, politely ask, "When may I expect to hear from you?"
- You may be asked to contact the employer later. Note the telephone number, the preferred time to call, and the contact person.
- You may be offered the job. You may be asked to decide right away whether you will take the job. If you are unsure, ask the interviewer if you may think about the offer. If this option is offered, be sure to follow up by responding promptly.
- You may not get the job. Do not be discouraged. The interviewer does not have to tell you why you are not being offered the job. Accept the decision gracefully.

After the Interview After each job interview, you have several responsibilities. First, send the interviewer a letter thanking him or her for the interview. Do this even if you have been turned down for the job. Be sure the correct address, contact information, and the correct postage are on the envelope.

If you have been asked to contact the employer, do so at the specified time. Send or deliver any materials or information, such as references, you have agreed to supply. If the employer has promised to contact you, wait the specified amount of time. If this time passes, telephone the employer and politely ask about the status of your application. You may be asked to go through a second interview.

Right after the interview, go over the session in your mind. Make notes on anything that you could do to improve. Note any key information, such as employer expectations and job responsibilities. List any questions you still have about the job.

Responding to a Job Offer

When you receive an offer of employment, you have three options. First, you can accept the offer. The employer will then give you information on when you will begin work. You may be asked to attend an employee orientation or a training session. You will be given specific details on pay, schedules, and other factors.

Second, you can ask for time to consider the offer. This is the time to bring up any unanswered questions that might affect your decision. Come to an agreement on when you will notify the employer of your decision. Do not be late.

Third, you can turn down the job offer. Perhaps the job is not right for you. You may have been offered a better job in the meantime. Whatever the case, if you do not intend to take the job, say so. You do not need to give reasons. Simply say, "Thank you for considering me, but I have decided not to take the job."

Reading Check

Recall *When should you send a thank-you letter?*

On the Job
What is ethical behavior?

When you accept a job, you enter into a relationship in which both parties have rights and responsibilities. In this section, you will learn about your rights as an employee and your responsibilities to your employer. You will become familiar with wages, taxes, and benefits. You will practice skills for getting along with others on the job. You will also identify some of the qualities required for advancement.

Rights and Responsibilities

Your employer will explain company rules and expectations when you begin your job. Your main responsibility is to do the best job possible. Here are some general guidelines:

- Use time responsibly. Be on time for work. Return promptly from authorized breaks and meal periods. Stay at work for your full shift, or specified hours of employment. Keep busy on the job. Avoid using company time or resources for personal business without permission.

- Respect the rules. Learn and follow your employer's rules and policies. You may be given an employee handbook. If you are in doubt about a company policy, ask your supervisor. Avoid drug and alcohol use, especially on the job.

- Work safely. Familiarize yourself with the safety requirements of your job. Learn how to operate and maintain equipment safely. Report any unsafe conditions or practices to your supervisor immediately.

- Earn your pay. Complete each task you are assigned to the best of your ability. Keep your work area neat and organized. Respect the value of the equipment and materials you work with. Use company resources responsibly.

Ethical Behavior Your employer has the right to expect ethical behavior from you. Your **ethics** are your inner guidelines for telling right from wrong. Ethical behavior consists of doing what is right.

Much of the time, it is easy to recognize the ethical course of action or the right decision to make. However, some choices are more difficult. When deciding between choices that may appear equally right or equally wrong, ask yourself the following questions:

- Does the choice comply with the law?
- Is the choice fair to those involved?
- Does the choice harm anyone?
- Has the choice been communicated honestly?
- Can I live with the choice without embarrassment or guilt?

On the Job

Hands-On Training On-the-job training (OJT) provides firsthand experience in learning important skills.

Employer Responsibilities

Your employer has responsibilities to you, too. Your employer must make sure that you are paid fairly for your work. You must also be given what you need to do your job. This often includes on-the-job training. Your employer must also provide safe working conditions and make sure you are treated fairly.

Safe Working Conditions Federal, state, and local regulations require your employer to provide you with safe working conditions. This includes:

- Eliminating recognized health and safety hazards. Injury prevention is part of this responsibility. For example, employers have supported research into repetitive stress injuries. These are injuries that develop among workers who perform the same motions repeatedly. They can affect a person's employability.

- Informing you when conditions or materials pose dangers to health and safety.

JOB SAFETY

WHAT IS NIOSH? NIOSH is the National Institute for Occupational Safety and Health. It is a federal agency that researches and makes recommendations for preventing work-related disease and injury. For example, if carpenters were to suffer from carpal tunnel syndrome (a painful wrist condition), NIOSH would study the problem to determine its cause and prevention. It would investigate working conditions that might lead to the problem. It would then make recommendations about preventing the problem and help train safety and health professionals to deal with the problem. NIOSH's sister agency is OSHA, the Occupational Safety and Health Administration, which administers laws related to workplace health and safety.

asiseeit/E+/Getty Images

- Maintaining records of job-related illnesses and injuries.
- Providing equipment and materials necessary to do the job safely.
- Complying with environmental protection policies for safely disposing of waste materials.
- Contributing to workers' compensation. If you are injured on the job and cannot work, your employer has a legal responsibility to provide financial help. This is called workers' compensation and it covers medical expenses and lost wages.

Fair Labor Practices Your employer has a legal responsibility to protect you from unfair treatment on the job. U.S. labor laws protect the following rights of employees:

- The right to have an equal opportunity to obtain and keep employment.
- The right to be paid a fair wage.
- The right to be considered fairly for a job or for a promotion.
- The right to be protected in times of personal and economic change.

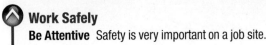

Work Safely
Be Attentive Safety is very important on a job site.

Equal opportunity, or *equity,* is part of the American workplace. Equity means fair treatment for everyone. Workers are chosen for jobs based on their skills and abilities, and not based on for how they look or who they are.

Employers must also pay their employees at least the federal minimum wage. This is the lowest hourly amount a worker can earn. Employers must give employees who work overtime extra pay or time off. This time off is called *compensatory time,* or comp time.

American workers are guaranteed the right to join a *labor union,* which is an organization of workers in a similar field. Labor unions act as the voice of their members in collective bargaining. Collective bargaining includes negotiating for working conditions, contracts, and other job benefits for a group of workers.

Employers must also protect their employees from *discrimination,* which is unfair treatment based on age, gender, race, ethnicity, religion, physical appearance, disability, or other factors. In addition, sexual harassment, any unwelcome behavior of a sexual nature, is prohibited in the workplace. Sexual harassment can include jokes, gestures, repeated or threatening requests for dates, and unwanted touching. Both males and females can be the victims of sexual harassment.

If you are the victim of discrimination or harassment, here are some suggestions:
- Immediately tell the person to stop. Be clear and direct.
- Write down what happened, noting the date, time, and place. Include the names of any witnesses. You should also include any comments about how the harassment or discrimination affected your work.
- Inform a trusted supervisor and follow up.
- If the issue is not resolved, you can get help from your local human rights group or the office of the U.S. Equal Employment Opportunity Commission (EEOC).

Wages and Benefits When you agree to take a job, you trade your skills and efforts for pay. Your pay is determined by a number of factors. These include your level of experience, the difficulty of the work, and the number of people competing for the same job. Pay periods differ from employer to employer. Some employers pay weekly, others every two weeks, still others once a month. Others pay a set amount for particular jobs.

If you earn an hourly wage, you are paid a certain amount for each hour you work. Your pay varies depending on how many hours you work all together. If you receive a salary, your employer pays you a set amount of money per year, regardless of the hours worked. This amount is divided up and paid at regular intervals.

Deductions

The amount of money you receive before deductions is known as your *gross pay*. The amount of money you actually receive is called your *net pay*, or take-home pay. *Deductions* are the amounts withheld from your gross pay for taxes, insurance, and other fees. Ask your employer to explain the deductions that will be taken from your pay. Some common deductions are shown in the callouts for the pay stub are shown below. These deductions include FICA (Federal Insurance Contributions Act), federal, state, and local income tax, gross pay, net pay, and other deductions, such as employee contributions to retirement plans or medical, dental, and life insurance.

Federal Income Tax
A personal income tax you pay on the amount of income you receive. This is the main source of revenue for the federal government.

FICA
Your Social Security taxes are paid on the money you earn and are withheld by your employer. Social Security taxes are withheld in two parts. The first part goes toward pension benefits; the second part covers Medicare benefits. FICA stands for the Federal Insurance Contributions Act.

State Income Tax
A personal income tax you pay on the amount of income you receive. This amount of state income tax varies by state. Some states have no income tax. Some cities may also have income tax.

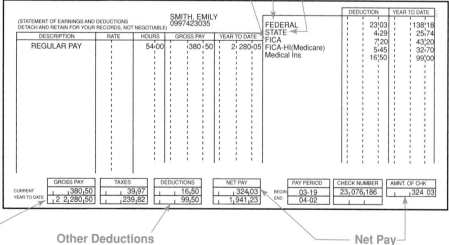

Gross Pay
The total amount of your earnings before taxes and other deductions.

Other Deductions
Other withholdings that are taken out of your paycheck. These might include employee contributions to medical, dental, or life insurance or retirement savings.

Net Pay
Your take-home pay, or the amount of your earnings left after all deductions are taken out.

 Pay Stub
Your Paycheck A pay stub shows you the amount of each deduction taken from your gross pay.

Benefits

In addition to giving you wages or a salary, your employer may offer benefits. Benefits are extras that workers receive on the job. Not all companies offer the same benefits. Some possible benefits include:

- health insurance
- dental insurance
- accident insurance
- vision insurance
- paid holiday days
- paid sick days
- paid vacation days
- life insurance
- disability insurance, which helps pay your expenses if you become disabled and can no longer work
- tuition reimbursement, which is full or partial repayment of tuition and fees you pay for education directly related to your career
- savings and investment plans, such as a 401k, to help you save money for retirement.

Be sure to figure in any benefits when calculating your job compensation. A high wage may make up for few benefits. A good range of benefits, on the other hand, can make up for a lower wage.

Working With Coworkers and Customers

When you take a job, you also enter into a relationship with your coworkers and customers. Every person is an individual, with his or her own personality, strengths, and weaknesses. You will need to get along and work together with a variety of different people.

There are many ways to work together effectively. Some strategies include:

- Keep a positive attitude. An upbeat, positive outlook contributes to team spirit. Complaining can decrease morale and affect your job performance.

- Respect yourself. You demonstrate self-respect when you accept responsibility for your actions, learn from your mistakes, and take care of your appearance, your tools, and your workspace.

- Respect others. Try to empathize, or to see things from another's point of view. This will help you get along with others.

- Show respect for the property, time, and beliefs of others as well.

Resolving Conflicts No matter how well you and your coworkers get along, you will not always agree. Conflicts are part of team interaction.

You may encounter some conflicts that cannot be resolved. Exercise self-control. If you are criticized, focus on the problem, not the personalities involved. While conflict can be unpleasant, you can learn to resolve conflicts respectfully.

Respond to customer complaints in a professional manner. Do not take them personally. If you cannot solve the problem yourself, ask your supervisor to step in.

Mathematics: Calculation

Calculate Pay Differences Mia is deciding between two four-week carpentry apprenticeships. Apprenticeship A pays $7.00 an hour for 50 hours of work a week and includes health insurance at no cost to Mia. Apprenticeship B pays $9.50 for 40 hours of work a week and does not include health insurance benefits. Mia thinks that she will need health insurance for her job in carpentry in case she becomes injured on the job, so she has decided to choose Apprenticeship A. How much less or more money will Mia make if she chooses Apprenticeship A instead of Apprenticeship B?

Starting Hint First, calculate how much Mia would earn for each apprenticeship. Then use subtraction to find the difference.

Advancing on the Job Advancement may involve a promotion. It may also mean staying at the same job level but with more or different responsibilities at a higher rate of pay.

Showing initiative can help you advance. *Initiative* is the quality of doing what needs to be done without having to be told to do it. Examples of initiative might include taking on new tasks and levels of responsibility. Workers with initiative do not have to be told what to do next.

Another way to advance your career is to continue your education or training through formal classes, workshops, or independent study. This shows a desire to learn and add to your skills and knowledge.

Leaving the Job

At some point, you may seek a new job. Perhaps work is slow and you have been laid off. Perhaps you think you would be happier with a different employer. In either case, behave in a professional way when you leave. If leaving is your choice, give your employer at least two weeks' notice. Thank your employer for the opportunities you have been given.

Do not criticize your past employers during job interviews. If you are asked why you are leaving, you might say something like, "I am looking for better opportunities."

Section 1.2 Assessment

After You Read: Self-Check

1. What basic academic skills are needed for any job?
2. How can technology resources help your career?
3. What is the purpose of a résumé?
4. What is the difference between a salary and an hourly wage?

Academic Integration: Mathematics

5. **Decimals and Percentages** Cuevas Construction has offered you a job as a carpenter's assistant. Your employer has told you that the tax on your weekly earnings will be 15% of your pay. Your pay is $80.00 per week. How much will you pay in taxes each month? Assume that one month = 4 weeks.

Math Concept A percentage is an amount that represents the part of a whole, where the whole is 100%. It is a relative comparison. A percentage is easily converted to a decimal, which makes it easier to perform calculations with other numbers.

Step 1: Convert the percentage to a decimal by dividing by 100 and moving the decimal point two places to the left ($15\% = 15 \div 100 = 0.15$).

Step 2: Multiply your total weekly earnings by the tax rate.

Step 3: Multiply the tax each week by the number of weeks in a month.

Go to **connectED.mcgraw-hill.com** to check your answers.

Review and Assessment

Chapter Summary

Construction is the process of building. Construction jobs fall into craft, technical, and professional categories. Customer needs influence construction trends. Education and training for a construction career includes certificate programs, apprenticeships, associate's degree and bachelor's degree programs, military training, and on-the-job training. An entrepreneur creates and runs his or her own business. Entrepreneurs need a business plan.

A person's employability skills include academic, interactive, and thinking skills; a strong work ethic; leadership skills; and the ability to use and manage resources. The process of finding a job includes gathering information, applying for a job, the interview, and responding to a job offer. An application form, résumé, and cover letter provide information about the job seeker. Both employers and employees have responsibilities on the job. Safety is an important concern.

Review Content Vocabulary and Academic Vocabulary

1. Use each of these content vocabulary and academic vocabulary words in a sentence or diagram.

Content Vocabulary

- career clusters (p. 6)
- career pathways (p. 6)
- apprentice (p. 6)
- trend (p. 8)
- certification (p. 9)
- entrepreneur (p. 11)
- business plan (p. 12)

- free enterprise (p. 12)
- work ethic (p. 15)
- networking (p. 17)
- résumé (p. 18)
- job application (p. 18)
- interview (p. 22)
- ethics (p. 24)

Academic Vocabulary

- visualization (p. 6)
- features (p. 8)
- evaluate (p. 9)
- specific (p. 12)

Speak Like a Pro

Technical Terms

2. Work with a classmate to define the following terms used in the chapter: *residential construction* (p. 6), *commercial construction* (p. 6), *master construction* (p. 7), *remodeling* (p. 8), *job shadowing* (p. 10), *internship* (p. 10), *transferable skills* (p. 13), *interpersonal skills* (p. 14), *non-technical slang* (p. 14), *body language* (p. 14), *active listening* (p. 15), *references* (p. 20), *initiative* (p. 29).

Review Key Concepts

3. **Name** three different career specialties in construction.

4. **Identify** two ways to prepare for a career in construction.

5. **Describe** the role of a business plan for an entrepreneur.

6. **List** three employability skills.

7. **Identify** the components of a chronological résumé.

8. **Explain** what it means to have a strong work ethic.

Critical Thinking

9. Explain What is the difference between gross pay and net pay?

Academic and Workplace Applications

10. Rounding and Estimation Coray makes $10 per hour for regular hours and 1.5 times regular pay for overtime. He worked 69.5 regular hours and 22.25 overtime hours this month. Round to the tens place to estimate his gross pay for this month.

Math Concept When rounding numbers, look at the digit to the right of the place to which you are rounding. If the digit is 5 or greater, round up. If the digit is less than 5, round down.

Step 1: Calculate overtime pay ($10 × 1.5).

Step 2: Round Coray's hours (69.5 and 22.25) to the nearest tens place.

Step 3: Multiply the rounded numbers by the corresponding rates per hour. Add the two numbers together to find the gross pay.

21st Century Skills

11. Communication Skills Your employer may ask you to present technical information to new employees or even a customer. Prepare a three-minute oral presentation for your classmates that describes the content of a topic listed in this chapter. First, evaluate your audience. How much do they already know? Next, prepare an outline of the material you want to cover. If you are explaining a process, start with the first step and move in order through to the last. Give background information or an overview first. If possible, provide visual examples and give a demonstration. Then summarize the information in your presentation. When you have finished, ask for questions.

21st Century Skills

12. Information Literacy Use the Internet and other resources to research information on construction-related certificate and apprenticeship programs in your area. Locate information on the associate's and bachelor's degree programs available in your state. Then visit at least one military Web site to learn about education available. Create a one-page chart that compares all the programs, including duration (length of time it takes to complete the program), cost, and the certification achieved or the employment result.

Standardized TEST Practice

True/False

Directions Read each statement. Fill in the bubble marked **T** if the answer is true. Fill in the bubble marked **F** if the answer is false.

(T) (F) **13.** Construction industry workers are always self-employed.

(T) (F) **14.** Listening and speaking are examples of communication skills.

(T) (F) **15.** Remodeling a house is an example of residential construction.

TEST-TAKING TIP

Make sure you understand the full statement before you decide whether the statement is true or false. Remember that all parts of a statement must be correct in order for the statement to be true. Statements that contain extreme words such as all, none, never, and always, or that have unsupported opinions, are often false.

*These questions will help you practice for national certification assessment.

Building Codes & Planning

Section 2.1
Codes & Zoning

Section 2.2
Architectural Drawing

Section 2.3
Reading Architectural Plans

Section 2.4
Estimating & Scheduling

Chapter Objectives

After completing this chapter, you will be able to:

- **List** the steps in planning to build a house.
- **Name** three sources of house plans.
- **Explain** how to obtain financing for construction.
- **Identify** several elements used in architectural drawings and architectural plans.
- **Summarize** the advantages of computer-aided drafting and design.
- **Describe** the three basic types of cost estimates and give an example of a direct cost and an indirect cost.

Discuss the Photo

Plans Builders should consult an accurate set of plans frequently during every phase of construction. *What components might be found in a standard set of building plans?*

Writing Activity: Business Letter

Write a brief letter or an e-mail to your local building department asking them which general building code is used in your area. Use correct spelling and include your contact information so that you will receive a response. Share the response you received with the class.

Hill Street Studios/JupiterImages/Brand X/Alamy

Before You Read Preview

Building codes and architectural plans guide the construction of a new house. Choose a content vocabulary or academic vocabulary word that is new to you. When you find it in the text, write down the definition.

Content Vocabulary

- O building code
- O building permit
- O stock plan
- O floor plan
- O mortgage
- O architect's scale
- O plan view
- O elevation
- O schedule
- O specifications
- O bid
- O quantity takeoff
- O board foot
- O indirect cost

Academic Vocabulary

You will find these words in your reading and on your tests. Use the academic vocabulary glossary to look up their definitions if necessary.

- ■ exceeds
- ■ scale
- ■ derived
- ■ allocation

Graphic Organizer

As you read, use a chart like the one shown to organize information about content vocabulary words and their definitions, adding rows as needed.

Content Vocabulary	Definition
building code	Standard set of regulations that govern the procedures and details of construction

Go to **connectED.mcgraw-hill.com** to download this graphic organizer.

Building Codes

What are building codes?

A house is a complex structure. It is made up of many materials and parts that must fit together with precision. Unless construction is planned carefully, the house will not be safe, durable, functional, or comfortable. A **building code** is a standard set of regulations that govern the procedures and details of construction. Its purpose is to ensure that buildings are structurally sound and safe from fire and other hazards. Most communities follow one or more building codes.

Building codes establish minimum standards of quality and safety. A builder can construct a house that exceeds code requirements. Many builders who have a reputation for high-quality work do this. However, by law, a builder must not construct a house that does not meet the code requirements. During construction, a building inspector visits the project at various times to ensure that the building codes are being followed.

Model Building Codes

Local building codes are usually based on model building codes. A *model building code* is a set of regulations developed by an independent organization on which local governments can base their own building codes. For many years, all U.S. construction was guided by one of three major model U.S. building codes. They set minimum standards for residential construction. The *National Building Code* was used primarily by a narrow band of states ranging from Missouri to Pennsylvania, and in New England. The *Standard Building Code* was used primarily in southern states. The *Uniform Building Code* was used primarily in western and upper midwestern states. Canada relies on the *National Building Code of Canada*.

Other factors beyond regional concerns may determine the information in model building codes. For example, the Americans with Disabilities Act (ADA) and the Fair Housing Act (FHA) contain provisions regarding new construction. Public buildings and certain multi-family dwellings must comply with these provisions to ensure that the building is accessible to individuals who have disabilities. The most recent model building codes include standards for meeting these provisions.

Model building codes also cover work related to installing such utilities as electricity and plumbing. The *National Electrical Code* and the *Uniform Plumbing Code* are two examples.

Building codes cover standard types of construction. However, an architect or builder may decide to use unusual materials and techniques. For example, it is possible to build a house entirely out of concrete. The designer must then prove to local officials that the house would meet or exceed standard requirements and that it would be safe.

Alternative Construction
Something Unusual This concrete home is an example of an alternate construction method.

©Glow Images/SuperStock

Modifying Codes The building department, which is a part of town, city, county, or state government, may adopt all the provisions of a model code. If so, the department does not need to develop its own code. However, sometimes it will adopt only those parts of the model code that fit local conditions. Throughout each unit of this book, *Regional Concerns* have been highlighted for this reason.

The International Residential Code To help make building codes more uniform, several of the organizations that develop the codes jointly produced the *International Residential Code for One- and Two-Family Dwellings.* The International Residential Code (IRC) is a standardized building code that is designed to account for regional variations. The map on this page shows one example of a regional variation. The IRC covers detached one- and two-family dwellings and townhouses that are no more than three stories high. It is updated every three years and has been adopted by all 50 states. However, most building departments review and update their codes every three to five years. Many have decided to follow the IRC, but some may still choose to use one of the older model codes. Because the IRC provides minimum standards that are widely recognized, this code is referred to throughout this book. *For your own work, always follow the codes that have been adopted in your local area.*

Permits and Inspections

In areas covered by building codes, the builder must obtain a building permit before beginning construction of a house. A **building permit** is a formal, printed authorization for the builder to begin construction. To obtain a permit, the builder must submit a full set of working drawings, called plans, to the local building department. Plans show exactly how a house will be built. Builders must check with the local building department for permit requirements. There can be financial penalties for building without a building permit. In some cases, work not covered under a permit must be demolished.

REGIONAL **CONCERNS**

Local Climate U.S. building codes vary by climate. Codes in coastal areas exposed to hurricanes include regulations aimed at reducing damage caused by high winds and flooding. Structures in earthquake-prone areas must withstand earthquakes, while those in "Tornado Alley," an area in the Great Plains where tornadoes frequently occur, must be built to reduce damage caused by high winds and debris. The northernmost states in the United States have detailed energy conservation regulations, while some western states have water conservation regulations.

Because codes are modified to suit local conditions, a house in South Carolina might be built to a different code than a house in Minnesota. This is generally not a problem for local carpenters and builders. They learn what is required during their training. However, if a carpenter or builder moves to another region, he or she must learn how the codes differ.

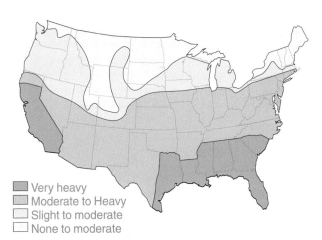

- Very heavy
- Moderate to Heavy
- Slight to moderate
- None to moderate

Note: Lines defining area are approximate only. Local conditions may be more or less severe than indicated by the region classification.

 Termite Infestation Probability
Regional Hazards The IRC addresses regional variations in hazards. This IRC code shows that the risk of termite damage varies from region to region. Codes in high-risk areas are stricter than codes in low-risk areas.

The building department examines the plans to make that they meet local codes. If they do, the builder is issued a permit, which must be posted on the building site throughout construction. The cost of a building permit is usually based on the estimated cost of construction.

At key points during construction, a city or county building inspector will visit the job site to examine the work. The inspections vary but often include an examination of footing trenches and foundation formwork, framing, wiring, plumbing, and insulation. The builder must contact the building department to schedule an inspection when each part of the job is complete, but before it is covered by other work. The inspector can require that work be done over if it does not meet building codes. If the work has passed inspection, the inspector will initial or sign off on the paperwork.

When the house is complete, one last inspection is made. If there are no problems, a *certificate of occupancy* (CO) is issued. This document states that the house is ready to live in.

Reading Check

Analyze *Why would a carpenter be familiar with multiple codes?*

Before Construction Starts

What is a floor plan?

A lot of planning occurs before a house can be built. This process involves more than just figuring out what the building codes are. The home has to fit the needs of the occupants, both at the present time and in the future. How construction will be paid for must also be determined. Sometimes a builder constructs a house before there is a buyer. This kind of house is called a speculative house, or spec house. A house built for a particular client is called a custom house. With a custom house, the builder and the

Good Communication
Planning Frequent discussions with homeowners can ensure that the house is completed on time and on budget.

future homeowners continue to communicate while the house is in the process of being built.

Lot or House Selection

When it comes to choosing a lot to build on, the following factors are important to consider.

Cost It is recommended that a homeowner spend not more than 25 percent of gross monthly income for housing. This amount of money includes mortgage payments, insurance, property taxes, utilities, and repairs.

Location How close is the property to jobs, schools, community services, and recreational facilities? Is the neighborhood likely to remain fairly stable, or will it change as the surrounding area grows? Are utilities such as water, sewer service, and electricity available?

Lot Shape and Contour Is the lot wide enough and deep enough for the desired house? Does it have unusual contours or other features that would make construction difficult or expensive? Will neighboring houses be so close that privacy will be a problem?

Special Conditions Examples may include: Is the lot on a flood plain? Are there underground springs in the area? Is there industrial contamination to consider?

Zoning Restrictions Most communities are divided into zones in which certain types of buildings are encouraged or restricted. Common zones are those for single-family dwellings, apartments and condominiums, commercial buildings, and industrial buildings. Many people prefer to reside (live) in a neighborhood that does not include commercial or industrial buildings.

Deed Restrictions Within any zone there may be deed restrictions on an individual lot. Deed restrictions might specify such things as the minimum-size house that can be built on the lot, requirements for certain architectural features, or setback distances. A setback distance is the minimum distance allowed from the house to adjacent features, such as other houses or the street. For example, a house near a stream might have to be located a certain distance above predicted flood levels. Additionally, a house built in a historic area might be limited to the use of certain exterior colors or roofing materials.

Legal Documents

Once the lot has been selected and the builder has been contracted, there are certain legal documents that must be processed. At least four legal documents are involved with the purchase of a house:

- The official survey, which shows the boundaries of the property.
- The deed, which is evidence of ownership.
- The abstract of title, which is a history of the deeds and other papers affecting the ownership of the property.
- The contract of sale (sometimes called a *sales contract*), which describes all the details relating to the purchase.

Before buying property, the purchaser should have the property surveyed to confirm that its dimensions match the dimensions noted on the sales contract. After the property is purchased, the buyer should retain a copy of the official survey, the deed, and the abstract of title. These are needed to secure financing and permits for new construction.

The contract of sale is required when a house or land is purchased. For an existing house, the contract contains such information as its exact location, the sales price, and a listing of anything inside the house that was not part of the sale. For example, the current owner might want to take the laundry appliances to another house. If a builder is hired to build a new house, the contract specifies such things as when the work will begin and when it will end. It also includes a great deal of detail regarding the house. For example, it might specify the level of quality the construction must reach. You can read more about specifications in Section 2.2.

To be valid, a contract of sale must have these features:

- It must be written (not verbal).
- It must clearly state the terms of the agreement, so there will be no disagreement about what is being purchased.
- It must include the price and the terms of payment.
- It must be dated.
- It must be signed by both the buyer and the seller. Both parties must be competent and old enough to sign legal documents.

House Plans

Every type of house, whether it is a spec house or a custom house, needs to have a house plan. House plans can be obtained in several ways. The buyer can purchase a stock plan. A **stock plan** is a standard house plan that can be adapted to fit many different lots. Companies that sell stock plans usually provide floor plans and other drawings to show what the finished house will look like. A **floor plan** is a scale drawing showing the size and location of rooms on a given floor. Once a suitable house plan has been chosen, complete working drawings and materials lists can be purchased. Stock plans are also available on the Internet, from plan books, and from some magazines.

House plans can also be obtained from local builders. Often, a builder specializes in a certain type of house. The builder develops one or two basic plans and then adapts them as needed to suit various buyers. This is why houses in a neighborhood often have similar features.

A third way to obtain house plans is to hire an architect or building designer to develop them to the buyer's specifications. A design fee may be based on a percentage of construction costs or it may be a flat fee. A percentage fee can range from five percent to ten percent. Some people prefer to use an architect or designer because the design can be tailored exactly to their needs. The architect may be hired to design the house as well as supervise its construction.

Financing

After a builder has been chosen, financing must be obtained from a lender, usually from a bank or a savings and loan company. The loan is for a certain percentage of the total cost. This percentage varies, but 80 percent is common. The borrower provides the balance as a down payment.

Typically, a borrower starts with a construction loan. A construction loan is a short-term loan used during construction. The lender provides money periodically as

Climate and Design The design of a house is often influenced by climate, available materials, and local traditions. For example, houses in New England must withstand heavy snowfalls, so eaves (roof overhangs) are short or nonexistent. In the Northwest, however, houses often have deep eaves to shield walls and windows from frequent rains. In the Southwest, houses are often built of dense materials, such as adobe and masonry (bricks or stone), which keep house interiors cool.

the work progresses. These sums of money are called draws, or advances.

After construction is finished, the construction loan is converted to a mortgage. A **mortgage** is a long-term (15 to 30 years) loan that is secured by the property. It allows the lender to claim the property if the borrower does not make the mortgage payments. The borrower pays interest and principal over the life of the loan.

Loan providers, as well as some federal agencies, may establish certain requirements for home construction. These requirements are not the same as building codes. They are

A

B

Regional Styles
Weather Resistant The roofing materials shown here are appropriate to the different climates in which these houses are found. **A.** Clay-tile roofs, such as those in the Southwest. **B.** Cedar roof shingles on a log house in the Northwest.

Estimating and Planning

This estimating and planning exercise will prepare you for national competitive events with organizations such as SkillsUSA and the Home Builder's Institute.

Construction Costs

Average Cost per Square Foot

To determine house construction costs, builders and designers often start with the average cost per square foot for residential buildings in their area. They multiply this figure by the number of square feet in the plans to estimate total costs.

It is important to understand that actual costs can be higher because of special features, unusual materials, custom products, or special building requirements. For example, kiln-dried lumber is more expensive than air-dried lumber. Housing standards for areas that have high risk of earthquakes or flooding may have different code requirements.

The best way to get an accurate price estimate is to ask subcontractors to bid on the project. They will study the plans to calculate exact costs. It is important to get several written bids and to talk to subcontractors in detail about what their bid includes.

The table below shows sample costs for a 1,040 ft^2 house in different geographic areas.

Calculating Average Cost

To calculate average cost by square foot, first determine the square footage of the house. This information is often available on architectural plans and drawings. A simple house plan is shown at the top of the next column. If the information is not available, you can use the measurements on plans to determine the number of square feet in the house.

1. Say that a two-story house has 1,040 square feet on the first floor and

1,040 SQ. FT.

900 square feet on the second floor. Add the two floors together to get the total size of the house.

$$1,040 \text{ ft}^2 + 900 \text{ ft}^2 = 1,940 \text{ ft}^2$$

2. Multiply the total square footage of the house by the cost of construction per square foot. You can determine this number by averaging the total costs of all the materials and labor. It is good to use a calculator when you are using large numbers. Using the costs provided in the table, you can calculate that the cost for a 1,940 ft^2 house in Charleston, SC would be $97,000.

$$1,940 \text{ ft}^2 \times \$50/\text{ft}^2 = \$97,000$$

Estimating on the Job

A contractor has given you the average cost of constructing a house in Euless, TX at $38 per square foot. Estimate the cost of building three houses with the following square footages: 1,500 ft^2, 3,422 ft^2, and 4,689 ft^2.

Sample Location	House Size	Cost of Construction per sq. ft.	Total Estimated Cost
New Orleans, LA	1,040 ft^2	$65	$67,600
Charleston, SC	1,040 ft^2	$50	$52,000
Chino, CA	1,040 ft^2	$120	$124,800
Minnville, KY	1,040 ft^2	$27	$28,080

standards that a builder must meet to obtain a certain type of mortgage. Builders must be aware of these additional requirements before construction begins.

Once financing has been arranged, contracts are signed for the construction. From then on, it is the responsibility of the builder and/or architect to make sure that the building goes as planned. A loan officer at the bank may also require progress reports to ensure that money loaned by the bank is being used properly.

Builders are usually paid a certain portion of the construction costs before work is started. They are then paid additional amounts at certain stages, such as after the roof is installed. Again, these are the draws from the construction loan. Final payment is made after the client and lender have inspected and approved the work.

Section 2.1 Assessment

After You Read: Self-Check

1. What is a building code? What is its purpose?
2. What is a building permit, and what must you provide to apply for one?
3. Describe the document that indicates that a house is ready to live in.
4. What is a mortgage?

Academic Integration: Mathematics

5. **Calculate Area** You have been given the following plan for the foundation of a one-story rectangular house. Calculate the surface area of the floor.

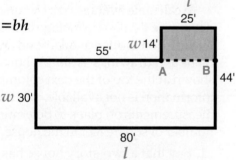

Math Concept Surface area is the sum of all of the areas of the shapes that cover the surface of the object. The area of a flat rectangular surface can be calculated using the following formula: **area = length × width**, or $A = lw$

Area can also be calculated as **area = base × height**, or $A = bh$

Step 1: The foundation is made up of two rectangles. A rectangle is a four-sided figure in which all four of its angles are right angles. This means that the sides opposite one another are of equal length. Draw a dotted line between point A and point B to help you see the two rectangles more clearly.

Step 2: Multiply length by width to calculate the areas of the larger rectangle (80' × 30') and the smaller rectangle (25' × 14'). Add both areas to find the total area of the foundation.

Go to connectED.mcgraw-hill.com to check your answers.

Section 2.2
Architectural Drawing

Drawing & Measurement
Why are plans important?

The ability to read and understand a set of construction plans is an essential skill. Plans tell you everything you need to know about how to build something. The use of lines, symbols, and words make the ideas of the designer or architect clear to those individuals who must work on a construction project. These individuals include carpenters and other tradespeople, materials suppliers, and building inspectors. To be successful in most construction-related jobs, you must be able to interpret the plans correctly. You must also be able to measure accurately and read a tape measure.

Types of Drawings

Construction plans consist of drawings of the structure that show how it should be assembled. A *sketch* of something is a quick and informal drawing. An architect might make a sketch to capture an original idea. He or she might also sketch a house to show clients how it would look. A master carpenter might sketch a framing connection as a way of showing an apprentice how to make it.

Builder's Tip

TRACKING CHANGES Various versions of the plans are produced during the design development of a house. Before construction begins, make sure you have reviewed the latest set of the plans. If you have any questions, contact the project manager or designer. Make sure any subcontractors that will be working for you also have the most recent set of plans.

Architectural plans, sometimes called construction drawings or working drawings, are a set of more formal drawings. They provide an organized and precise way of showing how an entire structure should be built and can be consulted at any time. Copies of original plans are called prints. Many people refer to prints as *blueprints*. This is because plans used to be printed as white lines on a dark blue background. Modern plans typically have blue lines on a white background. An architect, an architectural designer, or a drafter usually creates the original drawing from which prints are made.

 Reading Plans
Understanding Details Plans function as a universal language that even those who do not speak the same language can understand.

Measuring Systems

Two basic systems of measurement are used worldwide. The United States currently uses the customary system (Standard system) of measurement. Most other nations use the metric system. Some industries in the United States have changed over to the metric system. In construction, however, the customary system is still used. Some products, such as paint, are often labeled using both systems of measurement.

In the customary system, lengths are given in inches, feet, yards, and miles. In the metric system, lengths are given in millimeters, centimeters, meters, and kilometers. A meter, which is the basic metric unit of length, is slightly longer than a yard (39.37"). One inch is equal to 25.4 millimeters. The metric system is based on units of ten. The millimeter is equal to 1/1000 of a meter, and a centimeter is 1/100 of a meter. A kilometer consists of 1,000 meters.

The two most common length measures in the customary system used for residential construction are the inch and the foot. In the metric system, the two length measurements most commonly used in construction are the millimeter and the meter. A customary/metric conversion table titled "Metric Conversion Factors" can be found in the **Ready Reference Appendix** in the back of this book.

In the customary system, liquids are measured in quarts and gallons. In the metric system, they are measured in liters. A liter is about 5 percent larger than a quart. Liquid finishing materials, including paints, are normally given in liters, half-liters, and quarter-liters. Weight in the customary system is given in pounds. In the metric system it is given in kilograms. A kilogram is approximately 2.2 pounds.

In some parts of the world that use metric measures, particularly Great Britain, a standard unit of measurement is 300 millimeters—which is very close to one foot.

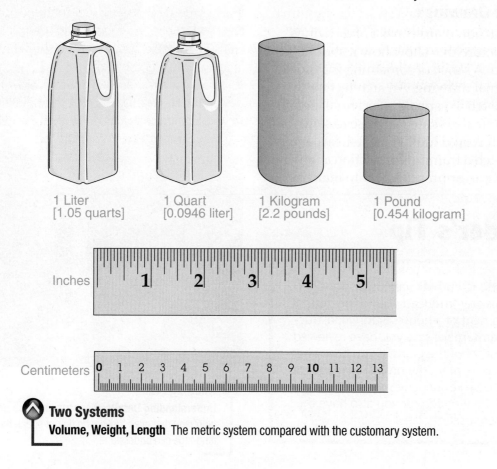

1 Liter
[1.05 quarts]

1 Quart
[0.0946 liter]

1 Kilogram
[2.2 pounds]

1 Pound
[0.454 kilogram]

Inches   1   2   3   4   5

Centimeters   0 1 2 3 4 5 6 7 8 9 10 11 12 13

Two Systems
Volume, Weight, Length The metric system compared with the customary system.

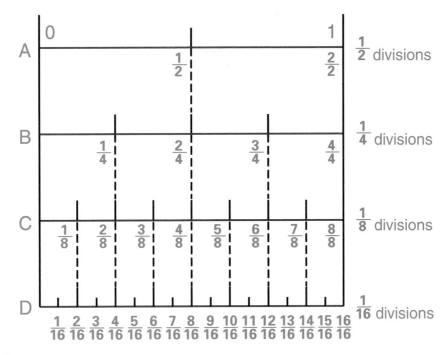

⬥ **Divisions of an Inch**
Common Fractions Carpenters rarely need to measure in increments smaller than ¹⁄₁₆ inch. For ease in reading, this drawing shows an inch larger than it really is.

On architectural plans, all building dimensions are given in millimeters. All site measurements appear in meters and millimeters. In the United States, building dimensions are given in feet, inches, and fractions of an inch. All site dimensions are given in feet and inches.

Reading a Customary Rule Measuring devices used in construction are based on multiples of 12 inches (one foot). The skills for measuring are the same, whether using a 1' rule or a 100' layout tape measure:

Take a look at the drawing, Divisions of an Inch. The distance between 0 and 1 represents 1". At *A*, the inch is divided into two equal parts. Each half represents ½". At *B*, the inch is divided into four equal parts. The first marker indicates ¼", the second marker ²⁄₄" (½"), and the third marker ¾". At *C*, the inch is divided into eight equal parts; each small division is ⅛". Two of these divisions make ²⁄₈" (¼"). Four make ⁴⁄₈" (½"). At *D*, the inch is divided into 16 parts. Notice that ⁴⁄₁₆" is equal to ¼". One mark past ¼" indicates ⁵⁄₁₆". Notice on your own rule that between one inch mark and the next, the ½" mark is the longest. The ¼" mark is the next longest, then the ⅛" mark. The ¹⁄₁₆" mark is the shortest.

To read a fraction of an inch, count the number of small divisions beyond the inch mark. For example, in Measuring an Object, you will find that it is 2" plus four ¹⁄₁₆" segments. This is 2 ⁴⁄₁₆", which is the same as 2¼".

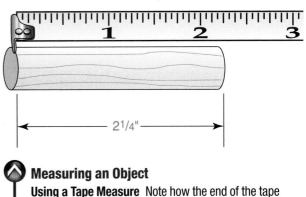

2¼"

⬥ **Measuring an Object**
Using a Tape Measure Note how the end of the tape measure is hooked on the object being measured. Use a rule to measure the distance indicated by the arrows.

Table 2-1: Customary Scales Used for Drawings		
Measurements based on ...	Scale of drawing	Scale as ratio
1' = 1'	full size	1:1
6" = 1'	half size	1:2
3" = 1'	one-forth size	1:4
1½" = 1'	one-eighth size	1:8
1" = 1'	one-twelth size	1:12
¾" = 1'	one-sixteenth size	1:16
½" = 1'	one-twenty-fourth size	1:24
⅜" = 1'	one-thirty-second size	1:32
¼" = 1'	one-forty-eighth size	1:48
3⁄16" = 1'	one-sixty-forth size	1:64
⅛" = 1'	one-ninety-sixth size	1:96

Understanding Scale

In order to represent large objects on small sheets of paper or computer screens, architectural plans are drawn to scale. Scale is the ratio between the size of the object as it is represented and the actual size of the object. Drawings that are done to scale are drawn with the same proportions as the objects they represent, but at a different size. In the case of architectural plans, the drawings are much smaller than the actual buildings and other objects they represent.

An architect can represent a building of any size on a small piece of paper by drawing it to a certain scale. However, it is important to understand that scale is not a unit of measurement. If the drawn object is exactly the same size as the real object, it is called a full-size or full-scale drawing. If the object is reduced, it will probably be drawn to one of the common scales shown in **Table 2-1**. Look at the scale in which 3" = 1'. It is called one-fourth size because there are four sets of 3" lengths in 1'. What size is a drawing in which ½" = 1'?

Many architects use computers to draw plans, but others use traditional tools. The tool that architects use when making scale drawings is called an **architect's scale**. It allows the measurements in reduced-scale drawings to be measured as if they were in actual feet and inches. A traditional architect's scale is shown below. Several sets of markings representing various scales can be displayed on its triangular shape. Some of these markings read left to right, while others read right to left. Architect's scales are also available in other shapes, including a flat scale that resembles a ruler.

A scale of ¼" = 1' 0" is most often used for drawing houses. A distance of ¼" on the drawing represents a distance of 1' 0" on the actual house. For example, if you used a tape measure to measure a window on a house, it might be 4' high and 3' wide. If you drew that window to a scale of ¼" = 1' 0", its size on the paper would be 1" high and ¾" wide. If you wrote its

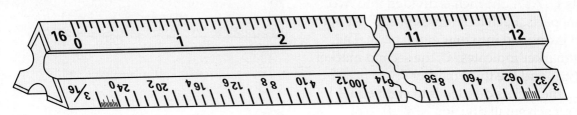

An Architect's Scale
Many Scales on One Tool The architect's scale is often used when making scale drawings. It is also useful for taking measurements from drawings that were made using a computer.

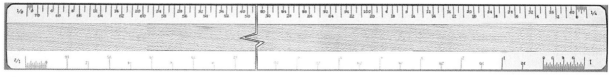

A Flat Scale

Another Shape This flat scale is easier for beginners to use than ones with a triangular shape.

dimensions next to the window on the drawing, however, you would write the size of the real window.

Using an Architect's Scale An architect's scale can be confusing for a beginner to use. Look at the flat architect's scale shown in Figure 2-10. Its left end is labeled ⅛, meaning that a ⅛" = 1' 0" scale starts at this end. This scale reads from left to right beginning at the zero mark. The other end of the instrument is labeled ¼, meaning that a ¼" = 1' 0" scale starts from that end. It reads from right to left beginning at the zero mark.

In A Flat Scale, a distance of 8' 6" is shown on the ¼" = 1' 0" scale. A distance of 14' 0" is shown on the ⅛" = 1' 0" scale. Can you see how these distances were determined? With practice, you will be able to draw lines of any distance using a scale.

Reading Check

Explain What is an architect's scale used for?

Making an Architectural Drawing

What are the elements of an architectural drawing?

Architectural drawings have traditionally been done by hand, and many are still done this way. However, computers are becoming the primary method for creating drawings. This is partly because relatively inexpensive computers are now powerful enough to run sophisticated software for drawing and design. The same basic elements are used whether drawings are made by hand or computer.

Elements of a Drawing

An architectural drawing consists of lines, dimensions, symbols, and notes. All four of these elements are very important to understanding, or reading, the drawing.

Lines Lines show the shape of an object and are used for many other purposes as well.

- *Centerlines* are used to indicate the center of an object. They are composed of long and short dashes, alternately and evenly spaced. At intersections, the short dashes cross. Very short centerlines may be broken if they will not be confused with other lines.

- *Dimension lines* indicate the start and end point of a particular dimension. They have arrowheads at each end. The dimension is written as a break in the middle of the line.

- *Leader lines* connect a note or a reference to part of the drawing. They usually end in an arrowhead or a large, circular dot. Arrowheads should always end at a line. Dots should be within the outline of an object. Leaders should end at any suitable portion of the note, reference, or dimension.

Science: Measurement

Metric Conversions What are the dimensions of a 12" × 2" × 1" plank in millimeters?

Starting Hint One inch = 25.4 mm.

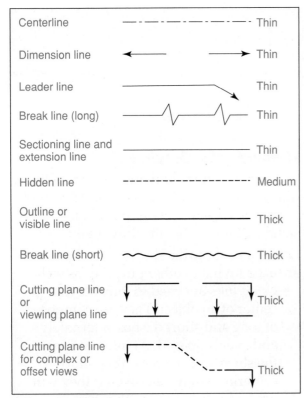

Centerline	—————·—————·—————·—————·	Thin
Dimension line	← ————————————— →	Thin
Leader line	———————————————→	Thin
Break line (long)	——————⋀⋁—————	Thin
Sectioning line and extension line	———————————————	Thin
Hidden line	- - - - - - - - - - - - - - - -	Medium
Outline or visible line	———————————————	Thick
Break line (short)	∿∿∿∿∿∿∿∿∿∿∿∿∿	Thick
Cutting plane line or viewing plane line		Thick
Cutting plane line for complex or offset views		Thick

 Line Meanings

Common Lines These are the lines most commonly found on architectural plans. The thickness, length, and shape of a line help to convey its meaning.

- *Break lines* may be solid, freehand lines that indicate short breaks. Full, ruled lines with freehand zigzags are used for long breaks.
- *Sectioning lines* indicate the exposed surfaces of an object in a sectional view. They are generally full, thin lines, but they may vary with the kind of material shown.
- *Extension lines* mark the end points of a dimension and should not touch the outline of the object.
- *Hidden lines* are short, evenly spaced dashes that show the hidden features of a part of the drawing. They always begin with a dash in contact with the line from which they start. However, a space is added when such a dash would form the continuation of a full line. Dashes touch at corners.

- *Outlines or Visible lines* represent those edges of the object that can actually be seen.
- *Cutting plane lines* or viewing plane lines show sections that would otherwise be hidden. A section is a view that shows an object as if part of it were cut away to expose the inside.

Dimensions Dimensions are numbers that tell the size of something. The dimension of a particular feature on a plan can be determined by using an architect's scale. However, some dimensions are also written on plans. Carpenters and other tradespeople must follow the written dimensions when laying out the framing of a structure.

Plans are dimensioned both outside and inside the building lines. Outside dimensions describe openings and other changes in the exterior wall, in addition to its overall dimension. Inside dimensions locate walls relative to each other and to exterior walls. All horizontal dimensions are shown in the plan (top) view. All vertical dimensions are shown in elevation (side) view. Some dimensions may not be shown. These can be **derived** by adding or subtracting other dimensions in the drawing.

Symbols Symbols are used to represent things that would be impractical to show in some types of drawings. For example, they are often used to represent doors, windows, electrical receptacles, plumbing fixtures, and heating equipment.

Many symbols appear on plans. Some of them are easy to interpret, such as plumbing symbols. Symbol keys are sometimes found on the plans to explain less obvious symbols. Just as a written language is composed of letters grouped into words, symbols are grouped in various ways to make them easier to interpret. Many electrical symbols have similar shapes. Some symbols are used to indicate objects, while others are used to indicate materials. You can see examples of these types of symbols on page 47.

Different Symbols

Symbol Keys Symbols can help workers understand the components of a building. *Why might symbols be useful for workers who might not speak the same language?*

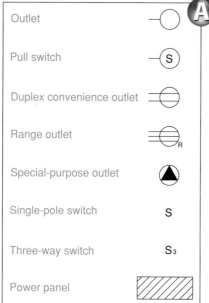

Outlet	
Pull switch	
Duplex convenience outlet	
Range outlet	
Special-purpose outlet	
Single-pole switch	S
Three-way switch	S₃
Power panel	

A Electrical
Common wiring symbols

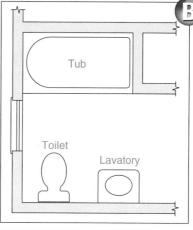

Tub

Toilet

Lavatory

B Plumbing
Common fixtures

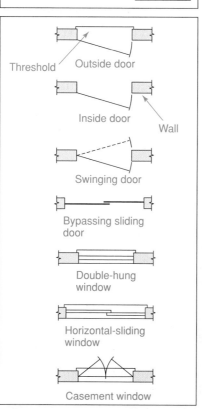

Threshold — Outside door

Inside door — Wall

Swinging door

Bypassing sliding door

Double-hung window

Horizontal-sliding window

Casement window

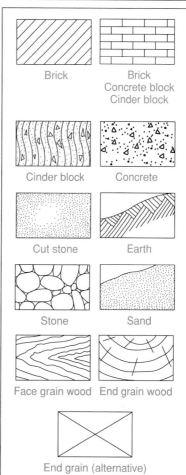

Brick

Brick
Concrete block
Cinder block

Cinder block   Concrete

Cut stone   Earth

Stone   Sand

Face grain wood   End grain wood

End grain (alternative)

C Doors and Windows
Note that the symbols show the direction of door swings and the way windows open.

D Building Materials
The way building materials are depicted on plans often represents what the actual material looks like.

Notes Notes are short, written explanations of some feature that might not be clear from the drawing or that requires extra emphasis. Notes give information about such matters as materials, construction, and finish. They are included wherever necessary. There are two kinds of notes: specific and general.

Specific Notes Specific notes might clarify dimensions or suggest a construction technique. For example, a note might be included telling the carpenters to be sure to check the dimensions of an unusually shaped bathtub before framing the bathroom walls.

General Notes General notes can be added that refer to many or all drawings in a set of plans. An example of a general note might be: "All dimensions are given from stud face to stud face." General notes may be underlined to attract attention.

To save space in notes, abbreviations are often used. Standard abbreviations for use on construction drawings can be found in the **Ready Reference Appendix** table called "Architectural Abbreviations," which is located in the back of this book.

Computer-Aided Drafting and Design

Architectural plans are often drawn by hand using pencil or ink. Plastic templates speed up the drawing of elements that are used over and over again, such as circles and arcs or symbols for plumbing fixtures and electrical components. However, computers are increasingly used for drawing architectural plans. Drawing on a computer is much faster than drawing by hand.

Computer-aided drafting and design (CADD or CAD) software programs can be used to create site plans, floor plans, elevation drawings, and even perspective (realistic) drawings of a structure. One advantage of using CADD software to prepare architectural plans is that details can be drawn once and reused on subsequent projects. However, software incompatibilities may prevent these drawings from being widely shared.

JOB SAFETY

AVOIDING STRESS HAZARDS AND EYESTRAIN
Anyone who prepares architectural plans may spend many hours in front of a computer. Be sure to arrange the monitor, keyboard, and electronic drawing tablets to minimize the risk of repetitive stress hazards. Eyestrain is another risk with computer use. Adjust lighting conditions accordingly, particularly to reduce glare.

The future establishment of uniform CADD standards will reduce or eliminate the problem of these incompatibilities. In the meantime, you must verify compatibility with your software when you receive files from outside sources.

Electronic drawing files can be printed out in the architect's office. They can also be given to a local printing company by uploading the files over the Internet or by delivering a CD-ROM. Printing companies have high-quality, high-volume color printers that can process many drawings quickly. Drawings can also be sent as e-mail attachments.

Using a computer offers many advantages. The drawings can be revised with ease. Estimating software can be combined with CADD software to produce a list of materials directly from the drawings. With a computer, symbols are easy to add, delete, or move. Computer-based symbol libraries are also available. When an architect wants to use a particular type of window in a house, its symbol can be obtained and inserted into the drawing with the click of a mouse. Manufacturers often provide symbol libraries that include specifications for their products.

Section 2.2 Assessment

After You Read: Self-Check

1. What does a centerline look like, and what does it represent?
2. What are extension lines?
3. Name two elements in plans that are commonly represented by symbols.
4. What does CADD software do, and how is it used in the construction industry?

Academic Integration: Mathematics

5. **Drawing to Scale** Create a scale drawing to represent a rectangular flat surface that is eighteen feet long and 1.5 feet wide. Use a scale of 1 yard = 1 inch. Label the dimensions of your drawing.

   Math Concept Scale is the ratio between the size of a representation of an object and the size of the actual object. A ratio is a comparison of two numbers. Ratios can be expressed with colons (1:1) or as fractions ($\frac{1}{1}$).

   Step 1: Convert the length from feet to yards. There are 3 ft in 1 yard. (18 feet ÷ 3 ft/yd = 6 yards). Convert the width from feet to yards (1.5 ft ÷ 3 ft/yd = 0.5 yards).

   Step 2: Use the scale of 1 yard = 1 inch to convert yards to inches.

   Step 3: Use a ruler to create your drawing. Label the length and the width of the rectangle.

Go to **connectED.mcgraw-hill.com** to check your answers.

Section 2.3 Reading Architectural Plans

Using Plans

What information is on a typical foundation plan?

A complete set of architectural plans consists of various views of the site and the building. Taken together, these views provide the information a carpenter, subcontractor, or builder would need in order to do their work. Knowing which view is likely to contain certain information is an important part of using plans.

The views of a building include general drawings and detail drawings. General drawings consist of plan views and elevations. Their purpose is to show large portions of the building. Details are shown with section views and detail drawings. They provide information about how parts fit together. Additional information is often given in the form of schedules and specifications. Other workers, such as estimators, depend on them.

Two-Story House with Triple Garage

House on Flat Lot Building plans for the house shown here appear in the following pages.

Plan Views

Seen above is a photograph of a house. Many of the architectural plans shown in the rest of this chapter are for this house.

A **plan view** is a top view. It is also known as a bird's-eye view. It allows you to see the width, length, and location of objects as if you were standing on a platform high above them and looking down. It is not possible to see the height of objects in a plan view. Several types of plan views are commonly used.

A *site plan*, or *plot plan*, shows the building lot with boundaries, contours, existing roads, utilities, and other details such as existing trees and nearby buildings. The basic elements of a site plan are drawn from notes and sketches based upon a survey. This plan shows where the driveway will be located. The outline of the building is often superimposed on the site plan, and corners are located by reference to natural objects, other buildings, and/or

survey markers. The excavation contractor relies on this plan.

A *foundation plan* is a top view of the footings and foundation walls. A foundation plan is shown on page 51. It also shows the location of posts and other elements,

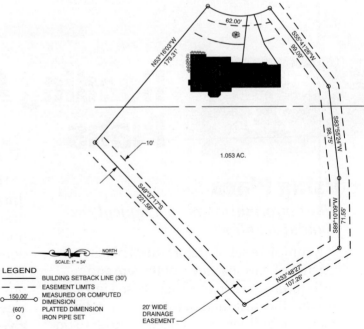

A Site Plan

House and Lot A typical site plan shows where the building will be placed on the lot.

such as pads needed to support an exterior deck. All openings in foundation walls are labeled and dimensioned. The type and location of foundation anchor bolts are identified. This plan is used by foundation contractors.

Floor plans such as those shown on page 52 are included for each level of the building. They are drawn as if the house were sliced horizontally at a level that would include all doors and window openings. This imaginary slicing is referred to as a *cutting plane*. A floor plan shows the outside shape of the building; the arrangement, size, and shape of rooms; the types of materials used; the thickness of walls; and the types, sizes, and locations of doors and windows. A floor plan may also include details of the structure, although these are usually shown on separate drawings called framing plans. Many tradespeople rely on floor plans.

Reflected ceiling plans are drawn as the ceiling would appear in a mirror placed on the floor below it. Reflected plans are used to show complex designs, such as tray ceilings (a type of decorative ceiling construction), or to show the locations of multiple lighting fixtures. They are not always included in a set of plans.

Framing plans show the size, number, and spacing of structural elements. Separate framing plans may be drawn for the floors and the roof. The floor framing plan must specify

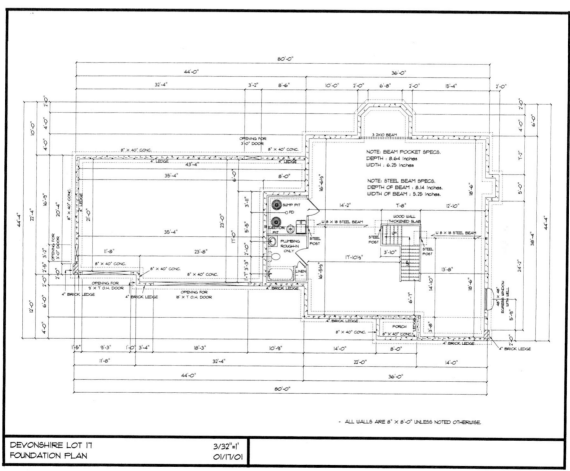

DEVONSHIRE LOT 17
FOUNDATION PLAN

3/32"=1'
01/17/01

Site Plan

The Foundation Mistakes made in reading the foundation plan can translate to mistakes in the layout of the slab, footings, or basement. This will have an impact on the wall framing, and the wall dimensions will need to be modified. Changing wall dimensions will affect doors, windows, and electrical systems.

Builder's Tip

VERIFY DIMENSIONS When building from architectural plans, always verify that the door and window schedules match dimensions given on the floor plans. It is far easier to correct mistakes before framing begins than after the house has been framed and sheathed.

the sizes and spacing of joists, girders, and columns used to support the floor. Doubled framing around openings and beneath bathroom fixtures is also shown. Detail drawings are added, if necessary, to show the methods of anchoring joists and girders to the

foundation walls. Roof framing plans show the size and spacing of rafters, as well as information about the roof slope and sheathing. Carpenters rely on these plans.

The *electrical plan,* which is drawn like a simplified floor plan, shows the location and type of every electrical feature of the building. These features include switches, ceiling lights, receptacles, and the service panel. The plan also indicates a schematic view of the electrical wiring that connects individual features to each other. The electrician relies on this plan.

The *mechanical plan* shows the arrangement and location of plumbing and heating features. Plumbers and mechanical contractors rely on this plan. A carpenter should also consult it to see if any special framing details might be required for these systems.

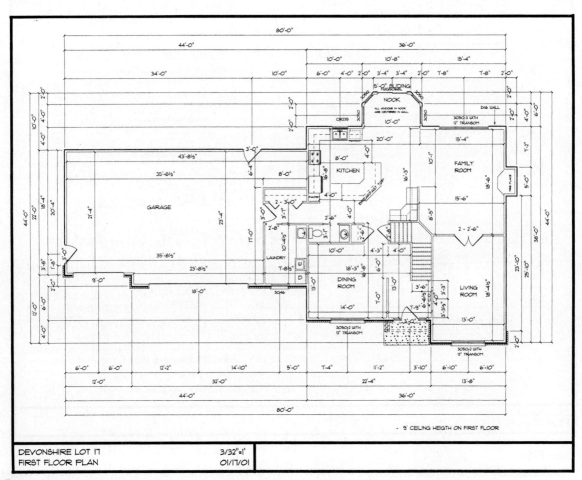

Floor Plan

First Floor This plan shows the first floor of the house and is similar to the foundation plan in overall shape.

Arnold & Brown

This is an example of how various types of plans often can be useful to more than one trade.

A *landscaping plan* shows the location of existing features such as trees and streams and provides information about new landscaping features that will be added later. These features include plants, trees, walkways, and irrigation systems. Decks are also included on landscaping plans (see Chapter 36). A landscaping plan is often developed by a landscape designer after the house is complete. Plants are represented by drawings that indicate the mature size of each one.

Elevations

An **elevation** is a side view that allows you to see the height and width of objects. An interior elevation shows a wall inside the building, as if you were in the room looking straight at it. This view shows information such as ceiling heights and the layout of interior trim and cabinets.

An exterior elevation shows one side of the building's exterior as shown in Elevation Views. Four elevation views are usually enough to show all sides of a house. However, if some of the walls do not meet at 90°, additional elevation views will be necessary to show them. Exterior materials (such as siding) are shown on all elevations. The size of windows and doors is sometimes given.

Reading Check

Contrast *What is the difference between a plan view and an elevation view?*

DEVONSHIRE LOT 17
FRONT ELEVATION   3/32"=1'   01/17/01

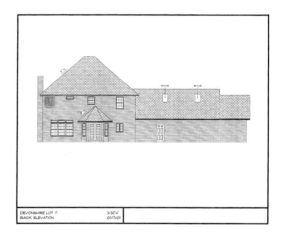

DEVONSHIRE LOT 17
BACK ELEVATION   3/32"=1'   01/17/01

DEVONSHIRE LOT 17
RIGHT ELEVATION   1/8"=1'   01/17/01

DEVONSHIRE LOT 17
LEFT ELEVATION   1/8"=1'   01/17/01

Elevation Views

Exterior Elevations These views show the front, back, right side, and left side of the house. A complicated house may require additional elevation views.

(t to b)Arnold & Brown

Section Views

Section views provide important information about materials, fastening and support systems, and concealed features. They show how an object looks when "cut" vertically by an imaginary cutting plane, as shown in Section View. The cut is not necessarily continuous but may be staggered to include as much construction information as possible. Section views are similar to elevation views because they allow you to see the actual shape of objects as shown from one side. They are extremely useful to many trades. For example, all the details of wall framing cannot be shown in a framing plan so they are usually shown in an elevation.

A wall-framing elevation would show the locations of studs, plates, sills, and bracing in a particular wall.

Where a section view is used to give more information about a larger drawing, the cutting plane is shown on the larger drawing by thick lines (see the cutting plane line in Line Meanings on page 46). These cutting plane lines, which sometimes have an arrow at each end, are identified with letters, numbers, or both. These labels help the reader understand exactly what portion of the house the section view represents. For example, a cutting plane line labeled "B-4" might be drawn on a floor plan or foundation plan. The section view that relates to this

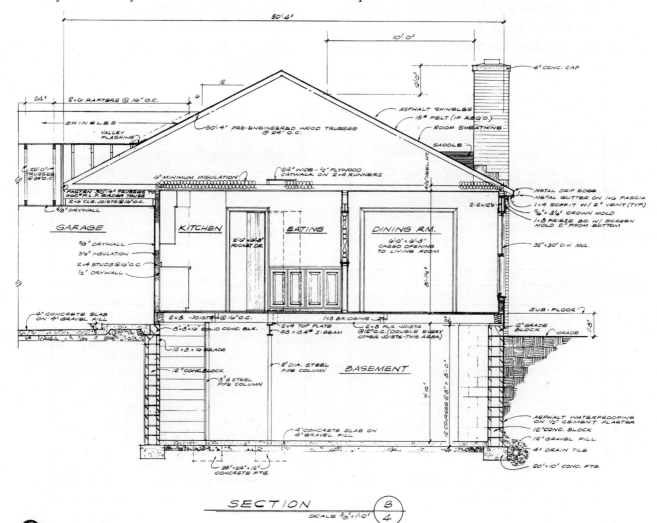

 **Section View**

A Horizontal Slice A section view shows the relationship of many details of a house. A complex house often requires several section views.

Arnold & Brown

slice through the house would be labeled "Section B-4."

Sections may be classified as typical or specific. *Typical sections* show construction features that are repeated many times throughout a structure. They are labeled "TYP," which is an abbreviation (shortened form) of "Typical."

When a feature occurs only once and is not shown clearly elsewhere, it is called a *specific section.* These features are generally not labeled.

Detail Drawings

When precise information is needed about a small or complex portion of the building, a *detail drawing* is made. The window detail shown in Trim Detail is a simple example of a detail drawing. Such drawings are used whenever the information given in elevations, plans, and sections is not clear enough. The construction at doors, windows, and eaves is often shown in detail drawings. Details are drawn at larger scales than plan views, such as ½" = 1' 0", ¾" = 1' 0", 1" = 1' 0", or ¼" = 1' 0". Some detail drawings may even be drawn at full size. Detail drawings are usually grouped so that references may be made easily from other drawings. They are often located on or near section drawings because they show a particular part of the section.

Detail drawings are sometimes made as isometric drawings. *Isometric drawings* are constructed around three basic lines that form 120° angles to one another. They sometimes illustrate an assembly detail, such as an interlocking joint in a timber-framed house. The isometric technique gives the detail a three-dimensional look.

Engineering Drawings

Many parts of a house are built using components that are manufactured elsewhere and then delivered to the job site. When these components are part of the house structure, *engineering drawings* may be required.

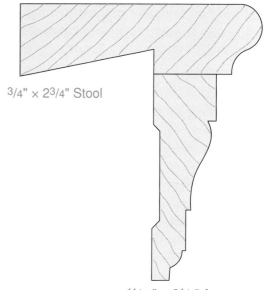

3/4" × 23/4" Stool

11/16" × 21/4" Apron

 Trim Detail
Up Close This simple detail drawing shows how the window trim parts fit together and gives their exact dimensions.

Engineering drawings are sometimes prepared by a civil or structural engineer hired specifically to solve a particular problem related to the house or its site. For example, an engineer might provide the design for an unusually tall retaining wall. In other cases, the drawings are made by engineers who work for the manufacturer of a product. For example, a roof truss manufacturer would employ engineers to design the trusses.

Engineering drawings show that each component has been evaluated by an engineer and is safe for its intended use. One way to recognize engineering drawings is to look for the engineer's official stamp. This stamp identifies the engineer who did the design and shows that engineer's license number for the state in which he or she is authorized to practice.

It is important to understand that a product or connection shown in engineering drawings must not be altered (changed) in any way without written approval of the engineer. For example, if the drawings show a connection held together with six bolts, all six bolts must be installed using the exact size and type of bolt called for in the drawings. Floor and roof trusses are always covered by engineering drawings. Any alteration, even a small notch, might cause the truss to fail.

Renderings

A *rendering* is sometimes called a presentation drawing. It is more like a picture of the structure than any other type of architectural drawing, as shown in Architectural Rendering. Its purpose is most often to show the exterior of a house as it would look when completed. A rendering often includes such elements as plants, shadowing, and other features that add to a sense of reality.

Renderings are sometimes created for clients to help them visualize what the house would look like after all the landscaping has been done. Renderings are often in color. Computer software makes it easy to change colors and other details shown in renderings. In some cases, an architect might even start the design process with a computer-rendered drawing and then develop other drawings based on the original drawing.

Reading Check

Summarize *What is the purpose of a rendering?*

Schedules & Specifications

What type of information is in a schedule?

A set of architectural plans consists of many drawings. This is their most obvious feature. However, plans also include additional elements that help the builder to understand exactly what to build. These elements relate directly to the drawings, but are presented as charts, lists, or text.

Schedules

A **schedule** is a list or a chart. For example, the window schedule shown in Window and Door Schedule on page 57 lists all the windows that will be used in the building. It contains information about the sizes of rough openings, glazing, finish, trim, manufacturer's name, and window type and size, among other things. It is information that would not necessarily appear in the drawings. Each window on the list is keyed to the floor plans with a letter or number. This ensures that windows will be put in the proper locations. A door schedule contains similar information.

Architectural Rendering
Finished View This is a rendering of the house shown in the preceding drawings.

SCHEDULES							
Windows				Doors			
Mark	Size	Type	Remarks	Mark	Size	Type	Remarks
W1	14-3'-4" × 5'-5"	D.H. VINYL		A	9'-0" × 6'-8"	OVERHEAD	GARAGE
W2	3-3'-4" × 4'-9"	D.H. VINYL		B	18'-0" × 6'-8"	OVERHEAD	GARAGE
W3	1-3'-4" × 4'-6"	D.H. VINYL		C	5'-0" × 6'-8"	FWD GLD PAT	NOOK
W4	2-2'-0" × 5'-5"	D.H. VINYL		D	3'-0" × 6'-8"	1/2 LIGHT	MN ENTR
W5	1-3'-4" × 2'-9"	D.H. VINYL		E	2-1'-0" × 6'-8"	1/2 LT SIDE LT	
W6	1-3'-0" × 3'-0"	D.H. VINYL		F	3-3'-0" × 6'-8"	6 PANEL	
W7	8-3'-0" × 1'-0"	D.H. VINYL	TRANSOMS	G	2-6'-0" × 6'-8"	BYPASS	
W8	2-4'-0" × 4'-0"	D.H. VINYL	FIXED PICT.	H	2-2'-6" × 6'-8"	BIFOLD	
W9	1-2'-2" × 1'-3"	D.H. VINYL	CIRCLE TOP	I	2-3'-0" × 6'-8"	BIFOLD	
W10	5'-0" × 1'-0"	D.H. VINYL	TRANSOM	J	2'-6" × 6'-8"	SGL POCKET	
				K	2'-4" × 6'-8"	SGL POCKET	
				L	2'-8" × 6'-8"	SGL POCKET	
				M	2-2'-6" × 6'-8"	FRENCH	
				N	2-2'-8" × 6'-8"	FRENCH	
				O	1'-8" × 6'-8"	ST 4 PANEL	
				P	2-2'-6" × 6'-8"	ST 4 PANEL	
				Q	4-2'-8" × 6'-8"	ST 4 PANEL	

Window and Door Schedule

A List of Sizes This schedule lists all the doors and windows for a house. It indicates their size and includes identification marks that identify the products on other drawings.

A *room-finish schedule* identifies the materials and finishes to be used for floors, walls, and ceilings for each room, including hallways. For example, a living room on the schedule might have strip-oak flooring, pine baseboard and window trim, painted drywall wall/ceiling surfaces, and wood paneling on two walls. In some cases, finish details may be included such as the color of the paint.

Specifications

Specifications are written notes that may be arranged in list form. They give instructions about materials and methods of work, especially those having to do with quality standards. Specifications such as those shown in Specifications on page 58 may explain the level of quality expected of tradespeople and give the minimum quality for materials and finishes. In commercial construction, complex projects often require full-time specification

writers. In residential construction, the specifications are often provided by the architect. Whether they are written by a specification writer or an architect, specifications should always be clear and brief.

Fire Ratings In many communities, fire hazards are a particular concern. In such cases the specifications for a house might include requirements for fire-resistant materials. Ratings of fire-resistant materials are often based on standardized tests that determine *flame-spread* ratings. Flame spread refers to how quickly flames will engulf the surface of a material. Materials are rated Class A (most resistant), Class B (less resistant) or Class C (least resistant). Another important rating is called *smoke density.* Smoke density is a measure of smoke created when a material is burning. For example, some foam plastics create dense, choking smoke when they burn and must be covered by nonflammable

SPECIFICATIONS

The house is to be built for_____ Owner,

residing at (Number)————————————(Street) ————————————

(City or Town)	(County)	(State)

and is to be built upon the Owner's property located as described below:

_____

_____

LOCATION OF HOUSE ON LOT - The location of the house shall be as shown and dimensioned on the Plot Plan included in the Working Drawings.

GENERAL CONDITIONS OF THE SPECIFICATIONS

GENERAL DESCRIPTION OF THE WORK -The Contractor shall supply all labor, material, transportation, temporary heat, fuel, light, equipment, scaffolding, tools and services required for the complete and proper shaping of the work in strict conformity with the Drawings and Specifications. All work of all trades included in the Specifications shall be Performed in a neat and workmanlike manner equal to the best in current shop and field practice.

BIDS-in receiving bids for the work specified herein, the Owner incurs no obligations to any bidder and reserves the right to reject any and all bids.

CONTRACT DOCUMENTS-The Contract Documents consist of the Drawings, Specifications, Plot Plan and the Agreement. The Contract Documents are complementary and what is called for by one shall be as binding as if called for by all. The intent and purpose of the Contract Documents is to include all labor, material, equipment, transportation and handling neccesary for the complete and proper execution of the work.

CLEANING-The Contractor shall at all times keep the premises free from accumulations of waste materials and rubbish, and at the completion of the work all rooms and spaces shall be left broom clean.

WORK NOT INCLUDED-The following items of work are excluded from the Contract, however, may be included if noted under "Special Items Included."

Blasting	Furniture and Furnishings
Sub-soil Drain	Venetian Blinds
Waterproofing	Window Shades
Driveways and Walks	Refrigerator
Finished Grading, Planting	Cooking Range
and Landscaping	Bathroom Accessories
Fences	Weatherstripping

EXCAVATION AND GRADING

Specifications

Detailed Instructions Specifications for a large house may be several pages long. This image shows only a portion of a specification sheet.

materials when used in a house. Building codes also include restrictions on the use of some construction materials based on their flame spread and smoke density performance.

Codes also include another type of fire rating called *fire resistance*. A fire-resistance rating is based on standardized tests of how long an assembly or product will withstand a fire on one side without letting it pass through to the other side. For example, a wall with a 1-hour rating will prevent flames from burning through the assembly for at least one hour. The wall's fire rating might be improved in various ways. For example, one side could be covered with materials that have a higher fire rating. Products such as exterior doors are often required to have a minimum fire-resistance rating.

Section 2.3 Assessment

After You Read: Self-Check

1. Name the six types of views commonly included in a set of architectural plans.
2. What is a typical section and how is it often identified?
3. Give three pieces of information that you would expect to find on a window schedule.
4. What is found on a room-finish schedule?

Academic Integration: English Language Arts

5. **Architectural Abbreviations** An abbreviation is a shortened form of a written word or phrase. Abbreviations are used to save space on architectural drawings. For example, the architectural abbreviation for "Typical" is TYP. Go to the **Ready Reference Appendix** in the back of this book and find the section titled "Architectural Abbreviations." Read the information at the top of the page. Then, find the following abbreviations and list the words that they stand for: **OC, BF, FDN**, and **H**.

 Go to **connectED.mcgraw-hill.com** to check your answers.

Estimating & Scheduling

Estimating

Why is it important to make accurate estimates?

General contractors and subcontractors alike regularly make estimates. *Estimating* determines the costs of building a house, particularly the costs of labor and materials. This information is important because it relates directly to how profitable the contractor's business is. If construction costs are underestimated all the time, a contractor may not be able to make a profit. If estimates are overestimated too often, the contractor may lose jobs to contractors with lower, more accurate bids. A **bid** is a signed proposal to do work and/or supply a material for a specified price. The ability to make accurate estimates for bids is one of the most important skills any builder or contractor can develop.

When creating estimates, a builder often contacts suppliers of materials and labor and asks them to bid on the project. This process is called competitive bidding. The builder often requests bids from several sources for each portion of the project, such as excavation, framing, and roofing. After receiving all of the bids, the builder chooses one. For example, the goal might be to complete the house quickly. In this case, the builder might choose the company that promises the earliest start date. If the goal is to keep costs as low as possible, the builder might choose the company that submitted the lowest bid price.

Large construction companies employ workers who specialize in preparing quantity and cost estimates. Many building material retailers also have at least one estimator on staff. In a small construction company, however, it is the owner who generally prepares the estimates. Anyone who prepares estimates must:

- Be able to read and measure building plans accurately.
- Have an excellent understanding of the materials and techniques used to build houses.
- Have an excellent understanding of local building codes.
- Be precise in assembling and computing numerical data.

To prepare an accurate estimate, a builder or contractor needs only a pencil, paper, and a calculator. However, computers and estimating software, as shown in Estimating Materials, make the estimating process much faster. This is especially true for large and complicated projects. The Internet is another aspect of technology that is rapidly changing the process of estimating. For example, it is now possible for an estimator to check a manufacturer's actual inventory to determine if there is enough material in stock to complete a project. Some manufacturers and suppliers even feature online estimating features on their Web sites.

Estimating Materials
Checking the Plans Estimators rely on accurate plans to help them determine the quantity of materials.

Types of Estimates

Estimates are used at many different times during the construction process. Some are informal and approximate while others are detailed and precise. The earliest estimates are often the least accurate. Accuracy increases as estimates are refined to reflect new information. Three types of estimates are pre-design estimate, quantity takeoff, and unit-cost feature.

Pre-Design Estimate During the early stages of working with a client, a builder is often asked how much a new house will cost. Providing an accurate answer is impossible without spending many hours studying a set of plans. Instead, the builder may multiply the square footage of the house by the approximate construction costs per square foot in that community. This type of estimate is called a *pre-design estimate*. It may also be called a preliminary estimate, ballpark estimate, or conceptual estimate. It is an estimate made before the exact features of the house are known.

Costs for materials and labor vary considerably in different parts of the country. An experienced builder usually knows the range of overall construction costs in his or her area. For example, a house built with modest materials might cost $100 per square foot in one area or $110 per square foot in another. A more complex house in which high-quality materials are used might run $135 per square foot. Therefore, the pre-design estimate for an 1,800-square-foot house could range from $180,000 to $243,000. The cost of land is added to determine a pre-design total cost for the project. This figure would enable the client to determine whether or not new construction was affordable.

An architect may use pre-design estimates to figure his or her design fee. Insurance companies also use pre-design estimates to establish what the approximate replacement costs would be if damage occurred.

Quantity Takeoff For a detailed understanding of costs, a builder or contractor develops a quantity takeoff. A **quantity takeoff**, is a cost estimate in which every piece of material required to build the house is counted and priced. It is also sometimes called a complete construction cost estimate or a quantity survey. A quantity takeoff is time consuming to create, but, once complete, it has other uses. For example, the builder may refer to the quantity takeoff when ordering materials.

		Basement Stair			
Part	Unit	Material	Length	Unit Cost	Total Cost
Stringer	LF	2 × 12 pine	12'		
Treads	PC				
Risers	PC				
Handrail	LF				
Balusters	PC				
Brackets	PC				
Other					

A Simple Quantity Takeoff
Tracking Costs In this example, the unit cost for the stringer (a length of lumber that supports steps) would be the cost per lineal foot. The total cost in this case would be the unit cost times 12, as the stringer is 12 feet long.

Special computer software can also be used to prepare quantity takeoffs. Some are linked to databases that can track fluctuations in the costs of materials. In some cases, an estimator may use a digitizing pen in order to measure and record dimensions directly from the plans and then store the information in a computer.

A complete set of building plans is necessary to prepare a quantity takeoff. The estimator must also review the project specifications carefully. There can be a big difference in cost between one grade of material and another, even though the number of pieces does not change. If a quantity takeoff is precise, there will be little or no difference between the estimate and actual construction costs.

Estimators must use the proper measure of quantity when preparing a quantity takeoff. For example, concrete is generally measured by the cubic yard. Framing lumber is measured by the lineal foot, by the piece, or by thousand-board-foot quantities. Carpeting is measured by the square yard and, increasingly, by the square foot. **Table 2-2** shows common abbreviations used in estimating.

Unit-Cost Estimate Another detailed estimate of construction costs is a *unit-cost estimate* or *component-cost estimate*. In making a unit-cost estimate the estimator divides the house into components, such as walls or roof. Estimates are made of the cost for each component. This is faster than the quantity takeoff method.

The "unit" of a unit-cost estimate depends on the component. The unit for walls is typically lineal feet (LF). Lineal feet is a measurement of length. The unit for floors and roofs is square feet (SF). For example, the 8' high interior partition shown in A Completed Partition Wall on page 62 would be measured in lineal feet. The cost of the wall would include the costs of every part, including the following:

- plates
- studs and 16d nails
- drywall on both sides of the wall
- drywall screws
- joint tape and compound
- drywall primer
- interior paint (two coats)
- baseboard trim.

Table 2-2: Common Abbreviations Used in Estimating		
APPR (approximate)	LBS/HR (pounds per hour)	OZ (ounce)
BDL (bundle)	LF (lineal foot))	PC (piece)
BF (board foot)	LH (left hand)	PR (pair)
CF (cubic foot)	M2 (square meter)	QT (quart)
CY (cubic yard)	M3 (cubic meter)	R/L (random lengths)
EA (each)	MH (man-hour)	RH (right hand)
GA (gauge)	MISC (miscellaneous)	SF (square foot)
GAL (gallon)	NA (not applicable)	SQ (square)
HR (hour)	NAT (natural)	UNF (unfinished)
LB (pound)	OA (overall)	YD (yard)

Note: These abbreviations are usually capitalized but may also be seen in other forms. For example, "each" may be written **Ea**. Some of these abbreviations may also be followed by periods. Some of these abbreviations differ from those used in this text.

If the wall were 12' long, then the total cost for the wall would be divided by 12 to determine the unit cost (the cost per lineal foot).

After a unit cost for this partition is figured, the estimator can quickly determine the total lineal footage of all partition walls in the whole house by measuring the plans. This figure is then multiplied by the unit cost to determine total costs for all the partition walls. It is important to note that this cost estimate would *not* cover exterior walls.

Depending upon the estimator's needs, the unit cost may be for materials only, or it may also include the cost of labor. A unit cost for labor for this particular project would include the labor of a carpenter, a drywall installer, and a painter.

Reading Check

Recall *What is a quantity takeoff?*

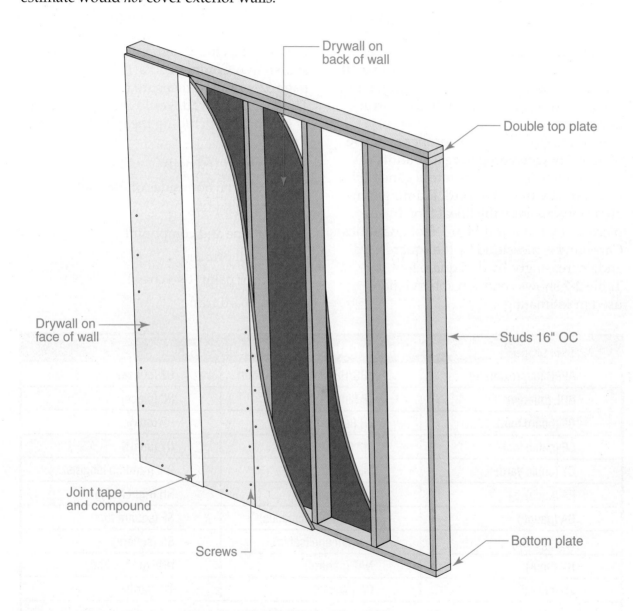

Drywall on back of wall

Double top plate

Drywall on face of wall

Studs 16" OC

Joint tape and compound

Bottom plate

Screws

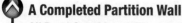
A Completed Partition Wall
All Parts Considered The unit measurement of this wall would be in lineal feet (LF). The unit cost would include every component of the wall from framing to finish work, including painting.

Calculating Board Feet

A **board foot** is a unit of measure that represents a piece of lumber having a flat surface area of 1 sq. ft. and a thickness of 1" nominal size, as shown in A Board Foot. The number of board feet can be found by using simple arithmetic or by referring to **Table 2-3**.

To determine the number of board feet in one or more pieces of lumber, use the following formula:

$$\frac{\text{Number of Pieces} \times \text{Thickness (in.)} \times \text{Width (in.)} \times \text{Length (ft.)}}{12}$$

Example 1: Find the number of board feet in a piece of lumber 2" thick, 10" wide, and 6' long. See Figuring Board Feet.

$$\frac{2 \times 10 \times 6}{12} = 10 \text{ bd. ft.}$$

Example 2: Find the number of board feet in 10 pieces of lumber 2" thick, 10" wide, and 6' long.

$$\frac{10 \times 2 \times 10 \times 6}{2} = 100 \text{ bd. ft.}$$

Table 2-3: Rules for Estimating Board Feet		
Width (in.)	Thickness (in.)	Board Feet
3	1 or less	¼ of the length
4	1 or less	⅓ of the length
6	1 or less	½ of the length
9	1 or less	¾ of the length
12	1 or less	Same as the length
15	1 or less	1¼ of the length

If all three dimensions are expressed in inches, the same formula applies. However, the divisor is changed to 144 (12^2).

Example 3: Find the number of board feet in one piece of lumber 2" thick, 10" wide, and 18" long.

$$\frac{1 \times 2 \times 10 \times 18}{144} = 2\frac{1}{2} \text{ bd. ft.}$$

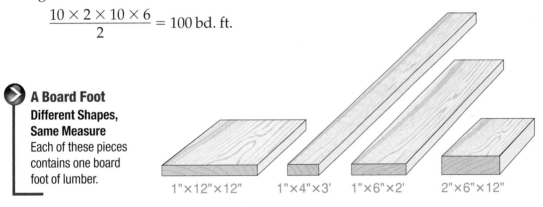

A Board Foot
Different Shapes, Same Measure
Each of these pieces contains one board foot of lumber.

1"×12"×12" 1"×4"×3' 1"×6"×2' 2"×6"×12"

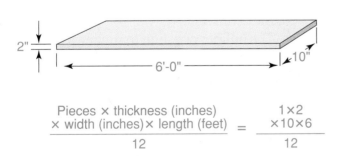

$$\frac{\text{Pieces} \times \text{thickness (inches)} \times \text{width (inches)} \times \text{length (feet)}}{12} = \frac{1 \times 2 \times 10 \times 6}{12}$$

Figuring Board Feet
Calculation Number of pieces × thickness (in.) × width (in.) × length (ft.) ÷ 12 = bd. ft.

Allowances

When an estimate is being prepared for a custom house, some costs may not be known until late in the building process. These costs are usually related to products that the client must choose, such as lighting fixtures, floor coverings, and cabinetry. To account for these items, the estimator includes an *allowance* in the estimate. An allowance is a dollar figure representing the cost of products that have not yet been chosen when a detailed estimate is made. For example, $3,500 might be the builder's allowance for interior and exterior lighting fixtures. If the client later chooses fixtures that cost more, the client must pay the difference.

The advantage of allowances is that the builder can provide an early cost estimate without forcing the client to make difficult product choices. The hazard of allowances is that they can be unrealistically low. The client might be surprised by large expenses late in the process.

Money is usually set aside in a builder's construction budget to cover the costs of unforeseen situations. One example would be an unexpected stretch of wet weather. Such weather might require the builder to use pumps to keep the excavation dry. This allocation of money is called a *contingency* allowance. If the money is not required, the builder either adds it to the profit or refunds it to the client.

Types of Costs

A builder must account for two types of expenses in order to turn a profit. Direct costs, or project costs, are related to a certain house. They include such costs as labor, materials, building permits, temporary power hookups, and some types of insurance. Most direct costs can be determined based on a review of the building plans. The estimating techniques noted throughout this book are those that determine direct costs.

Indirect costs, or overhead, are not related to a particular house. These costs relate to the organization and supervision of the project. They include the cost of office equipment and supplies, construction tools and equipment, office payroll, and taxes.

Some charges may include both direct and indirect costs. For example, the monthly access charge for the phone in a builder's truck would be an indirect cost. The costs of both framing lumber and long-distance calls made to a supplier for lumber used on a project are examples of direct costs.

A percentage for profit must be added to every estimate. This percentage may range from 10 to 40 percent, depending on the job size and the amount and type of work.

Checklists

The heart of an estimate is a checklist that identifies every piece of material used at each stage of construction. The purpose of a checklist is to ensure that nothing is left out of the estimate.

The builder or estimator reviews each item on the checklist and determines which ones apply to the project. For each item, the estimator calculates the dimensions and quantity of each item needed as well as the cost.

Construction-Order Checklist Many builders prefer items on a checklist to be in the same order as the building tasks that need to be done. Thus, the first items on a construction-order checklist relate to excavation, the next

Builder's Tip

CHANGE ORDERS A client may wish to change some aspect of the house after construction has begun. In such a case, the builder and client sign a document that describes the changes and then estimates the cost for the extra work. This document is called a change order. A change order helps the builder keep track of extra costs. It also reduces the chances for misunderstandings between builder and client.

items relate to building the foundation, and so on. One advantage of this approach is that it encourages the estimator and the builder to think logically about how the house will be built.

CSI MasterFormat Checklist The Construction Specifications Institute (CSI) is a professional association that develops standards for writing specifications. CSI has developed a system of organizing specifications for various aspects of commercial construction. This system is called *MasterFormat*. The MasterFormat system organizes all aspects of construction into 16 main categories. Each category consists of many subcategories. Although the system was developed for commercial construction, some residential builders have adapted it for use in developing residential estimating checklists. You can see an example of a MasterFormat checklist on this book's OLC through connectED.mcgraw-hill.com.

Sources for Cost Information

Gathering information about the exact cost of materials and labor can be time-consuming. The estimator must check various sources.

Material Suppliers The estimator can contact local material suppliers to get the cost of materials. Generally this is done after the quantity is determined. This is because a large quantity of flooring, for example, generally costs less per square foot than a small quantity. Several suppliers may then make bids. For example, several suppliers may bid on supplying all the lumber and sheathing for a project. A builder generally accepts the lowest bid for a given quality and quantity.

Prior Bids If the builder has recently completed another house in a similar area using similar materials, some of the cost information can be obtained from the previous estimate.

Pricing Guides Books are available that contain detailed listings of the prices of materials and of the labor to supply them. These books are published annually or semi-annually.

Builder's Tip

TIME-SENSITIVE BIDS When obtaining bids from materials suppliers, note that such bids may be valid for only a certain period of time. After that time expires, the bid price is likely to change. An estimate must take this into account.

Online Databases Increasingly, estimators use the Internet to check current prices. Manufacturers and distributors of materials and products will post prices on their Web sites. This information is usually more current than information in printed pricing guides.

Reading Check

Explain What is the difference between direct costs and indirect costs?

Scheduling

What two elements are scheduled during the building of a home?

Estimating and scheduling are two separate tasks. However, these tasks are often linked because both organize the materials and labor needed to build a house. Scheduling organizes the construction process so the contractor can make the most efficient use of resources. It also enables the builder to determine when the project is ahead of or behind schedule.

Two elements must be scheduled when building a home: materials and activities. It is the responsibility of the general contractor to set up and monitor these schedules. Naturally, the size of the building project affects the complexity of the scheduling.

A contractor who is building only a few houses each year with a small crew often works part-time as part of the crew on the job site. The rest of the contractor's time is spent coordinating the delivery of materials and the work of the subcontractors. Contractors with many projects may spend all their time on these matters or may hire supervisors to do it.

In residential construction, builders may use several different employment strategies. A builder may do any of the following:

- Build entirely with his or her own employees
- Build with a team composed of employees and subcontractors
- Build entirely with subcontractors.

It is important that all work be completed on time so as not to delay overall progress. Especially important is coordinating the work of subcontractors, because they work on different projects and for various builders.

Material Scheduling

Proper scheduling of material deliveries can have a major impact on the efficiency and quality of construction. The builder is responsible for ensuring access to the site. The builder must also make sure that delivered materials can be properly stored. For example, if a large load of oak flooring is delivered before the house is enclosed, it has to be stored outdoors. There, it is exposed to possible weather damage. In addition, the bundles of oak are in the way of workers, which may reduce worker efficiency.

Material scheduling must be coordinated by builders or general contractors. They work with suppliers to ensure that materials will be available for delivery when needed. Materials that are normally kept in stock, such as framing lumber, can usually be ordered on short notice. A non-stock item, such as a custom cabinet or Italian granite of an unusual color, may require a lead time of weeks or months.

Deliveries vary depending on the type and size of the project. They also depend on the number of people working on it and the time set for completion. Generally, material deliveries are made in the following order:

- *First Load:* all items needed to complete the house up to and including the subfloor.
- *Second Load:* wall framing and ceiling joists.
- *Third Load:* roof framing materials and roof coverings. If roof trusses are used, these will be shipped to the site on a special truck. This truck sometimes has a crane to lift trusses into position.
- *Fourth Load:* exterior doors, windows, exterior trim, and siding, as needed. After the house has been enclosed with doors and windows and can be locked up, the interior wall finish is applied. If the walls are plastered, adequate drying time must be allowed before additional material shipments are made.
- *Fifth Load:* hardwood flooring and underlayment materials.
- *Sixth Load:* interior doors, trim, and built-in cabinet materials.

Materials that are to be delivered to the job site are placed in the truck in the sequence in which they are to be used. When the materials are unloaded and stacked at the site, those materials that are needed first will be on top of the pile.

It is the general contractor's responsibility to check the delivered materials against the original order. If materials are damaged or missing, the supplier should be contacted immediately. If this is not done, construction may be delayed. If materials are left over, the supplier may accept them back for credit but may charge a restocking fee.

The supplier keeps a running tally of the materials shipped to the job site, as well as any credits for returns. The general contractor is expected to pay for materials on a certain time schedule, such as every week or month. In the case of special orders, full or partial payment may be required when the order is placed.

Activity Scheduling

The general contractor is responsible for scheduling subcontractors and other labor and keeps things moving smoothly. Careful scheduling can limit delays caused by subcontractors whose work needs to be done before other subcontractors can begin. For example, if a builder does not arrange for foundation subcontractors far enough in advance, the entire project may be delayed until a foundation subcontractor can fit the project into his or her schedule.

Following is a list of general steps in house construction. It is the general contractor's responsibility to see that these steps are carried out. However, sometimes jobs must be started ahead of schedule or delayed. Therefore, the steps may not always occur in this order. Note that required inspections must be scheduled at the appropriate time.

1. *Survey.* The job site is surveyed and the abstract of title (a record of ownership of the property) is brought up-to-date so that application for title insurance can be made. The abstract of title is the history of the ownership of a property.

2. *Permit.* A building permit is obtained from proper authorities so that work can begin.

JOB SAFETY

CALL BEFORE YOU DIG Before any excavation or trenching is done, the local gas, electric, water, and phone utility companies must be contacted. They then mark the location of utility lines. This "call before you dig" precaution helps to prevent the accidental cutting of buried lines that cross the property. Cutting into a power line is a serious safety hazard! Be sure you know where utilities are located before you begin an excavation.

Builder's Tip

SOIL TREATMENTS In areas of the country where termites are a particular problem (see page 35), the soil is sometimes treated with chemicals to prevent infestations. A termite control specialist is sometimes called to treat the soil before the foundation slab is poured. Be sure to coordinate this process with the concrete delivery company.

3. *Excavation.* The excavator brings in power equipment and strips the topsoil away, piling it in one corner of the lot for future use. If the building will have a basement, it is excavated at this time.

4. *Temporary Power.* The electric company must be contacted to set up a temporary power pole on the building site and hook it up. The electricity is needed for operating power tools.

5. *Temporary Water.* On some job sites, the plumber makes the temporary water hookup, which must be coordinated with the city utilities. In existing neighborhoods, water can sometimes be obtained from a neighbor. In this case, the permanent hookup for water to the building is not made until the foundation walls have been installed.

6. *Foundation.* Footings and foundation walls are installed by the foundation subcontractor.

7. *Plumbing.* Pipelines for the water supply are installed in trenches by the plumbing contractor. If the house will be served by a well, it may be drilled at this time.

8. *Slabs.* If the house has a basement, the concrete floor is poured after the rough plumbing is installed and before the interior finish work. The concrete must cure thoroughly. The garage floor is put in anytime after the backfill is completed.

Often this is done at the same time as the basement floor. The concrete is delivered to the site.

9. *Framing.* The carpenters can now frame and sheathe the floors, walls, and roof.

10. *Backfilling.* At some point after the foundation formwork has been removed, the excavation area must be filled in with dirt. This is called backfilling. Before any backfilling can be completed, the exterior walls of the foundation must be moisture-proofed and foundation drainage must be in place. In addition, backfilling should not be done until the floor system is framed or the foundation walls are otherwise braced. This allows time for the concrete to gain strength and reduces the chance that the pressure of backfill will damage the walls.

11. *Mechanicals.* At this point, a number of activities may be carried out at the same time, or at least in rapid succession. These include plumbing, heating, and electrical work. All mechanical subcontractors must work in two stages: rough-in work and finish work. For example, when the framing is complete, the electrician begins to do the rough wiring. This includes installing the main circuit panel and outlet boxes and feeding all the wires through the framing. This is the rough-in portion of the work. Later, after the interior walls are completed, the electrician comes back to install the switches, receptacles, and light fixtures. A plumber installs bathtubs during the rough-in phase, because tubs are a built-in feature of the house.

12. *Windows and Doors.* While the mechanical subcontractors are doing the rough-in work, the carpenters install exterior doors and windows and complete any remaining details of the framing.

Builder's Tip

REDUCING DAMAGE Windows and doors should be installed as soon as possible after they are delivered to the site. This reduces the chance of damage to glass or frames. It also minimizes the number of times doors and windows must be handled. Early installation also keeps these materials out of the way of workers.

13. *Roofing and Siding.* To weatherize and protect the house, contractors install roofing and siding. Generally the roofing is installed first.

14. *Insulation.* After all rough-in work is done, insulation is installed in the walls and, as needed, in the ceiling.

15. *Interior and Exterior Finishes.* Most interiors are finished with drywall or plaster. Plastering must be done immediately. At the same time, carpenters can work on the exterior of the building installing siding, exterior trim, and the garage door. Normally, plaster is applied in two stages. Often a week or ten days must be allowed for drying after each stage before proceeding with other interior work. With drywall, the drying period is much shorter since the only wet application is taping the joints and covering nail heads.

JOB SAFETY

BACKFILL TIMING If backfilling is delayed until framing is completed, workers have to work around a large excavation. This can be unsafe because the workers must carry materials on planks over the excavation. However, if backfilling is done *before* framing begins, the foundation walls must be braced from the inside to prevent them from being caved-in by backfill.

16. *Finish Carpentry.* At this stage the carpenters are ready to do the interior finishing, provided the plaster and concrete are thoroughly dry. Be careful not to store interior trim in a house with high humidity due to wet plaster and concrete, as the wood will absorb the moisture and swell. Later it will dry out and crack. Wood floors may be installed at this stage but are usually finished later. Built-in shelves, interior doors, and cabinets come next. Finally, the interior moldings are applied, including base, shoe, ceiling, window, and door trim.

17. *Exterior Painting.* While carpenters are working on the inside of the house, painters can be finishing the exterior. The ideal arrangement is for the painters to work closely behind the carpenters so that the wood is properly sealed. If exterior trim has been pre-primed at the factory, timing is not as important.

18. *Finish Grading.* While the carpenters are completing the interior of the house, the exterior finish grading is done. This includes preparation work for sidewalks and driveways.

19. *Concrete Driveways and Sidewalks.* After finish grading, at the very last stages of construction, concrete driveways and sidewalks are installed.

20. *Landscaping.* Landscaping is the final step in completing the exterior of the house.

21. *Interior Painting.* After the carpenters have completed the interior of the house, the painting is done.

22. *Floor Coverings.* After the paint is dry, floor tile and resilient flooring are installed.

23. *Finish Electrical.* At this point the electricians can return to add switches, outlets, and light fixtures.

24. *Finish Plumbing.* The plumbing fixtures are now installed by the plumbing contractor.

25. *Wood Flooring.* One of the last jobs on the interior of the house is to finish the wood flooring. Many homes are completely covered by carpeting and require no floor finishing. However, if hardwood floors are used, sanding should be done after the interior painting to remove any paint drops or spillage. The actual finishing is done as one of the last jobs so that traffic does not raise dust while the finish dries. Hardwood flooring can also be purchased prefinished, which greatly simplifies this part of the job.

26. *Carpeting.* After the wood floors are finished, carpeting is laid.

27. *Cleanup.* The general contractor is responsible for the final cleanup. A responsible contractor will make sure that the windows are washed and all waste materials are removed.

28. *Punch List.* After the entire house has been completed, the general contractor or builder walks through the house with the new owner. This is a chance for the owner to make sure everything has been done to his or her satisfaction. Often, the owner will spot such things as scuffed paint, cracked woodwork, or light fixtures that do not work properly. The contractor then makes a punch list. This list identifies all the repairs that must be completed before the house is acceptable to the owner.

Builder's Tip

MAINTAIN GOOD COMMUNICATION General contractors must stay in contact with subcontractors as they work on the house. However, it is also important to maintain contact with subcontractors well ahead of the time when their work is to begin. This will often give the general contractor advance warning of any delays that could affect the building schedule.

Staying on Schedule

A complex project may involve hundreds of individual tasks. Keeping track of scheduled jobs is the key to successful construction management. Many builders use graphic methods to do this, including bar charts and diagrams. Computer software makes it fairly easy to set these up and revise them as needed.

Bar Charts A bar chart is an easy way to keep track of a project. It shows how long each task will take and when each task will start and end. A calendar format displays the entire job over time. A simple bar chart such as the one below can be used to track small jobs, such as the addition of a bedroom to a house. Expanded versions of this type of bar chart can track the construction of an entire house.

The value of a bar chart is in its ability to show an overview of the entire project. It also shows how various tasks overlap. However, its simplicity limits its usefulness. A bar chart cannot show every complex relationship between various parts of the project, as shown in the construction sequence of Building a House on page 71. That is why it is used primarily as a general planning tool.

Location _____ Lot# _____
Model _____ Start _____ Finish _____

Calendar Days
Working Days: 1 2 3 4 5 6 7 8 9 10 11 12 13 14 15 16 17 18 19 20 21 22 23 24 25 26 27 28 29 30 31 32 33 34 35 36 37 38 39 40 41 42

ACTIVITY

- Permits
- Stakeout
- Excavation
- Footing
- Foundation
- Sewer lines
- Water lines
- Framing
- Roof framing/sheathing
- Roofing
- Windows/doors/stairs
- Rough HVAC
- Rough electrical
- Rough plumbing
- Insulation
- Brickwork
- Exterior trim
- Siding/gutters
- Exterior painting
- Backfill/grading
- Sidewalks/driveway
- Landscaping
- Drywall & finish
- Finish HVAC
- Finish electrical
- Finish plumbing
- Interior trim
- Interior painting
- Resilient flooring
- Carpeting
- Touch-up
- Housecleaning

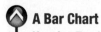

A Bar Chart

Keeping Track of Many Tasks A bar chart shows how long tasks take and presents them in chronological order.

Building a House

From Start to Finish A construction sequence shows how the various aspects of a project occur in relation to each other.

 Footings and Basement Walls

- The basement is excavated and then the footings are placed.
- The foundation walls are either poured or constructed of concrete block. The fill around the basement is omitted here to show footings and walls.
- The supporting columns and the center beam are installed.
- The exterior surface of the walls is moisture-proofed up to the finish grade level.
- The front porch excavation may be filled with sand and is ready for the porch and basement floors to be placed.

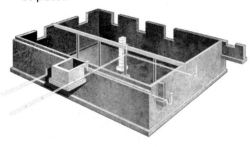

 First Floor Construction

- The rough grading is shown leveled off four inches below the finished grade line.
- The joists are installed. There are double joists at the stairwell and under inside partition walls.
- The double joists are separated with solid bridging. Metal or wood bridging is installed midway between the ends of other joists and the supporting beam.
- The plywood subfloor is laid.

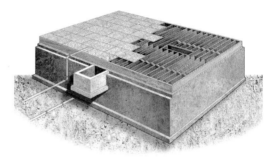

 Wall Framing Wall framing is completed. If the house has a second story, floor joists and additional wall framing will be added on top of the walls of the first story.

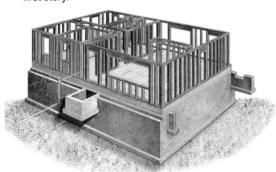

D Complete Framing Cutaway view of the house showing some finished plumbing and heating. The actual placement of the fixtures and appliances would not be done until the house has been closed in.

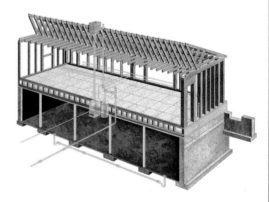

E Rendering of Finished Home The exterior of the home as it will look when it is completed.

Critical Path Method Diagrams The *Critical Path Method (CPM)* of scheduling shows the relationships among tasks as well as how long they take. It is the most common type of scheduling used in residential construction. The relationships are shown in a CPM diagram.

A CPM diagram's value is in identifying the tasks that are most important to the success of the project. Taken together, these tasks indicate the minimum amount of time needed. In other words, critical path tasks such as those shown in A CPM Diagram are those in which any delay will automatically delay completion of the entire project.

The CPM diagram was originally developed to keep track of maintenance work in an oil and chemical refinery. A similar diagram, called Program Evaluation and Review Technique (PERT), was developed about the same time to track the construction of nuclear submarines and is used by some builders.

To develop a CPM schedule, list all the work that has to be done. The list of tasks under "Activity Scheduling" in this chapter is an example of a list of all of the work that must be done to complete a project. The following three questions should be answered for each task:

1. What tasks come before this one? All tasks have a logical order in which they are performed. For example, the drywall must be taped before it can be painted.

2. What tasks cannot start until this one is complete? For example, rough plumbing cannot be installed until the framing is in place. Wall framing for a second story cannot be built until the first floor walls are in place.

3. What tasks can be worked on at the same time? Building jobs move more quickly when various tradespeople are working at the same time. For example, electricians can be working inside the house while other tasks are taking place outside the house.

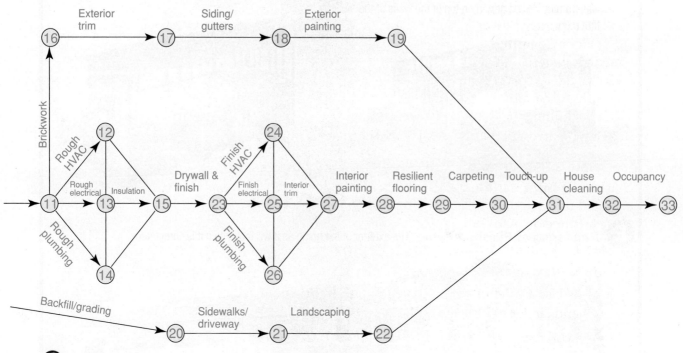

A CPM Diagram

Making Connections Obvious A CPM diagram arranges tasks based on their relationship to one another. The timeline would be placed at the bottom of the diagram and would depend on the schedule for the project.

After these questions have been answered, the tasks can be plotted on a CPM diagram. The arrows on the diagram indicate tasks. The tail of the arrow represents the start of a task, and the head represents the end. Boxes or circles, called *nodes,* represent events that can are key achievements in the project, such as "Roof trusses plumbed and braced." Arrows that follow one another along the same path indicate tasks that must end before the next task can begin. Parallel arrows represent tasks that go on at the same time. Arrows can be straight or curved.

Builder's Tip

PROJECT MANAGEMENT Project management software, which includes scheduling functions, makes schedule updates easy. If computers are *networked,* or linked together, workers can update their part of the project and inform others of changes simultaneously. Programs include graphic functions that can create charts and schedules that display the information visually.

Section **2.4** Assessment

After You Read: Self-Check

1. Why is it important that cost estimates be as accurate as possible?
2. Name two sources of information that can help a builder develop an estimating checklist.
3. Arrange the following tasks in chronological order, starting with the task that occurs first:
   A. Interior Painting
   B. Temporary Power
   C. Floor Coverings
   D. Insulation
   E. Finish Carpentry
   F. Floor Framing
   G. Backfilling
4. What do arrows and nodes represent on a CPM diagram?

Academic Integration: English Language Arts

5. **Manage Information** Create a Critical Path Method (CPM) diagram for the first 25 steps noted under "Activity Scheduling" on pages 67–69. Assume that each of the steps listed takes three days to complete.

   Step 1: Begin by identifying the steps that form the critical path.

   Step 2: Next, identify the steps that may proceed at the same time as the critical tasks.

   Step 3: Use a ruler or computer software to create your graph. Use arrows to indicate tasks and circles to indicate nodes.

   When you complete your CPM diagram, compare it with another student's CPM diagram and discuss the differences in class.

Go to **connectED.mcgraw-hill.com** to check your answers.

Review and Assessment

Section
2.1

Section
2.2

Section
2.3

Section
2.4

Chapter Summary

Building codes establish minimum standards of quality and safety in housing. A building department can develop its own codes or adopt model codes developed elsewhere. Financing can be obtained in the form of a construction loan and a mortgage.

Construction plans consist of drawings of each part of the structure as well as its measurements. Objects on plans are generally drawn to scale with an architect's scale. The elements of a drawing include lines, dimensions, symbols, and notes. Computer-aided drafting and design enables architects to create plans and other drawings electronically.

Plan views include site plans, foundation plans, floor plans, framing plans, electrical plans, and mechanical plans. Elevations, section views, schedules, engineering plans, and specifications give information that may not appear in plan views. Renderings show the finished building.

Accurate estimates of materials and costs help builders ensure a profit. Types of estimates include the pre-design estimate, the quantity takeoff, and the unit-cost estimate. A schedule helps builders keep track of project materials and activities.

Review Content Vocabulary and Academic Vocabulary

1. Use each of these content vocabulary and academic vocabulary terms in a sentence or diagram.

Content Vocabulary

- building code (p. 34)
- building permit (p. 35)
- stock plan (p. 37)
- floor plan (p. 37)
- mortgage (p. 40)
- architect's scale (p. 44)
- plan view (p. 50)
- elevation (p. 53)
- schedule (p. 56)
- specifications (p. 57)
- bid (p. 59)
- quantity takeoff (p. 60)
- board foot (p. 63)
- indirect cost (p. 64)

Academic Vocabulary

- exceeds (p. 34)
- scale (p. 44)
- derived (p. 46)
- allocation (p. 64)

Speak Like a Pro

Technical Terms

2. Work with a classmate to define the following terms used in the chapter: *model building code* (p. 34), *blueprints* (p. 41), *cutting plane* (p. 51), *detail drawing* (p. 55), *rendering* (p. 56), *change order* (p. 64), *punch list* (p. 69), *Critical Path Method (CPM)* (p. 72), *nodes* (p. 73).

Review Key Concepts

3. Identify the steps of planning to build a house.

4. Compare three sources of house plans.

5. Summarize how to obtain financing for construction.

6. Describe three elements used in architectural drawings and plans.

7. Restate the advantages of computer-aided drafting and design.

8. Define direct cost and indirect cost.

Critical Thinking

9. Analyze Describe the different types of cost estimates and how they differ from one another.

Academic and Workplace Applications

STEM Mathematics

10. Finding Volume A set of architectural plans shows an enclosed storage room. It will be 9' long × 9' wide × 8' high. The builders need to know the volume of the room so that they can install a dehumidifier to keep it dry. What is the volume of the storage room?

Math Concept The formula for finding the volume of a regular rectangular prism (a three-dimensional shape) is

$$V = lwh$$

where *V* stands for volume, *l* for length, *w* for width, and *h* for height. To solve this problem, insert the given length, width, and height into the formula. Solve for *V*. Use cubic units to express the volume of the room.

21st Century Skills

11. Economic Literacy: Purchasing Property The process for purchasing property differs from region to region. In some states, the attorney for the seller develops a contract of sale, and the attorney for the buyer does a title search through local records to make sure the title is free of any other claims on it. In other states, a realtor draws up a standardized contract and the title search is done by a title company.

A title company specializes in tracing the ownership of property through legal documents. Investigate the purchasing property process in your state. Summarize your findings in a two-paragraph report.

21st Century Skills

12. Financial Literarcy The total cost of a house is $450,000. The length of the mortgage is 30 years. What is the monthly payment?

Starting Hint This problem has many steps. First, multiply the number of years in the mortgage (30) by the number of months in a year (12) to calculate the total number of monthly payments needed. Then divide the total cost of the house ($450,000) by the total number of payments to calculate the average monthly payment.

Standardized TEST Practice

Multiple Choice

Directions Read the following questions. Then read the answer choices and choose the best answer.

13. Which of the following is NOT a legal document that must be processed to purchase a house?

   a. official survey
   b. deed
   c. elevation plan
   d. abstract of title

14. What is the ratio between the size of the object as drawn and its actual size called?

   a. customary rule
   b. scale
   c. dimension
   d. elevation

15. What are the two elements that must be scheduled when building a home?

   a. construction-order checklist and CSI MasterFormat checklist
   b. estimating and scheduling
   c. materials and activities
   d. prior bids and pricing guides

TEST-TAKING TIP

If you have time at the end of the test, reread the questions to make sure you have understood them.

*These questions will help you practice for national certification assessment.

CHAPTER 3

Construction Safety & Health

Section 3.1
Job Site Safety

Section 3.2
Personal Safety & Health

Chapter Objectives

After completing this chapter, you will be able to:

- **Describe** the purpose of OSHA.
- **Identify** practices for keeping a work site safe.
- **Apply** your knowledge of construction site safety to react appropriately in an emergency.
- **Identify** hazards on the job site.
- **Identify** the hazards associated with different types of tools.
- **Describe** protective clothing and personal protective equipment that suit various weather and job site conditions.

Jupiterimages/Getty Images

Discuss the Photo

Safety First The construction industry has the highest accident rate of any industry. *List at least two reasons why you think this might be.*

Writing Activity: Create a List

Create a list of five safety rules and reminders you think might be important for maintaining a safe work environment on a job site. Revise your list after you have read this chapter.

Before You Read Preview

In construction, working conditions vary widely and can be difficult to control. Workers and supervisors must take care to guard their own safety as well as the safety of others. Choose a content vocabulary or academic vocabulary word that is new to you. When you find it in the text, write down the definition.

Content Vocabulary

- Occupational Safety and Health Administration (OSHA)
- excavation
- first aid
- conductor
- electrical circuit
- grounding
- musculoskeletal disorder (MSD)
- ergonomics
- repetitive stress injury (RSI)
- wind chill

Academic Vocabulary

You will find these words in your reading and on your tests. Use the academic vocabulary glossary to look up their definitions if necessary.

- physical
- chemicals
- components

Graphic Organizer

As you read, use a chart like the one shown to organize key ideas about construction safety and health.

How to Keep the Work Site Safe	How to Work Safely on the Job

Go to **connectED.mcgraw-hill.com** to download this graphic organizer.

Job Site Safety

Safety Regulations

What is OSHA?

Most injuries can be prevented. Everyone on a job site has some responsibility for keeping the site safe. The best way to avoid accidents is to develop a safe attitude:

- Think about the consequences of your actions, not only for yourself, but also for those around you.

- Always follow safe procedures, obey the rules, and act responsibly.

- Keep in mind that you cannot rely on luck to protect you from accidents.

- Make safety a habit as you develop your construction skills.

- Learn good housekeeping habits because a clean job site is a safer job site.

Many builders and contractors feel that there is a direct relationship between quality work practices and safety. Workers who take the time to do a job carefully are safer workers. One reason for this is that it takes a lot of concentration to do a job right. A worker who concentrates is less likely to be distracted by other workers and is more alert to possible hazards. Another reason that careful workers are safer workers is that a high degree of craftsmanship slows the pace of construction. Injuries are more likely to occur when workers rush to complete a job.

OSHA

The purpose of the U.S. Occupational Safety and Health Act is "to assure so far as possible every working man and woman in the Nation safe and healthful working

Supervising Safety
Making Inspections OSHA Inspectors may check construction sites for safety violations.

Juice Images/Glow Images

conditions and to preserve our human resources." The Act is administered by the **Occupational Safety and Health Administration (OSHA)**, an organization that issues standards and rules for safe and healthful working conditions, tools, equipment, facilities, and processes. OSHA also conducts workplace *inspections* to ensure the standards are followed. It enforces those standards and assesses fines against violators.

Everyone has the right to a safe workplace. Under OSHA regulations, an employer must:

- Make sure the workplace is free from recognized hazards that are likely to cause death or serious physical harm.
- Provide proper training for all hazardous work and the use of personal protection equipment.
- Comply with OSHA standards, including record-keeping standards and accident reporting standards.

If an employee believes that OSHA standards are not being followed, he or she has the right to contact OSHA and request an inspection. Employees also have responsibilities. They must:

- Comply with OSHA regulations.
- Comply with other occupational safety and health standards that apply.

Workers Compensation

Even after every effort has been made to ensure a safe workplace, an employee may be injured. Each state has a type of insurance program that pays benefits for work-related injuries and illnesses. These programs are called workers compensation programs. The programs pay for reasonable and necessary medical care if an employee is injured or becomes ill on the job. If an employee is killed on the job, benefits go to his or her family.

Programs in each state vary, but they are designed to handle claims in a prompt and fair way. Workers compensation laws are complex, but they generally include the following information:

- The employee must be given prompt first aid, as well as necessary medical, surgical, rehabilitative, and hospital care.
- If a work-related injury prevents an employee from working for a certain number of days, he or she must be paid disability benefits during that period.
- If an employer denies compensation to a worker, the worker may appeal to state authorities, usually through the workers compensation program.

If you are injured at work, report your injury to your employer as soon as possible. Afterward, your employer must report the injury to the workers compensation insurance carrier. You should keep receipts for all expenses associated with your accident. You should also maintain records related to your care and recovery.

Reading Check

Recall *What is workers compensation?*

Keeping the Job Site Safe
What is good sanitation?

Younger workers are more likely to be injured than older workers. This might be partly due to their inexperience. In addition, new employees have a higher accident rate than long-time employees. This could be because they are unfamiliar with the job. Most apprentices are both young *and* new on the job.

Housekeeping and Sanitation

A hammer left balanced on a ceiling joist, a bucket left on the stairs, and oily rags thrown in a corner are instances of small acts of carelessness that can result in major injuries or property damage. Always practice good housekeeping on the job.

- Keep walkways clear of tools, materials, and clutter.
- Whenever you see protruding nails, remove them or bend them down.

- To prevent fires and reduce hazards, dispose of scraps and rubbish daily. Put oily rags and other highly flammable (able to catch fire easily) waste in approved metal containers.
- When working above other people, place tools and materials where they will not fall and cause injuries.

Good sanitation (cleanliness) helps prevent the spread of disease. Various OSHA and local regulations apply to sanitation. They cover such things as the supply of drinking water, food service and eating facilities, washing facilities, and toilets.

Signs, Tags, and Barricades

Several kinds of signs and tags are used at construction sites. Danger signs warn people about immediate hazards, such as open stairwells. Caution signs, such as those in Safety Signs, warn about potential hazards or unsafe practices. For example, a caution sign might be used to warn workers entering an area where laser equipment is in use.

If a tool or piece of equipment is defective, a temporary tag should be placed on it to warn people that it should not be used. The tags shown in Safety Tags are examples. A

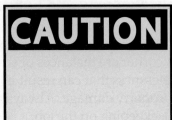

 Safety Signs
Identify Danger Learn to recognize safety signs and colors. The color of a sign relates to its meaning.

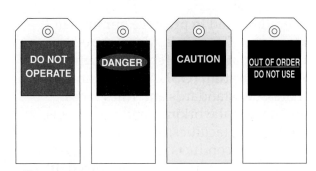

 Safety Tags
Identify Danger Tags are placed on tools to warn of faulty operation or maintenance issues. They should be connected with light wire or some other material that is not likely to come loose.

lockout/tagout procedure may be required. *Lockout/tagout* is the use of lockout devices and/or tags to prevent accidental machine startup or release of stored energy.

Barricades, or barriers, can be set up to prevent people from entering a dangerous area. This is often done to guard the outside edges of holes dug for foundations. It is also done to prevent workers from falling off the edge of flat roofs.

The colors used on signs, tags, and barricades have specific meanings. For example, red means "danger" and yellow means "caution." See **Table 3-1** for an explanation of the safety color codes.

Never use yellow caution tape to identify a dangerous area because yellow means "caution, proceed with care." Always use red tape to mark a dangerous area because red means "DANGER KEEP OUT."

Table 3-1: Color Safety Codes	
Color	**Meaning**
Red	Danger or emergency
Orange	Be on guard
Yellow	Caution
White	Storage or boundaries
Green	First aid
Blue	Information

Fire Prevention and Equipment

Flammable materials are often found at construction sites. These materials include wood, fuel for generators and other equipment, and paint and solvents. Weeds, grass, and *debris* (scraps) around the structure can also catch on fire. To prevent fires:

- Dispose of scraps and rubbish daily.

- Keep the area cleared of weeds and grass.

- Make sure electrical wiring and equipment are properly installed and working correctly.

- Maintain clearance around lights and heaters so that they do not set materials on fire.

- Store materials properly. For example, flammable liquids must be stored in approved, closed containers.

- If a flammable liquid spills or leaks, clean it up promptly and safely.

- Never smoke on the job site, particularly near flammable materials.

- Do not block exits with materials, equipment, or debris. In case of fire, people need to be able to exit quickly.

Construction sites are required to have firefighting equipment on hand. Some fires can be put out with water, so there is usually a water supply available as well as fire pails or hoses. Fire extinguishers are also needed, especially for fires that cannot be put out with water. The four types of fire extinguishers are show below. Note that three of the types are suitable for more than one kind of fire. Class D extinguishers are suited only to one special type of fire.

Using a Fire Extinguisher Various fire extinguishers might be found on a job site, including models that rely on dry **chemicals**, Halon, water, or carbon dioxide. However, the basic method for using any of these extinguishers is

Types of Fire Extinguishers		
Class of Fire	**Type of Flammable Material**	**Type of Fire Extinguisher to Use**
Class A	Wood, paper, cloth, plastic	Class A Class A:B
Class B	Grease, oil, chemicals	Class A:B Class A:B:C
Class C	Electrical cords, switches, wiring	Class A:C Class B:C
Class D	Combustible metals	Class D

 Types of Fire Extinguishers
Know Your Icons Newer fire extinguishers have pictorial icons that explain what types of fire they should be used on. A diagonal red slash through the icon means that the extinguisher should not be used for this type of fire. *Which class of fire is associated with combustible metals?*

similar. Remember this acronym: **PASS** (Pull, **A**im, **S**queeze, **S**weep).

- P = **Pull** the pin at the top of the extinguisher.
- A = **Aim** the nozzle toward the base of the fire. Stand about 8' from the fire.
- S = **Squeeze** the handle to discharge the extinguisher.
- S = **Sweep** the nozzle back and forth, aiming at the base of the fire.

Reading Check

Define *What does the acronym PASS stand for?*

Excavation Safety

An **excavation** is a cut, cavity, trench, or depression made by removing earth. Excavations are dug to prepare the site for footings and foundations. They are also required when installing pipes for site drainage. Workers near an excavation must be careful not to fall in. For those who are working within the excavation, the dangers include cave-ins and the possibility of equipment falling into the excavation. Excavations should be inspected daily for signs of soil movement or other problems. If an excavation is deeper than 5 feet, OSHA requires that workers be protected from cave-ins by shoring (reinforcing the walls), trench boxes, or by shaping the sides of the excavation to

minimize the hazard. Two basic ways to shape an excavation are called simple slope and benched slope, shown in Types of Excavations.

In benching, the soil is excavated to form one or more horizontal levels, or steps. The surfaces between levels are vertical or nearly vertical. Simple slope or benched slope excavations deeper than 20' must be designed by a registered engineer.

- Before starting an excavation, local utilities must be contacted in order to determine the location of existing underground utility lines. *Always call before you dig!*

- To minimize the danger of cave-ins, the sides of the excavation should not be too steep. The proper slope depends on the soil type.

- The soil that has been removed must be piled at least 2' away from the edge of the excavation. Any equipment must also be kept at least 2' from the edge to minimize the danger of falling in.

- Workers should not stand or work directly underneath the loads being handled by digging or lifting equipment.

- There must be a means for workers to get out of the excavation, such as a ladder or a ramp. The ladder or ramp should be no more than 25' away from any worker.

- Standing water must be pumped out of the excavation.

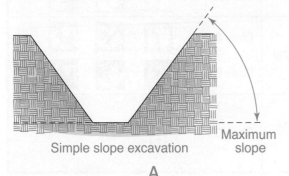

Simple slope excavation — Maximum slope

A

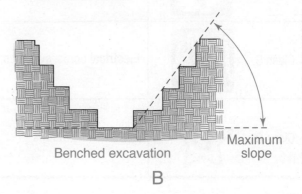

Benched excavation — Maximum slope

B

 Types of Excavations
The Right Slope The maximum safe slope for the sides of an excavation is determined by the soil type. **A.** A simple slope excavation is quite common on residential job sites. **B.** A benched slope excavation may also be suitable. *Why is it important to shore up the walls of a trench that is deeper than 5 feet?*

After You Read: Self-Check

1. What is the purpose of the Occupational Safety and Health Act?
2. What are an employer's basic responsibilities under OSHA?
3. What is the purpose of a barricade?
4. Why is it important to keep exits clear of materials, equipment, and debris?

Academic Integration: Mathematics

5. **Finding Perimeter** A 25' by 40' rectangular hole has been excavated for the foundation of a new building. An additional 6' of work area around all sides of the hole has been marked off. What is the perimeter of the work area?

   Math Concept The perimeter is the distance around a geometric figure. The perimeter can be calculated using the following formula:

   $$\text{Perimeter} = (\text{length} + \text{width}) \times 2, \text{ or } P = 2(l + w)$$

   Step 1: Draw a picture of the excavation site. Indicate the measurements of the hole.

   Step 2: Extend the length and width of the hole by 6 feet on both sides. Use the formula to find the perimeter.

Go to **connectED.mcgraw-hill.com** to check your answers.

Section 3.2 Personal Safety & Health

Responding to Emergencies

Who should administer first aid?

First aid is the initial help and care given to an injured person on the job site. If a construction site is not close to a hospital or other source of medical help, OSHA requires that a person who is certified to give first aid be at the job site. A first-aid kit must also be available on the job site.

Even if you are not certified to give first aid, you should always respond to an accident or injury. Make sure that you:

- Report all injuries immediately to your instructor or supervisor.
- Know how to summon help. Your instructor or supervisor can give you this information. Many workplaces require this information to be posted.
- Know the location of first-aid equipment. Even if you are not giving first aid, you may be asked to bring the materials to someone else.
- Do not give first aid unless you have been trained to do so. Untrained people can make an injury worse or endanger themselves.

- Use all suitable safety equipment.
- Get into the habit of scanning the work area for potential problems. Take immediate steps to prevent any problem from causing injury. For example, spilled liquid on a smooth surface or a loose railing could cause a person to slip or lose his or her balance.
- Always observe the entire area surrounding the injured person. Identify any life-threatening conditions, such as downed power lines or chemical spills, and react accordingly. For example, a person attempting to give aid to a person who has been electrocuted by a live wire could become a victim also if the injured person is still in contact with a live wire. Fumes from a toxic chemical spill could affect your breathing and eyesight.

Hazards on the Job Site

Why are unprotected openings dangerous?

Falls are the most common cause of major injury on a construction site. Other common injuries include cuts, puncture wounds, hearing damage, and injuries from falling objects or contact with toxic (poisonous) substances.

Falls and Falling Materials

Because falling is the most common cause of injury on a construction site, stay alert for possible hazards. Always wear proper footwear, which will help you to maintain traction. Always walk; do not run. Anyone who works at a height, such as carpenters, masons, roofers, and painters, should take extra care. Safety harnesses may be required. OSHA standards on fall protection require that any person working at a height above six feet must be protected by guardrails or wear fall protection and be properly secured.

Stairwells must have a safety railing around them at all times. Unprotected or poorly protected stair openings can lead to serious falls. However, not every railing

JOB SAFETY

BASICS OF FIRST AID First aid is the initial help and care given to an injured person on the job site.

- Always perform an initial assessment of the ABCs: **A**irway, **B**reathing, and **C**irculation.
- First aid providers must recognize the signs and symptoms of heart attack and stroke and perform CPR accordingly.
- Controlling bleeding is the first priority in treating an open wound.
- To perform first aid safely, you must recognize the signs and symptoms of job site injuries and emergencies.

is safe, as shown in A Stairway Hazard on page 86. When constructing a safety railing, try to imagine how it would perform if someone carrying tools suddenly slipped on a step. Would the railing prevent the person from toppling over the edge?

All connections in a railing system must be strong enough to withstand sudden stresses. Screwed connections are generally better than nailed connections. However, be sure the screws are suitable for the work. For example, drywall screws may snap when subjected to sudden shear forces. Temporary openings in a roof, such as skylight framing, should be covered with a thick sheet of plywood when not being worked on. The plywood should be nailed or screwed in place to prevent it from slipping. Covers over openings must be labeled using red lettering to signify danger to ensure that they are not removed accidentally or left in place longer than necessary. Roof jacks or other scaffolding should be used whenever necessary to prevent falls (see Chapter 7, "Ladders, Scaffolds, & Other Support").

Many building materials become slippery when wet or frosty. This is particularly true of plastic, plywood, and housewraps. Keep

A Stairway Hazard

Danger Zone The safety rail at the top of this opening should extend to the end of the opening and should have a full mid-rail. The handrail on the stairway should also include a mid-rail. *What portions of this railing are the weakest?*

workplace floors as clean and dry as possible to prevent accidental slips.

Be sure to dispose of debris properly. Never step on any material that is not nailed down.

Whenever construction debris must be dropped from a height of 20' or more to any point outside the building, an enclosed chute must be used. This reduces the danger of falling objects to workers below. When debris is dropped through a hole in the floor without the use of chutes, the spot where it lands must be completely enclosed with barricades.

Reading Check

Explain *What is one way to prevent falling when working on a roof?*

Electrical Hazards

Electric tools are common on the job site. Electrical safety is important because even a small jolt can injure or kill a worker. A

BLOOD-BORNE PATHOGENS If you provide first aid to an injury on the job site, you may encounter blood. Blood and other body fluids can contain microorganisms (viruses and bacteria) that are hazardous to your health. Take precautions to avoid direct contact with blood and other body fluids, and be especially careful to protect your eyes, mouth, and nose from these fluids.

conductor is a material that electricity readily flows through. Conductors include most metals. Materials that do not conduct electricity readily are called insulators. Rubber is a good insulator. However, moisture can cause some materials to conduct electricity, even if they are poor conductors when dry. This is particularly true of wood.

When electricity flows from a point of origin and returns to that point of origin through a conductor, it makes an **electrical circuit**. Most conductors used in circuits are metal wires insulated with a plastic or rubber casing. The insulation protects the user from the electricity. However, if the insulation is frayed or broken, the conductor is exposed. A person can become part of a circuit by touching both wires in the circuit or by touching the "hot" wire and a ground wire. (The hot wire is the one bringing electrical power to a device.) A person can also become part of the circuit if the tool being used comes in contact with a wire carrying electricity. This could be a hand tool, such as a screwdriver, or a power tool. When the body becomes part of the circuit, the electricity that flows through it can cause serious burns, injury, or even death.

For safety, all power tools must be grounded. **Grounding** provides a path for the electricity to flow safely from the tool to the earth. If a person accidentally becomes part of the circuit, electricity will flow through the ground wire, not through the person using the tool. Any break in the grounding system makes it useless, so it is very important to keep the system working properly.

Electrical Tool Safety Electric tools typically rely on one or more of these safety systems:

- A three-wire cord with a grounding prong. The grounding prong must never be removed from the plug. When an adapter is used on a two-hole receptacle, the adapter wire must be attached to a known ground.

- A two-wire cord and an insulating plastic shell around the tool's electrical **components**. The tool is then referred to as a double-insulated tool.

- A Ground-Fault Circuit Interrupter (GFCI). A GFCI is a fast-acting circuit breaker that can protect people from electrical shock. One version of this device is shown in A Portable GFCI. A GFCI can be attached to the cord supplying electricity to a power tool. It is particularly important whenever moisture is present. Employers are required to provide GFCIs for all temporary 120-volt, 15- and 20-ampere receptacle outlets on a construction site. A GFCI should also protect any permanent wiring used during the construction process. OSHA requires that GFCIs be used in addition to, not as a substitute for, grounding devices on tools.

GFCI protection is required by OSHA for any use of electrical tools on a job site. This can be accomplished either at the source by using a GFCI breaker in the electrical panel, using a GFCI outlet, or adding a GFCI pigtail (power cord adaptor) between the power chord and the receptacle.

Preventing Tool Injuries

Using hand and power tools can create many hazards. However, keeping five basic safety rules in mind can help to prevent many injuries:

1. Use the proper personal protective equipment, including protective eyewear, in the correct way.

A Portable GFCI
Safety Precaution A ground-fault circuit interruptor (GFCI) attached to an extension cord protects workers against electrical shock. *In what kind of weather conditions would a GFCI be particularly important?*

2. Use the right tool for the job.
3. Examine each tool for damage before use. Do not use damaged tools.
4. Operate tools according to the manufacturer's instructions.
5. Keep all tools in good condition by performing regular maintenance.

General Power Tool Safety Here are some basic safety guidelines to follow when you are using power tools:

- Avoid damaging cords or hoses. Never carry a tool by the cord or hose. Never yank the cord to disconnect it from a receptacle.

JOB SAFETY

POWER TOOLS This section includes general safety rules for power tools. We strongly advise you to check the manufacturer's manual for each tool for any special safety instructions. In addition, follow the suggestions for power tools given in other chapters of this book.

- Avoid accidental starts. Do not hold your fingers on the on/off switch while carrying a plugged-in tool. In addition, be sure the tool's switch is in the "off" position before plugging the tool in.

- Unplug or disable tools when not using them, before servicing or cleaning them, and when changing or adjusting accessories, such as blades, bits, and cutters.

- Always check lumber for knots, splits, nails, and other defects before machining it. Defects can cause a tool to move unpredictably.

- Use guards on power equipment and be sure they are installed correctly. Standard safety guards must *never* be removed when using a portable power tool.

- Keep your fingers away from the cutting edges of tools. For example, do not try to hold small stock while it is being cut, shaped, or drilled. Secure the stock with clamps or in a vise whenever possible. This frees both hands to operate the tool.

- Pay attention to the job. Always keep your eyes focused on where the cutting action is taking place. Do not talk with others while you are working with power tools. Never talk to or interrupt anyone else who is using a power tool.

- Always use a brush, not your hand, to clean sawdust away from a power tool.

- Be sure to keep good footing and maintain good balance when operating power tools.

- Always use the recommended extension cord size when using portable power tools. **Table 3-2** on page 89 provides recommended cord sizes.

- Loose clothing or jewelry can become caught in moving parts. Wear proper apparel for the task.

- Hair can become caught in moving parts. Tie back long hair.

- Keep work areas well lighted.

- Take steps to ensure that power cords will not create a tripping hazard.

- Never leave tools or materials on any piece of equipment while it is in use. This is especially important with table saws.

- When finished with a power tool, wait until the blade or cutter has come to a complete stop before walking away. Make all adjustments with the power off and the machine at a dead stop.

- Report strange noises or faulty operation of machines to your instructor or supervisor.

- Remove all damaged tools from use and tag them: "Do not use."

- Store electric tools in a dry place when not in use. Do not use them in damp or wet locations unless they are approved for those conditions.

Power Saws Kickback occurs when the stock being cut or the saw itself is thrown back at the operator at high speed. Because so many workers encounter kickback, the subject is discussed in detail in Chapter 5, "Power Saws." Always keep your hands clear of the cutting line. Never make adjustments while the saw is running. Disconnect the power source before changing a blade.

Always use a push stick to guide any piece of wood through the saw if it is too small to guide safely by hand. Also, never use a saw that does not have a working guard and anti-kickback device. A piece of wood is too small if your hand or fingers are in danger of coming in contact with the saw blade. Never wedge the guard or anti-kickback device out of the way.

Reading Check

Explain *Why is it important to keep your hands clear of the cutting line when operating a power saw?*

Builder's Tip

Portable Tools With Abrasive Wheels Portable abrasive grinding, cutting, polishing, and wire buffing wheels may throw off flying fragments. They must be equipped with suitable guards.

- Never stand directly in front of the wheel as it accelerates to full speed. An abrasive wheel may disintegrate during start-up.
- Ensure that the operating speed of the tool does not exceed the maximum speed marked on the wheel.
- Allow the tool to reach operating speed before you use it.
- Never clamp a hand-held grinder in a vise.

Gasoline-Powered Tools Various construction tools are sometimes fueled with gasoline. These include air compressors, chain saws, and portable concrete-cutters. Fuel vapors can burn or explode, and engines create dangerous exhaust fumes.

- Transport and store fuel only in approved containers for flammable liquids.
- Shut down an engine and allow it to cool before filling its fuel tank. This helps prevent vapors from igniting.
- Never fill the fuel tank of a piece of equipment in an enclosed area. Always clean up any spilled fuel immediately to prevent a sudden fire.
- Provide suitable ventilation when the tool is used in an enclosed area.

This lessens your exposure to carbon monoxide.

- Keep a fire extinguisher nearby.

Powder-Actuated Tools Powder-actuated tools are used to install various types of fasteners into steel, concrete, and masonry. They operate like a firearm and use gunpowder to propel the fastener into the material with great speed.

These tools must be treated with extreme caution. They should be operated only by employees certified in their proper use. OSHA requires that any person operating powder-actuated tools be trained and licensed. Different tools from different manufacturers require specific training for each. Workers using powder-actuated tools should do the following:

- Wear suitable ear and eye protection.
- Select a powder level that is right for the tool and that is able to do the work without excessive force.

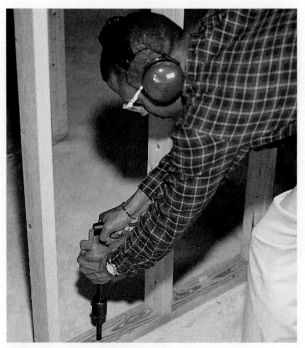

Powder-Actuated Tools
Proper Protection Powder-actuated tools are powerful. Hearing and eye protection are essential when using them. Never use this tool unless you have been trained in its use.

Jon P. Muzzarelli

Determine the Size of Extension Cord It is important to match a power tool to a suitable extension cord. A power tool connected to an improperly sized extension cord will not work properly. For example, if a tool draws more current than the cord can handle, the cord and the tool may overheat. This could cause a fire.

Wires inside an extension cord are rated by their gauge number. The gauge is a ratio of size: the *larger* the wire diameter, the *smaller* the gauge number: 10 gauge wire is larger than a 14 gauge wire. Use Table 3-2 to choose an extension cord based on the current draw of the tool and the length of extension cord needed. The numbers in the chart represent the different gauges of wire used in extension cords. The thicker the wire, the smaller the gauge.

Step 1 Check the nameplate on the tool to determine how much current it draws. The *amperage* of a tool is how much current it can carry. This is measured in *amperes*. An *ampere* is a measurement of electrical current or flow. It is often abbreviated as "amps" or "A."

Step 2 Find that number in the "Nameplate Amperes" column on the chart. For example, a standard electric drill might draw 4 amps of current.

Step 3 Determine how long the desired extension cord must be. In other words, how far will the tool be from an electrical receptacle? As an example, assume that you wish to use an electric drill 75' away from the receptacle. In the shaded row under "Cord Length in Feet," find 75.

Step 4 Follow the column down under 75 to a number that corresponds to the ampere number. This will be the minimum gauge of wire required for this application. In this example, if a 4 amp electric drill is on a 75' long extension cord, the cord should be 16 gauge or heavier.

Table 3-2: Recommended Extension Cord Sizes for Use with Portable Electric Tools

Nameplate Amperes	Cord Length in Feet							
	25	50	75	100	125	150	175	200
1	16	16	16	16	16	16	16	16
2	16	16	16	16	16	16	16	16
3	16	16	16	16	16	16	14	14
4	16	16	16	16	16	14	14	12
5	16	16	16	16	14	14	12	12
6	16	16	16	14	14	12	12	12
7	16	16	14	14	12	12	12	10
8	14	14	14	14	12	12	10	10
9	14	14	14	12	12	10	10	10
10	14	14	14	12	12	10	10	10
11	12	12	12	12	10	10	10	8
12	12	12	12	12	10	10	8	8
13	12	12	12	12	10	10	8	8
14	10	10	10	10	10	10	8	8
15	10	10	10	10	10	8	8	8
16	10	10	10	10	10	8	8	8
17	10	10	10	10	10	8	8	8
18	8	8	8	8	8	8	8	8
19	8	8	8	8	8	8	8	8
20	8	8	8	8	8	8	8	8

Notes: Wire sizes are for 3-CDR Cords, one CDR of which is used to provide a continuous grounding circuit from tool housing to receptacle. Wire sizes shown are A.W.G. (American Wire Gauge). Based on 115V power supply; Ambient Temp. of 30°C, 86°F.

- Faulty cartridges can explode unexpectedly. If a powder-actuated tool misfires, hold it in the operating position for at least 30 seconds before trying to fire it again. If the tool still does not fire, hold it in the operating position for another 30 seconds before removing the cartridge according to the manufacturer's instructions. Immerse the cartridge in water immediately after removal.
- Do not load the tool unless it is to be used immediately.
- Do not leave a loaded tool and cartridges unattended, especially where they might be found by someone not trained in their use.
- Never point the tool at anyone.
- Do not fire fasteners into a material they might pass completely through.
- Do not drive fasteners into very hard or brittle material that might chip, splatter, or make the fasteners ricochet.

Hand Tools The greatest hazards posed by hand tools result from misuse, carelessness, and poor maintenance. Because of the great variety of hand tools, it is impossible to provide a complete guide to their safe use. However, the following examples indicate the types of hazards you may encounter. You will also find safety rules in other chapters of this book where specific hand tools are discussed.
- Keep your fingers away from cutting edges.
- Do not hold small stock in your hand while it is being cut, shaped, or drilled.
- Do not use a chisel as a screwdriver. The tip of the chisel may break and fly off, hitting someone.
- If the wooden handle on a tool such as a hammer or an ax is loose, splintered, or cracked, the head of the tool may fly off and strike someone. Do not use tools in need of repair.
- Make proper adjustments to wrenches. If the jaws of a wrench are not adjusted properly, the wrench might slip.

- If impact tools, such as brick chisels, have a mushroomed head, the head can shatter, sending sharp fragments flying. Be sure heads are properly ground.
- Be sure cutting tools are sharp. If they are not, they require more force to use and are more likely to slip.
- When using a utility knife, consider wearing protective leather gloves. Always cut away from yourself. This simple tool is the source of many injuries.
- Never strike the head of a hammer with another hammer in order to drive the claws under the head of a nail to remove the nail from a piece of wood. The head of a hammer can shatter and strike a person with sharp slivers of metal.
- Always wear safety glasses when working with hand tools to avoid accidental damage to your eyes from flying objects.

Musculoskeletal Disorders

Lifting, fastening materials, and other tasks can cause musculoskeletal disorders (MSDs). A **musculoskeletal disorder (MSD)** is a disorder of the muscles, tendons, ligaments, joints, cartilage, or spinal discs. **Ergonomics** is the science of designing and arranging things to suit the needs of the human body. Using ergonomically designed tools can lessen the risks of MSDs. Tool features such as cushioned grips and properly angled handles can improve the safety of a tool as well as minimize physical stress and fatigue. Many hand tools and power tools are now available in versions with ergonomic improvements, such as those tools shown in Ergonomic Tools on page 91.

Repetitive Stress Injuries A worker's body can be damaged gradually by years of impact or motion. When a task done over and over causes minor irritation to nerves and tissues, the damage is called a **repetitive stress injury (RSI).** Carpal tunnel syndrome is one type of RSI involving the hand. Repetitive stress can cause permanent damage. A carpenter, for example, can develop RSI in a wrist after long periods of using a hammer. Hammering

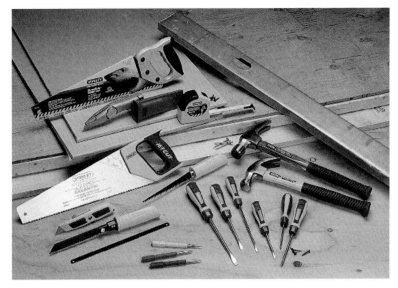

 Ergonomic Tools
Design for Safety The ergonomic design of tools increases their usability. Some tools shown here have cushioned handles while others are shaped in a way that reduces wrist strain. *Why do you think the shape of a tool can make a hand tool safer or easier to use?*

with a minimum of wrist movement can help carpenters avoid this problem. Pneumatic nailers also help, although they may stress a body in other ways.

Other workers, such as trim carpenters and tilesetters, work for long periods on their knees. Special kneepads can protect knees from repetitive stress injuries. In general, whenever work calls for a repeated motion, steps should be taken to lessen its negative effects.

Many protective devices, such as knee pads, feel awkward to use at first. This sometimes causes workers to stop using them. Be patient. Adjust the device for maximum comfort and use it consistently. Eventually, you will not even notice that you are wearing it.

Back Injuries Proper handling of materials makes any job site safer. Back injuries when handling materials are among the most common injuries suffered by carpenters and masons. Lifting that occurs with objects held below the knee or above the shoulders is more strenuous than lifting from the body's center.

Back injuries can be avoided or made less severe if you remember this rule: Lift with your legs, not with your back. This means that you should bend your knees and keep your back straight to prevent back muscles from being strained, as shown in Lifting Heavy Objects on page 92.

Obtain help for lifting loads heavier than you can manage. Whenever possible, stacks of material should be lifted by equipment designed for this task. This relieves workers from having to do tiring work, and it also shortens the delivery time.

Back muscles can also be injured if they are twisted while carrying a long piece of material. Long pieces of material should always be carried by two people. Proper support while cutting materials is also important. Whenever possible, the material should be supported securely at a comfortable height. Portable sawhorses, such as those shown in Folding Sawhorses on page 92, can be moved around a job site for this purpose. Never stand on sawhorses or use them for scaffolding. They are only meant to support material that you are working on.

Hazardous Materials

Repeated contact with such things as chemicals or dust is often ignored as a threat to a worker's health. However, these materials may affect a worker's health over time. When a worker finally realizes the problem, his or her health may already have suffered permanent damage. Hazardous and toxic substances common on construction sites include dust, fuels, solvents, and chemicals and must be labeled as such, as shown in Labels for Hazards on page 93.

OSHA currently regulates many hazardous substances. OSHA's Hazard Communication Standard, also known as the

Jon P. Muzzarelli

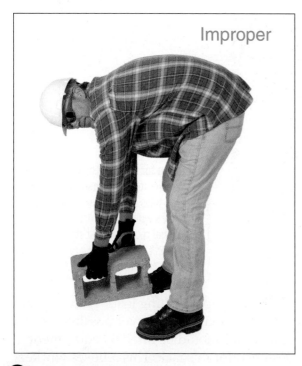
Improper

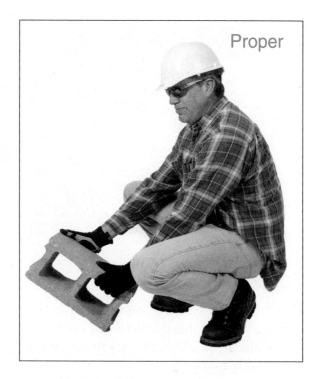
Proper

Lifting Heavy Objects
Bend Your Knees Lifting properly will reduce the possibility of back injury. *Can you name at least two small but heavy construction materials other than concrete?*

Folding Sawhorses
Strong Support Sawhorses can be used to position materials at the right height for cutting. Some have legs that can be adjusted to various lengths.

Right-to-Know Law, has led to better information about possible hazards. Manufacturers are required to provide Material Safety Data Sheets (MSDS) with shipments of their products. These sheets explain the hazards associated with the products and should be readily available. Many materials that are not typically thought of as hazardous are also covered by an MSDS. The sheets include information on physical data, like a material's melting point or flash point; levels of toxicity; recommendations for first aid, storage and disposal, protective equipment; and what to do if a spill or leak occurs. If an MSDS is not available on site, you may be able to locate it on the manufacturer's Web site. MSDS are easy to download.

Every worker must have access to MSDS. Regulations requiring that workers be notified of hazards are often called right-to-know laws. There are many different kinds of hazardous materials that workers may encounter on the job site. Some of the most common of these

Sandra Mesrine/McGraw-Hill Education

Warning Labels

Labels for Hazards Warnings OSHA requires clear labeling of all hazardous and toxic materials on construction sites.

substances include crystalline silica, formaldehyde, asbestos, lead, and arsenic. Have MSDS readily available for the use of emergency medical personnel when they are responding to an accident. The information contained in the sheets could save the life of the injured person.

Asbestos Asbestos is a once widely used, mineral-based material that resists heat and destructive chemicals. It is no longer used in most building products, but it is likely to be found in materials manufactured before it was banned. Asbestos is a fibrous material with a whitish appearance. It may release its fibers, which range in texture from coarse to silky, into the air. The fibers may be too small to see with the naked eye. When asbestos fibers are inhaled or swallowed, they can cause *asbestosis* (a scarring of the lungs) and various types of cancer.

Asbestos is highly regulated by OSHA and other federal agencies. It should be handled or removed only by professionals specially trained for asbestos mitigation work. *Mitigation* refers to any process that makes

a harmful material less dangerous. In the home building industry, workers involved in remodeling and demolition projects are the most likely to encounter asbestos.

Crystalline Silica When materials such as concrete, masonry, and rock are cut or ground, the dust may contain extremely fine particles of a mineral called crystalline silica. Breathing in these particles can produce permanent lung damage called silicosis. To reduce exposure to crystalline silica, follow these guidelines:

* Always wear a respirator designed to protect against fine, airborne particles. However, do not depend on a respirator as the primary method of protection.
* When sawing concrete or masonry, use saws that spray water on the blade.
* Use a dust collection system whenever possible.

Formaldehyde Formaldehyde is found in the urea, phenol, and melamine *resins* used in some construction materials and adhesives. Studies indicate that it is a potential cancer-causing agent. When working with liquid products that contain formaldehyde, take care to prevent contact with your skin. When cutting products that contain formaldehyde, wear appropriate breathing protection. Many manufacturers have taken steps to reduce or eliminate the amount of formaldehyde in their products.

JOB SAFETY

YOUR RIGHT TO KNOW The OSHA standard on Hazard Communication requires employers to inform their workers about dangerous chemicals they may encounter on the job. This requirement is often referred to as the workers' right to know. Many states also have their own right-to-know (RTK) laws.

Lead Many materials containing lead have been eliminated from use in residential building construction because lead has long been recognized as a health hazard. This includes lead-based solders (once used to solder copper-pipe joints) and lead-based paint. However, lead-based materials may still be encountered during remodeling and demolition work.

The removal of lead-based paint can create large quantities of lead-containing dust. Some states require that materials containing lead be removed only by certified lead-abatement contractors.

Arsenic Arsenic is a poisonous element that was used for many years to protect certain types of wood products against attack by wood-destroying insects. It was commonly found in preservative-treated lumber used to build decks, fences, and other outdoor structures. Various studies indicated that arsenic could leach out of wood over time and contaminate soil and groundwater. The U.S. Environmental Protection Agency (EPA) eventually banned the use of arsenic-treated wood for use in residential construction. As of January 1, 2004, no such wood could be sold. For more on this topic, see Chapter 35, "Decks & Porches."

Reading Check

Identify What are some hazardous or toxic substances commonly found on construction sites?

Personal Protective Equipment

Throughout this book, you will find references to the need for personal protective equipment. Generally, this refers to hearing protectors, hardhats, safety glasses, hard-toe footwear, and dust masks or respirators. Do not think of these devices as inconveniences. They provide a barrier between you and a possible hazard. *They are essential tools that help you get the job done safely.* The protective equipment you choose should be approved for a given use by the American National Standards Institute (ANSI).

Hearing Protection Construction workers are regularly exposed to noise. High levels of noise can cause hearing loss. Too much noise may also cause other harmful health effects, such as impaired balance and elevated blood pressure.

Temporary hearing loss results from short-term exposure to noise, with normal hearing returning after a period of rest. Permanent hearing loss is generally due to long exposure to high noise levels over a period of time. Hearing protectors or ear protectors, such as those in Hearing Protection are designed to reduce the noise level. Because there are so many types of hearing protectors, you should not have any problem finding a model that is both comfortable to use and appropriate for the type of work you do. Many safety-conscious builders own several pairs so that they are always available. For example, one pair might be near the table saw, and another might be in the case housing a powder-actuated fastening system. If hearing protection is readily available, a person is more likely to use it. Damage to your hearing from repetitive loud noises accumulates over time. You may not realize the damage is being done because

Hearing Protection
Noise Blocker There are many types of hearing protectors. This type fits over the ear. Others fit into the ear canal to block noise.

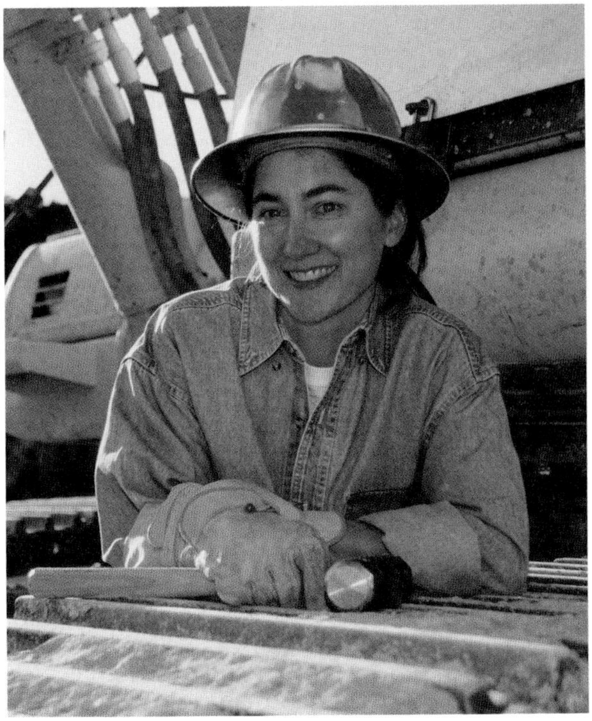

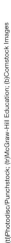

Head Protection

Heads Up Hardhats are often a requirement on job sites.

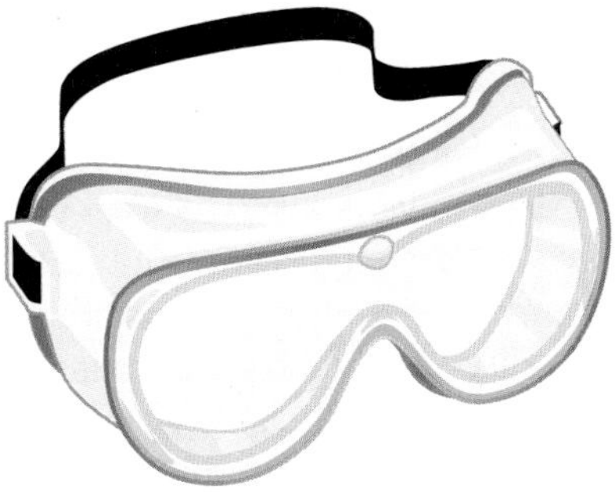

Eye Gear

Essential Eyewear Always wear adequate eye protection for the job you are performing. Note the side shields on this pair of safety glasses. They provide additional protection against debris that might be deflected toward a worker from an unexpected angle.

it happens a little at a time. Then, when you finally realize that damage has been done, it may be too late to reverse.

Head Protection If there is an overhead hazard at a job site, OSHA requires the use of *hardhats* (see Head Protection). An overhead hazard includes something as common as a falling tool. Though many situations do not expose workers to overhead hazards, wearing a hardhat all the time is a reasonable precaution to take. On some job sites, the project manager requires everyone who enters the area to wear a hardhat.

Eye Protection Eye hazards are particularly common in construction for workers in all trades. Many eye injuries are caused by sparks, by flying particles smaller than a pinhead, or by chemicals or liquids that splash into a worker's face.

Because you never know when you might be exposed to an eye hazard, you should wear eye protection, such as the pair of safety glasses shown in Eye Gear, whenever you are on a job site. To be effective, the eyewear must be the right type for the hazard associated with the job. It must fit properly and be in good condition. For example, scratched and dirty lenses reduce visibility, cause glare, and may even contribute to accidents. Sometimes it is necessary to wear a full face shield for operations that generate large quantities of sawdust. There are now many sources of good-fitting, attractive, and comfortable safety glasses.

Respiratory Protection A standard dust mask can filter out large particles, such as the sawdust created by a circular saw. It cannot protect you against very fine dust. For very fine dust, use a *dust respirator* or a full

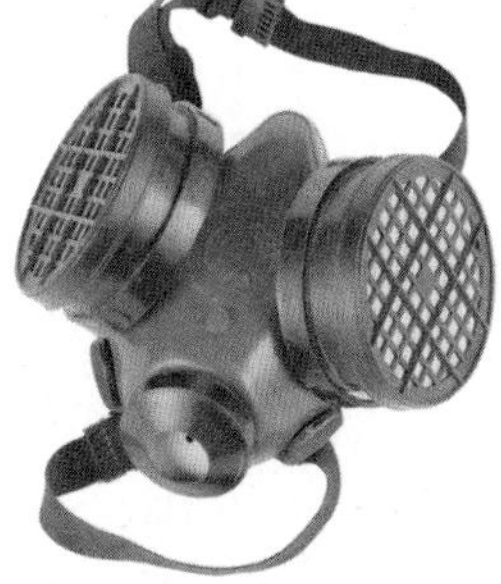

A Cartridge-Type Respirator

Extra Protection This partial-face respirator is fitted with replaceable cartridges that protect against specific hazards. Cartridges are often color coded to ensure proper use.

(tl)Photodisc/Punchstock; (tr)McGraw-Hill Education; (b)Comstock Images

Disposing of Oily Rags Oily rags soaked with solvents used to stain wood or clean equipment are common around a construction site. If these rags are not properly disposed of, they can pile up in an enclosed area such as a corner or an uncovered bucket. When the oils in the rags evaporate, heat can sometimes build up to the ignition point of the material used to make the rags. If enough oxygen is present, the rags can catch fire without any spark or flame to set them off. Cotton rags will catch fire if the temperature reaches about 430 degrees Celsius. What is the equivalent temperature using the Fahrenheit scale?

Starting Hint Solve for *F* using the following formula:

$$F = \frac{9}{5}C + 32$$

respirator. Respirators should be approved by the National Institute for Occupational Safety and Health (NIOSH). Respirators that contain replaceable filter cartridges, such as the one in A Cartridge-Type Respirator on page 95, can be used to protect against fumes. Be sure the cartridges you are using are designed to protect against the specific hazard you face. Many tools can also be equipped with a dust collection bag or connected to a vacuum system. Both catch dust before it becomes airborne, a great advantage in enclosed workplaces.

Suitable Clothing What you wear depends partly on the weather. However, you should choose suitable clothing for construction work, just as you would for any job.

- Avoid wearing pants or overalls that are too long or baggy. They tend to catch on your heels and cause falls. Avoid cuffs because they can collect debris such as sawdust and metal shavings.

- Upper-body clothing should not be so loose that it catches on nails or dangles near power tools. Keep the sleeves of shirts or jackets buttoned, or roll them up.

- Keep long hair tied back or cut it short.

- To protect feet from protruding nails, wear work boots with thick, sturdy soles. When working on a roof, wear boots with slip-resistant soles. Boots with safety toe caps protect feet from injuries caused by falling tools and materials.

- Rings, wristwatches, neck chains, and other jewelry can catch on tools or materials. Avoid wearing them.

Protection from the Environment

Most construction workers work at least part of the time outdoors in a variety of climates and weather conditions. These conditions may vary a great deal from region to region. It is important to take precautions to protect your health and comfort.

Heat Four factors affect the amount of stress a worker faces in hot weather: overall temperature, humidity, radiant heat (such as from the sun or a furnace), and air movement. The body reacts to high temperatures by giving off excess heat through the skin and by sweating. If a person cannot dispose of excess heat, various problems such as heat rash, cramps, exhaustion, and heatstroke may then occur.

Heat Rash also known as prickly heat, may occur in hot and humid environments where sweat does not easily evaporate. When extensive, heat rash can affect performance or cause temporary disability. It can be prevented by resting in a cool place and allowing the skin to dry.

Heat Cramps are painful spasms of the muscles caused when a worker drinks large quantities of water but fails to replace salt in the body. Tired muscles are usually the ones most susceptible to cramps.

Heat Exhaustion results from loss of fluid through sweating when a worker has failed to drink enough fluids, take in enough salt, or both. A worker who has heat exhaustion still sweats but experiences extreme weakness or fatigue, giddiness, nausea, or headache. The victim's skin will be clammy and moist and the body temperature will be normal or slightly elevated. The victim should rest in a cool place and drink an electrolyte beverage. Electrolyte beverages are those used by athletes to quickly restore potassium, calcium, and magnesium salts.

Heatstroke is the most serious heat-related health problem. It occurs when the body can no longer regulate its core temperature. Sweating stops and the body cannot rid itself of excess heat. Signs of heatstroke include the following:

- Mental confusion or delirium
- Convulsions
- Loss of consciousness or coma; a body temperature of 106°F (41°C) or higher
- Hot, dry skin that may be red, mottled, or bluish.

Heatstroke can be deadly. Summon medical help and move the victim to a cool area while waiting for help to arrive.

Most heat-related health problems can be prevented or the risk of developing them reduced. Follow these basic precautions:

- Wear suitable clothing.
- Install portable fans when working indoors.
- Drink plenty of water. You may need as much as a quart per hour.
- Alternate work with rest periods in a cool area. If possible, schedule heavy work during the cooler parts of the day.
- Get used to the heat for short periods. New employees and workers returning from a lengthy absence may need five days or more to adjust to the heat.

Sunlight Sunlight is a source of ultraviolet (UV) radiation known to cause skin cancer and various eye problems. People who have fair skin and light-colored hair are more sensitive to UV radiation than others.

Skin cancers detected early can almost always be cured. The most important warning sign for skin cancer is a spot that changes in size, shape, or color over a period of one month to two years. The most common skin cancers often take the form of a pale, waxlike, pearly nodule; a red, scaly, sharply outlined patch; or a sore that does not heal. The most serious type of skin cancer, called melanoma, often starts as a small, mole-like growth.

When working outdoors, take the following steps to protect yourself from UV radiation:

- Cover up. Wear clothing that protects as much of your skin as possible.
- Use sunscreen. A sunscreen with a Sun Protection Factor (SPF) of at least 15 blocks out 93 percent of the burning UV rays (UVB rays). The SPF number represents the level of sunburn protection. The higher the number is, the stronger the protection from burning rays will be.
- Wear a hat. A wide-brim hat is ideal because it protects the neck, ears, eyes, forehead, nose, and scalp. A baseball cap provides some protection for the front and top of the head, but not for the back of the neck or the ears, where skin cancers commonly develop.

JOB SAFETY

CPR BASICS If you come across an unconscious person on the job, cardiopulmonary resuscitation (CPR) may be necessary. CPR is a procedure that is used to restore normal breathing and heart rhythm after cardiac arrest. It includes the clearance of air passages, the mouth-to-mouth method of artificial respiration, and heart massage done by exerting pressure on the chest.

- Wear sunglasses. UV-absorbent sunglasses can block 99 to 100 percent of UV radiation. The UV protection comes from an invisible chemical applied to the lenses, not from their color or darkness. Choose a lens that is suitable for the work you do.
- Limit direct sun exposure. UV rays are most intense when the sun is high in the sky, between 10 a.m. and 4 p.m.

Cold When body temperature drops even a few degrees below 98.6°F, a worker is exposed to much stress. The four factors that cause cold-related stress are low temperatures, high/cool winds, dampness, and cold water. **Wind chill** is a combination of temperature and wind speed and increases the chilling effect. For example, when the actual air temperature is 40°F (4°C), a wind of 35 mph creates conditions equivalent to a temperature of 11°F (-12°C).

One of the most extreme conditions to result from exposure to cold is frostbite. When someone has frostbite, his or her skin tissue actually freezes. Initial effects include uncomfortable feelings of cold and tingling, stinging, or aching in the exposed area followed by numbness. Ears, fingers, toes, cheeks, and noses are commonly affected. Frostbitten areas appear pale and cold to the touch. If you suspect frostbite, seek medical assistance immediately.

Hypothermia occurs when body temperature falls to a level where normal muscle and brain functions are impaired. The first symptoms are shivering, an inability to make complex movements, lack of energy, and mild confusion. In severe cases, the person seems dazed and fails to complete even simple tasks. Speech becomes slurred and behavior may become irrational. Treatment of hypothermia calls for conserving the victim's remaining body heat and

Estimating and Planning

This estimating and planning exercise will prepare you for national competitive events with organizations such as SkillsUSA and the Home Builders Institute.

Clothing Costs

Protective Clothing

If you are a builder who lives in an area in which weather extremes are common, you will have to invest in the appropriate work clothing. In warm climates, the investment may be small. Outfitting for cold weather, however, can be expensive. For example, in the Midwest, if you work outdoors, your winter clothing may include the garments shown in the table in addition to your everyday garments.

Estimating on the Job

What is the climate like in your area or in an area in which you would like to work? Determine the most extreme weather conditions for that area. Then, using a catalog that features

Item	Approximate Cost*
Steel-toe insulated boots	$75 – $150
Thermal top and pants	$50 – $75
Thermal socks	$7 – $10
Flannel-lined jacket with hood	$85 – $125
Wind- and water-resistant insulated pants	$60 – $80
Insulated bib overalls	$100 – $150
*Costs are for moderately priced items.	

work clothing or by visiting a local store, estimate the cost of the clothing you would need on a day of bad weather.

providing additional heat sources. Seek medical assistance immediately.

The proper use of protective clothing as shown in Beating the Cold is important in avoiding cold-weather health problems. Pay special attention to protecting feet, hands, face, and head. Wear at least three layers of clothing:

- An outer layer to keep out the wind yet allow some ventilation.
- A middle layer of wool or synthetic fabric to absorb sweat but insulate against dampness.
- An inner layer of cotton or synthetic weave to allow ventilation.

>> **Beating the Cold**
Insulating Layers Multiple layers of clothing provide the best cold-weather protection. Choose clothing that will not interfere with the work you have to do. *What kinds of cold-weather clothing would be unsuitable for workers on a job site?*

Section 3.2 Assessment

After You Read: Self-Check
1. What is the most common cause of injury on a construction site?
2. How does a conductor differ from an insulator?
3. What is an MSDS, and why is it important?
4. List at least three symptoms of heat exhaustion.

Academic Integration: Science
5. **Protective Gear** Overexposure to the sun's ultraviolet rays can cause different types of skin damage, from sunburn to premature wrinkling to skin cancer. Research sun-protective clothing that can minimize exposure to these harmful rays. Write one or two paragraphs describing the design features and materials used in this type of clothing that makes it protective.

Go to **connectED.mcgraw-hill.com** to check your answers.

Jon P. Muzzarelli

Review and Assessment

CHAPTER 3

Chapter Summary

Section 3.1

The purpose of OSHA is to assure safe and healthful working conditions in the workplace. Workers must take responsibility for safety and know how to handle emergencies. Workers compensation helps workers financially after an injury. Keeping a work site safe includes good housekeeping and sanitation; proper use of signs, tags, and barricades; knowing about fire prevention and equipment; and following proper excavation procedures.

Section 3.2

To protect personal safety and health, workers need to take precautions against falls and falling materials, electrical shock, injuries from tools, development of musculoskeletal disorders, and harm from hazardous materials. They should wear appropriate clothing and personal protective equipment, and they should protect themselves from environmental conditions such as heat, sunlight, and cold.

Review Content Vocabulary and Academic Vocabulary

1. Use each of these content vocabulary and academic vocabulary words in a sentence or diagram.

Content Vocabulary

- Occupational Safety and Health Administration (OSHA) (p. 79)
- excavation (p. 82)
- first aid (p. 83)
- conductor (p. 85)
- electrical circuit (p. 85)
- grounding (p. 85)
- musculoskeletal disorder (MSD) (p. 90)
- ergonomics (p. 90)
- repetitive stress injury (RSI) (p. 90)
- wind chill (p. 98)

Academic Vocabulary

- physical (p. 78)
- chemicals (p. 81)
- components (p. 86)

Speak Like a Pro

Technical Terms

2. Work with a classmate to define the following terms used in the chapter: *inspections* (p. 78), *lockout/tagout* (p. 80), *barricades* (p. 80), *debris* (p. 81), *slope* (p. 82), *first aid* (p. 83), *ground-fault circuit* (p. 86), *asbestos* (p. 93), *mitigation* (p. 93), *resins* (p. 93), *hardhats* (p. 94), *dust respirator* (p. 95).

Review Key Concepts

3. Explain the rights of employers and employees under OSHA.

4. Describe two methods used to keep a work site safe.

5. Define first aid.

6. Recall the most common cause of major injury on construction sites.

7. List the five basic safety rules for hand and power tools.

8. Identify five types of personal protective equipment.

Critical Thinking

9. Apply What could you do as you work with gasoline-powered tools to avoid carbon monoxide poisoning?

Academic and Workplace Applications

STEM Mathematics

10. Line Graphs The chart below shows wind chill, the effect of wind on how we experience temperature. For example, when the wind is blowing at 10 mph and the temperature is 30°F, it feels like 21°F. Create a line graph to display the information in the charts.

Wind Chill Effect with a 10 mph Wind					
Temperature °F	0	10	20	30	40
Windchill °F	-16	-4	9	21	34
Wind Chill Effect with a 30 mph Wind					
Temperature °F	0	10	20	30	40
Windchill °F	-26	-12	1	15	28

Math Concept **Line Graphs** A line graph is a good way to show how two pieces of information are related and vary depending on each other.

Step 1: Draw and label a coordinate grid with an *x*-axis and a *y*-axis. Mark a scale that fits the data you are going to plot. Hint: mark off every 10 degrees.

Step 2: Plot the points and draw lines. Use the same coordinate grid to plot lines for both tables.

21st Century Skills

11. Media Literacy OSHA makes available posters that can be displayed at construction worksites to promote safety. The safety messages that follow are taken from actual OSHA posters. Which do you think would be most effective in encouraging worker safety on the job? Which is least effective? Write a few sentences explaining your choices. Write a message of your own that could be used on an OSHA poster.

1. Uncle Sam Wants You to Be Safe
2. Be a part of safety…Together Everyone Accomplishes More
3. Don't Be a Fool! Being Safe Is Being Cool.
4. Safety's Intention Is Accident Prevention
5. It wouldn't "Kill" you to attend a safety meeting . . . Would It?
6. A Clean Environment Is a Safe Environment
7. When there is a question. . . Wear your eye protection.

STEM Science

12. Heat Exposure Prolonged exposure to heat can cause heatstroke and heat exhaustion. Research these two conditions. Find out what causes them, how people working in construction can protect themselves from these conditions, and how to treat people who develop these conditions. Summarize your findings in a one-page report.

Standardized TEST Practice

Multiple Choice

Directions Choose the word or phrase that best answers the question.

13. Which of these materials is considered a good insulator?

   a. water   **c.** wire

   b. a GFCI   **d.** rubber

14. Which colors are used most often on caution signs?

   a. red and blue

   b. yellow and black

   c. orange and black

   d. black and white

15. What information about a material would not be found on an MSDS?

   a. levels of toxicity

   b. melting point

   c. price

   d. recommendations for disposal

TEST-TAKING TIP

Look for key words in test directions. For example, choose, describe, compare and contrast, *and* write a sentence *are key words and phrases that tell you how to answer the questions.*

*These questions will help you practice for national certification assessment.

UNIT 1
Hands-On Math Project

Green Construction in the Community

Your Project Assignment

For this project, you will estimate costs for green and standard construction.

- Find a local organization, such as YouthBuild U.S.A., that builds or renovates homes in your community.

- **Build It Green** **Research** green alternatives to standard home construction methods.

- **Prepare** the basic activity schedule for a home construction project by your organization.

- **Calculate** the cost differences in construction labor and materials for using a minimum of three green alternatives.

- **Create** a three- to five-minute presentation.

Applied Skills

Some skills you might use include:

- **List** the steps in planning to build a house.

- **Research** the factors involved in making construction cost estimates.

- **Define** and select three green alternatives you recommend to build this house. (Tip: Green alternatives can be found at nearly any stage of construction—be creative!)

- **Describe** the steps you use to make your basic cost estimate. (Tip: Don't forget to include your measurements!)

- **Compare** the costs for your green and standard construction operations.

The Math Behind the Project

The traditional math skills for this project are estimation and measurement. Remember these key concepts:

Estimation

For quick estimates, round measurements up for lengths and quantities. For example, if you were calculating the area of a concrete slab that was 24' 4" × 15' 9", you could use the numbers 25' × 16' for your estimate.

Percentage

To calculate percentage change in cost for materials, subtract the cost of green product from the cost of the standard product. Divide the difference by the cost of the standard product. For example, if wood studs cost $123.50 and steel studs cost $145.87, to find the percentage difference, use the following steps:

1. Subtract the costs.	$123.50 − $145.87 = $-22.37
2. Divide the difference by the standard product price.	$-22.37 ÷ $123.50 = $-0.181
3. Convert the decimal number to percentage.	-0.181 ÷ 100 = -18%

This means that the green alternative (steel) costs 18% more than the wood stud construction.

YouthBuild U.S.A. *Mission:* Unleash the intelligence and positive energy of low-income youth to rebuild their communities and their lives.

Project Steps

Step 1 Research

- Contact your organization to find out the average cost for their standard home construction projects.

- **Build It Green** List several green methods that could be used in place of standard construction methods.
- Select a minimum of three green methods you recommend be used.
- Determine the factors that influence price differences between your green methods and the standard ones.
- Confirm the cost and installation differences between green and standard construction.

Step 2 Plan

- Choose the basic design plan for your home.
- Measure the proposed dimensions of your home and calculate the total square footage.
- Determine the basic materials and supplies your organization will use to construct the home.
- List the types of skilled labor this construction will require.

Step 3 Apply

- Use your research to create a cost comparison of the two methods of construction.
- Create a material take-off for green methods and standards and calculate the cost differences.
- Make a CPM flow chart showing the construction process. Use general system terms such as rough plumbing, foundation, and wall framing.

Step 4 Present

Prepare a presentation combining your research and cost calculations using the checklist below.

PRESENTATION CHECKLIST
Did you remember to…
✓ Organize your research by step and topic?
✓ Use a calculator or spreadsheet for the cost calculations?
✓ Create visual aids
✓ Create notes you might need for your presentation?
✓ Focus on your recommendations for green methods?
✓ Explain the cost differences between green and standard methods?

Step 5 Technical and Academic Evaluation

Assess yourself before and after your presentation.

1. Did you plan your steps carefully?
2. Is your research thorough?
3. Were your cost estimates accurate?
4. Were your green method alternatives realistic?
5. Was your presentation creative and effective?

Go to **connectED.mcgraw-hill.com** for a Hands-On Math Project rubric.

UNIT 2
Tools &
Equipment

In this Unit:

Hands-On Math
Project Preview

Starting Your Own Carpentry Business

After completing this unit, you will research
the cost of the tools and equipment you will
need to start your own carpentry business.
You will also research terms and interest rates
for small-business loans and determine the
monthly payment, including interest, for a
loan to start your business.

Project Checklist

As you read the chapters in this unit, use this
checklist to prepare for the unit project:

✓ List the tools and equipment that every
  carpenter needs.

✓ Identify the skills that an independent
  contractor needs to be successful.

✓ Think about ways high-quality tools and
  equipment can save time and money, and
  help the environment.

  Go to **connectED.mcgraw-hill.com** for
the Unit 2 Web Quest activity.

Profile A construction and building inspector examines houses, buildings, streets, water systems, and other structures. The inspector ensures that construction, remodeling, or repair complies with building codes and other regulations.

Academic Skills and Abilities

- algebra
- geometry
- verbal and written communication skills
- interpersonal skills
- drafting
- blueprint reading

Career Path

- on-the-job training
- certificate or associate's degree in building inspection technology
- community college courses
- bachelor's degree in engineering or architecture
- apprenticeship programs
- certification

Explore the Photo

Tools for the Job Selecting the appropriate tool is vital to completing any job with accuracy and efficiency. *What features of this tool might help the carpenter work efficiently?*

©Hero/age fotostock

CHAPTER 4

Hand Tools

Section 4.1
Measuring & Layout Tools

Section 4.2
Cutting & Shaping Tools

Section 4.3
Tools for Assembling & Disassembling

Chapter Objectives

After completing this chapter, you will be able to:

- **Identify** hand tools used for measuring and layout.
- **Identify** hand tools used for cutting and shaping.
- **Identify** hand tools used for assembling and disassembling.
- **Explain** what various hand tools are used for.
- **Describe** common mathematical problems carpenters solve using hand tools.

Discuss the Photo

High-Quality Tools Most professionals recommend purchasing tools of the highest quality you can afford. *What characteristics do you think a good tool should have?*

Writing Activity: Create a Description

A description is a group of words that gives a mental picture of something. Select a common carpentry tool. Write three to five sentences describing the tool. Include details such as the shape, color, and size of the tool. Share your description with another person to see if he or she can identify the tool that you have described.

Chapter 4 Reading Guide

Before You Read Preview

Scan the chapter and pick out a picture of a tool you find. Brainstorm what tasks you think that tool is used for. When you find the tool in the text, check to see if your definition is correct. Change your definition as needed.

Content Vocabulary

- square
- level
- wrench
- pliers

Academic Vocabulary

You will find these words in your reading and on your tests. Use the academic vocabulary glossary to look up their definitions if necessary.

- angles
- adjacent
- arc
- horizontal
- vertical

Graphic Organizer

As you read, use a chart like the one shown to organize information about hand tools, adding rows as needed.

Type of Tool	Used for...

Go to **connectED.mcgraw-hill.com** to download this graphic organizer.

Measuring & Layout Tools

Tape Measures, Rules, & Other Tools

What are some common measuring tools?

A variety of hand tools is necessary for building a house, so a good selection of tools is essential. Some tools, such as a steel tape measure, are used in all building trades. Others are used primarily within one trade. For example, a framing square is used most often by carpenters. A **square** is a tool that is used primarily to measure or check angles.

Tools and Tasks

Members of all trades must measure, mark, and lay out projects many times a day. Carpenters must determine the correct **angles** and cuts for rafters. Plumbers must cut and assemble pipes to provide proper drainage.

Tool Safety

Before using any measuring and layout tools, review the safety material in Chapter 3.

Layout Tape Measure

Layout Tape Measure A steel or fiberglass tape in a rust-resistant case with a reel-in crank. Comes in lengths of 50', 100', and 200'. In situations that require greater accuracy, some prefer to use a steel tape. Steel tape does not display the stretch under tension that occurs in a fiberglass tape. It is used for:

- Laying out foundations.
- Locating site features, such as walkways and driveways.

Steel Tape Measure

Steel Tape Measure A steel tape with a sliding hook on the end is accurate whether it is used for inside measurements or outside measurements. The case has a belt clip and a tape-locking button. Comes in various widths and in lengths from 6' to 33'. The tape most commonly used by carpenters is 25'. The wider the tape, the better it is at measuring long lengths unsupported. It is used for:

- Performing general measuring tasks.
- Making accurate inside measurements. The measurement is read by adding a specified amount to the reading on the tape. This amount is the length of the case.

Chalk Line A chalk line is powdered chalk and a reel of string in a steel or plastic case. A hook is used to secure one end of the string. It is used for:
- Quickly creating a straight layout or cutting line over long lengths, as on panel products.
- Indicating the position of walls on a subfloor.

Chalk Line

Folding Rule A rigid wood rule, 6' or 8' long, that folds into a compact size. A metal slide in one end can be used for measuring depth. Uses:
- Making inside measurements and measuring the depth of a mortise or channel.
- Measuring plumbing runs.

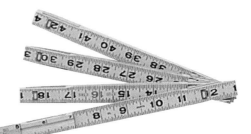

Folding Rule

Carpenter's Pencil A sturdy, thick pencil with a wood casing. The thick lead resists breakage and can be sharpened with a utility knife. Uses:
- All-purpose marking and layout where great precision is not necessary.

Carpenter's Pencil

Try Square

Try Square A fixed-blade square with a metal blade and a wood, plastic, or metal handle. The blade may be graduated in inches and is positioned at 90° to the handle. Uses:
- Checking **adjacent** surfaces for *squareness*.
- Making layout lines across the face or edge of stock.

Builder's Tip

CARING FOR TOOLS Good tools require proper care. Keep them in a toolbox when you are not using them. Do not leave tools outdoors in wet or damp weather. Inspect hand tools regularly for signs of wear or damage. Keep cutting edges sharp. If a worn or damaged tool cannot be repaired, replace it.

 **Combination Square**

Combination Square A blade that slides along its handle or head. The handle may contain a leveling vial. The removable blade can be used as a straightedge. Uses:
- Checking adjacent surfaces for a correct angle of 45° or 90°.
- Making layout lines at 45° and 90° across the face or edge of stock.
- Measuring the depth of a mortise or channel.
- Roughly leveling or plumbing a surface.

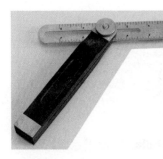

Sliding T-Bevel

Sliding T-Bevel A sliding metal blade that can be set at an angle to the handle and then locked into place. Uses:
- Transferring an angle, which means measuring an angle between 0° and 180° and duplicating it somewhere else.
- Checking or testing a miter cut at other than 45°.

Framing Square Sometimes called a carpenter's square. A large metal square consisting of a blade, or body, and a tongue. Identical in general appearance is the rafter square, which has tables printed on it that contain the solutions to various rafter layout problems. Uses:
- Checking for squareness.
- Using as a *straightedge*.
- Determining angle cuts on rafters and stair parts.

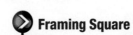 **Framing Square**

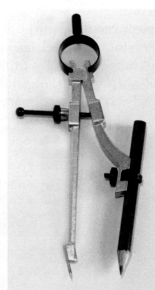

Dividers/Compass

Mathematics: Geometry

Right Angles An angle is a figure created by two lines that meet at a common endpoint, or vertex. Angles are measured in degrees. A *right angle* is formed when two lines are perpendicular to one another. A right angle can also be defined as one quarter of a circle. How many degrees are in right angle?

Starting Hint A full circle has 360 degrees. 1 degree is $\frac{1}{360}$th of a full circle.

Dividers/Compass A measuring tool with two metal legs. On some, one leg can be removed and replaced with a pencil, forming a compass (shown here). Uses:
- *Stepping off* measurements.
- Laying out an **arc** or circle.
- *Scribing* (fitting) a part such as a countertop to a surface such as a wall that may not be perfectly straight.

Triangular Framing Square A square marked with degrees for fast layouts. It also typically has marks to indicate plumb cuts for both common and hip/valley rafters. Its small size makes it easy to keep in a tool pouch. It is more durable than a standard framing square. Uses:

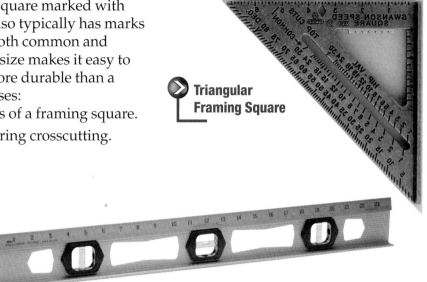

- Performing the functions of a framing square.
- Guiding power saws during crosscutting.

Triangular Framing Square

Carpenter's Level

Carpenter's Level A **level** is a long wood, metal, or fiberglass instrument with several glass leveling vials that measures the levelness or plumbness of a surface. Typical lengths are 24", 48", 72", and 96", with 48" being the most common. It is sometimes called a spirit level (after the liquid inside the vials). Uses:
- Determining whether or not a surface is level (horizontal) or plumb (vertical).

Torpedo Level

Torpedo Level A small spirit level approximately 9" long. It has leveling vials for plumb, level, and 45° angles. Sometimes called a pocket level. Uses:
- Leveling or plumbing small surfaces, such as pipes, or surfaces that are difficult to reach.

Laser Level A laser level uses a highly focused beam of light to project a straight, level line across a surface or through the air. An internal device levels the beam automatically. Laser levels come in a variety of shapes and sizes. Some models can be used to establish plumb lines. Uses:
- Establishing a level line across long distances or on intersecting walls.
- Installing shelving.
- Transferring layout marks from floor to ceiling.

Scratch Awl A pointed metal marking tool. Uses:
- Scribing a line accurately, particularly on metal.
- Starting a hole before drilling into wood.
- As an anchoring pin for chalk or masons' lines.

⬥ **Scratch Awl**

》 **Plumb Bob and String Line**

Plumb Bob and String Line A metal weight with a pointed tip. The tip may be replaced if damaged. The top of the weight has a hole for attaching the string line. Uses:
- Locating the corners of buildings during foundation layout.
- Establishing a true vertical line.

◀ **Construction Calculator**

Construction Calculator A portable calculator that computes measurements directly in feet and inches. Uses:
- Calculating volumes and areas.
- Converting decimals to fractions.
- Solving various roof framing layout problems.
- Solving stairway rise/run per step equations.

Section 4.1 Assessment

After You Read: Self-Check

1. What is a layout tape measure used for?
2. What tool can be used to indicate the position of walls on a subfloor?
3. Which type of square can be used to check 45° angles?
4. Which tool would you use to scribe a line on metal?

Academic Integration: Science

5. **Exploring Gravity** Carpenters use a plumb bob to determine if walls, doors, and other structural elements are vertical. A simple plumb bob is a weight that hangs at the end of a string. The gravity of the Earth pulls the plumb bob toward the center of the earth and the string represents a vertical line. *Gravity* is a force that draws objects toward one another. Use a plumb bob to test articles in your classroom to determine if they are truly vertical.

✈ Go to **connectED.mcgraw-hill.com** to check your answers.

Cutting & Shaping Tools

Precision Cutting & Shaping

What tasks are suitable for handsaws?

A wide variety of saws is available for use in construction. A handsaw can be useful for many sawing tasks, even though power saws are best for many jobs. Cutting, shaping, and drilling tools remove small chips from the workpiece. This makes a high degree of precision possible.

Tool Safety

Before using any cutting and shaping tools, review the safety material in Chapter 3 for hand tools on page 90.

Utility Drywall Saw

Utility Drywall Saw A slender saw with a pointed tip, a stiff blade, and large, sharp teeth. The blade can be pushed through the drywall surface and does not need an access hole. Uses:
- Making internal cuts in drywall
- Cutting drywall to irregular shapes.

Backsaw

Backsaw A fine-tooth crosscut saw with a heavy metal band across the back that strengthens the blade. A long model (24" to 28") is called a miter box saw. Uses:
- Making fine crosscuts trim. Often used in a miter box.
- Cutting miters in molding.

Handsaw A saw with a wide blade in lengths from 20" to 28". The teeth on crosscut models (A) cut across the grain. Rip models (B) cut with the grain. These cuts are shown above. Uses:
- Crosscutting and rip cutting wood.
- Completing cuts made by power saws.

Handsaw

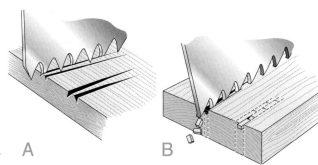

A

B

Hacksaw A saw with a U-shaped steel frame fitted with replaceable metal-cutting blades. Standard and high-tension models are available. Uses:
- Cutting all types of metal fasteners and hardware.
- Cutting plastic.

▲ **Hacksaw**

Keyhole Saw A saw with a narrow 10" replaceable blade with fine teeth. The blade tapers to a sharp point. A compass saw is similar but has slightly bigger teeth and a longer blade. Uses:
- Cutting curves.
- Enlarging holes.

◀ **Keyhole Saw**

Pull Saw A pull saw is used primarily for wood. It has teeth that are angled toward the handle so that they cut on the pull stroke. It is sometimes called a Japanese saw. The blade is unusually thin and the teeth have no set (deviation from a straight line). Uses:
- Cutting overhead, where use of a heavier saw would be tiring.
- Making fine cuts, as when mitering trim or cutting dovetails.
- Cutting the base of door jambs when laying finish flooring.

◀ **Pull Saw**

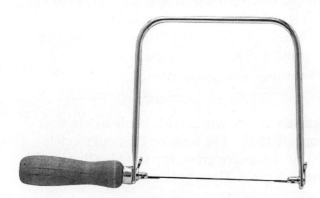

Coping Saw A saw with a U-shaped frame having a deep throat. The replaceable ⅛" wide blade with tiny teeth can be rotated to cut at various angles. Uses:
- Cutting curves in wood molding and trim.
- Shaping the end of molding for joints.
- Cutting scroll work.

◀ **Coping Saw**

Dovetail Saw This type of saw is similar to a backsaw, but the blade is narrower and thinner, and has very fine teeth. Uses:
- Making short, straight cuts of superior smoothness and accuracy.
- Cutting dovetails and other joints.

▲ **Dovetail Saw**

Block Plane A small (about 6" long) plane with a blade that cuts bevel-side up. Standard models have a blade angled 20° to the sole (bottom). Low-angle models are angled at about 12°. Uses:

- Planing end grain.
- Fitting doors.
- Beveling an edge.

 **Block Plane**

 ◄ **Jack Plane**

Jack Plane A 12" to 15" long plane with a blade that cuts bevel-side down. It is also called a *jointer plane*. Although there are no absolute definitions, a jack (or "#5") plane is typically 12" to 15" long. The jointer plane is longer, at 20" to 24". Both have blades that cut bevel side down. Uses:

- Smoothing and flattening edges for making a close-fitting joint.
- Planing long workpieces, such as the edges of doors.

 **Wood Chisel**

Wood Chisel A wood chisel has a steel blade sharpened to a fine edge at one end, with a wood or plastic handle at the other. Many blade types and shapes are available. A standard set of chisels usually includes blade widths from ¼" to 1½". Uses:

- Trimming and shaping wood.
- Clearing mortises.

 ◄ **Cold Chisel**

Cold Chisel A tool-steel chisel with a hardened and tempered edge for cutting metal. The angle of the beveled cutting edge is about 60°. Uses:

- Cutting off a rivet or nail.
- Getting a tight or rusted nut started.

 **Wood Rasp**

Wood Rasp A wood rasp is similar to a metal file but has raised teeth. On some models, the teeth are located on a replaceable, thin metal plate that is attached to the handle. In others, the teeth and handle are formed from a solid piece. Uses:

- Rounding corners and edges.
- Quickly removing stock where a smooth edge is not required.

Metal File A solid metal bar with patterned cutting ridges formed in one or more surfaces and edges. Single-cut and double-cut models are most common. Uses:

- Sharpening tools.
- Smoothing the edges of metal products and rounding metal corners.
- Trimming the edges of plastic laminate.

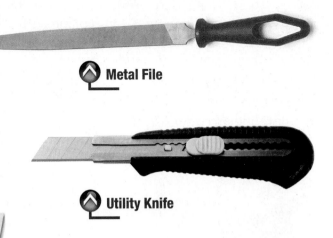

Metal File

Utility Knife

Utility Knife An all-purpose knife with extremely sharp, replaceable blades. On some models, the blade is retractable. Uses:

- Cutting roof shingles, tar paper, batt insulation, wood veneer, and many other materials.
- Cutting carpeting when a hooked blade is attached.
- Sharpening a carpenter's pencil.

Section 4.2 Assessment

After You Read: Self-Check

1. What is the difference in cutting action between a block plane and a jointer plane?
2. How does a wood rasp differ from a file?
3. What is a utility knife used to cut?
4. What material is cut using a hooked blade on a utility knife?

Academic Integration: Mathematics

5. **Fractions** A carpenter is using a plane to trim the edge of an antique door from 33⅞ inches to 33¾ inches. How much material should the carpenter remove?

Math Concept Fractions are a way of expressing an equal number of parts of an object. Fractions have two parts: the *numerator*, which gives the number of parts, and the *denominator*, which is the number of parts in a whole. To add or subtract fractions with different denominators, convert them to fractions with the same denominators. Then add or subtract their numerators to get your answer.

Step 1: Look at the two denominators and find the least common denominator (LCD). Since 4 divides evenly into 8, the LCD is 8.

Step 2: Convert ¾ to an equivalent fraction with 8 as the denominator by multiplying both the numerator and the denominator by 2: ¾ = 6/8

Step 3: Subtract the numerators: ⅞ − 6/8 = ?

Go to **connectED.mcgraw-hill.com** to check your answers.

Tools for Assembling & Disassembling

Clamps, Hammers, & Other Tools

What safety precaution might be necessary for these types of tools?

Clamps and hammers can be used to assemble objects and fasten things together. Other tools, such as pry bars, can help to take things apart.

Tool Safety

Before using any assembling and disassembling tools, review the safety material in Chapter 3 for hand tools on page 90.

Bar Clamp

Bar Clamp A clamp with a stationary head, a sliding tailstop, and an adjustable screw assembly, all mounted on a flat bar. When a pipe is used instead of a bar, the tool is called a pipe clamp. Its length ranges from 16" to 5' or more. Uses:

- Holding materials together as they are being glued.
- Compressing materials temporarily.
- Assembling cabinetry.

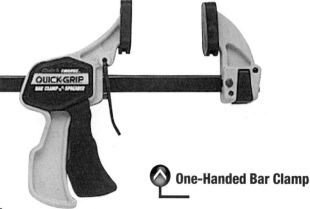

One-Handed Bar Clamp

Builder's Tip

USING CLAMPS Because clamps are often used to hold materials being glued, glue can build up on their surfaces. To prevent this, apply a coat of paste wax to the bars of the clamp. The wax will also help prevent rust.

One-Handed Bar Clamp A type of bar clamp with a trigger-handle tailstop that allows quick, one-handed adjustments. Uses:

- Securing wood that is being cut with a power saw.
- Clamping for general light-duty needs.

(t & b)Kevin May Corporation

 **Claw Hammer**

Claw Hammer A hammer with a curved claw. Heads weigh from 8 to 20 oz. A 16 oz. head is suitable for general construction. An 8 oz. head is a good finishing hammer. The face is slightly crowned, with beveled edges. Handles may be of hickory, steel, or fiberglass. Uses:
- Driving nails.
- Removing nails.

 **Rip Hammer**

Rip Hammer Also called a straight-claw hammer, a rip hammer has a wedge-shaped claw. Heads weigh from 13 to 25 oz. The handle may be made of hickory, steel, or fiberglass. Models used by framing carpenters have longer handles and checkered faces to reduce the chance of glancing blows and flying nails. Uses:
- Driving and removing nails.
- Prying apart pieces that have been nailed together.

 **Warrington Hammer**

Warrington Hammer A hammer with a flattened *peen* instead of a claw. Heads weigh from 3½ to 10 oz. Uses:
- Driving finishing nails into molding and trim.
- Starting brads.

Hand Sledge

Hand Sledge A hammer with a two-faced head weighing between 2 and 4 lbs. It has a wood or steel handle. Also called a hand drilling hammer. Uses:
- Striking steel tools such as cold chisels, brick chisels, and punches.
- Driving stakes during site layout.

Mallet

Mallet A hammerlike tool with two separate faces that may be rubber or plastic. A traditional mallet has a wood head. Uses:
- Striking blows where steel hammers would mar or damage the surface, as when assembling wood joints or driving wood pegs.
- Striking other tools, such as chisels, where a metal-headed hammer may damage the tool being struck.

Ripping Bar

Ripping Bar A bar with flat claws at each end. Available in lengths up to 8'. A 3' bar is suited for general use. Uses:
- Pulling large nails and spikes.
- Prying off old materials during renovation.
- Demolishing built work.

Cat's Paw A small pry bar/nail puller with a head that is aggressively scooped and is at a 90° degree angle to the handle. A hammer is used to drive the head (jaws) down into the wood to extract deeply imbedded nails.

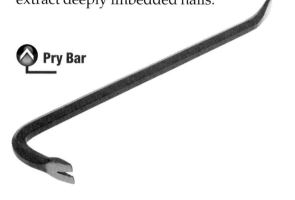
Pry Bar

Pry Bar A steel bar 6" to 14" long, with a nail-removing claw at one or both ends. Some models have a wide, flattened end for prying molding from a wall. Uses:
- Prying nailed lumber and trim apart.
- Pulling nails.

Nail Set

Nail Set A steel shank 4" long with a concave tip, a *knurled* body, and a square striking surface. Comes in a set of four that are sized from $\frac{1}{32}$" to $\frac{1}{8}$" at the tip of each shank. Use:
- Driving finishing nails below the surface of wood. Nail holes can then be filled.

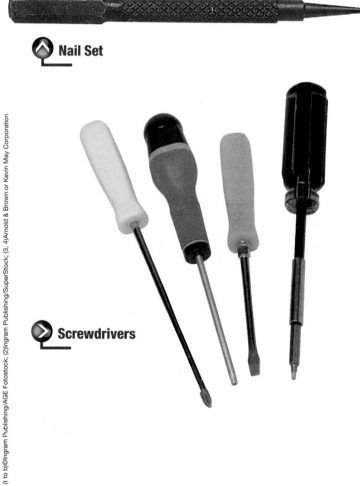
Screwdrivers

Screwdrivers A screwdriver is a steel shank of various lengths with a wood or plastic handle and a tip formed to fit a particular type of screw. A standard slotted head widens from tip to shank. A cabinet-slotted head has a uniform width to reach recessed screws. A Phillips head has the shape of an X at the tip to reduce slippage. Square-drive and Torx heads also prevent slippage. Uses:
- Driving and removing screws.

Stapler A heavy-duty stapler with a spring driven plunger that drives up to 9/16" staples. Uses:
- Attaching ceiling tile, screening, and other soft or thin materials.

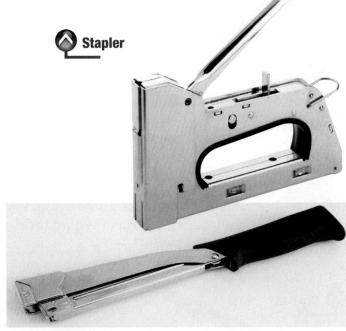

Stapler

Hammer Tacker A slender stapler with a handle at one end. It is used with a striking motion to quickly drive staples. Used for light-duty, high-volume fastening. Uses:
- Attaching insulation.
- Attaching roofing felt.
- Installing building paper.

Hammer Tacker

Wrenches

A **wrench** is a hand tool designed for turning a fastener, such as a bolt or a nut.

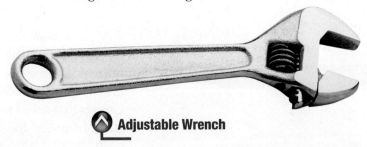
Adjustable Wrench

Adjustable Wrench A steel tool with one adjustable jaw. This wrench exerts its greatest strength when hand pressure is applied to the side with the fixed jaw. Uses:
- Turning nuts and bolts where there is plenty of clearance.

Open-End Wrench

Open-End Wrench A nonadjustable wrench with accurately machined openings on either end. For a variety of tasks, a complete set is needed. Sets are available in metric and standard sizes. Uses:
- Turning nuts and bolts in difficult-to-reach areas.

Box Wrench

Box Wrench A metal wrench with two enclosed ends. Heads are offset from 15° to 45°. It is available in metric and standard sizes. Uses:

- Making adjustments where there is limited space for movement.
- Turning nuts and bolts when a secure grip is essential.

Socket Wrench Set

Socket Wrench Set A series of metal sockets that fit onto a handle containing a *ratcheting* mechanism. Sets come in metric and standard sizes. A basic set contains ten sockets, a ratcheting handle, and a non-ratcheting handle. Uses:

- Installing and removing nuts, bolts, and lag screws quickly.

Locking Pliers An all-purpose tool with double-lever action that locks the jaws to clamp a workpiece. Uses:

- Substitutes for a vise, clamp, pipe wrench, fixed wrench, or adjustable wrench.

Locking Pliers

Pipe Wrench

Pipe Wrench A pipe wrench has hardened, cut teeth on the jaws. The jaws tighten as the wrench is turned. Uses:

- Tightening or removing pipes. It should not be used on nuts or bolts.

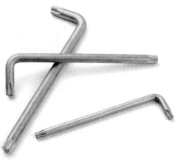

Allen Wrench A hexagonal steel bar with a bent end. Either end fits the hexagonal recess in the top of some screws. Sometimes called a hex key. Uses:
- Tightening and loosening set screws, some of which secure pulleys and wheels on power tools and equipment.

Allen Wrench

Pliers

Pliers are a hand tools with opposing jaws that are designed to hold things. They should not be used to turn nuts or bolts because they can damage them. Use a wrench instead.

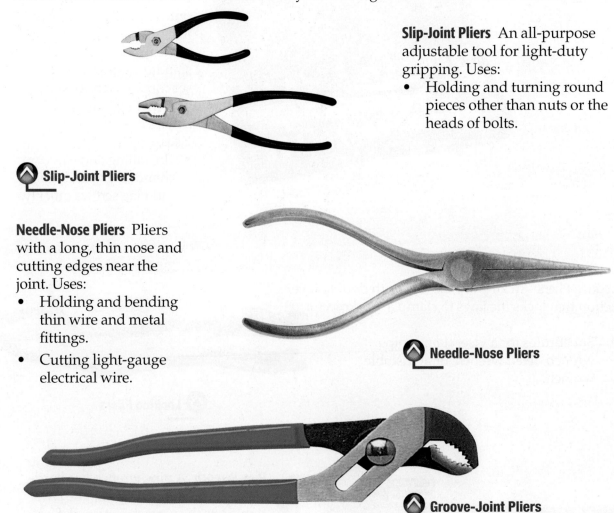

Slip-Joint Pliers An all-purpose adjustable tool for light-duty gripping. Uses:
- Holding and turning round pieces other than nuts or the heads of bolts.

Slip-Joint Pliers

Needle-Nose Pliers Pliers with a long, thin nose and cutting edges near the joint. Uses:
- Holding and bending thin wire and metal fittings.
- Cutting light-gauge electrical wire.

Needle-Nose Pliers

Groove-Joint Pliers

Groove-Joint Pliers Large pliers with a slip joint that can be set more than two positions. Sometimes called box-joint pliers or waterpump pliers. Uses:
- Holding and turning large, round parts.
- General gripping and turning.

Lineman's Pliers

Lineman's Pliers Pliers with stout, flattened jaws and long, slightly curved handles. Cutting edges are formed into one side of the jaws. Sometimes called side cutters. Uses:
- Cutting electrical and other wire.
- Twisting and grasping wire.

Metal Snips Tool with scissors-like handles for cutting metal. Metal snips are sometimes called tin snips. Those with straight blades are primarily used for making straight cuts. Duckbill blades on some models can make curved or straight cuts. Compound-action snips (sometimes called aviation snips) can cut thicker stock. Uses:
- Cutting *light-gauge* metal sheet stock and ducts.

 **Metal Snips**

Section 4.3 Assessment

After You Read: Self-Check

1. Which is the heaviest hammer described in this section?
2. Which tools are used for pulling nails?
3. Which general type of tool should be used to turn nuts or bolts?
4. What other terms are used for metal snips?

Academic Integration: Science

5. **Simple Machines** A *simple machine* is a machine that has no moving parts or few moving parts. A small amount of force is used to apply a greater force or to move a heavier weight. Simple machines can help make work easier, but they do not decrease the amount of work done. A lever is a simple machine made up of a stiff bar and a pivot point, called a *fulcrum*. By applying force to one side of the fulcrum, you increase the force on the other side and can lift heavier objects. Which tools in this section are levers? Explain.

   Starting Hint A claw hammer functions as a lever when removing nails.

🔎 Go to **connectED.mcgraw-hill.com** to check your answers.

Chapter Summary

Section 4.1
Workers in all construction trades need to measure, mark, and lay out projects. They need a variety of tools for measuring and layout. These include tape measures and folding rules; squares and sliding T-bevels for angular measurements; levels and plumb bobs for straight lines; and chalk lines, carpenter's pencils, and scratch awls for marking.

Section 4.2
Hand tools used for cutting and shaping can help workers achieve a high level of precision. These types of tools include utility drywall saws, backsaws, handsaws, hacksaws, keyhole saws, pull saws, coping saws, dovetail saws, block planes, jointer planes, wood chisels, cold chisels, wood rasps, metal files, and utility knives.

Section 4.3
Assembly, fastening, and disassembly tools help put things together and take them apart. These types of tools include clamps, hammers, mallets, bars, nail sets, screwdrivers, staplers, and tackers; wrenches such as adjustable, open-end, box, pipe, and Allen; different types of pliers such as slip joint, needle-nose, groove-joint, and lineman's pliers; and metal snips.

Review Content Vocabulary and Academic Vocabulary

1. Use each of these content vocabulary and academic vocabulary words in a sentence or diagram.

Content Vocabulary
- square (p. 108)
- level (p. 111)
- wrench (p. 120)
- pliers (p. 122)

Academic Vocabulary
- angles (p. 108)
- adjacent (p. 109)
- arc (p. 110)
- horizontal (p. 111)
- vertical (p. 111)

Speak Like a Pro

Technical Terms

2. Work with a classmate to define the following terms used in the chapter: *squareness* (p. 109), *straightedge* (p. 110), *stepping off* (p. 110), *scribing* (p. 110), *jointer plane* (p. 115), *peen* (p. 118), *knurled* (p. 119), *ratcheting* (p. 121), *light-gauge* (p. 123).

Review Key Concepts

3. Describe three hand tools that are used for measuring and layout.

4. Explain how a saw and a plane are used to cut and shape wood.

5. Demonstrate how one tool can be used for both assembling and disassembling.

6. Summarize the uses of a try square, coping saw, and a mallet.

7. Explain which tools carpenters can use to calculate angles.

Critical Thinking

8. Analyze Why are hand tools preferred for some carpentry tasks?

9. Apply In what situation might a carpenter find a one-handed bar clamp more useful than a bar clamp?

Academic and Workplace Applications

STEM Science

10. Work In physics, *work* is defined as force multiplied by distance (**W = fd**). You are doing work when you use a force (a push or a pull) to cause something to move. When calculating work, force, or distance, use *meters* as the unit of distance. Work is measured in units called *joules;* force is measured in units called *Newtons.*

   If you apply a force of 1N to a push a nail 2 centimeters into a piece of wood, what is the amount of work being done?

Step 1: Convert centimeters to meters. (2 centimeters = 0.02 meters)

Step 2: Multiply the force (1) by the distance (0.02) to find the amount of work done.

STEM Mathematics

11. Diameter and Circumference A carpenter used a compass to draw a circle on a piece of drywall that has a diameter of 5 inches. What is the circumference of the circle?

   Math Concept The *diameter* of a circle is the distance from one side of the circle to the other. The distance around a circle is called the *circumference.* Use a math calculator or construction calculator to find the circumference using the formula

$$C = d\pi$$

where **C** is the circumference and *d* is the diameter. If you do not have a calculator, use 3.14 in place of π.

21st Century Skills

12. Creativity and Innovation Choose three tools that you have read about in this chapter that are new to you or that interest you. Create an informational poster that includes a picture of each tool, a description of each tool, and the uses of each tool. You can draw the tool, copy a drawing, or cut out a picture from an advertisement.

Standardized TEST Practice

Multiple Choice

Directions Write down the letter next to the correct answer for each question.

13. Which type of saw is used for crosscutting and rip cutting wood?

   a. keyhole saw

   b. hacksaw

   c. handsaw

   d. none of the above

14. What is a name for a nonadjustable wrench with accurately machined openings on either end?

   a. tire wrench

   b. open-end wrench

   c. box wrench

   d. all of the above

15. What is the removable blade of a combination square used for?

   a. roughly leveling or plumbing a surface

   b. checking adjacent surfaces for the correct angle of 45° and 90°

   c. making layout lines at 45° and 90° across the face or edge of stock

   d. all of the above

TEST-TAKING TIP

In some multiple-choice tests, more than one answer may be possible. Read through all of the choices. Check for answers such as "all of the above" or "none of the above" before deciding on an answer.

*These questions will help you practice for national certification assessment.

Power Saws

Chapter Objectives

After completing this chapter, you will be able to:

- **Identify** circular, table, miter, radial-arm, and reciprocating saws as well as jigsaws.

- **Describe** the safety rules that apply to each type of power saw.

- **Explain** the causes of kickback on circular saws and table saws.

- **Construct** a rip cut with a circular saw, a table saw, and a jigsaw.

- **Demonstrate** how to maintain a circular saw and a table saw.

Discuss the Photo

Saws Safety should always be the number one consideration when using any type of power saw. *Why is it important to understand the differences between types of power saws?*

Writing Activity: Research and Report

Use the Internet to locate buying-guide articles about any saw listed in this chapter. Write a one-page report noting the saw features that are most highly recommended.

Before You Read Preview

The most common use of power saws on a job site is to cut solid wood or wood products. However, carpenters and other tradespeople also rely on power saws to cut many different kinds of materials. Choose a content vocabulary or academic vocabulary word that is new to you. When you find it in the text, write down the definition.

Content Vocabulary

- ○ crosscut
- ○ kickback
- ○ torque
- ○ kerf
- ○ miter cut
- ○ bevel cut
- ○ offcut
- ○ rip cut
- ○ feed rate
- ○ compound miter saw
- ○ internal cut

Academic Vocabulary

You will find these words in your reading and on your tests. Use the academic vocabulary glossary to look up their definitions if necessary.

- ■ diameter
- ■ precise

Graphic Organizer

As you read, use a chart like the one shown to organize information about content vocabulary words and their definitions, adding rows as needed.

Content Vocabulary	Definition
crosscut	a cut made across the grain of a piece of lumber and at a 90° angle to the edge

Go to **connectED.mcgraw-hill.com** to download this graphic organizer.

Circular Saws

Understanding Circular Saws

What are the two types of circular saws?

In the past, most carpenters had to depend on handsaws to cut framing lumber. A power saw is faster and cuts a wider variety of materials. Like many other power tools, power saws are available in cordless, as well as corded, models.

Circular Saws

A circular saw is one of the most important tools on a job site. It is especially useful for cutting panel products such as plywood and oriented-strand board (OSB), and for crosscutting framing lumber, as shown in A 7¼" Circular Saw. A **crosscut** is a cut made across the grain of a piece of lumber, and at a 90° angle to the edge. It is sometimes called a square cut. With the proper blade, even materials such as plastic laminate, masonry, fiber-cement, and nonferrous metals can be cut. *Nonferrous metals* are metals that do not contain iron.

Circular saws vary considerably in size, shape, and power. They are generally classified according to the **diameter** of the blade. A 6½" saw, for example, has a blade with a diameter of 6½". Blades can range in size from 4½" to 12" or more. The smallest saws are used for trim work and for cutting panel products such as sheathing. The largest saws can cut through large beams in one pass. The most common circular saw is the 7¼" model. It is very maneuverable, yet powerful enough to cut a great variety of materials.

Cordless saws that operate on rechargeable batteries are becoming increasingly popular. They may not have the power of a corded saw, but they can operate anywhere.

This makes them particularly useful on a roof or for installing wood fencing far from the nearest electrical outlet.

A 7¼" Circular Saw
Under Control Always maintain control of the saw. Use both handles if necessary. *What safety equipment do you notice in this photo?*

Arnold & Brown

On/off switch

Trigger lock

Blade-adjustment knob

Stationary blade guard

Blade guard lever

Bevel adjustment knob

Arbor

Baseplate (shoe)

Adjustable blade guard

 Parts of a Circular Saw

Circular Saws This contractor's saw has an electric brake. The blade is mounted on the right side of the motor. Some saws have a left-mounted blade instead.

Parts of a Circular Saw

The main parts of a circular saw include a motor, a handle, a *baseplate* or shoe, a fixed guard, an adjustable guard, a blade-guard lever, a blade, a blade-adjustment knob, and an on/off switch, as shown in Parts of a Circular Saw. Blades are described later in this section.

Types of Circular Saws

The two basic types of circular saws are the contractor's saw and the worm-drive saw. The motor on a contractor's saw is perpendicular to the blade. The blade is usually mounted on the right side of the motor and is driven directly by the motor shaft, called an *arbor*. A contractor's saw is sometimes called a *sidewinder*. A contractor's saw with a right-mounted blade as shown in Parts of a Circular Saw.

Kickback is a particular hazard with circular saws. **Kickback** is a reaction that occurs when a spinning blade encounters something that slows or stops it while the saw is under full power. Consequently, the

saw is violently "kicked back" at the operator. You can read more about preventing kickback on page 134.

The motor on a worm-drive saw is parallel to the blade, as shown in A Worm-Drive Saw. The blade is on the motor's left side.

 A Worm-Drive Saw

In Control A worm-drive saw has two large handles. This feature improves operator control.

Power from the motor is transferred to the blade through two gears mounted at right angles to each other. This arrangement generally results in slower blade speed but higher torque. **Torque** is a twisting force that produces rotation. The other parts of a worm-drive saw are similar to those of a contractor's saw.

Circular Saw Blades

Circular saw blades can cut a wide variety of materials. Blades that cut wood, plastic, and metal have teeth formed in the perimeter of the blade. Blades that cut masonry, ceramic tile, and other hard-to-cut materials have a grinding edge that is coated with an abrasive. Blades for circular saws are designed to withstand rough job-site conditions, such as cutting through green lumber or hitting an occasional nail. They are intended for general-purpose crosscutting and ripping in soft woods and panel products, such as plywood.

Reading Check

Describe What are some characteristics of circular saw blades used for cutting wood, plastic, and metal?

Standard Blades The toothed saw blades used on circular saws are similar to those used on table saws, radial-arm saws, and miter saws. Thus, the following information about blades also applies to blades for those tools.

The circular blade for a saw can be altered by the manufacturer to cut different materials or to increase the effectiveness of the saw, as shown in Circular Saw Blade. The key parts of a circular blade are the arbor hole, body, teeth or grinding edge, gullet, and shoulder.

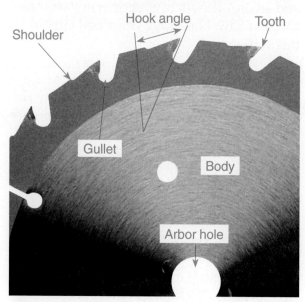

Circular Saw Blade
Parts of a Blade The main parts of a circular blade. This blade has carbide-tipped teeth.

JOB SAFETY

GENERAL POWER SAW SAFETY Specific safety tips for various types of power saws may be found in the sections dealing with those saws. The following safety tips apply when using any power saw. We strongly advise you to check the manufacturer's manual for any special safety instructions.

- Review the tool-related safety guidelines in Chapter 3, "Construction Safety & Health."
- Always wear safety glasses.
- Wear hearing protection, particularly when using miter saws.
- Avoid wearing loose clothing and jewelry because it could get caught in a blade.
- Make sure that the guard is installed and working properly.
- Never cut while off balance.
- If a loose piece of wood contacts a spinning blade, the wood can be violently ejected from the saw. For this reason, never stand in a direct line with the blade.
- When cutting certain materials, such as fiber-cement, it is important to collect the dust created so it is not inhaled. Most saws can be fitted to collect sawdust. Saws can also be connected to a vacuum hose. Wear a respirator that will protect your lungs.

Arbor hole This is where the blade is mounted on the saw. The hole is round on most power saws used in residential construction, but some arbor holes have a diamond shape. Be sure to select a blade with the correct arbor hole for your saw. Never install a blade with an arbor hole that is too large.

Body This is the flat disk that serves as a base for the teeth. It is also sometimes called the plate. Thin, straight slots at the edge of the body prevent the blade from warping as it heats up. Wider, curved slots between the arbor hole and the teeth are called body vents. They reduce the noise of the blade and help to cool it. Body vents also reduce blade vibration, which makes for cleaner cuts.

Teeth or grinding edge Like tiny chisels, teeth slice through the material being cut. Some blades have a grinding edge instead, coated with an industrial abrasive such as carbide or diamond grit.

Gullet This pocket in front of each tooth helps remove sawdust during the cut.

Shoulder The raised portion behind each tooth supports the tooth. A large shoulder is needed on blades that might encounter a nail. Some shoulders are raised to limit the amount of material each tooth can remove. A raised shoulder helps prevent kickback.

The cutting surface of saw blade teeth may be steel or carbide. Blades with steel teeth are inexpensive and can be sharpened on site by the user. However, most carpenters and builders now use carbide-tipped blades. Each tooth is a chunk of tungsten carbide that has been permanently fastened to a shoulder. Tungsten carbide is a metal alloy that maintains a sharp edge much longer than steel. These blades cut a wide variety of materials, but they must be taken to a professional for sharpening.

Differences in the angle, shape, size, and number of teeth greatly affect a blade's cutting ability, as shown in Carbide Tooth Patterns. The number of teeth also affects a blade's cost. More teeth usually means a more expensive blade. Choosing the right blade for a particular use can be difficult. It requires a compromise between cost and various performance features. In general, however:

- More teeth make a smoother but slower cut.
- Fewer teeth make a faster but rougher cut.
- Blades with less than 40 teeth are suitable for ripping.
- Blades with more than 40 teeth are good for making smooth crosscuts.

 Carbide Tooth Patterns

Flat Top Grind (FTG). These blades are particularly suited to making rip cuts. Each tooth, sometimes called a raker, removes a uniform chip of wood. FTG blades are intended for heavy-duty cutting.

Alternate Top Bevel (ATB). Pointed tooth surfaces sever wood fibers, which makes this pattern especially suitable for crosscutting.

Alternate Top Bevel with Rakers (ATB w/R). This arrangement combines two patterns. FTG teeth rake the bottom of the kerf while ATB teeth sever the fibers. One set of teeth cuts while the other set cleans up.

Triple-Chip Grind (TCG). This blade alternates raker teeth with others that have beveled edges. This arrangement excels at cutting hard materials.

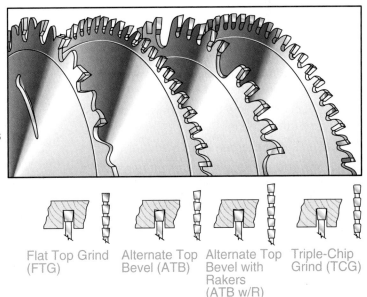

Flat Top Grind (FTG) Alternate Top Bevel (ATB) Alternate Top Bevel with Rakers (ATB w/R) Triple-Chip Grind (TCG)

- Teeth with a steep forward angle cut fast but leave a rough edge.
- Teeth with a shallow forward angle make slower, smoother cuts.

Thin-Kerf Blades The amount of power required to cut through a material depends partly on the thickness of the blade and the kerf. The **kerf** is the width of the cut. A thin-kerf blade has thinner teeth and a thinner body. It requires less power during cutting than a standard-kerf blade, and is particularly suitable for cordless circular saws. Thin-kerf blades are also often found on portable table saws.

Grinding-Edge Blades Grinding-edge blades can be classified by whether they cut wet or dry. They can also be classified by the type of abrasive that does the grinding: diamond, aluminum oxide, or silicon carbide.

Builder's Tip

KNOW YOUR SAW The owner's manual is a booklet packaged with a power tool. It contains specific instructions on how to use the tool safely and how to maintain it. Keep this manual for reference. If you no longer have the printed manual, you can usually download a copy from the saw manufacturer's Web site.

Wet-cutting blades must be sprayed with water during use. Water cools the blade and prevents it from clogging. Wet-cutting blades are also used to reduce the amount of dust caused by cutting materials such as concrete block and stone.

Dry-cutting blades are used without water, and they create a large amount of fine dust. When using them, wear a dust mask or respirator, depending on the material being cut. An inexpensive type of dry-cutting blade has a resinlike body of silicon carbide reinforced with fiberglass and crusted with abrasive. It is used to cut masonry materials such as concrete, brick, and cement block. It can also be used to cut nonferrous metals.

Care of Saw Blades

Any saw blade can be damaged by careless handling. Blades with carbide-tipped teeth are especially vulnerable to damage because carbide is a hard but brittle material. Teeth can be chipped if they come into contact with another metal, such as a nail embedded in the wood. Chipping might also occur when a blade is being changed.

Whenever you store a blade, be sure to protect its teeth. Do not stack blades in a way that allows their teeth to touch. Also, be careful when handling a blade. Sharp teeth can easily cut fingers. The teeth of any blade will eventually dull. A dull blade cuts poorly and is likely to warp. Dulling occurs most quickly when cutting abrasive materials, such as cement board or particleboard.

Carbide-tipped blades are also dulled by heat and corrosion. Always inspect a blade before using it to see if the teeth are sharp. Look closely at the teeth at different points along the blade. If they have chipped corners or rounded edges, the blade should be sharpened. A high-quality blade can be sharpened many times, but lesser-quality blades are often considered disposable.

Frequently inspect abrasive blades that have a resinlike body. Never use an abrasive blade that is cracked or damaged. It can shatter in use, spraying pieces away from the saw at high speed. Damaged or worn abrasive blades of this type should be discarded.

Circular Saw Maintenance

A saw in good working order is safer and will cut more accurately than a damaged saw. Blade guards are particularly prone to damage.

- Inspect blade guards frequently to ensure that they are working correctly. Never use a saw with a damaged or missing blade guard.

- Check that the baseplate is straight. Check along the bottom of the plate at the beginning of each day and if the saw has been dropped. If a baseplate cannot be bent back into alignment, replace it.

- Make sure the baseplate is perpendicular to the blade. Lower the blade as far as it will go, then rotate the blade guard out of the way and place a square against the blade and the baseplate.

- Check the alignment of the plate to ensure that it is perpendicular to the blade and make adjustments as needed.

- Worm-drive saws require a special lubricant to protect the internal gears. Drain the old lubricant periodically and replace it with

JOB SAFETY

CIRCULAR SAW SAFETY The following are general circular saw safety rules. Always check the manufacturer's manual for any special safety instructions.

- Make sure the teeth of the blade are sharp.
- Make sure that the stock you are going to cut is free of nails.
- When working outside or near moisture, make sure that the power source is protected with a GFCI (ground-fault circuit interrupter).
- If you cannot hold a piece of stock safely, clamp it to a sturdy work surface.
- Never make an adjustment on a saw while it is running.
- Be careful when making adjustments when a saw is plugged in but not running. Some adjustments are safe when the saw is plugged in, such as changing blade height or angle.
- Do not stand directly in line with the saw blade. If the blade binds, it may kick the saw back out of the cut. If this happens, turn off the switch immediately.
- Always keep the guard in place and the blade adjusted for the correct depth of cut.
- Use the correct blade for the work to be done.
- Make sure the power cord is clear of the blade.
- Unplug the saw before changing a blade. Make sure the teeth are pointed in the direction of blade rotation and the arbor nut is properly tightened.
- Always allow the saw to reach full speed before starting a cut.
- Always keep your hands clear of the cutting line.
- When finished with a cut, release the switch. Wait until the blade comes to a stop before setting the saw down.
- Always properly support the material being cut so it does not pinch the blade.
- Make sure the blade is the correct diameter and that the arbor hole is the right size and shape for the saw.

new lubricant as recommended by the manufacturer.

- Inspect the power cord. A saw's power cord can be damaged by the blade or by other abuse. If the damage extends through the cord's outer casing, the cord should be replaced.

Reading Check

Recall *List key parts of a circular blade.*

Using a Circular Saw

How do you prevent kickback?

If used improperly, a circular saw is much more likely to injure a worker than a handsaw. Its whirling blade can cause a serious injury before the user can react. For this reason, it is very important that you learn the proper use of this saw. Pay close attention to all the safety recommendations noted here.

Preventing Circular Saw Kickback

Though kickback can occur when using any type of saw with a circular blade, it is particularly dangerous with a circular saw. After kickback occurs, the saw operator can lose control of the saw when the blade starts spinning again. This can lead to serious injury or death. The following situations can result in kickback:

- The saw may be twisted to the side during the cut or pulled backwards. This causes the blade to bind.
- The material on one or both sides of the cut may bend inward, pinching the saw blade.
- The saw may encounter a large knot, which suddenly slows the blade.

To avoid injury from kickback, follow these precautions:
- Cut in a straight line.
- Support the wood in a way that prevents the cut pieces from pinching the blade.

Supporting Plywood
Strong Support The wood grid of this panel-cutting table supports plywood, OSB, and other panel products as it is being cut. The blade should extend no more than ⅛" below the wood being cut.

- Grip the saw firmly.
- Always use a sharp blade. Dull blades are more likely to bind or stall.
- If cutting through a knot is unavoidable, push the saw through at a slower rate.
- Never stand in a direct line with the blade.

Basic Cutting Techniques

Carpenters who cut a lot of plywood can work more safely and quickly by using a panel-cutting table, as shown in Supporting Plywood on page 134. This table is supported by sawhorses and provides a flat surface for cutting plywood and other large panels. Most importantly, it prevents the material being cut from pinching the saw blade. A pinched blade is a major cause of saw kickback. A panel-cutting table can be made on the job site.

Crosscuts and Rip Cuts The most common cut made with a circular saw is the crosscut. When using a circular saw to make a crosscut, place the work over sawhorses or support it securely in some other way. Make sure that the area beneath the cutting line is clear.

Because the blade of a circular saw cuts as its teeth enter from the bottom of the material, it will leave a smoother cut at the bottom than at the top, as shown in How a Circular Saw Blade Cuts. This is why plywood used for cabinetry or siding should be cut with the good side facing down. It is not important to do this when cutting sheathing.

To use a saw, loosen the blade adjustment lever to adjust the depth of cut. Only about ⅛" of the blade should show below the stock. Be sure to tighten the blade adjustment lever before making any cuts. Rest the front of the baseplate flat on the work, but do not let the blade touch the wood yet. Pull the trigger switch and allow the blade to reach full speed. Guide the saw across the board firmly but without force.

You can make a long rip cut in much the same way. A rip cut is made along the direction of the grain. If the cut does not need to be perfectly straight, you can follow a layout line freehand. For the straightest cuts, however, guide the saw with a rip guide such as the one shown in Using a Rip Blade. The rip guide is a metal guide that slips into slots in the saw's baseplate. A circular saw can also be guided by sliding it along a stiff straightedge that has been clamped to the *workpiece*. When ripping a long board, stop the saw before you overreach, then pull it back in the kerf slightly. Stand in a new position, then start the saw again to continue the cut.

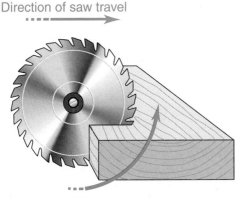

Direction of saw travel

Direction of blade rotation

 How a Circular Saw Blade Cuts
Cutting Action The cutting action of a circular saw blade. The blade cuts from the bottom up.

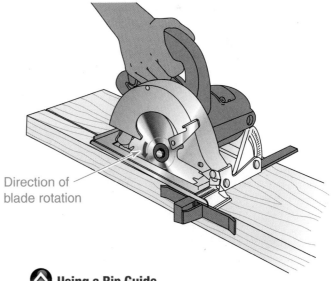

Direction of blade rotation

Using a Rip Guide
Ripping Straight A rip guide can be used to ensure a straight cut.

Miter Cuts

A **miter cut** is similar to a crosscut, except that it is made across the grain of a board at an angle other than 90°. Miter cuts can be made freehand, or they can be guided by a saw protractor such as the one in Making a Miter Cut. A saw protractor is a metal guide with two arms that can be adjusted to various angles. It is marked in degrees. To use it, hold the protractor firmly against the workpiece with one hand. Use your other hand to guide the saw's baseplate along the projecting arm. A triangular framing square can be used in a similar way to guide a saw for crosscuts.

> **Reading Check**
>
> ***Explain*** *Which hand tool may be used to help ensure a perfectly straight crosscut?*

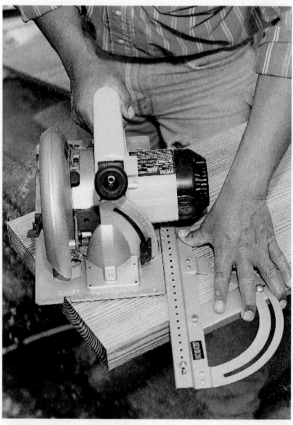

 Making a Miter Cut
Crosscutting Straight Use a saw protractor as a guide if perfectly straight cuts are necessary. It can be adjusted to any angle to make miter cuts.

Bevel Cuts

A **bevel cut** is a type of miter cut that is made at an angle through the thickness of a board, as shown in Cutting a Bevel. When making a bevel cut, the saw cannot cut as deeply as it can when making a crosscut. Most saws can be adjusted to make bevel cuts at angles between 45° and 90°.

To make a bevel cut, loosen the wing nut or adjusting lever and tilt the saw to the desired angle. Then retighten the wing nut or lever. Adjust the saw for the correct depth of cut. Retract the blade guard until the blade engages the wood, then allow it to slide back into position. Otherwise the guard may interfere with the cut. Make the bevel cut freehand or guide it with a saw protractor. Sometimes the blade guard will prevent a bevel cut from starting easily. In that case, using the saw blade lever, raise the guard slightly as you start the cut. Then release the lever.

A compound-bevel cut (which is also called a compound-miter cut) can be made when a bevel cut is combined with a miter cut. It can be created by tilting the blade *and* using a saw protractor to guide the saw.

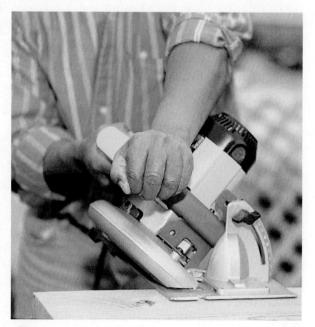

 Cutting a Bevel
Beveled End When making a bevel cut, it is important to keep good control of the saw by using both hands.

After You Read: Self-Check

1. When using a circular saw, what should you do with a 2×4 that cannot be held safely?
2. What special maintenance does a worm-drive saw require?
3. When making a crosscut with a circular saw, how much of the saw blade should show below the stock?
4. Describe how a bevel cut is made.

Academic Integration: English Language Arts

5. **Create an Outline** You have been hired by a major saw manufacturer to create a circular saw owner's manual. Use Section 5.1 to create an outline of the information you would include in the owner's manual. The outline should be no more than one page long.

Go to **connectED.mcgraw-hill.com** to check your answers.

Table Saws

Section 5.2

Understanding Table Saws

What purpose does a fence serve?

A table saw is one of the most versatile tools on a job site. The primary use of a table saw is to rip stock to width, though it can also be used to crosscut stock. It can rip cut with greater precision than a circular saw, which is why it is often used for ripping framing lumber and wood panel products. With the proper blade, a table saw can also be used to cut materials such as rigid sheet plastic. Though very similar to the type of table saw found in woodworking shops, table saws for building construction tend to be lighter and more portable.

Parts of a Table Saw

The main parts of a table saw are the table, the motor, the base, and the on/off switch, as shown in Parts of a Table Saw. For safety purposes, the on/off switch should be large and easy to reach. A miter gauge or a rip fence guides wood through the blade, and a guard prevents accidental contact. A splitter directly behind the blade prevents the stock from pinching the blade as it is being cut.

An opening in the center of the table allows the blade to be raised above the table surface. The opening is partially covered by a throat plate (the red area around the blade in Figure 5-11). This plate is an important safety feature that prevents small pieces of wood from being wedged against the blade.

The miter gauge is used during crosscut operations and can be adjusted for various angles. It slides forward and back in one of two grooves in the top of the saw table. Both grooves are parallel to the saw blade. When not in use, the miter gauge should be removed from the table.

Miter gauge slot

Blade guard with splitter

Throat plate

Table

Rip fence

Extension table
(outfeed table)

Miter gauge

On/off switch

Fence clamp

Blade height wheel

Pointer scale

Base

Parts of a Table Saw

Open Base The main parts of an open-base saw, or contractor's table saw. This saw is fitted with an extension table to increase cutting capacity.

The rip fence is used for all ripping operations. It is adjusted by sliding it left or right across the table. When it is ready for use, it is locked in place. If not being used, the fence should be moved well away from the blade and locked, or removed entirely. Also available are table extensions that can be fastened to the sides of the saw table. These are especially convenient when cutting a large panel, such as a sheet of plywood. The blade guard should be kept in place whenever possible. If a cut cannot be made with the

standard guard in place, some other safety method should be used instead.

Types of Table Saws

A table saw is generally classified according to the diameter of the largest blade it will accept. The most common table saw on a job site is the 10" saw. Table saws for the job site have the same features as models found in woodworking shops, but are easier to transport. Most table saws have a fixed, horizontal table and a blade that can be raised, lowered, and angled. This type is called a tilting-arbor saw. In contrast, a few saws feature a table that tilts; the blade can only be raised or lowered. This type has limitations on a job site and is rarely used.

Two types of tilting-arbor table saws are commonly used in building construction. Contractor table saws, sometimes called open-base saws or builder's saws, rest on a metal framework. This makes the saw light enough for two people to carry and move from site to site. Portable table saws, also called *benchtop saws,* are even lighter,

JOB SAFETY

UNEXPECTED MOVEMENT When using the fence to rip stock, always remove the miter gauge and store it elsewhere. This will prevent it from interfering with the cut or vibrating loose.

and can be carried by one person. These are easy to move around the job site, but are not generally used for heavy duty cutting. Many include folding stands with built-in wheels that make the saws even easier to transport.

Table Saw Blades

A table saw uses a circular blade that has the same basic features as a circular saw blade (see "Circular Saw Blades" on page 130). The main difference is that table saw blades are usually at least 10" in diameter. Also, the teeth may be shaped to provide a higher quality cut. Blades often feature a combination of ripping and crosscut teeth that make it suitable for a variety of cutting jobs. Several styles of teeth are available, and each has a particular application.

The blade of a table saw cuts as its teeth enter the top of the material, as shown in How a Table Saw Blade Cuts. This means it will leave a smoother cut at the top than at the bottom. Material should be cut with the good side facing up.

Table Saw Maintenance

A table saw requires less maintenance than some other saws. Refer to the manufacturer's instructions for specific recommendations. The following are general guidelines only.

- Keep the table clean and smooth. Remove patches of rust with metal

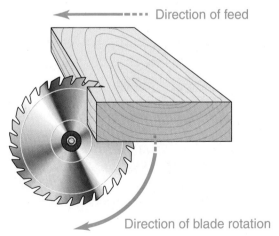

 How a Table Saw Blade Cuts
Good Side Up A table saw blade cuts from the top surface down.

polish. Scrape off paint drips and glue to prevent them from interfering with the movement of stock. After cleaning, some builders coat the top with paste wax. This makes it easier to slide stock over the surface. After cleaning, the table to may be coated with a paste wax as a lubricant so that the material cut may be fed smoothly towards the blade.

- Unplug the saw, then gently brush or blow accumulated sawdust from the blade-raising-and-tilting mechanism beneath the table. Lubricate the mechanism as recommended by the manufacturer.

- Some saws are driven by a V-belt that links the motor to the arbor. Replace the belt if it becomes worn or damaged.

Reading Check

Explain *What is the advantage of a benchtop saw?*

Using a Table Saw

Why is a push stick important?

Many of the same safety precautions that you would take with a circular saw also apply to table saws. However, because a table saw blade cuts in a different fashion, you must also know some safety rules that are specific to this tool.

Adjusting the Saw

To operate a table saw safely, you must carefully adjust the blade, the rip fence, and the miter gauge. Never make adjustments while the saw is running.

Setting the Blade Height To raise the blade, hold the workpiece near the side of the blade and carefully turn the blade-height wheel until the top saw tooth is at the correct height. For most cutting, the top of the blade should extend no more than $\frac{1}{8}$" above the stock. When making some types of cuts, such as grooves, the blade must be set for the exact depth of cut.

Tilting the Blade A blade-angle wheel on the side of the machine tilts the blade. A pointer or scale on the front of the saw indicates the degree of tilt. Turn the wheel until the blade is tilted to the desired degree. Lock the wheel to maintain the angle.

Adjusting the Rip Fence The rip fence is usually placed to the right of the blade. For accurate and safe cutting, set the rip fence exactly parallel to the blade. To check it, align it with one of the miter gauge slots in the table. Then lock the fence. Check alignment again. If the edge of the fence is not parallel to the slot, unlock the fence and readjust it.

To adjust the fence to a specific distance, first move it to the approximate location. Holding a rule or tape measure at a right angle to the fence, carefully measure the distance from the fence to the nearest edge of one tooth, as shown in Adjusting the Fence. Then lock the fence. Make a small test cut on the workpiece. Measure it to double-check the setup before making the complete cut.

Adjusting the Miter Gauge The miter gauge, which is used for crosscutting operations, can be used in either groove on the table. There is a pointer and scale on the miter gauge for setting it to any angle to the right or left. Most gauges have automatic stop positions at 30°, 45°, 60°, and 90°.

Before relying on the accuracy of a miter gauge, square it to the blade. Unplug the saw and use a framing square to see if the gauge and the blade form a 90° angle. If they do not, adjust the gauge.

Builder's Tip

BLADE INSTALLATION The arbor nut that holds a table saw blade in place always loosens when turned in the direction in which the blade's teeth are pointing. This is the direction of blade rotation.

Preventing Table Saw Kickback

Kickback from a table saw can be as dangerous as it is from a circular saw. However, on a table saw, the workpiece is kicked back, not the saw itself. A piece of wood ejected from the saw can become a deadly missile. Kickback can occur:

- When the workpiece is twisted to the side during the cut. This causes the blade to bind.
- When the stock on one or both sides of the cut bends, pinching the saw blade.
- When the rip fence is not parallel to the blade. This causes the blade to bind.

 Adjusting the Fence
Measure Carefully Measure from the fence to the blade tooth, then double-check the distance after locking the fence.

Arnold & Brown

- When wood is being crosscut with a miter gauge and a rip fence. A waste piece, called an **offcut**, can become wedged between the blade and the rip fence. This can produce kickback.

Follow these guidelines to avoid kickback:
- Be sure that one edge of the workpiece is always held firmly against the rip fence. Always keep a firm grip on the stock, and hold it flat against the table. Use a push stick when necessary.

JOB SAFETY

GENERAL TABLE SAW SAFETY The following are general table saw safety rules. Always check the manufacturer's manual for any special safety instructions.
- Remove rings, watches, and other items that might catch in the saw. Wear garments with short or tight sleeves. Keep long hair tied back.
- Wear proper eye protection.
- Always keep the area around the table saw clean and uncluttered.
- Never stand directly behind the blade.
- Never cut freehand.
- Do not cut warped or bowed material on the table saw.
- Make sure the stock is free of nails.
- When ripping stock that cannot be fed safely by hand, always push the stock past the blade with a push stick.
- Make all rip fence adjustments after the saw has been turned off and the blade is no longer spinning. Be certain the rip fence is clamped securely while cutting.
- Use the proper saw blade for the operation being performed. Always cut with a sharp blade.
- Unplug the saw before changing blades.

- Always adjust the saw blade so that it protrudes just enough above the stock to cut completely through.
- Never reach over a spinning saw blade. Instead, reach around the side of the machine.
- Keep your fingers away from the saw blade at all times.
- Always keep the guard and splitter in place. If the cut you are making does not permit use of the guard, use a featherboard or a special guard. A featherboard is a piece of stock with a series of long saw cuts on one end. It is used to hold narrow stock against the rip fence when making a rip cut with a table saw.
- When crosscutting with the miter gauge, never use the fence as a stop unless a clearance block is used. A clearance block prevents the wood from becoming trapped between the fence and the blade.
- When ripping, place the jointed or straightest edge of the stock against the fence.
- Keep the saw table clean. Remove all scraps with a brush or push stick, never with your fingers.
- Always hold the stock firmly against the miter gauge when crosscutting and against the rip fence when ripping. A helper should not pull the stock. The helper only supports the stock.

- Always support the wood in a way that prevents the cut pieces from pinching the blade. This might require a table or roller stands placed under the stock where it leaves the table.

- Always work with a sharp blade. Dull blades are more likely to bind or stall.

- When using the miter gauge to crosscut stock, make sure any offcuts cannot become trapped between the blade and the rip fence. To do this, remove the fence or push it well clear of the cutting area.

Reading Check

Explain *Why do sharp blades help prevent saw kickback?*

Basic Cutting Techniques

The most common cuts made on a table saw are rip cuts and cross cuts. Of the two, rip cuts are by far the most common. A **rip cut** is a cut made in the direction of the wood grain. It is usually parallel to the edge of a board or panel.

Ripping

It is essential to make sure the rip fence is properly adjusted when ripping stock. Whenever possible, place the widest part of the stock between the blade and the fence. The **feed rate** is the speed at which stock is pushed through the saw blade. For ripping thick stock or hardwoods, the feed rate should be slowed. The person on the infeed side controls the feed rate and the person on the outfeed side helps by supporting the material.

Wide Stock Stock with a width of 6" or more is considered wide. To rip wide stock:

1. Adjust the rip fence to the appropriate width and lock it.

2. Turn on the saw. Place the end of the stock flat against the table, as shown in Making a Rip Cut.

3. Push the wood against the fence with your left hand, and push it forward with the right. If the board is long, use one or more roller stands to support it as it leaves the table. Set up the stands before you begin the cut.

 Making a Rip Cut
Slow and Steady Rest the leading edge of the panel on the front of the saw table. A roller stand can support long stock as it exits a table saw. The guard should be in place for all tool operations.

Arnold & Brown

4. Feed the stock into the blade at a slow, steady speed. Hold your right hand close to the fence as you work. Be careful not to overload the saw.

5. To avoid kickback, push the stock completely past the blade to complete the cut.

Narrow Stock Narrow stock is generally considered anything with a width less than 6". Keep in mind that the narrower the stock, the more difficult it will be to rip safely. Always try to position the stock so that the widest portion is between the blade and the fence. To rip narrow stock:

1. Observe the same general practices as in starting the cut on wide stock.

2. As the uncut end of the board reaches the front of the table, use a push stick instead of your hand to guide the board between the blade and the fence, as shown in Ripping Narrow Stock.

3. Use a *featherboard* if necessary. Again, a featherboard is a safety device that can be attached to the saw. It helps to hold stock firmly against the saw fence.

Bevels On a table saw, a bevel cut is an angled cut made in the direction of the grain. To cut a bevel:

1. Tilt the blade to the correct angle for the bevel.

2. Place the fence on the table so the top of the blade is tilting away from the fence.

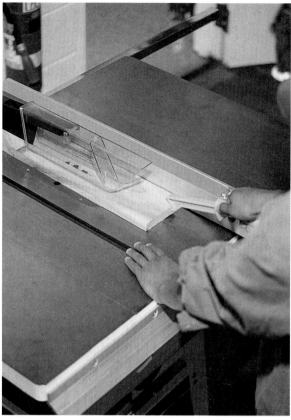

 Ripping Narrow Stock
Use a Push Stick Note that the push stick is used between the blade and the fence, not on the outside of the blade. The guard should be in place for all tool operations.

3. Adjust the fence for the correct width of cut.

4. Adjust the height of the blade to clear the top of the board slightly.

JOB SAFETY

EASY SAFETY: PUSH STICK Never under any circumstances cut narrow stock without a push stick. It is good practice to hang the push stick conveniently at the side of the saw so that you are not tempted to make a cut without it.

Builder's Tip

KNOWING WHEN STOCK IS FLAT Make sure that the material to be cut is flat against the saw table as cutting starts. Before cutting large pieces of stock, rock the wood forward and back on the front surface of the saw table. You will feel it "slap" the table when it is flat.

5. Hold the work firmly against the fence as the cut is made. Bevel Cutting shows the blade entering the stock.

Crosscutting

Use the miter gauge for most crosscutting operations. Always remove the rip fence or slide it well out of the way. In some cases the rip fence can be used during crosscutting operations. However, this is an advanced technique that must not be used without supervision by your instructor. To crosscut:

1. Hold the stock firmly against the gauge and advance it slowly into the blade, as shown in Crosscutting Narrow Stock.

2. When the cut is complete, slide the stock away from the blade and remove it. Never drag the cut edge back across the blade. Never put your fingers near the blade to remove the offcut.

Cutting Panels

Because of their size, panels such as plywood and oriented-strand board present cutting problems. The workpiece is often too large to fit conveniently on the saw table. The panel and any waste should be fully supported as they leave the top of the table saw. This can be done with a roller stand or an outfeed table. To cut panels:

1. Adjust the blade to the proper height. Adjust the fence for the width of cut.

2. Make sure the panel edge that will ride against the rip fence is flat and straight. If it is not, the cut will not be accurate. Place a different edge against the fence if necessary.

3. Turn on the saw, and rest the leading edge of the panel on the front of the saw table. The good side should be facing up, as shown on page 142.

4. Push the panel steadily forward, while at the same time applying pressure to keep its edge firmly against the rip fence. Do not allow the panel to move sideways or it will bind on the blade.

5. As you complete the cut, be sure to push the panel and the waste piece completely past the blade.

 Bevel Cutting
Angled Cut The guard is not shown here. This allows the operation to be shown more clearly. The guard should be in place for all tool operations.

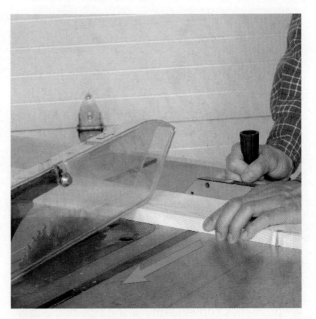

Crosscutting Narrow Stock
Feed Direction The arrow indicates the feed direction. Note how the guard slides over the stock as the blade cuts through the wood.

(l & r)Arnold & Brown

Installing a Table-Saw Blade Sometimes a blade must be removed for sharpening. In other cases, it must be replaced by another blade when cutting a different material. Knowing how to change the blade safely is important.

Step 1 Unplug the saw to prevent accidental startup.

Step 2 Remove the throat plate. This usually snaps in and out of position.

Step 3 Select a wrench to fit the arbor nut. On most saws, the arbor has a left-hand thread. The nut must be turned clockwise to loosen it. Hold a piece of scrap wood against the blade to keep the arbor from turning as you work, as shown in the figure.

Step 4 Remove the nut and the collar. Hold them securely or they may drop into the sawdust below and be difficult to find. Take off the old blade, being careful not to bump its teeth against the table.

Step 5 Slip the new blade onto the arbor. The teeth should point in the direction of blade rotation (toward the operator). Replace the collar and nut. Tighten the nut firmly, but not too tight. The nut will tighten further as the saw is used.

Step 6 Replace the throat plate.

Section 5.2 Assessment

After You Read: Self-Check

1. When you are cutting wood with a table saw, should the good side be facing up or down?
2. Name the two devices that are used to guide stock when cutting with a table saw.
3. Explain two ways of supporting long stock for ripping on a table saw.
4. List the required steps in installing a table-saw blade.

Academic Integration: English Language Arts

5. **Research** Research table saw blades for rip cutting and crosscutting. Describe the two types of blades. Write one paragraph about the differences in teeth as well as the type of cut produced by each.

Go to **connectED.mcgraw-hill.com** to check your answers.

Arnold & Brown

Miter & Radial-Arm Saws

Understanding Miter Saws

How are the various miter saws similar?

The miter saw, or chop saw, has become a very important tool on the job site. Frame carpenters may use it for making quick, repetitive crosscuts of framing lumber. Finish carpenters rely on it for accurate cuts on molding and trim, especially crown molding.

The miter saw makes a clean cut that is very accurate. It is portable, easy to set up, and maintains its accuracy, which is why it is often preferred over a radial-arm saw. However, it can be used only for crosscutting, not ripping. Some miter saws can make compound cuts in one pass. Fitted with a suitable blade, the miter saw can even cut metal pipe and light-gauge metal framing. Also, miter saws can easily be fitted with dust-collection equipment such as a dust bag.

The size of a miter saw is determined by its blade diameter. Most miter saws used for residential work range from 8¼" to 12" models.

Parts of a Miter Saw

A miter saw looks something like a circular saw mounted on a pivoting frame, as shown in Parts of a Miter Saw. The saw head may be mounted on a pivoting mechanism or on metal rails. The saw table has a miter scale. Most miter-saw scales contain positive (locked) stops at 90° and 45°. Positive stops allow the saw to be set quickly to commonly used angles. Some saw tables also contain positive stops at 15°, 22.5°, and 30°. On compound saws, the head may also contain positive stops.

Miter saws have a split fence that allows the blade to pass through. Miter saws also have retractable blade guards. The guard automatically pivots to shield whatever portions of the blade are not being used in a cut. This safety feature should never be disabled. All miter saws are equipped with a dust bag, a vacuum port, or both.

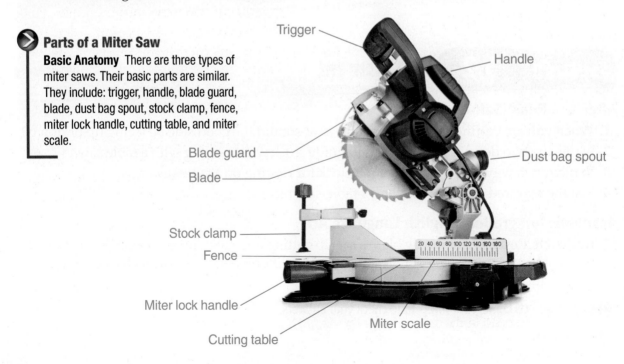

> **Parts of a Miter Saw**
> **Basic Anatomy** There are three types of miter saws. Their basic parts are similar. They include: trigger, handle, blade guard, blade, dust bag spout, stock clamp, fence, miter lock handle, cutting table, and miter scale.

Trigger

Handle

Blade guard

Blade

Dust bag spout

Stock clamp

Fence

Miter lock handle

Miter scale

Cutting table

Naumold/iStock/360/Getty Images

⌃ **A Sliding Compound-Miter Saw**

Large Capacity The blade of a sliding compound-miter saw can make a miter cut and a bevel cut at the same time, even across a wide board.

Types of Miter Saws

The three basic types of miter saws include conventional miter saws, compound-miter saws, and sliding compound-miter saws. Conventional miter saws were the first to become available. A **compound miter saw** is a saw in which the head of the saw pivots up and down and from side to side. Conventional models are typically the simplest and lightest. They are suitable for crosscuts and miter cuts.

A compound-miter saw has the same range of motion as conventional models. In addition, the head of the saw can also be tilted at an angle to one side or the other to make a bevel cut. This allows the saw to make compound cuts in one pass. It is often used when cutting crown moldings, handrails, and other trims that require complex fitting.

A sliding compound-miter saw, shown in A Sliding Compound-Miter Saw, is similar to a compound-miter model, except that it can cut wider stock. The head of the saw slides back and forth on one or two metal rails, and also pivots up and down.

With all three types of miter saws, the direction of blade travel is away from the operator, as shown in How a Miter Saw Cuts on page 148. In other words, the exposed teeth of the saw point toward the fence.

Miter Saw Blades The basic features of miter saw blades are the same as those for circular saw blades and table saw blades. However, these blades are sometimes referred to as crosscut blades or cutoff blades because they are used for trimming stock to length. A 40-tooth blade is a good general purpose miter saw blade because it provides a good-quality cut across the grain of a board. A blade with 60 teeth or more would be used for fine cuts on expensive molding.

Maintenance Check the positive stops for the table and the head (on compound models) regularly for accuracy. Adjust them as needed. The rails of a sliding-compound miter saw do not normally require attention other than brushing off excess sawdust. However, the sliding mechanism itself might

MITER-SAW SAFETY The following are general miter-saw safety rules. Check the manufacturer's manual for any special safety instructions.

- Remove rings, watches, and other items that might catch in the saw. Wear garments with short or tight sleeves. Tie back long hair.

- Wear proper eye protection.

- Use the correct cutting motion. The various types of miter saws have different requirements.

- Make sure that the stock is held firmly against the fence.

- Never disable the blade guard.

- Make any adjustments to the saw only after the blade has stopped moving.

- Unplug the saw before changing blades.

- Support the stock to be cut along its entire length. Never cut stock that is too short to hold safely.

- Most miter saws operate with a high-pitched whine. Wear hearing protection.

- Do not lift stock into the blade to complete a cut.

require lubrication. Consult the owner's manual for specific recommendations.

Brush off the table periodically to remove sawdust and small bits of wood that can interfere with table movement. In addition, remove any sawdust or small pieces of wood that might prevent the blade guard from operating properly. Empty the dust bag frequently. A buildup of sawdust in the bag encourages the release of fine dust into the air. This can lead to respiratory problems for anyone in the area.

Using a Miter Saw

The various types of miter saws differ in operation. For safety, be sure you understand the differences.

All Miter Saws If the saw has a manually operated blade brake, push down on the brake button when the cut is complete. If the saw has an automatic blade brake, it will engage as soon as the trigger switch is released. The blade will stop quickly.

Conventional Miter and Compound-Miter Saws To turn the saw on, pull the trigger switch. Make the cut by pivoting the saw head down into the wood. As soon as the cut is completed, release the trigger when the blade stops, then return the saw to its starting position (See A in How Miter Saw Cut.)

Sliding Compound-Miter Saws Slide the saw head outward, past the stock to be cut (see B in How Miter Saw Cut). Turn the saw on

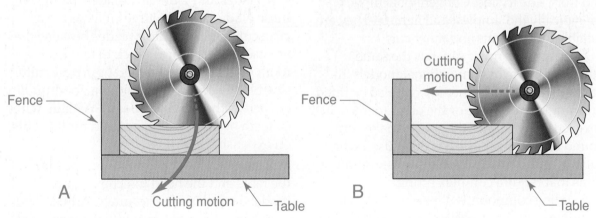

 How Miter Saws Cut

Cutting Motions **A.** The cutting motion of a conventional miter saw is downward and slightly toward the fence.
B. The cutting motion of a sliding compound-miter saw is directly toward the fence.

Installing a Miter-Saw Blade Replace a blade whenever the teeth become dull. A sharp blade improves safety and the quality of a cut. To replace a blade:

Step 1 Unplug the miter saw from its power source.

Step 2 Remove the blade guard and, if necessary, the blade housing from the saw. On some saws, the guard may simply pivot out of the way.

Step 3 Place a hex wrench in the depression in the end of the arbor to hold the arbor stationary. (Many saws have an arbor-lock button instead.)

Step 4 Loosen the arbor nut with an open-end wrench by applying pressure on the wrench in the direction of blade rotation. Remove the nut and the collar.

Step 5 Slide the old blade off the arbor.

Step 6 Slide a new blade onto the arbor. Make certain the teeth at the bottom are pointing away from you and toward the fence.

Step 7 With the recessed side against the blade, replace the collar. Replace and securely tighten the nut. Reinstall the blade housing and blade guard.

Step 8 Plug in the saw and turn it on and off several times without making a cut. This will help you to determine if the blade is properly secured.

---

using the trigger switch. To make the cut, pivot the saw head downward all the way and push it forward through the wood. When the blade is completely clear of the stock, release the trigger switch and lift the saw head.

Basic Cutting Techniques

Before making any cuts with a miter saw, make sure the tool is securely fastened to its stand. Also be sure that the stand rests on a solid surface and does not tip as the saw is used. The saw should be equipped with a blade brake. A blade brake allows the blade to be stopped quickly. This reduces the chances of hand injuries caused when the blade spins freely.

Crosscutting When cutting flat pieces, first check to see if the material is slightly *bowed*. If it is, make sure the material is positioned on the table as shown in A in Cutting Bowed Stock. If the material is positioned as shown in B in Cutting Bowed Stock, it will pinch the blade near the completion of the cut. This could result in kickback. The stock in these drawings is bowed to an extreme in order to make the drawing easier to understand. However, neither piece of wood could be safely cut on a miter saw if bowed this much.

Hold the stock flat against the table and tightly against the fence with one hand. Keep it well clear of the blade. Then make the cut by pulling the saw downward.

To make angle cuts, release the table lock. Move the indicator to the angle to be cut and engage the table lock. When making cuts at 45° or 90°, release the table lock and move the handle until the positive stop makes contact. Then engage the table lock.

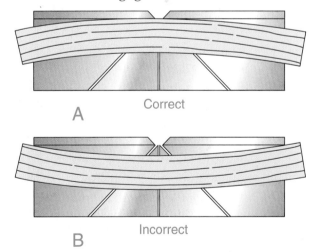

A Correct

B Incorrect

 Cutting Bowed Stock
Bow Caution A. If the material is slightly bowed, position it as shown. **B.** If positioned as shown, the stock may kick back. Bowing here is exaggerated.

Radial-Arm Saws

What is this type of saw used for?

The radial-arm saw, shown in Parts of a Radial-Arm Saw, can be used for ripping, *dadoing* (cutting a groove), and various combination cuts. However, it is best suited to crosscutting. For instance, a long board can be cut into shorter lengths easily because the board remains stationary on the table while the saw is pulled forward through the stock. Its primary advantage over miter saws is its ability to crosscut unusually wide or thick stock. However, this feature is seldom needed on residential job sites and the saw is awkward to transport, so this tool is much less common on residential job sites than it once was.

Like other saws with circular blades, the radial-arm saw is generally classified according to the largest diameter blade it will accept. Heavy-duty saws used in commercial construction might have a 14" blade.

The most common size used in residential construction is the 10" model.

Parts of a Radial-Arm Saw

A radial-arm saw is essentially a circular saw held by a yoke that allows it to slide forward and back on a track in the arm. The arm is attached to a vertical post that can raise or lower it. The saw takes its name from the fact that the arm moves in an arc from side to side (in other words, radially). Other parts of the radial-arm saw are shown in Parts of a Radial-Arm Saw.

Using a Radial-Arm Saw

A radial-arm saw tends to feed itself into the work. This is due to the position of the blade and the feed direction, as shown in How a Radial-Arms Saw Cuts on page 151. Though a radial-arm saw can make many types of cuts, its use should be limited to crosscutting.

To prevent the saw from lurching forward, hold it securely before turning it on and

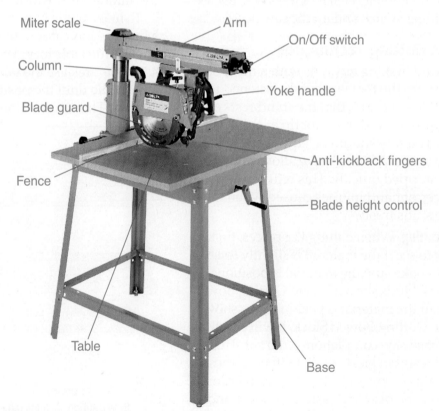

 Parts of a Radial-Arm Saw
Saw and Stand A radial-arm saw is mounted on a post located at the back of the saw stand. Infeed and outfeed tables support long stock.

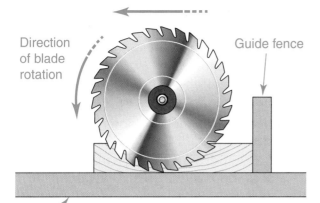

Direction of saw travel

Direction of blade rotation

Guide fence

Table

 How a Radial-Arm Saw Cuts
Cutting Path The teeth of the saw blade point away from the operator and cut from the top of the stock.

while cutting, as in Hand Position. Hold the stock to be cut tightly against the fence with one hand. With the other hand, grasp the motor yoke handle. For crosscutting and similar operations, pull the saw into the work. Return the saw to the rear of the table after each cut. Do not move the stock until the saw blade is behind the fence.

Mathematics: Geometry

Preparing to Cut Angles You are building a hexagon-shaped frame. You ask your teacher how to figure the angles at which your frame pieces will need to be mitered. The teacher draws the sketch below and challenges you to use it to figure out the angles. Write a few sentences explaining your reasoning and indicating the angle at which to cut the miters.

Starting Hint Figure out the interior angles of one of the triangles in the sketch. Notice that the triangles are congruent and isosceles.

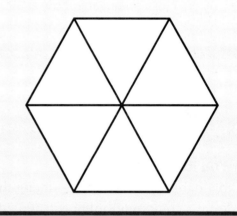

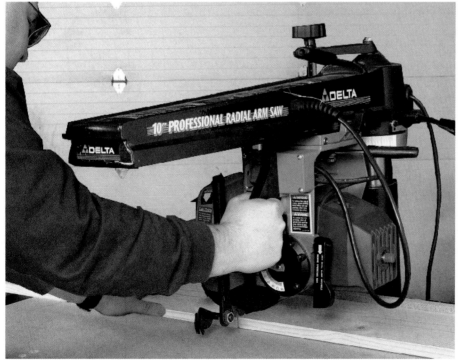

 Hand Position
Keep the Saw from Lurching The proper hand position for crosscutting. Notice the position of the anti-kickback fingers. The guard should be in place for all tool operations.

JOB SAFETY

RADIAL-ARM SAW SAFETY The following are general radial-arm saw safety rules. Check the manufacturer's manual for specific safety instructions.

- Remove rings, watches, and other items that might catch in the saw. Wear garments with short or tight sleeves. Tie back long hair.
- Wear proper eye protection.
- Always keep the safety guard and the anti-kickback device in position unless you are changing the blade.
- Before crosscutting, adjust the anti-kickback device (sometimes called anti-kickback fingers) to clear the top of the work by about ⅛".
- Make sure the clamps and locking handles are tight.
- Never place your hand closer than 6" to the blade.
- Use a brush or stick to keep the table clear of all scraps and sawdust. Don't brush the table with your hand.
- When finished with a cut, return the saw head all the way back past the fence.

Section 5.3 Assessment

After You Read: Self-Check

1. Explain the basic differences among the three types of miter saws.
2. What is a compound-miter cut, and what is it used for?
3. Why is a blade brake important?
4. What is the primary advantage of a radial-arm saw over a miter saw?

Academic Integration: Mathematics

5. **Complementary Angles** A miter saw makes angled crosscuts across the grain of a board. To cut a board at 90°, the saw is set at 0°. What is the saw setting for a miter joint for a 126° wall?

   Math Concept To solve this problem, you will need to determine the complementary angle you would set your saw to make the cut. Complementary angles are two angles that equal 90° when added together. For example, if one angle is 60°, its complementary angle would be 30°. To cut a board at 60°, you would set your saw to 30°.

   Starting Hint First divide the wall angle by 2, because you are measuring for a miter joint where two pieces of wood would meet at 126°. Next, determine your complementary angle and the appropriate miter saw setting.

Go to **connectED.mcgraw-hill.com** to check your answers.

Jigsaws & Reciprocating Saws

Jigsaws

Which part of a jigsaw is adjustable?

The jigsaw (sometimes called a saber saw or bayonet saw) is the best power saw for making curved or irregular cuts. Carpenters use a jigsaw to notch wood deck boards to fit around a post. Cabinet installers use it for fitting cabinets and for cutting sink holes in countertops. Siding installers use a jigsaw to cut siding to fit around arched windows.

Unlike the saws discussed in previous sections, the jigsaw has a straight blade. Teeth are formed into one edge, and, instead of spinning, the blade moves up and down. Fitted with a suitable blade, a jigsaw will cut metal, wood, plastic, and many other materials. It is often used instead of a hacksaw to cut angle iron and various kinds of metal or plastic pipe. The saws cannot be adjusted for depth of cut. Because the teeth of a jigsaw point upward, the good surface of a material should face down as it is cut.

Parts of a Jigsaw

A typical jigsaw has a baseplate, a housing/handle, and a blade-locking mechanism, as shown in Parts of a Jigsaw. The baseplate on many models can be tilted to one side or the other. This allows the saw to make bevel cuts. Most jigsaws have a variable-speed control. This enables them to cut a wide variety of materials.

The blade of a jigsaw is often secured to the shaft by a clamp or mounting screw. In order to change the blade, older jigsaws require the use of a screwdriver or a hex key. Some jigsaws are fitted with a convenient blade-locking mechanism that does not require a special tool. The cutting motion of a jigsaw is shown in How a Jigsaw Cuts.

Types of Jigsaws

Two types of jigsaws are commonly used in residential construction. Top-handled models are the most common. The handle contains the trigger switch and is used to guide the tool. Some people prefer the second type: barrel-grip models. Instead of a handle, the user grasps the body of the saw to guide it. Because the hand is held low on the tool and directly behind the blade, some people find this model easier to control. Mechanically, however, the two types are the same.

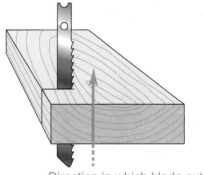

Direction in which blade cuts

How a Jigsaw Cuts
Cutting Motion The teeth of most jigsaw blades point upward, so the blade cuts only on the upstroke.

Parts of a Jigsaw

On/off switch
Handle
Chip shield
Motor
Blade-locking mechanism
Baseplate
Blade

Parts of a Jigsaw
Vertical Cutting A jigsaw is a simple power tool. Few adjustments are required.

Arnold & Brown

On some jigsaws, the blade moves straight up and down. On others, a setting on the tool allows the blade to move with a slight orbital motion as well. Jigsaws with an orbital setting cut much faster through wood. However, orbital motion is a disadvantage when cutting a material such as sheet metal. It causes the metal to vibrate and does not provide as smooth a cut.

Using a Jigsaw

The blade and the specific cutting technique used with a jigsaw should be adapted to suit the material. The following techniques are commonly used when cutting wood. Safety tips are much the same as those for reciprocating saws (see the Job Safety feature on page 157).

Straight and Irregular Cutting Though jigsaws can make straight cuts with some accuracy, the tool excels at curved cuts. For either type of cutting, proceed as follows:

1. Mark a layout line on the wood. It must be dark enough to be seen beneath the fine layer of sawdust that will be created around the cutting line.
2. Hold the wood tightly against a work surface, or clamp it in place as in Cutting a Curve. Make sure the area beneath the cutting line is unobstructed.

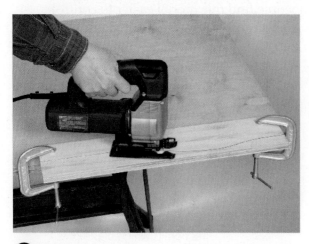

 Cutting a Curve
Clamping When cutting small pieces of stock, clamps may have to be repositioned in order to complete a cut. *Which clamp had to be repositioned first?*

3. Rest the front edge of the baseplate on the wood. Start the saw and allow it to reach full speed.
4. Move the saw blade slowly into the wood. Do not force it. Use only enough pressure to keep the saw cutting at all times.
5. When the cut is complete, stop the saw. Let the blade come to a complete stop before lifting it away from the wood.

Internal Cuts Sometimes an internal cut must be made. An **internal cut** is the technique of cutting a large hole in a material without starting at the edge. This is often required prior to installing a sink in a countertop. A jigsaw is an easy way to make an internal cut. There are two ways to make internal cuts.

Method #1

1. Mark the outline of the cut on the workpiece.
2. Using a ⅝" spade bit on an electric drill, drill through the material just inside the layout line, as shown in Internal Cut. This will be the starting point for the saw.
3. Proceed as in steps 2 through 5 in "Straight and Irregular Cutting."

Method #2

A plunge cut is an internal cut that is made without first drilling a hole. This is sometimes done for convenience, particularly with thin stock such as plywood and other sheet goods.

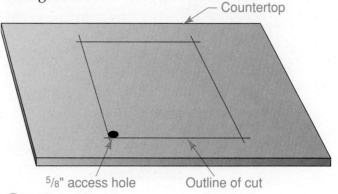

— Countertop

5/8" access hole   Outline of cut

 Internal Cut
Access Hole Once an access hole has been cut, slip the blade into it and start cutting to the inside of the layout lines.

1. Mark the outline of the cut on the workpiece.
2. Choose a convenient starting place inside the waste stock. Tip the saw forward with the baseplate resting on the surface of the material and the top of the blade clear of the work surface, as in Plunge Cutting.
3. Turn on the saw. When the blade reaches full speed, slowly lower the back of the saw until the blade begins to cut through the material. It is important to hold the saw firmly.
4. Continue cutting, using light pressure on the saw blade. When the blade cuts completely through the material, straighten the saw and cut normally.

Bevel Cuts The baseplate of a jigsaw can be adjusted from 0° to 45° for bevel cutting. A bevel cut made with a jigsaw will not be as straight or as smooth as a bevel cut made

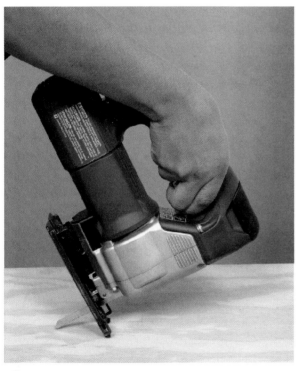

Plunge Cutting

Tip Up Tip the saw up on the front of the baseplate, then turn it on and gradually lower the blade into the wood.

Jon P. Muzzarelli

Builder's Tip

CHIP ELIMINATOR When cutting plastic laminate, draw cutting lines on masking tape. The tape will prevent the blade from chipping the countertop surface.

with a table saw or circular saw. However, a jigsaw can make a bevel cut that follows a curved layout line. After adjusting the baseplate to the bevel angle, follow the instructions for straight cutting.

Reading Check

Recall *What type of cut is best for cutting quickly through wood?*

Reciprocating Saws

What is the risk of using a very long blade?

A reciprocating saw has a straight blade with teeth along one edge. It has many of the characteristics of a jigsaw. It cuts with either a straight or an orbital motion. It can be used to make straight or curved cuts and often has variable-speed capability. Fitted with the appropriate blade, a reciprocating saw will cut metal, wood, plastic, and many other materials, as shown in A Reciprocating Saw on page 156. A reciprocating saw is most commonly used in remodeling and demolition work where a smooth finished cut is not required. When more **precise** control is needed, a jigsaw is often used instead.

Using a Reciprocating Saw

When operating a reciprocating saw, be sure to follow all safety rules. Be sure to wear proper eye protection to protect against dust, debris, and shards from snapped blades. Here are some general safety rules for this tool. These rules are slightly different from the rules for a jigsaw.

A Reciprocating Saw
Hand Position The saw should be controlled with two hands. Hold the shoe (the baseplate) firmly against the material being cut.

A reciprocating saw is a powerful tool. It is designed to be held with two hands, as shown in A Reciprocating Saw above. One hand should be on the handle, where it controls the speed of the tool and provides leverage for cutting. The other hand should be placed near the nose of the tool, just behind the blade. The operator should hold the saw's baseplate or shoe firmly against the material being cut. Otherwise, the saw may vibrate excessively.

Because the saw can accept blades as long as 12" long, it is important to check the area behind and to either side of the cutting line before starting the saw. This will reduce the chance of cutting something accidentally. Many electrical wires and water pipes have been severed by a reciprocating saw that was used carelessly. Before cutting into walls during remodeling work, use a keyhole saw to cut a small access panel. This will enable

you to look into the wall cavity and locate hidden pipes, wires, or other obstructions. Avoid using a blade that is longer than necessary.

Straight Blades
Why do some blades have very small teeth?

Straight blades with teeth on one edge are used in both jigsaws and reciprocating saws, as shown in Straight Blades. The type and size of the blade should be matched to the saw as well as to the material being cut. They are made from a flat piece of steel that has teeth or abrasive grit on one edge.

Choosing a Straight Blade

The main features of blades for jigsaws and reciprocating saws are essentially the same. However, reciprocating saw blades tend to be longer and wider than jigsaw blades.

Shank This is the end of the blade that is inserted into the saw. A shank made for one brand of saw may not fit another.

Body The body carries the teeth. It is made of various metals or combinations of metals that offer different cutting characteristics.

Gullet This is a pocket at the base of each tooth that helps to remove sawdust from the cut.

Straight Blades
Various Sizes The wide blades are for reciprocating saws. The narrow blades are for jigsaws.

Teeth Teeth are made of steel that is sometimes hardened for greater durability. Teeth point upward, toward the saw. This helps to hold the workpiece against the saw's baseplate.

A blade body made of high-speed steel (HSS) is best for cutting ferrous metals such as steel and iron up to ⅜" thick and nonferrous metals (brass, copper, and aluminum) up to ¼" thick. Blade bodies made with carbon steel will cut wood, plastics, and plastic laminates. Bimetal blades are versatile and durable, which makes them very popular in construction. They have flexible spring-steel bodies and hardened tool-steel teeth and are less likely to snap if they hit a nail. Bimetal blades can cut ferrous and nonferrous metals. However, they are also good for cutting wood and plastic. That makes them an especially good choice for reciprocating saw blades used in demolition work.

Blades are often classified by the number of teeth per inch (TPI). Wood-cutting blades usually range between 6 TPI and 12 TPI. Metal-cutting blades range between 12 TPI and 36 TPI. In general, jigsaw and reciprocating saw blades with fewer teeth cut fastest. Those blades with more teeth cut smoothest.

JOB SAFETY

JIGSAWS AND RECIPROCATING SAWS Jigsaws and reciprocating saws operate in similar fashion. The following safety rules apply generally to both tools. Check the manufacturer's manual for any specific safety instructions.

- Select the correct blade for the work and properly secure it in the saw.
- Be certain the material to be cut is properly clamped.
- Before starting a cut, look under the workpiece to make sure there are no wires or other obstructions near the line of cut.
- Keep the cutting pressure constant, but if you meet resistance, do not force the cut.
- Hold the baseplate firmly against the workpiece when cutting.
- When finished, turn off the power switch and allow the saw to come to a stop before pulling the blade from the cut and putting the saw down.

Section 5.4 Assessment

After You Read: Self-Check

1. By what other names is the jigsaw sometimes known?
2. How much pressure should you use when cutting with the jigsaw?
3. What other power tool is required before using a jigsaw to make an internal cut?
4. You have finished using a reciprocating saw. What should you do before pulling the blade from the cut and setting the saw down?

Academic Integration: English Language Arts

5. **Give Examples** Write a paragraph describing the two ways to make an internal cut using a jigsaw. Give at least two examples of situations requiring an internal cut.

Go to **connectED.mcgraw-hill.com** to check your answers.

Review and Assessment

Chapter Summary

Circular saws are used for crosscutting. The two basic types of circular saws are the contractor's saw and the worm-drive saw. Kickback can occur when a spinning blade hits something that stops it while under full power.

Table saws are used primarily for rip cuts. Tilting-arbor saws are commonly used on construction sites. The feed rate is the speed at which the stock is pushed through the saw blade.

Radial-arm saws can be used for crosscutting, ripping, dadoing, and combination cuts. Miter saws are used for quick crosscuts and accurate cuts on molding and trim. The three basic types include conventional miter saws, compound-miter saws, and sliding compound-miter saws.

Jigsaws are used for making curved, irregular, and internal cuts. Reciprocating saws are similar to jigsaws.

Review Content Vocabulary and Academic Vocabulary

1. Use each of these content vocabulary and academic vocabulary words in a sentence or diagram.

Content Vocabulary

- crosscut (p. 128)
- kickback (p. 129)
- torque (p. 130)
- kerf (p. 132)
- miter cut (p. 136)
- bevel cut (p. 136)
- offcut (p. 141)
- rip cut (p. 142)
- feed rate (p. 142)
- compound miter saw (p. 147)
- internal cut (p. 154)

Academic Vocabulary

- diameter (p. 128)
- precise (p. 155)

Speak Like a Pro

Technical Terms

2. Work with a classmate to define the following terms used in the chapter: *nonferrous metals* (p. 128), *baseplate* (p. 129), *arbor* (p. 129), *sidewinder* (p. 129), *workpiece* (p. 135), *benchtop saws* (p. 138), *featherboard* (p. 143), *bowed* (p. 149), *dadoing* (p. 150).

Review Key Concepts

3. Describe the saws listed in this chapter.

4. Summarize the safety rules for each type of power saw.

5. Describe how kickback occurs on circular saws and table saws.

6. Describe how to construct a rip cut with two types of saws.

7. List two guidelines for maintaining power saws.

Critical Thinking

8. **Analyze** Why is a thin-kerf blade best suited for cordless circular saws?

9. **Describe** Why are panels such as plywood and oriented-strand board difficult to cut? What can be done to ease this difficulty?

Academic and Workplace Applications

21st Century Skills

10. **Organizing Resources** Tools that you might use to help you make common saw cuts include a framing square, a carpenter's pencil, a permanent marker, sawhorses, an extension cord, and clamps. Describe which of these tools you would use to make a crosscut with a circular saw. List your process in a step-by-step format. Your format can be visual, written, or a combination of both.

Engineering

11. **Blade Manufacturing** The life of a saw blade depends mostly on what is being cut, and how often it is used. Other factors, however, affect how long a saw blade will last. The rubbing of one object against another, which is known as friction, can cause heat buildup. Heat buildup can damage a blade. Research and find one way in which blade manufacturers have designed blades to resist heat buildup and avoid damage. List your discoveries in a two-paragraph report.

21st Century Skills

12. **Media Literacy Skills** Consult a wide variety of media sources to locate as many product advertisements as you can for the various types of power saws. Use the Internet, local newspapers, or magazines to find your advertisements. Examine the ads you have found. Who is the intended audience for these ads? What features and benefits of the saw are brought to the attention of potential buyers? Construct a one-page chart or spreadsheet that compares pricing, features, benefits, and additional observations for at least three power saws.

> **Standardized TEST Practice**

Multiple Choice

Directions Choose the phrase that best answers the question.

13. Which phrase accurately describes an arbor hole?

    a. the raised portion behind each tooth that supports it

    b. the flat disk that serves as a base for the teeth

    c. the place where the blade is mounted on the saw

    d. the pocket in front of each tooth that helps remove sawdust during the cut

14. What is the primary use of the table saw?

    a. to make crosscuts

    b. to rip stock to width

    c. to make miter cuts

    d. to make bevel cuts

15. Which of the following is *not* one of the three types of miter saws?

    a. conventional miter saw

    b. compound miter saw

    c. sliding miter saw

    d. circular miter saw

TEST-TAKING TIP

When answering a multiple-choice question, do not linger on any one question. Mark your best guess and move on, returning later if you have enough time.

*These questions will help you practice for national certification assessment.

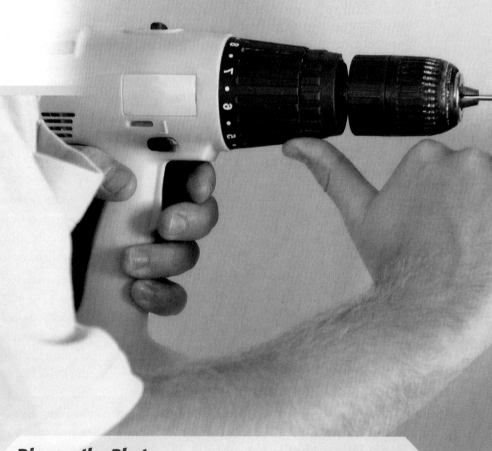

CHAPTER 6

Other Power & Pneumatic Tools

Section 6.1
Drills & Drivers

Section 6.2
Routers

Section 6.3
Sanders & Surfacing Tools

Section 6.4
Plate Joiners

Section 6.5
Power Nailers & Staplers

Chapter Objectives

After completing this chapter, you will be able to:

- **Describe** the uses of different types of drill bits.
- **Demonstrate** how to drill holes in wood and metal.
- **Identify** common uses for routers, sanders, planers, and jointers.
- **Describe** how a plate joiner works.
- **Identify** the uses of fasteners used with pneumatic tools.

Discuss the Photo

Electric Drill An electric drill is one of the most important power tools on a job site. *What activities might you perform with an electric drill?*

Writing Activity: Taking Notes

Taking notes helps you remember important information. Take notes when watching an explanation or demonstration of a tool in this chapter. Focus on examples, keywords, and facts. Avoid trying to write down every word. Use abbreviations to save time. Write in short phrases or very short sentences.

Corbis/Superstock

Before You Read Preview

Power and pneumatic tools improve the speed and accuracy of many tasks. Choose a content vocabulary or academic vocabulary word that is new to you. When you find it in the text, write down the definition.

Content Vocabulary

amperage	chamfer	pneumatic tool
countersink	template	regulator
pilot hole	cutterhead	collated fasteners
collet	biscuit	

Academic Vocabulary

You will find these words in your reading and on your tests. Use the academic vocabulary glossary to look up their definitions if necessary.

■ ranges   ■ approximate   ■ versatile

Graphic Organizer

As you read, use a chart like the one shown to organize information about power tools and their uses, adding rows as needed.

Power Tool	Used for...
right-angle drill	drilling in tight spaces

Go to **connectED.mcgraw-hill.com** to download this graphic organizer.

Drills

What is a grounding plug used for?

The portable electric drill is a versatile power tool. With the right bit, it can be used to drill holes in nearly any material. Fitted with various accessories, it can be used to install screws, cut holes, mix paints, and do many other jobs.

A typical builder or contractor will often have several electric drills on the job site. One will usually serve as a general-duty tool for drilling small and medium-size holes. The others might include a heavy-duty model for jobs requiring extra power, a drill/driver for installing screws, and a hammer drill for drilling into masonry. One or more of these drills may be battery-powered drills.

The major parts of an electric drill are shown in Parts of a Drill. The most common sizes of electric drills used in construction are ⅜" and ½". These dimensions refer to the diameter of the shank for the largest drill bit the chuck can hold. The *shank* is the end of the drill bit that fits into the chuck. The *chuck* is a device that has three jaws that can be tightened around the shank of a drill bit to hold it securely. Many drills, particularly heavy-duty drills, have a key-type chuck.

Some drills have a keyless chuck, which can be conveniently tightened by hand. Most drills have a pistol-grip handle.

Types of Drills

The two basic types of electric drills are corded drills and cordless drills. Older drills have a housing made of metal. Their power cord is fitted with a three-prong grounding plug to reduce the danger of electrical shock. Drills with a plastic housing typically have a two-prong plug because the housing itself insulates the operator against shock. Plastic housings are sometimes referred to as *double-insulated* housings. Many other types of power tools have plastic housings as well.

Most electric drills can be operated at various speeds. Speed control is important when drilling metal and when using the drill to start screws. The speed of a drill is controlled by finger pressure on the drill's trigger. Drill speed is rated in rpm (revolutions per minute), and typically **ranges** from 0 to 1,200 rpm. Speeds up to 900 rpm usually indicate a heavy-duty drill capable of providing great torque. *Torque* is a twisting force that produces rotation.

Parts of a Drill

Main Parts The main parts of a corded electric drill. Cordless drills include a rechargeable battery instead of a power cord. *What is the advantage of having a cordless drill?*

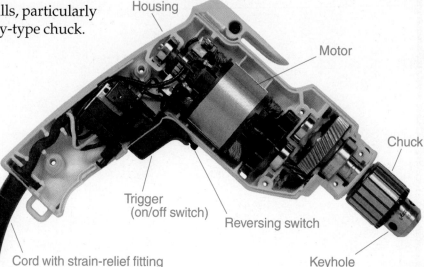

Housing

Motor

Chuck

Trigger (on/off switch)

Reversing switch

Keyhole

Cord with strain-relief fitting

Arnold & Brown

Some drills have an adjustable internal clutch, and are called *drill/drivers.* The clutch allows the tool to operate at 2 torque settings, one for drilling and one for driving screws. This allows it to drive screws more effectively than standard variable-speed drills.

Corded Drills A standard corded electric drill should be plugged into a properly grounded electrical outlet. These drills are best for drilling small and large holes, for drilling through difficult materials such as steel or concrete, and for drilling many holes in a short period of time.

Amperage is the strength of an electric current expressed in *amperes,* or *amps.* The amperage of an electric drill is an **approximate** measure of its power. The amperage rating is on a small metal specification plate permanently attached to the tool. Corded drills generally range from 3 to 8 amps. High-amperage drills are used for heavy-duty work.

There are various types of specialized corded drills. *Hammer drills* such as the one in Hammer Drill are used to drill holes in masonry. While the chuck revolves, the drill creates a rapid, hammerlike, reciprocating action. This helps to drive a masonry bit into the material. For drilling larger holes (⅜" and up), an SDS type drill (a heavy duty hammer drill using a specialized chuck and bits) is more effective.

JOB SAFETY

ELECTRIC DRILLS The following are general safety rules for electric drills. Check the owner's manual for any special safety instructions.

- Wear proper eye protection.
- Tie back long hair.
- When operating larger drills, use both hands and, if necessary, an auxiliary handle.
- Disconnect the power plug or remove the battery pack before installing or removing drill bits.
- Center the drill bit in the chuck and tighten the chuck securely. Make certain the drill bit is held securely in the chuck.
- Never use a bit with a square, tapered tang in an electric drill. The drill's chuck will not hold this type of bit securely.
- Be sure the chuck key has been removed before starting the drill.
- Do not force the drill into any material. Use an even, steady pressure.
- Never drill through cloth. It will twist around the bit.
- Do not hold small pieces of material with your fingers. Clamp them down to prevent them from spinning as they are being drilled.
- Put the drill down with the drill bit facing away from you. When laying down the drill, always point the drill bit away from you, even when it is coasting to a stop.
- Keep loose clothing or long hair away from the spinning bit, as they may become entangled very quickly.

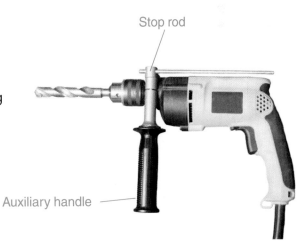

Stop rod

Auxiliary handle

Hammer Drill

Drilling Masonry The auxiliary handle allows better control of the tool. The stop rod limits the depth of the hole.

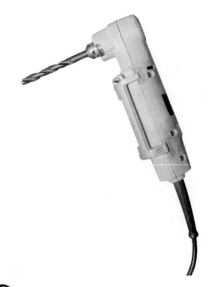

 A Right-Angle Drill
Cramped Quarters The position of the chuck makes this drill able to reach into places other drills cannot. *What parts are similar to a standard drill?*

Right-angle drills such as the one above are often used by electricians and plumbers. On such drills, the chuck is at 90° to the drill body. This allows drilling in tight spaces, such as through the sides of studs.

Cordless Drills Cordless electric drills such as the one below are powered by a rechargeable battery. They are especially useful where a

A Cordless Drill/Driver
Portable Power Advanced battery technology makes this tool lighter yet more powerful than previous models with the same battery voltage.

long extension cord would be undesirable or where electrical power is not available. The voltage of the battery roughly indicates the tool's power. Batteries typically range from 9.6 to 18 volts, but some models go up to 28 volts. Batteries with a higher voltage can operate longer between charges. However, the added capacity also makes the battery heavier. Recent improvements in battery technology have made cordless tools even more effective and useful.

The batteries are sealed within a plastic case inside or at the end of the drill's handle. The case and batteries form a unit called the battery pack. To charge the batteries, the battery pack is removed from the drill and placed in a charger. Full battery strength can be restored in one hour or less. Most builders keep two or more battery packs on hand. While one is recharging, the other is in use.

Build It Green The battery in any cordless power tool must be replaced eventually. Many batteries contain toxic chemicals that can enter the water supply if disposed of improperly. Always recycle old batteries. Check your owner's manual or contact the tool manufacturer to locate a recycling center.

Drill/Drivers A drill/driver is the best type of drill for driving screws, though a variable speed drill can also do the job. A drill/driver can be adjusted to two different speed ranges. The slower range is for driving screws while the faster range is for drilling. The adjustable clutch inside the tool reduces the chance that a screw will be driven too deeply. The clutch automatically disengages the drive mechanism at a preset level of resistance. Drill/drivers are particularly useful when many screws must be driven. They are most often used to drive Phillips-head and square-drive screws.

A screwdriving bit that is too small or too large for the screw head will spin out, or cam out when power is applied to the drill. The tip must fit snugly, with no sloppiness in the fit.

Drive screws as follows:

1. Start the screw at a slow speed.
2. Increase speed as the screw moves into the stock.
3. Stop the drill when the screw reaches the correct depth. Some drills have adjustable settings that prevent you from driving the screw too deep. Do not disengage the driving bit while it is spinning.
4. For precise control, drive the screw nearly all the way in and then finish with a screwdriver.

Drill Bits

The versatility of an electric drill comes from the great number and variety of bits, cutters, and other accessories that are available. Knowing when to use a particular bit is the key to getting the best results when using a drill. Common types of bits are shown in Drill Bits. Twist bits and spade bits are used most often on a job site.

- A *twist bit* is a **versatile** bit that can be used to drill holes in wood, plastic, or metal. Common diameters range from

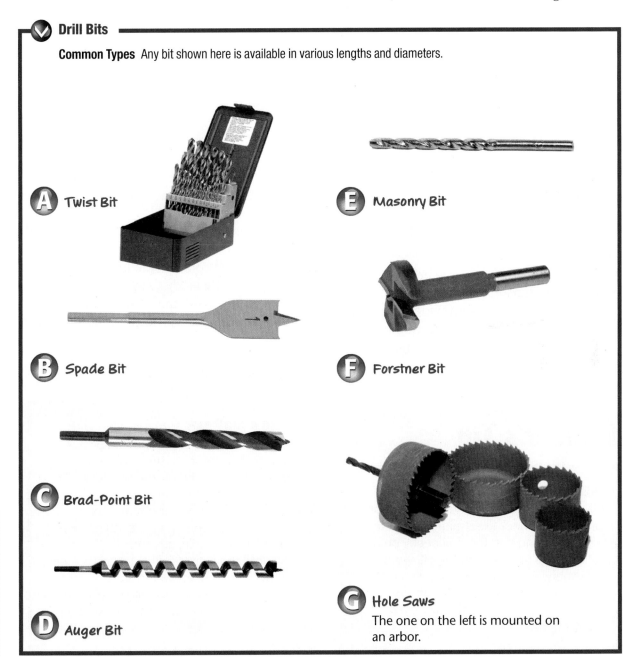

Drill Bits

Common Types Any bit shown here is available in various lengths and diameters.

A Twist Bit

B Spade Bit

C Brad-Point Bit

D Auger Bit

E Masonry Bit

F Forstner Bit

G Hole Saws
The one on the left is mounted on an arbor.

$\frac{1}{16}$" to $\frac{1}{2}$", in increments of $\frac{1}{64}$". Twist bits with a 118° tip are general utility bits used for drilling holes in wood, metal, and plastic. They have a cylindrical shank, spiral flutes (grooves), and a beveled tip. Twist bits made from high-speed steel (HSS) are particularly suited to drilling metal. A case that holds a group of twist bits in various sizes is called a drill index.

- A *spade bit* is used to bore holes in wood. The holes range in diameter from $\frac{3}{8}$" to $1\frac{1}{2}$". A large point guides the bit. Its horizontal cutting surfaces remove stock. A hexagonal shank reduces slippage.

- A *brad-point bit* has a small center point called a *brad point*. This prevents the bit from wandering as the hole is started. Sharp cutting edges cut very smooth, clean holes in wood.

- An *auger bit* is designed to cut deep holes quickly through wood. A screw point pulls the bit through the stock. Such an auger bit is called a self-feed bit. The wide, deep flutes remove chips efficiently. This bit is often used by electricians.

- A *masonry bit* is for use on brick, concrete, and other masonry materials. It has a beveled carbide tip and wide flutes that carry grit and dust away from the cut. If using a masonry bit in a hammer drill, be sure the bit is designed for use in a hammer drill.

- A *Forstner bit* has a brad point and a sharpened rim. It is excellent for boring smooth holes with flat bottoms in wood. Forstner bits can bore through end grain with ease. They are used primarily in cabinetmaking.

- A *hole saw* is a cylindrical metal sleeve with a sawtooth edge. It is commonly used by plumbers to cut large holes in wood framing for drain and vent piping. It is also used when installing a door lockset. A twist bit at the center of a hole saw centers the hole.

In many cases, the head of a screw must be flush with the surface of the wood. The screw head must be recessed as shown in Countersinking a Screw on page 167. A **countersink** is a special bit with beveled cutting edges. It creates a funnel shape called a countersink at the top of a drilled hole. This funnel shape allows the head of a wood screw to be flush with the wood surface. An 82° countersink is suitable for use with wood screws.

A *combination bit* is another convenient tool for countersinking wood screws. It will drill a pilot hole and countersink in one operation. A **pilot hole** is a hole drilled in wood to start and guide a screw. Combination bits are available in most of the common wood

Builder's Tip

CHECKING BITS A twist bit that is bent will cut an irregular hole and may not be safe to use. Check the straightness of the bit by spinning it in the drill briefly before drilling. The tip of a bent bit will wobble noticeably. A bit may also be rolled on a flat surface to check for straightness. Always discard drill bits that are even slightly bent.

Builder's Tip

USING A HOLE SAW When drilling with a hole saw, all the teeth should contact the wood at the same time. This ensures that the hole will be perpendicular to the surface of the wood. A hole saw may cause considerable tear out as it exits the back of the material. If both surfaces require a clean hole, advance the hole saw only until the center (pilot) bit exits, then withdraw the hole saw and start drilling from the other side, placing the pilot bit in the same hole to ensure that the finished holes line up.

screw sizes. For example, if a 1" #8 wood screw is used, a 1" #8 combination bit should be used.

Drilling Techniques

Drilling a hole properly is one of the most important skills a builder can develop. Though it might seem easy, drilling a hole that is perpendicular to the wood surface is a skill that takes practice. The most important factor in correct drilling is to choose a suitable bit for the material and the job to be done. An unsuitable bit will deliver poor results, could be damaged, and may be unsafe.

Installing and Removing a Bit To install or remove a bit, follow this basic procedure:

1. Unplug the drill or remove its battery pack.
2. Determine if the shank of the chosen drill bit will fit into the chuck. (A ⅜" chuck, for example, will not accept a ½" shank.)
3. Open the jaws of the chuck by twisting its collar.
4. Insert the shank of the bit as far as possible. Then turn the collar by hand to close the jaws. Check that the shank is centered between the jaws. If not, open the jaws and center it.
5. Tighten the chuck by inserting the chuck key in each of the three keyholes in succession. Remove the chuck key.
6. If the drill has a keyless chuck, twist the two portions of the sleeve in opposite directions until the jaws are tight.

Countersinking a Screw
Making a Recess A countersink is designed to cut a funnel-shaped opening.

Builder's Tip

TIGHTENING A CHUCK In most cases, a keyed chuck can be tightened by inserting the key in one hole. For maximum holding power, manufacturers recommend inserting the key in each hole and tightening it. This improves the gripping power of the chuck jaws.

7. The friction of drilling creates heat in a bit. Allow a bit to cool before removing it from the drill. To remove a bit, unplug the drill or remove the battery. Then open the chuck.

Drilling a Hole Hold the drill at a right angle to the work when starting a hole, as shown in Drilling a Hole. If the workpiece is too small to hold, clamp it down. Make sure you are using a sharp bit, and that it is centered securely in the chuck. Apply just enough pressure on the drill to keep the bit cutting.

Drilling a Hole
Drilling in Flat Stock Be sure to hold the drill at a right angle to the workpiece. If necessary, use a clamp or vise to hold the wood securely.

Too little pressure will dull the bit; too much pressure may break it. Do not move the drill from side to side as this may snap the bit. When the hole is complete, withdraw the drill with bit still spinning. When drilling a deep hole, periodically withdraw the bit and keep it spinning to clear shavings from the hole.

Drilling in Wood If it is important to prevent the underside of wood from splintering as the bit breaks through, clamp a piece of scrap wood behind the workpiece. Use this technique when drilling through hardwoods or cabinet-grade plywood.

When installing screws, drill a pilot hole first unless you are using self-drilling screws. Drill the correct-size pilot hole for the screw you are using. If the hole is too small, the screw will be hard to drive and may snap as it is driven. If the hole is too large, the screw will not hold.

When drilling hardwood, it is good practice to bore the pilot hole the same size as the root diameter of the screw. Diameters for common wood screws are shown in **Table 6-1**. In softwood, drill the pilot hole slightly smaller. Self-drilling screws, such as drywall screws, do not usually require a pilot hole when used in softwoods.

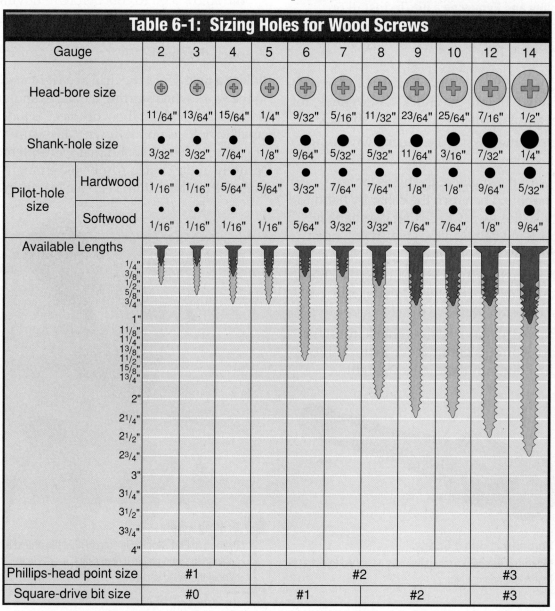

Table 6-1: Sizing Holes for Wood Screws

Gauge	2	3	4	5	6	7	8	9	10	12	14
Head-bore size	11/64"	13/64"	15/64"	1/4"	9/32"	5/16"	11/32"	23/64"	25/64"	7/16"	1/2"
Shank-hole size	3/32"	3/32"	7/64"	1/8"	9/64"	5/32"	5/32"	11/64"	3/16"	7/32"	1/4"
Pilot-hole size — Hardwood	1/16"	1/16"	5/64"	5/64"	3/32"	7/64"	7/64"	1/8"	1/8"	9/64"	5/32"
Pilot-hole size — Softwood	1/16"	1/16"	1/16"	1/16"	5/64"	3/32"	3/32"	7/64"	7/64"	1/8"	9/64"

Available Lengths: 1/4", 3/8", 1/2", 5/8", 3/4", 1", 1 1/8", 1 1/4", 1 3/8", 1 1/2", 1 5/8", 1 3/4", 2", 2 1/4", 2 1/2", 2 3/4", 3", 3 1/4", 3 1/2", 3 3/4", 4"

| Phillips-head point size | #1 | | | #2 | | | | | | #3 | |
| Square-drive bit size | #0 | | #1 | | | #2 | | | | #3 | |

When fastening two pieces of wood together, drill the pilot hole through the uppermost piece and into the bottom piece to the desired depth. Then drill a slightly larger hole through the uppermost piece. This hole is for the screw's shank (fully threaded screws do not require this second hole). Countersink the top of the shank hole.

Drilling in Metal A twist bit with a 135° split point is best for drilling metals. Drilling into stainless steel, cast iron, and some types of aluminum will quickly dull a twist bit. A cutting lubricant such as light-weight oil will cool the bit and extend its life. When drilling without lubricant, reduce the drill's rpm to prevent the bit from overheating.

Push firmly on the drill as it cuts, but do not force it. Just before the bit emerges from the metal, slow the feed rate to prevent the bit from catching on burrs.

Reading Check

Recall What is the correct angle to hold the drill to the workpiece when drilling a hole?

Impact Drivers

What are impact drivers best used for?

A cordless impact driver is a relatively new tool for builders, but it has quickly become very popular. The tool shown in An Impact Driver looks like a drill, but instead of an adjustable chuck it has a hex chuck. The chuck accepts only bits and drivers with

Builder's Tip

MAKING STRONG CONNECTIONS The strength of a connection made with a screw is improved more by increasing the length of the screw than by increasing its diameter.

Mathematics: Calculating Torque

Working with Formulas Torque is the force that causes an object to rotate. It consists of force acting on distance. Torque is measured in pound-feet (lb-ft) and may exist even if no movement occurs. Calculate the torque produced by a 65-pound force pushing on a 3" lever.

Starting Hint Torque is calculated by multiplying force by distance:

$$T = F \times D$$

hexagonal shanks. Though the tool can be used to drill holes, its primary purpose is to drive and remove screws and bolts. Deck builders use impact drivers to drive the lag screws that hold beams and posts together. Carpenters use them to drive many kinds of screws, particularly those that are long or

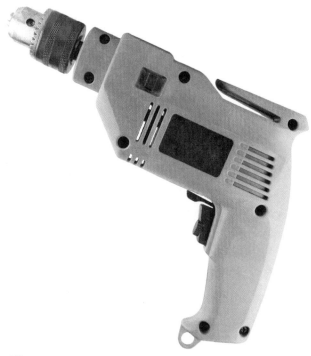

An Impact Driver
Good for Driving Impact drivers are extremely effective at driving and removing screws and hex-head fasteners such as lag screws.

large. Impact drivers are also ideal for driving masonry screws. The tool is most often fitted with driving bits for Phillips head and square-head screws, masonry bits, and nutsetter bits.

The chuck of an impact driver spins just like the chuck of a drill. However, as the chuck spins, the drive mechanism applies a rapid series of blows to the back of the driver bit. This combination dramatically reduces the amount of force required to drive the screw. In fact, most workers using the tool for the first time use too much force.

Section 6.1 Assessment

After You Read: Self-Check

1. In what units is the power of an electric drill measured?
2. What characteristic of masonry bits serves a similar purpose on auger bits?
3. For what type of drilling is a Forstner bit best suited?
4. What is the purpose of a countersink?

Academic Integration: Science

5. **Ohm's Law** Electrical engineers must calculate the amount of electrical current flowing through the wires of tools they design. To do this, they use an equation called Ohm's Law. Ohm's law is $V = IR$, where V = voltage (measured in volts), I = current (measured in amperes), and R = resistance (measured in ohms). To understand this equation, think of electricity like water in a hose: The *voltage* is the water pressure in the hose, the *current* is the rate at which the water flows, and the *resistance* is the size of the hose. What is the current of a 30 volt machine that is operated across a wire that has a resistance of 10 ohms?

Go to connectED.mcgraw-hill.com to check your answers.

Section 6.2 Routers

Understanding Routers

What work is a router used for?

The router is a portable electric tool designed to turn a sharpened cutter, called a *bit*, at high speed. The router is used primarily for finishing work, such as for cutting joints and shaping the surfaces and edges of stock. With accessories, a router can also be used to trim plastic laminate and cut openings in panel products.

Parts of a Router

There are two basic types of router. The motor and base of a *fixed-base router* always remain stationary during a cut. However, prior to cutting, the depth of cut can be adjusted by raising or lowering the base. A *plunge router*, shown on page 171, has a motor that is mounted on vertical metal posts. The motor assembly slides up and down on the posts. This allows the spinning

Motor housing

Power cord

Handle

Depth guide

Collet

Post

Turret

Chip shield

Base

A Plunge Router

Motor on Posts A plunge router is one type of router. It is similar to a fixed-base router but is better suited to some types of work, such as making stop cuts and repetitive cuts at several different depths. This one has an electric brake as a safety feature.

bit to be "plunged" into the workpiece and lifted away when the work is complete.

On some routers, the on/off switch is on the body of the tool. On other routers, it is on one of the handles. A handle-mounted switch makes it possible to turn off the tool without letting go of the handles. This is a good safety feature.

The **collet** is the part of the router that holds the bit. Routers are classified by the diameter of their collet. Thus, a ½" router is a router with a ½" collet. This determines the size of the bit shank the router will accept. Some routers come with interchangeable collets. These allow the routers to accept different sizes of bits.

A router motor is designed to deliver high speed rather than high torque. Router speeds range from 10,000 to over 25,000 rpm. (In contrast, an electric drill may have a top speed of only 1,200 rpm.) A variable-speed router allows the user to adjust the speed to suit the material being cut and the size of the

bit. Bits over 1" in diameter should be used at slower speeds. This improves safety and reduces the chance of damaging the stock.

Router Bits The cutting edges of most router bits are on the sides, rather than on the end. One important exception is the plunge-cutting bit. This bit has cutting edges on its sides and on its end. With a plunge-cutting bit in place, a router can drill a pilot hole and then cut or trim material starting from that hole. Router bits are available either with high-speed-steel cutting edges or with carbide-tipped edges. Carbide-tipped bits are generally more expensive and fragile, but remain sharp longer.

Router bits come in many shapes for doing grooved or decorative work on the surface or on the edge of stock. Some of the most common bits are shown in Router Bits on page 172, including straight, roundover, beading, cove, and chamfer bits. A **chamfer** is a beveled edge. The shank of a router bit commonly has a diameter of ¼", ⅜", or ½".

Arnold & Brown

Common Profiles These are a few of the most common router bits available.

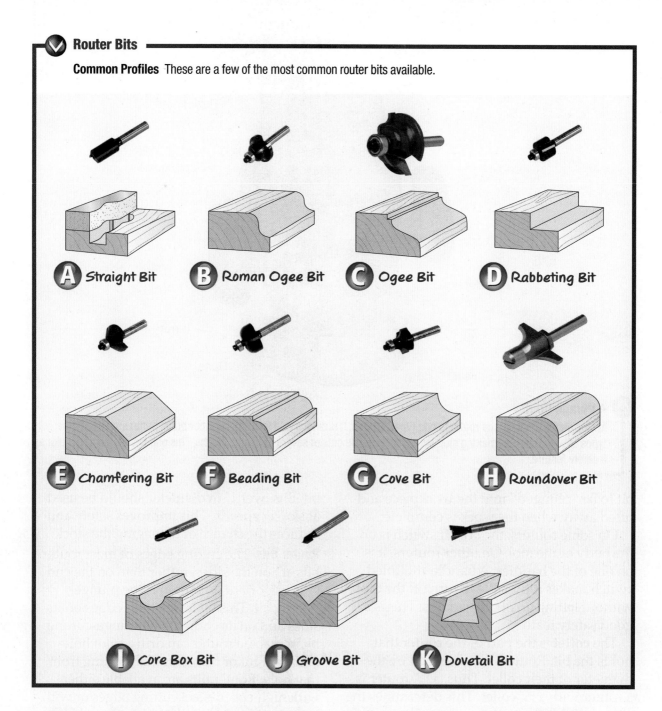

A Straight Bit

B Roman Ogee Bit

C Ogee Bit

D Rabbeting Bit

E Chamfering Bit

F Beading Bit

G Cove Bit

H Roundover Bit

I Core Box Bit

J Groove Bit

K Dovetail Bit

Bits that accept a bearing can be fitted with bearings of different sizes. It is common for many builders to keep several different diameters of bearings on hand. The bearing can be used to change the profile, width, or depth of the cut, as shown in Router Bit Bearings.

If a bearing is mounted on the side of the bit, the combination is called a *bearing-over bit*. If the bearing is mounted on the end, the combination is called a *bearing-under bit*. One example of a bearing-under bit is the flush-trimming bit shown in A Bearing-Under

Bit on page 173. It is used to trim plastic laminate.

Accessories

A number of router accessories are available. An edge guide rides against the edge of the stock, enabling the router to make a cut exactly parallel to the edge. A **template** is a guide made from metal or thin wood. Templates enable the router to quickly and accurately cut *mortises* (holes, grooves, or

(l to r, t to b)(1, 2, 4–7, 9–11)McGraw-Hill Education; (3, 8)Arnold & Brown

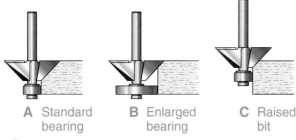

A Standard bearing **B** Enlarged bearing **C** Raised bit

 Router Bit Bearings
Effects of a Bearing A. A bearing controls the depth of cut by holding the bit away from the wood. **B.** Changing the bearing's size changes the size of the cut. **C.** The size can also be changed by raising or lowering the bit.

slots) for door hinges and other shapes. A router can also be controlled using a circle guide or a straightedge, such as the one shown in Guiding a Router.

A dovetail template, shown in A Fixed-Base Router on page 174, allows the user to cut a dovetail joint. A dovetail joint has interlocking pieces. Dovetail joints are used to assemble the drawers in high-quality cabinetry.

Some carpenters mount a router upside-down beneath a sheet of plywood or particleboard. The bit extends through a hole in the board. When a simple fence is added, this device is called a *router table*. In this case, the stock is moved past the cutter, rather than the cutter being moved along the stock.

A dust collection hose, connected to a shop vacuum cleaner, removes chips and dust. This allows the operator to see the cut more clearly. It also helps prevent chips from flying at the operator, which increases safety.

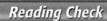
Reading Check

Define *What is a template?*

 A Bearing-Under Bit
Flush Trimming Bit This type of bit is often used when building countertops. As the bearing rides against the substrate, the bit trims off excess plastic laminate.

 Guiding a Router
Using a Straightedge A wood or metal straightedge clamped to the workpiece can be used to guide a router.

Using a Router
What causes a bit to overheat?

A router bit turns clockwise. Always feed against the direction of bit rotation, as shown in Bit Rotation on page 175. This reduces the tendency of the bit to "grab" the stock, which can make the router difficult to control. In addition, moving the router in the proper way improves the quality of the cut.

The speed at which the best cut is made will depend on the depth and width of the cut and on the hardness of the wood. If you move the router too quickly, the motor will slow down too much, making a poor cut. If you move the router too slowly, the bit may overheat. This can draw the temper from the cutting edge

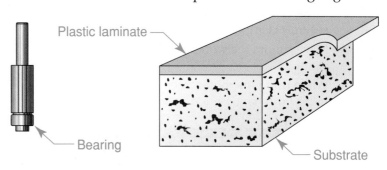

Plastic laminate

Bearing

Substrate

Arnold & Brown

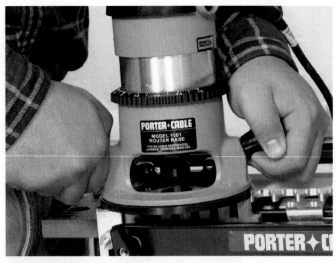

⬆ **A Fixed-Base Router**

Cutting Dovetails This router is being used with a dovetail template to shape the end of a drawer side. *Would this router be better for this joint than a plunge router? Why?*

or burn the wood. Do not force the cut. Allow the bit to cut freely. Listen to the motor for an indication of whether it is working at its most efficient speed.

Always make deep cuts in several passes. This is when a plunge router becomes especially useful. It can be quickly reset to several depths.

JOB SAFETY

ROUTER SAFETY The following are general safety rules for using routers. Check the owner's manual for any special safety instructions.

- Always wear proper eye protection.
- Wear hearing protection and a dust mask. Routers generate a lot of noise and sawdust.
- Be certain the power switch is off before plugging the router into an outlet. Always hold on to the router when turning it on.
- Make certain any fence or guide is securely clamped.
- Make certain the workpiece is securely clamped.
- When using the router, keep a firm grip, using both hands when appropriate.

- Make adjustments only when the bit is at a dead stop. When installing or removing bits, be sure the router is unplugged.
- Feed the router in the correct direction. Feeding the router generally means moving in a counter-clockwise direction around the outside perimeter of the piece being routed, and moving clockwise around anything being routed on the interior. Feeding the router is explained further on this page.
- When putting the router down, point the bit away from you. Be aware of a bit that is still moving.
- When using large bits, remove the stock with two or more passes.
- Never use a dull or damaged bit.
- Bring the router to full speed before cutting. Turn off the router after making the cut.

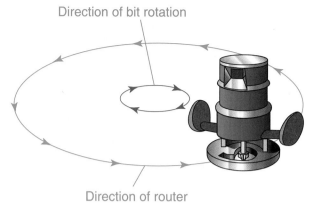

Direction of bit rotation

Direction of router

Bit Rotation

Feed Direction The router bit spins clockwise. Move the router counterclockwise when cutting outside edges. Move the router clockwise when routing inside edges.

Installing a Bit

The shank of a bit is held by the router collet. When installing a router bit, take care not to cut yourself with its sharp edges.

1. Disconnect the power cord.

2. Turn the router upside down. Depending on the kind of router, either lock the shaft or hold it with a wrench.

3. Slide the bit's shank all the way into the collet. Then back it off slightly and tighten the collet. A bit resting against the bottom of the collet will vibrate and loosen. It can also be difficult to remove.

4. Tighten the collet firmly with a wrench.

Builder's Tip

PILOT TIP BITS Some router bits used for edging have a solid pilot tip that rides against the uncut edge of the wood, instead of a bearing. Always keep the router moving when using a bit with a pilot tip. Otherwise, the heat generated from friction as the tip spins against the edge of the wood may scorch the wood.

Cutting a Decorative Edge Different decorative edges can be created using the many different bits (see Router Bits on page 172).

1. Install the required bit.

2. Adjust the bit to the approximate depth of cut.

3. Plug in the router and turn it on. Resist the starting torque of the motor by holding onto the router with both hands. Otherwise, it can twist out of your grip.

4. Make a test cut in a scrap piece of the same stock.

5. Adjust the depth of cut until the correct profile is obtained.

6. Make the final cut in the correct direction and at the appropriate feed rate.

Section 6.2 Assessment

After You Read: Self-Check

1. How is a plunge router different from a fixed-base router?
2. What bit is used to trim plastic laminate to size?
3. What accessory might be used for cutting a groove parallel to the edge of a plywood panel?
4. Why should you hold the router firmly when turning it on?

Academic Integration: Science

5. **Friction** Friction is the force that comes from two surfaces moving against one another. When these surfaces move against one another, the friction between the surfaces changes the energy of movement (*kinetic energy*) into heat (*thermal energy*). Give an example of friction found in this section.

Go to **connectED.mcgraw-hill.com** to check your answers.

Sanders & Surfacing Tools

Sanders

What is a belt sander used for?

Portable electric sanders are used for tasks ranging from heavy stock removal to delicate finish sanding of woodwork. The most common types are the belt sander, the pad or orbital sander, and the random-orbit sander.

Belt Sanders

The portable belt sander drives a revolving abrasive belt to remove stock quickly. The machine, shown in Parts of a Belt Sander, is classified by the width and length of its belt. For example, a small machine with a belt that is 3" wide and 18" in circumference would be referred to as a 3"×18" sander. Other sizes include 3"×21", 3"×24", and 4"×24". Each sander must be fitted with a belt of matching size. To reduce the amount of dust in the air, most belt sanders have a dust collection bag. The sander can also be connected to a vacuum system.

Installing the Belt Many sanding belts have a lap seam. If installed improperly, this type of belt can rip apart during use. There is an arrow on the inside surface of this kind of belt. The belt must be installed so that the arrow points in the same direction as the arrow on the side of the sander. "Seamless" belts are constructed differently. They can be installed in either direction. To install a new belt:

1. Unplug the sander.
2. Disengage the belt-release lever and remove the old belt.
3. If you are using a lap-seam belt, be sure the new one is turned in the right direction.
4. Slip the new belt onto the rollers and engage the belt-release lever.
5. Plug in the sander and turn it on. If the belt slides to one side or the other, correct this by turning the belt-tracking knob slightly while the sander is running.

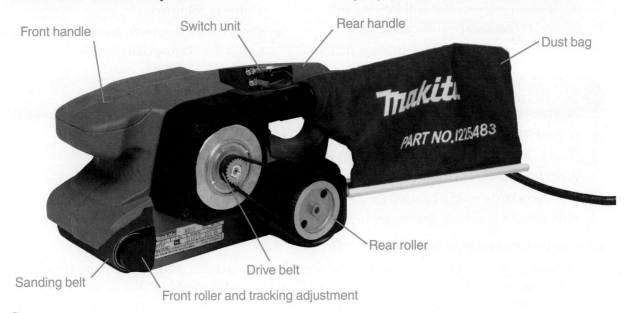

Front handle · Switch unit · Rear handle · Dust bag · PART NO. 1225483 · Rear roller · Drive belt · Sanding belt · Front roller and tracking adjustment

 Parts of a Belt Sander
Belt Drive Model This cutaway view of a belt sander shows how the rear roller is driven.

Arnold & Brown

Belt Sanding Techniques The portable belt sander is the most powerful of all portable sanders and should be used with care. Used carelessly, it can easily gouge the wood. To prevent this, always keep the tool moving when the belt is in contact with the workpiece. Be sure that the power cord is out of the way before starting the sander. The spinning belt can cut through a cord almost as quickly as a saw-blade can. This tool is generally used to sand in the direction of the wood grain.

1. Hold the sander with both hands and turn it on, as in Using a Belt Sander.

2. Slowly lower the sander onto the wood, letting the heel (rear portion) of the belt touch first.

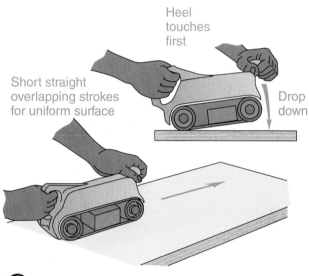

Heel touches first

Short straight overlapping strokes for uniform surface

Drop down

🔺 **Using a Belt Sander**
Feed Direction Lower the sander slowly onto the surface, then immediately move the machine in overlapping strokes.

Builder's Tip

CLEANING A BELT A clogged abrasive belt on a belt sander will not work well. Clean a clogged belt by running it against an inexpensive block of crepe rubber designed for this purpose.

3. Immediately move the sander either forward and back or from side to side. Never hold it in one place or it will gouge the workpiece. Some belt sanders have the option of using a *surfacing frame.* A surfacing frame is an accessory that surrounds the tool and limits the depth of cut, helping to reduce or eliminate the incidence of gouges.

Orbital Sanders

An orbital sander uses a sheet of abrasive paper instead of a belt. The paper

JOB SAFETY

POWER SANDER SAFETY The following are general safety rules for using sanders. Check the manufacturer's manual for any special safety instructions.

• Always wear proper eye protection.

• Always wear the proper dust mask or respirator when using sanding equipment.

• Be sure the sander's abrasive belt, disc, or pad is in good condition and that its grit is correct for the work being done.

• Be sure there are no nicks or tears in the edge of a disc or belt. An abrasive belt must be installed with the correct tension. Be sure it is tracking (aligned on the rollers) properly.

• Do not let go of the handles until the belt stops moving.

• Avoid nails and screws when sanding.

• Disconnect the power cord when changing abrasives.

• Make certain the tool's switch is in the off position before plugging in the power cord.

• Never touch a sanding belt or disc while it is moving.

• Do not use a sander to remove paint containing lead (see Chapter 33, "Exterior and Interior Paint").

is held in place by paper-locking levers such as those visible in Parts of an Orbital Sander. Some sanders use hook-and-loop or pressure-sensitive adhesive (PSA) systems to hold the sandpaper in place. The sanding pad moves with an orbital (circular) motion. Because orbital sanders are most often used to smooth a surface prior to painting or finishing, they are sometimes called finishing sanders. They are also called *pad sanders* because a rubber pad cushions the abrasive paper.

Orbital sanders are generally classified by the size of the pad, which may be square or rectangular. The pad size is based on standard-size abrasive sheets. Thus, there are *one-quarter-sheet* sanders, *one-third-sheet* sanders, and *one-half-sheet* sanders.

Orbital Sanding Techniques Rest an orbital sander evenly on the stock. Apply moderate pressure and move the sander back and forth, working from one side to the other. When using a standard orbital sander, move the sander in the direction of the wood grain to minimize cross-grain scratching.

Random-Orbit Sanders Some builders prefer a type of orbital sander called a *random-orbit sander*. This versatile tool, shown in

JOB SAFETY

DUST REMOVAL A "tool-triggered" shop vacuum can serve as a portable dust collection system when using any sander that has a dust collection hose. When a sander is plugged into the vacuum's on-board electrical receptacle, the vacuum will start and stop whenever the sander starts or stops.

A Random-Orbit Sander, can be used for fine finishing work as well as aggressive stock removal. It usually has a round sanding pad instead of a square one. As the pad spins, it also moves side-to-side. The combination of these two motions reduces visible scratches in the wood surface. You can use a random-orbit sander either with the grain or against it. A similar looking sander, called a *disk sander*, has a pad that spins but does not move side to side. It should not be used where a fine finish is required.

The abrasive paper of a random-orbit sander is attached to the pad by a

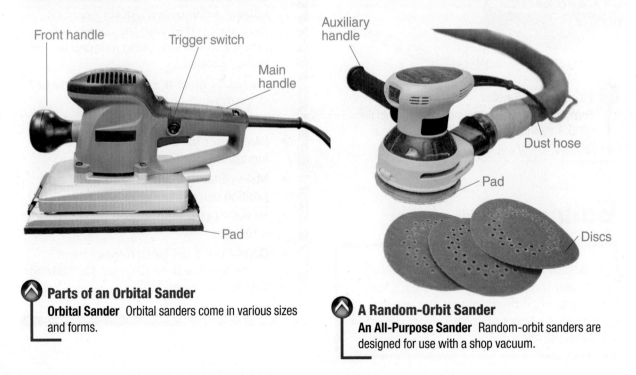

Parts of an Orbital Sander
Orbital Sander Orbital sanders come in various sizes and forms.

A Random-Orbit Sander
An All-Purpose Sander Random-orbit sanders are designed for use with a shop vacuum.

(l & r)McGraw-Hill Education

pressure-sensitive adhesive (PSA) backing or a hook-and-loop backing. PSA paper cannot be reused once removed but hook-and-loop paper can be reinstalled as often as necessary. This is an advantage when changing back and forth between grits. Holes in the sanding pad and matching holes in the abrasive paper make dust removal easier when the sander is connected to a vacuum system. Breathing quantities of fine sawdust can be hazardous to your health.

Reading Check

Recall *Which portable sander is the most powerful?*

Surfacing Tools

What are surfacing tools used for?

To meet the needs of the construction industry, manufacturers have designed small, portable versions of the surfacing tools common in woodworking shops. These tools, including jointers and planers, are easy to bring to the job site. In addition, builders find that turning rough stock into finished stock on site is sometimes less costly than buying finished stock from a lumberyard.

 A Jointer
Jointing an Edge A jointer is used to ensure a square edge.

McGraw-Hill Education

Jointers

A *jointer* is a power tool used to remove saw marks from stock and ensure a square edge. A jointer is most likely to be used at late stages of house construction, when cabinets and interior woodwork are being installed. The most common use for the jointer on a job site is for jointing an edge, as shown in A Jointer. An edge is said to be jointed when the edge forms a right angle with the face of the board along its entire length, shown in A Jointed Edge on page 180. A board is sometimes jointed after being cut to width on a table saw. The jointer shown in A Jointer is a model common in small shops. It might also be

JOB SAFETY

SURFACING TOOLS The following are general safety rules for using surfacing tools. Check the manufacturer's manual for any special safety instructions.

- Always wear proper eye protection.
- Wear hearing protection, especially when using routers, belt sanders, electric planes, and planers.
- Avoid wearing loose clothing or jewelry that could get caught in the tool. Tie back long hair.
- Protect yourself from inhaling dust by wearing the proper dust mask or respirator. Make sure the tool's dust bag is properly attached. Connect the tool to a vacuum system if possible.
- Always unplug the tool before changing bits, cutters, or belts.
- Make sure bits and cutters are sharp. Be careful when changing or adjusting them. They can cause serious cuts.
- Clamp the workpiece securely to prevent it from vibrating loose or being forcefully ejected.

A Jointed Edge

Flat and Straight A properly jointed edge is straight along its entire length and forms a 90° angle with the board's face.

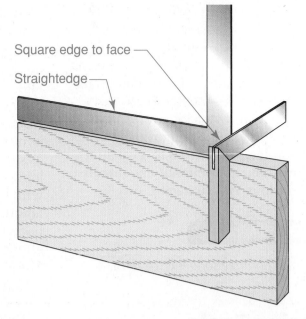

Square edge to face

Straightedge

used on a residential job site if the project calls for large quantities of custom wood-work. Portable jointers are also available for use on smaller projects.

A jointer has a **cutterhead**, a solid metal cylinder on which three or four cutting knives are mounted. The cutterhead is mounted below the bed of the machine. As the cutterhead spins, the knives shear off small chips of wood, producing a smooth surface. A guard covers the cutterhead but swings out of the way to enable stock to pass. Basic jointing operations should always be done with the guard in place.

JOB SAFETY

JOINTER SAFETY The following are general safety rules for using jointers. Check the manufacturer's manual for any special safety instructions.

- Wear proper eye protection.
- Be sure that portable jointers will not tip over during use. They should be secured temporarily to a structure's subfloor or to a workbench.
- Make sure that the guard is in place and operating easily.
- Check the stock for knots, splits, and other imperfections before jointing. Defective stock may break up or be thrown from the jointer.
- Always keep the knives of the jointer sharp. Dull knives tend to cause kickback. They also result in a poor cut.
- Never adjust the fence or the depth of cut while the jointer is running.
- Because of the danger of kickback, always stand to the side of the jointer, never directly behind it.
- Always allow the machine to come to full speed before using it.
- Always cut with the grain. Always use a pushstick or push block to move stock past the cutterhead. Do not make cuts too deep.
- Do not joint short pieces of wood.
- Use a brush to remove shavings from the table. Never use your hands.

A fence guides the stock. The size of a jointer is indicated by its maximum width of cut. A 6" or 8" jointer is common. The length of its bed also affects its usefulness. A longer bed provides better support for jointing longer pieces.

Portable Planers

A *planer* is used to reduce the thickness of a board, smooth its surface, and make one face parallel to another. For example, it might be used to square up stock for stair balusters and other finish work. A planer will not square stock that is not square to start with. The material must first be squared on two sides using a jointer or a table saw. Then the planer can be used to reduce the dimensions of the stock to the final finished size. Large planers are essential in woodworking shops. Portable planers such as the one in Portable Planer are sometimes used on a construction site, especially when much custom woodworking is required.

Like a jointer, a planer has a cylindrical cutterhead fitted with two or more knives.

Portable Planer
Board Smoother This 12" portable planer can be moved easily around the job site.

JOB SAFETY

PLANER SAFETY The following are general safety rules for using a portable planer. Check the manufacturer's manual for any special safety instructions.

- Wear proper eye protection.
- Because of the danger of kickback, always stand to the side of the planer, never directly behind it. Never look into the planer when it is running.
- Always provide support at the outfeed side of the planer to support long boards.
- Portable planers are light in weight. Before using one, make sure it is securely fastened to a work surface and will not tip during use.
- Check each board for loose or large knots, warped surfaces, and other flaws that might cause a problem.
- Avoid running used lumber through the planer. The blades can be damaged if they hit a nail or staple. Repairs are time-consuming and expensive.
- Do not force the stock; let the infeed roller pull the stock through. Do not pull stock out of the planer. Support it on the tips of your fingers or on an outfeed table as it leaves the machine.
- Take a series of shallow cuts rather than one deep cut. This is most important when planing hardwoods. A cut that is too deep can damage the stock and overload the planer.
- To be cut safely, a board must engage both the infeed and the outfeed rollers. Therefore, it must be at least several inches longer than the distance between them.

PORTABLE ELECTRIC PLANE SAFETY The following are general safety rules. Check the manufacturer's manual for any special safety instructions.

- Wear suitable protection for your eyes and ears.
- Be sure that the blades are sharp. Dull blades result in a poor cut that can be difficult to control.
- Do not allow the workpiece to move or vibrate. Secure it with clamps or in some type of holding device. Do not try to hold it freehand.

- Make adjustments to the plane only when the cord has been disconnected from the power source.
- Use two hands to guide the plane. Stand so you can guide the tool with an uninterrupted cutting motion.
- Do not put an electric plane down until the motor has come to a complete stop.
- As with a manual plane, a power plane should be laid to rest on its side (or upside down) to prevent the sharp blades from coming into contact with anything that could potentially dull them.

As the cutterhead rotates, the knives make many small cuts in the surface of a board. This brings the board to a uniform thickness. The cutterhead is mounted above the bed of the machine, as shown in How a Planer Works. A powered *infeed roller* moves the stock into the cutterhead. (The *infeed* end of the tool is where stock enters.) Between the infeed roller and the cutterhead is a *chip breaker.* The chip breaker keeps the stock firmly pressed against the bed and prevents tears and splinters. Just beyond the cutterhead is the *pressure bar.* This holds the stock firmly against the bed after the cut is made. An unpowered *outfeed roller* presses against the wood as it exits the machine. (The *outfeed* end of the tool is where stock exits.)

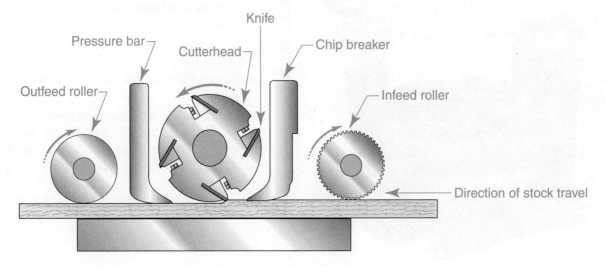

Knife

Pressure bar

Cutterhead

Chip breaker

Outfeed roller

Infeed roller

Direction of stock travel

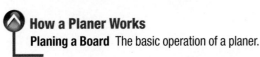

How a Planer Works
Planing a Board The basic operation of a planer.

The type and number of controls on a planer vary with its size. All machines, however, have a *handwheel* that moves the bed up and down to control the depth of cut. The size of a planer indicates the size of its bed and the widest board that it can surface. A 12" model can handle boards up to 12" wide.

Portable Electric Plane A portable electric plane, sometimes called a *power plane*, is shown in Electric Plane. It reduces the time and labor needed to plane by hand. It is used to trim or square an edge. Because it makes a smooth and accurate cut, it is useful for installing and trimming doors and paneling. It can also straighten lumber, trim siding, and surface large timbers.

The portable electric plane has a cylindrical cutterhead mounted above the fence and protected by a housing. In many cases, the cutterhead is fitted with three straight blades. Some cutterheads have curved blades mounted in a spiral pattern. These are more difficult to sharpen than straight

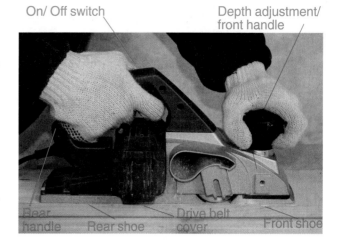

On/ Off switch   Depth adjustment/ front handle

Rear handle   Rear shoe   Drive belt cover   Front shoe

 Electric Plane
Portable Smoother The main parts of a portable electric plane include: on/off switch, depth adjustment/ front handle, rear shoe, drive belt cover, and front shoe.

blades, but they make a very smooth cut. In both cases, the cutterhead revolves toward the front of the tool.

Section 6.3 Assessment

After You Read: Self-Check

1. How is the size of a portable belt sander determined?
2. Describe the proper technique for using a portable belt sander.
3. How is the size of a jointer indicated?
4. How does the position of a cutterhead on a planer compare to its position on a jointer?

Academic Integration: Science

5. **Health Literacy** Mechanical abrasion of wood (cutting, drilling, sanding, and shaping) can release particles of preservative toxins such as arsenic into the air. These toxins can be inhaled, ingested, or absorbed into your skin. Because of the danger of poisoning, arsenic-based preservatives are no longer used in residential carpentry in the United States. Choose one of the safety guidelines in this chapter about reducing sawdust or protecting breathing in the workplace. Create a poster to hang in the shop or in the classroom.

Go to **connectED.mcgraw-hill.com** to check your answers.

©Konstantin Gushcha/Alamy

Plate Joiner Basics

What is a plate joiner used for?

A *plate joiner*, or *biscuit joiner*, is shown in A Plate Joiner. It is a portable power tool that cuts crescent-shaped grooves into the edge of a workpiece. Trim carpenters use the tool for such tasks as assembling molding and joining shelves to cabinetry. The tool can also be used for butt-joining custom wood flooring that is not end-matched. It can also be used to strengthen the joints in molding or trim.

A **biscuit**, or plate, is a small, flat piece of compressed wood. After the plate joiner cuts into the wood, biscuits are glued into the crescent-shaped grooves as shown in Biscuit Joinery. The workpiece can then form a joint with another workpiece in which matching grooves have been cut. Biscuits strengthen the joint and help to align the pieces accurately.

The 4" diameter blade of a plate joiner has carbide-tipped teeth. It is powered directly by the motor or by a flexible drive belt. At the front of the tool is a metal

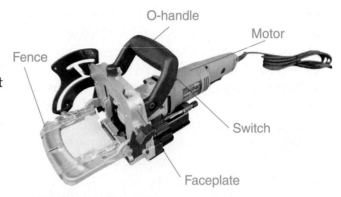

O-handle

Motor

Fence

Switch

Faceplate

A Plate Joiner
Tool for Strong Joints The main parts of a plate joiner.

faceplate. Small metal anti-kickback pins or rubber pads on the lower portion of the faceplate help to keep the tool from sliding during use. An adjustable fence positions the tool against the workpiece. The fence moves up and down and can be angled as well.

Biscuits

The small, thin, oval wood biscuits used in plate joinery are die-cut from beech blanks. The grain of each biscuit runs diagonally to

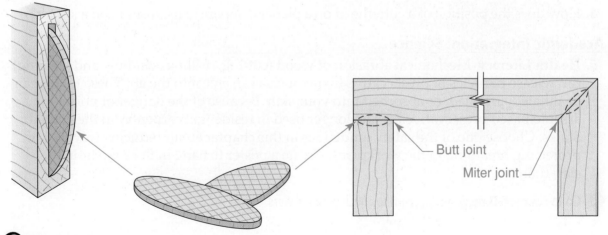

Butt joint

Miter joint

Biscuit Joinery
Using Biscuits Biscuits can be used to strengthen the joints in molding and trim.

its width. This helps it to resist shear forces across the completed joint.

The biscuits are compressed during manufacture. When one is placed in a glued joint, it absorbs moisture from the glue and expands slightly. This makes it fit the joint tightly as the glue dries. Use either white glue or carpenter's glue. Biscuits come in three standard sizes:

- #0 (approximately ⅝" by 1¾")
- #10 (approximately ¾" by 2⅛")
- #20 (approximately 1" by 2½")

Plastic biscuits are available for joining synthetic countertop materials such as Corian. Because plastic biscuits do not absorb moisture from adhesives, they will not expand within the joint. Plastic biscuits are used primarily to speed assembly and to strengthen the joint.

Using a Plate Joiner

The procedure for using a plate joiner is fairly simple. As an example, suppose that two 1×6 boards must be edge-joined to create 10"-wide stock for a closet shelf. You would follow these steps:

1. Place the boards edge to edge.
2. Draw short layout lines across the joint with a pencil. The lines should be 8" to 10" on center.
3. Adjust the joiner's depth of cut for the size of biscuit you wish to use.
4. Adjust the joiner's fence to center the cut in the edge of the board. In the case of 1× stock that is ¾" thick, the center of the cut will be approximately ⅜" from either surface.
5. Clamp one board to the workbench.
6. Use the centerline guide on the tool to align the faceplate with the board's layout marks.
7. Turn on the plate joiner. Bring it to full speed, and push it toward the board. This will plunge the blade into the stock.
8. When the cut is complete, pull the joiner away from the stock and line it up with

JOB SAFETY

PLATE JOINER SAFETY The following are general safety rules for using plate joiners. Check the manufacturer's manual for any special safety instructions.

- A plate joiner ejects dust and chips at a high rate of speed. Keep your face away from the dust ejection chute. Wear safety glasses at all times.
- Wear ear protection. Most plate joiners are noisy.
- Unplug the power cord when changing blades or performing routine maintenance.
- Any workpiece that is likely to move during the cut should be clamped.
- Be sure that the blades are sharp. Sharp blades improve the cutting action and minimize the possibilities for kickback.
- Check the operation of the guard before using the tool. It should close smoothly over the blade.
- Do not disable the anti-kickback points on the faceplate. Make sure the points engage the workpiece.
- Keep hands away from the blade area when making cuts.
- Never hold a workpiece in your hand while cutting.
- Retract the blade fully after a cut. Failure to retract the blade may allow it to contact the workpiece too soon during the next cut. This can cause kickback.

the next layout mark. Continue to make cuts in this manner.

9. After turning off the tool, clamp the second board in place and repeat Steps 6–8.

In order for the boards to register (line up on the surface) correctly, it is important that all biscuit joint cuts be made from the same surface on adjoining boards: Cuts must be made all from the top or all from the bottom surface, to allow for the fact that the cutter may not be perfectly centered in the thickness of the board.

To assemble the boards edge to edge, brush glue into the biscuit grooves of one board. Insert the biscuits. Then apply additional glue to the exposed portions of the biscuits and to the edges of both boards. Press the boards together, using the penciled layout lines to ensure precise alignment. Clamp the boards together until the glue dries. Then rip the stock to final width if necessary, using a table saw.

Section 6.4 Assessment

After You Read: Self-Check

1. Name two applications for plate joinery in residential construction.
2. What wood is used to make biscuits?
3. Why does the grain of a biscuit run diagonally?
4. What are the three standard sizes of biscuits?

Academic Integration: Mathematics

5. **Equidistant Points** Jao has marked 4 equidistant points along a 30" board to cut slots for biscuits. The first and last positions are 3" in from the ends of the board. How far apart are each of the biscuit slots? You can draw a picture to help you with this problem.

Math Concept The word *equidistant* means the same (equal) amount apart. For example, if three equidistant points are placed on a line, each of the line segments between the points is the same length.

Step 1: Add the inches at the end of the board together. (3" + 3" = 6")

Step 2: Subtract this sum from the length of the board to find the length of the board that includes the biscuit slots. (30" − 6" = 24")

Step 3: Divide the length of the board that includes the biscuit slots (24") by the number of equidistant points (4) to determine how far apart each biscuit slot should be.

Go to **connectED.mcgraw-hill.com** to check your answers.

Power Nailers & Staplers

Nailing & Stapling Systems

What does "pneumatic" mean?

Power nailers drive many types of fasteners, including framing nails, finish nails, roofing nails, drywall nails, brads, and corrugated fasteners. Power staplers are used primarily for installing sheathing, subflooring, and roofing. However, they can also be used to fasten framing, trim, and wood flooring. These tools allow carpenters to install fasteners more quickly and with less fatigue. In addition, they are useful in confined work spaces where it is difficult or impossible to use other tools.

Power nailers and staplers are either pneumatic or cordless. A **pneumatic tool**, such as the one shown in Pneumatic Nailer, is a tool powered by compressed air. It must be connected to an air compressor with a flexible air hose. The cordless types are sometimes driven by a rechargeable battery, but more often by an internal combustion engine and compressed gas. In this book, the terms nailer and stapler refer to both pneumatic and cordless models.

Pneumatic Tools

Compressed air is fed to a pneumatic tool through a high-pressure hose connected to an air compressor. The head of the tool or sometimes the handle holds the air. Most nailers and staplers operate on pressures of 60 to 120 psi (pounds per square inch). If the pressure is too low, the fastener may not be driven completely into the workpiece. If the pressure is too high, the fastener may be driven too deep. Excess pressure is also hard on the tool. The operating pressure appropriate for each tool can be found in the owner's manual.

Nailers and staplers are available in a variety of sizes that will drive a narrow range and type of fastener. For example, a nailer designed to drive 16d nails cannot be used to drive brads. When choosing a tool for a particular application, first determine the type and size of fastener needed. Then find a tool that will drive that fastener. If the tool is not cordless, you will need to then find an air compressor that will work with it.

How a Nailer Works Pulling the trigger on the tool releases the compressed air, which moves a piston in the head of the tool. This piston is attached to a driver blade. When the piston is forced downward, the driver blade strikes a fastener and pushes it into the workpiece at high speed. After the fastener has been driven, the piston retracts, pulling the driver blade with it. When this sequence is complete, another fastener is pushed into place, ready for the next pull of the trigger.

Newer nailers and staplers have a two-step firing sequence. This is an important safety feature. The trigger must be pulled *and* the

Head

Handle

Trigger

Nosepiece

Magazine

Pneumatic Nailer

Pneumatic Strip Nailer The main parts of a pneumatic strip nailer. This one has an angled magazine.

Arnold & Brown

JOB SAFETY

NAILERS AND STAPLERS The following are general safety rules for using nailers and staplers. Check the manufacturer's manual for any special safety instructions.

- Keep bystanders away from the work area. Power-driven fasteners sometimes ricochet and can injure anyone nearby.

- Always wear proper eye and hearing protection when using a nailer or stapler.

- Never carry a nailer or stapler while keeping your finger on the trigger. If you were to bring the nosepiece of the tool into contact with a person or object, a fastener could be fired accidentally.

- Never attempt to override the safety mechanism.

- Never use bottled gases to power the tool. The driver blade of a nailer or stapler sometimes makes a spark when it hits the fastener. Thus, running the tool on oxygen, for example, could cause an explosion. In addition, carbon dioxide and other gases are bottled at pressures that are unsafe for use by nailers and staplers.

- Never operate a nailer or stapler at a pressure higher than it was designed to handle. Check the pressure gauge of the air compressor periodically.

- If you are using a belt-driven air compressor, make sure that the belts are protected by a cover.

- Before transporting an air compressor, release the pressure in the air-storage tank. Secure the air compressor so it does not roll around in the back of the vehicle.

- Make sure the tool is pointed at the ground when you connect a pressurized air hose to it. The sudden entrance of pressurized air into the tool can cause it to fire.

- Check the hoses connected to a tool to make sure they are in good condition. Never step on a hose. This causes it to wear prematurely.

- Pay particular attention to hoses while using pneumatic tools on a roof. Hoses are easy to trip over. They can also sweep tools off the roof. Secure the hose to a point near the place where you are working. Do not work while moving backwards.

- Never fire the tool until the nosepiece is in contact with the workpiece.

- Never try to clear a jammed tool while it is still connected to an air supply or power source. Disconnect the tool before performing any maintenance on it.

- Nails do not always go in straight and may fishhook out one side of the wood. Keep your hand at least 6" to one side of the impact point.

---

nosepiece of the tool must be pressed against the workpiece before the tool can be fired. This helps to prevent the tool from being fired accidentally. A *sequential trip* nail gun prevents the tool from firing unless the nose is pressed against the work piece *before* the trigger is pulled. *Contact trip* nail guns do not have this additional safety feature.

Maintaining Pneumatic Tools Pneumatic nailers and staplers are used for high-volume and high-speed installation of fasteners. They must be given regular care. Otherwise,

fasteners will become jammed in the tool or be set improperly. Periodic maintenance also makes the tools safer to use.

- Store the tool at room temperature.

- Lubricate the gaskets on a regular basis in unusually cold weather. A gasket is a piece of flexible material that prevents air or liquid from moving between parts of a tool. The various gaskets on a pneumatic tool prevent air leaks. To lubricate the gaskets, place a few drops of tool oil into the air intake of the nailer just before connecting the hose. Check the owner's manual for the recommended frequency for lubrication. Another method is to attach a line lubricator to the air compressor. A line lubricator automatically adds small amounts of lubricant to the air in the hose, which then conveys the lubricant to the tool. Be sure to check the owner's manual for lubrication requirements.

- Check the magazine, which holds the fasteners. It can become clogged by dirt and sawdust. Spray the magazine with a lightweight lubricant recommended by the manufacturer. Then wipe it clean.

Compressors and Hoses

An air compressor squeezes air into an air-storage tank. The air can then be released through a hose to power tools. The main parts of an air compressor are shown in Air Compressor.

The **regulator** is a valve that controls the air pressure reaching a nailer or stapler. Inside the pump, one or more pistons compress the air into a small chamber. Most pumps on portable compressors are single-stage pumps. This means they pump the compressed air directly into the air-storage tank. Portable air compressors are often powered by electric motors. The motor usually drives the air pump by means of a belt connected to a flywheel. Large air compressors can be powered by gas engines.

The air-storage tank is usually a cylinder holding 1 to 10 gallons of air. The advantage of a larger tank is that it makes more air available to the tool at any given moment. The disadvantage is that a larger tank makes the air compressor heavier, which makes it harder to move. A pressure gauge measures air pressure within the air-storage tank. A line-pressure gauge monitors the pressure in the hose leading to the tool. This is important because the pressure in the air-storage tank may be different from the pressure in the hose.

Air compressors used on job sites are portable. They can easily be moved around the site to be close to the work being done. The smaller, more portable air compressors are not intended for high-volume applications.

Air Compressor Maintenance The following are general guidelines for air compressor maintenance:

- Maintain the proper oil level in the pump.
- Release the air in the air-storage tank at the end of each workday. This helps to clear any moisture from the tank, which can rust. Removing moisture also keeps the airlines from freezing in cold temperatures and keeps them from blasting moist air into your power tools. The air should be released by opening the drain petcock(s) at the bottom of the tank(s), so as to allow water that has accumulated in the tank to escape.
- Clean the air intake filter on the pump regularly. This filter traps dirt, moisture, and other contaminants. If these contaminants reach the tool, they can cause excessive wear.
- The vibration of an air compressor can loosen fittings over time. Check all the fittings periodically. Tighten them as needed.
- Check the drive belt. Replace the belt if it is worn or damaged.

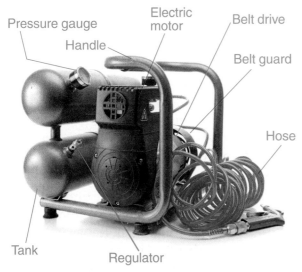

Pressure gauge
Handle
Electric motor
Belt drive
Belt guard
Hose
Tank
Regulator

Air Compressor

A Portable Model The main parts of an air compressor include: air pump, belt drive, belt guard, electric motor, pressure gauge, regulator, tank, hose fittings, and two handles. Handles allow two people to move the tool.

- Check the pressure gauge on the regulator periodically. Improper pressure will prevent the tool from setting fasteners completely.

Air Compressor Capacity The amount of air needed by various types and sizes of pneumatic tools varies. The rate and frequency of tool use will determine air consumption. To drive a framing nail, for example, can require 15 times the amount of air needed to drive a finish nail. Fastening subflooring is repetitive work that calls for many fasteners to be driven quickly. A nailer used for this work would require more air than one used for framing.

An air compressor must provide a steady air supply. Signs of a low air supply include air leakage from the tool, fasteners that are not set at the proper depth, and skipped shots. The volume of air is measured in cubic feet per minute (cfm) as it is delivered to the tool at a particular pressure. Pressure is measured in pounds per square inch (psi). For example, a framing nailer might operate best at 3 cfm and 90 psi. A brad nailer might require 2 cfm and 70 psi. Some carpenters find that they can fire fasteners more rapidly than an air compressor can supply air. This usually means that the air compressor is undersized.

Air Hoses The hose that is supplying air to a tool should have a minimum working-pressure rating that is 50 percent higher than the maximum pressure delivered by the compressor. This allows a margin of safety in case of malfunctions. Do not use hoses longer than 100'. The movement of air through a hose is slowed by friction. The longer the hose, the harder the air compressor must work to overcome friction.

Keep the outside of hoses clean. This helps to avoid premature wear. Keep the snap-on fittings at each end of the hose out of the dirt. Dirt and sawdust can clog the fittings, making it difficult to attach them to a nailer or stapler. Dirt-caked fittings can allow dirt into the tool.

Cordless Nailers

A cordless nailer or cordless stapler resembles a pneumatic model but operates differently. Fasteners are driven by a small internal combustion engine in the head of the tool. Fuel for the engine is liquefied gas compressed into disposable canisters. This gas is injected into a chamber above the piston. The gas is then ignited by a spark from a rechargeable battery in the tool's handle.

Cordless tools are self-contained and do not require a hose and air compressor. This makes them useful in remote locations or where air hoses and an air compressor would be awkward to use. Cordless models are sometimes referred to as gas, horseless, or portable nailers or staplers.

Maintaining Cordless Tools A cordless nailer requires different maintenance. Follow instructions in the owner's manual or have the tool serviced professionally.

General maintenance rules are as follows:
- Charge the battery and replace the fuel cylinders as needed. Be sure to use the correct fuel cylinder. They are sometimes color-coded.
- If a combustion chamber filter is present, clean it frequently. The filter prevents dust and debris from being drawn into the combustion chamber. Replace the filter if it cannot be cleaned.
- Periodically clean the combustion chamber with an aerosol degreaser.

JOB SAFETY

COMPRESSOR NOISE Air compressors are often noisy, and should be operated at a distance. Otherwise, workers may not hear someone approaching and may be startled or may swing the nailer around unexpectedly, causing an accident.

- Periodically clean the nosepiece. This is important when the tool is used to install roofing, because the asphalt from shingles will foul it. Use a putty knife to remove the asphalt. Although solvents are sometimes used for cleaning, they can damage O-rings in some nailers. Check the owner's manual for instructions. If the nosepiece cannot be cleaned or if it is worn, replace it.

Nails and Staples

Fasteners used in pneumatic or cordless tools are often purchased from the manufacturer of the tools. This is because fasteners made by one manufacturer may not fit another manufacturer's tools. When purchasing a stapler or nailer, be sure that you have access to a steady supply of suitable fasteners. Use of a fastener not recommended by the manufacturer may void the tool's warranty.

Nails Nailers must be loaded with nails that are collated. **Collated fasteners** are arranged into strips or rolls, with each fastener connected to the fasteners on either side. The nails are joined by plastic or paper strips or by fine wire, as shown in Collated Nails. This enables nails to be fed through the tool automatically. The plastic or wire falls away as the nails are driven.

Collated nails are available in a variety of metals, including galvanized.

Nail Head Shapes Nails are classified by the shape of the nail head, the type of shank, and the length, as shown in Types of Nails on page 192. The D-head, or clipped head, nail is used only with nailers. Part of the head has been removed, giving it a D shape. This allows the nails to be packed closely together. The disadvantage of a D-head nail is that it may not hold as well as a nail with a round head. Do not use D-head nails where building codes restrict their use. Such applications could include fastening shear walls and structures located where severe weather or earthquakes are common.

One option is to use *full round head* nails. The heads of these nails are not "clipped," but are instead the same size as a regular nail head. These nails are engineered to have the same holding power as a regular nail, and will meet codes in most areas. Always check with a local building official regarding the suitability of nail gun nail types. This is particularly important in regions that experience high winds.

Nail Shanks The shank of a nail is the portion below the head. The type of shank determines how well the nail will hold in various woods. Nails with several shank designs are available for pneumatic nailers. The following are among the more common types:

- Smooth-shank nails are used for general construction. The smooth shank provides good holding power in a variety of woods. Most framing and roofing nails are smooth-shank nails.

- Screw-shank nails have more holding power than smooth-shank nails. They have a spiral shape that is useful for nailing hardwoods.

Collated Nails
Paper or Wire Nails in **A.** coil form and **B.** strip form The nails shown here have a D-head.

A

B

(t)McGraw-Hill Education; (b)Arnold & Brown

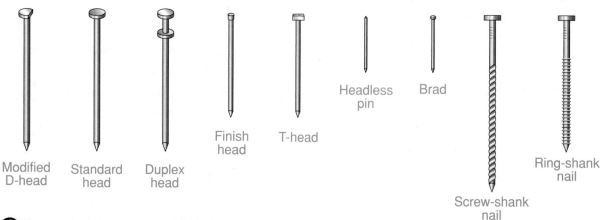

Modified D-head Standard head Duplex head Finish head T-head Headless pin Brad Screw-shank nail Ring-shank nail

 Types of Nails

Heads and Shanks Always choose a nail suitable for the type of work you are doing.

- Ring-shank nails have a series of ridges or rings running from the point nearly to the head. They are best for applications that require extra holding power, such as for nailing wood that has a high moisture content. Ring-shank nails are sometimes used to nail subfloors because they can reduce the occurrence of squeaks.

Staples Staples are made of various metals, including steel, galvanized steel, stainless steel, aluminum, and bronze. They are classified by leg length, width of crown, wire size, and type of point. Basic staple anatomy is shown in A Staple. Staples are generally available in lengths from ⅛" to 2½". The width of the crown is the overall width of the staple, including both legs. Crown width can be narrow, intermediate, or wide. Wire size is either heavy or fine.

Choosing the correct staple depends on the work to be done. For example, a fine-wire, narrow-crown staple is used where the staple must not show, as in fastening trim. Heavy-wire, wide-crown staples are used for attaching asphalt roofing and for other applications where extra holding power is needed. Common staple points are shown in Common Staple Points.

Fastener Magazines

The magazine is the container on a tool that holds a ready supply of fasteners. Fasteners are held in one of two types of magazine.

Strip-loaded tools, or strip tools, hold a straight row of nails or staples in a spring-loaded magazine. The magazine is sometimes angled toward the tool's handle. The angle helps the tool reach into confined areas.

Coil-loaded (or *coil*) tools are shorter and wider than strip-loaded models because they hold nails in a circular magazine (this magazine style is not available with staplers). Up to 300 nails can be loaded into a coil nailer at one time.

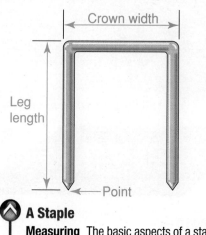

Crown width

Leg length

Point

A Staple

Measuring The basic aspects of a staple.

Builder's Tip

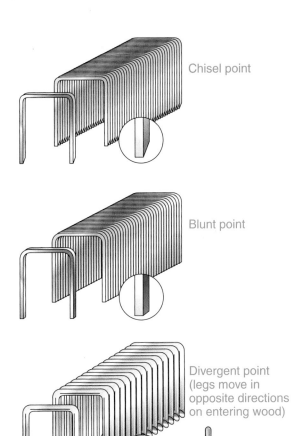

Chisel point

Blunt point

Divergent point
(legs move in
opposite directions
on entering wood)

Common Staple Points
Details Matter The point of a staple is an important factor in how it performs.

> **DRIVING STAPLES** The crown of the staple should be usually kept perpendicular to the grain of the wood. This "locks" it into the wood fibers. Inspect installed staples periodically to be sure that they are being driven properly. Staples driven into the wood at an angle do not hold as well. The exception is with staples used to fasten decorative trim.

Strip-loaded tools are more common on job sites than coil-loaded tools. Some builders prefer strip tools because they are narrow and can be used in tight spaces. For example, toenailing studs is easier with a strip nailer. In addition, spare strips of nails are much easier to carry in a tool pouch than are nail coils.

Section 6.5 Assessment

After You Read: Self-Check
1. What is the firing sequence of a nailer or stapler?
2. What is the best way to choose a pneumatic nailer or stapler for a particular use?
3. What are collated fasteners?
4. What is a disadvantage of a D-head nail?

Academic Integration: Science
5. **Pressure** Pressure is the force per unit area applied on a surface in a direction perpendicular to that surface. What is the pressure of a nail gun that exerts 950 lbs of pressure over 10 in²? Use the following formula to determine your answer:

$$\text{Pressure (psi)} = \frac{\text{Force (pounds)}}{\text{Area (square inches)}}$$

Go to **connectED.mcgraw-hill.com** to check your answers.

Review and Assessment

Chapter Summary

A corded electric drill is best for drilling large holes. Cordless drills are useful where a long extension cord would be undesirable or where electrical power is not available. A pilot hole is required for starting screws, except when using self-drilling screws.

A router is a portable tool that is used primarily for finishing work once a structure is enclosed. It is used for shaping the surfaces and edges of stock and for cutting joints.

Portable electric sanders are used for tasks ranging from heavy stock removal to delicate finish sanding. The most common sanders are the belt sander and the finishing sander. Surfacing tools can be used to convert rough stock into finished stock on site.

A plate joiner is used by finish carpenters to strengthen the joints in wood molding and for many other applications. The biscuits made for use with this tool are made of pressed wood.

A pneumatic nailer or stapler uses compressed air to drive fasteners. A cordless nailer or stapler uses a small internal combustion engine and fuel to drive fasteners. An air compressor must be chosen to suit the type and size of nail being driven. Fasteners for nailers and staplers are collated to fit the tool.

Review Content Vocabulary and Academic Vocabulary

1. Use each of these content vocabulary and academic vocabulary words in a sentence or diagram.

Content Vocabulary

- amperage (p. 163)
- countersink (p. 166)
- pilot hole (p. 166)
- collet (p. 171)
- chamfer (p. 171)
- template (p. 172)
- cutterhead (p. 180)
- biscuit (p. 184)
- pneumatic tool (p. 187)
- regulator (p. 189)
- collated fasteners (p. 191)

Academic Vocabulary

- ranges (p. 162)
- approximate (p. 163)
- versatile (p. 165)

Speak Like a Pro

Technical Terms

2. Work with a classmate to define the following terms used in the chapter: *shank* (p. 162), *chuck* (p. 162), *twist bit* (p. 165), *combination bit* (p. 166), *fixed-base router* (p. 170), *bearing-over bit* (p. 172), *bearing-under bit* (p. 172), *mortises* (p. 172) *infeed* (p. 182), *outfeed* (p. 182).

Review Key Concepts

3. Organize the different types of drill bits and their uses in a list or grid.

4. Explain how to drill holes in wood and in metal.

5. List one use each for the following types of tools: routers, sanders, planers, and jointers.

6. Explain how a biscuit is used with a plate joiner.

7. Describe the different types of fasteners used with pneumatic tools.

Critical Thinking

8. Infer Why might areas that experience high winds require nails with stronger holding power?

9. Analyze Why is a jointer often used at the late stages of house construction?

Academic and Workplace Applications

STEM Mathematics

10. Like Fractions Drill bit sizes are given in fractions. These fractions signify (refer to) the diameter of the bit. Place the following drill bit sizes in order from smallest to largest: ¼", ⅜", ³⁄₁₆", ¹³⁄₃₂", ½", ¹⁵⁄₆₄".

Math Concept You can compare fractions by converting them to like fractions. Like fractions have the same denominator. In fractions, the denominator refers to all possible parts of a whole. It is the number on the bottom of a fraction.

Step 1: Look at the denominators and find the least common multiple (LCM) of the denominators. The least common multiple of two numbers is the smallest number that is a multiple of both numbers.

Step 2: Convert each fraction to a like fraction with the LCM as the denominator by multiplying both the numerator and the denominator by the fraction form of 1 that will result in the LCM.

For example, take ⅜" bit: $\frac{3}{8} \times \frac{8}{8} = \frac{?}{64}$

$8 \times 8 = 64$, so use $\frac{8}{8}$ as the fraction form of 1 for conversion.

Step 3: Compare and order the fractions.

STEM Science

11. RPM *RPM* (revolutions per minute) is a measure of the rotational speed of the router bit, while the cutting speed measures the speed at which the router cuts through a material. The formula to determine RPM is:

$$RPM = \frac{\text{Cutting Speed}}{\text{Circumference}}$$

The cutting speed of a router bit increases as the diameter of the bit increases. What is the cutting speed of a ½" diameter router that turns at 25,000 rpm?

Starting Hint Calculate the circumference of the bit using the formula π times the diameter. Then multiply the circumference by the revolutions per minute.

21st Century Skills

12. Career Skills: Initiative and Self-Direction Research the job of *trim carpenter*. Write a one-paragraph job description. Include tools and equipment you would need to start a trim carpentry business.

Standardized **TEST** Practice

CERTIFICATION PREP

Short Response

Directions Write a phrase or sentence to answer each question. Use a separate piece of paper.

13. List at least two general safety rules for electric drills.

14. What is the best type of drill for driving screws and why?

15. In terms of sanding technique for wood, what is the difference between using an orbital sander and a random-orbit sander?

TEST-TAKING TIP

If a short response question is asking for facts, do not give your personal opinion on the topic.

*These questions will help you practice for national certification assessment.

CHAPTER 7

Ladders, Scaffolds, & Other Support

Section 7.1
Using Ladders & Scaffolds

Section 7.2
Other Support Equipment

Chapter Objectives

After completing this chapter, you will be able to:

- **Identify** various types of ladders and scaffolds.
- **Describe** how to follow the safety rules for working with ladders.
- **Demonstrate** how to set up a stepladder and a straight ladder.
- **Compare** manufactured metal scaffolding to wood scaffolding.
- **Identify** brackets, a pump jack, and a lifeline.

Discuss the Photo
Using a Scaffold A scaffold allows workers to safely reach heights not accessible with a stepladder. *Why is standard construction lumber not recommended for use in the planks?*

Writing Activity: Research and Summarize
Choose a trade and research the advantages and disadvantages of that trade using aluminum, fiberglass, or wood ladders. Summarize your findings in list form.

Chapter 7 Reading Guide

Before You Read Preview

Carpenters and other tradespeople depend on scaffolds and ladders to work in areas that would otherwise be out of reach. Choose a content vocabulary or academic vocabulary word that is new to you. When you find it in the text, write down the definition.

Content Vocabulary

- ladder
- rails
- stepladder
- spreader
- scaffold
- scaffold planks
- competent person
- trestle
- pump jack
- lifeline

Academic Vocabulary

You will find these words in your reading and on your tests. Use the academic vocabulary glossary to look up their definitions if necessary.

- ratio
- injure

Graphic Organizer

As you read, use a chart like the one shown to organize information about content vocabulary words and their definitions, adding rows as needed.

Content Vocabulary	Definition
ladder	a structure made up of two long side pieces joined by multiple crosspieces on which you can step

Go to connectED.mcgraw-hill.com to download this graphic organizer.

Section 7.1

Using Ladders & Scaffolds

Ladders

Which ladder is self-supporting?

Falls cause the greatest number of deaths in the construction industry. The proper use of ladders and scaffolds reduces the risk of falling. In addition, the use of guardrails and other safety devices can prevent deaths and injuries from falls. Before using ladders and scaffolds, it is very important to read and follow the safety guidelines in this chapter. You should also follow the instructions that come with the ladder or scaffold. They may include special safety instructions. Safety information may also be present on stickers attached to equipment.

A **ladder** is a structure made up of two long side pieces joined by multiple *crosspieces* on which you can step. Ladders are used to climb up and down. Ladders are fast and easy to set up and take down, but they must be moved frequently. They come in lengths from 3' to 50'. The basic parts of a ladder are the rungs, or steps, and the rails. Rungs and steps are the horizontal members that a worker climbs on. **Rails** are the vertical supports to which the rungs or steps are attached.

Types of Ladders

The two basic types of ladders are folding ladders and straight ladders. There are also folding articulated ladders. These ladders are all shown in Types of Ladders. Folding ladders

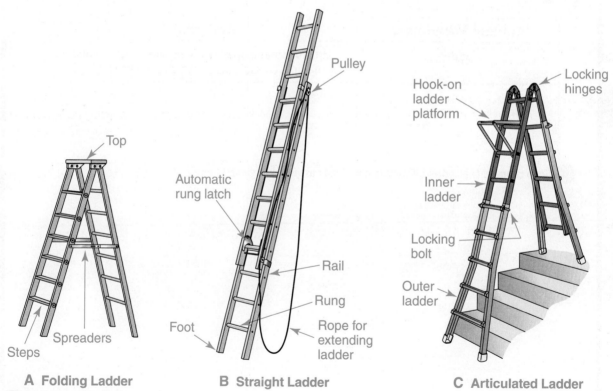

A Folding Ladder   **B** Straight Ladder   **C** Articulated Ladder

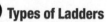 **Types of Ladders**
Folding, Straight, and Articulated A. The type of folding ladder shown is a stepladder. **B.** An extension ladder is a type of straight ladder. **C.** An articulated, or multipurpose, ladder.

are self-supporting and are used primarily indoors for reaching low and intermediate heights. They are the most portable type of ladder. A **stepladder** is a common type of folding ladder that has flattened steps instead of rungs. A *folding articulated ladder,* also called a multipurpose ladder, can be adjusted to fit into such spots as stairwells. Each side of the ladder can be adjusted to different lengths.

Straight ladders are used primarily outdoors where greater heights must be reached. They are not self-supporting. Instead, they must be leaned against a wall or some other object. The ladder's working length is the distance from the ground to the top support, measured along the ladder. An extension ladder is a common type of straight ladder that can be adjusted to various heights. It consists of two straight sections joined together. One of the sections can be extended beyond the other by way of a rope and pulley system.

Folding ladders and straight ladders are available in various grades or duty ratings, as shown in **Table 7-1**. The grade indicates the maximum load the ladder can support.

The materials used most often to make ladders are wood, aluminum, and reinforced fiberglass. **Table 7-2** lists the advantages and disadvantages of each.

Table 7-1: Grades of Ladders

Type	Duty Rating (lbs.)	Typical Uses
Household, Type III	200	Light duty. For household use.
Commercial, Type II	225	Medium duty. For painters and light-construction workers.
Industrial, Type I	250	Heavy duty. For contractors and maintenance workers.
Industrial, Type IA	300	Extra heavy duty. For rugged industrial and construction use.

Note: The user's weight, plus any tools, jacks, planks, and materials, must not exceed the duty rating.

Ladder Safety

The following are general safety rules for ladders. They are followed by rules specifically for stepladders and for extension ladders. As always, we strongly advise you to check the manufacturer's manual for any special safety instructions.

- Inspect ladders carefully. Keep nuts, bolts, and other fasteners tight.

- Do not try to repair a damaged ladder. Replace it or have it professionally serviced.

Table 7-2: Ladder Materials

Material	Advantages	Disadvantages
Wood	• Does not conduct electricity when clean and dry • Weight improves stability	• Less durable than other materials • Heavy • Susceptible to weather damage
Fiberglass	• Does not conduct electricity when clean and dry • High strength-to-weight ratio	• Long-term exposure to sunlight can lead to deterioration • Heavier than aluminum • Damage cannot be repaired • Expensive
Aluminum	• Lightweight • Durable • High strength-to-weight ratio • Weather resistant	• Conducts electricity • Damage cannot be repaired

- Keep steps and rungs free of oil, grease, paint, and other slippery substances. Wood ladders should not be coated with any opaque finish. Such a finish will hide cracks.
- Place the ladder on a firm, level surface. Make sure that it has non-slip safety feet.
- Never place a ladder in front of a door or other opening unless the ladder is secured to prevent an accident, or a barricade is used to keep traffic away.
- Never use a ladder as a scaffolding plank.
- Never use ladders after they have been soaked in water for a long time or been exposed to fire, chemicals, or fumes that could affect their strength.
- Always place the ladder close enough to the work to avoid a long, dangerous reach.
- Face the ladder when climbing up or down.
- Keep your weight centered between both side rails.
- Do not use any ladder where direct contact with a live power source is possible. Metal ladders and any ladder that is wet are especially hazardous.
- Do not overload a ladder. A ladder is designed to carry only one person at a time.

Safety with Stepladders
- During use, be sure that stepladders are fully open and the spreader is locked.
- Make sure all locking devices are secure.

- Never step on the stabilizing bars of a stepladder. These are the horizontal bars between the back rails. They are not designed to support a load.
- Never lay tools on the top step.

Safety with Straight Ladders
- Store straight ladders horizontally in a dry, ventilated place, as shown in Storing a Ladder.
- Make sure that the working length of the ladder will reach the proper height. The ladder should extend at least 3' above a roof or other elevated platform you wish to reach. Never stand on the top three rungs.
- For safety, the foot of the ladder should be a distance equal to one-fourth its working length from the building or other support. The angle created should be approximately 75°.
- To keep the legs from slipping when outdoors, drive a strong stake into the ground behind the ladder. Then tie the bottom of the ladder to the stake with rope.

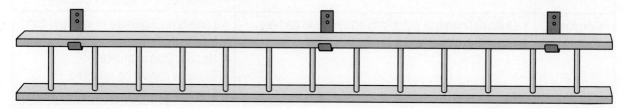

 Storing a Ladder
Dry and Straight Store a straight ladder horizontally on supports to prevent sagging. Never store ladders where they will be exposed to weather or near a source of heat.

- Do not tie or fasten ladders together to create longer sections. Use an extension ladder instead.
- Always make sure that both side rails are fully supported at top and bottom, as shown in Extension Ladder Setup.
- Tie the upper part of a straight ladder to an immovable object to prevent it from shifting.
- Before using an extension ladder, make sure all safety locks are securely hooked over their corresponding rungs.
- Do not adjust the height of an extension ladder while you are standing on it.

Using a Ladder

Ladders are frequently moved around a job site. To avoid accidents, be sure you know how to move a ladder safely. Position yourself at the center of the ladder before lifting it, as shown in Carrying a Ladder. Carry the ladder in a horizontal position, never a vertical one. This will prevent accidental contact with electrical power lines and other objects. In addition, you can control a horizontal ladder more easily and will be less likely to hurt your back.

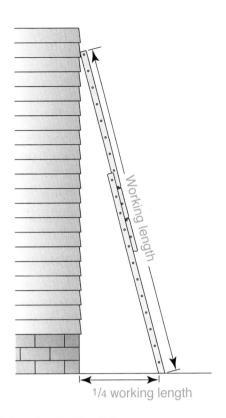

1/4 working length

Working length

Extension Ladder Setup
Proper Pitch and Distance For safety, the *pitch*, or angle, should be approximately 75°. The horizontal distance from the foot of the ladder to the support structure should be one-fourth of the ladder's working length.

Carrying a Ladder
Horizontal Movement Carry a ladder horizontally. If you carry it in a vertical position it could accidentally touch electrical power lines.

Setting Up a Straight Ladder Set up a straight ladder as follows:

Step 1 Brace the lower end against a step or other object so the ladder cannot slide.

Step 2 Grasp a rung at the upper end with both hands.

Step 3 Lift the upper end and walk forward under the ladder, grasping other rungs in turn as you proceed, as shown in the figure.

Step 4 When the ladder is erect, lean it forward into the desired position.

Step 5 Check the angle, height, and stability at top and bottom (see Extension Ladder Setup on page 201).

 Go to **connectED.mcgraw-hill.com** to check your answers.

Before using a stepladder, always be certain that its feet are firmly supported and that its spreader is locked into position. The **spreader** holds the ladder open and prevents it from closing accidentally. Never stand on the top of the ladder because your weight can easily unbalance the ladder and tip it over. Do not step on the bucket tray on the back of some stepladders. It is for holding tools and materials, not people.

When going up or down a straight ladder, grip the ladder firmly and place your feet squarely on the rungs. Make certain your

 Roof Safety

Extend the Top The top of the ladder should extend above the edge of the roof at least 3'. If the ladder is used when conditions are slippery, it must be tied off (secured) to the building.

work boots and the rungs are free of mud and grease. When using a ladder to access a roof, make sure the ladder extends above the edge of the roof by at least 3', as shown in Roof Safety on page 202.

There are limits to how far an extension ladder should be raised. The two sliding sections must overlap each other a certain amount or the ladder will be weak. It will also feel more wobbly when you are on it. The correct amount of overlap is based on the ladder's extended length. Overlap the sections by at least the following amounts:

- 3' for total extended lengths up to 32'.
- 4' for total lengths of 32' to 35'.
- 5' for total lengths of 36' to 47'.

Certain accessories make extension ladders safer to use. A ladder stabilizer, shown in Stabilizer, can be bolted to the top of the ladder. It has arms 4' apart that steady the top of the ladder and prevent it from slipping. Leg levelers can be attached to the feet of the ladder. They support the ladder on uneven ground.

Reading Check

List *What are the basic parts of any ladder?*

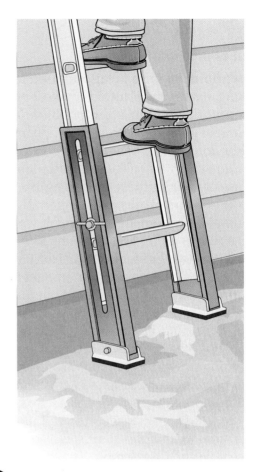

⌃ Levelers
Adjustable Legs Leg levelers adjust to uneven ground and help to prevent the ladder from tipping.

❯ Stabilizer
Wide Arms A ladder stabilizer prevents the ladder from sliding sideways or twisting.

Arnold & Brown

Scaffolds

What is pipe scaffolding?

A **scaffold** is a raised platform used for working at a height. Scaffolds make it possible to work safely, in a comfortable and convenient position, with both hands free. The horizontal parts of a scaffold on which a worker stands are called **scaffold planks**. Because of their stiffness and strength, laminated wood planks made especially for scaffolds are sometimes used. However, some builders prefer aluminum planks because they are lighter in weight and more durable. Standard construction lumber is not recommended for use as scaffold planks.

Commercial metal scaffolding, shown in Metal Scaffolding, is sometimes called *pipe scaffolding*. It has many advantages over scaffolding made from wood. Metal scaffolding is engineered and tested to withstand specific loads. It can be rented as needed, takes up less space than wood scaffolding, and is more weather resistant. It is also easier to assemble and disassemble. The end frames can be put together in a staggered position, making it possible to work from a stairway.

Scaffolds vary in design. For scaffold assembly, the manufacturer's instructions must be carefully followed. When a metal

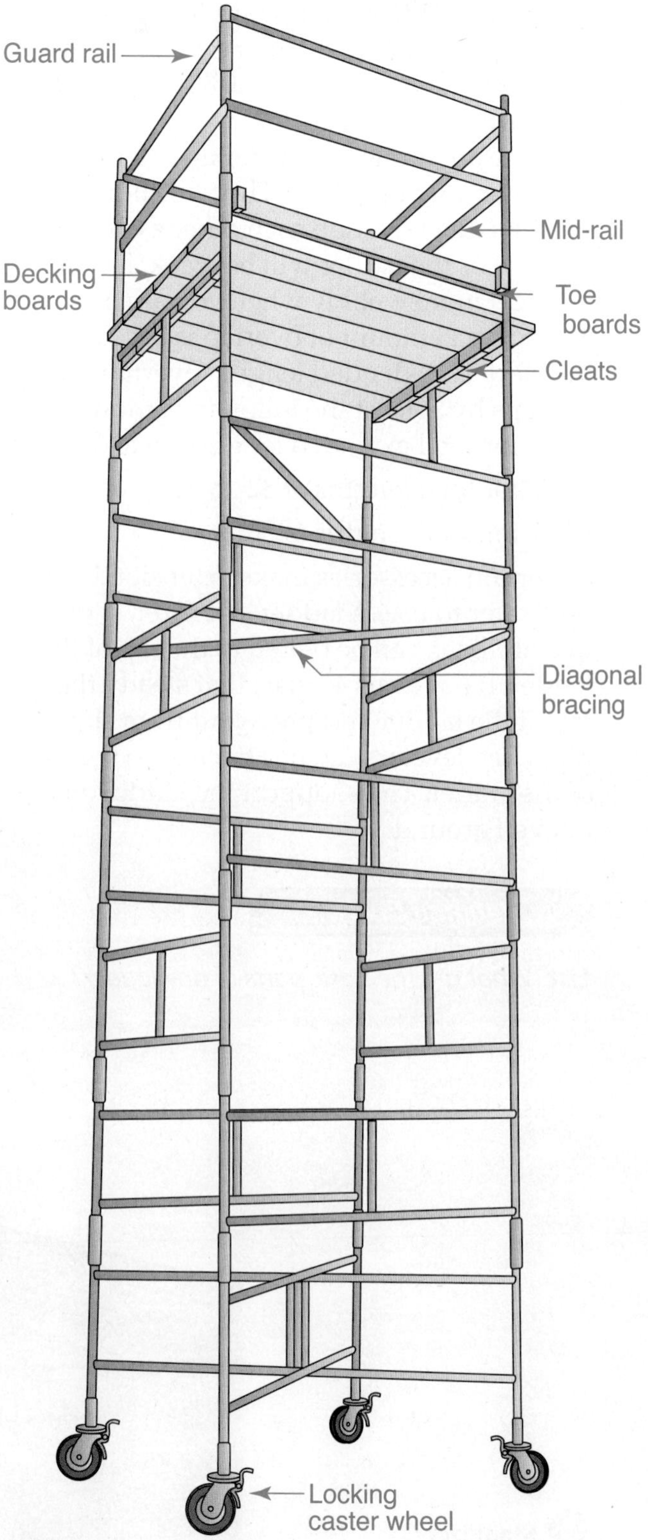

Guard rail →
Mid-rail
Decking boards
Toe boards
Cleats
Diagonal bracing
Locking caster wheel

Metal Scaffolding
Portable Scaffolding Commercial metal scaffolding assembled for interior use.

JOB SAFETY

SCAFFOLD STABILITY As a general rule, a freestanding scaffold should be restrained from tipping once its height is three times its minimum base dimension (a 3:1 ratio). The minimum base dimension is the shortest leg of the rectangle formed by connecting the points of each of the four legs of the scaffold. This 3:1 ratio holds true even if the scaffold is plumb, level, and square. Some codes allow a less strict 4:1 ratio.

JOB SAFETY

SCAFFOLDING SAFETY The following are general safety rules for scaffolding. We strongly advise you to check the manufacturer's manual for any special safety instructions.

- Always follow the manufacturer's instructions when erecting or removing scaffolding.

- Make sure all scaffolding is plumb and level. Use adjusting screws, not blocks, to compensate for uneven ground.

- Some scaffolds have casters or wheels. Make sure these are locked before using the scaffold.

- Provide adequate support for scaffolds. Use base plates, making sure that they rest firmly on the ground. Secure free-standing scaffold towers by attaching guy ropes or wires or by other means.

- Fasten all braces securely.

- Do not climb cross braces. Access to scaffolds should be by stairs or fixed ladders only.

- Provide proper guardrails. Add toe boards when required on planks. A toe board is a lip on the edge of the scaffold platform that prevents a worker's foot from sliding off of the platform. See examples of a toe board in Pump Jacks on page 208.

- Never use ladders on top of a scaffold.

- Never overload a scaffold. Follow manufacturer's recommendations regarding load limits.

- Inspect the scaffolding regularly and tighten any loose connections.

- Inspect lumber used for scaffold planks. It should be graded for that purpose. Both ends of wood planks should have cleats to prevent the planks from sliding off the supports.

---

scaffold is used indoors, it is sometimes equipped with locking caster wheels. With wheels, the scaffold can be rolled easily from place to place. Never move a scaffold while tools or workers are still on it.

To ensure scaffolding stability, be sure to follow the recommended **ratio** between the base of a scaffold and its height.

Scaffolding should include guardrails to protect workers from falls. For all scaffolding manufactured after January 1, 2000, the Occupational Safety and Health Administration (OSHA) requires that a guardrail be placed between 38" and 45" above the scaffold planks. A mid-rail should be placed about halfway between the guardrail and the planks.

Any part of a scaffold that has been damaged or weakened may not meet OSHA strength requirements. The part must either be repaired, replaced, or reinforced. Otherwise, the scaffold must be removed from

service. The scaffolding should be inspected frequently by a competent person. According to OSHA, a **competent person** is someone who has been trained to identify existing and predictable hazards on the job site. This person must also have the authority to take corrective actions to eliminate the hazard.

Dismantling Scaffolds

When dismantling a scaffold, work from the top down. As you remove parts, store them immediately at ground level. Do not stack parts on scaffolding planks because they could easily fall and **injure** someone below. If the scaffolding has been tied or braced to reduce tipping, do not remove the ties or braces until the scaffold has been dismantled to that level. Never throw scaffolding components off the scaffold. This could injure someone below, and it might also damage the component and make it unsafe for future use.

After You Read: Self-Check

1. Name and describe the two basic types of ladders.
2. What should you do if you encounter a ladder that is damaged?
3. What two types of planks are most suitable for use as scaffold planks?
4. What are the advantages of metal scaffolding over wood scaffolding?

Academic Integration: Mathematics

5. **Height-to-Base Ratio** In many codes, the height-to-base ratio for a freestanding (unsecured) scaffold support is 3:1. This means that the height of the scaffold should be at least 3 times the length of smallest base dimension. If not, the scaffold will need to be secured. Colin's scaffold has a base of 2' × 5'. What is the maximum height his scaffold's working platform can be before it has to be secured?

   Math Concept A *ratio* is a comparison of two numbers. Ratios can be expressed with colons (3:1). In the workplace, you may hear the ratio 3:1 expressed as "a three to one ratio."

   Starting Hint Multiply the smallest base dimension by 3 to find the maximum height.

Go to **connectED.mcgraw-hill.com** to check your answers.

Section **7.2** Other Support Equipment

Special Supports

What is a pump jack?

It is not always possible or desirable to work from scaffolding and ladders. This is often the case in cramped work areas or when working atop a roof. Other means must then be found for ensuring a safe way to work.

Brackets

Special brackets are available that can be attached to the frame of a structure. Scaffold planks are then laid on the brackets to form a platform. Some brackets are nailed to side-wall studs while others are bolted to them, as in

Wall Brackets. Nail-attached wall and corner brackets are secured with 20d nails driven into the stud at an angle through the tapered holes in the bracket. This allows the brackets to be easily removed without pulling the nails. Any nails remaining after the brackets have been removed can be driven flush.

Brackets for working on a roof are attached by nailing through them and into the rafters. They can be removed without pulling the nails. There are several types of roof brackets, as shown in Roof Brackets. One style holds a 2×4 or 2×6 against the roof. Another style can be adjusted to various roof pitches, from 90° to level.

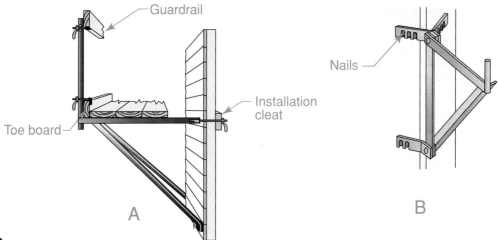

Wall Brackets

Types **A.** Bolt-attached brackets. Note the guardrail. This rail can also be used with other types of brackets.
B. A nail-attached corner bracket.

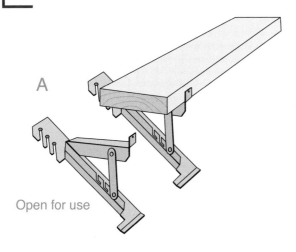

Trestles

A **trestle** is a portable metal frame with rungs that is used to support scaffold planks at various heights, as shown in A Trestle. Trestles are sometimes used by contractors for working on ceilings. They are available in a wide range of sizes and some are adjustable in height. Trestles accept the same types of planks used on standard scaffolds.

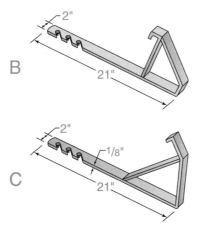

Roof Brackets

Types **A.** The folding roof bracket adjusts to various roof pitches. **B.** This bracket supports lumber at a right angle to the roof. **C.** This bracket positions lumber to provide a level walkway.

A Trestle

Portable Model A folding trestle has crossbars to hold scaffolding planks at various heights.

Though a trestle may resemble a saw-horse, it is designed specifically to support scaffolding. Sawhorses are not intended for this purpose and should not be used to support scaffolding.

Pump Jacks

A **pump jack** is a metal device with a foot pedal that a worker pumps to make it slide up and down on a wood or aluminum post. One type of pump jack is shown in Pump Jacks. Two or more jacks in a row support planks that a worker can use as a scaffold. Pump jacks are commonly used to reach the side walls of a house during siding or painting operations. To lower the jack, the worker turns a hand crank.

The wood posts are often created by two 2×4s fastened together. The lumber should be solid, knot-free, and no more than 30' long. For solid support, each post must rest on a wood or steel pad. It must also be anchored to the structure at least every 10' by metal stand-offs nailed or screwed into the studs.

Reading Check

Explain *How are brackets different from trestles and pump jacks?*

Stand-off screwed to house

Posts made of double 2×4s

Tool shelf

Guardrail

Pump jack

2×8 pad

Pump Jacks
Adjustable Height Pump jacks can be raised or lowered to any height along a wall.

Lifelines

A **lifeline** is a rope intended to prevent a worker from falling. The lifeline is fastened at one end to a secure point on the structure and at the other end to a harness worn by the worker. One type of lifeline is shown in Lifeline.

According to OSHA regulations, a lifeline must be secured to a structural member above the worker that is capable of supporting a minimum deadweight of 5,400 lbs. This amount accounts for the weight of the worker plus the force of the fall.

Lifelines that are used where they may be subjected to cutting or *abrasion* must have a wire core. For all other lifelines, manila rope or its equivalent, with a diameter of at least ¾" and a minimum breaking strength of 5,400 lbs., must be used. A lifeline should only be long enough to ensure that a worker can fall no more than 6'.

 Lifeline

Well Secured Harnesses and lifelines must be fastened securely. *Why is a lifeline often secured to the ridge of the roof?*

Section 7.2 Assessment

After You Read: Self-Check

1. In what kinds of situations might a worker prefer to set up planks on a bracket or pump jack instead of using standard scaffolding?
2. What might be the advantage of using a pump jack instead of a wall bracket when installing siding?
3. Which type of support would you use if you had a roof with many different kinds of pitches?
4. What is a lifeline, and what is its purpose?

Academic Integration: English Language Arts

5. **Create a Description** Write a one-paragraph description of a form of support found in this section. For each object, include what it looks like, what it is used for, what it is made of, and any other details you can think of. Be sure to include any important safety information.

🔝 Go to **connectED.mcgraw-hill.com** to check your answers.

Review and Assessment

Chapter Summary

The two types of ladders are folding ladders and straight ladders. They come in many sizes and are commonly made of wood, aluminum, or fiberglass. The proper use of ladders and scaffolds reduces the risk of falling. Before using any ladders or scaffolding, it is extremely important to read and follow the safety rules presented in this chapter. Read and follow the instructions in any manual that accompanies the ladder or scaffolding.

Brackets, trestles, and pump jacks also can be used for support. Brackets are attached to the structure. Trestles and pump jacks are freestanding and support scaffold planks. Lifelines are intended to prevent workers from falling more than 6'. They are attached to the structure and to the worker's harness.

Review Content Vocabulary and Academic Vocabulary

1. Use each of these content vocabulary and academic vocabulary words in a sentence or diagram.

Content Vocabulary

- ladder (p. 198)
- rails (p. 198)
- stepladder (p. 199)
- spreader (p. 202)
- scaffold (p. 204)

- scaffold planks (p. 204)
- competent person (p. 205)
- trestle (p. 207)
- pump jack (p. 208)
- lifeline (p. 209)

Academic Vocabulary

- ratio (p. 205)
- injure (p. 205)

Speak Like a Pro

Technical Terms

2. Work with a classmate to define the following terms used in the chapter: *crosspieces* (p. 198), *folding articulated ladder* (p. 199), *pipe scaffolding* (p. 204), *abrasion* (p. 209).

Review Key Concepts

3. Identify the basic parts of a ladder and the basic parts of a scaffold.

4. Summarize the safety rules for using an extension ladder.

5. Explain how to set up a stepladder and a straight ladder.

6. Describe the uses of metal scaffolding and wood scaffolding.

7. Summarize how brackets, pump jacks, and lifelines are used in the construction industry.

Critical Thinking

8. Synthesize A small interior painting company has been asked to paint a structure that has walls and ceilings up to 20 feet high. What types of supports should the company use?

9. Analyze Why would a fiberglass ladder be safer than a metal ladder for someone working close to electrical wires?

Academic and Workplace Applications

STEM Mathematics

10. Right Triangles A ladder is leaning up against a house. The top of the ladder is 24' off the ground. The distance from the foot of the house to the foot of the ladder is 7' from the house. What is the working length of the ladder?

Math Concept If you know the lengths of two sides of a right triangle, you can determine the length of the third using this formula, known as the Pythagorean Theorem: $a^2 + b^2 = c^2$, where a and b are the lengths of the two short sides of a right triangle, and c is the length of the side opposite the right angle.

Step 1: Identify the known values: $a = 24, b = 7$.

Step 2: Plug the known values into the formula: $24^2 + 7^2 = c^2$ Solve for c.

STEM Engineering

11. Drawing to Scale The most common scale used in drawing houses is ¼" = 1' (1:48). You are painting an attic vent in the gable of a two-story house. The vent, which is 3' square, is located 24' above the ground. Use ¼" = 1' as your scale and create a side view sketch to show how you would set up an extension ladder to perform this task. Label the important dimensions. Be sure your sketch includes the horizontal distance between the house and the foot of the ladder, the length of overlap needed for the extension section, and the length of the stabilizing arms required.

21st Century Skills

12. Problem Solving One online store has a 4' by 6' base scaffold with wood planks and swivel casters, brakes, and guardrails for $543. It weighs 146 pounds. The sideguard (midrail) for it costs another $75 and weighs another 30 pounds. Another online store has the same size scaffold in aluminum with brakes and casters for $443. It weighs 79 pounds. The Web site offers a two-piece guardrail system for $164 that includes the guardrails and mid-rails, which total 17 pounds. The shipping service will charge $60 for packages up to 100 pounds, $75 for packages up to 200 pounds, and $85 for packages more than 200 pounds. Create a table that organizes the information in the problem to help you determine which scaffold package is least expensive.

Component	Scaffold #1		Scaffold #2	
	cost	weight	cost	weight
Base Scaffold				
Sideguard				
Shipping				
Totals				

Standardized TEST Practice — CERTIFICATION PREP

Multiple Choice

Directions Choose the word or phrase that best answers the question.

13. Which item would not be used with a lifeline?

   a. rope   **c.** harness

   b. stabilizer   **d.** structural support

14. Which type of support might be used on a rooftop?

   a. stepladder   **c.** bracket

   b. sawhorse   **d.** trestle

15. Which type of ladder is best for a maintenance worker?

   a. Household, Type III

   b. Commercial, Type II

   c. Industrial, Type I

   d. Industrial, Type IA

TEST-TAKING TIP

Predict the answer before you read the different multiple-choice options. Then, read each option and choose the one that is closest to your prediction.

*These questions will help you practice for national certification assessment.

Starting Your Own Carpentry Business

Your Project Assignment

You are ready to start your own carpentry business. You will need to calculate how much money you will need for tools and supplies, and then apply for a business loan that meets your needs and budget.

- **Research** the cost of the tools and equipment you will need to start your business.

- 🏠 *Build It Green* **Identify** tools and equipment that can help reduce impact on landfills.

- **Estimate** the total cost of tools and supplies for your first year of business.

- **Research** terms and interest rates for small-business loans.

- **Determine** the size of the loan you will need and calculate the monthly payment, including interest.

- **Create** a three- to five-minute presentation.

Applied Skills

Practice your skills as you do the following:

- **Create** a complete list of tools you will need.

- **Identify** three tools on your list that can help reduce impact on the environment. (Tip: Cordless tools that have replaceable batteries are more environmentally friendly than cordless tools with fixed batteries.)

- **Research** the prices and features of at least two brands and models of each tool on your list.

- **Determine** which tools are best suited to your needs and the best value for the money.

- **Calculate** the total cost of the tools, including any sales tax.

The Math Behind the Project

The traditional math skills for this project are computation and algebra. Remember these key concepts:

Order of Operations

When solving an equation, first do all operations inside parentheses. Next, calculate any exponents.

An exponent shows the number of times you need to multiply a number by itself. For example, 2^3 means to multiply the number 2 three times ($2 \times 2 \times 2$). Then, working from left to right, do any multiplication and division followed by any addition and subtraction.

Compounding

Interest is the cost for borrowing money. Business loans have compound interest. This means that you pay interest not only on the principal (the amount of the loan), but also on the interest that accrues (grows) on the principal. This process is called compounding. Calculate compound interest using this formula: $A = P(1 + r)^n$. A is the total amount of money you will owe. P is the principal, r is the annual percentage interest rate expressed as a decimal, and n is the number of years you have to repay the loan. Solve for A and then divide this total into monthly payments. For example, if you borrow $10,000 for five years at 8% annual interest, calculate your monthly payment using the following steps.

1. Convert the percentage to a decimal and set up the equation.	$A = \$10,000 \times (1 + .08)^5$
2. Add the numbers in parentheses.	$A = \$10,000 \times (1.08)^5$
3. Use a calculator to calculate the exponent. Round to two decimal places.	$A = \$10,000 \times 1.47$
4. Multiply the two numbers.	$A = \$14,700$
5. Find your monthly payment by dividing this amount by the number of months in the term of the loan.	$\$14,700 \div 12 = \$245$

U.S. Small Business Association

Mission: To aid, counsel, assist, and protect the interests of small business concerns, to preserve free competitive enterprise, and to maintain and strengthen the overall economy of the nation.

Project Steps

Step 1 Research

- Research necessary tools and equipment for a general carpenter, and any other supplies you will need to start your business. Be sure to consider tools and equipment that will reduce impact on the environment.
- Compare brands and models of tools. Look on Web sites, in catalogs, and in local stores. Compare features, prices, warranty protection, and brand reliability. Also research cost to rent versus buy.
- Interview at least two contractors or carpenters in your area, asking for their recommendations and for any cost-saving tips.
- Contact the local field office of the Small Business Administration to gather information on rates and terms of loans for new businesses.

Step 2 Plan

- Itemize the tools and supplies you will need for the first year of your new business. List model numbers and prices.
- Create a cost comparison between standard and higher-quality tools. List the pros and cons of each.
- Determine the principal and term of the business loan you will need.

Step 3 Apply

- Determine the total cost of your tool and supply list, including tax.
- Calculate the total cost of the loan over its term based on an 8% annual percentage rate.
- Calculate your monthly payment. If the payment is not affordable, revise your tool list and desired loan.

Step 4 Present

Prepare a presentation combining your research and cost calculations using the checklist below.

PRESENTATION CHECKLIST
Did you remember to…
✓ Demonstrate the research you conducted to make your tool selections?
✓ Thoroughly explain the tool choices you made?
✓ Show how you determined the amount and term of the loan?
✓ Use and present a chart showing your calculations?
✓ Use a presentation program for your slides?

Step 5 Technical and Academic Evaluation

Assess yourself before and after your presentation.

1. Was your research thorough?
2. Was your tool list complete?
3. Were your cost calculations correct?
4. Were the interest calculations and monthly payments correct?
5. Was your presentation creative and effective?

Go to **connectED.mcgraw-hill.com** for a Hands-On Math Project rubric.

Zero Creatives/©Image Source/Alamy

UNIT 3
Building Foundations

Hands-On Math Project Preview

Checking for Square

After completing this unit, you will research the regulation size for a basketball court, volleyball court, or soccer field. You will then use the 3-4-5 rule to check a court for square.

Project Checklist

As you read the chapters in this unit, use this checklist to prepare for the unit project:

✓ Identify the different methods for checking the square.

✓ Describe the importance of laying foundations square.

✓ Think about what can happen to a house if the foundation is not square.

➤ Go to **connectED.mcgraw-hill.com** for the Unit 3 Web Quest activity.

Ana Abejon/Getty Images

Construction Careers Cement Mason

Profile Cement masons are structural workers who place concrete. They also set and align the forms that hold concrete. They may also make concrete beams, columns, and panels.

Academic Skills and Abilities ....

- mathematics
- blueprint reading
- geometry
- interpersonal skills
- mechanical drawing

Career Path ....

- on-the-job carpentry training
- apprenticeship programs
- trade and technical school courses
- certification

Explore the Photo

A Firm Foundation Concrete is often used as a building material. *What structures have you seen that are made with concrete?*

CHAPTER 8

Concrete as a Building Material

Section 8.1
Concrete Basics

Section 8.2
Working with Concrete

Chapter Objectives

After completing this chapter, you will be able to:

- **List** the characteristics of concrete that make it a useful construction material.
- **Restate** the basic ingredients of concrete.
- **Identify** the five basic types of cement.
- **Describe** how to mix a small batch of concrete from a pre-mix.
- **Describe** how to place concrete.
- **Name** the two basic types of steel reinforcement.

Discuss the Photo

Work Quickly When various concrete ingredients are mixed with water, the material begins to harden and should be placed in forms right away. *What might concrete be used for?*

Writing Activity: Writing Clear Questions
Locate a local supplier of Portland cement. Contact the company via e-mail or regular mail and ask what types of concrete admixtures are most frequently requested in your area. Be sure that your question includes the specific information you are looking for. Restate your question and the answer you found in a two-sentence summary.

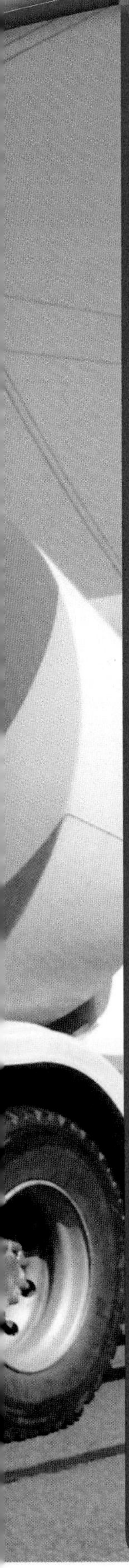

Chapter 8 Reading Guide

Before You Read Preview

Concrete is an extremely versatile and strong material that is used in nearly every residential construction project. Choose a content vocabulary or academic vocabulary word that is new to you. When you find it in the text, write down the definition.

Content Vocabulary

- ○ concrete
- ○ hydration
- ○ Portland cement
- ○ admixture
- ○ crazing
- ○ efflorescence
- ○ slump test
- ○ consolidation
- ○ chair

Academic Vocabulary

You will find these words in your reading and on your tests. Use the academic vocabulary glossary to look up their definitions if necessary.

- ■ techniques
- ■ series
- ■ minimum

Graphic Organizer

As you read, use a chart like the one shown to list the characteristics of concrete. Add rows as needed.

Characteristics of Concrete
1.
2.
3.
4.
5.

Go to connectED.mcgraw-hill.com to download this graphic organizer.

Concrete Basics

Understanding Concrete

What are the components of concrete?

Concrete is a hard, strong building material that is made by mixing cement, coarse aggregate (usually gravel or crushed stone), fine aggregate (a granular material, such as sand), and water in the proper proportions. When these materials, shown in Concrete Ingredients, are combined, a chemical reaction called hydration takes place, which causes the concrete to harden. **Hydration** is a chemical reaction that occurs when water combines with cement. This chemical reaction generates heat as the concrete cures (hardens).

Builders can alter concrete's characteristics by changing the proportion or type of ingredients or by adding other materials. The strength and usefulness of concrete depend on the quality and type of materials used in the mix. The strength of concrete is also affected by the curing methods and the curing time.

The fine and coarse aggregates in a concrete mix are its inert (inactive) ingredients, while cement and water are its active ingredients. Material is not considered concrete unless all four of these ingredients are present. For example, if the coarse aggregate is missing, the resulting material is called *mortar* or *grout*.

Advantages of Concrete

Concrete has been used as a building material for thousands of years. For example, you can see concrete in Roman architecture. Concrete has the following positive characteristics:

- It has tremendous *compressive strength,* which is the ability to withstand pushing forces.
- It is resistant to chemicals.
- It will not rot or be damaged by insects.
- It hardens even under water.
- When properly cured, it withstands extreme heat and cold.
- It can be formed into almost any shape.
- It is widely available and fairly inexpensive.

In residential construction, concrete is used primarily as a foundation material. This use takes advantage of concrete's compressive strength. It is also used for sidewalks, driveways, entry steps, floors, and even kitchen countertops.

Portland Cement

Roman builders obtained natural cement from pumice, a mineral deposited on the slopes of volcanoes. When mixed with

Concrete Ingredients
Basic Mix The basic ingredients of concrete are Portland cement, fine aggregate, coarse aggregate, and clean water.

mrKob/iStock/Getty Images

used for constructing buildings. Various types of Portland cement have different strength characteristics.

Manufacturing Portland cement consists of compounds of lime (calcium oxide) mixed with silica (silicon dioxide) and alumina (aluminum oxide). The lime comes from raw materials such as limestone, chalk, and even coral or seashell deposits.

To make Portland cement, the raw materials are crushed and then ground to a powder. They are then mixed in various proportions, based on the desired characteristics of the end product. The mixture is heated in a large kiln (oven) to approximately 2,700°F (1,482°C) or more. Heating changes the chemical composition of the ingredients and they form small lumps called *clinker*. A small amount of gypsum (no more than 5 percent) is added to the clinker. The resulting mixture is then pounded into the fine powder we call cement.

Basic Types There are five basic types of Portland cement. They are standardized in the United States by the American Society for Testing and Materials (ASTM). The basic types of Portland cement are shown in **Table 8-1**.

water it formed a hard, durable substance. The cement used in modern concrete is called **Portland cement**. It is a manufactured substance that is created using heat. It got its name from being similar in color to Portland stone, an English limestone

Table 8-1: Basic Types of Portland Cement	
Type and Use	**Characteristics**
Type I (standard) Most general construction purposes	Economical, with a long setting time
Type II (modified) Most general construction purposes	Generates less heat during hydration than Type I Resists breaking down when exposed to sulfates
Type III (high-strength) Used where forms must be removed quickly or concrete must be put in service quickly	Gains strength faster than other types of cement
Type IV (low heat) Used only on very large concrete projects, such as dams	Unusually low heat generated by hydration prevents cracking caused by wide ranges of temperature
Type V (sulfate-resistant) Used where concrete will be exposed to highly alkaline conditions and sulfates	Resists alkalines and sulfates

Specialty Cements In addition to the five basic types of cement, specialty cements are also avaliable. The following are used in new construction or remodeling:

- Self-leveling cement flows like thin syrup. It is often poured over a floor to cover tubes used in radiant heating systems. It is also used in remodeling work to level uneven subfloors.

- Hydraulic cement expands when mixed with water and hardens within minutes. It is used to plug holes and cracks in foundations.

- Anchor cement is fast-setting. It is used to secure railings and hardware in holes drilled in a concrete surface. It has a higher compressive strength than standard cement.

- Resurfacing cement is used to repair damaged concrete surfaces. Its fine aggregate allows it to be spread in thin layers.

Reading Check

List *What are the benefits of using concrete?*

Aggregates

Aggregate is granular material such as sand, gravel, or crushed stone. Fine aggregate consists of washed sand or other suitable materials up to ¼" in diameter. Coarse aggregate consists of pea gravel, crushed stone, or other suitable material larger than ¼" (see Concrete Ingredients on page 218). Large aggregate pieces used in concrete should be solid. Layered material such as shale must be avoided. All aggregates must be clean and free of dirt, clay, or vegetable matter, which reduces the strength of the concrete.

The size of the aggregate varies depending on the kind of work for which the concrete is being used. In walls, the largest pieces of aggregate should not be more than ⅕ the thickness of the finished wall section. For slabs, the pieces should not be more than ⅓ the thickness of the slab. Never use aggre-gate that is larger than ¾ the width of the narrowest space through which the concrete will be required to pass during placement.

A large percentage of finished concrete consists of aggregate. For this reason, aggregate quality can have a significant impact on the strength of the concrete. Contaminants, such as dirt and organic material, can generally be removed by washing the aggregate with clean water before it is mixed with other materials.

Water

The water used to mix concrete must be clean and free from oil, alkali (base), or acid. A good rule to follow is to use water that is suitable for drinking. Other contaminants must be avoided as well. For example, sugar prevents concrete from hardening. Sugar might accidentally be introduced if ingredients are mixed in a container that was once used for food products.

The ratio of water to cement is an extremely important factor in the strength of concrete. As more water is added, compressive and tensile strength decrease. *Tensile strength* is resistance to forces that bend and pull.

Hydration

The chemical reaction that occurs when cement is mixed with water is called hydra-

Science: Chemical Reactions

Hydration Portland cement reacts with water in a crystallization process called *hydration,* in which water combines with the compounds in the cement. The concrete continues to cure as long as water and unhydrated compounds are present in the cement. *Dehydration* is a reaction in which water is removed from a substance. What everyday or school tool can you think of that uses hydration or dehydration to work?

Starting Hint For hydration, think of materials that need water added in order to work correctly. For dehydration, think of materials that begin as a liquid and dry hard.

tion. Understanding hydration is the key to mixing and using concrete. Make sure the aggregate and other inert ingredients are thoroughly mixed with the cement. When the water is added, hydration between the water and the cement begins. This reaction causes the concrete to harden. Anything that slows hydration also slows the hardening process.

Notice the difference between the terms *hydration* and *dehydration*. In *dehydration*, a drying out takes place. Concrete does not dry out when it hardens, rather, a chemical reaction occurs. Concrete hardens just as well under water as in air. During the early stages of hydration, concrete must be kept as moist as possible. Premature drying causes the water content to drop below the amount needed for hydration to occur.

After a reaction stage, the initial hydration process comes to a stop. This *dormant* (inactive) period is what allows cement trucks to carry mixed concrete to the job site. Dormancy can last several hours, after which the concrete begins to harden.

Moist-Curing Moist-curing improves the strength of concrete. The surface is kept moist for at least several days after placement, if possible. This can be done by delaying the removal of formwork. It can also be done by covering the concrete with a material that retains moisture or by spraying it lightly with water or with chemicals that slow evaporation.

Concrete gains most of its strength in the 28-day period after it has been placed. However, concrete continues to gain strength for many years afterward.

Reading Check

Compare *What is the difference between hydration and dehydration?*

Admixtures

Ingredients called admixtures are sometimes added to concrete. An **admixture** is an ingredient other than cement, aggregate, or water that is added to a concrete mix to change its physical or chemical characteristics. For example, different admixtures can make concrete more workable or increase its strength, and can be added before or during the mixing process. The following are common:

Air-Entraining Admixtures These introduce tiny bubbles into the concrete, as shown in Air-Entrained Concrete. The bubbles increase the concrete's durability when it is exposed to moisture and frequent freeze/thaw cycles. Air-entraining admixtures are commonly added to concrete used in cold-weather climates. They also improve the material's workability.

Retarding Admixtures These make the concrete set up at a slower rate. This is useful in hot weather or when it is difficult to finish placement before the concrete normally sets up.

Accelerating Admixtures These increase the rate at which concrete gains strength. This can be important if the concrete must be put into service quickly. Calcium chloride is one type of accelerator. It is added to the mixing water in liquid form, rather than powdered form, to avoid problems caused by undissolved material.

Water-Reducing Admixtures These make it possible to reduce the amount of mixing water without reducing the workability of the concrete. This makes the concrete stronger.

Super-Plasticizing Admixtures These generally can do one of two things. They can make the concrete flow very easily, or they can significantly increase its strength.

Colorants

Color or pigment is sometimes added to concrete that will be used as a finished surface. An alternative method is to place a standard, uncolored layer of concrete and then immediately add a colored layer over it. A third method is to dust powdered colorant over the surface of wet concrete. As the surface is troweled flat and smooth, the colorant is absorbed into the surface. A *trowel* is a metal tool with a wide, flat blade that is used for shaping and spreading substances.

Troubleshooting Concrete

Concrete that has been mixed fully using the proper ingredients results in a durable and relatively trouble-free material. However, problems sometimes occur that can affect the strength, durability, or appearance of concrete. They are often the result of improper finishing techniques. Common problems include the following:

- **Crazing** is the appearance of fine cracks that appear in irregular patterns over the surface of the concrete. They typically appear within a week after the concrete has been placed and do not affect the strength of the concrete. Crazing is often caused by excessive floating or by spraying water on the concrete during finishing. Improper curing is another cause, particularly in hot or dry weather.

- *Plastic shrinkage cracks* occur mostly in concrete slabs. They appear as a series of shallow, parallel cracks in the surface. They are caused by the too-rapid drying of the concrete surface. The cracks rarely affect the strength or durability of the concrete.

- **Efflorescence** is a whitish crystalline deposit that sometimes appears on the surface of concrete or mortar. It is sometimes caused when salts in the concrete mix with water or moisture vapor and rise to the surface. It can occur at any time after the concrete cures substantially. It can also be caused when soluble (dissolvable in water) compounds in the soil are drawn into the concrete. Efflorescence is a cosmetic problem that generally does not affect strength or durability.

- *Cracks* that extend through the concrete can significantly reduce its strength and long-term durability. They can have many causes, including improperly compacted subgrades, excessive water in the concrete mix (high-slump), and the lack of expansion joints in large slabs. Concrete may also crack if there is not enough concrete cover over steel reinforcing. This allows moisture to reach the steel, causing rust deposits that expand.

- *Chalking* is a term that describes the formation of loose powder on the surface of hardened concrete. It is sometimes called *dusting*. The surface of the concrete is so weak that it can be crushed by surface traffic or even scratched with light pressure. It is sometimes the result of finishing the concrete before surface water (bleed water) has disappeared. It can also be caused by placing concrete directly over non-absorbative materials such as polyethylene sheet plastic, or by placing concrete in unusually cold weather.

- *Scale* is a term that describes widespread flaking of a hardened concrete surface. It is often the result of exposing the concrete to freeze/thaw extremes.

Section 8.1 Assessment

After You Read: Self-Check

1. If a material contains fine aggregate, cement, and water, but not coarse aggregate, what is the material called?
2. What is *clinker*?
3. How are plastic shrinkage cracks formed?
4. What is an admixture?

Academic Integration: Science

5. **Concrete Strength** When a material reaches the limit of its *compressive strength*, it is crushed. When a material reaches the limit of its *tensile strength*, it bends to the point that it breaks or is deformed. By itself, concrete has a comparatively low tensile strength, but a high compressive strength. With a partner, brainstorm one or two ways you could measure the compressive strength or tensile strength of a material.

Go to **connectED.mcgraw-hill.com** to check your answers.

Section 8.2 Working with Concrete

Placement Techniques

Why is a slump test important?

The word *pour* is often used to describe the process of putting wet concrete into position. However, the term favored by the industry is *place*. An example might read: "The concrete is placed in foundation forms."

Concrete is measured by the cubic yard. Builders often shorten this to "yard." One cubic yard contains 27 cubic feet.

Mixing

Concrete can be mixed on the job from raw materials or by adding water to bags of pre-mixed dry ingredients. It can also be delivered pre-mixed. This is sometimes called a "transit" mix because the concrete ingredients are mixed in transit by a truck traveling to the job site. Another term for this is *ready-mix*. A third method is to have the concrete ingredients delivered by a small-batch truck, sometimes called a "short load" truck. Instead of mixing all the ingredients in transit, the truck carries all the ingredients in separate containers, including water. The truck operator can than blend the ingredients as needed once the truck reaches the job site.

Strength, durability, watertightness, and wear resistance are controlled by the amount of water in proportion to the amount of cement. The lower the proportion of water, the stronger the cement. However, low levels of water also make the concrete stiffer. This can make it more difficult to place.

Mixing small amounts of concrete is often done in a wheelbarrow, but any similar container may be used. A mixing hoe with holes in the blade is often used. In general, the dry ingredients are mixed together first. Water is then poured into the dry ingredients. This reduces the formation of lumps. When mixing concrete by hand, do not add the water all at once. Instead, pour in about half the amount required and thoroughly mix it with the dry ingredients. Then pour in another quarter or so and mix it evenly. Add the remaining water gradually as you mix it in. This allows you to judge the consistency of the concrete as you work, reducing the chance of adding too much water.

Using Pre-Mixed Materials When small amounts of concrete are needed, the pre-mixed dry ingredients are most often purchased in 60-lb., 80-lb., or 94-lb. sacks. When mixed with water, a 60-lb. sack yields 1 cubic foot (cu. ft.) of concrete. However, if the job requires more than 12 sacks of pre-mix, it is generally more efficient and less expensive to obtain concrete in other ways.

Because water triggers the hydration process, take care to store sacks of pre-mix under dry conditions. Small amounts of moisture can cause the cement to become lumpy. Lumps that cannot be broken up by squeezing in your hand mean the pre-mix should not be used.

It is best to store the sacks indoors. If this is not possible, they must be covered with a waterproof tarp. Stack the sacks off the ground and arrange them tightly to limit air circulation. Material in sacks that have been stacked for a long time may seem hard. This is called warehouse pack. It can be loosened simply by rolling the sack back and forth.

Mixing on Site It was once common to mix concrete on site, using separate quantities of cement and aggregate. Most builders now rely on ready-mixed concrete, but there are still times when mixing on site is preferable. This might include occasions when sites cannot be reached by a ready-mix truck.

Table 8-2: Proportions for Various Mixes of Content				
Proportions		**Cement Bags**[b]	**Aggregates**	
			Fine (cubic feet)	**Coarse (cubic feet)**
With ¾" maximum size aggregate	Mixture for 1 bag trial batch[a]	1	2	2¼
	Materials per cu. yd. of concrete	7¾	17 (1,550 lbs.)	19.5 (1,950 lbs.)
With 1" maximum size aggregate	Mixture for 1 bag trial batch	1	2¼	3
	Materials per cu. yd. of concrete	6¼	15.5 (1,400 lbs.)	21 (2,100 lbs.)
With 1½" maximum size aggregate (preferred mix)	Mixture for 1 bag trial batch	1	2½	3½
	Materials per cu. yd. of concrete	6	16.5 (1,500 lbs.)	23 (2,300 lbs.)
With 1½" maximum size aggregate (alternate mix)	Mixture for 1 bag trial batch	1	3	4
	Materials per cu. yd. of concrete	5	16.5 (1,500 lbs.)	22 (2,200 lbs.)

[a]Mix proportions will vary slightly depending on gradation of aggregates. A 10 percent allowance for normal waste has been included in the above figures for fine and coarse aggregate.
[b]One bag of cement equals 1 cu. ft.

When concrete is mixed on the job site, the quantities of cement and aggregate must be figured separately for each cubic yard needed. **Table 8-2** shows the number of bags of Portland cement and the cubic feet of aggregates required to produce 1 cubic yard (27 cu. ft.) of mixed concrete for several mixes. **Tables 8-3** (below) and **8-4** (on page 226) show the amount of water to use in various mixes. These proportions are approximate. Always adjust the mix as needed to suit the job.

For accurate proportions, a bottomless measuring box may be used. This is a four-sided container with no top and no bottom. It has a capacity of 1, 2, 3, or 4 cu. ft. The box should be marked on the inside to show volume levels, such as 1 cu. ft., 2 cu. ft., or less. Handles on the side of the box make it easier to lift after the material has been measured.

To measure the materials, the box is placed on a mixing platform and filled with the required amount of material. The box is then lifted and the material remains on the platform.

Pails can also be used to measure proportion materials. For example, a batch of concrete could be measured by using one pail of Portland cement, two pails of sand, and three pails of gravel or crushed stone.

JOB SAFETY

PROTECT YOUR SKIN Because fresh concrete is highly alkaline, it can irritate the skin and eyes. Wear gloves, rubber boots, and eye protection when mixing and placing concrete.

This would be called a *1:2:3 batch.* The ingredients would be added directly to a portable drum mixer, as shown in A Drum Mixer on page 226. Measuring can also be done with shovels or wheelbarrows, depending on the amount required. However, these methods are less precise. Ingredients should be blended until all materials are uniformly distributed.

Using Ready-Mix Most concrete is supplied to job sites by ready-mix plants. Proper amounts of cement, fine and coarse aggregates, and water are poured into the rotating drum of a truck-mounted concrete mixer. The concrete is mixed as the truck travels to the site. At the site, the concrete slides down metal chutes as it is placed into forms. This method is most economical when at least

Table 8-3: Mix Proportions for Sand of Various Moisture Content							
Trial Mix Aggregate Size	**Gallons of Water Added to 1-Bag Batch if Sand is:**				**Suggested Mixture for 1-Bag Trial Batches(d)**		
	Dry	**Damp(a)**	**Wet(b)**	**Very Wet(c)**	**Bags of Cement (cu. ft.)**	**Aggregates (cu. ft.)**	
						Fine	**Coarse**
For mild exposure: 1½" max. size aggregate	7	6¼	5½	4¾	1	3	4
For normal exposure: 1" max. size aggregate	6	5½	5	4¼	1	2¼	3
For severe exposure: 1" max. size aggregate	5	4½	4	3½	1	2	2¼

(a)"Damp" describes sand that will fall apart after being squeezed in the palm of the hand.
(b)"Wet" describes sand that will ball in the hand when squeezed but leave no moisture on the palm.
(c)"Very wet" describes sand that has been subjected to a recent rain or been recently pumped.
(d)Mix proportions will vary slightly depending on gradation of aggregates.

Table 8-4: Water Proportions for Mixing Small Batches of Concrete				
Size of Batch	**Pints of Mixing Water to Add**			
	Very Wet Sand	**Wet Sand**	**Damp Sand**	**Dry Sand**
5 Gal. Water per Whole Sack of Cement				
½ sack	14	16	18	20
¼ sack	7	8	9	10
⅕ sack (18.8 lbs.)	5⅗	6⅖	7⅕	8
¹⁄₁₀ sack (9.4 lbs.)	2⅘	3⅕	3⅗	4
6 Gal. Water per Whole Sack of Cement				
½ sack	17	20	22	24
¼ sack	8½	10	11	12
⅕ sack	6⅘	8	8⅘	9⅗
¹⁄₁₀ sack	3⅖	4	4⅖	4⅘

A Drum Mixer
Portable Mixer A portable drum mixer may be used for mixing concrete on site. *Why would this tool save time and reduce labor costs?*

two cubic yards of concrete are ordered. The ready-mix company usually charges a premium for smaller volumes. In those cases, a small-batch truck may be more cost-effective.

Ready-mix concrete is ordered by the number of bags of cement required per cubic yard of concrete. Five-bag mix (that is, five bags per cubic yard) is considered to be the **minimum** amount required for most work. Where high strength is needed or where steel reinforcement is used, six-bag mix is commonly specified. Another

JOB SAFETY

RESPIRATOR USE Mixing concrete is a very dusty operation. Be sure to wear a suitable respirator, particularly when mixing the dry ingredients. The respirator should be specifically designed for protection when mixing these ingredients.

way of ordering ready-mix is by its compressive strength. Building plans often specify compressive strength, such as 2,500 or 3,500 psi. Ingredients are then blended to meet this requirement.

Where concrete will be exposed to moderate or severe weathering, building codes generally require stronger and more durable concrete. Such concrete may be used in sidewalks, exposed basement walls, porch slabs, carport slabs, and garage slabs. Codes may also require that the concrete be air-entrained. This concrete is better able to withstand temperature extremes and the chemicals sometimes used for melting ice and snow.

Slump Testing

After the ingredients have been mixed, a slump test is sometimes done at the job site or at the ready-mix plant. A **slump test** is a test to measure the consistency of concrete. The test should be done whenever the consistency of the concrete is of critical importance. It is often required in commercial construction and sometimes in

Tim Fuller

residential construction. In a slump test, concrete straight from the mixer is placed into a small sheet metal cone of specific dimensions, shown in Slump Test Cone.

After the concrete has been repeatedly speared with a rod to remove air pockets, the cone is lifted and removed. A measurement is then taken of how much the unsupported mass of concrete *slumps*, or loses its conical shape, as shown in Testing Slump. The greater the slump, the wetter the concrete. Concrete used in paving and floor slabs might have a minimum slump of 1" and a maximum of 4". However, concrete used for columns and walls might have a slump ranging from 4" to 8". The greater slump would make it easier for the concrete to flow into narrow forms.

Placement

Concrete should be placed continuously, or all at once, whenever possible. It should also be kept fairly level throughout the area. For best results, the concrete should be consolidated. **Consolidation** is a process that removes air pockets and forces the concrete into all parts of the forms. It causes fine particles of the mix to migrate toward the forms and heavier aggregate to move away, which leaves a smooth finish on the walls when the forms are stripped. In a concrete slab, consolidation makes it easier to put a smooth finish on the top of the slab. Consolidation also helps the concrete to flow around steel reinforcing. Concrete can be consolidated with a concrete vibrator as shown in Vibrating Concrete on page 228. This portable tool is powered by a small electric motor. It is used when large quantities of concrete must be placed, particularly in wall and column forms. On smaller jobs, concrete can be consolidated by tamping it or by repeatedly spearing it with a spade, or by tapping the forms repeatedly with a hammer to create a vibration.

After placement, concrete must cure properly to gain full strength. Rapid drying reduces its strength and may damage the exposed surfaces of sidewalks and drives. If maximum strength is critical, cover the concrete with a material that will slow its loss of moisture, such as polyethylene sheets, wet burlap, or wet straw.

In hot weather, make sure the concrete is covered for at least several days after placement. In some cases the concrete can be misted with water, though it is important not to use excessive amounts. Another

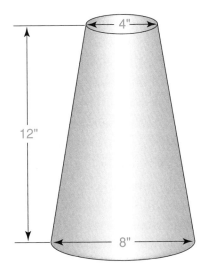

Slump Test Cone
Testing Concrete The dimensions of a test cone are critical to ensuring consistent tests.

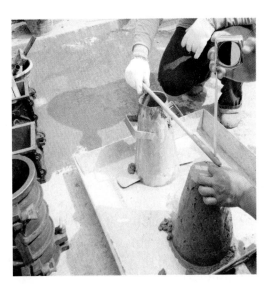

Testing Slump
Reading Slump When the cone is removed from the concrete, the resulting slump is determined by measuring from a rod to the top of the concrete.

 Vibrating Concrete

Consolidation A concrete vibrator helps to ensure that air pockets have been removed.

Concrete and the Weather In some geographic areas, extremely hot weather is common during the construction season. Elsewhere, work must continue during extremely cold weather. Various ingredients can be added to concrete in order to counteract the effects of temperature extremes. Check with your concrete supplier several days in advance of the delivery date.

protect the concrete from the cold. Heaters can be used to pump warm air into the framework until the concrete has set.

Once the dry concrete ingredients are mixed with water, there is not much time in which to place the concrete before hydration makes it difficult or impossible to work with. When the ready-mix truck shows up, all preparations for placement must be complete and all placement workers must be ready.

Pre-Placement Checklist Supervisors often fill out a *pre-placement checklist* to make sure no detail has been forgotten. This is a standardized form that lists the materials that must be on hand and the tasks that must be complete before the concrete is placed. Here are some of the questions that might appear on a pre-placement checklist:

- Is all wall formwork straight and plumb?
- Are scaffolding planks properly secured?
- Is all horizontal and vertical rebar in place?
- Is a concrete vibrator available? Is it in working order?
- Are anchor bolts and hold-downs on site?
- Are there enough shovels, rakes, and other placement tools for the crew?

Concrete can usually be placed directly by chute from the concrete truck. However, on a steep or heavily wooded site, it is sometimes impossible for the truck to get near enough to deliver the concrete by chute. In such cases, it

way to protect concrete in hot weather is to spray it with a curing compound. This is a liquid product that forms a transparent film on the concrete to reduce moisture loss.

In very cold weather, keep the temperature of the concrete above freezing until it has set. The rate at which concrete sets is affected by temperature, and is much slower at 40°F (4°C) and below than at higher temperatures. In cold weather, the use of heated water and heated aggregate during mixing is good practice. In severely cold weather, the concrete should be covered by waterproof insulating blankets. In some cases, a temporary framework of lumber and plastic sheeting can be built to

Tim Fuller

can be pumped through long, flexible pipes. The concrete is delivered from the ready-mix truck directly into the hopper of a pump truck. From there, it is pumped into the forms.

Reading Check

Explain What effect can rapid drying have on concrete?

Concrete Reinforcement
When should concrete be reinforced?

Both steel and synthetic fibers can be added to concrete to improve its qualities. Concrete has great strength in compression, which means that it can support huge loads placed directly upon it. Steel has excellent *tensile strength,* which is resistance to forces that bend and pull. When steel is embedded in concrete, the resulting material, called reinforced concrete, has some characteristics of both materials. Reinforced concrete has excellent compression strength and good tensile strength. Concrete footings, slabs, and walls are nearly always reinforced with steel.

Reinforcing Bar

Reinforcing steel can be purchased in the form of reinforcing bars, called *rebar,* or in the form of welded-wire fabric. Rebar, shown in types of Rebar, has a circular cross-section. It has a patterned, lugged, or otherwise "deformed" surface that helps the concrete grip the steel. Most rebar for residential

A Rebar Shear
Cutting Leverage The long handle of this rebar shear provides enough leverage for smooth cuts.

construction is made from uncoated steel. However, epoxy-coated rebar is sometimes used where conditions are especially corrosive. Rebar is used most often in footings and walls, while welded-wire fabric is used mostly in slabs.

Rebar comes in 20' lengths that can be cut or bent on the job site. A hacksaw or a cutting torch can cut rebar, but a rebar shear such as the one shown in A Rebar Shear makes the job easier. A rebar shear is sometimes called a rebar cutter.

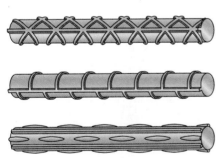

Types of Rebar
Tensile Strength Several types of rebar. The patterns help the concrete grip the bar.

JOB SAFETY

EXPOSED REBAR ENDS Rebar ends will sometimes stick out from an area of concrete that has already been placed. If they cannot be cut off immediately, the exposed ends of rebar must always be protected with a cap, shield, or some other device that will prevent accidental impalement injuries.

The diameter of the rebar needed for concrete varies according to the amount of tensile strength required for the concrete in the pour. **Table 8-5** shows common rebar sizes. Rebar size follows the diameter of the bar. The bar numbers represent multiples of ⅛". A number 3 rebar would be ⅜" diameter while a number 8 rebar would be 1" diameter. Diameters of residential rebar usually range from ⅜" to ⅝". Large construction projects could use a rebar size up to bar number 20, or 2½" diameter.

To be effective, any steel reinforcement must be covered by enough concrete. See **Table 8-6** for coverage guidelines. A suitable amount of concrete cover also protects the steel from rusting. If the steel reinforcement rusts, the concrete will be damaged and the strength of the installation may be weakened.

Wet concrete is dense and heavy, so workers must take steps to prevent the rebar from being dislodged as the concrete is placed. Spacers can be placed in vertical forms to prevent the rebar from shifting. In slab construction, rebar is often placed on small support devices called chairs, shown in Rebar Chair. A **chair** is made of non-corrosive plastic or metal and remains permanently in place after the concrete cures. The height and shape of the chair must suit the particular arrangement and size of rebar in the project.

Table 8-6: Concrete Cover for Reinforcing Steel

Location	Minimum Concrete Protection (inches)
Rebar in footings	3
Rebar in concrete surface exposed to weather	2 for bars larger than No. 5; 1½ for No. 5 bars and smaller
Rebar in slabs and walls	¾
Beams and girders	1½
American Concrete Institute ACI 318, *Building Code Requirements for Reinforced Concrete*.	

Chairs should be spaced as needed to prevent the rebar from sagging. Single-directional chairs support one length of rebar. Dual-directional chairs hold two lengths of rebar where they intersect. In some cases, rebar must be secured with light-gauge tie wire to prevent it from shifting off the chair when the concrete is placed. Do not use chunks of broken brick or similar materials to support rebar. These materials can draw moisture from the soil and cause the rebar to corrode over time.

Reinforcing Mesh

Welded-wire fabric is not really a fabric. It is an open mesh of wires running perpendicular to each other, as shown in Reinforcement Mesh. The most common welded-wire fabric used on a residential job site has wires spaced 6" apart in two directions. This type is referred to as 6 × 6 welded-wire reinforcement. When used to reinforce a slab, the wire

Table 8-5: Size and Weight of Reinforcing Bars

Bar Number[a]	Bar Diameter (inches)	Approximate Weight of 100 Ft.
2	¼	17
3	⅜	38
4	½	67
5	⅝	104
6	¾	150
7	⅞	204
8	1	267
[a]Bar numbers are multiples of ⅛"		

Rebar Chair
This rebar chair is a single-directional model.

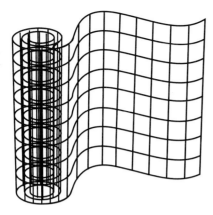

Fiber Reinforcement Short synthetic fibers are sometimes mixed with concrete to reinforce it. These fibers, however, are not a substitute for steel reinforcement. Instead, they help to reduce the shrinkage cracking that occurs sometimes as concrete cures. These fibers also increase concrete's resistance to impact and abrasion, and are often added to floor slabs.

Reinforcement Mesh
Support A roll of welded-wire fabric.

REGIONAL **CONCERNS**

is first unrolled. It is then pulled up into the concrete as the slab is being poured. A more precise technique is to support the welded-wire fabric on *wire chairs* so that the concrete can flow under it. Wire chairs for this purpose are usually between 2" and 4" in height.

Coastal Concrete Where concrete will be exposed to salt spray or other highly corrosive elements, such as in an ocean coast location, use corrosion-resistant rebar instead of regular rebar.

Section 8.2 Assessment

After You Read: Self-Check

1. What is a slump test?
2. What precaution should be taken when placing concrete in hot weather, and why?
3. What is the purpose of the patterned surface on rebar?
4. Identify two types of reinforcement and how they are used.

Academic Integration: Mathematics

5. **Equations** How many bags of five-bag mix of ready-mix concrete are needed to fill a foundation of 25 cubic yards?

Math Concept An equation shows the equal relationship between two expressions. A variable is used for the missing information.

Step 1: Let x represent the amount of bags needed.

Step 2: Select the operation you need to solve the problem. Look for words such as "how many."

Step 3: Create an equation to represent the problem ($5 \times x = 25$) and solve for x.

Go to **connectED.mcgraw-hill.com** to check your answers.

Review and Assessment

CHAPTER 8

Chapter Summary

Section 8.1

Concrete is one of the most common and important construction materials. It is made by mixing cement, fine aggregate (usually sand), coarse aggregate (usually gravel or crushed stone), and water in the proper proportions. Concrete hardens through a chemical process called hydration.

Section 8.2

The proportion of water to cement in a batch of concrete is extremely important in determining its strength. Concrete can be made on site from pre-mixed dry ingredients or from ingredients bought separately. It can also be ordered ready-mixed from a concrete supplier. When steel reinforcing is added to concrete, the resulting material combines compressive strength with tensile strength. Reinforcing materials come in several forms, including steel rods, rebar, welded-wire fabric, and synthetic fibers.

Review Content Vocabulary and Academic Vocabulary

1. Use each of these content vocabulary and academic vocabulary words in a sentence or diagram.

Content Vocabulary

- concrete (p. 218)
- hydration (p. 218)
- Portland cement (p. 219)
- admixture (p. 221)
- crazing (p. 222)
- efflorescence (p. 222)
- slump test (p. 226)
- consolidation (p. 227)
- chair (p. 230)

Academic Vocabulary

- techniques (p. 222)
- series (p. 222)
- minimum (p. 226)

Speak Like a Pro

Technical Terms

2. Work with a classmate to define the following terms used in the chapter: *compressive strength* (p. 218), *clinker* (p. 219), *tensile strength* (p. 220), *dehydration* (p. 221), *dormant* (p. 221), *trowel* (p. 222), *chalking* (p. 222), *scale* (p. 222), *place* (p. 223), *ready-mix* (p. 223), *slumps* (p. 227), *pre-placement checklist* (p. 228), *rebar* (p. 229), *wire chairs* (p. 231).

Review Key Concepts

3. Describe the characteristics of concrete make it useful for creating building foundations.

4. Identify the basic ingredients of concrete.

5. Name the five basic types of cement.

6. Explain how to mix a small batch of concrete from a pre-mix.

7. List the characteristics of properly cured concrete.

8. Describe the characteristics of rebar and welded-wire fabric.

Critical Thinking

9. Analyze Why is concrete used primarily as foundation material in residential construction?

Academic and Workplace Applications

10. Ratios Use the 1:2:3 mixture to calculate how many of pounds of fine and coarse aggregate are required for 300 lbs. of cement to produce one batch of concrete.

Math Concept A *ratio* is a comparison of two numbers that shows how the proportions are related.

Step 1: Multiply the amount of cement by 2 to find the correct ratio of fine aggregate.

Step 2: Multiply the amount of cement by 3 to find the correct ratio of coarse aggregate.

21st Century Skills

11. Communication Skills The terms *cement* and *concrete* are often used interchangeably. However, these two terms have different meanings. Define each term in one or two sentences. Then, write two to three sentences explaining why the phrases *cement sidewalk* and *cement mixer* are inaccurate.

21st Century Skills

12. Information Literacy: Regional Role-Play You are in charge of deciding what concrete should be used to build a parking garage project. The project is located in Detroit, Michigan, and the client's scheduler has determined that the concrete will be placed in January. Research what temperatures you can expect during the project. List an advantage or disadvantage of the weather.

What steps would you take to prevent problems with the concrete and to ensure the safety of workers during the placing of the concrete? Explain your reasoning to the client in a two-paragraph response.

Standardized TEST Practice

Multiple Choice

Directions Choose the word or phrase that best completes the statement.

13. When concrete is delivered to the job site pre-mixed, it is called a _____.

    a. short load mix

    b. small-batch mix

    c. transit mix

    d. site mix

14. The process that removes air pockets and forces the concrete into all parts of the forms is called _____.

    a. hydration

    b. consolidation

    c. admixture

    d. slump testing

15. _____ is widespread flaking of a hardened concrete surface.

    a. Crazing

    b. Efflorescence

    c. Chalking

    d. Scale

TEST-TAKING TIP

In a multiple-choice test, the answers should be specific and precise. Read the question first, then read all the answer choices. Eliminate answers that you know are incorrect.

*These questions will help you practice for national certification assessment.

Locating the House on the Building Site

Section 9.1
Basic Site Layout

Section 9.2
Establishing Lines & Grades

Chapter Objectives

After completing this chapter, you will be able to:

- **Create** a simple building layout, working from an existing reference line.
- **Identify** the basic types of surveying instruments and list their limitations.
- **Use** a builder's level to lay out a right angle.
- **Describe** how to set up batter boards.
- **Explain** how to measure a difference in elevation between two points using a level or transit.
- **Calculate** the volume of soil excavated for a house foundation.

Discuss the Photo
Surveying Many surveyors use a theodolite. *When is a theodolite preferable?*

Writing Activity: Step-by-Step Directions
Many locating tasks have step-by-step directions. Preview the step-by-step activities in this chapter. Then, create step-by-step directions for a simple task, such as hammering a nail into a board.

Jeff Vanuga, USDA Natural Resources Conservation Service

Chapter 9 Reading Guide

Before You Read Preview

Before a house can be built, its location on the plot of land must be accurately established. Choose a content vocabulary or academic vocabulary word that is new to you. When you find it in the text, write down the definition.

Content Vocabulary

- ○ plot plan
- ○ site layout
- ○ theodolite
- ○ bench mark
- ○ station mark
- ○ batter board
- ○ differential leveling
- ○ bearing capacity
- ○ porosity
- ○ overdig

Academic Vocabulary

You will find these words in your reading and on your tests. Use the academic vocabulary glossary to look up their definitions if necessary.

- ■ methods
- ■ locate
- ■ visible

Graphic Organizer

As you read, use a chart like the one shown to organize content vocabulary words and their definitions, adding rows as needed.

Content Vocabulary	Definition
theodolite	a transit that reads horizontal and vertical angles electronically

Go to **connectED.mcgraw-hill.com** to download this graphic organizer.

Site Layout

Basic Techniques

What is a reference line?

The **plot plan** is the part of the house plans that shows the location of the building on the lot, along with related land elevations (see Section 2.3). The location of the building must then be marked out on the land itself, in a process referred to as **site layout**. This may be done by a surveyor or by a builder familiar with basic surveying methods.

The purpose of site layout is to position the house correctly on the lot. The position of the house on the site must meet local building and zoning codes. Other factors also apply, such as the placement of utilities. The position may have been chosen to take advantage of views, to increase privacy, or to avoid a site feature such as a ledge (area of rock below grade) or a stream.

 Build It Green The position of the house may also have been chosen to maximize solar heat gain for increased energy efficiency, as shown in Solar Electricity.

Solar Electricity
Using the Sun This house uses photovoltaic panels to generate electricity from the sun. For these to work efficiently, the house must be placed properly on the lot.

Two basic methods can accurately determine the location of a proposed building on a property:
- Measuring from an existing reference line.
- Using a surveying instrument such as a level or a transit.

Measuring From a Reference Line

A building may be planned so that it is parallel to an existing line, such as a street or marked property line. Such a line can then be used as a guide, or *reference line*. This makes it possible to stake out the site without using a surveying instrument. When working in this way, it is best to make a drawing of the property first. Such a drawing is shown in Staking a Building. Rectangle ABCD represents the property lines, and boundary AB is the reference line.

Using Surveying Instruments

When a building cannot be laid out by working from reference lines, the builder or surveyor can use one of several kinds of surveying instruments. These instruments work with either optical or laser technology, though some are based on GPS (global positioning system) technology.

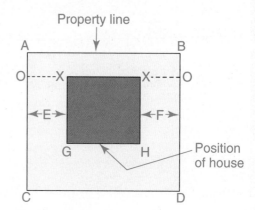

Staking a Building
Simple Layout Staking out a rectangular building without the use of a surveying instrument.

Users of optical instruments rely on line-of-site observations to determine position. The user must look through the instrument's telescope to spot a target or leveling rod held by an assistant. A *leveling rod* is a slender, straight rod marked with graduations in feet and fractions of a foot. The rod is held in a vertical position during use.

Laser instruments do not have a telescope. Instead, a highly focused beam of light is aimed at the target. The user does not need to look through the instrument, so an additional person is not required. One person can set up the laser and then move to another portion of the site and determine level by holding an electronic detector against a leveling rod. When the detector senses the reference plane projected by the laser, it signals the user with a light or sound. The disadvantage of a laser instrument is that its light can be difficult to see in bright daylight.

GPS instruments are the newest development in surveying systems. GPS stands for *global positioning system*. GPS consists of two parts, handheld instruments and 24 satellites evenly distributed around Earth's orbit. Each of these satellites transmits a unique signal, and when a GPS instrument receives several signals, it is able to **locate** its position on Earth's surface with great accuracy. These systems are much more expensive than optical or laser surveying instruments, but they are being used increasingly for construction site layouts.

Types of Instruments

The two basic types of surveying instruments commonly used in residential construction are *levels* and *transits*. Both sit atop a tripod.

Levels The telescope of a level is fixed in a horizontal plane. It can be used only for measuring horizontal angles because it cannot be tilted up or down. It can only be turned from side to side. This is sufficient for most building layouts. There are three types of levels.

A *builder's level* is sometimes called a *dumpy level*. It is the least expensive surveying instrument. It must be set up carefully to ensure accurate results. A good quality model is usually accurate to within ±¼" at a distance of 75'.

An *automatic level* automatically adjusts for variations in setup. It takes less time and effort to set up than a builder's level. It is also more accurate, usually to ±¼" at 100'. Some are accurate to ±¹⁄₁₆" at 200'.

A *laser level* on page 238, does not have a telescope, but is set up on a tripod like other levels. It projects its intense beam of light along a horizontal plane. The light shows up as a small red dot at great distances. Some models continually rotate atop the tripod. They project what appears to be a solid, level line around the job site. Other laser levels can be rotated as needed using a remote control. Some laser levels use

Builder's Level
Least Expensive A builder's level is suitable for small or simple layout tasks.

Automatic Level
Improved Accuracy An automatic level is more accurate than a builder's level.

Laser Level
Remote Control A laser level makes it possible for one person to determine elevation points.

an audible sound to signal a certain level point.

Transits The transit, shown below, is sometimes called a transit level. Unlike the telescope of a level, the telescope of the transit can be moved up and down as well as from side to side. A transit can measure horizontal angles and vertical angles as well. It can also be used to determine if a post or wall is plumb.

Transit
Plumb and Level A transit can be used for site layout but can also determine if parts of the building are plumb.

A transit is classified by the smallest increment that can be read on its *vernier scale*. (A vernier scale slides along a longer scale.) The graduations on a vernier scale are in minutes or seconds. A transit may be referred to as a five-minute transit, a one-minute transit, or a twenty-second transit. Measurements made in seconds are more accurate. A **theodolite**, as shown below, is a transit that reads horizontal and vertical angles electronically. It needs no vernier scale. It displays the measurements on an LCD (liquid-crystal display) screen. It is used when extremely accurate measurements are required. A device called an electronic field book can be attached to a theodolite to store information. A theodolite is easy to set up.

Reading Check

Recall What are the two tools used to determine the location of a building on the site?

Basic Layout with a Level or Transit

Most of the basic layouts described on the following pages can be made with a level.

Theodolite
Tracking Information This surveyor is entering information into a theodolite, or electronic transit. *What information might be entered?*

Step-by-Step Application

Leveling a Transit Head After you have chosen a position for the transit or level, level the instrument's head:

Step 1 Loosen the horizontal clamp screw and turn the telescope until the bubble is in line with one set of opposing leveling screws. Grip the screws, one in each hand, using a thumb and forefinger. Loosen one screw as you tighten the other to center the bubble. Keep the screws snug on the foot plate but do not overtighten them. Continue to adjust the screws until the bubble is centered. See Leveling.

Step 2 Rotate the telescope 90° so that it is over the second set of leveling screws and repeat the process described in Step 1.

Step 3 Return the telescope to the first position. Check the bubble to be sure it is centered, and readjust if necessary. Recheck the second position.

Step 4 Continue to check the bubble in both positions until it is within one graduation on either side of center in the bubble tube.

All of them can be made using a transit. The same concepts apply to the use of other surveying instruments, although layout methods will differ.

Locating a Bench Mark To lay out a building using a transit or a level, you must first have a basic starting point. This starting point, from which measurements can be made, is usually called the **bench mark**, or *point of reference (POR)*. A bench mark may be a mark on the foundation of a nearby building. More often it is a stone or concrete marker placed on the ground at a certain location. The location of the bench mark may appear on the architect's drawings. If so, the plans will usually be oriented to that point.

Setting Up a Transit or Level Set up the transit or level in a position outside the expected flow of activities on a job site. This will prevent it from being accidentally disturbed.

The point over which the level is directly centered is called the **station mark**. The layout is sighted from this point. The station mark may be a bench mark or a corner of the lot, but it should be where the lot can be conveniently sighted. (If the bench mark and the station mark are not the same, be sure you can sight the bench mark easily.)

A plumb bob is suspended from a hook beneath the head of the instrument. Use this plumb bob to center the level or transit directly over the station mark. Adjust the tripod so that it rests firmly on the ground, with the telescope at eye level.

Once a level or transit has been properly set up, be careful not to move or jar the tripod. If this occurs, the instrument must be readjusted. Some or all of the sightings may have to be redone.

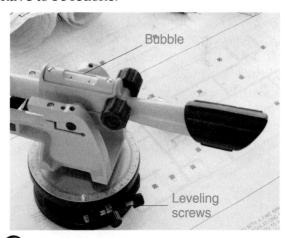

Bubble

Leveling screws

Leveling Controls
Level with Care The level of the head of the instrument can be adjusted by using the leveling screws and checking the bubble.

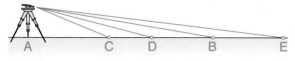

Points on a Line
Using a Transit Establishing points on a line.

Surveying measurements are based on degrees of a circle. A circle is divided into 360 degrees. To increase the accuracy of surveying measurements, degrees are further subdivided into minutes and seconds. A minute is $\frac{1}{60}$ of a degree. A second is $\frac{1}{60}$ of a minute.

Establishing Points Along a Line To establish points along a line, you must use a transit. Use the plumb bob to center the transit accurately over one point on the line. Then level the instrument. Sight the telescope on the most distant **visible** point along the same line. Lock the horizontal clamp screw to hold the telescope in position. Then adjust the instrument to place the vertical crosshair exactly over the distant point. Now, by tipping the telescope up or down, you can determine the exact location of any number of points on the line, as shown in Points on a Line.

Laying Out a Right Angle Using a plumb bob, set up the level or a transit directly over the point where the right angle is to be located. This is Point A in Right Angle. Sight a reference point along a base line and set the 360° scale to zero. The reference point is Point B, and the base line is $\overline{AB}$. Turn the telescope until the scale indicates that an arc of 90° has been completed.

Next, place a leveling rod in a vertical position along this second line at the desired distance, as shown in Leveling Rod. This is shown as Point D in Right Angle. A line

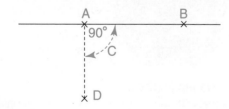

Right Angle
Swing an Arc Laying out a right angle with a transit or a level.

drawn from Point A to Point D will be perpendicular to the base line. Thus a right angle is formed where the lines intersect at Point A.

For accurate work, a spirit level (carpenter's level) may be attached to the leveling rod to check if the rod is being held plumb. The rod can also be kept plumb by aligning it with the vertical crosshair in the telescope. The person at the telescope can tell the rod holder which way to move the top of the rod in order to make it plumb. The rod holder should move the target up or down until the crossline on the target aligns with the crosshair sights in the telescope.

Laying Out a Simple Rectangle To perform this operation you must work from an existing line such as a road, street, or property line. In Rectangle Layout: Method #1, this reference line is shown as $\overline{AB}$. Locate the point (C) that represents the side limit for a front corner of the building. Set up the level or transit at Point C. Lay out a simple rectangle parallel to the existing line as follows:

1. Set the telescope at Point C to sight down line $\overline{AB}$.
2. Measure from Point C to establish Point D. This dimension will be found on the

Leveling Rod
Locating the Target The target on the leveling rod can be moved up or down. The rod is divided into whole feet, tenths, and hundredths of a foot.

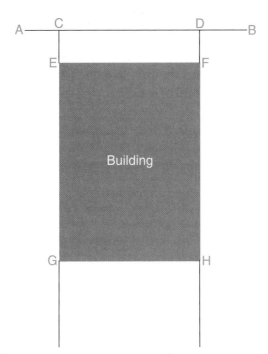

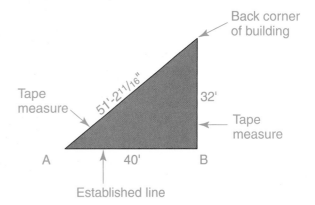

Rectangle Layout: Method #2
Using Two Tapes Measure from each end of the established line AB. The back corner will be where the diagonal measurement (51'-2^{11}/$_{16}$") and the end measurement (32') intersect.

Rectangle Layout: Method #1
Basic Layout Laying out a square or rectangular building with a level or transit.

building plans. It represents the width of the building.

3. Turn the telescope 90° and establish Points E and G by measurement.

4. Move the level or transit to Point D and sight through the telescope to Point C.

5. Turn the telescope 90° and establish Points F and H by measurement.

6. Points E and F represent the front corners of the building. To check your accuracy, measure lines $\overline{EF}$ and $\overline{GH}$. They should be the same length. Check the layout for square by measuring diagonally from Point E to point H, and from Point F to Point G. If these two measurements are identical, the layout is square.

Another method of laying out a rectangular project is to figure the diagonal of the rectangle, as shown in Rectangle Layout: Method #2. After one side of the project is established ($\overline{AB}$), usually the front setback line, the rectangle can be laid out accurately by using two long tape measures. The side and end measurements are input as *rise* and *run* into

a construction calculator, and the diagonal of the rectangle is figured automatically. Using one tape measure, measure the desired distance from one end of line AB. Then, with a second tape measure, measure the diagonal. The lines will intersect precisely at one of the corners. Do this from both ends of the established line and a perfect rectangle is established. The example in Rectangle Layout: Method #2 shows a house that is 40' long and 32' deep. It has a diagonal of 51'-2^{11}/$_{16}$".

Laying Out an Irregularly Shaped Building When the building is not regular in shape, it is usually best to start by laying out a large rectangle that will take in all or most of the building. This is shown in An Irregular Shape as QHOP. After the large rectangle is

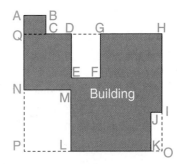

An Irregular Shape
Multiple Shapes Laying out an irregularly shaped building made of a series of squares and rectangles.

established, the remaining portion of the layout will consist of small rectangles. Each small rectangle can be laid out and proved separately. These rectangles are shown as NMLP, ABCQ, DGFE, and JIOK.

Even if the outline of a building is not a rectangle, the process for establishing each point is basically the same. More points must be located, and the final check is more likely to reveal a small error.

Step-by-Step Application

Staking Out the Site To stake out the site in the drawing, proceed as follows:

Step 1 Check the plot plan to find the setback distance. The *setback distance* is the minimum space allowed by local codes between a house and the property lines. Codes also specify setback distances between the house and utility lines, streams, and ponds. Along boundaries $\overline{AC}$ and $\overline{BD}$, measure this distance back from front line $\overline{AB}$. In the drawing, the setback is shown by segments $\overline{AO}$ and $\overline{BO}$.

Step 2 Stretch a line tightly between the points marked O. The front corners of the building will be located on line $\overline{OO}$. There are two ways to locate them. You can obtain the measurement from the plot plan to see how far the corners should be from the side boundaries. Then, along line $\overline{OO}$, measure the indicated distances in from $\overline{AC}$ and $\overline{BD}$. Xs represent the front corners of the building in Staking a Building on page 236. If the building is to be centered between the side boundaries, you need not refer to the plot plan. Instead, subtract the length of the building from the length of $\overline{OO}$. Then measure half this distance in from each end of $\overline{OO}$. Measure the distance between the two points marked X and check this distance with the plans. The distance $\overline{XX}$ represents the length of the building. It must be accurately measured and match the length on the plans.

Step 3 Check the plans to determine the width of the building (how far back it will extend from the front corners). Mark off the width by extending lines back from the two points marked X. If the boundary lines of the lot form a 90° angle at the corners, these lines should be parallel to $\overline{AC}$ and $\overline{BD}$. Note that E is the same as $\overline{OX}$, and F is the same as $\overline{XO}$. Thus, E and F show the distance between the sides of the building and the side boundary lines of

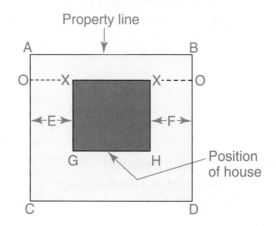

Property line

A
B
O----X
X----O
←E→
←F→
G
H
Position of house
C
D

the lot. Points G and H represent the rear corners of the building.

Step 4 The boundary lines of the lot may not be at right angles to each other. If this is the case, you will need to first establish the corner of the building that will be closest to a boundary line. Next, make certain that the minimum front and side yard requirements are established from this point. Then lay out the building by using the methods described on page 240, "Laying Out a Right Angle" and "Laying Out a Simple Rectangle." Establish a line to indicate the rear of the building. This is shown by $\overline{GH}$ in Staking a Building on page 236.

Step 5 If the building is complicated, divide it into smaller rectangles. Establish more lines such as $\overline{OO}$ to indicate the front of each rectangle. You can get the necessary information from the plans. Then carry out the same steps to establish the rest of each rectangle. The result will be a group of adjoining rectangles that will show the total outline of the building.

After You Read: Self-Check

1. What is a plot plan?
2. Name one advantage of surveying instruments based on laser technology.
3. What is the difference between a level and a transit?
4. What is the purpose of the hook under a level or transit?

Academic Integration: Mathematics

5. **Area of Irregular Shapes** Structures are not always regular shapes such as quadrilaterals. For example, some structures are L-shaped or U-shaped. The floor plan for the structure shown below is drawn to scale. Some of the dimensions on the floor plan are missing. Find all of the missing dimensions, then calculate the area of the structure.

Math Concept Area measures the size of a surface. The area (A) of a regular quadrilateral is calculated using the formula $A = length \times width$. You can calculate the area of an irregular shape below by dividing the shape into multiple rectangles.

Step 1: Divide the shape into a number of regular quadrilaterals.

Step 2: Calculate the sum of the areas of each shape that makes up the structure.

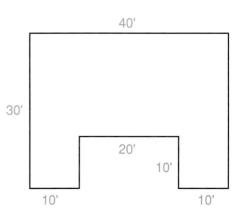

Go to **connectED.mcgraw-hill.com** to check your answers.

Section 9.2 Establishing Lines & Grades

Lines & Grades

What is an excavation?

After its location and alignment have been determined, a rectangle showing the outer dimensions of a structure is staked out. If the building is to form a simple rectangle, the staked-out area will follow the foundation line exactly.

Grade

The grade refers to the level of the ground where it will meet the foundation of the completed building. The grade is found on the building plans. Grade must be established accurately because it is used for making important measurements. From the grade, you can find the depth of

Setting Up Batter Boards After the corners of the building have been located, set up the batter boards as follows:

Step 1 Drive three 2×4 stakes at each corner of the building, at least 4' beyond the foundation lines.

Step 2 Nail 1×6 or 1×8 boards to the stakes horizontally. The top of each board must be level. All boards must be level with each other. Use a level or transit to locate the height of the boards.

Step 3 Stretch string across the tops of boards at opposite corners and align it exactly over the nails in the corner stakes. A plumb bob is handy for setting the lines.

Step 4 Make a shallow saw kerf where the string overlaps the boards. This will keep the string from sliding. It will also make it easy to reposition the string if it must be temporarily removed to let equipment or machinery pass. Make similar cuts in all eight batter boards.

Step 5 Tie the string to the boards, using the saw kerfs to position it (see page 245). This establishes the lines of the house.

Step 6 Check the diagonals again to make sure the corners are square. (The area for an L-shaped building can be divided into rectangles and the diagonal measurement of each rectangle can be checked.)

the excavation. An excavation is a cut, cavity, trench, or depression made by removing earth.

Sometimes the bench mark is used as a reference point for establishing the grade. At other times, the grade may be located in relation to the level of an existing street or curb. The grade is indicated on a stake driven into the ground outside the excavation area.

You can also use the grade to establish other points, such as the height of a foundation or the elevation of the first floor.

Batter Boards

After the grade and the corners of the house have been established, building lines must be laid out as aids in keeping the work level and true. A **batter board** is a board

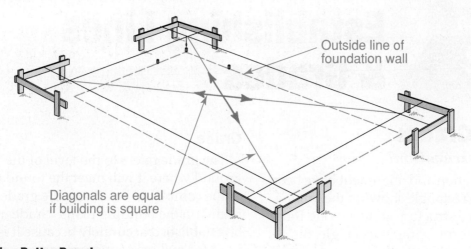

Outside line of foundation wall

Diagonals are equal if building is square

Using Batter Boards
Building Outline Using batter boards to establish the outline of the foundation wall. The top edge of a batter board represents the height of the foundation wall.

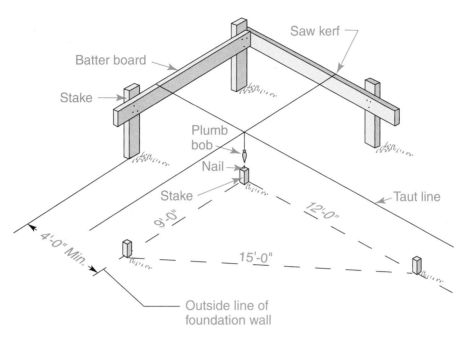

Saw kerf

Batter board

Stake

Plumb bob

Nail

Stake

Taut line

12'-0"

9'-0"

4'-0" Min.

15'-0"

Outside line of foundation wall

Batter Board Detail
Squaring a Corner You can check that the corners of a layout are square by using the 3-4-5 system.

fastened horizontally to stakes placed to the outside of where the corners of the building will be located. These boards and the string tied between them locate and mark the outline of the building as shown in Using Batter Boards. The height of the boards may be the height of the foundation wall. Before setting up batter boards, locate the corners of the building precisely by one of the methods already discussed. Drive nails into the tops of the stakes that indicate the outside edge of the foundation walls.

Setting up batter boards is explained in the Step-by-Step Application.

To be certain that the corners are square, use the following procedure:
1. Measure the diagonals of the completed layout to see if they are the same length, as shown in Using Batter Boards.
2. The corners can also be squared by using the *3-4-5 method* shown in Batter Board Detail. This is done by measuring a distance along one side in 3' increments, such as 6, 9, or 12.

3. Measure along the adjoining side in the same number of 4' increments (8, 12, or 16). The diagonal drawn between the end points will then measure in an equal number of 5' units (10, 15, or 20) when the unit is square. Thus, a 9' distance on one side and a 12' distance on the other should result in a 15' diagonal measurement if the corner is a true 90°.

Land Elevations

Check the architect's plan to see where elevations are to be determined. Set up a level or transit. Be certain that the locations can be seen through the telescope. Place a leveling rod upright on any location to be checked. Then sight through the telescope at the leveling rod. Take a reading of where the horizontal crosshair in the telescope crosses the rod. Then move the rod to the second location to be established. Raise or lower the rod until the reading is the same as for the first point. The bottom of the rod is then at the same elevation as the original point.

Elevation Difference #1
Points Visible Obtaining the difference in elevation between two points that are visible from an intermediate point.

Measuring a Difference in Elevation To learn the difference in elevation between two points, such as A and B above, set up a transit or level at an intermediate point. With the measuring rod held at point A, note the mark where the horizontal crosshair in the telescope crosses the rod. Then with the rod held at point B, sight the rod and note the point where the horizontal crosshair crosses the rod. In our example, the difference between the reading at A (5' 0") and the reading at B (7' 0") is the difference in elevation between A and B. Thus the ground at point B is 2' lower than the ground at point A. This technique can be used to position foundation footings and other points below grade.

Sometimes it is not possible to sight two points from a single point between them. A high mound can cause this difficulty. To solve this problem, one or more additional intermediate points, such as C and D shown below, must be used for setting up the

instrument. The process of determining differences in elevation between points that are remote from each other is called **differential leveling**.

Reading Check

Recall What points can be established using the grade?

Foundation Wall Height
The foundation wall footings must rest on firm soil below the frost line. Footings are continuous concrete pads that support a foundation wall. The depth of the frost line varies according to the local climate and is given in local building codes. This determines how deep the excavation must be. To determine this depth, it is common practice to use the highest point on the perimeter of the excavation as the reference point, as shown in

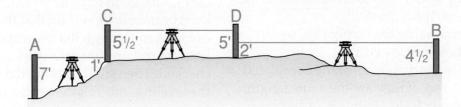

Elevation Difference #2
Points Not Visible Obtaining the difference in elevation between two points not visible from a single intermediate point.

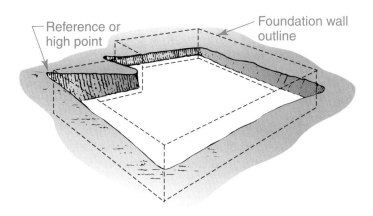

Determining Excavation Depth
Highest Point To establish the depth of the excavation, use the highest point around the excavation as the reference point.

Determining Excavation Depth. This is true for both graded and ungraded sites.

Good drainage is ensured if sufficient foundation height is allowed for the sloping of the finish grade, as shown in Proper Drainage. *Finish grade* is the level of the ground when grading is completed. Foundation walls at least 7'-4" high are necessary for full basements, and 8' walls are common.

The top of the foundation wall must be at or above a minimum height set by code. This distance is often at least 8" above the finished grade, but this may vary with local code. This allows the siding and framing members to be protected from soil moisture and places them well above the grass line. In termite-infested areas, this makes it possible to observe signs of any termite activity between the soil and the wood. Protective measures can then be taken before the wood is damaged.

Crawl spaces should have enough height to permit inspections for termites and to allow plastic soil covers to be installed. Soil covers reduce the effect of ground moisture on framing members. Ordinarily, there should be at least 18" between the undersides of the joists and the highest point of ground enclosed by the foundation walls.

If the ground beneath the structure is excavated or is otherwise lower than the outside finished grade, measures must be taken to assure good drainage. The finished grade should always slope away from the house, as shown in Figure 9-20. Below-surface drainage systems might also be required.

The Excavation

Any variation from standard construction practices increases the cost of the foundation and footings. This might influence the design of the house. Before excavating for a new home, subsoil conditions should be determined by test borings or by obtaining soil samples in some other way. This has several purposes. It helps the excavators to determine whether any special precautions will be necessary. Unusually soft or hard soil conditions may call for special excavation techniques or equipment. It is also important to know the bearing capacity of the soil. **Bearing capacity** is a measure of how well the soil can support the weight of a house. Another purpose of soil testing

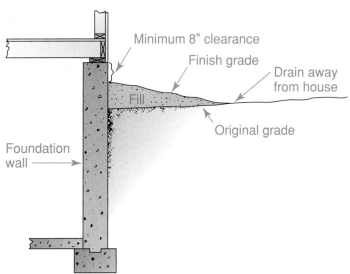

Proper Drainage
Finished Grade Added fill brings the finish grade above the original grade and ensures drainage away from the house.

is to determine its porosity. **Porosity** is the measurement of the ability of water to flow through the soil. If the soil does not drain well, additional material may have to be excavated and replaced with better soil.

Test borings will also identify problems that could significantly affect the cost of the house. A rock ledge may require costly removal. A high water table may require design changes from a full basement to a crawl space or concrete slab construction. If there has been a previous excavation on the site, the soil may have been disturbed and may not have sufficient bearing strength for footings.

Several types of earth-moving equipment, such as a power shovel or backhoe, can be used for basement excavations. Topsoil is often stockpiled by a bulldozer, a front-end loader, or a grader for future use, as in Grading the Site. Power trenchers are often used in excavating slab footings or shallow foundation walls.

It is best to excavate only to the top of the footings or the bottom of the basement floor. The soil must be stable enough to prevent cave-ins, and some soil becomes soft upon

exposure to air or water. Thus, unless form boards are to be used, it is advisable not to make the final excavation for footings until it is nearly time to pour the concrete.

The excavation must be wide enough to provide space to work, as shown in Excavation Detail on page 249. For example, there must be enough room to install and remove concrete forms, to lay up block, to waterproof the exterior surfaces of the walls, and to install foundation drainage. **Overdig** is the term used to describe the additional excavation needed to provide clearance for work. The sides of the excavation may have

Grading the Site
Stockpiling Topsoil A grader being used to strip the topsoil from a building site in preparation for excavating the basement.

ewg3D/Getty Images

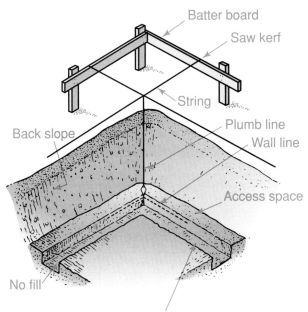

Batter board
Saw kerf
String
Back slope
Plumb line
Wall line
Access space
No fill
Footing trench. Make sharp cut or use form boards.

 Excavation Detail

Back Slope The basement excavation is back-sloped to prevent cave-ins. Note the use of the plumb line for accurately locating the foundation wall line.

JOB SAFETY

ESCAPE ROUTES Any excavation or trench that is more than 4' deep must be provided with a stairway, ladder, ramp, or other safe means of escape. This includes foundation excavations. The escape devices must be within 25' of a worker.

to be sloped to prevent them from caving in. This is sometimes called *back slope*. The steepness of the back slope of the excavation is determined by the type of subsoil.

Once soil is excavated, the sides of the excavation are exposed to the weather. They will dry out and may become unstable. Allow sufficient overdig and back slope to reduce the hazards. Follow OSHA regulations for slopes and check local building codes for information regarding soil types.

When excavating for basements, some contractors *rough-stake* only the outline of the building to indicate where earth must be removed. When the proper elevation has been achieved for the basement floor, the footings are laid out and the final excavation dug. After the footings have been placed and the concrete has hardened somewhat, the building wall outline is

established on the footings and marked to indicate the position of the formwork or concrete block wall.

Excavation costs are based on the total cubic yards of earth to be removed. To determine the volume of material, multiply the length of the excavation times the width times the depth. Calculating is usually done in decimals. See **Table 9-1** for conversion figures and **Table 9-2** on page 250 for excavating features for standard depths.

Table 9-1: Converting Inches to Decimal Fractions of a Foot	
Inches	Feet
1	0.083
2	0.167
3	0.250
4	0.333
5	0.417
6	0.500
7	0.583
8	0.667
9	0.750
10	0.833
11	0.916
12	1

Estimating and Planning

CERTIFICATION PREP

This estimating and planning exercise will prepare you for national competitive events with organizations such as SkillsUSA and the Home Builder's Institute.

Excavation Volume

Excavation Costs

Multiply the length of the excavation times the width times the depth to determine the volume of the material to be removed.

1. For the house shown in the floor plan below, multiply 7' (depth of excavation) times 30' (26' width of house plus 2' clearance at each end, between the excavation and the outside of the foundation wall) times 44' (40' length of house plus 2' clearance at each end). The answer is 9,240 cubic feet.

2. There are 27 cubic feet in 1 cubic yard. To convert cubic feet to cubic yards, divide by 27:

$$9,240 \div 27 = 342.2$$

Rounded off, approximately 342 cubic yards of material will have to be excavated.

There is another method to determine the cubic yards of material to be removed. It can be used if the excavation depth is a standard one as shown on Table 9-2.

1. Refer again to the floor plan. The excavation needed for this house is 30' wide, 44' long, and 7' deep. Multiply the width by the length to find the area of the excavation.

2. Refer to Table 9-2. For an excavation 7' deep, 0.259 cubic yards of material are removed for each square foot of area. Multiply the area of the excavation by 0.259 to find how many cubic yards of material will be removed. Round up.

Table 9-2: Excavating Factors for Standard Depths			
Depth per Square Foot	Cubic Yards Removed	Depth per Square Foot	Cubic Yards Removed
2"	0.006	4'-6"	0.167
4"	0.012	5'-0"	0.185
6"	0.018	5'-6"	0.204
8"	0.025	6'-0"	0.222
10"	0.031	6'-6"	0.241
1'-0"	0.037	7'-0"	0.259
1'-6"	0.056	7'-6"	0.278
2'-0"	0.074	8'-0"	0.298
2'-6"	0.093	8'-6"	0.314
3'-0"	0.111	9'-0"	0.332
3'-6"	0.130	9'-6"	0.350
4'	0.148	10'-0"	0.369

Note: To find the factor, first locate the excavation depth. The factor is in the column to the right.

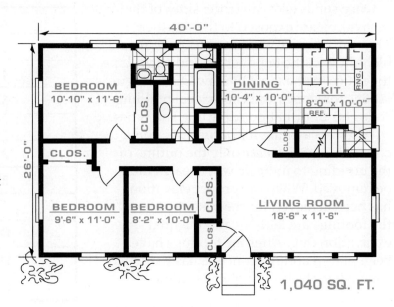

Trenches The amount of material that will be removed from trenches, such as those trenches that might be dug for utilities, can be calculated by using **Table 9-3**. For example, if a trench is to be 42" deep and 18" wide, the table shows that 19.4 cubic yards of material will be removed for every 100 lineal feet. Such a trench might be dug for a house with a 30' setback. To determine how much material would be removed, multiply 19.4 by 0.30 (because 30' is about 0.30 of 100'). The answer is that 5.82 cubic yards of material would be removed.

Mathematics: Measurement

Volume How many cubic yards of material will be removed for a trench that is 24" deep, 24" wide, and 100 lineal feet long?

Starting Hint Consult Table 9-3.

Table 9-3: Cubic Yard Content of Trenches per 100 Lineal Feet

Depth in Inches	Trench Width in Inches						
	12	18	24	30	36	42	48
6	1.9	2.8	3.7	4.6	5.6	6.6	7.4
12	3.7	5.6	7.4	9.3	11.1	13.0	14.8
18	5.6	8.3	11.1	13.9	16.7	19.4	22.3
24	7.4	11.1	14.8	18.5	22.2	26.0	29.6
30	9.3	13.8	18.5	23.2	27.8	32.4	37.0
36	11.1	16.6	22.2	27.8	33.3	38.9	44.5
42	13.0	19.4	25.9	32.4	38.9	45.4	52.0
48	14.8	22.2	29.6	37.0	44.5	52.0	59.2
54	16.7	25.0	33.3	41.6	50.0	58.4	66.7
60	18.6	27.8	37.0	46.3	55.5	64.9	74.1

Section 9.2 Assessment

After You Read: Self-Check

1. What are three possible reference points for establishing a grade?
2. What is a batter board and what is its purpose?
3. Briefly describe the process of identifying the difference in elevation between two points.
4. What is overdig?

Academic Integration: Science

5. **The Water Table** During excavation, soil samples must be obtained to identify any potential problems that could significantly affect the cost of construction. For example, a high water table may require design changes from a full basement to a crawl space or concrete slab construction. Find out more about the science of the water table. Write a few sentences about your findings. Be sure to define *water table*, then explain the difference between the saturated zone and the unsaturated zone.

Go to **connectED.mcgraw-hill.com** to check your answers.

Chapter Summary

Section 9.1

A house can be positioned or located on a piece of property either by measuring from an established reference line or by using an instrument such as a level or a transit. Most layouts begin at a bench mark. Surveying instruments include levels and transits. The point at which they are set up is called the station mark.

Section 9.2

String stretched between batter boards is used to establish the outline of a house. Once this has been done, the excavation can proceed. The excavation must be wide enough to provide space to work. Soil taken from an excavation is measured in cubic yards.

Review Content Vocabulary and Academic Vocabulary

1. Use each of these content vocabulary and academic vocabulary words in a sentence or diagram.

Content Vocabulary

- plot plan (p. 236)
- site layout (p. 236)
- theodolite (p. 228)
- bench mark (p. 239)
- station mark (p. 239)

- batter board (p. 244)
- differential leveling (p. 246)
- bearing capacity (p. 247)
- porosity (p. 248)
- overdig (p. 248)

Academic Vocabulary

- methods (p. 236)
- locate (p. 237)
- visible (p. 240)

Speak Like a Pro

Technical Terms

2. Work with a classmate to define the following terms used in the chapter: *reference line* (p. 236), *leveling rod* (p. 237), *builder's level* (p. 237), *dumpy level* (p. 237), *automatic level* (p. 237), *laser level* (p. 237), *vernier scale* (p. 238), *3-4-5 method* (p. 245), *finish grade* (p. 247), *back slope* (p. 249), *rough-stake* (p. 249).

Review Key Concepts

3. Describe how to work from an existing reference line to establish a simple building layout.

4. Describe the different types of surveying instruments.

5. Demonstrate how to use a builder's level to lay out a right angle.

6. Demonstrate how to set up batter boards.

7. Explain how to use a transit or level to measure a difference in elevation between two points.

8. Explain how to determine the depth of an excavation for a house foundation.

Critical Thinking

9. Infer How could you best keep the cost of foundations and footings for a building at a minimum?

Academic and Workplace Applications

10. Finding Volume A rectangular excavation 5' deep is needed to build a building that is to be 26' by 32'. How many cubic yards of material should be removed if a 2' clearance is needed outside the foundation walls?

Math Concept The formula for finding the volume of a rectangular prism is $V = lwh$. To determine the actual dimensions of the excavation, draw a picture and label the length and width to account for the building and the clearance all around.

Step 1: Draw a picture of the dimensions of the building. Then add the clearance.

Step 2: Multiply the length by the width. Then multiply the result by the depth of the excavation.

Step 3: Use cubic units to express the volume of the material. Convert cubic feet to cubic yards.

STEM Engineering

11. Measurement Accuracy A surveyor makes a measurement of 60½' using a laser measure. It is accurate to within ±¼". What is the range within which the actual length could fall?

Starting Hint All measurements fall within a range of accuracy depending on the size of the smallest mark on the measuring device used and the quality of the device. Determine the lower end of the range by subtracting the accuracy factor.

21st Century Skills

12. Career Skills: Problem Solving You are a member of a surveying crew that has been hired to survey a building site this Saturday. On Friday evening, the other two members of your crew call to inform you that they will not be able to work on Saturday. The job needs to be completed before Monday. You were planning on working with your crew members and have all of the optical instruments ready to use. Now you will have to determine how you will complete the surveying job alone. Describe what changes you can make in order to complete the job yourself in a one-page summary. Identify the equipment you will need and explain why this equipment will be suitable for the situation.

Standardized TEST Practice

Multiple Choice

Directions Choose the phrase that best completes the following statements.

13. The top edge of a batter board represents the _____.
   a. depth of the foundation wall
   b. outside edge of the foundation wall
   c. height of the foundation wall
   d. width of the foundation wall

14. A measure of how well the soil can support the weight of a house is called _____.
   a. a bench mark
   b. the bearing capacity
   c. an overdig
   d. the station mark

15. The purpose of a site layout is to _____.
   a. position a house correctly on the lot
   b. determine the depth of an excavation
   c. locate a bench mark
   d. maximize solar heat gain

TEST-TAKING TIP

Skipping a question when you do not know the answer can waste valuable time. A better strategy is to mark the answer you believe to be correct and come back to the question after you finish the test.

*These questions will help you practice for national certification assessment.

Foundation Walls

Section 10.1
Footings

Section 10.2
Concrete Foundation Walls

Section 10.3
Concrete Block Walls

Chapter Objectives

After completing this chapter, you will be able to:

- **Explain** the purpose of footings.
- **Determine** the exact location of the footing.
- **Explain** how concrete foundation walls are formed.
- **Identify** the two types of foundation walls.
- **Describe** the process of laying a concrete block wall.
- **Identify** the types of concrete block used in concrete block walls.

Discuss the Photo

Foundation Walls Foundation walls can be made of concrete, concrete masonry units, or pressure-treated lumber and plywood. *Why do you think most foundation walls are made of concrete?*

Writing Activity: Make Predictions

A footing is a base that distributes weight over a wide area of soil. All foundation walls are supported by footings. A foundation wall carries the weight of a house down to the footings. Write a paragraph describing what you think will happen if a house is built on poorly constructed footings.

James Hardy/Photo Alto

Chapter 10 Reading Guide

Before You Read Preview

A foundation anchors a house to the earth and provides a solid, level base for framing. Choose a content vocabulary or academic vocabulary word that is new to you. When you find it in the text, write down the definition.

Content Vocabulary

- footing
- wales
- cold joint
- head joint
- bed joint
- story pole
- control joint
- parging
- radon

Academic Vocabulary

You will find these words in your reading and on your tests. Use the academic vocabulary glossary to look up their definitions if necessary.

- framework
- incorporate

Graphic Organizer

As you read, use a chart like the one shown to organize information about content vocabulary words and their definitions, adding rows as needed.

Content Vocabulary	Definition
footing	a base that provides a surface that distributes weight over a wide area of soil

Go to connectED.mcgraw-hill.com to download this graphic organizer.

Footings

Wall Footings

Why do you think the base of a foundation wall is called a footing?

All foundation walls are supported on a widened base called a footing. A **footing** is a base that provides a surface that distributes weight over a wide area of soil. Footings are generally made of concrete. In some cases, the footing is placed first and the foundation wall is built on top of it later, as shown in Footing Forms. In other cases, the footing and the wall are built as a single unit, sometimes called a *monolithic wall*.

The size and shape of a footing are specified on the building plans. The depth and width are based on factors such as the weight the footing must bear, the bearing capacity of the soil, and local building codes. If a building official thinks the bearing capacity of the soil may be less than 1,500 lbs./sq. ft., he or she may require a soil test to determine the soil's actual bearing capacity.

Footings must always rest on undisturbed soil (soil that has not been dug up previously). This lowers the chance of the foundation settling unevenly. It is especially important where the building site has been raised by adding compacted fill. If the site for a footing has been dug too deep, it should never be filled with soil. It should be filled with concrete to make the foundation more stable.

Reading Check

Describe *What factors determine the depth and width of a footing?*

Footing Forms
Shaping Concrete Wood or steel forms prevent the concrete from spreading out as it is placed.

Table 10-1: Minimum Width of Concrete or Masonry Footings in Inches				
Load-Bearing Value of Soil (psf)	1,500	2,000	3,000	≥4,000
Conventional light-frame construction				
1-story	12	12	12	12
2-story	15	12	12	12
4-inch brick veneer over light frame or 8-inch concrete masonry				
1-story	12	12	12	12
2-story	21	16	12	12
8-inch solid or fully grouted masonry				
1-story	16	12	12	12
2-story	29	21	14	12

Footing Design

Local building codes specify the strength of the footing concrete and the minimum footing depth. Footings should be placed at least 12" below grade. In cold climates the footings should be far enough below finished grade to protect them from frost.

The width of a footing depends on the bearing capacity of the soil, specified in building code charts such as **Table 10-1**. On standard soils, footing size is generally based on the thickness of the foundation wall. The width of the footing should be twice the thickness of the foundation wall. The footing would project out one-half the thickness of the foundation wall on each side, as in

Typical Footing Dimensions. If the load-bearing capacity of the soil is low, wider and thicker footings may be needed.

Footing Reinforcement The strength of a footing is greatly improved when reinforcing bar, or rebar, is embedded in it. Reinforcement often consists of two lengths of ½" diameter (#4) rebar. The rebar must be at least 3" above the bottom of the footing.

Footing Forms

The exact location of footings is determined by plumb bobs hung from the foundation batter boards. Once the location is known, the shape and size of the footing can be established. The shape of the footing

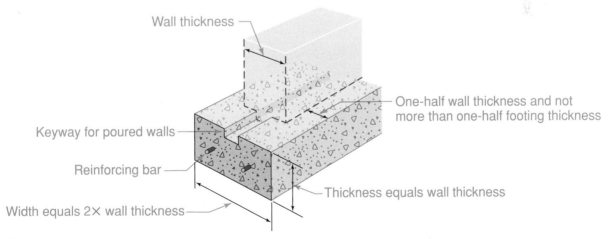

Wall thickness

One-half wall thickness and not more than one-half footing thickness

Keyway for poured walls

Reinforcing bar

Thickness equals wall thickness

Width equals 2× wall thickness

Typical Footing Dimensions

Standard Footing The basic proportions noted here are guidelines only. Always refer to the building plans.

Ground Conditions Local building codes usually specify the depth frost penetrates into the soil. In the northern United States, this may be 48" or more, which means the footings have to be placed deeper. In areas prone to earthquakes, such as California, local codes have additional requirements for footings, such as steel reinforcement.

is created by pouring it into a form. A form is any **framework** designed to contain wet concrete. Forms can be made of steel, lumber, or a combination of lumber and plywood. A common type of wall footing form is shown in Footing Forms. The sides are formed by 2×4 lumber and braced to prevent them from being spread apart by the wet concrete. These boards are sometimes called *haunch boards. Spreaders,* or *form brackets,* are the boards that hold apart the sides of the forms. Lumber formwork is often assembled with duplex head nails to make disassembly easy later on.

A keyway should be formed in the top of the footing, as shown in Typical Footing Dimensions on page 258. The keyway locks the foundation walls to the footing. This prevents moisture from seeping between the wall and the footing. A keyway is usually 3½" wide and 1½" deep because a 2×4 is often used to form it. After the concrete has been placed, lengths of 2×4 can be pressed into the footings directly below where the foundation wall will be. They are removed after the footings have cured.

Some builders form the key by sliding a short length of 2×4 along the top of the footing before the concrete starts to stiffen. After the rebar has been positioned and the footings have been poured, the top of the footing should be troweled smooth.

Other Types of Footings

Other load-bearing parts of a structure, such as columns and chimneys, must also be supported by footings. Their exact size and location are specified on the building plans.

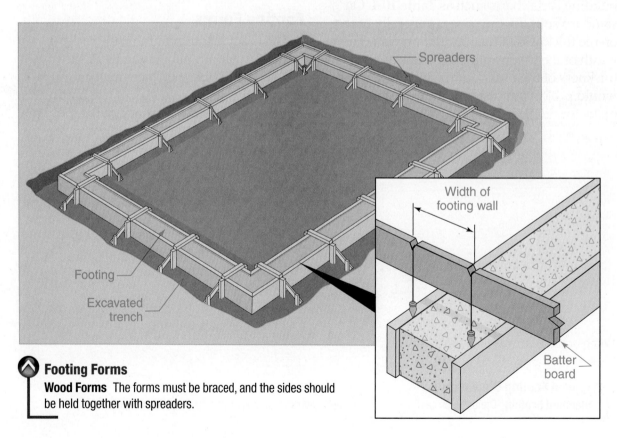

Footing Forms
Wood Forms The forms must be braced, and the sides should be held together with spreaders.

Pier Footings A footing for a pier can be round or square. A *pier* is a block or column of concrete separate from the main foundation. It is often used to support girder floor systems or exterior decks. A steel pin or a metal bracket is sometimes anchored in a pedestal above the footing, as shown in Pier Footing. The pin or bracket will secure a wood post. A pedestal should be about 3" above the finished basement floor and at least 12" above finished grade in crawl-space foundations.

When steel posts are used, they are sometimes set directly on the footing and concrete floor is then poured around them. If a concrete column is poured on top of a footing, rebar placed vertically in the footing will keep the column in position.

Stepped Footings *Stepped footings* are often used on a lot that slopes. Instead of being set at the same height around the entire foundation, the footings "step" down the sloped

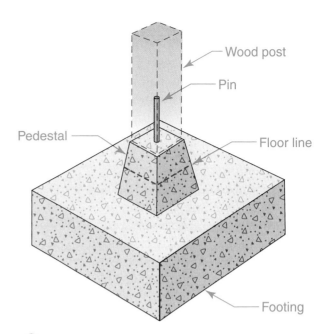

Labels: Wood post, Pin, Pedestal, Floor line, Footing

Pier Footing
Post Support A steel pin can be used to anchor a post, but many builders use a metal bracket instead.

Estimating and Planning

This estimating and planning exercise will prepare you for national competitive events with organizations such as SkillsUSA and the Home Builder's Institute.

Reinforcing Bar for Footings

Working with Rebar

Footings do not always have to be reinforced. Where steel reinforcing bar (rebar) is required, the building plans should identify the size, number, and placement of the rebar.

Step 1 Identify the combined length of all the footings. Suppose a foundation measures 42' by 24'. The total length of the footings is 132' ((42' + 24') × 2).

Step 2 Determine the exact number of lineal feet (l.f.) of rebar. Multiply the number of bars by the length of the footings. If a house has 132 l.f. of footings and two bars are required for each, it will need 264 l.f. of rebar (132 × 2).

Step 3 Add an amount for overlaps. Each length of rebar must overlap the connecting length by an amount determined by local codes. In most residential projects it is enough

to add 10% to the total lineal footage. Using the example above:

$$264 \times .10 = 26.4$$
$$264 + 26.4 = 290.4$$

Step 4 Rebar typically comes in lengths of 20 ft. To figure the number of lengths needed, divide the total lineal footage by 20.

$$290.4 \div 20 = 14.52, \text{ rounded up to } 15$$

15 lengths of 20' rebar will be needed for the footings.

Estimating on the Job

Determine how many lengths of 20' long rebar will be needed for a house with a foundation that measures 58' by 38'. Local codes call for 3 reinforcing bars in the footing.

site (see A Stepped Footing). The vertical step should be poured at the same time as the rest of the footing. If the foundation wall is built with concrete block, the height of the step should be in multiples of 8". This is the height of a block with a standard ⅜" mortar joint.

The bottom of the footing is always placed on undisturbed soil below the frost line. Each run of the footing should be level. A *run* is a horizontal section between two vertical sections. With a concrete block foundation, the runs should be calculated so that the horizontal spacing of the block can be maintained across the step.

The vertical step should be at least 6" thick and be the same width as the rest of the footing. On steep slopes, more than one step may be required. It is good practice, when possible, to limit the vertical step to 2' in height. This results in a stronger wall and makes finish grading much easier.

Table 10-2 provides estimated material and labor for footings.

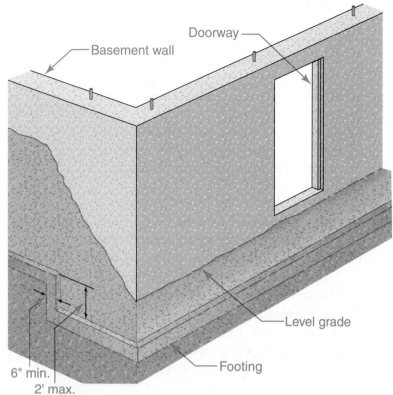

A Stepped Footing

Footing Details The details of a stepped footing, such as its height and thickness, will be shown on the plans.

Footing Size	Material			Labor	
	Cubic Feet of Concrete Per Lineal Foot	Cubic Feet of Concrete Per 100 Lineal Feet	Cubic Yards of Concrete Per 100 Lineal Feet	Excavation Hours per 100 Lineal Feet[a]	Placement Hours per Cubic Yard[b]
6×12	0.50	50.00	1.9	3.8	2.3
8×12	0.67	66.67	2.5	5.0	2.3
8×16	0.89	88.89	3.3	6.4	2.3
8×18	1.00	100.00	3.7	7.2	2.3
10×12	0.83	83.33	3.1	6.1	2.0
10×16	1.11	111.11	4.1	8.1	2.0
10×18	1.25	125.00	4.6	9.1	2.0
12×12	1.00	100.00	3.7	7.2	2.0
12×16	1.33	133.33	4.9	9.8	2.0
12×20	1.67	166.67	6.1	12.1	1.8
12×24	2.00	200	7.4	15.8	1.8

Table 10-2: Estimating Material and Labor for Footings

[a]Reduce hours by ¼ for sand or loam. Increase hours by ¼ for heavy clay soil.
[b]Placement labor based on ready-mixed concrete.

Estimating and Planning

This estimating and planning exercise will prepare you for national competitive events with organizations such as SkillsUSA and the Home Builder's Institute.

Concrete and Labor for Footings

Concrete Details

To determine the amounts of concrete required for footings, calculate the total length of the footings and the volume. Another way to do this is to refer to a volume table such as Table 10-2.

1. Note that the foundation in the house plan below measures 42' × 24'. The perimeter (the total length of the four sides) is 132 lineal feet.

2. Suppose that the footing size is 8" × 16". Table 10-2 shows that for 8" × 16" footings, 3.3 cu. yds. of concrete are needed for 100 lineal feet. The footings in the example are longer than 100', so additional calculations are needed. If 3.3 cu. yds. of concrete will fill 100', how many cubic yards will be needed for 132'? Divide 132 by 100. The answer is 1.32. This indicates that the house footings are 1.32 times longer than 100'. Therefore, multiply 1.32 × 3.3 to find the total amount of concrete needed. The answer is 4.36 cu. yds. (Note: A cubic yard is a measure of volume that represents a cube measuring 3' by 3' by 3', or 27 cubic ft.

Labor for Excavation

The hours of labor required for excavation of the footings can also be determined using Table 10-2.

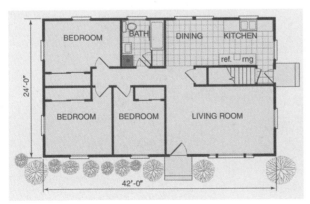

1. For 8" × 16" footings, Table 10-2 shows that it will take 6.4 hours to excavate 100 lineal feet. Because the perimeter is 132 lineal feet, allowance must be made for the excess over 100. You again divide 132 by 100 and get 1.32.

2. Multiply 1.32 by 6.4 for an answer of 8.448 hours of labor, rounded off to 8.5. In other words, if it takes 6.4 hours to excavate 100 lineal feet, it will take 8.5 hours to excavate 132 lineal feet.

Labor for Placing Concrete

Table 10-2 can also be used to determine the hours of labor needed to place the concrete in the footing forms. Note: This figure is based on the use of a ready-mixed concrete.

1. The house required 3.3 cu. yds. of concrete per lineal foot. The table shows that for 8" × 16" footings, it will take 2.3 hours to place one cubic yard of concrete in the forms.

2. To calculate the total time, multiply 2.3 (placement hours per cubic yard) by 3.3 (yards of concrete to be placed). The answer is 7.59 hours, rounded off to 7.6 hours of labor. This also includes the time for forming the footings. Estimates may have to be corrected to account for differing soil conditions, as noted at the bottom of the table.

Estimating on the Job

Using Table 10-2, estimate the cubic yards of concrete needed and the hours required for excavation and placement of 10" × 12" footings for a foundation that measures 62' × 38'.

Footing Drains

Why should holes in a drainpipe be on the bottom of the pipe and not the top?

If water builds up on one side of a foundation wall, the pressure created may force moisture through the concrete and into any joints. This is called *hydrostatic pressure.* Footing drains, sometimes called *foundation drains* or *perimeter drains,* relieve pressure by allowing water to drain away. As shown in Foundation Drains, they are located near the outside face of the footing.

Drains are generally required for full-height foundation walls. They are also required where a house is located near the bottom of a long slope that is subject to heavy runoff. The drains direct subsurface water away from the foundation. This helps to prevent damp basement walls and wet floors. Many builders install drains even when they are not required to do so by code.

REGIONAL CONCERNS

Ordering Gravel What you get when you order "gravel" varies in different parts of the United States. In some areas gravel means crushed stone; in other areas it means crushed stone with added fines. The latter is not suitable to cover drainpipe. Always specify the use for the gravel you order. For example, ask for drainage gravel instead of driveway gravel.

Most new houses use a network of plastic pipes as footing drains. These 4" diameter pipes are placed alongside the base of the footing. (See Drainpipe on page 297 of Chapter 11, "Concrete Flatwork," for more on foundation drainpipes.) They are usually connected to storm sewers but may run

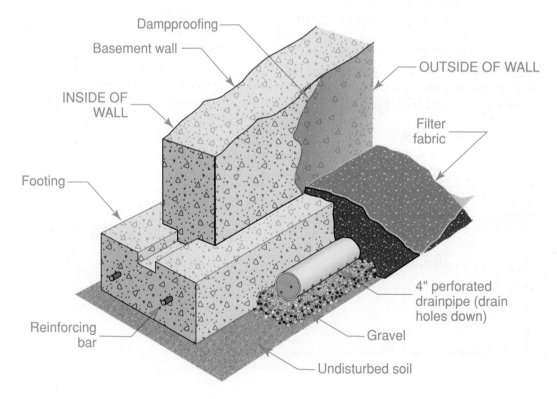

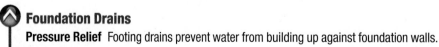

Foundation Drains
Pressure Relief Footing drains prevent water from building up against foundation walls.

to daylight. This means that they lead to a low portion of the site and the end of the pipe is exposed at that point. The piping can also drain into subsurface drain fields if permitted by code, but it must not empty into the drain field of a septic system. This is restricted by code because large amounts of water can damage the septic system.

Plastic drainpipe is different from other types of drainpipe. There are many small holes along the bottom edge of the pipes. The pipes should be placed with the holes facing down. In this position, water is carried away from the house as soon as it rises into the pipes. To keep the water moving, the pipes should be sloped toward the drain at least ⅛" per foot. After the pipes are in place, the drainage area should be covered with *filter fabric* (also called *geotextile* or *landscaping fabric*). This fabric is made of polyester or polypropylene. It allows water to pass through but prevents tiny particles of soil (called *fines*) from getting into the drainage system and clogging it. The filter fabric is backfilled with more drainage gravel. The foundation is then backfilled up to rough grade with dirt.

Section 10.1 Assessment

After You Read: Self-Check

1. What is the purpose of a footing?
2. What type of reinforcement is commonly added to strengthen a footing?
3. When are stepped footings required?
4. What is a footing drain, and why is it important?

Academic Integration: Mathematics

5. **Calculating Volume** Concrete footings are poured prior to laying the block for the foundation wall. Footings must be at least twice the width of the block wall above and as deep as the wall is thick. Calculate the cubic yards of concrete needed for the footings under an 8" block foundation that measures 10' × 10'.

   Math Concept To calculate accurately, measurements should be converted to like units before adding, subtracting, multiplying, and dividing.

   Step 1 Determine the size of the footings for an 8" block foundation.

   Step 2 Calculate the perimeter (in inches) of the foundation.

   Step 3 Calculate the volume by multiplying the perimeter times the width and depth of the footings.

   Step 4 Convert cubic inches to cubic yards. There are 36 × 36 × 36 cubic inches in a cubic yard. Round up to the nearest quarter of a cubic yard of concrete.

Go to **connectED.mcgraw-hill.com** to check your answers.

Concrete Foundation Walls

Types of Foundation Walls

Why is it important for forms to be properly braced?

There are two types of foundation walls: *full-height foundation walls* and *crawl-space foundation walls*. Full-height foundation walls are tall enough to make room for a basement. Crawl-space foundation walls are shorter, typically less than five feet in height. They do not create enough space for a basement, but allow access to pipes and wiring beneath the first floor.

A foundation carries all the loads of a house and transmits them to the ground. Most foundation walls are made of concrete or concrete masonry units (see Section 10.3). Some walls are made of pressure-treated lumber and plywood, but concrete foundation walls are the most common. They are durable and water-resistant. They can be installed on most building sites and can support any type of house. Some houses can be built using precast concrete panels. Walls used to support and contain concrete slabs are called *stem walls* (see Chapter 11, "Concrete Flatwork").

In residential construction, solid foundation walls usually range from 8" to 10" in thickness. The minimum compressive strength for such walls is 2,500 psi (pounds per square inch). Many foundation contractors pour walls that are 8' high above the footings. This provides a clearance of 7'-9½" from the top of the finished concrete floor to the bottom of the first-floor joists.

Reading Check

Contrast What is the difference between full-height foundation walls and crawl-space foundation walls?

Full-Height Walls

Forms must be installed for each concrete foundation wall. Reusable forms are the most cost effective when a contractor does this work regularly. Forms can also be built on site. In any case, the forms must be accurately constructed and properly braced, as shown in Foundation Wall Formwork. This enables them to withstand the forces of the placing and vibrating operations. When concrete is first placed, it puts considerable pressure on the forms. The higher the form, the greater the pressure. If forms are not constructed properly and braced well, the pressure can cause forms to "blow out." This failure allows concrete to spread over the job site. One or more horizontal members, called **wales**, are usually required to brace forms.

Wall Form Details Wall forms may be made from wood or metal, depending on how durable they must be. Many are made from plywood and lumber. Although any exterior-grade plywood can be used, special form-grade plywood is available. Form-grade plywood made by member mills of APA, the Engineered Wood Association, is referred to as *Plyform*. An overlaid surface material is bonded to both sides

JOB SAFETY

WEAR PROPER CLOTHING When placing or finishing concrete, take care to prevent excessive or prolonged contact of concrete with your skin. It is good practice to wear gloves, safety goggles, long pants, and high rubber boots.

of the plywood under high heat and pressure to make it stronger. Medium-density overlay (MDO) has a smooth surface and can be reused many times. High-density overlay (HDO) offers the smoothest finish and can be reused the most. Mill-oiled plywood, another type of plywood, has a sanded veneer surface that is coated with a release agent at the mill. This coating prevents the forms from sticking to the concrete walls. The coating must be reapplied periodically.

Forms built on site may be taken apart after the concrete hardens. The lumber can then be reused elsewhere in the project. It is generally more cost effective and efficient to use reusable forms.

Before any concrete is placed, the sides of each form are fastened together with clips or other ties. Thin metal rods called *snap-ties* are commonly used, as shown in Form Ties. The rods extend through the foundation. Metal brackets attached to the rods prevent the forms from spreading. After the concrete is placed and the forms are stripped (removed), the protruding ends of each rod are snapped off. The end of each rod on the exterior side of the foundation must be sealed with grout to prevent the rods from acting as a conduit for water to enter the interior of the foundation wall. Forms should be left in place for three to seven days before being stripped. This slows the curing process and results in stronger walls.

Insulating Wall Forms Another type of formwork, shown in Insulating Concrete Forms on page 266, is made of rigid foam insulation, usually expanded or extruded polystyrene. These products are referred to as *insulating concrete forms (ICFs)*. Rather than being

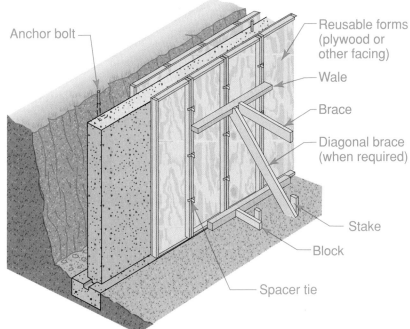

Anchor bolt · Reusable forms (plywood or other facing) · Wale · Brace · Diagonal brace (when required) · Stake · Block · Spacer tie

Foundation Wall Formwork
Stiff and Solid One method of constructing a form for a concrete foundation wall.

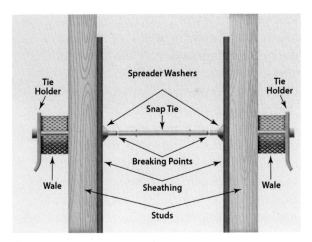

Tie Holder · Spreader Washers · Tie Holder · Snap Tie · Breaking Points · Wale · Sheathing · Wale · Studs

Form Ties
Preventing Spread Snap-ties prevent the two sides of a concrete form from spreading apart when concrete is placed.

The basic components of an ICF can be planks, sheets, or hollow blocks. In many cases, the two sides of the form are held together with plastic or steel connectors that remain within the finished wall. Depending on the product, the concrete placed for the foundation may form a flat surface or a waffle surface, as shown in Types of ICF Forms.

Flat Wall The concrete forms a solid wall identical to a wall poured between traditional concrete forms. See A in Types of ICF Forms.

Grid Wall The concrete forms a wafflelike grid and varies in thickness at different places. This type uses less concrete than other ICF foundations. See B in Types of ICF Forms.

Placement Concrete should be poured continuously, without interruption. This prevents a cold joint. A **cold joint** occurs where fresh concrete is poured on top of or next to concrete that has already begun to cure. A cold joint is more likely to leak and is weaker than the surrounding wall.

The water content of concrete is very important. Although it is tempting to add extra water to make the concrete flow better into the forms, this weakens the finished walls and encourages cracking. Concrete should always be as stiff as is practical.

stripped off after the concrete cures, they are left in place permanently. This eliminates the need to strip and store the formwork. It also greatly increases the insulating level of the foundation walls, an advantage when living space is located below grade. Because the forms are very light in weight, they are easy to install. For the same reason, they must be braced with care. Always consult manufacturer's instructions for bracing any concrete placement. Check local building codes for any special requirements for placement of reinforcing bar, and make sure that local codes permit use of the type of ICF you are considering. Care must also be taken to ensure that termites and other insects cannot reach the foam. Insects do not eat rigid foam insulation, but they will tunnel through it to reach wood. In areas where termite infestation is considered "very heavy" by code, ICF construction is not permitted.

Mathematics: Converting Units

Working with Cubic Units Concrete is frequently purchased in cubic yards, a measure of volume. When calculating the volume of a footing, you will probably work with measurements in feet or inches instead of yards. Create a chart or list showing the number of cubic inches in one cubic foot and one cubic yard, and the number of cubic feet in one cubic yard.

Starting Hint Volume is found by multiplying length by width by height. To find the number of cubic inches in a cubic foot, multiply $12 \times 12 \times 12$.

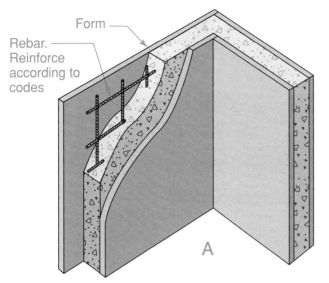

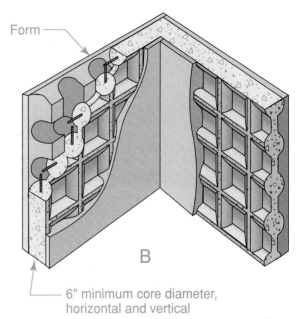

6" minimum core diameter, horizontal and vertical

 Types of ICF Forms

Basic Types A. Standard ICF walls create a channel for concrete that is similar to traditional formwork. **B.** A waffle-like grid wall uses less concrete than other ICF foundations.

Concrete should always be placed as close as possible to where it is needed. This reduces the need to push it around with shovels.

Concrete is normally delivered by ready-mix trucks. As it is placed in the forms, it should be worked to remove air pockets and to help it flow. The most basic technique is to jab it repeatedly with a shovel or pipe as it is being poured. This is generally enough for residential foundations. However, a *concrete vibrator*, sometimes called a *stinger*, is more effective. It is commonly used in commercial construction.

Crawl-Space Walls

Crawl-space foundations are common in mild climates. A crawl-space foundation is shown on page 268. One of the main advantages of the crawl-space foundation is reduced cost. Little or no excavation or grading is required except for the footings and walls. For crawl-space foundations:

- Soil beneath the house must be covered with a material to block moisture vapor from reaching the floor structure. This material, called a *vapor retarder,* is often 6-mil or 8-mil thick plastic sheeting. The best products are reinforced to minimize tearing.
- The crawl space usually must be ventilated. Check local codes.
- The floor framing above the crawl space should be insulated to reduce heat loss.

Poured-concrete or concrete-block piers are often used to support floor girders in crawl-space houses. They should be no closer than 12" to the ground. To prevent ground moisture from reaching floor framing, bare dirt should be covered with 6-mil plastic sheeting. Otherwise, the floor framing may absorb enough moisture to encourage fungi. When temperatures favorable for fungus growth are reached, much decay may result. To protect

Flood-Resistant Foundations The building code now requires special construction details for foundations built where flooding is likely. Foundations must be designed to resist "flotation, collapse, or permanent lateral movement" due to stresses caused by flood waters.

 Crawl-Space Foundation
Crawl-Space Walls A crawl-space foundation is poured with anchor bolts to secure the sill plate and the vapor retarder on the floor may be protected with pea gravel.

the plastic from damage, some builders cover it with a layer of pea (rounded) gravel.

Reinforcement for Walls

In some parts of the country, it is not required to reinforce concrete foundation walls with steel. Some builders add vertical and horizontal rebar anyway to provide extra strength. The rebar should be centered in the wall. Window and door openings in the wall call for special attention. The concrete over these openings should contain rebar according to local codes. In earthquake hazard zones, the code requires that foundation walls be reinforced. Always check local codes for reinforcing requirements.

Where concrete work includes a connecting porch or garage wall not poured with the main basement wall, rebar ties must be provided. The rebar is placed as the main wall is poured. Keyways may also be used to resist sideways movement by forming a lock between the walls. Connecting walls should extend below the normal frost line and be supported by undisturbed soil.

Sill-Plate Anchors

In wood-frame construction, the sill plate must be securely fastened to the foundation. In areas exposed to high winds or earthquakes, well-anchored plates are especially important. For this reason, anchorage requirements are more stringent in such areas. Bolts or other anchoring devices may have to be placed closer together. In some cases they must be connected to reinforcing bar within the wall.

Most builders use ½" diameter L-shaped bolts called *anchor bolts* (shown in A in Sill-Plate Anchors). These are embedded in the concrete immediately after the top of the foundation walls have been smoothed. They should be spaced no more than 6' apart and no more than 12" from the ends of any plate

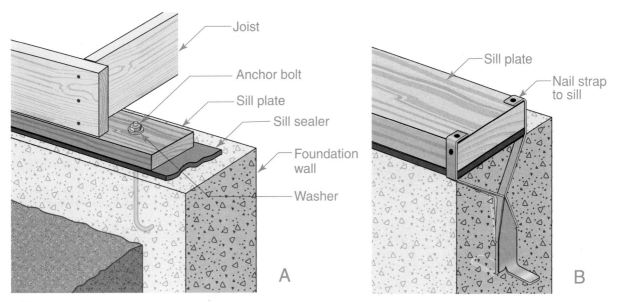

A

B

 Sill-Plate Anchors
Secure Connection A. Anchor bolts embedded in the foundation wall are used to secure the sill plate. **B.** Some builders use metal straps embedded in the concrete.

Labels for image A: Joist, Anchor bolt, Sill plate, Sill sealer, Foundation wall, Washer

Labels for image B: Sill plate, Nail strap to sill

section or wall corner. Each bolt must extend at least 7" into the concrete. A large flat washer must be used beneath the nut that holds the sill plate in place.

Another type of anchor is a metal strap that is embedded in the concrete, as shown in B in Sill-Plate Anchors. The legs of the strap wrap around the sill plate.

A sill sealer is often placed under the sill plate on poured walls to smooth any uneven spots that might have occurred during placement. If termite shields are used, they should be installed under the plate and sill sealer. A termite shield is typically a continuous length of galvanized sheet metal with the edges bent slightly downward.

Foundation Wall Details

A foundation wall must often have special details, such as brick-veneer siding or utility sleeves. These details must be accounted for in the design of the foundation.

Masonry Ledges Brick or stone veneer is often used for the outside finish over wood-frame walls. In such cases, the foundation must include a *masonry ledge,* a supporting ledge or an offset about 5" wide, as shown in

A Masonry Ledge on page 270. Including a masonry ledge results in a space of about 1" between the masonry and the sheathing that is needed for ease in laying the brick. A *base flashing* is used at the brick layer below the bottom of the sheathing and framing. The flashing should be lapped with sheathing paper. *Weep holes* (to provide drainage) are also located at this course and are formed by omitting the mortar in a vertical joint. (Brick-veneer walls are discussed in Chapter 24, "Brick Masonry & Siding.")

Utility Sleeves It is often necessary for pipes, such as the main drain to the sewer or septic system, to pass through the foundation walls. Other pipes may also need to pass through the foundation, including water supply pipes and electrical conduits. It is easier to provide space for these pipes as the forms are being placed rather than drill large holes in the foundation later. Where a pass-through is required, a tight-fitting foam block is placed within the formwork and secured with nails. A short length of plastic pipe can also be used. These barriers prevent concrete from flowing into these areas, creating a hole in the wall at that point.

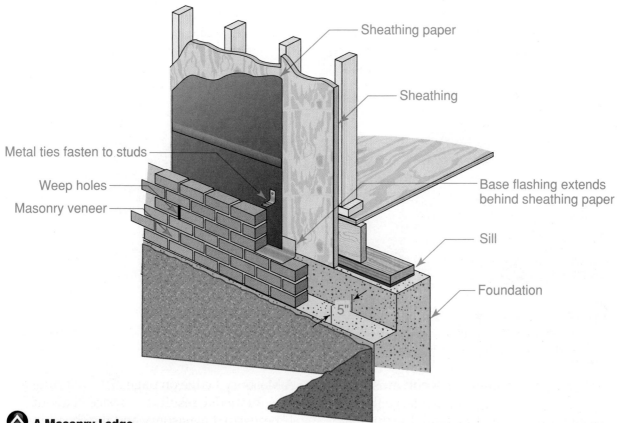

Sheathing paper

Sheathing

Metal ties fasten to studs

Weep holes

Masonry veneer

Base flashing extends
behind sheathing paper

Sill

Foundation

5"

A Masonry Ledge

Brick or Stone Support A foundation wall with a masonry ledge. Note that weep holes in the brick veneer are located just above the base flashing.

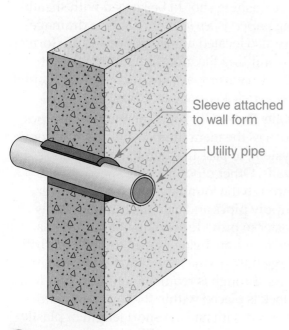

Sleeve attached
to wall form

Utility pipe

A Pipe Sleeve

Hole Through Foundation A sleeve provides openings through the basement wall for utilities.

After the forms have been stripped, the block is removed. Later, pipes can be routed through the hole, as shown in A Pipe Sleeve. Any space around them can be sealed with hydraulic cement and waterproofed.

Foundation Vents and Windows In crawl-space foundations, metal or wood-framed vents are sometimes installed within the forms before the concrete is poured. Vents are visible in Crawl-Space Foundation on page 268. In full-height foundation walls, frames for small, grade-level windows may also be placed in the forms. The rust-resistant steel frame of these windows will then be locked securely to the foundation. Where larger openings are required, wood frames may have to be inserted in the forms. In this case, the wood is sometimes left in place after the forms are stripped away. This wood must be pressure treated. Small anchor bolts should be inserted into pre-drilled holes in the wood frame before the concrete is poured.

Beam Pockets A wall notch, or beam pocket, is needed for basement beams or girders. The notch allows the top of the girder to be flush with the top of the sill plate. It should be large enough to allow at least ½" of clearance at the sides and ends of the beam. This clearance is for ventilation. If wood beams and girders are so tightly set in wall notches that moisture cannot readily escape, they may decay. A waterproof membrane, such as roll roofing, is applied under the end of the beam to reduce moisture absorption.

Stripping and Maintaining Forms

Forms should not be removed until the concrete has enough strength to support the loads of early construction. Leaving the formwork in place also slows the loss of moisture, which improves the strength of the concrete. At least two days (preferably longer) are required before forms can be stripped in temperatures above freezing. A week may be required when outside temperatures are below freezing.

Metal prybars should not be used when stripping wood forms. They can easily damage the edges and faces of the panels. Use wood wedges to pry panels away from the concrete. As soon as the forms have been removed, they should be cleaned, inspected for damage, and repaired if necessary. Concrete residue and scaling can be removed by scraping the surfaces with a hardwood wedge and brushing them with a stiff bristle brush. Do not use a wire brush because it can damage the wood surface. Before the forms are used again, recoat them with a *form-release agent*. This is a liquid that prevents concrete from sticking to the forms.

Moisture Protection

Before full-height concrete foundation walls are backfilled, steps must be taken to protect them from ground moisture. Where the walls will be exposed to standard soil conditions and no unusual drainage problems, this may be done by *dampproofing* them. Dampproofing is now required by code for all foundation walls, not just those enclosing habitable space. The walls are coated with a material that protects against ordinary seepage, such as seepage that may occur after a rainstorm. The coating should extend from the top of the footings to the finished grade level. It should not be applied until the surface of the concrete has dried enough. Otherwise, it may not stick. Various materials can be used. The most common is a black bituminous coating that is either sprayed or brushed over the walls.

Where the soil drains poorly, where the water table is high, or where living spaces will be located below grade, greater efforts must be made to protect the foundation. They often involve applying a waterproofing membrane to the foundation walls. The membrane should extend from the top of the footings to the finished grade level. All joints in the waterproofing membrane must be overlapped and sealed with an adhesive suitable for the membrane material. Various materials can serve as a waterproofing membrane:

1/2" clearance, sides, and end

4" minimum bearing

Sill plate

Asphalt bearing pad

Location of beam

A Beam Pocket
Solid Bearing A beam pocket is a notch created to support the end of a beam.

- 2-ply hot-mopped felts
- 55-lb. rolled roofing
- 6-mil PVC or polyethylene sheeting
- 40-mil polymer-modified asphalt
- 60-mil flexible polymer cement
- ⅛" cement-based fiber-reinforced waterproof coating
- 60-mil solvent-free liquid-applied synthetic rubber

In some cases, a multilayer combination of rigid insulation board, drainage media, and spray- or sheet-membranes may also be approved for use.

Backfilling

Backfilling, is the process of filling in the excavated area around a foundation with soil. This brings the area around the house up to rough grade.

 Backfilling
Careful Work To avoid damage, only an experienced operator should use heavy equipment to backfill a foundation.

Estimating and Planning

Concrete Foundation Walls

Formwork

Refer to the floor plan on page 273, which measures 40' × 26'.

1. To determine the total foundation wall area, assume that the wall is 8' high. Multiply 8' × 132' (perimeter of the building). The answer is 1,056 sq. ft.

2. Assume the wall thickness is 8". Refer to the **Table 10-3** on page 273. Read down the column headed "Wall Thickness" to 8". Then read across to the column titled "Forming." Remember, the wall is to be 8' high. The table shows that the wall will require 7.75 hours per 100 sq. ft. of wall area.

3. Next, calculate the total time for installing the forms. Since you know it will take 7.75 hours for each 100 sq. ft., divide the total number of square feet by 100 and multiply by 7.75.

$$1{,}056 \div 100 = 10.56$$
$$10.56 \times 7.75 = 81.84$$

It will take about 82 hours to install the forms.

4. Next, calculate the time needed to remove the forms. According to the table, between 2 and 3 hours are needed to remove forms for 100 sq. ft. of an 8' wall. Using the larger number as an example, the calculation would be:

$$10.56 \times 3 = 31.68$$

It will take about 31⅔ hours total labor time for removing the forms.

Concrete

You can also calculate the amount of material needed using the table. In our example, the wall is 8" thick and has a total area of 1,056 sq. ft.

1. Find the 8" thickness in the column at left. Reading across, under "Material" you find that 2.47 cu. yds. of concrete are needed for every 100 sq. ft. of wall. Therefore you must again divide the total area by 100 to

Table 10-3: Estimating Concrete Foundation Walls

Walls	Material		Forming			Concrete Placement
	Per 100 Feet of Wall		Hours per 100 Square Feet of Wall			Hours per Cubic Yard
Wall Thickness (inches)	Cubic Feet Required	Cubic Yards Required	Place		Remove	
			0' to 4'	4' to 8'		
4	33.3	1.24	4.7	7.13	2.0	Average 3.25 Hours
6	50.0	1.85	4.7	7.75	Varies as to Height	
8	66.7	2.47	5.0	7.75		
10	83.3	3.09	5.0	7.90		
12	100.0	3.70	5.0	7.90	3.0	

find how many hundreds of square feet there are.

$$1{,}056 \div 100 = 10.56$$

2. Then multiply by 2.47.

$$10.56 \times 2.47 = 26.08$$

Round your answer to the next larger ¼ cu. yd. Thus, a total of 26.25 cu. yds. of concrete are needed.

Labor

To estimate placement of concrete in the forms for the wall, again use the table.

1. Under "Concrete Placement," it says that 1 cu. yd. takes an average of 3.25 hours.
2. Multiply the total cubic yards by the time required to pour 1 cu. yd. This will tell you the total time required. In our example, 26.25 cu. yds. of concrete are required. Therefore:

$$26.25 \times 3.25 \text{ (hours)} = 85.31$$

Rounded off, this comes to 85⅓ hours of labor.

Estimating on the Job

Using the table, estimate the time to install and remove forms, the cubic yards of concrete needed, and the time required to place the concrete for a foundation that measures 43' × 27'. The wall will be 8' high and 10" thick.

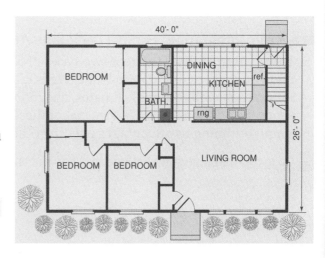

Backfilling should be done as soon as possible for safety. Backfilling also makes it easier to transport materials to and from the house. A foundation must not be backfilled too soon. The weight of the earth can damage walls that are not yet strong enough to withstand the pressure. During backfilling, the vertical portions of drainpipes should be temporarily capped to prevent soil from getting into the drain system.

All foundation drainage, dampproofing, and waterproofing must be complete before backfilling begins. Under ideal conditions, the floor framing (or floor slab) is also in place. This braces the tops of the foundation walls. In cases where the wall must be backfilled before floor framing is in place, the walls can be temporarily braced from inside the excavation. This can be done using framing lumber.

Follow the best local building practices when choosing backfill material. Do not use materials that expand and drain poorly, such

as clay. Layer gravel into the excavation as needed to ensure proper drainage. Backfill 6" to 8" at a time, and compact the soil to prevent it from settling too much later. Do not allow wood debris, such as lumber scraps and tree limbs, to be included in backfill. This encourages insects and uneven settlement when limbs and/or wood debris decompose.

JOB SAFETY

FALLS AND CAVE-INS The area around a new foundation wall is an open excavation. Care must be taken when working around this area to prevent falls. To prevent cave-ins, keep trucks and other equipment well away from the perimeter.

Section 10.2 Assessment

After You Read: Self-Check

1. What is a wale?
2. Why might ICFs be used to form foundation walls for a house that was designed to include a basement recreation room?
3. Why must concrete always be poured in as stiff a mix as is practical?
4. Name three important aspects of crawl-space foundation construction.

Academic Integration: English Language Arts

5. **Foundation Walls** There are two types of foundation walls: full-height foundation walls and crawl-space foundation walls. Walls tall enough to create a basement are full-height foundation walls. In mild climates, it is more common to find shorter walls called crawl-space foundation walls. Choose one of these types of foundation walls and write a paragraph outlining its advantages over the other type of foundation wall.

Go to connectED.mcgraw-hill.com to check your answers.

Concrete Block Walls

Block Basics

What is a CMU?

Concrete block is popular for building foundation walls (see Concrete Block Walls). This is because the walls do not require formwork and the blocks are fairly inexpensive. Unlike work on a solid concrete foundation, which must be done all at once, work on a block foundation can start and stop as needed.

Full-height foundation walls are often constructed of eleven *courses* (rows) of block above the footings, with a 4" solid cap block. The cap block seals the cores of the foundation walls. The cores, or *cells*, are the hollow part of the block. This results in about 7'-4" of headroom between the joists and the basement floor. A wall with 12 courses of block would add another 8" of headroom.

Concrete Block Walls
Building a Wall Concrete blocks should be stacked near where they will be used. *Why might this be?*

Mel Stoutsenberger/iStock/360/Getty Images

Strengthening Walls Sometimes a block wall must be strengthened with rebar. If this is required by the building's designer, #4 to #7 rebar is inserted into the vertical channels created in the wall by successive block cores. Each core containing rebar is then filled with concrete. This creates a reinforced column within the wall. These columns should be spaced as required by local codes, depending on the height of the wall and the local soil type. Generally, columns in a 12" thick wall will be spaced no more than 72" OC (on center).

Adding pilasters to the wall is another way to strengthen it. *Pilasters* are projections resembling columns that may be used to strengthen a wall under a beam or girder. Pilasters are placed on the interior side of the wall and are constructed as high as the bottom of the beam or girder they support. Basement door and window frames should be keyed to the foundation for rigidity and to prevent air leakage.

Types of Block

Any hollow masonry unit is called a concrete block, or *concrete masonry unit (CMU)*. The most common type is made with Portland cement, a fine aggregate, and water. Concrete blocks come in many shapes and sizes for a large variety of applications. Some of the most common sizes are shown in Concrete Masonry Units on page 276. The most widely used sizes are 8", 10", and 12" wide (nominal dimension). The nominal dimensions allow for the thickness and width of a standard ⅜" mortar joint. Thus, the actual dimensions of a standard block are usually 7⅝" high by 15⅝" long. This results in assemblies that measure 8" high and 16" long from centerline to centerline of the mortar joints. A vertical mortar joint is called a **head joint**. A horizontal joint is called a **bed joint**.

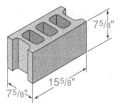
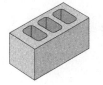

| Stretcher (3 core) | Corner | Double corner or pier | Bull nose | Jamb |

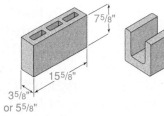

| Full cut header | Half cut header | Solid top | Stretcher (2 core) | 4" or 6" partition | Beam or lintel |

 Concrete Masonry Units

Common Shapes Typical shapes and sizes of concrete masonry units. Half-length sizes are usually available for most of the units shown above.

Specialty blocks are made for specific purposes. Split-face blocks have one rough face that looks something like stone, as shown in Split-Face Block. Insulated blocks come in various forms, including some that contain inserts of polystyrene. Heat transfer from one surface to another is reduced when blocks contain rigid insulation. Another type of block has one or more glazed surfaces. It can be used as a structural as well as a finish material.

Bond Patterns

Block courses are laid in a *common bond*. A common bond is the overlapping arrangement (see A in Bond Patterns). Joints should be tooled smooth to seal them against water seepage. Mortar should be spread fully on all contact surfaces of the block. Such spreading is called a *full bedding*.

When exposed block foundation is used as a finished wall for basement rooms, the *stack bond* pattern can give a pleasing effect. In a stack bond, blocks are placed directly above one another, resulting in continuous vertical joints, as shown in B in Bond Patterns. It is

Split-Face Block

Rough Surface Split-face block is sometimes used for decorative effect in exposed locations.

necessary to add some type of joint reinforcement at every second course. This usually consists of steel rods arranged in a grid pattern.

Protecting Block Walls Freshly laid block walls should be protected in temperatures below 32°F (0°C). Freezing of the mortar before it has set will often result in low adhesion, low strength, and joint failure. Care must be taken to keep blocks dry on the job. They should be stored on planks or other supports so the edges do not touch the ground. They should be covered for protection against moisture. Concrete block must not get wet just before or during installation.

Block walls should not be backfilled until they have gained sufficient strength. Follow the precautions noted in Section 10.2 regarding backfilling.

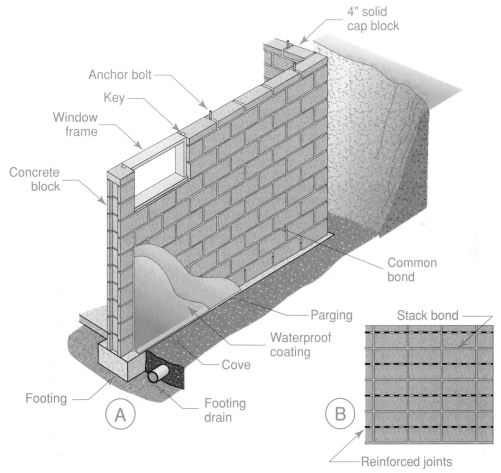

 Bond Patterns
Foundation Walls A. Block walls are usually laid in a common bond, as shown here. **B.** In some cases, walls are laid in a stack bond, with horizontal reinforcing as needed.

Cutting Block Blocks are usually available in half-length as well as full-length units. It is sometimes necessary to cut a block to fit it into place. This can be done in two ways. The traditional method is to use a brick hammer and chisel to cut block, as shown in A in Methods for Cutting Block. The block is scored on both sides with the chisel to make a clean break. A faster method to cut block is to use a masonry saw, as shown in B in Methods for Cutting Block. The saw leaves a smooth, uniform edge but takes time to set up. It is also possible to cut block with a standard circular saw fitted with a dry-cutting masonry blade, but this method creates clouds of dust and should be avoided whenever possible.

Mortar

Good mortar is essential for a strong, solid wall. The strength of the mortar bond depends on:

- The type and quantity of mortar.
- The workability, or *plasticity,* of the mortar.
- The surface texture of the mortar bedding areas.
- The rate at which the masonry units absorb moisture from the mortar.
- The water retention of the mortar.
- The skill of the person laying the block.

Mortar Mixtures Mortar is a mixture of Portland cement, hydrated lime, sand, and water. The individual ingredients are often mixed together on site using a mechanical drum mixer. Mortar mixes for various purposes are shown in **Table 10-4**. Varying the ingredients yields mortar with different characteristics. A relatively high proportion of Portland cement improves strength. Lime reduces compressive strength but increases flexibility and makes the mortar "stickier." Sand reduces shrinkage as the mortar cures.

Mortar can also be made from prepackaged mortar mix or masonry cement. These products must be mixed with water on site. Mortar mix contains all the dry ingredients, including sand. Masonry cement contains all the dry ingredients *except* sand. Various masonry cement mixes are shown in Table 10-4. The following types of mortar are the most common:

- Type N mortar has average strength for most general masonry work above grade. It has only moderate compressive strength.

A

B

Methods for Cutting Block

Hand or Power Saw A. To cut by hand, score the blocks along both sides with a chisel. **B.** A masonry saw is fast and precise.

Table 10-4: Proportions of Mortar Ingredients by Volume					
Mortar type	Portland or blended cement	Masonry cement type			Fine aggregate
		M	S	N	
M	1	–	–	1	4½ to 6
	–	1	–	–	2¼ to 3
S	½	–	–	1	3⅜ to 4½
	–	–	1	–	2¼ to 3
N	–	–	–	1	2¼ to 3
O	–	–	–	1	2¼ to 3

- Type M mortar has high compressive strength and is particularly durable. This makes it good for heavily loaded or below-grade foundation walls.

- Type S mortar has a high tensile strength as well as high compressive strength. This makes it suitable for regions exposed to earthquakes or high winds.

- Type O is a low-compressive-strength mortar used primarily for interior walls.

Mixing and Placing Mortar To ensure that ingredients are well blended, mortar should be mixed in power mixers. For very small jobs, it may be mixed in a wheelbarrow or a mortar tray.

Mortar will stiffen after being mixed because of evaporation and hydration. Evaporation occurs when moisture is lost from the mixture. In that case, water can be added and mixed in to restore the mortar's workability. Mortar stiffened by hydration should be thrown away. It is not easy to tell whether evaporation or hydration is the cause. A judgment can usually be made on the basis of how much time has passed since initial mixing. Mortar should be used within two-and-a-half hours when the air temperature is 80°F (27°C) or higher, and within three-and-a-half hours when air temperature is below 80°F (27°C). If more time has passed, assume that any stiffness is caused by hydration.

Mortar must be sticky so that it will cling to the concrete block. Mortar is taken from the mortar board with the trowel and then "set" on the trowel with a quick vertical snap of the wrist. The excess will fall from the trowel with the snap of the wrist.

Laying Block Foundation Walls

What is a mason's line used for?

Laying block foundation walls is a job for skilled masons. The following section is intended primarily as an overview of the process.

Concrete block is heavy. It should always be stacked close to the work area as to minimize the need to carry it, as shown in Concrete Block Walls on page 275. Boards, building paper, or tarpaulins should be used to cover the tops of unfinished block walls at the end of the day's work. This prevents water from entering the cores.

Building the Corners

The corners of the wall are built first, usually four or five courses high. After locating the outside corners of the wall, use a chalked line to mark the footing and help align the first block accurately. A full mortar bed should then be spread with a trowel. The corner block should be laid first and carefully positioned.

JOB SAFETY

PRECAUTIONS WITH MORTAR Prolonged contact with wet mortar is harmful to your skin. Wear protective clothing, including gloves, to minimize contact with the material. If skin does come into contact with mortar, wash off the mortar as soon as possible. Change clothing that has become saturated with mortar.

The first course of the corner should be laid with great care to make sure it is properly aligned, leveled, and plumbed. This will ensure a straight, true wall. After three or four blocks have been laid, use the mason's level as a straightedge to ensure correct alignment. Make blocks plumb by tapping them with the trowel handle.

After the first course is laid, apply mortar to the top of the face shells. A *face shell* is the side wall of a concrete block (Bedding the Face Shell). In some cases, a full mortar bed may be specified. Mortar for the vertical joints can be applied to the ends of the next block or to the ends of the block previously laid. There is the danger of building up the joint size when applying mortar to the ends of both blocks and it is also a wasted motion (multiplied over 300 or 400 blocks set per man per day). Buttering one end of the block is usually sufficient.

As each course is laid at the corner, check it with a level for alignment, for levelness, and for plumb. Check each block carefully with a level or straightedge to make certain that the faces of the blocks are all in the same

Full Mortar Bed
No Bare Spots This is what a full mortar bed should look like.

Bedding the Face Shell
Mortar Bed Mortar bedding the face shell in preparation for laying up additional courses.

Laying the Block
Block Pressure The weight of the block will usually be sufficient pressure to embed it in the mortar.

plane. Check the horizontal spacing by placing the level diagonally across the corners of the blocks (Check Alignment). A **story pole**, or *course pole*, is a board with markings 8" apart. It can be used to gauge the top of the masonry for each course.

Check for Level
Level Check the alignment of the blocks frequently.

Check Alignment
Alignment If the blocks have been positioned correctly, the alignment can be checked by holding a level or straightedge diagonally across the corners of the block.

Check for Plumb
Plumb After the corners have been built up, be sure to check the corner for plumb before continuing.

A Story Pole
Height A wood story pole marked with the course levels should be used to maintain the proper height of the courses.

 Laying to a Line
Mason's Line After the corners have been built up, stretch a mason's line from corner to corner for each course. Between the corners, set the blocks so their top edges align with the mason's line.

Filling In Between Corners

When filling in the wall between the corners, a mason's line is stretched from corner to corner for each course, as shown in Laying a Line. The top, outside edge of each block is laid to this line.

Handling or gripping the block correctly is important and is learned with practice. Roll the block slightly to a vertical position and shove it against the adjacent block. Final positioning of the block must be done while the mortar is soft and plastic. Any attempt to move or shift the block after the mortar has stiffened will break the mortar bond and allow water to seep into the completed installation. "Dead" mortar that has been picked up from the scaffold or from the floor should not be used.

To assure a good bond, mortar should not be spread too far ahead of actual laying of the block or it will stiffen. As each block is laid, excess mortar at the joints is cut off with the trowel. Applying mortar to the vertical joints of the block already in the wall and to the block being set results in well-filled joints.

The block that fills the final gap in a course between corners, shown in Fitting the Closure Block, is called the *closure block*. To install this block, spread mortar on all edges of the opening and all four vertical edges of the block itself. The closure block should be carefully lowered into place.

 Fitting the Closure Block
Last Block The closure block is carefully placed in position to complete a course.

Intersections

Load bearing walls built of intersecting concrete blocks should not be tied together in a masonry bond, except at the corners. Instead, one wall should end at the face of the other wall, with a control joint at that point. A **control joint** is a joint that controls movement caused by stress in the wall. The joints are built into the wall in a way that permits slight movement without cracking the masonry. They are continuous from the top of the wall to the bottom. They are the same thickness as the other mortar joints.

Control joints should be placed at the junctions of bearing as well as nonbearing walls, at places where walls join columns and pilasters, and in walls weakened by openings.

For sideways support, bearing walls are tied together with a metal reinforcing bar called *tie bar*, shown in below. The bends at the ends of tie bars are embedded in cores filled with mortar or concrete. Pieces of metal lath placed under the cores support the concrete or mortar filling.

For tying nonbearing block walls to other walls, strips of metal lath or ¼" mesh galvanized hardware cloth are placed across the joint, as shown in Metal Lath. The metal strips are placed in alternate courses. When one wall is constructed first, the metal strips are built into the first wall and later tied into the mortar joint of the second wall. Another type of reinforcement is called *ladder-reinforcement*. Lengths of ladder-shaped metal wire can be set lengthwise into horizontal mortar joints. In some cases, this is required in every third course of a block wall. Ladder-reinforcing significantly strengthens the wall.

Tie Bar
Reinforcing a Corner A reinforcing bar can be placed in the mortar joint to tie intersecting walls together.

Metal Lath
Reinforcing a Joint Metal lath placed across the joint is used to tie a non-bearing intersecting wall to the main wall.

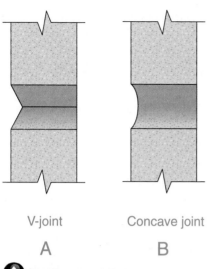

V-joint
A

Concave joint
B

 Tooling the Joints
Bed Joints Joints are usually tooled with a jointer shaped to create either **A.** a V-shaped joint or **B.** a concave joint

Tooling the Joints

Weathertight joints and a neat appearance depend on proper tooling. This is done after the mortar has become "thumbprint hard" (the thumb makes no indentation when pressed into the mortar). Tooling is a process of using shaped metal bars to compact the mortar and force it tightly against the masonry on each side of the joint. Proper tooling produces joints of uniform appearance, with sharp, clean lines. Unless otherwise specified on the plans, all joints should be tooled in either a concave or V-shape.

Tooling of the head joints should be done first, using a small S-shaped jointer. Tooling of the bed joints should follow, as shown in Tooling the Joints. The horizontal

Estimating and Planning

Block Walls

Estimating Materials

When estimating materials for block walls, you must consider several factors. These include the number of blocks, the amount of mortar, and the cost of labor.

Block

The number of blocks needed for a foundation can be determined by the area of each wall to be built.

1. Nine 8" × 8" × 16" blocks will make eight square feet of wall area. Therefore, take the total number of square feet in the wall and divide it by eight. Multiply the result

by nine. You will then have a good estimate of the number of blocks needed for the wall.

For example, consider a house with a 25' × 40' foundation that is 7' high. The simplest way to find the total square footage is to multiply the perimeter times the height. The perimeter (the total length of the four sides) is 130'. Multiply this by 7' to find the total area of the four basement walls, which is 910 sq. ft. Now apply the formula:

$$910 \div 8 = 113.75$$
$$113.75 \times 9 = 1,023.75$$

Rounded off, your answer would be 1,024.

2. Next, because the courses overlap or interlock at the corners, subtract one-half block for each corner of each course. The wall in the example would be 11 blocks high; therefore, subtract 5½ blocks for each corner, or 22 blocks altogether.

$$1,024 - 22 = 1,002$$

A total of 1,002 blocks would be needed. This number would be reduced even more to allow for windows or other openings.

3. The number of concrete blocks necessary for a wall can also be determined by referring to the table below. In the left column, find the size of the block used. If you select an 8" × 8" × 16" block, the table indicates 110 concrete blocks for each 100 sq. ft. of wall. The walls in our example have an area of 910 sq. ft. Divide this by 100 to find the number of square feet expressed in hundreds: 910 ÷ 100 = 9.1.

The table shows that 110 blocks are needed for each 100 sq. ft., so multiply 9.1 by 110 to find the total number of blocks needed: 9.1 × 110 = 1,001 total blocks.

Some adjustment may still be necessary if there are openings in the wall. The table allows for the overlapping of blocks at the corners, so it is not necessary to subtract for this as in the previous example. Note also that the answer is not precisely the same as when calculated by the first method. However, the estimates are very close, and both methods are reliable.

Mortar

The number of cubic feet of mortar needed for a block wall can also be determined from the table.

1. For the walls in our example, the table shows that 3.25 cu. ft. of mortar would be needed for every 100 sq. ft. of wall area.

2. There are 9.1 hundreds of square feet in the walls. By multiplying 9.1 by 3.25 you find the total amount of mortar needed.

9.1 × 3.25 = 29.575 cu. ft. of mortar, rounded off to 29.6

Labor

To determine labor costs, again consult the table.

1. You will see that 8" × 8" × 16" blocks are laid at a rate of 18 per hour.

2. Using the figure 1,001 for the total number of blocks, divide by 18 to learn the number of hours needed:

$$1,001 ÷ 18 = 55.6$$

3. Multiply the hours needed by the hourly rate of pay to find the labor cost.

Estimating on the Job

Using the table, determine the number of blocks, amount of mortar, and labor hours needed to make a 22' × 38' foundation that is 7' high. Concrete blocks that measure 8" × 8" × 16" need to be used.

Estimating Table for Masonry Blocks			
Lightweight Block	Material for 100 Sq. Ft. of Wall		Labor
Size	Number of Units	Mortar (cu. ft.)	Blocks per Hour
8×4×12	146	4.0	24
8×4×16	110	3.25	22
12×4×12	100	3.25	30
8×6×16	110	3.25	21
Concrete Block	Material for 100 Sq. Ft. of Wall		Labor
Size	Number of Units	Mortar (cu. ft.)	Blocks per Hour
8×8×16	110	3.25	18
8×10×16	110	3.25	16
8×12×16	110	3.25	13

Note: Mortar quantities based on ⅜" mortar joints, plus 25% waste. For ½" joints add 25%.

Filling Cores

Top Course With metal lath in place, the cores of the top-course blocks can be filled and troweled smooth.

joint should appear continuous. A jointer for tooling horizontal joints is upturned on one end to prevent gouging the mortar. For concave joints, a tool made from a ⅝" round bar is fine. For V-shaped joints, a tool made from a ½" square bar is generally used. After the joints have been tooled, a trowel or stiff brush is used to trim mortar burrs flush with the wall face.

Completing the Walls

Foundation walls of hollow concrete block must be capped with a course of solid masonry to distribute the loads from the floor beams and to act as a termite barrier. *Solid-top blocks,* in which the top 4" is of solid concrete, can be used to accomplish this (see Concrete Masonry Units on page 276). The course can also be covered with a solid 4" thick block called a *cap block.* A third method is to use stretcher (standard) blocks and then fill all cores with concrete or mortar. In this case, a strip of metal lath wide enough to cover the

core spaces is placed in the joints under the top course. The cores are then filled and troweled smooth, as shown in Filling Cores.

Subterranean termites can crawl through hidden cracks in a wall to the wood in the building above. Installing metal termite shields on top of the block walls prevents this.

Installing Anchor Bolts

The house framing rests on preservative-treated wood sill-plates that are fastened to the top of the foundation walls. This is done by means of anchor bolts ½" in diameter and 18" long, spaced not more than 8' apart. These anchor bolts are placed at least 16" deep in the cores of the top two courses of block, and the cores are filled with concrete or mortar. The threaded end of the bolt should extend above the top of the wall, as shown in Anchor Bolt. Pieces of metal lath are placed in the second horizontal joint from the top of the wall and under the cores to be filled. The lath supports the concrete or mortar filling.

Cleaning Block Walls

Any mortar droppings that stick to the block wall should be allowed to dry slightly before removal with a trowel. The mortar may smear if removed while too soft. When dry and hard, most of the remaining mortar can be removed by rubbing it with a small piece of concrete block and then brushing.

Anchor Bolt

Position Carefully Fill cores with concrete or mortar and insert the bolt so that the threads extend above the top of the wall. The bolt should be plumb.

Additional Techniques

What is radon?

After the walls are complete, additional steps should be taken to ensure that they are able to resist moisture. In some cases, block walls can be assembled using surface-bonding techniques.

Moisture Protection

Like solid concrete walls, block walls must either be dampproofed or water-proofed. Block walls are sometimes parged as part of this process. **Parging** is the process of spreading mortar or cement plaster over the block, as shown in below. A cove should be formed where the wall joins with the footing, as shown in Bond Patterns on page 277. The parging should be at least ⅜" thick. When the parging is dry, a coating

 Parging

Moisture Protection Parging a block wall blocks water infiltration. It can be applied to above-grade as well as below-grade walls.

of asphalt is applied to the exterior of the wall. This, along with a properly designed footing drain, will normally ensure a dry basement.

Sometimes added protection is needed, as when soil is often wet. In such cases, the entire wall should be waterproofed like a solid concrete foundation (see Section 10.2).

Lintels and Bond Beams

In some situations, concrete is added to a block wall to span an opening or to provide additional strength. Where openings occur in the foundation, a *lintel* must be installed over the opening to provide support for the masonry above it. A lintel is a horizontal member that supports the weight of the wall above. A lintel in a masonry wall is like a header in a wood-frame wall. It directs loads around the opening. One way to create a lintel is to use L-shaped steel angles. One leg of the L fits under the masonry to support it over the opening.

Another type of lintel is made of rein-forced pre-cast concrete. It is a manufactured product that is delivered to the job site in finished form. It is placed over an opening just as a wood header would be placed.

A third way to create a lintel is to use lintel blocks (see Concrete Maonry Units on page 276). Lintel blocks are temporarily supported over the opening by a wood framework. The open portions of the blocks are then filled with concrete and reinforced with rebar. When the concrete has cured, the wood framework can be removed.

Building codes where earthquakes are a hazard may require that masonry walls be strengthened with a *bond beam*. A bond beam is a course of reinforced concrete or reinforced lintel block. It is sometimes called a *collar beam*. It is often positioned as the top course of a wall. In some cases it is placed in more than one location. For example, a bond beam might be placed at every fourth course in the wall to stiffen the wall. Bond beams can be created by a continuous course of reinforced lintel

blocks. Another method is to secure metal or plywood forms to the top of the wall. Once the bond beam cures, the forms can be removed and construction can continue.

Surface Bonding

Mortared block walls are the most common type of concrete block wall. Another technique called *surface bonding,* or *dry-stacking,* is also used. It starts out similar to parging. The first course of block is bedded in mortar as usual. Additional courses are stacked dry, with no mortar. Fiberglass-reinforced mortar, or *surface-bonding mortar,* is then troweled over both sides of the walls in a layer at least ⅛" thick, as shown in Surface Bonding. The ½" long fibers improve the tensile strength of the mortar much like steel mesh reinforces concrete. Because individual joints are not mortared, walls are built more

quickly and are easier for unskilled workers to install. The coating of surface-bonding mortar provides water resistance.

Radon

Radon is a colorless and odorless radio-active gas that travels through soil. According to the U.S. Environmental Protection Agency, radon can be extremely toxic to humans if it builds up inside a house. Long-term exposure to radon has been linked to an increased risk of lung cancer. All types of house foundations, including concrete slabs, should be designed to reduce penetration by radon.

Because house foundations are in direct contact with the soil, they are a common entry point for radon. Radon enters through floor and wall cracks, expansion joints, gaps around pipes, and even through the pores in concrete. Because radon is soluble in water, it can also enter a basement through water seepage and through water vapor.

In addition, radon is nine times heavier than air, so it tends to accumulate in basements. Air circulation and other forces help to distribute radon throughout a house.

Radon-Resistant Construction Radon can be found in every area of the country, but it does not affect every region or every house equally. Therefore, steps should be taken during the foundation construction of every house to minimize radon problems. The following protective features are common:

Gas-permeable layer This is a 4" thick layer of drainage gravel directly beneath the floor slab. It allows radon to move freely beneath the house. A 4" thick layer of sand, topped with geotextile fabric, is an alternative.

Soil-gas retarder Polyethylene sheeting 6-mil thick is placed on top of the gas-permeable layer. This prevents radon from moving through the slab.

Sealants All openings and joints in the foundation floor are sealed to reduce radon entry. Sealant techniques include the use of high-performance caulks as well as plastic covers over sump pits.

Vent pipe A 3" or 4" diameter PVC pipe is connected to the gas-permeable layer. It leads to the roof. The pipe acts as an exhaust to safely vent radon outside the house.

Cap course Concrete block foundation walls must **incorporate** either a continuous course of solid masonry, a continuous course of concrete, or one course of masonry-grouted solid. This prevents radon from moving through the hollow cores of the block.

Building codes in some parts of the country require the use of radon-resistant foundation techniques. Always check local codes for specific construction requirements. For more on radon-resistant construction, see Chapter 11, "Concrete Flatwork," page 299.

Section 10.3 Assessment

After You Read: Self-Check

1. List three advantages of concrete block foundation walls as compared to solid concrete walls.
2. Name the type of packaged mortar that is most suitable for regions where earthquakes occur.
3. What is a story pole and how is it used when laying concrete block?
4. What is parging and what is its purpose?

Academic Integration: Mathematics

5. **Estimating Block** Estimate the number of 8" blocks you would need for a foundation wall that measures 20' in length and 8' in height. Use these facts to guide your estimation:
   - One block, including a ⅜" mortar joint, measures 8" in height. Therefore, 3 blocks will stack 2' high.
   - One block, including a ⅜" mortar joint, measures 16" in length. Therefore, 3 blocks will lay out 4' in length.

   Math Concept Many estimating problems can be solved using basic addition, subtraction, multiplication, and division operations. It helps to visualize a problem if you are using mental math.

   Step 1 Since 3 blocks lay out 4' in length, you must multiply the length of the wall (20') by ¾ to determine the number of blocks needed for every course. Another way to multiply by ¾ is to multiply by 3, then divide by 4.

   Step 2 Since 3 blocks stack 2' high, multiply the height of the wall (8') by ½ to determine the number of courses.

   Step 3 Multiply the number of blocks in one course by the number of courses to determine the number of 8" blocks that will be needed for the wall. Round up to the nearest whole number.

Go to connectED.mcgraw-hill.com to check your answers.

Chapter Summary

Section 10.1
Footings provide a base that supports a foundation wall, a pier, or a post. Footings can be reinforced with rebar. Footing drains help prevent damp basements.

Section 10.2
Foundation walls may be full-height for basements or shorter for crawl spaces. Poured foundations require forms. Rebar can be added for extra strength. Dampproofing and waterproofing must be done where moisture is a potential problem.

Section 10.3
Concrete block walls do not require formwork. The blocks are fairly inexpensive. Mortar holds the blocks together. Block walls require protection from moisture.

Review Content Vocabulary and Academic Vocabulary

1. Use each of these content vocabulary and academic vocabulary words in a sentence or diagram.

Content Vocabulary

- footing (p. 256)
- wales (p. 264)
- cold joint (p. 266)
- head joint (p. 275)
- bed joint (p. 275)

- story pole (p. 281)
- control joint (p. 283)
- parging (p. 287)
- radon (p. 288)

Academic Vocabulary

- framework (p. 258)
- incorporate (p. 289)

Speak Like a Pro

Technical Terms

2. Work with a classmate to define the following terms used in the chapter: *monolithic wall* (p. 256), *haunch boards* (p. 258), *form brackets* (p. 258), *stepped footings* (p. 259), *hydrostatic pressure* (p. 262), *perimeter drains* (p. 262), *filter fabric* (p. 263), *fines* (p. 263), *full-height foundation* (p. 264), *crawl-space foundation* (p. 264), *stem walls* (p. 264), *snap-ties* (p. 265), *insulating concrete forms* (ICFs) (p. 265), *stinger* (p. 267), *vapor retarder* (p. 267), *anchor bolt* (p. 268), *masonry ledge* (p. 269), *weep hole* (p. 269), *dampproofing* (p. 271), *course* (p. 275), *cell* (p. 275), *pilaster* (p. 275), *common bond* (p. 276), *full bedding* (p. 276), *stack bond* (p. 276), *face shell* (p. 280), *course pole* (p. 281), *tie bar* (p. 283), *cap block* (p. 286), *surface bonding* (p. 288).

Review Key Concepts

3. Summarize the function of the footing.

4. Describe the exact location of the footing.

5. Explain the process for forming concrete foundation walls.

6. Describe two types of foundation walls.

7. List the steps in laying a concrete block wall.

8. Describe the types of concrete block used in concrete block walls.

Critical Thinking

9. Explain Which type of foundation would you recommend to an individual planning to build a house in the southern United States who wished to conserve costs as much as possible? Explain the reasons for your recommendation.

Academic and Workplace Applications

STEM Mathematics

10. Estimating Labor Costs On average, it takes between 4.7 and 5.2 man-hours per 100 blocks to lay, clean, and joint $8 \times 8 \times 16$ concrete block. How much time would it take 4 workers to build a foundation estimated to require 1,200 such blocks?

Math Concept A man-hour is the amount of work one individual can do per hour.

Step 1: Determine how many man-hours it will take to complete 1,200 blocks given that it will take one man between 4.7 and 5.2 hours to complete 100 blocks.

Step 2: Divide the number of man-hours by the number of workers on the job.

STEM Science

11. Soil Drainage Where the soil drains poorly, where the water table is high, or where living spaces will be located below grade, greater efforts must be made to protect the foundation of a house. Find out more about soil drainage. Write a paragraph explaining why soil in some environments drains poorly.

21st Century Skills

12. Career Skills Rumblestone Concrete will be undertaking a large concrete project this winter. They will begin work in December, when temperatures are likely to be near or below 32° F. Their new foreman, Jim, has only conducted work in warm climates. You must help Jim ensure that the project is a successful one by outlining some strategies to protect the freshly placed concrete from freezing temperatures. Write a one-page letter to Jim in which you give him advice on how to best protect the concrete.

> **Standardized TEST Practice** — CERTIFICATION PREP

Short Answer

Directions Write one or two sentences to answer the following questions.

13. How far below grade should footings be placed?

_____

_____

_____

_____

14. What risks are present when freshly laid concrete block walls are exposed to temperatures at or below 32°F?

_____

_____

_____

_____

15. How do you prevent radon from moving through the hollow cores of block in a concrete block foundation wall?

_____

_____

_____

_____

> **TEST-TAKING TIP**
>
> *If each item on a test is worth the same number of points, do not spend too much time on questions that are confusing. If you do not know the answer to a question, make a note and move on. Come back to that question later, after you have answered all the other questions.*

*These questions will help you practice for national certification assessment.

Concrete Flatwork

Section 11.1
Foundation Slabs

Section 11.2
Finishing Flatwork

Chapter Objectives

After completing this chapter, you will be able to:

- **Identify** the two types of foundation slabs.
- **Describe** foundation slab details and reinforcement.
- **Explain** how to place a slab.
- **List** forms of flatwork other than foundations.
- **Demonstrate** the steps in finishing flatwork.
- **Explain** how temperature extremes affect fresh concrete.

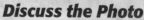

Discuss the Photo
Concrete Foundation The foundation of a building is placed in the first stage of construction. *Why do you think concrete is used for building foundations?*

Writing Activity: Descriptive Writing
Look around your school and a neighborhood that interests you. What types of structures do you see that are made with concrete? Write a one-paragraph description of at least two of these structures.

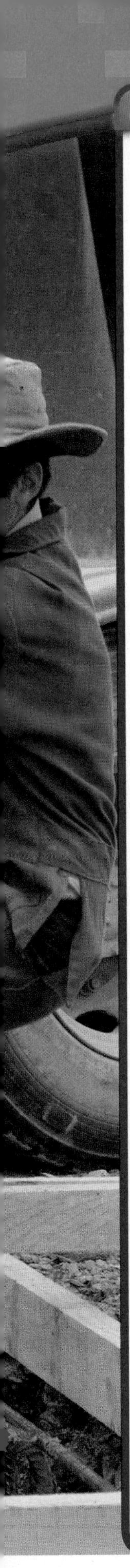

Before You Read Preview

Concrete slabs are widely used as the foundation for houses in mild climates. Choose a content vocabulary or academic vocabulary word that is new to you. When you find it in the text, write down the definition.

Content Vocabulary

- concrete flatwork
- frost depth
- monolithic slab
- independent slab
- subgrade
- lift
- fines
- screed
- bull float
- kneeboard

Academic Vocabulary

You will find these words in your reading and on your tests. Use the academic vocabulary glossary to look up their definitions if necessary.

- layer
- occurs
- adjusted

Graphic Organizer

As you read, use a chart like the one shown to organize terms and their definitions. Keep notes on how each product or process is used, adding rows as needed.

Term	Definition	Use
monolithic slab	A footing and a floor slab that are formed in one continuous pour	Used in warm climates where frost is not a problem Used in areas where termite infestations are common
kneeboard	Board measuring about 12" by 24" placed on the concrete to support the weight of the finisher	Allows the finisher to move from one area to another without stepping onto the fresh concrete

Go to **connectED.mcgraw-hill.com** to download this graphic organizer.

Foundation Slabs

Foundation Slab Basics

What is a monolithic slab?

Concrete flatwork consists of flat, horizontal areas of concrete that are usually 5" or less in thickness. Flatwork is placed either directly on the ground or over compacted gravel or sand. Examples of flatwork include foundation slabs, basement floors, driveways, and sidewalks. Concrete flatwork must be contained by forms until it is strong enough to hold its shape without forms.

Once a slab has been placed, its top surface must be finished. This means that the surface must be smoothed, textured, or otherwise worked using a combination of hand and power tools. Then, steps must be taken to ensure that the concrete cures properly. Improper curing reduces the strength of the concrete and can cause various other problems, such as cracking.

Concrete flatwork is generally installed by subcontractors who specialize in this work. This chapter will provide a general introduction to the topic.

Flatwork is commonly used in residential construction for foundation slabs in houses built without a basement or a crawl space. In mild or warm climates, a foundation slab has these advantages:

- Excavation costs are reduced because very little earth must be removed.
- Extensive or complex formwork is not required.
- A concrete slab eliminates the need for a separate subfloor.
- Construction costs are lower.
- The concrete provides a solid base for concrete block walls, which are sometimes used in warm climates as the exterior walls of the house.

A foundation slab also has some disadvantages. The primary disadvantage is that utilities must be planned carefully and roughed-in in advance, as shown in A Slab Under Construction. Changes are very difficult once the slab has been placed.

Frostline, Moisture, and Soil

The foundation of a house must always be deep enough so that it rests on soil that is *below* the local frost depth. The **frost depth** is the depth in any climate below which the soil does not freeze. Frost depth is also called *frostline* or *freezing depth*. If moisture is in the soil beneath a foundation, it can actually lift the foundation when it freezes and expands. When the soil thaws, it will return to its

 A Slab Under Construction
Planning Ahead Pipes for the rough plumbing are cast into the floor as the concrete is placed during pouring.

United States Department of Agriculture - Forest Service, Northern Region previous volume. This cycle of expansion and contraction is an important consideration in foundation design because the movement, called *frost heave,* can cause a foundation to crack.

The type of soil is another factor that can affect foundation design. Various types of soil, particularly those containing clay, tend to hold moisture. These are called *expansive soils.* Other soils, particularly those with a high sand content, do not hold moisture. They are sometimes called *free-draining soils.* Because expansive soils hold water and are unstable, they should be avoided under and around a foundation. The best soil to have under a foundation is one that drains freely.

Types of Foundation Slabs

There are two types of foundation slabs: monolithic and independent. The choice to use one or the other depends largely on climate and local custom. In both cases, the minimum thickness required by code is 3½". A thickness of 4" is more common.

A **monolithic slab** consists of a footing and floor slab that are formed in one continuous pour. It is also referred to as a unified slab, a thickened-edge slab, or a slab with a turned-down footing. The perimeter of the slab is thicker than the main area, as shown in A in Types of Slab Foundations on page 296. It is strengthened with rebar at the edges. The bottom of the footing should be at least 1' below the natural grade line and supported by solid, unfilled, well-drained soil. A monolithic slab is useful in warm climates where frost penetration is not a problem and where soil conditions are favorable. It is also preferred in areas where termite infestations are common.

An **independent slab**, also called a *ground-support slab,* is a slab that is used in areas where the ground freezes fairly deep during winter. The house is supported by foundation walls that extend to solid bearing below the frostline. The slab is then poured between the foundation walls. (Foundation

walls are discussed in Chapter 10, "Foundation Walls.") After the foundation wall formwork has been removed, the slab can be placed. One method for laying this type of slab is shown in B in Types of Slab Foundations on page 296.

Slab Details

Slab details include support, formwork, drainage, and reinforcement. Always check local codes for requirements that relate to concrete slabs. This is especially important where earthquake hazards must be considered.

Support for Bearing Walls Exterior bearing walls are supported either by the thickened edge of the slab or by foundation walls. Beneath interior bearing walls, the slab may be thickened to provide the necessary support, as shown in Support for Bearing Walls on page 297. This thickened area is like a footing. It should be strengthened with rebar. Unlike a standard footing, the thickened area is formed by a trench, not by formwork.

Formwork A slab foundation does not require much formwork. For a monolithic slab, builders often use lumber to form the slab edges. Foundation contractors may prefer reusable metal or wood forms. In any case, the outer edges of the forms must be braced to resist the pressure of the wet concrete.

REGIONAL **CONCERNS**

Slab or Basement? Where winter temperatures are fairly mild, slab foundations are often more cost-effective than other types of foundations. This is because a slab does not require the deeper extensive excavation that is needed for basement foundations. Full basements are not common in mild climates for this reason.

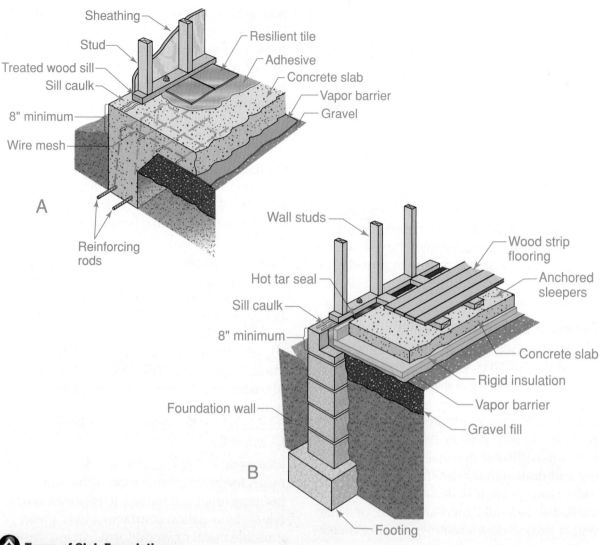

Sheathing
Stud
Treated wood sill
Sill caulk
8" minimum
Wire mesh
Resilient tile
Adhesive
Concrete slab
Vapor barrier
Gravel

A

Reinforcing rods

Wall studs
Hot tar seal
Sill caulk
8" minimum
Foundation wall
Wood strip flooring
Anchored sleepers
Concrete slab
Rigid insulation
Vapor barrier
Gravel fill

B

Footing

Types of Slab Foundations
Two Versions In a monolithic slab **A.**, the footing and the slab are placed at the same time. In an independent slab **B.**, the slab is placed after the foundation wall has been placed. *What are the main factors in choosing which type of slab foundation to use?*

After the concrete has partially cured, the forms can be removed.

Drainage The finish-floor level should be high enough above the natural ground level that finish grade around the house can be sloped away for good drainage. The top of the slab should be no less than 8" above the ground and the siding no less than 6".

A perimeter drain should be placed around the outside edge of the exterior wall footings, as shown in Drainpipe. The drain helps keep ground moisture from wicking into the slab. Drain lines are not always required by code where the floor is located

on fairly high ground, where subsoil is well drained, or in a very dry climate. Placement of the perimeter drain is similar to that for a full-height foundation (see Foundation Drains on page 262 in Chapter 10).

Reinforcement Metal reinforcement is often placed in a concrete slab to increase its tensile strength and reduce cracking. It also helps to prevent slabs from separating if they do crack. Welded-wire mesh keeps slabs from separating once the concrete has cracked. This reinforcement can consist of either rebar or welded-wire mesh fabric. Wire fabric is a grid of horizontal and vertical wires. It comes

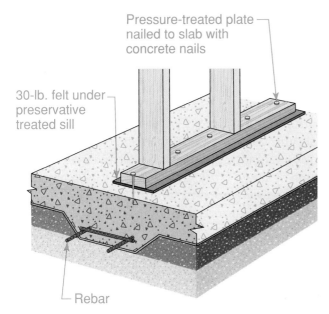

Pressure-treated plate nailed to slab with concrete nails

30-lb. felt under preservative treated sill

Rebar

Support for Bearing Walls
Thickened Slab A foundation slab should be thickened beneath bearing walls to support the walls and prevent cracking.

Drainpipe
Perforated The holes in a drainpipe should face downward. They allow water in the pipe to be drained away.

in rolls and is cut to size at the site. Any reinforcement should be in place before the concrete is poured.

Code requirements for rebar vary, depending on the type of slab and its location. However, with a monolithic slab, either one #5 or two #4 bars should be located in the middle third of the footing's depth. Vertical lengths of rebar are sometimes added to reinforce the thickened portion of the slab. In earthquake areas, reinforcement is important. Local codes should be followed carefully.

Wire fabric should be placed near the center of the slab thickness. Contractors sometimes roll out the fabric over the excavation. They then use a rake or hook to pull it up into the concrete during the pour. However, this method makes it difficult to tell the exact position of the reinforcement. A more precise method is to support it on chairs. A *chair* is a small metal or plastic device that supports the wire fabric at a particular height (see Rebar Chair on page 230). Chairs are left in place as the concrete is poured. The IRC code requires the use of chairs when a slab is reinforced. The chair must support the

reinforcement so that it will stay in place at or slightly above the center of the slab when the concrete is placed.

Insulation

In some climates, a foundation slab can feel uncomfortably cold if it is not insulated. The best insulation for slabs is rigid, nonabsorbent boards or sheets, such as extruded or expanded polystyrene. It can be placed around the perimeter of a monolithic slab, where the concrete is exposed to colder temperatures. For independent slabs, rigid insulation can be placed between the foundation walls and the edge of the floor slab, as shown in B in Types of Slab Foundations on page 296. Studies have shown that this edge insulation is important in reducing the amount of heat lost by conduction. Insulation may also be placed below the slab. This is especially important when radiant heating tubes are built into the slab (see Chapter 30, "Mechanicals").

Termite Protection Some areas of the United States have problems with termites. In these areas, special care must be taken to prevent

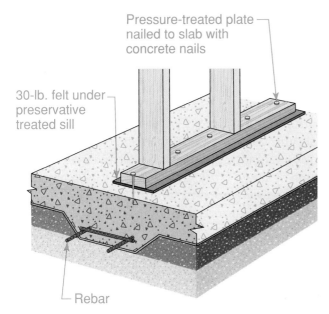panic_attack/Getty Images

termites from getting into the wood framing above the concrete. One method is to chemically treat the soil before placing a slab. The chemicals, their strength, and the application methods are determined by local and state building officials. Their guidelines should be followed carefully.

Physical barriers, such as metal termite shields, should also be included. On monolithic slabs, shields should be located between the slab and the wall plate. For independent slabs, this barrier is continued to cover the gap between the slab and the foundation wall.

Where the chance of termite infestation is very high, the IRC building code prohibits the use of foam plastic insulation on the outside of foundation walls and beneath slabs below grade. This is because insects may tunnel through the insulation to reach wood framing.

Installing a Concrete Slab

A foundation slab is installed by foundation subcontractors. The following outlines the basic steps they use.

Preparing the Subgrade The **subgrade** is the earth below the slab. The subgrade must be uniformly compacted (pressed down) to prevent any uneven settlement of the floor slab. Uneven settlement is a common cause of cracks in concrete.

All organic matter, such as sod and roots, should first be removed and the ground leveled off. Any holes or cracks in the subgrade should be filled and compacted. Material for fill should be uniform. It should not contain large lumps, stones, or material that will rot. Any fill should be compacted in lifts no more than 6" deep. A **lift** is a uniform and fairly shallow **layer** of material. If fill is compacted in thicker layers, it may appear to be firm on the surface but it will not be uniformly firm. A *power tamper*, or plate tamper, is often used to compact fill. This tool is powered by

a gasoline engine and guided by hand over the area. Areas that are difficult to reach with this equipment can be compacted with a hand tamper.

After any holes are filled, the entire subgrade should be thoroughly compacted by tamping or rolling. The finished subgrade should then be carefully checked for height and levelness. Any variations can create a slab of uneven thickness. When this **occurs**, the slab may not cure evenly, which is another common cause of cracking.

A Hand Tamper
Tight Spaces Compact coarse fill with a hand tamper.

Soil cannot be properly compacted if it is too wet or too dry. You can get a rough idea of the proper moisture content of ordinary soils, except very sandy ones, by squeezing some in your hand. With proper moisture content, the soil will cling together but will not be *plastic* (clay-like) or muddy. If the soil is too dry, it should be sprinkled with water before compacting. If the soil is too wet, it must be allowed to dry.

Providing for Other Trades Electrical conduit, ducts for heating systems, and plumbing supply and waste lines can be placed in trenches cut in the subgrade. Care should be taken to protect water supply lines from freezing if the building will not be occupied during cold weather. Careful planning ensures that connections to these utilities can be made where specified on the building plans.

After a monolithic slab has been poured and partially finished, anchor bolts are inserted around the perimeter. Carpenters will use these to secure wall framing to the slab.

Preparing the Subbase Coarse fill should be placed over the compacted subgrade to form the subbase, or base course. This fill should consist of coarse slag, gravel, or crushed stone no more than 2" in diameter. The fill particles should be of uniform size to prevent them from packing together tightly. If necessary, the material should be sifted through a screen to remove any fines. **Fines** are finely crushed or powdered materials. The subbase, along with drainage pipes at the perimeter of the foundation, helps to drain water that might collect under the slab. When the slab is below grade, the subbase must be at least 4" thick. The fill should be brought to the desired grade and then thoroughly compacted.

Installing Vapor Retarder Water will not penetrate good quality concrete unless the water is driven by pressure (see Chapter 10, "Foundation Walls"). A properly constructed drainage system will prevent pressures from building up beneath a foundation slab. However, concrete can be penetrated by water vapor. If vapor passes through a slab, moisture can cause problems inside the house. For example, flooring surfaces glued to the slab may loosen. To prevent this, a 6-mil polyethylene vapor retarder with joints lapped at least 6" must be placed between the slab and the subbase. Standard polyethylene may be used. However, cross-laminated polyethylene is more durable. In either case, workers should be warned not to puncture the membrane when placing the concrete.

One disadvantage of a vapor retarder is that it forces moisture in the fresh concrete to escape through the exposed top surface. This can cause shrinkage cracks in the slab surface, as well as other problems. For this reason, it is sometimes recommended that concrete not be placed directly on the vapor barrier. Instead, a 3" thick layer of sand can be spread over the vapor barrier and compacted. The concrete can then be placed over the sand.

Radon Control Radon is a colorless, odorless radioactive gas given off by some soils and rocks. In some parts of the United States, the seepage of radon into houses poses a health threat (see Chapter 10).

Houses built in areas with radon seepage must be built to resist radon entry. In some houses, the combination of a granular subbase and a carefully installed vapor barrier is enough. However, where concentrations are very high, stronger methods are needed.

REGIONAL CONCERNS

Radon Hazard Areas The Environmental Protection Agency (EPA) has developed a map showing which areas of the country have the highest radon risk. This map is in the IRC. Broadly, states located in a band from the Southeast through Texas are least likely to have radon problems. Alaska, Hawaii, and portions of the Pacific Northwest are also considered low-risk areas. However, homes with elevated levels of radon have been found in all states. Homes should be tested regardless of geographic location or zone designation.

One such method is the sub-slab ventilation system shown in Radon Control System. This is called a *passive venting system* because no fans are involved. An *active venting system* uses one or more fans to move the air.

Placing the Concrete Concrete for the floor slab and bearing-wall footings should be made with durable, well-graded aggregate. It must have a compressive strength of at least 2,500 psi. The concrete should be workable so it can be placed without developing large air pockets (honeycombing) or excess water on the surface. However, too much water should not be added to the concrete just to make it easier to place. This reduces its strength. If necessary, the proportion of fine and coarse aggregate should be **adjusted** to obtain a more workable mix. Another way to increase workability is to add an admixture

to the concrete (see Chapter 8, "Concrete as a Building Material").

Concrete should not drop more than approximately 4' to the ground as it is delivered by the ready-mix truck, as shown in Placing Concrete. A greater drop can cause large aggregate to settle unevenly,

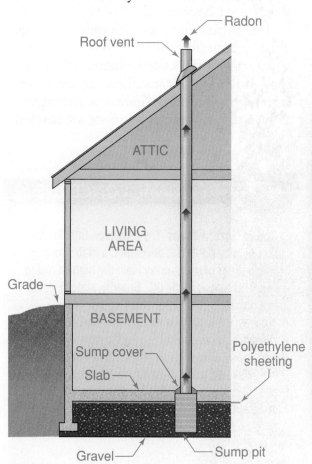

Radon Control System
Passive Venting One method of venting radon gas from beneath a slab.

Placing Concrete
Maximum Chute Height The end of the concrete chute should be within approximately 4' of ground level.

James Hardy/Photo Alto

weakening the concrete. Extension chutes, temporary ramps, or methods such as pumping prevent this problem. After placing, the concrete should be made to settle by vibrating, tamping, or spading. Then it should be finished. The steps in finishing the surface will depend upon the floor finish specified. You will learn more about this in Section 11.2.

Reading Check

Explain *When placing the concrete, why is it important not to add too much water?*

Other Types of Flatwork
What is concrete used for?

There are other types of concrete slabs besides ones that are used as a foundation. Basements typically include a concrete floor slab. The slab is placed after all improvements, such as sewer and water lines, have been connected. Concrete is also often used for walks and driveways, especially where snow removal is important.

Basement Floors

Basement floor slabs should be no less than 3½" thick and should slope toward a floor drain. There should be at least one drain in a basement floor. For large floors, two drains may be required.

When concrete is placed in an enclosed area, such as a basement, the foundation walls serve as forms. However, the concrete still must be leveled to the correct thickness. This is done by means of rail-like devices on which a screed rides, as shown in Screeding Concrete. A **screed** is a long, straight length of metal or wood that is used to "strike off" (level) the concrete. The screed is pulled by hand across the top of the rails. At the same time, the screed is moved back and forth in a sawing motion. The rails are made of sections of 1" pipe set on stakes driven into the sub-grade. The pipes used as rails are called screed strips. The stakes are driven deep enough so that when the pipes are set on them, the tops of the pipes will be at the level desired for the surface of the slab. After screeding, the pipes and stakes are removed. A float is then used to pack concrete into any gaps. You will learn more about this in Section 11.2.

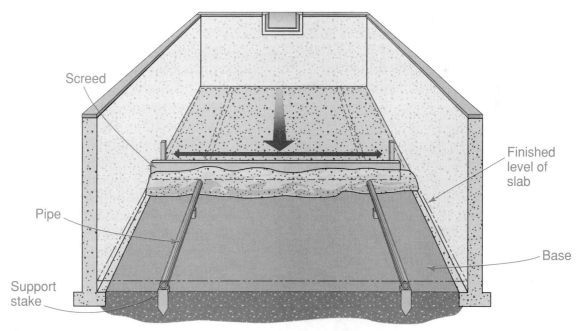

Screeding Concrete
Maintaining Thickness Metal pipes may be used to guide the screed when leveling a basement floor.

Driveways

The grade, width, and radius of curves in a driveway are important when establishing a safe entry to the garage. Driveways that have a grade of more than 7 percent (7' rise in 100 lineal feet) should not be covered with gravel because the gravel will gradually wash away. Concrete or asphalt is often used instead. A common type of concrete driveway is the full-width slab. The concrete should be given a broom finish to prevent both cars and people from slipping. It should also be slightly crowned. This means it should be slightly higher at the center than at the edges. This allows it to drain properly.

A gravel base is not ordinarily required on sandy, undisturbed soil. If the area has been recently filled, the fill, preferably gravel, should settle first and be well tamped. A gravel base should be used on all other soils. The concrete should be about 5" thick, and a vapor barrier is not required. Side forms are often built of 2×6 boards. These members establish the elevation and alignment of the driveway. They are also used to support the screed used to strike off the concrete.

Though not required, the addition of 6×6 wire fabric reinforcing reduces pavement cracking. Expansion joints with asphalt-saturated felt strips inserted should be used where the driveway joins the public walk or curb and at the garage slab. They should also be used about every 40' on long driveways. Concrete containing an air-entraining agent should be used in areas having severe winter climates. Air-entraining produces tiny air bubbles that help the concrete resist damage during freeze/thaw cycles.

Sidewalks and Walkways

Concrete sidewalks are constructed in much the same way as concrete driveways. They should not be poured over filled areas unless the fill has settled and is well tamped. This is especially true of areas near the house after basement excavation backfill has been completed.

Minimum thickness of concrete over normal undisturbed soil is usually 4". Control joints should be used and spaced on 4' centers. A *control joint* is a joint that helps to minimize random cracks in a concrete slab. By creating a slightly weakened area, it encourages a crack to form in a straight line, rather than across the concrete in an irregular line. Note that it does not prevent the concrete from cracking. It simply controls the location of the crack. An expansion joint may also be required for sidewalks. An *expansion joint* is a gap between portions of concrete that is filled with a flexible material. The concrete is thus able to expand and contract without damage to itself or to adjacent surfaces.

When slopes to the house are greater than 5 percent, a walkway should have steps, as shown in A Steep Walkway. Steps make the walkway easier to use. The riser (vertical portion) of each step is sometimes formed by 2×6 lumber. In effect, each platform of a stepped walkway is formed like an individual slab.

5 percent maximum slope

A Steep Walkway
Steps for Safety Steps in a steep walkway make it safer and more comfortable to use.

Estimating and Planning

This estimating and planning exercise will prepare you for national competitive events with organizations such as SkillsUSA and the Home Builder's Institute.

Concrete for Flatwork

Calculating Volume

The volume of concrete needed for flatwork is based on the size of the area covered and the thickness of the slab. The result is expressed in cubic yards.

1. To calculate the amount of material required for the basement floor of the home shown in the floor plan below, first figure the area of the slab. The house measures 26' × 40', so the total area of the basement slab will be 1,040 square feet.

2. The floor will be 4" thick. **Table 11-1** shows that at a thickness of 4", one cubic yard of concrete covers 81 sq. ft. To calculate the total amount of concrete required, divide the total slab area (1,040 sq. ft.) by the number of square feet covered by one cubic yard (81):

$$1,040 \div 81 = 12.39$$

If you round this off, you will need 13 cu. yds. of concrete. When estimating, you should always round up.

Another way to calculate the volume of concrete is to use the following formula:

length in feet × width in feet × thickness in feet ÷ 27 = cubic yards

To use the formula, each dimension must be in the same unit. For example, if you calculate the volume of concrete for a slab that is 20'-6" long, 10' wide, and 4" thick, convert all dimensions to feet, using decimal equivalents:

$$20.5' \times 10' \times 0.33' = 67.65 \text{ cu. ft.}$$

Then divide by 27 to obtain cubic yards:

$$67.65 \div 27 = 2.50 \text{ cu. yds., or } 2\tfrac{1}{2} \text{ cu. yds.}$$

Estimating on the Job

Using Table 11-1, estimate the number of cubic yards of concrete needed for a basement floor that measures 20' by 30' and is 4" thick. Assume that 1 cubic yard of concrete can be placed in .43 hours. Estimate the time it would take to place the concrete.

Table 11-1: Estimating Materials for Concrete Slab	
Material	
Thickness (inches)	Square Feet from One Cubic Yard
2	162
3	108
4	81
5	65
6	54

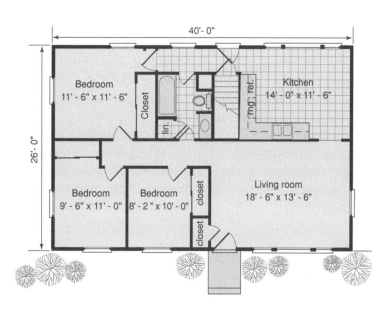

After You Read: Self-Check

1. What is concrete flatwork?
2. Why is it important to compact the subgrade in lifts?
3. What is the purpose of the subbase?
4. What is the formula for calculating cubic yards of concrete?

Academic Integration: Mathematics

5. **Using Formulas** Calculate the cubic yards of concrete required for the basement floor for a home that is 36' wide, 45' long, and 6" thick.

> **Math Concept** A formula is an equation that gives the relationships between the numbers in a group of numbers. A letter or word is often given to represent the value. In this chapter, you are given the following formula to calculate volume in cubic yards:
> Volume = (length in feet × width in feet × thickness in feet) ÷ 27 = cubic yards.

Go to **connectED.mcgraw-hill.com** to check your answers.

Section 11.2 Finishing Flatwork

Finishing a Slab

What is the first step in finishing flatwork?

A flatwork surface, such as a floor, driveway, or sidewalk, must be finished. Finishing is a multiple-step process that levels the concrete and gives it a surface that will suit the intended use. A basement or foundation slab, for example, is given a very smooth finish. This makes the concrete easy to clean and allows finished floors to be attached directly to the surface. A broomed, non-slippery finish is best for sidewalks, while a somewhat coarser surface suits driveways. Various decorative finishes are also available.

Steps in Finishing

Finishing is a multiple-step process that may includes screeding, bullfloating, edging, jointing, floating, and troweling. Some steps, such as jointing, are not required on every job.

Screeding The first step in finishing any flatwork is screeding. The concrete is *struck off* just after it is placed in the forms. The screed rides on the edges of the side forms or on wood or metal strips set up for that purpose. Two people move the screed along the slab, using a sawing motion. Screeding may also be done with mechanical equipment. It leaves a level surface with a coarse finish.

⬆ **Screeding**
Team Effort Screeding a concrete slab for an office building

Bullfloating Bullfloating makes the concrete surface more even with no high or low spots. A **bull float** is a wide, flat metal or wood pad that is pushed back and forth over the concrete to make the surface even. A long handle enables the worker to reach every area of the slab. A similar tool, called a *darby,* has a somewhat shorter handle.

Bullfloating is done shortly after screeding, while the concrete is still wet enough to allow a slight paste of mortar to be brought to the surface. However, there must be no water visible on the concrete. Otherwise, an excess amount of fines and moisture will also come to the surface. This is one of the principal causes of defects. Among other problems, it causes fine hairline cracks (*crazing*) or a powdery material (*dusting*) on the surface. Only enough bullfloating should be done to remove defects and to bring enough mortar to the surface of the slab to produce the desired finish.

Edging and Jointing When the sheen has left the surface and the concrete has started to stiffen, other finishing operations can be done. Edging produces a rounded edge on the slab to prevent chipping or damage, as shown in Edging a Slab on page 306. The edger is run back and forth, covering coarse aggregate particles.

Immediately following edging, larger slabs are jointed, or grooved. Sometimes shrinkage stresses are present in the slab as a result of temperature changes or dryness. These stresses can cause the concrete to crack. Joints reduce the thickness of the slab and cracks are then likely to occur only at these weakened points. When the concrete shrinks, these joints open slightly, preventing other uneven and unsightly cracks. A jointing tool is used to cut the control joints about ¾" deep in the slab. The joints should be perpendicular to the slab's edge. To ensure straight joints, it is good practice to guide the jointer with a straight 1×8 or 1×10 board, as shown in Jointing a Slab. A crooked joint detracts from the appearance of the finished slab.

Hand Floating In some cases, an additional floating step is done, using wood or metal floating trowels, or floats. Hand floating further evens the surface of the concrete. It also compacts the surface mortar in preparation for the next finishing steps. It produces a very even surface with a light texture. Hand floating also removes any ridges left by jointing tools. Aluminum or magnesium floats must be used when hand floating air-entrained concrete because wood floats stick to the concrete surface. If floating is the last step in finishing, it may be necessary to float the surface a second time after the concrete has hardened slightly.

Troweling For a dense, smooth finish, floating is followed by troweling with a steel trowel. For large areas, a power trowel is used instead of a hand trowel, as in Power Troweling on page 307. Troweling cannot be started until the concrete has hardened enough to prevent fine material and water from working to the surface. In fact, troweling should be planned carefully.

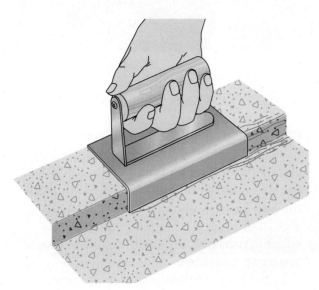

 Edging a Slab
Rounded Edge An edger smoothes the edge of a slab.

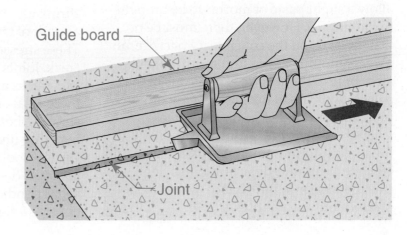

Guide board

Joint

 Jointing a Slab
Stress Relief A jointer is guided by a board or some other straight edge.

A surface that is troweled too early will not be durable. A surface that is troweled too late will be difficult to finish properly.

Hand or power troweling should leave the surface smooth, even, and free of marks and ripples. For a fine-textured surface, the first troweling is immediately followed with a second. In this second operation, the trowel is held flat and passed lightly over the concrete with a circular motion.

For a hard steel-troweled finish, the second troweling should be delayed until the concrete has become hard enough to make a ringing sound under the trowel. In hard steel-troweling, the trowel is tilted slightly. Heavy pressure is applied to compact the surface.

Another type of finishing tool is called a *fresno*, a steel trowel that it is attached to a long handle. It enables concrete finishers to

Hand Troweling
Smoothing a Surface Hand troweling a concrete slab.

Power Troweling
Super Smoother Power troweling a concrete slab.

Builder's Tip

Stamped Concrete
Brick Pattern This concrete was colored and stamped with a pattern shortly after it was placed.

trowel the concrete without having to use kneeboards. A **kneeboard** distributes the weight of a finisher over the surface of the concrete, as shown in Using a Kneepad.

However, the finisher cannot put as much pressure on a fresno as on a hand trowel. Therefore, a concrete surface finished with a fresno will not be as dense as concrete finished by hand or power troweling.

Special Finishes

Concrete can be given a color, a pattern, or a texture. This is done after the concrete has been placed but before it has cured. Patterns come from metal stamps that are pressed into the wet concrete, as shown in Stamped Concrete. Colors come from various types of masonry dyes. In some cases, small, smooth pebbles can be scattered into the fresh concrete after bullfloating. They

are pressed into place during later finishing operations.

A *broom finish* roughens the surface of the concrete slightly so that it will be more slip-resistant. To create the finish, a stiff bristle broom can be dragged over the surface in parallel strokes. This should be done after bleed water is completely gone from the surface. Brush strokes should be perpendicular to the grade. A *rubbed finish* creates the opposite effect. Shortly after the concrete has hardened and the forms have been removed, the surface is rubbed with abrasives to create a uniform surface that is very smooth. A *salt finish* is created by broadcasting rock salt (a type of very coarse salt) onto fresh concrete. A roller is then used to press the salt into the surface. The salt should be washed away after the concrete has set. This leaves a surface filled with shallow, angular indentations.

Curing Flatwork

After finishing, the concrete should be kept moist for at least two days. This ensures that hydration will continue. It also improves the concrete's strength. When the finished floor is to be exposed concrete, at least five days of moist curing are required. Burlap, canvas, or a waterproof concrete curing paper may be used to cover the floor slab during this period. If burlap or canvas is used, it should be kept wet by sprinkling it with water. Curing should begin as soon as the concrete is hard enough to make damage unlikely. Chemicals can also be used to coat concrete for curing. By slowing the rate at which moisture leaves the concrete, they help to increase its strength.

When the concrete has cured enough to withstand foot traffic, wall plates can be laid out and construction can continue.

Temperature Extremes Problems can be caused in concrete by temperature extremes that occur in the early stages of curing. This is because so much surface area is exposed.

Placing concrete in unusually hot weather can reduce its strength. Hot conditions also encourage workers to add more water to the mix, which further reduces its strength. In hot weather, the water and aggregates should be kept as cool as possible before being mixed with the cement. Forms, rebar, and the subgrade should be cooled by sprinkling them with water just before the concrete is placed. In some cases, it may be wise to place the concrete early in the morning, or even at night, to avoid very hot temperatures. Moist curing is particularly important under these conditions and should be started as soon as possible.

In moderately cold weather, the heat of hydration is usually enough to prevent damage. However, concrete placed in temperatures below freezing can suffer a loss of strength unless protected. In fact, if concrete is frozen shortly after being placed, it can lose up to 50 percent of its strength. Concrete can be protected by placing it in insulated forms or by covering it temporarily with insulation. High-early strength, air-entrained, and low-slump concrete can also be used to counteract such conditions.

Section 11.2 Assessment

After You Read: Self-Check

1. What is the purpose of a bull float?
2. What defects can be caused by bullfloating too soon?
3. When is a concrete slab ready for jointing and edging?
4. What are kneeboards used for?

Academic Integration: English Language Arts

5. **Imperative Mood** In English verbs, the *indicative mood* is used to show an act, state, or occurrence is actually happening, while the *imperative mood* is used to give commands. The unstated subject of an imperative sentence is "you." The street sign "STOP" and the phrases "Come here" or "Don't do that" are examples of imperative sentences. Directions are often given in the imperative mood. Use your own words to write a brief description of two steps of the finishing process. Use the imperative mood.

Example:
Indicative mood: "The concrete should be kept moist for at least two days."
Imperative mood: "Keep the concrete moist for at least two days."

Go to connectED.mcgraw-hill.com to check your answers.

Review and Assessment

Chapter Summary

Section 11.1

Concrete foundation slabs are often used in warm or mild climates to provide a foundation and a subfloor for houses. Proper preparation of the subgrade and subbase are important to a high-quality job. Slabs are more prone than foundation walls to being damaged by temperature extremes.

Section 11.2

Finishing a slab is a process of preparing the surface for various end uses. After the concrete is placed, the excess is screeded off the slab. After any water has disappeared, bullfloating, edging, jointing, floating, and troweling can take place.

Review Content Vocabulary and Academic Vocabulary

1. Use each of these content vocabulary and academic vocabulary words in a sentence or diagram.

Content Vocabulary

- concrete flatwork (p. 294)
- frost depth (p. 294)
- monolithic slab (p. 295)
- independent slab (p. 295)
- lift (p. 298)

- subgrade (p. 298)
- fines (p. 299)
- screed (p. 301)
- bull float (p. 305)
- kneeboard (p. 308)

Academic Vocabulary

- layer (p. 298)
- occurs (p. 298)
- adjusted (p. 300)

Speak Like a Pro

Technical Terms

2. Work with a classmate to define the following terms used in the chapter: *frost heave* (p. 295), *power tamper* (p. 298), *plastic* (p. 299), *struck off* (p. 304), *fresno* (p. 307), *broom finish* (p. 308), *rubbed finish* (p. 308), *salt finish* (p. 308).

Review Key Concepts

3. Define the two types of foundation slabs.

4. Describe various types of slab reinforcement.

5. Summarize the steps in placing a slab.

6. List two forms of flatwork other than foundations.

7. Identify the tools needed to finish flatwork.

8. Illustrate how temperature extremes affect fresh concrete.

Critical Thinking

9. Synthesize How does metal reinforcement strengthen a concrete slab?

Academic and Workplace Applications

STEM Mathematics

10. Slope/Intercept Form Calculate the slope of a regular concrete wheelchair ramp that is 10 feet long. The bottom of the ramp is horizontal to the ground and the top of the ramp is 4 feet from the ground.

Math Concept Slope is the measure of a line's slant. Slope is given by the ratio of rise (vertical movement) to run (horizontal movement).

$$\text{Slope} = \frac{\text{rise}}{\text{run}}$$

Slope is often represented as a fraction. The fraction can be reduced to lowest terms. A horizontal line and a vertical line both have no slope (the rise over run is $\frac{0}{0}$). If a ramp is three feet high after you have gone six feet up the ramp, the slope of the ramp is $\frac{3}{6}$. Reduced to lowest terms, the slope is $\frac{1}{2}$.

Step 1: Substitute the appropriate numbers for rise and run.

Step 2: Reduce the fraction to lowest terms.

21st Century Skills

11. Communication Skills: Flow Charts A flow chart is a type of visual aid that represents a system or a process. Flow charts often use symbols, pictures, or other graphic elements to represent actual steps or things. For example, you might use the arrow symbol to indicate that one task should be completed after another task. You might use a picture of a tool or a color to indicate who might be performing a particular task. Imagine that you are supervising a team of three finishers, Jude, Frederick, and Sita, who all speak different languages. Use a graphics program or art materials to create a one-page flow chart that shows the steps they must follow when finishing a basement floor with a fine-textured surface. Assign each worker specific tasks.

STEM Science

12. Frostline The colder the climate, the deeper the frost depth. However, different factors, including geothermal energy (energy from the earth) can affect the frostline. What is the frost depth in your area?

Starting Hint Consult local building codes.

Standardized TEST Practice

Multiple Choice

Directions Select the word, phrase, or amount that best answers the question.

13. According the Environmental Protection Agency, which state is considered a low-risk state for radon risk?

  a. Texas

  b. Florida

  c. Hawaii

  d. Georgia

14. What is the minimum possible thickness that is recommended for basement floor slabs?

  a. 1"

  b. 3"

  c. 1'

  d. 3'

15. Which type of finish is best for sidewalks?

  a. salt finish

  b. broomed finish

  c. rubbed finish

  d. fresno finish

TEST-TAKING TIP

Reread all questions and answers containing measurements, abbreviations, or symbols. Make sure you have selected the answer with the correct symbol.

*These questions will help you practice for national certification assessment.

Checking for Square

Your Project Assignment

One of a carpenter's most important jobs is to lay foundations square. In this project you will determine if a basketball court, volleyball court, or soccer field near your home is laid square.

- **Research** the regulation size for a basketball court, volleyball court, or soccer field.
- **Interview** a local carpenter about tools and processes for measuring and checking for square.
- **Measure** the length and width of the court or field, then predict and measure the diagonal.
- **Use** the 3-4-5 method to check the square area.

Applied Skills

Your success in carpentry will depend on your skills. Some skills you might use include:

- **Research** the standard sizes for athletics courts and fields.
- **Locate** and interview a local carpenter about measuring and checking for square.
- **Explain** the purpose of the 3-4-5 method.
- **Compare** calculated length with measured length.
- **Create** a sketch of a court or field with clearly labeled dimensions.

The Math Behind the Project

The traditional math skills for this project are geometry and measurement. Remember these key concepts:

Geometry

The Pythagorean theorem states that the sum of the squares of the sides of a right triangle is equal to the square of the hypotenuse. This is expressed as the equation $a^2 + b^2 = c^2$. To find the length of the hypotenuse (c), we take the square root of both sides of the equation:

$$\sqrt{a^2 + b^2} = c$$

The diagonal of basketball court divides it into two right triangles. The diagonal is the hypotenuse of the triangles.

Units of Measurement

When we solve for the equation above using measurements in feet, our answer will often contain a decimal. We need to convert the decimal to inches and fractions of an inch. First multiply the decimal by 12 to convert the decimal to inches. Then multiply the decimal from this calculation by 16 to convert it to $1/16$ of an inch.

For example, if a court is 24' wide and 54' long, we can calculate the length of the diagonal in feet, inches, and fractions of inches using the following steps:

1. Use the Pythagorean theorem to calculate the diagonal.	$\sqrt{24^2 + 54^2} = 59.093$ ft
2. Multiply the decimal by 12 to convert it to inches.	$12 \times .093$ ft. $= 1.116$ in.
3. Multiply the decimal from the above step by 16 to convert it to $1/16$ of an inch. Round.	$16 \times .116 = 1.856 \approx 2$
5. Write the total as inches and feet. Reduce the fraction if possible.	59 ft., $1^2/_{16}$ in. = 59 ft., $1^1/_8$ in.

To convert inches and fractions of inches to a decimal, reverse the steps shown above.

Project Steps

Step 1 Research

- Research the regulation size of a basketball court, volleyball court, and soccer field.
- Determine if the courts or field at your school or in your neighborhood are regulation.
- Contact the local of the United Brotherhood of Carpenters and Joiners of America. Locate a local carpenter who is available for an interview.
- Write interview questions about methods of checking for square.

Step 2 Plan

- Select the court or field you will check. If no court is available, choose another rectangular area, such as a meeting room or parking lot.
- Sketch the area and indicate which dimensions you will measure.
- Choose a measuring tool, such as a measuring tape, a rigid ruler, or a scrap-lumber measuring stick.
- Interview the carpenter you located. Take notes.

Step 3 Apply

- Measure the length and width of the area to the nearest $1/16$ of an inch. Mark these on your sketch.
- Use the Pythagorean theorem to predict what the diagonal must be if the area is a perfect rectangle. Convert fractions of inches to decimals. Then use the equation $\sqrt{a^2 + b^2} = c$ to solve for the diagonal (c). Convert the decimal in your answer to sixteenths of an inch.
- Measure the diagonal. If it is longer or shorter than your prediction, the corners are not exactly 90°.
- Check the court for square again, this time using the 3-4-5 method. Start in one corner. Measure down one side 3 feet and down the other side 4 feet. Then measure the diagonal line between these two points. If the court is square, how long should the diagonal be?
- Compare the results of your two methods of checking for squareness. Did they yield the same result—square or not square? If not, recheck your measurements and calculations.

United Brotherhood of Carpenters and Joiners of America

Mission: To represent and offer training to North America's carpenters, cabinetmakers, millwrights, piledrivers, lathers, framers, floorlayers, roofers, drywallers, and workers in forest-products and related industries.

Step 4 Present

Prepare a report combining your research, calculations, and measurements using the checklist below.

PRESENTATION CHECKLIST
Did you remember to…
✓ Demonstrate the research you conducted? .
✓ Discuss the tools and methods you used to measure?
✓ Explain the differences between measured length and calculated length?
✓ Use and present your sketch?
✓ Explain how you determined whether the court was square?

Step 5 Technical and Academic Evaluation

Assess yourself before and after your presentation.

1. Did you prepare well for the interview?
2. Did you calculate the diagonal correctly?
3. Was your sketch legible and neat?
4. Were your measurements accurate and precise?
5. Was your presentation clear and creative?

Go to **connectED.mcgraw-hill.com** for a Hands-On Math Project rubric.

UNIT 4

Wood Frame Construction

In this Unit:

Hands-On Math Project Preview

Construction Calculations

After completing this unit, you will use trigonometry to calculate the rafter lengths for various design options and create a table to present the options. You will also calculate the total rise of specific roof angles and spans.

Project Checklist

As you read the chapters in this unit, use this checklist to prepare for the unit project:

✓ List the materials used in roof framing.

✓ Describe the different styles of roof design.

✓ Identify how to calculate rafter length and rise.

➤ Go to connectED.mcgraw-hill.com for the Unit 4 Web Quest activity.

©Huntstock, Inc/Alamy

Construction Careers Framing Carpenter

Profile A framing carpenter is a type of structural worker. Framing carpenters assemble the basic structural features that are later finished by other specialty carpenters.

Academic Skills and Abilities ....

- mathematics
- blueprint reading
- geometry
- interpersonal skills
- mechanical drawing

Career Path ....

- on-the-job training
- apprenticeship programs
- trade and technical school courses
- certification

Explore the Photo

Wood Frame Construction Framing carpenters construct the frame of the building. *Why do you think teamwork and communication are an important to a framing carpenter?*

CHAPTER 12

Wood as a Building Material

Section 12.1
Wood Basics

Section 12.2
Protecting Wood

Chapter Objectives

After completing this chapter, you will be able to:

- **List** the advantages of wood as a building material.
- **Define** hardwoods and softwoods.
- **Describe** flat-sawn and quarter-sawn boards.
- **Explain** how the moisture content of wood is controlled.
- **Identify** common defects in lumber.
- **Summarize** conditions and factors that lead to wood damage.

Discuss the Photo

Natural Resources Wood is a renewable resource. This means that wood that has been consumed or used can be replaced by reproduction. *Why should human beings help renew natural resources such as forests?*

Writing Activity: Collaborative Writing

Team up with two other classmates. Brainstorm a list of ideas about how lumber is used in structures. Then, work with your team to write one or two paragraphs summarizing your ideas.

© moodboard/Corbis

Chapter 12 Reading Guide

Before You Read Preview

Wood is used more than any other material in the construction of a house. It is easy to work with and extremely versatile. Choose a content vocabulary or academic vocabulary word that is new to you. When you find it in the text, write down the definition.

Content Vocabulary

- cambium
- deciduous tree
- coniferous tree
- lumber
- moisture content (MC)
- fiber-saturation point
- seasoning
- kiln
- grade
- grade stamp
- warp
- nominal dimension
- dry rot
- wood preservative

Academic Vocabulary

You will find these words in your reading and on your tests. Use the academic vocabulary glossary to look up their definitions if necessary.

- species
- ecosystems
- photosynthesis
- equilibrium

Graphic Organizer

As you read, use a two-column chart like the one shown to organize types of trees.

Hardwood	Softwood

Go to **connectED.mcgraw-hill.com** to download this graphic organizer.

Wood Basics

The Value of Wood

Why must we take care of our forests?

For all-around utility, wood has no equal as a building material. It can be used to form most of the house's frame and many interior and exterior surfaces. It is also a key element in the construction of doors, windows, cabinetry, stairs, and other features. Wood is versatile and readily available. It can be cut to different sizes and formed into many different shapes.

Because wood has been used for thousands of years, carpenters, builders, architects, and others know a great deal about how it performs. Even so, new types of wood and wood-based materials are being developed all the time. For more on these materials, see Chapter 13. This chapter will focus on solid wood.

Solid wood is used for many types of construction for several reasons:

- Wood is strong. Certain common framing woods are as strong and rigid as some types of steel.
- Wood is easily fastened with nails, staples, bolts, connectors, screws, or glue.
- Wooden buildings are easily altered or repaired. Openings can be cut and additions made without difficulty.
- Wood has low heat conductivity, which helps to reduce heat loss.
- Wood accepts decorative coatings such as paint and stains.
- Wood resists acids, saltwater, and other corrosive agents better than many other structural materials.
- Wood is a renewable resource.

Our Forest Resources

Most of the wood used in the United States is harvested from millions of acres of forestland spread across North America. Many years ago, trees were cut without regard for the effect cutting would have on the forest itself and the surrounding areas. However, many people now understand that we must take greater care of forests across the globe. The benefits include:

- Continued ability to harvest lumber
- Protection of important water resources
- Soil conservation
- Production of oxygen and absorption of carbon dioxide
- Preservation of wildlife and plant habitats (environments) and endangered species (types of organisms)
- Maintenance of scenic areas

Build It Green If forests are managed properly, there will be no shortage of wood. Individual trees may grow back, but forest ecosystems are fragile and may require thousands of years to reach a mature stage.

How Trees Grow

The growing, "working" parts of a tree are shown in How a Tree Grows on page 319. They include the tips of its roots, the buds, the leaves, and a thin layer of cells just inside the bark called the cambium. The **cambium** is a layer of living tissue that produces new wood, called *sapwood*, along its inner surface. New bark is created along the cambium's outer surface. Sapwood enables water and nutrients from the tree's roots to reach its leaves. As successive layers of sapwood build up around the tree, the layers nearest the center gradually turn into a nonliving material called *heartwood*. Heartwood does not contribute to growth, but it gives strength and rigidity to the tree.

Oxygen

Carbon dioxide

Water

Food

Heartwood

Cambium

Outer bark

Sapwood

Inner bark

How a Tree Grows
A Growth System A tree depends on all its elements in order to grow.

Water from the soil enters a tree through its roots. The water travels upward through the sapwood into the leaves. Through the process of **photosynthesis**, carbon dioxide in the atmosphere and water are combined in the presence of chlorophyll and sunlight. This provides food to nourish the whole tree. This food is carried from the leaves to the rest of the tree through the inner bark. Oxygen is released through the leaves as a byproduct of photosynthesis.

After a tree has been harvested, its life story can be read in the *annual rings* (growth rings) of the stump. In temperate climates, the tree adds one annual ring during each year of growth. Most annual rings consist of a light band formed in the spring (early wood) and a dark band formed in the summer (late wood). When growth conditions are good and food and water abundant, the rings are wide. When long dry spells or other adverse conditions occur, growth slows and the rings are narrow. Annual rings are visible in the end grain of lumber, as shown in Growth Rings. Experienced builders sometimes study these rings to determine the suitability of the lumber for various uses.

Growth Rings
History Growth rings are visible on the ends of boards. *When are growth rings narrow?*

Hardwoods and Softwoods

The terms *hardwood* and *softwood* identify woods based on the two main types of trees. The terms do not indicate actual softness or hardness of the wood. In fact, some hardwoods, such as balsa wood, are softer and less dense than many softwoods. Some softwoods, such as yew, are harder than some hardwoods. Hardwoods are cut from broad-leaved, deciduous trees. A **deciduous tree** is a tree that sheds its leaves annually, during cold or very dry seasons. In the United States and other countries with temperate climates, most broad-leaved trees are deciduous and lose their leaves in the fall. Some common hardwoods are walnut, mahogany, maple, birch, cherry, oak, and ash. Softwoods are those that come from coniferous trees. A **coniferous tree** is a tree that produces seeds in cones and has needle-like or scalelike leaves. Common examples of coniferous trees are pine, hemlock, fir, cedar, and redwood trees.

Wood for construction has traditionally been used in the general region in which the logs were harvested. **Table 12-1** lists the North American softwoods that are used most for wood products. However, the global economy and improvements in shipping have made wood products from one region available in other regions. For example, tropical hardwoods from Central and South America are found increasingly in North American markets. Only those tropical woods purposely grown on tree plantations (farms) can be considered renewable resources.

Reading Check

Recall *How long might it take to replace a forest ecosystem?*

Processing Lumber

What is a nominal dimension?

When a tree has been cut down and its limbs have been removed, the result is called

Table 12-1: Principal Commercial Softwoods

Common Commercial Names	Alternate Names
Cedar	
Alaska cedar Eastern red cedar Incense cedar Northern white cedar Port Orford cedar Southern white cedar Western red cedar	Southern red cedar Atlantic white cedar
Cypress	
Cypress	Bald cypress Pond cypress
Fir	
Balsam fir Douglas fir Noble fir White fir	Fraser fir Subalpine fir California red fir Grand fir Pacific silver fir
Hemlock	
Eastern hemlock Mountain hemlock West Coast hemlock	Carolina hemlock Western hemlock
Juniper	
Western juniper	Alligator juniper Rocky Mountain juniper Utah juniper
Larch	
Western larch	None
Pine	
Jack pine Lodgepole pine Norway pine Ponderosa pine Sugar pine Idaho white pine Northern white pine Longleaf yellow pine Southern yellow pine	Red pine Western white pine Eastern white pine Longleaf pine Slash pine Loblolly pine Longleaf pine Pitch pine Shortleaf pine
Redwood	
Spruce	
Eastern spruce Engelmann spruce Sitka spruce	Black spruce Red spruce White spruce Blue spruce
Tamarack	
Yew	
Pacific yew	None

a log. Logs are sawn lengthwise into smaller pieces at a mill. These pieces of wood have a uniform thickness and width and are referred to as **lumber**.

Cutting Boards From Logs

The way in which a board is cut from the log can affect its appearance and performance. Two methods are commonly used: flat-sawing and quarter-sawing.

Flat-Sawn Lumber Most construction lumber is *flat-sawn* lumber, as shown on page 322. At the mill, a log is squared up lengthwise, and then sawn into boards. As you look at the end grain of a flat-sawn board, you can see that the growth rings run across the board's width. As you look at the face, you can see a distinctive archlike pattern. Flat-sawn lumber is relatively inexpensive. Flat sawing produces boards of greater width than other cutting methods. However, such boards are more likely to shrink and warp.

Quarter-Sawn Lumber *Quarter-sawn* lumber is a premium wood. At the mill, a log is first sawn lengthwise into quarters. Boards are then cut from the faces of each quarter, as shown on page 322. Looking at the end grain of a quarter-sawn board, you can see that the growth rings run across the thickness of the board. These growth rings generally form angles of 60° to 90° to the board's surface. Quarter-sawn boards with end grain at angles between 30° and 60° are referred to as rift-sawn boards.

Quarter-sawn boards have a low tendency to warp, shrink, or swell. They also provide a more durable surface than flat-sawn lumber. They do not tend to twist or cup. They hold paints and finishes better. However, quarter-sawn lumber is more expensive and less plentiful than flat-sawn lumber and is not generally carried by home centers. In addition, it is generally not carried by do-it-yourself building supply outlets.

Controlling Moisture Content

The amount of water wood contains is referred to as its **moisture content (MC)**. It is expressed as a percentage of what the wood would weigh if it were completely dry. For example, assume that a block of wood that has just been cut from a tree weighs 60 lbs. After being dried in an oven, it weighs only 50 lbs. Thus the original piece contained 10 lbs. of water. That is 20 percent of the wood's dry weight ($10 \div 50 = .20$). The lower the percentage, the drier the wood.

Fiber-Saturation Point A living tree takes in a lot of water. The tree stores water first in the cell walls. When a tree's cell walls have absorbed all the water they can hold, the wood is at the **fiber-saturation point**. For most woods, the fiber-saturation point occurs when the wood contains about 28 percent moisture, though this number can vary. If the tree takes in additional water, it stores that water in the cambium cell cavities.

Removal of water from the cell cavities of harvested wood has no apparent effect upon its properties except to reduce its weight. For this reason, drying the wood until its moisture content is roughly 28 percent does not result in shrinkage. However, reducing moisture to less than 28 percent will remove water from the cell walls, causing the wood to shrink in all directions.

Seasoning The process of drying wood is called **seasoning**. There are two methods of seasoning wood: air drying and kiln drying.

In *air drying,* the rough lumber is stacked outdoors in layers separated by thin wooden cross-pieces called *stickers.* The lumber remains stacked from one to three months or longer. After air drying, the lumber has an average moisture content of 19 percent or less.

In *kiln drying,* the lumber is also stacked in layers with stickers between. It is then placed in a kiln. A **kiln** is an oven in which moisture, airflow, and temperature are carefully controlled. Properly kiln-dried (KD) lumber has less than 10 percent moisture content. In a kiln, drying may take less than four days.

Even though lumber is dried to a certain moisture content, it continues to absorb or give off water, depending on the humidity

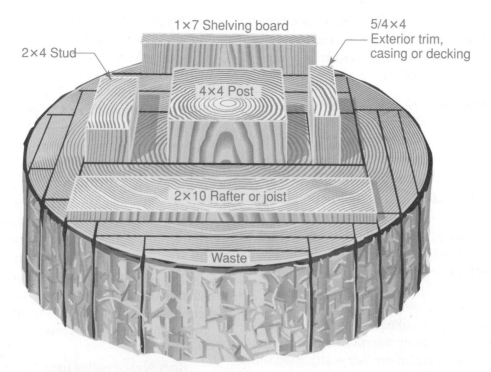

Flat-Sawn Lumber

Rings across the Width Notice the orientation of growth rings on the ends.

2×4 Stud

1×7 Shelving board

5/4×4 Exterior trim, casing or decking

4×4 Post

2×10 Rafter or joist

Waste

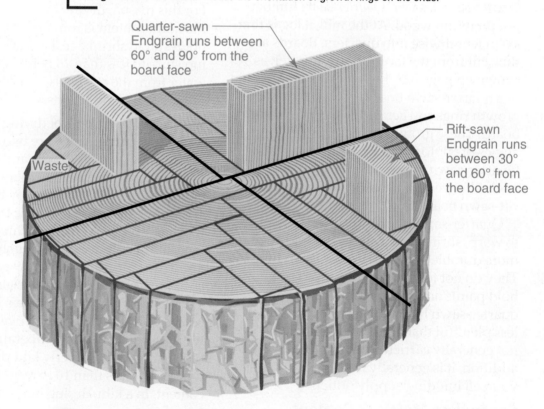

Quarter-sawn Endgrain runs between 60° and 90° from the board face

Rift-sawn Endgrain runs between 30° and 60° from the board face

Waste

Quarter-Sawn Lumber

Rings Across the Thickness Compare the direction of grain on the ends of these boards with the directions on those above.

of the surrounding air. If the air is damp, dry wood absorbs moisture and swells. If the air is very dry, the wood shrinks. If the air alternates between moist in the summer and dry in the winter, as it does in many parts of North America, wood expands and contracts. This is sometimes referred to as *seasonal expansion.*

Lumber shrinks in both width and length. Shrinkage in length is usually so small that it is not considered a problem. Shrinkage across the width of the board, however, can be more troublesome. The shrinkage of studs, for example, can cause drywall nails to pop. The shrinkage of floorboards can cause gaps to appear in the flooring, as shown in Wood Shrinkage. In wood maintained at a consistent moisture content, swelling and shrinkage are kept to a minimum. This may require the builder to temporarily condition a home once it has been dried-in to reduce the cracking and shrinking of any installed finished millwork.

The size of a board will vary about 1 percent for each 4 percent change in moisture content. When the moisture content of the wood is in balance with the humidity of the surrounding air, it neither gains nor loses moisture. At that point, the moisture content is said to have reached **equilibrium**. A portable moisture meter, is the most common instrument for checking the moisture content of wood.

On-Site Storage Lumber kept outside at the job site should be supported off the ground and stacked with stickers (thin pieces of wood) between layers. It should be covered loosely with waterproof material.

Lumber kept indoors will absorb or lose moisture until it reaches a balance with the moisture of the air in the room. Flooring and wood paneling should be delivered at least several days in advance of installation to allow the wood to reach equilibrium. Storing the materials in the room prior to installation is called *conditioning.*

Grading Lumber

Lumber is graded according to various characteristics of the wood. **Grade** is a general indication of the quality and strength of a piece of lumber. Being able to identify grades is an important skill for carpenters and other building professionals.

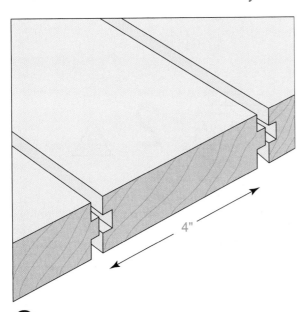

Wood Shrinkage
The Cause of Gaps If the moisture content is excessive in floor boards, gaps will appear between the boards as the wood dries.

A Moisture Meter
Checking Moisture An electronic moisture meter can be used to check the moisture content of wood.

Hardwood Grades Hardwood is used where beauty or durability is important, such as for door and window casings, stair treads, balusters, handrails, and cabinetry. Hardwoods are available in three common grades, *firsts and seconds (FAS)*, *select*, and *No. 1 common*. Each kind of hardwood lumber is graded by a slightly different standard. Generally, firsts and seconds are used for built-ins, fine casework, and paneling.

Softwood Grades Grading standards for softwoods have been developed by governmental agencies in cooperation with producers, distributors, and users of softwoods. According to these standards, softwood lumber is divided into two basic groups:

- Green (unseasoned) lumber with a moisture content of more than 19 percent.
- Dry (seasoned) lumber with a moisture content of 19 percent or less.

Each major lumber trade association has developed a complete set of grading standards. These rules are extensive and vary somewhat with each association. The builder who uses lumber primarily from one section of the country should understand the grading used in that area. For example, if most of the lumber used comes from the western states, then the standards published by the Western Wood Products Association would apply. If the lumber comes from southern states, standards set by the Southern Forest Products Association would apply. Familiarity with common lumber abbreviations will simplify the selection and specifications of softwood lumber. For information, refer to the **Ready Reference Appendix** table "Lumber Abbreviations".

After a softwood board has been graded, it is marked with a grade stamp, as shown below. A **grade stamp** is a permanent

(A) WWPA Certification Mark: Certifies Association quality supervision. Ⓦ is a registered trademark.

(B) Grade Designation: Grade name, number, or abbreviation.

(C) Species Identification: Indicates species by individual species or species combination.

(D) Mill Identification: Firm name, brand, or assigned mill number. WWPA can be contacted to identify an individual mill whenever necessary.

(E) Condition of Seasoning: Indicates condition of seasoning at the time of surfacing.
S-GRN. Over 19 percent moisture content (unseasoned).
MC15 or KD15. 15 percent maximum moisture content.
S-DRY or KD. 19 percent maximum moisture content.

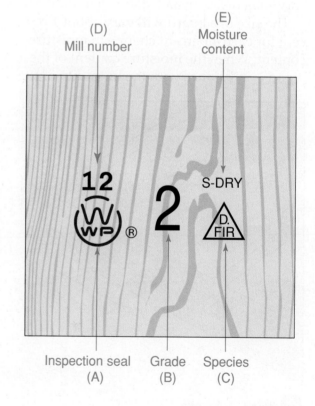

 A Grade Stamp

Knowing the Board Most grade stamps contain information similar to the information in this stamp from the Western Wood Products Association.

mark that identifies the board's species, quality, mill source, and gives a general indication of strength. Grade stamps let building inspectors know if a house is being built with lumber of suitable quality. Information on the specific grades of framing lumber used in residential construction can be found in Chapter 14.

Lumber Defects

A lumber defect is a flaw that detracts from the quality of the wood in either appearance or usefulness. About 25 characteristics and conditions are considered when wood is graded. They are described in any set of grading rules. Some of the more common defects are shown and described in Lumber Defects on pages 326–327. The term **warp** is a general description of any variation from a flat surface. It includes *bow, crook,* and *cup,* or any combination of these.

When selecting lumber, the carpenter must examine each board for defects. However, boards do not always have to be perfect to be useable. It is important to consider the exact use required of the wood. For example, a minor defect along one edge of a board would not be a problem if the board was to be ripped to a slightly narrower width. Likewise, a large knot near the end of a 12' board would not be a problem unless the carpenter needed the full length of the board. Sometimes a flawed board can be set aside for cutting into shorter lengths. For example, carpenters often cut flawed 2×4 stock into blocking.

Standard Sizes of Lumber

The width and thickness of lumber are given by two types of measurements: *nominal dimensions* and *actual dimensions*. The **nominal dimension**, such as 2" by 4" (2×4), is the size of the board, in inches, as originally cut. Nominal dimensions, or nominal sizes, refer to the width and thickness of rough-sawn lumber (not its length).

After the board has been surfaced and seasoned at the mill, its *actual dimension* becomes less than its nominal dimension. Building codes use actual dimensions when they address maximum allowable sizes for drilling holes and notching. *Dressed sizes* apply after the wood has shrunk and been surfaced with a planing machine. The width and thickness of dressed lumber are less than its nominal width and thickness. For example, a 2×4 stud actually measures about 1½ × 3½. For the differences between nominal and actual sizes, see the **Ready Reference Appendix** table "Standard Sizes for Framing Lumber, Nominal and Dressed" on the OLC.

Sizes of lumber used in building construction have been standardized for convenience in ordering and handling. Softwood lumber is readily available in actual lengths of 8', 10', 12', 14', and 16'. Common nominal dimensions are 2", 4", 6", 8", 10", and 12" in width; and 1", 2", 4", and 6" in thickness.

Hardwoods are not standardized for length or width. They run ¼", ½", 1", 1¼", 1½", 2", 2½", 3", and 4" in thickness.

Metric Sizes The United States is the only major lumber-producing country that does not exclusively use the metric system of measurement. However, wood intended for export to other countries may be sized to metric measurements. In addition, wood produced in other countries and imported to the United States for certain uses may be sized to metric measurements. Thickness and width are given in millimeters (mm), length is given in meters (m).

Science: Life Science

Photosynthesis Green plants make their own food. Through photosynthesis, glucose, a form of sugar, is formed. Oxygen is released into the atmosphere as a byproduct. Where does the glucose go?

Starting Hint Sugar is food.

Looking for Trouble A piece of lumber could have one or more of these defects.

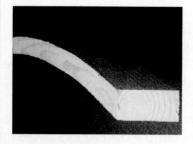

Bow A flatwise deviation (bend) along the grain from a straight, true surface. Bow is measured at the point of greatest deviation.

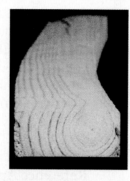

Crook An edgewise deviation (bend) from a straight, true surface. Crook is measured at the point of greatest deviation.

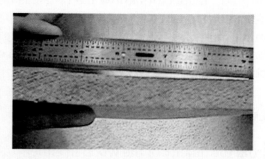

Cup A flatwise deviation (bend) across the grain from a straight, true surface. Cup is measured at the point of greatest deviation.

Check A small crack that runs across the growth rings, parallel with the grain. Check usually occurs as a result of seasoning.

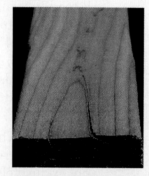

Shake A lengthwise grain separation between or through the growth rings. It may be further classified as *ring shake* or *pitch shake*.

Split A lengthwise separation extending from one surface through the piece of lumber to the opposite side or an adjoining surface.

Decay Disintegration of wood due to the action of wood-destroying fungi. It also may be called *dote*, *rot*, or *unsound wood*.

Knot Place where a branch once grew. Knots are classified according to size, quality, and occurrence. To determine the size of a knot, average the maximum length and maximum width, unless otherwise specified. A sound encased knot and a sound intergrown knot are shown above.

(l to r, t to b)(1–5, 8)Jon Muzzarelli; (6 & 7)Denise McCullough; (9)Design Pics/Ken Welsh

Knothole Hole left by the removal of an embedded knot.

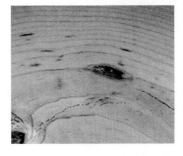

Pitch Accumulation of resin in the wood cells in a more or less irregular patch.

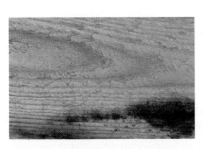

Stain Discoloration on or in lumber other than its natural color.

Torn grain Torn spot created as the board is machined to size.

Section 12.1 Assessment

After You Read: Self-Check

1. What is the difference between hardwoods and softwoods?
2. Compare the advantages of quarter-sawn lumber with those of flat-sawn lumber.
3. What is KD lumber?
4. What causes the moisture content of lumber to reach a point of equilibrium?

Academic Integration: Science

5. **Measurement Systems** Common lumber sizes are almost identical in both the customary and metric systems. For example, the basic thicknesses are 25 mm and 1" (25.4 mm), and the basic widths are 100 mm and 4". Metric lengths range from 1.8 m (about 6') to 6.3 m in increments of 300 mm. Note that 300 mm is close to, but slightly shorter than, 1'. Estimate the nominal dimensions in millimeters of a 2" × 2" piece of softwood lumber.

Go to **connectED.mcgraw-hill.com** to check your answers

Protecting Wood

Section 12.2

Causes of Wood Damage

Which insects damage wood?

Wood must be protected from decay and from being damaged by certain insects. Both problems are more likely to occur when wood becomes wet.

Rain is the most obvious source for the introduction of water. A house must be designed to encourage rainwater to drain freely away. Water vapor is a less obvious source. Water vapor is given off during cooking, washing, and other household activities. This vapor can pass through walls and ceilings. When it reaches a cold surface, such as sheathing or studs in the winter, it condenses into water droplets. Leaking pipes are a third source of moisture. They should be fixed immediately to prevent damage to the house. Gaps around exterior wood trim or between a chimney and the siding can allow water to seep into walls. Such problems should be eliminated with regular maintenance, such as caulking all gaps.

Wood Decay

Wood used where it will always be dry, or even where it may be wetted briefly and promptly dried, will not decay. However, wood will decay if kept wet for long periods at temperatures favorable to the growth of decay organisms. Wood decay, or rot, is caused by *fungi* (such as mildew and mold) that use wood for food. These fungi require air, warmth, food, and moisture for growth. Damp wood provides an ideal environment for them.

In early stages, decay caused by fungi may show up as a discoloration of the wood. Paint also may become discolored where the underlying wood is rotting. Advanced surface decay is more easily recognized. Generally, the affected wood is brown and crumbly. Brown, crumbly decay is sometimes called **dry rot**. This is a misnomer because wood must be damp for rotting to occur. At other times the wood may be rather white and spongy. Decay inside the wood is often indicated by sunken areas on the surface or by a hollow sound when the wood is tapped with a hammer. Where the surrounding air is very damp, the decay fungus may grow out on the surface, appearing as white or brownish growths in patches or strands, as shown in Decay Fungus. In some cases, decay fungus has a vinelike structure. The presence of fungus stains or mold is a warning that conditions are or have been suitable for decay. Affected lumber should always be examined for decay damage before installation.

Preventing Decay Fungi grow most rapidly at temperatures between 70°F and 85°F (21°C and 29°C). High temperatures, such as those used in kiln-drying of lumber, kill fungi. Low temperatures, even far below 0°F (-18°C), merely cause them to become dormant. The best way to prevent fungi attack is to keep wood dry. Wood-destroying fungi cannot grow in dry wood. A moisture

 Decay Fungus
Damp Wood Damp conditions are essential for the growth of fungi on wood, such as this lichen.

Perry Mastrovito/Image Source

content of 20 percent or less is generally dry enough to prevent or stop growth.

Construction lumber that is improperly seasoned may be infected with one or more fungi and should be avoided. Such wood may contribute to serious decay in both the structure and exterior parts of buildings. You may see signs of infection in wood that is improperly seasoned. You can also determine the moisture content of such wood with an electronic moisture meter.

Decay-Resistant Woods When untreated, the sapwood of all common native woods has low resistance to decay. This gives it a short life under decay-producing conditions. The natural decay resistance of native woods lies in the heartwood. Of the species commonly used in house construction, the heartwood of bald cypress, redwood, and various cedars is highest in decay resistance. However, lumber made entirely from heartwood is becoming more and more difficult to obtain. This is because increasing amounts of timber are cut from the smaller trees of second-growth stands in which little heartwood has developed. In general, when decay resistance is needed in load-bearing members that would be difficult or expensive to replace, preservative-treated wood is used (see Chapter 35 for more information).

Reading Check

Recall Which wood species resist decay?

Insects

Under certain conditions, wood can be damaged by insects such as termites, carpenter ants, and beetles. Some of the conditions that encourage fungi, such as too much moisture, can also encourage wood-infesting insects.

Termites *Termites,* shown above, are the most destructive of the insects that infest wood. The best time to protect against them is during the planning and construction of the building. Remove all woody debris, such as stumps and lumber scraps, from the soil at the building site before and after construc-

 Termites
A Hazard to Wood Termites are the most destructive insect that affects wood.

tion. No wood member of the structure should be in contact with the soil. Forms of protection include chemical termiticide (termite-killing) treatments, physical barriers, natural termite resistant woods, or pressure-preservative treated wood.

Termites can be grouped into two main classes: subterranean and dry-wood termites. *Subterranean termites* account for about 95 percent of all termite damage. They eat the interior of the wood and can cause much damage before they are discovered. They honeycomb the wood with tunnels separated by thin layers of sound wood. They are common throughout Hawaii and the southern two-thirds of the United States, except in mountainous and extremely dry areas. The map on page 330 shows termite infestation probability for the contiguous United States.

Subterranean termites thrive in moist, warm soil containing a large supply of food in the form of wood or other material containing cellulose (plant fibers). In their search for additional food, they build shelter tubes over foundation walls, in cracks, or on pipes or supports leading from the soil into the house. These flattened tubes, from ¼" to ½" or more in width, protect the termites in their travels. Metal or masonry barriers such as those shown in Termite Shields on page 330 should be installed in areas where these termites are common.

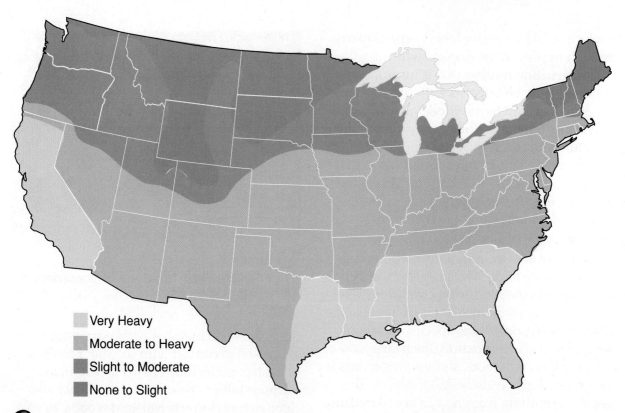

	Very Heavy
	Moderate to Heavy
	Slight to Moderate
	None to Slight

Termite Probability Map

Areas of Greatest Hazard Lines separating areas are approximate. Local conditions may be more severe than shown here.

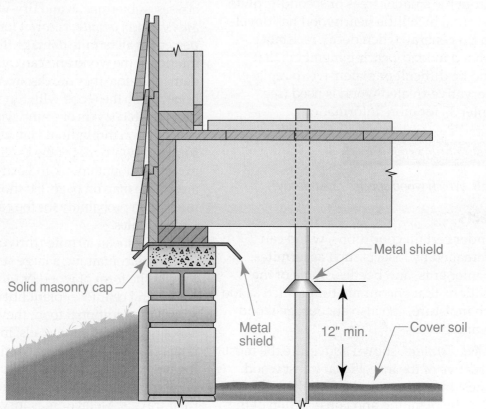

Solid masonry cap

Metal shield

Metal shield

12" min.

Cover soil

Termite Shields

Metal Blockers Two types of shields used to block termite access to wood.

An especially destructive type of subterranean termite is the *Formosan termite*. This termite was introduced into the United States after World War II, and it is now found in most of the southern states as well as California and Hawaii. Native species feed on dead trees and processed wood. Formosan termites will eat these plus anything else that contains wood fiber, including live trees and many plants. Because they are aggressive, live in very large colonies, and can survive on many different food sources, Formosan termites are more destructive than native species.

Dry-wood termites fly directly to the wood instead of building tunnels from the ground. They chew across the grain of the wood, creating broad pockets, or chambers. These chambers are connected by narrow tunnels.

The termites remain hidden in the wood and are seldom seen, except when they take flight. They are more difficult to control, but they cause less serious damage than subterranean termites. Dry-wood termites are common in the tropics. They have also been found in the United States along the Atlantic Coast from Virginia to the Florida Keys, westward along the coast of the Gulf of Mexico, and up the Pacific Coast as far as northern California.

Other Insects The insects shown in Other Destructive Insects can also cause wood damage.

Carpenter Ants *Carpenter ants* are a problem primarily in the Northeast, Midwest, and Northwest, though they can be found throughout the country. They nest in the ground as well as in dead trees, firewood, and houses. Carpenter ants do not eat wood.

▽ Other Destructive Insects

Wood Insects These insects can cause wood damage.

Carpenter Ants Destructive tunnelers that burrow into wood to create nests.

Carpenter Bee Expert drillers that commonly infest fascia boards and window trim.

Beetles The powderpost beetle, shown at left, and the deathwatch beetle, shown at right.

They eat plant juices, insects, honey, and food particles found inside a house. The damage they cause comes from the irregular tunnels they create in wood for their nests. Small sawdust piles may indicate their presence. Controlling carpenter ants can be difficult. To be most effective, chemical treatments must be applied to the nest itself.

Beetles Two types of beetles commonly infest wood: the powderpost beetle and the deathwatch beetle. The *powderpost beetle* is more common. It is second only to the termite in the amount of damage it causes. These beetles commonly enter the house via already-infested wood such as firewood, rough-sawn timbers, and barn wood. They attack hardwoods only, preferring ash, oak, mahogany, hickory, maple, and walnut. Most of the infestation occurs in sapwood that has a moisture content between 10 and 20 percent.

One common sign of a powderpost beetle infestation is a tiny pile of fine, flourlike powder. This material is pushed out of the wood as the beetles emerge. Infested wood should either be removed or professionally treated. Treating only the surface of the infested wood with an insecticide will not kill beetles deep within the wood.

Deathwatch beetles are larger beetles. They infest the sapwood of both hardwoods and softwoods. They are most likely to be found in wood with a high moisture content, such as wood used in unheated housing and damp crawl spaces. Treatment is similar to that required for powderpost beetles. In addition, it is very important to cut off the supply of moisture.

Carpenter Bees *Carpenter bees* resemble bumblebees in shape. The bee has a metallic blue-black body covered with yellow or orange hairs. They cut a ½" diameter hole in bare or untreated wood. They then build their nests by boring a tunnel parallel to the surface of the wood. Sawdust seen around a small, symmetrical hole indicates their presence. New generations of bees will return to the nest annually. Carpenter bees can be controlled with insecticides. However, one of the best ways to prevent infestation is to paint the wood.

Preservative Treatments

Lumber can be treated with liquid wood preservatives soon after being milled. This can increase the lumber's resistance to decay, insects, and fire. A **wood preservative** is a chemical that protects wood.

Decay-Resistant and Insect-Resistant Wood Various woods can be successfully treated, but fir, spruce, and pine are the most common. Preservative-treated wood is generally used outdoors or where it is in contact with concrete or masonry. In one treatment process, the wood is dipped into chemicals and then air-dried. In another process, the chemicals are forced deep into the wood under pressure. In a third process, the chemicals are injected into the wood. Depending on the process and the chemicals used, treated wood will be various shades of green or brown. Some treated wood can be painted or stained.

The amount of preservative used can be adjusted to provide different levels of protection. Wood that is in direct contact with the ground (such as fence posts) should have the

JOB SAFETY

TREATED WOOD The chemicals in treated wood can be toxic. Wear gloves to protect your hands when you are handling the material frequently. When sawing or machining, wear a dust mask to prevent nose, throat, and lung irritation. Cut treated lumber outdoors to avoid indoor accumulations of airborne sawdust. Wear eye protection. After handling, wash your hands thoroughly, particularly before eating. Do not burn scraps, as this may result in toxic fumes. Always dispose of treated wood according to local regulations.

highest level available. Preservative-treated lumber is graded and stamped to indicate its suitability for various uses. Special fasteners may be required when assembling pressure-treated wood products because of the tendency of preservatives to accelerate metal corrosion. Check with the manufacturer for installation instructions. For more on this topic, see Chapter 35.

Fire Retardant Treated Wood Another type of preservative is used to make FRT (fire retardant treated) wood. FRT wood is important in wildfire-prone areas and areas where there are limited firefighting services or water supplies. The preservatives do not prevent wood from burning, but they can significantly reduce the ability of flames to spread. This can give firefighters more time to put out a fire that has started.

Fire retardant chemicals are forced into the lumber under pressure, then the wood is re-dried to a moisture content of 19 percent or less. Plywood can also be treated. It is dried after treatment to a moisture content of 15 percent or less.

FRT lumber can be cut to length, drilled, and lightly sanded without affecting its performance. You do not have to apply additional preservative to the cut ends of boards. However, ripping, surfacing, and milling operations are only permitted in certain cases. Always check manufacturer's installation requirements when working with FRT lumber. It should be used only in interior applications protected from the weather. It is not intended for exterior use. Check local building codes for any other restrictions on the use of FRT wood products. Though FRT chemicals are less toxic than other preservatives, you should follow the same precautions for handling and cutting FRT wood.

Section 12.2 Assessment

After You Read: Self-Check

1. Decay fungi grow best within what temperature range?
2. What condition creates the best environment for decay?
3. Wood from what part of the tree is most decay-resistant?
4. Explain the differences between the two basic types of termites.

Academic Integration: Science

5. **Local Ecosystem** Find your state on the map on page 330 if possible. Research which types of termites, if any, are in your state using your local Cooperative Extension office. Locate information about safe ways to prevent local termite damage. For example, are any special building materials recommended? Are there any non-toxic ways to destroy termites that are offered in your state? Report your findings in a one-page summary.

Go to **connectED.mcgraw-hill.com** to check your answers.

Review and Assessment

Chapter Summary

The suitability of lumber for construction depends on the species of wood, how the lumber is manufactured, and on various measurements of its strength and stiffness. Wood may be either hardwood or softwood. Softwoods are more common in construction. Defects can reduce the utility of lumber. Grading standards take defects and other characteristics of wood into account.

Wood can be damaged by decay and by wood-infesting insects, such as termites and beetles. This damage can be prevented by reducing the moisture content of lumber and ensuring that wood is not allowed to remain wet. Decay-resistant or preservative-treated woods can prevent problems.

Review Content Vocabulary and Academic Vocabulary

1. Use each of these content vocabulary and academic vocabulary words in a sentence or diagram.

Content Vocabulary

- cambium (p. 318)
- deciduous tree (p. 320)
- coniferous tree (p. 320)
- lumber (p. 321)
- moisture content (MC) (p. 321)
- fiber-saturation point (p. 321)
- seasoning (p. 321)
- kiln (p. 321)
- grade (p. 323)
- grade stamp (p. 324)
- warp (p. 325)
- nominal dimension (p. 325)
- dry rot (p. 328)
- wood preservative (p. 332)

Academic Vocabulary

- species (p. 318)
- ecosystems (p. 318)
- photosynthesis (p. 319)
- equilibrium (p. 323)

Speak Like a Pro

Technical Terms

2. Work with a classmate to define the following terms used in the chapter: *sapwood* (p. 318), *heartwood* (p. 319), *annual rings* (p. 319), *flat-sawn* (p. 321), *quarter-sawn* (p. 321), *air drying* (p. 321), *stickers* (p. 321), *kiln drying* (p. 321), *seasonal expansion* (p. 323), *conditioning* (p. 323), *firsts and seconds (FAS)* (p. 324), *select* (p. 324), *No. 1 common* (p. 324), *actual dimension* (p. 325), *dressed sizes* (p. 325), *termites* (p. 329), *carpenter ants* (p. 331), *powderpost beetle* (p. 332), *deathwatch beetle* (p. 332), *carpenter bees* (p. 332).

Review Key Concepts

3. Classify two advantages and disadvantages of using wood for building.

4. Give an example of a hardwood and a softwood.

5. Explain the difference between a flat-sawn board and a quarter-sawn board.

6. Summarize how to calculate the fiber-saturation point of lumber.

7. Describe these three closely related lumber defects: bow, crook, and cup.

8. Identify two ways to prevent wood damage.

Critical Thinking

9. Synthesize Why is it important to maintain forests responsibly?

Academic and Workplace Applications

STEM Mathematics

10. Percentages Moisture content is the weight of water held in the wood expressed as a percentage of the weight of the oven-dry wood. It can be calculated using the following formula:

$$\frac{\text{initial (wet) weight} - \text{oven-dry weight}}{\text{oven-dry weight} \times 100\%}$$

Calculate the moisture content of a piece of wood that weighs 56 grams wet and 48 grams when oven-dry. Then state whether the MC is above or below the fiber-saturation point.

Math Concept Percentage is a proportion that means *per hundred*. For example, 37% means 37 out of 100.

Step 1: Subtract the oven-dry weight from the initial wet weight (56g − 46g).

Step 2: Multiply the oven-dry weight by 100% (48g × 100%).

Step 3: Divide the first number by the percentage to determine the moisture content and compare it to the fiber-saturation point percentage.

STEM Science

11. Metric Conversions You can convert feet to meters using the following conversion formula:

1 foot = 0.3048 meters

You are a U.S. lumber exporter and you are sending a shipment to a country which uses the IS (the metric system). Your customer has asked you to convert the total lineal footage of their order from feet to meters on the invoice. The total lineal footage is 244. How many meters is this? Round up to the nearest tenth.

21st Century Skills

12. Communication Skills A brochure is a printed document containing both text and visual information. For example, various organizations publish brochures about design values, spans, standard sizes, grades, and the properties of wood. Examples of organizations might include the American Wood Council, American Forest and Paper Association, the Canadian Wood Council, the Southern Forest Products Association, and Western Wood Products Association. With a partner, create a brochure for either the organization or for one aspect of wood that you learned in this chapter, such as a brochure about sustainable forest products.

- Use software or art materials to design your brochure.
- Choose art and text to identify the main details about the organization on the interior panels.

Standardized TEST Practice

Multiple Choice

Directions Choose the word or phrase that best completes the following statements.

13. The process of drying wood is called _____.
   - **a.** saturation
   - **b.** controlling moisture
   - **c.** seasoning
   - **d.** expansion

14. The insect that represents the greatest risk to wood is the _____.
   - **a.** carpenter bee   **c.** beetle
   - **b.** termite   **d.** moth

15. The average moisture content of kiln-dried lumber is _____.
   - **a.** 19% or less   **c.** 10% or less
   - **b.** 19% or more   **d.** 10% or more

TEST-TAKING TIP

Pay attention to key words in the question and in each answer choice. For example, in question 14, the key words are **greatest risk.** *The word* **greatest** *helps you know that the question is a comparison.*

*These questions will help you practice for national certification assessment.

Engineered Wood

Section 13.1
Plywood

Section 13.2
Composite Panels, Laminated Veneer, & I-Joists

Section 13.3
Other Types of Engineered Lumber

Chapter Objectives

After completing this chapter, you will be able to:

- **Explain** how the use of engineered lumber helps conserve wood resources.
- **Explain** the grading system for plywood.
- **Discuss** how to store, handle, and install LVL I-joists.
- **Describe** the differences among various types of engineered lumber.
- **Employ** safety rules when handling or machining engineered panels.

Discuss the Photo

Engineered Wood Engineered, or manufactured, wood products are often used in place of solid lumber. *How might engineered wood products conserve resources?*

Writing Activity: Persuasive Paragraph

After you have read the chapter, write a persuasive paragraph to support the use of engineered wood for a specific purpose, such as for subfloor. Be sure to state your position clearly, and use facts to support your position.

Before You Read Preview

Solid wood was once the only wood product used in residential construction. The introduction of engineered wood products has dramatically changed residential construction techniques. Choose a content vocabulary or academic vocabulary word that is new to you. When you find it in the text, write down the definition.

Content Vocabulary

- engineered panel
- plywood
- plies
- veneer match
- composite panel product
- oriented-strand board (OSB)
- medium-density fiberboard (MDF)
- engineered lumber
- laminated-veneer lumber (LVL)
- glulam
- camber
- finger joint

Academic Vocabulary

You will find these words in your reading and on your tests. Use the academic vocabulary glossary to look up their definitions if necessary.

- assemble
- components
- temporary
- retain

Graphic Organizer

As you read, use a chart like the one shown to organize content vocabulary words and their definitions, adding rows as needed.

Content Vocabulary	Definition
engineered panel	any manufactured sheet product, including plywood that is made of wood or wood pieces bonded with a natural or synthetic adhesive

Go to **connectED.mcgraw-hill.com** to download this graphic organizer.

Plywood Basics

What are panels?

For many years, solid wood was the only wood product used in residential construction. Today, both solid wood and engineered wood are used in residential and commercial construction. The increasing use of engineered wood changed many residential construction techniques.

In this book, the term **engineered panel** refers to any manufactured sheet product, including plywood, which is made of wood or wood pieces bonded with a natural or synthetic adhesive. **Plywood** is a building material that consists of layers of wood veneer and sometimes other materials that have been glued together.

Unlike solid lumber, plywood comes in large sheets, called *panels*, that can be installed quickly. Several other panel products are now common on residential, as well as commercial, job sites. Some are made of wood flakes, wood dust, or wood fibers mixed with *adhesives*. Others are made of wood fiber mixed with Portland cement. All of these manufactured panels, including plywood, share the following characteristics:

- They are engineered for the efficient use of wood resources. They are often made of wood that would otherwise be unused or wasted.
- They are manufactured using various natural or synthetic adhesives.
- Their performance is highly predictable.

Plywood is a versatile building material. It is relatively light but very strong, and comes in a variety of thicknesses. It is important at almost every stage of home building. For example:

- Plywood made into foundation forms provides a stiff, uniform surface for forming concrete.

- Plywood floor, wall, and roof sheathing stiffens and strengthens the structure of the house, as shown in Plywood.
- Plywood soffits provide a smooth, easy-to-paint surface.
- As underlayment, plywood makes a smooth substrate for finish flooring.
- Cabinets and built-ins are often made of plywood because it holds fasteners well and accommodates various types of joinery.

Types of Plywood

Plywood panels can be divided into two basic categories: structural plywood (once called softwood plywood or construction plywood) and hardwood plywood. Hardwood plywood is used for paneling, cabinets, built-ins, and other interior features. This chapter deals primarily with structural plywood because that is the type most widely used in building construction. Structural plywood is used for sheathing, subflooring, concrete forms, and other uses where strength is more important than looks.

The Composition of Plywood As mentioned previously, plywood consists of layers of wood veneer and sometimes other materials that have been glued together. Wood veneer is a very thin, pliable sheet of wood that has been sawed, peeled, or sliced from a log. When used in plywood, these thin sheets are called **plies**. Plywood is manufactured with an odd number of layers of these veneers. The grain of each layer runs at a right angle to that of the neighboring layers. Construction and industrial plywoods may have three, five, or seven layers.

The grain of the outermost plies always runs in the same direction relative to each other. This is usually along the length of

 Plywood

Structural Sheathing Plywood sheathing stiffens the house structure and provides a solid base for siding, roofing, and other materials.

the panel, as shown in Structural Plywood. The outermost plies are called *face plies* (A). The face ply of best quality is called the *front face*, or *face*. The other is called the *back face*, or *back*. The plies between the two face plies make up what is called the core (B). Plies that are arranged at a 90° angle to the face plies are called *crossbands*.

Hardwood plywoods may be made entirely of veneers or of veneers bonded to a core of glued-up lumber, as shown in Hardwood Plywood on page 340. The latter is called *lumber-core plywood*. Some types of hardwood plywood have a particleboard core. Particleboard is made of very small particles of wood bonded together by adhesives.

How Plywood Is Manufactured

After a tree is cut and the log is trucked to a plywood mill, it goes into a pond for storage. Only select logs qualify as plywood *peeler* logs, from which veneers will be cut. A chosen log is lifted from the pond and cut to length. As it moves into the mill, it passes through a debarker that uses high-pressure jets of water to blast off the bark. An overhead crane lifts the stripped log into a lathe. The log is then spun against a long, razor-sharp steel blade that slices a continuous strip of thin wood veneer from the log. As the veneer is sliced from the log, it moves over conveyors to the clipping machine, where giant knives cut it to a specific width. Then the veneer sheets pass into the dryers to reduce the moisture content of the wood.

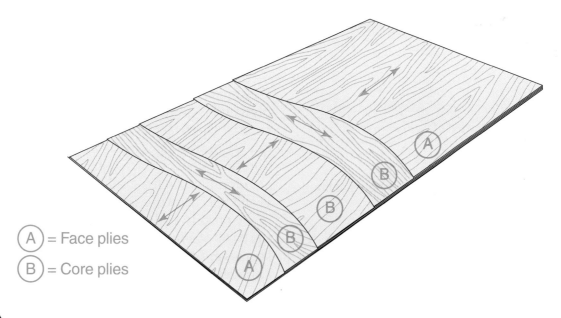

(A) = Face plies
(B) = Core plies

 Structural Plywood

Veneer Layers This is an example of a plywood panel with five layers, or *plies.* The arrows indicate grain direction.

Next, natural defects in the veneer sheets are cut out and the holes are patched with solid wood or synthetic patching material. Some sheets then go through the glue spreader, where large rollers cover both sides with adhesive. Glue-covered sheets

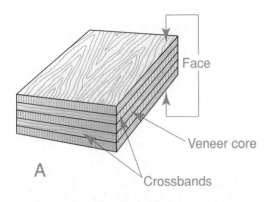

A

Face

Veneer core

Crossbands

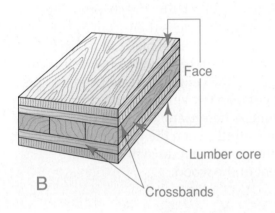

B

Face

Lumber core

Crossbands

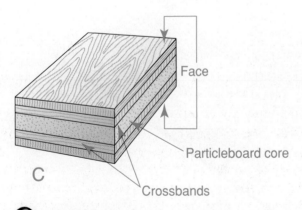

C

Face

Particleboard core

Crossbands

Hardwood Plywood

Mixed Layers Three types of cores found in hardwood plywood. **A.** A veneer core is made of thick wood veneer. **B.** A lumber core consists of strips of lumber bonded together. **C.** A particleboard core is commonly used for cabinet doors because it is very stable.

are stacked alternately with unglued sheets to make up panels of the desired thickness. This is called the *lay-up process*. After lay-up, panels go into presses where the wood and glue are bonded together using pressure and, in some cases, heat. This creates a product that is strong, rigid, stable, and able to resist great impact. The panels are trimmed to exact length and width, sanded to final thickness, and inspected. Blemishes (imperfections) in the face plies are repaired. The panels are then graded and bundled.

How Plywood Is Graded

Grade marks are stamped on the back face of structural plywood. Typical grade marks are shown in Plywood Grade Stamps. The purpose of a grade mark is to identify all the features of a panel. This enables builders to select the right panel for the job. The grade mark also enables building inspectors to verify that the correct materials have been used.

Structural plywood is graded by one of two methods. In the *prescriptive* method, a panel is graded by the quality of its veneer, the adhesives used to **assemble** it, the number of its layers, and its dimensions. In the *performance* method, the panel is rated according to its suitability for a particular use. Hardwood plywoods are graded primarily by veneer quality, species, and the arrangement of the face plies.

Prescriptive Grading Under this system, plywood is graded according to a variety of factors. These factors relate to the type and quality of wood it is made from, its construction, and other characteristics. These factors are identified on the grade stamp.

Wood Species Most structural plywood is made of softwoods such as Douglas fir. On the basis of stiffness and other factors, these species are divided into five groups. The strongest woods are found in Group 1, listed in **Table 13-1** on page 342.

Veneer Quality The quality of the veneer is specified by a letter, ranging from A (highest) to D (lowest). Plywood often has faces of

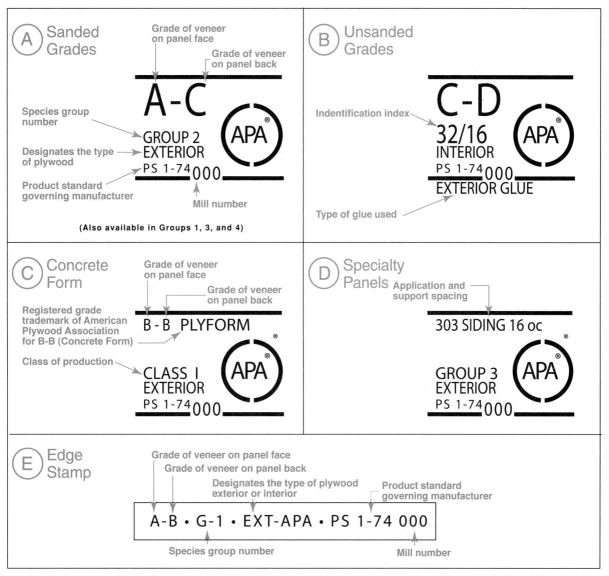

A Sanded Grades

Grade of veneer on panel face
Grade of veneer on panel back

A-C

Species group number
GROUP 2
Designates the type of plywood
EXTERIOR
Product standard governing manufacturer
PS 1-74 000

APA®

Mill number

(Also available in Groups 1, 3, and 4)

B Unsanded Grades

Indentification index
C-D
32/16
INTERIOR
PS 1-74 000
EXTERIOR GLUE

Type of glue used

APA®

C Concrete Form

Grade of veneer on panel face
Grade of veneer on panel back

Registered grade trademark of American Plywood Association for B-B (Concrete Form)
B - B PLYFORM

Class of production
CLASS I
EXTERIOR
PS 1-74 000

APA®

D Specialty Panels

Application and support spacing

303 SIDING 16 oc

GROUP 3
EXTERIOR
PS 1-74 000

APA®

E Edge Stamp

Grade of veneer on panel face
Grade of veneer on panel back
Designates the type of plywood exterior or interior
Product standard governing manufacturer

A-B • G-1 • EXT-APA • PS 1-74 000

Species group number
Mill number

Plywood Grade Stamps

Panel Identification A grade stamp may be located on the back face of a panel or on one edge.

differing quality. Because smoothness is not as important on the side installed against the framing, it is typically a lesser grade. Veneer quality grades are listed in **Table 13-2** on page 343. By mixing veneers, mills can conserve quality woods and produce panels more cost effectively.

Adhesives The type and durability of adhesive used to assemble the panel affects its resistance to weathering and moisture. Exterior-grade plywood is designed for long-term exposure to the weather. It is made with water-resistant glue. The veneer-quality marking for a panel made with water-resistant adhesive is followed by an *X* (for example, CDX). Panels made with adhesives that are not water resistant should be used indoors only.

Construction The number of layers in a plywood panel usually corresponds to the number of plies. In some cases, two of the inner plies may be glued together with their grain running parallel, forming a single inner layer. The two plies count as one layer. Most structural plywood has 3, 5, or 7 plies.

Table 13-1: Classification of Wood Species Used in Plywood[a]

Group 1	Group 2		Group 3	Group 4	Group 5
North American Species—Applicable to trees grown in North America					
Beech, American	Cedar, Port Orford	Pine	Adler, Red	Aspen	Basswood
Birch	Cypress	Pond	Birch, Paper	Bigtooth	Poplar, Balsam
Sweet	Douglas fir[b]	Red	Cedar, Alaska	Quaking	
Yellow	Fir	Virginia	Fir, Subalpine	Cedar	
Douglas fir [b]	Balsam	Western White	Hemlock, Eastern	Incense	
Larch, Western	California	Spruce	Maple, Bigleaf	Western Red	
Maple, Sugar Pine,	Red	Black	Pine Jack	Cottonwood	
Southern	Grand	Red	Lodgepole	Eastern	
Loblolly	Noble	Sitka	Ponderosa	Black (W. Poplar)	
Longleaf	Pacific Silver	Sweetgum	Spruce	Pine	
Shortleaf	White	Tamarack	Redwood	Eastern White	
Slash	Hemlock, Western	Yellow Poplar	Spruce	Sugar	
Tanoak	Maple, Black		Engelmann		
			White		
Non-North American Species					
Apitong[c][d]	Lauan	Mengkulang[c]		Cativo	
Kapur[c]	Almon	Meranti, Red[c][e]			
Keruing[c][d]	Baglikan	Mersawa[c]			
Pine	Mayapis				
Caribbean	Red Lauan				
Ocote	Tangile				
	White Lauan				

[a]Table 1 species classified in accordance with ASTM D 2555.

[b]Douglas fir from trees grown in the states of Washington, Oregon, California, Idaho, Montana, Wyoming, and the Canadian Provinces of Alberta and British Columbia shall be classified as Group 1 Douglas fir. Douglas fir trees grown in the states of Nevada, Utah, Colorado, Arizona, and New Mexico shall be classified as Group 2 Douglas fir.

[c]Each of these names represents a trade group of woods consisting of a number of closely related species.

[d]Species from the genus Dipterocarpus marketed collectively: Apitong if originating in the Philippines, Keruing if originating in Malaysia or Indonesia.

[e]Red Meranti shall be limited to species having a specific gravity of 0.41 or more based on green volume and oven dry weight.

Dimension The thickness of structural plywood ranges from ¼" to 1¼" or more. Plywood is most commonly available in panels that are 4' wide and from 8' to 12' long. The most common sheet size used in residential construction is 4' × 8'. When referring to the size of panel products, remember that the width is always given first. Thus, a 4' × 10' sheet is 4' wide and 10' long. The grain of the face plies runs along the length.

Performance Grading Some structural plywood panels, sometimes called *span-rated* panels or *performance-rated* panels, are graded using a performance-based standard. Instead of identifying how the panel is made, the standard identifies how it will perform. Performance-rated panels are typically used for single-layer subflooring, exterior siding, and sheathing. Each of these categories is subdivided into further categories based on how resistant it is to moisture. A grade stamp developed by APA—The Engineered Wood Association is shown in Performance Grade Stamp on page 343.

Specialty Plywood

Plywoods are often adapted for special uses. For example, foundation-grade plywood includes panels that have been treated with

Table 13-2: Veneer Quality	
N	Intended for natural finish. Selected all heartwood or all sapwood. Free of open defects. Allows some repairs.
A	Smooth and paintable. Neatly made repairs permissible. Also used for natural finish in less demanding applications.
B	Solid surface veneer. Repair plugs and tight knots permitted.
C	Sanding defects permitted that will not impair the strength or serviceability of the panel. Knotholes to 1½" and splits to ½" permitted under certain conditions.
C plugged	Improved C veneer with closer limits on knotholes and splits. C-plugged veneers are fully sanded.
D	Used only for inner plies and backs. Permits knots and knotholes to 2½" maximum dimension and ½" larger under certain specified limits. Limited splits permitted.

preservative chemicals. These are used where the wood will be permanently installed in contact with the earth. Tongue-and-groove plywood is used in the construction of single-layer flooring systems. Concrete formwork plywood has a finish that resists damage caused by contact with cement, and the edges may be painted to increase the panel's durability.

Hardwood Plywood

Hardwood plywood is more expensive than structural plywood of a similar thickness and size. Hardwood plywoods are commonly found in 4' × 8' × ¾" sheets, but other sizes and thicknesses are available. Birch, maple, and oak are frequently used for the face plies. Panels made from other domestic and exotic woods can be obtained from specialty supply sources.

The appearance of a hardwood plywood panel depends upon the species of wood used and the way the face plies are milled and applied. Hardwood plywood is used for finish work in which appearance is very important.

Mills provide many options for the manufacture and arrangement of face plies. There are four basic milling methods, called *cuts*.

THE ENGINEERED WOOD ASSOCIATION

RATED STURD-I-FLOOR
24 OC 23/32 INCH
SIZED FOR SPACING
T&G NET WIDTH 47- 1/2
EXPOSURE 1
— **000** —
PS 1-95 UNDERLAYMENT
PRP-108

 Performance Grade Stamp
Rated for Use This grade stamp is for a subflooring panel. *What is the maximum spacing of joists allowed?*

They are shown in Cuts of Plywood on page 344. Two logs of the same species, with their veneers cut differently, will have an entirely different appearance even though their colors are similar.

Face plies are applied to the panels using one of five basic arrangements. The arrangement of pieces of veneer is called a **veneer match**. This creates different patterns and effects, as shown in Veneer Matching on page 345. An unusual combination of cut and veneer match may have to be special ordered from the plywood supplier. Hardwood plywood veneers are precision matched and available in a variety of veneer types.

Reading Check

Recall *List two prescriptive grading factors.*

How Plywood Is Used
How should plywood be stored?

Working with plywood involves knowing how to store, cut, shape, and fasten it. All of these factors relate to each other. For example, if plywood is not stored properly, it will not perform as intended.

 **Cuts of Plywood**

Different Slices The manner in which veneers are cut is an important factor in producing various effects.

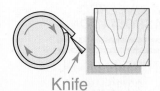

Knife

A **Rotary Cutting**

The log is mounted centrally in the lathe and turned against a razor-sharp blade. Since the cut follows the log's annual growth rings, a bold, irregular grain marking is produced. Eighty to 90 percent of all veneer is cut by the rotary lathe method.

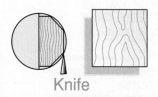

Knife

B **Plain or Flat Slicing**

The half log, or flitch, is mounted with the heart side flat against the guide plate of the slicer. The slicing is done parallel to a line through the center of the log. This produces an irregular figure that is similar to that of sawn lumber.

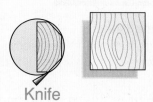

Knife

C **Half-Round Slicing**

With this method, log segments are mounted off-center in the lathe. This results in a cut slightly across the annual growth rings, and shows characteristics of both rotary and plain sliced veneers. This method is often used on red and white oak.

Knife

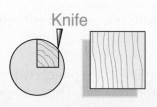

D **Rift Cutting**

Rift-cut veneer is produced from species of oak. Oak has medullary ray cells that radiate from the center of the log like the curved spokes of a wheel. The rift, or comb, grain effect is obtained by cutting perpendicularly to these medullary rays on either the lathe or the slicer.

Storage

Store plywood sheets flat whenever possible. This reduces the chance that sheets will warp or tip over. If flat storage is not possible, you can store sheets on edge for short periods. Take precautions to prevent them from tipping over.

Plywood should be stored indoors until it is ready for use and is delivered to the job site. It can be stored outdoors as long as it is protected from the weather by a waterproof tarp. The sheets should be stacked atop wood spacers that support the pile at least 1½" above the ground. Use enough spacers to prevent the sheets from bowing.

Cutting and Shaping

Like any panel product, plywood must always be supported firmly as it is cut, as shown in Safe Support for Plywood on page 345. This prevents dangerous kickback when using power saws. It also prevents binding when using a handsaw. Sawhorses are often used to support plywood sheets. Some builders make a simple cutting table to improve safety and convenience.

Plywood should be supported on each side of the cut line, at each end of the piece to be kept, and at the offcut. However, when the offcut is less than 1' wide, it can be difficult to support. In that case, it is generally left unsupported and allowed to fall away.

When hand-sawing plywood, always place the best face up and use a saw that has at least 10 to 15 points to the inch. Hold the saw at a low angle. If you must cut cabinet-quality plywood with a circular saw, place

Veneer Matching

Different Arrangements Veneer matching refers to the alignment of veneer strips on a panel.

Book Match
This is accomplished by turning over every other piece of veneer peeled in sequence from the same log. The finished face resembles the opened pages of a book, with opposite patterns identical.

Book Match

Slip Match
Pieces of veneer cut in sequence are joined side by side, same sides up. The result is a grain pattern more uniform than book match.

Slip Match

Whole-Piece
This method uses a single piece of veneer to expose a continuous grain characteristic across the entire panel.

Whole-Piece

Pleasing Match
The face veneer is matched for color at the veneer joint but not necessarily matched for grain characteristics.

Pleasing Match or Unmatched

Unmatched
The veneer is assembled with no regard for color, pattern, or grain uniformity. This method is usually used for panel backs.

Mismatch or Random Match
Veneers are joined with the intention of creating a casual, unmatched effect, such as for pre-finished wall panels. Veneers from several logs may be used in the manufacture of these panels.

Mismatch or Random Match

Safe Support for Plywood
Preventing Kickback When cutting plywood or any other sheet product, always support it on each side of the cut line. This will prevent it from binding on the saw blade.

©Hero/age fotostock

the best face down to minimize splintering. When cutting plywood on a table saw, always place the best face up. With either saw, adjust the blade so that its teeth just clear the top of the plywood. This reduces blade exposure and increases safety. Scoring the veneer before cutting helps reduce splintering.

Plywood can be shaped with various tools. A belt sander can be used to round off corners or shape edges to fit an irregular surface. Large holes can be made in plywood using a jigsaw or a large-diameter, toothed drill bit called a *hole saw*. Routers can be very useful for cutting large holes in a panel once it is in place.

Fastening

Plywood can be fastened to other materials using wood adhesives, nails, and screws. However, no fastener works as well on the edges of plywood as it does on the face plies. It is important to remember this, especially when attaching hinges.

When driving screws into plywood by hand, a hole should be pre-drilled. Holes should also be pre-drilled when inserting screws close to the edge of a sheet to prevent the edge from splitting. Screws with sharp points, such as drywall and decking screws,

Builder's Tip

PREVENTING SPLINTERING When you are drilling through plywood, the panel may splinter on the back where the bit emerges. To prevent this, clamp a scrap piece of plywood beneath the area to be drilled. Then drill through the top piece and into the scrap. The scrap minimizes splintering by supporting the edges of the hole.

can be driven directly into plywood with a screw gun, impact driver, or an electric drill fitted with a screwdriving tip.

When nails or screws are not enough to hold plywood in place, adhesives should also be used. The combination, called glue-nailing, produces a particularly strong bond. Many builders glue and nail sheathing plywood to the floor framing to produce an extra stiff floor.

Where finish nails are used, conceal the nail holes with wood putty. Press the putty into the hole, level it off with a putty knife, and then sand the putty after it is dry. When using screws to fasten plywood, set the heads flush with the plywood.

Section 13.1 Assessment

After You Read: Self-Check

1. Name at least three general uses for plywood in residential building construction.
2. Describe the three types of cores found in hardwood plywood.
3. What is the purpose of veneer matching of hardwood plywood?
4. What is the technique used to minimize splintering when drilling through plywood?

Academic Integration: English Language Arts

5. **Specialty Plywood** Plywood is often adapted for special uses. Research one type of specialty plywood. Write three to five sentences about your findings. State the name of the specialty plywood and explain what it is used for.

Go to connectED.mcgraw-hill.com to check your answers.

Composite Panels, Laminated Veneer, & I-Joists

Composite Panel Basics

What is a composite?

Unlike plywood, which is made entirely from layers of wood veneer, a **composite panel product** is made from pieces of wood in a variety of forms, including veneer, chips, and fibers. The wood is mixed with adhesive and formed into panels. These panel products often share many of the same characteristics as plywood. Many can be used for sheathing, subflooring, cabinetry, and paneling. The most common types of composite panels are oriented-strand board (OSB), hardboard, medium-density fiberboard (MDF), and particleboard, as shown in Composite Panels. Various other names are often used to describe certain engineered panels. For example, *chipboard*, *waferboard*, and *flakeboard* are all names for panels that are essentially the same or similar to oriented-strand board. *Com-ply* is a panel developed by APA—The Engineered Wood Association. It consists of a core made of compressed wood strands and has faces made of wood veneer.

Composite Panels
Wood Bits and Glue Common types of composite panels include oriented-strand board (OSB), particleboard, medium-density fiberboard, hardboard, and tongue-and-groove oriented-strand board.

Build It Green Composite panels are often made from wood that would otherwise be unused or wasted. This includes trees that are unsuitable for veneers, trees that are too small for lumber production, and trees that grow too quickly to produce sound lumber. Portions of trees that would otherwise go unused, such as stumps and limbs, can also be used.

The Structure of Composites

The wood pieces used in composite panels range in size from small fibers, as shown in Composite Panel Fibers on page 348, to large flakes. Two strong influences on the properties of composite panels are:

- The size and shape of individual flakes and particles.
- The ratio of adhesive to particles.

Particle shape and adhesive content can be controlled to create a given set of physical properties. The size, type, and position of the particles also will affect a panel's surface smoothness.

Adhesives A critical ingredient of a composite panel is the adhesive used to bond the wood pieces or particles. Adhesives may be water-resistant or waterproof. They are generally synthetic resins such as urea formaldehyde. Urea-formaldehyde adhesives are highly water resistant and have excellent bond strength. There is a trend toward developing other types of adhesives because of health concerns about off-gassing

McGraw-Hill Education

in formaldehyde-based products. Off-gassing is the gradual release of gaseous materials from a solid product. Some people are very sensitive to formaldehyde vapors. The chemical can cause eyes to water or lead to breathing problems. When products containing formaldehyde are used in tightly sealed structures, vapors can accumulate to unhealthy levels. Urea-formaldehyde resins are a particular problem in this regard. Urea resins release formaldehyde vapors. Because of health concerns, the amount of formaldehyde has been greatly reduced in various panel products. The composite panel industry is also developing new adhesive products that will reduce or eliminate off-gassing.

Additives Other ingredients may be added to the mix of adhesives and wood to change panel characteristics. Small amounts of wax (0.5 percent to 1 percent) reduce a panel's tendency to absorb moisture and make it more suitable for such uses as sheathing. Fire-retardant chemicals can be added to make the panel more suited to areas where wildfires occur. Preservatives can be added to make the panel last longer.

Making a Composite Panel

The difference in manufacturing between plywood panels and composite panels rests in the way the wood is processed. In composite products, logs or wood scraps are processed mechanically to create fairly small, uniform pieces. In some cases, the pieces are chemically treated to make them even smaller. The pieces are mixed with adhesives and other additives. The mixture is then formed into a thick layer called a *mat*, as shown in A Composite Panel Mat. The mat is squeezed under heat and pressure into sheets of a specified thickness.

Composite panels can be given special finishes or treatments at the mill. Surfaces may be filled or primed for easy painting, embossed or textured for a decorative surface, or covered with a vinyl overlay. The edges of panels may be banded with lumber, machined for tongue-and-groove joints, or given special sanding or overlays. Panels can be laminated together to make unusually thick panels. They also can be glued edge to edge to create large panels. The most common panel size is 4' × 8', but panels up to 8' × 16' are available.

The Future of Composites

As our forests are managed with ever more care, scientists and wood technologists continue to develop new composite products. The goal is to make the most efficient use of our forest resources. Various materials are being studied in addition to those noted in this chapter. They include recycled wood products, compressed natural grasses, and wood fiber mixed with thermoplastics.

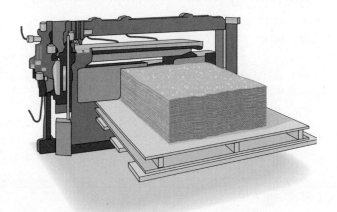

A Composite Panel Mat
Big Squeeze When this composite panel mat leaves the press, it will be only about 0.12" thick.

 Cutting Composites
Good Edges Use blades with carbide teeth when cutting composite panels.

 Oriented-Strand Board
Interlocking Strands In OSB, the strands are directionally oriented.

Each new product must be thoroughly tested before it is approved for construction use. Many composite products incorporate chemicals and other additives that are not in solid wood. Therefore it is always important to learn as much as you can about a new product before using it. This would include gathering information about its long-term structural performance as well as its impact on the environment and human health.

Always review the material safety data sheet (MSDS) for any composite product you are not familiar with. It will suggest suitable precautions.

Working with Composites

Composite panels are free from cracks and other imperfections commonly found in solid and veneered wood. They present none of the problems related to grain in wood. Generally, they can be worked with standard woodworking tools. Because composite panels are made to exact thicknesses at the mill, there is little need for further surface preparation, such as sanding. They can be sawed, routed, shaped, and drilled cleanly, with crisp edges and corners. Carbide teeth are generally recommended for circular saw blades, as shown in Cutting composites.

All types of joints for casework or architectural assemblies are readily made with composites. Architectural panels may

be butted or splined. In cabinetry, miter, lock-miter, doweled, mortise-and-tenon, and tongue-and-groove joints are common. The absence of voids gives these products a full, uniform contact surface for gluing. This means strong butt joints. Short lengths can be glued into longer sections for a minimum of waste. Check the manufacturer's literature for advice on the best adhesives to use.

Oriented-Strand Board Oriented-strand board (OSB) is made from wood strands bonded with adhesive under heat and pressure. It has been available since the early 1980s. It is now considered similar to plywood in strength and usefulness, especially for sheathing and subflooring. In many parts of the country it has replaced plywood as the most common sheathing product.

Though a variety of wood species can be used, most OSB is made from aspen, southern pine, and various medium-density hardwoods. The strands in each layer of an OSB panel are oriented (positioned) so they run more or less in one direction, as shown in Oriented-Strand Board. Then layers are placed perpendicular to each other. Panels usually have three or five layers.

OSB is available with square or tongue-and-groove edges. Thicknesses range from ⅜" to 1⅛". Though OSB is generally made with a waterproof adhesive, the panels are not suited for long-term exposure to the weather. OSB sheathing should be covered as soon as is practical. To increase its moisture resistance, the edges are often coated at the factory with a sealant.

Like most wood products, OSB will shrink or swell slightly with changes in humidity. It is more likely to change in thickness than plywood. Sheathing and subflooring should be installed with a ⅛" gap between the ends of adjacent panels and ¼" at the sides. If the edge seal is damaged during storage or installation, moisture can wick into the panel and swell its edges.

Panels at a job site should not be stored directly on the ground. Instead, they should be stacked on a level platform supported by 4×4 stringers or other blocking. They should be covered loosely with a waterproof tarp as soon as possible. The tarp should be arranged so as not to trap ground moisture beneath it. The steel banding that secures the panels during delivery should be cut right away to prevent the edges from being damaged if swelling occurs.

Fiberboard The term fiberboard generally includes panel products such as hardboard and medium-density fiberboard. To make fiberboard, logs are chipped into small pieces of wood that are then reduced to fibers by steam or mechanical processes. These fibers are refined and mixed with an adhesive. They are then compressed under heat and pressure to produce panels. A fiberboard product will not split, crack, or splinter. It is dense, with extremely smooth front and back surfaces, and has superior wear resistance.

Hardboard *Hardboard* is a high-density fiberboard that is often used for such things as interior paneling, flooring underlayment, and cabinet back panels. The manufacturing process is similar to that used for paper. A slurry (a mixture of fibers and water) is formed into a mat. The mat is then compressed in several stages under pressure and steam heat until the final panel is formed. In some cases, binders or adhesives do not have to be added to the mat. The lignin in the wood fiber, when heated and pressed, serves as a natural adhesive.

Hardboard is generally available in standard and tempered grades. *Standard hardboard* is given no additional treatment after manufacture. It has high strength and good water resistance. It is sometimes used in cabinetwork because it has a very smooth surface. *Tempered hardboard* is standard hardboard to which linseed oil or tung oil has been added prior to pressing. This process improves stiffness as well as scratch and water resistance.

Hardboard is manufactured with one or both sides smooth. Hardboard with one side smooth is known as *S1S*. Hardboard with two sides smooth is *S2S*. It is available in thicknesses from ⅛" to ⅜". The standard panel size is 4' × 8', but widths up to 6' and lengths to 16' are also available. *Perforated hardboard* has very closely spaced holes punched or drilled into it. The holes can be fitted with metal hooks, holders, supports, or similar fittings, as shown in Perforated Hardboard on page 351. Wood-grain hardboard is printed to match the color and texture of oak, walnut, mahogany, and many other woods. It is popular for interior paneling.

Medium-Density Fiberboard (MDF) Medium-density fiberboard (MDF) is made of compressed wood fibers mixed with

JOB SAFETY

ADHESIVE DUST When machining or sanding a composite product or the edges of plywood, wear a dust mask. This will prevent you from inhaling dust from the adhesives used in the panel's manufacture. Excessive exposure to these adhesives is a health hazard. This is especially important with products that generate very fine dust, such as particleboard, MDF, and fiber-cement board.

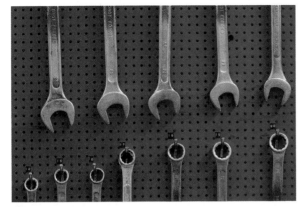

Perforated Hardboard
Storage Board Perforated hardboard is often used with metal hooks to store tools.

ureaformaldehyde adhesive. Because of the uniformity of the fibers used in the manufacturing process, MDF panels have a uniform thickness and an extremely smooth surface. This makes them ideal for use where the end product will be painted, such as door panels and cabinetry. Painted cabinet doors that appear to be frame-and-panel doors are often made of MDF, for example. Though considered a type of fiber-board, MDF is manufactured in a way similar to that used for particleboard. MDF can be worked with standard cutting tools, but carbide-tipped saw blades give the best results.

Particleboard Particleboard is used indoors wherever a smooth and relatively inexpensive surface is required. It is often used as a cabinet carcase material and as a substrate to which other materials are applied. It is also used as a substrate for plastic laminate countertops because of its unusually smooth surface. Construction particleboard is made by combining wood particles or flakes with adhesives and hot-pressing them into panels. The particles near the top and bottom surfaces are relatively fine. Somewhat coarser particles are located at the core. This construction is not obvious, however. Because the particles do not interlock the way fibers do, particleboard is not as strong as MDF and OSB. When stressed beyond certain limits, fasteners may pull out of the material. As with OSB, small amounts of

wax may be added to the material during manufacture to improve its water resistance.

During manufacturing, density, adhesives, and moisture content can be controlled to produce a variety of products, as shown in **Table 13-3** on page 352. Stock panels range in density from 24 to 62 pounds per cubic foot. Ordinarily, a high-density panel will also have greater strength and a smoother, tighter edge than a low-density panel. Panels range from ¼" to 1½" thick, from 3' to 8' wide, and up to 24' in length. They come in ten different grades and three different densities (high, medium, and low).

Fiber-Cement Board *Fiber-cement board* differs from other engineered panel products. Instead of formaldehyde-based adhesives, its cellulose fibers are bound together with a mixture of Portland cement, ground sand, additives, and water. The panels are as efficient, uniform, and predictable as the other panel products discussed in this chapter. They are considered non-combustible and rot proof.

The standard fiber-cement panel is ⁵⁄₁₆" thick and very dense. It comes in sheets 4' wide, and 8', 9', 10', or 12' long. One side is typically very smooth, while the other is rougher, but finishes of various textures are available. Its smooth finish, along with its stability, make fiber-cement board desirable for use as siding. It is available as lap siding as well as in shingle siding patterns.

Table 13-3: Uses for Particleboard

Type	Composition	Uses
Corestock	Flakes or particles bonded with urea-formaldehyde or phenolic resins; has various densities and related properties	Furniture, casework, architectural paneling, doors, and laminated components
Wood-veneered particleboard	Corestock overlaid at the mill with various wood veneers	Furniture, panels, wainscots, dividers, cabinets, etc.
Overlaid particleboard	Particleboard faced with impregnated fiber sheets, hardboard, or decorative plastic sheets	Furniture doors, wall paneling, sink tops, cabinetry, and store fixtures
Embossed particleboard	Surfaces are heavily textured in decorative patterns by branding with a heated roller	Doors, architectural paneling, wainscots, display units, and cabinet panels
Filled particleboard	Particleboard surface-filled and sanded; ready for painting	Painted end-products requiring firm, flat, true surfaces
Primed or undercoated particleboard	Factory-painted base coat on either filled or regular board; exterior or interior	Any painted products
Floor underlayment	Panels specifically engineered for floor underlayment	Underlay for carpets or resilient floor coverings

Handling and Cutting Fiber-cement panels should be stored under cover on a dry, level surface. Steps should be taken to protect the edges and corners because they can be damaged if struck. The panels can be awkward to transport because they are thin, somewhat flexible, and quite heavy. For comfort and safety, each panel should be carried on edge by two people.

Panels can be cut with special shears but carbide-tipped circular saw blades are more often used. However, cutting with a blade generates a great deal of very fine dust that cannot be contained by standard dust bags. Fiber cement contains silica. Breathing excessive amounts of silica dust can lead to an illness called *silicosis*.

Engineered Lumber Basics
Where should engineered lumber not be used?

Engineered lumber is any manufactured product made of solid wood, wood veneer, wood pieces, or wood fibers in which the components have been bonded together with adhesives. Engineered lumber products are stiff, strong, dependable, and versatile. They are increasingly popular as substitutes for solid lumber (see Engineered Wood Products on page 353). Engineered lumber products include:

- Laminated-veneer headers and I-joists.
- Glue-laminated beams.
- Finger-jointed studs.
- Laminated-strand lumber posts.
- Oriented-strand lumber window framing stock.

Engineered products are often used in combination with conventional materials. For example, a house might be built with

JOB SAFETY

DUST MASKS Always wear a dust mask when cutting fiber-cement products. If large quantities of dust are generated, a respirator may be necessary. Special dust collection accessories are available for circular saws.

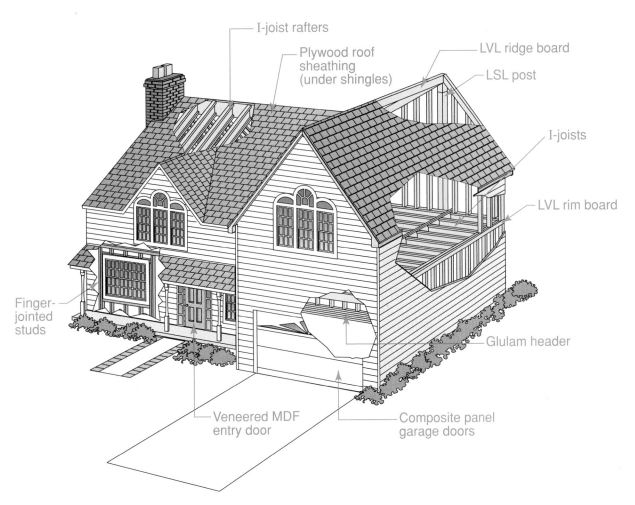

I-joist rafters

Plywood roof
sheathing
(under shingles)

LVL ridge board

LSL post

I-joists

LVL rim board

Finger-
jointed
studs

Glulam header

Veneered MDF
entry door

Composite panel
garage doors

Engineered Wood Products

Around the House Engineered wood products have many structural and finish uses.

engineered lumber floor joists and solid lumber rafters. Engineered lumber has the following advantages:

- It uses wood that might otherwise be wasted.
- Its performance is highly predictable.
- It is available in a wide variety of dimensions and in unusually long lengths.
- It is free of the defects often found in solid lumber.

Engineered lumber is not suitable for all purposes. It should not be used where it will be permanently exposed to the weather. In addition, engineered lumber should not be installed in direct contact with concrete or masonry.

Engineered lumber is made by various manufacturers. Always consult the manufacturer's instructions for handling, care, and usage information. Check local building codes for any restrictions on usage.

Laminated-Veneer Lumber

What does the "I" in I-joist represent?

Laminated-veneer lumber (LVL) is a family of lumber products made with wood veneer as the basic element. LVL products are lightweight, rigid, and available in lengths of up to 66'. To make LVL, layers of veneer are glued together in various ways. In this

respect, LVL products are similar to plywood. However, LVL products are used for beams, headers, joists, and rafters instead of as panel products, as shown in Laminated-Veneer Lumber. LVL is now common in both residential and commercial construction.

Manufacturing Methods

Laminated-veneer lumber products are made in plants in the United States and Canada. Any species of wood may be used, depending on availability and the manufacturer's preference. Sheets of veneer are moved through large, open-ended drying ovens. They are dried until they have a moisture content of about 8 percent. As each sheet leaves the drying oven, it is checked for quality and graded.

The veneer is then fed into an automatic glue spreader. The spreader coats the top of each sheet with an exterior-grade adhesive. As the glued sheets are assembled, the highest-grade veneers are placed at the top and bottom of the stack. This increases the overall strength of the finished product. The sheets are then fed into a machine that uses heat and pressure to cure the adhesive. The resulting material, called a *billet*, may be up to 80' long. Billets are then cut into stock shapes and sizes for headers, beams, or portions of wood I-joists.

Finished LVL products are typically shipped to materials dealers in 60' lengths. When a builder orders LVL products, the dealer cuts them to common lengths in increments of 2' and delivers them to the job site. The builder then cuts them to shorter lengths if necessary.

Performance

A cross section of laminated-veneer lumber looks similar to a cross section of plywood, as shown in Cross-Section of LVL Stock. There is, however, an important difference. Plywood is cross laminated. This means that the grain of each layer runs perpendicular to the grain of adjoining layers. In contrast, the grain of every layer in LVL runs in the same direction. This is called *parallel lamination,* and it produces a material that is more uniform. It also means that the end grain of each veneer layer is exposed only at the ends of the product.

Pieces of solid lumber may shrink or swell in slightly different degrees after leaving the mill. However, the qualities of each piece of LVL stock can be controlled. This means that LVL products are very predictable in their performance. Each piece of LVL behaves exactly like the other pieces in a load. For example, if there is any swelling, all the LVL pieces will swell by the same percentage. As a result, architects and engineers can better control the structural soundness of buildings. LVL is stronger and stiffer than solid lumber of the same size. It replaces solid lumber in many building code requirements.

Care and Handling Laminated-veneer products are produced using waterproof adhesives. They can withstand normal exposure to moisture during the construction process but should not be exposed to moisture needlessly. Most LVL products are wrapped in protective material for transport to the job site. Do not remove this wrapping until you are ready to install the materials.

Do not store LVL products in direct contact with the ground. Rest the bundles on stickers to encourage air circulation and to prevent contact with ground moisture, as shown in Storing LVL Products.

 Storing LVL Products

High and Dry LVL products should be protected from moisture until it is time to install them.

I-Joists and Rim Boards

One of the most common laminated-veneer lumber products is the *I-joist* (as seen in Figure 13-16 on page 354). A *joist* is a type of beam used to support a floor, ceiling, or roof. Seen on end, an I-joist is shaped like the letter *I*. Its vertical member is called the *web*. The two horizontal members are called *flanges*. I-joists are most often used in floor construction to support subflooring. They can also be used in place of rafters in roof construction, as shown in I-Joist Roof Framing on page 356.

The construction of an I-joist varies according to the manufacturer. The top and bottom flanges may be made of solid lumber or laminated-veneer lumber. I-joist flanges range from 1½" to 3½" in width. The web may be made from sections of structural ⅜" plywood or oriented-strand board (OSB). Waterproof adhesive is used to attach webs to flanges. No nails or staples are used. LVL I-joists are commonly available in depths of 9½", 11⅞", 14", and 16".

I-joists have several advantages over solid lumber joists. Because they are available in long lengths, a single I-joist can run the entire width of a house. This removes the need to overlap joists at the center of the floor system. It also reduces the number of separate pieces that must be handled. The flooring system can be installed faster. An I-joist is lighter in weight than an equal length of solid lumber. This makes it easier to carry. For example, an I-joist 26' long and 9½" deep weighs about 50 lbs. A piece of 2×10 solid lumber of the same length weighs about 96 lbs.

Installation I-joists are made by a number of manufacturers. Each manufacturer provides span tables and recommended installation details. A *span table* is a list of distances that a particular structural product can span between structural supports, such as walls or columns. In addition, the trade association of engineered wood manufacturers provides span tables for "performance rated" products that it

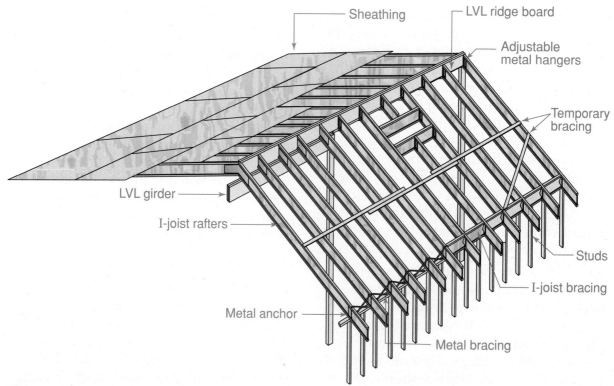

Sheathing

LVL ridge board

Adjustable metal hangers

Temporary bracing

LVL girder

I-joist rafters

Studs

I-joist bracing

Metal anchor

Metal bracing

 I-Joist Roof Framing

Roof Details When I-joists are used to frame a roof, they must be braced until the sheathing can be applied. This prevents them from tipping.

promotes. Be sure to use the span tables for the specific product you intend to use. The following installation methods are typical, but do not apply to every I-joist product.

The web of an I-joist is not in the same plane as its flanges, so take care when cutting the product. For crosscutting, the easiest method is to use a radial-arm saw or a large slide-type miter saw (see Chapter 5, Section 5.3 for information about these tools). To cut I-joists safely with a circular saw, you must prevent the shoe of the saw from lodging against a flange during the cut. You can do this by placing a wood block against the web and between the flanges, as shown in I-Joist Saw Guide. You can then cut the I-joist with ease. The I-joist manufacturer may also supply a cutting guide.

In general, each I-joist should be nailed into place as it is installed, then braced temporarily if necessary. Permanent bracing is provided by the sheathing, by the rim board, and by cross-bracing. If **temporary** bracing

is required before the sheathing is in place, install it as shown in I-Joist Roof Framing. The method is as follows:

1. Use stock at least 1' × 4' in size.
2. Braces should be at least 8' long. Space them no more than 8' on center.
3. Secure each brace into the top of each I-joist using two 8d nails.
4. Nail the bracing to a lateral restraint, such as an existing subfloor or a braced end wall, at the end of each bay.
5. Lap the ends of adjoining bracing over at least two I-joists.

I-joists used in floor construction are installed in a way similar to solid lumber joists. They can be nailed to the plate by toe-nailing through the lower flange or secured by metal joist hangers. (For more on joist hangers and similar metal connectors, see Chapter 14, "Structural Systems".) They can be braced with solid blocking, I-joist blocking, or metal cross-bracing.

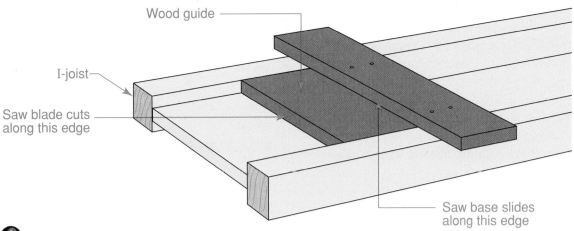
Wood guide

I-joist

Saw blade cuts
along this edge

Saw base slides
along this edge

I-Joist Saw Guide

Cutting Guide A wood guide will prevent a circular saw's shoe from lodging against a flange.

When using joist hangers, keep several factors in mind. Joist hangers are generally nailed to the I-joist with 10d common nails. Never drive nails sideways (parallel to the layers) into an I-joist flange. This tends to split the layers, reducing the strength of the joist. Instead, drive nails into the flange at a 45° angle, as shown in Joist Nailing.

In some cases, you must permanently install thin blocks of wood against both sides of the web. These are called *web stiffeners, bearing blocks,* or *squash blocks.* They reinforce the web and prevent it from buckling at points of high stress. For example, stiffeners are often placed where an I-joist crosses a mid-span support, such as the girder shown

in Web Stiffener. A stiffener used in this way is called a *bearing stiffener.* In other cases, stiffeners are installed where a concentrated load is expected from above, such as from a load-bearing post. This would be called a *load stiffener.* Stiffeners should be at least 2¹⁵⁄₁₆" wide. Their thickness varies according to the width of the I-joist flange. For example:

- A flange 1½" wide calls for stiffeners at least ¹⁵⁄₁₆" thick.
- A flange 2¹⁵⁄₁₆" wide calls for stiffeners at least 1" thick.
- A flange 3½" wide calls for stiffeners at least 1½" thick.

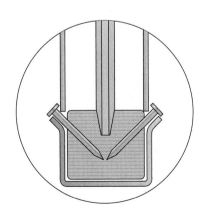

 I-Joist Nailing

Angle Nails Drive nails through an I-joist flange from both sides at a 45° angle to avoid splitting the stock.

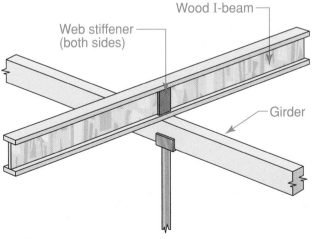
Wood I-beam

Web stiffener
(both sides)

Girder

 Web Stiffener

Stress Relief Web stiffeners should be installed on both sides of the web where the I-joist crosses a support.

The web of a wood I-joist often has pre-scored knockouts located about 12" on center (OC) along its entire length. You can punch these out with a hammer to create passages for plumbing and electrical lines. This removes the need for time-consuming drilling or notching. Additional holes can be cut along the length of the web, but *only* according to the manufacturer's instructions.

Rim Boards When a floor is framed with conventional lumber, the ends of the floor joists are connected with solid lumber of the same size. This lumber is called a *rim joist*. Solid-lumber rim joists do not work with flooring systems framed with I-joists. This is because the two products expand and shrink differently. A rim board is used instead. A rim board is a length of engineered wood stock that has the same depth as the I-joists, as shown in Rim Board. Rim boards are often made from laminated-veneer lumber. They may also be made from plywood, OSB, or laminated-strand lumber. They are available in thicknesses that range from 1" to 1½".

Care and Handling It is important to store and carry I-joists on edge because they are fairly weak in lateral strength. Storing or carrying

an I-joist on its side or allowing it to flex back and forth could break the glued butt joints that join individual sections of web. This would severely weaken the product. Two people should work as a team to carry unusually long I-joists, as shown in Carrying an I-Joist

When I-joists are first delivered to the job site, do not open the protective covering around each bundle. Instead, open or remove it shortly before the I-joists are to be installed. The covering protects them from weather and reduces the chances of damage. If bundles are stacked, separate them with stickers. Before installing an I-joist, inspect it to make sure it has not been damaged during storage.

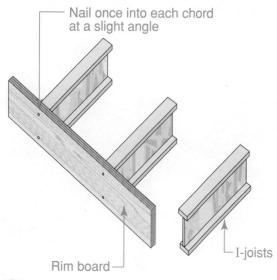

Rim Board
End Cap An LVL rim board is used to tie the ends of I-joists together.

Storing an I-Joist
On Its Edge I-joists are weak in lateral strength so they should be stored on their edges.

McGraw-Hill Education

I-Joist Performance Ratings Instead of carrying a standard grade stamp, I-joists are stamped to identify a specific performance standard it meets. In construction, a *performance standard* is a standard that defines the required behavior of a specified building component. The position of this stamp varies by manufacturer but is often located on the flange. In some cases, the stamp may include additional information such as the maximum on-center spacing permitted for the product.

LVL Headers and Beams

Laminated-veneer lumber can be used like solid lumber for many purposes. It is most often used to replace solid-wood or built-up wood headers and beams. LVL header and beam stock comes in various thicknesses. Stock that is 1¾" thick is most common in residential work. When two pieces are combined, they equal the thickness of a standard 2×4 wall. LVL headers and beams commonly range in depth from 5½" to 18", as shown in **Table 13-4**.

Cutting and nailing LVL stock requires the same tools as those used with solid lumber. When nailing LVL headers face to face, use three rows of 16d nails spaced 12" on center. Some builders have found that the nail-holding ability of LVL headers and beams is greater than that of solid lumber.

LVL headers can be cut to length on site, using carbide-tipped blades. However, holes should not be cut in LVL headers or beams without consulting the manufacturer's recommendations.

Table 13-4: Common LVL Headers and Beams	
Depth (inches)	**Thickness (inches)**
5½, 7¼, 9¼, 9½, 11¼, 11⅞, 14, 16[a], 18[a]	1¾, 3½
9½, 11⅞, 14, 16[a], 18[a]	5¼
[a]16" and 18" beams should be used only in multiple thicknesses.	

Section 13.2 Assessment

After You Read: Self-Check

1. How does a composite panel differ from a plywood panel?
2. Why is fiber-cement board used as a siding material?
3. What are some common uses of LVL products?
4. What are two uses of I-joists?

Academic Integration: Mathematics

5. **Using a Construction Calculator** Use a construction calculator to calculate the number of board feet in 24 pieces of lumber that measure 2" × 10" × 22'.

   Math Concept The board foot is the basic unit of lumber measurement. One board foot is equal to a piece of lumber that measures 1" × 12" × 12". You can use the Bd. Ft. function on a construction calculator to convert the product to board feet.

Go to **connectED.mcgraw-hill.com** to check your answers.

Other Types of Engineered Lumber

Glue-Laminated Lumber

What is "glulam" short for?

When layers of lumber are glued together, their strength and stiffness are greater than that of solid lumber of an equal dimension. This is the principle behind the glue-laminated beam, shown in I-Glue-Laminated Beams. A glue-laminated beam is often called a **glulam**. A glulam is stronger than a steel beam. Glulams can take various forms and span great distances. These characteristics make them useful in both residential and commercial construction. They are used for garage door headers, patio door headers, carrying beams, window headers, and even exposed stair stringers. Glulam posts are also available. Glulams are very fire resistant and do not ignite easily. If they do catch fire, they burn slowly. In some fires where an unprotected steel beam fails completely, a glulam will **retain** much of its strength.

Manufacturing Methods

Glulams are made by gluing lengths of dimension lumber together. The individual layers are adhered face to face, clamped together, and allowed to cure at room temperature. The grain of all layers is parallel along the length of the beam. Each layer is generally no more than 1½" thick. The woods most commonly used for glulams are southern yellow pine and Douglas fir.

The best quality material is used in the top and bottom layers, as shown in Layers of a Glulam. This improves the

A

B

Glue-Laminated Beams

Long Spans Glulam beams can be manufactured as curved beams (**A**) or as straight beams (**B**).

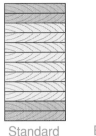

Standard beam

Extra strength beam

One extra tension lamination

 Layers of a Glulam
Improving a Beam Precise layering of high-quality wood improves the strength and fire resistance of a glulam.

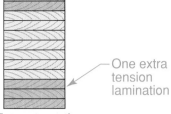
strength of a glulam by nearly 100 percent as compared to random layering. When a one-hour fire rating is required, additional layers of high-quality lumber are placed on the bottom of the beam.

Glulams are sometimes manufactured with a slight upward curve, as shown in Glulam Camber. This curve is called **camber**. The beam must be installed with the curve oriented up. When the beam is in place and fully loaded, the curve straightens out. The amount of camber in any glulam varies according to its length. Camber is measured in the following two ways:

- *Inches of camber* The actual amount of camber is measured in inches at the center of the beam. It is the amount the beam curves above a flat surface.

- *Radius of curvature* The camber of a glulam represents a segment of a huge circle. Stock beams used in residential construction are cambered based on a radius of 3,500'. In commercial construction, this radius may be 1,600' or 2,000'.

Grades

Four grades of glulams are generally available. The differences among the grades are based on appearance. There are no differences in strength. As the appearance of the beam becomes more important, its cost increases.

- *Framing grade glulams* are intended for use where they will not be seen. They are available in widths that fit flush with 2×4 or 2×6 framing.

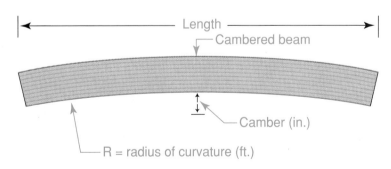

Length

Cambered beam

Camber (in.)

R = radius of curvature (ft.)

 Glulam Camber
Upward Curve In this drawing, the curve (camber) of the glulam has been exaggerated.

- *Industrial grade glulams* are for areas in which appearance is not of much importance. Voids (gaps) may appear on exposed edges. Beams are surfaced only on the sides.

- *Architectural grade glulams* are used where appearance is more important. Some voids are permitted, but any over ¾" in diameter will be filled. All exposed faces are surfaced, and the exposed edges are eased (slightly rounded over).

- *Premium grade glulams* are available only as a custom order. They are used where appearance is very important. All knotholes and voids are filled. All exposed faces are surfaced, and the exposed edges are eased.

Specifying Glulams When glulams are specified or ordered, width and depth are the most important factors. Stock beams are manufactured in widths of 3⅛", 3½", 5⅛", 5½", and 6¾". Depths range from 9" to 36". It is also important to specify whether any camber is necessary or whether the glulam should instead be flat. Custom glulams can be ordered if an unusual strength, length, shape, or degree of camber is required.

Builder's Tip

FRP Glulams A new type of glulam beam is being used in light commercial construction. Though it is not yet used in residential construction, the technology may eventually find applications there. The addition of fiber-reinforced polymers (FRP) to a glulam significantly improves its bending strength and its stiffness. This also reduces the cost of the beams and requires less wood than an unreinforced glulam of the same size. The FRP layers are glued into place as the beam is manufactured.

Storage and Installation

Glulams generally leave the mill with a moisture content of 12 percent. Before shipping, they are sealed, primed, or wrapped in water-resistant paper. The ends are sealed to limit moisture penetration. When a beam must be stored on site before installation, support it off the ground. It can be stored on edge, but laying it flat reduces the chance that it will tip over accidentally. If the beam is wrapped in protective paper, make small slits in the underside to allow moisture to drain and to encourage air circulation.

Take care to install glulams properly. They must not be notched or drilled in any way unless this has been accounted for in their design. If a beam must be cut to length on site, the cut ends should be sealed according to the maker's specifications. Heavy-gauge metal framing connectors are often used to support glulams.

Other Engineered Lumber Products

Why is it becoming difficult to find lumber of consistent quality?

In addition to LVL and glulam products, other engineered structural products are available. They include finger-jointed lumber and a family of products based on a mix of wood strands and adhesive. For installation details on any of these products, review the product literature provided by each manufacturer. Find out if the products are accepted by the building codes in your area.

Finger-Jointed Lumber

It is increasingly difficult to find framing lumber of the consistent quality that was once available. This is partly due to the heavy demand for wood products, but it is also a result of decreasing forest resources. To help meet demand, *finger-jointed lumber*, sometimes referred to as *structural end-jointed lumber*, has been developed. Lengths of solid wood are joined end to end. A **finger joint** is a closely spaced series of wedge-shaped cuts made in the mating surfaces of lumber. These cuts create a large surface area that results in a strong glue bond between the two parts, as shown in Finger-Jointed Lumber.

Finger-jointed lumber has several important characteristics:

- It is always straight.
- It can be sawed and nailed exactly like solid lumber.
- It makes use of short pieces of wood that might be otherwise wasted.
- It is available in longer lengths than standard lumber.

Finger-jointed products are available for use as wall framing lumber, paint-grade interior molding, and exterior trim.

Finger-Jointed Studs The grading agencies that supervise the manufacture of standard lumber also supervise finger-jointed lumber. Each piece of finger-jointed lumber should be marked with a grade stamp, as shown in Grade Markings for Finger-Jointed Lumber. The stamp indicates the grade of wood, its species, and the type of glue used in the joints. Building codes generally treat lumber with *certified exterior joints* as if it were standard lumber. It can be used interchangeably with standard lumber of the same size, species, and grade. This type of finger-jointed lumber uses exterior-grade adhesives and the fingers are 7/8" to 1 1/8" long.

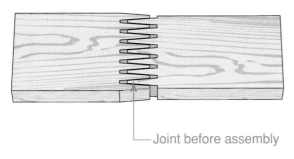

Joint before assembly

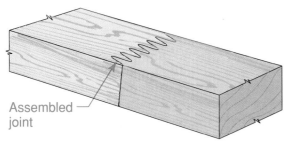

Assembled joint

 Finger-Jointed Lumber
Interlocking Joints Finger-jointed lumber consists of many short pieces of solid wood joined together.

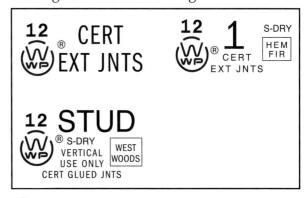

 Grade Markings for Finger-Jointed Lumber
Board Information These three grade stamps are typical of those found on finger-jointed lumber.

Lumber with *certified glued joints* is suited to vertical use only. It should not be used in any horizontal application. Finger-jointed lumber of this type is sometimes stamped "STUD USE ONLY" or "VERTICAL USE ONLY." This type of lumber does *not* use exterior-grade adhesives. Its fingers are ⅜" to ⅝" long. You should never store such lumber where water might collect in the stack for prolonged periods. Store this product indoors or under cover until you are ready to use it.

Finger-jointed studs are manufactured in sizes of 2×2, 2×3, 2×4 (the most common size), 3×4, and 2×6. They may be up to 12' long. Some manufacturers assemble finger-jointed lumber from oversize blocks of wood and then saw the studs to standard dimensions. Thus a 2×4 will measure 1½"

Laminated-Strand Lumber

Short Strands The surface of LSL has an appearance similar to oriented-strand board panels.

by 3½". Other manufacturers assemble the lumber from short pieces of 2×4 lumber and then plane ⅟₃₂" off both edges to ensure smooth surfaces. A finger-jointed 2×4 made using this method would measure 1½" by 3⁷⁄₁₆". This smaller size must be indicated on the grade mark.

Laminated-Strand Lumber

Laminated-strand lumber (LSL) is made of wood strands glued together and cut to uniform dimensions, as shown in Laminated-Strand Lumber. Its performance is predictable. It can be used for rim boards, studs, plates, headers, beams, and columns.

An LSL product is typically made from fast-growing aspen or yellow poplar, though small amounts of other hardwoods may be included. After debarking, logs are processed into wood strands from 0.03" to 0.05" thick, 1" wide, and about 12" long. The strands are dried and sorted to remove waste and pieces that are too short. The dried strands are coated with a blend of adhesive and wax. The coated strands are then formed into a thick mat. The mat is cut to a 35' or 48' length and formed under heat and pressure into a billet. A billet is a long block of manufactured wood that will be cut into smaller pieces. When cooled, the billet is sanded and cut to lengths of up to 22'. The finished LSL lumber is then marked with a grade stamp.

Parallel-Strand Lumber

Parallel-strand lumber (PSL), shown on page 365, is a product that can be used for columns, studs, and beams. The wood comes from Douglas fir, western hemlock, southern pine, or yellow poplar logs. The logs are cut into 8' lengths, then rotary-peeled into veneer with a thickness of ⅟₁₀" or ⅛". The veneer is then cut into large pieces, dried, and sliced into ribbon-like strands about 1" wide and up to 8' long. Unwanted short strands and other waste are removed.

The strands are then coated with an adhesive mixed with a small amount of wax. The wax helps to keep out moisture. Strands are laid parallel and built up into a mat. The mat is then compressed and the adhesive is cured using microwave energy. The resulting billet may be as much as 11" thick. When cool, it is cut to size and sanded. The pieces are grade stamped and coated with a sealant to further slow moisture intake.

Oriented-Strand Lumber

A relatively new engineered lumber product is called *oriented-strand lumber (OSL)*. It is similar to oriented-strand board (OSB) except that the strands are aligned along the length of the product. It is also similar to laminated-strand lumber except that the strands are shorter. Its uses are still being developed but the material has been used to form the core of entry doors, the structural frame of windows, and the framework of upholstered furniture.

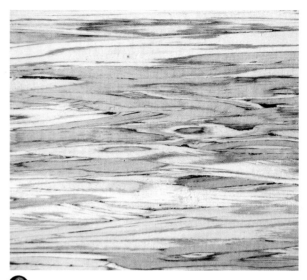

Parallel-Strand Lumber
Veneer Strands PSL lumber is often used for beams, headers, and columns.

Section 13.3 Assessment

After You Read: Self-Check

1. What are some common uses of glulams in residential and commercial construction?
2. What are four important characteristics of finger-jointed lumber?
3. Name the two basic types of finger-jointed lumber.
4. What types of woods are used to make laminated-strand lumber and parallel-strand lumber?

Academic Integration: English Language Arts

5. **Building Codes** Building codes regulating the use of finger-jointed studs vary. Research your local building codes to find out if finger-jointed studs are approved for use in your area. Make note of any restrictions. Record your findings in a one-page summary.

Go to **connectED.mcgraw-hill.com** to check your answers.

Review and Assessment

Chapter Summary

Plywood is made from layers of wood veneer called plies. Construction plywood grades are based on type of adhesives, veneer quality, wood species, construction, size, performance, and special characteristics. Plywood may be made from either softwood or hardwood.

Composite panels are made from pieces of wood mixed with adhesives and joined under heat and pressure. Composites include oriented-strand board, fiberboard, particleboard, and fiber-cement board. Some panels are given special finishes. Engineered-lumber products are often used in combination with conventional materials. Laminated-veneer lumber (LVL) is a family of engineered products made with wood veneer. One common LVL product is the I-joist, used in floor and roof framing.

Glulam beams are made of layers of lumber glued together. A finger joint is a way of joining solid wood end to end. Finger-jointed lumber can be sawed and nailed like solid lumber. Laminated-strand lumber (LSL) is made from strands of wood glued together. Parallel-strand lumber (PSL) is made from ribbons of wood veneer glued together.

Review Content Vocabulary and Academic Vocabulary

1. Use each of these content vocabulary and academic vocabulary words in a sentence or diagram.

Content Vocabulary

- engineered panel (p. 338)
- plywood (p. 338)
- plies (p. 338)
- veneer match (p. 343)
- composite panel product (p. 347)
- oriented-strand board (OSB) (p. 349)
- medium-density fiberboard (MDF) (p. 350)

- engineered lumber (p. 352)
- laminated-veneer lumber (LVL) (p. 352)
- glulam (p. 360)
- camber (p. 361)
- finger joint (p. 363)

Academic Vocabulary

- assemble (p. 340)
- components (p. 352)
- temporary (p. 356)
- retain (p. 360)

Speak Like a Pro

Technical Terms

2. Work with a classmate to define the following terms used in the chapter: *panels* (p. 338), *adhesives* (p. 338), *face plies* (p. 339), *front face* (p. 339), *back face* (p. 339), *prescriptive* (p. 340), *performance-rated* (p. 342), *fiber-cement board* (p. 351), *billet* (p. 354), *parallel lamination* (p. 354), *flanges* (p. 355), *web stiffeners* (p. 357), *bearing blocks* (p. 357), *inches of camber* (p. 361),

radius of curvature (p. 361), *framing grade glulams* (p. 361), *architectural grade glulams* (p. 362), *certified exterior joints* (p. 363), *certified glued joints* (p. 364), *laminated-strand lumber (LSL)* (p. 364), *parallel-strand lumber (PSL)* (p. 364).

Review Key Concepts

3. Discuss how the use of engineered lumber helps conserve wood resources.

4. Describe the grading system for plywood.

5. Explain how to store LVL I-joists.

6. List the differences between composite panel and plywood.

7. List two safety rules for machining engineered panels.

Critical Thinking

8. Synthesize Describe one advantage of the availability of I-joists at multiple lengths.

9. Explain How does laminated-veneer differ from plywood?

Academic and Workplace Applications

STEM Science

10. Adhesives A critical ingredient of a compound panel is the adhesive used to bond the wood pieces or particles. Most adhesives work by either chemically bonding materials or by mechanically bonding them together. Use your school library or the Internet to research these two types of bonds. Then write a one-page explanation of the difference between a chemical bond and a mechanical bond.

STEM Science

11. Expansion and Contraction When a floor is framed with conventional lumber, the ends of the floor joists are connected with solid lumber of the same size. This lumber is called a rim joist. Solid lumber rim joists do not work with flooring systems framed with I-joists. This is because the two products expand and shrink differently. Do you know why they expand and shrink differently? Find out more about expansion and contraction of wood products. Then write three to five sentences describing why various wood products expand and contract.

STEM Mathematics

12. Estimate Sheathing Engineered panel products are usually purchased in 4×8 panels, but are also available in 4×10 and 4×12 panels. A 4×8 panel covers 32 sq. ft. (4 × 8 = 32 sq. ft.) of space; a 4 × 10 panel covers 40 sq. ft. A 4×12 panel covers 48 sq. ft. Determine how many 4×8 panels of plywood are actually needed for the walls of a room that measures 10' × 12' and has 8' high ceilings.

Step 1: Calculate the perimeter of the rectangular room.

Step 2: Multiply the perimeter times the wall height to determine the area of the walls to be covered.

Step 3: Determine the square footage (area) of one panel of plywood.

Step 4: Divide the number of square feet to be covered by the area of one sheet of plywood to determine the actual number of sheets needed.

> **Standardized TEST Practice**

CERTIFICATION PREP

Multiple Choice

Directions Choose the term or number that best answers the following questions.

13. Which can be used to fasten plywood to other materials?

  a. ropes **c.** screws
  b. staples **d.** stakes

14. What is the moisture content of laminated-veneer products after they are dried?

  a. 20% **c.** 15%
  b. 8% **d.** 35%

15. What can excessive exposure to sunlight cause the wood in a glulam beam to do?

  a. break down **c.** shrink
  b. fade in color **d.** expand

TEST-TAKING TIP

If you skip a question, make a mark next to it. Go back to the questions you skipped after you have reached the end of the test.

*These questions will help you practice for national certification assessment.

CHAPTER 14

Structural Systems

Section 14.1
Framing Systems & Structural Design

Section 14.2
Nails & Connectors

Chapter Objectives

After completing this chapter, you will be able to:

- **Describe** the differences between platform-frame construction and balloon-frame construction.
- **Name** the stresses that structural wood must resist.
- **List** the advantages of structural insulated panels.
- **Demonstrate** how to read a span table.
- **Explain** the difference between a live load and a dead load.
- **Identify** various nails and connectors.

Discuss the Photo

Well Kept The Paul Revere House in Boston is one of the oldest buildings in North America. It is proof of wood's strength and durability if maintained properly. *Why is wood a common framing material?*

Writing Activity: Create a Survey

Plan to interview a builder in your community about what type of framing he or she uses to build structures. Write at least three grammatically correct questions. Report your findings in a one-paragraph summary.

Before You Read Preview

The structural design of a house is based on an understanding of how materials and fastening systems interact. Choose a content vocabulary or academic vocabulary word that is new to you. When you find it in the text, write down the definition.

Content Vocabulary

- balloon-frame construction
- platform-frame construction
- in-line framing
- post-and-beam framing
- structural insulated panel
- spline
- shear wall
- load
- design value
- span table
- on center (OC)
- dead load
- live load
- framing connector

Academic Vocabulary

You will find these words in your reading and on your tests. Use the academic vocabulary glossary to look up their definitions if necessary.

- crucial
- function

Graphic Organizer

As you read, use a chart like the one shown to organize the characteristics and process of each type of framing. Add rows as needed.

Type of Framing	Characteristics	Process
balloon-frame construction	Studs run from the sill plate to the top plate of the second floor.	The frame is constructed as one piece. The second floor joists are connected to the same studs as the first floor joists.

Go to connectED.mcgraw-hill.com to download this graphic organizer.

Framing Systems & Structural Design

Structural Materials & Framing

What is balloon framing?

Most homes in the United States and Canada have a structural frame made of wood or wood products. Metal-frame houses are structurally similar.

Wood-frame houses have several important features. They often cost less than houses built using other structural systems. They are easily insulated, which reduces heating and air-conditioning costs. They can support a wide variety of exteriors. This flexibility allows architects and builders to produce nearly any architectural style. In addition, a well-built and properly maintained wood-frame home is very durable.

Conventional Framing

Most houses are built using wood framing that consists of many individual pieces. The main pieces are joists, studs, beams, and rafters, as shown in Conventional Framing. These pieces are spaced at regular intervals. They are fastened together in a way that enables them to support and strengthen the house. In this way, every piece supports part of the load.

Wood panels, called *sheathing*, are fastened to the wood framing to give it more strength and stiffness (see Chapter 13, "Engineered Wood" for more on sheathing products). Together, the framing and sheathing form the basic structure of a house. The two main types of conventional framing are balloon-frame construction and platform-frame construction, shown in Balloon-Frame Construction on the next page and Platform-Frame Construction on page 372.

Balloon Framing In **balloon-frame construction**, also called *balloon framing*, the studs run from the sill plate to the top plate of the second

floor, as shown in Balloon-Frame Construction. The first-floor joists also rest on this sill. The second-floor joists bear on 1×4 ribbons (sometimes called *ribands*) cut into the inside edges of the studs. Wood expands and contracts *across* the grain but is relatively stable *with* the grain. Because less cross-grain framing is used, balloon-frame construction is less affected by expansion and contraction.

Balloon framing is not used often to frame modern houses. This is partly because long, straight lengths of lumber are no longer readily available. Remodeling and restoration contractors see this system primarily when they remodel and repair old houses. With the development of finger-jointed lumber, however, portions of new houses may be constructed using this system.

Platform Framing In **platform-frame construction**, also called *platform framing*, each level of the house is constructed separately. The floor is a platform built independently

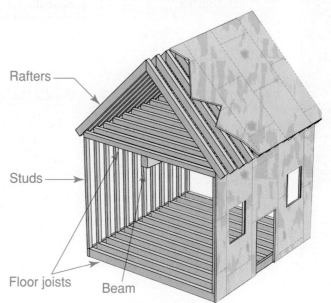

Rafters

Studs

Floor joists

Beam

Conventional Framing
Four Major Parts Joists, studs, beams, and rafters.

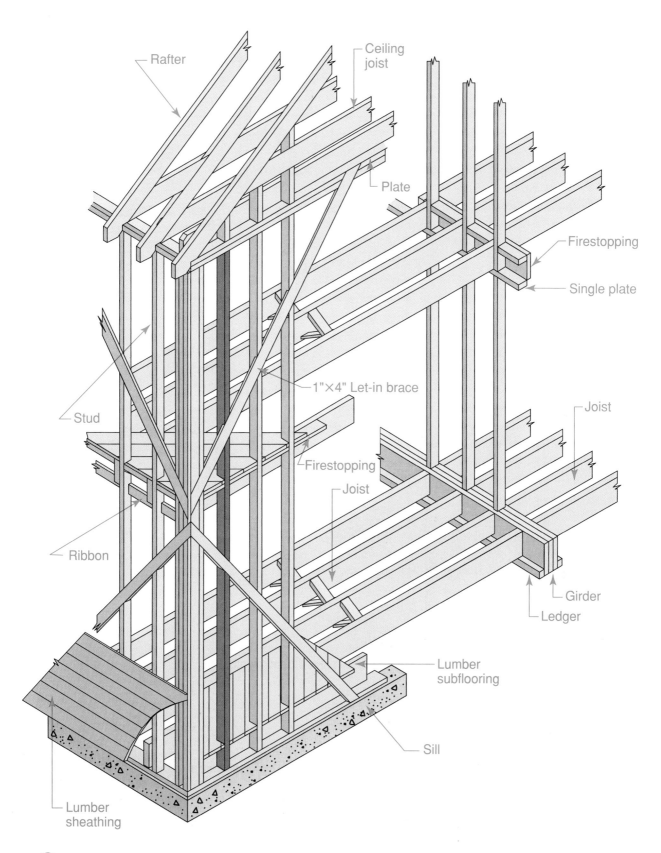

Rafter

Ceiling joist

Plate

Firestopping

Single plate

1"×4" Let-in brace

Joist

Stud

Firestopping

Joist

Ribbon

Girder

Ledger

Lumber subflooring

Sill

Lumber sheathing

Balloon-Frame Construction

Long Lengths Wall studs extend in continuous lengths from one story to another. All joints are nailed.

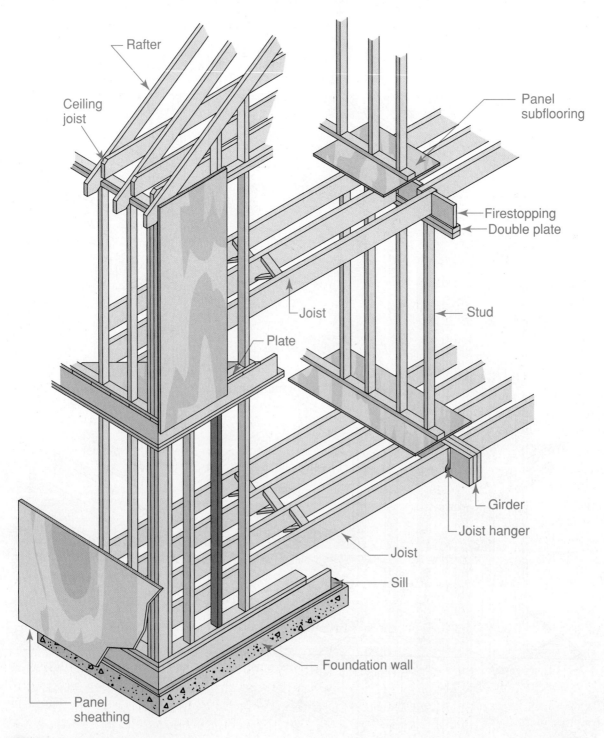

Rafter

Ceiling
joist

Panel
subflooring

Firestopping
Double plate

Joist

Stud

Plate

Girder

Joist hanger

Joist

Sill

Foundation wall

Panel
sheathing

Platform-Frame Construction
Short Lengths Wall studs are only long enough to form the walls of one story. Joints are nailed and may be held with framing connectors.

of the walls, as shown in above. The top surface of this platform is called the subfloor. It extends to the outside edges of the building. Each wall is usually assembled flat on top of a subfloor, and then tilted into place.

The general sequence of constructing a multi-story house using platform framing starts with the installation of a foundation. The first level floor joists are installed next and then sheathed. The first level exterior and interior walls follow. The exterior walls should be sheathed before the second level is built, in order to provide rigidity to the frame. The second level floor joists can then be installed and sheathed. Then, the second level walls are framed and sheathed. At this point ceiling joists and rafters are installed, and the roof is sheathed.

Because the floors, walls, and roof are all separate parts, the connections between them can fail if not made properly. This is particularly true in areas affected by severe weather or earthquakes. Most builders, especially in seismic and high-wind zones, improve the strength of connections by using metal framing connectors (see Section 14.2 for more information). However, if its parts are securely connected, a platform frame will be strong and rigid.

Platform-frame construction is easily adapted to prefabrication. Walls can be built elsewhere and then lifted into place on the subfloor. Another advantage of platform-frame construction is that it does not require unusually long lengths of lumber. Building techniques in most parts of the United States have developed almost entirely around platform-frame construction. Because it is the most common method used for one- and two-story houses, this book will focus on its techniques.

Spacing Variations In standard platform-frame construction, wall studs are commonly spaced 16" apart, measured from the center of one stud to the center of the next. This is called 16" on center, or 16" OC. On center is discussed further in Section 14.2. However, floor joists might be spaced at intervals of 12", 16", 19.2", or 24" OC. The result is that structural loads are not always passed directly from one framing member to another. This is not a problem because wall plates distribute the loads. In addition, the number of studs in a typical wall makes up for slight irregularities in the load distribution. This helps to even out the load.

One variation on this system is called in-line framing. In **in-line framing**, all joists,

studs, and rafters are given the same spacing, as shown in In-Line Framing. This spacing is usually 16" or 24" OC. It creates a direct path for loads, from the rafters right down to the foundation wall. This increases the load-bearing efficiency of the frame and reduces the amount of lumber needed for a house. For example, double-top plates are not required. A single wall plate is adequate.

In-line framing is one element of a more comprehensive system designed to reduce the amount of lumber required to build a house. The system is sometimes called *advanced framing,* but is also known as Optimum Value Engineering (OVE). In addition to in-line framing, the system calls for planning the house in 2' modules and includes details such as two-stud corners.

Post-and-Beam Framing **Post-and-beam framing** is a framing system that relies on fewer but larger pieces of framing members, as shown in Post-and Beam Framing on page 374. The framing members are spaced

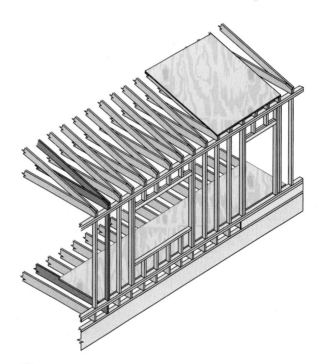

In-Line Framing
Parallel Rows Note how rafters, joists, and studs are lined up with each other. Joints are nailed or made with metal framing connectors.

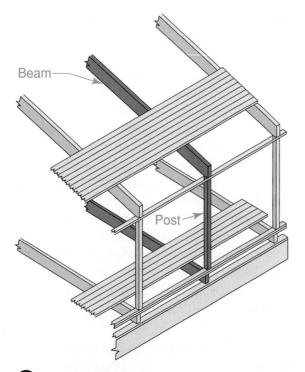

Post-and-Beam Framing
Large Dimensions Note the similarities between this system and in-line framing.

farther apart than those used in conventional framing. Subfloors and roofs are supported by a series of beams spaced up to 8' apart. The ends of the beams are supported by structural timber posts. (*Structural timber* is lumber that is 5×5 or larger. It is used mainly for posts and columns.) The roof sheathing and the subfloor may consist of planks, usually with a 2" nominal thickness, or structural tongue-and-groove (T&G) plywood that is 1⅛" thick. Spaces between posts are framed as needed for attaching exterior and interior finish.

One advantage of post-and-beam framing is the architectural effect provided by the exposed framing in the ceiling. Thick roof planking serves as the finished ceiling as well as the structural support for the roofing. Generally, the planks are selected for appearance. No further ceiling treatment is required.

A variation on this system is often used in portions of the Pacific Northwest and in other areas of mild weather. The first floor is framed using post-and-beam techniques, but the rest of the house is framed using platform framing. This is a cost-efficient method to use when building a house with a crawlspace foundation.

Timber Framing A *timber frame* is a type of post-and-beam frame that rests on a foundation, as shown in Timber-Framing. The supporting members are fairly far apart. They are made from either hardwood or softwood timbers. The timbers are surfaced and then connected with interlocking joinery, which are often secured with wooden pegs. This requires a high degree of woodworking skill. Some joints are quite complex, but most are a variation of the mortise and tenon joint.

Construction of a post-and-beam structure starts with a foundation. The entire structural frame is assembled next, including the roof. The frame is self-bracing, which means

Timber Framing
Timbers and Joinery The individual pieces of a timber frame are connected with interlocking joints. *What disadvantage does this system have compared to conventional framing?*

Huntstock/Getty Images

that it is rigid enough to stand without requiring sheathing. When the frame is complete, it is often covered using structural insulated panels (SIPs). At that point construction follows the same sequence as standard construction, with the installation of doors, windows, mechanicals, and interior finishing.

One of the most desirable aspects of a modern timber frame is the structure, which is typically exposed on the inside of the house, as in An Exposed Frame. The use of timbers to frame buildings is a technique with a very long history. There has been a revival of interest in timber framing, particularly where nearby forests can provide the timber stock of suitable dimensions.

Structural Insulated Panels Structural insulated panels (SIPs) are used with increasing frequency to form the walls, floors, and roof of a house, such as the one shown in SIP Framing on page 376. A **structural insulated panel** consists of 3½" thick expanded polystyrene (EPS) foam insulation between sheets of exterior plywood or oriented-strand board (OSB). Because their interiors are made of foam insulation, SIPs are sometimes referred to as *foam-core panels.* SIPs are sometimes used along with conventional framing.

Panels usually range in size from 4' × 8' to 8' × 28'. Larger panels are also available. These load-bearing panels are built in a factory and delivered to the job site. There they are fastened together using a system of 2×4 or 2×6 splines, as shown in Installing SIP on page 376. A **spline** is a thin strip of wood used to reinforce a joint. Depending on how they are designed, panel walls may be structural or non-structural. When applied to a timber frame house, for example, they are non-structural because the frame is carrying all the loads.

An Exposed Frame
Visible Structure The timber frame is typically exposed to the interior. It has a structural purpose as well as a decorative role.

SIP Framing
Panel Walls The use of structural insulated panels speeds the construction process.

Panel

Spline

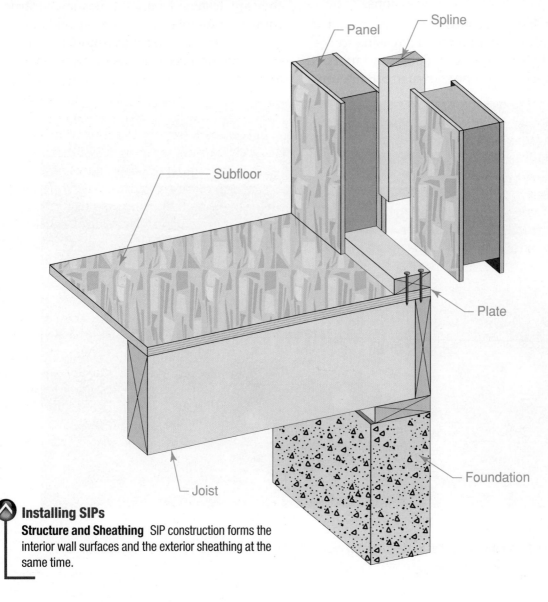

Subfloor

Plate

Foundation

Joist

Installing SIPs
Structure and Sheathing SIP construction forms the interior wall surfaces and the exterior sheathing at the same time.

©Streeter Photography/Alamy

Building a house with structural insulated panels has several advantages. First, the shell of the house can be erected very quickly. The house is very strong because there is wood sheathing on the inside as well as on the outside. Some panels also come with an inside skin of drywall. This saves time and work in completing the interior of the house. The panels are energy efficient because they allow very little cold air to leak into the house. Also, the foam within each panel is a very effective insulator. On the negative side, it can be difficult to run wiring through the panels. Plumbing and wiring plans should be developed with this in mind. Also, heavier panels require a crane to lift them into place.

Manufactured Housing

Many houses today are built as *manufactured housing*. This means that the houses are built completely or partially on factory assembly lines. The resulting building is then trucked to the job side and installed. This system is very efficient. Few materials are wasted, and delays caused by weather are minimized.

Manufactured housing has evolved over time. Prefabricated roof trusses became common in the 1950s, and were a precursor to prefabricated houses. The next step was the development of factory-built wall panels. These were used to speed construction of the house shell. Later, entire portions of a house could be built in the factory, and then trucked to the job site for assembly on top of a foundation.

The structural aspects of a manufactured house are often similar to those of site-built houses. There may be a wood frame, for example, as well as structural elements that serve as joists and rafters. However, the assembly methods and the actual materials are often quite different from those found on a job site.

Special Framing Techniques

Areas that have hazards such as severe weather and earthquakes have stricter building codes than other areas. Carpenters and builders must take extra steps to strengthen house framing. This is true no matter what type of framing system is used.

Wind Resistance Many U.S. regions, such as the Gulf Coast, are exposed to hurricanes and high winds, as shown in Wind Zones on page 378. In these areas, building codes require special techniques, such as using metal straps to secure the roof framing to the wall framing, as shown in Hurricane Straps and Hold-Down Anchors on page 379. Gable roofs are more easily damaged by high winds than hip roofs or flat roofs. Strengthening a gable end may require special structural bracing. Follow local building codes carefully when building in a region subject to high winds.

Earthquake Resistance The ground moves violently during an earthquake. This may deform the structure of a house or even push it off its foundation. After the ground stops moving, the house will continue to move for a short time. This movement puts great strain on

JOB SAFETY

LIFTING WITH A CRANE Building components such as wood timbers and SIPs are often too heavy to lift into place by hand. Small cranes may be used instead. Ensure that any object being lifted by crane is properly balanced and secured. Attach a guide rope to the object to prevent it from swinging out of control.

Science: Earth & Space Science

Hurricanes A hurricane is a severe tropical cyclone. What are the characteristics of a hurricane? Do you live in a hurricane-prone area?

Starting Hint U.S. government Web sites may be a good place to find the answer.

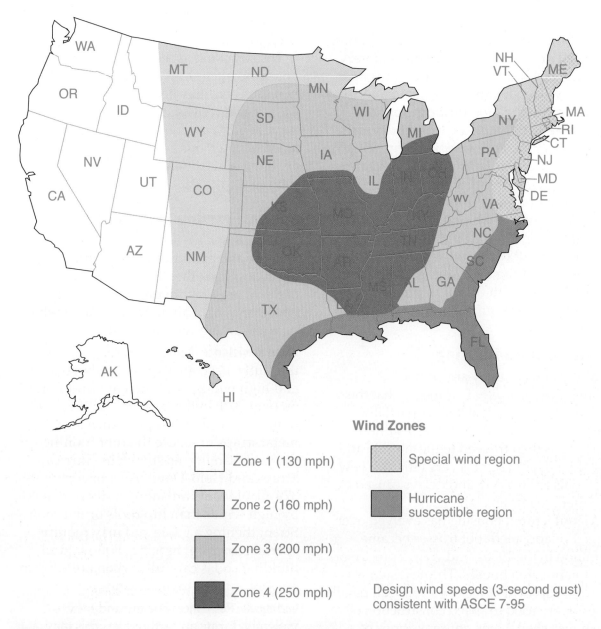

Wind Zones

□ Zone 1 (130 mph)		▨ Special wind region	
□ Zone 2 (160 mph)		■ Hurricane susceptible region	
▢ Zone 3 (200 mph)			
■ Zone 4 (250 mph)		Design wind speeds (3-second gust) consistent with ASCE 7-95	

Wind Zones

Danger Areas This is a general guide to wind hazard areas. The International Residential Code book contains more detailed maps.

structural materials and connections. In earthquake areas (seismic zones), building codes require additional construction features. These are designed to make the structure more rigid and to hold it securely to the foundation.

Shear walls are a type of wall commonly used to strengthen houses in seismic zones. A **shear wall** is a wall designed to resist lateral (sideways) forces. Using a specific nail spacing to attach sheathing is one way

to create a shear wall. Most exterior walls of a house can be designed as shear walls. It is **crucial** to provide shear strength at the corners of a house. The top of a shear wall must be fastened to the second-floor framing. The bottom of the wall must be fastened to the sill plate. The sill plate must be bolted securely to the foundation. A sheathing panel should be set in place vertically in order to be able to make these connections. If

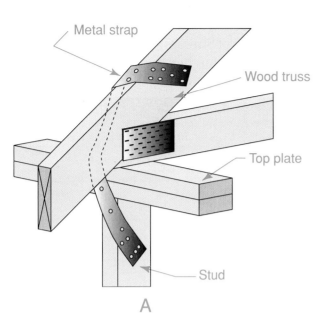

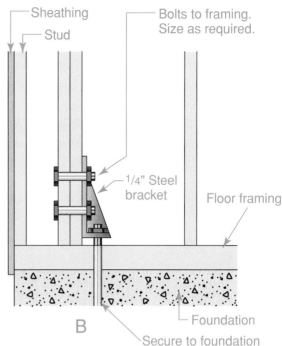

A

B

 Hurricane Straps and Hold-Down Anchors
Strong Connections A. Metal hurricane straps can be used to reinforce the connection between roof and wall framing.
B. Hold-down anchors are steel brackets that prevent walls from toppling.

a 4×8 panel will not reach, a 4×9 or 4×10 panel or solid blocking at joints may be required. The panel must be nailed to the framing according to local codes.

To provide added security, seismic connectors called *hold-down anchors* can be installed at each corner of the house, or as required by local codes.These steel brackets, shown above, prevent walls from tipping over. They are attached to the foundation with anchor bolts, and to the framing with lag screws or machine bolts.

Reading Check

Recall *What is sheathing?*

Structural Design
What is a design value?

When a load is placed on lumber, stresses are created inside the wood. A **load** is a force that creates stresses on a structure. Weight is one type of load, but wind is also a load. Wind creates stress when it pushes against

the walls. The size and spacing of joists, studs, and rafters are determined by the way wood responds to stress.

Design Values

How wood will behave can be calculated once its species and grade are known. The results of laboratory stress tests on wood are summarized in tables of design values. A **design value** is a number assigned to how well a particular wood resists stresses. Part of a design value table is shown in **Table 14-1** on page 380. This is shown only as an example. *Always read the footnotes that accompany the table.*

Carpenters and builders should have a basic understanding of what design values mean. Design values are based on the following stress factors shown in Types of Stresses on page 381.

Extreme Fiber Stress in Bending (F_b) When a load is applied to a joist, header, or beam, it bends. This produces tension stresses in the wood farthest from the load and compression stresses closest to the load (A in Types of Stresses).

Tension Parallel to Grain (F_t) When the ends of a piece of wood are pulled in opposite directions, tension along the grain results. (B in Types of Stresses). This might occur in a floor joist attached to two walls that are bowing outward.

Horizontal Shear (F_v) Shear stresses occur where two portions of the wood are trying to slide past each other in opposite directions. (C in Types of Stresses). A deep, heavily loaded beam might experience shear stresses near the centerline of the wood.

Table 14-1: Design Values for Beams and Stringers[a]

5" and thicker, width more than 2" greater than thickness[b] Grades described in sections 53.00 and 70.00 of *Western Lumber Grading Rules*

Species or Group	Grade	Extreme Fiber Stress in Bending Single Member F_b	Tension Parallel to Grain F_t	Horizontal Shear[c] F_v	Compression		Modulus of Elasticity E
					Perpendicular $F_{c\perp}$	Parallel to Grain F_c	
Douglas Fir-Larch	Dense Select Structural	1,900	1,100	170	730	1,300	1,700,000
	Dense No.1	1,550	775	170	730	1,100	1,700,000
	Dense No. 2	1,000	500	170	730	700	1,400,000
	Select Structural	1,600	950	170	625	1,100	1,600,000
	No. 1	1,350	675	170	625	925	1,600,000
	No. 2	875	425	170	625	600	1,300,000
Douglas Fir-South	Select Structural	1,550	900	165	520	1,000	1,200,000
	No. 1	1,300	625	165	520	850	1,200,000
	No. 2	825	425	165	520	550	1,000,000
Hemlock-Fir	Select Structural	1,300	750	140	405	925	1,300,000
	No. 1	1,050	525	140	405	750	1,300,000
	No. 2	675	350	140	405	500	1,100,000
Mountain Hemlock	Select Structural	1,350	775	170	570	875	1,100,000
	No. 1	1,100	550	170	570	725	1,100,000
	No. 2	725	375	170	570	475	900,000
Sitka Spruce	Select Structural	1,200	675	140	435	825	1,300,000
	No. 1	1,000	500	140	435	675	1,300,000
	No. 2	650	325	140	435	450	1,100,000
Spruce-Pine-Fir (South)	Select Structural	1,050	625	65	335	675	1,200,000
	No. 1	900	450	65	335	575	1,200,000
	No. 2	575	300	65	335	350	1,000,000
Western Cedars	Select Structural	1,150	700	70	425	875	1,000,000
	No. 1	975	475	70	425	725	1,000,000
	No. 2	625	325	70	425	475	800,000
Western Hemlock	Select Structural	1,400	825	170	410	1,000	1,400,000
	No. 1	1,150	575	170	410	850	1,400,000
	No. 2	750	375	170	410	550	1,100,000
Western Woods (and White Woods	Select Structural	1,050	625	65	335	675	1,100,000
	No. 1	900	450	65	335	575	1,100,000
	No. 2	575	300	65	335	350	900,000

[a]Design Values in pounds per square inch. See Sections 100.00 through 180.00 in the *Western Lumber Grading Rules*.
[b]When the depth of a sawn lumber member exceeds 12", the design value for extreme fiber stress in bending (F_b) shall be multiplied by a size factor in Table J.
[c]All horizontal shear values are assigned in accordance with ASTM standards, which include a reduction to compensate for any degree of shake, check, or split that might develop in a piece.

Compression Perpendicular to the Grain (F$_{c\perp}$)
This occurs when the wood rests on supports. An example would be a joist (D in Types of Stresses). Any load on the wood tends to crush wood fibers at the bearing points. This problem can be reduced by increasing the bearing area.

Compression Parallel to the Grain (F$_c$) This occurs when loads are supported on the ends of the wood (E in Types of Stresses). This is typical of studs, posts, and columns. The resulting stresses affect the wood fibers uniformly along the full length of the wood.

Modulus of Elasticity (E) This is the ratio showing the amount that wood will bend in proportion to its load. The actual amount of bending is called *deflection*. An example would be how "springy" a floor is when walked on, as in F in Types of Stresses.

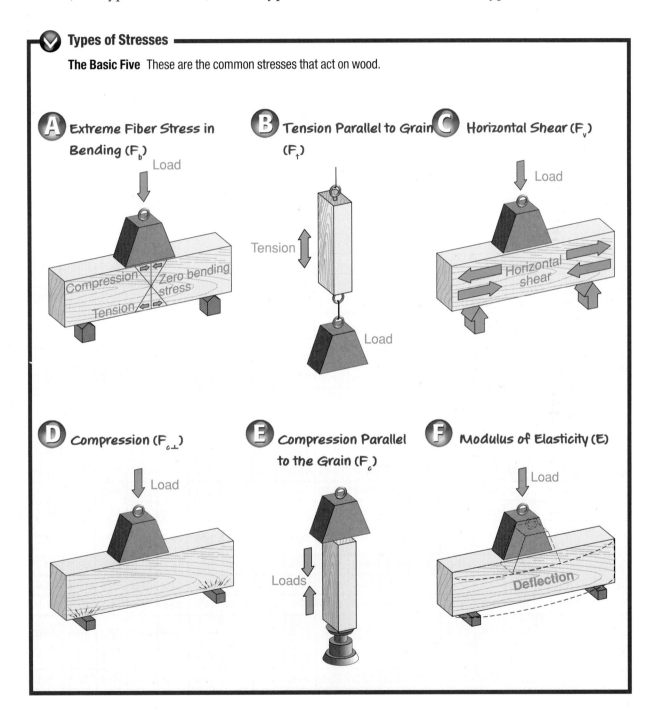

Types of Stresses

The Basic Five These are the common stresses that act on wood.

A Extreme Fiber Stress in Bending (F$_b$)

B Tension Parallel to Grain (F$_t$)

C Horizontal Shear (F$_v$)

D Compression (F$_{c\perp}$)

E Compression Parallel to the Grain (F$_c$)

F Modulus of Elasticity (E)

Span Tables

Carpenters do not need to figure out how a certain wood will behave in a floor or ceiling. Instead, they can refer to a span table, such as the one for floor joists in **Table 14-2**. *This table is a sample only; always check your local building code book for current span tables.* A **span table** lists the maximum spacing allowed between different sizes of joists or rafters. This type of spacing is referred to as on center spacing. **On center (OC)** refers to the distance from the centerline of one structural member to the centerline of the next closest member.

Using span tables, a carpenter can quickly find the right spacing for the species, grade, and dimensions of wood being used. Span tables are included in building code books and in literature from the major lumber trade associations. More complete examples

Table 14-2: Floor Joist Spans

40 lbs. LIVE LOAD. 10 lbs. DEAD LOAD. L/360
Design Criteria: *Strength*– 10 lbs. per sq. ft. dead load plus 40 lbs. per sq. ft. live load.
Deflection– Limited in span in inches divided by 360 for live load only.

Species or Group	Grade	2×8				2×10				2×12				2×14			
		12"	16"	19.2"	24"	12"	16"	19.2"	24"	12"	16"	19.2"	24"	12"	16"	19.2"	24"
Douglas Fir-Larch	Sel. Struc.	15-0	13-7	12-10	11-11	19-1	17-4	16-4	15-2	23-3	21-1	19-10	18-6	27-4	24-10	23-5	21-4
	1 & Btr.	14-8	13-4	12-7	11-8	18-9	17-0	16-0	14-9	22-10	20-9	19-1	17-1	26-10	23-4	21-4	19-1
	No. 1	14-5	13-1	12-4	11-0	18-5	16-5	15-0	13-5	22-0	19-1	17-5	15-7	24-7	21-4	19-5	17-5
	No. 2	14-2	12-9	11-8	10-5	18-0	15-7	14-3	12-9	20-11	18-1	16-1	14-9	23-4	20-3	18-5	16-6
	No. 3	11-3	9-9	8-11	8-0	13-9	11-11	10-11	9-9	16-0	13-10	12-7	11-3	17-10	15-5	14-1	12-7
Douglas Fir-South	Sel. Struc.	13-6	12-3	11-7	10-9	17-3	15-8	14-9	13-8	21-0	19-1	17-11	16-8	24-8	22-5	21-1	19-7
	No .1	13-2	12-0	11-3	10-6	16-10	15-3	14-5	12-11	20-6	18-4	16-9	15-0	23-8	20-6	18-9	16-9
	No. 2	12-10	11-8	11-0	10-2	16-5	14-11	13-10	12-5	19-11	17-7	16-1	14-4	22-8	19-8	17-11	16-1
	No. 3	11-0	9-6	8-8	7-9	13-5	11-8	10-7	9-6	15-7	13-6	12-4	11-0	17-5	15-1	13-9	12-4
Hemlock-Fir	Sel. Struc.	14-2	12-10	12-1	11-3	18-0	16-5	15-5	14-4	21-11	19-11	18-9	17-5	25-10	23-6	22-1	20-6
	1 & Btr.	13-10	12-7	11-10	11-0	17-8	16-0	15-1	14-0	21-6	19-6	18-3	16-4	25-3	22-4	20-5	18-3
	No. 1	13-10	12-7	11-10	10-10	17-8	16-0	14-10	13-3	21-6	18-10	17-2	15-5	24-4	21-1	19-3	17-2
	No. 2	13-2	12-0	11-3	10-2	16-10	15-2	13-10	12-5	20-4	17-7	16-1	14-4	22-8	19-8	17-11	16-1
	No. 3	11-0	9-6	8-8	7-9	13-5	11-8	10-7	9-6	15-7	13-6	12-4	11-0	17-5	15-1	13-9	12-4
Spruce-Pine-Fir (South)	Sel. Struc.	13-2	12-0	11-3	10-6	16-10	15-3	14-5	13-4	20-6	18-7	17-6	16-3	24-1	21-11	20-7	19-2
	No. 1	12-10	11-8	11-0	10-2	16-5	14-11	14-0	12-7	19-11	17-10	16-3	14-7	23-0	19-11	18-2	16-3
	No. 2	12-6	11-4	10-8	9-8	15-11	14-6	13-3	11-10	19-4	16-10	15-4	13-9	21-8	18-9	17-2	15-4
	No. 3	10-5	9-0	8-3	7-5	12-9	11-0	10-1	9-0	14-9	12-10	11-8	10-5	16-6	14-4	13-1	11-8
Western Woods	Sel. Struc.	12-10	11-8	11-0	10-2	16-5	14-11	14-0	12-9	19-11	18-1	16-6	14-9	23-4	20-3	18-5	16-6
	No. 1	12-6	11-1	10-1	9-0	15-7	13-6	12-4	11-0	18-1	15-8	14-4	12-10	20-3	17-6	16-0	14-4
	No. 2	12-1	11-0	1-01	9-0	15-5	13-6	12-4	11-0	18-1	15-8	14-4	12-10	20-3	17-6	16-0	14-4
	No. 3	9-6	8-3	7-6	6-9	11-8	10-1	9-2	8-3	13-6	11-8	10-8	9-6	15-1	13-1	11-11	10-8

of span tables are shown in the **Ready Reference Appendix**. Steps in reading a span table are given below.

To simplify span tables, loads on a structure are divided into two types, dead loads and live loads. A **dead load** is the total weight of the building. This includes the structural frame and anything permanently attached, such as wall coverings. A **live load** is weight that is not permanently attached. Examples of live loads include furniture and people. Live loads are determined based on the use of the building.

Step-by-Step Application

Reading a Span Table Use a span table to determine what dimension of floor joist is suitable for use over a span of 18'.

Step 1 Determine the live load category of the building (30 or 40 lbs. per square foot). The designer determines the live load based on the usage of the building. For example, refer to **Table 14-2**, which deals with structures with a live load of 40 psf (pounds per square foot).

Step 2 Locate the "Species or Group" column. Identify the species of wood being considered.

Step 3 Refer to the "Grade" column. Identify the wood grade.

Step 4 Follow the row to the right until you find 18-0 or greater. This is the span you are looking for, in feet and inches.

Step 5 Now, follow the column directly upward to the "Spacing on Center" row. The numbers there will tell you how far apart the joists must be spaced.

Step 6 In the row above that are the lumber dimensions. For example, Hem-Fir (hemlock-fir) joists graded No. 1 would have to be 2×12s in order to span 18'. They could be spaced either 12" or 16" on center (OC), but no further apart.

Section 14.1 Assessment

After You Read: Self-Check

1. What are the basic pieces that form the structure of a wood-frame house?
2. What are the two basic types of conventional framing?
3. What do design value tables show and what are they used for?
4. What is a *live load*?

Academic Integration: Science

5. **Stress Factors** Stress factors include extreme fiber stress in bending, tension parallel to grain, horizontal shear, compression, compression parallel to the grain, and modulus of elasticity. A floor joist has been attached to two outside walls that are bowing outward. Which type of stress is being applied to the floor joist?

   Starting Hint Draw a simple sketch with arrows indicating the direction of the force.

Go to connectED.mcgraw-hill.com to check your answers.

Nails & Connectors

Construction Nails

What does the "d" in nail sizing stand for?

Creating a strong structural frame for a house depends on good materials and correct structural design. Good connections are also important. Without them, even the best materials would not **function** properly. Screws, staples, and adhesives are often used to assemble the framework of a house. However, *nailing* is still the primary connection method, though builders are increasingly relying on metal framing connectors as well. These two connection methods will be discussed in this section.

There are literally hundreds of different types of nails. They are made from many different materials, but nails made of mild steel are used most in building construction. The most commonly used types are shown in below.

Nail Anatomy

The basic parts of a nail are the head, shank, and point. Each of these parts can vary according to the type of use the nail is intended for.

Head The head of a nail determines how easily it can be pulled through the material. Nails typically have a round head, although nails driven by pneumatic tools may be shaped in other ways. Nails with large heads, such as common nails and roofing nails, will not pull through under normal conditions. These nails are intended for structural uses where the visibility of the head is not objectionable. Brads and finish nails have small heads, and can easily be pulled through most material. They do not have much holding power, and are primarily used in finish work.

Shank The shank of a nail is the portion below the head. The type of shank determines how well the nail will resist withdrawal. *Withdrawal* refers to forces that can pull the nail headfirst out of the wood. These forces are applied parallel to the nail shank.

There are two basic types of nail shanks. *Smooth-shank* nails are used for general construction. The shank provides good holding power in a variety of woods. Most framing and roofing nails are smooth-shank nails. *Deformed-shank* nails are best for

 Types of Nails

Nail Group The basic types of nails used most often in construction: **A.** Wire brad. **B.** Finish nail. **C.** Box nail. **D.** Roofing nail. **E.** Common nail. **F.** Ring-shank spike

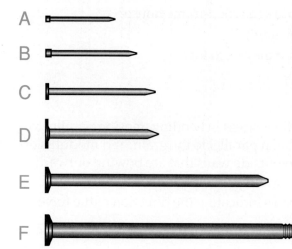

A

B

C

D

E

F

applications that require extra holding power. The shank features a series of ridges or a spiral, which increases withdrawal resistance. One new type of nail has a spiral pattern near the head, a ridged pattern near the point, and a smooth shank in between. The combination increases withdrawal resistance without reducing the nail's strength. A shank's diameter determines how well the nail will resist shear. *Shear* refers to forces acting perpendicular to the nail. They cause the nail to bend. The larger the nail's diameter, the greater its resistance to shear.

Point The point of a nail determines how likely it will be to split the wood and how easy it will be to drive. Nails with sharp points are easy to drive but are more likely to split the wood. Nails with relatively blunt points are not as likely to split the wood.

A blunt point tends to crush or sever the wood fibers as it enters, where a sharper point tends to wedge them apart. A compromise between sharp and blunt is a four-sided tapered cut called a *diamond point*. This is the point found on most nails used for framing.

Nail Sizes

Nails are classified by pennyweight, abbreviated "d." For example, a 16d nail is pronounced "sixteen-penny." The origins of this system are unclear but it is generally thought to have begun in England. The "d" is the abbreviation for the Latin word *denarius*, a small Roman coin similar to a modern penny. Note that the penny number of a nail refers to its length, not its cost or its weight. The larger the number, the longer the nail, as shown in Relative Nail Sizes. Also, the longer the nail, the larger its diameter.

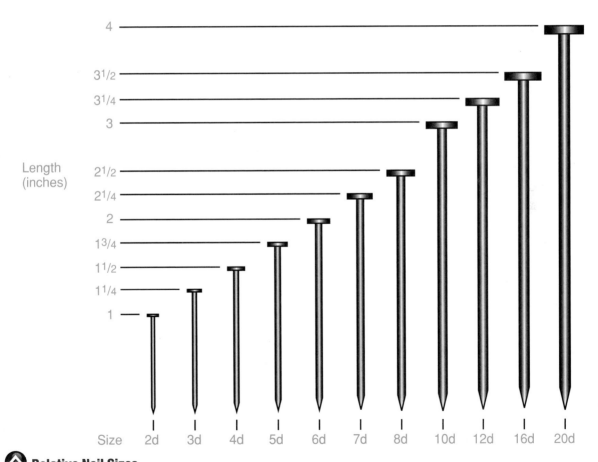

Relative Nail Sizes
Penny Is Length These are the lengths of nails most often used in construction.

Builder's Tip

Using Nails In order to hold a connection properly, the correct size and type of nail must be used. Carpenters must also be sure to use the correct number of nails. Using fewer nails than specified results in a weakened connection. However, in many cases the same problem occurs when too many nails are used. This is because the additional nails tend to create weakened areas of wood that will eventually split. The number and type of nail used for each type of connection in a house is specified in the building code. **Table 14-3** shows some of the more common nailing requirements. Consult tables in the code book called "nailing schedules" for more information about nailing.

Reading Check

Recall What is a shank?

Metal Framing Connectors
How should a joist hanger be installed?

At one time in modern platform framing, all wood-to-wood connections were secured by nails alone. Because of differences in the skills of carpenters, the strength of the connections varied. In order to strengthen connections and make them more uniform, metal framing connectors were developed. A metal **framing connector** is a formed metal bracket that is installed at framing connections using nails. Connectors are often used to install engineered lumber. Some engineered products, such as I-joists, are difficult or impossible to secure without the use of

Table 14-3: Nail Details for Framing	
Description of Building Elements	**Number and Type of Fastener***
Joist to sill or girder, toe nail	3-8d
1" × 6" subfloor or less to each joist, face nail	2-8d
2" subfloor to joist or girder, blind and face nail	2-16d
Sole plate to joist or blocking, face nail 16" OC	16d
Top or sole plate to stud, end nail	2-16d
Stud to sole plate, toe nail	3-8d or 2-16d
Double studs, face nail 24" OC	10d
Double top plates, face nail 24" OC	10d
Sole plate to joist or blocking at braced wall panels 16" OC	3-16d
Double top plates, minimum 24-inch offset of end joints, face nail in lapped area	8-16d (3½" × 0.135")
Blocking between joists or rafters to top plate, toe nail	3-8d
Rim joist to top plate, toe nail 6" OC	8d
Top plates, laps at corners and intersections, face nail	2-10d
Built-up header, two pieces with ½" spacer 16" OC along each edge	16d
Continued header, two pieces 16" OC along each edge	16d
Ceiling joists to plate, tow nail	3-8d
Continuous header to stud, toe nail	4-8d
Ceiling joist, laps over partitions, face nail	3-10d
Ceiling joist to parallel rafters, face nail	3-10d
Rafter to plate, toe nail	2-16d
Built-up corner studs 24" OC	10d
Roof rafters to ridge, valley or hip rafters: toe nail face nail	4-16d 3-16d
Rafter ties to rafters, face nail	3-8d
Collar tie to rafter, face nail, or 1¼" × 20 gauge ridge strap	3-10d

*All nails are smooth-common, box, or deformed shank.

metal framing connectors. Connectors are very important where severe conditions can be expected, such as high winds or earthquakes. For example, the connector shown in Hurricane Tie, ties the wall framing to the roof framing. It is aptly called a *tie*.

Manufacture

A metal framing connector not only makes wood-to-wood junctions stronger, it makes wood-to-masonry and wood-to-concrete connections stronger as well. Some ornamental connectors are meant to be exposed. However, most will never be seen after the building is completed. These are made from various gauges of galvanized steel, as shown in **Table 14-4**.

Metal connectors are galvanized after they have been formed into a specific shape. Galvanizing deposits a layer of zinc on all sides. This protects the metal by slowing corrosion (the formation of rust). The thickness of the zinc coating is indicated by a code. The standard zinc coating is G 60. This means that the zinc is 0.005" thick on each side of the steel. A connector with a G 90 coating would have a layer of zinc one-and-one-half times as thick as one with a G 60 coating.

Table 14-4: Thickness of Galvanized Steel Used for Framing Connectors			
Gauge	In Decimal Inches	In Millimeters	In Approximate Fractions of an Inch
7	0.186	4.8	3/16
10	0.138	3.5	1/8
11	0.123	3.1	1/8
12	0.108	2.7	3/32
14	0.078	2.0	3/32
16	0.063	1.6	1/16
18	0.052	1.3	1/16
20	0.040	1.0	1/32
22	0.034	0.8	1/32

Note: Actual steel dimensions will vary from nominal dimensions according to industry tolerances.

Some connectors have a G 185 coating. If even greater corrosion resistance is required, stainless steel connectors should be considered. They must be installed using stainless steel nails because other nails would rust.

Types of Connectors

A wide variety of connectors is available. The best source of information about a connector is the manufacturer. Framing connectors are widely accepted by building codes. However, you should always make sure that the use you are planning for is approved by the manufacturer and by local codes.

Hurricane Tie
Stronger Joints This connector ties walls to roof framing. It can be used where a stud is directly beneath a rafter.

Andy Kass

REGIONAL **CONCERNS**

Corrosion-Resistant Connectors Salt spray from the ocean increases the speed of corrosion. The amount of salt spray in the air may be a significant factor as far as 3,000' inland. Builders in these areas should use metal connectors that have extra corrosion resistance. Check local codes to see how far inland the "corrosion zone" extends.

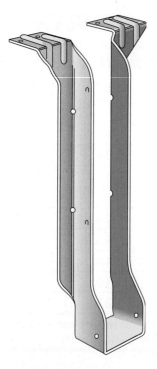

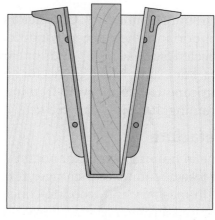

A

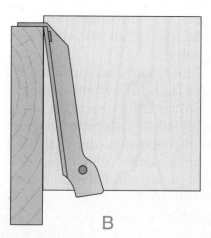

B

 A Joist Hanger

Joist Support A joist hanger secures a joist to an intersecting framing member.

Standard Joist Hangers Perhaps the most common metal connector is the joist hanger, shown in above. These sturdy brackets are used where floor or ceiling joists meet another framing member, such as a beam. Standard hangers are made from 18-gauge galvanized metal and are intended for use with solid lumber joists. They are typically installed with 10d common nails. However, always follow the manufacturer's recommendations for the type and size of nail. Special *joist-hanger nails* may also be supplied by the manufacturer. They are the same diameter as a 10d nail but shorter.

To install a joist hanger, first nail it to a beam. Then slip the joist into the hanger. Finally, hammer nails through the holes in the hanger and into the joist. Proper installation is important. If the sides of the hanger are spread too wide, the joist will be raised slightly, as shown in Improper Installation. This can cause a lump in the floor sheathing. If the seat of the hanger is "kicked out" from the beam, settling later on may cause the floor to squeak.

 Improper Installation

Avoid This An overspread hanger **A.** raises the height of the joist. If the hanger is "kicked out" from the header **B.** but not overspread, the floor may squeak.

The most common mistake made when installing joist hangers is to use too few nails. The connection depends on nails for shear strength. Undernailing can cause the connection to fail when loads are placed on

Builder's Tip

POSITIONING JOISTS To ensure that the tops of all floor joists are in exactly the same plane, a joist hanger can be nailed to the end of each joist first. Hold the joist in the desired position, then secure the hanger to the beam.

Jon P. Muzzarelli

Flush Edges

Alignment A joist hanger must be positioned so that the top of the joist is flush with other joists. The position of the joist's lower edge is less critical. *Why is the top edge more important?*

it. Check the manufacturer's instructions carefully. Use all the nails recommended.

Joist hangers are available in sizes to fit most common framing situations. Headers, for example, may be hung from a pair of joist hangers. Even large glulam beams and solid timbers may be secured with joist hangers. They can be used when the top edge of a joist must be at the same level as the top edge of an intersecting beam. They are also used when the bottom edges of the intersecting members must be flush, as in Flush Edges.

I-Joist Hangers Because the shape of I-joists might allow them to tip from side to side in a standard joist hanger, choose hangers

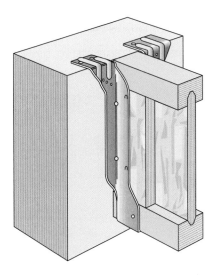

Web Stiffener

Extra Support Web stiffeners fit between the joist hanger and the I-joist. They prevent the web from buckling at stress points.

with care. If the sides of the hanger do not extend at least ⅜" up the sides of the I-joist's top flange, install a web stiffener. This will prevent the I-joist from tipping from side to side, as shown in Web Stiffeners. I-joist hangers are designed specifically for use with I-joists, and web stiffeners may not be needed.

When I-joists are attached to an intersecting I-joist, the joist hangers should be *backed*. To back joist hangers means placing backer blocks against the web, between the flanges of the supporting I-joist. Nails are driven into the blocks for extra support.

Builder's Tip

ENSURING MAXIMUM STRENGTH Metal brackets, anchors, and straps increase a structure's ability to resist severe weather and earthquakes. However, to be effective they must be installed *exactly* according to the manufacturer's specifications. It is particularly important to use the correct number and size of nails.

Reinforcement Metal framing ties connect and reinforce materials.

Tie Plate
Metal ties are being used here to connect various parts of a truss.

Railing Ties
Railing ties are first connected to the post. Galvanized screws may be used instead of nails because the connection is not load-bearing.

Post Base
A post base holds the wood slightly above the top of the pier to reduce the chance of rot.

Ties and Straps *Metal framing ties* are also used to hold pieces of wood together or to reinforce a joint. The most common form of tie is a flat strap, shown in Tie Plate in Connectors. The straps are perforated so they can be nailed in place without pre-drilling. They may also be bent to fit various angles.

A tie in an angular shape can be used to join wood members at right angles. Such ties do not carry structural loads. They simply hold the pieces of wood together. An example is the tie that connects deck railings to deck posts. The post holds the railing up. The tie simply holds them together. An advantage of this connection is that it removes the need for surface-nailing. It also reduces the chance that water will penetrate the area around the nails.

Other Metal Connectors The wide variety of metal connectors makes them useful from foundation to roof. Metal post bases can be embedded in concrete slabs or piers, as shown in Connectors. The base holds the wood post slightly above the level of the concrete. This reduces the possibility of rot. When the post is bolted to the post base, it is securely tied to its foundation.

Various types of metal clips and bracket can be used to tie rafters and trusses to a top plate. Metal connectors such as these are sometimes required in areas affected by earthquakes or severe weather. For example, metal hurricane clips that connect top plates to rafters prevent a roof from lifting off in high winds.

Nails for Framing Connectors

Most structural framing connectors are used in situations where the nails that fasten them are exposed to shear stresses. It is very important that the nails used are able to withstand this shear stress. Otherwise, the connection may fail even if the connector itself does not. For this reason, drywall screws should never be used because they do not have the necessary shear strength that nails have. Use only nails in most framing connectors. In some instances, bolts are also appropriate.

The length of nail required varies with the type of connection. The manufacturer's instructions include nail schedules. However, when 16d nails are specified, this generally refers to common nails, not 16d sinkers. (*Sinkers* are nails that are slightly thinner and shorter than common nails.) Some manufacturers provide special nails, sometimes called joist-hanger nails, for use with their connectors. Their larger diameter, as compared with standard nails of similar length, improves their shear strength.

Pneumatic nailers can be used to fasten metal connectors into place. However, you must be careful to place the nail through existing holes in the connector. A nail that pierces the metal elsewhere reduces the connector's strength. Some pneumatic nailers are specially designed for use when installing framing connectors. They are sometimes called *hardware framing nailers*. A hardened metal probe in the nose of the tool locates the connector hole and guides the nail into it as it is driven.

Some connectors have angled holes for nails, which increases the strength of the connection. However, they must be installed by driving the nails with a hammer, not a pneumatic nailer.

Section 14.2 Assessment

After You Read: Self-Check

1. How are a common nail and a finish nail different?
2. What is the purpose of a deformed nail shank?
3. What is galvanizing and what is its purpose?
4. Name the types of metal framing connectors.

Academic Integration: Science

5. **Matching Materials to Regional Needs** The proper use of metal framing connectors is very important in regions where earthquake hazards and high winds are common. Summarize the reason for this in one or two sentences. Then, contact a company that makes metal framing connectors and gather information about connectors designed for one of these two hazards. Determine which types of connections are used in your region and why. Summarize your findings in a one-page report.

Go to **connectED.mcgraw-hill.com** to check your answers.

Review and Assessment

CHAPTER 14

Chapter Summary

Section 14.1

Balloon-frame and platform-frame construction make use of many individual pieces of wood. Platform-frame construction is the most common system used to build houses. Other framing methods include post-and-beam framing, timber framing, and the use of structural insulated panels. The design of wood framing is based on laboratory tests of wood samples. The wood is given a rating based on its resistance to stresses. Test results are summarized in tables of design values.

Section 14.2

Choosing the correct nail is very important when assembling wood framing. Metal framing connectors improve the strength of joints. Types of connectors include standard joist hangers, I-joist hangers, and ties and straps. Always use the type and size of nail recommended for the installation of framing connectors.

Review Content Vocabulary and Academic Vocabulary

1. Use each of these content vocabulary and academic vocabulary words in a sentence or diagram.

Content Vocabulary

- balloon-frame construction (p. 370)
- platform-frame construction (p. 370)
- in-line framing (p. 373)
- post-and-beam framing (p. 373)
- structural insulated panel (p. 375)
- spline (p. 375)
- shear wall (p. 378)
- load (p. 379)
- design value (p. 379)
- span table (p. 382)
- on center (OC) (p. 382)
- dead load (p. 383)
- live load (p. 383)
- framing connector (p. 386)

Academic Vocabulary

- crucial (p. 378)
- function (p. 384)

Speak Like a Pro

Technical Terms

2. Work with a classmate to define the following terms used in the chapter: *sheathing* (p. 370), *ribands* (p. 370), *advanced framing* (p. 373), *structural timber* (p. 374), *manufactured housing* (p. 377), *hold-down anchors* (p. 379), *smooth-shank* (p. 384), *deformed-shank* (p. 384), *diamond point* (p. 385), *tie* (p. 387), *joist hanger* (p. 388), *backed* (p. 389), *sinkers* (p. 391), *hardware framing nailers* (p. 391).

Review Key Concepts

3. Summarize the process of balloon-frame construction.

4. Identify the six stresses structural lumber must resist.

5. Restate how SIPs are used in framing.

6. Explain what information is found on a span table.

7. List examples of a live load and a dead load.

8. Explain when a joist hanger is used during framing.

Critical Thinking

9. Compare and Contrast Describe the difference between in-line framing and standard platform frame construction.

Academic and Workplace Applications

STEM Mathematics

10. Problem Solving A rectangular building built with platform framing uses 24 joists set 19.2" O.C. What is the length in feet of the floor of the building?

Math Concept To solve this problem, you need to multiply the distance between each joist by the *number of spaces between joists,* not the number of joists. For help solving this problem, sketch the layout of the floor to figure out the number of spaces.

Step 1: Multiply the distance between each joist by the number of spaces between joists.

Step 2: Convert inches to feet.

21st Century Skills

11. Information Literacy Obtain a copy of the building code for your area. The reference section of your local library may have a copy. Locate references to the fire safety of foam plastic insulation, such as the type of foam plastic insulation used in structural insulated panels (SIP). What is this material made of? What safety precautions must be taken when using this material during construction? Summarize your findings in a one-page list or report.

21st Century Skills

12. Problem Solving Use the six-step problem solving process to describe how to prevent one type of framing damage caused by severe weather. Create a graphic organizer or list and fill in the steps of the problem-solving process.

1. **State the problem clearly.** This helps define what needs to be done.
2. **Collect information.** What is causing the problem? What resources are available?
3. **Develop possible solutions.** Consider several ideas.
4. **Select the best solution.** Look at the advantages and disadvantages of each.
5. **Test what appears to be the best solution.** This will reveal its strengths and weaknesses.
6. **Evaluate the solution.** Is it effective? If not, select another possibility and test that one until you find one that works.

Standardized TEST Practice

Short Answer

Directions Write one or two sentences to answer the following questions. Use a separate piece of paper to record your answers.

13. What are two natural hazards that can cause damage to housing frames?

_____

_____

14. What two features make flat straps the most common form of tie?

_____

_____

15. How does galvanizing protect nails?

_____

_____

TEST-TAKING TIP

Underline key words in short answer questions. This can help you make sure that you understand the question.

*These questions will help you practice for national certification assessment.

Floor Framing

Section 15.1
Floor Framing Basics

Section 15.2
Framing with Joists & Girders

Section 15.3
Subfloors

Chapter Objectives

After completing this chapter, you will be able to:

- **Identify** the basic floor-framing components and explain the purpose of each.
- **Construct** posts and girders.
- **Explain** sill plates and the layout of basic joist spacing.
- **Recognize** cases where special framing details may be required, such as beneath a bearing wall.
- **Construct** a panel subfloor.
- **Recognize** other types of framing systems and products, such as those using trusses and girders.

Discuss the Photo

Subflooring The layer of material that goes over a floor frame is called the subfloor. *What might go over a subfloor?*

Writing Activity: Observe and Record

Examine the floor system at your house or the house of a friend. Can you determine which trades might be affected by the work of the floor-framing carpenter? With a classmate, try to identify other parts of the house that could be affected by the quality of the floor framing. Summarize your thoughts in a bulleted list.

pojoslaw/Getty Images

Chapter 15 Reading Guide

Before You Read Preview

Floor framing depends on a precise arrangement of joists, beams, and panel products in order to perform properly. Choose a content vocabulary or academic vocabulary word that is new to you. When you find it in the text, write down the definition.

Content Vocabulary

- ○ girder
- ○ box sill
- ○ subflooring
- ○ crown
- ○ bridging
- ○ bearing wall
- ○ header
- ○ trimmer joist
- ○ tail joist
- ○ cantilever

Academic Vocabulary

You will find these words in your reading and on your tests. Use the academic vocabulary glossary to look up their definitions if necessary.

- ■ criteria
- ■ perpendicular
- ■ offset

Graphic Organizer

As you read, use a chart like the one shown to organize information about content vocabulary and their definitions, adding rows as needed.

Content Vocabulary	Definition
girder	a large principal horizontal number used to support floor joists

Go to connectED.mcgraw-hill.com to download this graphic organizer.

Posts, Girders, & Trusses

What components go into floor framing?

When building a single-level house on a concrete slab, the slab itself serves as the floor system. However, houses that have a basement or crawl-space foundation have a floor system that is assembled from various types of wood framing and panel products. Floor framing consists of posts, girders, sill plates, joists or trusses, and subflooring. These framing members are fastened together to form a strong platform that supports the house. In first-floor framing, joists rest on the sill plate. They may also rest on girders or be attached to them. In second-floor framing, they rest on a double top plate. Once covered with sheathing, floor joists distribute loads to the foundation walls and provide a solid base on which to walk.

Nominal 2" lumber or laminated-veneer lumber (LVL) I-joists are generally used for floor framing. Floor trusses can be used either as girders or as floor joists. Engineered materials offer the advantages of light weight, consistent strength, and long spans (see Chapter 13, "Engineered Wood"). This chapter describes floor framing using conventional lumber. However, framing with LVL I-joists calls for similar techniques. Where they differ, the differences will be noted.

In the average house, the distance between opposite foundation walls is too great for a single floor joist to span. A *floor joist* is any light beam that supports a floor. Pairs of joists are often used to span the distance. The outer ends of the paired joists are supported by the foundation walls, while the inner ends are supported by a girder. A **girder** is a large principal horizontal member used to support floor joists. The ends of a girder are supported by the foundation walls. A *post* is a wood or steel vertical member that provides intermediate support for a girder. Girders and posts can be made out of solid lumber, engineered lumber, or steel. Posts are sometimes made from concrete or brick.

Posts

Posts are often used in basements beneath the main girder. They may also be used in a garage to support ceiling girders. They are generally spaced 8" to 10" on center (OC). The exact spacing depends on the size of the load.

A wood post must be solid and not less than a 4×4. It is often a 6×6. Its ends must be flat and securely fastened. The loads it carries are transferred to a fairly small area of the foundation. Because of this, a portion of the concrete slab directly below a post is made thicker to provide greater bearing capacity, as shown in **Thickened Slab**. This must be done when the slab is poured. In some cases, a wood post is supported by a heavy-gauge metal bracket that lifts it clear of any moisture that might cause rot.

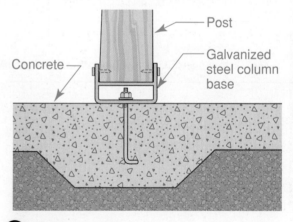

Thickened Slab
Extra Support A concrete slab floor must be made thicker to support the load from a post. The thickened portion is called a footing.

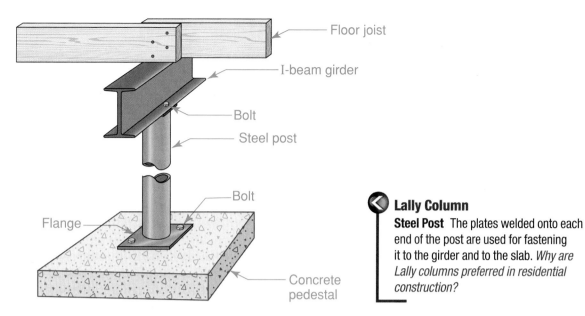

Floor joist

I-beam girder

Bolt

Steel post

Bolt

Flange

Concrete pedestal

Lally Column
Steel Post The plates welded onto each end of the post are used for fastening it to the girder and to the slab. *Why are Lally columns preferred in residential construction?*

Steel posts, sometimes called *Lally columns,* are often preferred in residential construction. They are strong, easy to handle, and take less space than solid wood posts. They must be at least 3" in diameter and protected against rust, such as with rust-resistant paint. Steel posts have steel bearing plates at each end. In some cases the top plate has metal straps that can be used to secure it to a steel girder. Some posts are adjustable to various lengths. Steel posts are sometimes filled with concrete for extra strength.

Girders

Girders are generally placed halfway between the longest foundation walls and parallel to them. They may be made of wood or steel. Steel does not shrink as solid wood does. However, wood girders are lighter and therefore easier to install. It is also easier to connect floor joists to wood girders.

In one framing method, the floor joists rest on top of the girder as shown in **Joists on Top**. However, the top of the girder must be at the same level as the top of the sill plate. (The *sill plate* is the horizontal framing member anchored to the foundation wall.) If more clearance is needed, the girder can be installed so that its top surface is level with the top of the floor joists as shown in **Joists Flush** on page 398. In this case, metal framing connectors attached to the girder support the ends of the joists.

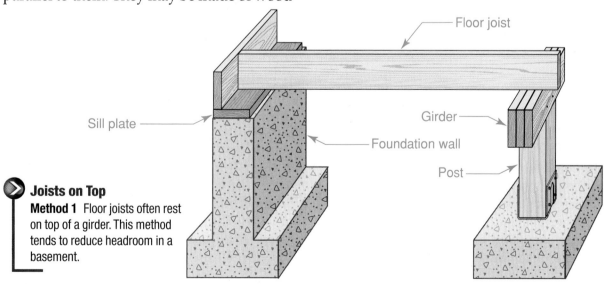

Floor joist

Sill plate

Girder

Foundation wall

Post

Joists on Top
Method 1 Floor joists often rest on top of a girder. This method tends to reduce headroom in a basement.

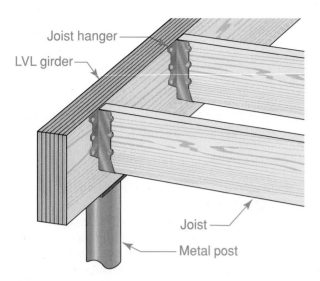

Joist hanger

LVL girder

Joist

Metal post

◆ Joists Flush
Method 2 Joists can also be supported by joist hangers nailed to the side of the girder.

Wood Girders Wood girders are available in several forms. Solid wood girders were once common, but solid wood of suitable size and quality is no longer readily available. Another approach is to use built-up wood girders. These are built on-site and consist of three or four pieces of solid lumber nailed face to face, as shown in **Built-Up Girder on page 399**. To determine material requirements, refer to **Table 15-1**.

To make a built-up girder, face-nail each layer with 10d nails as follows:

- Stagger the nails 32" OC at top and bottom.
- Nail two or three times at the end of every board (depending on size), including splices.
- Stagger the joints.

Glue-laminated beams and laminated-veneer lumber can also be used as girders. They offer the dimensional stability of steel and the easy installation of wood.

Steel Girders A steel beam has great strength and is often used in floor systems as a girder. Like a wood I-joist (see Section 13.2), a steel beam consists of horizontal flanges separated by a vertical element called a web. The depth of the web and the

Builder's Tip

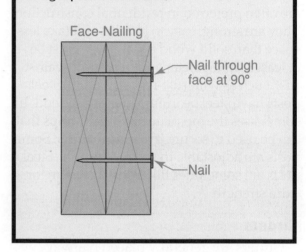

Face-Nailing

Nail through face at 90°

Nail

thickness of the beam's steel determine how strong the beam is, not the size of the flanges. Steel girders come in two basic shapes. An I-beam looks like a capital I, with the web wider than the flanges. A W-beam, sometimes called a wide-flange beam, is squatter in appearance because the web and flanges are the same width.

Steel beams are ordered based on two **criteria**: height and weight. Height is measured from flange to flange. Weight is based on the weight of a one-foot length of beam. For example, a steel beam might be specified this way: W12 × 45#. This would refer to a wide flange beam (W)

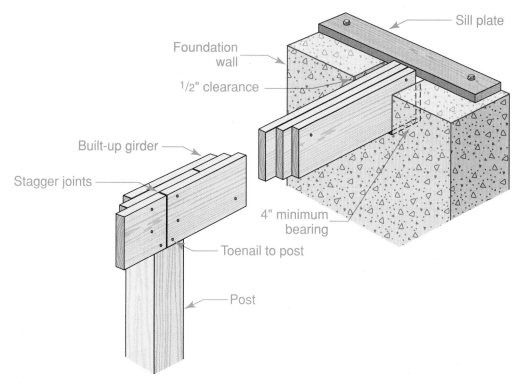

Sill plate

Foundation wall

1/2" clearance

Built-up girder

Stagger joints

4" minimum bearing

Toenail to post

Post

 Built-Up Girder
Layers of Lumber A built-up girder consists of layers of solid wood. This girder fits into a notch in the foundation wall called a *beam pocket*.

Table 15-1: Estimating for Built-Up Girders[a]		
Size of Girder	Board Feet per Lineal Foot	Nails per 1,000 Board Feet
4×6	2.15	53
4×8	2.85	40
4×10	3.58	32
4×12	4.28	26
6×6	3.21	43
6×8	4.28	32
6×10	5.35	26
6×12	6.42	22
8×8	5.71	30
8×10	7.13	24
8×12	8.56	20

[a] A 4×6 girder 20" long contains 4" board feet of lumber (20 × 2.15 = 43).

with flanges that are 12" wide. The beam would weigh 45 lbs. per foot. An I-beam would be designated with an S instead of a W. If it is to be used as a girder, a wood bearing plate must be attached to the top of the beam. The plate will enable you to toenail floor joists to the beam, as shown in Bearing Plate on page 400.

Wood bearing plates can be fastened to steel with steel pins. The pins are driven with a powder-actuated fastening tool. An explosive gunpowder charge, called a *load*, drives the hardened steel pins into steel or concrete. Attaching the plate to a girder with pins may be easier to do before the girder is lifted into place. As an alternative, the steel fabricator can weld short lengths of threaded steel rod to the top of the girder. The wood plate can then be drilled to match and be secured with nuts and washers.

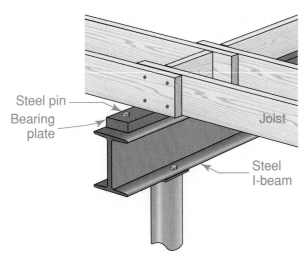

Bearing Plate
Attachment Aid Fasten a wood bearing plate to a steel girder so that joists can be nailed to the plate. This girder is an I-beam.

Floor Trusses

A floor truss is another structural product that can be used as a girder. These trusses are made in a factory to the specifications of the job. Floor trusses can also be used in place of lumber joists where long spans are required, as shown in Parallel-Chord Trusses. A floor truss has only three basic parts: chords, webs,

and connector plates. The open webs allow heating ducts, water lines, drain lines, and other items to be passed through with ease. The most common type of floor truss in residential construction is the *parallel-chord floor truss.* The top and bottom chords are parallel to each other over the length of the truss.

It is important to brace floor trusses as they are being installed. This increases the safety of workers installing the subfloor. It also prevents the trusses from being damaged by sideways movement before the subfloor is laid. The truss fabricator can provide detailed bracing instructions.

Installing Posts and Girders

Posts and girders are installed after the foundation walls are complete and the forms have been stripped. In houses that will have a basement, the basement floor slab may also be in place. Posts should be located only where the slab has been made thicker to distribute the expected loads. To locate these points, metal anchors are sometimes placed when the slab is poured. The posts can then be plumbed, braced temporarily, and bolted to the anchors. If there are no anchors to indicate the location of posts, consult the plans for the location.

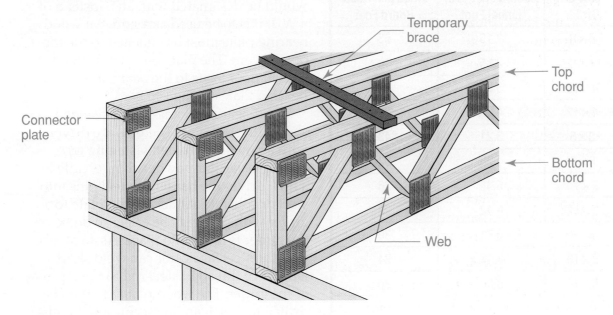

Parallel-Chord Trusses
Engineered Support Floor trusses should be braced temporarily with 2×4 stock until they are sheathed.

After the posts are in place, install the girder and brace it if necessary. Steel and glulam girders are placed by a small crane or a special forklift. Solid wood and built-up girders can sometimes be lifted into position by hand. The ends of wood girders should bear at least 4" on masonry walls. This will reduce the risk of crushing the wood fibers. A ½" clearance should be provided at each end and at each side of a wood girder framed into the masonry (see Built-Up Girder on page 399). This will prevent the wood from absorbing moisture from the masonry. The bearing details for a steel girder are specified on the plans.

To determine the height of a post, stretch a string line tightly across the foundation. Measure from the floor to the line. Then calculate the length of the post based on deducting the depth of the girder. Be sure to account for the thickness of any other elements that will be attached to the post, such as bearing plates. For a prefabricated steel post, the height from the floor to the bottom of the girder is all that is needed. The post should be centered on the girder's width.

JOB SAFETY

POWDER-ACTUATED TOOLS Because a powder-actuated fastening tool can be dangerous if used improperly, special training is required. Only qualified users should operate this tool. It is also essential that users wear hearing and eye protection. Others in the area should be warned when the tool is to be used. See Chapter 3 for more on safety requirements for powder-actuated tools.

Section 15.1 Assessment

After You Read: Self-Check

1. What is a girder, and what supports it?
2. What is a Lally column, and why are Lally columns often preferred in residential construction?
3. How is a built-up wood girder constructed?
4. What must be done to a steel beam before it is ready to use as a girder?

Academic Integration: Mathematics

5. **Word Problems** A 56' long girder will be supported at each end by foundation walls. It will be supported along its length by posts spaced 8' OC. How many posts are needed to support the girder?

   Math Concept Many word problems can be solved using the four basic operations: addition, subtraction, multiplication, and division. Identify what information is required and choose the operations that you need to solve the problem.

   Starting Hint Divide the length of the girder in feet by the OC spacing of the posts in feet.

🚀 Go to connectED.mcgraw-hill.com to check your answers.

Framing with Joists & Girders

Framing with Joists

What is the lowest member of a wood frame?

Solid lumber was once the only material used for floor joists but modern builders now use laminated-veneer lumber (LVL) as well. The most common LVL product is the I-joist, shown in **below**. The information that follows relates to solid lumber as well as to LVL floor framing. Any differences between the two will be noted.

Box sill construction is used for platform-framed floors. A box sill consists of a sill plate (also called a *mudsill,* or just the *sill*) and a rim (or *band*) joist. The sill plate is anchored to the foundation wall and the rim joist is toenailed to the sill plate. Floor joists and subflooring complete the floor system. **Subflooring** consists of engineered wood sheets such as plywood or oriented-strand board (OSB). The elements of box sill construction are shown in Box Sill Construction.

Installing the Sill Plate

The sill plate is the lowest member of the wood frame and provides a smooth bearing surface for the floor joists. It is made of 2×4 or 2×6 preservative-treated lumber. Preservatives protect the wood against moisture damage and insect attack. Joists are typically toenailed to the plate but may be secured with metal framing anchors where required by code.

Sill plates establish the quality of all the framing that will follow. If they are not level, the entire floor system will not be level. If they are poorly secured to the foundation, the house may not survive very severe weather or an earthquake. Follow local building codes carefully.

Anchor bolts or straps for the sill plates are installed when the foundation is placed. However, always verify that the bolts have been placed correctly. Sill plates should be anchored to the foundation with at least two ½" bolts in each plate. According to the building codes, the bolts must be spaced no more than 6' OC and there must be a bolt within 12" of each end of every plate.

To prevent cold air from leaking into the house, the plate should be set on top of a foam or fiberglass *sill sealer.* This product comes in

I-Joist
Floor Framing Option This I-joist is typical of those used for floor framing.

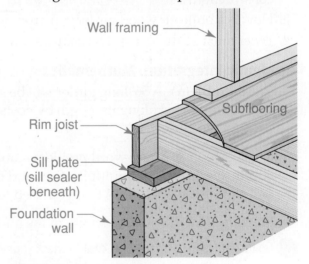

Wall framing

Subflooring

Rim joist

Sill plate
(sill sealer
beneath)

Foundation
wall

Box Sill Construction
Key Elements First-floor framing at the exterior wall using box-sill construction.

David R. Frazier Photolibrary, Inc.

a roll approximately 6" wide. It fills any gaps between the irregular surface of the concrete and the smooth surface of the sill plate.

Steps for installing sill plates are given on page 404.

Installing Lumber Joists

Joists are usually placed perpendicular to girders on 16" or 24" centers. However, check the house plans for the exact size, spacing, and direction of the joists. If the sizes for joists are not specified, refer to **Table 15-2**.

Plans will also specify a lumber grade. The **Ready Reference Appendix** table

REGIONAL **CONCERNS**

Termite Shields Where termites are a significant problem, a metal termite shield should be installed beneath the sill plate. This shield makes it difficult for termites to reach wood members directly from the foundation walls without being noticed.

Nonstress-Graded Lumber provides information about grades.

Joist Spacing (inches)	Species and Grade		Dead Load = 10 psf				Dead Load = 20 psf			
			2×6	2×8	2×10	2×12	2×6	2×8	2×10	2×12
			Maximum floor joist spans							
			ft.-in.	ft.-in.	ft.-in.	ft.-in.	ft.-in.	ft.-in.	ft.-in.	ft.-in.
16	Douglas fir-larch	SS	11-4	15-0	19-1	23-3	11-4	15-0	19-1	23-0
	Douglas fir-larch	#1	10-11	14-5	18-5	21-4	10-8	13-6	16-5	19-1
	Douglas fir-larch	#2	10-9	14-1	17-2	19-11	9-11	12-7	15-5	17-10
	Douglas fir-larch	#3	8-5	10-8	13-0	15-1	7-6	9-6	11-8	13-6
	Hemlock-fir	SS	10-9	14-2	18-0	21-11	10-9	14-2	18-0	21-11
	Hemlock-fir	#1	10-6	13-10	17-8	20-9	10-4	13-1	16-0	18-7
	Hemlock-fir	#2	10-0	13-2	16-10	19-8	9-10	12-5	15-2	17-7
	Hemlock-fir	#3	8-5	10-8	13-0	15-1	7-6	9-6	11-8	13-6
	Southern pine	SS	11-2	14-8	18-9	22-10	11-2	14-8	18-9	22-10
	Southern pine	#1	10-11	14-5	18-5	22-5	10-11	14-5	17-11	21-4
	Southern pine	#2	10-9	14-2	18-0	21-1	10-5	13-6	16-1	18-10
	Southern pine	#3	9-0	11-6	13-7	16-2	8-1	10-3	12-2	14-6
	Spruce-pine-fir	SS	10-6	13-10	17-8	21-6	10-6	13-10	17-8	21-4
	Spruce-pine-fir	#1	10-3	13-6	17-2	19-11	9-11	12-7	15-5	17-10
	Spruce-pine-fir	#2	10-3	13-6	17-2	19-11	9-11	12-7	15-5	17-10
	Spruce-pine-fir	#3	8-5	10-8	13-0	15-1	7-6	9-6	11-8	13-6
24	Douglas fir-larch	SS	9-11	13-1	16-8	20-3	9-11	13-1	16-2	18-9
	Douglas fir-larch	#1	9-7	12-4	15-0	17-5	8-8	11-0	13-5	15-7
	Douglas fir-larch	#2	9-1	11-6	14-1	16-3	8-1	10-3	12-7	14-7
	Douglas fir-larch	#3	6-10	8-8	10-7	12-4	6-2	7-9	9-6	11-0
	Hemlock-fir	SS	9-4	12-4	15-9	19-2	9-4	12-4	15-9	18-5
	Hemlock-fir	#1	9-2	12-0	14-8	17-0	8-6	10-9	13-1	15-2
	Hemlock-fir	#2	8-9	11-4	13-10	16-1	8-0	10-2	12-5	14-4
	Hemlock-fir	#3	6-10	8-8	10-7	12-4	6-2	7-9	9-6	11-0
	Southern pine	SS	9-9	12-10	16-5	19-11	9-9	12-10	16-5	19-11
	Southern pine	#1	9-7	12-7	16-1	19-6	9-7	12-4	14-7	17-5
	Southern pine	#2	9-4	12-4	14-8	17-2	8-6	11-0	13-1	15-5
	Southern pine	#3	7-4	9-5	11-1	13-2	6-7	8-5	9-11	11-10
	Spruce-pine-fir	SS	9-2	12-1	15-5	18-9	9-2	12-1	15-0	17-5
	Spruce-pine-fir	#1	8-11	11-6	14-1	16-3	8-1	10-3	12-7	14-7
	Spruce-pine-fir	#2	8-11	11-6	14-1	16-3	8-1	10-3	12-7	14-7
	Spruce-pine-fir	#3	6-10	8-8	10-7	12-4	6-2	7-9	9-6	11-0

Table 15-2: Floor Joist Spans for Common Lumber Species

Source: 2006, International Code Council, Inc., Falls Church, Virginia. 2006 International Residential Code. Reprinted with permission of the author. All rights reserved.

Installing Sill Plates Sill plates on opposite walls should be installed so that they are parallel to each other.

Step 1 Check to see that foundation is level and square. To check a simple rectangular foundation, measure diagonally from corner to corner. To check a more complex foundation, use the 3-4-5 method for squaring walls (see Chapter 16, Wall Framing & Sheathing). Adjust the plates as needed so they will be square.

Step 2 Establish the location of the sill plate. From the outside edge of the foundation wall, measure back a distance equal to the width of the sill plate. If the outside of the wall sheathing will be flush with the outside edge of the foundation wall, measure back the width of the sill plate plus the thickness of the sheathing.

Step 3 Place sill plate stock around the foundation. Use only straight, flat lumber that has been preservative-treated. Any wood placed against concrete should be preservative-treated to resist rot. Place the edge of each piece against the foundation anchor bolts and mark the centerline of the bolts on the plate. Using a square, extend these marks across the width of the sill plate.

Step 4 Measure from the center of each bolt to the chalk line on the foundation. Measure the same distance on the plate, starting from the edge that is resting against the bolt. Mark the bolt centerline at this point.

Step 5 Using a ⅝" spade bit, bore holes through the plate at each marked point. (If termite shields will be used, bore holes at the same locations in them, using a suitable drill bit.)

Step 6 Roll out sill sealer over the top of the foundation walls and press it into place. The weight of the building presses the sill sealer against the foundation wall to stop drafts.

Step 7 Slip the sill plate over the anchor bolts. Start at the high point of the foundation wall and check to see that the sill plate is level, using a builder's level. An even more accurate technique would be to check the plate with a laser level set up in a location that would allow you to check the entire foundation. Shim beneath the sill plate with cement grout or with preservative-treated wood shims, as needed, to make it level.

Step 8 Place a flat washer and a nut on each foundation bolt. Use a wrench to tighten each nut securely.

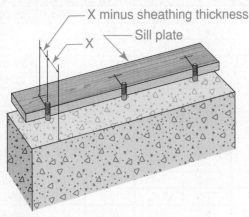

Sill Plate Layout

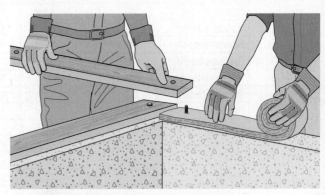

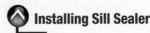
Installing Sill Sealer

Installing a Joist Hanger

Secure Connection Joist hangers strengthen the connection between floor joists and the rim joist.

Joists are often nailed into place. However, metal connectors can be used to replace many of the nailed connections described in this chapter. The most common metal connector used in floor framing is the joist hanger, shown being installed above.

Joist Layout

Joists are spaced evenly from one end of the house to the other. However, certain factors may interrupt this spacing, such as the need to provide a stairwell opening or extra room for plumbing drain lines. Always consult the plans to identify these issues before you begin the joist layout.

Use a tape measure to lay out the desired joist spacing on the sill or wall plate. Begin the layout by measuring from the corner of the sill plate. Make a mark 15¼" from the outside edge of the sill plate. This will be the location to the edge of the first joist. Mark an X on the side of the line where the joist will be, as shown in Locating the First Joist. That will ensure that the joist will not be placed on the wrong side of the layout line. From that point on, mark every 16" to indicate the positions of all the joists on that plate, as shown in Layout of Remaining Joists.

When you have marked the position of all the joists, double-check to be sure that a joist is centered every 4'. When joists are laid out in this way, the edges of the floor sheathing panels will always fall along the centerline of a joist.

If the joists span the distance from one foundation wall to the opposite wall, as I-joists often do, the layout on both sill plates

Locating the First Joist

Initial Layout The X next to each layout mark indicates which side of the line the joist will be on.

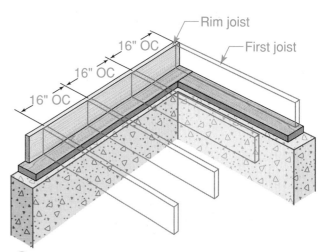

Layout of Remaining Joists

Important Location Notice that the edge of the first joist will be 15¼" from the outside edge of the sill plate.

will be identical. However, when joists will be overlapped at a girder, the layouts will differ as shown in Layout of Lapped Joists. In this case, first mark the layout on one sill plate. Mark the same layout on the girder. On the opposite wall, **offset** the position of joists by 1½" (the thickness of a joist) to ensure that the joists overlap at least 3". In this case, the floor sheathing panels will slightly overhang the offset joists.

Select floor joist lumber carefully so the floor will be flat and strong. Any joists having a slight edgewise bow should always be placed with the crown on top. The **crown** is the outermost curve of the bow. Mark the joist with an arrow pointing to the crown. A crowned joist will tend to straighten out when subfloor and normal floor loads are placed on it. Be sure that knots in the joist

are on its compression (top) side. They are less likely to cause failure in this location. A large knot on the tension side of a joist can be pulled apart, weakening the joist as shown in The Effect of Lumber Defects.

Alternate Layout Method Some carpenters lay out the spacing of the floor joists by marking the sill plates. Others prefer to mark their layout on the top edge of the rim joists. This is done after the rim joist has been toenailed to the plate.

Installing I-Joists

The specific installation details for I-joists vary according to the product's manufacturer. Always read the instructions for the specific I-joist you plan to use. You must

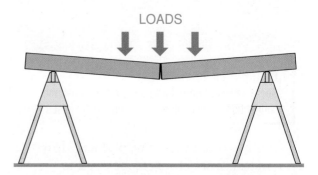

Saw cut at bottom of board opens up

A

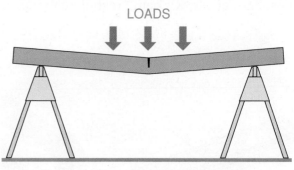

Saw cut at top of board will compress

B

Layout of Lapped Joists
Layout Offset Lapped joists call for a different layout on opposite plates to compensate for the thickness of the joists.

The Effect of Lumber Defects
Choose Joists Carefully A knot or other defect in a joist will have the same results as a saw cut. **A.** With a saw cut at the bottom, the joist opens up and breaks. **B.** A saw cut at the top closes up (compresses), so the joist retains more strength.

Mathematics: Measurement

Squaring Floors Floors need to be "squared up" as they are being constructed. All four angles of a rectangular wall or floor must be right angles. The diagonals running from opposite corners of a rectangular wall must also be equal. It is important to know the length of each of these diagonals. Calculate the length of the diagonals of a floor that measures 24' × 32'.

Starting Hint The Pythagorean Theorem states that the square of the hypotenuse of a right triangle is equal to the sum of the squares of the other two sides.

$$a^2 + b^2 = c^2$$
$$a = \text{altitude}; \; b = \text{base}; \; c = \text{hypotenuse}$$

also check the span tables for that product, not the span tables for lumber joists. Do not assume that the span listed for one line of engineered lumber products will be the same as the span for another line. In general, however, an I-joist floor system will have details such as those shown in An I-Joist Floor System. Layout of I-joists on the plates is largely the same as layout for lumber joists. The difference is that I-joists are not generally lapped over a girder because they are available in long lengths.

The following details are generally used when installing I-joists.

Backing and Blocking I-joists are nearly always supported by metal joist hangers. The width of the hanger should match the width of

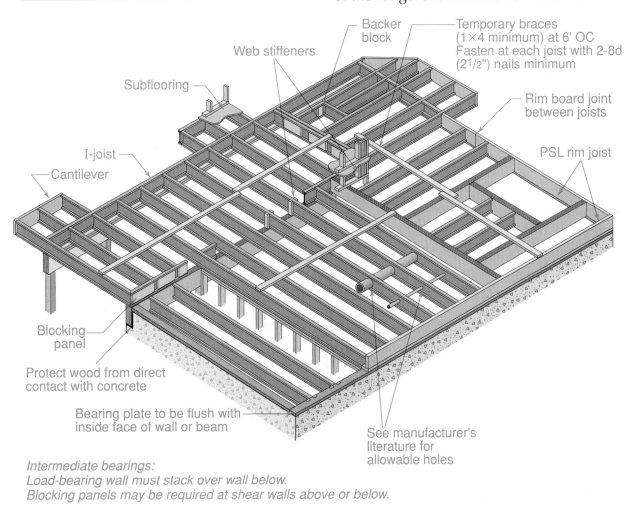

Backer block

Temporary braces (1×4 minimum) at 6' OC Fasten at each joist with 2-8d (2¹/₂") nails minimum

Web stiffeners

Subflooring

Rim board joint between joists

PSL rim joist

I-joist

Cantilever

Blocking panel

Protect wood from direct contact with concrete

Bearing plate to be flush with inside face of wall or beam

See manufacturer's literature for allowable holes

Intermediate bearings:
Load-bearing wall must stack over wall below.
Blocking panels may be required at shear walls above or below.

An I-Joist Floor System
Typical Details The overall arrangement of I-joists is similar to a solid lumber floor system but many details differ.

the I-joist. A backing block can be nailed to both sides of the I-joist to improve the fit as shown in **Backing Blocks**.

Where an I-joist runs continuously over a support (a girder, for example), web stiffeners should be nailed to both sides of the web. They will improve load-bearing ability of the I-joist. The stiffeners can also provide additional bearing surface for lumber or I-joist blocking. Depending on the dimension of the I-joist, web stiffeners may be made of ½", ⅝", or 1" thick plywood or OSB-rated sheathing. I-joists with unusually wide flanges may even require a web stiffener made from nominal 2" lumber. Stiffeners are ⅛" shorter than the exposed portion of the web so that they do not force the upper flange out of position. One method of installing a stiffener is shown in Web Stiffeners. A method recommended by another manufacturer calls for

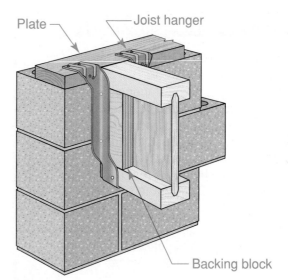

Plate — Joist hanger

Backing block

Backing Blocks
Improves Fit A backing block prevents an I-joist from moving in the joist hanger.

Step-by-Step Application

Installing Lumber Joists Inspect each piece for usability before carrying it to the foundation. Set aside unusable lumber to cut up later for blocking.

Step 1 Toenail the rim joists to the sill plates, using 16d nails every 16" OC. Be sure that the outside face of the joist is in the same plane as the outside edge of the sill plate.

Step 2 Place the joists over the layout marks, laying them flat for now. Add extra joists or leave out joists where large openings will be located.

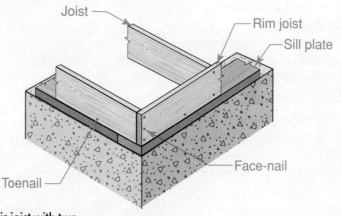

Joist — Rim joist — Sill plate — Face-nail — Toenail

Step 3 Tip the outermost joist up on edge and align one end with the end of a rim joist. Nail through the rim joist and into this joist with two 16d nails as shown in the figure below. Toenail this joist to the plate with 16d nails spaced 16" OC.

Step 4 Proceeding from one end of the house to the other, tip each joist on edge, crown up, and align it with the layout marks. This process is sometimes called rolling the joists. Nail through the rim joist and into each floor joist with two 16d nails. Toenail each joist to the sill plate with three 8d nails.

Step 5 Toenail each joist to the girder with 8d nails.

Step 6 Face-nail overlapping joists to each other with at least three 10d nails.

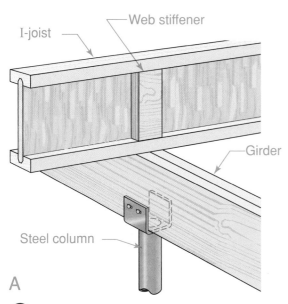

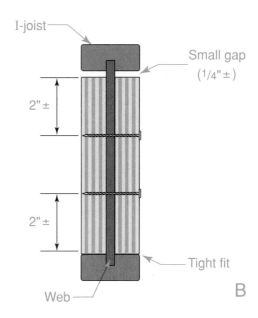

I-joist

Web stiffener

Girder

Steel column

A

I-joist

Small gap
(1/4" ±)

2" ±

2" ±

Tight fit

Web

B

 Web Stiffeners

Installing Stiffeners Web stiffeners should be installed on both sides of an I-joist where it crosses a support, as shown in **A.** One method of installing stiffeners is shown in **B.** In this case, spiral-shank nails were used.

installing the stiffeners using 10d *clinched* nails. A clinched nail is a nail of any size or type whose point has been bent over with a hammer where it exits the wood. This prevents the nail from pulling out. Always follow manufacturer's nailing details when installing stiffeners.

Where solid-wood blocking would be required on a solid-wood joist system, similar blocking may be required for an I-joist floor. Short lengths of I-joist stock may be used instead of solid lumber. Where I-joist blocks intersect I-joists, a backing block or web stiffener may be required.

Rim Boards A solid-wood rim joist is not suitable for use with an I-joist floor. Instead, an engineered product such as LVL is used. This piece, shown in Rim Board, is called a *rim board.* A rim board ties the ends of the I-joists together. It does not shrink as much as solid lumber and comes in lengths of up to 24'. It may be up to 1⅛" thick and must be the same depth as the I-joists.

To install a rim board, toenail it to the sill plate with 8d common or box nails spaced 6" OC, or as recommended by the manufacturer. When nailing through the rim board and into an I-joist, make sure one 8d

common or box nail penetrates the center of each flange.

Cutting and Notching Floor Joists

During installation of ducts, plumbing pipes, or wiring, solid-wood floor joists must sometimes be notched or drilled. A *notch* is a saw cut made in the end or edge

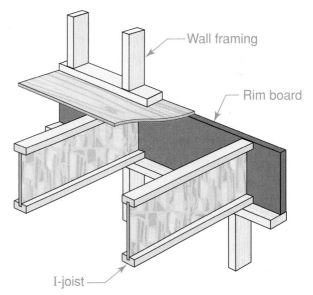

Wall framing

Rim board

I-joist

 Rim Board

End Tie A rim board ties the ends of I-joists together. Here it is shown on second-story in-line floor framing.

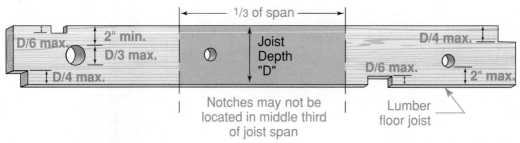

1/3 of span

D/6 max.

2" min.
D/3 max.

D/4 max.

Joist
Depth
"D"

D/4 max.

D/6 max.

2" max.

Notches may not be
located in middle third
of joist span

Lumber
floor joist

 Notching Lumber Joists
Code Restrictions The location and size of notches permitted in a solid-wood floor joist are based partly on the depth (D) of the floor joist.

of lumber. In some cases a notch might also be made when the joists are being installed. Careless notching or drilling will reduce the strength of a joist. For this reason, building codes restrict notching, as shown in Notching Lumber Joists. For example, joists must never be notched in the middle one-third of their span or on the bottom edge. Codes also restrict the size and position of holes. For example, holes must be at least 2" away from the top and bottom of a joist.

The situation with I-joists is more straightforward. Except when cutting an I-joist to length, the flanges must *never* be cut, drilled, or notched. The existing knockouts in the web of the I-joist are used for utility access. These pre-scored holes can be punched out by striking them with a hammer. One such hole is

visible on p. 402. If holes of any other size are needed, consult the manufacturer's literature.

Bridging Long joists should be stiffened with bridging. **Bridging** is a method of bracing between joists. It is done to distribute loads, prevent the joists from twisting, and add stability and stiffness.

There are two types of bridging. *Solid bridging* is made of solid lumber that is the same dimension as the joists, as shown in A in Joist Bridging. *Cross bridging* (also called *diagonal bridging*) is more common because it is very effective and requires less material. Precut 1×3 or 2×2 lumber is sometimes used for cross bridging, as in B in Joist Bridging. Metal-strap cross bridging with nailing flanges may also be used, as in C in Joist Bridging.

Builder's Tip

INSTALLING BRIDGING Before installing bridging, snap a chalk line across the tops of the joists as a guide. When installing solid-wood cross bridging, leave the bottom ends loose until the subfloor has been laid. This permits the joists to adjust themselves to their final positions. Then, complete the nailing later.

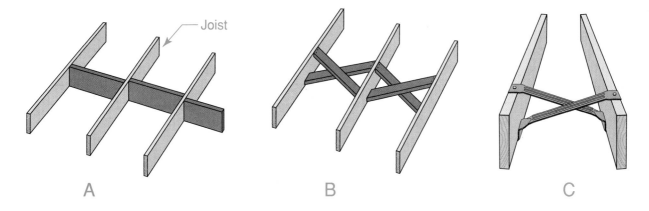

A   B   C

Joist

 Joist Bridging

Three Types Bridging comes in various forms. **A.** Solid bridging. Offsetting the blocks allows them to be end nailed. **B.** Wood cross bridging. **C.** Metal-strap cross bridging.

Bridging is not generally required by code unless joists exceed 2×12 in depth. However, bridging is a cost-effective and efficient way to stiffen a floor. Many builders add it even though it is not required. If the joists are over 8' long, install one row of bridging at the center of the joist span. For joists 16' and longer, install two rows of bridging equally spaced on the joist span.

Special Floor Framing Conditions

Often a carpenter must adjust the layout of a floor system to accommodate special conditions. These conditions should be identified before the joist layout begins.

Framing Under Bearing Walls Joists should be doubled under each load-bearing wall that is parallel to the joists. If needed, a double joist hanger or two joist hangers could be used as in Joists Under Bearing Walls. A **bearing wall** is a wall that supports loads in addition to its own weight. If the wall will contain plumbing pipes or heating ducts, the joists can be separated by blocking. The blocking must be cut from the same size stock as the floor joists. Blocking should be spaced not more than 4' OC.

Framing Large Openings It is often necessary to create large openings in the floor system, such as for framing around stairwells and

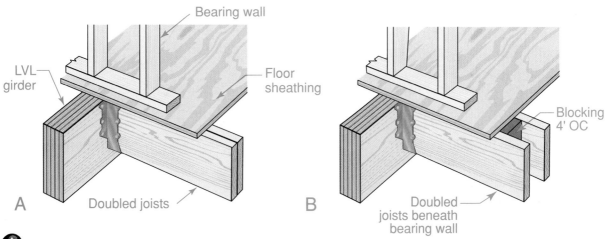

Bearing wall

LVL girder

Floor sheathing

A   Doubled joists

Blocking 4' OC

B   Doubled joists beneath bearing wall

 Joists Under Bearing Walls

Added Support Floor joists should be doubled under bearing walls, as in **A.** If clearance space is needed for plumbing drains, the joists can be installed as in **B.**

chimneys. In these cases, the joists framing the opening should be doubled, and the interrupted joists must be supported by headers as in Joist Headers. A **header** is a horizontal member that carries loads from other members and directs them around an opening. In a floor system, a header is supported by trimmer joists. A **trimmer joist** is used to form the sides of a large opening. A **tail joist** is a floor joist interrupted by a header.

The exact layout of headers is based on the flooring loads, as well as the size and shape of the opening. However, doubled lengths of joist stock are generally used when the header must span more than 4'. In the case of an I-joist floor, an LVL header is often used. Consult the plans for framing details as well as for rough opening sizes.

The difficulty of framing a stairwell depends on whether the opening runs parallel to the floor joists, as in Stairwell Parallel to Joists, or perpendicular to them, as in Stairwell Perpendicular to Joists. A parallel opening is easier to frame. In either case, the rough opening must be at least 37" wide. If ½" drywall will be used to cover the walls of the stairwell, this will leave a finished opening of 36", as required by code. The length of a stairwell opening is specified on the plans. (For further information about stairwells, see Chapter 25, "Stairways.") Steps in framing an opening are given on page 415.

Cantilevered Floor Framing The framing for a bay window or similar projection is often arranged so that the floor joists extend beyond the foundation wall. A **cantilever** is a supporting member that projects into space and is itself supported at only one end. The joists carry the necessary loads, and there is no need for separate foundation walls. This cantilevered extension should normally not exceed 2'.

The joists forming each side of the bay, as well as the header, should be doubled. How this is done depends on whether the floor joists run parallel to the cantilevered section, as in Cantilever Framing on page 414, or perpendicular to them, as in Another Type of Cantilever Framing on page 414.

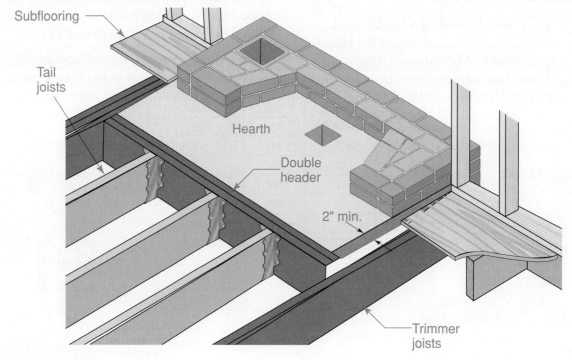

Joist Headers
Transferring Loads A header is often found in the floor framing near a fireplace.

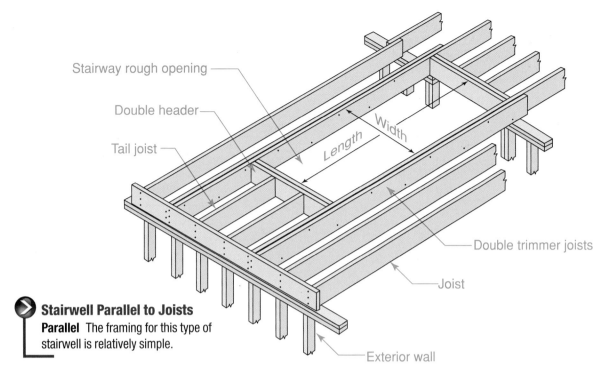

Stairway rough opening

Double header

Tail joist

Length

Width

Double trimmer joists

Joist

Exterior wall

Stairwell Parallel to Joists
Parallel The framing for this type of stairwell is relatively simple.

Nailing, in general, should conform to that for floor openings. The subflooring is extended to the outer framing member and sawed flush with that member. Check local building codes for other rules related to cantilevered framing. For example, preservative-treated lumber may be required.

Bathroom Floor Framing The weight and drainage requirements of plumbing fixtures involve special framing. Bathroom floor joists that support a tub or shower should be arranged so that no cutting is necessary when connecting the drainpipe, as Framing Under a Bathtub on page 414.

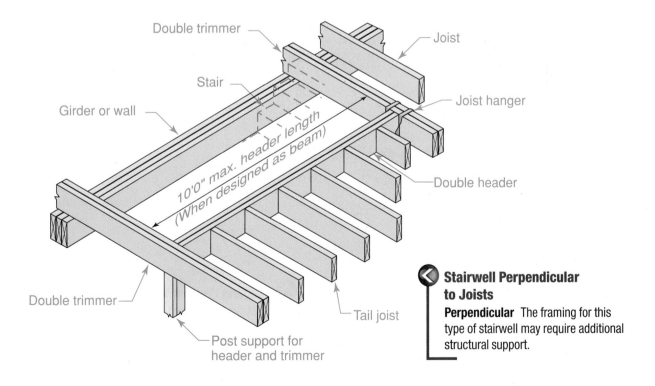

Double trimmer

Stair

Girder or wall

Joist

10'0" max. header length (When designed as beam)

Joist hanger

Double header

Double trimmer

Post support for header and trimmer

Tail joist

Stairwell Perpendicular to Joists
Perpendicular The framing for this type of stairwell may require additional structural support.

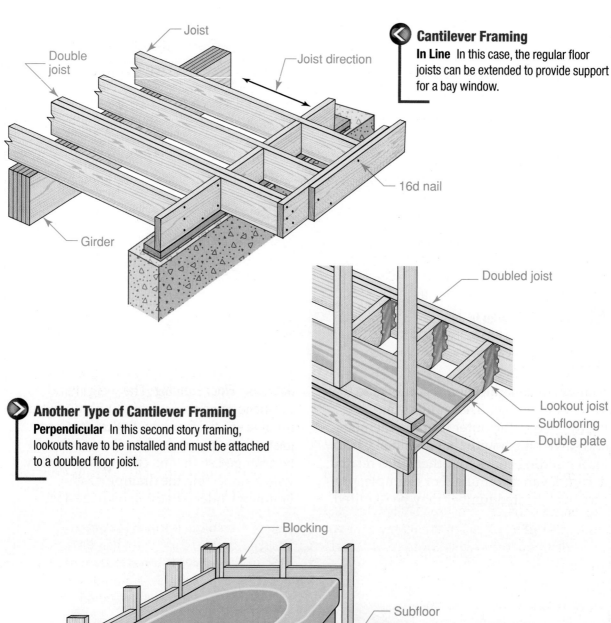

Double joist

Joist

Joist direction

Cantilever Framing
In Line In this case, the regular floor joists can be extended to provide support for a bay window.

16d nail

Girder

Doubled joist

Another Type of Cantilever Framing
Perpendicular In this second story framing, lookouts have to be installed and must be attached to a doubled floor joist.

Lookout joist

Subflooring

Double plate

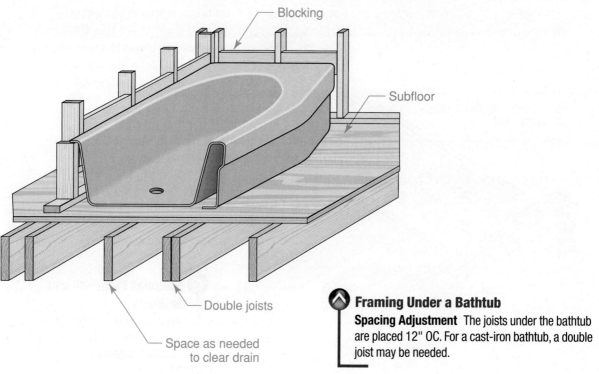

Blocking

Subfloor

Double joists

Space as needed to clear drain

Framing Under a Bathtub
Spacing Adjustment The joists under the bathtub are placed 12" OC. For a cast-iron bathtub, a double joist may be needed.

Framing an Opening The opening frame can be assembled entirely with nails. However, it is easier to use metal framing connectors, as described here.

Step 1 Make sure the trimmer joists have been doubled. Double-check the width of the opening against the plans.

Step 2 Use a square to lay out the position of both headers. The dimensions of the rough opening will be noted on the plans.

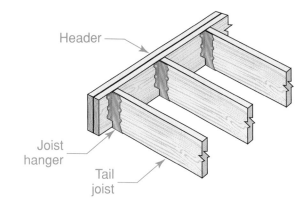

Header

Joist hanger

Tail joist

Step 3 Attach framing connectors to the sides of the trimmer joists (see page 405). Framing connectors are required by code if the header spans more than 6'. Nail with 16d common nails or as specified by the connector's manufacturer.

Step 4 Cut header stock to fit snugly between the trimmers. Insert the stock into the connectors. Nail into the header with 10d common nails or with joist-hanger nails.

Step 5 Install the tail joists. Support the tail joists on joist hangers nailed as in Step 4. According to the building code, tail joists over 12' long must be supported by framing connectors or on 2" square (or larger) ledger strips.

This may require only a small adjustment in spacing the joists. When joists are parallel to the length of a tub, they are usually doubled under its outer edge. Unusually large tubs and whirlpool tubs may require additional support.

Second-Story Framing The layout and installation of second-story floor joists is basically the same as for the first story. However, instead of resting on a sill plate, the joists rest on the double top plate of the first-story walls. It is also important to remember that finish ceiling materials will be nailed to the underside of the second-story floor joists. This calls for some special framing details.

At the junction of a wall and ceiling, doubled joists provide a nailing surface for the ceiling and interior wall finish, as shown in Providing a Nailing Surface on page 416. Another method of providing nailing at the ceiling line is to install solid blocking as in Ceiling Finish Blocking on page 416. This

may be required where a wall runs perpendicular to the floor joists. If 2×6 blocks are centered over a 2×4 partition wall, they will provide a nailing surface of approximately 1" on each side of the wall. Blocks should be firmly secured with 16d common nails so that they will not be hammered out of position when the drywall is installed.

Reading Check

Recall What is a cantilever?

Framing with Girders

What is the difference between a girder floor and a joist floor?

The floor of most houses is built using some type of floor joist. However, there are other ways to build floor systems. For example, girder floor framing is

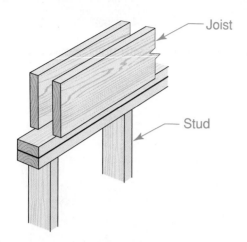

 Providing a Nailing Surface
Extra Joist Joists can be added or spaced differently to provide a nailing surface for interior wall finish or ceilings.

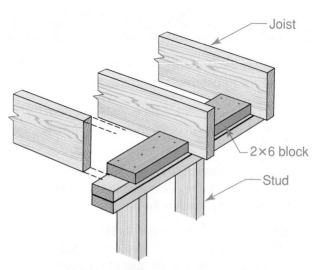

 Ceiling Finish Blocking
Extra Stock Horizontal blocks can be installed as needed to support the edges of ceiling finishes.

common in mild climates. It is a form of post-and-beam framing. A girder in this system serves a similar role to the girders noted in Section 15.1. However, in a girder floor system the girders are used like joists because they directly support a subfloor.

Components of the Floor

The girder method of floor framing is sometimes used where homes are built over a crawl space instead of a basement. A system of posts and girders, instead of joists, supports the subfloor. Frequently, 4×6 girders are used and spaced 4' OC. They are supported by 4×6 posts spaced no more than 5' OC. Sometimes girders are combined with box-sill framing, as shown in Girder Floor Framing. In other cases, the box sill is replaced by a plate. Asphalt roofing material or metal post anchors keep the wood from contacting the pier. If square-edge subflooring panels are used instead of T&G (tongue-and-groove) panels, blocking is required at unsupported edges.

Installation Details After the foundation walls are in place, locations for concrete piers are laid out and holes can be dug for the pier footings. The piers should be set in a

reasonably straight line. Their height is not critical because the posts will be cut to length as needed. The sill is then cut to size and bolted in place.

The bearing posts must be cut to length accurately to provide a level floor. A string is pulled tight from opposite sill plates over the piers. Then the distance is measured from the line to the top of each pier and recorded. This process is repeated for each line of piers until the height of each bearing post has been determined and recorded. Posts can be cut to length with a circular saw or radial-arm saw.

Each post is attached to a pier. Then the girders are cut to length and toenailed to the posts. If a low house profile is desired or if the finished floor is to have a step-down area, the tops of the girders in the step-down area are set flush with the top of the sill. A special metal hanger is used to support the girder.

Working space under the girders is limited, so plumbing and heating lines are roughed in before the subfloor is laid. The subfloor is then cut and nailed in place.

Laying the Subfloor The subfloor is usually of 1⅛" or thicker tongue-and-groove plywood. Some local building codes permit the use of 2×6 tongue-and-groove plank subflooring

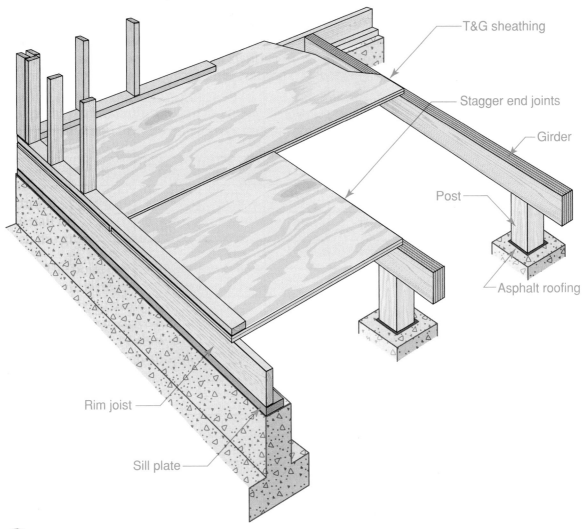

Girder Floor Framing

Posts and Girders Girder construction with box-sill framing.

Labels in figure: T&G sheathing, Stagger end joints, Girder, Post, Asphalt roofing, Rim joist, Sill plate

over girder floor framing. The subflooring is cut even with the outside of the framing and nailed to the top of the girders, as shown below. 16d nails are used to toenail at the tongue and to face-nail at a joint on all girders. After the subfloor has been cut and nailed in place, the surface is ready for the layout and erection of the walls.

Plank Subflooring

Lots of Lumber Wall framing on top of girder construction using 2×6 T&G plank subflooring.

Arnold & Brown

Table 15-3: Board Feet in Standard Lumber

Width and Depth (inches)	Length (feet)							
	10	12	14	16	18	20	22	24
1×2	$1\frac{2}{3}$	2	$2\frac{1}{3}$	$2\frac{2}{3}$	3	$3\frac{1}{3}$	$3\frac{2}{3}$	4
1×3	$2\frac{1}{2}$	3	3	4	4	5	5	6
1×4	$3\frac{1}{3}$	4	$4\frac{2}{3}$	$5\frac{1}{3}$	6	$6\frac{2}{3}$	$7\frac{1}{3}$	8
1×5	$4\frac{1}{6}$	5	$5\frac{5}{6}$	$6\frac{2}{3}$	$7\frac{1}{2}$	$8\frac{1}{3}$	$9\frac{1}{6}$	10
1×6	5	6	7	8	9	10	11	12
1×8	$6\frac{2}{3}$	8	9	$10\frac{2}{3}$	12	13	$14\frac{2}{3}$	16
1×10	$8\frac{1}{3}$	10	$11\frac{2}{3}$	$13\frac{1}{3}$	15	$16\frac{2}{3}$	$18\frac{1}{3}$	20
1×12	10	12	14	16	18	20	22	24
$1\frac{1}{4}$×4	$4\frac{1}{6}$	5	$5\frac{5}{6}$	$6\frac{2}{3}$	7	$8\frac{1}{3}$	$9\frac{1}{6}$	10
$1\frac{1}{4}$×6	$6\frac{1}{4}$	$7\frac{1}{2}$	$8\frac{3}{4}$	10	$11\frac{1}{4}$	$12\frac{1}{2}$	13	15
$1\frac{1}{4}$×8	8	10	$11\frac{2}{3}$	$13\frac{1}{3}$	15	$16\frac{2}{3}$	$18\frac{1}{3}$	20
$1\frac{1}{4}$×10	10	12	$14\frac{1}{2}$	$16\frac{2}{3}$	$18\frac{2}{3}$	$20\frac{5}{6}$	$22\frac{5}{6}$	25
$1\frac{1}{4}$×12	$12\frac{1}{2}$	15	$17\frac{1}{2}$	20	$22\frac{1}{2}$	25	27	30
$1\frac{1}{2}$×4	5	6	7	8	9	10	11	12
$1\frac{1}{2}$×6	$7\frac{1}{2}$	9	$10\frac{1}{2}$	12	$13\frac{1}{2}$	15	16	18
$1\frac{1}{2}$×8	10	12	14	16	18	20	22	24
$1\frac{1}{2}$×10	12	15	17	20	22	25	27	30
$1\frac{1}{2}$×12	15	18	21	24	27	30	33	36
2×4	$6\frac{2}{3}$	8	$9\frac{1}{3}$	$10\frac{2}{3}$	12	$13\frac{1}{3}$	$14\frac{2}{3}$	16
2×6	10	12	14	16	18	20	22	24
2×8	$13\frac{1}{3}$	16	$18\frac{2}{3}$	$21\frac{1}{3}$	24	$26\frac{2}{3}$	$29\frac{1}{3}$	32
2×10	$16\frac{2}{3}$	20	$23\frac{1}{3}$	$26\frac{2}{3}$	30	$33\frac{1}{3}$	$36\frac{2}{3}$	40
2×12	20	24	28	32	36	40	44	48
2×14	$23\frac{1}{3}$	28	$32\frac{2}{3}$	$37\frac{1}{3}$	42	$46\frac{2}{3}$	$51\frac{1}{3}$	56
4×4	$13\frac{1}{3}$	16	$18\frac{2}{3}$	21	24	$26\frac{2}{3}$	$29\frac{1}{3}$	32
4×6	20	24	28	32	36	40	44	48
4×8	$26\frac{2}{3}$	32	$17\frac{1}{3}$	$42\frac{2}{3}$	48	$53\frac{1}{3}$	$58\frac{2}{3}$	64
4×10	$33\frac{1}{3}$	40	$46\frac{2}{3}$	$53\frac{1}{3}$	60	$66\frac{2}{3}$	$73\frac{1}{3}$	80
4×12	40	48	56	64	72	80	88	96

This estimating and planning exercise will prepare you for national competitive events with organizations such as SkillsUSA and the Home Builder's Institute.

CERTIFICATION PREP

Estimating and Planning

Floor Framing

Number of Joists

Estimating methods are similar for both solid-lumber joists and I-joists.

Step 1 To find the number of joists needed for a house, first divide the length of the floor (in feet) by the joist spacing (in feet). Conventional joist spacing is 16" (1.33') on center. Dividing by 1.33 is the same as multiplying by 0.75. Therefore, for joists 16" on center, simply take three-fourths of the length of the building. For example, for a building that is 40' long, multiply 0.75 by 40. The answer is 30.

Step 2 Add 1 for the end joist, which gives a total of 31 joists.

Step 3 I-joists may extend from wall to wall, but solid lumber joists usually do not. More are needed. Suppose that the building in Step 1 is 20' wide and that you are using 10' joists. The joists will extend only from one wall to a center girder. Another 31 joists will be needed to cover the span from the girder to the opposite wall, for a total of 62 joists.

Step 4 Add one extra joist for each wall for which double joists are specified.

Material Costs

An accurate cost estimate of materials is made by multiplying the number of joists required by the cost per joist. However, a rough estimate can be made without knowing the exact number of pieces needed.

Step 1 Find the area of the floor by multiplying the length times the width for each level. For example, a one-story building 20' wide and 40' long has a floor area of 800 sq. ft.:

$$20 \times 40 = 800$$

Step 2 The number of board feet required for joists can be found by referring to **Table 15-4** on page 420. According to the table, using 2×6 joists placed 16" OC, 102 board feet of lumber are needed for each 100 sq. ft. of floor surface area.

Step 3 Divide the total floor area by 100, and multiply by the number of board feet that you obtained from the table. The answer to the example problem is 816 board feet:

$$800 \div 100 = 8, \text{ and } 8 \times 102 = 816$$

Step 4 By multiplying the cost per board foot of lumber by the number of board feet required, you can obtain a rough cost estimate.

Step 5 Table 15-4 also helps determine the number of nails needed. For 2×6 joists, 10 lbs. of nails are needed for each 1,000 board feet. Since our floor has only 800 board feet, it will require about 8 lbs. of nails.

Step 6 Multiply the number of pounds needed by the cost per pound to find the total cost of the nails.

Labor Costs

To determine the labor cost for framing a floor, you must know the joist size.

Step 1 In our example, the joists are 2" × 6" × 10'. Refer to Table 15-3, which shows that 2" × 6" × 10' boards contain 10 board feet. For a building with 62 joists, as in our example, there would be a total of 620 board feet of joists:

$$10 \times 62 = 620$$

Step 2 Refer to Table 15-14 on page 420 to find the labor rate. One worker in one hour can frame 65 board feet of 2×6 joist material.

Step 3 To find the total hours needed, divide the total board feet by the number of board feet framed in one hour. The answer is 9.5 hours (620 ÷ 65 = 9.5).

Step 4 Multiply the number of hours by the hourly rate.

Table 15-4: Estimating Board Feet, Nails, and Labor

Size of Joist	Joists				Nails	Labor
	Board Feet Required for 100 Sq. Ft. of Surface Area				Per 1,000 Bd. Ft. (pounds)	Bd. Ft. (per Hour)
	12" OC	16" OC	20" OC	24" OC		
2×6	128	102	88	78	10	65
2×8	171	136	117	103	8	65
2×10	214	171	148	130	6	70
2×12	256	205	177	156	5	70

Section 15.2 Assessment

After You Read: Self-Check

1. What is a sill plate, and what is its purpose?
2. What size and type of bit should be used when boring holes in the sill plate for foundation anchor bolts that are ½" in diameter?
3. What does the term *crown* refer to, and why is it important?
4. What is the purpose of a web stiffener?

Academic Integration: Mathematics

5. **Estimate Floor Joists** Normally floor joists are placed 16" or 24" OC. Estimate the number of 2' × 8' × 22" joists laid 16" OC needed for a 42' × 66'-6" structure. A girder supported by posts runs the entire length of the building.

   Math Concept When estimating, be sure that all your calculations are in the correct units (such as feet, meters, or inches).

   Step 1: Divide 12" by the OC spacing in inches (16") to convert to a decimal factor.

   Step 2: Convert the length of the structure (66'-6") to decimal feet.

   Step 3: Multiply the length of the structure (66.5') times the decimal factor (0.75), and round up.

   Step 4: Add 1 starter joist.

   Step 5: Multiply the joists by 2 to account for both sides of the floor system. (Joists are 22' long and the building width is 42'. The joists overlap over the girder.)

   Step 6: To determine the extra stock needed for rim joists, multiply the length (66.5') by 2, and divide by 22'. Add the quotient to the number of joists.

   Go to connectED.mcgraw-hill.com to check your answers.

Subfloors

Installing Subflooring

Are there any disadvantages to both gluing and nailing a subfloor?

The layer of material directly over the floor joists is called *floor sheathing*. It forms what is called the *subfloor*. Sheathing serves several important purposes. It lends bracing strength to the building. It provides a solid base for the finish floor. By acting as a barrier to cold and dampness, it helps keep the building warmer and drier in winter. In addition, it provides a safe working surface for building the house.

Many years ago, floor sheathing was made of solid 1× boards, which were nailed diagonally across the floor joists. Today, solid lumber sheathing has been replaced by 4' × 8' engineered panel products such as plywood and OSB. These panels are easier to install and create a stiffer subfloor that is less likely to squeak.

A typical subfloor consists of sheathing panels fastened directly to joists or girders, as shown in Subflooring.

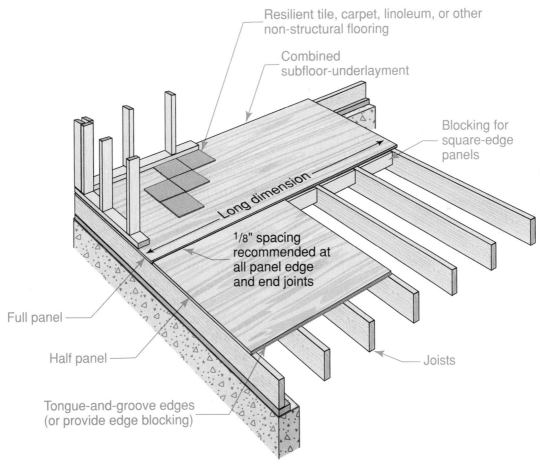

Resilient tile, carpet, linoleum, or other non-structural flooring

Combined subfloor-underlayment

Blocking for square-edge panels

Long dimension

1/8" spacing recommended at all panel edge and end joints

Full panel

Half panel

Joists

Tongue-and-groove edges (or provide edge blocking)

Subflooring

Single Layer Subflooring can be laid over supports spaced up to 48" apart, depending on the design of the floor system and the thickness of the panel.

Table 15-5: Fastening Schedule for Subfloors[a]

Span Rating (Maximum Joist Spacing) (inches)	Panel Thickness (inches)[b]	Fastening: Glue-Nailed[c]			Fastening: Nailed Only		
		Nail Size and Type	Spacing (inches)		Nail Size and Type	Spacing (inches)	
			Supported Panel Edges	Intermediate Supports		Supported Panel Edges	Intermediate Supports
16	¾" or less	6d ring- or screw-shank[d]	12	12	6d ring- or screw-shank	6	12
20			12	12		6	12
24			12	12		6	12
24	⅞", 1"	8d ring- or screw-shank[e]	6	12	8d ring- or screw-shank	6	12
32			6	12		6	12
48	1⅛"		6	(e)		6	(e)

[a] APA Rated Sturd-I-Floor. Special conditions may impose heavy traffic and concentrated loads that require construction in excess of the minimums shown.

[b] Panels in a given thickness may be manufactured in more than one span rating. Panels with a span rating greater than the actual joist spacing may be substituted for panels of the same thickness with a span rating matching the actual joist spacing.

[c] Use only adhesives conforming to APA Specification AFG-01, applied in accordance with the manufacturer's recommendations. If non-veneered panels with sealed surfaces and edges are to be used, use only solvent-backed glues; check with panel manufacturer.

[d] 8d common nails may be substituted for ring- or screw-shank nails if ring- or screw-shank nails are not available and 10d common nails may be substituted with 1⅛" panels if supports are well seasoned.

[e] Space nails 6" for 48" spans and 12" for 32" spans.

If the sheathing has a smooth surface, finish flooring such as carpeting, resilient tile, and sheet flooring can be applied directly to it. Sheathing rated for this purpose is identified by a grade mark stamped on the back of the panel.

Reading Check

Contrast Why are 4×8 engineered panel products used in place of solid lumber for floor sheathing?

Fastening Details

For fastening details, refer to **Table 15-5**. Common nails are often used to install floor sheathing. When panels are glued *and* nailed, fewer nails are required, as compared to nailing only. This is called *glue-nailing* and it results in a very strong and stiff floor system. The adhesive is a type of construction mastic applied to the top of each joist with a caulking gun. Only a single bead of adhesive is required. Panels should be installed immediately after the adhesive has been applied in a small area. If the adhesive is applied too long in advance, it will dry out or harden and will not hold. Hand nailing a subfloor is time consuming and physically difficult. Using pneumatic nailers is much more efficient. This has become the standard method.

For builders who prefer to screw the sheathing to the joists, special attachments enable electric drills to drive a great number of screws quickly. These attachments use coils of screws, automatically feeding each one to the tip of a drill such as the one on page 423. Some models are fitted with a long extension so that the carpenter does not have to lean over during installation.

Underlayment

In some cases, an extra layer of material is installed over the subfloor. This layer is called underlayment. The underlayment covers any minor construction damage to the subfloor and provides a smooth substrate for finish flooring such as sheet vinyl. It should not be installed until just before the

Laying a Panel Subfloor The general method for installing a subfloor is the same for plywood and OSB panels. However, you should always consult the manufacturer's instructions.

Step 1 Measure 48" along the side of the foundation from the starting corner and mark this point. Repeat the process on the opposite side of the foundation. Snap a chalk line between these points. This serves as an alignment guide for the sheathing.

Step 2 Place a full panel even with one of the outside corners of the floor joists. Align the edge with the chalk line. The grain of the plywood should run at right angles to the joists. If the subfloor will be glue-nailed, spread a bead of construction adhesive on the joists just before installing each sheet.

Step 3 Drive just enough nails to hold the panel in place.

Step 4 Place the next full panel in position at the end of the first panel. Be sure the joint is centered over the joist, and leave about ⅛" space between panels.

Install it as in Step 2. *Tip: Rather than measure this space, use a 10d box nail as a spacer. It is approximately ⅛" in diameter.*

Step 5 Begin the second row of panels at the end of the building, alongside the first panel laid. Cut a panel in half, lay the end flush with the outside of the floor framing, and nail the half panel to the joists, as shown on page 421. Continue to lay and nail full panels in this row.

Step 6 Start the next (third) row with a full panel. This alternating method will stagger the joints used for support and provide the strongest floor. Continue to lay panels, driving just enough nails in each to hold it in position until all panels are laid. Then snap chalk lines to indicate the location of floor joists. Complete the nailing as required.

Self-Feeding Screw Gun
Efficient Fastening This tool increases the speed at which screws can be installed in a subfloor.

Arnold & Brown

finish flooring is ready for installation. This will prevent the underlayment from being damaged by other construction activities. For fastening details refer to **Table 15-6**.

Underlayment is made of plywood with a *touch-sanded surface*. This means that it is sanded at the mill just enough to ensure uniform thickness. The inner plies of underlayment-grade plywood resist dents and punctures from heavy loads, such as furniture. To improve the stiffness of the floor, the face grain of the underlayment should be placed perpendicular to supports. The edges should be offset at least 2" from the edges of the subfloor panels. This is usually done automatically because the subfloor extends beneath the wall plates, while the underlayment does not.

Blocking

The edges of square-edged subflooring panels should be supported. They must rest either on a joist or on blocking laid between the joists. This blocking is cut from nominal 2" lumber.

Sheathing with a tongue-and-groove (T&G) edge is shown in on page 425. It does not need to be blocked between supports. The surface of the panel is a full 4'×8' with an allowance for the tongue. Tongue-and-groove subflooring should be started with the tongue toward the outside of the building. Any pounding required to close the joints between the panels can then be done on a scrap block held against the groove.

Table 15-6: APA Plywood Underlayment Fastener Schedule					
Plywood Grades[a]	Application	Minimum Plywood Thickness (inches)	Fastener Size and Type[b]	Fastener Spacing (inches)[c]	
				Panel Edges	Intermediate
APA UNDERLAYMENT	Over smooth subfloor	¼	3d ring-shank nails[d]	3	6 each way
APA C-C Plugged EXT APA RATED Sturd-I-Floor (¹⁹⁄₃₂" or thicker)	Over lumber subfloor or other uneven surfaces	¹¹⁄₃₂	3d ring-shank nails	6	8 each way
Same grades as above but species Group 1 only	Over lumber floor up to 4" wide; face grain must be perpendicular to boards	¼	3d ring-shank nails[d]	3	6 each way

[a] In areas to be finished with thin floor coverings such as tile or sheet vinyl, specify Underlayment C-C Plugged or Sturd-I-Floor with "sanded face." Underlayment A-C, Underlayment B-C, Marine EXT, or sanded plywood grades marked "Plugged Crossbands Under Face," "Plugged Crossbands (or Core)," "Plugged Inner Plies," or "Meets Underlayment Requirements" may also be used under thin floor coverings.

[b] Other code-approved fasteners may be used.

[c] Space fasteners so they do not penetrate framing.

[d] Use 3d ring-shank nails for ½" panels and 4d ring-shank nails for ⅝" or ¾" panels.

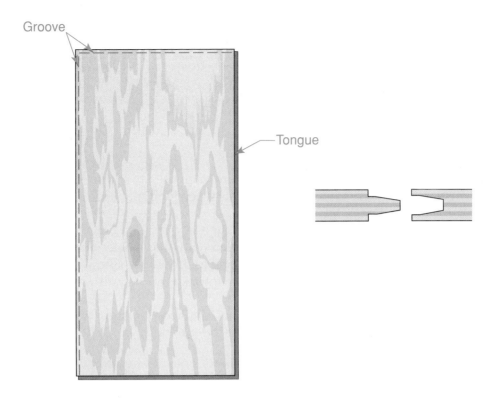

Groove

Tongue

 T&G Subflooring
Solid Edges This subflooring panel is 1⅛" thick with a groove on one end and one edge and a tongue on the other end and edge.

Section **15.3** Assessment

After You Read: Self-Check

1. When glue-nailing a subfloor, when should the panels be installed?
2. List four fasteners used to attach sheathing to the floor joists.
3. What is underlayment and why is it used?
4. What can be used to gauge the required ⅛" separation between subflooring panels?

Academic Integration: English Language Arts

5. **Adhesives and Subfloor** Construction adhesives can dramatically improve the performance of a subfloor system. However, improvements in adhesive technology are expected in coming years. Using the most current sources of information available, locate at least three adhesives that are suitable for subfloors. Is the product suitable for use on wet lumber? Is the product suitable for use on preservative-treated lumber? What features of the product do you think carpenters would find most important? Summarize your results in a one-page table or a spreadsheet.

Go to **connectED.mcgraw-hill.com** to check your answers.

Review and Assessment

Section
15.1

Section
15.2

Section
15.3

Chapter Summary

Girders, supported by wood or metal posts, are often used to support floor systems. Floor trusses can be used as girders, or as joists when long spans are required.

Floor joists are commonly made of solid lumber or I-joists. Rim boards should be used instead of rim joists when the floor is assembled with I-joists. Special framing techniques must be used for large openings. Cross bridging improves the stiffness of joists. Girder floor framing is often used with a crawl-space foundation.

The subfloor consists of plywood or OSB panels applied to joists or girders. In some cases, underlayment plywood is applied over the subfloor.

Review Content Vocabulary and Academic Vocabulary

1. Use each of these content vocabulary and academic vocabulary words in a sentence or diagram.

Content Vocabulary

- girder (p. 396)
- box sill (p. 402)
- subflooring (p. 402)
- crown (p. 406)
- bridging (p. 410)

- bearing wall (p. 411)
- header (p. 412)
- trimmer joist (p. 412)
- tail joist (p. 412)
- cantilever (p. 412)

Academic Vocabulary

- criteria (p. 398)
- perpendicular (p. 403)
- offset (p. 406)

Speak Like a Pro

Technical Terms

2. Work with a classmate to define the following terms used in the chapter: *floor joist* (p. 396), *post* (p. 396), *Lally columns* (p. 397), *sill plate* (p. 397), *face-nailing* (p. 398), *toenailing* (p. 398), *beam pocket* (p. 399), *load* (p. 399), *parallel-chord floor truss* (p. 400), *mudsill* (p. 402), *sill* (p. 402), *band* (p. 402), *sill sealer* (p. 402), *clinched* (p. 409), *rim board* (p. 409), *notch* (p. 409), *solid bridging* (p. 410), *cross bridging* (p. 410), *diagonal bridging* (p. 410), *glue-nailing* (p. 422), *touch-sanded surface* (p. 424).

Review Key Concepts

3. List the purpose of each basic floor-framing component.

4. Explain how girders are constructed.

5. Describe the basics of joist spacing.

6. Identify one case in which special framing details may be required.

7. Explain the procedure for laying a panel subfloor.

8. Identify two other types of framing systems and products that are used in building construction.

Critical Thinking

9. **Analyze** Why is it important that sill plates are level? Under which conditions may sill plates that are not level be most problematic?

Academic and Workplace Applications

STEM Mathematics

10. **Estimate Area** Estimate the number of 4×8 plywood sheets needed to place a subfloor in a one-story building measuring 32' × 52'.

Math Concept Think of the area of a rectangular space as the number of 1 foot by 1 foot squares it would take to completely cover the space with no overlapping squares.

Step 1: Find the area of the floor in square feet.

Step 2: Figure the number of square feet in 1 sheet of plywood.

Step 3: Divide the square footage of floor area by the square feet in 1 sheet of plywood.

STEM Science

11. **Rust** Lally columns are often preferred in residential construction. They are strong, easy to handle, and take less space than solid wood posts. They must be at least 3" in diameter and protected against rust, such as with rust-resistant paint. Rust can cause a severe amount of damage to steel posts. Use the Internet to find out more about the rusting process. Write a few sentences describing what rust is and how it is caused.

21st Century Skills

12. **Information Literacy** Research solid bridging and cross bridging. List the pros and cons of each method of bracing between joists. Contact a local contractor or builder and arrange a brief interview, asking about floor framing methods. Determine whether they prefer solid bridging or cross bridging when bracing between joists. Compare your research with the information gained in the interview. Write an essay comparing the information you researched, and the information from the interview.

Standardized TEST Practice

CERTIFICATION PREP

Multiple Choice

Directions Choose the best answer to each of the following questions.

13. Which two criteria are the ordering of steel beams based on?

    a. length and weight
    b. weight and color
    c. height and weight
    d. width and length

14. What prevents an I-joist from moving in the joist hanger?

    a. a backing block
    b. a crown
    c. a rim board
    d. bridging

15. Under which circumstance is the girder method of floor framing used?

    a. when homes are built in extreme climates
    b. when homes are built over a basement
    c. when materials are in short supply
    d. when homes are built over a crawl space

TEST-TAKING TIP

Right before taking the test, review key vocabulary words.

*These questions will help you practice for national certification assessment.

CHAPTER 16
Wall Framing & Sheathing

Section 16.1
Wall-Framing Materials

Section 16.2
Wall Layout

Section 16.3
Assembling & Erecting Walls

Section 16.4
Special Framing Details

Chapter Objectives

After completing this chapter, you will be able to:

- **Identify** the basic parts of a framed wall.
- **Describe** how to apply sheathing.
- **Estimate** materials for wall framing and sheathing.
- **Explain** how to lay out and frame a wall.
- **Assemble** and erect a wall.
- **Identify** situations that require special framing.

Discuss the Photo
Walls Walls support the roof of a structure and also enclose the structure. *What kinds of tools do you think are used in framing walls?*

Writing Activity: Research and Summarize
Contact the U.S. Forest Products Laboratory or similar organization and research new wood-based wall framing materials they are developing. Write a two-paragraph summary of your findings.

Chapter 16 Reading Guide

Before You Read Preview

Wall framing can begin once the floor system is complete and the subfloor is in place. Choose a content vocabulary or academic vocabulary word that is new to you. When you find it in the text, write down the definition.

Content Vocabulary

- sheathing
- stud
- plate
- header
- rough sill
- cripple stud
- trimmer stud
- rough opening (RO)
- corner post
- temporary bracing
- soffit

Academic Vocabulary

You will find these words in your reading and on your tests. Use the academic vocabulary glossary to look up their definitions if necessary.

- primary
- dimensions

Graphic Organizer

As you read, use a chart like the one shown to organize the components of wall framing. Add rows as necessary.

component	purpose

Go to connectED.mcgraw-hill.com to download this graphic organizer.

Wall-Framing Materials

Wall-Framing Basics

How does a bearing wall differ from a non-bearing wall?

The framing and sheathing for the first floor of a house create a flat and level platform on which to build the walls. Carpenters can then lay out, assemble, and erect the walls. This work must be done properly. Mistakes at this stage will make it more difficult to complete the rest of the house. For example, cabinets will not fit well if walls are not plumb.

After the walls are framed, sheathing must be attached to them. **Sheathing** consists of rigid 4'×8' or larger panels that are attached to the outside surface of the exterior wall framing. Sheathing adds great stiffness and strength to the walls. This is important because the walls of a house will provide a framework for attaching interior and exterior coverings such as siding and drywall. A wall that also supports weight from portions of the house above, such as the roof, is called a *load-bearing wall*, or simply, a *bearing wall*.

Exterior walls are nearly always load-bearing walls. Interior walls, also called *partition walls* or *partitions*, are sometimes load-bearing walls. If they carry only their own weight and the weight of wall coverings, they are not considered load-bearing.

When roof trusses span the entire width of a house, the exterior walls support the weight of both the roof and ceiling loads. Interior walls then serve mainly as room dividers. When a roof is framed using ceiling joists and rafters instead of roof trusses, however, interior walls usually carry some of the ceiling load, as shown in Bearing Wall. It is important for a carpenter to understand the difference between bearing walls and non-bearing walls. Each type of wall is framed differently.

The standards and specifications found in this chapter and the following chapters are based on the 2006 IRC. However, you should always follow the codes used in your area. Even if your town or state has adopted the 2006 IRC as its model code, variations in the code may be required to address local conditions.

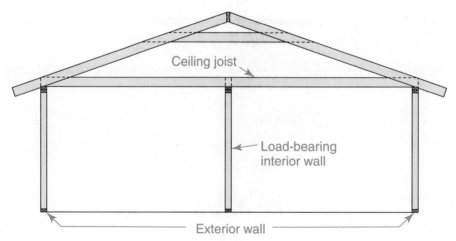

Ceiling joist

Load-bearing interior wall

Exterior wall

 **Bearing Walls**

Span Support In most structures, ceiling joists cannot span the entire width of the house. This requires an interior load-bearing wall or beam to support the ends of the joists not supported by an exterior wall.

Choosing Lumber

Wall-framing lumber should be stiff, free from warpage and twist, and have good nail-holding capability. Bottom plates should be made of preservative-treated lumber when installed on a concrete slab. The woods used for wall-framing members are, in general, the same species as those used for floor framing. Common species include Douglas fir and Southern yellow pine.

Lengths of general framing lumber are available in increments of 2'. However, wall studs are usually precut to a particular length. Standard precut studs are usually 92⅝" long for an 8' wall, as shown in Precut Studs.

There are many different types and grades of lumber. The building code considers finger-jointed lumber to be no different

Stud Length Precut studs are supplied by local lumberyards. Standard lengths vary in different parts of the United States. In some regions, precut studs are 92¼" or 93" long, while in other regions they are 92⅝" long. There is no particular reason for this other than local preferences. Most standard precut studs are now 92⅝".

than solid-sawn lumber of the same species and grade. Wall studs (vertical members) should be at least No. 3, standard, or stud-grade lumber. Lumber used for wall plates is sometimes a higher grade or different species than that used for studs due to

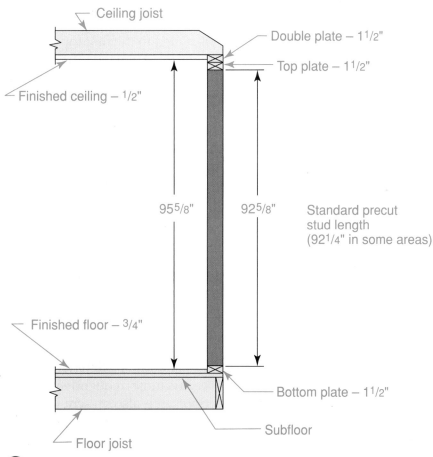

Ceiling joist

Double plate – 1½"

Top plate – 1½"

Finished ceiling – ½"

95⅝"

92⅝"

Standard precut stud length (92¼" in some areas)

Finished floor – ¾"

Bottom plate – 1½"

Subfloor

Floor joist

Precut Studs
Determining Ceiling Height A standard precut stud is 92⅝" long.

the need for long, straight lengths. Wall plates must be straight or else the wall will not be straight. High-quality construction may require kiln-dried (KD) lumber to minimize problems associated with lumber shrinkage.

Wall-Framing Members

The **primary** framing members for walls are *studs* and *plates*. Other framing members, including headers, sills, cripple studs, and trimmer studs, are physically connected to one or both of these framing members. These elements are all shown in A Framed Wall. on the next page.

Studs A **stud** is a vertical framing member. Conventional construction commonly uses 2×4 studs spaced 16" on center (OC). The full-length stud on either side of an opening is sometimes referred to as a *king stud*. The use of 2×6 studs for exterior walls is increasingly popular. The extra thickness of the resulting wall allows space for additional insulation. These 2×6 studs may be placed 16" or 24" OC. Some interior walls are also framed with 2×6 or larger studs, particularly those that will contain the main drainpipes for plumbing fixtures (see A Sound Wall on p. 458). However, most interior walls are framed with 2×4 lumber. Spacing for interior-wall studs is normally 16" OC.

The ends of studs must be cut square so they bear evenly on the plates. Cutting may be done with a circular saw, power miter saw, or radial-arm saw. Precut studs require no additional trimming.

Plates A **plate** is a horizontal framing member used to tie together interior and exterior wall framing. The width of the plates determines the thickness of the wall. In a 2×6 wall, for example, the plates would be made from 2×6 lumber. Each wall has three plates: a bottom plate and two top plates.

The bottom plate, also called the *sole plate*, ties the bottom ends of the studs together. It also provides a nailing surface for the bottom edge of wall coverings and wall sheathing. The bottom plate is secured to

Framing Terminology There are two plates at the top of a load-bearing wall. The terms can be confusing, partly because the first plate is called the *top plate.* The plate above the top plate may be called a *double plate,* a *doubler,* or a *rafter plate.* Together, the two plates are sometimes called a *double plate* or *doubled plate.*

the bottom of the studs with nails. It is also nailed to the subfloor.

The top plate is nailed to the top ends of the studs and ties them together. It also provides a nailing surface for wall coverings and sheathing. The top plate has the same dimensions as the bottom plate.

The second top plate, called a *double plate,* is nailed to the first top plate after the walls have been erected. The second top plate has four purposes:

- It adds strength and rigidity to the top of the wall.

- It supports the ends of joists and the bottom ends of rafters.

- It helps distribute structural loads that do not fall directly over studs.

- It ties intersecting walls together.

The double plate can be omitted in some cases, such as when an in-line framing system is used (see Chapter 14). If the double plate is omitted, intersecting walls must be tied together with a steel plate.

Headers Wherever an opening in a wall is wider than the stud spacing, parts or all of some studs will have to be left out. This occurs most frequently with windows and doors. It is also present with fireplaces and pass-throughs. To prevent the wall from being weakened at this point, a header is installed.

A **header** or *lintel* is a wood beam placed at the top of an opening. The header supports structural loads above the opening and transfers them to framing on each side of the opening. Headers are sometimes made of solid lumber. They are also built up from two or more pieces of 2× lumber laid on edge with spacer blocks to match the thickness of the wall. They are also made from engineered lumber, such as laminated-veneer lumber (LVL).

Step-by-Step Application

Framing a Wall The following is a basic technique for wall framing that can be used to frame nearly any straight wall.

Step 1 Cut the bottom plate and top plate to length. If the plates will require more than one piece, make sure the break falls where a stud can support the end of both pieces, usually at 16" OC.

Step 2 Align the bottom plate with the top plate. Tack them together temporarily.

Step 3 Lay out the location of the windows, partitions, studs, and other components on the edges of both plates simultaneously.

Step 4 Spread the top and bottom plates apart.

Step 5 Place the required number of precut studs between the plates.

Step 6 Install cripple studs, headers, and other parts of the framing. Nail the bottom and top plates to the studs by driving nails through the plates and into the end of each stud. At this stage, some carpenters also install the double top plate. If this technique is chosen, spaces must be left where the double plate of intersecting walls will tie in.

Step 7 Square the wall and brace the corners.

Step 8 Install panel sheathing (plywood or oriented-strand board).

Step 9 Tilt the wall into place.

Step 10 Make sure the wall is plumb, then install temporary braces to prevent it from tipping over.

Step 11 Repeat the process for adjacent walls. As walls are erected, the double top plate can be installed. It overlaps the top plate of intersecting walls to tie walls together.

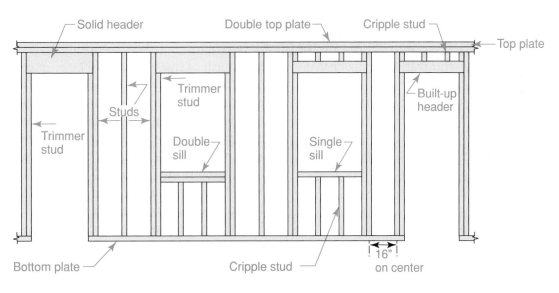

A Framed Wall

Rough Sill A **rough sill** is a horizontal member placed at the bottom of a window opening to support the window. It connects the upper ends of the cripple studs that are below the window. The rough sill does not need the same strength as a header because it supports only the window, not structural loads. It is made from lumber having the same dimensions as the studs. It may be a single piece of lumber (single sill) or two pieces (double sill) if extra strength is required. Note that a rough sill is a framing member, while a windowsill is a part of the window itself.

Cripple Studs A **cripple stud**, or *cripple,* is a stud that does not extend all the way from the bottom plate to the top plate of a wall because of an opening such as framing for a window. Cripple studs are installed above headers and below rough sills. They are located where a full-length precut stud would be placed if there were no opening.

Cripple studs provide a nailing surface for the sheathing (outside) and for wall covering (inside). To conserve lumber, they are often cut from stock that is too short for other purposes.

Trimmer Studs A **trimmer stud**, or simply *trimmer* or *jack stud,* supports a header over a window or door opening. A trimmer stud is shorter than a standard stud but longer than a cripple stud. It is cut to fit beneath the header. A trimmer stud transfers structural loads from the header to the bottom plate. For wide openings, additional trimmer studs may be needed. Check local building codes.

Estimating Studs, Plates, and Headers

Wall framing requires many individual pieces of lumber. At the design and planning stages, you may need to know only the approximate number of pieces necessary. Later, you will need a more accurate calculation.

For exterior walls with studs spaced at 16" OC, figure one stud for every lineal foot of wall. For example, if you are building a 16-foot long wall, buy 16 studs. This allows for the extra framing required around openings and at corner posts. To determine the number of studs needed for a partition, refer to the table "Partition Studs Needed" in the **Ready Reference Appendix**.

To determine the number of lineal feet of top and bottom plates for walls having double top plates, multiply the length of the wall by three. Add materials for such items as gable-end studs, corner braces, fireblocking, and wall blocking. Sometimes a builder may wish to get a rough idea of how much lumber will be required for a house. A rough estimate can be made as follows:

1. Figure the total length of the outside walls, and then double it. This will give you the approximate combined total length of all the walls, interior and exterior.

2. Multiply the total length of all walls by five. This will give you the approximate total lineal footage of plate material plus the additional miscellaneous framing for gables, bracing, and blocking.

The dimensions of each header will sometimes be found on the building plan. The length of a header is generally 3" longer than the rough opening width. This assumes that the header will be supported by one trimmer stud on each end. This is generally the case with openings less than 6' wide. Headers over larger openings should be supported by two trimmers at each end. In that case, the length of the header will be 6" greater than the rough opening width. Make a list of headers and their dimensions for use as a cutting list during construction. Review the plans to identify any areas requiring special framing. Add this material to the estimate.

Framing lumber is usually sold by the lineal foot. To estimate total cost, multiply the number of lineal feet by the cost per lineal foot. Add to that the cost of the precut studs. Another way to do this would be to determine the total cost by multiplying the total number of board feet by the cost of one board foot of lumber.

Reading Check

Explain What is a plate?

Wall Sheathing

What is the purpose of wall sheathing?

Wall sheathing is a panel product nailed to the outside surface of exterior walls, as shown in below.

Wall sheathing has several functions:

- It strengthens and braces wall framing, and adds great rigidity to a house.
- It forms a solid nailing base for the siding.
- It helps to seal a house by reducing air infiltration.
- It ties wall framing to floor framing. A solid connection here is especially important in areas prone to high winds and earthquakes.

The use of diagonal bracing, also called corner bracing or wind bracing, is another way to strengthen a wall. It can be used with or without sheathing. In one bracing system, the studs are notched to receive 1×4 pieces that are let into the notches. This is known as let-in bracing. Another system uses metal angle strips that are nailed to the outside edges of the studs.

In mild climates, plywood sheet siding is sometimes applied directly to the outside of the wall studs (see Chapter 23). In this case, it serves both as siding and as sheathing. No separate sheathing is required. In this case, grades, thicknesses, and types of plywood vary from standard sheathing requirements. For example, the plywood must be thicker and of a higher grade in order to maintain wall strength.

Installing Sheathing

The most common sheathings used in residential construction are square-edged 4×8 panels made of plywood or oriented-strand board (OSB). Panel thickness ranges from $\frac{5}{16}$" to 1". Walls with stud spacing 16" OC must have sheathing that is at least $\frac{5}{16}$" thick, although $\frac{1}{2}$" is more common. When finish

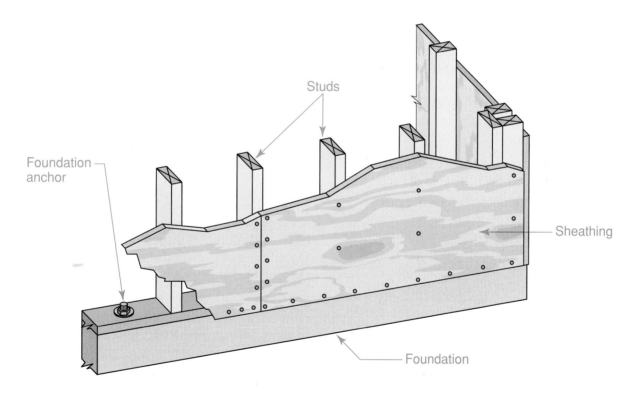

Studs

Foundation anchor

Sheathing

Foundation

Wall Sheathing

Strong Panels Wall sheathing may be made of plywood or oriented-strand board (OSB). Sheathing started at the foundation wall may require blocking behind the uppermost joint.

siding requires nailing between studs (as with wood shingles), the sheathing should be at least ⅜" thick.

Though OSB is different from plywood in terms of manufacture, it is considered the same by codes in terms of sheathing. Nailing and installation details for OSB sheathing are similar to those for plywood.

Walls that have been covered with sheathing provide a more solid support for the ceiling and roof members. That is why carpenters apply the sheathing as soon as possible. However, it may be applied at either of the following stages:

- Sheathing may be applied when the wall frame is lying on the subfloor, completely framed and squared. The advantage in applying the sheathing at this time is that it can be nailed in place while the wall sections are lying flat. This eliminates

Estimating and Planning

This estimating and planning exercise will prepare you for national competitive events with organizations such as SkillsUSA and the Home Builder's Institute.

Estimating Sheathing

Number of Sheets Needed

Using the following example will allow you to determine how many sheets of sheathing will be needed to cover the walls of a house. For example, a one-story house has four walls and a gable roof. (A gable is the triangle formed in a wall by the sloping ends of a roof.) Two of the walls are 8' high and 28' long. The other two walls are 8' high and 36' long. The bottom of each gable is 28' long. Make a sketch of this arrangement to help you understand the situation.

Step 1 Multiply the height of each wall by its width to determine the total area of each wall.

$$8' \times 28' = 224 \text{ sq. ft.}$$
$$224 \text{ sq. ft.} \times 2 = 448 \text{ sq. ft.}$$
$$8' \times 36' = 288 \text{ sq. ft.}$$
$$288 \text{ sq. ft.} \times 2 = 576 \text{ sq. ft.}$$

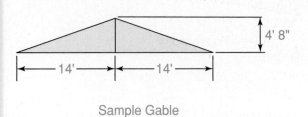

Sample Gable

Step 2 For a house with a gable roof, find the area of one gable. To find the area of a gable, multiply the height of the gable by one-half its width at the bottom, as shown in the sample gable below. Multiply the result by the number of gables to determine the total gable area. (For ease in calculating this example, change 4' 8" to a decimal, or 4.66'.)

$$4.66' \times 14' = 65.24 \text{ sq. ft.}$$
$$65.24 \text{ sq. ft.} \times 2 = 130.48 \text{ sq. ft.}$$

Step 3 Add the total area of each wall and the gables to determine the total wall area of the house.

$$448 + 576 + 130.48 = 1,154.48 \text{ sq. ft.}$$

Step 4 There are 32 square feet in a 4×8 sheet of plywood. Divide the total wall area by 32 and round off the result. This will give the number of plywood sheets required to sheathe the house.

$$1,154.48 \div 32 = 36.08$$

However, a waste factor of 10% should be added to allow for cut pieces that cannot be used.

$$36.08 \times 10\% = 3.608.$$

Rounded up, the house would require 40 sheets of sheathing.

the need for ladders or scaffolding. The disadvantage is the added weight that must be lifted when erecting the walls.

- Sheathing can also be added after the wall frame has been erected, plumbed, and braced, and the ceiling joists have been installed.

Orientation and Fastening Sheathing is usually applied vertically, using perimeter nailing with no additional blocking. If a panel does not extend to the top of the wall, its top edge should be nailed to blocking, as in Blocking on page 435. An alternative would be to use longer sheathing panels. Local building codes may require that sheathing be applied only vertically near the corners of a building. This provides additional rigidity to the structure. Plywood can also be applied horizontally, although the horizontal joints between panels should be supported by solid blocking as a base for nailing.

Building codes sometimes allow sheathing to be fastened by stapling. However, nailing is more common. The spacing and gauge of fasteners is important in creating a solid connection between sheathing and framing. Nailing and stapling requirements for plywood wall sheathing are listed in the **Ready Reference Appendix**. See the tables titled "Stapling Schedule" and "Plywood Wall Sheathing Application Details."

Where earthquakes or high winds are a common hazard, nailing requirements are stringent. This is especially true around window and door openings. Extra nailing strengthens a sheathed wall to the point that it can be considered a shear wall, sometimes called a braced-wall. A shear wall resists severe forces that tend to separate the sheathing from the framing. To create a shear wall, the sheathing must be nailed to all studs, blocking, and sills 3" OC, using 8d common or galvanized box nails. Another way to create a shear wall is to use a prefabricated wall product made of a light-gauge steel panel and wood framing. The steel resists shear forces and is bolted to the foundation. Engineering requirements may also require a moment frame. This is a load-bearing assembly that resists bending or twisting forces.

Section 16.1 Assessment

After You Read: Self-Check

1. What stud spacing is commonly used in conventional construction?
2. What is the purpose of a header?
3. What problems might occur if a carpenter did not install cripple studs under a window opening?
4. What are the most common sheathings for residential construction?

Academic Integration: Mathematics

5. **Calculate Bulk Savings** Professional builders may be offered a lower price per sheathing panel because they usually order in bulk. Contact a building supplier to find out the price of one 4×8 sheet of ½" OSB wall sheathing. Ask the supplier what sort of a discount you would get if you purchased 50 sheets at one time. Calculate the difference in price per sheet and compare this to the price for one sheet. Report your findings.

Go to **connectED.mcgraw-hill.com** to check your answers.

Wall Layout

Responsibility for Layout

What is wall layout?

The most experienced carpenter, called the lead carpenter, is entrusted with reading the plans and translating them into a series of lines and symbols marked on the subfloor. These marks are called a layout. The layout shows other carpenters exactly where to install the walls. Wall layout involves two main steps:

- Marking the location of walls on the subfloor.
- Marking the location of studs, windows, and doors on the wall plates.

In layout, accuracy is more important than speed. The lead carpenter must thoroughly understand the building plans before layout begins. In addition, he or she must be aware of special framing requirements for work to be done by other skilled workers, such as plumbers and electricians. Before the layout is started, the subfloor must be swept clean. This makes it easier to snap chalk lines on the surface. All objects that might be in the way, such as sawhorses, must be removed.

Builder's Tip

USING A CONSTRUCTION CALCULATOR

You can use a construction calculator to determine whether intersecting lines are square. Using the calculator's roof-framing setting, enter one line length as the Rise. Enter the other line length as the Run. The answer will be the length of the diagonal line connecting the two end points.

The layout is often done by a carpenter and an apprentice. The carpenter is responsible for measuring and marking. The apprentice observes and learns while holding one end of a chalk line or tape measure. A carpenter doing the layout alone might use a nail or an awl to hold the end of the tape or chalk line.

In some parts of the country, it is common to build houses on a slab foundation instead of on foundation walls. All the wood-framed walls of such a house must be bolted directly to the foundation. However, the principles for laying out and assembling walls are essentially the same as described here.

Laying Out Wall Locations

The first step in wall framing is to lay out the location of two intersecting exterior walls. Carpenters usually start with two long walls that meet at a corner. Measurements taken from these two walls can then be used to locate other walls. Once the exterior walls have been located, layout proceeds to the interior walls.

Exterior Walls The outside edge of an exterior wall's bottom plate should be flush with the outside edge of the subfloor. To begin layout, measure 3½" in from the edge of the sheathing (or 5½" for a 2×6 wall) and snap a chalk line parallel to the edge. Repeat the process for the intersecting wall.

Check the two chalk lines to make sure they form a 90° angle. Good carpenters never assume that the floor framing is perfectly square. To check for squareness, measure from the chalk line intersection exactly 3' along one line and 4' along the other, as shown in Using the 3-4-5 Rule. If the diagonal measurement between these two points is exactly 5', the corner is square. This is an example of the 3-4-5 rule. Any multiples of these numbers that preserve the

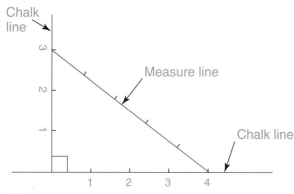

Using the 3-4-5 Rule

Useful Geometry If two legs of a triangle measure 3' and 4', they form a 90° angle if the diagonal between the ends of the legs measures exactly 5'.

same ratio, such as 9, 12, and 15, will work in the equation. The larger the triangle formed using the 3-4-5 rule, the more accurately the building will be checked for square.

If the corner is not square, adjust one of the layout lines until it is. After the first two layout lines are correct, locate and mark the position of the remaining exterior walls.

Interior Walls Consulting the plans, locate the position of interior walls by measuring from the chalk lines that indicate exterior walls. Pull a chalk line taut and snap it to indicate the exact location of one edge of each partition's bottom plate. To prevent confusion, mark an X on the subfloor to show the side of the line on which the plate will be located. This is where new carpenters often make a mistake. If the X is marked on the wrong side of the layout line, the wall will be built

3½" from its correct position. Check partition walls during layout to ensure that they are square with intersecting walls.

It is important to identify special partitions as layout proceeds. Special partitions will contain plumbing drains or other features. They may have to be thicker than standard walls. This should be noted on the subfloor.

Cutting the Plates

After the layout for the exterior and interior walls has been snapped out on the subfloor or the slab, cut the top and bottom plates to fit the layout. Some carpenters use a tape measure to measure the length of plates. Others mark plates using the subfloor layout marks as guides.

Plates for the exterior walls are cut first. Before cutting, you must decide which exterior walls are by-walls and which are butt-walls. A *by-wall* runs from the outside edge of the subfloor at one end of the building to the outside edge of the subfloor at the opposite end, as shown in Wall Identification. By-walls are framed first and then erected into position. A *butt-wall* fits between the by-walls. Butt-walls are framed after by-walls. The double plate (rafter plate) will later tie butt-walls and by-walls together. The top plates of a given wall will be the same length.

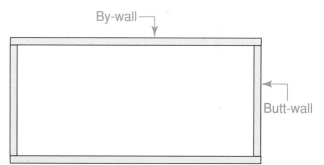

Wall Identification

First Decision To begin layout of the plates, decide which walls will be by-walls and which will be butt-walls. Usually, the longest walls will be by-walls, then the shorter walls are butt-walls. Sometimes the butt-walls have butt-walls that attach to them. For example, hall walls would normally be by-walls with the bedroom walls being butt-walls. Then closet walls would be butt walls to the bedroom walls.

Layout on a Concrete Slab In mild or warm climates, houses are often built on concrete slabs. Layout on a slab can be difficult because of the rough plumbing drains and electrical conduits that protrude through the slab. The layout may have to be adjusted if pipes are slightly out of position.

Cut the exterior plates to length, making sure that the ends of the plates for long walls break on 16" OC marks. Then stack the plates on the subfloor and align them with the chalk lines. After you have cut the exterior wall plates, cut the interior wall plates. Place these plates on the X side of the chalk lines.

Laying Out the Plates

Plate layout identifies the location of each stud in a wall, as well as the location of doors and windows. These locations are marked on the edges or the sides of the plates using a carpenter's pencil. Start by tacking the top and bottom plates together with two or three 8d nails. This prevents them from shifting during layout. If the edges of the plates are marked, then the edges should face up, as shown in Lining Up the Plates If the sides of the plates are marked, then the sides should face up.

The procedure for laying out the plates depends partly on how walls, windows, and doors are dimensioned on the plans. If they are dimensioned in reference to centerlines, use the procedure described under "Openings." If another dimensioning system is used, adapt the technique in some way. For example, you might measure between the faces of the plates, instead of to the centerlines of the plates.

Openings

Refer to the building plans to find the distance from one corner of the building to the center of the first opening. Measure this distance and square a line across both plates at this point. Mark the line with a centerline symbol and an identification letter or number. This can be used for reference when cutting other parts for this opening.

Continue to lay out and mark openings on the remaining exterior wall plates. Use a letter or symbol to distinguish between door and window centerlines. When laying out plates on a slab foundation, pay attention to the location of foundation anchor bolts during the layout process. There should be an anchor bolt within 12" of the end of each section of plate.

As they mark centerlines, many carpenters also "detail" the openings: this means to mark the rough opening as well as the location of trimmer studs and king studs. The **rough opening (RO)** is the space into which a door or window will fit, as shown in Layout Symbols. It allows room for the door or window and its frame. It also provides space for leveling and plumbing the frame. Note in Framing for a Rough Opening that the centerline of a door opening has been marked 7'-7" from the outside corner. The rough opening (RO) measures 34½". One-half the RO (17¼") is laid out on each side of the centerline. The thickness of the trimmer stud is laid out on each side of this. The header (37½" long) will rest on top of the trimmer studs and between the king studs (marked with an X).

Building plans include window and door *schedules,* which are charts that provide rough-opening sizes. Maximum spans for

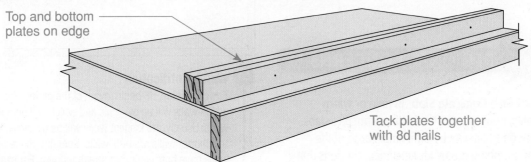

Top and bottom plates on edge

Tack plates together with 8d nails

Lining Up the Plates
Preparing for Layout Place top and bottom plates together on edge and tack them together. In this case, the layout will be marked on the edges.

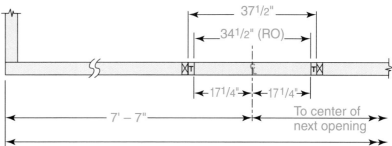

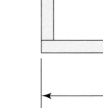
Layout Symbols
Identifying Marks Lines and symbols identify the location of all wall elements.

headers and estimates of material needed for studs are listed in the **Ready Reference Appendix**. Refer to the tables titled "Maximum Spans for Headers," "Exterior Wall Studs, Including Corner Bracing," and "Partition Studs, Including Top and Bottom Plates."

When the rough opening size for windows is not provided, it can be obtained from the window manufacturer's catalog. The rough opening sizes vary somewhat among manufacturers. Each catalog typically contains tables showing four width and height measurements for each window: masonry openings, rough openings, frame size, and glass size. Some may also list the sash size. Rough openings usually allow ½" on each side of the window or door and ½" at the top of the unit to allow adjustment of the unit for plumb and level installation.

After you mark a window or door centerline on the plates, measure from each side of the centerline a distance equal to one-half the rough opening, as shown in below. Square a line at this point. This line represents the inside face of the trimmer stud. Now mark the plate to locate the position of each king stud. The face of the king stud is 1½" away from the inside face of the trimmer stud.

Wall Intersections Mark the exterior plates to indicate the centerlines of all intersecting interior walls. Again, start from one corner of the building. Mark the place where the interior wall would intersect with a *P*, as in Marking Intersections on page 442.

Exterior Corner Posts A **corner post** is an assembly of full-length studs at the corner of a building. An *exterior* corner post is one that

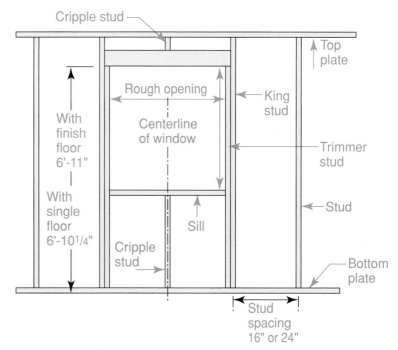

Framing for a Rough Opening
Window Framing Some carpenters add cripple studs under the ends of the sill for increased strength.

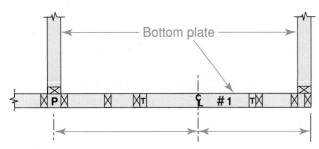

Marking Intersections
Where Walls Meet The location of the interior wall has been marked with a *P*. Start the layout from a corner.

forms an inside corner and an outside corner. The inside corner provides nailing surfaces for interior wall coverings. The outside corner provides nailing surfaces for sheathing. (Sheathing is discussed in Section 16.1.)

Corner posts are usually built from three or more studs to provide greater strength. They may be made in several ways. Two of the more common methods are shown in Corner Post #1 and Corner Post #2. Whatever method you choose, remember to mark the arrangement of studs on the exterior wall plates.

Partition Corner Posts A *partition corner post* is a particular type of post required where a partition meets another wall. A partition corner post is sometimes called a *channel*, a *partition-T*, or a *T-post*. In the type shown in In

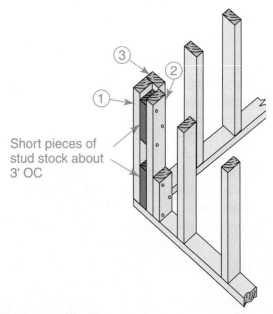

Short pieces of stud stock about 3' OC

Corner Post #2
Another Method The pieces numbered 1, 2, and 3 are studs. The short blocks are usually 10" or 12" long. Nail Studs 1 and 2 to the blocks with 10d nails. Stud 3 is part of the butt-wall.

Partition Corner #1, the regular spacing of the outside wall studs is interrupted by double studs where the partition ties in. The double studs are set 3" apart. This interval allows the partition's end stud to lap the others just enough to permit nailing. It leaves most of the inner edges of the other studs clear to serve

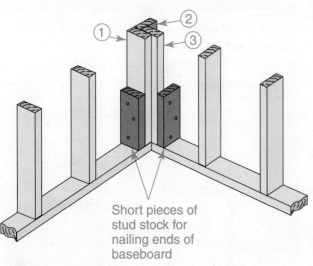

Short pieces of stud stock for nailing ends of baseboard

Corner Post #1
Simple Corner Post The pieces numbered 1, 2, and 3 are studs. This is an energy-efficient post because it allows insulation to reach the outer edge of the building.

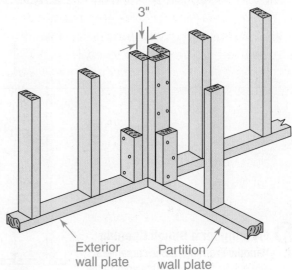

3"

Exterior wall plate

Partition wall plate

Partition Corner #1
One Option A double-stud partition corner assembly.

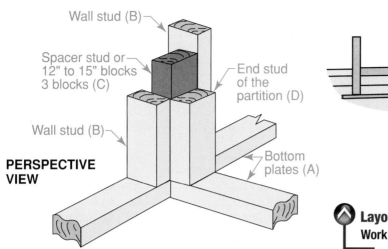

Wall stud (B)

Spacer stud or
12" to 15" blocks
3 blocks (C)

End stud
of the
partition (D)

Wall stud (B)

Bottom
plates (A)

**PERSPECTIVE
VIEW**

Partition Corner #2

A More Popular Option This method allows more nailing space on the inside corners than the corner post shown in Partition Corner #1. Nail the end stud of the partition to these studs and spacers when the partition is

Layout Template

Work Aid Using a template speeds the layout process.

as nailing bases for inside wall covering. The variation in Partition Corner #2 gives more nailing surface for the inside wall covering.

Many carpenters detail the arrangement of partition corner posts as they locate partition centerlines. This is sometimes done with the aid of a site-built jig made of two blocks of framing lumber nailed together.

Stud Locations Mark all the exterior plates and all the partition plates for the location of wall studs and cripple studs. There are various ways to do this. Tape measures marked with special symbols at intervals of 16" and 24" may be used. Some carpenters find it faster and easier to use a layout template, as shown in Layout Template. A *layout template*, or *layout stick*, is an aluminum bar with 1½" wide "fingers" that correspond to the particular stud spacing being used. The template is 4' long, the standard width of wall sheathing. It has four fingers, each 1½" wide and spaced 16" OC, representing four stud markings.

When laying out stud locations, you should be thinking ahead. Remember that sheathing and wallboard come in standard 4' widths. There must always be a stud where two panels will meet in a vertical joint. The panels are fastened to the studs at these joints.

Begin the layout on the plates by measuring from the corner of a by-wall. Make a mark 15¼" from the end of the plate. This will be the location to the edge of the first stud. Mark an X on the side of the line where the stud will be. This will ensure the stud will not be placed on the wrong side of the layout line. From that point on, mark every 16" to indicate the edge of each stud on that wall, as shown in Layout of a By-Wall on page 444. Double-check the layout by measuring along the plate to see that a stud will always be located where there will be a vertical joint between 4' wide sheathing panels.

Before laying out the studs on a butt-wall, check the plans to determine the thickness of the wall sheathing. The spacing of the butt-wall studs must account for the thickness of the intersecting by-wall and its sheathing. This is needed because the butt-wall sheathing will overlap the by-wall and its sheathing. Measure 15¼" in from the outside edge of the sheathing as shown in Layout of a Butt-Wall on page 444. This ensures that the edges of the sheathing on the butt-wall will be properly supported at 4' intervals.

Where the stud layout is interrupted by a window or door opening, lay out cripple studs above and below the opening. Maintain the 16" or 24" OC spacing that you used for the full-length studs. Mark on the plates the position of each cripple stud. The position of a cripple stud is usually marked with a C, rather than an X.

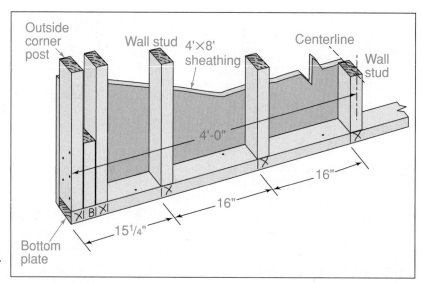

 Layout of a By-Wall
Account for the Sheathing Check the layout carefully to make sure sheathing will be properly supported.

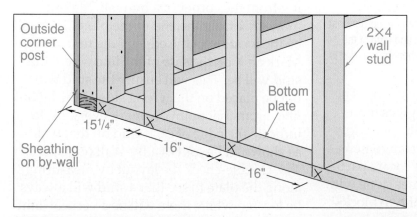

Layout of a Butt-Wall
Check the Plans The spacing depends on the thickness of the sheathing.

Section 16.2 Assessment

After You Read: Self-Check

1. What is the 3-4-5 rule used for?
2. What is plate layout?
3. Explain the difference between a by-wall and a butt-wall.
4. What is a rough opening?

Academic Integration: Mathematics

5. **3-4-5 Rule** With a partner, use a 25' tape measure to measure the diagonals of at least four large rectangles in a structure to see whether they are square. Examples of measurable spaces might be the walls of a room or a rectangular deck. Start in one corner of the layout. Record the results of your investigation for each rectangle, providing the length of the diagonal and whether or not the structure is square.

Go to connectED.mcgraw-hill.com to check your answers.

Assembling & Erecting Walls

Knowing Procedures

When would you prefer to erect a wall before you sheathed it?

When plate layout is complete, the various parts of the wall framing can be cut to length, assembled on the subfloor, nailed together, and lifted into place, as shown in Erecting a Wall. Several procedures can be used to assemble walls. Some carpenters prefer to tip framed walls into place and sheathe them later. Other carpenters install sheathing, windows, and sometimes siding on exterior walls before lifting them into place.

 Erecting a Wall

Team Effort In this case, the sheathing will be applied after the wall is in place. This reduces the weight of the wall and makes it easier to tilt up.

In either case, the walls should be squared. A common technique is to sheathe the walls on the floor deck but install windows and siding after the walls have been erected.

Choosing a Strategy

The method chosen depends in part on the length and weight of the walls. Whichever method is used, the order in which the exterior walls are to be assembled and erected must first be determined. The by-walls are usually erected first. The butt-walls are erected next.

In areas of the country where severe weather or earthquakes are a risk, buildings require the use of metal straps and anchors to strengthen the connections between framing members. In some cases, steel straps must be used to tie wall framing to the roof framing. Be sure to follow local codes. For more information, see Chapter 17.

Preparing Components

In the most common method of assembling walls, the cripple studs, trimmer studs, and headers are precut to length on the job site. Then they are distributed to the area on the subfloor where they will be assembled.

Builder's Tip

MAKING A CUT LIST Experienced carpenters always try to save time on a project without sacrificing quality. One technique is to make a cut list. This is a written list showing the length and dimension of all components in a wall. With the list in hand, one carpenter can quickly cut all the pieces without having to go back and forth to make measurements.

Making a Story Pole If there are several different heights to mark off above the subfloor, you may need to make additional story poles. These should be labeled for the various rooms or areas. To make a story pole:

Step 1 Select a straight length of framing lumber.

Step 2 Nail a 2×4 block to the bottom of the stud. This block represents the bottom plate of a wall.

Step 3 Examine the plans for standard window- and door-opening dimensions.

Step 4 Transfer these dimensions to the pattern. You can now use the story pole as a reference for cutting cripple studs and trimmer studs.

Cutting is usually done with a circular saw. However, a radial-arm saw or power miter saw ensures square cuts. Such a saw is often easier to use when making repetitive cuts to a standard length.

Using a Story Pole The length of each wall component can be determined from the plans. Because the height of wall openings is standardized, carpenters often lay out a story pole. A *story pole* is a piece of framing lumber that represents the wall from the top of the subfloor to the bottom of a ceiling joist. It includes information about the location and size of the window headers, sills, and door headers. It also includes the heights of various openings above the subfloor.

Trimmer Studs and Cripple Studs Trimmer studs should be cut to fit snugly under the header so they will support it properly. If a header settles, cracks in the plaster or drywall may develop, and doors and windows may fit improperly. The trimmer studs also reinforce the door and window openings. In fact, it is very important for all framing members to fit against each other tightly. This will help to reduce shifts in framing that can result in cracked wall surfaces as well as poorly fitting doors and windows.

Determine the lengths of the cripple studs

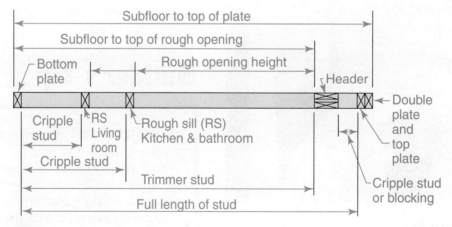

A Story Pole
Dimension List A story pole is an efficient method for laying out repetitive components of a wall.

by referring to the story pole. To determine how many are required, count the cripple-stud layout marks on the wall plates.

Headers The depth of a header (lintel) is determined by the length of the opening it must span. Longer openings require stronger, deeper headers. This information will be found in the building plans or local code requirements. Make sure that the header is long enough to bear on all of the trimmer studs.

Header lengths are obtained by measuring the top plate between layout marks for the king studs. In the case of the door opening shown in Framing for a Rough Opening on page 441, the header length would be 37½". Window and door headers are sometimes cut from solid pieces of 4×6 or larger stock. Commonly, however, the header is built up from 2×6 or wider framing stock. Side by side, two 2× members are only 3" thick. A ½" thick plywood spacer must be sandwiched between the two pieces to give the header the full 3½" thickness of the wall. The members are nailed with 16d nails staggered on 16" centers.

Various types of headers can be assembled using solid lumber, as shown in Headers. Many builders today use engineered lumber as header material. These products are generally stronger and save time because they do not need to be assembled.

A header will normally be supported by the trimmer studs. In some cases, as in remodeling, headers may be supported by metal framing brackets, as in Window Framing Brackets. In states where high winds are a threat, codes may require that metal straps be used to connect headers/plates and studs.

Rather than cut one header at a time, it is faster to first number the openings (such as windows, doors, and fireplaces) for

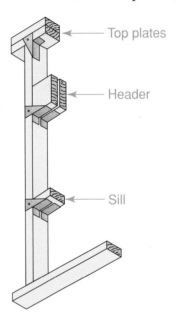

Top plates

Header

Sill

 Window Framing Brackets
Superior Strength Metal framing brackets can be used to support the studs, header, and sill. They also tie all the pieces together.

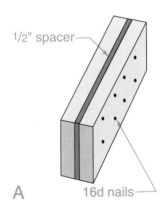

1/2" spacer

16d nails

A

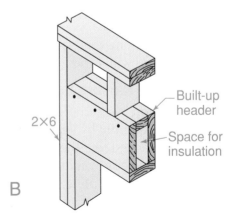

2×6

Built-up header

Space for insulation

B

 Headers
Two Common Types A header is often made from 2× lumber with a plywood spacer in **A**. The type of header shown in **B** is sometimes used because it can be insulated. The insulation must be inserted before the header is installed.

identification. Then you can make a cutting schedule for all headers. One person can cut these to length as another assembles them. Use 16d nails, two near each end. Stagger the others 16" apart along the length of the header. Do not forget to use ½" spacers between the 2× members. Place assembled headers at their locations on the subfloor in readiness for the assembly of the wall sections.

Assembling Corner Posts Because corner posts are made from precut studs, cutting is not required. Corner posts are nailed together with 10d and 16d nails. They are then taken to the place on the subfloor where they will be used for assembly of the wall sections. The short pieces of 2×4s at the base of the corner posts on page 442 are installed after the walls have been raised. They will provide places for nailing the ends of the baseboard. *Baseboard* is a type of finish trim installed at the base of a wall. In areas where energy efficiency is especially important, other corner-post assemblies can be used.

Reading Check

Recall *What is a story pole?*

Assembling & Raising Walls

Why is it important to align the edges of studs in a wall?

There are different methods for assembling and raising exterior walls and interior walls.

Assembling and Raising Exterior Walls

Each wall section is assembled on the subfloor. Begin the assembly by separating the plates that were tacked together for layout. Lay the top plate on edge on the subfloor about 8' from the bottom plate. Do not flip the top plate end-for-end as you move it. If the plate is flipped, the layout marks will not match those on the bottom plate. This is a common mistake.

Lay a full stud at each mark. Place the

Builder's Tip

ALIGNING STUDS When framing long walls such as those that form a hallway, inspect each stud before nailing it in place. Studs that are slightly bowed should be placed so that all the bows face the same side of the wall. This eliminates the wavy wall appearance that opposing bows create. If a stud's bow is significant, however, the stud should not be used for wall framing.

header so that the rough sill, cripple studs, and trimmer studs are in position. Place the preassembled exterior corners and partition corners at the marked locations. You can now nail the components together. The order of assembly depends on the preference of the carpenter. The following is one approach.

Beginning at one end of the top plate, drive two 16d nails through the plate into each stud at the correct location. Secure the bottom plate in the same way. Be careful to keep the edges of the framing members flush with each other. This is essential if sheathing is to fit correctly. If the wall contains a door or window, nail all those components into place. To keep the edges of studs and plates perfectly aligned during nailing, carpenters step on the intersection as they nail through the plate. This also keeps the wall from sliding if it is being hand nailed. This problem is eliminated with pneumatic nailing.

Framing a Window Opening Once all the full-length studs are in place, gather the components surrounding the window opening, as shown in Window Framing Brackets.

- The cripple studs at *A* are toenailed with four 8d nails, two on each side.

- The full stud is nailed to the header at *B* with four 16d nails and to the trimmer at *C* with 10d nails 16" OC.

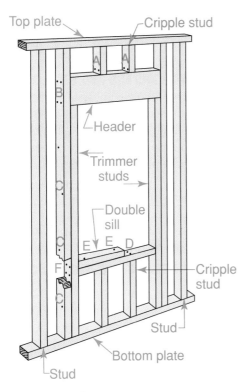

Window Framing

Basic Elements Assembly details for a window opening.

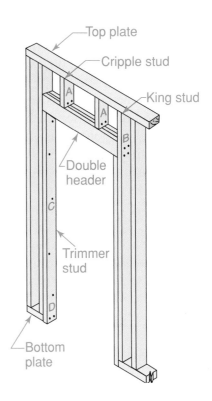

Door Framing

Basic Elements Assembly details for a door opening.

- The full studs are toenailed to the bottom plate or nailed through the bottom plate.
- The lower part of the double sill is nailed with two 10d nails into the ends of the cripples at *D*.
- The upper part of the sill (*E*) is nailed to the lower with 10d nails 8" OC and staggered.
- The ends of the sills are nailed through the trimmer studs with two 16d nails at each end (*F*).

Framing a Door Opening Assembly of a door opening such as the one shown in Door Framing should proceed as follows.

- The cripple studs (*A*) are toenailed with four 8d nails, two on each side.
- The full studs (*B*) are nailed to the header with four 16d nails on each side and toenailed to the bottom plate with two 8d nails. The full stud (*B*) could be nailed from the bottom up through the plate, with two 16d nails, if the plate is attached before the wall is erected.

- The trimmer (*C*) is nailed with 16d nails staggered 16" OC.
- Two 10d nails are driven into the end of the bottom plate at *D*.

General Details Trimmer studs fit under a window or door header. They are nailed to a king stud with 16d nails, spaced 16" apart and staggered. Notice that the trimmer stud for a door may extend from the header to the bottom plate. The portion of the plate within the door opening will be cut out

Builder's Tip

SPLIT-FREE NAILING When securing the studs at the ends of a wall, slightly blunt each nail by tapping its point with a hammer. The nail will then be less likely to split the plate.

later, after the wall is erected so that finish flooring can be laid. If this portion of the plate is cut out earlier, the wall will be more difficult to erect.

Nail through the king studs and into the header with 16d nails. Once the header is secure, insert the cripple studs above it. Nail their tops as if they were studs. Toenail the bottom end of each cripple to the header with two 8d nails on each side.

When you have fully assembled the wall on the subfloor, square it. Do this by running a tape measure across diagonally opposite corners, as shown in Squaring a Wall. If the diagonal measurements between all corners are the same, the wall is square. If the wall is not square, push the plates in opposite directions until it is.

As mentioned earlier, some carpenters apply sheathing at this point. Others apply it after the wall has been erected. If walls are framed *and* sheathed while flat on the subfloor, raising the completed wall can be awkward. This is because the sheathing makes it hard for carpenters to get a good grip on the wall.

 Squaring a Wall

Measure Diagonally Always check a wall for squareness before sheathing it. Measure diagonally across the corners. If both measurements are equal, the wall is square.

To make walls easier to raise, use a prybar to lift the top end of the wall off the deck. Slip in scrap 2×4 blocks at numerous locations beneath this. When the crew is ready to raise the wall, they will find it easier to get a solid grip.

Temporary Bracing Temporary bracing is bracing that has the following two purposes:

- It prevents walls from tipping as they are being erected.
- It holds walls in position after they have been plumbed and straightened.

Temporary bracing may consist of 2×4 or 2×6 members nailed to one face of a stud and to a 2×4 block nailed to the subfloor, as shown in Raising a Wall. The wall braces may also be nailed to wood stakes driven into the ground outside the perimeter of the foundation.

Take care not to let the ends of the temporary braces project above the top plate. Otherwise, the braces could interfere with ceiling and roof framing and would have to be removed. This would disturb the plumbed and straightened walls. Use enough nails to brace the wall securely, but do not drive the nails in all the way. Each nail head should project enough to allow easy withdrawal. Leave the temporary

FOTOGRAFIA INC./iStock/Getty Images Plus

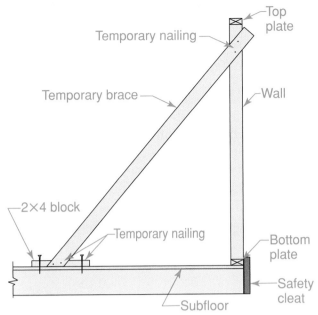

- Top plate
- Temporary nailing
- Temporary brace
- Wall
- 2×4 block
- Temporary nailing
- Bottom plate
- Safety cleat
- Subfloor

Raising a Wall
Slow Is Safe The temporary brace pivots into place as the wall is raised. The block nailed on the subfloor holds it in place. A safety cleat keeps the wall from slipping off the subfloor.

bracing in place until the ceiling and roof framing have been completed, and sheathing has been applied to the outside walls.

Raising the Wall As a wall is lifted into position, align the bottom plate with the chalk lines made earlier. This is a job for several carpenters. One carpenter is needed at each end of the wall. One or more may be needed in between, depending on the length of the wall. Have temporary bracing ready for use as soon as the wall is partially upright. As an extra measure of safety, nail 2×4 cleats to the outside of the rim joist to prevent an exterior wall from slipping off the subfloor as it is lifted. (A *rim joist* is a joist at the edge of the floor system.) In some cases, hand-cranked lifts are used to tilt a heavy wall into position.

When the wall is upright, fasten the bottom plate to the floor framing with 16d nails spaced 16" apart and staggered when practical. You can now plumb and brace the wall.

Plumbing the Wall To *plumb* a wall means to make sure it is perpendicular to the subfloor. Either a carpenter's level or a plumb bob

may be used to plumb wall sections. As noted earlier, framed exterior walls may be raised into position with or without sheathing already applied. In either case, the walls must be plumbed and straightened. This is done after all the framed walls are in position and temporarily braced.

Using a Plumb Bob To plumb a corner post with a plumb bob, attach the plumb line (string) securely to the top of the post, as shown in Plumb Bob Method. Make sure the line is long enough to allow the plumb bob to hang near the bottom of the post. Use two blocks of wood identical in thickness as gauge blocks. Tack one block near the top of the post between the plumb line and the post. Insert the second block between the plumb line and the bottom of the post. If the entire face of the second block makes contact with the line, the post is plumb.

Using a Carpenter's Level To plumb a corner with a carpenter's level, do not place the level directly against a stud. The face or edge of the stud may be bowed. Instead, place an 8' level against blocks nailed to the top and bottom plates. To increase accuracy when plumbing the corner, hold the level so that you can

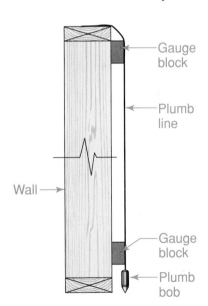

- Gauge block
- Plumb line
- Wall
- Gauge block
- Plumb bob

Plumb Bob Method
String and Weight Plumbing a wall using a plumb line and plumb bob.

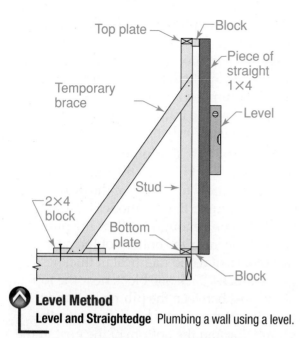

Top plate

Block

Piece of straight 1×4

Temporary brace

Level

Stud

2×4 block

Bottom plate

Block

▲ **Level Method**
Level and Straightedge Plumbing a wall using a level.

look straight in at the bubble. If a long level is not available, place a shorter level against a 1×4 straightedge, as in Level Method. The

blocks can be attached to the framing or to the straightedge. While one carpenter reads the level, another should be ready to move the braces as needed and to secure them as soon as the correct position is found.

Plumb outside corners by checking them on two adjacent surfaces, then brace them to prevent the wall from shifting. After you have plumbed and braced all exterior walls, plumb and brace the intersecting interior walls. This will also plumb the exterior wall at the point of intersection.

Straightening Walls To straighten walls, fasten a string line to the outside top of one of the corner posts. Stretch the line to the outside top of the corner post at the opposite end of the building. Fasten the line to this post in the same manner as for the first post. Place a ¾" wood block under each end of the line to give clearance, as shown in Straightening a Wall. Place

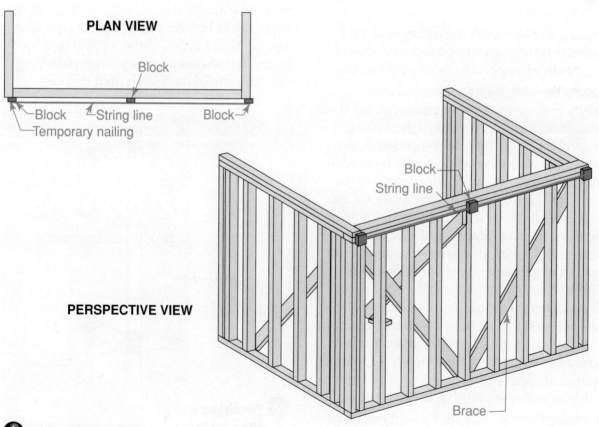

PLAN VIEW

Block

Block

String line

Block

Temporary nailing

Block

String line

PERSPECTIVE VIEW

Brace

▲ **Straightening a Wall**
The Line Is the Guide Use a string line and blocks as a reference to straighten the walls.

additional temporary braces at intervals close enough to hold the wall straight. When the wall is far enough away from the line to permit another ¾" block to slide between the line and the plate, nail the braces. This straightening procedure is sometimes called *lining the walls.* It is carried out for the entire perimeter of the building. Later, you should also straighten any long partitions in the same manner.

Assembling and Raising Interior Walls

After all the exterior walls are plumbed and braced, and the bottom plates securely nailed, assemble and erect the interior walls. Interior walls are easier to assemble than exterior walls because they usually do not require framing for windows. However, they do require framing for doors. Otherwise, they are assembled in the same manner as exterior walls. To determine the sizes of the various parts (such as headers, trimmer studs, and cripple studs), refer to the building plans and the story pole.

Careful planning of the order in which the interior walls are assembled and erected is very important. See the Step-by-Step Application on page 454.

Reading Check

Recall What safety precautions are important when raising walls?

JOB SAFETY

RAISING WALLS A wall that is being raised must be braced quickly to lessen the chance that it will topple. Carpenters often secure the tops of the braces to the studs before erecting the wall. A single nail at this point allows the brace to swing into position as the wall is being raised.

Installing the Double Plate
Why is the double plate necessary?

The double plate is nailed to the top plate after the exterior and interior walls have been erected. Use the same type and quality of material as for the top plate. Be sure to cut the pieces accurately to length. Cutting to length is sometimes done when the top and bottom plates are cut. However, it is most often done after the walls are erected because the double plates are a different length than the other two plates.

One of the main purposes of the double plate is to tie the walls together at the top. Therefore, it laps over the joint formed at a corner of intersecting exterior walls, as shown in below. It also laps the joint formed by intersecting partitions, as shown on page 454. On a long wall, joints in the double plate should be at least 4' from any joint in the top plate. Fasten the double plate with 10d nails spaced 16" OC and staggered. Nail end laps between adjoining plates with two 16d nails on each lap (see bold arrow in on page 454).

Order of Assembly on page 454 shows the order of assembly for assembling and erecting interior walls.

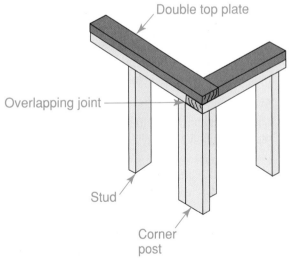

Double top plate

Overlapping joint

Stud

Corner post

Plate Lap at Corner
Important Overlap Double plates are joined by a lap joint.

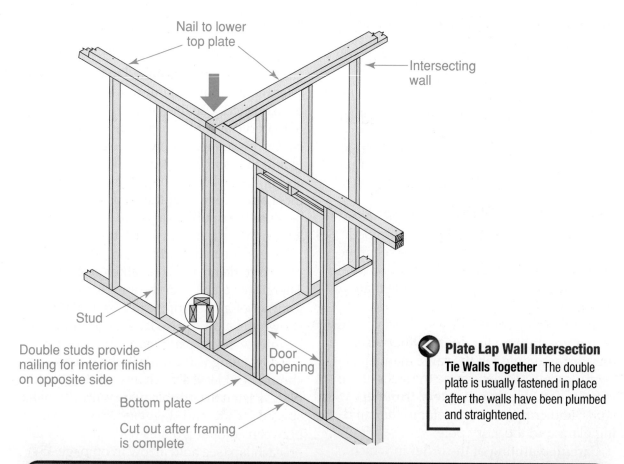

Nail to lower top plate

Intersecting wall

Stud

Double studs provide nailing for interior finish on opposite side

Door opening

Bottom plate

Cut out after framing is complete

◀ **Plate Lap Wall Intersection**
Tie Walls Together The double plate is usually fastened in place after the walls have been plumbed and straightened.

Step-by-Step Application

Assembling and Erecting Interior Walls The following is a general procedure.

Step 1 Raise, fasten, and temporarily brace the longest center partition (1). Work should then proceed from the center wall out to the exterior walls, and from one end of the building to the other. Complete operations in one area before moving to the next area.

Step 2 Note that partition 1, though interrupted by openings, is considered to be one piece. Erect and plumb partition 2 next. Note that it helps support partition 1 and connects it to the previously plumbed exterior wall. Partition 3 comes next, then 4, 5, and 6, etc.

Step 3 Continue erecting partitions that are at right angles to each other, all the way to the back of the building. This sequence is better than erecting two parallel partitions (such as 2 and 4) and then working in a confined area to erect the connecting partition (3).

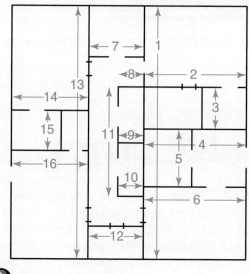

▼ **Order of Assembly**

Section 16.3 Assessment

After You Read: Self-Check

1. What is the purpose of a story pole?
2. What may happen if a trimmer stud does not properly support a header?
3. What determines the depth of a header?
4. What size nail is used to nail through the top and bottom plates and into the studs? How many nails are used?

Academic Integration: Mathematics

5. **Understand Tool Accuracy** Suppose a 2' carpenter's level is accurate enough to measure only within ¹⁄₁₆" of true level over 2'. If the level is used to plumb an 8' wall, by how much might the top of the wall vary from true plumb?

Go to **connectED.mcgraw-hill.com** to check your answers.

Section 16.4 Special Framing Details

Dealing with Special Conditions

When is special framing called for?

Special framing adds strength and quality to the construction. Its requirements are not always noted on the building plans. However, the carpenter should be familiar with them. Special framing details are required in various situations:

- For unusual architectural features.
- To provide openings for plumbing vents and fixtures.
- To provide openings for heating ducts.
- To add support for heavy items.
- To add blocking that supports the edges of interior wall coverings.
- To provide extra strength to houses built in earthquake or hurricane zones.
- In some cases, for fire safety.

A building must be enclosed quickly to protect it from the weather. However, special framing can be time-consuming. Therefore, it may not always be installed as the walls are being framed and erected. Instead, a builder may concentrate on sheathing the walls first and getting the house closed in. The builder may then install special framing as fill-in work during slack periods in later construction stages.

Unusual Walls

Most walls can be framed using standard precut studs and other components. In some situations, however, wall framing must be handled differently.

Gable Walls Walls that angle upward to meet the underside of the roof framing are called *gable walls* or *rake walls*. The spacing of the studs is the same as for surrounding walls, but precut studs cannot be used. Instead, each stud in the gable wall must be cut to a specific length. Its top end must be cut at an angle that matches the roof pitch (slope). The bottom plate of a gable wall is similar to that in a standard wall. The top plate slopes to follow the roof angle.

Gable walls are sometimes built and erected along with the rest of the wall framing. In other cases, the roof framing is installed first and the gable-wall framing is added afterwards. If the roof framing consists of trusses, the truss manufacturer supplies the gable-wall framing.

The cuts involved in building a gable wall are directly related to the slope of the roof. For this reason, gable-wall construction is discussed in Chapter 18.

Bay Windows Bay windows project outward from a wall or roof and have special framing requirements. The floor framing is provided during an earlier stage. The projecting walls of the bay may be framed and erected as if they were standard walls. Sometimes, however, an angled bay may call for framing that has been beveled with a circular saw or table saw. Note also that a bay window may require two headers. One header is over the opening in the main wall and one is over the window itself. See Chapter 15 for details of bay window floor framing.

There are various ways to frame a bay window, depending on its size, roof type, and the type of windows. The bay window in Bay Window Framing is one example.

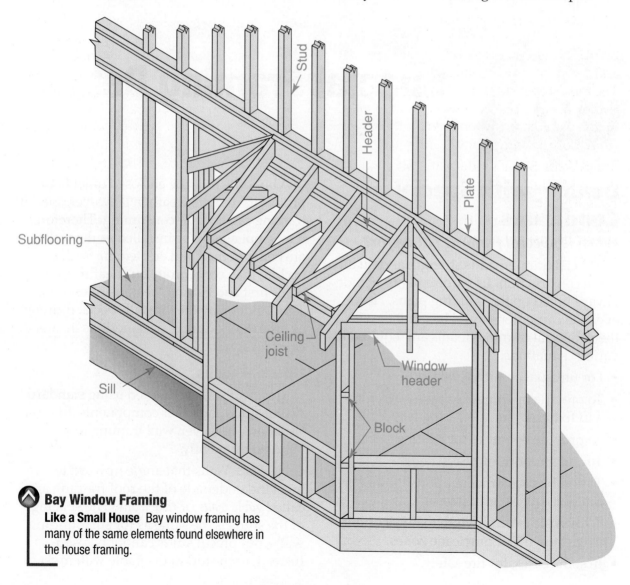

Bay Window Framing
Like a Small House Bay window framing has many of the same elements found elsewhere in the house framing.

The ceiling joists in the framing of this bay window are set on top of the window headers. The top of a bay window should be kept in line with the other windows and doors in the room. Therefore, the wall header will not be a standard header height. It will have to be raised so that its bottom is in line with the bottom of the bay ceiling joists, as shown.

Shear Walls A *shear wall* has been engineered to withstand unusual stresses. These walls are installed in order to make the structure of a house more rigid. They are often used in areas where earthquakes and severe storms are common. Such a wall may also be a feature of houses with unusual designs.

Shear walls may require extra nailing, hold-downs and/or special anchor bolts. Wider studs may be needed to accommodate close nailing patterns. Shear walls may also call for the use of construction adhesives and/or sheathing on both sides of the wall. The requirements for a shear wall are specified by an architect or engineer and detailed on the plans. The carpenter must not deviate from those specifications.

Radius Walls Some modern architectural designs include curved walls, called *radius walls*. The **dimensions** of the curve are detailed on the plans. The studs may be the same dimensions as in surrounding walls. However, the curved plates are cut from plywood instead of lumber. The ends of

blocking must be cut to an angle. The ability to build such unusual walls is the mark of an experienced framing carpenter.

Reading Check

Summarize *How is a bay window like a miniature house?*

Other Special Conditions

Carpenters often encounter special framing requirements that relate to providing framing for the work of other tradespeople. It is important for the carpenters to understand these requirements while performing their own trade.

Plumbing Needs Plumbing vents are usually installed in a 2×6 or thicker wall, as shown in Framing a Plumbing Wall. This is sometimes called a *plumbing wall*. It provides the needed wall thickness for the bell (large end) of a 4" cast-iron soil pipe, which is larger than the thickness of a 2×4 stud wall.

In some cases, builders stagger wall studs in a way that reduces sound transmission through the wall. Fiberglass batt insulation or sound-deadening batts are then added to form a continuous, unbroken layer, as shown

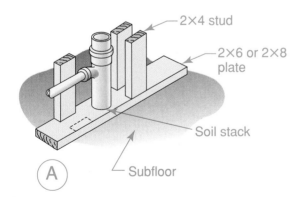

A

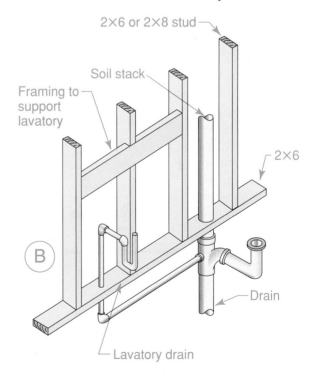

B

Framing a Plumbing Wall
Thick Wall To create a thick wall, **A.** 2×4 studs can be installed sideways on a 2×6 plate, or **B.** 2×6 studs can be installed as studs.

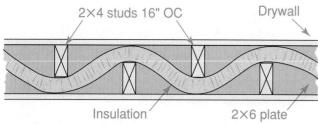

2×4 studs 16" OC   Drywall

Insulation   2×6 plate

 A Sound Wall

Muffling Noise An extra thick wall stuffed with insulation helps to block sound transmission generated by drain pipes. *Where might these walls be found?*

in A Sound Wall. To further reduce sound transmission, the wall can be covered with sound-resistant drywall (see Chapter 32, "Wall & Ceiling Surfaces"). This type of construction is sometimes called a *sound wall*. It helps to muffle the sound of water rushing through drain pipes.

Some plumbing fixtures may require extra backing prior to installation. This extra backing is sometimes called *blocking*. Backing is usually made from short lengths of 2× lumber but in some cases ¾" plywood can be used. This framing is sometimes noted on the plans and if so, it should be installed by the carpenter. Blocking can also be installed by the plumber. See the Step-By-Step Application below for this procedure.

Step-by-Step Application

Blocking for Plumbing Fixtures Blocking helps to support plumbing fixtures. It should be installed shortly after the walls are framed.

Step 1 Determine the height of the fixture and mark the location.

Step 2 Nail a block on the side of the stud. Set it back from the edge a distance equal to the thickness of the backing material.

Step 3 Cut the backing to fit between the studs and nail it in place. The backing material can also be notched into the studs at the correct height and face-nailed with 10d nails.

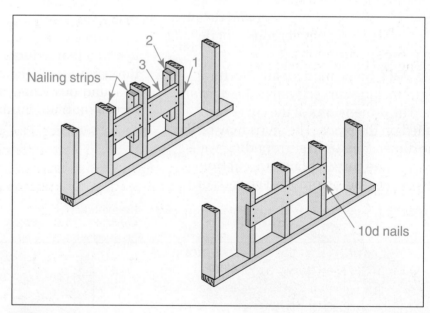

Nailing strips

10d nails

 Backing

One use of blocking might be that where a bathtub is enclosed by walls, support for its edges must be provided. (See page 415.) Backing must also be provided for the shower-arm fitting and for any grab rails that will be installed in the tub area.

Sometimes a hatch will be built on the back of the shower wall. This provides access to the tub drain and overflow riser. Suitable backing should be installed to support the edges of the hatch. Wall-mounted sinks and toilets require special blocking to ensure a secure attachment to the wall.

Cabinets Special support and blocking must be provided for inset cabinets that are to fit between studs and be flush with the wall covering. These cabinets are usually designed to be fastened directly to the faces of the studs or to blocking. Backing for nailing the wall covering must be provided at the top and bottom of the cabinets.

Bathroom vanities, kitchen cabinets, and other cabinets must be securely fastened to the wall. A good framing job includes special blocking for this purpose. The location of the blocking can be determined by studying the building plans. One method of blocking for an upper cabinet is shown in Cabinet Blocking. In some cases it is advisable to install an extra full-length stud or even a 2×6 flat in the wall. This is especially true in kitchens where upper and lower cabinets will be installed. To install the blocking:

1. Mark on the studs the top and bottom of the cabinets.
2. Blocks for attaching the cabinet backing are then fastened between the studs. These blocks must be back from the edge of the studs a distance equal to the thickness of the cabinet backing.
3. Mark the position of the cabinet backing onto the blocks.
4. Fasten the cabinet backing to the blocks at the location marks.

There are other ways to provide support for cabinets. Some carpenters inset ¾" plywood panels into the framing in the area of

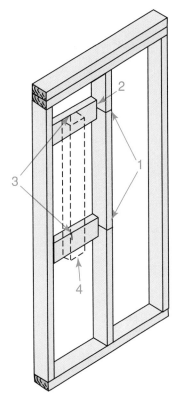

Cabinet Blocking
One Method Hanging cabinets require solid support within the wall.

the cabinetry. Others inset continuous strips of 2× lumber where the top and bottom of the cabinets will be.

Drywall Blocking Drywall must be nailed or screwed around the entire perimeter of each sheet. This means that the edges of drywall must fall on solid wood. Though the carpenter does not install the drywall, he or she should know what type of support it needs. Added blocking or even whole studs that support drywall is much easier to install while the framing is being assembled. However, blocking requirements are not typically noted on plans.

Trim Blocking The installation of baseboard, chair rail, crown, and other moldings is made easier if blocking is provided for nailing at the ends. Without it, the nails must be driven very near the ends of the molding and usually at a slight angle to reach the corner posts. This often results in splitting the ends of the molding. Blocking,

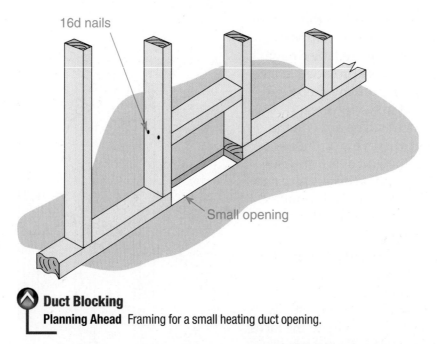

Duct Blocking
Planning Ahead Framing for a small heating duct opening.

such as that shown on page 442, will minimize this problem. These small blocks are made from scrap framing stock that would otherwise be discarded. Trim carpenters appreciate this detail but do not often encounter it. Trim backing is often left out at the framing stage unless specifically noted on the plans.

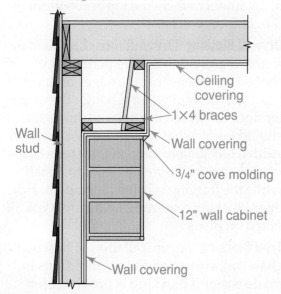

Soffit Framing
Filling Space One method for constructing a cabinet soffit.

Heating Ducts Heating ducts require openings in the ceiling, floor, or wall. Backing must therefore be provided for fastening the wall-covering material. An opening in the wall larger than the distance between studs requires cutting off one or more studs. It also requires a header to support the bottom end of the shortened stud and serve as a nailing surface for the wall covering.

Soffits A **soffit**, sometimes called a *bulkhead*, is a lowered portion of the ceiling. It sometimes contains lighting. More commonly it is used where upper cabinets do not extend to the ceiling, as shown in Soffit Framing. A soffit is usually about 2" deeper than the cabinets, so that molding may be installed at the cabinet top. Some carpenters will assemble a soffit from framing lumber and lift it into place as a unit. Others may install it in place. Whatever the method, the soffit must be level and securely attached to the surrounding framing. This makes cabinet installation easier.

The bottom of the soffit is usually about 84" from the finished floor. It is assembled from 2×2 and 2×4 lumber nailed together with 16d and 10d nails. It should be fastened directly to the wall and ceiling framing. After the wall covering has been applied and painted, the cabinets are attached to the wall and to the bottom of the soffit. A piece of cove or quarter-round molding may be used to close the joint between the cabinet and the soffit.

Chases A *chase* is a framed passageway that contains drainage pipes, wiring, or other features that must be concealed. In this respect, it is similar to a soffit. A chase may be vertical or horizontal. It typically does not carry structural loads.

Fireblocking

Building codes may require *fireblocking* in walls that are over 10' high. Fireblocking is meant to slow the passage of flames through wall cavities. It also strengthens the walls. It is made from short lengths of 2× framing lumber installed crosswise between studs. The blocking must be the same width as the surrounding framing and should fit snugly. The blocking can

be staggered to make end nailing easier. In standard-height walls, the top and bottom plates are considered sufficient fireblocking.

Mathematics: Estimating

Estimate Sheathing A one-story house has four walls and two gables. Two of the walls are 38' long. The other two walls are 50' long. The walls are 10' high. Each gable has a height of 12'. The gables are above the shorter walls. How many sheets of 4'×8' wall sheathing will be needed? Round to the nearest whole number.

Starting Hint To find the area of a gable, multiply the height of the gable by one-half its width at the bottom. Multiply the result by the number of gables to determine the total gable area.

Section 16.4 Assessment

After You Read: Self-Check

1. What is a shear wall? When might it be used?
2. Name three types of special framing that might be required in a bathroom.
3. What is the purpose of a cabinet soffit?
4. What is the purpose of fireblocking?

Academic Integration: Science

5. **Continuous Blocking** In cabinet blocking, blocking must be installed between studs in continuous runs. This means that some will have to be toenailed to the studs. To avoid this, some carpenters alternate 2×4 blocks with 2×6 blocks (or 2×6 blocks with 2×8 blocks). First, the narrower blocks are installed in every other stud bay and end-nailed through the studs. Then the wider blocks are installed in the remaining stud bays. The extra width of the second set of blocks allows them to be end-nailed as well. Why is end-nailing used rather than toenailing?

Go to connectED.mcgraw-hill.com to check your answers.

Chapter Summary

Exterior walls are nearly always load-bearing walls. Wall framing members include studs, plates, headers, cripple studs, sills, and trimmer studs. Sheathing may be applied before or after a wall is raised. The spacing and gauge of sheathing fasteners is important for establishing a solid connection.

Wall framing begins with laying out the location of two intersecting exterior walls. Plate layout identifies the location of each stud in a wall, as well as the location of doors and windows.

A story pole is useful as a quick layout reference. Exterior walls are set up, plumbed, and braced before interior walls.

Special framing adds strength and quality to a structure. Special framing is often done after the building is enclosed.

Review Content Vocabulary and Academic Vocabulary

1. Use each of these content vocabulary and academic vocabulary words in a sentence or diagram.

Content Vocabulary

- sheathing (p. 430)
- stud (p. 432)
- plate (p. 432)
- header (p. 432)
- rough sill (p. 434)
- cripple stud (p. 434)
- trimmer stud (p. 434)
- rough opening (RO) (p. 440)
- corner post (p. 441)
- temporary bracing (p. 450)
- soffit (p. 460)

Academic Vocabulary

- primary (p. 432)
- dimensions (p. 457)

Speak Like a Pro

Technical Terms

2. Work with a classmate to define the following terms used in the chapter: *load-bearing wall* (p. 430), *bearing wall* (p. 430), *partition walls* or *partitions* (p. 430), *sole plate* (p. 432), *double plate* (p. 432), *lintel* (p. 432), *trimmer* or *jack stud* (p. 434), *by-wall* (p. 439), *butt-wall* (p. 439), *schedules* (p. 440), *partition corner post* (p. 442), *channel* (p. 442), *partition-T* (p. 442), *T-post* (p. 442), *layout template* (p. 442), *layout stick* (p. 443), *baseboard* (p. 447), *lining the walls* (p. 453), *shear wall* (p. 457), *sound wall* (p. 458), *bulkhead* (p. 460).

Review Key Concepts

3. List the components of a framed wall.

4. Summarize how to apply sheathing.

5. Demonstrate how to estimate sheathing materials.

6. List the two main steps in wall layout.

7. Explain how to frame a window.

8. Describe two situations in which special framing might be used.

Critical Thinking

9. Synthesize What special framing techniques, if any, might be required in your local area because of climate or weather? Explain.

Academic and Workplace Applications

STEM Mathematics

10. Measuring Diagonals Find the length of the diagonal of a rectangular floor that measures 24' by 36'. Draw a sketch to help you. Round your answer to the nearest tenth.

Math Concept A diagonal separates a rectangle into two congruent (identical) right triangles. The lengths of the sides of a right triangle are related according to the Pythagorean theorem, which states that the sum of the square of the altitude and the square of the base of a right triangle equals the square of the hypotenuse. To square a number means to multiply it by itself.

Step 1: Find the sum of the squares of 24' and 36'.

Step 2: Use a calculator to find the square root of that sum.

STEM Engineering

11. 3-4-5 Rule The 3-4-5 rule relies on proportions, or ratios, to work. The 3-4-5 rule states that if two legs of a triangle measure 3' and 4', they form a 90° angle if the diagonal between the ends of the legs measures exactly 5'. Multiples of these numbers that preserve the 3:4:5 ratio (such as 9, 12, and 15) will also work in a 3-4-5 equation:

$$3 \times 3 = 9, 3 \times 4 = 12, 3 \times 5 = 15.$$

Use the 3-4-5 rule to fill in the missing number in the following proportional measurements where the two legs of a right triangle are 21' and the diagonal is 35':

$$21\text{-} \underline{\ ?\ } \text{-}35$$

12. Career Skills: Communicating Information Imagine that you work for a construction company that builds 20 houses each year. You are responsible for quality control. Create a one-page checklist that will help the builders review each set of plans for the special framing requirements. Include a paragraph that summarizes how the checklist will be used.

Standardized TEST Practice

CERTIFICATION PREP

Short Answer

Directions Answer each question below with a complete sentence.

13. What are the primary wall-framing members?

14. What is a bearing wall?

15. What are three things good wall-framing lumber should be?

TEST-TAKING TIP

In some short-answer tests, a short phrase may be an acceptable format for your answer. For other short-answer tests, a complete sentence may be required in order for you to receive full credit for your response. Be sure to reread the directions to make sure that your answer is in the correct format. A complete sentence contains a subject and a predicate (a verb phrase).

*These questions will help you practice for national certification assessment.

Basic Roof Framing

CHAPTER 17

Section 17.1
Planning a Roof

Section 17.2
Roof Framing with Common Rafters

Section 17.3
Ceiling Framing

Section 17.4
Roof Trusses

Chapter Objectives

After completing this chapter, you will be able to:

- **Identify** the basic roof styles.
- **Understand** the basic terms relating to roof-framing carpentry.
- **Explain** the layout of a common rafter, using at least one of the four basic methods.
- **Describe** the layout of ceiling joists.
- **Name** the three basic parts of a roof truss.
- **Demonstrate** how to install roof trusses.

Discuss the Photo

Building a Roof Roofs are constructed in many different shapes. *What are some things to consider in determining the shape of a roof?*

Writing Activity: Research and Summarize

Contact a manufacturer of residential wood trusses. Find answers to the following questions. Then summarize your findings in a one-page report.

1. How long does it take to design a truss, and how far in advance should a builder order them?
2. How are computers used in the design of trusses?

© iStockphoto/James Brey

Before You Read Preview

Roof framing is the last major framing activity in the construction of a house. Choose a content vocabulary and academic vocabulary word that is new to you. When you find it in the text, write down the definition.

Content Vocabulary

rafter	unit run	pitch
ridge board	total rise	chord
span	unit rise	web
total run	slope	

Academic Vocabulary

You will find these words in your reading and on your tests. Use the academic vocabulary glossary to look up their definitions if necessary.

■ benefit   ■ distributed   ■ access   ■ specify

Graphic Organizer

As you read, use a two-column chart like the one shown to organize content vocabulary words and their definitions.

Content Vocabulary	Definition
rafter	A rafter is an inclined framing member that supports the roof.

Go to connectED.mcgraw-hill.com to download this graphic organizer.

Planning a Roof

Roof Styles

Which roof styles are most common where you live?

Roof framing is considered the most complicated frame carpentry in a house because of all the angles involved. It may also seem difficult to learn because of the many special terms. However, it is important to understand that even the most complex roofs are based on a few standard designs.

Roof framing begins after the house walls have been framed. In most cases, the walls have also been sheathed to increase their strength and stiffness. Nominal 2" lumber is generally used for roof framing but I-joists are increasingly common (see Chapter 13,

"Engineered Wood"). This and following chapters describe roof framing using conventional lumber. Using I-joists requires the same basic understanding of rafter layout and roof design. Truss roof framing is covered in Section 17.4.

The main purpose of a roof is to protect the house in all types of weather with a minimum of maintenance. The roof must be appropriate for the climate in which the house is being built. A roof must be strong to withstand snow and wind loads. A roof should provide a continuous downward slope to shed rain water and snow melt. The parts must be securely fastened to each other to prevent them from coming apart in high winds or collapsing under a heavy load of snow.

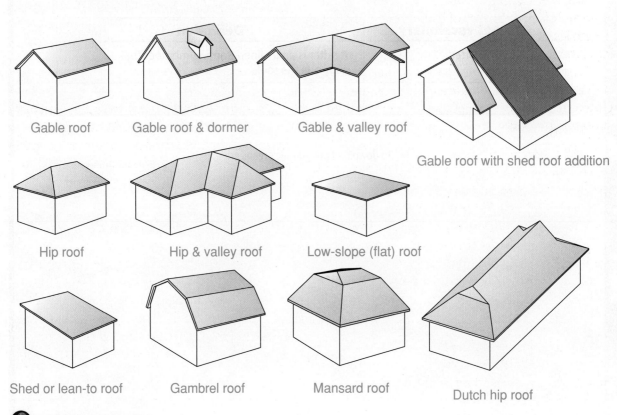

Gable roof   Gable roof & dormer   Gable & valley roof   Gable roof with shed roof addition

Hip roof   Hip & valley roof   Low-slope (flat) roof

Shed or lean-to roof   Gambrel roof   Mansard roof   Dutch hip roof

Common Roof Styles
Shape Determines Style Each style of roof has advantages and disadvantages.

Two Styles of Roof
Multiple Styles Gable roofs and hip roofs are common styles in homebuilding. *Which roof styles do you see on this house?*

Arnold & Brown

A carpenter must understand and be able to frame roofs in different styles. The basic roof styles used for homes and small buildings are gable, hip, low-slope, and shed, as shown in Common Roof Styles. Variations are associated with architectural styles of different regions or countries. Some of these include the gambrel roof, the mansard roof, and the Dutch hip roof.

A roof should also add to the attractiveness of the home. Roof styles are used to create different effects. A house may have more than one style of roof.

Gable Roof

The *gable roof* has two sloping sides that meet at the top to form a gable at each end. A *gable* is the triangular wall enclosed by the sloping ends of the roof. A gable roof may include *dormers* (upright window projections) that add light and ventilation to second-floor rooms or the attic. The gable roof is the most common type of roof.

Hip Roof

A *hip roof* slopes at the ends of the building as well as at the two sides. The slope on all sides results in an even overhang all around the building and gives a low appearance. Because there is no siding above the overhang, maintenance needs are reduced. The hip is also a very strong roof and is often found in regions where severe storms are common.

Low-Slope Roof

Sometimes called a *flat roof*, a *low-slope roof* is not perfectly flat. Instead, the rafters are laid at a slight angle to encourage water to drain. A **rafter** is an inclined framing member that supports the roof. Sheathing and roofing are applied to the top of the rafters. The ceiling material is applied to the underside of the rafters. Because a flat roof can be difficult to waterproof, it is found most often in dry climates.

Shed Roof

Sometimes called a *lean-to roof*, the *shed roof* slopes in one direction only. A shed roof is often used for an addition to an existing structure. In this case, the roof may be attached to the side of the structure or to the roof.

Gambrel Roof

The *gambrel roof* is a variation of the gable roof. It has a steep slope on two sides. A second slope begins partway up and continues to the top. It is commonly used on barns. A gambrel roof allows for more useable attic space than a typical gable or hip roof. This space can also be used as a second floor.

Mansard Roof

The *mansard roof* is a variation of the hip roof. It has steep slopes on all four sides. Partway up, a shallow second slope is developed and continues to the top where it meets the slopes from the other sides. The mansard roof style was brought to North America by the French who settled in Quebec, Canada.

Dutch Hip Roof

A *Dutch hip roof* is related to both the gable roof and the hip roof. Basically, it is a hip roof with a small gable at each end near the top. Like a hip roof, it has an even overhang around the entire building. This protects the walls from rain. Like a gable roof, portions of a Dutch hip roof are formed by two slopes that meet at the top.

Roof Framing Basics

How does the slope of a roof relate to local weather patterns?

Mastering the special terms used in roof framing will make it easier for you to understand roofing concepts. Become familiar with the framing square, which you will be using. Its short leg is called the *tongue*. Its long leg is called the *blade*, or *body*.

Parts of a Roof

A basic, conventional roof consists of rafters, ceiling joists, and a ridge board, as shown in Conventional Roof Framing. More complex roofs include other elements such as braces, valley rafters and jack rafters. (See Chapter 18, "Hip, Valley, and Jack Rafters.")

A *rafter* is an inclined member of the roof framework. Rafters serve the same purpose in the roof as joists in the floor or studs in the wall. They are usually spaced 16" or 24" apart. Rafters vary in depth depending on their length, the distance they are spaced apart, their slope, and the kind of roof covering to be used. A *collar tie* is a horizontal tie that connects opposite pairs of rafters to help stiffen the roof. A **ridge board**, or *ridge*, is the horizontal piece that connects the upper ends of the rafters. It extends the full length of the house.

Rafters often extend beyond the exterior walls to form *eaves* (also called *overhangs*) that protect the sides of the house from sun and water damage. The tail is the portion of the rafter that extends beyond the wall of the building to form the eave. Unsupported eaves

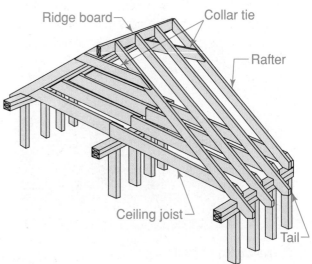

Ridge board — Collar tie

Rafter

Ceiling joist —

Tail —

Conventional Roof Framing
Basic Elements The basic parts of a roof frame.

commonly range from 6" to 24" in depth. Deeper eaves should be supported by posts.

The types of rafters shown in on page 469 can be used to frame various styles of roof. Not every type of rafter will be needed for each roof.

- *Common rafters* extend from the top plate to the ridge board at 90° to both.
- *Hip rafters* extend diagonally from the corners formed by the top plates to the ridge board.
- *Valley rafters* extend diagonally from the top plates to the ridge board along the lines where two roofs intersect.
- *Jack rafters*, also called *jacks*, never extend the full distance from the top plate to the ridge board. There are three kinds of jack rafters. *Hip jack rafters* extend from the top plate to a hip rafter. *Valley jack rafters* extend from the ridge board to a valley rafter. *Cripple jack rafters* extend between a hip rafter and a valley rafter or between two valley rafters.

Calculating Roof Slope

The slope of a roof must be calculated before construction can begin. It depends upon several factors, including the roof's span, run, and rise (see Terms Used in Roof Framing).

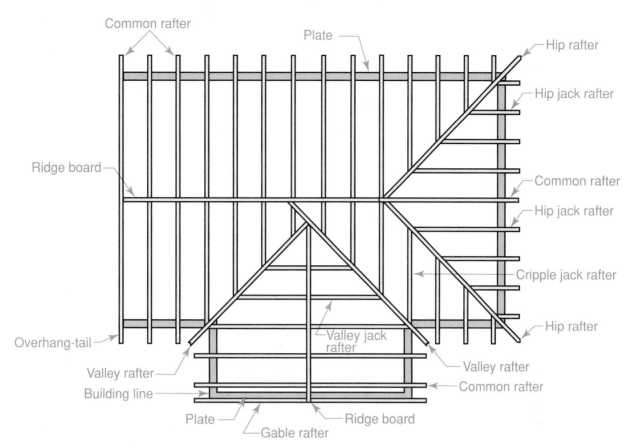

Types of Rafters

Plan View This is a plan view of a roof that contains various types of rafters.

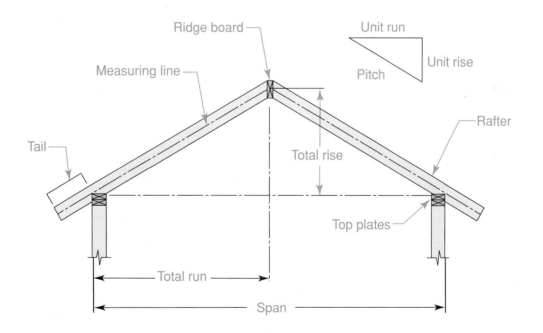

Terms Used in Roof Framing

Key Terms This drawing shows the basic elements a builder must be familiar with in order to frame a roof.

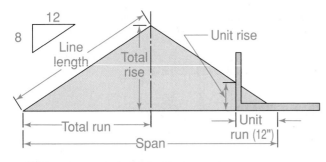

Unit Run and Unit Rise

Understanding Run The unit run of a common rafter is always 12". The rise in inches is variable, depending on the slope of the roof. In the example shown here, there are 8 inches of rise per unit of run.

The **span** is the distance between the outer edges of the top plates. It is measured at right angles to the ridge board. The **total run** is one-half the span (except when the slope of the roof is irregular). The *unit run*, or *unit of run*, is a set length that is used to figure the slope of rafters. The **unit run** for a rafter that is at a 90° angle to the ridge (a common rafter) is always 12". The unit run for a rafter that is at a 45° angle to the ridge is 17". The *measuring line* is an imaginary line running from the outside wall to the top of the ridge. The **total rise** is the vertical distance from the top of the top plate to the upper end of the measuring line. The **unit rise** is the number of inches that a roof rises for every 12" of run

(the unit run). As the unit rise varies, the slope of the roof changes, as shown in Unit Run and Unit Rise.

Many carpenters, and even some construction dictionaries, use the terms *slope* and *pitch* as if they are synonymous. However, they do not mean the same thing. **Slope** refers to a ratio of rise to run. **Pitch** refers to a ratio of rise to span. Either term can be used to describe the inclination (slant) of roofs and rafters, but slope is the term most suited to roof framing.

The triangular symbol above the roof shows the slope visually. When the slope is written out in words, the unit rise is separated from the unit run by a slash mark. For example, a roof may have a unit rise of 6" and a unit run of 12". The information would be written "$^6/_{12}$ slope" and pronounced "six twelve slope" or sometimes "six in twelve slope." See Describing Slope for a visual description of this concept. The slope of a roof can also be given in degrees. For example, a roof with a $^{12}/_{12}$ slope forms a 45° angle, as shown in Another Way to Describe Slope. However, referring to degrees is not a common practice.

Plumb lines and level lines refer to the direction of a line on a rafter, not to any particular rafter cut. Any line that is vertical when the rafter is in its proper position is called a

Describing Slope

By Rise and Run One way to understand the relationship of run to rise is to look at a framing square.

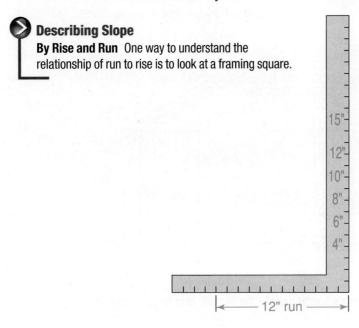

15" rise per 12" run ($^{15}/_{12}$ slope)

12" rise per 12" run ($^{12}/_{12}$ slope)

10" rise per 12" run ($^{10}/_{12}$ slope)

8" rise per 12" run ($^{8}/_{12}$ slope)

6" rise per 12" run ($^{6}/_{12}$ slope)

4" rise per 12" run ($^{4}/_{12}$ slope)

12" run

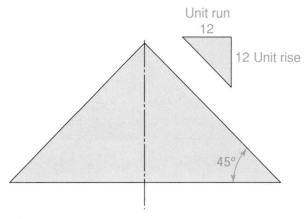

Unit run
12

12 Unit rise

45°

Another Way to Describe Slope
By Degrees The slope of a roof can also be described by giving the angle formed by the roof.

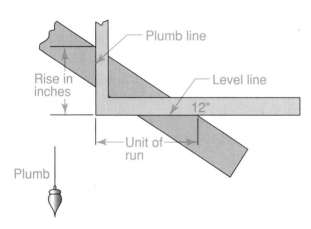

Plumb line

Rise in inches

Level line

12"

Unit of run

Plumb

Plumb and Level Lines
Using a Square A plumb line is drawn along the tongue of the square. A level line is drawn along the body, or blade.

plumb line. Any line that is horizontal when the rafter is in its proper position is called a *level line.* A framing square is used to lay out plumb and level lines on a rafter, as in Pumb and Level Lines.

Reading Check

Recall *State the purposes of the collar tie and the ridge.*

Laying Out a Roof Framing Plan

Before cutting rafters, the carpenter must determine what kinds of rafters are needed to frame the roof. A roof framing plan may be included in the set of building plans to help in this regard. If it is not included, you must lay one out for yourself as shown in the following illustrations.

Gable Roof The gable roof framing plan is the simplest to develop.

1. Lay out the outline of the building (A).
2. Determine the direction in which the rafters will run.
3. Draw the centerline at right angles to this direction (B). The centerline determines the location of the ridge line (C). This corresponds to the location of the ridge board.
4. Determine the distance between the rafters and lay out the roof frame plan (D).

Gable and Valley Roof A gable and valley roof is simply two gable roofs that intersect. In the majority of cases, they intersect at a 90° angle.

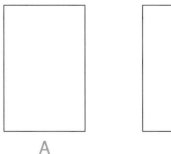

A

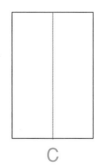

B

C

D

Framing Plan for a Gable Roof
A Simple Plan The framing plan for a shed roof would be one-half of this.

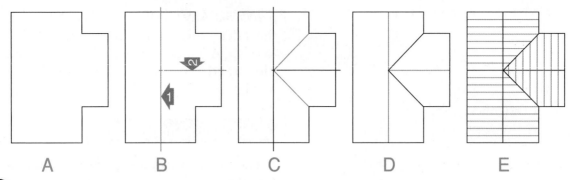

A   B   C   D   E

Framing Plan for a Gable and Valley Roof
T-Shaped Plan The rafters of an addition will be perpendicular to rafters of the main roof.

The intersection creates two valleys.

1. Lay out the outline of the building (A).
2. Draw the centerline of the larger rectangle (B, arrow 1).
3. Draw the centerline of the smaller rectangle (B, arrow 2).
4. Draw 45° lines from the interior corners of the building to where the centerlines intersect (C).
5. Draw the ridge lines (D).
6. Determine the distance between the rafters and lay them out on the roof framing plan (E).

Hip Roof The angle at which the hip extends from each corner is usually 45°. However, other angles are possible. A 45° angle will insure that each surface of the roof will have the same slope. If the angle is something different than 45°, the slopes will not be the same.

1. Lay out the outline of the building (A).
2. Locate and draw a centerline (B).
3. Starting at each corner, draw a 45° line from the corner to the centerline (C). This establishes the location of the hip rafters.
4. Draw the ridge line between the intersecting points of the hip rafters (D).
5. Determine the distance between the rafters and lay them out on the roof framing plan (E).

Hip and Valley Roof A hip and valley roof can be quite complex. It is created when one or more hip roofs intersect at 90° angles, as shown in on page 473.

1. Lay out the outline of the building (A).
2. Outline the largest rectangle inside the building outline (B).

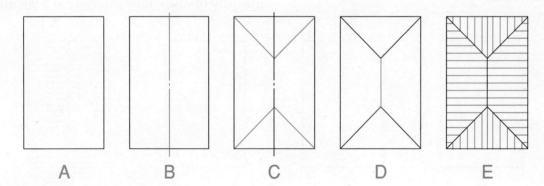

A   B   C   D   E

Framing Plan for a Hip Roof
Hip Roof Like a gable roof, the framing plan for a hip roof is symmetrical.

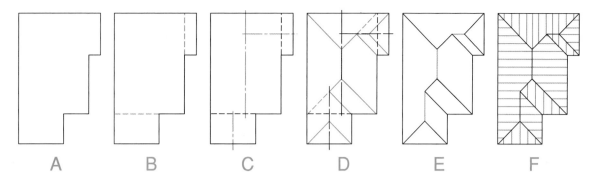

A   B   C   D   E   F

 Framing Plan for a Hip and Valley Roof
Complex Plan Note how the roof is first divided into rectangular sections.

3. Draw centerlines for every rectangle formed inside the building outline (C).

4. Draw a 45° line from each inside and outside corner. Extend these lines to intersect with the centerlines (D). The lines indicate the location of the hip rafters on outside corners and valley rafters on inside corners.

5. The centerlines drawn in (C) connect the hip and valley rafters. Draw these as solid lines where the ridges will be located (E).

6. Figure the distance between the rafters and lay them out on the roof framing plan (F).

JOB SAFETY

LIFTING RAFTERS During the layout and assembly of a roof frame, a great volume of lumber is handled. Due to their size and length, rafters are much heavier than wall studs and far more unwieldy. Organize the work to minimize back strain. For example, have lumber delivered close to where it will be cut. Work with a helper to lift rafters into position.

Section 17.1 Assessment

After You Read: Self-Check

1. Name the four basic roof styles used for homes.
2. The gambrel roof is a variation of which basic roof style?
3. Explain the difference between a plumb line and a level line.
4. What is the purpose of a roof framing plan?

Academic Integration: Mathematics

5. **Slope and Pitch** The terms slope and pitch are often used synonymously. However, slope and pitch are two different calculations. What is the slope of a roof if the unit run is 12" and the pitch is ⅓?

   Math Concept Pitch is a ratio of total rise to span. Slope is a ratio of unit rise to unit run. To find the slope, multiply the pitch times 24. Then place the product over the unit run.

Go to connectED.mcgraw-hill.com to check your answers.

Laying Out Common Rafters

What is a seat cut?

In conventional roof construction, carpenters assemble the roof from individual ceiling joists and rafters. The rafters should not be erected until the ceiling joists have been fastened in place (see Section 17.3). The ceiling joists act as a tie to prevent the rafters from pushing the exterior walls outward. The roof framing methods that follow are used for a gable roof. Variations apply to gambrel, shed, and flat roofs.

The rafters form the skeleton of the roof. They must be carefully made and fitted if they are to support the roof's weight. The top of the rafter rests against the ridge board, as shown in Visualizing Rafter Cuts. The cut made in the rafter so it fits against the ridge

is called a *plumb cut*. The bottom of the rafter rests on the plate. The cut made here is called a *level cut*, or *seat cut*.

- **Plumb cut** A line for the plumb cut can be drawn using a framing square as a guide, as shown in A in Making Plumb and Seat Cuts. The unit run (12" mark) on the blade of the square is aligned with the edge of the rafter. The unit rise on the tongue of the square will correspond to the pitch of the roof. The unit rise is aligned on the same edge of the rafter. The line for the plumb cut is then drawn along the edge of the tongue.

- **Seat cut** A line for the seat cut is drawn with the square in the same orientation on the rafter but with the square in a different location. The line is drawn along the body of the framing square as shown in B in Making Plumb and Seat Cuts.

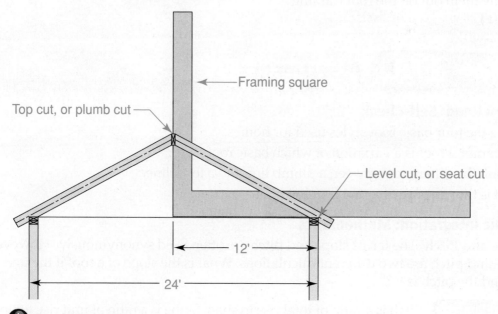

Framing square

Top cut, or plumb cut

Level cut, or seat cut

12'

24'

Visualizing Rafter Cuts
Plumb and Seat Cuts A framing square was enlarged to show its relationship to the roof and to the top and bottom cuts.

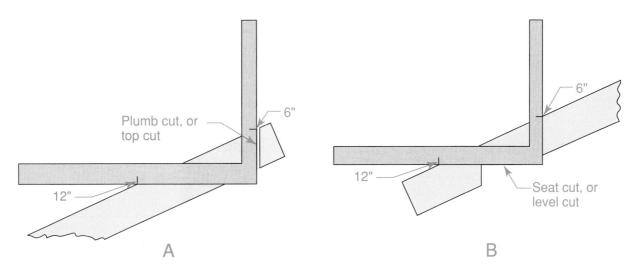

Making Plumb and Seat Cuts

The Square Shifts A. A plumb line has been drawn for the plumb cut on a roof with a 6" unit rise (¼ pitch). **B.** A level line was drawn for the seat cut. This cut was made for a roof with a 6" unit rise.

Look at Theoretical Rafter Length and Actual Rafter Length. The *theoretical length* of a common rafter is the shortest distance between the outer edge of the plate (A) and a point where the measuring line of the rafter meets the ridge line (B). This length is found along the measuring line. It may be calculated in the following ways:

- By using the Pythagorean theorem.
- By using the unit length obtained from the rafter table on the framing square.

- By stepping off the length with the framing square.
- By entering the rise and run into a calculator designed for solving construction problems.

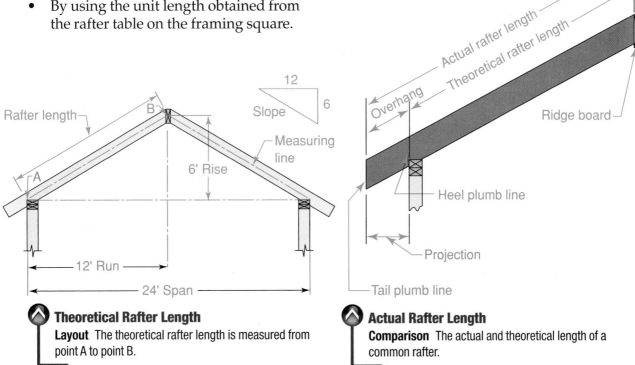

Theoretical Rafter Length

Layout The theoretical rafter length is measured from point A to point B.

Actual Rafter Length

Comparison The actual and theoretical length of a common rafter.

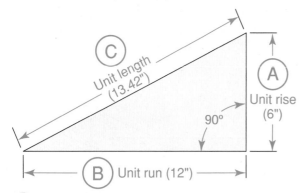

Unit Length

Hypotenuse The unit length of the rafter is represented by the line C. The length of this line can be found in the rafter table on the framing square.

Builder's Tip

ANOTHER TYPE OF FRAMING SQUARE Some carpenters find a triangular framing square easier to use and more convenient to carry. Another advantage is that the thickened lip along one side allows it to be used to guide a circular saw when crosscutting lumber. One brand of triangular framing square is called a Speed Square™. To make a plumb cut with the triangular framing square (shown below), hold the square's pivot point against one edge of the rafter stock. Pivot the square until the appropriate rise number on the "common" scale of the square lines up with the same rafter edge. As shown here, the rise is 6-in-12. Mark the rafter from the pivot point along the top edge of the square. This is the plumb line.

Pythagorean Theorem Method

The Pythagorean theorem states that the square of the hypotenuse of a right triangle is equal to the sum of the squares of the other two sides (see Unit Length):

$$A^2 + B^2 = C^2$$

The length of the hypotenuse (C) will be the square root of the sum of the square of the other two sides.

$$C = \sqrt{A^2 + B^2}$$

The rise, the run, and the rafter of a roof form a right triangle. Therefore the hypotenuse can be used to represent the rafter. The length of the rafter (C) can thus be calculated from the rise (A) and the run (B).

Unit-Length Method

The unit-length method uses the rafter table on a framing square. *Unit length* is the length of a rafter per foot of run. It can be expressed as the hypotenuse of a right triangle. The unit run (12") is the base, and the unit rise (in inches per foot of run) is the altitude, as in Unit Length. Look at the rafter table on the framing square in Finding the Unit Length. The top line of the table reads: "Length Common Rafters per Foot Run." The inch markings along the top represent unit rise. For example, if you follow across the top line to the figure under 6 (for a unit rise of 6"), you will find the number 13.42. This is the unit length for a roof triangle with a unit run of 12" and a unit rise of 6".

Let us figure the total length of a rafter for a small building with a unit rise of 5", a span of 6', and a run of 3', as shown in Sample Problem. Look at the rafter table to obtain the unit length. For a unit rise of 5", the unit length is 13" per unit run. The total length is the unit length times the total run. The total run of the building in this example is 3'. Therefore the total length of the common rafters is 39". Remember that the overhang must be added to the calculated rafter length, and half of the ridge width must be subtracted to determine the true total length.

Enlarged portion of framing square rafter table:

LENGTH	COMMON	RAFTERS	PER FOOT	RUN	21	63	20	81
"	HIP OR	VALLEY	"	"	24	74	24	02
DIFF	IN LENGTH	OF JACKS	16 INCHES CENTERS		28	7/8	27	3/4
"	"	"	2 FEET	"	43	3/4	41	3/8
SIDE	CUT	OF	JACKS	USE	6	11/16	6	15/16
"	"	HIP OR	VALLEY	"	8	1/4	8	1/2

Finding the Unit Length

Using the Tables To find the unit length of common rafters, check the rafter table on the face of the framing square. An enlarged portion is shown here.

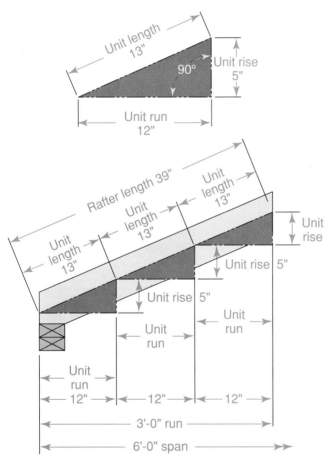

Sample Problem

Finding the Length The total theoretical length of a rafter is the total run times the unit length. In this example, the total run is 3' and the unit length is 13". Therefore, the length of the rafter is 39".

Step-Off Method

A third method for finding the theoretical rafter length is by using the framing square to "step off" the length, as shown on page 478. Place the square on the rafter with the tongue along the plumb cut. Step off the length of the unit run on the rafter stock as many times as there are feet in the total run. In this case, it would be three times.

Often the total run of a building will not come out in even feet. For example, the run might be 3'-4". To handle that, put the square at the first position, then draw a line along the edge of the tongue to represent the plumb cut at the ridge board, as in Uneven Runs on page 478. At the 4" mark on the blade, make a mark on the rafter along the level line—not along the edge. Then, starting at this mark, step off the unit run three times, for a total run of 3'-4". This is the theoretical length of the rafter.

Calculator Method

Small, easy-to-use construction calculators, such as the one on page 478, are common on job sites. Measurements can be entered into the calculator in feet and inches, including fractions. If you know the rise and the run, you can easily determine the length of a common rafter by entering these figures into the calculator. You can also use the tool to calculate cuts for hip rafters and valley rafters.

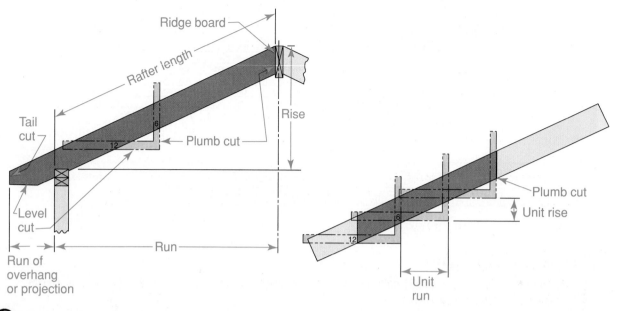

Stepping Off the Unit Length

Using the Square This is another way to find the length of a common rafter.

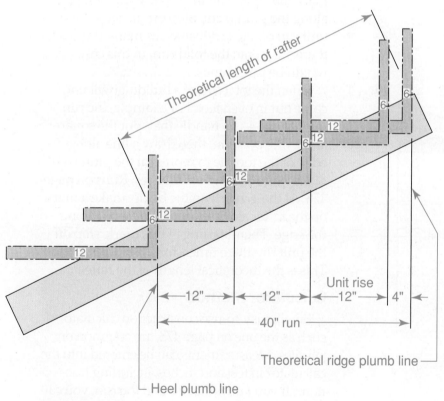

Uneven Runs

Using the Square Stepping off the rafter when the total run is not an even number of feet.

A Construction Calculator

Time Saver A construction calculator is often used to solve roof framing problems.

Completing the Layout

What are some advantages of having a roof with an overhang?

After the basic rafter layout is complete, additional work is required. The ridge allowance, rafter overhang, and bird's mouth must be considered. (Layout of the ridge board itself will be discussed in Chapter 19, "Roof Assembly & Sheathing").

Common Rafter Ridge Allowance

The theoretical rafter length does not take into account the thickness of the ridge board or the length of the overhang, if there is one. To cut a rafter without an overhang to its actual length, you must deduct one-half the thickness of the ridge board from the ridge end, as shown in Understanding Ridge Allowance. For example, if 2× stock is used for the ridge board, its actual thickness is 1½". One-half of this is ¾". The ¾" is indicated along the level line, and the line for the actual ridge plumb cut is drawn, as shown in Marking the Ridge Allowance.

Common Rafter Overhang

A roof may or may not have an overhang, or eave. If not, the rafter must be cut so that its lower end is even with the outside of

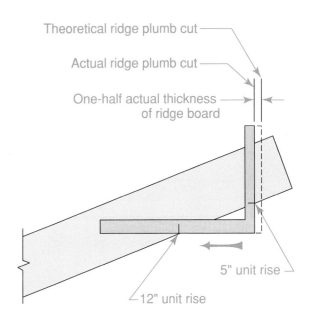

Marking the Ridge Allowance
Avoid the Edge Lay off one-half the thickness of the ridge board along the level line. Do not lay it off along the edge of the rafter.

the exterior wall. If the end is cut parallel to the ridge plumb cut, it is said to have a *heel*, as shown in Rafter Heel. The portion of the rafter that rests on the plate is called the *seat*. To lay out the seat, place the tongue of the framing square on the heel plumb

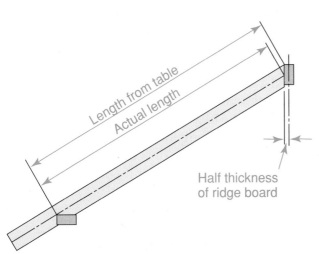

Understanding Ridge Allowance
Ridge Thickness Subtract one-half the *actual* thickness of the ridge board from the theoretical length of the rafter to obtain the rafter's actual length.

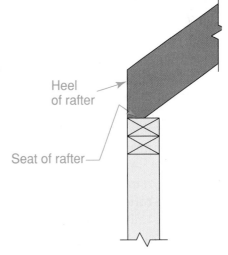

Rafter Heel
End of Rafter A rafter without an overhang rests on the exterior wall plate.

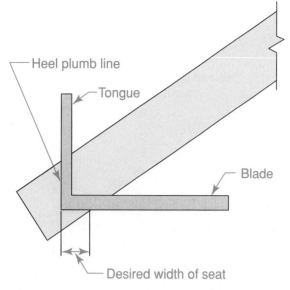

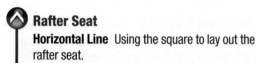

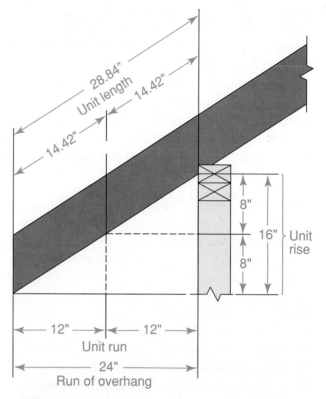

line. The rafter edge will intersect the correct seat width on the blade, as shown in Rafter Seat. Indicate the seat by drawing a line from the heel plumb line along the blade.

A roof with wide overhangs provides protection for side walls and end walls. Though it adds slightly to the initial cost, this type of roof extension saves on maintenance later.

If the roof does have an overhang, the overhanging part of the rafter is called the tail. Its length must be added to the length of the rafter in order to find the total length of the rafter. The length of the tail may be calculated as if it were a separate short rafter. Any of the methods used for finding rafter length may be used to find the length of the tail. For example, suppose the run of the overhang is 24" and the unit rise of the roof is 8", as shown in Calculating Overhang. Look at the rafter table on the framing square to find the unit length for a common rafter with a unit rise of 8". You will see the unit length is 14.42". Since the total run of the overhang is 24", the total length is 28.84", or 28 27/32":

14.42" (in. per unit run) × 2 (units of run) = 28.84"

Another way to lay out the overhang is with the framing square. Suppose the run of the overhang is 10". Start the layout by placing the tongue of the square along the heel plumb line and setting the square to the pitch of the roof. In Overhang Using a Square, the square is set to a unit rise of 8" and a unit run of 12". Move the square in the direction of the arrow, as shown, until the 10" mark of the blade is on the heel plumb line. Draw a line along the tongue. This will mark the tail cut. If fascia or soffits are to be added later, be sure to allow for them in figuring the length of the rafter tail.

Many carpenters do not cut the tail to the finished length until after the rafters have been fastened in place. Instead, a sufficient amount of material is allowed for the overhang. After the rafters are fastened in place, a chalk line is snapped on the top edge of all the rafters. A tail plumb line is then drawn down from this chalk line on each rafter and the tail is cut along the line. The **benefit** of this method is that it results in perfectly aligned rafter ends even if the house walls are not perfectly straight or if there is a slight bow in the ridge beam. The disadvantage is that the carpenter must make plumb cuts with a circular-saw while on the roof or on a ladder, which can be awkward.

In some cases, the end of a rafter does not require a plumb cut. It can be square cut instead. A square is placed on a chalk mark snapped across the top edge of the rafters so that a line can be drawn 90° across each rafter. This results in a square cut. This can only be done when there is no requirement for gutters on the house. A square cut makes gutters very difficult to mount.

Laying Out a Bird's Mouth

A *bird's mouth* is a notch made in a rafter with an overhang so that the rafter will fit against a plate. The plumb cut for the bird's mouth, which bears against the side of the plate, is called the *heel cut*. The level cut, which bears on the top of the plate, is called the *seat cut*.

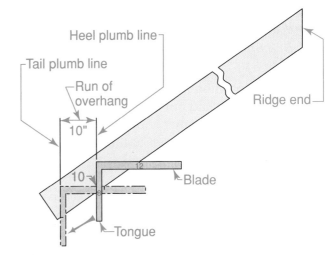

Overhang Using a Square
Step It Off Laying out the run of the overhang directly on the rafter using the framing square.

The size of the bird's mouth for a common rafter is usually stated in terms of the depth of the heel cut rather than the width of the seat cut. The bird's mouth is laid out much the same way as the seat cut for a rafter without an overhang. Measure off the depth of the heel on the heel plumb line, set the square, and draw the seat line along the blade as shown in Bird's Mouth Layout.

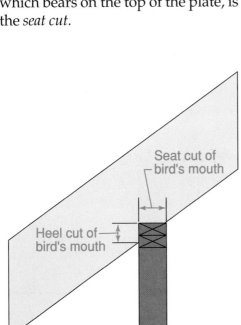

Bird's Mouth
A Notch for Bearing The bird's mouth on a rafter with an overhang.

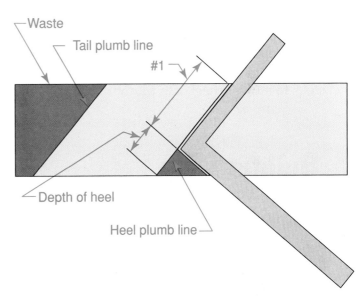

Bird's Mouth Layout
Layout of the Notch Using the square to lay out a bird's mouth. The line at arrow #1 is important when laying out the bird's mouth for a hip and valley rafter.

Cutting Rafters

When should a rafter have the ridge cut facing up?

To cut common rafters, the actual length of one rafter is laid out on a piece of stock. The crown of the rafter should be on the top edge. After the first rafter is cut, it is used as a pattern for cutting a second.

The two rafters are then tested on the building using the ridge board or a scrap piece of the same size material to see how the heel cut and the top cut fit. If they fit properly, one of these rafters can be used as a pattern to cut all the others needed.

Once cut, the rafters should be **distributed** to their locations along the building. The rafters are usually leaned against the building with the ridge cut up. The carpenters on the building can then pull them up as needed and fasten them in position.

In large developments, houses must be built quickly. Carpenters who use conventional roof framing methods instead of trusses, which are installed more quickly, must develop efficient work habits. One way in which they can speed their work without sacrificing quality is to gang-cut rafters. After the master rafter pattern has been established, many pieces of rafter stock are clamped together atop sawhorses. Lines are marked across the edges of the stock to indicate heel cuts, plumb cuts, and bird's-mouth cuts. Then the cuts are made on a group of rafters at the same time, sometimes using special saws.

Gambrel, Shed, & Low-Slope Roof Framing

Why are gambrel roofs common on barns?

Other kinds of roofs can be framed using variations of the same basic techniques used to build a gable roof. These roofs include gambrel roofs, shed roofs, and flat, or low-slope, roofs.

Framing a Gambrel Roof

The framing for a gambrel roof combines primary and secondary rafters. The lower (primary) rafter has a steep pitch, and the upper (secondary) rafter has a low pitch. If the pitches are known, the rafters may be laid out in the same manner as any common rafter.

The roof may also be laid out full size on the subfloor, as shown in Gambrel Roof. Use the run of the building (AB) as a radius and draw a semicircle. Draw a perpendicular line from point A to intersect the semicircle at E. This locates the ridge line. Find the height of the walls from the plans.

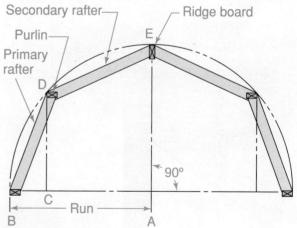

 Gambrel Roof
Full-Size Layout The patterns for the rafters in the gambrel roof may be made by laying out the full-size roof on the subfloor.

Builder's Tip

CUTTING THE BIRD'S MOUTH Rafter cutting is usually done with a circular saw. It is important not to overcut when making the bird's mouth. This will weaken the rafter. It is best to stop the heel cut and the seat cut short of each intersecting layout line. In other words, make partial cuts. Then finish the cut with a handsaw or jigsaw.

Draw a perpendicular line (CD) to this length between the plate and the semicircle. Connect points B and D and points D and E. This gives the location and pitch of primary rafter BD and secondary rafter DE. From this layout the rafter patterns can be made and cut for test fitting on the building.

Framing a Shed Roof

A shed roof is basically one-half of a gable roof. The full-length rafters in a shed roof are common rafters. The total rise is the difference in height between the walls on which the rafter will bear. The total run is equal to the span of the building minus the width of the top plate on the higher wall, as shown in Shed Roof. The run of the overhang on the higher wall is measured from the inner edge of the top plate. With these exceptions, shed roof common rafters are laid out like gable roof common rafters. A shed roof common rafter has two bird's mouths. They are laid out just like the bird's mouth on a gable roof common rafter.

Framing a Low-Slope Roof

A low-slope roof is any roof with a slope of $3/12$ or less. This includes flat roofs, which actually have a slight slope to encourage water to drain off. Low-slope roofs generally require larger rafters than roofs with steeper slopes, but the total amount of framing lumber is usually less because the rafters also serve as ceiling joists. Thus, their size is based on both

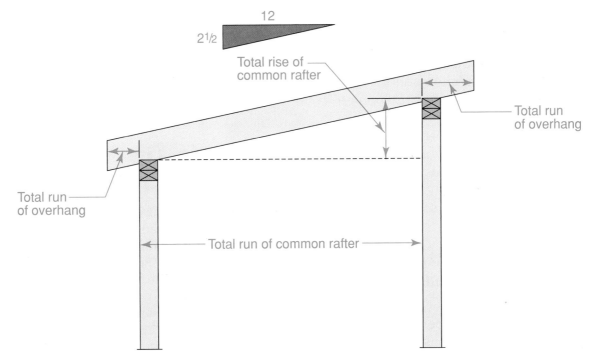

Shed Roof
Single Slope A shed roof rafter is like a rafter on just one side of a gable roof, except that it has two bird's mouth cuts.

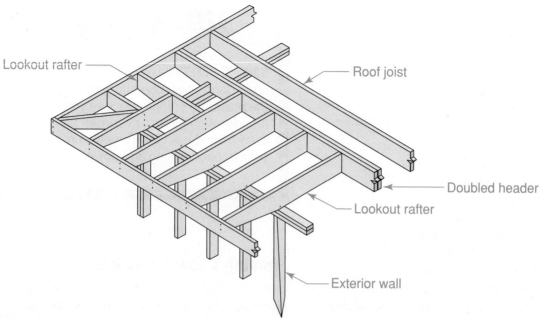

Lookout rafter

Roof joist

Doubled header

Lookout rafter

Exterior wall

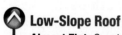

Low-Slope Roof

Almost Flat Construction of a low-pitched roof with an overhang.

roof and ceiling loads. The size is given on the plans or determined from rafter span tables.

When there is an overhang on all sides of the house, *lookout rafters* (sometimes called *outriggers*) are ordinarily used on two of the sides. Lookout rafters project beyond the walls of the house, usually at 90° to the common rafters, as shown in Low-Slope Roof. Where they run perpendicular to common rafters, they are nailed to a double header and toenailed to the wall plate. The distance from the double header to the wall line is usually twice the overhang. Rafter ends may be capped with a barge rafter, which will serve as a nailing surface for the fascia.

Using I-Joist Rafters

I-joists can be a good substitution for what?

Laminated-veneer lumber (LVL) can be used to build many portions of the house frame, including the roof. (For more on laminated-veneer lumber, see Chapter 13). It is also used in commercial wood framing. LVL I-joists can be used in place of lumber

rafters. Various companies make I-joists and you must follow the manufacturer's instructions for the product you are using.

The details shown in LVL Framing Details provide a general introduction to the subject of I-joist roof framing. Always follow manufacturer's installation instructions for installing these products. Be sure to check the building code for additional information.

Builder's Tip

NAILING I-JOIST RAFTERS The flanges of I-joist rafters are made of LVL stock. Improperly nailing through this material seriously weakens it. The correct way to nail is shown here. The other methods will split the material.

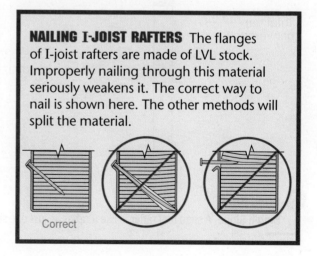

Correct

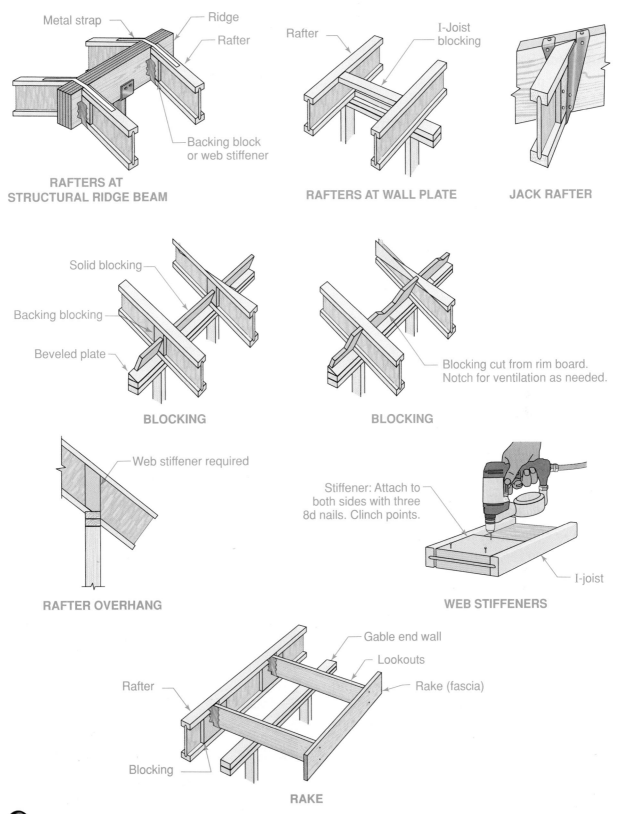

RAFTERS AT STRUCTURAL RIDGE BEAM

Metal strap
Ridge
Rafter
Backing block or web stiffener

RAFTERS AT WALL PLATE

Rafter
I-Joist blocking

JACK RAFTER

BLOCKING

Solid blocking
Backing blocking
Beveled plate

BLOCKING

Blocking cut from rim board. Notch for ventilation as needed.

RAFTER OVERHANG

Web stiffener required

WEB STIFFENERS

Stiffener: Attach to both sides with three 8d nails. Clinch points.
I-joist

RAKE

Gable end wall
Lookouts
Rake (fascia)
Rafter
Blocking

LVL Framing Details

Suggested Details These illustrations show various installation details for I-joist rafters. These are suggestions only. Always consult the manufacturer's instructions before installation.

After You Read: Self-Check

1. What prevents rafters from spreading and pushing out on the exterior walls?
2. State the Pythagorean theorem.
3. Name the parts of a bird's mouth and tell where they bear.
4. When making cuts for a bird's mouth, what mistake should you avoid?

Academic Integration: English Language Arts

5. **Writing Instructions** The theoretical length of a common rafter is the shortest distance between the outer edge of the plate and a point where the measuring line of the rafter meets the ridge line. There are four ways to find the theoretical length of a rafter: the Pythagorean theorem method, the unit-length method, the step-off method, or the calculator method. Write a set of instructions for an individual wishing to find the theoretical length of a rafter using the step-off method.

 Go to **connectED.mcgraw-hill.com** to check your answers.

Section 17.3 — Ceiling Framing

Ceiling Joists

How is the size of a ceiling joist determined?

Ceiling joists are the framing members that support ceiling loads. In the first story of a two-story house, the same framing serves as both ceiling and floor, as in Ceiling Framing. In other words, if you stood downstairs and looked up, you would refer to it as the ceiling framing. However, if you stood upstairs and looked down, you'd call it the floor framing. The floor framing for the first story of a house is covered in Chapter 15, "Floor Framing."

The ceiling framing discussed in this chapter is directly related to the roof framing. It prevents walls from bowing outward by tying the lower ends of the rafters together. At the same time, it ties the walls of the house together and forms the floor of the attic.

Ceiling Framing
Two Roles Ceiling framing often supports a second floor.

Tim Fuller

Ceiling framing for the top level of a house usually proceeds at the same time as roof framing. While the rafters are being laid out and cut, other carpenters cut and install the ceiling joists. Like floor joists, ceiling joists may be supported by girders or by bearing walls.

Sizing Ceiling Joists

The size of the ceiling joists is determined by the distance they must span and the load they must carry, as shown in **Table 17-1**. The species and grade of wood must also be considered. The correct size for the joists will be found on the building plans. Spacing and span limitations must comply with local building codes.

Estimating

The methods for estimating the number of ceiling joists, as well as the material cost, are the same as for estimating floor joists. Refer to Chapter 15, "Floor Framing."

Layout

The layout for ceiling joists is determined as one lays out the rafters. Rafter spacing and placement are determined first. Ceiling joist spacing and placement are determined second.

Ceiling joists are usually placed across the width of the building and parallel to the rafters. The ends of the joists that rest on the exterior wall plates next to the rafters will usually project above the top edge of the rafter, as shown in Trimming Joist Corners. If left untrimmed they would interfere with

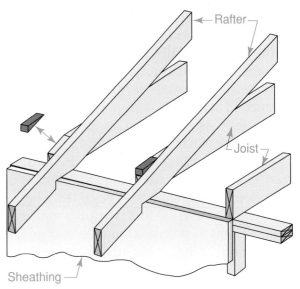

 Trimming Joist Corners
Trim for Clearance The upper corners of the ceiling joists must be cut off at an angle to match the angle of the rafters.

the installation of roof sheathing. These ends are cut off at an angle to match the angle of the rafters. The cuts should be about ⅛" below the rafter's top edge. This is best done before the joists are installed.

REGIONAL **CONCERNS**

Wind Uplift When houses are built in areas with high winds, extra precautions such as metal framing anchors should be taken to secure framing joints.

Table 17-1: Allowable Spans for Ceiling Joists Using Non-Stress-Graded Lumber					
Size of Ceiling Joists (inches)	Spacing of Ceiling Joists (inches)	Maximum Allowable Span (feet and inches)			
		Group I	Group II	Group III	Group IV
2×4	12	11-16	11-0	9-6	5-6
	16	10-6	10-0	8-6	5-0
2×6	12	18-0	16-6	15-6	12-6
	16	16-0	15-0	14-6	11-0
2×8	12	24-0	22-6	21-0	19-0
	16	21-6	20-6	19-0	16-6

Installation begins at one end of the house and continues to the other end. The spacing of the joists is usually 16" or 24" OC. Extra joists are placed, as needed, without altering the spacing. For example, a ceiling joist will be needed at the inside edge of the plate on an end wall. This provides an edge nailing surface for the ceiling finish as shown in Adding a Ceiling Joist. An extra joist is sometimes located over the studs in a partition wall, as shown on page 416. The distance between the first two joists at this location will then be less than the normal OC spacing but each succeeding joist is spaced 16" or 24" on center. The ends of all ceiling joists must have at least 1½" of bearing on a wood framed wall. They must have at least 3" of bearing on a masonry wall.

Lumber ceiling joists meet other ceiling joists at the center of the building. The joists are normally offset 1½" on the two outside walls so that they lap each other when they meet over the bearing partition wall, as shown in A in Ceiling Joist Spacing. Code requires that the joist lap be at least 3" long. This lap is face-nailed with three 10d nails.

The joists are toenailed to the bearing wall plate with three 8d nails. In an alternate method, the end of the joists butt against each other over a bearing wall as shown in B in Ceiling Joist Spacing. The ends that

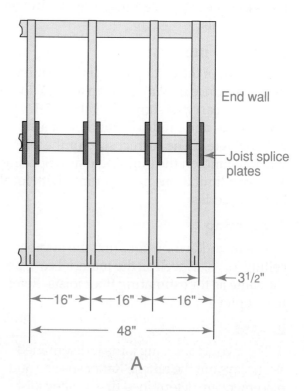

A

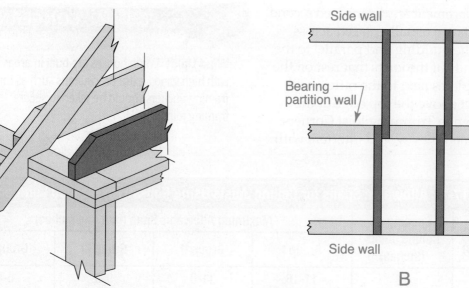

B

Adding a Ceiling Joist
Extra Joist A ceiling joist should be set on the inside edge of the end wall to permit nailing of the material for the ceiling surface.

Ceiling Joist Spacing
Two Methods Ceiling joists are usually lapped at a bearing wall (A). However, it is also possible to have the joists meet in a butt joint as long as the joint is reinforced (B).

butt will have to be squared and cut off to length. Each joist end will be resting on just half of the wall plate. A wood or metal splice plate must be nailed securely to both sides of the joists to hold them together. Note that the distance between the first two pairs of joists is less than 16".

Nonbearing partitions that run parallel to the ceiling joists are nailed to blocks installed between the joists, as shown in Blocking at a Nonbearing Wall. Blocking or manufactured spacers can also be used in the middle of long spans to prevent the ceiling joists from twisting or bowing. Blocking can be removed if heating or air-conditioning ducts must run through the spaces between joists.

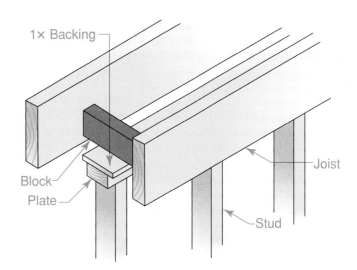

 Blocking at a Nonbearing Wall
Important Details Notice the 1× backing that has been attached to the top of the wall for nailing the ceiling material.

Step-by-Step Application

Installing Ceiling Joists The locations for ceiling joists are laid out like the locations for floor joists (see Chapter 15, "Floor Framing," page 405). The spacing of the joists will be found on the building plans. Mark the plates for the correct spacing. Then install the joists as shown in the figure.

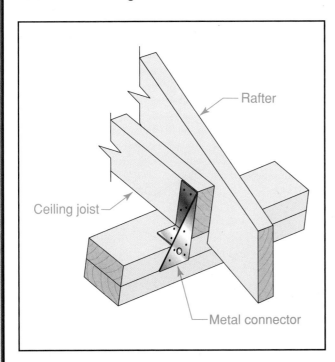

Step 1 Cut each joist to length. Sight down the edge of the joist to determine where the crown is. Trim off the corners that will extend above the rafters.

Step 2 Distribute the joists around the building so that they can be lifted into place.

Step 3 Place each joist with the crown up. Align the end of the ceiling joist with the outside edge of the exterior wall plate.

Step 4 At one end, toenail three 8d nails through the joist and into the plate or use a metal framing connector.

Step 5 Toenail the other end of the joist to a girder or bearing wall.

Step 6 Nail lapped joists to each other with three 10d nails.

Special Ceiling Framing

What does a flush girder do?

Ceiling framing must often accommodate interruptions in the regular spacing of joists. Before beginning layout and installation, the carpenter should check the plans to identify these special situations.

Hip Roofs

In the framing for a hip roof with a shallow slope, the first ceiling joist will interfere with the bottom edge of the rafters. *Stub joists* (short joists) installed at right angles to the regular joists will correct this situation, as shown in Stub Joists. Space the stubs 16" OC for attaching the finished ceiling. Locate them so that the rafters, when installed, may be nailed directly to their sides.

Reading Check

Recall *How are stub joists used in hip roofs?*

Ceiling Openings

Openings in the ceiling may be required for a chimney or for **access** to the attic. These openings are often larger than the spacing between the joists and will require the cutting of one or more joists. Such joists must be supported and framed as described in the section titled Framing Large Openings in Chapter 15.

Building codes require that any framing, including ceiling framing, be kept at least 2" from the front and sides of masonry fireplaces, and at least 4" from the back.

Framing Flush Ceilings

In the past, homes usually had many small rooms. Today, however, homeowners often prefer larger and more open living spaces. A combined kitchen and family room is common, for example. To visually tie the rooms together, *flush ceilings* (the two

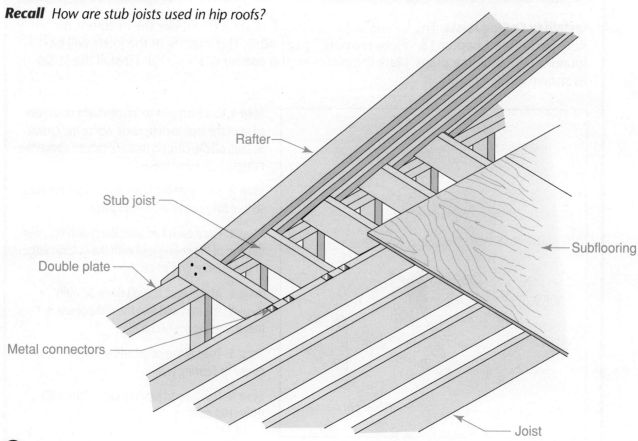

Stub Joists

Plan Ahead Stub joists are securely anchored to the regular joists with metal framing connectors.

ceilings flow together as one) are desirable. Because there is no partition, a girder is often needed to support the interior ends of the ceiling joists.

This support can be provided by a *flush girder*. A flush girder is usually built up from the same stock used to frame the rest of the ceiling. It can also be a glulam or LVL beam. Instead of resting on top of the girder, ceiling joists are fastened to the side with joist hangers as shown in A Flush Girder. Joist hangers are nailed to the girder with 10d or larger nails and to the joist with joist hanger nails. It is

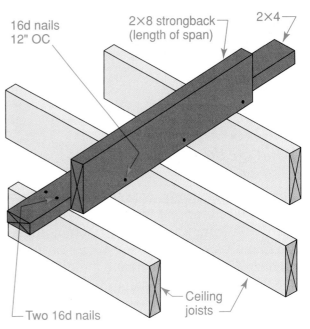

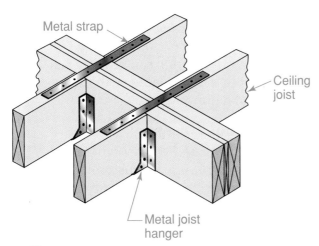

 A Strongback
Hidden Strength A strongback is used to give long joists additional support.

often easiest to fasten the hangers to the ends of the joists before raising the joists into place.

Another approach that can sometimes be used with shorter spans is to stiffen the ceiling joists with a member called a *strongback*. A strongback, shown in above, should be nailed to the tops of the ceiling joists. However, this method will not eliminate the need for some kind of header.

A Flush Girder
Flush Gains Headroom Ceiling joists are fastened to a flush girder with joist hangers. The joints can be reinforced with metal straps.

Section 17.3 Assessment

After You Read: Self-Check

1. What factors determine the size of ceiling joists?
2. How are ceiling joists arranged in relation to the building and rafters?
3. How much space must be left between ceiling joists and the front and sides of a masonry chimney?
4. How are joists attached to a flush girder?

Academic Integration: English Language Arts

5. **Flush Ceilings** Flush ceilings are often used to visually tie two rooms in a home together. In this case, a girder is often needed to support the interior ends of the ceiling joists. Write a paragraph describing what a flush girder is constructed from and how it is installed.

Go to **connectED.mcgraw-hill.com** to check your answers.

Roof Truss Basics

What are the chord and the web?

A roof truss is an assembly of members forming a rigid framework of triangular shapes. It can support loads over long spans. Many residential and commercial buildings are framed with roof trusses. Trusses are made in factories and delivered to the job site by truck. One advantage of using trusses is that they save money on materials and on-site labor. Materials may cost about 30 percent less than those used for traditional roof framing because they utilize smaller dimension lumber.

The basic parts of a roof truss are shown in below. A **chord** is the top or bottom outer member of the truss. The **web** is the framework between the chords. It creates a rigid assembly. Chords and webs are often connected at the joints by rectangular connector plates. A *connector plate* is a pre-punched metal plate with many teeth. It is pressed into the wood under hydraulic

pressure. Most trusses are built from 2×4 stock. They may also be built of other materials or combinations of materials, including metal and timber.

Trusses can be erected quickly. They are usually designed to span from one exterior wall to the other with lengths of 20' to 32' or more. They are usually concealed by interior finish materials but can be left exposed. Because no interior bearing walls are required, interior design can be more flexible. Partitions can be placed without regard to structural requirements.

Types of Roof Trusses

Though the overall shape of most common trusses is triangular, trusses come in a wide variety of shapes to solve nearly any problem. Span and load requirements (for snow, wind, etc.) govern the type of truss to be used. King-post, Fink, and scissors trusses are most commonly used for houses and are shown in Three Basic Trusses. These and similar trusses are most adaptable to

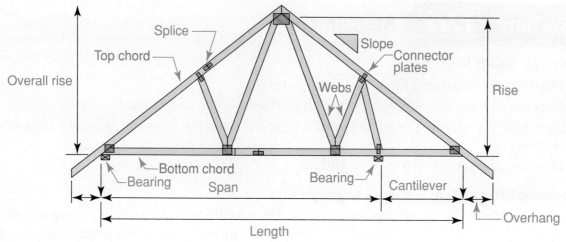

Parts of a Roof Truss

Three Elements The main parts of a truss are the chords, webs, and connector plates (also known as gusset plates).

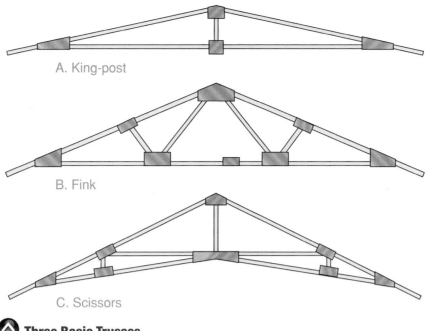

A. King-post

B. Fink

C. Scissors

⌂ **Three Basic Trusses**
The Big Three These trusses are the ones most often used in residential roof framing.

rectangular houses because the uniform width requires only one type of truss. However, trusses can also be used for L-shaped houses. For hip roofs, hip trusses can be provided for each hip and valley area.

King-Post Truss The *king-post truss* has upper and bottom chords and a single vertical post in the center, as in above. This vertical post is sometimes called a strut. It is the simplest form of truss used for houses. For short and medium spans, the king-post truss is probably more economical than other types. It has fewer pieces and can be fabricated faster. However, because so much of the upper chord is unsupported, allowable spans are somewhat shorter than for the Fink truss when the same size members are used.

Fink Truss The *Fink truss*, also called a *W-truss*, uses three more supporting members than the king-post truss. Distances between connections are shorter, as shown in B, above. This usually allows the use of lower-grade lumber and somewhat longer spans for the same member size. It is perhaps the most popular and most widely used of the light-wood trusses.

Scissors Truss The *scissors truss* features sloped top and bottom chords, as shown in C, above. It is more complicated than the Fink truss. It is used for houses with cathedral ceilings, where it provides a savings in materials over conventional framing.

Truss Design

The design of a truss depends not only on the loads it must carry but also on the weight and slope of the roof itself. Generally, the flatter the slope, the greater the stresses. Therefore, flatter roofs require trusses with larger members and stronger connections.

Many lumber dealers can provide the builder with completed trusses ready for installation. Often, the builder orders trusses directly from a truss manufacturer. To order a series of trusses, the builder must supply a precise description of what is needed. Much of the information is on the set of building plans. Ordering information includes the following:

- *Nominal span* Generally, the nominal span is the length of the bottom chord.

- *Overhang length* The overhang length is the horizontal distance from the end of

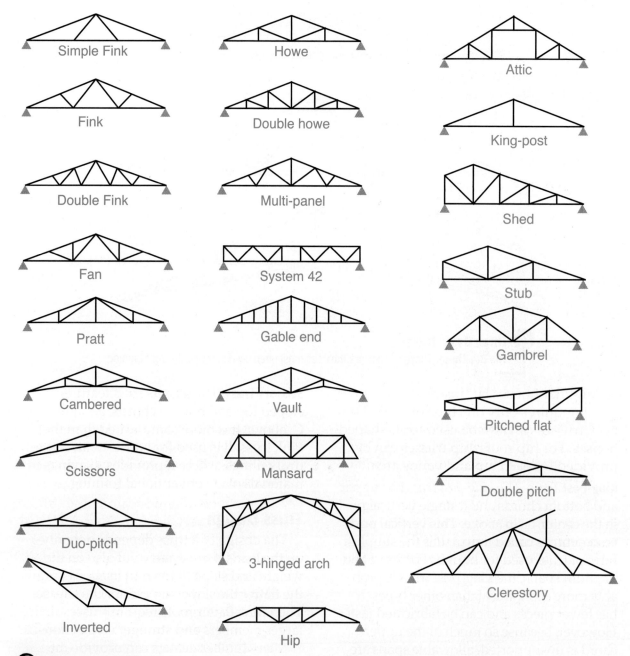

Simple Fink Howe Attic

Fink Double howe King-post

Double Fink Multi-panel Shed

Fan System 42 Stub

Pratt Gable end Gambrel

Cambered Vault Pitched flat

Scissors Mansard Double pitch

Duo-pitch 3-hinged arch

Inverted Hip Clerestory

Types of Trusses

Many Shapes Some of the many types of roof trusses. The solid triangles indicate bearing points.

the bottom chord to the bottom edge of the rafter (extension of the top chord).

- *End cut of rafter* Will the ends be plumb cut, square cut, or untrimmed?

- *Roof pitch* State the vertical unit rise per 12" of run. Strictly speaking, this is the roof's slope, but the term *pitch* is often used instead.

- *Type of truss* See Types of Trusses above for different types of trusses.

- *Quantity* The number of trusses required, including gable-end trusses, must be specified.

- *Design parameters* These would include information about the expected loads, particularly anything unusual, such as very heavy snow.

⊙ **Roof Trusses**
An Open Floorplan Because these trusses would be exposed to the living space, the truss builder took extra care to make them visually appealing.

Robin Matthews/Ingram Publishing

- *Special requirements* These include anything unusual about the use of the truss, such as the need for it to cantilever past a wall.

When the trusses have been designed, drawings made by the manufacturer must be approved by local building officials. These drawings are then made a part of the approved set of building plans.

Machine Stress-Rated Lumber *Stress-rated lumber* is structural lumber that has been graded electronically and stamped to indicate the specific load it will support. Because the grading is done with a machine, lumber rated in this way is called machine stress-rated lumber (MSR). When strength properties are critical, an architect or engineer may specify machine stress-rated lumber of a certain species and grade. Manufacturers of roof trusses often use MSR lumber since trusses are so important to the strength of the house.

Sheathing Requirements Trusses are commonly designed for 24" OC spacing. This spacing requires thicker interior and exterior sheathing or finish material than is needed for conventional 16" OC rafter spacing. The interior trusses in Roof Trusses have been left exposed.

Costs A manufacturer delivers trusses to the job site and sets them on the rafter plate with a crane. They are then ready for workers to tip them into place for less than the cost of the material alone in conventional framing. To estimate exact costs, the truss fabricator reviews the building plans. When a bid is provided, the builder should know exactly what it covers. For example, is setting the trusses in place on the house included in the cost?

Installing Trusses
What weather conditions might make it unsafe to install trusses?

When handling and storing trusses, avoid placing unusual stresses on them. They are designed to carry roof loads in a vertical position. For this reason, they must be lifted and stored upright. However, it is important to prevent them from tipping and possibly injuring nearby workers. If they must be handled or stored in a flat position, they should be supported along their length to lessen bending. Never support the trusses only at the center or only at each end when they are in a flat position.

If the trusses will be stored outdoors before being installed, they should be supported above the ground to protect them from dampness or water. A tarp should cover them to prevent rain damage. The bands around bundles of trusses should not be cut or removed until just before the trusses are ready to be erected. Because every part of the truss has been specifically designed for a particular job, a truss must never be cut or altered.

Raising Trusses

Completed trusses can be raised into place by hand or by crane. The truss fabricator may bring a truck-mounted crane when delivering the trusses. Because a large truss can be heavy and awkward to handle, a crane is the preferred method. However, take great care to secure the truss properly as the crane is lifting it. A guide rope must be attached so that a worker on the ground can keep the truss from swinging out of control. For more on rigging techiques, see the **Ready Reference Appendix**.

When cranes and other equipment are used to lift heavy loads such as trusses, the operator often relies on someone on the

JOB SAFETY

ground to help place the load in the right spot. Because voices cannot be heard over loud engine sounds and other noises, hand signals are often used to communicate. For example, a raised thumb is the signal for raising the load, and a closed fist is the signal to stop. For more on hand signals, see the **Ready Reference Appendix**.

When raising small trusses by hand, care is needed to avoid damage to the trusses and accidental injury to workers. One by one, the trusses should be laid across the building and swung up into place by two or more workers. Another worker should be at the roof level to brace the trusses as soon as they are tipped into place.

Bracing Trusses

Trusses must be braced temporarily as they are being installed. This helps to maintain precise spacing during installation. Bracing is also important for safety because it prevents trusses from tipping like dominoes. The manufacturer's drawings will accompany the trusses when they are delivered to the job site. These drawings should be studied for information about proper bracing.

Permanent bracing may also be required. This type of bracing stiffens the entire roof assembly and is meant to be left in place after the structure is complete.

Temporary Bracing The gable-end truss is the first truss on the building and the most important to brace. It should be braced with lumber standoffs anchored to stakes driven into the ground, as shown in Ground Bracing.

As each additional truss is put into place, it is braced temporarily to the adjacent trusses with a length of nominal 2" lumber. The lumber is secured diagonally to the top chord with two 16d nails at every intersection, as shown in Diagonal Bracing. Such diagonal bracing is sometimes called sway bracing or cross bracing. Lateral bracing may also be required. Finally, a second gable-end truss is placed at the other end of the building. Top-chord bracing can be removed as the roof sheathing is installed.

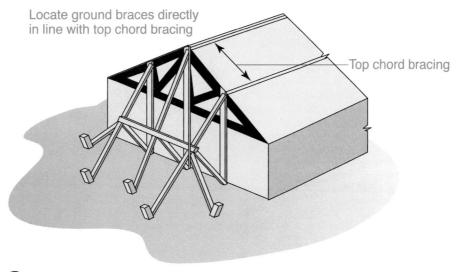

Locate ground braces directly in line with top chord bracing

Top chord bracing

 Ground Bracing

Bracing for Safety The gable-end truss should be securely braced. This method is sometimes called ground bracing.

Permanent Bracing There are several types of bracing that are designed to remain in place once installed. Permanent lateral metal bracing ties the individual trusses into a rigid structural system. It also ensures precise spacing. However, it does not eliminate the need for temporary gable-end and diagonal bracing with nominal 2" lumber. See Truss Bracing on page 498.

Permanent continuous lateral bracing consists of 2×4 or wider stock that is nailed to the web or to the bottom chord of each truss. The exact location of the lateral bracing is usually specified in the truss design.

On trusses with long spans, permanent bracing prevents the bottom chords from moving as the ceiling finish is applied. Nominal 2" lumber is nailed to the top edge of the bottom chord and runs the length of the building. After the permanent bracing is in place, the roof should be sheathed as soon as possible. Roof sheathing prevent trusses from collapsing sideways and strengthens the entire roof system.

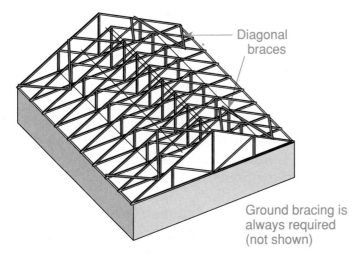

Diagonal braces

Ground bracing is always required (not shown)

 Diagonal Bracing

Diagonal Ties Diagonal bracing helps to prevent the trusses from shifting until the roof sheathing can be applied.

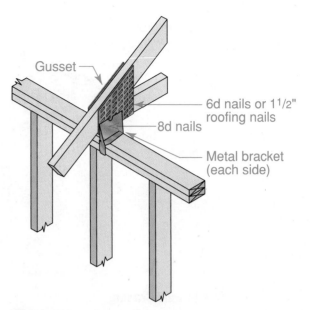

Gusset

6d nails or 1¹/₂"
roofing nails

8d nails

Metal bracket
(each side)

 Securing a Truss
Secure Connection Fastening trusses to the wall plate using metal brackets.

 Truss Bracing
Double Duty Prefabricated metal spacers help to brace trusses and ensure exact spacing.

Fastening Trusses

Trusses are fastened to the outside walls with nails or metal framing anchors. Resistance to wind uplift stresses and thrust must be considered. A ring-shank nail provides a simple connection that resists modest uplift. Toenailing is also sometimes done, but this is not always satisfactory because metal truss plates are located at the wall plate and make toenailing difficult.

A better system uses a metal bracket, such as the one in Securing a Truss. These brackets are available commercially in a variety of shapes. They are nailed to the top and sides of the wall plate and to the bottom chord of the truss. They provide superior resistance against wind uplift and may be required by building codes.

Truss Layout The location of each roof truss is marked on the top plate, in the same way as standard rafters (see Section 19.2, Roof Assembly). Unlike rafter framing, however, there is no need to incorporate ceiling framing. This is because the bottom chord of the truss serves as the ceiling framing.

As you mark the plates, be sure to double check the overall layout as you work.

Interior Partitions

Sometimes partitions run parallel to and between the bottom truss chords. When these partitions are erected before the ceiling finish is applied, install 2×4 blocking between the bottom chords of the truss. This blocking should be spaced not more than 4' OC. To provide nailing for lath or drywall, nail a 1×6 or 2×6 continuous backer to the blocking.

When these partitions are erected after the ceiling finish is applied, 2×4 blocking is placed with its bottom edge level with the bottom of the truss chords. The blocking is fastened with two 16d nails in each end.

Metal brackets, called *clips*, should also be used to align and fasten the bottom chord of a truss where it passes over nonbearing partitions. These clips, shown in Partition Clip, prevent the chords from moving from side to side. They also allow the bottom chord to flex upward slightly when the truss is loaded. This is important in order to prevent the walls from interfering with truss movement.

Problems with Framing

Unlike conventional framing, roof trusses are manufactured to precise tolerances before being delivered to the job site. For this reason, carpenters must take extra care when framing a house with a trussed roof. The trusses cannot be altered to make up for minor errors in framing.

If an interior partition wall is too high, it will prevent trusses from seating properly. A similar problem occurs when the top plate of one or more of the partition walls is not level. These problems might also put dangerous stresses on the bottom chord of the truss, which could cause the truss to fail. The stresses could also cause interior wall finishes to crack later.

Wall framing problems are sometimes caused by errors in reading the plans, or by mistakes made in measuring or assembly. These problems are fairly easy to correct. However, problems caused by a poorly constructed foundation or an unlevel floor system are much more troublesome.

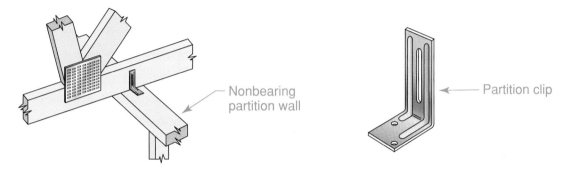

Nonbearing partition wall

Partition clip

Partition Clip
Simple Stabilizer Leave a 1/16" gap between the nailheads and the clip to help prevent squeaking.

Section 17.4 Assessment

After You Read: Self-Check

1. What precautions should be taken when handling and storing roof trusses?
2. What two methods can be used to raise roof trusses into place, and which one is preferable?
3. What is the first thing that should be done when a truss has been raised into place?
4. What is the procedure for temporary bracing of roof trusses?

Academic Integration: Science

5. **Truss Safety** Trusses provide a strong and efficient way to frame a roof. However, many accidents have occurred due to improper installation and bracing. Review the latest safety guidelines provided by OSHA regarding roof truss installation. In addition, locate at least one manufacturer of metal truss braces for further safety guidance. Prepare a three-minute presentation in which you discuss common mistakes that lead to improper installation, how to avoid these mistakes, and how to work safely with trusses.

Go to connectED.mcgraw-hill.com to check your answers.

Review and Assessment

Chapter Summary

Section 17.1

Planning a roof calls for an understanding of architectural styles as well as an understanding of how the individual pieces are assembled. A roof plan must be developed before any framing can begin.

Section 17.2

There are four basic methods for laying out the cuts required for a common rafter. They include the Pythagorean theorem method, the unit-length method, the step-off method, and the calculator method.

Section 17.3

Ceiling framing is much like floor framing. However, it is often considered to be a part of roof framing.

Section 17.4

Roof trusses are strong, efficient, and cost less than traditional framing. Common truss shapes include the king-post, the Fink (W-truss), and the scissors. The design of a truss depends on the loads it must carry and the weight and slope of the roof. Each truss should be solidly braced immediately after being lifted into place.

Review Content Vocabulary and Academic Vocabulary

1. Use each of these content vocabulary and academic vocabulary words in a sentence or diagram.

Content Vocabulary

- rafter (p. 467)
- ridge board (p. 468)
- span (p. 470)
- total run (p. 470)
- unit run (p. 470)
- total rise (p. 470)
- unit rise (p. 470)
- slope (p. 470)
- pitch (p. 470)
- chord (p. 492)
- web (p. 492)

Academic Vocabulary

- benefit (p. 481)
- distributed (p. 482)
- access (p. 490)
- specify (p. 495)

Speak Like a Pro

Technical Terms

2. Work with a classmate to define the following terms used in the chapter: *dormers* (p. 467), *low-slope roof* (p. 467), *lean-to roof* (p. 467), *collar tie* (p. 468), *eaves* (p. 468), *overhangs* (p. 468), *hip jack rafters* (p. 468), *valley jack rafters* (p. 468), *cripple jack rafters* (p. 468), *unit of run* (p. 470), *measuring line* (p. 470), *plumb line* (p. 471), *level line* (p. 471), *theoretical length* (p. 475), *bird's mouth* (p. 481), *outriggers* (p. 484), *stub joists* (p. 490), *flush girder* (p. 491), *strongback* (p. 491), *king-post truss* (p. 493), *fink truss* (p. 493), *W-truss* (p. 493), *scissors truss* (p. 493).

Review Key Concepts

3. Name the four basic roof types.

4. Recall common terms used in roof-framing carpentry.

5. Use one of the four basic methods to lay out a common rafter.

6. Explain how ceiling joists are laid out.

7. Identify the three basic parts of a roof truss.

8. Describe how roof trusses are installed.

Critical Thinking

9. Compare What is the difference between common rafters and hip rafters?

Academic and Workplace Applications

STEM Mathematics

10. Fractions and Decimals Small, easy-to-use construction calculators are now common on job sites. Measurements can be entered into the calculator in feet and inches, including fractions. Use a construction calculator to make the following calculation: Add 14'-6⅜" to 23'-9¹¹⁄₁₆".

Math Concept Fractional measurements have decimal equivalents. For example, ¹⁄₁₆" = 0.0625"; ⅛" = 0.125"; ³⁄₁₆" = 0.1875"

Step 1: Input the following: 14 [feet] 6 [Inch] 3 [/] 8 [+] 23 [Feet] 9 [Inch] 11 [/] 16 [=]

Step 2: Use the correct keys to convert to decimal feet, then to decimal inches, then to fractional inches.

STEM Science

11. Fungi Certain precautions must be taken in order to prevent damage to lumber products such as trusses. If the trusses will be stored outdoors before being installed, they should be supported above the ground to protect them from dampness or water that can cause the wood to rot. Fungi are a direct cause of wood rot. Find out more about wood rot caused by fungi and summarize your findings in a one-page report. State what fungi are and how they interact with their environment to cause wood rot.

21st Century Skills

12. Critical Thinking: Research Regional Styles Find out more about roof styles such as gable, hip, flat, and shed roofing. List the purposes of each type and classify them according to the regions of the country for which they are appropriate. Observe residential homes in your area and become expert at identifying roofing styles. Assess the neighborhood you live in and determine which roofing styles have been used. Write a paragraph stating which types you have observed and state why these types have been chosen for this region.

Standardized TEST Practice

Multiple Choice

Directions Choose the phrase that best completes the following statements.

13. The angle at which the hip extends from each corner is usually _____ .

   a. 30°   **b.** 90°   **c.** 45°   **d.** 180°

14. To obtain the actual length of a rafter it is necessary to _____ .

   a. add one-half the actual thickness of the ridge board from the theoretical length

   b. subtract one-half the actual thickness of the ridge board from the theoretical length

   c. subtract one-quarter of the actual thickness of the ridge board from the theoretical length

   d. add one-quarter of the actual thickness of the ridge board from the theoretical length

15. On trusses with long spans, _____ prevents the bottom chords from moving as the ceiling finish is applied.

   a. temporary bracing

   b. ceiling joists

   c. permanent bracing

   d. sheathing

TEST-TAKING TIP

When you first get a test, look at the total number of questions and sections before answering any questions. Once you have scanned the test, estimate what pace you should maintain in order to finish the test in the time allotted.

*These questions will help you practice for national certification assessment.

Hip, Valley, & Jack Rafters

Section 18.1
Hip Rafters

Section 18.2
Valley Rafters

Section 18.3
Jack Rafters

Chapter Objectives

After completing this chapter, you will be able to:

- **Explain** how to lay out a hip rafter for a given roof.

- **Explain** how to lay out a valley rafter for a given roof.

- **Determine** the rafter overhang for a hip or valley rafter.

- **Define** a dormer.

- **Explain** how to lay out a jack rafter for a given roof.

- **Summarize** why the intersection of two roofs calls for more complex framing.

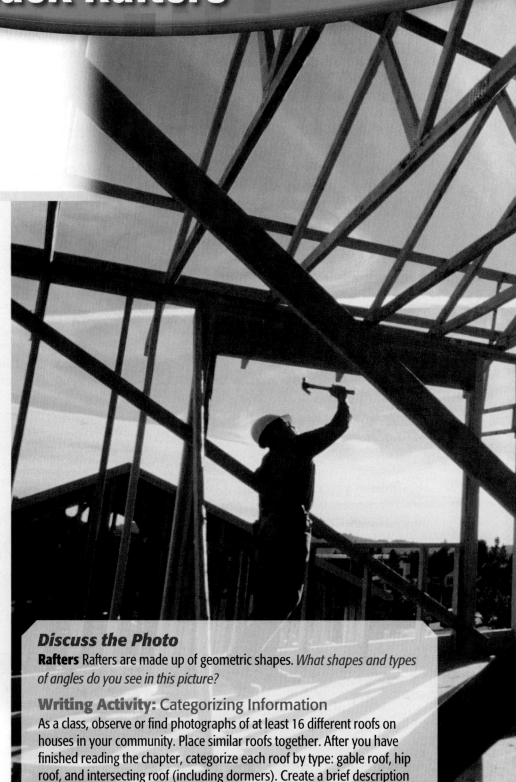

Discuss the Photo

Rafters Rafters are made up of geometric shapes. *What shapes and types of angles do you see in this picture?*

Writing Activity: Categorizing Information

As a class, observe or find photographs of at least 16 different roofs on houses in your community. Place similar roofs together. After you have finished reading the chapter, categorize each roof by type: gable roof, hip roof, and intersecting roof (including dormers). Create a brief description of each type of roof.

Ingram Publishing

Chapter 18 Reading Guide

Before You Read Preview

Roof framing with hip, valley, and jack rafters is more complex than framing entirely with common rafters. Choose a content vocabulary or academic vocabulary word that is new to you. When you find it in the text, write down the definition.

Content Vocabulary

○ hip rafter
○ valley rafter
○ jack rafter

○ seat cut
○ backing the hip
○ dropping the hip

○ addition
○ dormer
○ doghouse dormer

Academic Vocabulary

You will find these words in your reading and on your tests. Use the academic vocabulary glossary to look up their definitions if necessary.

■ hypotenuse
■ significant
■ ensure

Graphic Organizer

As you read, use a chart like the one shown to organize information about the three types of rafters.

hip rafter	valley rafter	jack rafter
forms a raised area, or hip, usually extending from the corner of the building diagonally upwards to the ridge		

Go to connectED.mcgraw-hill.com to download this graphic organizer.

Understanding Complex Roofs

When is a hip rafter called for?

A simple gable roof can be built entirely with common rafters. However, a carpenter must also know how to lay out and cut hip, valley, and jack rafters. These rafters, are required when framing complex roofs, such as hip roofs and intersecting gable roofs (for more on roof types, see Chapter 17, "Basic Roof Framing"). A **hip rafter** forms a raised area, or *hip*, usually extending from the corner of the building diagonally upwards to the ridge. A **valley rafter** forms a depression in the roof instead of a hip. Like the hip rafter, it extends diagonally from the top plate to the ridge. A hip rafter is called for only when framing a hip roof, but a valley rafter is needed on both hip and gable roofs whenever roof planes intersect. A **jack rafter** is a shortened common rafter that may be framed to a hip rafter, a valley rafter, or both. Thus, there are *hip jack rafters* and *valley jack rafters*.

The total rise of hip and valley rafters is the same as that of common rafters. They are also the same thickness as common rafters. However, they should be 2" wider in their nominal dimension. For example, if you use 2×6 common rafters, use 2×8 hip rafters to provide full bearing for the end of intersecting jack rafters, as shown in Width of Hip and Valley Rafters.

A mastery of roof framing with hip, valley, and jack rafters is what distinguishes the true

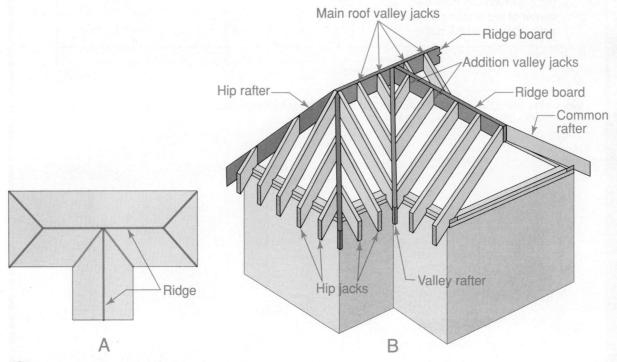

Hip, Jack, and Valley Rafters
Roof Anatomy **A.** The roof framing plan. **B.** The general arrangement of rafters shown in the larger drawing.

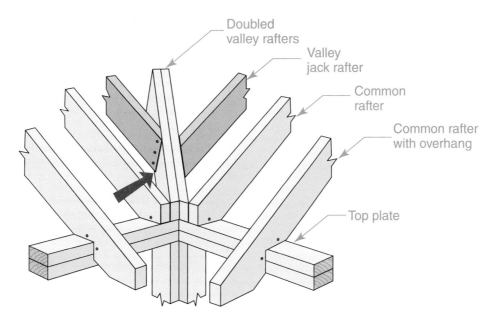

Doubled valley rafters

Valley jack rafter

Common rafter

Common rafter with overhang

Top plate

Width of Hip and Valley Rafters
How the Parts Fit The doubled valley rafter in this drawing has been cut off at the top plate. Normally it is extended to become part of the overhang. Doubled valleys are sometimes used to provide more bearing for the roof sheathing.

professional from the casual carpenter. This chapter describes how to figure rafter layouts manually using a standard framing square. On the job, construction calculators and triangular framing squares are often used for this purpose. A calculator works quickly and with great precision. This makes it invaluable when laying out hip, valley, and jack rafters. Most construction calculators have built-in functions to make roof calculations even easier.

Reading Check

Recall *What tools are used to figure layouts?*

Hip Rafter Layout
What is a hypotenuse?

Any of the methods for determining the length of a common rafter may be used for determining the length of a hip rafter (see Chapter 17, "Basic Roof Framing"). However, some of the basic data used is different.

Part of a framing plan for a hip roof is shown in Hip Roof Framing Plan. Remember that a line on the framing plan indicating a rafter represents the total run of the rafter, but not its actual length. On a hip roof framing plan, the lines that indicate the hip rafters (EC, AC, KG, and IG) form 45° angles with the edges of the building. You can see from the plan that the total run of a hip rafter is the **hypotenuse** of a right triangle. The two shorter legs of this triangle are each equal to the total run of a common rafter, or half the span of the roof.

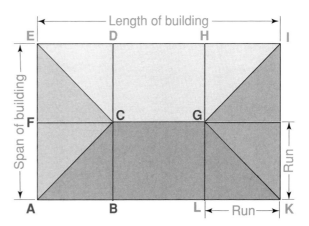

Hip Roof Framing Plan
Framing Plan This is the framing plan for a small rectangular building with a hip roof.

In Comparison of Hip and Common Rafters, one corner of the roof framing plan (ABCF) has been drawn in perspective. This shows the relative position of the hip rafter to the common rafter.

The unit run of a hip rafter is the hypotenuse of a right triangle with the shorter sides each equal to the unit run of a common rafter, as shown in Comparing Unit Runs. The unit run of a common rafter is 12". Using the Pythagorean theorem, $a^2 + b^2 = c^2$, the unit run of a hip rafter is the square root of 144 + 144 which is 16.97" (which can be rounded up to 17"), as shown in A in Unit Run and Unit Length.

Like the unit length of a common rafter, the unit length of a hip rafter may be obtained from the rafter table on the framing square. In B in Unit Run and Unit Length, the second row in the table is headed "Length Hip or Valley per Foot Run." This means "for every 12" of a common rafter in the same roof." Another way to state this would be "per 16.97" run of hip or valley rafter." For example, the unit length for a unit rise of 8" is 18.76". To calculate the length of a hip rafter, multiply the unit length by the number of feet in the total run of a common rafter.

Look again at Comparing Unit Runs on page 507, which shows the corner of the building in Hip Roof Framing Plan on page 505. In this example the total run of a common rafter is 5'. The unit rise is 8" and the unit length of the hip rafter for this unit rise is 18.76". The unit length multiplied by the total run in feet is the length of the hip rafter in inches (18.76" × 5 = 93.8", or 7'-9¹³⁄₁₆"). As in the case of common rafters, this is the theoretical length. To obtain the actual length, the ridge board shortening allowance and the rafter tail length will have to be calculated and laid out.

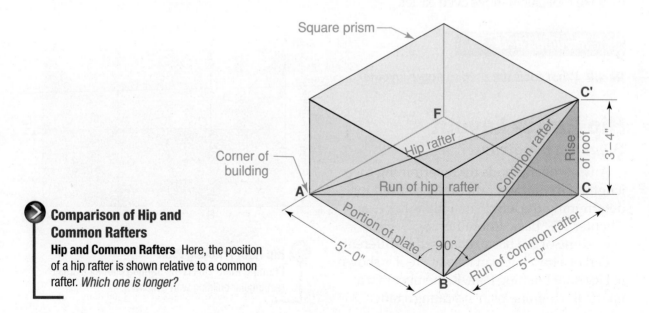

Comparison of Hip and Common Rafters
Hip and Common Rafters Here, the position of a hip rafter is shown relative to a common rafter. *Which one is longer?*

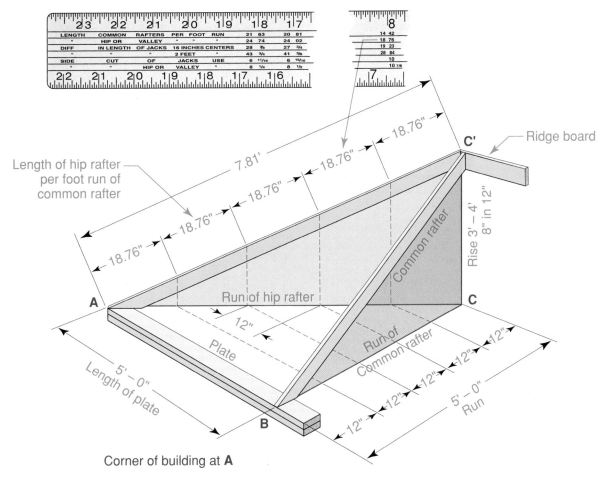

Comparing Unit Runs

Visualizing Unit Run The relationship between the unit run of a hip rafter and the unit run of a common rafter.

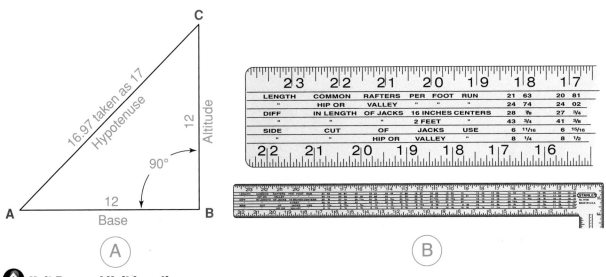

Unit Run and Unit Length

Finding Unit Run and Unit Length A. The hypotenuse of a right triangle, the shorter sides of which each equal 12", is 16.97". This can be rounded up to 17". **B.** Unit length can be obtained from the framing square.

Pythagorean Theorem Find the run of a hip rafter if the run of the common rafter is 10 feet. Round to the nearest tenth.

Starting Hint The run of the hip rafter is the hypotenuse of a right isosceles triangle. The length of the legs is the length of the common rafter.

REGIONAL CONCERNS

Roof Styles A roof is a very important element in the architectural style of a house. The style of a house can change dramatically by changing features as basic as roof pitch. For example, a hip roof with a steep pitch would typically indicate a French style of architecture that is often seen in the Gulf Coast region, particularly in Louisiana. A low-pitched hip roof characterizes Prairie style houses often found in the Midwest. The type of roof framing that carpenters become familiar with depends on the style of house most common where they work.

Plumb and Level Lines

Cuts made in a hip or valley rafter are made either along *plumb lines* (plumb cuts) or along *level lines* (level cuts), as shown in Plumb and Level. To lay out the plumb and level cuts of the hip or valley rafters, set off 17" on the *blade* (the long leg) of the framing square. On the *tongue* of the square (the short leg), set off the rise per foot of common rafter run. A line drawn along the tongue then indicates the plumb cut. A line drawn along the blade indicates the level cut. When the completed rafter is to rest on its level cut, the level cut is sometimes referred to as the **seat cut**.

Shortening Allowance

The theoretical length of a hip rafter does not take into account the thickness of the ridge board. This must be allowed for by deducting the shortening allowance. The *shortening allowance* for a hip rafter depends on the way the rafter is cut to fit against the other structural members. Some carpenters make a single side cut, as in Single Side Cut. Other carpenters prefer a double side cut, as in Double Side Cut.

> **Plumb and Level**
> **Same Square Position** Marking the plumb cut and the level (seat) cut on a hip rafter. *Along what lines are cuts made in hip and valley rafters?*

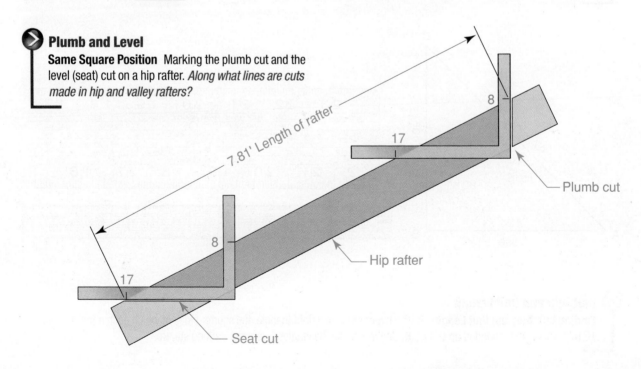

7.81' Length of rafter

8

17

Plumb cut

Hip rafter

8

17

Seat cut

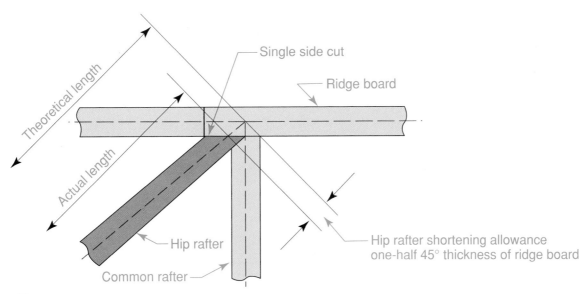

Single Side Cut

One Cut A hip rafter framed against the ridge board calls for a single side cut. Depending on the layout of common rafters, the end common rafter may require a 45° plumb cut so that it will fit against the side of the hip rafter.

If the ridge board is a different thickness than the rafters, the shortening allowance must take this into account. If the hip rafter is framed against the ridge board, using a single side cut, the shortening allowance is one-half the 45° thickness of the ridge board. (The 45° thickness is the length of a line laid at 45° across the thickness of the board.) However, if the hip rafter is framed against

the common rafters, using a double side cut, the shortening allowance is one-half the 45° thickness of a common rafter.

To lay out the shortening allowance, set the tongue of the framing square along the rafter's plumb line. Measure the shortening allowance along the blade and mark this point, as shown in Step 1 on page 510. Then slide the square sideways until the tongue

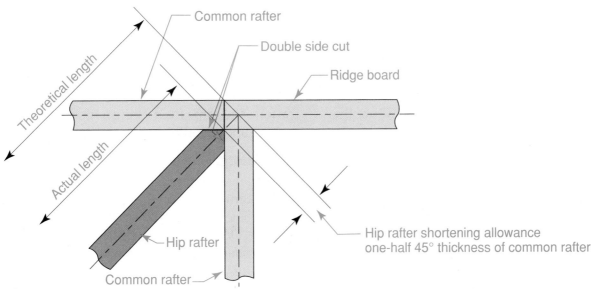

Double Side Cut

Two Cuts A hip rafter framed against the ridge-end common rafters requires a double side cut.

⌄ **Shortening Allowance for Hip Rafter**

Laying Out Shortening Allowance Start by setting the tongue along the plumb line.

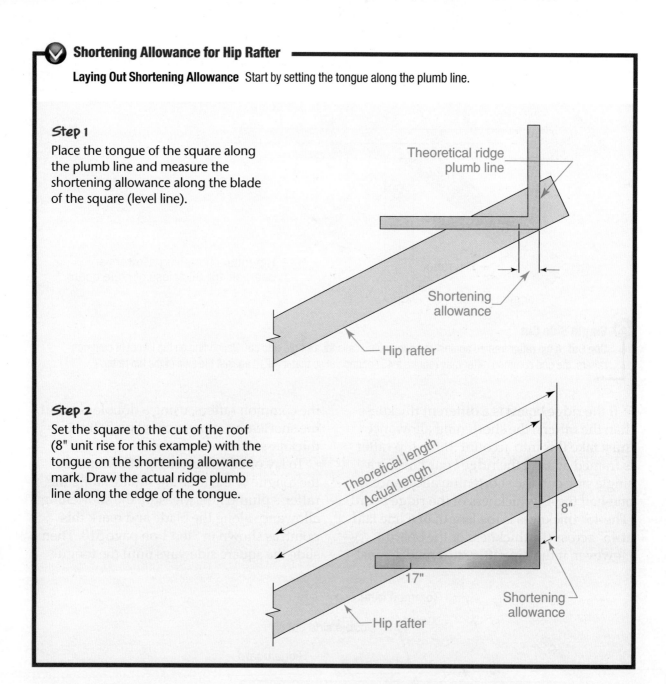

Step 1

Place the tongue of the square along the plumb line and measure the shortening allowance along the blade of the square (level line).

Theoretical ridge plumb line

Shortening allowance

Hip rafter

Step 2

Set the square to the cut of the roof (8" unit rise for this example) with the tongue on the shortening allowance mark. Draw the actual ridge plumb line along the edge of the tongue.

Theoretical length

Actual length

8"

17"

Shortening allowance

Hip rafter

is lined up with the mark and draw another plumb line, as shown in Step 2. This line marks the actual plumb cut for the rafter.

Hip Rafter Tail Cut

A common rafter tail has a single plumb cut at its lower end. A hip rafter tail, however, will butt against the corner of two intersecting planes, and requires two cuts. See Hip Tail Cut.

The face of each cut should be in the same plane as cuts on the ends of adjacent common rafters. The steps for making a hip rafter tail cut are shown on page 511 and on page 512.

Overhang

The amount of rafter overhang has a **significant** impact on the appearance of a house. The amount is often related to the climate. Deep overhangs protect walls from rain or shade them from intense sun. Shallow overhangs help to prevent ice dams caused when melted snow refreezes at the overhang. Deep overhangs are also typical of certain

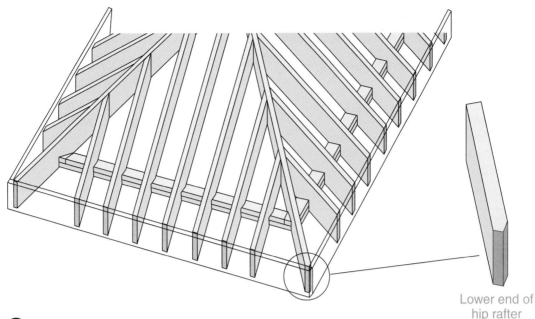

Lower end of
hip rafter

 Hip Tail Cut

Fit for Fascia The end of the hip rafter has a double side cut so that the fascia (installed later) will fit properly.

Step-by-Step Application

Making Hip Rafter Side Cuts The end of a hip rafter joins the ridge board (or the ends of the common rafters) at an angle. The cut is called a *side cut* or sometimes a *cheek cut* (see page 509). The side cut may be laid out in one of two ways.

Method 1

Step 1 Place the tongue of the framing square along the actual ridge board plumb cut line. Measure one-half the thickness of the hip rafter along the blade (level line) and place a mark.

Step 2 Shift the tongue to the mark, set the square to the cut of the rafter (17" and 8" in this example), and draw a plumb line (A).

Step 3 Turn the rafter on edge and draw a centerline along its edge, indicated by the red arrow on page 512.

Step 4 Extend the plumb lines from the face of the rafter to intersect the centerline at 90°. The side cut line is drawn from line A through the intersection of the centerline and the actual ridge-end plumb line.
Note: A hip rafter that will be framed against the ridge board has only a single side cut. A hip rafter framed against the ends of the common rafters requires a double side cut.

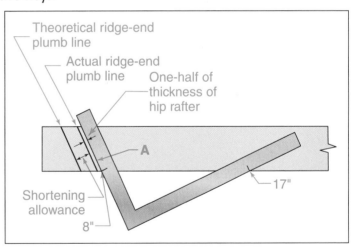

Theoretical ridge-end plumb line
Actual ridge-end plumb line
One-half of thickness of hip rafter
A
Shortening allowance
8"
17"

 Drawing a Plumb Line

Step 5 The tail of the rafter must have a double side cut at the same angle, but in the reverse direction, to allow attachment of the fascia board.

(continued)

Method 2 For this method, refer to the rafter table on the framing square.

Step 1 On the framing square, the bottom line of the table is headed "Side Cut Hip or Valley Use" (see page 507). Follow this line over to the column under the number 8 (for a unit rise of 8"). The number shown is 10⅞.

Step 2 Place the framing square face up on the rafter edge, with the tongue on the ridge-end plumb cut line.

Step 3 Set the square to a cut of 10⅞" on the blade and 12" on the tongue. Draw the side cut angle along the tongue.

To determine the overhang, see the Step-by-Step Application on page 513.

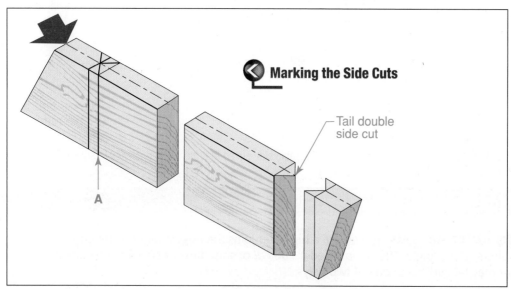

Marking the Side Cuts

Tail double side cut

A

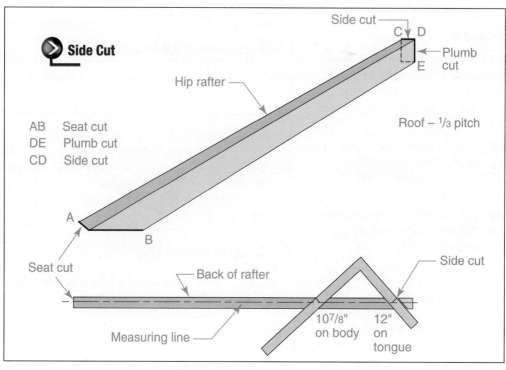

Side Cut

Side cut

Plumb cut

Hip rafter

Roof – ⅓ pitch

AB   Seat cut
DE   Plumb cut
CD   Side cut

Seat cut

Back of rafter

Side cut

Measuring line

10⅞" 12"
on body on tongue

Determining the Overhang As with a common rafter overhang, a hip or valley rafter overhang is figured as if it were a separate rafter. The run of this overhang, however, is not the same as the run of a common rafter overhang in the same roof. Instead, the run of a hip or valley overhang is the hypotenuse of a right triangle whose shorter sides are each equal to the run of a common rafter overhang. If the run of the common rafter overhang is 2' for a roof with an 8" unit rise, the length of the hip or valley rafter tail is figured as follows.

Step 1 Find the unit length of the hip or valley rafter on the framing square. For this roof, the unit length is 18.76".

Step 2 Multiply the unit length of the hip or valley rafter by the run of the common rafter overhang: 18.76" (unit length of hip or valley rafter) × 2 (feet of run in common rafter overhang) = 37.52", or 37½".

Step 3 Add this product to the theoretical rafter length.

The overhang may also be stepped off as described in Chapter 17 for a common rafter. When stepping off the length of the overhang, set the 17" mark on the blade even with the edge of the rafter. Set the unit rise, whatever it might be, on the tongue, even with the same rafter edge.

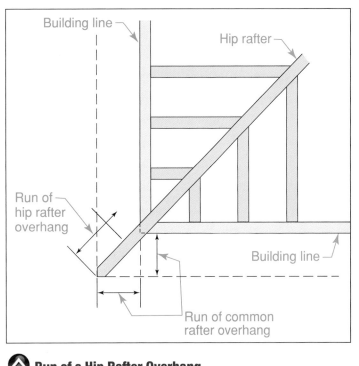

Run of a Hip Rafter Overhang

architectural styles, such as Arts and Crafts, Italianate, or Mission styles.

The parts of a hip rafter overhang are shown in Run of a Hip Rafter Overhang.

Bird's Mouth

Laying out the bird's mouth for a hip rafter is much the same as for a common rafter. However, there are a couple of things to remember. When you lay out the plumb (heel cut) and level (seat cut) lines on a hip rafter, set the body of the square at 17" and the tongue to the unit rise (depending on the roof pitch; see Plumb and Level on page 508). When laying out the depth of the heel, measure along the heel plumb line down from the top edge of the rafter, as shown in Layout of a Bird's Mouth

on page 514. This must be done because the hip rafters are usually wider than common rafters, and the distance should be the same on both. An additional step must also be taken to **ensure** that the top edge of a hip rafter will be in alignment with jack rafters. In this step, the hip rafter must either be *backed* or *dropped*.

Backing or Dropping a Hip Rafter

If the top edge of the hip rafter extends slightly above the upper ends of the jack rafters, it will interfere with the sheathing. **Backing the hip** means to bevel the upper edge of the hip rafter, as shown in Backing or Dropping a Hip on page 514. This allows the roof sheathing to be installed without hitting the corners of the hip rafter.

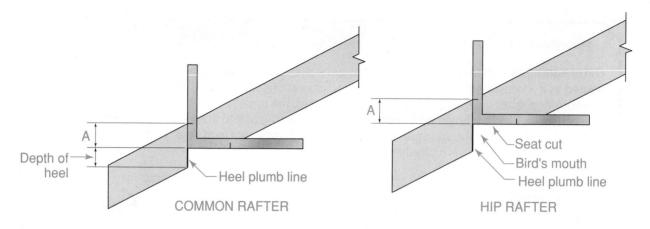

Depth of → heel

A

— Heel plumb line

COMMON RAFTER

A

— Seat cut
— Bird's mouth
— Heel plumb line

HIP RAFTER

 Layout of the Bird's Mouth

Measure Down When laying out the bird's mouth on a hip rafter, measure down from the top edge. Dimension A must be the same for both common and hip rafters so that the tops of all the rafters will be level for the application of sheathing.

Dropping the hip means to deepen the bird's mouth so as to bring the top edge of the hip rafter in line with the upper ends of the jacks.

The amount of backing or drop required is calculated as shown in A in Determining Backing or Drop. Set the framing square to the cut of the rafter (8" and 17" in this example) on the upper edge. Measure off one-half the thickness of the rafter from the edge along the blade. For backing, a line drawn through this mark and parallel to the edge will indicate the bevel angle, as in B in Determining Backing or Drop. For dropping, the perpendicular distance between the line and the edge of the rafter will be the amount of drop. This is the amount by which the depth of the hip rafter bird's mouth should exceed the depth of the common rafter bird's mouth, as in C in Determining Backing or Drop.

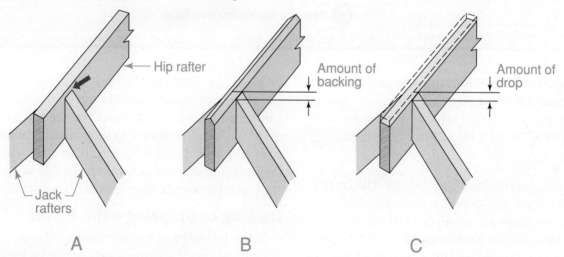

— Hip rafter

Amount of ↓ backing ↑

Amount of ↓ drop ↑

Jack rafters

A

B

C

 Backing or Dropping a Hip

Two Solutions A. The top of a hip rafter may extend above the upper ends of the jack rafters. **B.** Backing the hip rafter. **C.** Dropping the hip rafter.

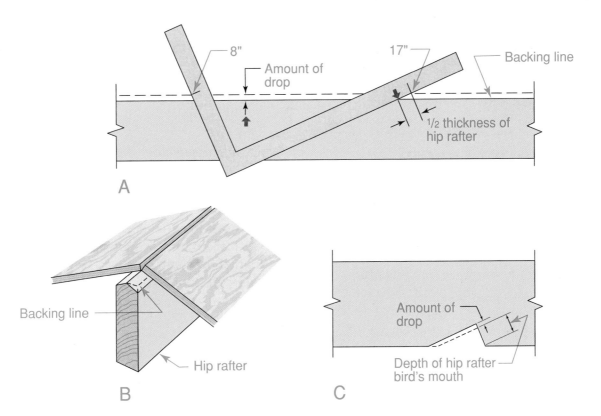

A

Backing line — Hip rafter

B

Amount of drop — Depth of hip rafter bird's mouth

C

 Determining Backing or Drop
Fine-Tuning the Rafter **A.** Determining the amount of backing or drop. **B.** Bevel angle for backing the rafter. **C.** Deepening the bird's mouth for dropping the rafter.

Section 18.1 Assessment

After You Read: Self-Check

1. What is the shortening allowance for a hip rafter when the ridge end is framed against the ridge board?
2. What is the amount of overhang often related to?
3. What is the main difference between a hip rafter and valley rafter?
4. What does *backing the hip* mean?

Academic Integration: Mathematics

5. **Explaining Unit Run** The unit run of a hip rafter is 17" and the unit run of a common rafter is 12". Explain how to calculate the hip rafter's unit run using an equation.

   Math Concept The unit run of a rafter is also the hypotenuse of a right triangle. The Pythagorean theorem is an equation that shows the relationships of the lengths of the sides of a right triangle.

   Step 1: Review the Pythagorean theorem.

   Step 2: Substitute known values into the formula.

   Step 3: State the formula and tell how to use the Pythagorean theorem to find the unit run.

Go to connectED.mcgraw-hill.com to check your answers.

Valley Rafters

Valley Rafter Layout

Could an addition roof have a span and a pitch different from the main roof?

A valley rafter is necessary where two roofs intersect. It is also needed at the intersection of a dormer roof with the main roof. Most intersecting roofs that contain valley rafters each have the same pitch. The valley rafters always run at a 45° angle to the building perimeter and the ridge boards.

Equal and Unequal Spans

A roof that intersects the main roof is sometimes referred to as an **addition**. This is because the main roof is generally framed first, and the intersecting roof is then added. Another reason is that a common method for expanding an existing house is to build an addition that intersects the main house.

Equal-Span Roof In equal-span framing, the span, or width, of the addition is the same as the span of the main roof, as shown in An Equal-Span Roof. When the pitch of the addition's roof is the same as the pitch of the main roof, the ridges of both roofs are at the same height.

The total run of a valley rafter (indicated by AB or AC in Ridge-End Shortening Allowance) is the hypotenuse of a right triangle. Each shorter side of the triangle is equal to the total run of a common rafter in the main roof. The unit run of a valley rafter is therefore 16.97", the same as the unit run for a hip rafter. Figuring the length of an equal-span valley rafter is thus the same as figuring the length of a hip rafter.

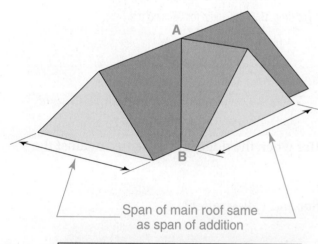

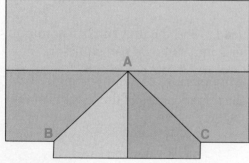

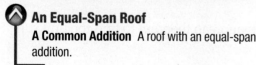

An Equal-Span Roof
A Common Addition A roof with an equal-span addition.

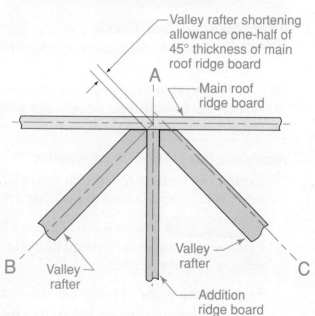

Ridge-End Shortening Allowance
Allow for the Ridge Ridge-end shortening allowance for an equal-span addition valley.

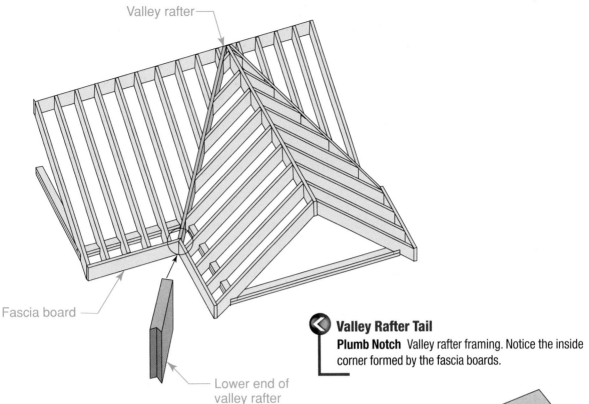

Valley rafter

Fascia board

Lower end of
valley rafter

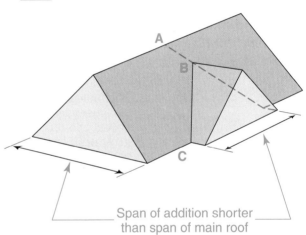

Valley Rafter Tail
Plumb Notch Valley rafter framing. Notice the inside corner formed by the fascia boards.

The shortening allowance for an equal-span addition valley rafter is one-half the 45° thickness of the ridge board. Side cuts are laid out as they are for a hip rafter. The valley rafter tail has a double side cut, like the hip rafter tail, but running in the opposite direction. This is because the tail cut must form an inside rather than an outside corner. The bird's mouth and the overhang, if any, are figured just as they are for a hip rafter.

Unequal-Span Roof A single full-length valley rafter (AD in An Unequal-Span Roof) is framed between the top plate and the ridge board. A shorter valley rafter (BC) is then framed to the longer one at a 90° angle. The total run of the longer valley rafter is the hypotenuse of a right triangle, the shorter sides of which are each equal to the total run of a common rafter in the main roof. The total run of the shorter valley rafter is the hypotenuse of a right triangle with shorter sides each equal to the total run of a common rafter in the addition. The total run of a common rafter

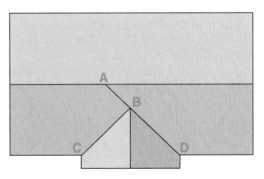

Span of addition shorter
than span of main roof

An Unequal-Span Roof
One Method An addition with a span shorter than the main roof span. This addition is formed with a long and a short valley rafter.

in the main roof is equal to one-half the span of the main roof. The total run of a common rafter in the addition is equal to one-half the span of the addition.

Determining the Length of a Valley Rafter

When the total run of any rafter is known, the theoretical length can be found by multiplying the total run by the unit length. Suppose, for example, that the addition shown in An Unequal-Span Roof has a span of 30' and that the unit rise of a common rafter in the addition is 9". The rafter table on page 507 shows that the unit length for a valley rafter in a roof with a common rafter unit rise of 9" is 19.21".

To find the theoretical length of the valley rafter, multiply its unit length by the total run of a common rafter in the roof to which it belongs. (The total run of a common rafter is equal to one-half the span.) Therefore, the length of the longer valley rafter on page 517 would be 19.21" times one-half the span of the

main roof. The length of the shorter valley rafter would be 19.21" times one-half the span of the addition. Because one-half the span of the addition is 15', the length of the shorter valley rafter is 19.21" × 15, or 288.15". Converted to feet, this is 24.01'.

The shortening allowances for the long and short valley rafters are shown in Long and Short Valley Rafter Shortening Allowances. Note that the long valley rafter has a single side cut for framing to the main roof ridge board. The short valley rafter is cut square for framing to the long valley rafter.

Explain *How can you determine the theoretical length of a valley rafter?*

Framing Dormers
What do dormers do?

Dormers are often added to a roof. A **dormer** is a roofed projection from a slanted roof. In addition to a roof, dormers typically include a window. They add architectural interest, allow natural light to reach the top floor, and provide more headroom beneath steep slopes. In many respects, framing some kinds of dormers is like framing a small house with a small roof.

Dormers Without Side Walls

When constructing a gable dormer without side walls, the dormer ridge board is fastened to a header. The header is supported on each end by doubled common rafters in the main roof. The valley rafters are framed between this header and a lower header. The total run of a valley rafter is the hypotenuse of a right triangle, the shorter sides of which are each equal to the total run of a common rafter in the dormer.

The arrangement and names of framing members in this type of dormer framing are shown in Parts of a Dormer Without Side Walls. Note that the upper edges of the headers must be beveled to the pitch of the main roof.

Shortening allowance of longer valley rafter equals one-half of 45° thickness of main roof ridge board

Main roof ridge board

Shortening allowance of shorter rafter equals one-half of thickness of longer valley rafter

Addition ridge board

Long and Short Valley Rafter Shortening Allowances
Long and Short Long and short valley rafter shortening allowances.

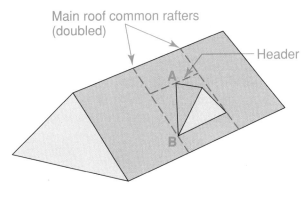

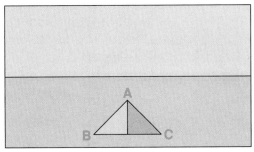

A Dormer Without Side Walls

All Roof Framing a dormer without side walls.

In this method, the shortening allowance for the upper end of a valley rafter is one-half the 45° thickness of the inside member (the member closest to the dormer) in the doubled upper header. For example, see page 520. The shortening allowance for the lower end is one-half the 45° thickness of the inside member in the doubled common rafter. Each valley rafter has a double side cut at the upper and lower ends.

Dormers With Side Walls

A method of framing a gable dormer with side walls is shown on page 520. This type of dormer is sometimes referred to as a **dog-house dormer** because of its shape. The total run of the valley rafter is the hypotenuse of a right triangle. The shorter sides of the triangle are each equal to the total run of a common rafter in the dormer.

Figure the lengths of the dormer corner posts and side studs just as you would the lengths of gable-end studs (see Chapter 19,

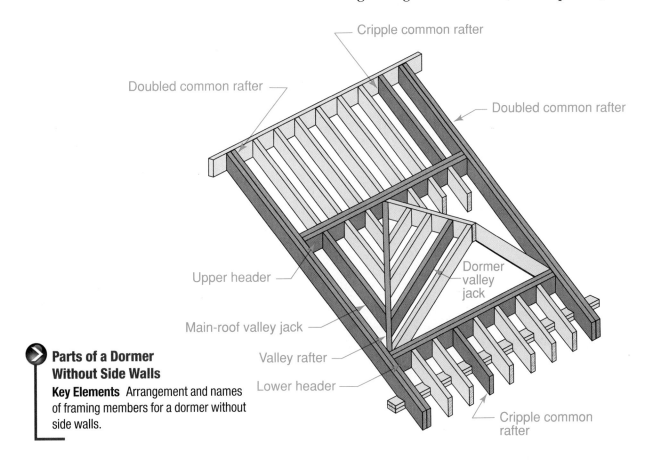

Parts of a Dormer Without Side Walls

Key Elements Arrangement and names of framing members for a dormer without side walls.

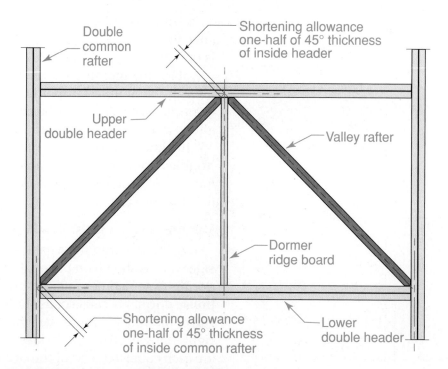

Double common rafter

Shortening allowance one-half of 45° thickness of inside header

Upper double header

Valley rafter

Dormer ridge board

Shortening allowance one-half of 45° thickness of inside common rafter

Lower double header

 Valley Rafter Shortening: Dormer With Side Walls
A Rigid Structure Valley rafter shortening allowances for a dormer without side walls.

"Roof Assembly & Sheathing"). Lay out the lower-end cutoff angle by setting the square to the pitch of the main roof. The valley rafter shortening allowances for this method of framing are shown in Valley Rafter Shortening.

Another type of dormer with side walls is the *shed dormer*. This type is usually tied into a gable roof. This dormer is discussed in detail in Chapter 19, "Roof Assembly & Sheathing."

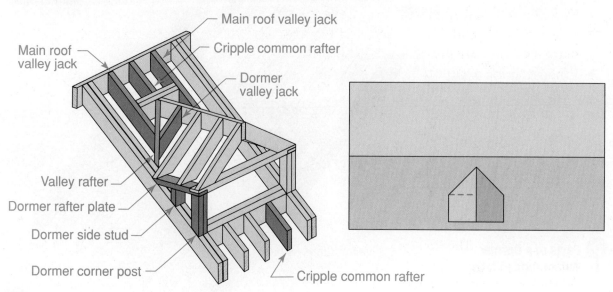

Main roof valley jack

Cripple common rafter

Main roof valley jack

Dormer valley jack

Valley rafter

Dormer rafter plate

Dormer side stud

Dormer corner post

Cripple common rafter

 A Gable Dormer With Side Walls
Key Elements Framing a gable dormer with side walls.

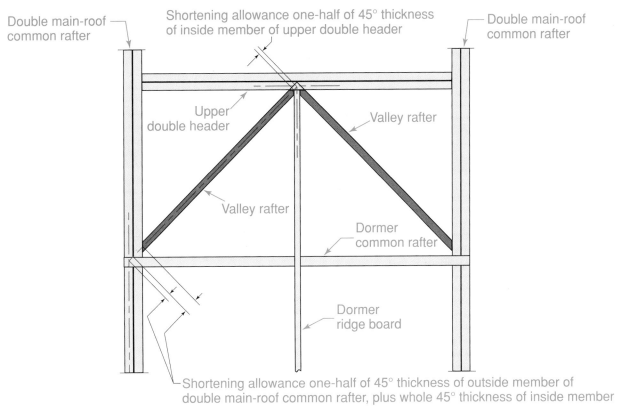

Double main-roof common rafter

Shortening allowance one-half of 45° thickness of inside member of upper double header

Double main-roof common rafter

Upper double header

Valley rafter

Valley rafter

Dormer common rafter

Dormer ridge board

Shortening allowance one-half of 45° thickness of outside member of double main-roof common rafter, plus whole 45° thickness of inside member

 Valley Rafter Shortening
Extra Steps Valley rafter shortening allowances for a dormer with side walls.

Section 18.2 Assessment

After You Read: Self-Check

1. Describe an equal-span roof.
2. When the pitch and the span of an addition roof are the same as the pitch and span of the main roof, how are the ridge boards positioned in relation to each other?
3. When framing a gable dormer without side walls, what is the dormer's ridge board attached to?
4. When framing a doghouse dormer, how is the run of a valley rafter determined?

Academic Integration: Mathematics

5. **Using Tables** Use the framing square to find the length of the rafter for the two problems below. Round answers to the nearest $\frac{1}{16}$".
   A. Run = 13'; Slope = $\frac{4}{12}$
   B. Run = 15'; Slope = $\frac{5}{12}$

   Math Concept The first line on a framing square will give you the length of a common rafter. For example, for a slope that rises 9" for every foot of run, the first line of the table tells you that the length of the common rafter for every foot of run is 15. You would multiply this by the total run to find the length of the common rafter (the hypotenuse).

Go to **connectED.mcgraw-hill.com** to check your answers.

Jack Rafters

Jack Rafter Layout

How does knowing the common difference save a carpenter time?

A *jack rafter* is a shortened common rafter that may be framed to a hip rafter, a valley rafter, or both. This means that in an equal-span framing situation, the unit rise of a jack rafter is always the same as the unit rise of a common rafter.

There are several types of jack rafters. A *hip jack rafter* extends from a hip rafter to a rafter plate. A *valley jack rafter* extends from a valley rafter to a ridge board. A *cripple jack rafter* does not contact either a plate or a ridge board. There are two kinds of cripple jack rafters. The *valley cripple jack* extends between two valley rafters in the long-and-short-valley-rafter method of addition framing. The *hip valley cripple jack* extends from a hip rafter to a valley rafter.

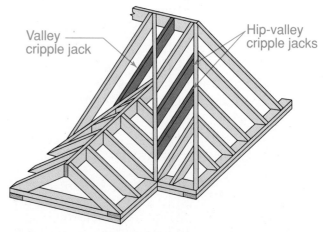

Other Types of Jack Rafters
Basic Elements Valley cripple jack and hip-valley cripple jacks.

Lengths of Hip Jack Rafters

A roof framing plan for a series of hip jack rafters is shown in Framing Plan. The jacks are always on the same spacing as the common rafters. The spacing in this instance is

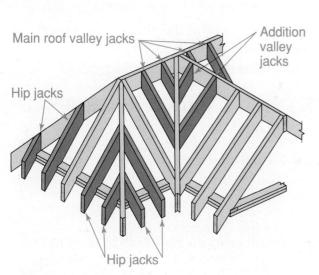

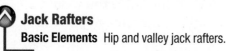

Jack Rafters
Basic Elements Hip and valley jack rafters.

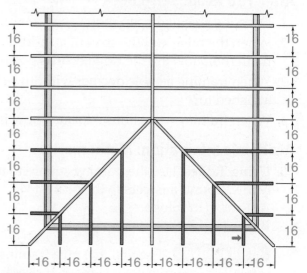

Framing Plan
How the Parts Fit A framing plan for a roof with hip jack rafters.

16" on center. You can see in the lower-right part of the plan that the total run of the shortest jack is also 16".

Suppose the unit rise of a common rafter in this roof is 8" per 12" of run. The hip jacks have the same unit rise as a common rafter. The unit length of a common rafter is the hypotenuse of a right triangle with the unit run as base and the unit rise as altitude. The unit length of a hip jack rafter in the example is therefore the square root of 144 + 64, or 14.42. This means that a hip jack is 14.42" long for every 12" of run.

The theoretical total length of the shortest jack rafter (X) can now be calculated using this formula:

$$\frac{12\text{" (unit run)}}{14.42\text{" (unit length)}} = \frac{16\text{" (total run)}}{X \text{ (total length)}}$$
$$X = 19.23"$$

This is the length of the shortest hip jack when the jacks are spaced 16" on center and the unit rise is 8". It is also the *common difference* in length between one jack and the next. This means that the next hip jack will be 2 × 19.23" long, the one after that 3 × 19.23" long, and so on.

The common difference for hip jacks spaced 16" on center and for hip jacks spaced 24" on center can also be found in the rafter table on a framing square (see page 507).

Builder's Tip

MAKING REPETITIVE CUTS When making repetitive angled cuts on roof framing lumber, a radial-arm saw or compound-miter saw can improve the speed and accuracy of your work. Once you have determined the proper angle, set a stop at one end of the saw's outfeed table. All stock resting against this stop will then be cut to the exact same length. Do not let sawdust collect around the stop. It will affect the cut length.

For example, the third row of the table reads "Difference in Length of Jacks 16 Inches Centers." Follow this row to the column headed 8 (for a unit rise of 8") to find the length of the first jack rafter and the common difference, which is 19¼.

Lengths of Valley Jacks and Cripple Jacks

The best way to figure the total lengths of valley jacks and cripple jacks is to lay out a roof framing plan. Part of a framing plan for a main hip roof with a long-and-short-valley-rafter gable addition is shown in Framing Plan with Gable Addition.

By studying the plan, you can figure the total lengths of the valley jacks and cripple jacks as follows:

- The run of valley jack No. 1 is the same as the run of hip jack No. 8, which is the shortest hip jack. The length of valley jack No. 1 is therefore equal to the common difference between jacks.

- The run of valley jack No. 2 is the same as the run of hip jack No. 7. The length is therefore twice the common difference between jacks.

- The run of valley jack No. 3 is the same as the run of hip jack No. 6. The length

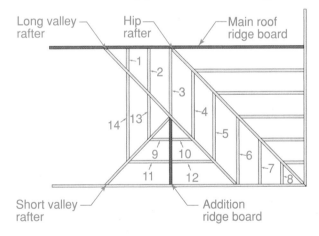

 Framing Plan with Gable Addition
Jack Layout Jack rafter framing plan for a hip roof with a gable addition.

is therefore three times the common difference between jacks.

- The run of hip-valley cripples No. 4 and No. 5 is the same as the run of valley jack No. 3. The length of these rafters is thus the same as the length of No. 3.

- The run of valley jacks No. 9 and No. 10 is equal to the spacing of jacks on center. Therefore, the length of each of these jacks is equal to the common difference between jacks. The run of valley jacks No. 11 and No. 12 is twice the run of valley jacks No. 9 and No. 10. The length of each of these jacks is therefore twice the common difference between jacks.

- The run of valley cripple No. 13 is twice the spacing of jacks on center, and the length is therefore twice the common difference between jacks. The run of valley cripple No. 14 is twice the run of valley cripple No. 13, so the length is twice the common difference between jacks.

Shortening Allowances A hip jack rafter has a shortening allowance at the upper end equal to one-half the 45° thickness of the hip rafter. A valley jack rafter has a shortening allowance at the upper end equal to one-half the thickness of the ridge board. It also has a shortening allowance at the lower end equal to one-half the 45° thickness of the valley rafter. A hip-valley cripple has a shortening allowance at the upper end equal to one-half the 45° thickness of the hip rafter, and another at the lower end equal to one-half the 45° thickness of the valley rafter. A valley cripple has a shortening allowance at the upper end equal to one-half the 45° thickness of the long valley rafter. At the lower end, the allowance is equal to one-half the 45° thickness of the short valley rafter.

Reading Check

Explain *What are three types of jack rafters?*

Side Cuts

The side cut on a jack rafter can be laid out by the method shown on pages 511–512 for laying out the side cut on a hip rafter.

Another method is to use the rafter table on the framing square (see page 507). Find the row headed "Side Cut of Jacks Use" and read across to the figure under the unit rise. For a unit rise of 8", the figure given is 10. To lay out the side cut on a jack with this unit rise, set the square face-up on the edge of the rafter to 12" (the unit run) on the tongue and 10" on the blade. Draw the side cut line along the tongue (see page 512).

A jack rafter pattern can also be used to save time.

Bird's Mouth and Overhang

A jack rafter is a shortened common rafter. Consequently, the bird's mouth and overhang are laid out just as they are on a common rafter (see Chapter 17, "Basic Roof Framing").

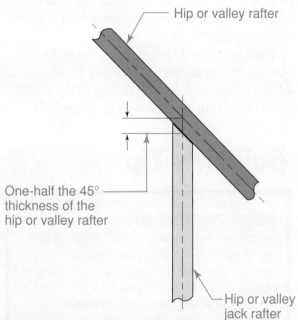

Hip or valley rafter

One-half the 45° thickness of the hip or valley rafter

Hip or valley jack rafter

Shortening Allowance for Hip or Valley Jack Rafters
Plan View The shortening allowance for the upper end of a hip jack or the lower end of a valley jack rafter.

Cutting a Jack Rafter Pattern Rather than lay out and mark each jack rafter individually, a pattern is used to save time. When all the rafters have been cut, the rafter used as a pattern becomes part of the roof frame.

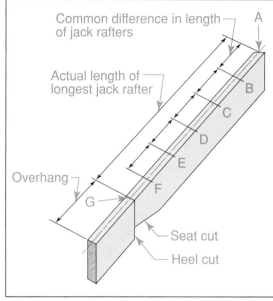

Common difference in length of jack rafters — A

Actual length of longest jack rafter

B

C

D

E

F

Overhang

G

Seat cut

Heel cut

Jack Rafter Pattern

Step 1 Lay out and cut the longest jack rafter first, including the overhang, if there is one. Be careful to calculate and make all necessary allowances to determine the actual length.

Step 2 Set the rafter in place on the building and check the fit of all the cuts. See that the spacing between the centers of the rafters is correct.

Step 3 When everything is correct, use this rafter as a pattern. On the top edge of the rafter, measure down the center line from the ridge end a distance equal to the common difference measurement (found on the framing square rafter table). This is the length of the second-longest jack rafter.

Step 4 Continue to mark the common difference measurements along the top edge until the lengths of all the jacks have been laid out.

Step 5 Use the longest jack rafter (AG) as a pattern to lay out all the jack rafters. The second jack rafter is BG, the third jack rafter is CG, and so on.

Section 18.3 Assessment

After You Read: Self-Check

1. What is a jack rafter?
2. What is a valley jack rafter?
3. What is the best way to figure the total lengths of valley jacks and cripple jacks?
4. What is the purpose of a jack rafter pattern?

Academic Integration: Mathematics

5. **Lengths of Hip Jack Rafters** For a house with a hip roof, the run of a common rafter is 14', the pitch is ⁶⁄₁₂, there is a 2' overhang, and the rafters are 16" OC. Figure the length of the shortest hip jack rafter.

    Math Concept The run, rise, and length of a hip jack rafter are like the base, altitude, and hypotenuse of a right triangle.

    Step 1: Use the Pythagorean Theorem to find the unit length.

    Step 2: Set up a proportion comparing the ratio of unit run to unit length with the ratio of total run to total length.

Go to **connectED.mcgraw-hill.com** to check your answers.

CHAPTER 18 Review and Assessment

Chapter Summary

Section 18.1
Three types of rafters are hip, valley, and jack rafters. The length of a hip rafter is calculated on the basis of the unit run and unit rise and/or the total run and total rise. Any of the methods previously described for determining the length of a common rafter may be used. However, some of the basic data for hip and valley rafters is different.

Section 18.2
The span of an addition roof may be equal or unequal to that of the main roof. Dormers are framed either with or without side walls. Those with side walls are called doghouse dormers.

Section 18.3
Jack rafters are shortened common rafters framed to a hip rafter, valley rafter, or both. The best way to figure the total lengths of valley jacks and cripple jacks is to lay out a framing plan. A hip jack rafter extends from a hip rafter to a plate. Hip jacks always have the same spacing as common rafters.

Review Content Vocabulary and Academic Vocabulary

1. Use each of these content vocabulary and academic vocabulary words in a sentence or diagram.

Content Vocabulary
- hip rafter (p. 504)
- valley rafter (p. 504)
- jack rafter (p. 504)
- seat cut (p. 506)
- backing the hip (p. 513)
- dropping the hip (p. 514)
- addition (p. 516)
- dormer (p. 518)
- doghouse dormer (p. 519)

Academic Vocabulary
- hypotenuse (p. 505)
- significant (p. 510)
- ensure (p. 513)

Speak Like a Pro

Technical Terms

2. Work with a classmate to define the following terms used in the chapter: *hip jack rafters* (p. 504), *valley jack rafters* (p. 504), *plumb lines* (p. 508), *level lines* (p. 508), *blade* (p. 508), *tongue* (p. 508), *backed* (p. 513), *dropped* (p. 513), *cripple jack rafter* (p. 522), *valley cripple jack* (p. 522), *hip valley cripple jack* (p. 522), *common difference* (p. 523).

Review Key Concepts

3. **Demonstrate** how to make hip rafter side cuts.

4. **Define** a valley rafter.

5. **Explain** how to calculate rafter overhang.

6. **List** at least two reasons to add a dormer to a roof.

7. **Demonstrate** how to construct a jack rafter.

8. **Explain** why the intersection of two roofs calls for more complex framing.

Critical Thinking

9. **Explain** Can the actual length of a rafter be taken from the framing plan? Explain your reasoning.

Academic and Workplace Applications

STEM Mathematics

10. Pythagorean Theorem The right triangle is the basis for many roof structures. Understanding how to use the Pythagorean theorem will help you in your construction career. Find the length of the hypotenuse, c, of $\triangle XYZ$.

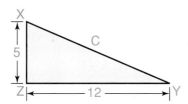

Math Concept The Pythagorean theorem states that the sum of the squares of the sides of a right triangle are equal to the square of the hypotenuse. The hypotenuse is the side of a right triangle opposite the right angle. The formula is $a^2 + b^2 = c^2$.

Step 1: Substitute known lengths for a and b.

Step 2: Square the two known lengths.

Step 3: Find the sum of the squares of the two legs (the base and the altitude).

Step 4: Take the square root of the sum to solve for c.

STEM Engineering

11. Triangles The triangle is used often in architecture and design. Unlike the shapes of a square or a rectangle, the shape of a triangle is rigid. This means that the shape of a triangle cannot be changed without changing the length of one of its sides or breaking one of its joints. A single truss between two diagonal corners strengthens a square or rectangle by turning it into two triangles. Find three instances of triangles used in design, such as in buildings, bridges, or other structures. You can use structures in your neighborhood or pictures of structures found elsewhere. Describe the structure. Try to determine if the triangles in the structure are used to decorate or to support the structure. Summarize your findings in a one-page report. Include pictures or drawings.

21st Century Skills

12. Career Skills: Investigate Roofing Careers Interview two carpenters with roofing experience. Ask them to recall their experiences as they were learning to frame roofs with hip and valley rafters. Ask how they learned to do this work accurately. How much practice did it take? What math skills did they use? What tools did they find helpful? Take notes during your interview. Summarize the interview in a one-page document.

Standardized TEST Practice

CERTIFICATION PREP

Multiple Choice

Directions Choose the best answer for each of the following questions:

13. The formula you would use to calculate the length of a hypotenuse is _____.

   a. $A = bh$ **c.** $A = 2\pi r$

   b. $a^2 + b^2 = c^2$ **d.** $y = ax + b$

14. In mathematics, a right angle is equal to _____.

   a. $60°$ **c.** $180°$

   b. $90°$ **d.** $45°$

15. The shortening allowance of a hip jack rafter at the upper end is _____.

   a. ½ the 45° thickness of the hip rafter

   b. ¼ the 45° thickness of the hip rafter

   c. ½ the thickness of the long rafter

   d. ¼ the thickness of the long rafter

TEST-TAKING TIP

Science and mathematics require you to memorize the relationships between groups of numbers as well as specific numbers. Using these relationships in real-world applications, such as measurement, may help you be able to recall formulas and special numbers.

*These questions will help you practice for national certification assessment.

Roof Assembly & Sheathing

Chapter Objectives

After completing this chapter, you will be able to:

- **Identify** the two basic types of ridges.
- **Calculate** ridge length.
- **Create** the ridge layout for gable roofs, hip roofs, addition roofs, and dormers.
- **Identify** different types of cornice construction and name the parts.
- **Assemble** a simple box cornice.
- **List** the basic requirements for the placement and nailing of panel roof sheathing.

Discuss the Photo
Roofing Safety must always be a concern when working on a roof. *What is the biggest danger?*

Writing Activity: Quick Write
Roof assembly includes laying out the rafters, erecting the ridge board, and erecting the rafters. Each step is usually done by two or more carpenters. Write a paragraph describing why you think most steps in roof assembly would require more than one person.

Terry J Alcorn/Getty Images

Chapter 19 Reading Guide

Before You Read Preview

Scan the section headings, subheadings, and illustrations in this chapter. Write a question that you expect to be answered within the chapter. When you find the answer to that question, write it down.

Content Vocabulary

- ridge
- ridge beam
- collar tie
- purlin
- brace
- common difference
- eaves
- cornice
- fascia
- soffit
- lookout
- rake

Academic Vocabulary

You will find these words in your reading and on your tests. Use the academic vocabulary glossary to look up their definitions if necessary.

- intermediate
- suspended
- version

Graphic Organizer

As you read, use a chart like the one shown to organize the steps in roof assembly and details about each step, adding rows as needed.

Steps in Roof Assembly	Details
Step 1:	
Step 2:	
Step 3:	
Step 4:	

Go to connectED.mcgraw-hill.com to download this graphic organizer.

Types of Ridges

How would a roof's strength change if it had no ridge?

A **ridge** is a roof framing member placed at the intersection of two upward-sloping surfaces. Carpenters may install a ridge in various ways. In most cases they will cut the rafters first. Laying out and cutting common rafters is discussed in Chapter 17.

There are two basic types of ridges: nonstructural and structural. The type of ridge is indicated on the building plans. A *nonstructural ridge* does not support the weight of the rafters or the roof. In fact, it is the rafter system that holds a non-structural ridge in place. This ridge serves as a bearing surface for the top ends of the rafters and helps to tie them into a rigid structure. In contrast, a *structural ridge* actually supports some of the weight of the roof system. The loads it carries must be transferred to the foundation, either by structural posts or by bearing walls.

The framing member that forms a nonstructural ridge is called a *ridge board.* This type of ridge is the most common. It is usually made of nominal 2" lumber that is slightly wider than the rafter stock. For example, the ridge board for a roof framed with 2×8 rafters would be a 2×10. The extra width ensures that angled cuts at the ends of the rafters will bear fully on the ridge board. However, code also allows nominal 1" lumber to be used as a ridge board. A ridge board can also be made from a continuous length of LVL stock. In any case, the thickness of the ridge stock must be accounted for when calculating the actual length of the rafters.

The framing member that forms a structural ridge is called a **ridge beam**. A ridge beam is made from LVL, glue-laminated lumber, or nominal 4" lumber. The rafters

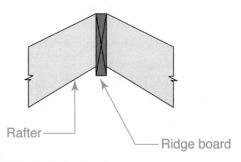

Rafter

Ridge board

 Ridge Board
One Size Up A ridge board must be wider than the rafters that support it.

rest on top of the ridge beam or are supported by metal brackets or hangers nailed to its side. The ends of the ridge beam are supported by posts or bearing walls. **Intermediate** support posts may also be needed. A structural ridge is commonly used when framing low-pitched or shed roofs or when the house is framed using posts and beams (see Chapter 14).

Whether installing a structural or nonstructural ridge, the stock should be as long and straight as possible. If the ridge is bowed, twisted, or warped, this will create a lot of extra work for the roof framing crew. When a lumber ridge board is used, it can

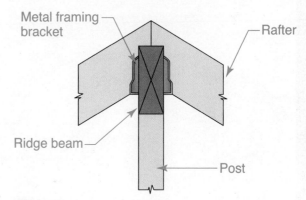

Metal framing bracket

Rafter

Ridge beam

Post

 Ridge Beam
Two Options A ridge beam supports rafters. They rest on top of the ridge or are notched to fit over it.

be the same grade of lumber as the rafters. An LVL ridge board is used when the roof is framed with engineered lumber. Seams between lengths of ridge board should occur only between opposing pairs of rafters. Seams between lengths of ridge beam should occur only over support posts.

Reading Check

Contrast *What is the difference between a nonstructural ridge and a structural ridge?*

Calculating Ridge Length

How do you determine ridge length?

The following text refers to solid-lumber ridge boards. However, the information also applies to engineered-lumber ridge boards and ridge beams.

Gable Roofs

Calculating the length of the ridge board for a gable main roof is easy. The theoretical length of the ridge board (or ridge beam) is equal to the length of the building, measured to the outside edge of the wall framing. The actual length of the ridge board includes any overhang.

Hip Roofs

For a main hip roof, the ridge board layout requires calculations. In an equal-pitch hip roof, the theoretical length of

the ridge board amounts to the length of the building minus twice the total run of a main roof common rafter. The actual length depends on the way in which the hip rafters are framed to the ridge.

The theoretical ends of the ridge board are at the points where the ridge centerline and the hip rafter centerline cross. If the hip rafter is framed against the ridge board, the actual length of the ridge board exceeds the theoretical length, at each end, by one-half the thickness of the ridge board plus one-half the 45° thickness of the hip rafter. If the hip rafter is framed between the common rafters, the actual length of the ridge board exceeds the theoretical length, at each end, by one-half the thickness of a common rafter.

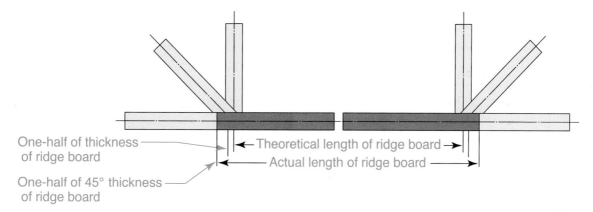

One-half of thickness of ridge board

One-half of 45° thickness of ridge board

←Theoretical length of ridge board→

←——— Actual length of ridge board ———→

Ridge Board Length

Against the Ridge Theoretical and actual lengths of a hip roof ridge board. In this case the hip rafter is framed against the ridge.

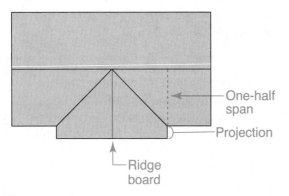

One-half span

Projection

Ridge board

 Equal Span
Note Shortening Allowance Determining the length of a ridge board for an equal-span addition.

Equal-Span Additions

For an equal-span addition, the length of the ridge board is equal to the distance that the addition projects beyond the building, plus one-half the span of the building, minus the shortening allowance at the main-roof ridge. The *shortening allowance* accounts for the thickness of the main-roof ridge board when determining the length of an intersecting ridge. It is different for different framing situations. For an equal-span addition, it equals one-half the thickness of the main-roof ridge board.

Reading Check

Explain How is the length of the ridge board for a gable main roof calculated?

Unequal-Span Additions

When the width of an addition is less than the width of the main portion of the house, their roof spans are unequal. The length of the ridge board for an unequal-span addition varies with the method of framing the ridge board. If the addition ridge board is **suspended** from the main roof ridge board, the length is equal to the distance the addition projects beyond the building, plus one-half the span of the main roof.

Builder's Tip

VERIFYING DIMENSIONS The length of the ridge board or ridge beam can be taken from the building plans. However, a carpenter should always confirm this dimension by measuring the actual framing. This will account for any minor differences between the house as planned and the house as built.

If the addition ridge board is framed by the long-and-short-valley-rafter method (see Addition Roofs on page 536), its length is equal to the distance the addition projects beyond the building, plus one-half the span of the addition, minus a shortening allowance. In this case, the shortening allowance is one-half the 45° thickness of the long valley rafter.

If the addition ridge board is framed to a double header set between a pair of doubled main-roof common rafters, the length of the ridge board is equal to the distance the addition projects beyond the building, plus one-half the span of the addition, minus a shortening allowance. This shortening allowance is one-half the thickness of the inner member of the double header.

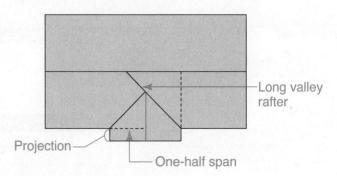

Long valley rafter

Projection

One-half span

 Unequal Span
Long Valley Determining the length of a ridge board for an unequal-span addition.

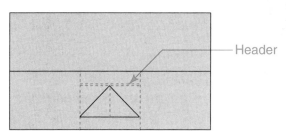

 Dormer Without Side Walls
Header Determining the length of a ridge board on a dormer without side walls.

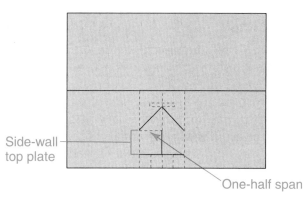

Side-wall top plate

One-half span

 Dormer With Side Walls
Top Plate Determining the length of the ridge board on a dormer with side walls.

Dormers

The length of the ridge board on a dormer without side walls is equal to one-half the span of the dormer, minus a shortening allowance. The shortening allowance is one-half the thickness of the inner member of the upper double header.

The length of the ridge board on a dormer with side walls is equal to the length of the dormer side-wall top plate, plus one-half the span of the dormer, minus a shortening allowance. The shortening allowance is one-half the thickness of the inner member of the upper double header.

Mathematics: Measurement

Ridge Board Length What is the length of a ridge board on a 12' dormer without sidewalls and a 2" shortening allowance?

Starting Hint Draw a diagram to help you visualize the measurements.

Section 19.1 Assessment

After You Read: Self-Check

1. What is the difference between a ridge board and a ridge beam?
2. Why must a ridge board be wider than the rafters?
3. Based on what you have read, why is the ridge board for a hip roof shorter in length than the ridge board for a gable roof?
4. How do you calculate the length of the ridge board for an equal-span addition?

Academic Integration: English Language Arts

5. **Length of a Ridge Board** The length of the ridge board on a dormer without a sidewall is different than the length of the ridge board on a similar dormer with sidewalls. Write a paragraph explaining why there is a difference in length.

Go to connectED.mcgraw-hill.com to check your answers.

Roof Assembly Steps

Who does roof assembly?

Roof assembly includes laying out the rafters, erecting the ridge board, and erecting the rafters. This is generally a job for two or more carpenters and at least one helper. The order of steps may vary, depending on the type and complexity of the roof. However, work generally proceeds in this order:

1. Install the common rafters and ridge boards.
2. Install hip and valley rafters, if any.
3. Install jack rafters, if any.
4. Frame special items such as gable ends and roof openings.
5. Install roof sheathing.

Laying Out Rafter Locations

Where can rafter spacing be found?

Laying out the locations of common rafters is much like laying out the locations of floor joists. However, other roof members may make the layout more complex.

The rafter spacing on the wall plates and ridge board is found on either the building plans or the roof framing plan (see Chapter 17). Rafter locations are laid out on plates, the ridge board, and other rafters with the same lines and Xs used to lay out stud and joist locations (see Chapter 16).

In some cases all the rafters are located next to the ceiling joists. The rafters can then be fastened to the side of the joists as well as to the plate in order to tie the building together. In most cases, however, some rafters will be next to joists and others will rest between the joists. This is because the on-center spacing of the joists is often different from the on-center spacing of the rafters.

Gable Roofs

For a gable roof, lay out the rafter locations on the top plates first. Transfer the locations to the ridge board by laying the ridge board on edge against a top plate and matching the marks.

The first rafters on each end are usually set even with the outside wall to provide a smooth, unbroken surface for the wall sheathing. Because the first ceiling joist is along the inside edge of the exterior wall, place a spacer block between the first rafter and the first ceiling joist. Fasten the other rafters to the side of the joists along the length of the building.

If the rafters are on 24" centers and the ceiling joists are on 16" centers, place the first rafter. The second rafter will rest on the plate between the second and third joists. Nail the third rafter to the side of the fourth joist. The rafters will continue to alternate in this fashion along the length of the building.

Sometimes the rafters and the ceiling joists will have the same on-center spacing.

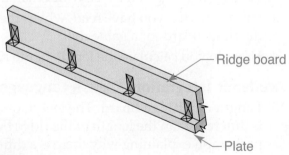

Ridge board

Plate

 Ridge Board Layout
Transfer the Layout Lay the ridge board on edge on the top plate and extend the layout lines from the plate onto the ridge board.

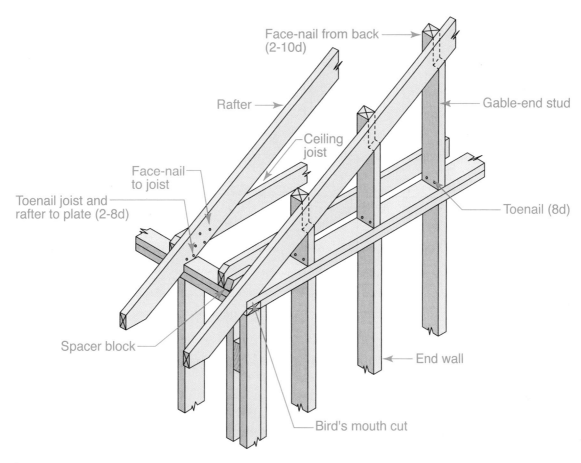

Face-nail from back (2-10d)

Rafter

Gable-end stud

Ceiling joist

Face-nail to joist

Toenail joist and rafter to plate (2-8d)

Toenail (8d)

Spacer block

End wall

Bird's mouth cut

 Rafter Positions

Gable End Details Note the spacer block. Sometimes the gable-end studs are cut all the way across, rather than notched. The gable-end studs are then toenailed to the rafter.

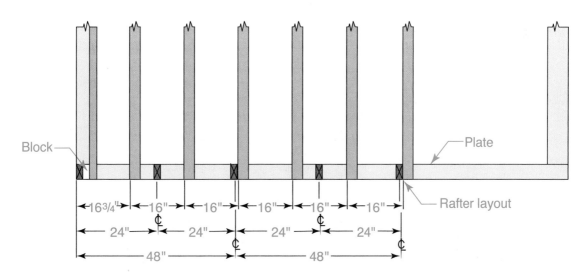

Block

Plate

16³/₄"   16"   16"   16"   16"   16"

24"   24"   24"   24"

48"   48"

Rafter layout

 Alternating Spacing

Rafters and Joists Layout of a building with the rafters on 24" centers and the ceiling joists on 16" centers.

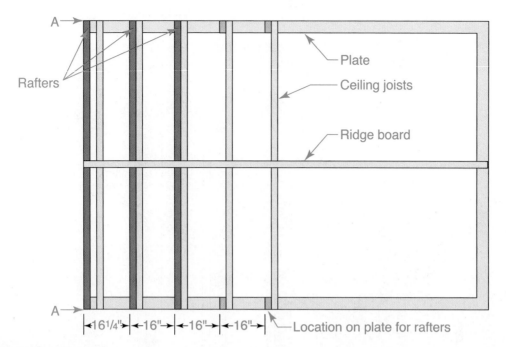

Matching Spacing

Accuracy Improves Always begin the layout of opposing rafters from the same end of the building. In this drawing, the layout for each phase began at arrow A on the same side wall.

In that case, the layout would be as shown in Matching Spacing. In any case, always begin the rafter layout for opposing plates from the same end of the building. This will ensure that the rafters butt against the ridge board directly opposite each other.

Hip Roofs

The ridge-end common rafters in an equal-pitch hip roof are located inward from the building corners at a distance equal to one-half the span (or the run of a main-roof common rafter). The locations of these ridge-end rafters and the common rafters lying between them can be transferred to the ridge board by matching the ridge board against the top plates.

Addition Roofs

An addition complicates the process of laying out the locations of the rafters and ridges. Study the following drawings carefully.

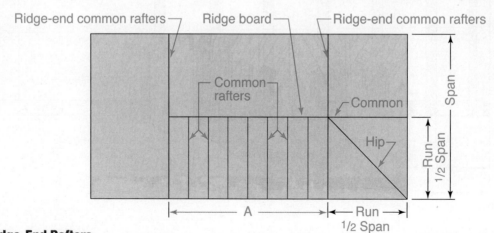

Ridge-End Rafters

Transfer the Layout The locations of the rafters in area A are transferred to the ridge board from the top plate.

Equal Spans For an equal-span addition, mark the main ridge board to indicate where it will be intersected by the addition ridge board. The top ends of the addition's valley rafters will rest on either side of this location. In Gable Roof Addition, the distance between the end of the main-roof ridge board and the point where it intersects the addition ridge board is equal to distance A plus distance B (distance B equals one-half the span of the addition). In Hip Roof Addition, the distance between the theoretical end of the main-roof ridge board and the point where it intersects the addition ridge board is the same as distance A.

Unequal Spans If framing is by the long-and-short-valley-rafter method, the distance from the end of the main-roof ridge board to the upper end of the longer valley rafter is equal to distance A plus distance B (distance B is one-half the span of the main roof). See Unequal Span Addition.

The intersection of the shorter valley rafter and the longer valley rafter can be located in the following way. Obtain the unit length of the longer valley rafter from the rafter table on the framing square shown in Framing Square. For example, suppose that the common rafter unit rise is 8". In that case, the unit length of a valley rafter is 18.76".

The total run of the longer valley rafter is the hypotenuse of a right triangle. The shorter sides of this triangle are each equal to the total run of a common rafter in the addition. The total run of a common rafter in the addition is one-half the span. If the addition is 20' wide, the run of a common rafter would be 10'. Refer to distance C in Unequal Span Addition.

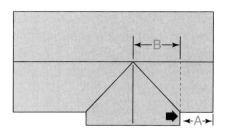

Gable Roof Addition
Mark the Intersection Ridge board location for an equal-span addition on a gable roof.

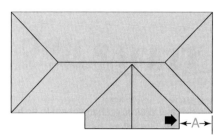

Hip Roof Addition
Check the Distance Ridge board location for an equal-span addition on a hip roof.

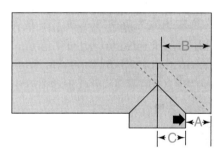

Unequal Span Addition
Long and Short Valleys Ridge board and valley rafter locations for an unequal-span addition.

Framing Square
Useful Table To find the unit length of the longer valley rafter, check the rafter table on the face of the framing square.

The valley rafter in our example is 18.76" long for every foot of common rafter run. The point where the inboard end of the shorter valley rafter intersects the longer valley rafter can be calculated as follows: 18.76 (in. per ft. of run) × 10 (ft. of run) = 187.6" 187.6" = 15.63' (15' 7⁹⁄₁₆"). This is the distance from the heel plumb cut line of the longer valley rafter to the intersecting point.

Assembling the Ridge

Why must the ridge be straight and level?

Many carpenters raise the ridge board and the gable-end rafters all at one time because the members support one another. Other carpenters prefer to put the ridge board in place before raising any rafters. To do this they support it with temporary framing. The framing rests on any partition top plates that are handy. The ridge board should also be braced along its length to prevent it from swaying.

Reading Check

Recall *Name two strategies used by carpenters for placing the ridge board.*

Raising the Rafters

What weather would make rafter installation hazardous?

Depending on the type and height of the roof, you may have to install the rafters while working from scaffolding. The scaffold planking should not be less than 4' below the level of the ridge board. In some cases, it may be possible to work from ladders instead. (See Chapter 7 for ladder safety guidelines.)

If the building has an addition, frame as much of the main roof as possible before starting the addition roof framing. All types of jack rafters are usually left until after the headers, hip rafters, valley rafters, and related ridge boards have been installed.

The following text describes standard assembly techniques. Other techniques may be required in areas of the country exposed to unusually high winds and seismic activity. In these areas, local building codes often require the addition of special metal anchors, straps and hold-downs to connect the roof framing to the wall framing. These anchors must be installed with care. Follow all code requirements regarding the type, spacing, and number of nails used to secure these devices.

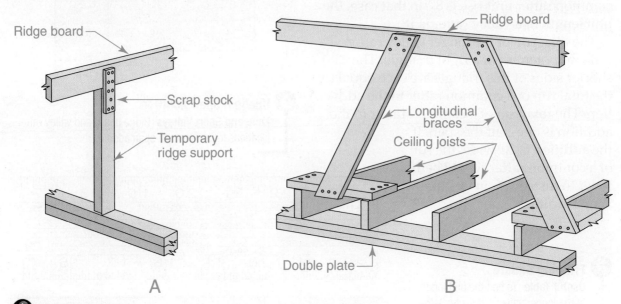

Ridge Supports
Temporary Support A. An upright (leg) supports the ridge board in position for erecting the rafters. **B.** Brace the ridge board as needed along its length.

Gable Roofs

For a gable roof, the two pairs of gable-end rafters and the ridge board are usually erected first. Two people, one at each end of the building, hold the ridge board in position. Meanwhile, a third person sets the gable-end rafters in place and toenails them at the top plate with 8d nails, two on one side and one on the other side. Nailing at the plate first prevents the rafter from slipping out of position as the ridge is being installed. Make certain the heel (plumb) cut of the bird's mouth is tight against the side of the building when the rafter is nailed at the plate. Otherwise, the ridge will not be set at the correct height.

The ridge board is then secured to one of the rafters with three 16d nails driven through the ridge board and into the end of the rafter. The opposing rafter is toe-nailed to the ridge board with four 16d nails, two on each side of the rafter. If the ridge board has not been previously erected and braced, temporary braces like those for a wall should be installed at the ridge ends. These will prevent the rafters from tipping from side to side. Ceiling-joist ends are nailed to adjacent rafters with three 10d nails, two to each side. Nailing continues in this fashion until all the rafters are in place.

Hip Roofs

On a hip roof, first install the ridge board and the common rafters extending from the ridge ends to the side walls. This is done in about the same manner as for a gable roof. Then fill in the intermediate common rafters. Next, install each common rafter that extends from the ridge end to the midpoint on the end wall. Do this for both end walls. These rafters are sometimes referred to as end rafters. Finally, install the hip rafters and hip jacks.

The common rafters in a hip roof do not have to be plumbed. If the hip rafters are correctly cut, installing the hip rafters and the common rafter that projects from the end of the ridge board to the end wall will make the common rafters plumb.

Toenail hip rafters to the plate with 10d nails, two to each side. At the ridge board, toenail hip rafters with four 8d nails. After the hip rafters are fastened in place, drive a nail partway into the top edge of the hip rafter at the ridge end and at the plate end. Pull a string taut between the nails as the hip jacks are nailed to the hip rafter. Keep it centered on the top edge of the hip rafter. This allows you to see if the hip rafter is being pushed out of alignment by the jacks and ensures a straight hip line.

The hip jacks should be nailed in pairs, one opposite the other. Do not nail all the jacks on one side of the hip first. This would push the hip out of alignment and cause it to bow. Toenail hip jacks to hip rafters with l0d nails, three to each jack, and to the plate with 10d nails, two to each side.

Nailing Rafters to the Ridge
The First Rafters Help Once the first pairs of rafters are nailed to the ridge board, the remaining rafters will be easier to position.

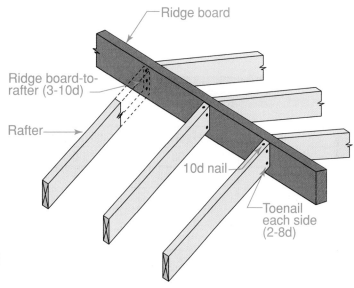

Ridge board

Ridge board-to-rafter (3-10d)

Rafter

10d nail

Toenail each side (2-8d)

Estimating and Planning

This estimating and planning exercise will prepare you for national competitive events with organizations such as SkillsUSA and the Home Builder's Institute.

Roofing Framing Materials

Estimating Materials

Estimating lumber and nails for a roof can be done in several ways.

Method 1

The number of rafters needed may be counted directly from the roof framing plan. For a gable roof, the number may also be estimated as follows:

Step 1 For rafters on 16" centers, take three-fourths of the building's length in feet, add one for the end rafter, and then double this figure. For example, if a rectangular building is 40' long, 31 rafters will be required for each of the longer sides:

$$\frac{3}{4} \times 40 = 30$$
$$30 + 1 = 31$$
$$31 \times 2 = 62$$

A total of 62 rafters would be needed.

Step 2 Add to this amount extra rafters for the required trimmers and any other special framing. An accurate cost estimate can then be figured by multiplying the number of rafters required by the cost per rafter.

Method 2

Sometimes a builder does not make up a complete bill of materials and needs only a rough cost estimate. This method will give an approximate cost for materials.

Step 1 Find the area of the roof. The area is the length times the width of each roof plane. Add the areas of each roof plane to find the total area of the roof.

Step 2 Refer to the table to determine the number of board feet needed for rafters, ridge board, and collar ties. For example, if the rafters are 2×6 spaced 16" OC, 102 bd. ft. of lumber are needed for each 100 sq. ft. of roof surface area.

Step 3 Divide the total roof area by 100 and multiply by the factor in the table. For our example:

$$867 \div 100 = 8.67$$
$$8.67 \times 102 = 884.3 \text{ bd. ft.}$$

Step 4 Multiply this figure by the cost of the lumber per board foot to find the total cost of lumber for the roof.

Step 5 The table can also be used to determine the number of nails needed. For the roof in the example, 12 lbs. of nails are needed for each 1,000 bd. ft. Since the roof in the example has only about 884 bd. ft., it will require about 10½ lbs. of nails:

$$884 \div 1,000 = 0.884$$
$$0.884 \times 12 = 10.6 \text{ or } 10\frac{1}{2} \text{ lbs. of nails}$$

Step 6 The cost of nails for roof framing is determined by multiplying the number of pounds needed by the cost per pound.

Estimating on the Job

Using Method 2, estimate the board feet and pounds of nails needed for a roof that measures 22' wide and 37' long and has a rise of 6". There is no overhang, and 2×8 rafters will be placed 16" OC.

Estimating Materials and Labor for Roof Framing					
MATERIALS					**LABOR**
	Board Feet Required for 100 Square Feet of Surface Area			**Nails per 1,000 Board Feet**	**Board Feet per Hour**
Rafters	**12" OC**	**16" OC**	**24" OC**		
2×4	89	71	53	17	Common-35
2×6	129	102	75	12	Hip-35
2×8	171	134	112	9	Jack-25
2×10	212	197	121	7	Valley-35
2×12	252	197	143	6	Ridge-35
					Collar-65

Note: Includes common rafters, hip and valley rafters, ridge boards, and collar ties.

Additions and Dormers

When there is an addition or dormer, the valley rafters are usually erected first. Toenail them to ridge boards and headers with three 10d nails. Install the ridge boards and ridge-end common rafters next. Then install the other addition common rafters and, last, the valley and cripple jacks. As with hip rafters, pull a string along the top edge of the valley rafter and nail the jacks in pairs. A valley jack calls for special attention as it is being nailed. When the jack has been properly positioned, the end of a straightedge laid along its top edge should contact the centerline of the valley rafter.

Using Roof Framing Brackets

Metal brackets may be used to attach common rafters to the plate. In parts of the country that are regularly exposed to high winds, the lower rafter brackets should extend from the rafter and connect to a wall stud wherever possible. This strengthens the connection between the roof system and the walls.

When using certain types of roof framing brackets, it is sometimes helpful to install them on the ridge board before it is lifted into place. This technique is also suitable when a ridge beam is being used. Brackets may be attached to the wall plates as well before the rafters are lifted into place. However, care must be taken not to bend or otherwise damage the brackets. Also, protect your hands. The edges of metal brackets can be sharp.

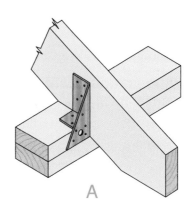

A

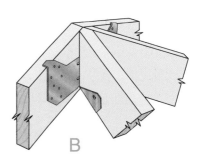

B

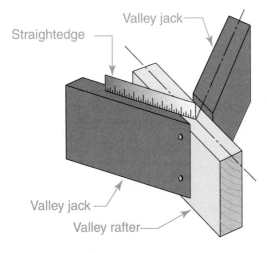

 Aligning the Jacks
Align to the Centerline Correct position for nailing a valley jack rafter.

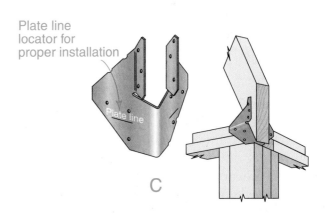

C

 Roof Framing Brackets
Strong Connections Metal brackets are often used to secure the rafters and are available in many shapes. **A.** A simple rafter bracket. **B.** An adjustable-pitch bracket. **C.** A bracket to anchor the end of a hip rafter.

After You Read: Self-Check

1. Describe the procedure for laying out rafter locations for a gable roof.
2. When nailing common rafters in place in a gable roof, why must the rafter be nailed at the plate first?
3. How should the ends of the ceiling joists be connected to the rafters in a gable roof?
4. Why must hip jack rafters be installed in pairs?

Academic Integration: Mathematics

5. **Converting Fractions to Decimals** Estimate the number of rafters on 16" centers needed for a gable roof on a rectangular building measuring 48' long.

   Math Concept It is often easier to perform mathematical operations if you convert fractions to decimals. Memorize the decimal equivalents of common fractions, such as ¾ = 0.75.

   Step 1: Figure ¾ the building's length in feet.
   Step 2: Add one for the end rafter.
   Step 3: Multiply by two to get the total number of rafters that will be needed.

    Go to connectED.mcgraw-hill.com to check your answers.

Section 19.3

Special Framing Details

Collar Ties

What is a collar tie used for?

Rafters in a gable roof are sometimes reinforced by collar ties. A **collar tie** is a horizontal framing member that prevents opposing rafter pairs from spreading apart. It also prevents the rafters from bowing inward when weight is placed upon them. In a finished attic, collar ties may also support the ceiling surfaces where the ceiling joists have been omitted, or where ceiling joists run perpendicular to the rafters. When ceiling joists tie opposite walls together, collar ties may not be required.

If the collar ties will support a ceiling, they should be installed at every rafter pair. Otherwise, attach a collar tie to every fourth rafter pair if the spacing is 16" OC and every third rafter pair if the spacing is 24" OC. Local codes may require a closer spacing.

Collar-Tie Length

A collar tie may be made of nominal 1" or 2" thick lumber. Check the building plans for the specified dimensions. The length of a collar tie can be found either by calculation or by measurement.

Calculation Method This method is used when collar tie framing must be done at a precise

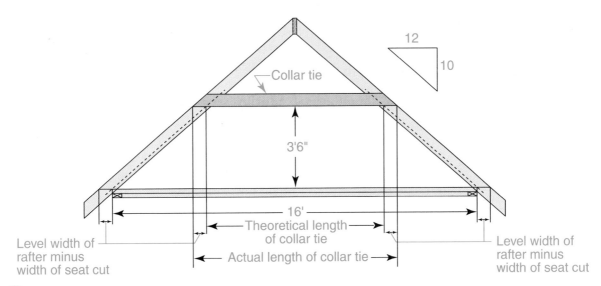

 Calculating Rafter Tie Length
Find the Height Laying out a collar tie based on calculations.

height. This is often the case when the collar ties will form the base for a finished ceiling surface. The length of a collar tie can be calculated based on its distance above the level of the side-wall top plates. The theoretical length of a tie in feet is found by dividing this distance in inches by the unit rise of a common rafter and subtracting twice the result from the span of the building.

In the roof shown in Calculating Rafter Tie Length, the collar tie is 3'6" (42") above the top plate. The unit rise of a common rafter in the roof is 10". Forty-two divided by 10 is 4.2, and twice 4.2 is 8.4. This number is subtracted from the span of the building: $16 - 8.4 = 7.6'$, or about 7'–7³⁄₁₆". This is the theoretical length of the tie.

To bring the ends of the collar tie flush with the upper edges of the common rafters, you must add to the theoretical length of the tie, at each end, an amount equal to the level width of a rafter minus the width of the rafter seat cut. One way to obtain the level width is to hold a framing square on the rafter set to the pitch of the roof. You then draw a level line from edge to edge and measure the line's length.

Measurement Method Collar ties are sometimes used only for structural purposes. In such cases, the length of the collar tie can

be easily determined by measuring. Simply measure between the rafters on a level line, starting from the height noted in the building plans. Cut one collar tie and check its fit before cutting all the collar ties to length.

After the overall length of a collar tie is determined, the ends must be cut to the pitch of the roof to prevent the tie from getting in the way of roof sheathing. Lay out the end cuts with a framing square set to the pitch of the roof. These cuts can be made with a circular saw, radial-arm saw, or slide-compound miter saw.

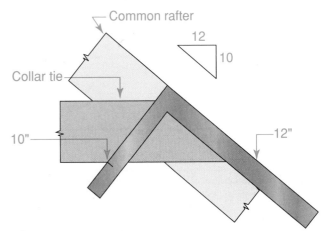

 Cutting Rafter Tie Ends
Determine the Angle Laying out the end cut on a collar tie for a roof with a unit rise of 10".

Installing Collar Ties

Collar ties must be aligned during installation to ensure that their lower edges are in the same plane. First, snap a chalk line across the rafters on one side of the house, indicating the desired height of the top or bottom edge of the collar tie. Then install one tie at each end of the house by aligning one end to the chalk line and using a level to align the other end. Nail the ties into place. Now stretch a string tightly between the ties. Align the remaining ties to the chalked line and to the string. Nail nominal 1" collar ties to the common rafters with four 8d nails in each end. Nail nominal 2" collar ties with three 16d nails at each end.

Purlins, Braces, & Gable Ends

What does a purlin do?

Carpenters may have to install various types of framing to provide extra support for rafters. At each end of a gable roof, framing must be installed to close in the gable end and provide a nailing surface for sheathing.

Purlins and Braces

To span a greater distance, a rafter must have a greater depth. However, deeper rafters are not always desirable or available. A system of purlins and braces can be used instead. A **purlin** is a horizontal structural member that supports roof loads and transfers them to structural supports. A **brace** is a member used to stiffen or support a structure.

Purlins should be no smaller than the rafters they support. They must be continuous between braces. Braces should be at least 2×4 stock. They should connect to bearing walls at no less than a 45° angle. They should be no longer than 8" and be spaced not more than 4' OC. Braces should not be connected to non-bearing walls.

Gable Ends

Wall studs must be installed at each end of a gable to support sheathing. These studs rest on the top plate and extend to the rafter line. There are various ways to install them. In the case of a gable end with no overhang as shown in Non-Structural Gable End, the gable end studs are not load bearing framing members. The gable end rafters are self supporting and would carry the roof loads whether or not the studs were in place. When the gable end will have an overhang, as in Structural Gable End, the gable end must support roof loads. The studs must be capped with a wall plate. In either case, gable-end studs should be installed like standard wall studs: one edge should be flush with the outside wall.

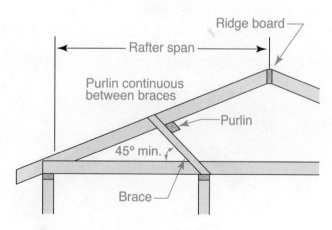

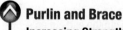
Purlin and Brace
Increasing Strength A system of purlins and braces can be used to provide additional support for rafters.

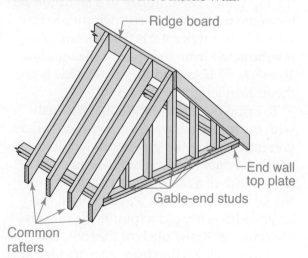

Non-Structural Gable End
Filler Studs The only purpose for these studs is to provide support for roof sheathing, so this is not a bearing wall.

Carpenters have different preferences for how the tops of non-structural gable-end studs connect to the gable-end rafters. Some notch the studs to fit around the rafters, as shown A in Gable-End Studs. This allows the studs to be toenailed as well as face nailed to the rafter. Another approach is to bevel the top ends of the studs to fit the slope of the rafters, as B in Gable-End Studs.

Layout and Installation To install studs of the type shown A in Gable-End Studs, proceed as follows. Similar techniques, without the notch, can be used to lay out gable-end studs with a beveled end.

Locate the first gable-end stud by making a mark on the double plate directly above the wall stud nearest the ridge line (see arrow A in Gable-End Stud Layout). Plumb the gable-end stud on this mark. Mark the pitch of the roof across the edge of the stud (see arrow B).

Now determine the length of the stud. (It must not extend above the top edge of the rafter.) Cut the stud to length and notch it to a depth matching the thickness of the rafter (see arrow C). Toenail it into place with three 8d or two 16d nails at each end. As you nail the studs into place, take care not to force a crown into the rafter.

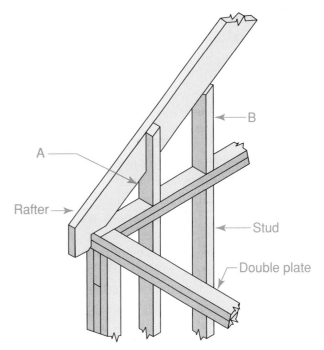

Gable-End Studs
Two Methods A. Some carpenters notch the studs to fit over the rafter. **B.** Others prefer to bevel the ends.

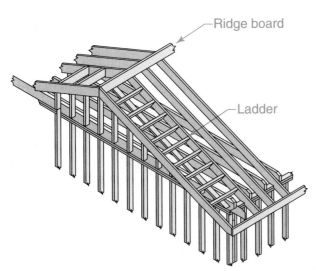

Structural Gable End
Bearing Studs These studs support roof loads, so the gable end is a bearing wall. They have the same spacing as studs in the wall below.

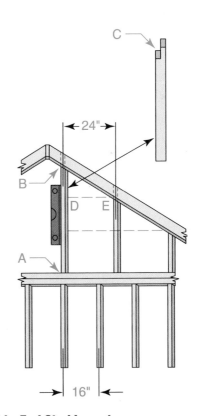

Gable-End Stud Layout
Making Them Fit Locating the gable-end studs and determining the common difference in length.

All remaining studs can be sized using this method. However, it is much easier to calculate stud lengths by using the common difference method. The basic calculation technique is the same, whether the studs are notched or just bevel cut.

Calculating the Common Difference

Gable-end studs have the same on-center spacing as standard wall studs. However, each stud is a different length than the studs on either side. Their differences in length are based on a single figure that depends on the pitch of the roof. This figure is called the **common difference**. After you have determined the length of the tallest gable-end stud, you can subtract the common difference to find the length of all the shorter

gable-end studs. This is faster than making individual measurements for each stud.

The common difference is calculated using the unit run and unit rise. For example, to find the common difference in the length of gable-end studs placed 24" OC:

$$24" \div 12" \text{ (unit run)} = 2$$
$$2 \times 6" \text{ (unit rise)} = 12"$$

A common difference of 12" means that the second stud will be 12" shorter than the first (tallest) stud. The third stud will be 12" shorter than the second stud, and so on. If the studs are spaced 16" OC for the same roof, the common difference is 8":

$$16" \div 12" \text{ (unit run)} = 1.333$$
$$1.333 \times 6" \text{ (unit rise)} = 8"$$

Steps for finding the common difference using a framing square are given below.

Step-by-Step Application

Figuring the Common Difference Using a Framing Square The common difference in the length of the gable-end studs may also be figured directly with the framing square.

Step 1 Place the framing square on the stud and set it to the unit rise and unit run of the roof (6 and 12 for this example). Draw a line along the blade at A, as shown in the figure below.

Step 2 Slide the blade along this line in the direction of the arrow at B until the spacing between the studs

(16 for this example) is at the intersection (C) of the line drawn at A and the edge of the stud.

Step 3 Read the dimension on the tongue where it meets the same edge of the stud. This is the common difference (8" for this example) for the gable-end studs.

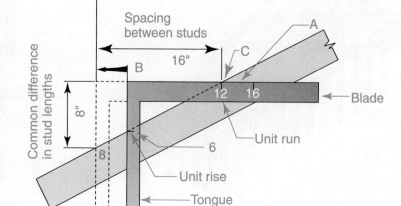

Roof Openings

What common features require roof openings?

Roof openings often interrupt the normal spacing or run of rafters. Openings may be required for a dormer, a chimney, or skylights. Roof openings, like floor openings, are framed by headers and trimmers as shown below. Single or double headers are used at right angles to the rafters. The rafters are set into the headers in the same manner as joists around a floor opening. Just as trimmers are double joists in floor construction, they are double rafters in roof openings.

There are two ways to frame roof openings, as shown Types of Roof Openings on page 548. The headers may be plumb, as shown in part A. This method is used to accommodate vertical objects that must pass through the framing, such as chimneys. In this method, the end of an intersecting rafter must be cut at an angle to fit against the header.

A second method is to keep the headers in the same plane as the surrounding roof framing, as shown in part B of on page 548. Such an opening is easier to install and is sometimes used for skylights. In this case, the end of an intersecting rafter must be cut square to fit against the header.

Reading Check

Contrast What are the two methods for framing roof openings? How are they different?

Shed Dormers

How does a shed dormer differ from a gable dormer?

Dormers are framed after all of the common rafters are in place and a roof opening has been created. The framing of a gable dormer was discussed in Chapter 18. Shed dormers (Framing a Shed Dormer on page 548) will be discussed here.

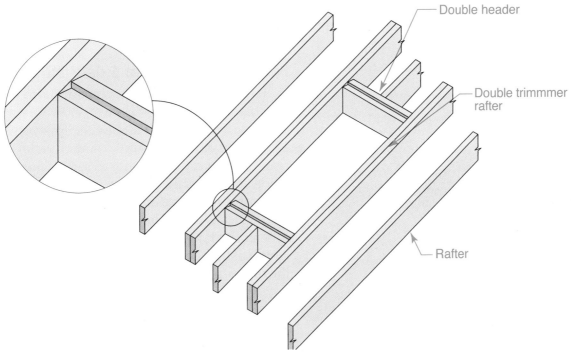

Double header

Double trimmmer rafter

Rafter

 Roof Framing Around a Chimney
Details Are Important The top edges of the headers are kept below the top edge of the rafter. The lower edges of the headers are kept even with the top edge of the rafter.

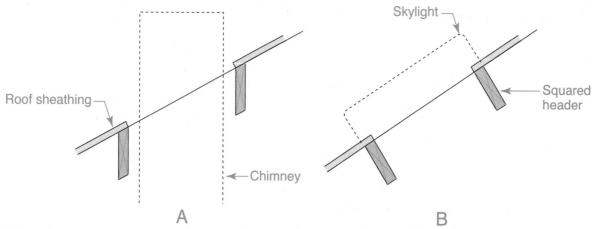

Roof sheathing

Chimney

Skylight

Squared header

A

B

 Types of Roof Openings

Two Methods The two basic methods for framing roof openings. The headers may be single or double, as needed.

Laying Out the Rafters

To determine the total run of a shed-dormer rafter, divide the height of the dormer end wall by the difference in inches between the unit rise of the dormer roof and the unit rise of the main roof. See Rafter Top Cut on page 549. Suppose the height of the dormer end wall is 9', or 108", as in A. The unit rise of the main roof is 8". The unit rise of the dormer roof is 2½". The difference between them is 5½". The total run of a dormer rafter is therefore 108" divided by 5½", which is 19.64". Knowing the total run

 Framing a Shed Dormer

Basic Elements This shows the framing for a small shed dormer. The dormer roof appears flat but is actually pitched to encourage drainage.

and the unit rise, you can figure the length of a dormer rafter by any of the methods already described.

The inboard ends of dormer rafters must be cut to fit the slope of the main roof, as in B. To get the angle of this cut, set a framing square on the rafter to the pitch of the main roof, as in C. Measure off the unit rise of the dormer roof along the tongue, starting at the heel. Make a mark at this point and draw the cut-off line through this mark starting at the 12" mark on the blade.

Finding the Length of Side-Wall Studs

To frame a shed dormer, you must also find the lengths of the side-wall studs. Suppose a dormer rafter rises 2½" for every 12" of run, and a main-roof common rafter rises 8" for every 12" of run, as in A. If the studs are spaced 12" OC, the length of the shortest stud is the difference between 8" and 2½", which is 5½". (This is also the common difference.) If the stud spacing is 16", the length of the shortest stud is the value of x in the proportional equation $12:5½ :: 16:x$. Thus $x = 7\frac{5}{16}$. The shortest stud will be $7\frac{5}{16}$" long. The next stud will be $2 \times 7\frac{5}{16}$" long, or $14\frac{5}{8}$", and so on.

A second method of determining the length of the shortest stud (the common difference) is to make the layout directly on a stud with the framing square, as shown in Direct Layout

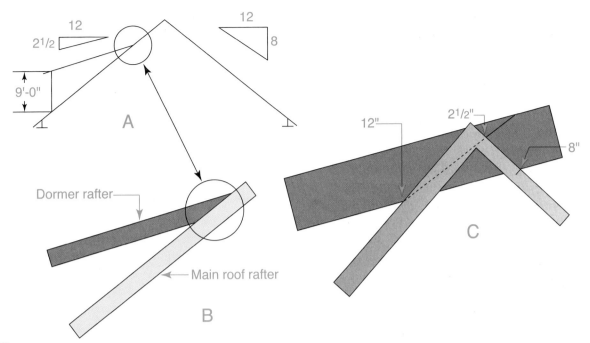

 Rafter Top Cut
Laying Out the Top Cut This diagram shows how to determine the top cut of the dormer rafters where they meet the main roof rafters.

Method. The difference in the rise of the two roofs is 5½". Find the 5½" mark on the tongue of the square and place it on the edge of a stud. Place the blade's 12" mark on the same edge of the stud. Draw a line on the stud along the blade. Slide the square along this line until the blade's 16" mark (the on-center spacing between the studs) is over where the 12" mark had been. Draw a line along the tongue of the square. This completes the layout for the shortest stud. The second stud will be longer by this measure (the common difference), and so on.

To get the lower-end cut-off angle for studs, set the square on the stud to the pitch of the main roof. To get the upper-end cut-off angle, set the square to the pitch of the dormer roof.

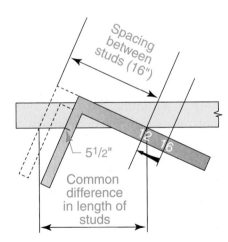

Direct Layout Method
Using the Square Determining the common difference in the length of dormer side-wall studs by direct layout.

Reading Check

Recall How are the lower-end cut-off angle and the upper-end cut-off angle for studs found?

Mathematics: Calculation

Calculate Dormer Side Walls A dormer has a sidewall height of 8' 7" (103") and a unit rise of 2½' on the dormer roof. If the main structure's roof has a unit rise of 10', what is the total run of the dormer rafter?

Starting Hint Find the difference between the unit rise of the main roof and the unit rise of the dormer roof.

Chimney Saddles

Where is the chimney saddle constructed?

A *chimney saddle*, or *cricket*, diverts water around a chimney and prevents ice from building up on the roof behind it. It is a fairly small piece of framing but it can be challenging to build.

The saddle may be constructed while carpenters are on the roof. However, if the chimney span and roof pitch are known, it can also be fabricated on the ground. The completed assembly can then be lifted into position and nailed to the roof framing. There are various methods for building chimney saddles. One method is shown here.

Valley strips for the saddle are 1×4 or 1×6 stock, as shown in Chimney Saddle. The distance across the widest part of the valley strips must be slightly less than the width of the chimney. This accounts for the distance that the saddle sheathing will project beyond the strips (see B). This distance should be estimated by the carpenter. It varies, depending on the slope of the saddle. The length is determined in the same way as for a valley rafter. Use the framing square. Lay out the top and bottom cuts along the tongue of the square. For the length of the strip, use the unit length of a common rafter from the roof on which the saddle is to be framed.

Suppose a roof with a unit rise of 5" has a unit length of 13". To lay out the valley strip, position the square with the tongue's 13" mark and the blade's 12" mark on the edge of the strip. Draw a line along the tongue for the top cut. Measure and lay out the length of the valley strip. With the square set the same as for the top cut, place the edge of the blade on the length mark and draw a line along the blade for the bottom cut.

The end of the saddle's ridge board rests on the valley strips (see A). This cut is the same as the seat cut for a common rafter in the main roof. Place the square on the ridge board for the pitch of the roof (in our example, 5" on the tongue and 12" on the

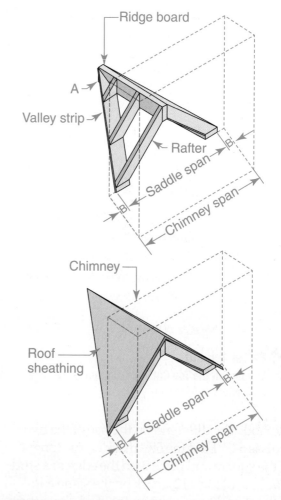

Chimney Saddle

Sheathing Projection When sheathing is attached to the framing, the saddle should be the same width as the chimney.

blade) and draw a line along the blade. The length of the ridge is equal to the run of the common rafter in the saddle's span minus the allowance for the drop of the ridge, which is approximately ¾".

To determine the theoretical length of the longest rafter, multiply the saddle's run (half the saddle's span) by the unit length of a common rafter. Deduct the ridge shortening allowance to obtain the actual length. The top and bottom cuts are the same as for a common rafter in the main roof. However, there is a side cut on the bottom where the rafter rests on the valley strip. This cut is the same as for regular valley jacks (see Section 18.3).

The cuts are the same for all the rafters in the chimney saddle. However, the rafter lengths differ. The difference in the length of the rafters can be found on the framing square's rafter table under Difference in Length of Jacks. For rafters 16" on center in a roof with a unit rise of 5", the second rafter will be 17⅝₁₆" shorter than the first rafter. The third rafter will be 34⅝" (2 × 17⅝₁₆) shorter than the first rafter, and so on. When the saddle framing is complete, nail the roof sheathing to it.

Section 19.3 Assessment

After You Read: Self-Check

1. What is a collar tie and what is its purpose?
2. What is the purpose of purlins and braces?
3. In gable-end framing, what is meant by "common difference"?
4. At what point in roof construction are dormers framed?

Academic Integration: Mathematics

5. **Calculate Common Difference** What is the common difference in length of gable-end studs placed 16" OC for a roof with a ⁵⁄₁₂ pitch?

Math Concept The common difference is calculated using the unit run and the unit rise.

Step 1: Divide the OC distance by the unit run.

Step 2: Multiply by the unit rise.

Go to connectED.mcgraw-hill.com to check your answers.

Section 19.4 Rakes & Cornices

Roof Edge Details

Does every style of roof have eaves?

The **eaves** are those portions of a roof that project beyond the walls. On a house with a hip roof, the eaves are the same width and height around the entire house. On a house with a gable roof, the eaves at the ends of the house follow the slope of the roof. They may be a different width than eaves at the sides of the house or may be omitted entirely. The upward slope of the eaves at a gable end is called the *rake angle*.

The roof-edge details can take many forms, depending on tradition and local preferences. They can include molding and decorative cuts in exposed rafter tails. Ventilation of the attic is another important factor to consider when planning and constructing roof-edge details.

The eaves themselves are formed by the rafter overhangs. For more on that topic see Chapters 17 and 18. The material in this section relates to details that are installed after the rafters are in place.

Cornices

What materials would not be good to use in cornice construction?

A **cornice** consists of a fascia, a soffit, and various types of molding. The **fascia** is a board that is nailed to the ends of the rafter tails. It protects the end grain of the rafters and serves as a mounting surface for gutters. Sometimes a sub-fascia made of 2× lumber will be nailed to the ends of the rafters to provide extra strength in this area. It is sometimes called a structural fascia. The sub-fascia is always covered later by a finished fascia, which is high-quality 1× stock. In some cases a finished fascia can be nailed directly to the rafter ends.

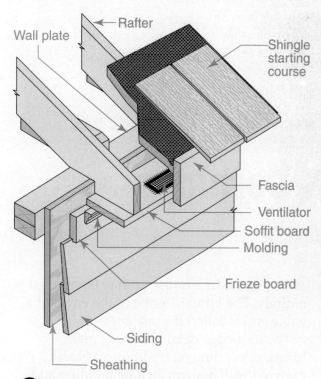

Basic Cornice Construction

Parts of a Cornice A cornice can take various forms. This illustration shows one common way to build a cornice.

The **soffit** is the underside of the eaves. It is sometimes enclosed with plywood, prefabricated vinyl panels, or aluminum sheets. It can also be left open, exposing the rafter tails.

Some cornice work may be done as soon as the roof has been framed. However, the cornice can also be built after the roof covering is in place. Cornice construction details are shown on the wall sections of the house plans. Detail drawings are usually included as well.

Types of Cornices

There are three basic types of cornices: open, box, and closed, in Types of Cornices. An *open cornice* consists of frieze blocks, molding, and a fascia. The underside of the roof sheathing and the rafters are exposed. It is simple to construct. One variation includes a continuous *frieze board*, which runs above the top course of siding, as shown in A. Unlike a frieze board, a frieze block is strictly functional. It is a short piece of 2× framing lumber nailed between the roof rafters to seal off the attic space.

A *box cornice* entirely encloses the rafter tails. It is built of roof sheathing, fascia, and a soffit. There are several ways of building a box cornice. The soffit can be nailed directly to the underside of the rafters. However, it is more often nailed to lookouts as in B. A **lookout** is a horizontal member that extends from a rafter end to a nailer or the face of the wall sheathing. Lookouts form a horizontal surface to which the soffit material is attached.

A *closed cornice* (C) appears on a house that has no rafter overhang. One **version** consists of a frieze board and one or more pieces of molding. This type of cornice is common on older houses in some parts of the United States. However, it is seldom used on newer houses because of the difficulty in providing attic ventilation.

Solid Wood Solid wood is the traditional material used for cornices. Because portions are exposed to the weather, rot-resistant

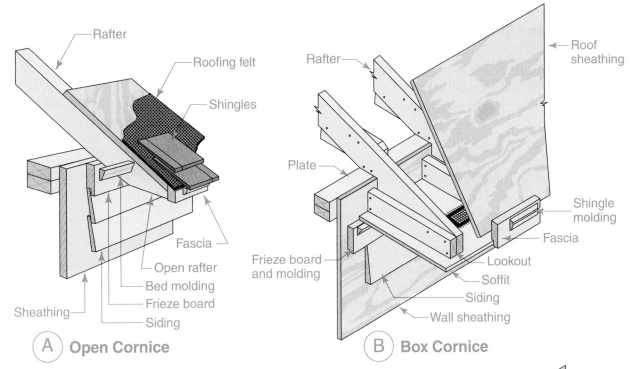

A **Open Cornice**

Labels: Rafter, Roofing felt, Shingles, Fascia, Open rafter, Bed molding, Frieze board, Siding, Sheathing

Labels (B): Rafter, Roof sheathing, Plate, Shingle molding, Fascia, Frieze board and molding, Lookout, Soffit, Siding, Wall sheathing

B **Box Cornice**

woods are preferred. The cornice is very visible, so top grades of lumber should be used. Avoid any board that contains sapwood, surface cracks, or loose knots. The fascia may be nominal 1× or 2× stock.

Other Materials The use of synthetic, composite, and engineered materials for cornice construction is now common. These materials come in many varieties, but some of the most common consist of rigid boards made from high-density polyurethane or from recycled wood fibers. Builders like using these materials because they are more uniform than solid lumber and free of defects. They are also available in long lengths. Many are pre-primed at the factory on all surfaces and edges. This saves labor at the job site and improves durability. Some of these products are composite blends of wood fiber and plastic. Others are made of high-grade PVC (polyvinyl chloride) resins and other polymers. Still others are made from finger-jointed lumber or laminated-veneer lumber with MDO (medium-density overlay) surfaces. With any of these materials, always follow the manufacturer's guidelines for installation. If the manufacturer's guidelines are not followed, any warranty on the product is void.

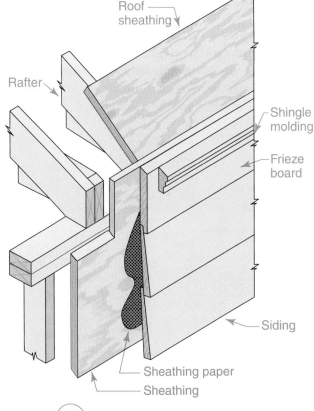

Labels (C): Roof sheathing, Rafter, Shingle molding, Frieze board, Siding, Sheathing paper, Sheathing

C **Closed Cornice**

Types of Cornices
Three Types **A.** An open cornice with a fascia board and frieze board. **B.** A box cornice with a flat soffit enclosure and lookouts. **C.** A closed cornice.

Building an Open Cornice

One method of constructing an open cornice is to install 2× frieze blocks between the rafters. The blocks are drilled for ventilation holes and circular, screened metal vents are fit into the holes. To install the blocks, nails are driven through the side of the rafter into the end of a block on one side. Nails have to be toenailed on the other side.

The frieze block is sometimes positioned at an angle to the walls. However, some carpenters prefer to install the block so that it is parallel to the walls and at an angle to the rafters. In this case, the block must be a size larger than the nominal dimension of the rafters. It may have to be cut to a width that will fit the space and beveled to the slope of the roof.

A disadvantage of an open cornice is that the underside of the roof sheathing is exposed. The material used for roof sheathing may have surface imperfections that are unattractive. To counter this problem, carpenters sometimes install a higher grade of plywood sheathing above an open cornice. Another solution is to install tongue-and-groove boards as sheathing in this area (see Section 19.5). Cornice workmanship is readily visible from the ground. For this reason, all joints in the construction of an open cornice should fit together tightly. Moldings should be mitered at outside corners and mitered or coped on inside corners.

Reading Check

Recall What are the two ways frieze blocks might be installed?

Building a Box Cornice

Before adding a box cornice, check the plumb cuts on the rafter tails to make certain they are all in line with one another. This check can be done by stretching a line along the top ends of the rafters from one corner of the building to the other. However, many carpenters do not make the plumb cut on the rafter tails when the rafter is cut. Instead, they install the rafters with the tails running longer than necessary. Then they snap a chalk line across the top of the tails to indicate the top of the cut. After drawing a plumb line downward from the chalk line on every rafter, they cut it to length. See pages 556–557 for steps in building and installing a box cornice.

Installing Sectional Soffits Several materials may be used for the soffit of a box cornice. Because wide overhangs are popular, materials available in large sheets are often used. The installation of exterior-grade plywood

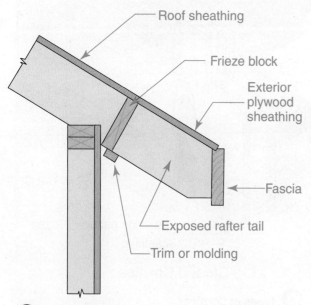

Frieze Blocks
Quick and Inexpensive This is a very simple type of open cornice.

Labels in figure: Roof sheathing, Frieze block, Exterior plywood sheathing, Fascia, Exposed rafter tail, Trim or molding

Builder's Tip

WHO INSTALLS SOFFITS? All contractors should review the plans and specifications regarding soffit finish materials. If a house is sided with wood, wood soffits might be installed by carpenters. However, if the house is sided with vinyl, vinyl siding contractors might install soffit panels along with the siding.

is described in the Step-by-Step Application on page 556. However, the use of vinyl soffit material is very popular, even on houses that do not have vinyl siding. This is because vinyl soffits require little maintenance. The material is light in weight and easy to install. It is entirely prefinished and available solid or perforated for ventilation. Aluminum soffit material can be used for many of the same reasons. When using either material, always follow the manufacturer's installation instructions. Installing Sectional Soffit Material shows one installation method.

Installing Sectional Soffit Material

Piece by Piece Installing soffit material.

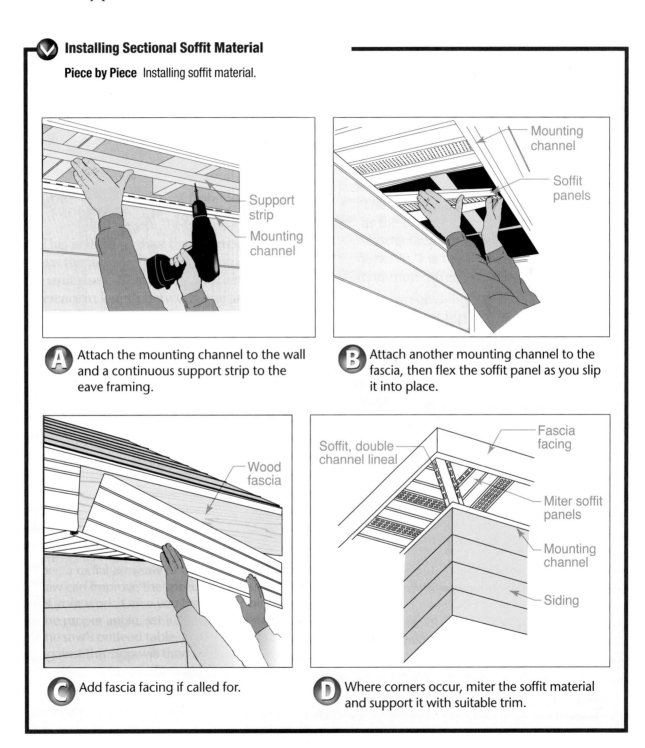

A Attach the mounting channel to the wall and a continuous support strip to the eave framing.

B Attach another mounting channel to the fascia, then flex the soffit panel as you slip it into place.

C Add fascia facing if called for.

D Where corners occur, miter the soffit material and support it with suitable trim.

Building a Box Cornice The general procedure is to install lookouts first. The fascia and soffit can then be installed.

Step 1 Use a piece of 1×4 material as a ledger. Temporarily nail it tight against the wall and against the rafters, and align it with the inside edge of the first rafter. The bottom edge of the ledger should be even with the bottom of the rafter tail. (A ledger is a horizontal length of lumber used to support other structural elements.) With a straightedge against the side of the rafter, draw a line on the ledger. Place an X on the side of the line away from the underside of the rafter to indicate the location of the lookout. Do this along the entire length of the building.

Step 2 Determine the length of the lookouts. Measure on a level line from the plumb cut on the rafter tail to the wall. Subtract ¾" from this measurement to allow for the thickness of the ledger. Subtract another ¾" to make sure that the lookouts do not project beyond the end of the rafters. Otherwise, any deviation in the wall would cause the lookout

to extend beyond the end of the rafter tail. This would interfere later with installation and alignment of the fascia.

Step 3 Lookouts are generally made from 2×4 lumber. After they have been cut to length, remove the ledger from its temporary position. Nail the lookouts to the ledger over the Xs. Nail through the back of the ledger into the end of each lookout with two 16d coated nails. Some carpenters toenail the ends of the lookouts to the ledger instead. Note that the end lookout is nailed into the end of the ledger strip. This means that the end lookout has to be of the same thickness as the rafter and longer than the rest of the lookouts. It will have to be cut to fit under the rafter tail.

Step 4 Locate the ledger on the wall by leveling from the rafter tail in toward the wall and placing a mark on the sheathing (point B in C). Do this at each end of the building. Snap a chalk line along the length of the building on the sheathing.

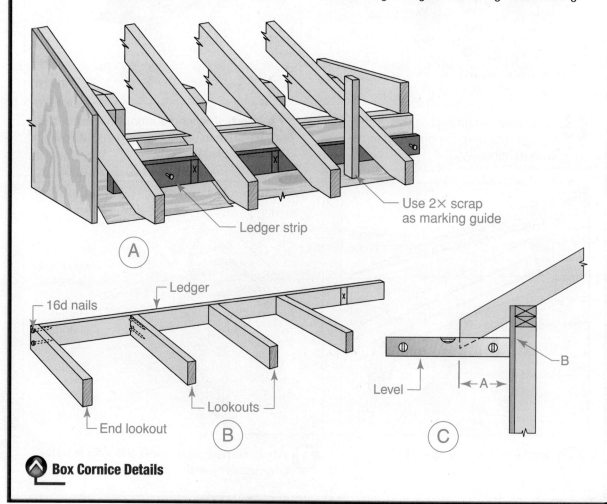

Use 2× scrap as marking guide

Ledger strip

Ⓐ

16d nails

Ledger

Lookouts

End lookout

Ⓑ

Level

◄─ A ─►

B

Ⓒ

▲ **Box Cornice Details**

Step 5 Place the bottom edge of the ledger on this line. Nail it through the sheathing and into the studs. Nail each lookout to the side of a rafter tail, except the end lookout. The end lookout should be cut to fit against the underside of the rafter. Level each lookout as it is nailed.

Step 6 If the soffit is narrow, as on page 552, the connection between the fascia and the soffit may be a butt joint. If the soffit is wide, as in D, one edge of the soffit material can fit into a groove cut in the back of the fascia. That method is described in the following steps.

Step 7 Rip the fascia stock to width if necessary. The top edge of the fascia may be beveled to the same angle as the pitch of the roof. If it is not, its outside top corner must be in line with the top edge of the rafter (E). Then cut a groove in it to receive the soffit. The groove should be located about ⅜" up from the bottom edge of the fascia board. This creates a drip edge that prevents water from being drawn into the joint.

Step 8 Nail the fascia to the ends of the rafter tails so that the top of the groove is even

with the bottom edge of the lookouts (D). Make certain that the fascia is straight along its length. If it is not, the soffit material will not fit properly. If the fascia must be spliced, the joint should be mitered and fall on the end of a rafter tail.

Step 9 Cut exterior-grade plywood soffit material to fit and slip it into place so that one edge fits into the groove in the fascia. Nail the soffit to each lookout and to the ledger strip with 4d galvanized nails spaced about 6" apart. As with the fascia, any joints in the soffit material must occur over solid backing (E). This joint is not mitered.

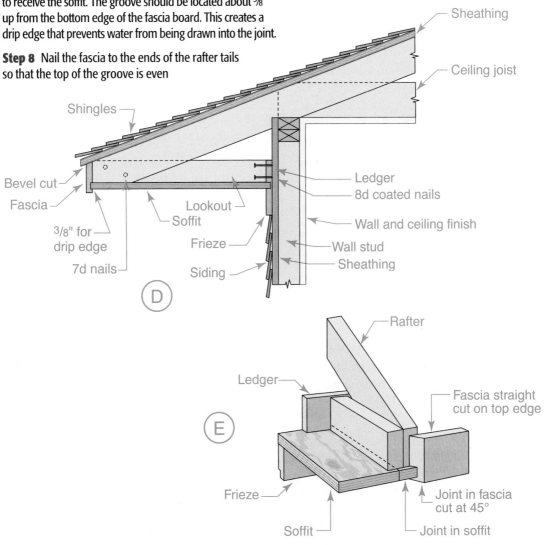

Estimating Cornice Materials

The materials for cornice framing, such as ledger strips, are estimated based on lineal foot measurements. This information is easily obtained from the building plans. Estimates for moldings and most other materials that are attached to the walls can be figured based on the perimeter of the house. The amount of material required for the fascia board and any molding attached to the fascia is figured by determining the perimeter of the roof at the rafter ends (not the perimeter of the walls).

Soffits The amount of soffit material required is based on the length and width of the soffit. These dimensions can be obtained from the house plans.

The method for determining the quantity of material needed depends on the material being used. With plywood, the estimate is based on the lineal footage of strips that can be ripped from a 4×8 panel. Aluminum and vinyl soffit material may be estimated by the square footage of soffit to be covered, or by referring to coverage charts provided by the manufacturer. Vinyl and aluminum soffit material comes in small sectional panels or in sheets that are 12' long.

Gable Rakes

What joint would you use where two fly rafters meet?

The part of a gable roof that extends beyond the end walls is called the **rake**. It is like a cornice in some respects so it is sometimes called a rake cornice. However, it has different installation requirements. A rake may be either closed or extended.

Closed Rake

A *closed rake* consists primarily of the frieze board and moldings. Some additional protection and overhang can be provided by using a 2×3 or 2×4 fascia block over the sheathing. This member acts as a frieze board. The siding can be butted against it. The fascia, often 1×6 stock, then serves as trim. Metal roof edging is used along the

rake to seal out water. Rakes with little or no overhang are inexpensive and simple to build. However, extending the rake helps to protect side walls from weathering, which reduces maintenance costs.

Extended Rake

An *extended rake* may be as narrow as 6" or as wide as 2'. An extended rake is shown in page 559. If the underside of the roof sheathing is exposed, it is called an *open rake*. If it is not exposed, it is called a *boxed rake*.

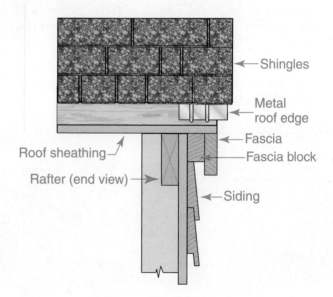

Shingles

Metal roof edge

Fascia

Fascia block

Roof sheathing

Rafter (end view)

Siding

Closed Rake
Some Protection A closed rake consists of various materials applied directly to the gable end of the house.

When the rake extension is only 6" to 8", the fascia and soffit can be nailed to a series of short lookout block (A). The fascia is further secured by nailing through the projecting roof sheathing. A frieze board and appropriate moldings complete the construction.

In a moderate overhang, both the sheathing and a fly rafter aid in supporting the rake section (B). A *fly rafter* extends from the ridge board to a structural fascia and is made of nominal 2" stock. The roof sheathing should extend from inner rafters to the end of the gable projection to provide rigidity and strength. It is nailed to the fly rafter and to the lookouts. The lookouts also serve as nailing surfaces for the soffit material. The assembly of lookouts, blocks, and fly rafter is sometimes called a *ladder*.

Wide rake extensions require rigid framing to prevent deflection. This is usually done by installing a series of lookout rafters that cantilever over the end walls, as shown on page 560. It may be constructed in place or built on the ground and hoisted into place. The lookouts are usually spaced 16" or 24" OC.

When the framing is preassembled, it is usually made with a header rafter on the inside and a fly rafter on the outside. Each is nailed to the ends of the lookouts that bear on the gable-end wall (rake wall). Lookouts are a type of purlin. In this use they are sometimes called an *overhanging purlin*.

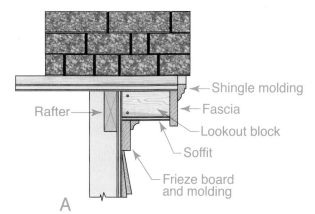

Rafter — Shingle molding — Fascia — Lookout block — Soffit — Frieze board and molding

A

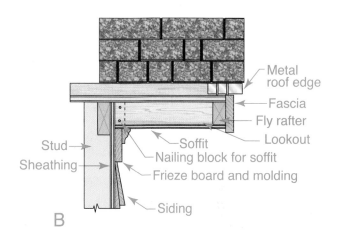

Stud — Sheathing — Metal roof edge — Fascia — Fly rafter — Lookout — Soffit — Nailing block for soffit — Frieze board and molding — Siding

B

 Extended Rakes
More Protection An extended rake with **A.** a narrow overhang and **B.** a moderate overhang.

When the header is the same size as the rafter, it should be cut just as a standard rafter, including the bird's mouth. The header rafter is face-nailed directly to the standard rafters with pairs of 12d nails spaced 16" to 20" apart. Each lookout should be toenailed to the rake wall plate.

The lineal footage for rake moldings is figured in the same way as the length of the gable-end rafter. The amount of material needed for the lookouts is obtained by multiplying the projection times the number of rafters.

Reading Check

Describe *What is the difference between an open rake and a boxed rake?*

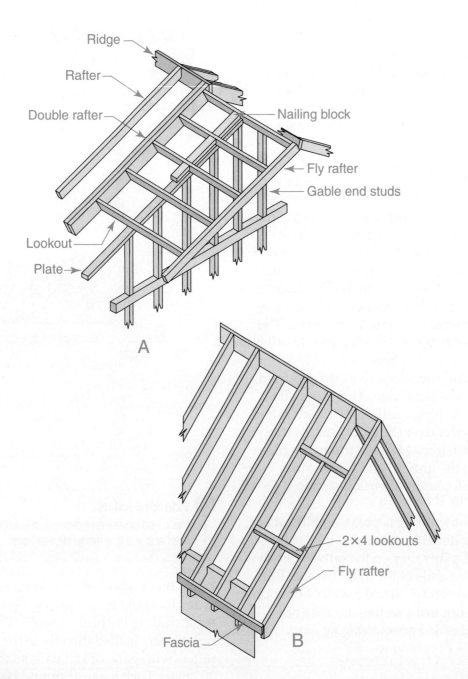

Ridge

Rafter

Double rafter

Nailing block

Fly rafter

Gable end studs

Lookout

Plate

A

2×4 lookouts

Fly rafter

Fascia

B

 Deep Rakes

Most Protection Lookouts can **A.** rest directly on the top plate of the gable wall or **B.** fit into notches cut into the end rafter.

Cornice Returns

A cornice return provides a transition between the rake and a cornice, as shown in Cornice Returns. How it is built depends on how the cornice is built and on how far the rake projects beyond the side walls.

When the cornice is boxed and there is some rake extension, the cornice return is also boxed, as in A. A boxed return is often used in houses of Cape Cod or Colonial design. The fascia board and shingle molding of the cornice are carried around the corner of the rake projection. When a house has open cornices, the cornice return is sometimes handled quite simply, as in B. A curved piece of wood can be attached to the underside of the rake trim. This piece is sometimes called a *pork chop*.

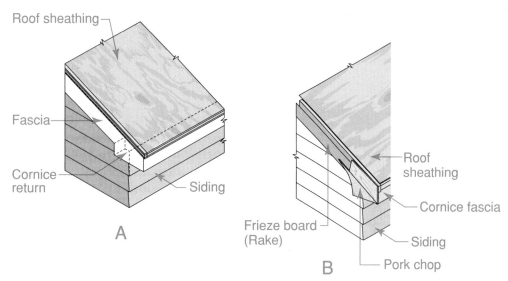

Roof sheathing

Fascia

Cornice return

Siding

A

Roof sheathing

Cornice fascia

Frieze board (Rake)

Siding

Pork chop

B

 Cornice Returns
Two Methods A. A narrow cornice with boxed return. **B.** An open cornice with a pork chop.

Section 19.4 Assessment

After You Read: Self-Check

1. Name the three basic types of cornice.
2. What is the disadvantage of open cornices?
3. What is a cornice return?
4. What is a pork chop?

Academic Integration: Mathematics

5. **Perimeter and Estimation** A house measures 20' by 35'. It has a hip roof, and the rafters project 18" from the walls. The plans call for a box cornice. Calculate how many 4' × 8' sheets of plywood will be required for the soffit.

   Math Concept Perimeter is the distance around a geometric figure, such as a rectangle. The formula for finding the perimeter of a rectangle is $P = 2l \times 2w$, where l is length, and w is width.

   Step 1: Figure the perimeter of the roof at the rafter ends.

   Step 2: Determine how many feet of plywood measuring 18" wide you can get from a 4' × 8' sheet of plywood.

   Step 3: Divide the perimeter by the number of feet of plywood you can get from one sheet.

Go to connectED.mcgraw-hill.com to check your answers.

Roof Sheathing

Panel Sheathing

How should roof panel sheathing be laid?

Sheathing provides a nailing base for the finish roof covering and gives rigidity and strength to the roof framing. Spaced boards are sometimes used to sheath roofs that will be covered with wood shingles or shakes (see Chapter 22, "Roofing & Gutters"). Other types of lumber sheathing can be used as well. Most roofs, however, are sheathed with panel products such as OSB or plywood. Though they are manufactured in different ways, they have about the same capabilities when used as roof sheathing.

The top surface of a sheathed roof is sometimes referred to as the roof deck.

Panel sheathing can be installed quickly over large areas. It provides a smooth, solid base with a minimum of joints. It can be used under almost any type of shingle or built-up roofing. Waste is minimal, which helps keep costs low.

Spans

Depending upon its thickness, panel roof sheathing can be used to span various distances. Many panels are performance rated. That means they are stamped to indicate their suitability for particular spans. The stamp consists of a pair of numbers

Roof Sheathing

A Base for Roofing The roof of this home was sheathed with plywood. *What are the advantages of panel sheathing?*

Lynn Betts/USDA Natural Resources Conservation Service

separated by a slash mark, such as 32/16 or 12/0. The number in front of the slash indicates the maximum spacing (in inches) of supports when the panel is used for roof sheathing. The number following the slash refers to the maximum spacing (in inches) of supports beneath panels used for subflooring. When one of the numbers is zero, the panel is unsuitable for that particular use. In this rating system, it is assumed that the long dimension of most panels will span at least three supports. Note that greater spans are generally allowed for roof sheathing than for floor sheathing.

Installation

Panel roof sheathing should be laid with the grain (the long dimension) perpendicular to the rafters. End joints should occur over rafters. The end joints of adjacent rows of panels should be staggered by using half-sheets. Unsupported edge joints can be strengthened with metal panel clips that tie them together. These joints can be supported by wood blocks instead, but blocks

are time-consuming to install and can limit ventilation beneath the roof.

Panels shrink or swell slightly as their moisture content changes. If panels are butted tightly during installation, they may buckle as they expand. To prevent buckling, allow ⅛" between panels or as recommended

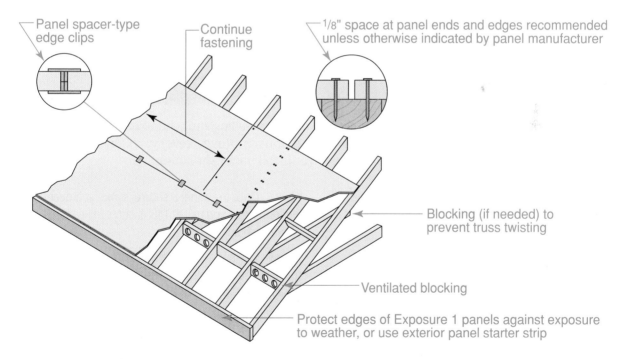

Panel spacer-type edge clips

Continue fastening

⅛" space at panel ends and edges recommended unless otherwise indicated by panel manufacturer

Blocking (if needed) to prevent truss twisting

Ventilated blocking

Protect edges of Exposure 1 panels against exposure to weather, or use exterior panel starter strip

Sheathing Details

Details Are Important Seemingly small details make a big difference in the strength of a roof sheathing installation.

Table 19-1: Minimum Fastening Schedule for APA Panel Roof Sheathing

Panel Thickness (inches)	Size	Nailing[a] Spacing (inches)		Leg Length (inches)	Stapling[b] Spacing (inches)[c]	
		Panel Edges	Intermediate		Panel Edges	Intermediate
5/16	6d	6	12	1¼	4	8
3/8	6d	6	12	1⅜	4	8
7/16, 15/32, ½	6d	6	12	1½	4	8
19/32, 5/8, 23/32, ¾, 7/8	8d	6	12[d]	–	–	–
1⅛, 1¼	8d or 10d	6	12[d]	–	–	–

[a] Use common smooth or deformed shank nails with panels to 1" thick. For 1⅛" and 1¼" panels, use 8d ring- or screw-shank or 10d common smooth-shank nails.

[b] Values are for 16-gal. galvanized wire staples with a minimum crown width of 3/8".

[c] For stapling asphalt shingles to 5/16" and thicker panels, use staples with a ¾" minimum crown width and a ¾" leg length. Space according to shingle manufacturer's recommendations.

[d] For spans 48" or greater, space nails 6" at all supports.

by the manufacturer. This spacing should be used at all edge joints and end joints. Some panel clips are constructed to automatically space panels the proper distance apart. No surface or edge should be permanently exposed to the weather.

Space nails according to **Table 19-1**. Use 6d common, ring-shank, or spiral-thread nails for plywood ½" thick or less. For plywood more than ½" thick, use 8d common nails. For additional holding power, use ring-shank or spiral-thread nails, or glue-nail the sheathing. Place nails approximately 3/8" in from panel ends and edges.

Lumber Sheathing

Why is closed sheathing rarely used?

Not all roofs are sheathed with panel products. For various reasons, a roof may be sheathed with solid wood instead. For example, plank sheathing is common where

Step-by-Step Application

Installing Panel Sheathing Always use caution when installing roof sheathing. If conditions are wet or windy, postpone installation until conditions improve. Never work on top of a panel that is not nailed securely to the rafters. To install panel sheathing:

Step 1 Position the panel on the rafters. If necessary, tack it (nail it temporarily) to the rafters to prevent it from shifting.

Step 2 Nail one end of the panel. Drive nails flush with the panel surface. Remove any temporary nails.

Step 3 Snap a chalk line across the panel to indicate the centerline of each rafter. Nail the panel across its width, starting at one end and working toward the other.

Step 4 Once the panel is secure, stand on it over the framing as you nail it. This ensures full contact between sheathing and rafter.

Step 5 When installing the next panel, allow a ⅛" space between panels (the width of a 10d box nail). Install edge clips as required by local building codes.

Step 6 Cover the sheathing with roofing felt as soon as possible to minimize exposure to the weather (see Chapter 22, Roofing & Gutters).

post-and-beam framing techniques are used (see Chapter 14). *Open sheathing* (also called *skip sheathing* or *spaced sheathing*) is common in regions where wood shingles and shakes are popular. However, in areas where wind-driven snow conditions prevail, panel sheathing is recommended under wood shingles and shakes instead.

Another type of sheathing, called *closed sheathing*, was common before it was replaced by panel sheathing. Nominal 1" boards with T&G or shiplap edges were installed over the entire roof surface. Closed sheathing is sometimes encountered during remodeling work.

Open Sheathing

Open sheathing consists of 1×4 square-edged boards. It is often called spaced sheathing. Spaces between sheathing boards promote ventilation around wood shingles and shakes, allowing them to dry out evenly. However, open sheathing is not suitable for use in earthquake regions. In addition, in regions where the average daily temperature in January is 25°F (-4°C) or less,

panel sheathing is required on portions of the roof that require an ice barrier. This area generally runs from the edge of the eaves to a point at least 24" inside the building line. Consult local building codes for specific requirements.

The boards of open sheathing are laid with on-center spacing equal to the amount of shingle exposed to the weather. The boards are laid perpendicular to the rafters. Open sheathing is nailed to each rafter with two 8d nails. Joints must be made on the rafters. Each board should bear on at least two rafters.

Reading Check

Recall *In which regions should open sheathing be avoided?*

Plank Sheathing

Plank sheathing with double tongue-and-groove edges (also called *roof decking*) provides a solid surface for roofing and an attractive ready-to-finish interior ceiling. It is available in several patterns and thicknesses.

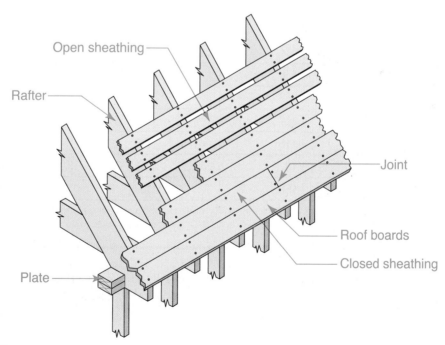

Spaced and Closed Sheathing
Old and New Installation of lumber roof sheathing, showing open (spaced) and closed types.

Estimating and Planning

This estimating and planning exercise will prepare you for national competitive events with organizations such as SkillsUSA and the Home Builder's Institute.

Sheathing Panels

Materials and Dimensions

To figure the area to be sheathed without actually getting on the roof and measuring, find the dimensions on the plans.

Step 1 Multiply the length of the roof times the width and include the overhang. For example, suppose that a home is 70' long and 30' wide, including the overhang. The roof has a rise of 5½":

$$70' \times 30' = 2{,}100 \text{ sq. ft.}$$

Step 2 Refer to the table. Multiply by the factor shown opposite the rise of the roof. The result will be the roof area. For a rise of 5½", the factor on the table is 1.100. The result will be the area of the roof.

$$2{,}100 \text{ sq. ft.} \times 1.100 = 2{,}310 \text{ sq. ft.}$$

Step 3 To estimate the number of sheathing panels required, divide the roof area by 32 (the number of square feet in one 4' × 8' sheet). For example, 2,310 ÷ 32 = 72.19.

Step 4 Add 5 percent for a trim and waste allowance:

$$72.19 \times .05 = 3.6$$

$$72.19 + 3.6 = 75.79 = 76 \text{ sheets (rounded up)}.$$

Estimating on the Job

How many sheets of panel sheathing would be required for a roof that measured 50' long and 27' wide, including the overhang, and that had a 4" rise?

Estimating Roof Sheathing from Plans			
Rise (inches)	Factor	Rise (inches)	Factor
3	1.031	8	1.202
3½	1.042	8½	1.225
4	1.054	9	1.250
4½	1.068	9½	1.275
5	1.083	10	1.302
5½	1.100	10½	1.329
6	1.118	11	1.357
6½	1.137	11½	1.385
7	1.158	12	1.414

Note: When a roof has to be figured from a plan only, and the roof pitch is known, the roof area may be fairly accurately computed from this table. The horizontal or plan area (including overhangs) should be multiplied by the factor shown opposite the rise, which is given in inches per horizontal foot. The result will be the roof area.

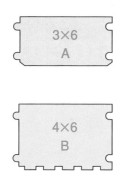

 Plank Sheathing
Limited Uses Planks come in nominal widths of 4" to 12" and in nominal thicknesses of 2" to 4".

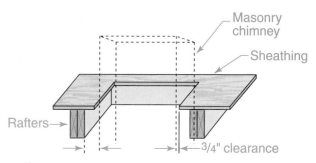

Masonry chimney

Sheathing

Rafters

3/4" clearance

 Sheathing Around a Chimney
Fire Safety Framing and sheathing around a masonry chimney should be installed so that they are not touching the masonry.

The patterned side of the plank sheathing faces down so that it will be exposed to the room below. Plank sheathing is no longer allowed in earthquake-prone areas.

Sheathing Details

Where gable ends have little or no extension other than the molding and trim, the roof sheathing is usually sawed flush with the outer face of the side-wall sheathing. Cuts should be even so that the trim and molding can be properly installed. The sheathing at the valleys and hips should be fitted to give a tight joint. It should be securely nailed to the valley or the hip rafter.

Roof sheathing should have a ¾" clearance from the finished masonry on all sides of a chimney opening. This gap is covered by flashing. Framing members should have a 2" clearance for fire protection. The sheathing should be securely nailed to the rafters and to the headers around the opening.

Section 19.5 Assessment

After You Read: Self-Check

1. A performance-rated panel is stamped ³²⁄₁₆. What does the number 32 represent?
2. What is the correct spacing for nails when installing roof sheathing panels?
3. What is the advantage of using open sheathing beneath wood shingles?
4. What clearance is recommended between roof sheathing and the finished masonry for a chimney?

Academic Integration: English Language Arts

5. **Roof Sheathing Safety** Sheathing provides a nailing base for the finish roof covering and gives rigidity and strength to the roof framing. Most roofs are sheathed with panel products. Use the Internet to locate a trade association responsible for maintaining standards for plywood roof sheathing. Search the association's Web site to find publications related to residential roof sheathing. Download and print out any publications that relate to safe working procedures during installation. After reading the publication, write a one-page summary of its contents.

Go to connectED.mcgraw-hill.com to check your answers.

Chapter Summary

Ridges can be structural or nonstructural. Calculating the length of a ridge requires actual measurements taken from the framed building. The ridges for dormers and additions can be calculated with the assistance of a sketch of the roof plan.

A careful rafter layout is important so that rafters will bear properly on the ridge board or ridge beam. Common rafters are generally installed first, then hip and valley rafters. Jack and hip jack rafters are installed last.

Special framing details include collar ties, purlins and braces, gable ends, roof openings, shed dormers, and chimney saddles.

Cornices can be constructed in various ways, based on the architectural style of the house and the climate. The three types of cornices are open, box, and closed. A rake is the part of a gable roof that extends beyond the end walls. Careful detailing is required at the rake, particularly at the cornice returns. The joints must be tight to prevent water from getting through.

Roofs may be sheathed with panels or lumber. Panel sheathing adds considerable strength to the roof system.

Review Content Vocabulary and Academic Vocabulary

1. Use each of these content vocabulary and academic vocabulary words in a sentence or diagram.

Content Vocabulary

- ridge (p. 530)
- ridge beam (p. 530)
- collar tie (p. 542)
- purlin (p. 544)
- brace (p. 544)
- common difference (p. 546)
- eaves (p. 551)
- cornice (p. 552)
- fascia (p. 552)
- soffit (p. 552)
- lookout (p. 552)
- rake (p. 558)

Academic Vocabulary

- intermediate (p. 530)
- suspended (p. 532)
- version (p. 552)

Speak Like a Pro

Technical Terms

2. Work with a classmate to define the following terms used in the chapter: *nonstructural ridge* (p. 530), *structural ridge* (p. 530), *shortening allowance* (p. 532), *cricket* (p. 550), *closed cornice* (p. 552), *open rake* (p. 558), *fly rafter* (p. 559), *pork chop* (p. 560), *skip sheathing* (p. 565), *roof decking* (p. 565).

Review Key Concepts

3. Name the two basic types of ridges.

4. Explain how to calculate ridge length.

5. Describe how ridge layouts are created for gable roofs, hip roofs, addition roofs, and dormers.

6. Name different types of cornice construction and name the parts.

7. Construct a simple box cornice.

8. Explain the basic requirements for the placement and nailing of panel roof sheathing.

Critical Thinking

9. Infer What might happen to panel roof sheathing if the panels are installed with no space between them?

Academic and Workplace Applications

STEM Mathematics

10. Calculating Area Estimate the number of 4×8 sheathing panels required for a roof that is 60' long and 25' wide, including overhang. The roof has a rise of 8 feet.

Math Concept Just as the length of a rafter is greater than its run, the area of a gable roof is greater than the footprint of the building it covers.

Step 1: Find the horizontal area, or plan area, of the roof in square feet.

Step 2: Refer to Table 19-1 on page 564 to obtain the rise factor. Multiply by the rise factor to obtain the area of the roof.

Step 3: Divide the area of the roof in square feet by the square footage of one sheathing panel.

Step 4: Add five percent for a trim and waste allowance.

STEM Engineering

11. Load In a post-and-beam house, the ridge beam bears the weight of other roof components, such as rafters and roof sheathing. If the roof has a low pitch and the house is well insulated, the beam may also have to bear the substantial weight of snow build-up. What are the characteristics of a ridge beam and what factors determine how much weight it can bear without sagging or failing? Write a list of as many factors as you can think of that affect the ability of a ridge beam to bear weight.

21st Century Skills

12. Information Literacy A lack of headroom often prevents attics from being converted to living space. One solution is to build a long shed dormer on one or both sides of a house. A back-to-back pair of long shed dormers are sometimes called saddlebag dormers. They increase the amount of usable attic floor space. Using library or Internet resources, locate examples of saddlebag dormers. Be sure to check sources of stock building plans. When you find at least one example, make a cross-section sketch of the house with and without the dormer. See if you can determine approximately how much usable floor space, in square feet, is lost if the dormer is removed.

Standardized TEST Practice

True/False

Directions Read each of the following statements carefully. Mark each statement as either true or false by filling in **T** or **F**.

(T) (F) **13.** An open cornice consists of frieze boards, molding, and a fascia.

(T) (F) **14.** When the width of an addition is less than the width of the main portion of the house, their roof spans are unequal.

(T) (F) **15.** When installing collar ties, nominal 1" collar ties should be nailed to the common rafters with three 16d nails at each end.

TEST-TAKING TIP

If any part of a true/false question is false, then the entire statement is false. However, just because part of a statement is true does not necessarily mean the entire statement is true.

*These questions will help you practice for national certification assessment.

Construction Calculations

Your Project Assignment

Your company is building new houses for people who have lost their homes in a natural disaster. You will use a shed roof design. You need to calculate the rafter lengths for various design options and present them in a table. You will pick roof angles in 5 degree increments, calculate the rafter length using trigonometry, and pick a material size for the rafters.

- **Research** the safe spans of common rafter materials such as saw lumber, PSL, and LVL.

- 🏠 *Build It Green* **Research** green roofing materials.

- **Calculate** the rafter lengths needed for specific roof angles and spans.

- **Calculate** the total rise of specific roof angles and spans.

- **Create** a three- to five- minute presentation.

Applied Skills

Your success in carpentry will depend on your skills. Some skills you might use include:

- **Calculate** rafter lengths using trigonometric functions.

- **Create** a span table for roof specifications given as angles.

- **Present** your results and demonstrate how to use your roof angle rafter table.

The Math Behind the Project

The traditional math skills for this project are trigonometry and measurement. Remember these key concepts:

Pythagorean Theorem

The rafter of a shed roof forms the hypotenuse of a right triangle. The legs of the triangle are the rise and run of the roof. You can use the following equation derived from the Pythagorean theorem, $a^2 + b^2 = c^2$, to calculate the length of the rafter:

$$\sqrt{a^2 + b^2} = c$$

Trigonometric Functions

You can determine the length of the rafter using trigonometric functions. The circles below show abbreviations for standard trigonometric terms—opposite, tangent, adjacent, sine, hypotenuse, and cosine. To solve for a term in the top half of any of these circles, multiply the terms in the two bottom quarters to get your answer. For example, to find the adjacent of a triangle, multiply the cosine by the hypotenuse. To solve for a term in the lower half of the circle, divide the quantity on the top by the other quantity on the bottom. For example, to solve for tangent, divide opposite by adjacent.

To find the angle of the triangle, take the result of your calculations and use the inverse or 2nd function key on a scientific calculator. Your answer will be a whole number and a decimal. Convert the decimal to a fraction. For example, to change 59.093 ft to feet and inches, use the following steps:

1. Multiply the decimal portion by 12.	.093 × 12 = 1.116
2. Multiply the decimal portion of the product by 16.	.116 × 16 = 1.856
3. Round to the nearest whole number.	2

This means that .116" is about $^2/_{16}$" or $^1/_8$". The final answer is 59 feet, $1^1/_8$ inch.

Project Steps

Step 1 Research

- Research the allowable spans for saw lumber, PSL, LVL, and I-joists. You can find this information at building centers and on manufacturers' Web sites.

- 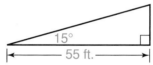 *Build It Green* Contact or use the online resources of the U.S. Green Building Council for guidance on selecting the most energy-efficient, sustainably produced materials suitable for roofing in your region.

- Decide on a material and material size for the rafters.

Step 2 Plan

- Specify the span length of your roof. (Remember that on a shed roof, span and run are the same.)

- Determine the roof angles you will use for your calculations. Pick three roof angles that differ in 5 degree increments.

- Sketch each of your three designs as in the example below. Label the span length and specify the roof angle.

15° — 55 ft.

Step 3 Apply

- Use trigonometry to calculate the rafter length for each of your three roof angle designs. Add this information to your sketches. Add a tail for eaves if desired.

- Calculate the total rise for each roof angle design. Add this information to your sketches.

- Create a chart for your three roof angles that lists the different rafter lengths needed for each.

- Write an instruction sheet that explains how to use the chart and how you calculated the data.

U.S. GREEN BUILDING COUNCIL

Mission: To transform the way buildings and communities are designed, built, and operated, enabling an environmentally and socially responsible, healthy, and prosperous environment that improves the quality of life.

Step 4 Present

Prepare a presentation combining your research and measurement calculations using the checklist below.

PRESENTATION CHECKLIST
Did you remember to…
✓ Show your calculations for each roof angle?
✓ Use and present your sketches and a worksheet with your calculations?
✓ Explain how you used trigonometry to calculate rafter length?
✓ Write notes you might need for your presentation?
✓ Explain how to use the rafter chart?

Step 5 Technical and Academic Evaluation

Assess yourself before and after your presentation.

1. Was your research thorough?
2. Were your green alternatives realistic?
3. Were your calculations correct?
4. Were your sketches accurate?
5. Was your presentation creative and effective?

Go to **connectED.mcgraw-hill.com** for a Hands-On Math Project rubric.

UNIT 5
Completing the Structure

In this Unit:

Hands-On Math Project Preview

Assembling Resources

After completing this unit, you will you will find a local organization, such as Rebuilding Together, that builds or repairs homes for low-income members of the community. You will also identify the dimensions and materials needed for a rectangular storage shed that could be part of one of the organization's building projects:

Project Checklist

As you read the chapters in this unit, use this checklist to prepare go the unit project:

✓ Examine the amount of labor involved in applying vinyl siding.

✓ Identify the methods for estimating exterior siding.

✓ Think about how the steps in building a shed would be similar building a house.

➤ Go to connectED.mcgraw-hill.com for the Unit 1 Web Quest activity.

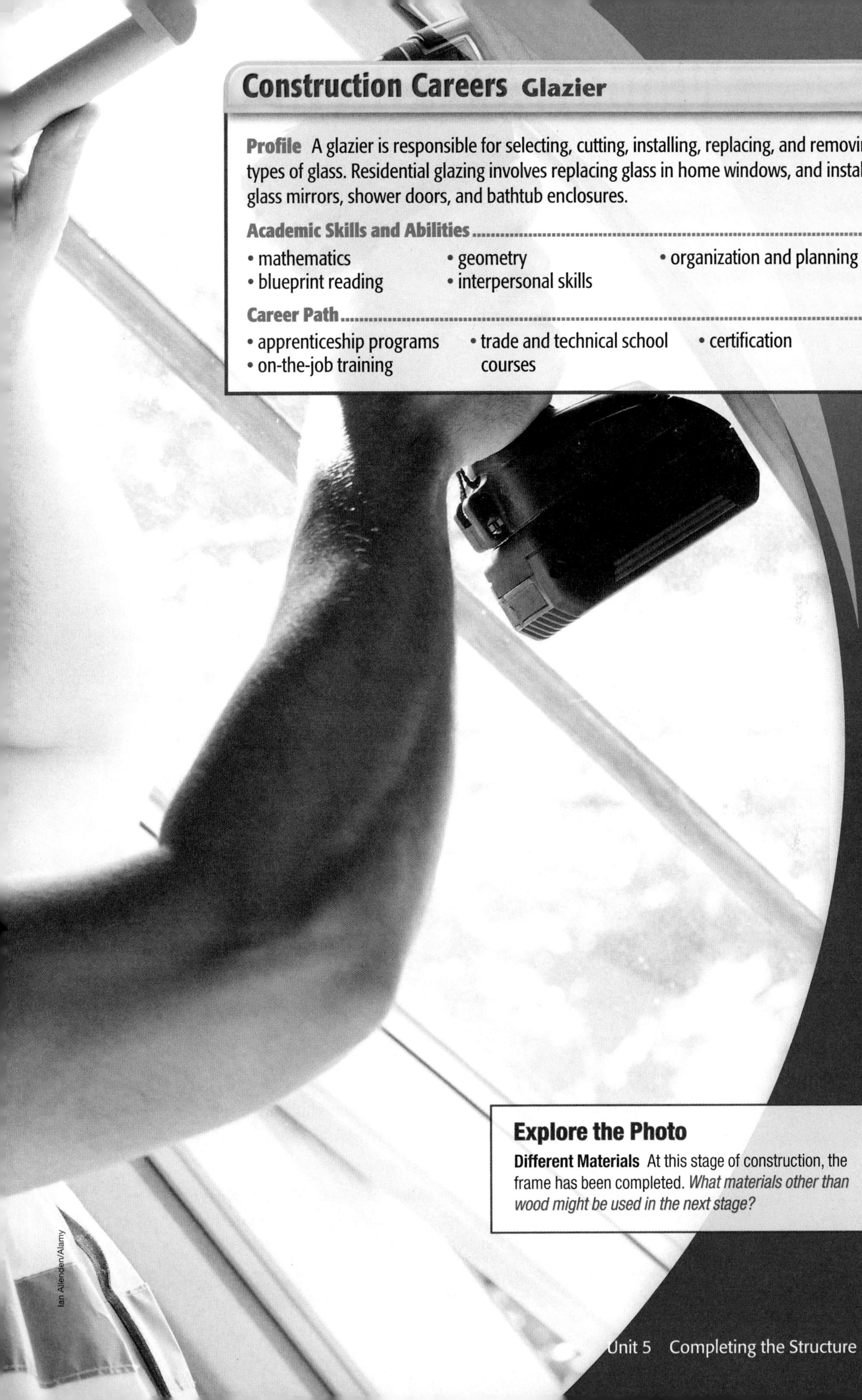

Construction Careers Glazier

Profile A glazier is responsible for selecting, cutting, installing, replacing, and removing all types of glass. Residential glazing involves replacing glass in home windows, and installing glass mirrors, shower doors, and bathtub enclosures.

Academic Skills and Abilities

- mathematics
- blueprint reading
- geometry
- interpersonal skills
- organization and planning skills

Career Path

- apprenticeship programs
- on-the-job training
- trade and technical school courses
- certification

Explore the Photo

Different Materials At this stage of construction, the frame has been completed. *What materials other than wood might be used in the next stage?*

Ian Allenden/Alamy

Windows & Skylights

Section 20.1
Types of Windows & Skylights

Section 20.2
Installing Windows & Skylights

Chapter Objectives

After completing this chapter, you will be able to:

- **Describe** the basic types of windows.
- **Identify** the ways in which windows are made energy efficient.
- **Use** a window schedule and a manufacturer's size table.
- **Show** how to install a standard double-glazed or casement window.
- **Calculate** an estimate for the cost of a window.

Discuss the Photo
Skylights Skylights admit more light into rooms but must be fitted properly. *Why are skylights prone to leaking?*

Writing Activity: Brainstorming
Windows may be installed for a variety of reasons. Different types of houses in various climates make use of very different window styles. Brainstorm the many functions that windows serve. List your ideas.

Ingram Publishing

Chapter 20 Reading Guide

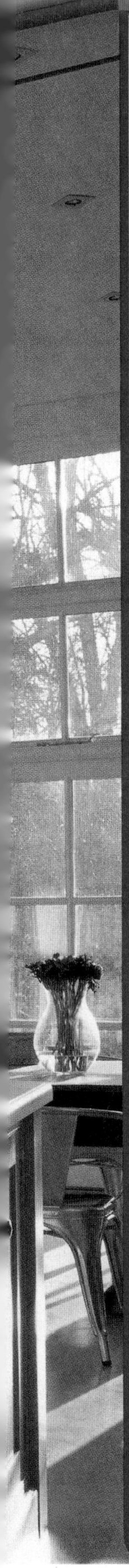

Before You Read Preview

Windows let light and air into a house and are also an important part of its architectural design. Before reading this chapter, take a few minutes to examine the windows at school and in your home. After you have read the chapter, write one or two sentences describing your findings.

Content Vocabulary

- glazing
- sash
- muntin
- window schedule
- unit dimension
- mullion strips

Academic Vocabulary

You will find these words in your reading and on your tests. Use the academic vocabulary glossary to look up their definitions if necessary.

- sufficient
- technology
- function

Graphic Organizer

As you read, use a chart like the one shown to compare the features of different types of windows.

Type of Window	Sash	Hardware
double-hung	upper and lower	metal sash locks
casement	side-hinged	rotary opener, hinge assembly, sash lock

Go to connectED.mcgraw-hill.com to download this graphic organizer.

Types of Windows & Skylights

Windows

What is glazing?

Windows let light and air into a house. They are also an important part of its architectural design. However, 20 to 30 percent of the heat lost from some houses is through the windows. This loss is due to air leaking around the window or by heat being radiated through the glass. In hot climates, cool indoor air can be lost in a similar fashion. As heating and cooling costs have climbed, manufacturers have greatly improved the energy efficiency of windows.

Windows should suit the style of the house. Different rooms within the house, however, will require different sizes and types of windows. For example, bedrooms and living rooms will often require larger windows to allow more light and ventilation. In a bathroom, however, smaller windows will provide greater privacy. Convenient window operation is another important consideration when selecting windows.

Design Requirements

The building code requires that the total area of window glass in a room should be not less than 8 percent of the floor area. This ensures **sufficient** natural light. The total window area that can be opened for ventilation should be not less than 4 percent of the floor area, unless mechanical air conditioning and ventilation are provided. Bathrooms must have no less than 3 sq. ft. of glazing, unless the room is ventilated with a fan. In the kitchen, windows should provide good ventilation of cooking odors.

 Windows and Style

Many Shapes The style, size, and shape of windows heavily influence the style of the house. *How many different window sizes and shapes do you see in the photo?*

©Dennis MacDonald/Alamy

Windows serve an additional and often overlooked purpose. They provide a way for rescuers to enter a room in an emergency. Windows also allow emergency exit from the room, particularly during a fire. According to building codes, every bedroom must have at least one window (or exterior door) that is suitable for *egress*, or emergency escape. It must have the following characteristics:

- Sill height no more than 44" above the floor.

- Height of opening no less than 24".

- Width of opening no less than 20".

- Unblocked open area no less than 5.7 sq. ft., except those on grade floor openings. These must have an open area of no less than 5.0 sq. ft.

(Note: A window with the minimum opening height and the minimum opening width will have an unblocked open area of only 3.3 sq. ft. This will not meet code. However, a window with the minimum opening height that is also 48" wide would have an unblocked open area of 8 sq. ft. That would be acceptable.)

Reading Check

Recall *What is one advantage of installing smaller windows in a bathroom?*

Types of Windows

The basic parts of a window are the glazing, the sash, and the frame. **Glazing** refers to the glass portions of a window. The glass within each section of window is also called a *pane* or a *lite*. The **sash** is the part that holds the glazing. The *frame* is the fixed part of the assembly that receives the sash. It consists of a *sill*, *side jambs*, and a *head jamb*.

At one time, the sash and the frame of a window were supplied separately. Today, windows are factory assembled as complete units, sometimes with the exterior casing in place. A completely assembled window

unit is sometimes called a *prehung window*. The carpenter installs this entire unit at once instead of installing the frame and sash separately.

There are six major types of windows: *double-hung*, *casement*, *stationary*, *awning*, *hopper*, and *horizontal-sliding* windows. The types that open are shown on page 578.

Most windows can be fitted with screens to keep insects out when the window is open. Screens used with double-hung, sliding, and hopper windows are installed on the outside of the window. For casement and awning windows that open out, the screen is installed on the inside.

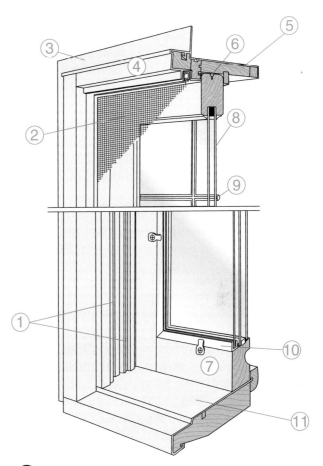

Parts of a Window Unit

Window System Parts of an assembled double-hung window: **1.** Tracks **2.** Screen **3.** Mounting flange **4.** Exterior casing **5.** Head jamb **6.** Weatherstripping **7.** Sash **8.** Glazing **9.** Muntins **10.** Removable storm panel **11.** Sill

Casement

Awning

Hopper

Horizontal-Sliding

Double-Hung

 Basic Window Styles
Windows That Open There are five basic window styles. Stationary windows do not open and are not shown here.

Double-Hung Windows The *double-hung* window consists of an upper and a lower sash that slide up and down in channels in the side jambs. A *jamb* is an exposed upright member on each side and at the top of the frame. Each sash has springs or balances that hold it in any position. Some types allow the sash to be removed or pivoted away for easy cleaning, painting, or repair. A *single-hung* window is a variation, where the upper sash is fixed in place, and only the lower sash slides up and down. Only half of a double-hung or single-hung window can be opened at one time.

A sash may have muntins. A **muntin** is a vertical or horizontal piece that holds a pane of glass. When glass was first used in

REGIONAL **CONCERNS**

Storm-Protected Windows Codes in areas with severe storms often require the installation of storm-resistant windows in new construction. These windows have reinforced glazing, reinforced frames, and strengthened locking mechanisms. These features reduce damage caused by wind-borne debris. They also reduce the chance that high winds will pull the window unit out of the framing. Folding or rolling shutters can be installed over the windows to increase protection from damage.

houses, only small panes were available. Muntins were used to hold a group of panes together to form a single window. Modern glass **technology** makes very large panes possible. However, muntins are often part of a window's design because they create a traditional appearance. Sometimes the muntins are purely decorative. For example, some manufacturers sell preassembled muntin grids that snap in place over a single lite. Although these are not true muntins, they make it look as if the glass were divided into six or more portions. Grids simplify painting and other maintenance. A window in which the muntins actually hold individual panes is called a *true divided lite* window.

Hardware for a double-hung window includes one or two metal sash locks. When engaged, these locks prevent the sash from being opened from the outside. They also draw the sash together at the meeting rails to reduce air infiltration.

Casement Windows Casement windows have a side-hinged sash that swings inward or outward. An outward-swinging sash does not get in the way of furniture. Also, wind tends to push an outward-swinging sash against the weatherstripping, making a stronger seal. One advantage of the casement window over the double-hung type is that the entire window area can be opened for ventilation. Hardware consists of a rotary opener, a hinge assembly, and a sash lock.

Stationary Windows Stationary windows, also called *fixed-glass windows,* are sometimes

Fixed and Awning Windows
Two Types of Windows The top row consists of stationary windows. The bottom row consists of stationary windows above smaller awning windows.

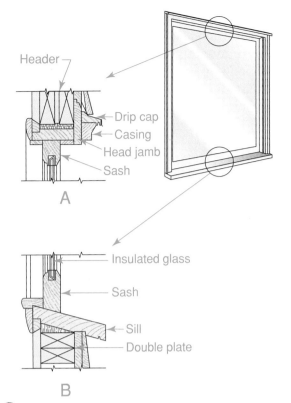

A

B

Header

Drip cap
Casing
Head jamb
Sash

Insulated glass

Sash

Sill
Double plate

Parts of a Stationary Window
Simple Construction This cross section shows the **A.** head jamb area and **B.** the sill area, as well as some of the wall framing.

used alone, but they are more commonly combined with other window types. They consist of a single pane of insulated glass fastened permanently into the frame and cannot be opened. Stationary windows can be installed with a sash, or without. When they are installed without a sash, the glass is set directly into the frame members and held in place with stops.

Glass block is a type of stationary window, although technically it is not a window at all. Glass block is typically sold by glass manufacturers, not window companies. It is installed by masons, not carpenters, because the individual units are held in place with mortar. However, assemblies of glass block, called *panels,* are often used to serve the same purposes as a window. They can be used to admit light in places where transparency and ventilation are not required. It is difficult to break glass block, so the material is also used where security is important.

Glass Block
Strong and Private Glass blocks can be used to admit light where ventilation is not required.

Awning and Hopper Windows The sash of an *awning window* swings outward at the bottom. Awning windows are sometimes grouped in pairs. A *hopper window* is similar to an awning window, except that the sash swings inward at the top. Both types provide protection from rain while open. They are sometimes combined with stationary windows. Hardware includes hinges, pivots, and sash-supporting arms.

Horizontal-Sliding Windows *Horizontal-sliding windows* resemble casement windows in appearance. However, the sashes (in pairs) slide horizontally in separate tracks, or *guides*, located on the sill and head jamb.

Frame & Sash Materials
What does cladding do?

Any of the basic types of windows can have sashes and frames made of wood, metal, vinyl, fiberglass, or wood composites. *Hybrid windows* are a combination of two or more materials.

Wood

Wood window frames and sashes should be made from a clear grade of all-heartwood stock. The wood should be decay-resistant or given a preservative treatment. Species commonly used include ponderosa and other pines, cedar, cypress, and spruce. All

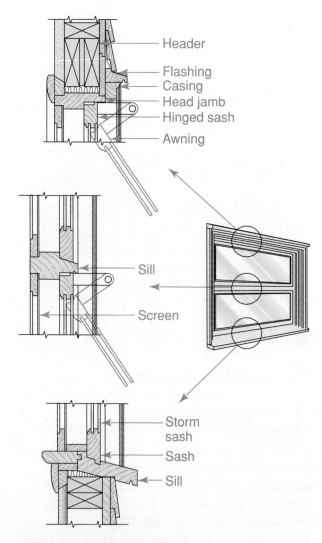

Header
Flashing
Casing
Head jamb
Hinged sash
Awning

Sill

Screen

Storm sash
Sash
Sill

Double Awning Window
Good Ventilation Cross section of a double awning window. This features two awning windows combined into a single unit.

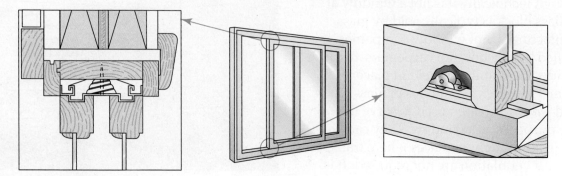

Sliding Window
Horizontal Slider The track at the top of the sash is spring loaded. This provides a weathertight seal and also permits lifting the sash out of the window frame. Along the sill, the sash travels on a nylon roller for easy operation.

wood components should also be treated with a water-repellent preservative at the factory.

The wood parts of a *clad-wood window* are covered, or clad, with vinyl or aluminum. The wood provides strength, and the cladding protects the wood. This type of wood window never needs painting.

Metal

Metal window frames and sashes are usually made of steel or aluminum. They are generally lighter and less costly than windows made of other materials. Frames and sashes are narrower than those of wood windows. This allows a larger glass area for a given rough opening. Unlike wood windows, metal windows are not subject to insect attack. They are available with a baked-on or anodized finish. Painting is not required.

Aluminum windows are uncommon where winters are cold. This is because heat loss through metal frames and sashes is much greater than through similar wood units. A related problem is that moisture-laden air inside a house can condense on metal surfaces exposed to cold outside air. Windows that have a thermal break reduce these problems. A *thermal break* is a material such as rubber or dense foam insulation that slows the transmission of heat and cold. The most energy-efficient metal windows have two-piece frames separated by a thermal break.

Most metal windows have a nailing flange on all sides. This makes them easy to install. Manufacturer's instructions should be followed carefully. However, the techniques generally follow those required for other flanged windows (see Step-by-Step Application on page 589).

Vinyl, Fiberglass, and Composites

Windows with structural PVC (polyvinyl chloride, or vinyl) sashes and frames are easy to maintain, and they resist heat loss. This material used for windows is sometimes referred to as *uPVC* (unplasticized PVC). This means it does not contain plasticizers, which are thought to be environmentally harmful.

The vinyl is colored all the way through, so it does not need painting. It also resists attack by insects. Vinyl window frames have hollow channels beneath the surface. In *insulated vinyl* windows, these cavities are filled with foam insulation for greater energy efficiency.

Window frames can also be made of fiberglass, which is a polyester-based material reinforced with very thin glass strands. Fiberglass window frames can be

Clad-Wood Window
Protective Layer A cutaway of a clad-wood window.

Structural uPVC Window
uPVC A cutaway view of a structural uPVC window.

(l & r)Joe Mallon

hollow or insulated, much like vinyl frames, though fiberglass is stiffer and stronger.

Another material used for windows consists of polymers (plastics) mixed with wood particles. This mixture is made into various shapes under pressure. Composites have properties similar to those of solid wood, but are more resistant to decay.

Reading Check

Synthesize *In areas prone to termites, which frame and sash material would be best?*

Energy Efficiency

The energy efficiency of a window depends on more than one component. For example, a window with the most energy-efficient glazing would still be inefficient if faulty weatherstripping allowed heat loss around the sash.

Ratings When choosing windows, compare independent ratings of overall performance. The most accurate ratings consider glazing, weatherstripping, materials, and construction. The National Fenestration Rating Council (NFRC) has developed a window rating system that considers solar heat gain, R-value (a measure of resistance to heat transfer), and air leakage. The rating numbers indicate the percentage of heating or cooling

energy the window saves compared to an inefficient window with single glazing and an aluminum frame. The higher the number, the greater the savings.

Heat transfer through windows is expressed in U-values, or U-factors. U-values are like the R-values for insulation, except they are reversed. A lower U-value indicates less heat loss or gain and greater insulating performance. Another useful efficiency rating is called the *solar-heat-gain coefficient,* or SHGC. It expresses the amount of solar heat that passes through a window. Its scale goes from 0 (zero), for none, to 1, for 100 percent of available solar heat. Numbers in between are given in decimals, such as 0.55. *Visible transmittance (VT)* is a number that indicates how easy a window is to see through and how well it admits daylight. For example, tinted glass may be rated as 15 percent VT and clear glass as 90 percent VT.

Glazing Most windows installed in new houses contain insulating glass. Sometimes called *double-glazed windows,* they are made with two or more sheets of glass separated

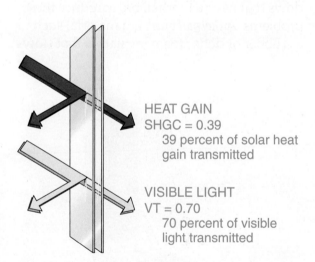

HEAT GAIN
SHGC = 0.39
39 percent of solar heat gain transmitted

VISIBLE LIGHT
VT = 0.70
70 percent of visible light transmitted

 High Efficiency Glazing
Managing Heat Transfer Efficiency rating for one type of high-efficiency window glazing. SHGC stands for solar heat gain coefficient, and VT stands for visible light transmittance.

Science: Condensation

Chemical Reactions One problem with using aluminum as a window framing material is its tendency to permit condensation. Explain how moisture-laden air inside a house can lead to condensation on metal window materials such as aluminum.

Starting Hint Condensation is the change of water vapor into liquid water. What would cause this change?

by an air space. The edges are sealed to trap the air between the sheets, which provides the insulation. This type of window has more resistance to heat loss than one with a single sheet of glass. Insulating glass is used in both stationary and movable-sash windows.

In very cold climates, windows may even be triple glazed. The added airspace improves the energy efficiency of the window. Because triple-glazing can be expensive, it is important to balance its cost against the energy savings it delivers.

The type of glass used also affects energy performance. The following are available:

Low-e Glazing Many window manufacturers offer low-emissivity, or *low-e,* glazing. *Low-emissivity* means that the glass radiates less heat to the outdoors than regular glass. For one common type, a special coating is applied directly to one of the glass surfaces facing the airspace. This coating reduces energy flow through the glazing by as much as 50 percent. Low-e glass can be useful in both warm and cool climates.

Heat-Absorbing Glazing This type of glass contains special tints (dyes) that enable it to absorb large amounts of solar energy. This is particularly helpful in cool climates.

Gas-Filled Glazing Energy efficiency is improved if the air between double-glazed panes is replaced with a denser gas that insulates better. Colorless gases such as argon and krypton are sometimes used.

Low-Conductance Spacers

In the 1960s and 1970s, manufacturers used aluminum spacers to separate the two panes of double-glazed windows. Because aluminum is a good conductor of heat, these windows were not very efficient. Modern double-glazed windows use materials such as silicone foam or thermoplastics. These materials are sometimes referred to as *warm edge spacers.* They conduct less heat and improve the overall efficiency of a window.

Weatherstripping

The main purpose of weatherstripping around a window is to prevent air from leaking between the sash and the frame. Weatherstripping is made of various flexible materials, including foam and fibrous pile. Over time, weatherstripping can lose its effectiveness due to wear. Worn or damaged weatherstripping should be replaced. Weatherstripping should not be painted or stained.

Reading Check

Discuss *What advantage do double-glazed windows offer?*

Skylights

What might make a skylight more energy efficient?

A skylight is essentially a window that is placed in a flat or pitched roof. The purpose of a skylight is primarily to admit natural light into a room, but some skylights can be opened for ventilation.

Skylights come in two basic types: fixed and ventilating. *Fixed skylights* cannot be opened. They are generally less expensive

▲ Insulating Glass Window
Double Glazed A cross section of an insulating glass window.

Joe Mallon

and easier to install. *Ventilating skylights* swing open on hinges. They can allow heated air to escape the house in hot weather. They can also funnel cooling breezes inside. Some units can be fitted with adjustable blinds or shades that can be used to control the amount of light admitted.

The glazing in a skylight can be either flat glass or plastic that is flat or domed. A skylight with flat glazing is sometimes called a *roof window*. Skylights should be double glazed to reduce heat loss. Some skylights include triple glazing or high-performance glazing with a low-e coating.

Tunnel Skylights In recent years, a new type of skylight has become available. Unlike skylights that offer natural light and ventilation as well a view of the sky, these units, called *tunnel skylights*, have one purpose only: to bring natural light into a room without reducing energy efficiency.

 Build It Green The outside portion of a tunnel skylight consists of a clear dome located on the roof. The dome is connected to a sealed tubular shaft that passes through the roof system and attic. At the other end of the shaft is a light diffuser mounted flush with the room ceiling. When sunlight enters the dome it is channeled downwards by light-reflecting surfaces on the inside of the shaft. The light is then spread into the room by the diffuser. This type of skylight is energy efficient because the shaft is sealed. Also, by bringing natural daylight into an otherwise dark room, lights do not need to be turned on.

Ventilating Skylight
Light and Air This skylight can be opened and closed from inside to let in light and air.

Tunnel Skylight
Easy Installation A tunnel skylight captures light and channels it to a room below.

Section 20.1 Assessment

After You Read: Self-Check

1. Why is it important for every bedroom to have at least one window or an exterior door?
2. Name the six basic types of windows.
3. What is a hybrid window?
4. What is the purpose of a tunnel skylight?

Academic Integration: English Language Arts

5. **Persuasive Paragraph** Double- and triple-glazed windows have greatly improved the energy efficiency of modern homes. They have also provided greater flexibility in how homeowners choose to try to regulate indoor temperatures and humidity levels. Write a paragraph to try to persuade homeowners to install double- or triple-glazed windows in their home.

Go to **connectED.mcgraw-hill.com** to check your answers.

Section 20.2 Installing Windows & Skylights

Sizing Windows

How does the work of a framing carpenter affect window installation?

A **window schedule** is a portion of the building plans that contains descriptions of the windows and the sizes for the glass, the sash, and sometimes the rough opening. The location of each window in a house is found by matching the number of the window in the schedule with the corresponding number on the house plan, as shown on page 586.

The width of a window's jambs must be the same dimension as the thickness of the wall, including the exterior sheathing and the interior finished wall covering. Window jambs are made of nominal 1" or thicker lumber. The sills are made from nominal 2" lumber and are sloped for good drainage. The sash is normally 1¾" thick.

Rough-Opening Sizes

The *rough opening* is the space left in the wall framing to receive a window. It is always larger than the window itself to allow for shimming and ease of installation. Information about the rough-opening sizes for all windows in a house is needed when the walls are framed.

When the dimensions of a window are specified, the width is the first dimension given and the height is the second. The number of lites and the window style may follow or precede these dimensions. For example, consider this specification: 28½" × 24", 2 lites D.H. This means that the glass itself is 28½" wide and 24" high and that there are two pieces of glass in a double-hung unit. Often the term *lite* is abbreviated L.T.

The rough-opening size of a wood window can be figured if the glass size is known. The rough opening should be about

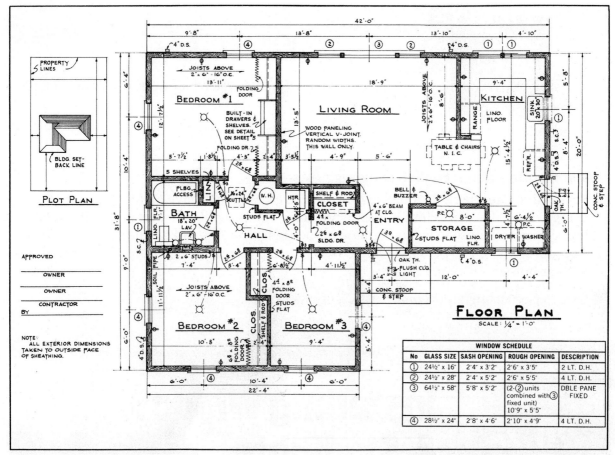

Planning for Windows

Determining Locations A house plan, including a basic window schedule. The letters D.H. stand for *double hung*.

The floor plan shows the following labeled areas and details:

PLOT PLAN
- PROPERTY LINES
- BLDG. SET-BACK LINE

APPROVED
- OWNER
- OWNER
- CONTRACTOR
- BY

NOTE:
ALL EXTERIOR DIMENSIONS TAKEN TO OUTSIDE FACE OF SHEATHING.

FLOOR PLAN
SCALE: ¼" = 1'-0"

Rooms: BEDROOM #1, LIVING ROOM, KITCHEN, BATH, HALL, CLOSET, ENTRY, STORAGE, BEDROOM #2, BEDROOM #3

WINDOW SCHEDULE				
No	GLASS SIZE	SASH OPENING	ROUGH OPENING	DESCRIPTION
①	24½" x 16"	2'4" x 3'2"	2'6" x 3'5"	2 LT. D.H.
②	24½" x 28"	2'4" x 5'2"	2'6" x 5'5"	4 LT. D.H.
③	64½" x 58"	5'8" x 5'2"	(2-② units combined with③ fixed unit) 10'9" x 5'5"	DBLE PANE FIXED
④	28½" x 24"	2'8" x 4'6"	2'10" x 4'9"	4 LT. D.H.

6" wider and 10" higher than the window glass size. To figure the rough-opening width for our example, add 6" to the glass width: 28½" + 6" = 34½", or 2'-10½". To obtain the rough-opening height, add the upper and lower glass heights. Then add 10": 24" + 24" + 10" = 58", or 4'-10". These allowances are fairly standard. They provide room for plumbing, squaring, and normal adjustments to the window position. However, when the window manufacturer is known, always use its recommended rough-opening sizes because these will be more precise.

If the rough-opening size is not on the plans, it can be obtained from the window manufacturer's catalog. Sizes vary somewhat among manufacturers. Each catalog contains tables showing four width and height measurements for each window:

masonry opening, rough opening, frame size, and glass size. Some may also list the sash size. Note the headings in the top left corner:

- *Masonry* refers to the masonry opening. This is the size of the opening that should be used if the house is built using brick or stone.

- *Rough* refers to the rough opening. This is the size that should be used in wood-frame houses with wood, vinyl, or metal siding.

- *Frame* refers to the frame size. This is the measurement of the window from edge to edge, excluding exterior casings. In some cases, the unit dimension may be listed instead of the frame size. The **unit dimension** is the overall size of the window, including casings.

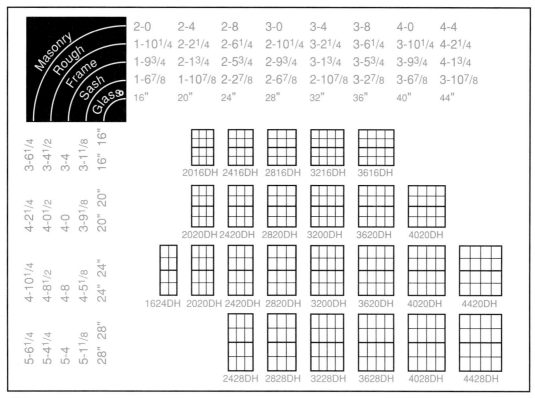

 Window Size Table

Key Framing Data An example of a manufacturer's size table for double-hung windows.

- *Sash* refers to the actual dimensions of the sash.
- *Glass* refers to the dimensions of the glass, both visible and covered, in a single sash.

Combination Windows

Many times, windows of various styles and sizes are combined to make up a larger unit for a particular room and use. These combined units are separated only by vertical wood pieces called **mullion strips**. Windows grouped in this way are sometimes referred to as *mulled windows*.

The rough opening for a combined unit is smaller than the rough openings for the individual units added together. For example, refer to the house plan on page 586. Note that the window schedule on the plan calls for a combination window unit in the living room. This unit consists of two No. 2 and one No. 3 window units. It would be important to consult the manufacturer's catalog to find the rough opening for this combined unit.

Installing Windows
Why must windows be plumb?

Careful installation is necessary for a window to function properly. It is also important because an improperly installed window can allow water to seep into the framing. This would eventually cause the framing to rot.

Before installation, apply a coat of primer to all wood portions of the window (including both sides of the jambs). This will increase durability. Do not apply primer to wood portions that will be varnished or similarly sealed later.

Windows are put in after the exterior walls have been sheathed but before the wood, vinyl, or metal siding has been installed. Many windows have an installation flange, often called a *nailing flange*, that makes installation easier. This is shown on page 588. The flange is either part of the window unit or made of separate pieces

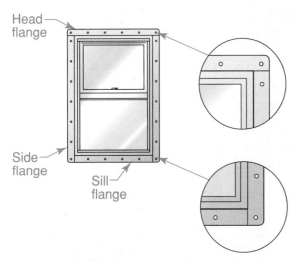

Head flange

Side flange

Sill flange

Window Installation Flange
Makes Mounting Easy Individual installation flanges on a double-hung window.

Standard Windows

Most window frames are installed in the same general way, regardless of style or manufacturer. However, you should always refer to the manufacturer's instructions for recommendations. For example, some manufacturers recommend that the sash be removed from the frame to prevent breakage and to allow for easier handling. Others specify not only that the sash be left in the frame but also that diagonal braces and, in some cases, reinforcing blocks, be left in place. This ensures that the frame remains square and in proper alignment during installation. If the sash is to be left in during installation, make sure the locks are engaged.

Another difference concerns the method for adding flashing to a window. The traditional technique is to use 8" wide splines of 15 lb. building paper. The bottom spline is installed first, then the sides, then the top piece. These are overlapped and stapled around the opening before the window is installed. However, many builders are now using strips of flexible, self-adhesive flashing instead. These strips are made of either

Builder's Tip

BRICK MOLDING During the manufacture of some windows, thick wood molding is permanently attached to the outer edges of the jambs. This molding is called *brick mold* or *brick molding*. Depending on the type of window, brick mold sometimes serves as a nailing flange, particularly on wood frame windows. Casing nails may be driven through the brick mold and into the sheathing. Later, siding will be installed so that it butts up to the outer edges of the brick mold.

butyl rubber or modified bitumen (*rubberized asphalt*). Check with the window supplier to see which method is recommended.

Once windows have been installed, they are surrounded with wood *trim* on the outside of the house and wood *casing* on the inside. For more on this subject see Chapter 26, "Molding & Trim."

Windows in Masonry Walls

Window manufacturers specify methods for installing their windows into masonry walls, such as those of brick veneer. A common method is to replace the installation flanges on the windows with metal jamb clips. The clips are screwed to the window jambs. They

Jamb Clips
Masonry Connection Metal jamb clips allow windows to be installed in a masonry wall.

inserted into grooves in the outer face of the jamb. See page 589 for steps in installing a window.

Tim Fuller

Installing a Flanged Window Window installation is a two- or three-person job. One installer works inside the house, while the others work outside. Always follow the window manufacturer's installation instructions, otherwise the manufacturer's warranty may be void. The following method assumes that housewrap (see Chapter 16) has already been installed over the sheathing.

Step 1 Prepare the window. Inspect the sash and frame for damage. If nailing flanges are separate pieces, insert them into their grooves on the window. Tap them into place with a wood block and hammer. The head flange should overlap the side flanges. The side flanges should overlap the sill flange.

Step 2 Cut the housewrap so the window can be placed in the rough opening. There should be a horizontal cut at the head jamb. Then slip the top flange of the window under the resulting flap. If there is no housewrap in place, some builders staple lengths of felt paper (9" to 12" wide) to the sheathing around the opening. These are called splines. The upper splines must overlap those that are lower. This helps to drain any moisture that might later get behind the siding.

Step 3 Insert the window. Large or heavy windows should be lifted by at least two people. Place the frame in the opening from the outside, allowing the subsill to rest in the rough opening. Hold the window in place against the sheathing. Center it from side to side in the opening.

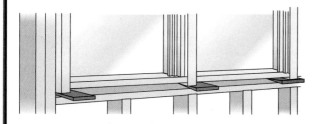

Step 4 Level the window sill by inserting blocks or tapered shims beneath it from inside the house. Place the shims or blocks under the legs near the corners of the window and at the center, as shown in the figure below. Check the window for plumb.

Step 5 When the window is plumb and level, nail through a flange at one corner, using a 1¾" roofing nail. Check the window again to be sure it is plumb and level. Check it for squareness by measuring diagonally across the corners. If the two measurements are the same, the window is square. If the window is not square, shim the side jambs as needed. Then recheck plumb and level. Measure across the window at the top, bottom, and center. Measurements should be equal.

Step 6 Nail each corner of the installation flange to secure the window. Check the window for easy operation. Then nail the entire perimeter of the installation flange. Space roofing nails every 6" to 8" or as recommended by the manufacturer. Remove any remaining packing on the window unit.

Step 7 From inside, fill gaps between the jamb and the framing with expanding foam sealant. Do not use too much insulation or the jambs will bow inward, and the window will not open properly.

are then nailed into furring strips or connected directly to masonry fasteners. There should be at least ½" clearance from the top of the masonry to the bottom of the sill, as shown on page 590.

Basement Windows

Basement window units are made of wood, plastic, or metal. In most cases, the sash is removed from the frame. The frame is set into the concrete forms for a poured wall. The wall is poured with the window frame in place. If the windows are to be set into a concrete block wall, special blocks are placed around the frame that accommodate the various frame types. The windowsills are usually installed after the basement floor framing is constructed.

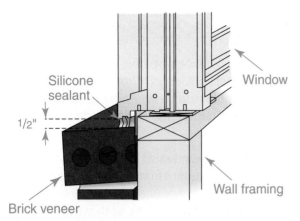

Silicone sealant

1/2"

Window

Wall framing

Brick veneer

 Installation Clearance
Gap Sealant Window installation in brick-veneer construction.

Installing Skylights

Why are skylights more likely to leak than windows?

Some skylights are complete units, ready to set into the roof. Others must rest on a lumber curb that lifts them above the level of the roof.

Because skylights are often high in a ceiling, they can be fitted with small motors that open and close them. The motors are controlled electronically from below. Rough wiring is put in before finished wall and ceiling surfaces are installed. Skylights without motors can be opened and closed with long poles or cranks.

Skylight Flashing Because skylights are located on roofs, they require a heavier duty flashing to prevent leaks than windows do. Great care must be taken to ensure this flashing is installed properly. The most durable flashing is made of copper, but aluminum flashing is more commonly used because it is much less expensive.

There are two methods of installing skylight flashing. *Step flashing* consists of small pieces of L-shaped metal that are interwoven with the roof shingles. The other option is *pan flashing*, where a one-piece metal assembly called a *pan* fits over the skylight curb. The pan must be fabricated by a sheet metal shop for a specific size of skylight. After pan flashing is in place, the roof shingles are installed. It is also possible to install a skylight in a tile roof. In such cases, flexible lead step flashing is often used.

Estimating Costs

The cost of an individual window unit depends on its quality, style, glass type, frame material, and any factory-applied finish. To determine an accurate cost, a complete list

 Curb-Type Skylight
Elevated from Roof Some curbs may be supplied by the skylight manufacturer, as shown here, or be constructed on site from 2x6 lumber.

 Skylight Flashing
Properly Flashed Pan flashing is placed before the roof shingles are installed.

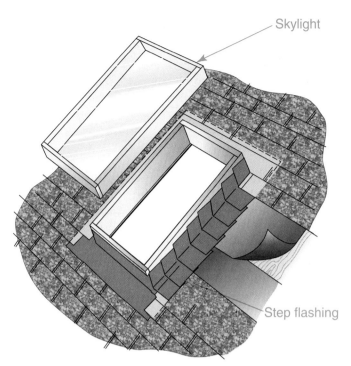

Skylight

Step flashing

Pan flashing

 Step Flashing
Many Pieces Like shingles, each piece of step flashing overlaps another to shed water.

of the windows should be submitted to the supplier for pricing.

The labor required for installing and setting windows depends on their size and style. The approximate time can be estimated as follows based on window glass size:

- Up to 10 sq. ft. of glass area: 1 hr.

 Pan Flashing
One Piece Seams in pan flashing should be soldered to prevent leakage.

- Up to 20 sq. ft. of glass area: 1½ hrs.
- Over 20 sq. ft. of glass area: 2 hrs.

These estimates include only the preparation of the opening and the actual installation of the window unit. They do not include installation of interior trim.

Section 20.2 Assessment

After You Read: Self-Check

1. Name two places where you can find dimensions for the rough opening of a window.
2. Describe the two types of installation flange.
3. Describe the traditional technique for adding flashing to a window.
4. Describe the difference between step flashing and pan flashing for skylights.

Academic Integration: Mathematics

5. **Problem Solving** Reno's Replacement Window Co. charges $50 per hour to install replacement windows. Figure the labor cost of installing four windows with a glass size of 30" by 48" windows, four windows with a glass size of 36" by 48", and two windows with a glass size of 72" by 48".

Go to connectED.mcgraw-hill.com to check your answers.

CHAPTER 20

Review and Assessment

Chapter Summary

Section 20.1
A single house has various types and sizes of windows. The six basic types include double-hung, casement, stationary, awning, hopper, and horizontal-sliding windows. The energy efficiency of a window depends on glazing, weatherstripping, materials, construction, and other features.

Section 20.2
A window schedule gives important specifications. Window installation requires two or more people. The care with which a window and its flashing are installed has a large impact on its ability to prevent air and water leaks.

Review Content Vocabulary and Academic Vocabulary

1. Use each of these content vocabulary and academic vocabulary words in a sentence or diagram.

Content Vocabulary
- glazing (p. 577)
- sash (p. 577)
- muntin (p. 578)
- window schedule (p. 585)
- unit dimension (p. 586)
- mullion strips (p. 587)

Academic Vocabulary
- sufficient (p. 576)
- technology (p. 578)

Speak Like a Pro

Technical Terms

2. Work with a classmate to define the following terms used in the chapter: *egress* (p. 577), *lite* (p. 577), *frame* (p. 577), *prehung window* (p. 577), *double-hung* (p. 578), *jamb* (p. 578), *single-hung* (p. 578), *true divided lite* (p. 578), *fixed-glass windows* (p. 578), *glass block* (p. 579), *panels* (p. 579), *awning* (p. 580), *hopper* (p. 580), *guides* (p. 580), *hybrid windows* (p. 580), *thermal break* (p. 581), *solar-heat-gain coefficient* (p. 582), *visual transmittance (VT)* (p. 582), *double-glazed windows* (p. 582), *low-emissivity* (p. 583), *warm edge spacers* (p. 583), *roof window* (p. 584), *rough opening* (p. 585), *nailing flange* (p. 587).

Review Key Concepts

3. List the types of windows used in residential construction.

4. Explain how windows are made energy efficient.

5. Describe how a window schedule and a manufacturer's size table are read.

6. Demonstrate how a standard double-glazed or casement window is installed.

7. Show how window costs are estimated.

Critical Thinking

8. Analyze With safety in mind, state which of the five basic window styles would be the least accessible for rescuers. Explain your reasoning.

9. Explain Which type of window material would you not recommend to an individual installing windows in cold climates? Explain your recommendation.

Academic and Workplace Applications

STEM Mathematics

10. Window Measurements The only window in a second-floor bedroom has a height of 24 inches. If building codes require an unblocked open area of 5.7 square feet, what is the minimum width of the window? Express your answer in inches. Round up to the nearest whole inch.

Math Concept Computing with measurements should be done using equivalent units of measure.

Step 1: To convert 5.7 square feet to square inches, multiply by the number of square inches in one square foot.

Step 2: Divide the total area in inches by the height of the window.

Step 3: Round up the quotient to the nearest whole inch.

STEM Technology

11. Energy Efficiency It is estimated that 20 to 30 percent of the heat lost from some houses is through the windows. This loss is due to air leaking around the window or by heat being radiated through the glass. As heating and cooling costs have climbed, manufacturers have greatly improved the energy efficiency of windows. Double- and triple-glazed windows are one example of technological advances that have lead to more efficient windows. Research the technological advances that have made double-glazed windows more efficient. Write a paragraph summarizing your findings.

21st Century Skills

12. Collaboration Working in groups of 3 or 4, use the diagram your instructor has handed out to label the parts of an assembled double-hung window. Discuss the diagram with your team members and correct any errors. Finally, elect a representative from your team to present your diagram to the class.

Standardized TEST Practice

Multiple Choice

Directions Choose the phrase or word that best answers the following questions.

13. Which type of window has a side-hinged sash that swings inward or outward?

  a. stationary window

  b. double-hung window

  c. casement window

  d. hopper window

14. Generally, the total area of window glass in a room should be no less than what percent of the floor area?

  a. 10

  b. 8

  c. 3

  d. 15

15. What term is used to describe a short vertical or horizontal piece of wood used to hold a pane of glass?

  a. mullion strip

  b. screen

  c. sash

  d. muntin

TEST-TAKING TIP

When answering multiple-choice questions, ask yourself if each option is true or false. This may help you find the best answer.

*These questions will help you practice for national certification assessment.

CHAPTER 21

Residential Doors

Section 21.1
Door Basics

Section 21.2
Exterior Doors & Frames

Section 21.3
Interior Doors & Frames

Chapter Objectives

After completing this chapter, you will be able to:

- **Identify** the various types of interior and exterior doors and door hardware.
- **Explain** how to handle a door properly at a job site.
- **Identify** the hand of any door.
- **List** at least three aspects of exterior door construction and installation that improve energy efficiency.
- **Identify** the correct clearances for installing exterior and interior doors.
- **Summarize** how to install an interior pre-hung door.

Discuss the Photo
Careful Fitting A door has many functions. It improves security, ensures privacy, protects against the weather, and lends beauty and character to the home's architecture. All of these functions require careful installation. *Why do most doors open into a room rather than outwards?*

Writing Activity: Descriptive Paragraph
Think about the doors in your home. Consider their locations, their sizes, and the materials they are made of. Choose one door. What is its function? Is it an interior door (for the inside of a house) or an exterior door (a door that leads to or from the outside)? Describe the door and its function in a descriptive paragraph.

Chris Rose/Getty Images

Chapter 21 Reading Guide

Before You Read Preview

There are many kinds of residential doors. Builders and carpenters can choose from many different designs, materials, assembly techniques, and sizes. Choose a content vocabulary or academic vocabulary word that is new to you. When you find it in the text, write down the definition.

Content Vocabulary

- passage door
- strike plate
- solid-core construction
- hollow-core construction
- stiles
- door frame
- gain
- lockset
- hand

Academic Vocabulary

You will find these words in your reading and on your tests. Use the academic vocabulary glossary to look up their definitions if necessary.

- dexterity
- accommodate
- consistent

Graphic Organizer

As you read, use a chart like this one to categorize types of doors. Add more rows if necessary.

Exterior Doors	Interior Doors
sliding-glass doors	
French doors	

Go to connectED.mcgraw-hill.com to download this graphic organizer.

Door Basics

Types of Doors

What is hollow-core construction?

All doors require mounting hardware, such as hinges and tracks. They also require hardware for operation, such as knobs or pulls. Beyond that, however, doors come in a variety of types. They can be categorized in the following ways:

- Location (exterior wall or interior wall)
- Material (wood, fiberglass, vinyl, metal, and others)
- Operation (hinged doors, sliding doors, folding doors)
- End use (new construction or replacement)
- Purpose (closet doors, garage doors, fire doors)
- Construction type (solid core, foam core, hollow core)

 A Patio Door
Door Features This sliding door can serve as an entrance, but it also allows views.

- Style (raised panel, flat panel, arched top)
- Installation method (pre-hung, hung on site)

Any door can be described using a combination of any of these characteristics. For example, a door may be a pre-hung, exterior, hinged, wood door having raised panels.

The most common type of door is a flat-panel or raised-panel passage door. A panel is a wide piece of solid wood or plywood. A **passage door** is a door that swings open and closed on two or more leaf hinges mounted along one side. It allows passage from one area into another. It has two doorknobs, a latching mechanism, and sometimes a locking mechanism. The *latch* slips into a hole in a **strike plate**, a metal plate which fits into an opening in the door jamb. The latch holds the door closed.

An exterior passage door is often surrounded by moldings that emphasize the architectural style of the house. This is particularly common on houses with traditional styling.

Flat-Panel Doors

Flat-panel doors are sometimes referred to as *flush doors* or *slab doors*. Their entire surface is flat. They are made with plywood, hardboard, metal, or some other suitable facing applied over a light framework and core. Cores are either solid or hollow, as shown in Core Construction.

Solid-Core Construction Solid-core construction consists of strips of wood, particleboard, rigid foam, or other core material covered with a thin outer material, such as wood veneer (A). Solid-core construction reduces warping and is generally preferred for exterior doors. It is also more fire-resistant than hollow-core doors.

zveiger alexandre/Getty Images

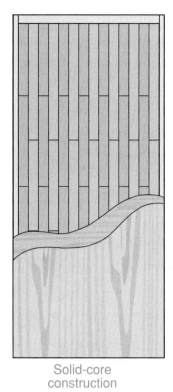

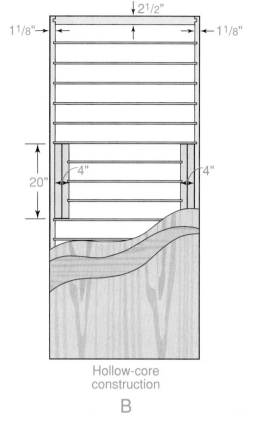

Solid-core
construction

A

Hollow-core
construction

B

⬢ **Core Construction**
Hidden Structure The core construction of a flat-panel door can be **A.** solid or **B.** hollow.

Hollow-Core Construction Hollow-core
construction consists of a light framework
of wood or corrugated cardboard faced with
thin plywood or hardboard (B). Plywood-
faced flush doors usually have surface
veneers of birch, oak, or mahogany. Most of
these are suitable for natural finishes. Other
wood veneers are usually painted, as are
hardboard-faced doors. Areas of a hollow-core
door are built up with solid wood, as shown
in B. This provides solid backing for the lock.
Solid wood around the edge allows a carpenter to trim the door to fit the opening.

Raised-Panel Doors

   Raised-panel wood doors have panels
that are thicker at the center than at the
edges. They consist of **stiles** (vertical side
members), rails (horizontal crosspieces), and
the raised panels that fill the spaces between.
Stiles and rails are generally made of solid
wood. The panels may be made of solid wood
or plywood.

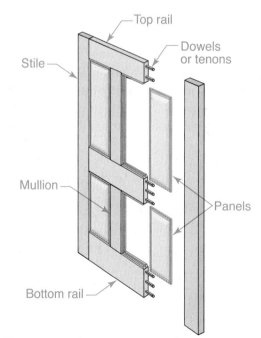

⬢ **Raised-Panel Wood Door**
Basic Anatomy The individual parts of a door are held
together with a combination of interlocking joinery and
adhesives. Nails and screws are not typically used.

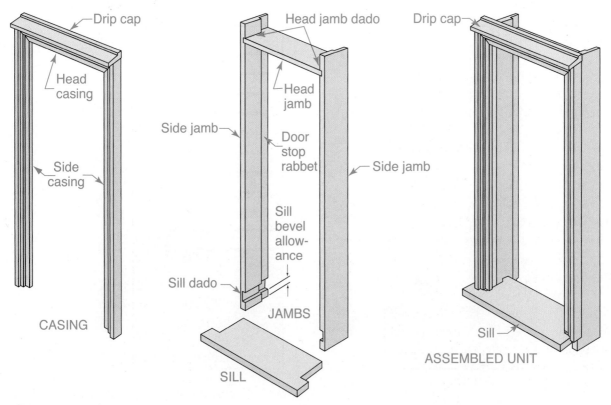

Labels on diagram:
Drip cap — Head casing — Side casing — CASING
Head jamb dado — Head jamb — Side jamb — Door stop rabbet — Sill bevel allowance — Sill dado — JAMBS — SILL
Drip cap — Side jamb — Sill — ASSEMBLED UNIT

Wood Door Frame
Door Frame Parts Parts of a door frame: sill, casing, and jamb assembly.

Door Frames

A door is mounted on a door frame, which is attached to the wall framing. The **door frame** consists of two side jambs and a head jamb, and is the surrounding assembly into which the door fits. Exterior door frames may also include a sill or threshold. Molding

is usually nailed to the door frame on the inside of the house after the frame is installed. Exterior molding and wood stops are sometimes attached by the door manufacturer. However, they are often installed on site.

The door jambs fit inside a rough opening in the wall framing. They are installed so that their outer edges will be aligned with the finished wall surfaces. For example, a standard interior jamb is 4⁹⁄₁₆" wide. Jambs are manufactured in standard widths to suit various wall thicknesses. However, wood jambs may be easily cut to fit. If the wall is unusually thick, strips of solid wood called *jamb extensions* can be nailed to the edges of the jamb to make it wider. Jambs may also be custom-made to any size.

Wood is the traditional material used for jambs. However, metal is not uncommon in a residential building, particularly when the door itself is metal. When exterior door frames include a sill, it is usually

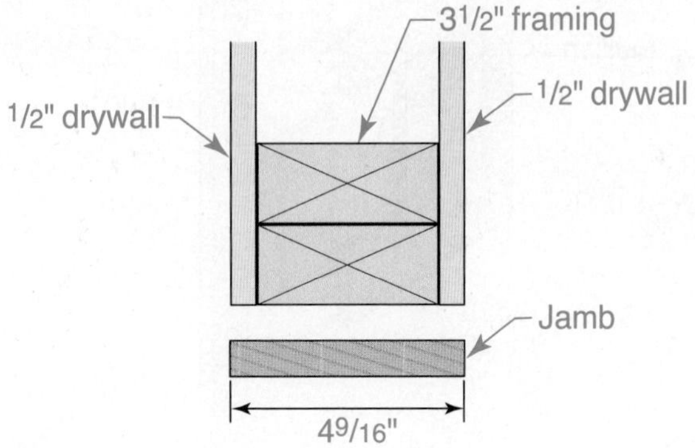

Diagram labels:
3¹⁄₂" framing — ¹⁄₂" drywall — ¹⁄₂" drywall — Jamb — 4⁹⁄₁₆"

Jamb Width
Sized to the Wall The width of an interior door jamb should match the overall thickness of the wall.

made of oak (for wear resistance), aluminum, or a combination of wood and aluminum.

Door frames may be purchased knocked down or preassembled with just the exterior casing or brick molding applied. Brick molding is a type of exterior casing. In some cases, door frames come preassembled with the door already hung in the opening. These are called *pre-hung doors.*

Door Frames for Masonry Preassembled metal door frames with welded corners are often specified for exterior walls made of concrete block. These frames are installed as the masonry is laid up. Metal tabs embedded in selected courses hold the frame in place. The hollow area between the frame and the masonry may be filled with masonry grout. If this is done, temporarily brace the frame until the grout hardens. This will prevent the frame from bowing outwards. Metal thresholds should be bedded in high-performance sealant or otherwise sealed according to the manufacturer's instructions.

KD Metal Door Frames Metal door frames are sometimes installed in residential construction because of their durability. Knock-down (KD) frames consist of three U-shaped channels sized to fit over the combined thickness of framing and drywall. The top channel, called a header, is installed first. The jamb channels are installed next. After the door has been hung, the frame should be plumbed, squared and secured by screws. Finally, metal casing can be snapped in place over the frame.

Reading Check

Recall *Which type of door construction is preferred for exterior doors?*

Door Hardware

What materials would not be suitable for door hardware?

Hardware for doors is made using various metals and finishes. Steel, brass, bronze, and nickel are perhaps the most common. The type of metal and its finish affect not only how the hardware looks, but also its durability and cost.

Hinges

The *loose-pin butt mortise* hinge is most often used for hanging residential doors. It has two rectangular *leaves* that pivot on a loose metal pin that can be removed. One leaf is screwed to the edge of the door. The other leaf is screwed to the door jamb. The pin connects the two. The door may be removed from the jamb simply by removing the hinge pins and lifting the door out. On exterior doors, hinges are mounted so the pins cannot be removed from the outside. This prevents an intruder from getting into the house simply by removing the hinge pins. The size of a loose-pin butt mortise hinge is the length of a leaf in inches.

The leaves fit into gains cut into the edge of the door and the jamb. A **gain** is a

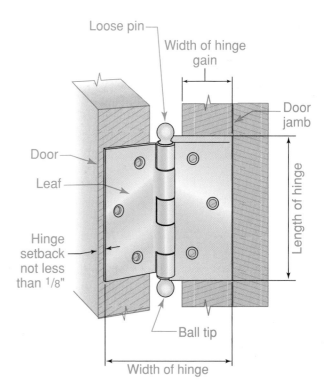

Typical Door Hinge
Basic Hinge A loose-pin butt mortise hinge.

mortise (notch) that has a depth equal to the thickness of a single hinge leaf.

Locksets

Passage doors are fitted with different types of opening hardware. The assembly of knobs, latch, and locking mechanism is called a **lockset**. In some cases, lever handles replace the knobs. Lever handles are easier to use for people with limited strength or **dexterity**.

Locksets are available in three basic types: entry, privacy, and passage. *Bored locks* fit into a large hole bored into the face of the door stile. *Mortise locks* fit into a large notch cut into the edge of the door stile.

Entry Locksets These are used on exterior doors and contain a locking mechanism and key. They may be of the bored or mortise type.

Privacy Locksets These light-duty locksets are used on doors leading to bedrooms and bathrooms. They can be locked from the inside by pressing or turning a button in the knob. There is usually a small slot or hole in the non-locking knob. A small screwdriver inserted into the hole pops the lock open. This is a safety feature that allows the door to be opened from the outside in an emergency. All privacy locksets are bored locks.

Passage Locksets These are sometimes referred to as *latchsets*. They are similar in design to privacy locksets but do not include a locking mechanism. They are often used on closet doors.

Storage and Handling

When should doors be delivered to a job site?

Proper care and finishing of a door will ensure the most service and satisfaction. Proper door care includes the following:

- Wood doors should not be delivered to the house until wet materials, such as plaster and concrete, have given up most of their moisture. Otherwise, the doors can swell and will not fit properly.

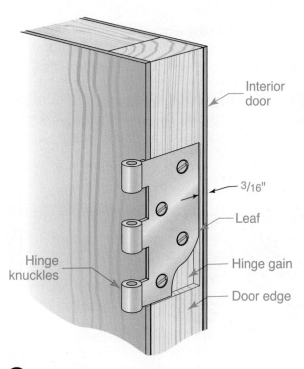

Interior door

3/16"

Leaf

Hinge knuckles

Hinge gain

Door edge

  Hinge Gain
Correct Depth The hinge gain, or mortise, recesses each hinge leaf so that its surface is flush with the surface of the door's edge.

Locksets
Different Grips Two locksets with lever handles and one with knobs. The tall model is a mortise-type lockset. *How do passage locksets differ from privacy locksets?*

- Keep all doors away from unusual heat or dryness. Sudden changes, such as heat forced into a building to dry it out, should be avoided.
- Store doors under cover in a clean, dry, well-ventilated area.
- Condition wood doors to the average local moisture content before hanging (see Chapter 12).
- Store doors on edge on a level surface.
- Handle doors wearing clean gloves. Bare hands leave finger marks and soil stains on surfaces that have not yet been sealed or painted. When moving a door, carry it. Do not drag it.
- Seal the top and bottom edges of wood doors immediately upon taking delivery to prevent moisture from reaching the end grain. Any warranty that covers the door may be void if the door is not sealed promptly and properly. Even doors that are ordered factory-primed may not have properly sealed edges.

Direction of Swing

What can happen if a door swings in the wrong direction?

When ordering and installing a door, it is important to know the direction in which the door should swing. A door that opens in the wrong direction may interfere with wall switches, cabinetry, and traffic patterns. The direction in which a door swings is called its **hand**. Doors are either left-hand doors or right-hand doors. Unfortunately, there are several methods commonly used to describe the hand of a door. Some of them contradict each other, and even door companies differ in the way they describe door hand.

Here are the two most common methods to determine the hand of a door:

- *Method 1* Stand with your back against the hinge jamb. A door that swings toward your right is a right-hand door. One that swings to your left is a left-hand door. This is shown in on page 602.
- *Method 2* Face the inside of a closed door. In general, the inside of the door is the side from which the hinge knuckles are visible when the door is closed. If the hinges are on the right side of the door, it is a right-hand door. If they are on the left, it is a left-hand door.

As you can see, each of these methods gives different results. It is therefore important to make sure you understand what is intended when ordering or installing a door. The hand of a door will generally be noted on the building plans and door schedules. The door swing is indicated by a curved line that visually indicates the direction.

Science: Earth Science

Humidity Humidity is the amount of water in the air. It is expressed as a percentage. You know that wood can shrink or swell as a result of humidity. What are some things or activities that could increase the humidity of the air in your home?

Starting Hint Think about household activities that involve water.

JOB SAFETY

CARRYING A DOOR All doors are unwieldy to carry, particularly pre-hung units. Solid-core doors are very heavy. Get help when moving a door. This will reduce the chance of strained muscles and any damage to the door or finished wall surfaces.

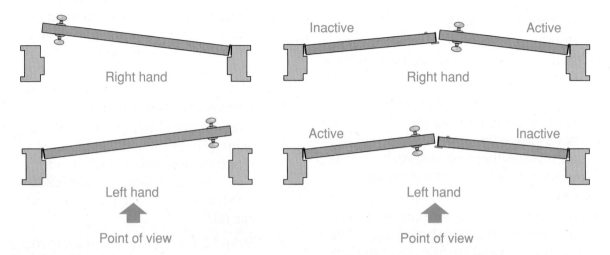

SINGLE DOOR

Hand of door may be determined by referring to sketches below. Door must always swing away from the point from which it is viewed.

Right hand

Left hand

Point of view

PAIRS OF DOORS

Hand of doors is determined by location of active leaf when doors swing away from point viewed.

Inactive   Active

Right hand

Active   Inactive

Left hand

Point of view

 Door Hand

Two Methods Two ways to determine the swing direction of a door. Double doors are divided into active and inactive sections.

Section 21.1 Assessment

After You Read: Self-Check

1. What is the most common type of door used in residential construction?
2. Describe a hollow-core door.
3. What is the standard width of an interior door jamb?
4. What type of hinge is most often used for hanging residential doors?

Academic Integration: English Language Arts

5. **Gathering and Displaying Information** A table organizes information into rows and columns. With a partner, make a table listing three different doors in the building where you live or in your school. Include the following information about each door: location, dimensions, type of door (exterior or interior), type of lockset, and the hand of the door. Record this information in the table. Use one column for each door.

	Location	Dimension	Type of Door	Type of Lockset	Hand
Door 1					
Door 2					
Door 3					

Go to **connectED.mcgraw-hill.com** to check your answers.

Exterior Doors & Frames

Exterior Doors

How does climate affect door choice?

Many combinations of door and entry designs, trim, and decorative elements are available, from modern to classical. Just as with windows, care should be taken to select a door that is correct for the architectural style of the house. Strength, durability, and energy efficiency are also important characteristics of exterior doors.

Exterior doors are sometimes installed along with glazed panels called sidelights and transoms. A sidelight is a tall glazed panel installed at one or both sides of a door. A transom is a horizontal glazed panel that fits above a door. Sidelights and transoms do not open. Their role is to allow natural light into the entry area.

Exterior doors are usually 1¾" thick and not less than 6'-8" high. The main entrance door is required by building codes to be at least 3' wide, but a side or rear service door may be 2'-8" wide. A hardwood or metal threshold helps to prevent water from getting under a door. It also serves as a base for weatherstripping.

Types of Exterior Doors

Types of exterior doors include combination, sliding-glass, French, entry, fire-rated, and garage doors.

Combination Doors Lightweight screen or storm doors are often attached to the frame of an exterior door. Screen doors allow extraventilation during hot weather. Storm doors protect the main door from weather damage and reduce heat loss in cold weather. Combination storm and screen doors are the most common. The screen panel and the storm panel may be exchanged as needed. Door Details on page 604 shows design details that include a combination door.

©George Hammerstein/Corbis/Glow Images

⌃ Entry Door
Elegant Entry This entry includes a solid wood door with glass inserts and glass transoms.

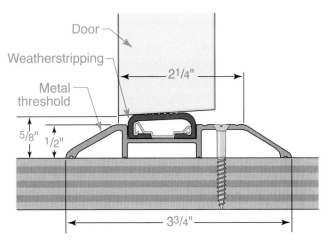

Door
Weatherstripping
Metal threshold
2¹/4"
5/8" 1/2"
3³/4"

⌃ Door Threshold
Weather Blocker A metal threshold with a vinyl weatherstripping insert. *What does this prevent?*

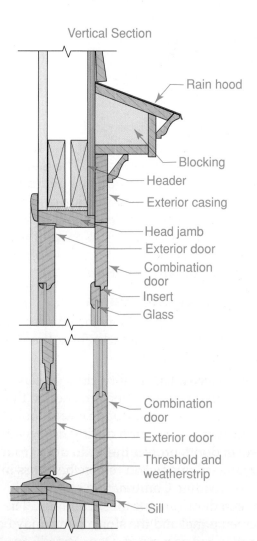

Vertical Section

Rain hood

Blocking

Header

Exterior casing

Head jamb

Exterior door

Combination door

Insert

Glass

Combination door

Exterior door

Threshold and weatherstrip

Sill

Door Details
Traditional Style Cross section details of a door assembly. Note how the combination door relates to the exterior door.

Sliding-Glass Doors *Sliding-glass doors*, or *patio doors*, are available with either wood, vinyl, or metal frames. Insulating glazing helps limit heat loss, but these doors are not as energy efficient as other types. One side of the door is usually stationary; the other side slides. Door operation may be right-hand or left-hand (as viewed from the outside). Doors are available in widths from 30" to 120".

French Doors *French doors* are hung in pairs on hinges located at each side of the door opening. The doors swing toward each other and meet at the center. French doors are usually fully glazed and often lead to a deck

Sliding-Glass Door
One Side Opens A wood sliding-glass door with muntins.

or patio. They can swing into the room or out. When both sides of the door are open, the area between the jambs is unobstructed and unusually wide.

Molding with a T-shaped profile is attached to the edge of one of the doors.

French Doors
Both Sides Open A pair of French doors with decorative muntins.

This provides a stop against which the other door closes. It is sometimes called an *astragal stop*. Because there is no framing in the middle of the opening, sealing exterior French doors against the weather can be difficult. Be sure to examine the weather-stripping carefully.

Entry Doors An entry door is mounted on three or more mortise hinges (see Section 21.1). The hinges are mounted so that the door swings inward. Entry doors may be solid or glazed. A glazed door is any door that contains glass. The glass in a door is called a *lite* (a panel of glass, like that used in window construction). The glazed portion is often decorative.

According to building codes, doors are a hazardous location for glazing. For this reason, glazing in any door must be shatter-resistant. Each lite must be permanently marked at the factory to indicate its suitability for use in hazardous locations. The mark is etched or sandblasted into a corner of the glass, where it will not be covered by muntins.

Entry Door
Matching Lites A glazed solid-wood entry door with two side lites.

Joe Mallon

Fire-Rated Doors *Fire-rated doors* are built to resist the passage of fire. These doors are not fireproof. They withstand fire only long enough for occupants to reach safety. Though more commonly used in commercial construction, fire-rated doors are suitable for some places in a house. For example, building codes require that any door between a house and an attached garage have one of the following characteristics:

- A 20-minute fire rating
- Solid wood construction not less than 1⅜" thick
- Solid-core or honeycomb-core steel construction not less than 1⅜" thick

Fire doors generally should not be trimmed in the field. This could affect their fire rating.

Garage Doors The unusually large openings in garages require special types of doors. The standard single garage door is 9' wide and 7' high. Double garage doors are usually 16' wide. The opening is lined with wood jambs, similar to the opening for other doors. However, the jambs are made from 2×6 or sometimes 5/4×6 lumber. The jambs are nailed directly to the rough opening.

The most widely used garage door in new construction is the overhead sectional door. For an overhead door, four or five sections are hinged together. The sections may be made of wood, steel, or fiberglass. Rollers on the ends of each section ride in a metal track at the side of the opening for the door. Mounting clearance required above the top of sectional overhead doors is usually about 12". However, low-headroom brackets are available when such clearance is not possible. Overhead doors are usually installed by the door supplier.

The bottom edge of a garage door should be scribed and cut to conform to the garage floor. Weatherstripping is recommended for the bottom rail. It seals any minor irregularities in the floor and acts as a cushion in closing. The header over the door opening may be a steel I-beam, built-up framing lumber, a glulam beam, or laminated-veneer lumber (LVL). Sizing charts for garage-door headers are found in building code books and in instructions provided by header manufacturers.

Sectional doors may be raised manually, but it is common to install an electric door opener. For safety, building codes require sensors on all newly installed garage doors fitted with electric openers. These sensors, one on each side of the opening, detect the presence of a person or object beneath the door. Then they will not allow the door to close.

Installing Exterior Doors

Installing an exterior door requires a high level of ability. In terms of installation technique, there are two basic types of doors: pre-hung doors and those hung on site. Hanging a door on site is described here. A pre-hung door is installed as an entire unit, much as a pre-hung window is installed as a unit. Follow manufacturer's instructions when installing a pre-hung door.

Estimating Materials and Time

The cost of materials for an exterior door and frame depends on the style and trim.

An accurate price should be obtained from a local supplier.

A conventional exterior frame and brick molding with an oak sill requires about two hours to assemble and install. It requires one extra hour to hang the door and half an hour to install the lockset. Combination storm and screen doors require about one hour for installation.

Installing the Door Frame

Before installing the frame, prepare the rough opening. The opening should be about 2" wider and 2" higher than the door. The sill should rest firmly on the floor framing, which must sometimes be sloped to **accommodate** solid wood sills. The top of the sill should be even with the finished floor surface. This will allow the threshold to bridge any gap between the sill and the flooring. The top of the threshold should be higher than the finished floor. This allows the door to open over entry rugs or mats.

Another method would be to use an adjustable metal sill. This type of sill has a built-in threshold. It often incorporates weatherstripping.

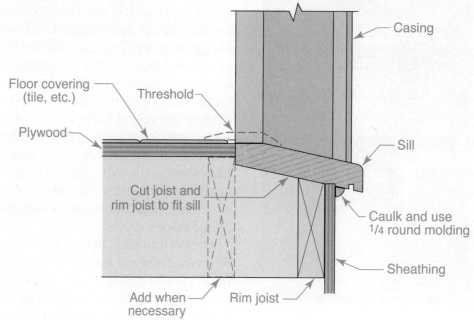

Floor covering (tile, etc.)
Threshold
Plywood
Casing
Sill
Cut joist and rim joist to fit sill
Caulk and use 1/4 round molding
Add when necessary
Rim joist
Sheathing

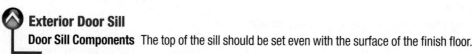

Exterior Door Sill
Door Sill Components The top of the sill should be set even with the surface of the finish floor.

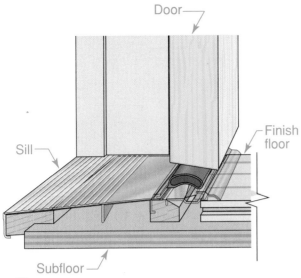

Adjustable Door Sill

Time Saver This preassembled door frame features a sill that is adjustable.

The trimmer studs should be set so that they are plumb and straight. This is conveniently done with a 6' long level or by measuring to the string of a plumb line hung from the head jamb. Once the trimmers are plumbed, they should be secured to the king studs in a way that prevents them from moving. Many carpenters clinch trimmers. They do this by driving a 16d nail partway into the trimmer at various points. They then bend it over horizontally so that the edge of the nail head is embedded in the edge of the king stud. Last, they drive a second nail partway into the king stud and bend it downwards to secure the first nail. Another method is to use wood shims to plumb each trimmer. The trimmer is then face-nailed to the king stud through the shims.

Line the rough opening with a strip of 15 lb. asphalt felt or flexible self-adhesive flashing that is 10" or 12" wide. Another way to protect this area from moisture is to install a prefabricated pan flashing system. This ensures that any moisture that gets under the sill cannot reach the subflooring or floor joists below.

To install the frame, tip the frame into place in the rough opening. Brace it loosely to keep it from falling out during adjustment.

Once it is in place, the frame must be plumbed carefully. This will ensure that the door will close properly and that the weatherstripping will be effective.

Plumb the frame by inserting wood shingles as wedges. Check the sill with a level, and wedge it up as necessary. Insert wedges on each side alternately between the side jambs and the trimmer studs until the space between them is exactly the same on both sides. Then drive a 16d casing nail through the side casing and into the trimmer studs on each side, near the bottom of the casing. This will hold the sill in position. Drive the nails in only part way. Do not drive any nails all the way in until all the nails have been placed and a final check has been made for level and plumb.

Next, place your level against one of the side jambs. Adjust the remaining wedges on that side until the jamb is perfectly true and plumb. Repeat on the other side. Make a final check for level and plumb. Fasten the frame in place with 16d casing nails driven through the casings into the trimmer studs

Installing a Door Frame

Straight and Plumb Installing and leveling the frame.

and the door header. Place nails ¾" from the outer edges of the casings and space them about 16" on center. Set all nails with a nail set.

Preparing the Door Check the door for imperfections as you remove any protective packaging. The door should be square and flat, with smooth and unmarked surfaces. You may notice a minor warp or bow caused by unequal moisture conditions on the two sides. The door will usually straighten when the moisture equalizes.

Determine from the floor plan which edge of the door is the hinge edge and which is the lock edge. Mark both door edges and the corresponding jambs accordingly.

Carefully measure the height of the finished opening on both side jambs and the width of the opening at top and bottom. The opening should be perfectly rectangular. When hung properly, the door should fit with an opening clearance of ¹⁄₁₆" at the sides and on top. If the door has a sill but no threshold, the bottom clearance should be ¹⁄₁₆" above the sill. If it has a threshold, the bottom clearance should be ⅛" above the threshold. Any sill or threshold should be in place before the door is hung. Lay out

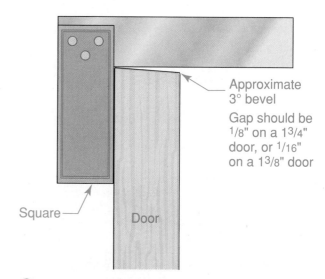

Approximate 3° bevel

Gap should be ¹⁄₈" on a 1³⁄₄" door, or ¹⁄₁₆" on a 1³⁄₈" door

Square

Door

Beveling the Lock Edge
Gradual Trimming Use a square to gauge the bevel while planing the edge of a door.

the dimensions of the finished opening, less clearance allowances, on the door.

Plane the door edges to the lines. Set the door in the opening frequently to check the fit. Bevel the lock edge so that the inside edge will clear the jamb. This angle is about 3°.

As an aid in fitting the door, you might want to build a *door jack,* or *door stand.* The

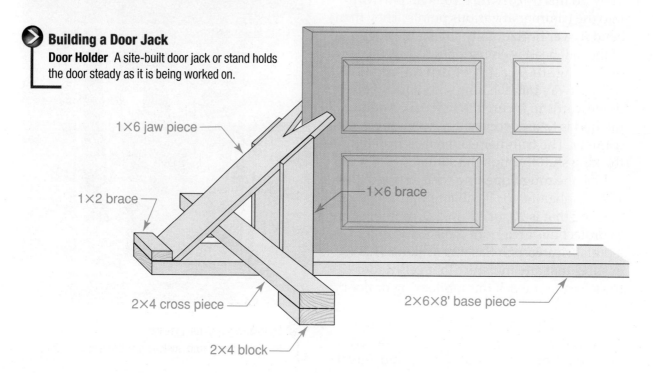
Building a Door Jack
Door Holder A site-built door jack or stand holds the door steady as it is being worked on.

1×6 jaw piece

1×2 brace

1×6 brace

2×4 cross piece

2×6×8' base piece

2×4 block

jack holds the door upright for planing edges and for the installation of hardware. Commercially made jacks are also available.

Some carpenters prefer to scribe a door to the jamb opening directly. This removes the need to move the door frequently to check its fit. To do this, the door is held against the opening and marked. A simple tool called a door hook, can be used to hold the door in position as it is marked.

Hanging the Door

After the door has been properly fitted, lay out the locations of the hinges on its edge and on the hinge jamb. Exterior doors usually have three hinges. If the exact positions are not specified, the measurements used are those shown in Laying Out Hinge Locations. For an exterior door, a 3½" or 4" hinge is recommended. Three or more hinges help to prevent the door from warping.

Set the door in the frame and force the hinge edge of the door against the hinge jamb with a wedge. See Wedge A. Then insert a 4d finish nail between the top of the door and the head jamb. Force the top of the door up against the nail with another wedge (Wedge B). Since a 4d finish nail has a diameter of ⅟₁₆" (the standard top clearance), the door is now at the correct height.

Measure the distance from the top of the door to the top of the upper hinge and from the floor up to the bottom of the lowest hinge. Mark these locations.

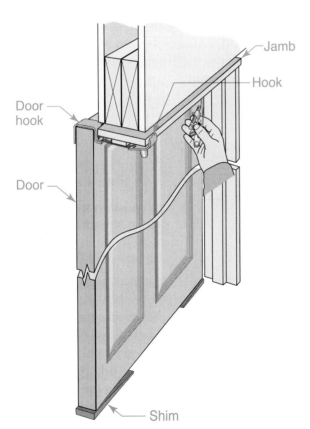

Using a Door Hook
Door Holder A door hook can be used to hold a door in position for scribing.

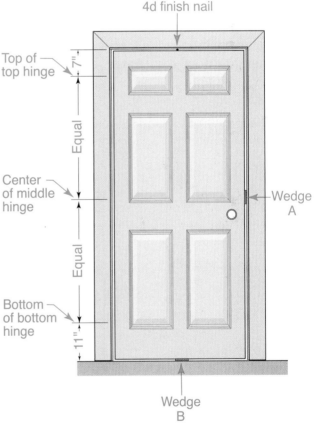

Laying Out Hinge Locations
Basic Layout Measurements commonly used in laying out hinge locations on the exterior door and door jambs.

Butt-Hinge Marking Gauge
Mark the Gains A butt-hinge marking gauge is struck with a hammer to create incised layout marks for a hinge gain.

When marking for the middle hinge, remember that the mark must locate the center of the hinge. If a 4" hinge is used, measure 2" from each side of the location line and make a mark. The gain will be cut between the marks.

Remove the door from the opening. Place it in a door jack and lay out the outlines of the gains on the edge of the door using a hinge leaf or a butt-hinge marking gauge as a marker. The door-edge hinge setback (see page 599), should not be less than ⅛". It is usually about ¼". Lay out gains of exactly the same size on the hinge jamb. Chisel out the gains to a depth equal to the thickness of a single hinge leaf.

Separate the hinge leaves by removing the pin. Screw the leaves into the gains on the door and the jamb. Make sure that the leaf into which the pin will be inserted is in the uppermost position when the door is hung in place.

Hang the door, insert the pins, and check the clearances at the side jambs. If the clearance along the hinge jamb is too large (more than ¹⁄₁₆") and that along the lock jamb is too small (less than ¹⁄₁₆"), remove the pins from the hinges and remove the door. Then remove the hinge leaves from the gains and deepen the gains slightly. If the clearance along the hinge jamb is too small and that along the lock jamb is too large, the gains are too deep. This can be corrected by shims of stiff paper stock (such as business cards) placed in the gains.

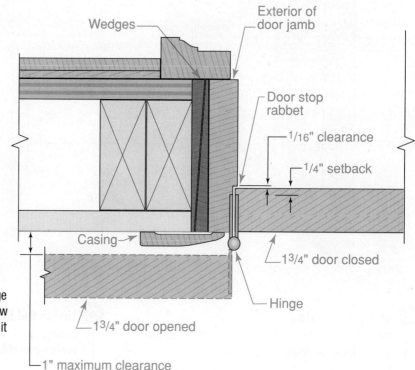

Door Swing Clearances
Important Details The door hinge should be set back enough to allow the door to clear the casing when it is swung wide open.

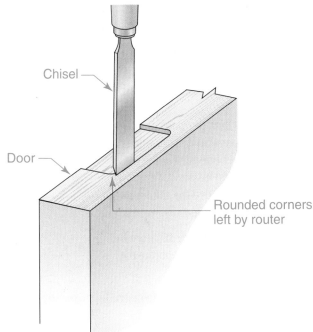

Chisel

Door

Rounded corners left by router

Butt-Hinge Template
Time Saver Use a butt-hinge template and a router to cut hinge gains on the edge of a door.

Routing for a Butt Hinge

A special metal template is available for locating the gains for butt hinges. The butt-hinge template may be adjusted for most common hinge spacings. It is easily mounted on the door by driving small pins into the door edge.

The template guides a router, and the gains are cut quickly and accurately. The template is then transferred to the door jamb for cutting the gains that match those on the door. Because router bits leave a rounded corner, the corners must be chiseled square. An alternative is to use hinges with leaves that have rounded corners. The radius of the corners may vary by brand, size, or type of hinge.

Installing an Entry Lockset

Entry locksets come in many styles, from very simple to very sophisticated. Sometimes a separate deadbolt is installed to increase security. A deadbolt can have a separate key from that of the lockset, or it can have a matching key. A bored-hole lockset is inexpensive and fairly easy to install. Steps for installing a lockset are given on pages 612–613. An electric drill fit with two different sizes of hole-saws, is required

Squaring the Hinge Gains
For a Good Fit If the hinge to be installed has square corners, use a sharp chisel to square up rounded corners left by a router.

for this work. Forstner bits can also be used. The smaller hole-saw (⅞") is for the latch assembly. The larger one (2⅛") is for the knob assembly.

Tools for Lockset Installation
Two Sizes Two holesaws, a drill, and a latch assembly.

Installing a Bored-Hole Lockset Methods for installing locksets vary with the manufacturer. Always refer to the instructions. A general procedure is explained here. The same method is used for locksets on interior doors.

Step 1 Open the door to a convenient position. Place wedges under the bottom near the outer edge to hold it steady. You may prefer to remove the door and place it on padded sawhorses.

Step 2 Measure up 36" from the bottom of the door to locate the lockset.

Step 3 Fold the marking template (which comes with the lockset) along the indicated lines. Place it on the beveled edge of the door. Using the template guides, mark the center of the door edge and the center of the hole. (If a special boring jig is used, no template is needed.) The distance from the door edge to the center of the hole is called the backset. It is typically $2\frac{1}{8}$" or $2\frac{1}{4}$".

Step 4 Using a $2\frac{1}{8}$" hole saw, bore a hole of the correct size in the face of the door. To prevent splitting of the faces, bore the hole on one side until the point of the pilot bit breaks through. Then complete the boring from the other side.

Step 5 Using a $\frac{7}{8}$" hole saw, bore a hole of the correct size in the center of the door edge for the latch.

Step 6 Insert the latch assembly into the hole. Keep the faceplate parallel to the edge of the door and mark its outline with a sharp pencil or utility knife.

Step 7 Remove the latch assembly. Chisel out the marked area so that the faceplate will be mounted flush with the edge of the door.

Step 8 Install the latch with its curved surface facing in the direction of the door closing. Insert and tighten the screws.

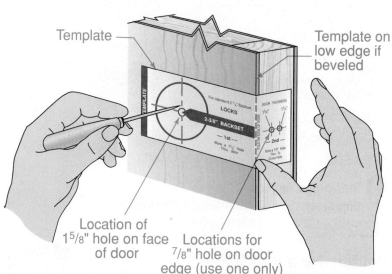

Template

Template on low edge if beveled

Location of $1\frac{5}{8}$" hole on face of door

Locations for $\frac{7}{8}$" hole on door edge (use one only)

Marking Template

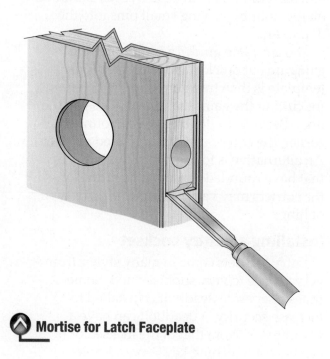
Mortise for Latch Faceplate

Step 9 Insert the exterior knob with the stems into the latch. Make certain that the stems are positioned correctly inside the latch holes.

Step 10 Install the interior knob by placing it over the stems and aligning the screw guides. Push the assembly flush with the door. Insert the screws and tighten them until the lockset is firm.

Step 11 To locate the strike plate, place it over the latch in the door. Then carefully close the door against the stops. The strike plate will hang on the latch in the clearance area between the door edge and the jamb. Push the strike plate in against the latch. With a pencil, mark its position on the jamb.

Step 12 Open the door and hold the strike plate in position against the jamb. Make sure that it is parallel to the edge of the jamb. Mark around the strike plate. Chisel out the marked area so that the strike plate will mount flush with the surface of the jamb.

Step 13 Drill a $1^5/_{16}$" clearance hole for the latch bolt $^1/_2$" deep in the door jamb. Locate it on the centerline of the screws (top to bottom). Install the strike plate and tighten the screws.

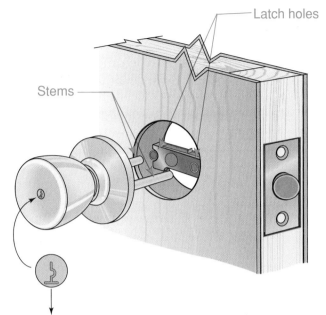

Latch holes

Stems

(Down) - Correct position of keyway when installing locksets

Installing the Knobs

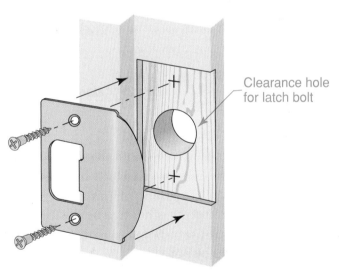

Clearance hole for latch bolt

Installing the Strike Plate

Weatherstripping

An exterior door provides a barrier to the weather. It must also prevent heated indoor air from being lost in the winter and cooled air from being lost in the summer. The construction of the door helps to accomplish both of these tasks. However, weatherstripping is equally important.

Weatherstripping consists of flexible lengths of rubber, vinyl, polypropylene pile, flexible metal, or other materials that are attached to the edge of the door or its frame. Magnetic weatherstripping is available for use with metal doors. Any weatherstripping material must close air gaps around the door without interfering with door operation.

Over time, weatherstripping wears down and becomes less effective. It should be replaced as a regular part of door maintenance. Weatherstripping should not be painted.

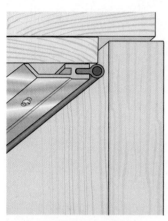
Head of door

Side of door

Door bottom

Weatherstripping

Two Types Many kinds of weatherstripping are available. Two types are shown here: bulb weatherstripping for the head and side of the door and sweep weatherstripping for the bottom.

Section 21.2 Assessment

After You Read: Self-Check

1. How much larger than an exterior door should the rough opening for the door be?
2. What is the difference between a butt-hinge gauge and a butt-hinge template?
3. Exterior doors should be hung on three hinges. One should be centered on the door's length. Where should the other two be positioned?
4. How many holes must be bored in a door to mount a typical entry lockset, and where should these holes be bored?

Academic Integration: English Language Arts

5. **Analyze Information** Obtain installation instructions for any type of bored-hole entry lockset for an exterior door. What types of drill bits does the manufacturer recommend for installation? Do you think these bits would be suitable for all types of exterior doors? Write a paragraph explaining your reasoning.

 Go to **connectED.mcgraw-hill.com** to check your answers.

Interior Doors

What are folding doors?

Most interior passage doors are 1⅜" thick. Standard interior door height is 6'-8". Common minimum widths for single doors are as follows:

- Bedrooms and other habitable rooms: 2'-6"

- Bathrooms: 2'-4"

- Small closets and linen closets: 2'

In most cases, jambs, a stop, and a casing are used to frame and finish the opening. Hinged doors should swing in the direction of natural entry, against a blank wall whenever possible. They should not be obstructed by other doors. For safety reasons, interior doors should never swing into a hallway. See page 602 for determining the hand, or swing, of a door.

Types of Interior Doors

In addition to basic flush or panel styles, other types of interior doors are available to solve special problems.

Louvered Doors A *louvered door* is used for closets because it provides some ventilation. It is essentially a standard passage door in which the panels have been replaced by louvers (angled slats).

Sliding Doors The bypass *sliding door* is designed for closets and storage walls with wide openings. It requires no open swinging area, so it does not interfere with furniture placement. Only half of the opening is exposed at one time. The exterior surface of sliding doors is sometimes covered with a shatter-resistant mirror.

Sliding doors hang from small roller wheels mounted in a metal track. The track is screwed to the underside of the door head

jamb and can be hidden from view with a piece of trim. The doors are guided at the bottom by a small piece of hardware screwed into the floor where the doors overlap.

Doors are available 1⅜" thick, 6'-8" or 7'-0" high, and in many widths. Most sliding door hardware will also adapt to 1¾" or 1⅛" door thicknesses. The rollers are adjustable so the door may be plumbed and aligned in the opening. Rough opening sizes for sliding doors differ among manufacturers. Be sure to consult the specifications.

Bifold Doors *Bifold doors* are often used to enclose a closet, pantry, or laundry area. The doors may be wood, metal, or a composite such as wood covered with vinyl. Unlike sliding doors, bifold doors can be opened so the entire opening is exposed at one time.

Louvered Bifold Doors
Compact When Folded Each side of a bifold door can be folded upon itself and stored against the door jamb.

The doors are available in 6'-8", 7'-6", and 8'-0" heights and in widths of 3', 4', 5', and 6'. Each individually hinged portion of the door is called a panel. If desired, the tracks may be cut in half and only two panels installed. For example, two panels from a 3'-0" door could be used for a 1'-6" linen closet opening.

A bifold door is usually installed in a conventional door frame. The frame may be trimmed with casing to match the trim in the rest of the house, or the jamb may be finished the same as the walls. The rough opening is framed in the same way as for a conventional swinging door. The finish opening size varies with the manufacturer.

To install a bifold door, install the top track first. The top track contains a fixed pivot near the jamb. It also serves as a guide for another pivot that is at the opposite side of the door. Then fasten the pivot mounting plates directly to the floor and against each side jamb. The doors pivot on pins inserted in their top and bottom edges. The pins fit into adjustable metal sockets in the tracks. Install the doors by inserting the bottom pivot pin into the bottom track socket. Then insert the upper pivot pin into the top track socket. Adjust the panels to the opening by adjusting the position of the sockets. To make the tops of the panels even, raise or lower the panels by adjusting the lower pins.

Folding Doors A *folding door* may be used as a room divider or to close off a laundry area, closet, or storage wall. They are made from wood, reinforced vinyl, or plastic-coated wood. Folding doors with a metal framework covered with vinyl are sometimes called *accordion-fold doors.*

Folding doors hang on nylon rollers that glide smoothly in a metal track that may be concealed with matching wood molding. Standard or stock doors are available 6'-8" high and 2'-4", 2'-8", 3', or 4' wide. They are installed in a standard frame and trimmed in a conventional manner. Doors are shipped from the factory in a package containing hardware, latch fittings, and installation instructions.

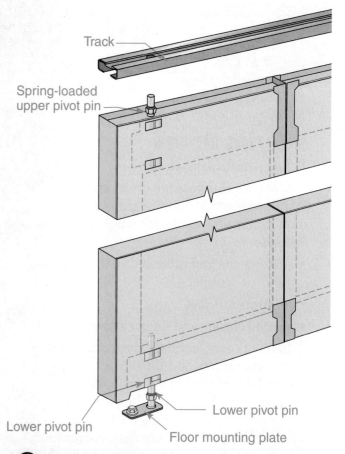

Bifold Door Mounting
Pivots The doors are supported on pivots. The top pivot fits in a track. The bottom pivot is attached to the floor.

Double-Acting Doors *Double-acting doors* are common in commercial construction, particularly in restaurants. They are typically full-height and are often used to separate restaurant kitchens from the dining room. However, in residential construction, they are called *café doors.* These doors are much lighter than commercial doors and do not cover the entire opening. They are usually only about 40" tall, and come in widths to fit standard openings. They are centered in the door opening. Café doors are often made much like interior wood shutters. They have a wood frame that is typically 1⅛" thick.

Each café door panel is connected to a jamb so that both doors meet in the middle of the door opening. Each door panel is hinged so that it can swing inward or outwards. This allows someone to push the door open from either side. Some hinges,

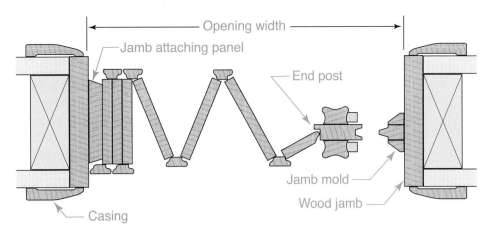

Jamb attaching panel

Opening width

End post

Jamb mold

Wood jamb

Casing

 Folding Door
Neatly Stacked A section view showing a folding door installed with wood jambs.

called *gravity hinges,* are designed so that the panel raises slightly as it is opened. The weight of the panel then causes the panel to swing closed. Other hinges contain a spring mechanism that automatically closes the door when someone passes through.

Café doors are easy to install. They should be trimmed to the proper width for the opening, with a clearance allowance for swing. Once the hinges are screwed to the door jambs, the doors can be mounted on the hinges.

Pocket Doors A *pocket door* slides into an opening or pocket inside the wall. When a pocket door is open, it is concealed except for one edge. A pocket door is convenient where there is not enough clearance space for a swinging door. One disadvantage is that a pocket door can be difficult to repair or adjust because the tracks are inaccessible.

Standard widths are 2'-0", 2'-4", 2'-6", 2'-8", and 3'-0". Any style of door with a thickness of 1⅜" can be installed in the pocket to match the other doors in the home. However, special hardware is required when the door itself is unusually heavy.

Framing carpenters must be made aware of special framing needs for pocket doors. Before the opening for the pocket unit is roughed in, the manufacturer should be consulted for specifications. The rough opening is usually 6'-11½" or 7' high and

twice the door width plus 2" or 2½". The wall header above the pocket must be adequate to support any weight on the wall so there will be no weight on the door itself. The pocket door mounting frame, a stiff metal assembly, comes complete and ready for installation into the rough wall opening. After the finish wall covering has been applied, hang the door and install the stops.

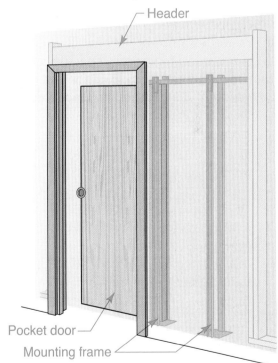

Header

Pocket door

Mounting frame

 Pocket Door
Hidden Door A pocket door is supported on a mounting frame that fits in the wall cavity.

Installing Interior Doors

Rough openings in the stud walls for interior doors are usually 2" higher than the door and 2" wider. This provides room for plumbing and leveling the frame in the opening.

Interior door frames are made of two side jambs, a head jamb, and stop moldings that the door closes against. They are available in 5¼" widths for plaster walls and 4⁹⁄₁₆" widths for walls with ½" drywall finish. Wider jambs are required if the walls are thicker. Two- and three-piece adjustable jambs (also called split jambs) are available. Their chief advantage is in being adaptable to different wall thicknesses. Some manufacturers produce interior door frames with the door fitted and pre-hung, ready for installation. Adding the casing completes the job.

Installing the Door Frame If a door unit is not pre-hung, the jambs must be cut to length, assembled, and installed. The side jambs should be nailed through the notch into the head jamb with three 7d or 8d coated nails. Cut a spreader to a length exactly equal to the distance between the jambs at the head jamb. A spreader is a temporary spacing device that ensures that the side jambs will be a **consistent** distance apart. Place the

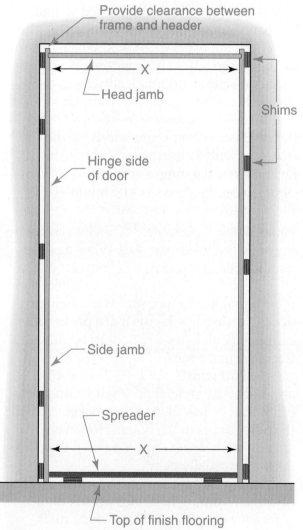

Provide clearance between frame and header

Head jamb

Shims

Hinge side of door

Side jamb

Spreader

Top of finish flooring

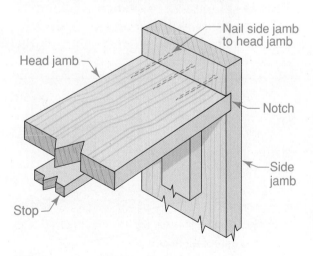

Head jamb

Nail side jamb to head jamb

Notch

Side jamb

Stop

Interior Door Frame Details

Door Support A door frame supports the door and provides a finished look to the rough opening.

Door Frame Details

Spacing Aid Cut the spreader equal to the distance (X) between the side jambs just below the head jamb.

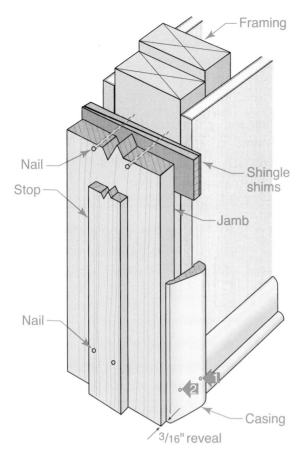

Framing

Nail

Stop

Nail

Shingle shims

Jamb

2 1

Casing

3/16" reveal

⬥ Installation Details

Frame and Casing Drive a 6d or 7d finish nail at arrow 1 to nail through the casing into the wall stud. At arrow 2, use a 4d or 5d finish nail to fasten the casing to the jamb.

spreader at the floor line so that the distance between jambs will be identical to the distance near the head jamb. This is more accurate than measuring the distance.

Plumb the assembled frame in the rough opening using pairs of shingle shims placed between the side jambs and the studs. One jamb, usually the hinge jamb, is plumbed using four or five sets of shims along the height of the frame. Two 8d finishing nails are installed at each wedged area, one driven so that the door stop will cover it.

Place the spreader in position at the floor line. Fasten the opposite jamb in place using shingle shims and finishing nails. Use the first jamb as a guide in keeping a uniform width. This can be done by using a second

precut spreader as a gauge, checking several points. It can also be done by carefully measuring at various points along the height of the door frame between the side jambs.

Installing a Pre-Hung Door If the door unit is pre-hung, installing the frame follows much the same process as for one that is not pre-hung. However, spreaders may not be necessary if the unit has already been squared and braced at the factory. If so, leave the braces on and fit the entire unit into the rough opening. Then shim the frame and nail it as before. When the frame is secure, remove the braces and double-check the fit of the door.

Hanging an Interior Door Interior doors are often hung with two 3½" by 3½" loose-pin butt hinges. However, three hinges will strengthen the door and help to prevent it from warping. Gains should be routed into the edge about ³⁄₁₆" from the door back (see page 600).

The door fits into the opening with the clearances shown on page 620. Clearances and locations for hinges, lockset, and doorknob may vary depending on the products being used.

The lock stile is the door stile on which the lock will be located. The door stile on which the hinges are located is sometimes called the hinge stile. The edge of the lock stile should be beveled slightly to permit the door to clear the jamb when swung open. This detail is shown on page 608. If the door is to swing across thick carpeting, the bottom clearance should be increased.

When fitting doors, the stops are usually temporarily nailed in place until after the door has been hung. Stops for doors in single-piece jambs are generally ⁷⁄₁₆" thick and may be ¾" to 2¼" wide. They are installed with mitered joints at the junction of the side and head jambs.

Many interior doors feature hollow-core construction and a wood-veneer surface. To prevent the veneer from splintering when

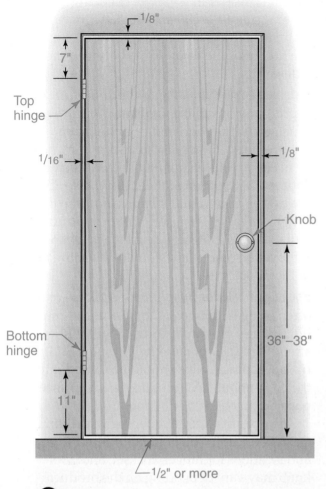

Fitting a Door
Proper Fit These dimensions and clearances are commonly used when installing interior doors.

the door is trimmed, some carpenters use a site-built cutting guide. The shoe (base) of a circular saw rides on top of the guide, and the shoe's edge is guided by the stop. The stop is fastened in place with glue and brads.

Door Stops and Trim After the door is in place, permanently nail the stops with 1½" finish nails. Nail the stop on the lock side first, setting it tightly against the door face while the door is latched. Space the nails 16" apart in pairs.

Nail the stop behind the hinge side next. Allow a 1/32" clearance from the door face to prevent scraping as the door is opened. Finally, nail the head-jamb stop in place. Note that when the doors and trim are painted, some of the clearances will be taken up.

Splinter Stoppers When trimming interior doors, some carpenters use a cutting guide. Use a blade specifically designed for crosscutting plywood to help reduce splintering even more. Another method is to score the cut line with a utility knife before trimming the door.

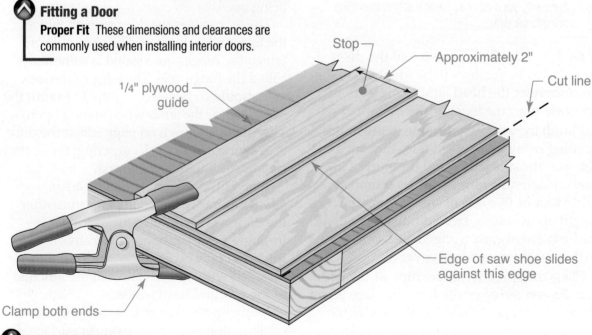

Door Cutting Guide
Carpenter-Made Guide A cutting guide is used when trimming doors to length. It must be clamped to the door.

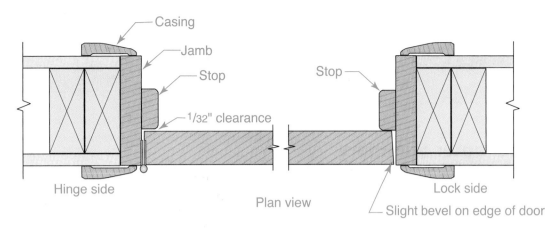

Casing

Jamb

Stop

Stop

¹/₃₂" clearance

Hinge side

Lock side

Plan view

Slight bevel on edge of door

 Door Stop Installation Details
Last Step The door stops prevent the door from swinging into the opening.

Door Trim Door trim, or casing, is nailed around openings and is also used to finish the room side of exterior door frames. Casings are nailed to both the jambs and the framing studs or header. When the casings have been installed, the door opening is complete except for fitting and securing the hardware. For more on installing door casing, see Chapter 28.

Estimating Materials and Time

The cost of interior doors and door frames varies a great deal with the type of door, the materials it is made of, and the quality of its manufacture. The number, type, and often the manufacturer are specified on the plans.

Check the door schedule and specifications. Then confirm the number and type of doors by checking the floor plans. The cost of each door must be established separately. Be sure to determine if the cost includes the door frame and mounting hardware.

The time needed to install an interior door depends on the type of door. A pre-hung door may take approximately 1½ hours. A sliding door may take as much as 3 hours. Pocket doors may take even longer.

Section 21.3 Assessment

After You Read: Self-Check

1. What is the standard thickness of an interior passage door?
2. What is a pocket door?
3. When installing an interior door frame, what is the purpose of a spreader?
4. What can be done to prevent the surface veneer of a door from splintering?

Academic Integration: Mathematics

5. **Calculate Dimensions** The maximum height for the rough opening of a pocket door is 7'. Determine the maximum size of the width of a rough opening for a 2'6" pocket door. Draw a sketch and label the appropriate measurements.

Math Concept Creating a sketch of a geometric object and labeling the measurements can help you visualize the missing measurements.

Go to **connectED.mcgraw-hill.com** to check your answers.

Chapter Summary

Section 21.1
Flat-panel or raised-panel passage doors are the most common types of doors in houses. They may be made of various materials, though wood is the most common material. Door quality depends upon the door's construction and hardware.

Section 21.2
An important feature of any exterior door is its energy efficiency, which depends on construction and weatherstripping. Exterior doors can be part of a preassembled package that includes the hinges and jambs. One of several types of keyed locksets may be used, depending on the level of security required.

Section 21.3
Various types of sliding and folding doors are available for closets, storage areas, and other locations. They are sometimes mounted on metal tracks instead of on hinges. Installation of an interior door is less complicated than the installation of an exterior door because it does not include a sill or weatherstripping.

Review Content Vocabulary and Academic Vocabulary

1. Use each of these content vocabulary and academic vocabulary words in a sentence or diagram.

Content Vocabulary

- passage door (p. 596)
- strike plate (p. 596)
- solid-core construction (p. 596)
- hollow-core construction (p. 597)
- stiles (p. 597)
- door frame (p. 598)
- gain (p. 599)
- lockset (p. 600)
- hand (p. 601)

Academic Vocabulary

- dexterity (p. 600)
- accommodate (p. 606)
- consistent (p. 618)

Speak Like a Pro

Technical Terms

2. Work with a classmate to define the following terms used in the chapter: *latch* (p. 596), *flat-panel doors* (p. 596), *flush doors* (p. 596), *slab doors* (p. 596), *jamb extensions* (p. 598), *pre-hung doors* (p. 599), *loose-pin butt mortise* (p. 599), *leaves* (p. 599), *bored locks* (p. 600), *mortise locks* (p. 600), *latchsets* (p. 600), *sliding-glass doors* (p. 604), *patio doors* (p. 604), *French doors* (p. 604), *astragal stop* (p. 605), *fire-rated doors* (p. 605), *door jack* (p. 608), *door stand* (p. 608), *weatherstripping* (p. 614), *louvered door* (p. 615), *sliding door* (p. 615), *bifold doors* (p. 615), *folding door* (p. 616), *accordion-fold doors* (p. 616), *double-acting doors* (p. 616), *café doors* (p. 616), *pocket door* (p. 617).

Review Key Concepts

3. Name three types of exterior doors and three types of interior doors.

4. List three things you can do to protect a door after it is delivered to a job site.

5. Describe one method used to determine the hand of a door.

6. List the features of an exterior door that would result in maximum energy efficiency.

7. Recall the standard height for exterior and interior doors.

8. Demonstrate how to install an interior pre-hung door.

Critical Thinking

9. Analyze Identify two types of doors that would not be good to have in a region with high winds. Explain your reasoning.

Academic and Workplace Applications

STEM Mathematics

10. Estimating Costs Tino has been hired to install 6 pre-hung doors and 2 sliding doors for a house remodeling project. He also needs to allow 5 hours to install a pocket door. The work needs to be completed in 2 days, and he will charge time-and-a-half overtime for any hours he needs to work beyond 8 hours a day. If he charges $20 per hour, how much will Tino earn for the job?

Math Concept Making a chart can help you keep track of different types of information in story problems. It can also help you to eliminate the information in the story problem that you do not need.

Starting Hint Organize the information you need in a chart like the one below.

type of door	number of doors	time per door	total time
pre-hung			
sliding			
pocket			

STEM Engineering

11. Customized Doors You have been asked to create a customized interior door for a specific user who uses wheelchairs and braces to get around. Brainstorm what modifications you might make to doors for this person so that the door would be easy for him or her to operate. Think about the materials the door will be made from, how the door will be used, and what locking mechanism might be used. Record your ideas in a bulleted list or a one-paragraph description. You might also include a sketch of your door that includes measurements.

21st Century Skills

12. Career Skills: Work Orders If you have been asked to make a repair to a newly completed house, you may be given a work order. Work orders describe work that needs to be done. They help maintenance departments schedule jobs and keep track of work in progress. They include the date the request is made, the date by which the work should be completed, where the problem or task is located, and a description of it. A list of necessary materials and parts may also be included, as well as any special safety warnings. Most work orders also require that you enter information on them, such as your name, the date you received the order, the date you finished the work, your diagnosis of the problem, what you did to correct it, and the hours it required. Obtain or create a sample blank work order. Print or make a copy of the work order. Assume that you have been asked to replace a door trim. Include the information from the requesting party, as well as the information you will need to fill in as the individual completing the order.

Standardized TEST Practice

True/False

Directions Decide if the following statements are true or false by filling in **T** or **F**.

(T) (F) **13.** Steel, brass, bronze, and nickel are the most common materials used for door hardware.

(T) (F) **14.** A conventional exterior frame and brick molding with an oak sill requires about six hours to assemble and install.

(T) (F) **15.** Folding doors with a metal framework covered with vinyl are sometimes called accordion-fold doors.

TEST-TAKING TIP

Be sure to reread true/false questions to make sure that you have not misread a false statement as a true statement by mistake.

*These questions will help you practice for national certification assessment.

Roofing & Gutters

Chapter Objectives

After completing this chapter, you will be able to:

- **Identify** different roofing products.

- **Demonstrate** how to install strip shingle roof covering.

- **Describe** the various types of flashing and explain where they are used.

- **Explain** the difference between shakes and wood shingles.

- **Identify** the main parts of a gutter system.

- **Describe** the two basic types of gutter systems.

Discuss the Photo
Roofing After framing is in place, roofers install one of a variety of roofing products over the framing. *What influences the type of material used in roofing?*

Writing Activity: Research and Summarize
Wood shingles and shakes are traditional roofing materials. However, asphalt and fiberglass shingles are much more common today. Research the history of roofing materials in the United States. Try to determine what materials were used on colonial houses. Summarize your findings in a one-page report.

Chapter 22 Reading Guide

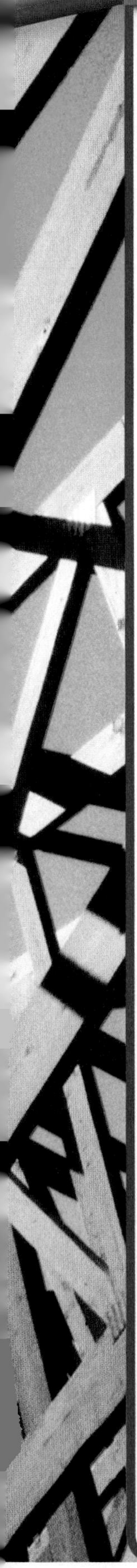

Before You Read Preview

A roof covering must provide long-lasting waterproof protection for a house. Rain that falls on a roof should be channeled away from the house by a gutter system. Create a list of possible problems that may occur because of a faulty roof or gutter system. Compare your list with the list of a classmate.

Content Vocabulary

- flashing
- square
- exposure
- butt edge
- top lap
- side lap
- underlayment
- ice dam
- open valley
- Boston ridge

Academic Vocabulary

You will find these words in your reading and on your tests. Use the academic vocabulary glossary to look up their definitions if necessary.

- precaution
- succeeding

Graphic Organizer

As you read, use a chart like the one shown to organize information, adding rows as needed.

Action	Result
carefully installing flashing	prevents water from getting behind the roofing itself

Go to connectED.mcgraw-hill.com to download this graphic organizer.

Roofing Basics

Roofing Terms & Concepts

Why is roofing sold by the square?

The choice of materials and how to apply them is influenced by many factors including cost, fire resistance, wind resistance, and climate. Because a roof is large and visible, appearance is also important. Codes in some areas of the United States are particularly stringent regarding the type of roofing allowed and the methods used to install it. This is due to concerns over damage caused by high winds.

The most common type of roof covering is the *shingle*. Shingles can be made from many materials, and all of them are applied to roof surfaces in an overlapping fashion in order to shed water. The installation of a roof covering includes more than just shingles. It also involves metal flashing at roof openings such as chimneys, skylights, and vents. **Flashing** is a piece of metal that protects against water seepage. It must be installed carefully to prevent water from getting behind the roofing itself.

Roofing is usually installed by roofing contractors. These specialists are familiar with the different types of products but often specialize in a particular type. For example, some contractors install only wood shingles. Others may install only metal roof panels.

Roof shingles and other types of roofing are estimated and sold by the *square*. One **square** of roofing is the amount of roofing required to cover 100 sq. ft. of roof surface. The amount of weather protection provided by the overlapping of shingles is called *coverage*. Coverage depends on the kind of shingle and the method

of application. Shingles may furnish one thickness (single coverage), two thicknesses (double coverage), or three thicknesses (triple coverage) of material over the surface of the roof.

Exposure is the amount of a shingle that shows after installation. The exposed edge of the shingle is called the **butt edge**. The portion of the shingle that is not exposed to the weather is the **top lap**. The *head lap* is the shortest distance from the lower edge of an overlapping shingle to the upper edge of the shingle in the second course below. The **side lap**, or *end lap*, is the amount that adjacent roofing sheets overlap each other horizontally. This applies primarily to rolled

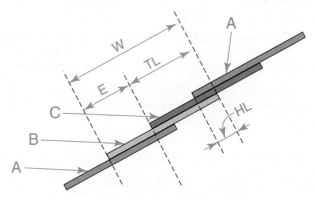

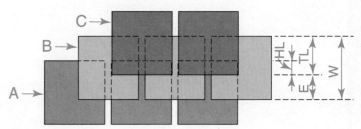

⌃ Shingle Terminology

Shingles Terms used to describe shingle placement: E = exposure, TL = top lap, HL = head lap, W = width of strip shingles or length of individual shingles. A, B, and C correspond to rows of shingles.

roofing and underlayment. **Underlayment** is a material, such as roofing felt, applied to the roof sheathing before shingles are installed.

Roof Slope

The slope of a roof is expressed as a ratio of vertical rise to horizontal run. It is written as a number "in 12." For example, a roof that rises at the rate of 4" for each foot (12") of run has a *4-in-12* slope. The triangular symbol above the roof conveys this information.

Slope is important because some roofing products are suited for use only on roofs with a slope great enough to provide proper drainage. In terms of building codes, there are three basic types of roofs based on the amount of slope. The three basic slope categories are:

- *Flat* Roof slope is below 2½-in-12.
- *Low slope* Roof slope is between 2½-in-12 and 4-in-12.
- *High slope* Roof slope is 4-in-12 or greater.

Most shingled roofs are *high-slope* roofs. Shingle manufacturers specify what roof slopes their product can be used on. Make sure the shingles you use are of the correct type for the slope of roof they will be applied to. The roof slope should never be less than the minimum specified by the shingle manufacturer, or the roof may leak.

Types of Roofing
What types of materials might make good roofing products?

In general, there are two basic types of roofing products: *single units* and *sheet products.* Single units, such as shingles and tiles, are applied to the roof in an overlapping fashion to keep water out. Sheet products cover much larger areas but, depending on the roof size, are sometimes overlapped as well. Sheet products include *roll roofing, built-up roofing, metal sheets,* and *single-ply roofing.* Single-ply and built-up roofing are used only on flat roofs. Roll roofing is sometimes used on pitched roofs and as underlayment for tile roofs. It can also be used as valley flashing under shingle roofs.

Reading Check

Recall What are the two basic types of roofing products?

Shingles

Shingles are the most common form of roof covering for residential homes. They are easy to transport and install, and are offered in a wide variety of colors and shapes. Shingles may be made of asphalt, fiberglass, wood, cement, or slate.

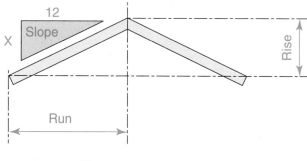

 Understanding Slope
Rise Over Run Slope is expressed as a ratio of rise to run.

Roofing Preferences The popularity of many roofing products varies across the United States. Strip shingles are common in every region of the United States. Metal sheet roofing is common in snowy regions such as the Rocky Mountains and Vermont. Slate roofing is common in Pennsylvania, where slate quarries are nearby. Wood shingles are common in the Northwest because of the nearness of vast forests. Tile roofing is most common in the Southwest and southern California.

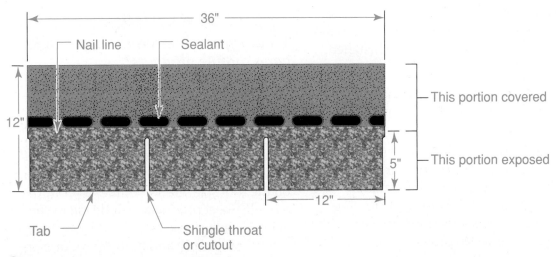

36"

Nail line   Sealant

12"

This portion covered

5"   This portion exposed

12"

Tab   Shingle throat or cutout

 Anatomy of a Shingle

Shingle Basics The dimensions shown are approximate and may vary by product and manufacturer.

Asphalt shingles are available in three styles: *strip*, *individual*, and *large individual*. Strip shingles are widely available and the most popular type of shingle. They are easy to apply and cover a roof fairly quickly. Individual shingles interlock or are stapled down. Large individual shingles are applied using either the American or Dutch lap methods.

One type of strip shingle that is particularly popular is the *architectural shingle*, sometimes called a *laminated shingle*. In this type, two layers of strip-type shingles are bonded together by the manufacturer. Segments of the top layer are cut out at random intervals, producing varying thicknesses in any one shingle. Architectural shingles give roofs a deeply textured appearance, and they should always be installed according to the manufacturer's instructions. Random wide cutouts in the top layer give a more textured appearance than that given by standard strip shingles. Architectural shingles should be installed according to manufacturer's instructions.

Interlocking (*lock-down*) asphalt shingles are designed to resist strong winds. They feature locking tabs that help to hold the shingle in place in high winds. Interlocking shingles do not require adhesives, although

roofing cement may be needed along rakes and eaves. However, the locking device on each shingle must be engaged properly and nails properly located. The shingle manufacturer's instructions specify where nails should be located for best results. Standard (non-interlocking) three-tab shingles that are used in high-wind areas must be installed using extra nails, usually a minimum of six.

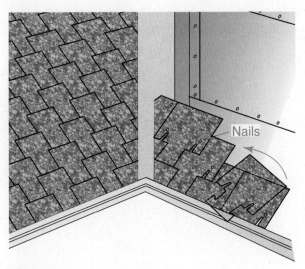

Nails

 Interlocking Shingles

Wind Resistant Interlocking asphalt shingles are joined by locking tabs.

Roll Roofing

Wide Sheets Application details for roll roofing.

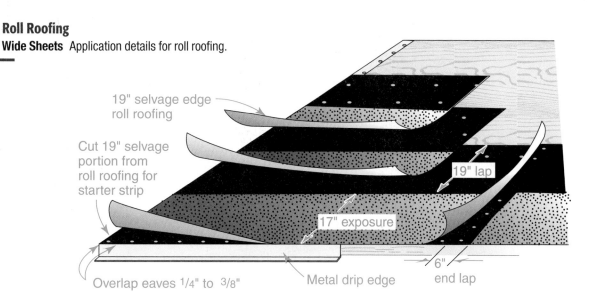

19" selvage edge roll roofing

Cut 19" selvage portion from roll roofing for starter strip

19" lap

17" exposure

Overlap eaves 1/4" to 3/8"

Metal drip edge

6" end lap

Roll Roofing

When roofing costs must be minimized, such as when roofing utility buildings, mineral-surfaced roll roofing is often suitable. It comes in 36" wide rolls that usually cover 100 sq. ft. Depending on the product, it typically weights 75 lbs. to 90 lbs. per roll. It can be applied quickly over large areas. Roll roofing should be installed over a double-coverage underlayment.

The first course of roll roofing is applied horizontally across the roof, beginning at the eaves. It should be attached using galvanized roofing nails. The remaining courses should then be lapped over previous courses 19", leaving just the mineral surface exposed. Once the roofing is in place, lift the mineral surface of each course. Apply a quick-setting lap cement to the underlying sheet to within 1/4" of the exposed edge. Finally, apply firm pressure over the entire cemented area.

Built-Up Roofing

Built-up roof coverings are installed by roofing companies that specialize in this field. Roofs of this type receive three, four, or five layers of roofer's felt, each mopped down with tar or asphalt. The final surface may be coated with asphalt and gravel embedded in tar, or it may be covered with a cap sheet. Gravel acts as a ballast that helps to hold the roofing in place. It also protects

the asphalt from sunlight. The *cap sheet* is a sheet of roll roofing or roofing felt that becomes the top layer.

Metal Sheets

Metal sheet roofing comes in widths up to 4' and lengths up to 22' and covers large areas quickly. It can be used on slopes as low as 4-in-12 or, if a single panel will cover from eave to ridge, as low as 3-in-12.

Unlike inexpensive, corrugated metal sheets, high-quality metal roofing has an enameled top surface that is available in a variety of colors. This type of roof on page 630, should be installed by a specialized contractor who has the tools needed for working and forming metal.

Single-Ply Roofing

Single-ply roofing is used primarily to cover the large, flat roofs of commercial buildings, but it is increasingly used in residential construction on flat or very low-slope roofs. Single-ply roofing is made of various materials, but EPDM (ethylene propylene diene monomer) is common. Such sheets are very wide and are generally 45 or 60 mils thick. Single-ply roofing is installed by specialized roofing contractors, particularly those who specialize in commercial construction.

 Metal Roofing
Raised Seams Metal sheet roofing is sometimes called *standing seam roofing* or *raised ridge roofing*. The raised seams prevent water from collecting at joints.

Tile

Clay and lightweight concrete tile roofing products are very common in southern California, Florida, and parts of the southwestern United States. Both materials are very durable, particularly against fire hazards. Tile comes in many styles, colors, and shapes. Tile is installed with nails and sometimes also with mortar. It can be attached to a roof sheathed with solid panel products, or to spaced structural sheathing boards.

Tools & Equipment

What might be some disadvantages of using pneumatic tools?

Regardless of the type of shingle to be installed, always check the instructions provided by the shingle manufacturer. This will ensure that you have the correct tools and materials. Instructions often carry valuable safety information as well.

 Tile Roofing
Heavy Roof Tile roofing is much heavier than other types of roofing. Roof framing must be designed to support the extra weight.

Hatchets and Nailers

Traditionally, the primary tool required for installing most types of shingles was the *shingling hatchet*. It is also called a *roofing hatchet* or *roofing hammer*. A sharp blade on one end is used for scoring or cutting shingles. (Many roofers, however, use a utility knife for this purpose.) The edge will eventually wear down but can be resharpened. A V-shaped notch on the underside of the blade is used to remove nails. Most shingling hatchets also have an adjustable gauge used to measure the exposure of shingle courses.

Many roofers now use pneumatic nailers instead of shingling hatchets. Nailers speed installation and reduce strain on the roofer's arm and hand. Various types of nailers are available. (For more on pneumatic nailers, see Section 6.5.)

> **Reading Check**

Explain *What is the traditional primary tool for installing shingles?*

 Shingling Hatchet
Nailing by Hand This shingling hatchet has an adjustable gauge, or *pin,* for measuring the amount of shingle exposure.

Ladders, Lifts, and Brackets

Working on a roof is one of the most dangerous occupations in residential construction. Many workers, particularly roofers, are killed or injured each year in falls. Falls from roofs tend to result in serious injury. One study found that over 85 percent of those who fell from a roof were fatally or seriously injured. Take every precaution to prevent falls. Always follow appropriate OSHA requirements for wearing safety harnesses. A harness must fit properly and be tied off securely. Falls from ladders are another danger faced by roofers. Make sure the ladder will not slip out of position while you are on it. Always keep your weight centered along the axis of the ladder. For more on ladder safety, see Chapter 7.

Whenever possible, use a hoist or a lift to transport shingles to the roof or have the shingles delivered directly to the roof. This speeds the work and minimizes back strain. Shingle suppliers often provide the service of delivering their products directly to the roof.

Roof brackets (sometimes called *roof jacks*) should be used to increase safety while working on the roof. These temporary

 Roofing Nailer
Time Saver The nails in this pneumatic roofing nailer are coiled in a circular magazine located below the operator's hand.

brackets are secured by nails driven through the sheathing and into the rafters. They hold one or more boards at a right angle to the roof surface. The boards prevent workers and tools from sliding off. For more on roof brackets, see Chapter 7.

JOB SAFETY

GUARD AGAINST FALLS Working on a roof presents some unexpected hazards. Loose sawdust and loose shingles scattered on the roof sheathing dramatically increase the chances that a worker will step on them and slip. Likewise, tools and pneumatic nailer hoses can be hazardous. Always keep the work area clean and as clear as possible.

Installation Materials

How might a fan be used to reduce the severity of an ice dam?

The outer roofing materials, such as shingles, are most exposed to weather and are the most visible part of a roof. However, a number of unseen interior materials are just as important in creating a durable, leak-free roof.

Underlayment

Underlayment is a layer of weather-resistant material that is applied to the roof sheathing before the final roofing material is installed. Underlayment is required for all types of shingles, as well as for some other roofing products. It can be applied to solid sheathing or spaced sheathing. Underlayment is typically a material with low vapor resistance such as asphalt-saturated felt. Felt comes in rolls and is installed easily and quickly. Roof underlayment generally has four purposes:

- It protects the sheathing and the house interior from moisture until the shingles can be applied. In some parts of the country, builders refer to this stage as drying-in the house.

- It provides a second layer of weather protection. If wind drives rain or snow under the shingles, the underlayment protects the sheathing.

- It prevents asphalt shingles from sticking to the sheathing, which can damage the shingles over time.

- It prevents condensation on the sheathing.

Eaves Protection

In cold climates, ice dams can occur. This causes water to leak into the house, damaging interior ceilings and walls. An **ice dam** is formed by melting snow that freezes at the eave line (eaves are the roof's lower overhang). As more snow melts, the water backs up behind the ice and seeps beneath the shingles.

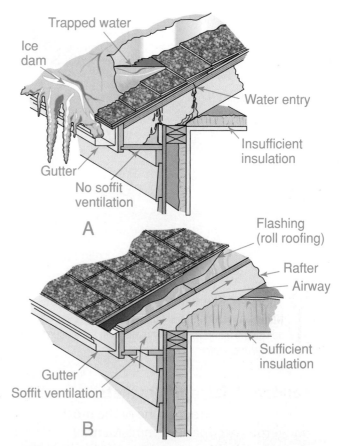

 Preventing Ice Dams
Poor Ventilation If warmth reaches a snowy roof, as in **A**, ice dams build up near the roof overhang. With proper ventilation, as in **B**, the warmth is carried away before it can melt the snow.

Proper ventilation, along with attic insulation, helps to prevent ice dams. The insulation slows heat loss and the ventilation carries away any heat that does escape the house. As an additional precaution, eaves flashing is added at the lower portion of the roof. Building codes require eaves flashing (sometimes called *eaves protection* or an *ice barrier*) in areas where the average daily temperature in January is 25°F (−4°C) or less. This flashing is usually provided by one of the following:

- At least two layers of underlayment cemented together.

- A single layer of self-adhering bitumen sheet that acts as an ice and water shield beneath the shingles.

- Exposed metal flashing sheets with soldered joints.

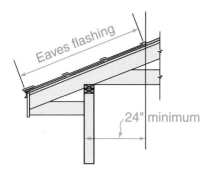

 Placement of Eaves Flashing
Water Blocker Eaves flashing protects the roof system from leaks caused by ice dams. *How else can ice dams be prevented?*

Eaves protection should extend from the end of the eaves to a point at least 22" inside the exterior wall line of the house.

Flashing

Flashing is a thin metal sheet or strip used to protect a building from water seepage. It comes in various forms, including rolls of various widths, plates, and angle stock. Flashing prevents water from entering roof joints. It is required wherever the roof covering intersects another surface, such as a wall, chimney, skylight, or vent pipe. It is sometimes used in roof valleys as well. Angled flashing called *drip edge* protects the edges of the roof.

Flashing must be installed so that it sheds water. Metal used for flashing must be corrosion resistant. Galvanized steel (at least 26 gauge), 0.019" thick aluminum, 16 oz. copper, or lead-coated copper can be used.

It is important not to use a variety of flashing metals on the same roof. This can cause a process called *electrolytic corrosion.* This is corrosion caused by electrolysis. *Electrolysis* is the creation of tiny electrical currents when different metals are in contact with each other and with water. Electrolysis can also occur where water washes first over one metal and then over another. For example, it might result when one type of metal is used for skylight flashing and another at the eaves.

Step Flashing Step flashing consists of small L-shaped pieces of metal that can be used wherever a sloped roof meets a vertical surface such as a wall or the sides of a chimney. The pieces are installed in an overlapping fashion as the roof shingles are being installed.

Stack Flashing This type of flashing is a manufactured product that helps to seal the area around vent pipes that penetrate the roofing. Stack flashing simply slips over a pipe when the shingles reach that point.

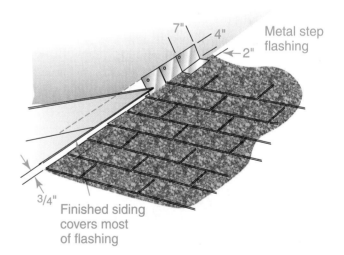

 Step Flashing
Overlapping Pieces Step flashing is often used at the intersection of a roof and a wall. Its top edge fits under the siding.

 Vent Stack Flashing
Covering Gaps Stack flashing fits over a vent pipe. The top edge will be covered by the next row of shingles to be applied. The bottom laps over the shingles already in place.

Subsequent courses of shingles cover the top of the flashing.

Valley Flashing Used in open valleys, this product consists of strips of copper or galvanized steel. It has a splash-diverting rib down the center. The rib reduces the tendency for water to pour down one side of the roof and splash up on the adjacent side. The flashing is held in place with metal cleats nailed to the roof. These cleats allow the flashing to expand and contract. They do not puncture it as nails would. Valley flashing should be made of at least 26-gauge

 Valley Flashing
Channeling Water The center of metal valley flashing is raised to prevent heavy volumes of water from flowing down one side and up the other.

galvanized metal or an equivalent noncorrosive metal.

Drip Edges *Drip edges* are designed and installed to protect the edges of the roof. They prevent leaks by causing water to drip free of underlying eave and cornice construction. A drip edge is recommended for most shingle roofs. It is applied to the sheathing and *under* the underlayment at the eaves, but *over* the underlayment at the rake.

Nails

No single step in applying roof shingles is more important than proper nailing. This depends on several factors, including:

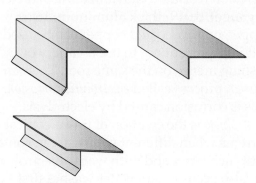

 Drip Edges
Edge Protector Various shapes of drip edge are used on a roof. The products come in 10' lengths.

- Selecting the correct nail for the kind of shingle and type of sheathing.
- Using the correct number of nails.
- Locating the nails in the shingle correctly.
- Choosing nails of a metal compatible with the metal used for flashing.

Specific recommendations for the type, size, number, and spacing of roofing nails are given in Sections 22.2 and 22.3.

Cements

Roofing cements include plastic asphalt cements, lap cements, quick-setting asphalt adhesives, roof coatings, and primers. They are used for installing eaves flashing, for flashing assemblies, for cementing tabs on asphalt shingles and laps in sheet material, and for roof repairs. The materials and methods used should be those recommended by the manufacturer of the roofing material.

Section 22.1 Assessment

After You Read: Self-Check

1. What term is used for the amount of roofing required to cover 100 sq. ft.?
2. Define shingle *coverage* and *exposure*.
3. What are two advantages of using a pneumatic nailer instead of a shingling hatchet?
4. Why is roofing one of the most dangerous occupations in residential construction?

Academic Integration: English Language Arts

5. **Proper Nailing** No single step in applying roof shingles is more important than proper nailing. Write a paragraph identifying the factors that a roofer must consider to ensure proper nailing.

Go to connectED.mcgraw-hill.com to check your answers.

Section 22.2 Strip Shingles

Installation Issues

Before installing strip shingles, what should already be installed?

Strip shingles, sometimes called *three-tab shingles,* are the most common roofing product used on houses. Before applying shingles make sure that:

- The underlayment, drip edge, and flashings are in place.

- The roof sheathing is tight and provides a suitable nailing base.
- The chimney is completed and the counter-flashing has been installed. *Counter-flashing* is metal flashing that covers the top edge of base or step flashing.
- Vents and other items requiring openings in the roof are in place, with counter-flashing where necessary.

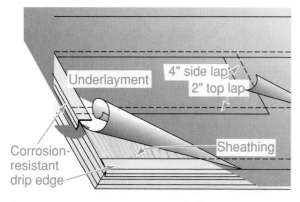

Single Coverage Underlayment

Good Protection Applying the underlayment for single coverage. Note the position of drip edge.

The most common strip shingles are made from asphalt or fiberglass. Both types are installed using the same basic methods. The following text relates to the installation of strip shingles on roofs with a slope of 4-in-12 or greater. Consult the shingle manufacturer's literature for installation advice on roofs with a lower slope.

Installing Underlayment

How many fasteners should be used to hold the underlayment?

The first step in strip shingle installation is to install the underlayment. Apply the underlayment as soon as the roof sheathing has been completed. For single coverage, start at the eave line with No. 15 felt. Roll the underlayment across the roof and work your way upwards. Make sure to create a top lap of at least 2" at all horizontal joints and a 4" side lap at all end joints. Lap the underlayment over all hips and ridges for 6" on each side. End laps should be spaced at least 6' apart.

Double coverage can be started with two layers at the eave line if desired. They should be flush with the fascia board or molding. For the remaining strips, allow 19" head laps and 17" exposures. Cover the entire roof in this manner.

Use only enough fasteners to hold the underlayment in place until the shingles can

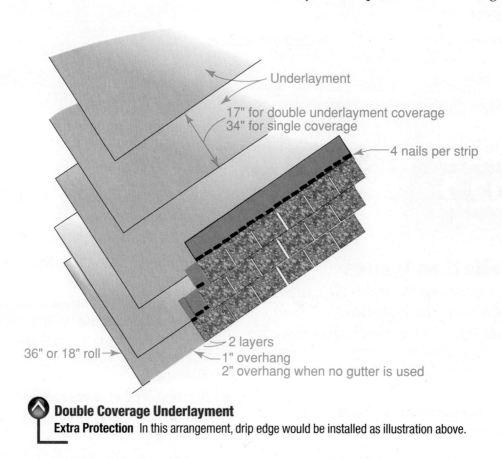

Double Coverage Underlayment

Extra Protection In this arrangement, drip edge would be installed as illustration above.

be applied. Fasteners may be nails or staples. Nails fitted with thin plastic washers, sometimes called *button caps,* can be used to prevent wind damage to the underlayment prior to shingle installation. Some roofers use a tool called a *hammer tacker* to attach roofing felt. Others use pneumatic equipment.

Laying Shingles

When should shingles not be applied?

Shingles are packaged in paper-wrapped bundles. A bundle contains about 25 shingles, more or less, and there are usually three bundles of shingles in a square of roofing. Bundles of shingles should be delivered to the roof and distributed evenly. This is sometimes called *stacking the roof* or *loading the roof.* Spreading the weight around prevents damage to the house structure. Bundles of shingles should be laid out for maximum safety and efficiency. They should not be in the way of the roofers' work.

Shingles should never be applied over wet underlayment. The sealant on the shingles will not adhere to a wet surface, and a wet surface is not safe to work on.

Strip shingles may be laid from either end of the roof, but many roofers prefer to work from right to left.

Apply the first course of shingles, called the *starter course*, over the eaves flashing. Roll roofing or inverted shingles can be used. This first course and starter strips should project just past the roof edge or edge metal. Fasten the starter strip with roofing nails placed about 3" or 4" above the eave edge and spaced so that the nail heads will not be exposed at the cutouts between the tabs on the first course. If strip shingles are used as a starter strip, cut 3" off the first starter course shingle to be laid at the rake. This ensures that the slots will be offset from the next course. Then lay the first course right side up, starting with a full shingle. A starter shingle is available from most roofing suppliers. The advantage of using a starter shingle is the proper location of the self-sealing adhesive.

Reading Check

Recall Why should shingles never be applied over wet underlayment?

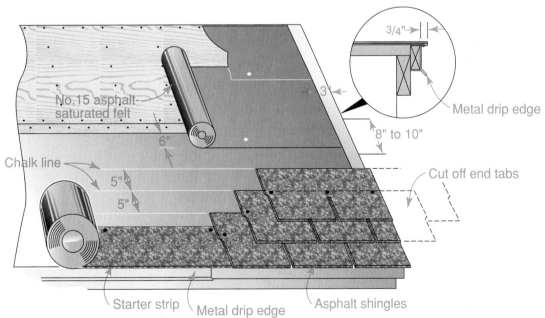

Laying Strip Shingles
Working Right to Left Basic details showing how to start the job.

Start **succeeding** courses with full or cut shingles. The choice depends on how the roofer wants to align the cutouts in the shingles. The choice affects the look of the roof but does not affect its function or durability. Two methods of alignment are described as follows:

- *Method 1* This method aligns the cutouts of every other course. Start the first course with a full shingle. For the second course, cut 6" (one-half of a tab) off the end of a full shingle and use the shingle to start the course. For the third course, cut 12" (a full tab) off the end of a full shingle and use the shingle to start the course. For the fourth course, cut 18" (1½ tabs) off the end of a full shingle and use the shingle to start the course. This process is sometimes called *breaking the joints on halves.*

- *Method 2* This method separates aligned cutouts by two courses. Start the first course with a full shingle. Start the second course with a full shingle cut 4" short (one-third of a tab). Start the third course with a full shingle cut 8" short (two-thirds of a tab). Start the fourth course with a full shingle. This is sometimes called *breaking the joints on thirds.*

Regardless of the method chosen, place each succeeding course of shingles so that the lower edges of the butt ends are aligned with the top of the cutouts on the underlying course. To ensure proper alignment of the shingle courses, snap a chalk line periodically from one end of the row to the other at the top of the cutouts. Align the courses with the chalk line.

Nailing

Nails should be made of hot-dipped galvanized steel, aluminum, or stainless steel. A roofing nail has a sharp point and a large, flat head at least ⅜" in diameter. Shanks should be 10- to 12-gauge wire. They may be smooth or threaded for increased holding power.

The number and the placement of nails are important. Nailing should start at the end of the shingle nearest the shingle last applied and proceed to the opposite end. To prevent buckling, be sure each shingle is in perfect alignment before driving any nails. Drive the nail straight to avoid cutting the shingle with the edge of the nail head. Do not sink the nail head below the surface of the shingle.

Three-tab shingles require four nails for each strip. Individual tabs or parts of tabs must be attached with at least two nails. In areas where wind hazards are high, local codes may require six nails per shingle. When the shingles are applied with a 5" exposure, the four nails are placed ⅝" above the top of the cutouts. The nails are located horizontally with one nail 1" back from each end of the shingle and one nail on the centerline of each cutout. To provide extra resistance to uplift in high wind areas, use six nails for each strip.

🔺 **Shingle Nails**
Three Types Standard and ring-shank roofing nails.

McGraw-Hill Education

Roofing Details

Why does air move in a passive vent system?

Installing shingles in uninterrupted rows can proceed quite quickly. This is particularly true if a roofer is nailing as an apprentice positions a steady supply of shingles. Work proceeds at a slower pace at hips, ridges, and in other locations around the roof.

Hips and Ridges

Hips and ridges may be finished by using *hip and ridge shingles* furnished by the manufacturer. They are sometimes called *caps*. You can also cut pieces at least 9" × 12" either from shingle strips or from mineral-surfaced roll roofing of a color to match the shingles. Apply these by gently bending each shingle lengthwise down the center and placing an equal amount on each side of the hip or ridge. To ensure proper align-

ment, snap a chalk line down one side of the ridge and align the edge of the shingle with it as you nail.

Apply the hip and ridge shingles by beginning at the bottom of a hip or one end of the ridge. Use a 5" exposure. Secure each shingle with one nail at each side 5½" from the exposed end and 1" away from each edge. When laying the shingles on the ridge, always lay the exposed edge away from the prevailing winds.

Ridge Vents Proper ventilation of the roof is very important. It removes hot air that could damage shingles from the attic. It also removes moisture that could damage the roof sheathing. There are various ventilation methods. One common method is the use of the full-length ventilating ridge. For a ridge vent to be effective, air must be drawn in through the eaves and soffit vents. As this relatively cool air is drawn in, warmer air is exhausted through the ridge vent. This

Hip and Ridge Shingles

Hip and Ridge Caps The hip and ridge shingles on this roof give it a neat, finished appearance and protect joints between roof planes.

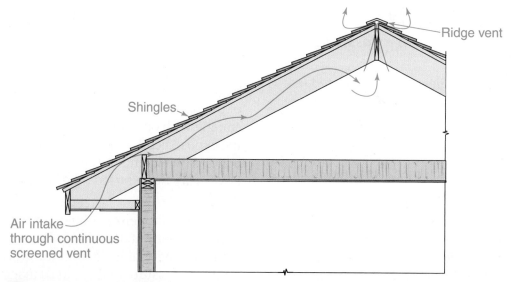

Ridge vent

Shingles

Air intake through continuous screened vent

 How a Ridge Vent Works
Steady Flow In passive venting, air is drawn in through vents in the soffit as warm attic air exits a vent at the ridge of the roof.

ensures a flow of air through the roof cavities or attic.

This method is called *passive venting* because it does not rely on fans. The roof sheathing is cut away on each side of the ridge board, creating a narrow opening along the entire length of the roof. One of various ridge-vent products is then nailed over the opening. Depending on the product, ridge shingles may be nailed to the top of the ridge vent. Ridge vents vary considerably in design. Always follow the manufacturer's instructions when installing them. This will ensure that the vent is weatherproof.

Vent Pipes

Vent pipes, sometimes called *soil stacks*, often penetrate the roof. These pipes must be flashed with vent stack flashing (page 634) to prevent water from getting past the shingles. Before the flashing is placed, the roof shingles are applied in courses up to the exposed pipe. At that point, the shingles are cut to fit around the pipe. Then the flashing sleeve is slipped over the stack. Shingling can now continue, with shingles overlapping the top and sides

of the flashing. They are cut to fit around the stack and pressed firmly into roofing cement. However, shingles should never cover the bottom of the flashing because this would prevent water from draining away freely.

Closed Valleys

When two different roof planes meet, they often form a valley. Valleys can be shingled in one of two ways: *closed* or *open*. For a closed valley (sometimes called a *woven valley*), strip shingles are interwoven

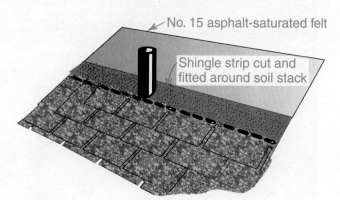

No. 15 asphalt-saturated felt

Shingle strip cut and fitted around soil stack

 Shingles at Vent Pipes
Good Fit Shingles are cut to fit around the pipe before the flashing is placed over the pipe.

 A Properly Flashed Vent
Waterproof Note that the bottom edge of the flashing is *not* covered by shingles.

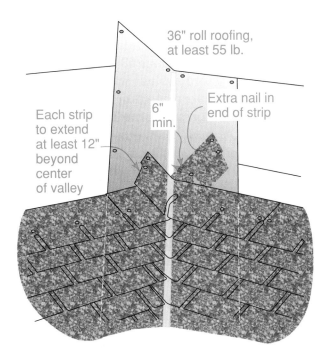

36" roll roofing, at least 55 lb.

Each strip to extend at least 12" beyond center of valley

6" min.

Extra nail in end of strip

 Closed Valley
Double Coverage Closed valley construction using woven strip shingles.

to protect the valley from seepage. The shingles are laid as shown in Closed Valley. A valley lining made from a 36" wide strip of 55 lb. (or heavier) roll roofing should be placed over the valley underlayment and centered in the valley. No metal flashing is necessary.

A closed valley results in a double coverage of shingles throughout the length of the valley. Valley shingles are laid over the lining by either of two methods. In the first method, they are applied on both roof surfaces at the same time, with each course in turn woven over the valley. In the second method, each surface is covered to a point approximately 36" from the center of the valley. The valley shingles are woven in place later.

In both methods, the first course at the valley is laid along the eaves of one surface over the valley lining. It is then extended along the adjoining roof surface for at least 12". The first course of the adjoining roof surface is then carried over the valley on top of the previously applied shingle. Succeeding courses are then laid alternately, the valley shingles being laid over each other.

Each shingle is cemented to the valley lining with asphalt cement and pressed down to ensure a tight seal. The shingles are then nailed in the usual manner, except that no nail should be located closer than 6"

to the valley centerline and no joints should occur within 12" of the valley centerline. Two nails are also used at the end of each terminal strip.

Closed-Cut Valley One variation of the closed valley is called a *closed-cut valley*. One side of the intersecting roofs, the shingles are applied across the valley, as with a standard closed valley. End joints should not occur within 6" of the center of the valley. A chalk line is then snapped on the shingles parallel to the center of the valley and on the unshingled side of the valley. Shingles from the adjacent roof area are then installed on top of the overlapping shingles and cut at the chalk line. A 3" wide strip of asphalt cement is applied under the cut ends and no fasteners are installed within 6" of the valley.

Open Valleys

In an open valley, flashing is used to cover the valley. Shingles overlap the edges of the flashing. An **open valley** is a type of roof valley in which shingles are not applied to the intersection of two roof surfaces. This leaves the underlying flashing exposed along the length of the valley. Open valley flashing can be strips of metal (see page 634) or a length of mineral-surfaced asphalt (roll roofing).

You can calculate roof area without measuring it by hand using **Table 22-1**.

Metal Each section of metal valley flashing should lap at least 4" over the next lowest section. The sides must extend at least 8" up each side of the intersecting roof. A 36" wide layer of underlayment should run the full length of the valley under the flashing. Special underlayment similar to that used for eaves protection may be required in areas where ice dams are often a problem.

Mineral-Surfaced Asphalt When mineral-surfaced roll roofing is used to flash a valley, it is matched to the color of the shingles. An 18" wide strip of roll roofing is placed over the underlayment, with the mineral-surfaced side down. When it is necessary to splice the material, the ends of the upper segments overlap the lower segments by 12". They are then secured with asphalt plastic cement.

Only enough nails are used in rows 1" in from each edge to hold the strip smoothly in place. Another strip, 36" wide, is then placed over the first strip. It is centered in the valley with the surfaced side up and secured with nails. If necessary, it is lapped the same way as the underlying strip. As shingles are applied later, they are cut at an angle to fit against chalked lines snapped on either side of the valley. No exposed nails should appear along the valley flashing. No nails should be placed within 6" of the valley's centerline.

Table 22-2 on page 644 provides nail and labor requirements for asphalt roofing products.

Vertical Intersections

Step flashing is necessary where a vertical surface, such as a chimney or a second-story wall, meets the roof. Step flashing consists of individual L-shaped pieces of metal.

Where the roof meets a wall, one leg of a piece of step flashing fits under a shingle. The other leg fits under the siding, as shown on page 633. Where the roof meets a chimney, the top edge of the step flashing fits under metal cap flashing. The cap flashing is inserted 1½" into the joints between bricks. The joints are then filled with mortar.

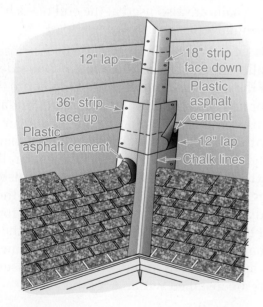

12" lap
18" strip face down
Plastic asphalt cement
36" strip face up
Plastic asphalt cement
12" lap
Chalk lines

Open Valley
Details Count Roll roofing used as open valley flashing. The upper corners of each shingle at the valley should be cut off to prevent them from obstructing water flow. *What is the difference between a closed valley and an open valley?*

Table 22-1: Determining Roof Area from a Plan			
Rise (inches)	Factor	Rise (inches)	Factor
2	1.014	7½	1.179
3	1.031	8	1.202
3½	1.042	8½	1.225
4	1.054	9	1.250
4½	1.068	9½	1.275
5	1.083	10	1.302
5½	1.100	10½	1.329
6	1.118	11	1.357
6½	1.137	11½	1.385
7	1.158	12	1.414

Estimating and Planning

This estimating and planning exercise will prepare you for national competitive events with organizations such as SkillsUSA and the Home Builder's Institute.

Strip Shingles and Underlayment

Estimate Area and Materials

To determine the amount of shingles and underlayment needed, you must first know the total area to be covered.

Gable Roofs

To figure the roof area without actually getting on the roof to measure it, use Table 22-1.

Step 1: Look at the building plans to find the roof rise. The rise is given on Table 22-1 in inches per horizontal foot.

Step 2: The area of the roof line (including the overhang) is then multiplied by the factor shown in the table. For example, if a home is 70' long and 30' wide, including the overhang, the roof line area is 2,100 sq. ft. Look at Table 22-1. If the rise of the roof is 5½", the factor needed is 1.100. The 2,100 sq. ft. multiplied by 1.100 results in a total roof area of 2,310 sq. ft.

Step 3: One square of shingles covers 100 sq. ft. of roof surface. To determine the number of squares needed to cover the roof, divide the total area by 100 and add 10 percent for waste and cutting.

$$2,310 \div 100 = 23.10$$
$$23.1 \times 0.10 = 2.31$$
$$23.1 + 2.31 = 25.41$$

25.41 = about 25 squares of shingles.

Step 4: Be sure to subtract for large openings in the roof such as skylights and chimneys. Subtract also for the area where a dormer intersects the roofline so that you do not include the dormer area twice.

Another method of figuring the area of a plain gable roof is to multiply the length of the ridge by the length of a rafter. This gives one-half the roof area. Then, multiply by 2 to obtain the total square feet of roof surface.

Hip Roofs

To find the area of a hip roof, multiply the length of the eaves by one-half the length of the common rafter at the end. Multiply this by 2 to obtain the area of both ends. To find the area of the sides, add the length of the eave to the length of the ridge and divide by 2. Multiply this by the length of the common rafter to obtain the area of one side of the roof. Multiply by 2 to find the number of square feet on both sides of the roof. Add this to the area of the two ends and divide the total area by 100 to get the number of squares.

To obtain the area of a plain hip roof running to a point at the top, multiply the length of the eaves at one end by one-half the length of the rafter. This gives the area of one end of the roof. To obtain the total area, multiply by 4.

Quantities of starter strips, eaves flashings, valley flashings, and ridge shingles all depend upon linear measurements along the hips, rakes, valleys, eaves, and ridge. Measurements for horizontal elements can be taken from the roof plan. The rakes, hips, and valleys run on a slope. The actual length of rakes, hips, and valleys must therefore be measured on the roof.

Nails

The number of nails needed for asphalt roofing can be determined from Table 22-2 on page 644.

Step 1: In the above example, 25 squares of three-tab shingles are required to cover the roof area. Read down the table from the heading Pounds per Square to the line "3 tab sq. butt on new deck."

Step 2: If 11-gauge nails are used, 1.44 lbs. are required for each square. The total number of pounds of nails needed would then be 36.

$$25 \times 1.44 = 36 \text{ lbs. of nails}$$

Estimating on the Job

Using Table 22-1, estimate the number of squares of shingles that would be required to cover a house that is 82' long and 28' wide, including overhang. The rise is 7". Round your answer to the nearest whole number.

Table 22-2: Nail and Labor Requirements for Asphalt Roofing Products

Type of Roofing	Shingles per Square	Nails per Shingle	Length of Nails[a] (inches)	Nails per Square	Pounds per Square (approximate)		Labor Hours per Square
					12 ga. by $\frac{7}{16}$" head	11 ga. by $\frac{7}{16}$" head	
Roll roofing on new deck	—	—	1	252[b]	0.73	1.12	1
Roll roofing over old roofing	—	—	1¾	252[b]	1.13	1.78	1¼
19" selvage over old shingle	—	—	1¾	181	0.83	1.07	1
3 tab sq. butt on new deck	80	4	1¼	336	1.22	1.44	1½
3 tab sq. butt reroofing	80	4	1¾	504	2.38	3.01	1⅚
Hex strip on new deck	86	4	1¼	361	1.28	1.68	1½

[a]Length of nail should always be sufficient to penetrate at least ¾" into sound wood. Nails should show little, if any, below underside of deck.
[b]This is the number of nails required when spaced 2" apart.

Section 22.2 Assessment

After You Read: Self-Check

1. What does *stacking the roof* mean?
2. What is often used as a starter course on shingled roofs?
3. What is the typical exposure of ridge shingles?
4. How is a closed valley protected from seepage?

Academic Integration: Mathematics

5. **Percent Increase** One square of shingles covers 100 sq. ft. of roof surface. Michael determined that he needs 22 squares of shingles to cover a roof with an area of 2,000 sq. ft. His calculation included a percent increase to allow for waste. What percent increase did Michael use in his calculation?

   Math Concept Percent increase is when you find a percentage of an amount and then add it to the amount. A percent increase is always based on an original amount.

   Step 1: Find the amount of squares needed without the percent increase. This is the original amount (x).

   Step 2: Divide the difference between 22 and the amount by the original amount $\left(\frac{22 - x}{22}\right)$. Convert the decimal to a percentage.

   Go to connectED.mcgraw-hill.com to check your answers.

Wood Shingles & Shakes

Installation Issues

Why is wood a good shingle material?

Wood shingles and shakes are generally laid over spaced sheathing boards. The spaced sheathing, sometimes called *open sheathing,* allows the wood to dry uniformly after exposure to rain. Shingles and shakes may also be laid over solid sheathing. This is sometimes required in regions where earthquakes are common. It is also recommended where wind-driven snow is common.

Underlayment protects the sheathing and reduces air infiltration. For underlayment, No. 15 asphalt-saturated felt may be used.

However, underlayment is not installed over the entire roof in one layer, as with strip shingles. Instead, it is overlapped with successive courses.

Types and Grades

What types of wood might not be good for shingles?

Shakes and shingles are graded according to the quality of the wood, considering any imperfections. The most common material is red cedar. Other types of cedar, along with oak and other woods, are sometimes used.

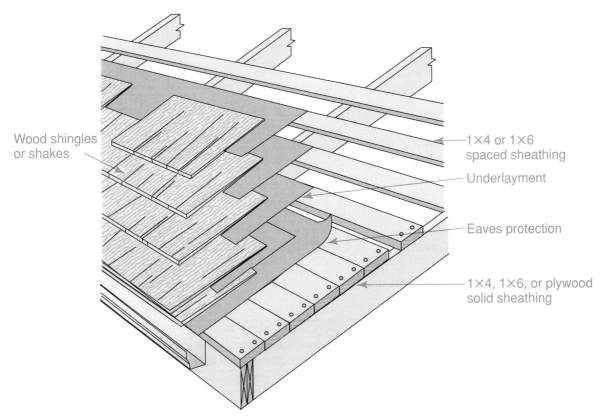

Wood shingles or shakes

1×4 or 1×6 spaced sheathing

Underlayment

Eaves protection

1×4, 1×6, or plywood solid sheathing

 Wood Shingle Roof System
Spaced Sheathing The manufacturer's installation requirements should be closely followed when installing wood shingles and shakes.

Shakes

A shake is a type of wood roofing product that is larger in size than a wood shingle and thicker at the butt edge. Shakes are split on one or two sides. This gives each one a much rougher, rustic texture compared to a wood shingle. There are three types of wood shakes:

- *Handsplit-and-resawed* shakes have split faces and sawed backs. Blanks or boards are split and then run diagonally through a band saw to produce two tapered shakes.

- *Tapersplit* shakes are produced largely by hand, using a sharp-bladed steel froe and a wooden mallet. (A *froe* is a hand tool with a sharp blade and a wood handle at a right angle to the blade.) The taper is achieved by reversing the block end-for-end with each split.

- *Straightsplit* shakes are similar to taper-split shakes. However, because they are split from the same end of the block, the shakes are not tapered.

Shakes are available in three lengths: 16", 18", and 22". The maximum exposure recommended for double coverage is 13" for 32" shakes, 10" for 22" shakes, and 7½" for 18" shakes. Triple coverage can be achieved by reducing these exposures to 10" for 32" shakes, 7½" for 22" shakes, and 5½" for 18" shakes. Shakes are not recommended for roofs with slopes of less than 4-in-12.

Wood Shingles

A wood shingle is manufactured by being sawn from a short length of log. This gives shingles a relatively smooth and uniform texture as compared to a wood shake. Wood shingles are manufactured in lengths of 22" (*Royals*), 18" (*Perfections*), and 16" (*Fivex*). They are available in No. 1 grade (sometimes called *blue label*), No. 2 grade (sometimes called *red label*), No. 3 grade (sometimes called *black label*), and undercourse grade used for underlying starter courses. The exposure of wood shingles depends on the slope of the roof, as shown in **Table 22-3**.

In addition, decorative *fancy-butt* shingles are available. The butt edge (the thickest part of the shingle) can be curved, beveled, or cut into any number of other shapes. Some wood shingles are pressure-treated with fire-retardant chemicals at the factory. These may be required by code where fire danger limits the use of untreated shingles. Generally, only No. 1 grade shingles are treated.

Wood shingles are thinner than wood shakes. The surface of either product may be sawed (for a relatively smooth appearance), or split (for a more rustic look). The installation methods are similar.

Reading Check

Recall *Why are shakes split on one or two sides?*

Installing Wood Shingles

Are ice dams less common on wood-shingle roofs?

The installation of wood shingles requires many of the same types of materials required for installing strip shingles. Some of them are installed in different ways.

	Table 22-3: Wood Shingle Exposure								
	Maximum Exposure Recommended for Roofs Length (inches)								
Slope	No. 1 Blue Label			No. 2 Red Label			No. 3 Black Label		
	16	18	22	16	18	22	16	18	22
3:12 to 4:12	3¾	4¼	5¾	3½	4	5½	3	3½	5
4:12 and steeper	5	5½	7½	4	4½	6½	3½	4	5½

Eaves Protection

In many cold climates, ice dams can form along the eaves and cause water to leak into the house (see page 634). In such areas, solid (plywood) sheathing should be applied above the eave line to a point at least 22" inside the interior wall line of the building. The solid sheathing should then be covered with a double layer of No. 15 asphalt-saturated felt. A comparable product such as *self-adhering bitumen sheet* can also be used.

Flashing

Galvanized metal flashing is commonly used with wood shingles. If copper flashing is used with wood shingles or shakes, take special precautions. Early deterioration of the copper may occur when the metal and wood are in direct contact in the presence of moisture.

On slopes up to 12-in-12, metal valley sheets should be wide enough to extend at least 10" on each side of the valley centerline. The open portion of the valley should be at least 4" wide.

Reading Check

Explain *What may occur if copper and wood are in direct contact in the presence of moisture?*

Mathematics: Measurement

Roof Slope The slope of a roof is the ratio comparing unit rise with unit run. The unit run is always 12". In other words, the slope of a roof is its rise per foot of run. What is the slope of a roof if the pitch is ⅛? Is a roof with this pitch a flat, low slope, or high slope roof?

Starting Hint When the roof pitch is known, multiply the pitch times 22 to find the unit rise. Place this number over the unit run to determine the roof slope.

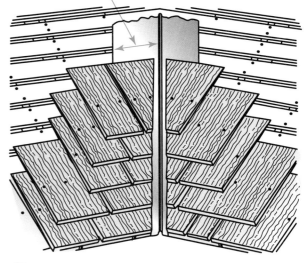

On slopes up to ¹²/₁₂, valley sheets should extend at least 10" from valley center

 Open Valley Flashing
Good Drainage Open valley flashing is the preferred valley technique for wood shingles and shakes.

Laying Shingles

As with strip shingles, the first course of wood shingles is a starter course. Double or triple the first course of shingles at the eaves. It should project 1" to 1½" beyond the eaves to provide a drip edge.

Nail the second layer of shingles in the first course over the first layer to provide a minimum side lap of at least 1½" between joints on page 648. A triple layer of shingles in the first course provides additional insurance against leaks at the cornice. Undercourse shingles or No. 3 grade shingles are frequently used for the starter course.

Space shingles at least ¼" apart to provide for expansion. Joints between shingles in any course should be separated not less than 1½" from joints in the adjacent course above or below. Joints in alternate courses should not be in direct alignment. When shingles are laid with the recommended exposure, triple coverage results. Some shingle grades contain small defects, such as knots. Shingles should be laid so that the edges of the nearest shingles in

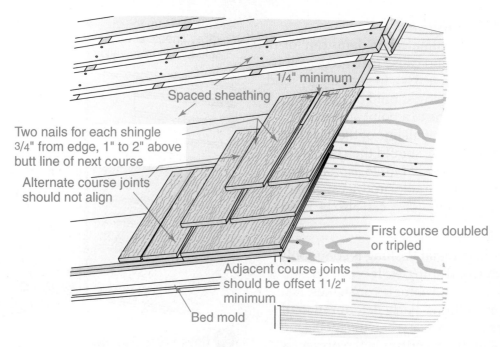

Spaced sheathing

1/4" minimum

Two nails for each shingle 3/4" from edge, 1" to 2" above butt line of next course

Alternate course joints should not align

First course doubled or tripled

Adjacent course joints should be offset 11/2" minimum

Bed mold

Nailing Details
Watch the Joints Details for applying wood shingles over spaced or solid sheathing.

the course above are at least 1½" away from the defect.

When the roof terminates in a valley, carefully cut the shingles for the valley to the proper miter at the exposed butts. Nail these shingles in place first so that the direction of

1 1/2"

Knot or similar defect

Spacing Details
Separate Flaws Shingles should be spaced so any slight flaws are separated as shown. This protects the flaws from exposure.

shingle application is away from the valley. This permits valley shingles to be carefully selected. It also ensures that shingle joints will not break over the valley flashing.

Nailing

To ensure that shingles will lie flat and give maximum service, use only two nails to secure each one. Place nails not more than ¾" from the side edge, at a distance of not more than 1" above the exposure line. Drive nails flush, but take care that the nail head does not crush the wood. The recommended nail sizes for the application of wood shingles are shown in **Table 22-4**.

Hips, Ridges, and Rakes

Hip and ridge shingles should overlap in a fashion sometimes referred to as a *modified Boston ridge*. A **Boston ridge** calls for alternating the joint position in each pair of intersecting cap shingles. The top edges of the cap shingles should be beveled for a neat appearance. This also eliminates projections that would encourage water to seep into the joint.

Table 22-4: Recommended Nail Sizes for Application of Wood Shingles and Shakes				
For 16" and 18" Material		For 22" Material	For 16" and 18" Material	For 22" Material
1¼" long	1¼" long 14½ gauge	1½" long 14 gauge	1¾" long 14 gauge	2" long 13 gauge
Approx. 376 nails per lb.	Approx. 515 nails per lb.	Approx. 382 nails per lb.	Approx. 310 nails per lb.	Approx. 220 nails per lb.

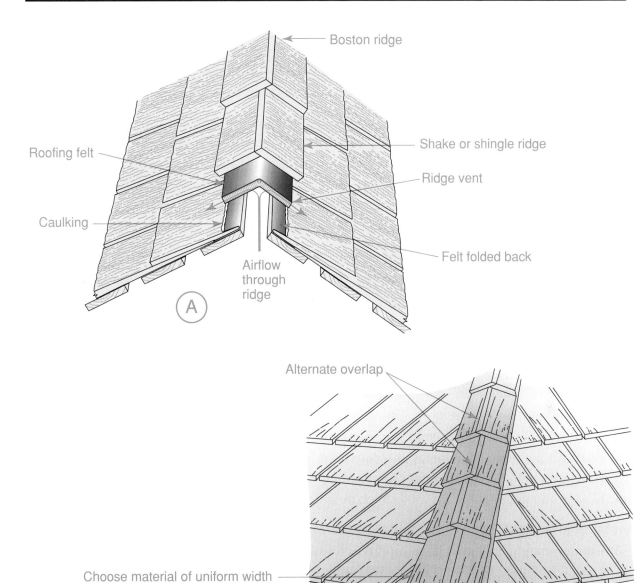

 **Hips and Ridges**

Alternating Laps Wood shingle construction. **A.** Boston ridge construction. **B.** Hip construction.

Nails that are at least two sizes larger than those that are used to apply the shingles are required for a Boston ridge. A continuous ridge vent may be installed on a wood shingle roof but is not required if other ventilation methods have been provided. Check local code requirements.

Hips and ridges should begin with a double starter course. Either site-applied or pre-formed factory-constructed hip and ridge units may be used. Shingles should project 1" to 1½" over the rake.

Estimating

The general area of a gable roof can be estimated by consulting Table 22-1 on page 642 and the instructions given in the Estimating and Planning feature on page 643. To estimate the number of wood shingles required, consult **Table 22-5**. A waste factor is always considered when estimating shingle quantities. This allows the roofer to cull (remove) shingles that have serious imperfections, and to replace shingles that split during installation. Flashing and other materials can be estimated in the same manner as with strip shingles.

Table 22-5: Materials and Labor for Wood Shingles			
Material and Labor	Material per 100 Sq. Ft. of Surface		
Wood Shingles Laid to Weather (inches)	Shingles	Shingles with 10% Waste	Labor Hours per 100 Sq. Ft.
4	900	990	3¾
5	792	792	3
6	600	660	2½

Note: Increase time factor 25% for hip roofs.

Section 22.3 Assessment

After You Read: Self-Check

1. What type of sheathing is most commonly used beneath wood shingles and shakes?
2. What grades of wood shingles are often used for starter courses?
3. How far should wood shingles be spaced from each other and why?
4. How many nails should be used to secure a shingle?

Academic Integration: English Language Arts

5. **Compare and Contrast** Shakes and wood shingles are common roof covering materials. Write two paragraphs describing how the two products are similar and how they differ.

Go to connectED.mcgraw-hill.com to check your answers.

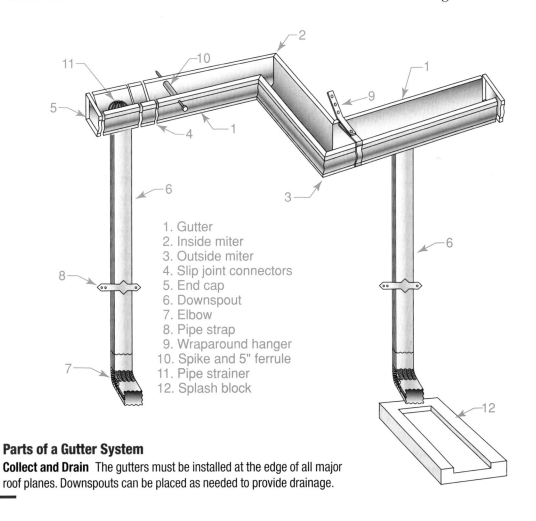

Gutter Basics

How does corrugating a downspout help prevent ice damage?

Because they are attached to the fascia of a house, gutters are often considered to be part of the cornice construction. However, gutters are not generally installed by carpenters. Instead, they are the work of contractors who specialize in fabricating and installing gutters.

Gutters are part of a system that collects water from the roof and drains it away from the house. The gutters themselves are horizontal members that collect the water

and channel it to vertical downspouts, or *leaders*. *Elbows* join the gutters to the downspouts. They are used at the bottom of the downspouts as well. *Splash blocks* direct downspout water away from the foundation of the house. This is important in preventing large quantities of water from collecting near the foundation, where it could cause damage.

Wooden gutters were once widely used but are very rare now. Today, most gutters are made of aluminum, copper, or vinyl. The two general types are the *formed-metal gutter* and the *half-round gutter*. The most common formed metal gutter is a box

1. Gutter
2. Inside miter
3. Outside miter
4. Slip joint connectors
5. End cap
6. Downspout
7. Elbow
8. Pipe strap
9. Wraparound hanger
10. Spike and 5" ferrule
11. Pipe strainer
12. Splash block

Parts of a Gutter System

Collect and Drain The gutters must be installed at the edge of all major roof planes. Downspouts can be placed as needed to provide drainage.

shape called a *K-style gutter* and is typically 5" wide. If unusually large amounts of water must be channeled away, 6" wide gutters should be used. Half-round gutters are generally about 1" wider than K-style gutters of an equivalent drainage capacity.

Downspouts can be round or rectangular. The round downspout is used most often with half-round gutters. Both types are usually corrugated for added strength. Corrugated downspouts are less likely to burst if plugged with ice.

Gutters can be purchased in 10' sections and joined with slip joint connectors. However, it is common for gutters to be fabricated on site using machines. Gutters of almost any length can then be formed from continuous coils of flat aluminum stock. These are sometimes called *seamless*, or *continuous*, gutters. They are less likely to leak because they are not assembled from several shorter lengths. In either case, the basic material is generally coated with a

baked-on finish in one of several common colors.

Reading Check

Explain *Why are gutters often considered part of the cornice construction?*

Installation
What other method could be used to determine the slope of gutters?

Metal gutters on a house appear to be level. However, they actually slope at least 1" every 16' toward the downspouts ($\frac{1}{16}$" per foot). This is essential for proper drainage. The maximum distance between the gutter's high point and the downspout should not ordinarily exceed 25'.

To ensure the correct slope, measure the distance in feet from one end of the fascia to the other. Round up to the nearest whole foot. Multiply this number by $\frac{1}{16}$". For example, a measurement of 20'-4" is rounded up to 21':

$$21 \times \frac{1}{16}" = 1'\text{-}\frac{5}{16}"$$

The answer is the difference in inches between the gutter's highest and lowest points. Locate these points on the fascia and snap a chalk line between them. Align the top of the gutter with this line. To prevent gutters from being damaged by sliding snow or ice, position them so that the outer edge is below the plane of the roof.

Gutters are held in place using one of several methods. One calls for flat metal hangers, called *wraparound hangers*, spaced

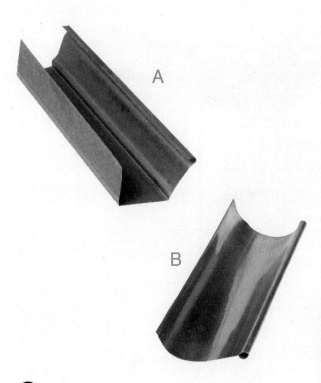

Shapes of Gutters
Basic Shapes Gutters: **A.** Formed. **B.** Half-round (this one is made of copper).

JOB SAFETY

AVOIDING SHARP EDGES The edges of metal gutters can be very sharp. Wear leather gloves when handling and cutting them.

3' to 4' on center (see page 651). Another is called the *spike-and-ferrule method*. *Ferrules* are short metal tubes placed between the inner and outer faces of the gutter. Aluminum spikes are then driven through the face of the gutter, through the ferrule, and into the fascia.

Downspouts are fastened to the wall by leader straps or hooks. These allow a space between the wall and the downspout. At least two straps should be used to secure an 8' length. An elbow directs the water to a splash block that carries it away from the foundation. The splash block should be at least 3' long. In final grading, the slope should ensure positive drainage of water away from the foundation walls.

Some builders eliminate the splash blocks and lower elbows. Instead, they connect the downspouts to a system of below-grade pipes that drain into a storm sewer. This

Builder's Tip

TIMING GUTTER INSTALLATION Some types of gutter systems are best installed before the roofing has been applied. Others are best installed after the roofing has been applied. However, the roofers and the gutter installers should never be working simultaneously. The general contractor supervising the project must coordinate the work of the two trades so that schedule conflicts can be avoided.

ensures that water will not seep into the soil around the house. However, building codes prohibit drainage into a septic system.

Section 22.4 Assessment

After You Read: Self-Check

1. Name the two general types of gutters.
2. Downspouts often have a corrugated shape. What purpose does this serve?
3. What is the minimum slope required to ensure that gutters drain properly?
4. What is a ferrule?

Academic Integration: Mathematics

5. **Proportions** Metal gutters on a house appear to be level. However, they actually drop toward the downspouts at least 1" for every 16'. How many inches must a metal gutter measuring 80 feet drop from one end to the other to allow for proper drainage?

> **Math Concept** A proportion is an equation that sets two ratios equal to one another.

Step 1: Use x to represent the unknown quantity, the total drop over 80'.

Step 2: Write two ratios, one describing the drop over 16', the other describing the drop over 80'. Set them equal to one another.

Step 3: Solve the proportion for the unknown quantity using cross-multiplication.

Go to **connectED.mcgraw-hill.com** to check your answers.

Chapter Summary

Section 22.1
Each type of roofing product has different characteristics that make it suitable for a particular house. The choice of a roofing material depends in part on the climate in which the house is located. Safety is of criti al importance when working on a roof. Underlayment and flashing prevent water from reaching the sheathing and causing damage.

Section 22.2
Installing strip shingles starts at the eaves of a roof and works up to the ridge. Shingles should be installed in an orderly fashion. Proper nailing is very important for the durability of the shingle and to maintain the manufacturer's warranties.

Section 22.3
Wood shingles and wood shakes can be installed in similar ways. Various grades and types of shingles and shakes are available. They are installed over spaced sheathing.

Section 22.4
Gutters and downspouts are part of a system that drains water away from the house. Gutters must be sloped toward downspouts. They are often installed by specialty contractors.

Review Content Vocabulary and Academic Vocabulary

1. Use each of these content vocabulary and academic vocabulary words in a sentence or diagram.

Content Vocabulary
- flashing (p. 626)
- square (p. 626)
- exposure (p. 626)
- butt edge (p. 626)
- top lap (p. 626)
- side lap (p. 626)
- underlayment (p. 627)
- ice dam (p. 632)
- open valley (p. 641)
- Boston ridge (p. 648)

Academic Vocabulary
- precaution (p. 631)
- succeeding (p. 638)

Speak Like a Pro

Technical Terms

2. Work with a classmate to define the following terms used in the chapter: *lock-down* (p. 628), *cap sheet* (p. 629), *shingling hatchet* (p. 630), *roof jacks* (p. 631), *ice barrier* (p. 632), *drip edge* (p. 633), *electrolytic corrosion* (p. 633), *three-tab shingles* (p. 635), *counter-flashing* (p. 635), *button caps* (p. 637), *hammer tacker* (p. 637), *stacking the roof* (p. 637), *starter course* (p. 637), *passive venting* (p. 640), *soil stacks* (p. 640), *open sheathing* (p. 645), *Royals* (p. 646), *perfections* (p. 646), *Fivex* (p. 646), *wraparound hangers* (p. 652), *ferrules* (p. 653).

Review Key Concepts

3. List the roofing products used in residential construction.

4. Name the steps for installing a strip shingle roof covering.

5. Explain the purpose of flashing.

6. List the properties of shakes and wood shingles.

7. Name the parts that make up a gutter system.

8. Explain how gutter systems work.

Critical Thinking

9. Explain A gutter system must be installed to slope at least 1 inch for every 16 feet. Explain why this is.

Academic and Workplace Applications

STEM Mathematics

10. Finding Perimeter Galvanized drip edge is purchased in 10' sections. Aluminum flashing is purchased in 10' × 10' rolls. Estimate the number of 10' sections of galvanized drip edge and 10' × 10' rolls of aluminum flashing needed for a gable roof with a 58' ridgeboard, 22' common rafters, and one double-flued 2'-8" × 4'-0" chimney.

Math Concept Perimeter is the distance around a figure or shape. The perimeter of a rectangle is two times the length plus the width.

Step 1: Calculate the perimeter of the roof and of the chimney opening.

Step 2: Convert the measurements to feet using decimal numbers, if necessary. Multiply by 110% to add a 10% allowance for waste.

Step 3: In each case, divide the length by 10', the number of feet in one section of drip edge and one roll of flashing. Round up to the nearest ¹⁄₁₆".

STEM Engineering

11. Stresses and Design Roofing materials vary in their composition and weight. For example, roofing tiles are made of clay or lightweight concrete and weigh more than wood or asphalt shingles. To accommodate roofing tiles, roof framing must be designed to support that extra weight. Research aspects of the materials and techniques used in roof framing must be modified in order to support roofing tiles. Summarize your findings in a one-page report.

21st Century Skills

12. Information Literacy The roof covering must provide long-lasting waterproof protection for a house. The choice of materials and methods is influenced in part by climate. Find out more about which roof covering materials are best suited for the climate in which you live. Use the Internet and library or contact a local roofing company to investigate roof covering materials. Write a paragraph stating which materials are best suited for your climate. Support your statement with facts from your research.

Standardized TEST Practice

Multiple Choice

Directions Read each of the following questions carefully. Select the best word or phrase to answer the question.

13. _____ is *not* a common material used for making shingles.

 a. Wood   c. Rubber
 b. Cement   d. Slate

14. Roof brackets should be used to increase

_____ .

 a. durability   c. drainage
 b. safety   d. weatherproofing

15. The first step in strip shingle installation is

_____ .

 a. nailing the shingles
 b. installing the underlayment
 c. stacking the roof
 d. measuring the roof

TEST-TAKING TIP

Eat well before taking a test. Have a good breakfast or lunch and avoid junk food. Studies show that you need good nutrition to concentrate and perform your best.

*These questions will help you practice for national certification assessment.

CHAPTER 23

Siding

Chapter Objectives

After completing this chapter, you will be able to:

- **Identify** common siding materials.

- **Describe** how to prevent moisture from seeping into or behind siding.

- **Install** plain-bevel wood siding.

- **Describe** the four coursing styles for wood shingles.

- **Describe** vinyl, plywood, stucco, and fiber cement siding.

- **Explain** proper safety techniques when installing siding.

Christine Glade/Getty Images

Discuss the Photo
Bevel Siding Wood bevel siding is the most common type of siding. *How might the beveled shape allow the boards to shed water effectively?*

Writing Activity: Write a Letter
Siding is available in a wide variety of materials, including solid wood, vinyl, plywood, steel, aluminum, stucco, and fiber cement. Write a one-page letter to a local building contractor. Request information about which type of material is best for a house in your region of the country.

Chapter 23 Reading Guide

Before You Read Preview

The siding of a house should be selected with great care. Siding has an effect on the overall appearance of a home, as well as on the proper ease of maintaining it. Look through the chapter, noting the photographs and reading their captions.

Content Vocabulary

- siding
- back-priming
- story pole
- undercourse
- J-channel
- drainage plane

Academic Vocabulary

You will find these words in your reading and on your tests. Use the academic vocabulary glossary to look up their definitions if necessary.

- traditional
- overall
- indicate

Graphic Organizer

As you read, use a chart like the one shown to describe problems and solutions associated with siding installation.

Problem	Solution
Moisture seeps into or behind the siding	Seal out water with proper detailing, flashing, and high-quality sealants

Go to connectED.mcgraw-hill.com to download this graphic organizer.

Siding Basics

Understanding Siding

Is siding waterproof?

Siding is the exterior wall covering of a house. Its purpose is to shed rain and protect the structural portions of walls. Depending on the material, siding is installed by either a carpenter or a siding specialty contractor. In either case, the process normally begins after windows and doors are in place and after the roofing has been installed.

Siding is available in a wide variety of materials, including solid wood, vinyl, plywood, steel, aluminum, stucco, and fiber cement. Many are available prefinished. This eliminates the need to paint or stain them after installation. Another siding material popular in some areas is brick, which is discussed in Chapter 24.

Wood Siding

The material most characteristic of North American houses is solid-wood siding. Many other siding materials, including fiber cement, engineered wood, and vinyl, are formed to look like wood siding.

Wood Quality Woods used for siding should have the ability to accept paint or stains, should be easy to work with, and should be dimensionally stable. These properties are present to a high degree in the cedars, eastern white pine, Western white pine, sugar pine, cypress, and redwood. They are present to a good degree in Western hemlock, ponderosa pine, spruce, and yellow poplar.

Exterior siding materials should be of a premium grade that is free from knots. The moisture content at the time of application should be the same that the wood will have during service. This is about 12 percent, except in the dry southwestern United States. There the moisture content should average about 9 percent.

Types of Wood Siding

The following basic types of wood siding are shown on page 659, along with some less common types.

Wood Siding
Traditional Material Wood lap siding can be found in every region of the country.

Builder's Tip

CODE TERMINOLOGY Building codes often use unfamiliar terminology. For example, the IRC refers to siding as "exterior wall covering." These materials provide weather resistance to the "exterior wall envelope." You will find it easier to locate information in the code book if you learn the code's language. Whenever you find an unfamiliar word, check the Definitions section to determine its meaning.

Name	Board		Bevel	Bungalow	Dolly Varden	Panel
Version	Board and Batten	Board on Board	Plain	Plain	Rabbeted Edge	
Description	Available surfaced or rough textured.		Plain bevel may be used with smooth face exposed or sawn face exposed for textured effect.	Thicker and wider than bevel siding. Sometimes called *colonial*. Plain bungalow may be used with smooth face exposed for textured effect.	Thicker than bevel siding. Rabbeted edge.	Typically 4'×8' or 4'×9' in size. May include decorative grooves.
Application and Nailing	Recommended 1" minimum overlap. Use 10d siding nails as shown. Installed vertically or horizontally.		Recommended 1" minimum overlap on plain bevel siding. Use 6d siding nails as shown. Installed horizontally.	Same as for bevel siding, but use 8d siding nails. Installed horizontally.	Same as for rabbeted bevel, but use 8d siding nails. Installed horizontally.	Exterior-grade plywood sheets are applied vertically directly to studs. Horizontal joints are flashed. Vertical joints may be covered with wood battens.

Name	Channel Rustic	Drop	Log Cabin	Tongue & Groove	Wood Shingle
Version	Board and Gap	Shiplap Patterns		Plain	
Description	Available in rough-sawn or smooth surface. Several thicknesses.	Available in various patterns.	1½" at thickest point.	Available in smooth surface or rough surface.	Available in random widths of 4" to 14". Lengths are 16", 18", and 24".
Application and Nailing	May be applied horizontally or vertically. Has ½" lap and 1¼" channel when installed. Use 8d siding nails as shown for 6" widths. Wider widths, nail twice per bearing. Installed horizontally or vertically.	Siding nails installed horizontally.	Nail 1½" up from lower edge of piece. Use 10d casing nails. Installed horizontally.	Use 6d finish nails as shown for 6" widths or less. Wider widths, face nail twice per bearing with 8d siding nails. Installed horizontally.	Applied in overlapping rows much like wood roof shingles.

 Types of Wood Siding

Many Shapes Wood siding comes in many forms, shapes, and sizes. *How is wood shingle siding installed?*

Vertical Solid wood boards with a uniform thickness are sometimes placed vertically over sheathing. The boards may be formed with a shiplap or tongue-and-groove edge to keep water out. When square-edge board siding is used, joints between boards must be covered with slender pieces of solid wood called *battens*. This is called *board-and-batten siding*.

Horizontal The most common type of solid wood siding is available as boards placed so that each piece overlaps the one below. A common type is plain-bevel siding, which is also called *clapboard siding*.

Panel Exterior-grade plywood panels may also be used for siding. The exposed surface is covered with a high-grade wood veneer that may have a smooth or rough-sawn surface and may have grooves cut into it. The edges of panels may be covered with battens or have a tongue-and-groove or shiplap joint to seal out water.

Shingle In some parts of the country, wood shingle siding is common. The individual shingles are nailed to the wall sheathing in much the same way as wood roof shingles are nailed to the roof sheathing.

Window Shutters

The installation of window shutters is sometimes coordinated with the installation of siding. Window shutters were first used for protection from fierce weather and other hazards. Shutters are still used on today's homes, but in many parts of the United States are considered purely decorative. They are screwed to the siding or held with special clips. Installation may fall to various trades. The siding contractor installs vinyl shutters. Wood shutters might be installed by the carpenter or by the painting contractor after the house is painted.

In regions where storms and high winds are common, homes are increasingly being fitted with working shutters that can be closed to protect windows and sliding glass doors. These shutters are designed to withstand a moderate level of impact caused by wind-borne debris.

Reading Check

Summarize *What is the main purpose of siding?*

Preventing Moisture Problems

As noted earlier, siding is simply the first barrier to water. However, it is not the only barrier. Unless a secondary barrier is installed properly, moisture will be able to reach the wall sheathing and will eventually cause it to rot. The following construction methods help to seal out water:

Proper Detailing The siding must be installed so that water is directed away from joints. This is often done by lapping one piece of siding over another. Also, it is a code requirement that the lowest edge of any wood siding must be at least 8" above grade level. This helps to keep it dry and reduces the chance of insect infestation.

Flashing Metal flashing is used to seal the joints where the siding meets a horizontal surface. This includes the areas over door and window frames. Siding cannot dry out quickly where there is a tight fit. Flashing should extend well under the siding and sufficiently over edges and ends of a well-sloped drip cap to prevent water from seeping in.

Builder's Tip

WHERE TO CAULK The vertical joints between lengths of wood siding should be caulked with a durable, paintable sealant. Caulk is also needed around all corner boards and around other exterior trim elements. However, the horizontal joints between courses of beveled siding should never be caulked. Caulking prevents moisture vapor from escaping from behind the siding, which can cause rot.

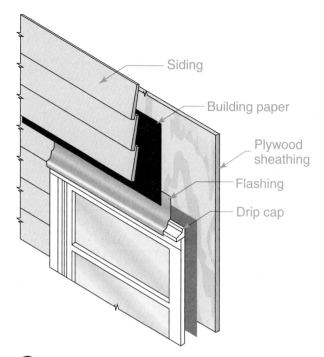

Siding

Building paper

Plywood sheathing

Flashing

Drip cap

Flashing
Proper Overlaps Flashing should be used above windows and doors. The upper leg fits beneath the siding and building paper. Note how each material overlaps another.

High-Quality Sealants Exterior-grade sealants (a type of high-quality caulking) can be used to seal minor joints. Sealants should remain permanently flexible. Vinyl or fiber cement sidings may require a special sealant recommended by the siding manufacturer.

Protecting the Sheathing The sheathing of a house is entirely covered by the siding and trim. However, wind-blown rain or water vapor can still reach the sheathing. That is why the sheathing should be covered by a barrier of building paper or housewrap. The IRC refers to these materials as a weather-resistive barrier. They are stapled to the sheathing just before the siding is installed. Additional layers are placed under trim, such as around windows and doors. Every type of siding should be installed over building paper or housewrap.

It is very important to understand that these barriers are *water-resistant* but *not moisture-vapor resistant*. This means that they block moisture that is in the form of water,

but they do not block moisture that is in the form of vapor. Water that would reach these materials would come from outside the house. Vapor that would reach these materials would come from inside the house. These barriers are designed to prevent water from getting in without preventing vapor from getting out. The reason for this is simple. If vapor cannot escape, it will be trapped against the sheathing and will eventually cause serious problems.

Building Paper The traditional method for protecting sheathing is to cover it with horizontal courses of asphaltic felt, often called *building paper* or *building felt*. This is a heavy-duty asphalt-based paper product that comes in rolls 36" wide. Building paper comes in two weights: No. 15 (often referred to as 15 lb. paper) and No. 30 (30 lb. paper). The lighter paper (No. 15) is the minimum required by code beneath siding. Some manufacturers make a fiberglass-reinforced building paper that is less likely than standard paper to pucker if it gets wet.

Building paper should be applied smoothly to the sheathing using staples. Succeeding layers should lap at least 2" over strips previously applied. Where vertical joints are necessary, the building paper should be lapped at least 6". Particular care should be taken around window and door openings. Builders install extra strips of paper around the rough openings. Extra strips of paper about 6" wide should be installed behind all exterior trim.

Housewrap *Housewraps* are made from high-density polyethylene fibers. These fibers interlock to allow water vapor to pass through, but not water. A housewrap on page 662, has some advantages over asphaltic felt. Because of its light weight, it comes in rolls 9' wide. This speeds installation and reduces the number of seams. A housewrap is difficult to rip, so it is less likely to be damaged during installation. Special adhesive tape called *housewrap seam tape* should be used to seal the seams. It should also be used to seal any punctures or tears in the housewrap.

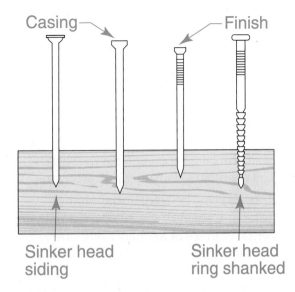

Casing — Finish

Sinker head siding   Sinker head ring shanked

Siding Nails
Nail Types Nails commonly used for the application of siding.

Housewrap should not be used beneath stucco sidings. The stucco can bond to the housewrap in a way that can cause it to lose effectiveness.

Siding Nails Nails used to install siding should be rust resistant and must not cause the siding to discolor or stain. Three types of nails with these characteristics are galvanized steel nails, stainless steel nails, and high-tensile-strength aluminum nails.

There are two kinds of galvanized nails: plated and hot-dipped. Mechanical plating provides a uniform coating, giving the nail predictable corrosion resistance. However, hot-dipped nails are generally more corrosion resistant.

Stainless steel nails are recommended when maximum weather resistance is required, as in areas affected by salt spray. They are also a good choice for installing cedar siding because they are not affected by natural substances in the wood that can cause corrosion in other types of metals. Aluminum nails can be difficult to apply by hand. However, they are readily installed by pneumatic nailers.

Nail heads are usually small so they will not show. Nail shanks may be smooth. However, ring-shank or spiral-threaded nails offer increased holding power.

Section 23.1 Assessment

After You Read: Self-Check

1. List three characteristics of wood that are important when it is used for siding.
2. Name the woods that have a high degree of the characteristics you noted for question 1.
3. What methods will help protect siding from moisture damage?
4. What is the primary purpose of building paper and housewrap?

Academic Integration: English Language Arts

5. **Wood Siding** The basic types of wood siding are vertical, horizontal, panel, and shingle. Create a brief description of each type of siding.

Go to **connectED.mcgraw-hill.com** to check your answers.

Alamy

Wood Bevel Siding

Preparation and Layout

What effects do various exposures have on siding?

Beveled siding is a very common type of horizontal wood siding. The beveled shape, allows the boards to be overlapped in horizontal courses that shed water effectively. There are several types of beveled siding. The most common is plain-bevel siding, sometimes called *clapboard siding* or *lap siding*. That is the type of siding that will be discussed in the following material. However, many of the installation details also apply to other types of beveled siding.

Sizes

Plain-bevel siding is made in nominal 4", 5", 6", 8", and 10" widths. The butt edge (the thickest part of the board) ranges from $7/16$" to $11/16$", depending on the board's width. The top edge is $3/16$" thick in all sizes. Plain-bevel siding generally comes in random lengths from 4' to 16'. One face of each board is rough-sawn, while the other face is smooth. The siding can be installed with either surface exposed.

Preparing the Wood

Before installing plain-bevel siding, allow it to reach a moisture content compatible with local conditions. This will prevent excessive shrinkage after the siding is in place.

Wood siding that will be painted should be primed before it is installed. This ensures that all surfaces are protected and improves the durability of the siding as well as the final finish. If the siding has not been primed on all surfaces by the manufacturer (this is called *pre-priming*), the back surface of the boards should be primed on site, which is called **back-priming**. The front surface can then be primed after installation.

Another protective option is to treat the boards with a water repellent before installation. This may be done by brushing or spraying the water repellent on all surfaces. The ends of boards cut during installation should also receive treatment with water repellent.

Reading Check

Recall *What is the purpose of the beveled shape of beveled siding?*

Determining Exposure

The first row of bevel siding is installed at the bottom of a wall and installation progresses upward from there. Each new course overlaps the top edge of the previous course on page 664.

The spacing for plain-bevel siding should be carefully laid out in advance. This ensures an even appearance. It also helps to avoid awkward details, such as having to make a deep notch in a siding board to make it fit over a window or door. Layout starts by determining the number of courses required to cover a wall.

Determine the number of courses by measuring from the underside of the soffit to a point 1" below the sheathing on page 664. Divide that distance by the maximum exposure of a single piece of siding. The *exposure* is the amount of surface exposed to the weather. To determine the exposure, deduct the minimum overlap (head lap) from the total width of the siding. The minimum overlap is 1" for 4" and 6" widths and 1¼" for widths over 6".

For example, if nominal 10" plain-bevel siding is used, its actual width is 9¼". A minimum overlap of 1¼" is required. Therefore, the maximum exposure is 8" (9¼" − 1¼" = 8").

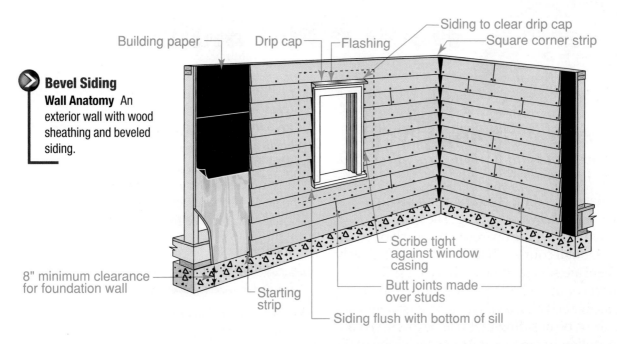

Bevel Siding

Wall Anatomy An exterior wall with wood sheathing and beveled siding.

Building paper

Drip cap

Flashing

Siding to clear drip cap

Square corner strip

8" minimum clearance for foundation wall

Starting strip

Scribe tight against window casing

Butt joints made over studs

Siding flush with bottom of sill

With a pair of dividers set at 8", make a trial layout on the wall. Begin at the bottom and "walk off" the height of the wall in 8" increments. The bottom of the piece of siding that passes over the top of the first-floor windows should meet with the top of the window. If it does not line up properly, adjust the exposure distance until it does. Note that in this case, 8" is the maximum exposure. Any adjustments must be to something less than 8". Another way to make the layout work is to raise or lower the first piece of siding slightly.

It is good practice to have a consistent exposure on each course of siding. However,

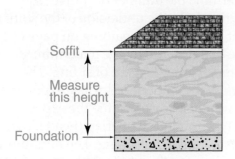

Soffit

Measure this height

Foundation

Determining the Number of Courses

Measure the Wall Measure the vertical distance to be covered by the siding to determine how many courses will fit.

it is also good practice to avoid having to notch siding to fit around the top and bottom of windows and doors. Sometimes it is not possible to follow both practices. Many carpenters think it is most important to avoid notching. In order to avoid this, they make slight corrections in the spacing of several courses. If the adjustments are small, they will not be noticeable. The appearance of a notch in siding is generally more noticeable.

Making a Story Pole Many carpenters rely on a story pole when installing siding. A **story pole** is a measuring device made on site to ensure a uniform layout all around the house. A story pole is shown in using a story pole.

The story pole should reach from the underside of the soffit beyond the bottom edge of the first piece of siding as shown from A to B. Hold the story pole in position against the building. Lay out the spacing on the story pole as shown at C. Check to be certain that the bottom edge of the siding over the window is even with the top of the window as shown at D.

To make a story pole, select a straight piece of 1×2 stock approximately 8' to 10' long. Determine the number of courses and the spacing as described above. Lay out the spacing on the story pole. Additional information,

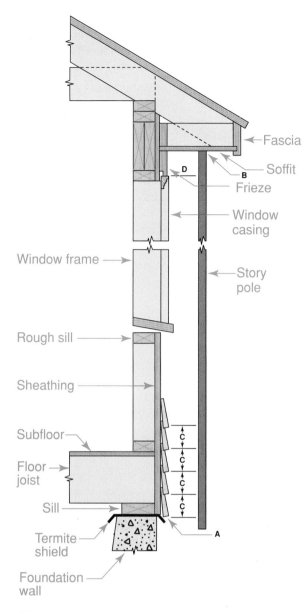

Fascia

Soffit

Frieze

Window casing

Window frame

Story pole

Rough sill

Sheathing

Subfloor

Floor joist

Sill

Termite shield

Foundation wall

Using a Story Pole
Siding Aid The story pole should reach from the underside of the soffit to a point beyond the bottom edge of the siding.

applied determines how level and uniform the succeeding courses will be.

First, install all trim around windows and doors. Install *corner strips* at inside corners and *corner boards* at outside corners. These are lengths of solid wood trim that are placed vertically over the sheathing where walls intersect. They protect the corner joint from water seepage. The size of this trim may be specified on the building plans. It varies according to the style of the house and the thickness of the siding. However, corner boards are generally made of 1¼" thick stock. This stock is sometimes called ⁵⁄₄ *(five quarter) stock*. Insider corner strips are approximately 1⅛" × 1⅛" in size, depending upon the thickness of the siding.

Assembling Corner Boards Outside corner boards are nailed together before being nailed to the house. This ensures a tight joint and is much easier than nailing each board to the house separately.

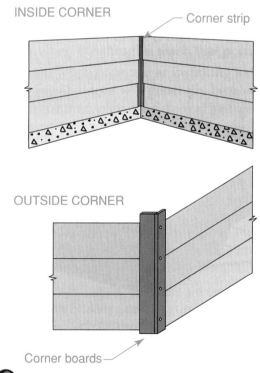

INSIDE CORNER

Corner strip

OUTSIDE CORNER

Corner boards

 Corner Boards
End Cap The ends of bevel siding fit against corner strips or corner boards.

such as the height of windowsills, can also be marked on the story pole. The story pole can now be used on each wall of the house to ensure consistent spacing of the siding.

Application

Why is ⁵⁄₄ stock used for corner boards?

Siding can be installed after the house has been wrapped with building paper or house-wrap. How well the first course of siding is

Installing Plain-Bevel Siding Before you begin, cut a piece of stock to serve as a story pole.

Step 1 Hold the story pole in position against the soffit. Transfer the marks from the story pole to the house on all corners and on all window and door casings. On a long wall, intermediate layout marks may also be needed. Make sure that the bottom marks are clearly visible on the foundation.

Step 2 Snap a chalk line between the lowest marks around the perimeter of the house. Nail a furring strip (sometimes called a *starting strip*) about ³⁄₈" above this line to provide support for the first course. The strip can be a board ripped at an angle or an extra siding board.

Step 3 Start the first course at one end of the wall and work toward the other end. Attach the first board to the bottom plate by placing nails just below each stud. This marks the nailing locations for the succeeding courses. Use 6d siding nails for standard beveled siding. Use 8d siding nails for thicker siding.

Step 4 On long walls, two or more lengths of siding may be required for each course. Cut the ends of adjoining boards square to create a tight-fitting butt joint. Joints should be staggered so that they do not line up with joints in the adjacent three or four courses (see page 664). Some carpenters prefer to use a scarf joint for extra weather tightness, but this is more time consuming. A *scarf joint* is formed by cutting an angle on the ends of boards so that they overlap.

Step 5 Continue to install additional courses by aligning the siding with your layout marks. To prevent splitting and to allow expansion clearance, it is generally best not to nail through the course underneath. However this is difficult to avoid with some exposures. Nails should be perpendicular to the face of the siding. Tap the nail head flush with the surface. Note that the nails do not pierce the siding in the course below.

Story pole

Story Pole

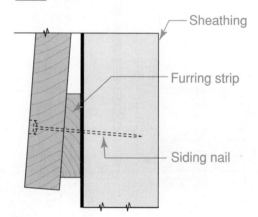
Sheathing

Furring strip

Siding nail

Starter Strip

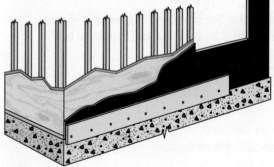

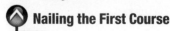
Nailing the First Course

Step 6 Where siding must be cut to fit against trim, hold a piece of siding in place. Use a small wood block gauge to accurately mark pieces that must fit against a vertical surface. The siding gauge can be made from a scrap of plywood. To use it, cut a siding board slightly long and hold it in position against the house. Hold the siding gauge over the board and against the side of the window casing. Mark the siding board to length.

Step 7 Siding that passes under a windowsill should be cut to fit into the groove in the bottom of the sill. Siding installed over doors and windows should stop slightly above the window flashing to encourage drainage.

Step 8 Where siding meets a roof, as on a dormer, allow a 2" clearance between the cut ends of the siding and the flashing. This ensures proper drainage. It also protects the vulnerable end grain of the siding from prolonged contact with water. See pages 667–668 for an explanation.

Nail Location

Step 9 Trim the last course of siding to fit under the eaves. Apply any molding or trim called for in the plans. After the face of the siding has been primed, but before it has been painted, caulk vertical joints between lengths of trim.

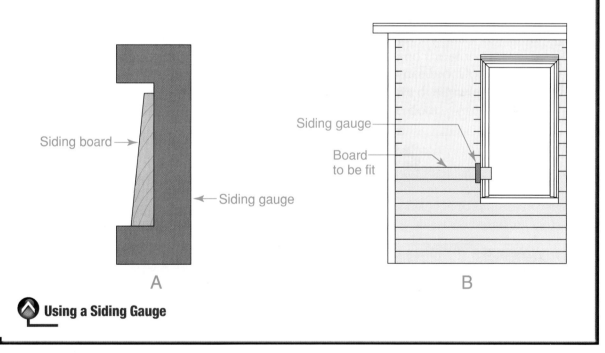

Using a Siding Gauge

Alternative Corner Treatment

At the outside corners of a house, plain-bevel siding is generally fitted snugly against the corner boards. However, the ends of the boards can be mitered instead. Mitered corners, sometimes used with thicker siding, should be cut with a compound-miter saw. They must fit tightly and smoothly for the full depth of the miter. Nail mitered ends to the sheathing, not to each other.

Estimating and Planning

This estimating and planning exercise will prepare you for national competitive events with organizations such as SkillsUSA and the Home Builder's Institute.

Beveled Siding Materials

Estimating Siding

To estimate the amount of siding needed, you must determine the area to be covered.

Step 1 Find the area by multiplying the perimeter of the house by the wall height. For a house with a gable roof, figure the gable-end area separately and add the result to the side wall area. Subtract any areas greater than 50 sq. ft. that will not be sided, such as garage doors. You do not need to subtract smaller areas, such as windows and doors. For example, consider a house 46' long and 26' wide with a hip roof. Its perimeter is 144' (46 + 46 + 26 + 26 = 144). With 8' high walls, the total area to be covered is 1,152 sq. ft. (144 × 8 = 1,152). If the garage door openings total 112 sq. ft., the total area to be sided is 1,040 sq. ft. (1,152 − 112 = 1,040).

Step 2 Multiply the square footage of walls by the appropriate area factor from **Table 23-1**. This table includes factors for calculating different dimensions of beveled siding. The factors are based on actual exposures, including an allowance for trimming and waste. For example, if nominal 1×10 beveled siding is to be used, 1,238.4 sq. ft. would be required (1,040 × 1.21 = 1,238.4).

Step 3 Round up the answer to the nearest hundred. In the example, that would be 1,300 sq. ft. Any siding left over after the project is complete should be left for the homeowner to use for future repairs.

Step 4 To determine the number of nails needed, refer to **Table 23-2**. For example, 1×10 siding requires ½ (0.5) lb. of nails per 100 sq. ft. In our example, about 1,300 sq. ft. of siding will be installed. Divide this by 100 (1,300 ÷ 100 = 13). Then multiply the resulting figure by the weight of nails required per 100 sq. ft. (13 × 0.5 = 6.5 lbs. of nails).

Estimating on the Job

A house is 58' long and 27' wide and has a hip roof. The walls are 8' high. How many square feet of 1×8 siding will be required and how many pounds of nails would be needed?

Table 23-1: Coverage Estimator for Beveled Siding			
Nominal Size	Width		Area Factor
	Dress	Face	
1×4	3½	3½	1.60
1×6	5½	5½	1.33
1×8	7¼	7¼	1.28
1×10	9¼	9¼	1.21
1×12	11¼	11¼	1.17

Table 23-2: Nails Required for Beveled Siding	
Size	Nails (per 100 sq. ft.)
1×4	1½ pounds
1×5	1½ pounds
1×6	1 pound
1×8	¾ pound
1×10	½ pound
1×12	½ pound

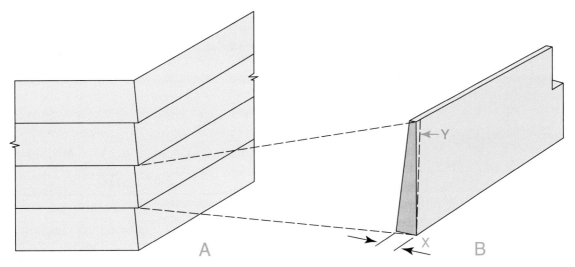

A   B

 Mitered Corners
Outside Corners Mitering plain-bevel siding.

Mitering siding at outside corners calls for careful workmanship to make a weather-resistant joint. To lay out and cut the joint, measure the butt edge thickness (X). Measure back along the top edge a distance equal to the butt edge thickness. Then connect these two points as shown by the dashed lines (Y). With the saw blade set at about a 47° angle, make the cut beginning at the butt end.

Section **23.2** Assessment

After You Read: Self-Check

1. What is back-priming and why is it important?
2. What other finishing method can be used to improve the durability of siding?
3. What is the best way to ensure consistent spacing of siding?
4. Explain the use of a small wood gauge when cutting siding.

Academic Integration: Mathematics

5. **Mental Math** Suppose that the distance from the top of the foundation to the top of the window in a garage is 6'-6". Beveled siding of 8" nominal width is to be applied. Beveled siding of 8" nominal width has an actual width of 7¼". The minimum overlap recommended for this siding is 1¼". How could you use mental math to determine how many courses are required to reach the top of the window?

   Math Concept To use mental math to solve a measurement problem, visualize the situation and look for convenient numbers to work with.

   Step 1: First think through how to find the width of exposure for each course.

   Step 2: Divide the total distance to the top of the window by the exposure width.

   Step 3: If the number of courses comes just to the top of the window, you have solved the problem. Otherwise, rethink how the courses should be overlapped.

Go to connectED.mcgraw-hill.com to check your answers.

Wood Shingle Siding

How are wall shingles similar to roof shingles?

When wood shingles are used as siding they are sometimes referred to as *side-wall shingles.* They are most commonly made from red cedar, but white cedar and other woods are also used.

Wood shingles are generally installed one by one. However, some companies manufacture a product that combines groups of shingles. The shingles are attached at the factory to a plywood backing. The resulting shingle panel is then nailed into place.

Grades and Sizes

Sidewall shingles are usually classified into four grades. The first grade includes clear, all-heartwood shingles. The second grade consists of shingles with a clear exposed area (butt) and allows defects in the part that will be covered in use. The third grade includes shingles that have defects other than those permitted in the second grade. The fourth is a utility grade used for undercourses on double-coursed side walls. An **undercourse** is a low-grade layer of shingles that will not be exposed to the weather.

Shingles come in lengths of 16", 18", and 24". They are packaged in bundles but the coverage of a single bundle will vary depending on how the shingles are installed. Shingles may be prefinished or finished after installation. They are made in random widths. In the first grade, they vary from 3" to 14" wide, with only a small proportion of the narrowest width permitted in each bundle. Shingles cut to uniform widths of 4", 5", or 6" are also available. They are known as *dimension shingles* or *rebutted-and-rejointed shingles.* Their edges are machine trimmed so as to be exactly parallel; exposed ends are trimmed at 90° angles. Dimension shingles are applied with tight-fitting joints to create an unbroken horizontal line.

Reading Check

State *What types of wood are most commonly used for wood shingle siding?*

Application

Shingles are generally spaced ⅛" to ¼" apart. This allows them to expand and prevents buckling. Maximum exposures vary. Spacing for the shingle courses is determined in the same way as for plain-bevel siding.

There are several methods for installing siding shingles. The method selected determines the **overall** look of the house.

Single Coursing In single coursing, each course of shingles overlaps the one below so that every part of the wall is covered with two layers. The same grade of shingle must be used throughout. Weather exposure is fairly small.

Double Coursing One purpose of this method is to obtain deep shadow lines for appearance. High-grade shingles are laid over undercourse-grade shingles. This method is less expensive than single-coursing because fewer high-grade shingles are used. Rebutted-and-rejointed shingles are sometimes used as the outer course for length.

Ribbon Coursing In this method, a double shadow line is created by raising the outer course slightly above the undercourse. Because the undercourse is partially exposed, however, it should be the same grade of shingle as the outer course.

Decorative Coursing A rustic effect can be created if the exposed end of every other shingle is placed at a slightly different distance from a horizontal layout line. This results in an edge line that appears staggered.

Types of Shingle Coursing

Three Methods Each different method for installing shingles results in a different overall appearance: **A.** Single coursing **B.** Double coursing **C.** Ribbon coursing.

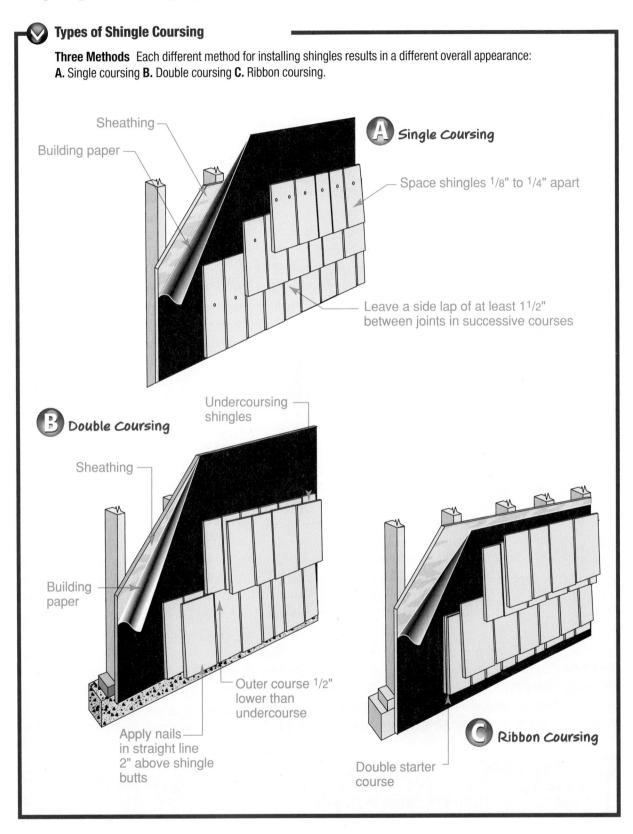

A Single Coursing

Sheathing

Building paper

Space shingles 1/8" to 1/4" apart

Leave a side lap of at least 1 1/2" between joints in successive courses

B Double Coursing

Undercoursing shingles

Sheathing

Building paper

Outer course 1/2" lower than undercourse

Apply nails in straight line 2" above shingle butts

C Ribbon Coursing

Double starter course

Nailing

For double coursing, secure each outer-course shingle with two small-head, rust-resistant, 5d nails driven about 2" above the butt edges and ¾" in from each side. Drive additional nails about 4" apart across the face of the shingle. Single coursing involves the same number of nails, but they can be shorter, such as 3d. Blind nail them not more than 1" above the butt edge of the next higher course. Never drive the nail so tight that its head crushes the wood.

Outside corners can be interlaced, mitered, or butted into corner boards. Laced ends overlap in an alternating pattern. Inside corners may be mitered over metal flashing or butted to a corner strip similar to that used when installing plain-bevel siding.

Estimating

To determine the number of shingles needed, figure the area to be covered plus a trim and waste allowance. For single coursing, one *square* (4 bundles) of 16" shingles with a 7½" exposure will cover 150 sq. ft. as shown in **Table 23-3**.

For example, suppose a house requires 902 sq. ft. of siding. Add 5 percent to this number for trim and waste. Then divide by 150 (the area covered by 1 square of shingles):

$$902 \times 0.05 = 45.1$$
$$902 + 45.1 = 947.1 \text{ sq. ft. to be covered}$$
$$947.1 \div 150 = 6.31, \text{ or } 6\tfrac{1}{2} \text{ squares}$$

Since there are 4 bundles in a square, 26 bundles will be required to shingle the house in this example ($4 \times 6.5 = 26$).

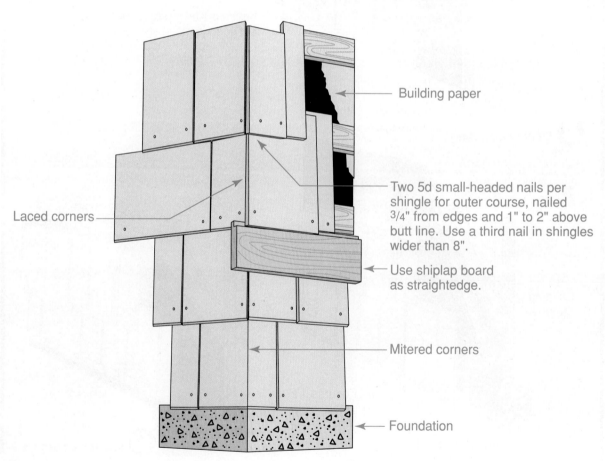

Laced corners

Building paper

Two 5d small-headed nails per shingle for outer course, nailed 3/4" from edges and 1" to 2" above butt line. Use a third nail in shingles wider than 8".

Use shiplap board as straightedge.

Mitered corners

Foundation

⌂ Installation Details

Nails and Alignment One method of finishing outside corners. Shingles may also be butted into vertical corner boards. A board can be used to keep the rows of shingles straight.

Table 23-3: Sidewall Shingle Exposure and Coverage

Length and Thickness (inches)	Approximate coverage in square feet of one square (4 bundles) of shingles based on the following exposures												
	3½"	4"	4½"	5"	5½"	6"	6½"	7"	7½"	8"	8½"	9"	9½"
16×5/2	70	80	90	100(a)	110	120	130	140	150(b)	160	170	180	190
18×5/2¼	–	72½	81½	90½	100(a)	109	118	127	136	145½	154½(b)	163½	172½
24×4/2	–	–	–	–	–	80	86½	93	100(a)	106½	113	120	126½
	10"	10½"	11"	11½"	12"	12½"	13"	13½"	14"	14½"	15"	15½"	16"
16×5/2	200	210	220	230	240(c)	–	–	–	–	–	–	–	–
18×5/2¼	181½	191	200	209	218	227	236	245½	234½	–	–	–	–
24×4/2	133	140	146½	153(b)	160	166½	173	180	186½	193	200	206½	213(c)

Note: The thickness dimension represents the total thickness of a number of shingles. For example, 5/2" means that 5 shingles, measured across the thickest portion, when green, measure 2 full inches.
(a)Maximum exposure recommended for roofs.
(b)Maximum exposure recommended for single coursing on side walls.
(c)Maximum exposure recommended for double coursing on side walls.

Section 23.3 Assessment

After You Read: Self-Check

1. Describe the top two grades of siding shingles.
2. What is an undercourse?
3. With ribbon coursing, why should the undercourse be the same grade of shingle as the outer course?
4. For single coursing, how many square feet will 1 square of 16" shingles with a 7½" exposure cover?

Academic Integration: Mathematics

5. **Estimating Siding** Suppose a house requires 860 sq. ft. of siding. How many bundles of 16" shingles with a 6" exposure will be needed to shingle the house if you add 5% for trim and waste?

Math Concept To add 5% to a number, you can use a calculator to multiply that number by 105%.

Step 1: Find the total number of square feet of siding you will need including trim and waste.

Step 2: Use the table to determine the approximate coverage of one square of shingles given the size and exposure of the shingles you are using.

Step 3: Multiply by 4, the number of bundles in a square.

Go to connectED.mcgraw-hill.com to check your answers.

Other Types of Siding

Vinyl Siding

What is a buttlock?

Vinyl siding is a manufactured product made primarily of polyvinyl chloride (PVC). It is colored throughout the thickness of the material, not just on the surface. This means that scratches are not noticeable. Vinyl siding can be installed more quickly than wood siding. If damaged, it can be easily removed and replaced. It will not rot and requires little maintenance. Various textures, colors, and grades of siding are available. The thickness of the vinyl itself ranges from 0.035" to 0.055".

Vinyl siding is applied in horizontal pieces, called *panels,* that consist of one or more courses. The panels are shaped to resemble wood lap siding. Vinyl siding should always be installed over a weather-resistive barrier such as housewrap. Though vinyl itself is waterproof, the joints of a siding installation are not. Under some circumstances, wind can drive water through the joints. This is why a weather-resistive barrier is required. The joints between panels allow moisture vapor to escape.

A slotted *nailing flange* (also called a *nail hem*) at the top of each panel is used when nailing the panel to sheathing. The bottom edge, called the *buttlock,* fits into the top lock in the panel below. The specific characteristics and durability of manufactured siding materials vary from manufacturer to manufacturer. Always follow the manufacturer's instructions for the product you are installing. This will ensure that the product warranty will cover the completed project.

Wood trim and corner pieces are not used when installing vinyl siding. Instead, various accessory parts are used to serve the same purpose.

Accessories include starter strips, corner posts, and window trim. Accessories must be ordered at the same time and from the same manufacturer as the siding. This ensures that the color will match.

Application

Vinyl siding is typically installed by contractors who specialize in one or more vinyl product lines. They sometimes

Vinyl Siding
Interlocking Joints The lower edge of a vinyl siding panel hooks onto the flange of a panel already installed.

Carol Gering/E+/Getty Images

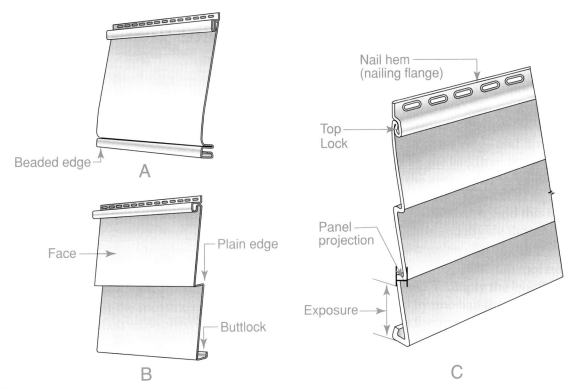

Beaded edge

A

Face — Plain edge

Buttlock

B

Nail hem
(nailing flange)

Top
Lock

Panel
projection

Exposure

C

 Vinyl Siding Panels

Panel Anatomy Types of vinyl siding panels. **A.** Single lap. **B.** Double lap. **C.** Triple lap. The terminology used here applies to any vinyl siding panel.

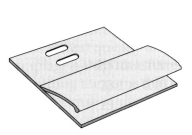

Starter strip

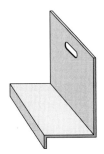

Window head flashing
(drip cap)

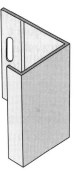

J-channel

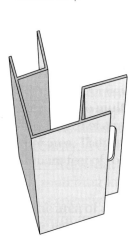

2" narrow face
outside corner post

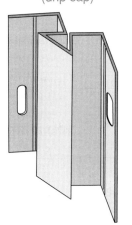

Inside corner post

 Siding Accessories

Important Details Specially shaped pieces, called *accessories,* must be used at corners and openings to ensure a weather-resistant installation.

undergo training programs sponsored by vinyl siding manufacturers. The IRC does not specify a particular installation method for vinyl siding. Instead, it simply says that vinyl siding must be installed according to the manufacturer's installation instructions. However, the methods described in the following pages are generally accepted.

Reading Check

Recall *Why is it important to include an additional weather-resistive barrier?*

Fastening Techniques Vinyl siding expands and contracts much more than wood. For this reason, using the correct fastening technique is extremely important. Vinyl siding is generally nailed through the sheathing and into studs using corrosion-resistant nails. Staples and pan-head screws may also be permitted, but check the manufacturer's installation literature for details. Nails should have heads at least 5⁄16" in diameter, with a 1⁄8" shank. They should be long enough to penetrate at least 3⁄4" into the framing.

Good nailing technique includes the following:

- Never drive nails tight, unless manufacturer's instructions specifically recommend otherwise. Leave approximately 1⁄32" (the thickness of a dime) between the underside of the head and the vinyl. This allows the vinyl to expand and contract and prevents it from buckling with changes in temperature.

- Always drive fasteners straight and level. This prevents the panel from distorting. Start in the center of the panel and work toward the ends. Nails should be centered in each slot. Steps for installing vinyl siding are given on page 677.

- Space fasteners no more than 16" apart on panels and 8" to 10" apart on accessories.

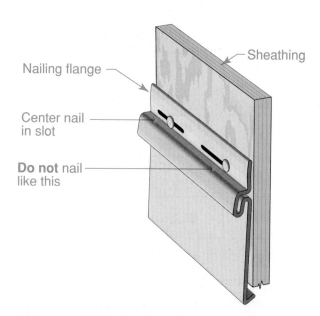

Nailing flange

Sheathing

Center nail in slot

Do not nail like this

Nailing Details
Always Leave Clearance The careful placement of nails will prevent panels from distorting as they expand and contract.

- Center all fasteners in the nailing slots, except those at the very top of a corner post. This allows the siding to expand and contract. Also, leave a 1⁄4" space at joints, channels, and corner posts for expansion and contraction.

- For best appearance, locate any overlaps on panels in a way that minimizes their appearance. Stagger end laps a minimum of 24" or as suggested by the manufacturer. End laps should be 1" wide.

Cutting Techniques Several methods can be used to cut vinyl siding. Vinyl siding is often cut using a power miter saw. The blade should be a fine-tooth plywood blade. However, it should be mounted in the saw so that the teeth point *backward*: they should point toward the operator, not toward the saw fence. This is the opposite direction from normal blade installation. This results in a smoother cut through vinyl, especially in cold weather. Vinyl siding can also be cut with tin snips. For best results, do not close the blades completely at the end of a stroke. A third way to cut vinyl siding is by scoring it with a sharp utility knife or vinyl scoring tool.

Installing Vinyl Siding The following instructions are a general guide. Instructions specific to a particular siding product must always be followed closely.

Step 1 Snap a level chalk line along the bottom of the sheathing. Align the starter strips with this line and nail them into place around the house. Leave room for corner posts at each inside and outside corner.

Step 2 The ends of the corner posts should be ¼" below the bottom of the starter strips. Plumb each post carefully. Then nail it to the sheathing through the uppermost nailing slot on each side. The post should "hang" on these nails. Continue nailing down the length of the post on both sides. Install all inside and outside corner posts in similar fashion.

Step 3 Apply housewrap or flexible self-adhesive flashing around windows, doors, exterior electrical boxes, and other openings. Overlap it. The flashing should be wide enough to extend past the nailing flange of any accessory.

Step 4 A **J-channel** is a plastic or metal channel shaped like a J that is used to support trim. Nail J-channel to the sheathing wherever necessary to hold the ends of siding panels. This would include areas around windows, at gable ends, and where dormer side walls meet a roof. Follow manufacturer's instructions carefully.

Step 5 Place the first panel in the starter strip and nail it into place. Subsequent panels can then be held in place, locked to the previous course, and nailed. The ends of panels fit into J-channel around windows and doors. Leave a ¼" gap between the siding and all corner posts and channels. (Increase the gap to ³⁄₈" when installing siding in temperatures below 40°F (4°C).)

Step 6 Cut the siding as needed to fit under windows. Be sure to include space for an expansion gap. Check the siding with a level after installing every five or six panels. Adjust the panel's position slightly as needed to maintain level.

Step 7 The top course of siding may have to be cut to fit beneath the soffit. In this case, use a special siding tool called a *snaplock punch* to create a nailing flange along the cut edge. Use vinyl trim to cap the top of the wall as needed.

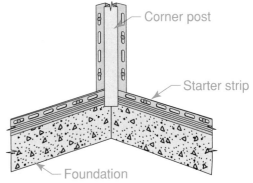

Corner post

Starter strip

Foundation

Corner Posts

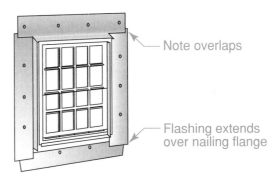

Note overlaps

Flashing extends over nailing flange

Window Flashing

J-channel

J-Channel

Vinyl Siding & Wind Always check local building codes for specific installation requirements for vinyl siding. In regions where high winds are common, local codes may include more stringent installation instructions.

CUTTING VINYL WITH A MITER SAW
Vinyl panels can be cut with a plywood blade mounted so that the teeth face the opposite direction from normal. However, never use this blade position to cut other materials. Also, never use this blade position with any other types of saw blades. Always wear suitable eye protection.

Cut with the siding panel facing upward, and guide the blade with a metal straight-edge such as a framing square. Cut part way through the panel, then snap the panel along the scored line.

Estimating Vinyl Siding

Most manufacturers of vinyl siding indicate the number of panels required to cover one square (100 sq. ft.) of wall area. For example, 12 pieces of 8" siding might cover one square.

Figure the area to be covered by multiplying the perimeter measurement by the wall height. Divide this total by 100 to find the number of squares. Multiply the number of squares by the number of pieces needed to cover one square (in our example, 12). For example, for a house having 947 square feet to be covered, you would need 114 pieces:

$$947 \div 100 = 9.47$$
$$9.47 \times 12 = 113.64, \text{ or } 114$$

Plywood Siding

When is primer used?

Plywood siding panels can also be used as an exterior wall covering. When this is done, sheathing is not required because the panels provide enough rigidity to the walls. Using plywood sheet siding has several advantages. It comes in many grades and surface textures. It covers large areas quickly, taking less time to install. Little material is wasted. By omitting sheathing, costs are reduced. Plywood siding panels are 4' wide. They come in standard

8' lengths as well as 12', 14', and 16' special lengths. Panels range in thickness from $^{11}/_{32}$" to ¾".

Preparation

The edges and ends of plywood siding panels should be sealed before installation. This prevents sudden changes in moisture

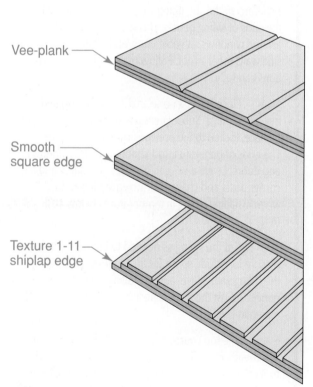

Vee-plank

Smooth square edge

Texture 1-11 shiplap edge

 Plywood Siding Profiles
Many Choices Three of the many plywood siding styles available.

Builder's Tip

ASSEMBLING CORNER BOARDS Outside corner boards are nailed together before being nailed to the house. This ensures a tight joint and is much easier than nailing each board to the house separately.

content caused by wet weather. Primer is used if the siding will be painted. A paintable, water-repellent sealer is used if the siding will be finished with a solid-color stain. To seal the panels, brush or roll the sealant over the edges of stacked panels. Any fresh edges created by cutting the panels during installation should also be sealed.

Application

Plywood siding is normally installed vertically, but it may be installed horizontally. All vertical edges should be supported by framing. Horizontal joints should be supported by blocking. To prevent staining of the siding, use galvanized, aluminum, or other non-corrosive nails.

Vertical joints between panels can be handled in two ways to keep out water. The joint must either be a shiplap joint or be a butt joint covered by a wood batten. Horizontal joints between panels should be lapped at least 1", shiplapped, or protected with Z-shaped flashing as in Figure 23-26. Flashing prevents water from entering the joint. Space panels ⅛" apart at sides and ends to allow for expansion and contraction.

Follow the nailing recommendations provided by the panel manufacturer or local codes. **Table 23-4** on page 680 provides some guidelines. Add corner boards and trim around windows after the siding is in place. Building paper or housewrap is usually not required by code. However, always check local codes to be sure.

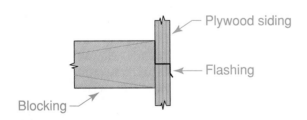

Plywood siding

Flashing

Blocking

Horizontal Joints
Weather Detailing Horizontal joints should always be supported by blocking. The joint should be flashed as shown.

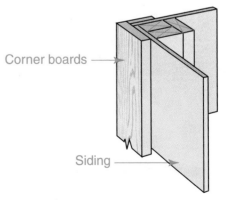

Corner boards →

Siding —

Wood outside corner

Corner Detail
Joint Coverage Outside corner detail for plywood panel siding.

Table 23-4: Framing and Nailing Schedule for Plywood Panel Siding						
Panel Siding Thickness	⁵⁄₁₆"	3.8"	½" grooved	½" flat	⅝" grooved	⅝" flat
Single-Wall Construction						
Maximum Stud Spacing		16" OC	16" OC	24" OC	16" OC	24" OC
Nail Size		6d	8d	8d	8d	8d
Over ³⁄₈" Sheathing						
Maximum Stud Spacing	24"	24"	24"	24"	24"	24"
Nail Size	6d	6d	6d	6d	8d	8d
Approximate Nail Spacing[a]						
Edges	6"	6"	6"	6"	6"	6"
Intermediate Members	12"	12"	12"	12"	12"	12"

[a]Use non-corrosive casing, siding, or box nails.

Stucco

In what type of climates is stucco popular?

In some parts of the country, a stucco finish is the most common type of siding. *Stucco* is a durable product that is applied by trowel in several layers over reinforcing wire called *lath*. The stucco itself is a mixture of clean water, bagged silica sand, Portland cement, and lime. It forms a hard coating that is all of one piece. It is particularly suitable for use in hot or mild climates. Stucco may have a natural cement color or be colored as desired.

If stucco is to be applied on houses more than one story high, balloon framing (see Chapter 14) or steel framing should

 Stucco Siding

Regional Favorite The stucco exterior of this house provides a durable finish.

be used for the exterior walls. This reduces expansion and contraction that could crack or otherwise damage the stucco. The framing should be sheathed according to local building codes. However, plywood is commonly used. Stucco can also be applied to masonry surfaces such as concrete block. In such cases, code allows for a two-layer application instead of the standard three-layer application required over wood framing.

Portland-cement stucco that has been commercially prepared should be mixed and applied according to the manufacturer's instructions. If the material is mixed on site, it is generally one part Portland cement, three parts sand, and a portion of hydrated lime equal to 10 percent of the Portland cement by volume. When describing masonry mortars, cement is the first ingredient listed followed by the lime content and then the proportion of sand.

Application

On a wood-framed house, stucco is typically applied over metal lath. Acceptable types of lath include:

Zinc-Coated or Galvanized Metal This may have large openings or small openings.

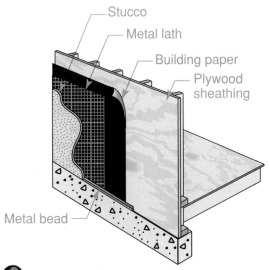

Installing Stucco
Wood Framing Stucco applied on lath over plywood sheathing.

Stucco

Metal lath

Building paper

Plywood sheathing

Metal bead

Galvanized Woven-Wire Fabric This material may be 18-gauge wire with 1" maximum mesh, 17-gauge wire with 1½" maximum mesh, or 16-gauge wire with a 2" maximum mesh.

Galvanized Welded-Wire Fabric This may be made of 16-gauge wire with 2" × 2" mesh and waterproof paper backing. It may also be made of 18-gauge wire with 1" × 1" mesh without paper backing.

The lath should be held at least ¼" away from the sheathing so that the lath will be embedded completely as the stucco is forced through it. Galvanized furring nails, metal furring strips, or self-furring lath are used for this spacing.

Nails used to attach the lath should be 1½" long with an 11 gauge shank and a ⁷⁄₁₆" diameter head. Staples can be used instead if permitted by local codes. They should be ⅞" long and 16 gauge in size.

Traditional stucco should be applied in three layers to a total thickness of about 1".

Scratch Coat The first layer is called the *scratch coat*. It should be forced through the lath and worked so as to embed the lath at all points. The surface should be scratched or scored to provide a rough surface that helps the next layer to hold.

Brown Coat The second layer is called the *brown coat*. It contains more sand than the scratch coat. This makes it easier to work with but it is not as strong.

Finish Coat The third layer is the *finish coat* (sometimes called the *top coat* or *color coat*). Colorants are added to this layer, which is only about ⅛" thick.

Builder's Tip

SAND FOR STUCCO Inexpensive sand that is sometimes called *yard sand* contains impurities that can cause stains to appear in the stucco. Always use silica sand when mixing stucco.

Newly applied stucco should be shaded and kept moist for three days. Do not apply stucco when the temperature is below 40°F (4°C). It sets very slowly, and it may freeze before it has set.

As with other siding materials, stucco should not be considered waterproof. Water can seep through the material itself or through tiny cracks in the surface that are sometimes created as the house settles. Always follow local codes for the installation of sheathing paper. This typically includes requirements for creating a drainage plane behind the stucco. A **drainage plane** is a gap or series of gaps behind the siding that allows water to drain freely. Without a drainage plane, the stucco would stick to the sheathing paper and block water draining. One way to create a drainage plane is to cover the sheathing with wrinkle wrap. This is a flexible sheet material with an irregular surface that promotes drainage.

Fiber Cement Siding

What methods could be used to contain cutting dust?

Fiber cement is a material that has become very popular for siding. It can be formed into various shapes, including sheets, planks, and even bevel siding. The substance is made of Portland cement, ground sand, cellulose fiber, additives, and water. The ingredients are mixed, formed into siding boards, and cured in an autoclave. An *autoclave* is a chamber filled with steam under high pressure. The result is a product that will not burn, rot, or split. It resists mold, mildew, fungus, salt spray, UV rays, and pests such as termites. It contains no defects that must be removed before installation.

Though fiber cement can be formed into shingles, it is more commonly sold as planks. These planks are 5⁄16" thick, 12' in length, and available in widths from 5¼" to 12". Planks are available pre-primed or unprimed.

 Fiber Cement Siding
Wood Texture Fiber cement siding planks, shown here in detail, have a grain pattern that mimics wood.

Working with Fiber Cement

Fiber cement products should be stored flat and kept dry. If the material is installed wet, butt joints will open up as the material shrinks. Individual planks are dense and fairly heavy, so take care when lifting them. Long planks should be carried on edge to prevent them from snapping across their width.

Cutting, drilling, or sanding fiber cement releases a cloud of fine dust that may contain silica. Inhaling this dust over time can cause silicosis, a disabling lung disease (see Chapter 3). A NIOSH-approved respirator is recommended if exposure to the dust will exceed normal limits. Consult the product's material safety data sheet (MSDS) for additional information.

Cutting Fiber Cement There are several methods for cutting fiber cement. The best methods are to score it with a blade and then snap

it, or to use pneumatic or handheld shears that slice through the material. Either of these methods minimizes the amount of fine dust that is created during cutting. It is also possible to cut the material with a circular saw. However, the blade should be approved by the siding manufacturer for use with this material, and the saw should be fitted with a special dust collection system. For low to moderate cutting volumes, the saw can be fitted with a dust guard. For better dust collection, the saw should also be connected to a shop vacuum fitted with a HEPA filter.

Application

Fiber cement planks are installed much like plain-bevel siding. Each plank overlaps the one below and is nailed to the sheathing. Building paper or housewrap may not be required by code, but fiber cement manufac-

Fiber Cement Dust Collection
Safety Feature A clear housing on this circular saw would help to contain the fine dust created when fiber cement products are cut.

turers generally recommend it. Joints between planks should be no more than ⅛" wide. Use a high-quality paintable caulk to seal butt joints and gaps between siding and trim.

Nails should be of 6d hot-dipped galvanized steel or stainless steel and have a blunt or diamond point. Local codes may allow the use of thinner siding nails in some cases. Nails must be long enough to penetrate at least 1" into the wood. When metal framing is used instead of wood, the siding is installed with corrosion-resistant, Phillips-type, bugle-head screws.

Section 23.4 Assessment

After You Read: Self-Check

1. List the advantages vinyl siding has over wood siding.
2. How much space should be left under the nail head when securing vinyl siding?
3. Which type of siding serves as both sheathing and exterior wall covering?
4. What materials are combined to form stucco?

Academic Integration: English Language Arts

5. **Comparison** Write a two-paragraph comparison of stucco and fiber cement as siding materials.

Go to **connectED.mcgraw-hill.com** to check your answers.

Chapter Summary

Siding is made from many materials. Solid-wood siding is the type of siding most commonly used on houses. It must be a high grade of material for maximum durability. Steps should be taken to prevent problems associated with water and water vapor.

Plain-bevel wood siding is installed in overlapping courses. The exposure of each course is important when determining the layout. Tight-fitting joints prevent water seepage.

Wood shingle siding may be installed in several ways that result in different looks. Various grades of shingles may be mixed, depending on the method used.

It is important with sidings such as vinyl, fiber cement, and plywood panels to follow manufacturer instructions closely. This ensures the most durable results. While cutting and handling fiber cement, use extra safety precautions to reduce dust exposure.

Review Content Vocabulary and Academic Vocabulary

1. Use each of these content vocabulary and academic vocabulary words in a sentence or diagram.

Content Vocabulary
- siding (p. 658)
- back-priming (p. 663)
- story pole (p. 664)
- undercourse (p. 670)
- J-channel (p. 677)
- drainage plane (p. 682)

Academic Vocabulary
- traditional (p. 661)
- overall (p. 670)
- indicate (p. 678)

Speak Like a Pro

Technical Terms

2. Work with a classmate to define the following terms used in the chapter: *building paper* (p. 661), *building felt* (p. 661), *house-wraps* (p. 661), *pre-priming* (p. 663), *exposure* (p. 663), *corner strips* (p. 665), *corner boards* (p. 665), *scarf joint* (p. 666), *side-wall shingles* (p. 670), *dimension shingles* (p. 670), *rebutted-and-rejointed shingles* (p. 670), *square* (p. 672), *nailing flange* (p. 674), *nail hem* (p. 674), *buttlock* (p. 674), *snaplock punch* (p. 677), *stucco* (p. 680), *lath* (p. 680), *zinc-coated* or *galvanized metal* (p. 681), *scratch coat* (p. 681), *brown coat* (p. 681), *finish coat* (p. 681), *fiber cement* (p. 682), *autoclave* (p. 682).

Review Key Concepts

3. List common siding materials.

4. Explain the procedures used to prevent moisture from seeping into or behind siding.

5. Summarize how to install bevel wood siding.

6. List the four coursing styles for wood shingles.

7. Identify other types of siding.

8. Describe the appropriate safety techniques for installing siding.

Critical Thinking

9. Discuss When installing vinyl siding, what consequence may result from driving nails too tight?

Academic and Workplace Applications

STEM Mathematics

10. Estimating Perimeter Aluminum or vinyl siding can be installed to the exterior of a building by first attaching it to a starter strip which runs around the perimeter of the building. When estimating starter strip needs, include an additional 10% for trim and waste. What is the number and cost of 10' lengths of starter strip at $2.99 per length for a rectangular 32' × 56' structure?

Math Concept To find the perimeter of a rectangle, add the length and width and multiply the sum by 2.

Step 1: Find the perimeter of the structure.

Step 2: Multiply the perimeter by 110% to figure a 10% trim and waste allowance.

Step 3: Divide the total by the length of a section of starter strip. Round up to the next whole section.

Step 4: Multiply the number of sections needed by the cost of a section.

STEM Engineering

11. Shingle Spacing Shingles are generally spaced ⅛" to ¼" apart. This allows them to expand and prevents buckling. Suppose a carpenter or siding installer were to decrease the spacing to ¹⁄₁₆" or ¹⁄₃₂" apart in order to increase protection against weather and moisture. What issues may occur because of this spacing adjustment? Write a one-paragraph response.

21st Century Skills

12. Cross-Cultural Skills: The Bilingual Workspace Think about working with on a job site where both English and Spanish are spoken. State some reasons why it would be helpful to know both languages in order to communicate. Discuss some of the possible difficulties that may result from a language barrier.

> **Standardized TEST Practice**

True/False

Directions Read each of the following statements carefully. Mark each statement as either true or false by filling in **T** or **F**.

Ⓣ Ⓕ **13.** Woods used for siding should have the ability to accept paint or stain, be easy to work with, and be dimensionally stable.

Ⓣ Ⓕ **14.** Wood siding that will be painted does not need to be primed before it is installed.

Ⓣ Ⓕ **15.** When installing vinyl siding, nails should have heads at least ⁸⁄₁₆" in diameter.

TEST-TAKING TIP

To help your concentration, try to minimize your anxiety when you are taking a test. For example, do not worry about whether you should have studied more.

*These questions will help you practice for national certification assessment.

CHAPTER 24
Brick Masonry & Siding

Section 24.1
Tools & Materials

Section 24.2
Building Brick-Veneer Walls

Section 24.3
Fireplaces & Chimneys

Chapter Objectives

After completing this chapter, you will be able to:

- **Identify** the tools used in working with brick and mixing mortar.
- **Use** a mason's hammer to cut brick.
- **Name** the type of mortar used most often for brick veneer.
- **Explain** why care should be taken when laying brick in cold weather.
- **Identify** the main parts of a chimney.
- **Identify** the main parts of a fireplace.

Discuss the Photo
Brick Veneer Brick veneer can add distinction to a home. *What advantages might brick veneer offer to the homeowner?*

Writing Activity: Create a List
Many of the tools used in brick construction are similar to those used to install other types of masonry, such as concrete block. Create a list of the tools that might be used in brick masonry and siding. Include a short phrase or sentence that describes how each tool might be used.

Chapter 24 Reading Guide

Before You Read Preview

Brick masonry is a popular building material that has many uses, including residential siding, chimneys, and fireplaces. Before reading the chapter, list some ways in which you think brick masonry is used as a building material in the construction of a new home.

Content Vocabulary

- jointer
- retempering
- weep hole
- lintel
- lead corner
- line block
- firebox
- makeup air
- refractory cement
- hearth
- draft
- flue
- corbel

Academic Vocabulary

You will find these words in your reading and on your tests. Use the academic vocabulary glossary to look up their definitions if necessary.

- initial
- transmission
- prolonged

Graphic Organizer

As you read, use a chart like the one shown to organize events in a sequence. For example, you could list the steps of how to cut a brick in half across its width with a hand tool.

1. Gather the bricks, a brick hammer, and a brick set.

↓

2. Line the bricks up evenly.

↓

3. Use a straightedge to scribe a line across the bricks.

Go to connectED.mcgraw-hill.com to download this graphic organizer.

Tools & Materials

Brick Masonry Tools

Are hand tools better than power tools for cutting brick?

Brick comes in many colors and shapes. On some houses, it covers the entire surface of every exterior wall, while on others it is used on just some of the walls. In every case, it should be applied by skilled brick masons using suitable tools. This ensures that the brick will properly protect the house.

Hand Tools

Many of the tools used in brick construction are similar to those used to install other types of masonry, such as concrete block (see Chapter 10).

Brick Trowel The brick mason's basic tool is the *brick trowel.* The trowel has a steel blade and a wood handle. The end of the blade is called the *toe,* or *point.* The wide portion is called the *heel.* Trowels are available in many sizes and in two main shapes:

Philadelphia Pattern A Philadelphia trowel has a blade with a squared-off heel. It is best suited to laying block because it holds more mortar.

London Pattern This trowel has a rounded heel and is the pattern used most for laying brick. The shape of the blade holds less mortar than a Philadelphia trowel, and mortar tends to be supported more toward the toe of the trowel.

The width of a trowel, measured at the heel, is often between 5" and 5½". The most popular length is 11". However, each mason should find a style and size of trowel that feels good in the hand. Some masons find that short, wide trowels are most comfortable to use because their weight, when loaded with mortar, is centered nearer the handle. This puts less strain on the mason's wrist. Wood handles and cushioned-wood handles are available.

Jointer A **jointer**, or jointing tool, is a simple metal bar with a shaped end. It is run over the joints to pack the mortar into them and give them a particular shape.

Mason's Level A good quality mason's level is another important tool for the brick mason. It should have horizontal and vertical leveling vials that can be read from both sides. The edges should be metal to withstand wear.

Mason's Rule A *mason's rule,* is used for measuring the height and spa-cing of brick courses (rows) as they are laid. A folding rule is preferred over a tape measure because the tape can be damaged by contact with mortar.

 Brick Trowel
Basic Anatomy The brick mason's trowel has a steel blade and a wooden handle.

© Ingram Publishing/Alamy

Jointer
Jointing Tool A jointer is used to form and compact mortar joints. Various types are available.

Two types of folding rule are available. The standard mason's rule is white. It is used for measuring standard, or modular, brick. An oversized mason's rule is yellow. It is easier to use when working with over-sized bricks.

Brick Tongs *Brick tongs* help masons carry small quantities of brick efficiently. The metal tool clamps over a row of six to eleven bricks. The lever action of the handle holds the bricks in place. When the handle is lowered, the bricks are released.

Brick Hammer A *brick hammer*, is used for splitting and rough-breaking bricks. It often has a hardwood handle. The head has a chisel blade (sometimes called the peen end) and a square face.

Brick Set A *brick set* is a chisel-like tool made of tempered steel. One edge is beveled. Striking the brick set with a brick hammer cuts a brick cleanly.

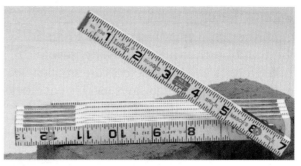

Mason's Rule
Brick Measurements The front side of the rule reads in feet and inches. The back side of the rule is marked with brick coursing dimensions.

Brick Tongs
Quality Control Tongs reduce the damage that might occur to bricks if they were just dumped into a wheelbarrow.

Brick Hammers
Essential Tools These brick hammers have wood handles and carbide chisel ends.

Power Tools

In some cases it is most efficient to cut brick with a tool such as a brick masonry saw (also called a brick saw). These tools are smaller and more portable than saws used to cut block and can cut to a depth of 3¼" or 5", depending on model. The saws typically include a small pump that sprays water to lubricate the blade and reduce dust. Brick is a harder but less abrasive material than block, so it should be cut with a diamond blade designed specifically for brick. Blades are available for cutting brick dry, but the process generates a great deal of dust that is hazardous to breathe.

Cutting Brick

Cutting brick with hand tools is often the easiest and quickest method. To cut brick with a brick hammer, make a cutting line

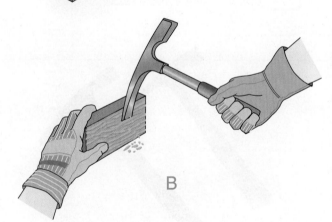

around the brick by striking it lightly and repeatedly with the square face. A sharp blow to one side of the completed cutting line will then split the brick along the line. Rough surfaces can be cleaned up somewhat with the chisel blade.

To cut a brick in half across its width, use a brick hammer and brick set. If several bricks must be cut to the same size, line them up evenly. Then use a straightedge to scribe a line across them. This limits variations in the sizes of the cut brick.

Brick Basics

Why are there so many sizes of brick?

Like concrete block and similar materials, brick is categorized as a unit masonry product. Unit masonry consists of individual pieces of material that can be assembled into larger structures using mortar.

Brick is produced in factories by crushing clay and shale, tempering it with water, forming it into bricks, and then drying the bricks in large kilns (ovens). Other ingredients may be added during manufacture to improve some qualities of the brick. The tempered material is formed into bricks in one of three basic ways:

 Brick set

Cutting Across the Length
Half Size Cutting brick in half lengthwise. **A.** Splitting the brick along the cutting line using a brick hammer. **B.** Trimming brick with the hammer's chisel blade.

Cutting Across the Width
Brick Set Using a brick hammer and brick set to cut brick.

- By extruding the material through dies and slicing it into individual bricks.
- By placing the material into individual molds.
- By dry-pressing the material under high pressure.

Types

The suitability of a brick for a particular use depends on several factors. These include the type and source of the clay and how the brick was manufactured. Many types of bricks are available. Following are the three basic types:

Building Brick Building brick is a strong, general-purpose brick. Its color varies from brick to brick. Sizes are somewhat inconsistent.

Facing Brick Facing brick is used primarily for exposed exterior surfaces such as veneer walls. Because manufacturing is carefully controlled, the resulting bricks are consistent in size, texture, and color.

Firebrick Firebrick is usually pale yellow or buff in color. It is used specifically for lining fireplaces and other heating units. It is sometimes referred to as *refractory brick.*

Bricks with holes (cores) through them are called *hollow bricks,* or sometimes *cored bricks.* Those without holes are called *solid bricks.* Both kinds of bricks are suitable for veneer walls. During firing, the cores enable the bricks to dry and harden more evenly. Firing is a process that heats the brick to temperatures as high as 2,400°F (1,316°C). Bricks with cores are also less expensive to produce, because they use less clay and the firing process uses less fuel. Because they weigh less, they are less expensive to ship and easier to lay. One other advantage is that when mortar oozes into the cores, it makes a very strong mechanical connection.

Some hollow bricks have three fairly large cores. Others have ten or more smaller cores. The exact arrangement is determined by the manufacturer. However, the overall area of coring must not be more than 60 percent of the brick's surface.

A depression in one bedding surface of a solid brick is called a *frog,* or sometimes a panel. A frog serves a purpose similar to that of a core. Frogs are limited to a specified depth and a specified distance from the brick's face. Bricks should always be laid with the frogged surface down.

Sizes

For many years, only three sizes of bricks were available: standard, Roman, and Norman. Today there are hundreds of sizes. When specifying the size of a brick, always list the dimensions in the following order: thickness by height by length.

All brick can be classified into two size groups: modular and nonmodular. The length of a modular brick is based on multiples of 4". It is a nominal size that includes an allowance for the thickness of a standard ⅜" mortar joint. The actual dimensions of a modular brick are therefore smaller than the nominal dimensions, just as the actual dimensions of a 2×4 stud are smaller than its nominal dimensions.

The size of nonmodular brick does not take the thickness of the mortar joint into account. The dimensions given for non-modular brick are the actual dimensions.

Builder's Tip

BRICK HEIGHT Regular modular and nonmodular building brick are different lengths but they are always the same height. Specialty brick may be different heights. This is because brick manufacturers who first introduced modular brick wanted the material to match the coursing of nonmodular brick walls. That way builders could use the new brick when adding on to houses built with the old brick. That is why reference tables that list course heights for brick walls apply to either type of brick.

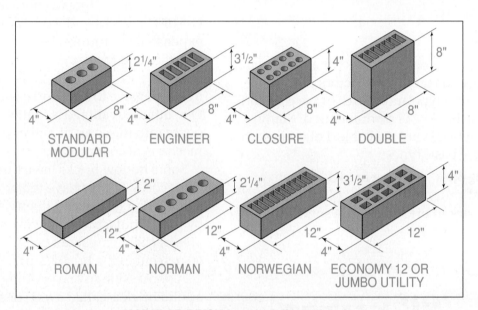

MODULAR BRICK (nominal dimensions)

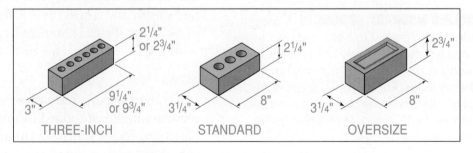

NONMODULAR BRICK (actual dimensions)

Sizes of Modular/Nonmodular Brick
Common Sizes Sizes of modular and nonmodular brick, shown with nominal and actual dimensions.

Textures

Along with color and size, bricks vary in texture. The texture of a brick can significantly affect its appearance. Some bricks have smooth surfaces. Others may have a wirecut (velour), brushed, or other type of rough texture.

Brick Positions and Bonds

A brick can be mortared into place in any one of six basic positions. Some of these positions are used to strengthen a wall, while others are used primarily for decorative purposes. A skilled mason should know how to install brick in each of these positions. A row of brick in one position is called a *course.* Thus a row of brick arranged as in A in Brick Positions would be called a *stretcher course.*

There are many ways to position the bricks in a wall to create various patterns. These patterns are called *bonds.* The bond used on a brick project affects the look of a wall as well as its cost, strength, and ease of construction. For example, if each brick was stacked directly above the brick below, this would be called a *stacked bond.* It would result in a grid-like pattern of mortar joints but would not create an especially strong wall. Another bond is called a *running bond* or a *stretcher bond.* In this method, each brick is centered directly over the joint between the two bricks below. In other words, each brick overlaps adjacent bricks by one-half its length. This is the method used most often in residential brick construction. All illustrations in this chapter show a running bond.

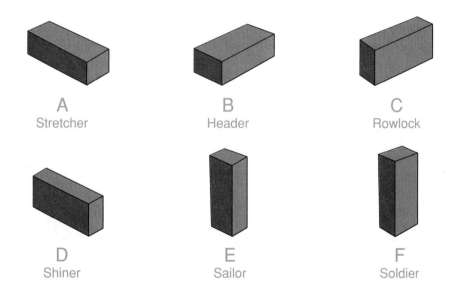

A
Stretcher

B
Header

C
Rowlock

D
Shiner

E
Sailor

F
Soldier

Brick Positions

Six Possibilities Bricks can be installed in these basic positions: **A.** Stretcher **B.** Header **C.** Rowlock **D.** Shiner **E.** Sailor **F.** Soldier. The darkened surface is the one that would be exposed in a wall.

Mortar Basics

What effect would hot weather have on mortar?

Mortar holds bricks together as they are laid. Its bond strength is its most important quality. Mortar must be durable and have enough bearing capacity to support the brick. It must also block the passage of water.

The basic dry ingredients of mortar are Portland cement, masonry cement, hydrated lime, and sand. When mixed with clean water, these ingredients form a durable and easily worked material.

Types of Mortar

Different types of mortar are needed for different types of brick construction. They vary as to ingredients and proportions. Following are the four basic types of mortar:

Type M This type is recommended specifically for masonry below grade and in contact with soil. Uses include foundations, retaining walls, and walkways.

Type S This type is recommended when the brick must resist high levels of lateral (sideways) force. It may be required in areas of high earthquake activity.

Type N This type is a general-purpose mortar for brick-veneer walls. It is suitable for general use in exposed masonry above grade. It is recommended specifically for exterior walls exposed to severe weather conditions.

Type O This type is used for load-bearing walls that will not be subjected to moisture and freeze/thaw cycles.

Mixing Mortar

Mixing proportions for these basic types of mortar are shown in **Table 24-1**. Measurements should be made carefully. A common mistake is to add too much sand. A finer sand requires a greater proportion of cement to produce a workable mix. Colorants may also be added to change the appearance of mortar. This is sometimes done to complement the color of the brick.

The proper amount of water to use in mixing mortar is often misunderstood. Mortar should not be confused with concrete. Concrete is mixed with the least amount of water possible in order to maximize its ability to bear heavy loads. Mortar, on the other hand, requires workability and high bond strength. Mortar is considered workable if it spreads easily and will readily stick to vertical surfaces. Another difference between the two materials is that concrete does not contain lime.

Unlike concrete, mixed mortar may be retempered. **Retempering** is the process of adding water to a batch of mortar that has become too stiff to work. Only enough water is added to replace that lost by evaporation. Also, mortar should be used within 2½ hours after **initial** mixing if the air temperature is 80°F (27°C) or higher. Cooler weather allows a somewhat longer working time.

If a large volume of mortar is required, it may be mixed in a mortar mixer (sometimes called a *power mixer* or *mechanical mixer*). It is similar to a concrete mixer, except that

Table 24-1: Mortar Proportions by Volume			
	Portland Cement-Lime Mortars		
Type	**Portland Cement**	**Hydrated Lime or Lime Putty**	**Sand**
M	1	¼	3
S	1	½	4½
N	1	1	6
O	1	2	6
	Masonry Cement Mortars		
Type	**Portland Cement**	**Masonry Cement Type II**	**Sand**
M	1	1	6
S	½	1	4½
N	—	1	3
O	—	1 (Type I or II)	3

Note: Numbers represent parts. For example, "1 part lime to 6 parts sand."

the mixing drum does not rotate. Instead, mixing paddles inside the drum are turned by a heavy-duty electric motor or a gas engine. Mortar ingredients are added to the drum and blended by the rotating paddles. All the dry ingredients should be measured and placed in the mixing drum and mixed for one minute before clean water is added. The mixer should continue to run until the ingredients are thoroughly blended.

A mortar mixer should be maintained regularly. This includes cleaning the mixing drum thoroughly after use. Any drive belts should be checked for proper tension, and the retractable guard over the mixing drum must work smoothly. Mortar should not be allowed to dry on the mixer's moving parts, particularly on the drive gears.

Small amounts of mortar may be mixed in a wheelbarrow. However, a steel mortar box is less likely to tip over. A mortar hoe is used to blend the ingredients. It has holes in the blade to make mixing easier. Blend all the dry ingredients, and then add clean water as needed. Do not let the hoe or any other mixing tools come into contact with dirt.

 Mortar Mixer
High-Volume Mixing Mortar mixers come in capacities of 7, 9, and 12 cubic feet.

Section 24.1 Assessment

After You Read: Self-Check

1. What is the term used to describe the wide portion of a brick trowel blade?
2. Name two safety precautions to be taken when cutting brick.
3. How should a mortar mixer be maintained?
4. What must be done before adding water to a mortar mixture?

Academic Integration: English Language Arts

5. **Brick Trowels** The brick mason's basic tool is the brick trowel. Write a one-paragraph description of the brick trowel. Name its components and uses. State how it is best used.

Go to connectED.mcgraw-hill.com to check your answers.

Building Brick-Veneer Walls

Construction Details

What supports a brick-veneer wall?

In residential construction, a layer of brick may be used for part or all of the exterior covering over standard wood-frame walls and sheathing. This is called *brick-veneer siding,* or more often, *brick veneer.* Brick is also sometimes used as a load-bearing structural material, rather than as veneer. These types of walls are called *cavity walls.* This is because the walls consist of two layers of brick separated by an air space, or cavity. This chapter will focus on brick veneer, because this is the most common use of brick wall in modern residential construction.

One advantage of brick veneer over wood materials is that it reduces the **transmission** of sound to the inside of the house. Another advantage is brick's natural fire resistance. Although fairly high in cost, brick has low maintenance requirements and a long life. In addition, it is versatile and suitable for a variety of architectural styles.

For brick veneer to protect the house, it must be installed with skill. All mortar joints must be fully filled because partially filled joints allow water to pass through. This can cause wood sheathing behind the wall to rot. It can also damage the wall when water freezes and expands in the joints. However, if too much mortar is used, it will squeeze out and stain the bricks below. All mortar joints should be filled so that some mortar is forced out of the joint when the brick is set. In this way a full bond is achieved. The mortar momentarily hangs on the brick until the mason cleans the excess off with a trowel. If done properly there is no mortar that falls on the ground. It is equally important to keep mortar out of areas intended to let moisture drain from behind the wall. Brick courses should be straight and level. In addition, vertical joints must fall in regular patterns that are plumb.

A brick-veneer wall must be supported by a masonry or concrete foundation and be tied into the framework of the house. The brick rests on a supporting ledge or shelf formed into the main house foundation. This ledge is often flashed. This ledge is approximately 5" wide, which allows the brick to be spaced at least 1" away from the sheathed wall. If moisture penetrates the wall, this air space will allow the water to drain out through gaps in the bottom of the wall. This space also prevents mortar from blocking the drainage plane.

The first course of brick is a leveling course intended to make up for any irregularities in the support ledge. Subsequent courses should be checked frequently for level and plumb. When a house has deep overhangs, they are often framed. At the top of the veneer wall, building code calls for a gap of at least ¾" between the brick and the underside of the soffit framing. Cove molding can be used to conceal the joint between the soffit material and the brick. The cove molding should be attached to the underside of the soffit material, not to the brick.

REGIONAL **CONCERNS**

Brick Masonry & Earthquakes In areas of the country where the risk of seismic activity is high, extra precautions must be taken to ensure that brick-veneer walls will not topple. These may include restricting the height of the wall, installing additional wall ties, and limiting the type of mortar allowable.

Another way to handle veneer at the soffit is to continue the brick veneer into the soffit cavity by adding one course of brick to the wall. The horizontal framing forming the soffit can then be butted into the side of the veneer. Cove molding would be used to conceal the joint. This method is suitable for relatively short overhangs but increases the amount of brick that must be used.

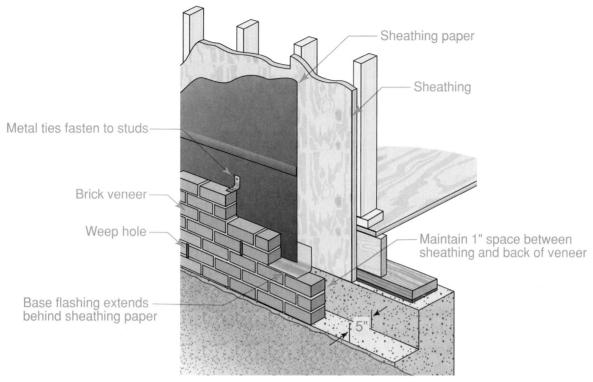

Sheathing paper

Sheathing

Metal ties fasten to studs

Brick veneer

Weep hole

Maintain 1" space between sheathing and back of veneer

Base flashing extends behind sheathing paper

5"

Brick Veneer at the Foundation
Support and Drainage Details of brick-veneer construction supported by a concrete foundation wall.

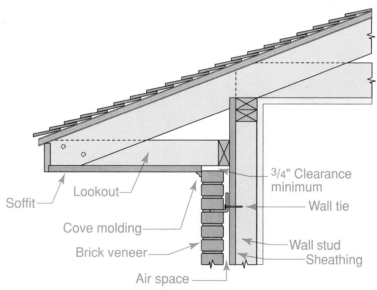

Soffit

Lookout

Cove molding

Brick veneer

Air space

3/4" Clearance minimum

Wall tie

Wall stud

Sheathing

Brick Veneer at the Soffit
Planning for Movement If soffits are framed in this manner, a gap is necessary between the brick and the wood framing to allow for movement.

Flashing and Drainage

It is important to remember that no siding is considered waterproof. Brick is no exception. Materials behind the brick should be installed on the assumption that water *will* reach this area eventually. For example, base flashing should be used at the brick course below the bottom of the sheathing and framing. The top edge of the flashing should be slipped behind the building paper or building felt already attached to the sheathed walls. This is an important detail that is often overlooked. Various flashing materials are acceptable, but copper is highly durable and is often preferred.

A series of weep holes must be built into the wall. A **weep hole** is a hole that provides drainage near the bottom of walls. Weep holes are often formed by omitting some of the mortar in a vertical joint every 18" to 24" along the wall. Moisture that builds up behind the veneer can then escape to the outside. Weep holes are located in the lowest mortar joints above grade. They may be formed by leaving out certain head joints, by inserting plastic tubing into the wall, or by creating a hole using a metal rod. Where insect infestations are a problem, plastic weep hole vents can be inserted into the

open mortar joint. The vents allow moisture to escape but prevent insects from getting behind the brick veneer.

Wall Ties

The veneer wall must be tied to the frame of the house with corrosion-resistant fasteners, called *wall ties*, secured with galvanized nails. Several types of wall ties are available. Corrugated metal straps ⅞" wide and 6" long are traditional for brick-veneer walls. However, wire-type ties are more corrosion resistant. Ties must be secured to the wall studs.

One side or leg of each tie is nailed or screwed through the sheathing and into the studs. The other portion is embedded in the bed joint mortar as the wall goes up. The ties should be embedded 2" into the mortar joint. Check local codes for the required spacing and type of ties. Generally, ties should be spaced no more than 24" on center horizontally and vertically. However, more stringent requirements may apply in areas of earthquake activity.

Window and Door Openings

Special care is required when installing brick around windows and doors. Bricks above an opening must be supported by a lintel that spans the opening. A **lintel** is a structural

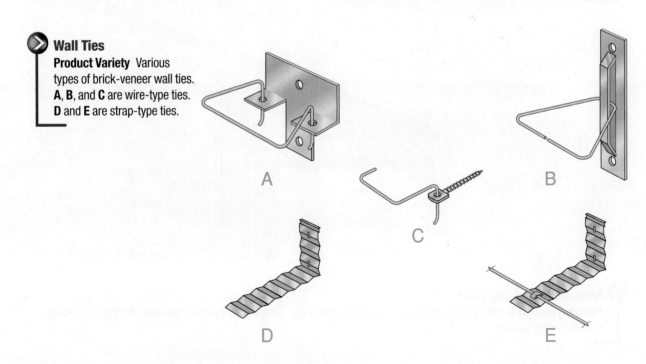

Wall Ties
Product Variety Various types of brick-veneer wall ties. **A**, **B**, and **C** are wire-type ties. **D** and **E** are strap-type ties.

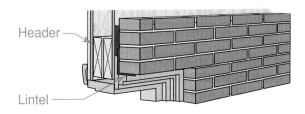

Header

Lintel

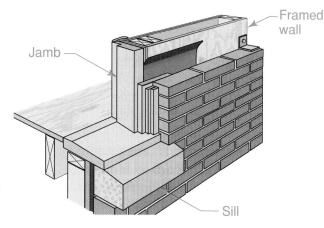

Framed wall

Jamb

Sill

 Wall Opening Details
Openings These details show a door frame installed in brick-veneer construction. Window installation would be similar.

support for masonry. It can be made of solid stone, pre-cast concrete, or an L-shaped length of ¼"-thick steel. In any case, a lintel must have at least 4" of bearing on each jamb. The sill below a window or door is sometimes made of a row of bricks slanted downward to shed water. This is called a *rowlock*. However, it is also common to install a sill made of solid stone or precast concrete.

Reading Check

Recall *What are some advantages of brick veneer over wood materials?*

Building a Veneer Wall
How would you create a raked joint?

Before actually laying brick, the mason should review the plans for the house. This will reveal any special details called for, such as arches or decorative coursing. The mason should also plan the courses of brick so that they line up with the top and bottom of window and door openings wherever possible. This reduces the need to cut brick and results in a better overall appearance. On a small project, a mason may mix mortar as well as lay brick. On larger projects, he or she may have a helper, called a *tender*. This person mixes mortar and provides a steady supply to the mason. The mason can then focus on laying brick. Steps for laying a brick appear on the next page.

Building Lead Corners

A **lead corner** is a partially constructed corner of brick. After lead corners are established at both ends of a wall, the remaining

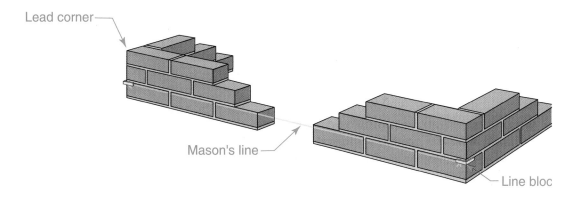

Lead corner

Mason's line

Line bloc

 Lead Corners
Keeping Courses Straight Line blocks hooked over each lead corner of a wall hold a mason's line in place to help keep the courses straight.

brick is laid between them. The first course of each corner is placed into a thick bed of mortar laid on the foundation wall. Additional courses are stepped upward toward the corner. Work then proceeds toward the center of the wall. This technique helps to maintain straight and level courses. Line blocks may be hooked onto the outside edges of each lead corner. In the case of an unusually long wall, an additional lead, called a *middle lead*, is built between the two corner leads.

Step-by-Step Application

Laying a Brick It is important for a mason to work efficiently. Each brick should be laid using the fewest possible motions. This reduces physical stress and increases speed. Following are the basic steps for laying brick for a wall that has already been started.

Step 1 Scoop up a portion of mortar with the trowel. Spread it over three to five bricks by sliding the mortar off the trowel with one smooth sweeping motion. This creates a mortar bed.

Step 2 Furrow the mortar by drawing the point of the trowel across it. This provides better coverage for the brick. However, it is important not to leave any voids in the bed joint. For this reason, some masons do not furrow the mortar.

Step 3 Pick up a brick with one hand and butter one end. Buttering means swiping a small amount of mortar over the end of the brick. This mortar will fill the vertical (head) joint. (An alternate method is to butter the end of a brick already in place.)

Step 4 Shove the brick into place so that the mortar squeezes out the top of the head joint.

Step 5 Using the end of the trowel handle, bed the brick by tapping the top of the brick into place. Make sure that it is level with the adjoining brick and

aligned with, but not touching, the mason's line used for alignment.

Step 6 Slide the edge of your trowel at an angle across the bed and head joints (across the surface of the wall) to cut off mortar that has squeezed out. This prevents the mortar from staining the bricks below. Mortar that is cut off can be reused. Continue to lay brick in this fashion until reaching the end of the mortar bed. Apply a new bed and repeat the process.

Step 7 After several complete courses of brick have been laid, use a jointer to compact bed and head joints. This is called *tooling the joints*. The head joints should always be tooled before the bed joints. Finally, brush the wall with a mason's brush to remove stray bits of mortar.

Laying Brick to a Line

To ensure that a course of brick is level and straight, masons align it with a string line that is tightly stretched across the wall (see page 699). The line identifies the top outside edge of each course. To begin a wall, a mason sometimes secures the line to stakes driven into the ground at each end of the wall. As the wall gains height, the line is then stretched between line blocks instead. A **line block** sometimes called a *corner block,* is a small L-shaped device made of wood or plastic. It hooks over the edge of a brick and is held in place by the tension of the string. Opposing pairs of line blocks are moved upward as the wall gets higher.

Another method of aligning a mason's line is to stretch it between corner poles (also called *adjustable masonry guides* or *corner story poles*). Corner poles come with fittings that allow them to be attached to either inside corners or outside corners. The fittings can be nailed to the sheathing or secured to the brick. They provide an accurate guide to course height and can be adjusted quickly. The pole itself is sometimes marked with brick coursing dimensions.

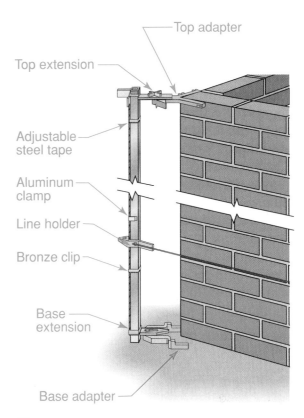

Using a Corner Pole
Quick Adjustment Corner poles can be used to align courses of brick.

Tooling the Joints

Properly finished mortar joints help to create a strong, weather-resistant bond between the mortar and the brick. This is especially important when the wall will be exposed to the weather. There are two basic types of joints:

Struck Joints This is a joint that has been finished using only a trowel. Struck joints are easy to make but not very weather-resistant.

Tooled Joints This is a joint that has been finished using a jointer (see page 689). Use of a jointer results in a denser and more weather-resistant joint. The joint should be tooled as soon as the mortar is thumbprint hard.

The shape of the joints have an impact on how a wall looks, shown on page 702. However, they also affect how the wall performs. Tooled concave joints and V-joints resist rain well. They are recommended in areas where heavy rains and high winds are common.

Builder's Tip

USING A TRIG On a long wall, a mason's line will sag at the middle. To provide support, masons often loop a short piece of line around the midpoint of the main line. The piece, sometimes called a *trig line,* is held in place simply by weighting the loose ends with a brick resting on top of the middle lead. This brick is sometimes called a *trig brick.* In some cases, a brick is used to weight a flat piece of metal that has a loop at one end to hold the line.

Common Types There are many brick joints but these joints are often used in construction: **A.** Concave. **B.** V-shaped. **C.** Weathered (also called weather-struck). **D.** Rough cut (also called flush). **E.** Raked.

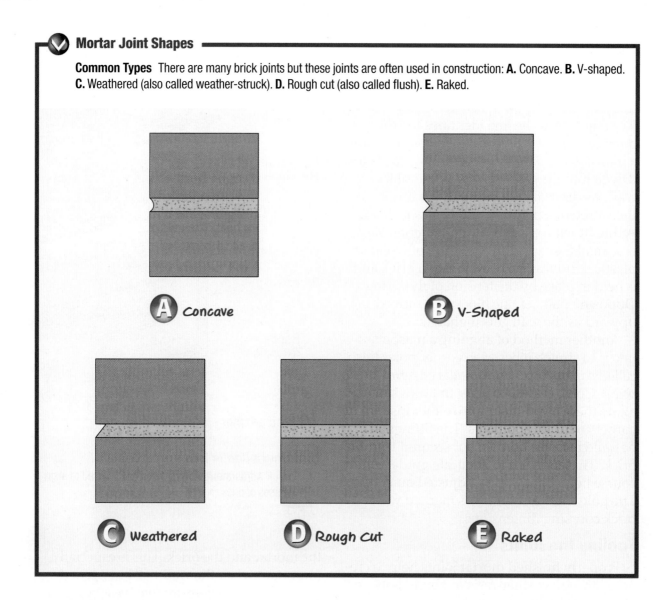

A Concave

B V-Shaped

C Weathered

D Rough Cut

E Raked

Working in Cold Weather

Cold weather slows the hydration process in mortar and can affect the strength of the wall. Building codes therefore include rules for the installation of masonry in cold weather. Bricks that are cold affect curing. When outdoor temperatures are below 20°F (-7°C), bricks should be warmed before use. However, it is rarely necessary to warm them to more than 40°F (4°C). If wet brick has become frozen, it must be thawed and dried completely before use.

When mixing mortar in cold weather, you may need to heat the sand and water before mixing them with the other ingredients. Mortar should never be allowed to freeze as

it is curing. This makes it less weather resistant. It can also reduce or destroy the bond between brick and mortar. For this reason, masons often work beneath protective tents in cold weather.

Estimating Brick

A rough estimate of bricks needed may be made based on the wall's square footage. Approximately seven standard bricks are needed for every square foot of a veneer wall. This includes a small allowance for waste. After calculating the square footage of walls, minus any openings, multiply this figure by 7 to get the number of bricks required. Another method is to consult a table such as **Table 24-2**.

Table 24-2: Modular Brick and Mortar Required for Single-Wythe Walls in Running Bond

Nominal Size of Bricks (inches) T H L	Number of Bricks per 100 Sq. Ft.	Cubic Feet of Mortar			
		Per 100 Sq. Ft.		Per 1,000 Bricks	
		⅜" Joints	½" Joints	⅜" Joints	½" Joints
4 × 2⅔ × 8	675	5.5	7.0	8.1	10.3
4 × 3⅕ × 8	563	4.8	6.1	8.6	10.9
4 × 4 × 8	450	4.2	5.3	9.2	11.7
4 × 5⅓ × 8	338	3.5	4.4	10.2	12.9
4 × 2 × 12	600	6.5	8.2	10.8	13.7
4 × 2⅔ × 12	450	5.1	6.5	11.3	14.4
4 × 3⅕ × 12	375	4.4	5.6	11.7	14.9
4 × 4 × 12	300	3.7	4.8	12.3	15.7
4 × 5⅓ × 12	225	3.0	3.9	13.4	17.1
6 × 2⅔ × 12	450	7.9	10.2	17.5	22.6
6 × 3⅕ × 12	375	6.8	8.8	18.1	23.4
6 × 4 × 12	300	5.6	7.4	19.1	24.7

Note: Running bond is a particular arrangement of brick. No allowances are made for breakage or waste.

Section 24.2 Assessment

After You Read: Self-Check

1. What supports a brick-veneer wall?
2. Why is a wire-type tie an improvement over the corrugated strap tie?
3. What is a lead corner?
4. Why do masons use a string line to align a course of brick?

Academic Integration: Mathematics

5. **Equations** Approximately 7 standard bricks are needed for every square foot of a veneer wall. Write an equation that could be used to solve the following problem: Determine how many standard bricks are needed to cover the four rectangular walls of a house that measures 60' × 25' and has a height of 9'.

   Math Concept An equation is a mathematical sentence stating that two things are equal. Equations often use symbols such as x to represent unknown quantities.

   Step 1: Use x to represent the number of bricks needed for the whole job.

   Step 2: List the steps needed to solve the problem. Use your own words.

   Step 3: Use your list and the numbers from the problem to write the equation.

   Go to connectED.mcgraw-hill.com to check your answers.

Fireplace Design & Planning

What is a firebox?

Fireplaces are sometimes installed as a prefabricated product made of metal, but brick fireplaces are a traditional favorite. Whatever the material, it is essential for safety to follow all local building codes when installing fireplaces and chimneys. This reduces fire hazards.

An ordinary fireplace has a heating efficiency of only about 10 percent. Its value as a heating unit is low compared with its decorative value. However, its heating efficiency can be increased with a factory-made metal unit that is built into the brick structure. This unit circulates the heated air throughout a room.

The design of a fireplace should harmonize in detail and proportion with the room, but safety and utility should not be sacrificed for appearance. Many fireplace designs are possible. In some cases, the fireplace is made entirely of brick. In other cases, it can be made of concrete or concrete block with brick facing or detailing. In this figure, decorative brick is used to surround the opening. It is supported by a steel lintel. Firebrick forms the back and sides of the firebox. The **firebox** is the area of the fireplace where burning takes place. *Firebrick*, also called *refractory brick*, is a brick made of a material that can withstand high temperatures.

Fireplace openings are usually from 2' to 6' wide. Their height commonly ranges from 18" to 28". In general, the wider the opening, the greater the fireplace depth. A deep opening holds larger, longer-burning logs. A shallow opening throws out more heat than a deep one, but it requires smaller pieces of wood. A minimum depth of 16" lessens the danger of firebrands falling out on the floor. However, for a special type of fireplace called a *Rumford*, it may be as narrow as 12". Suitable screens or glass doors should be placed in front of the opening to minimize the danger from brands and sparks.

Coordinating the Work

The installation of a masonry chimney and fireplace calls for close coordination among masons, carpenters, and roofing contractors. The work of each of these trades affects the others. Scheduling must be monitored carefully because masonry often proceeds in several stages separated by

Traditional Fireplace
Careful Construction Building codes regulate most details of fireplace construction, including the position and depth of mantles and hearths.

Joe Mallon

days or weeks. It is also affected by weather extremes. The various trades should contact each other directly and be cooperative. They should notify the general contractor or project supervisor when they have technical questions or scheduling problems that may affect the other trades.

Makeup Air

Combustion requires a source of air. In many cases, air for a fireplace is drawn from inside the house. That air is in turn replaced by outside air that is pulled in through gaps and cracks. In cold weather, this means that heated household air is drawn up the chimney and replaced by cold outside air. In houses built to reduce air infiltration, the fire may not burn well because it cannot get enough air, and the chimney may smoke.

To reduce this problem, makeup air should be drawn into the firebox directly from outdoors. **Makeup air** replaces air exhausted by a combustion appliance. Most local codes now require that all types of fireplaces be supplied with makeup air. The makeup air passageway (sometimes called

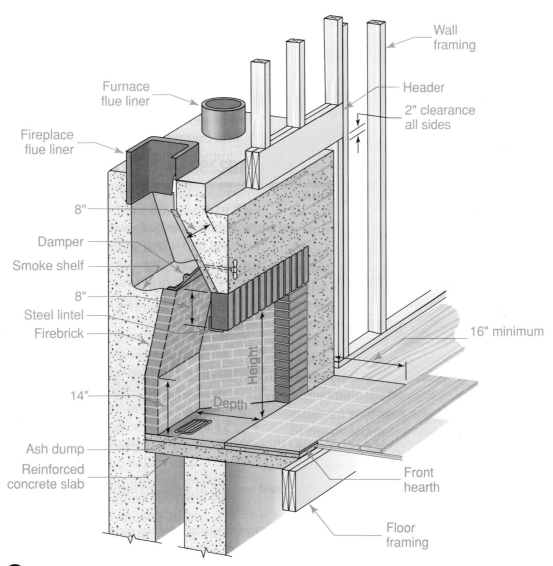

Furnace flue liner

Wall framing

Header

2" clearance all sides

Fireplace flue liner

8"

Damper

Smoke shelf

8"

Steel lintel

Firebrick

Height

Depth

16" minimum

14"

Ash dump

Reinforced concrete slab

Front hearth

Floor framing

⬣ Anatomy of a Fireplace
Construction Details Basic fireplace construction and framing details. The dimensions of the firebox determine its ability to function properly.

the *fresh air intake*) must have a cross section of at least 6 sq. in. and be covered with a corrosion-resistant screen. It may be located in the back or sides of the firebox, or within 24" of the firebox opening on or near the floor. It cannot be located in a basement or garage. The outlet must be closable. Be sure to check local codes for makeup air requirements.

Fireplace Construction

Does every fireplace need a hearth?

The relationships among the depth, height, and width of the firebox are important for proper operation of a fireplace. The cross-sectional size of the flue is also important. Building codes contain charts and graphs as an aid to fireplace design.

The main parts of a fireplace include the fire-box, hearth, lintel and throat, damper, smoke shelf, and smoke chamber. Building codes regulate the relative sizes of these elements as well as their relationship to each other. The following information should serve as a general construction guide. However, always consult local codes prior to constructing a fireplace.

Firebox

Building codes generally require that the backs and sides of fireboxes be constructed of solid masonry, stone, reinforced concrete, or hollow masonry units grouted solid. When lined with at least 2" of firebrick, the walls should be at least 8" thick. The firebrick must

Regional Heating Though many houses have chimneys, fewer have fireplaces. They are least popular where the climate is mild. However, a masonry or prefabricated fireplace is common in cold climates and in cool, rainy climates, such as in the Northwest. In rural areas, a woodstove may provide all the heating needs of an entire house.

be laid using refractory cement, with joints no greater than ¼" wide. **Refractory cement** is a cement resistant to high temperatures.

Hearth

The **hearth** is the floor of the firebox, plus the fireproof area in front of the fireplace. The hearth has two parts: the front hearth (sometimes called the *hearth extension* or the *finish hearth*) and the back hearth, under the fire. Because the back hearth must withstand intense heat, it is built of or lined with firebrick. It should be at least 4" thick. The front hearth protects against flying sparks. While it must be noncombustible, it does not have to resist intense **prolonged** heat. It should be at least 2" thick.

If the fireplace opening is less than 6 sq. ft. in area, the front hearth should extend at least 16" in front of it and at least 8" on both sides. If the fireplace opening is 6 sq. ft. or more, the hearth should extend 20" in front of the opening and at least 12" to either side. However, always be sure to check local codes because they may require somewhat different dimensions.

The hearth can be flush with the floor or it can be raised. Raising and lengthening the hearth is presently common practice, especially in contemporary design. If there is a basement, a convenient ash dump can be built under the back of the hearth.

In wood-framed buildings, the front and back hearths are sometimes supported by steel-reinforced concrete poured in place. If this method is used, any wood formwork must be removed after construction is complete. No combustible material can remain against the underside of the back or front hearths.

Lintel and Throat

Every standard masonry fireplace includes a lintel. A lintel is a length of steel angle iron installed across the top of the fire-box opening to support the masonry. It cannot be seen from the front of the fireplace. Angle iron measuring ¼" thick and having 3½" wide legs is commonly used.

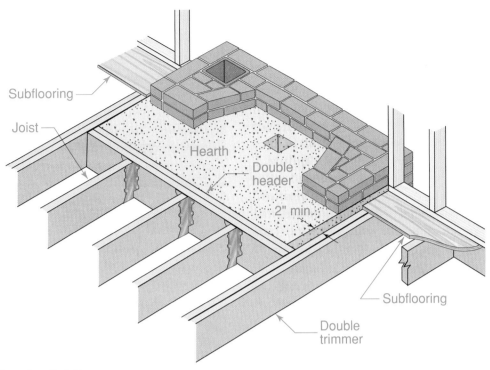

 Floor Framing Details
Fire Prevention The framing details around this brick fireplace are designed to transfer loads around the opening.

However, the actual dimensions depend on the width of the opening and the load to be supported.

Proper construction of the throat is essential for a satisfactory fireplace. The throat is the narrowest part of the firebox, where the damper is located. The sides of the firebox must be vertical up to the throat, which should be 8" or more above the bottom of the lintel. The area of the throat must not be smaller than that of the flue.

Damper

A *damper* consists of a cast iron frame with a hinged lid. It opens or closes to vary the size of the throat opening. In cold weather, closing the damper reduces heat loss when the chimney is not in use. In warm weather, a closed damper prevents insects and small animals from entering the house.

Dampers of various designs are available. However, it is important that the size of the damper opening equal the cross-sectional area of the flue.

Smoke Shelf and Smoke Chamber

The *smoke shelf* is on the back wall of the smoke chamber. It helps to prevent downdrafts from driving smoke back down into the firebox. It is made by setting the

 Chimney Throat Construction
Careful Detailing Approved installation details for the chimney throat and surrounding areas must be followed closely. This assures fire safety and improves the draw of the fireplace and chimney.

brickwork at the top of the throat back to the line of the flue wall for the full length of the throat. The depth of the shelf may be 6" to 12" or more, depending on the depth of the fireplace. The smoke shelf is concave to hold any slight amount of rain that may enter.

The *smoke chamber* is the area from the top of the throat to the bottom of the flue. Its sidewalls slope inward to meet the flue, and its front is formed by corbeled bricks. To make the surfaces of the smoke shelf and the smoke chamber walls smooth, they should be plastered with cement mortar at least ½" thick. If corbelled gradually, the chamber walls can be plastered with refractory cement that can be sponged smooth. This seals any voids in the walls and reduces the amount of creosote buildup on the walls.

Clearances

It is the front, or *surround,* of a fireplace that can have the greatest effect on the architectural style of a room. Many different materials can be used. These include brick, ceramic tile, wood molding and trim, and wood mantels. However, any combustible material must be installed with care and according to building codes. Use the following guidelines:

- No woodwork can be placed within 6" of the firebox opening.

- Woodwork, including any mantel located between 6" and 12" from the firebox opening, must not project forward more than ⅛" for every inch above the opening. For example, a mantel located 9" above the firebox opening can be no more than 1⅛" wide (9 × ⅛" = 1⅛").

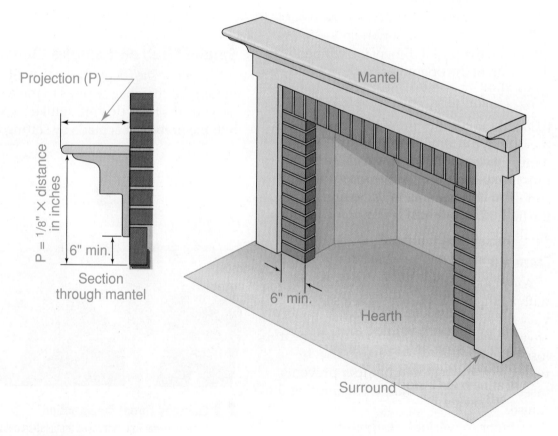

Projection (P)

$P = 1/8" \times$ distance in inches

6" min.

Section through mantel

Mantel

6" min.

Hearth

Surround

Clearance to Wood

Mimimum Clearances Maintaining a suitable distance between combustible trim and the fireplace opening helps to prevent trim from overheating.

Prefabricated Fireplaces

Some fireplaces are made with a heavy-gauge metal firebox. Some of these are designed to be concealed by brickwork or other materials. Others can be set into a wood-framed opening and connected to a prefabricated metal chimney system. In either case, a prefabricated fireplace includes all the essential parts: firebox, damper, throat, smoke shelf, and smoke chamber. The correctly designed and proportioned firebox provides a ready-made form for masonry. This reduces the chance of faulty construction.

Reading Check

Recall *Why is the back hearth built or lined with firebrick?*

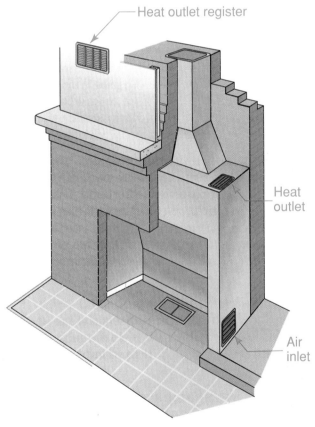

Heat outlet register

Heat outlet

Air inlet

Prefabricated Firebox
Efficient Heater Air from the room is drawn through the air inlet. The air is heated upon contact with the metal and discharged through the heat outlet.

Chimney Design & Planning

What design problems would a cylindrical chimney create?

Just as with fireplaces, chimneys may be built of various materials but brick is a popular choice. It is essential for safety to follow all local building codes when installing chimneys. This reduces fire hazards and ensures that dangerous combustion gases are not released into the house.

A chimney is required for any fuel-burning appliance, such as a fireplace, wood stove, or furnace. The purpose of a chimney is to produce sufficient draft. **Draft** is the upward movement of air within the chimney. Draft draws air into the appliance. This aids in combustion (burning) and expels smoke and harmful gases. The greater the difference in temperature between chimney gases and the outside air, the stronger the draft. An interior chimney will usually have a better draft than an exterior chimney of the same height because it is better able to retain heat.

Chimneys are generally constructed of brick or other masonry units supported by a concrete slab foundation. However, lightweight metal chimneys can be used when approved by local building codes. The design of a chimney is very important for its safety and effectiveness. It may be well constructed, but if it is not designed properly it will not work well. The following information is meant only as a general guide. Always consult local building codes.

Flue Size

The **flue** is the passage inside the chimney through which the air, gases, and smoke rise. Its dimensions, height, shape, and interior smoothness determine how effective the chimney is in creating enough draft. When a fuel-burning unit is connected to a chimney, consult the unit's specifications to establish the flue size.

Height

The height of a chimney above the roofline is usually based on its location in relation to the ridge. On a pitched roof, the top of the

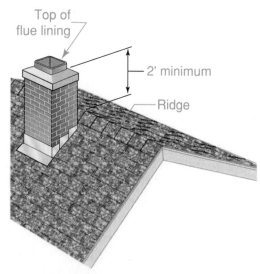

Top of flue lining

2' minimum

Ridge

Chimney Height
A Safe Height The top of the flue lining should be at least 2' above the ridge.

flue lining should be at least 2' above any portion of the roof that is within 10' (measured horizontally) of the chimney. One reason for this is to ensure that burning embers leaving the chimney will cool before they can ignite roofing materials. Another reason is to encourage proper draft. A chimney shielded from the wind by a nearby ridge may not draw well.

In addition, codes may require that the top of the chimney be above any operable window within a 20' radius. This helps to prevent smoke from being drawn into the house when the window is open.

Earthquake Protection

In many areas, chimneys need little in the way of reinforcement. However, where earthquake activity is a serious risk, building codes require masonry chimneys to be reinforced vertically and horizontally. The exact size and type of reinforcement depend on the size of the chimney and local codes. However, chimneys up to 40" wide must contain at least four continuous lengths of No. 4 reinforcing bar. Horizontal reinforcements must be located no farther apart than every 18" along the height of the chimney.

In addition, a masonry chimney located outside the house walls must be anchored to the framing. This must be done with two $\frac{3}{16}$" $\times$ 1" metal straps at each floor, ceiling, and roofline that is more than 6' above grade. The straps should be bolted to framing with ½" diameter bolts. These reinforcement details are not required if the chimney is located completely within the exterior walls.

Clearances

Any portion of a chimney located within the house must have a 2" clearance between its walls and the wood framing. Subflooring and finish flooring can be laid within ¾" of the masonry. If the chimney is located entirely outside the exterior walls, the gap must be at least 1". Exterior wood sheathing, siding, and trim can touch the side walls of the chimney, but only where they will be at least 12" from the nearest flue liner.

The space between wall and floor framing must be filled with a code-approved fire-stopping material. This prevents shavings or

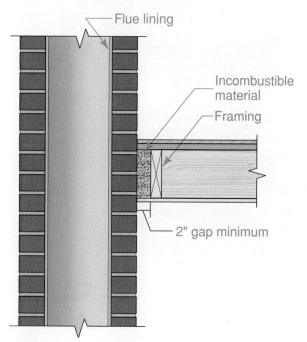

Flue lining

Incombustible material

Framing

2" gap minimum

Clearance to Framing
Heat Shield Wood floor joists must be protected from chimney heat with a suitable fire-stopping product.

other flammable material from building up in these areas. It also keeps combustible materials away from the masonry. Several types of materials may be used, including mineral wool. Unbacked fiberglass insulation is not recommended. Brickwork or standard mortar should not be used because they conduct heat. Fire-stopping materials should be added before the floor sheathing is installed.

Chimney Construction Details

The main parts of a chimney include the foundation, flue liners, walls, and cleanout opening.

Foundation The chimney is usually the heaviest part of a building. It must rest on a foundation and footings to prevent uneven settling and to avoid exceeding the load-bearing capacity of the soil. The foundation is usually built at the same time as the house foundation walls. It must extend at least 6" beyond the chimney on all sides and be at least 12" thick. Steel reinforcement should be added as required by local codes.

Flue Liners Building code now requires that all chimneys be lined during installation. A separate flue and flue liner is needed for each fireplace, furnace, and boiler. A *flue liner* is a fire-clay or stainless-steel pipe

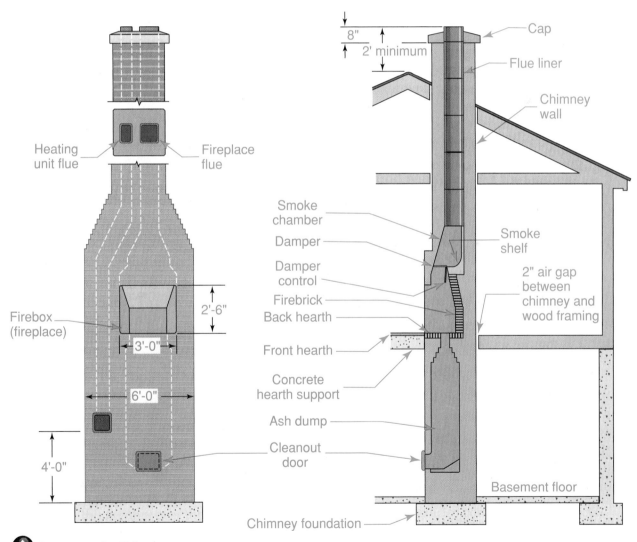

Anatomy of a Chimney

Dual Use This chimney is designed to exhaust gases from the furnace and from one fireplace.

assembled from individual sections that sit within the chimney brickwork. Without a liner, mortar and bricks directly exposed to heat and flue gases can crack. This creates a fire hazard.

Rectangular fire-clay flue liners or round glazed (vitrified) tile can normally be used in all chimneys. Glazed tile or a stainless steel liner is usually required for gas-burning equipment. Local codes outline specific requirements.

Each length of fire-clay or glazed-tile liner should be set in refractory cement with the joint struck smooth on the inside. The liner and brick are installed together as the chimney is built. In masonry chimneys with walls less than 8" thick, there should be an air space between the liner and the chimney walls. This allows for expansion and contraction, and improves heat dissipation. This space should not be blocked by mortar.

The flue liner above a fireplace starts at the top of the throat and extends to the top of the chimney. If a chimney contains three or more lined flues, each group of two must be separated from the others by a layer of brick called a *wythe*. Joints in the liners of two flues grouped together without a wythe between them should be staggered at least 7".

Chimney Walls

Walls of masonry chimneys must have a nominal thickness of at least 4", not including the lining. This is about the width of one brick. Greater thickness may be required if the chimney is unusually tall or located in an earthquake zone.

To strengthen the exposed portion of an interior chimney, the bricks are sometimes corbeled in the portion around the roofline. A **corbel** is a course of brick offset to extend past the course below it. A corbel must not project more than one-half the height of the brick or one-third the width of the mortar bed depth, whichever is less. The top of a chimney is also sometimes corbeled, though this is done purely for decoration. Corbeling can be used to alter the position

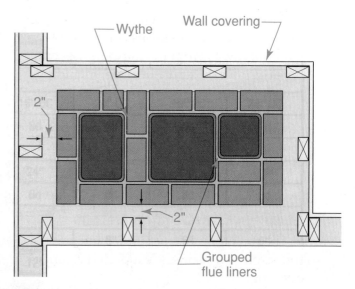

Three-Flue Chimney

Plan View The joints of successive courses of brick should be staggered. This strengthens the chimney. Wood framing should be at least 2" from the masonry.

of a chimney, as when avoiding framing that cannot be relocated. Local codes may specify other requirements.

The chimney wall and its flue liner must not change dimension within 6" above or below where the chimney passes through a floor, ceiling, or roof assembly. This reduces the risk of leaving gaps in the masonry that might allow sparks to reach the framing.

Joints Brickwork around chimney flues and fireplaces should be laid with cement mortar. It is more resistant to the action of heat and flue gases than lime mortar. All bricks and blocks require full, push-filled mortar joints having no gaps anywhere. A *push-filled joint* is one created by pushing the brick into a thick bed of mortar and then striking off the excess that squeezes out.

Cleanout Opening

Soot and ash that build up in a chimney can be removed through a small cleanout. A cleanout is required within 6" of the base of each flue. A single cleanout should serve only one flue. If two or more flues are connected to the same clean-out, air drawn from one to the other will affect the draft in all

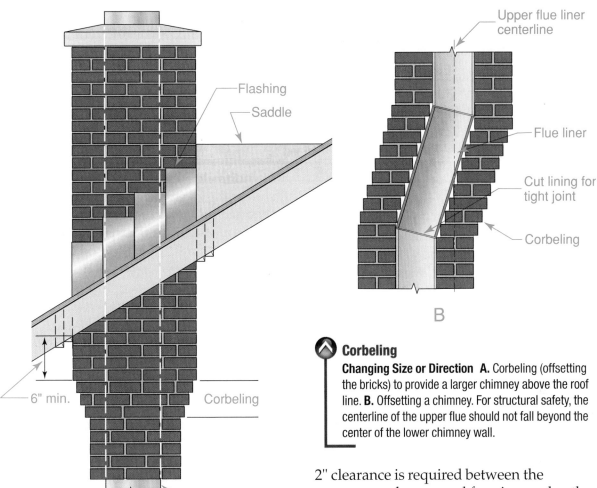

Flashing

Saddle

6" min.

Corbeling

Flue liner

A

Upper flue liner centerline

Flue liner

Cut lining for tight joint

Corbeling

B

Corbeling

Changing Size or Direction A. Corbeling (offsetting the bricks) to provide a larger chimney above the roof line. **B.** Offsetting a chimney. For structural safety, the centerline of the upper flue should not fall beyond the center of the lower chimney wall.

the flues. The cast-iron door on the cleanout should fit snugly and be kept tightly closed to keep air out.

Reading Check

Summarize *Why must the chimney rest on a foundation and footings?*

Roof Details

Care must be taken where the chimney meets the roofing system. The area must resist heat from the chimney, but it must also resist water leakage. Wood and masonry expand and contract at different rates, so construction must allow for this movement.

At the Roof Where any portion of the chimney is located within the exterior walls, a

2" clearance is required between the masonry and any wood framing or sheathing. This is done to reduce fire hazards. However, if the chimney is entirely outside the exterior walls, the minimum clearance is reduced to 1". This clearance also permits expansion due to temperature changes, settling, and slight movement of the chimney during heavy winds.

JOB SAFETY

WORKING ON A ROOF When completing the top of a chimney, work from scaffolding whenever possible. This is safer than trying to work standing on the roof itself. Scaffolding offers a flat, stable surface from which tools and materials will not slide. It is also a more comfortable surface from which to work.

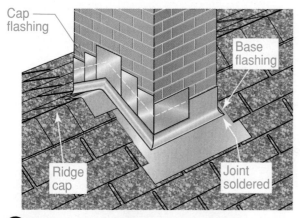

Chimney Flashing
Water Blocker A chimney located on a ridge calls for cap flashing that fits over base flashing.

Cap flashing

Base flashing

Ridge cap

Joint soldered

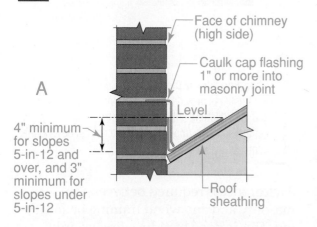

A

Face of chimney (high side)

Caulk cap flashing 1" or more into masonry joint

Level

4" minimum for slopes 5-in-12 and over, and 3" minimum for slopes under 5-in-12

Roof sheathing

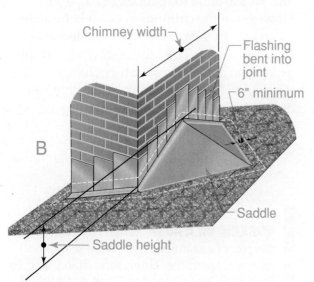

B

Chimney width

Flashing bent into joint

6" minimum

Saddle

Saddle height

Special Flashing Details
Preventing Leaks Flashing a chimney that rises through a sloped roof. **A.** A narrow chimney where no saddle is required. **B.** A wide chimney with a saddle.

Chimneys must be flashed and counter-flashed to make the junction with the roof watertight. Corrosion-resistant metal, such as copper, galvanized steel, or lead-coated copper, should be used for flashing.

When a chimney rises through a sloped roof, careful detailing is required to prevent water flowing down the roof from causing leaks at the chimney. There are two methods for handling this detail.

Overlapping Flashing Cap flashing embedded in the chimney mortar joints overlaps a wide piece of base flashing on the roof.

Chimney Saddle A *saddle* (sometimes called a *chimney cricket*) actually diverts water around the chimney. Building code requires a saddle when the chimney dimension parallel to the ridge line is greater than 30" and does not intersect the ridge. For installation information, see Section 19.3.

Above the Roof A mortar or precast concrete cap should be placed over the top course of brick to prevent moisture from seeping between the brick and flue liner. The cap should be sloped away from the flues on all sides to drain water away. The flue liner should extend at least 4" above the top course of brick. Any gaps between the flue liner and the cap should be sealed with mortar.

A metal or stone hood over the flue openings, help to keep rain out of the chimney.

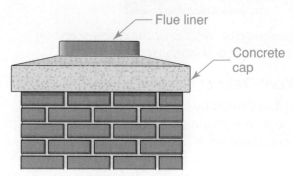

Flue liner

Concrete cap

Chimney Cap
Decorative and Functional A cap prevents water from seeping between the chimney walls and the flue.

It also prevents wind downdrafts from causing a fireplace to smoke. Spark arresters may be required around the hood when chimneys are on or near flammable roofs, or in areas where fire hazards are high. Spark arresters do not eliminate sparks, but they greatly reduce the hazard. A spark arrester also prevents birds and small animals from getting into the chimney.

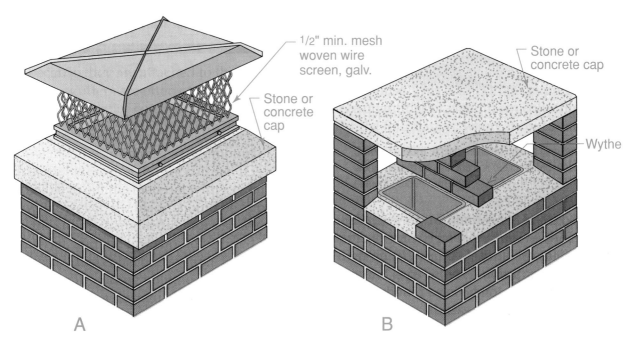

 Chimney Hoods

Blocking Downdrafts A metal or stone cap prevents downdrafts and keeps water out of the chimney. **A.** A metal hood with spark-arrester screening also keeps small animals out. **B.** A stone hood is often found on houses with traditional styling.

Section 24.3 Assessment

After You Read: Self-Check

1. What is makeup air?
2. Name the main parts of a fireplace.
3. Name at least two important characteristics of the flue that affect the draft of a chimney.
4. How much clearance should there be between the chimney walls and wood framing members if the chimney is located within the house?

Academic Integration: Mathematics

5. **Clearance** Woodwork, including any mantel located between 6" and 12" from the firebox opening, must not project forward more than ⅛" for every inch above the opening. What is the maximum width for a mantel located 8.5" above the firebox opening?

Math Concept Numbers expressed as a decimal can also be expressed as a fraction.

Go to **connectED.mcgraw-hill.com** to check your answers.

Review and Assessment

Section
24.1

Section
24.2

Section
24.3

Chapter Summary

Important brick-laying tools include the brick trowel and the jointer. Brick comes in three main types: building brick, facing brick, and firebrick. The four types of mortar are M, S, N, and O. Mortar may be mixed by hand or by machine.

Weep holes provide drainage for veneer walls. Wall ties secure the wall to the framing. Line blocks, corner poles, and mason's lines help masons align courses. A wall is begun at the corners. Joints are tooled for strength and appearance.

Chimneys may be made of masonry or prefabricated parts. It is extremely important to construct a chimney so that it does not create a fire hazard. Height above the roofline and clearance to wood framing must be considered. A chimney saddle diverts water around the chimney. A cap keeps moisture from seeping between the brick and the flue. The interior portions of a masonry fireplace must be properly designed and carefully constructed. This ensures that the fireplace will work properly. It also ensures that smoke and harmful gases are drawn out of the house efficiently. The main parts of a fireplace include the firebox, damper, throat, lintel, smoke shelf, smoke chamber, and hearth.

Review Content Vocabulary and Academic Vocabulary

1. Use each of these content vocabulary and academic vocabulary words in a sentence or diagram.

Content Vocabulary

- jointer (p. 688)
- retempering (p. 694)
- weep hole (p. 698)
- lintel (p. 699)
- lead corner (p. 699)

- line block (p. 701)
- firebox (p. 704)
- makeup air (p. 705)
- refractory cement (p. 706)
- hearth (p. 706)

- draft (p. 709)
- flue (p. 709)
- corbel (p. 712)

Academic Vocabulary

- initial (p. 694)
- transmission (p. 696)
- prolonged (p. 706)

Speak Like a Pro

Technical Terms

2. Work with a classmate to define the following terms used in the chapter: *brick trowel* (p. 688), *mason's rule* (p. 688), *brick tongs* (p. 689), *brick hammer* (p. 689), *brick set* (p. 690), *frog* (p. 691), *wall ties* (p. 698), *damper* (p. 707), *smoke shelf* (p. 707), *smoke chamber* (p. 708), *surround* (p. 708), *flue liner* (p. 711), *wythe* (p. 712), *push-filled joint* (p. 712), *saddle* (p. 714).

Review Key Concepts

3. Name three tools used in working with brick and mixing mortar.

4. List the steps in cutting brick with a mason's hammer.

5. Identify the type of mortar used most often for brick veneer.

6. Describe how cold may affect laying out brick.

7. List the main parts of a chimney.

8. List the main parts of a fireplace.

Critical Thinking

9. Summarize What may happen if building codes for installing masonry in the cold are not followed properly?

Academic and Workplace Applications

STEM Mathematics

10. Using Standard Units If a fireplace opening is less than 6 sq. ft. in area, the front hearth should extend at least 16" in front of it and at least 8" on both sides. If the fireplace opening is 6 sq. ft. or more, the hearth should extend 20" in front of the opening and at least 12" to either side. What is the minimum area of a front hearth for a fireplace that is 40" wide and 30" tall?

Math Concept When solving a measurement problem, convert all measurements to the same units.

Step 1: Determine the area of the fireplace opening in square feet.

Step 2: Determine the length and width of the front hearth.

Step 3: Calculate the area.

STEM Engineering

11. Heating Efficiency Describe the heating efficiency of a traditional fireplace. Then describe some ways in which you could improve its heating efficiency in order to fulfill your architectural desires, as well as your heating needs. Your description should be no more than one page and can include lists, diagrams, or bullets.

21st Century Skills

12. Communication Skills You are the foreman on a masonry project. You have just hired an inexperienced laborer to help complete the job. You have noticed that he is using a tape measure to measure the height and spacing of brick courses. Write a short paragraph explaining how you could explain to the laborer that a mason's rule would be better suited for this purpose. Explain what a mason's rule is and why this tool is better for this particular job than a tape measure is.

Standardized TEST Practice

Short Answer

Directions Write one or two sentences to answer the following questions. Use complete sentences.

13. What is a struck joint?

_____

_____

14. How are weep holes formed?

_____

_____

15. How is a brick hammer used to cut brick?

_____

_____

TEST-TAKING TIP

Before answering short-answer questions, write down key words or phrases from the material you remember before writing your sentences. Make sure your complete sentences include both a subject and a predicate.

*These questions will help you practice for national certification assessment.

Unit 5
Hands-On Math Project

Assembling Resources

Your Project Assignment

For a weekend volunteer project, you will design and assemble a storage shed. The shed will have a simple gable roof with a 4-in-12 slope. The exterior will have vinyl siding. Before you begin, you need to estimate and price all the materials you will need for the project.

- **Find** a local organization, such as Rebuilding Together, that builds or repairs homes for low-income members of the community.
- **Specify** the dimensions and materials for a rectangular storage shed that could be part of one of the organization's building projects.
- **Calculate** the quantity of materials required for the shed.
- **Create** a three- to five-minute presentation.

Applied Skills

Some skills you might use include:

- **Define** the dimensions for a shed based on its location and intended use.
- **Compare** shed materials and specify which one best fits your needs.
- **Research** the steps in building a shed.
- **Calculate** the total materials and costs for your project.
- **Identify** the amount of labor involved in building the shed.

The Math Behind the Project

The traditional math skills for this project are estimation and geometry. Remember these key concepts:

Estimation

Many construction materials, such as vinyl siding, come in standard sizes. To estimate how much of these materials you will need, first calculate the amount of the material you will need for the project. Then divide this total by the standard size of the material. For example, if your exterior walls measure 250 square feet and 12 pieces of 8' vinyl siding covers 100 square feet, you will need 30 pieces of siding.

Geometry

A 4-in-12 gable roof has 4 inches of rise for every 12 inches of run. To calculate the rise of a 4-in-12 gable roof, first divide the span in half to find the run. Then multiply the run by 4" to find the rise. Once you know the run and the rise, you can use the Pythagorean theorem to calculate the rafter length.

For example, if your shed is 8 feet wide, to find the run, rise, and rafter length, use the following steps:

1. Divide the span by two to find the run.	$8' \div 2 = 4'$
2. Multiply the run by 4" to find the rise.	$4 \times 4" = 16"$
3. Convert the run from feet to inches.	$4 \times 12 = 48"$
4. Use the Pythagorean theorem to calculate the rafter length. Round the result.	$\sqrt{(16)^2 + (48)^2} = 50.6"$

Project Steps

Step 1 Research

- Find a local organization that provides repairs or builds homes for members of the community.
- Contact your organization to find out the size, materials, and likely use of a shed that would be most suitable for its building projects.
- Compare the costs of various shed materials such as redwood lumber and enamel-coated steel.
- Research the steps involved in building a shed.
- Investigate what building permits may be required for a new shed.

Step 2 Plan

- Determine the main use for your shed. Sketch a suitable plan, labeling dimensions and indicating the placement and size of the door.
- Calculate the run, rise, and rafter length of the roof.
- Calculate total square footage of siding and roof materials.
- Estimate how long it will take to build the shed, working either alone or in a team.

Step 3 Apply

- Use your plan and measurements to calculate how much wall and roof material, siding, paint, and other materials you will need to build the shed.
- Create a quantity take-off for the project.
- Calculate the total cost of building the shed.

Rebuilding Together

Mission: To bring volunteers and communities together to improve the homes and lives of low-income homeowners by providing free repair services for those with the greatest need.

Step 4 Present

Prepare a presentation that describes your research, presents your design, and explains your cost calculations using the checklist below.

PRESENTATION CHECKLIST
Did you remember to…
✓ Explain and describe your shed design?
✓ Demonstrate how you calculated the run, rise, and rafter length of the roof?
✓ Include costs for all materials needed to build the shed?
✓ Describe the basic steps in assembling the shed?
✓ Use a standard format for your quantity take-off?

Step 5 Technical and Academic Evaluation

Assess yourself before and after your presentation.

1. Did you choose a realistic and practical shed design?
2. Was your research thorough?
3. Were your cost estimates accurate and realistic?
4. Were your measurements and calculations accurate?
5. Was your presentation creative and effective?

Go to connectED.mcgraw-hill.com for a Hands-On Math Project rubric.

UNIT 6

Finish Carpentry

In this Unit:

Hands-On Math Project Preview

Presenting a Professional Estimate

After completing this unit, you will you will create an estimate for remodeling a kitchen. You will also draw up and present a bid sheet.

Project Checklist

As you read the chapters in this unit, use this checklist to prepare go the unit project.

✓ Identify the methods for estimating materials, labor, and cost of a remodeling project.

✓ Research materials and appliances

✓ Calculate costs and check estimates.

🡒 Go to connectED.mcgraw-hill.com for the Unit 6 Web Quest activity.

Corbis/Superstock

Construction Careers Finish Carpenter

Profile Finish carpenters work on the detailed final stages of a project. They install stairs, wood floors, cabinets, countertops, doors and windows, and molding and trim..

Academic Skills and Abilities

- mathematics
- blueprint reading
- geometry
- interpersonal skills
- organization and planning skills

Career Path

- apprenticeship programs
- on-the-job training
- trade and technical school courses
- certification

Explore the Photo

Specialty Carpenters Finish carpenters usually pick up where framing carpenters leave off. *Would a remodeling business be likely to employ finish carpenters?*

Stairways

Section 25.1
Stair Basics

Section 25.2
Stair Construction

Chapter Objectives

After completing this chapter, you will be able to:

- **Identify** the method of construction used on any stairway.
- **Identify** the different parts of a stairway and the purposes of each part.
- **Understand** the building code requirements that apply to stairs.
- **Summarize** the steps of stair construction.
- **Explain** how to lay out a cut-stringer stairway.
- **Explain** how to install a cleat-stringer stairway.

Discuss the Photo

Interior Stair Construction of a stairway calls for detailed calculations to ensure that it meets building codes. *What are the two basic parts of a stairway?*

Writing Activity: Writing a Letter

Stairways are made up of a combination of structural lumber and millwork items such as treads, balusters, and trim. Write a one-page letter to a millwork company asking which types of wood are used for these common stairway items.

Before You Read Preview

Wood stairway construction is often considered the hallmark of fine finish carpentry. The work is entrusted to experienced carpenters because stairways must be durable and safe as well as attractive. Choose a content vocabulary or academic vocabulary word that is new to you. When you find it in the text, write down the definition.

Content Vocabulary

- stairwell
- treads
- stringer
- risers
- step
- stairway
- balusters
- flight
- winders
- headroom
- skirtboard
- rabbet
- kick plate

Academic Vocabulary

You will find these words in your reading and on your tests. Use the academic vocabulary glossary to look up their definitions if necessary.

- involves
- process
- parallel

Graphic Organizer

As you read, use a chart like the one shown to organize information about content vocabulary words and their definitions, adding rows as needed.

Content Vocabulary	Definition
stairwell	the vertical shaft inside of which a stairway is built; sometimes called well opening

Go to connectED.mcgraw-hill.com to download this graphic organizer.

Understanding Stairways

What features of a house might make stairway placement difficult for a designer?

Falls on stairs can result in many injuries. This is why building codes tightly regulate the design of stairways. To reduce the possibility of injury by people using a stairway, the carpenter must understand the local building codes for stair design and construction. In addition, the carpenter must work carefully, because even small differences in step height can make the stairs unsafe.

Stairways are typically built by carpenters on site. They use a combination of structural lumber and millwork items such as treads, balusters, and trim. However, stairs are sometimes assembled in a cabinet shop and delivered to the site for installation. This is particularly true of curved stairways.

This chapter covers primarily the construction of interior stairways. Many houses also have exterior stairways that lead to decks. Those are discussed both in this chapter and in Chapter 35.

Stairway Location

Determining the location of a stairway is one of the first and most important decisions made during the design of a house. There are several reasons for this. In addition to being a functional element of the house, a stairway can be a very important architectural feature. This is due in part to the size of the stairway. It is also due to the fact that a stairway is typically the only portion of a house that contains walls two or more stories high.

The stairway location also has an impact on the size and location of rooms on both floors. In fact, many architects start their floorplan sketches by determining the location of the stairway. The stairway is usually centrally located. This makes it equally accessible from various rooms.

Finally, the location of a stairway has an impact on the structure of the house. Because the stairwell interrupts the spacing of floor joists, structural reinforcement is required to channel loads to the foundation. A **stairwell** is the vertical shaft inside of which a stairway is built. It is sometimes called a *well opening.* Carpenters must provide framing details that are strong and unlikely to squeak as the stair is used.

Parts of a Stairway

Some houses have two interior stairways. The stairway that connects the main levels is called a *main stairway, primary stairway,* or *finish stairway.* A main stairway often provides a dramatic focal point for an entry hall.

A Main Stairway
Focal Point Stairs can be custom built or assembled from a collection of stock parts.

Library of Congress, Prints & Photographs Division, [HABS NY,11-KINHO,V,1--99 (CT)]

Inclined Planes In physics, *work* is the amount of force required to move an object over a distance (work = force × distance). A *simple machine* is a tool that makes work easier through the application of a force. However, a simple machine does not reduce the amount of work done. Any decrease in force is accompanied by a reciprocal increase in distance.

A stairway is an example of a simple machine called an inclined plane. An *inclined plane* makes it easier to raise or lower objects over a vertical height. It allows you to overcome the force of gravity by applying a smaller amount of force over a longer distance than the distance that the load is to be raised. Why is it easier to carry a 10 pound weight up a stairway than up a ladder?

Starting Hint Find a stairway and lean a ladder against it. Compare the length of the stairway to the length of the ladder.

The stairway that leads to the basement is called a *service stairway*. Either type of stairway may be built completely by hand or assembled from prefabricated parts purchased as stock millwork items.

The framing of the stairwell is explained in Chapter 15. Most stairways are built within framed walls that form a stairwell. Sometimes it completely encloses the stair from one floor to another. In other cases, it may be open on one or more sides. When the stair builder begins to work, carpenters should already have completed the framing for the floor opening and the stairwell. In some cases, temporary stairs may also be in place.

Though stairways vary in their construction, all stairways have three common elements. The **treads** are the parts of the stair upon which you step. The stringers support the treads, and a *handrail* is required for

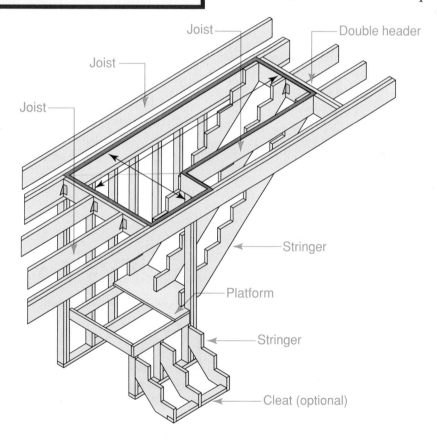

Basic Stair Framing
Stairwell and Stairway Framing for a stairwell and stairway. The top of the stairwell is outlined in red.

safety. A **stringer** is a long piece of 2× lumber that supports a stair. On some stairways, the spaces between treads are enclosed by vertical boards called **risers**.

In this book, the term **step** refers to a tread and a riser. The term **stairway** refers to a series of steps along with all the related elements, including structural elements such as stringers and finish elements such as handrails and balusters. **Balusters** are the slender vertical members that support the handrail.

Reading Check

Explain *What can a carpenter do to reduce the possibility of injury on stairways?*

Types of Stair Construction

There are three common types of stair construction. Each type is based on how the stringers are installed.

Cleat-Stringer Stairway The simplest stair construction has two stringers and a series of treads made from wood planks. It is called a *cleat-stringer stairway* because the treads are

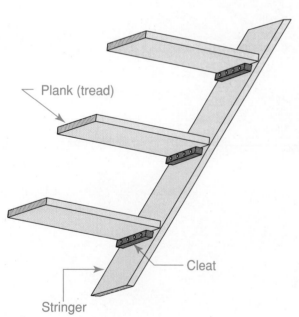

Plank (tread)

Cleat

Stringer

Cleat-Stringer Stairway
Quickly Built Cleat-stringer stairways are sometimes used as service stairways because they are inexpensive and easy to build.

supported by small pieces of wood called *cleats* that are attached to the sides of each stringer. Metal brackets are also available to take the place of wooden cleats. Cleat-stringer stairways are fairly easy to build and are usually made of inexpensive materials. They are sometimes used as service stairways. Note that the stringers are not notched to accept treads or risers.

Cut-Stringer Stairway A *cut-stringer stairway* is the most common type of stairway found in houses. The treads and risers are attached to notches sawn into the upper edge of each stringer. A cut-stringer stairway may have two or more stringers. Three stringers are common on main stairways, but unusually wide stairs may have four stringers.

Housed-Stringer Stairway *Housed stringers,* sometimes called *closed stringers* or boxed stair construction, have recesses in the sides. The treads and risers fit into the recesses and are held in place with wood wedges. This type of stair construction is often found in prefabricated stair systems.

Handrails and Balusters

In order to be used safely, all stairways require some type of handrailing. In some cases, the handrailing can be attached to the inside surfaces of the stairwell using special handrail brackets. With open stairs, which

Builder's Tip

FIRE SAFETY The space beneath a stair is often enclosed to provide storage. Building codes require that drywall ½" or thicker be installed on the inside of all walls. Drywall should also be attached to the underside of the stair stringers. The drywall provides a measure of fire protection in these spaces. This requirement should be taken into account when framing the platform walls.

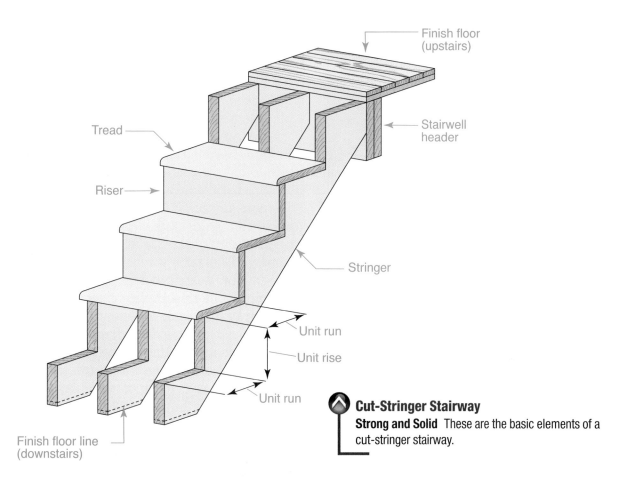

Tread

Riser

Finish floor
(upstairs)

Stairwell
header

Stringer

Unit run

Unit rise

Unit run

Finish floor line
(downstairs)

◆ **Cut-Stringer Stairway**
Strong and Solid These are the basic elements of a
cut-stringer stairway.

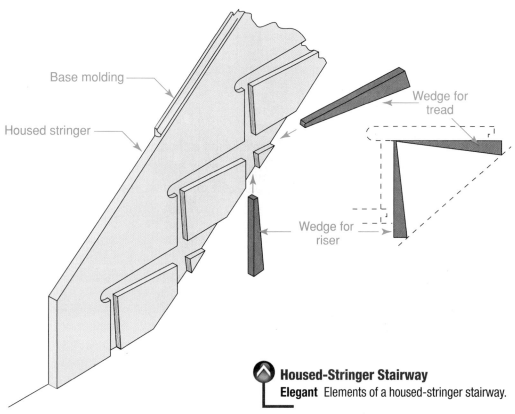

Base molding

Housed stringer

Wedge for
tread

Wedge for
riser

◆ **Housed-Stringer Stairway**
Elegant Elements of a housed-stringer stairway.

are not enclosed in a stairwell, the handrail must be supported by balusters. Newel posts support each end of the handrail. The entire system of balusters, handrails, and related support pieces is sometimes called a *balustrade*.

Baluster A vertical member that supports the handrail on open stairs.
Gooseneck The curved piece between the main handrail and a newel post.

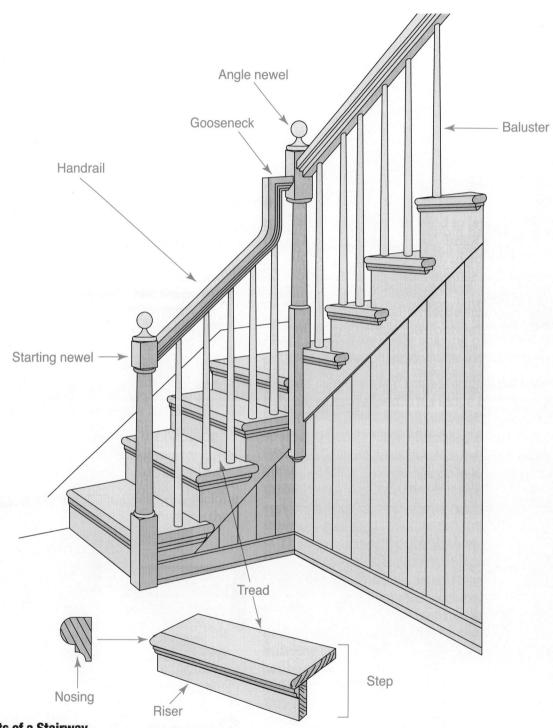

Parts of a Stairway
Finished Stair These parts can be found on many types of stairs.

Handrail The portion of the stair that is grasped when going up or down. It sometimes continues past the stairway to form a balcony handrail.

Newel A newel is a post that supports the handrail at the top and bottom. A *landing newel,* also called a *starting newel,* is located at the bottom of a stair. It is often more elaborate than other newels and can serve as an architectural feature. An *angle newel* supports the handrail at landings, particulary where the stair changes direction.

Nosing The part of a tread that projects beyond the face of the riser.

Riser The vertical portion of a step.

Step One tread or a tread and riser. The first step of a stair is often called the *starting step.* If the starting step is lengthened and rounded at one or both ends, it is referred to as a bullnose starting step.

Tread The horizontal portion of a step.

Riser The vertical portion of a step.

Stringer The piece of lumber that supports the treads and risers. A stringer may also be called a *string,* a *carriage,* or a *horse.*

Metal Stair Parts Most stairs are built entirely of wood products, but some include parts made of metal as well. This is often the case with balusters and handrails. For example, an ornate balustrade might be made of wrought iron. It could support a handrail made of wood or metal. It would also include wrought iron newel posts. Metal stair parts must be ordered well in advance of stair construction. They are sometimes shipped as a collection of pieces that must be installed according to the manufacturer's instructions. In other cases, portions of the railing system may be shipped as complete sections that can then be linked together and attached to the stair.

Reading Check

Explain *What type of stairway is sometimes used as a service stairway?*

Stairway Planning

How is headroom measured?

In most houses, a stairway makes a straight, continuous run. A straight run of stairs is called a **flight**. A flight goes from one landing to another. The *landing* is the floor area where a flight ends or begins. However, a stairway may make a turn to conserve space. Stairways that turn usually include a platform. This platform can also be considered a landing because it is the start of one flight and the end of another. The platform or landing must be at least as wide as the stairway itself.

The stair shown below is an example of an L-shaped stair. It contains two flights arranged at 90° to each other and separated by a platform. The platform is built with short joists covered by floor sheathing and

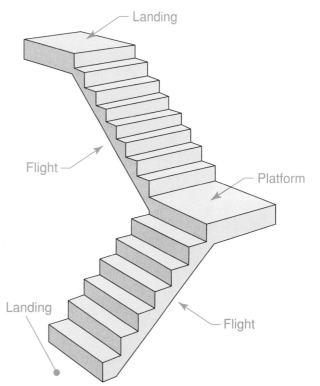

Flights and Landings
Stair Layout The arrangement of a stair is a combination of flights leading to landings (and sometimes to platforms).

is supported by framed walls. The enclosed sides of a platform may be nailed into one or more existing walls. Flights may also be arranged at 180° to each other, with a platform between.

Radiating treads called **winders** can be used instead of a platform when a stairway must change directions. Winders are often found in older houses, but they are not as safe as standard treads. This is because a winder is wedge-shaped and a portion of the tread is too narrow for safe use. Current building codes now restrict how narrow the point of the winder can be.

Stair Codes

The following factors are important to consider when designing and building a stair. The code dimensions found in this chapter are from the 2006 IRC. Before designing a stair, be sure to find out what building codes have been adopted in

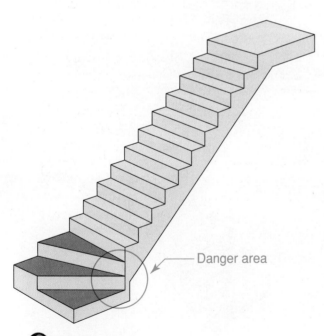

— Danger area

Unsafe Winders
Difficult to Use The winders are the shaded steps at the bottom of this stair. Building codes now prohibit winders that come to a point as shown here.

your area. The codes noted here apply to stairways with straight flights. Codes for spiral stairways and curved stairways are somewhat different. Temporary access stairs used during construction should conform to OSHA guidelines.

Headroom The clearance directly above a step is called **headroom**. Headroom is measured from the outermost edge of the nosing to the ceiling surface directly overhead. If a beam projects below the ceiling surface, headroom is measured to the underside of the beam. Headroom on a landing is measured from the floor surface to the ceiling. Although the minimum headroom required by code on stairs and landings is 6'-8", main stairways generally have between 7'-4" and 7'-7" of headroom. Sometimes two or more flights of stairs are arranged one above the other in the same stairwell. An example would be a basement stairway under the main stairway. Headroom on the lower stair must be carefully calculated because support beams for the first floor can make it difficult to provide suitable headroom.

Calculating headroom is critical in framing stages because the header height must be positioned to provide the minimum headroom for the finished stairway.

Stairway Width A stairway must be wide enough to allow two people to pass comfortably and to permit furniture to be carried up and down. The minimum width for a stairway is 36", measured between the finished walls of the stairwell. However, a width of 42" makes moving furniture easier. The *minimum clear width* of a stair is the distance between the handrails. If a stairway has two handrails, the minimum distance allowed between them is 25". If there is only one handrail, the minimum clear width is measured to the opposite wall surface and is 31½". There is also a minimum distance between balusters.

Risers and Treads The height of the riser and the depth of the tread determine the ease with which the stairs are used. If the risers

 Stair Comfort
Too Steep A stairway should be built with the proper height of risers and treads. On this stairway, the tread is too shallow and the riser is too high.

are too high, climbing the steps can be difficult. If the treads are too shallow, toes will bump the riser at each step. The building code allows a maximum riser height of 7¾" and a minimum tread depth of 10". Safety research indicates that the ideal riser is 7" high and the ideal tread is 11" deep. Winders are no longer allowed. Codes now specify that all winders must be at least 10" wide at a point measured 12" away from the wall. In addition, no part of a winder can be less than 6" wide. The general rule of thumb for stair design is that the riser and tread should add up to between 17 and 18 inches. In addition, the wider the tread, the lower the riser should be. This is both for safety and comfort when accessing the stairs.

Whatever dimensions are chosen for steps, the height and depth of each step must be uniform. Perfect uniformity is difficult, but the code allows a variation of no more than ⅜" between maximum and minimum riser heights. Likewise, the difference between the deepest and the shallowest tread must be no more than ⅜". Variations that are greater can cause someone to trip when using the stairs. Warped or poorly secured treads reduce uniformity. As a result, they are a tripping hazard.

Treads should have a slip-resistant surface. Materials such as polished stone and glazed tile can be dangerous on a stair and should be avoided.

Handrails One of the most important features of good stair design is the handrail. A solid, easily grasped handrail can prevent falls and serious injuries. A continuous handrail must be provided on at least one side of every flight that has four or more risers.

You should be able to curl your fingers around a handrail with ease. Handrails are easiest to grasp when they are made from metal tubing or solid wood shaped into a cylinder. The diameter of such handrails should be between 1½" and 2". Handrails with a larger diameter are more difficult to hold. Other shapes of handrails are permitted, but their shape and size is regulated by code in great detail. In general, however, the edges of the handrail must have a minimum radius of ⅛" for comfort. They should be rounded over and have no sharp edges. There should be at least 1½" of space between the handrail and the wall on which it is mounted.

A handrail should be 34" to 38" high. This distance is measured vertically from the upper edge of the nosing to the top of the handrail.

Balusters The purpose of balusters is to prevent anyone, particularly children, from slipping under the handrail and falling. Balusters should be spaced so that a sphere 4" in diameter cannot pass between them. One exception is at the triangular opening sometimes formed between adjacent steps. In that case, a sphere 6" in diameter should not be able to pass through.

After You Read: Self-Check

1. What is a stairwell?
2. What are risers?
3. Name the three basic types of stair construction.
4. What role do the dimensions of risers and treads play in the ease with which a stair can be used?

Academic Integration: English Language Arts

5. **Write a Description** Research online and print sources for photos or illustrations of open stairways. Select one. Write a description of how the stair on page 728 compares to the stairway you have selected. In your description, use the following terms: platform, nosing, risers, railing, and open and closed stringers.

 Go to **connectED.mcgraw-hill.com** to check your answers.

Section 25.2 Stair Construction

Basic Calculations

What are unit rise and unit run?

The construction of any stairway calls for detailed calculations to ensure that it meets building codes. The calculations for a basic stairway can be easily done with pencil and paper. However, some builders use a construction calculator instead. The handheld calculator can display results in fractions of an inch and is programmed to solve stair problems.

The basic calculations determine the dimensions of details. Use this illustration to study and familiarize yourself with the basic terminology needed to understand stairway calculations.

- *Stairwell Header* The doubled framing that forms the ends of the stairwell opening. It supports the floor joists that were cut to create the stairwell opening.

- *Total Rise* The vertical distance from the finished surface of one floor to the finished surface of the next floor.

- *Total Run* The horizontal length of the stairs.

- *Unit Rise* The vertical distance from the top of one tread to the top of the next highest tread.

- *Unit Run* The horizontal distance between the face of one riser and the face of the adjacent riser.

Unit Rise and Unit Run

The first task in stairway layout is to determine the unit rise and unit run per step. The unit rise is the height of one riser. It is based on the total rise of the stairway and the fact that the unit rise for stairs should be about 7". The unit run is the distance from

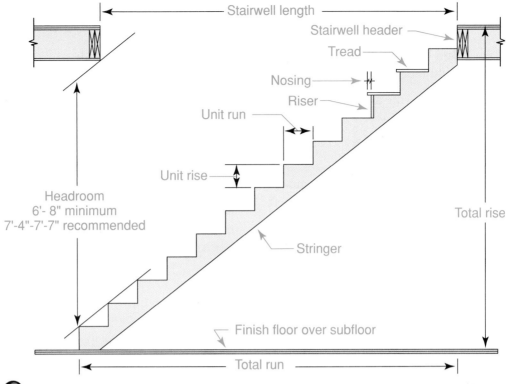

Stairwell length

Stairwell header

Tread

Nosing

Riser

Unit run

Unit rise

Headroom
6'- 8" minimum
7'-4"-7'-7" recommended

Total rise

Stringer

Finish floor over subfloor

Total run

Basic Layout of a Stairway
Critical Dimensions This illustration identifies the elements of a stair that must be calculated.

the face of one riser to the face of the next riser, or the depth of a tread. It does not include the nosing. The total rise is the vertical distance between the surface of the finish floor on one level and the surface of the finish floor on the next level. It is shown on the plans in elevations and wall sections.

It is important to actually measure the total rise before stair construction starts.

It may vary slightly from the distance specified on the plans. For example, suppose that you are building a service stairway from the basement, and the concrete basement floor is already finished. You must then add the thickness of the upper floor's finish flooring to your total rise measurement in order for it to match the measurement on the plans.

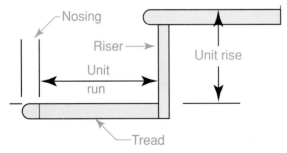

Nosing

Riser

Unit run

Unit rise

Tread

Unit Rise and Unit Run
Step Details The unit run is the distance from the face of one riser to the face of the next riser and does not include the nosing. The unit rise is the distance from the top of one tread to the top of the next tread.

Builder's Tip

TIGHT JOINTS Careful construction techniques improve the safety and durability of a stair. Be sure to make all cuts square and even. This will result in tight joints and will also prevent stair parts from wobbling or squeaking over time. Securing key parts with woodworking glue or construction adhesive helps to reduce wood movement that can result in squeaks.

Calculating Unit Rise and Unit Run Here is an example of how to calculate unit rise and unit run. Assume that the total rise for this stairway is 8'-11".

Step 1 Convert the total rise to inches. That number is 107" ($12 \times 8 = 96$; $96 + 11 = 107$).

Step 2 Divide 107" by 7", the ideal riser height. The result is 15.28 ($107 \div 7 = 15.28$).

Step 3 Round 15.28 to the nearest whole number, which is 15. This gives you the total number of risers in the stairway.

Step 4 To find the unit rise, divide the total rise (107) by the number of risers (15) for a result of $7\frac{1}{8}$" ($107 \div 15 = 7.13$, or $7\frac{1}{8}$").

Step 5 As a general rule, the sum of one riser and one tread should be between 17" and 18". If you subtract the unit rise from this sum, you can find the unit run. For example, if the sum of one riser and one tread is $17\frac{1}{2}$" and the unit rise is $7\frac{1}{8}$", the unit run will be $10\frac{3}{8}$" ($17\frac{1}{2}$" $- 7\frac{1}{8}$" $= 10\frac{3}{8}$").

Once you know the total rise for a stair, you can calculate the unit rise and unit run. For an example of how to do this, see the Step-by-Step Application on this page.

Calculating Total Run The *total run* is a measurement equal to the unit run times the number of treads in the stairway. The total number of treads depends on the manner in which the upper end of the stairway is anchored to the upper landing.

A complete tread at the top of the stairway is shown in A in Anchoring the Stringers. This tread requires a larger stairwell opening, yet it allows the stringer to bear solidly against the header. This means that the number of treads in the

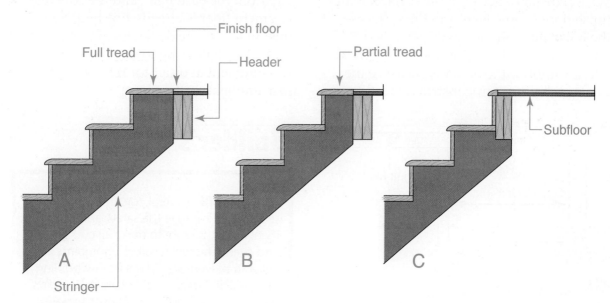

 Anchoring the Stringers
Three Methods In **A.**, the top of the stringer forms a full tread. In **B**, it forms a partial tread. In **C**, it stops short of the upstairs finish floor.

stairway is the same as the number of risers. If there are 15 treads and the unit run is 10⅜", the total run of the stairway is 12'-11⅝" (15 × 10⅜" = 155⅝", or 12'-11⅝").

Only part of a tread at the top of the stairway is shown B in Anchoring the Stringers. In this case, the number of complete treads is one less than the number of risers. Using the earlier example, that would mean 14 treads. The total run of the stairway would then be 14 × 10⅜", plus the run of the partial tread at the top. This partial tread may be dimensioned in detail on the plans. If not, you will have to estimate it as closely as possible. If we assume that it is about 7", the total run is 12'-8¼" (14 × 10⅜" = 145¼"; 145¼" + 7" = 152¼", or 12'-8¼").

In C in illustration on page 734, the upper finish flooring serves as the top tread. In this case the number of treads is one less than the number of risers. Using our example, the total run would be 12'-1¼" (14 × 10⅜" × 145¼", or 12'-1¼").

After you have figured the total run of the stairway, drop a plumb bob from the stairwell header to the floor below. Measure off along the floor the total run, starting at the plumb bob. This locates the anchoring point for the lower end of the stairway.

Sometimes there may not be enough room for a straight run. In this case, a landing and two or more sections of stairway would be needed.

Height Floor to Floor H	Number of Risers	Height of Risers R	Width of Tread T	Run		Run	
				Number of Risers	L	Number of Risers	L2
8'0"	13	7⅜" +	10"	11	8'4" + W	2	0'10" + W
8'6"	14	7⁵/₁₆" -	10"	12	9'2" + W	2	0'10" + W
9'0"	15	7³/₁₆" +	10"	13	10'0" + W	2	0'10" + W
9'6"	16	7⅛" -	10"	14	10'10" + W	2	0'10" + W

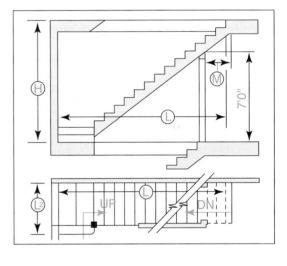

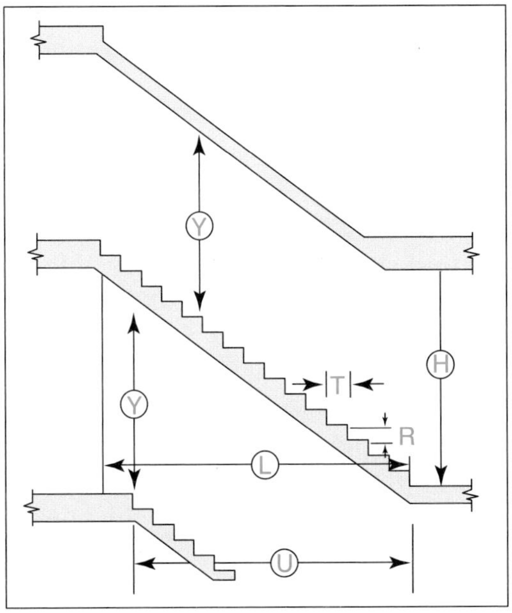

Height Floor to Floor H	Number of Risers	Height of Risers R	Width of Treads T	Total Run L	Minimum Head Rm. Y
8'0"	12	7³/₈" +	9"	8'3"	6'8"
	13	7³/₈" +	9¹/₂"	9'6"	6'8"
	13	7³/₈" +	10"	10'0"	6'8"
8'6"	13	7³/₈" +	9"	9'0"	6'8"
	14	7³/₈" +	9¹/₂"	10'3¹/₂"	6'8"
	14	7³/₈" +	10"	10'10"	6'8"
9'0"	14	7³/₈" +	9"	9'9"	6'8"
	15	7³/₈" +	9¹/₂"	11'1"	6'8"
	15	7³/₈" +	10"	11'8"	6'8"
9'6"	15	7³/₈" +	9"	10'6"	6'8"
	16	7³/₈" +	9¹/₂"	11'10¹/₂"	6'8"
	16	7³/₈" +	10"	12'6"	6'8"

Sample Stair Layout Dimensions

Various Possibilities Layout dimensions for some standard stairways.

Cut-Stringer Stairways

What factors might weaken a stringer?

A cut-stringer stairway is very versatile. It can be built of expensive materials or common lumber. It can have treads and risers, or just treads. (If it has no risers, it is called an *open-riser stairway*.) It can be a permanent part of the house or be used only during the construction phase. All of these variations rely on the same basic concepts.

The building of a stair **involves** a series of operations. These operations must be done carefully and accurately.

Laying Out the Stringers

The treads and risers are supported by two stringers that are solidly fastened in place. When the treads are less than 1⅛" thick, or if the stairs are more than 2'-6" wide, a third stringer should be installed in the middle of the stairs.

Cut stringers for main stairways are usually made from 2×12 stock. To lay out the stringer, you must first determine how long a piece of stock you will need. We will use the same figures used in the calculation examples. Assume that the method of anchoring the stair is the one shown in C on page 734. The total rise is 8'-11". The total run is 12'-1¼".

1. On the framing square twelfth scale, measure the distance between a little over 12¹⁄₁₂" on the blade and 8¹¹⁄₁₂" on the tongue. You will find that it comes to just about 15". Therefore, you will need a piece of stock at least 15' long. You should allow extra stock for waste (about 3' more in this case).

2. Select or cut a piece about 18' long. Lay out the stringer from the lower end.

3. Set the framing square to the unit run and unit rise. Draw line $\overline{AB}$ along the blade and line $\overline{BC}$ along the tongue. $\overline{AB}$ indicates the first tread, $\overline{BC}$ the second riser.

4. Reverse the square and starting at A, draw line $\overline{AD}$ perpendicular to $\overline{AB}$. It

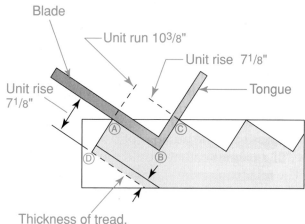

Blade
Unit run 10³⁄₈"
Unit rise 7¹⁄₈"
Tongue
Unit rise 7¹⁄₈"
Thickness of tread, or the thickness of tread less the thickness of finish floor

 Starting the Layout
Riser and Tread Laying out the lower end of a cut stringer.

should be equal in length to the unit rise. Line $\overline{AD}$ indicates the first riser.

5. The first riser has to be shortened, a **process** that is called *dropping the stringer*. In the completed stair, the unit rise is measured from the top of one tread to the top of the next. Assume that the bottom of the stairway is to be anchored to a finished floor, such as a concrete basement floor. If $\overline{AD}$ were cut equal to the unit rise, the first step would be too high when the first tread was put on. Its height would equal that of the unit rise plus the thickness of the tread. To make the height of the first step equal to just the unit rise, shorten $\overline{AD}$ by the thickness of a tread. If the bottom of the stringer is to be anchored on a subfloor to which finish flooring will be applied, shorten $\overline{AD}$ by the thickness of a tread minus the thickness of the finish flooring.

6. When you have shortened $\overline{AD}$ as required, proceed to step off the unit run and unit rise as many times as the stairway has treads. In our example,

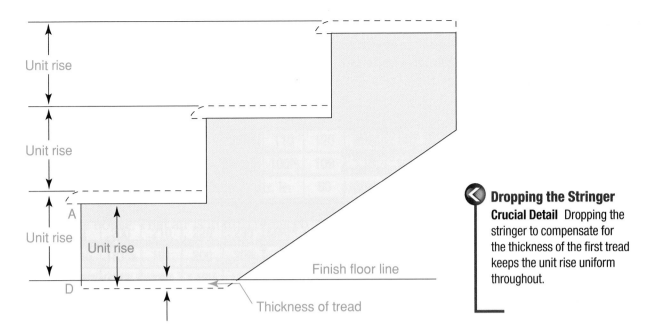

Dropping the Stringer
Crucial Detail Dropping the stringer to compensate for the thickness of the first tread keeps the unit rise uniform throughout.

that would be 14. To maintain the square in the same position as you slide it along the rafter, attach stair gauges to the framing square. They ride along the edge of the stock to prevent the square from slipping out of position.

7. Finish the layout at the upper end as shown on page 738. Remember, we are going to anchor the upper end by the method shown in C. First lay out

line $\overline{AB}$, which represents the last of the treads.

8. Lay out dotted line $\overline{BC}$, which indicates the face of the header.

9. Extend $\overline{BC}$ down to D, so that $\overline{BC}$ plus $\overline{BD}$ will equal the depth of the header.

10. To make the stringer fit close under the lower edge of the header, you must

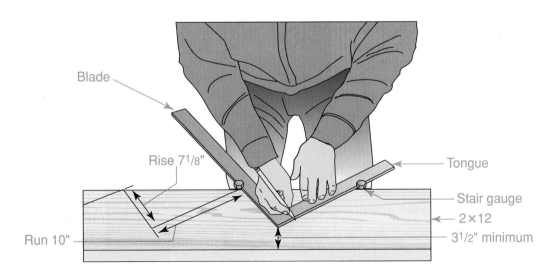

Continuing the Layout
Stair Gauges To complete layout on the stringer, use the framing square and stair gauges to "step off" the layout.

shorten $\overline{BD}$ by the amount the stringer was dropped. Draw $\overline{EF}$ equal in length to the thickness of the header. Draw line $\overline{FG}$ to square off the stringer.

11. Carefully cut out the first stringer with a circular saw. Do not overcut intersecting cuts, which will weaken the stringer.

Instead, finish with a handsaw or jigsaw. Then set the stringer in position, and check it. If it fits properly, use it as a layout pattern for the others.

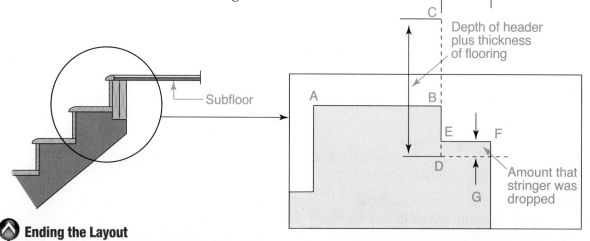

— Subfloor

Ending the Layout
Header Notch Laying out the upper end of a cut stringer.

A

B

Cutting the Stringer
Two Saws With the stringer resting on sawhorses, use a circular saw to cut just along the layout lines. **A.** To avoid overcutting, use a handsaw or jigsaw to finish the cut. **B.** *Why should overcutting be avoided?*

Installing Stringers

The methods used for framing stairways and securing stringers vary in different regions. The stair builder must determine which method will be used before laying out the stringers. Regardless of method, the goal should be to have a structurally strong, safe stairway.

The stringers are the first stairway members installed. Install stringers as follows:

1. Tack the stringers into position.

2. Check each stringer for plumb by holding the carpenter's level vertically against a riser cut.

3. Check if each stringer is level with the other stringers by setting a carpenter's level across the stringers on the tread cuts.

4. When the stringers are level and plumb, nail them into place. A stringer that lies against a trimmer joist should be nailed to the joist with at least three 16d nails.

5. The bottom of a stringer that rests on the subfloor should be toenailed with 10d nails, four to each side if possible. The nails should be driven into the subflooring and, if possible, into a joist below.

Installing Risers and Treads

After you have placed the stringers, you will install the treads and risers.

1. Cut the treads and risers to length.

2. Nail the bottom riser to each stringer with two 6d, 8d, or 10d nails, depending on the thickness of the stock.

3. If the first tread is $1\frac{1}{16}$" thick, nail it to each stringer with two 10d finish nails and to the riser below with at least two 10d finish nails. If the first tread is $1\frac{5}{8}$" thick, a 12d finish nail may be required. Use three nails at each stringer, but do not nail to the riser below. When using hardwood stock, nail holes should be pre-drilled to prevent the stock from splitting. Set all finish nails.

4. Proceed up the stair in this same manner.

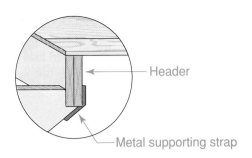
Header
Metal supporting strap

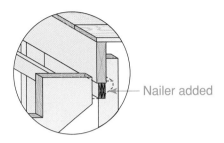
Nailer added

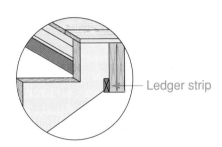
Ledger strip

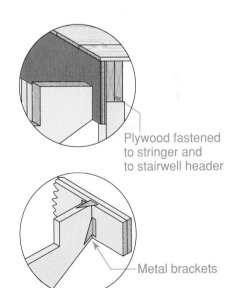
Plywood fastened to stringer and to stairwell header

Metal brackets

Anchoring a Stringer
Various Methods These are some of the ways in which the top of a stringer can be anchored to the framing.

Skirtboard

Sometimes, a skirtboard may be installed. A **skirtboard** is a finished board that is nailed to the wall before the stringers are installed. The risers and treads are nailed to the stringers and butted to the skirtboard. The skirtboard protects the wall from damage. It also provides a finished edge against the wall, which makes it easier to paint or wallpaper the adjacent areas.

Prefabricated Stairs

Many stairs are constructed either partly or entirely using manufactured parts. The parts include risers, treads, stringers, and anything else needed to assemble the stair. Parts can be purchased from a local millwork company or ordered online from any number of suppliers. In some cases, a carpenter will cut and install stringers on site, then finish the stair by using prefabricated treads and risers. In other cases, such as housed stairs, the outer stringers are manufactured parts routed out to the profile of the tread, riser, and nosing (see page 727).

Sometimes the treads and risers are assembled with interlocking joinery. This is often the case when the stairway is being assembled from manufactured parts. The top of the riser is rabbeted to fit into a groove in the bottom front of the tread. The back of the tread is rabbeted to fit into a groove in the bottom of the next riser. A **rabbet** is a cut or groove along or near the edge of a piece of wood. It allows another piece to fit into it to form a joint. The treads and risers are fitted together and slipped into place. They are then tightened by driving and gluing wood wedges behind them. With this method, the outer stringers are visible above the steps, so the stringer must be of very good quality.

Prefabricated stairs should be assembled following instructions supplied by the manufacturer. It is extremely important to

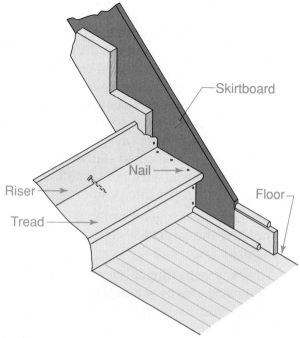

Using a Skirtboard
Finished Detail A skirtboard and a rough stringer nailed in place.

verify the total run and total rise where the stair will be installed. Once the stair has been ordered and arrives on the job site, changes are difficult.

Completing the Stair

The handrail, newels, and balusters complete a stair and make it safe to use. These parts may be plain or elaborate, but they should be in keeping with the style of the house. Building codes regulate the placement of these elements.

Handrail For closed stairways, the handrail is typically attached to the wall with adjustable metal brackets. Locating the position of the brackets is not difficult. After the height of the handrail has been decided upon, a chalk line is snapped on the wall and the brackets are aligned with it. They should be screwed to the stairwell framing. A bracket should be located near the top of the railing, near the bottom, and at various locations in between.

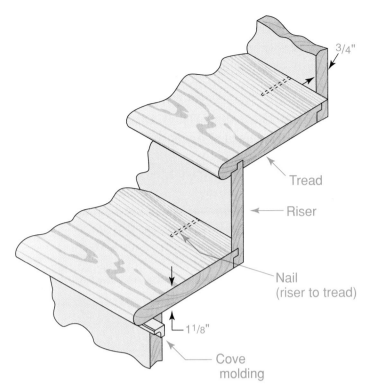

Tread

Riser

Nail
(riser to tread)

3/4"

1 1/8"

Cove
molding

 Interlocking Joinery

Close Fit Interlocking joinery provides a tight fit between treads and risers. Cove molding can be added under the nosing to conceal the joint there.

Some manufacturers recommend that brackets be spaced no more than 48" apart, but closer spacing provides extra support. Make

sure that that the handrail is **parallel** to the stringer.

For open stairways, the handrail and the balusters are assembled together. If the top of a baluster is cylindrical, it often fits into a hole drilled into the underside of the handrail. If the baluster is rectangular, it may be toenailed to the underside of the handrail with finishing nails.

The balusters are doweled or dovetailed into the treads. For the dovetail method, a strip called a *nosing return* is cut to fit the end of the tread. Dovetails on the lower ends of the balusters fit into dovetail recesses in the end of the tread. The dovetails are glued into the recesses. The nosing return is then nailed into place to conceal the dovetails. If a baluster later breaks, the return can be removed in order to replace the damaged baluster. Balusters that are dowelled into the treads cannot be replaced as easily.

It is important to understand that balusters are not the primary support for a handrail. That is the function of the newel posts at each end. For this reason, the newel posts should be firmly anchored. Where half-newels are attached to a wall, blocking should be provided at the time the wall is framed.

Baluster Attachment

Two Methods Balusters are attached to the treads with either dowels or dovetails.

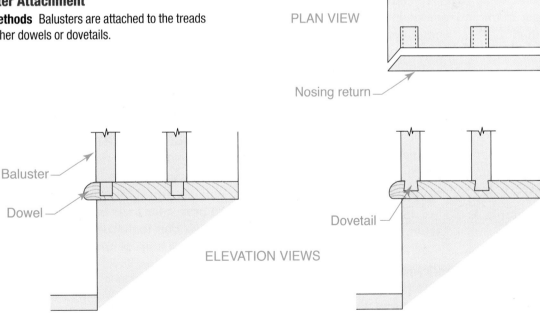

PLAN VIEW

Nosing return

Baluster

Dowel

Dovetail

ELEVATION VIEWS

Other Types of Stairways

There are many types of stairways in addition to those mentioned previously. Some, such as circular stairs, are very complex to build. Others, such as cleat-stringer stairways, are easy to build. Others, such as disappearing stairs, are used to solve particular access problems.

Cleat-Stringer Stairways For a cleat-stringer stairway, the stringers are not cut in order to support treads. This stairway does not normally have risers, and the treads are usually made of thick softwood planks.

Specialty Stairways One kind of specialty stairway is the spiral stairway. Spiral stairways are prefabricated units that often incorporate a steel support system. They are sometimes used to reach lofts and other secondary areas. Their advantage over other types of stairs is that they take up little floor space. Steps and railings are shipped in pieces for easy handling at the job site. Stairwell dimensions and other framing requirements are provided by the stair manufacturer. However, spiral stairways should not be used as primary stairs. Before ordering a spiral stair, check local building codes for minimum requirements.

Estimating and Planning

This estimating and planning exercise will prepare you for national competitive events with organizations such as SkillsUSA and the Home Builder's Institute.

Stairway Materials and Labor

Materials

Estimating the quantity of materials for a stairway is done on a piece-by-piece basis. Once the design of the stairs has been determined, make a detailed materials list showing the quantity and quality of the individual pieces required, such as treads, risers, and balustrade parts. Total the cost of the materials. In the case of a stairway built primarily of manufactured components, the manufacturer may furnish a package price.

Labor

Labor costs can be only roughly estimated because of the many variables that will affect construction time. The suggestions below are very rough approximations. All estimates depend on the carpenter's experience, the style of the stairway, and the type of wood.

Example A: For an open stairway less than 12' long and 42" wide:

Step 1 Estimate construction time at 8¾ hours. This includes rough-cutting the stringers and framing and installing the stringers, treads, and risers.

Step 2 Add 3 hours if there is a turn in the stairway involving a platform or landing.

Step 3 Add one hour to install a handrail.

Step 4 Add 2½ hours for the installation of the newel posts, rails, and balusters.

Example B: For a pre-cut stairway less than 12' long and 42" wide:

Step 1 Estimate 6 hours for assembly.

Step 2 If this stairway has a turn that includes a platform, add about 3 more hours.

Step 3 Add 1 hour for a handrail.

Step 4 For an open stairway with newel posts, rails, and balusters, add 2½ more hours.

Estimating on the Job

Suppose your company is to build an open stairway that is 10½' long and 40" wide. It will have one platform. Estimate the time and labor cost for this open stairway if you hire one worker to do the job and pay him or her $22.50 an hour. Round your answer to the nearest dollar.

Installing a Cleat-Stringer Stairway Basic steps for laying out a cleat-stringer stairway. Note the distance from A to C is the same as the distance from C to D and is equal to the riser height. However, the distance between the floor and line A is less than the riser height to allow for the thickness of the first tread.

Step 1 Determine the total rise and run. Divide the rise by 7. If this does not result in even spacing, adjust the divisor until equal spacings are obtained. Try to keep this spacing between 6½" and 7½".

Step 2 Use a square to lay out a suitable angle at the bottom of the stair. Set a T-bevel to this angle. Then cut each stringer along the layout line using a circular saw.

Step 3 To locate the position of the first cleat, measure up from the bottom of the stringer. Mark off a distance equal to the riser height minus the thickness of the tread. Use the T-bevel to draw a line parallel to the bottom of the stringer at this point. This line represents the top of the cleat and the bottom of the tread.

Step 4 Measure up from line A a distance equal to the riser height and establish point C. Position the T-bevel at point C and mark another line across the stringer. Continue this operation until all tread positions have been located.

Step 5 Lay out and cut the top of the stringer according to whatever method is used to support it.

Step 6 Cut the cleats for each stringer from 1 × 2 or heavier stock. Screw them in position at each line using suitable woodscrews. Place the stringers in the stairwell and nail them in place.

Step 7 Cut the treads to length. Starting with the bottom tread, place each tread in position. Nail it securely to the cleat.

Step 8 Install a railing system to ensure that the stair can be used safely.

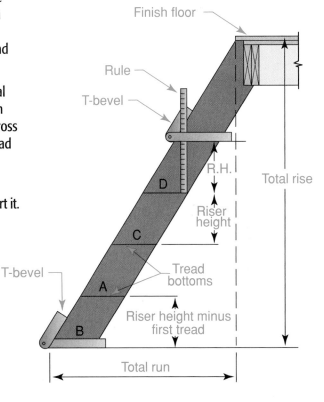

 Cleat-Stringer Stair Layout

 Spiral Stair
Space Saver Note the use of wood for the balusters and treads on this spiral stairway.

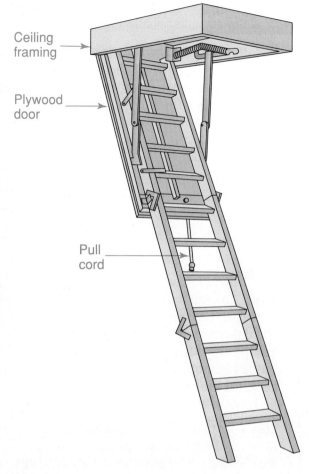

Ceiling framing

Plywood door

Pull cord

 Disappearing Stairs
Basic Access When the stairs are in the stored position, the plywood door is barely noticeable because it is painted to match the ceiling.

Hinged stairs or *disappearing stairs* are often used for access to an attic. They may be made of wood or aluminum. They are installed as a complete unit, fitting into a framed opening in the ceiling. Disappearing stairs swing up into the attic space when not in use. They are suitable only for occasional attic access, not as access to a living area.

Basement stairs can be built much like any other stair and must follow the same building codes. However, basement stairs often incorporate one unusual detail: the lower end of the stairway rests on concrete so they cannot be secured by toenailing. To prevent the stringers from moving, a kick plate is screwed or nailed to the concrete. The stringers can then be toenailed to the kick plate. A **kick plate** is a short piece of framing lumber that is used to anchor the bottom of a stair.

The stairs in residential construction are usually made of wood. In commercial construction, some stairs are built of wood but fire-resistant metal service stairs may be required by code. Technical advances in metal forming and fabrication have improved the quality of metal service stairs. Such stairs may be classified as fire stairs.

During construction of a house, it is often necessary to build temporary service stairs to enable workers to reach upper levels. These stairs are not made with finished materials because they will only be used during construction. However, any temporary stair should be built with safety in mind. Because it can be built quickly and

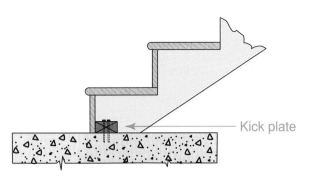

inexpensively, a cleat-stringer stair is often used as a temporary service stair. Later on, the stair is removed and a new, permanent stair can be built. Any temporary service stair should be fitted with a sturdy railing to ensure its safe use. Another way to provide a service stair is to install an open riser stair. This would have cut stringers and uses 2× stock as for treads. There are no risers. Layout and construction of this type of stair is much the same as with cut-stringer stairs.

 Basement Stairs

Kick Plate The lower ends of the stringers should be anchored against a kick plate that has been bolted or nailed to the concrete floor.

— Kick plate

Section **25.2** Assessment

After You Read: Self-Check

1. The total rise of a stairway is the vertical distance between which two points?
2. What is the first task in stairway layout?
3. What precaution should you take when cutting out a stringer for a cut-stringer stairway?
4. On a closed stairway, what supports the handrail?

Academic Integration: Mathematics

5. **Calculating Total Rise and Total Run** Use a calculator to find the total rise and total run of a stairway with 14 treads that are 11" deep, have nosing of ⅜", and have risers measuring 7". A top step has a tread measuring 7".

   Math Concept When using a calculator, convert fractions to decimals. It is useful to memorize common conversions, such as ⅜ = 0.375.

   Step 1: Calculate the unit run. Subtract the nosing from the depth of the tread. Multiply by the number of treads.

   Step 2: Add the depths of any treads that were shortened at the top of the flight.

   Step 3: Calculate the unit rise. Multiply the height of the riser by the number of risers.

   Step 4: Convert your answers to feet, inches, and fractions of an inch.

Go to connectED.mcgraw-hill.com to check your answers.

Review and Assessment

Chapter Summary

The three main parts of a stairway include the treads, the stringers, and a handrail. The treads on cleat-stringer stairways are supported by cleats attached to the stringers. The treads on cut-stringer stairways are supported by notches cut into the stringers. In stairs with more than one flight, the flights are separated by landings. Stair designers must consider headroom, width, riser and tread dimensions, handrails, and balusters. All are important to consider in making a stairway safe and easy to use. In most cases, building codes specify maximum and minimum dimensions.

The first step in stairway construction is to calculate the unit rise and unit run. Total rise is given on the plans. Total run is based on the unit run. The next step is to lay out the stringers and install them. The third step is to install treads and risers (if any). Finally, the handrail and any balusters are put in place.

Review Content Vocabulary and Academic Vocabulary

1. Use each of these content vocabulary and academic vocabulary words in a sentence or diagram.

Content Vocabulary

- stairwell (p. 724)
- treads (p. 725)
- stringer (p. 726)
- risers (p. 726)
- step (p. 726)
- stairway (p. 726)
- balusters (p. 726)
- flight (p. 729)
- winders (p. 730)
- headroom (p. 730)
- skirtboard (p. 740)
- rabbet (p. 740)
- kick plate (p. 744)

Academic Vocabulary

- involves (p. 736)
- process (p. 736)
- parallel (p. 741)

Speak Like a Pro

Technical Terms

2. Work with a classmate to define the following terms used in the chapter: *main stairway* (p. 724), *service stairway* (p. 725), *handrail* (p. 726), *cleat-stringer stairway* (p. 726), *cleats* (p. 726), *cut-stringer stairway* (p. 726), *housed stringers* (p. 726), *balustrade* (p. 728), *angle newel* (p. 729), *landing* (p. 729), *minimum clear width* (p. 730), *total run* (p. 734), *open-riser stairway* (p. 736), *nosing return* (p. 741).

Review Key Concepts

3. Explain the basics methods used to build any stairway.

4. Identify the components of a stairway.

5. Explain how building code requirement affect stairway construction.

6. List the steps of stair construction.

7. Describe how to lay out a cut-stringer stairway.

8. Describe how to install a cleat-stringer stairway.

Critical Thinking

9. Analyze A homeowner has suggested that a spiral staircase be installed as the main stairway in their new home. Is this a good choice? Why or why not?

Academic and Workplace Applications

STEM Mathematics

10. Unit Rise and Unit Run What is an appropriate unit rise and unit run for a stairway whose total rise is 12' and total run is 18'?

Math Concept Sometimes a problem has more than one solution. The final answer requires making a judgment.

Step 1: Convert the total rise to inches. That number is 144" ($12 \times 12 = 144$).

Step 2: Divide 144" by 7", the ideal riser height. The result is 20.57 ($144 \div 7 = 20.57$).

Step 3: Round 20.57 up to the nearest whole number, which is 21. This gives you the total number of risers, and steps, in the stairway.

Step 4: To find the unit rise, divide the total rise by the number of risers. To find the unit run, divide the total run by the number of steps.

Step 5: Since the unit rise comes out to be a bit under 7", the ideal, refigure the plan using 20 steps instead of 21. Compare the two plans, and explain which plan you think is best.

STEM Engineering

11. Architectural Design Evaluate the riser-tread dimensions for a stairway in your school or home. Decide whether the stairway is too steep, just right, or too shallow. Measure the handrail height. How does it compare with the code requirements noted in this chapter? Sketch the basic dimensions of the stairway.

21st Century Skills

12. Career Skills Finish carpenters must undergo a great deal of training and on-the-job instruction in order to learn their trade. Write a paragraph explaining how an apprenticeship that includes on-the-job training would be beneficial to you on your path to becoming an expert finish carpenter. For example, you might list what skills outside of carpentry you would expect to learn through an apprenticeship.

Standardized TEST Practice

Multiple Choice

Directions Choose the best answer to each question.

13. What is the term for a stairway that leads to a basement?

   a. secondary stairway

   b. primary stairway

   c. service stairway

   d. finish stairway

14. What is the term for a long piece of 2× lumber that supports a stair?

   a. step

   b. stringer

   c. tread

   d. riser

15. When laying out stringers, what should your goal be?

   a. to do quick, efficient work

   b. to save money on materials

   c. to follow the plans closely

   d. to have a structurally strong, safe stairway

TEST-TAKING TIP

If you are unsure about a question on a written test, place a check mark next to the question in pencil so that you remember to go back to it. Be sure to erase any stray marks before you turn in the standardized test. Stray marks might cause your test to be graded incorrectly.

*These questions will help you practice for national certification assessment.

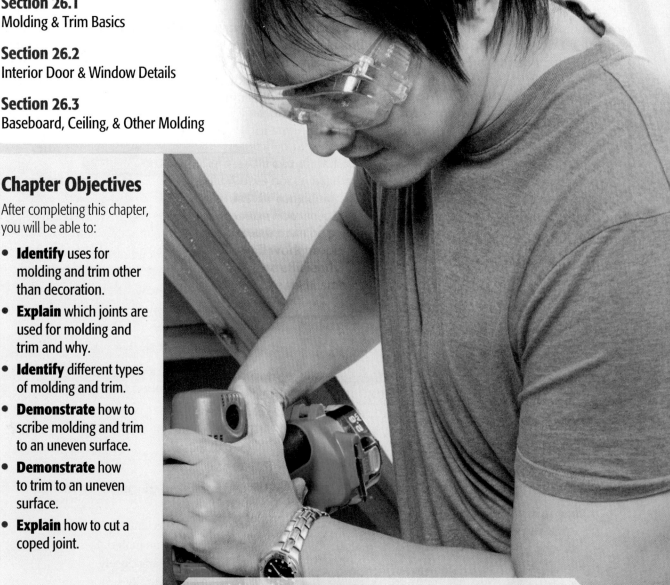

CHAPTER 26

Molding & Trim

Section 26.1
Molding & Trim Basics

Section 26.2
Interior Door & Window Details

Section 26.3
Baseboard, Ceiling, & Other Molding

Chapter Objectives

After completing this chapter, you will be able to:

- **Identify** uses for molding and trim other than decoration.
- **Explain** which joints are used for molding and trim and why.
- **Identify** different types of molding and trim.
- **Demonstrate** how to scribe molding and trim to an uneven surface.
- **Demonstrate** how to trim to an uneven surface.
- **Explain** how to cut a coped joint.

Discuss the Photo

Trim Carpentry A trim carpenter is a specialty carpenter who installs many types of interior woodwork. *What are some examples of interior woodwork you have seen?*

Writing Activity: Job Advertisement

Use the *Occupational Outlook Handbook* and other resources to investigate the job of finish carpenter. Create a 200-word job advertisement for a finish carpenter. Include information about training, tool use, and basic skills required.

Purestock/SuperStock

Chapter 26 Reading Guide

Before You Read Preview

Careful trim carpentry is one characteristic of high-quality building construction. Choose a content vocabulary or academic vocabulary word that is new to you. When you find it in the text, write down the definition.

Content Vocabulary

- molding
- trim
- casing
- side casing
- head casing
- reveal
- return
- baseboard
- coping
- crown molding
- springing angle
- backing
- wall standard

Academic Vocabulary

You will find these words in your reading and on your tests. Use the academic vocabulary glossary to look up their definitions if necessary.

- reinforces
- absorbs
- accurate

Graphic Organizer

As you read, use a chart like the one shown to organize information about content vocabulary words and their definitions, adding rows as needed.

Content Vocabulary	Definition
molding	narrow lengths of wood shaped to a profile
trim	a length of wood with square edges that is surfaced on 4 sides (S4S)

Go to connectED.mcgraw-hill.com to download this graphic organizer.

Molding & Trim Basics

Types of Millwork

Why are molding and trim often combined?

Houses are made of many different materials, but no material is used more extensively than wood. In Unit 4, many fundamental principles of wood growth and usage were discussed, including frame construction techniques using lumber. Lumber is one category of wood used in construction. Millwork is another category. In the broadest sense, millwork includes doors and door frames, window frames, stair parts, cabinetry, trim, molding, and any wood product with a finished surface. Unlike lumber, millwork is meant for use where it will be visible once the house is complete.

Installing millwork is the job of the trim carpenter. Generally, trim carpentry involves all the woodwork that is installed inside a building, with the exception of wood flooring. This chapter will cover molding and trim.

 Using Millwork
All Millwork This photo shows several types of millwork, including molding and trim.

Molding and Trim

The term **molding** usually refers to narrow lengths of wood with a shaped profile. The term **trim** refers more often to a length of wood with square edges that is *surfaced on four sides (S4S)*. However, the two terms are often used interchangeably. *Trim* is also used as a verb. For example, a builder might *trim out* a window (attach molding and trim to it).

Although they are used decoratively, molding and trim often have practical purposes as well. For example, window molding reinforces the window jambs and conceals the large gap between the jambs and the surrounding framing. Baseboard molding protects the lower portion of a wall from damage when the floors are cleaned. (For this reason, it was once known as *mop board*.)

Molding and trim can be made on site from rough stock. This is sometimes done when

REGIONAL CONCERNS

Wood Species The woods used for interior trim that will not be painted can vary from region to region. For example, in the Pacific Northwest, high-quality grades of Douglas fir are sometimes used for baseboard as well as window and door trim. In Massachusetts, however, oak would be more common.

unusual patterns and profiles are required. For example, a trim carpenter could plane the surfaces of rough stock to convert an unusual hardwood into trim. Another example would be to use a custom-made router bit to shape the edge of trim stock, then rip the shaped edge off on a table saw to create molding. If large quantities of a unique molding are needed, they can be fabricated at a millwork company. However, most trim and molding is purchased from local sources. Molding and trim can be used individually or combined to form many interesting designs.

Standard molding patterns and shapes are readily available. Likewise, standard sizes of trim can be purchased. Typical molding profiles are shown on pages 752–753.

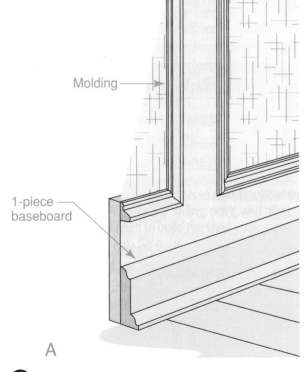

Molding

1-piece baseboard

A

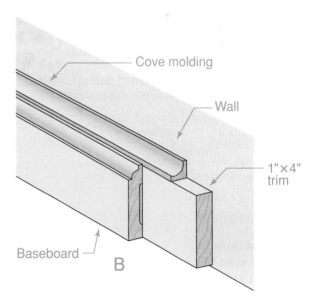

Cove molding

Wall

1" × 4" trim

Baseboard

B

Using Molding and Trim
Design Flexibility Various types of moldings can be used separately, as in **A**, or they can be combined with trim. For example, the assembly of molding and trim in **B** would be called a *built-up baseboard*.

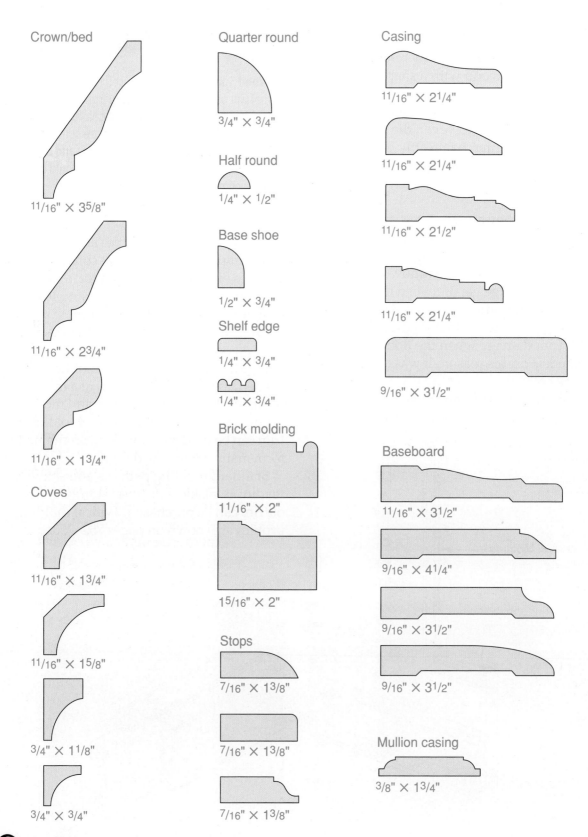

Crown/bed

11/16" × 35/8"

11/16" × 23/4"

11/16" × 13/4"

Coves

11/16" × 13/4"

11/16" × 15/8"

3/4" × 11/8"

3/4" × 3/4"

Quarter round

3/4" × 3/4"

Half round

1/4" × 1/2"

Base shoe

1/2" × 3/4"

Shelf edge

1/4" × 3/4"

1/4" × 3/4"

Brick molding

11/16" × 2"

15/16" × 2"

Stops

7/16" × 13/8"

7/16" × 13/8"

7/16" × 13/8"

Casing

11/16" × 21/4"

11/16" × 21/4"

11/16" × 21/2"

11/16" × 21/4"

9/16" × 31/2"

Baseboard

11/16" × 31/2"

9/16" × 41/4"

9/16" × 31/2"

9/16" × 31/2"

Mullion casing

3/8" × 13/4"

Typical Molding Profiles

Many Shapes These moldings are shown in cross section, called a *molding profile*. The profiles shown here are among the most common. The actual dimensions listed here are only a sample of what is available. Moldings come in many sizes.

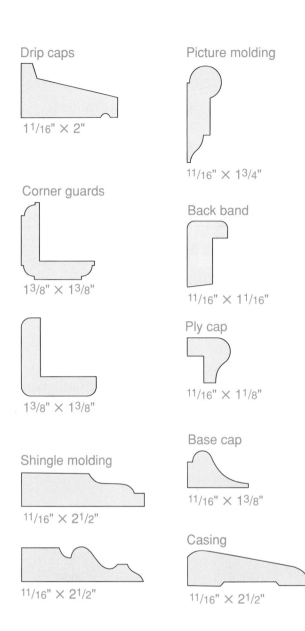

Drip caps

1¹/₁₆" × 2"

Corner guards

1³/₈" × 1³/₈"

1³/₈" × 1³/₈"

Shingle molding

1¹/₁₆" × 2¹/₂"

1¹/₁₆" × 2¹/₂"

Picture molding

1¹/₁₆" × 1³/₄"

Back band

1¹/₁₆" × 1¹/₁₆"

Ply cap

1¹/₁₆" × 1¹/₈"

Base cap

1¹/₁₆" × 1³/₈"

Casing

1¹/₁₆" × 2¹/₂"

Stools

1¹/₁₆" × 3¹/₄"

1¹/₁₆" × 3¹/₄"

Planning for Molding and Trim Installation of molding and trim is one of the last steps in construction, but planning needs to start early. Orders for materials should be placed at least six to eight weeks in advance. Everyone on the building team (such as the contractor, finish carpenters, and so on) should have drawings and specifications at the earliest possible stage. Any changes need to be communicated because they will affect the final installation as well as the budget.

Materials

Most molding and trim used in residential construction are made of solid wood, but other materials are becoming quite common.

Solid Wood Most products are made of pine, hemlock, poplar, or fir. Hardwoods such as oak and maple are common, and other hardwoods are available by special order.

Finger-Jointed Stock Some manufacturers make molding and trim from short lengths of solid wood that have been finger-jointed and glued together. This utilizes wood that would otherwise be wasted. Finger-jointed wood is considered paint-grade product. This means that it is suitable for finished use if it will be painted, but not if it will be stained. Paint conceals the finger joints.

Veneered Stock Good quality wood is increasingly difficult to find, so molding and trim is sometimes made from a base material that is then covered with wood veneer. The base material could be finger-jointed solid

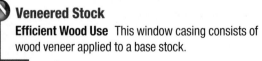
Veneered Stock
Efficient Wood Use This window casing consists of wood veneer applied to a base stock.

Arnold & Brown

wood, a composite material, or a synthetic material. The veneer can be prefinished or finished on site.

Synthetic Stock Some manufacturers produce trim made from synthetic materials such as polyurethane that are much lighter than wood. However, synthetic stock can be cut and nailed with standard woodworking tools, and it is decay resistant. This makes it especially useful for exterior use. These products come with a primer coat because they are always painted. Always consult the manufacturer's literature for joinery details. Some types of joints used for wood molding are not suitable for use on synthetics. Also, special adhesives may be required.

The cost of interior trim varies a great deal with wood species and styles. For example, pine used for door and window frames may cost half as much as some hardwood trims. The choice of materials is therefore based on where the trim is located and how it will be finished. For example, oak crown molding with a stain finish might be specified for a living room and dining room. Simple cove moldings made of pine might be specified for bedrooms. They would also be painted. Paint-grade molding can often be purchased in lower grades because painting will cover minor imperfections of color and figure. Such

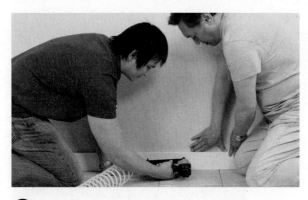

Synthetic Stock
Easy Installation Synthetic materials are light in weight and can be installed with standard woodworking tools.

Science: Materials

Moisture Content What is the recommended average moisture content for wood in your region?

Starting Hint Begin by identifying your state if the map on page 755 includes your state. Use other resources if necessary.

details are covered in the building plans and specifications.

The moisture content of wood determines how stable it will be once installed. The recommended moisture content for interior wood trim varies from 6 to 11 percent, depending on climate. The averages for various parts of the contiguous United States are shown in Recommended Moisture Content. In Canada the recommended moisture contents for the four major geographical areas are: Vancouver, 11 percent; Saskatoon, 7 percent; Ottawa, 8 percent; Halifax, 9 percent.

Finishes

Interior molding and trim may be painted, stained, or given a clear finish such as varnish or polyurethane. In some cases, the wood is stained before it is coated with a clear finish. The type of finish desired often determines the species of wood to be used.

Woodwork to be painted should be smooth, close-grained, and free from pitch streaks. Two woods having these qualities in a high degree include northern white pine and yellow poplar. When the finish is to be clear, or natural, the wood should have a pleasing grain and uniform color. Woods with these qualities include ash, birch, cherry, maple, oak, and walnut.

Curved Trim and Molding

Most trim and molding is used in straight lengths. However, there are times when the installation of curved stock is necessary. For example, a room with curved walls will require baseboard that follows that

Purestock/SuperStock

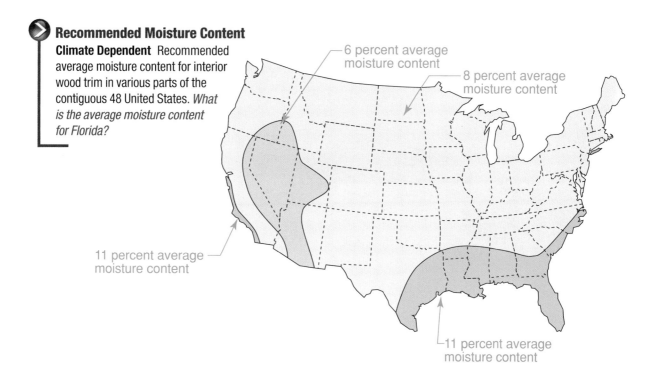

Recommended Moisture Content

Climate Dependent Recommended average moisture content for interior wood trim in various parts of the contiguous 48 United States. *What is the average moisture content for Florida?*

6 percent average moisture content

8 percent average moisture content

11 percent average moisture content

11 percent average moisture content

curve. Window trim around the top of a *Palladian window* also must be curved. (A Palladian window is a tall window with a curved top, flanked on each side by a shorter, rectangular window.) In some cases, such as with curved-head windows, the window manufacturer will supply curved trim. However, it is also possible to create curved trim.

Direction of Curve In some cases, a piece of straight, solid-softwood stock can be forced into a gentle curve. However, the thicker the stock, the less likely this method will succeed. Excessive bending will cause the wood to snap or crack. Also, bending is only possible across the thickness of the wood, not across its width.

Reading Check

Recall What does baseboard molding protect against?

Making Curved Stock Curved molding and trim can be made in various ways. Some methods are shown on page 756. The method chosen depends on what material will be used, the degree of curve necessary,

Palladian Window

Curves Required This type of window requires trim that follows the curve of the uppermost section of the window.

Bendable Four methods for creating curved molding or trim using solid wood.

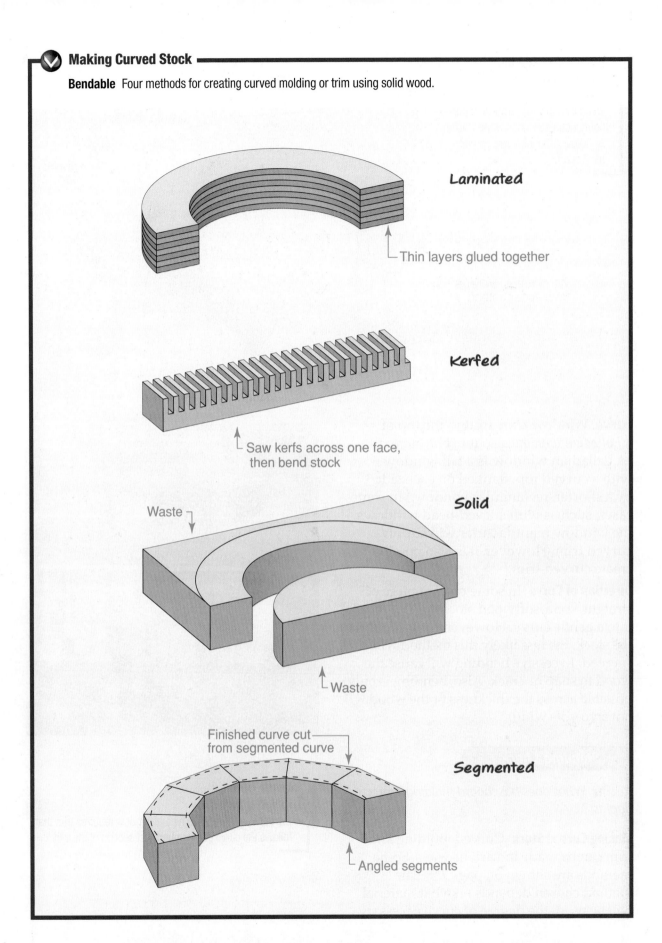

Laminated

Thin layers glued together

Kerfed

Saw kerfs across one face, then bend stock

Solid

Waste

Waste

Segmented

Finished curve cut from segmented curve

Angled segments

and how the material will be finished. Here are the basic methods:

Laminating Thin layers of wood can be glued together in curved forms. When the glue dries, the layered assembly will retain the shape of the forms.

Kerfing A shallow saw cut is called a *kerf*. When one side of a piece of wood is repeatedly kerfed, the wood will be weakened enough to bend. The spacing and depth of kerfs will affect how much of a bend is possible.

Solid Stock A curved piece can be cut from a single piece of solid wood. However, this method has limited use, because the starting stock must be wide enough to contain the entire curve. Also, much of the wood will be wasted.

Segmental Stock Instead of cutting one large curved piece, many smaller pieces can be cut from straight stock and then glued together end to end. This creates a large, angular piece in the approximate shape of the curve desired. This piece is then cut or shaped to create the final curved stock. This method wastes much less wood than the solid stock method.

Steam Bending Subjecting wood to hot steam relaxes wood fibers and enables the wood to bend somewhat. This method is labor intensive because the wood to be bent must be contained in a custom-built steambox.

Use of Synthetics Molding and trim made of flexible polymers can be bent into a wide range of curved shapes. These products can be painted or stained.

Cold Bending The wood is soaked in water and then bent. This is only possible with thin strips of wood. Also, the bend is not permanent, so the bent wood must be contained in some sort of frame.

Chemical Bending Certain chemicals will make wood more pliable. This method should only be used by professionals.

Section 26.1 Assessment

After You Read: Self-Check

1. Define trim carpentry.
2. What is the meaning of the abbreviation *S4S*?
3. What advantages does synthetic trim have compared to solid wood?
4. What qualities are desired in wood that is to be painted?

Academic Integration: Science

5. **Create a Diagram** Refer to page 751 and pages 752–753. Using any of these molding profiles and any dimension of trim, design a built-up baseboard. Make a sketch of your design using CAD software if possible. Describe your design and how it might be installed in a bulleted or one-paragraph summary.

Go to connectED.mcgraw-hill.com to check your answers.

Interior Door & Window Details

Basic Skills

What tools do you think are most important to a trim carpenter?

The molding or trim around a window or door is called the **casing**. When installing casing in a room, door and window frames are usually trimmed first because baseboard and some other moldings must fit against them.

Preparing the Room

Before any molding or trim can be installed, the interior wall covering must be complete (except for paint). Cabinets, built-in bookcases, rough fireplace mantels and similar features should be in place.

The finish floor may or may not be in place at this stage. If it is, precautions should be taken to protect it from damage. If it is not, spacers can be used to approximate the thickness of the finished flooring. The subfloor should be scraped clean and be free of any irregularities. Lightly mark the location of all wall studs on the floor or lower portion of the wall. The marks will be covered later by trim.

Basic Joinery

The two most common corner joints that are used for casing are the butt joint and the miter joint. The joints connect side casings with the head casing. A **side casing** is a vertical piece at the side of a door or window. A **head casing** is the horizontal piece at the top of a door or window.

Butt Joint A butt joint is a quick and easy joint to make. It is made using cuts made at 90° across the face of the stock. Casing that is made primarily of trim stock often uses this joint.

Miter Joint Casing with a shaped surface is mitered at a 45° angle. Mitering ensures that the shape will be continuous from side casings to head casing. As the wood dries, a mitered joint may open slightly at its outer edge. Nailing across the joint after pre-drilling the hole and gluing the joint help to hold the joint together.

Many trim carpenters use compressed-wood biscuits to hold trim joinery together. This technique can be used on butt joints or mitered joints. A biscuit joiner (see Chapter 6,

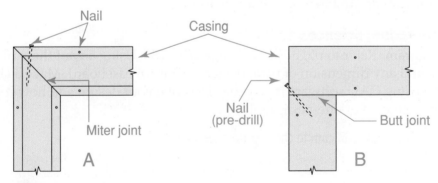

Casing Joints
Joining Casing A. Casing with a molded shape must have a mitered joint at the corner. **B.** Square-edged casing may be joined with a miter joint or a butt joint. In both cases, the casing is nailed to the wall and the joints may be reinforced by nailing at the locations shown by arrows.

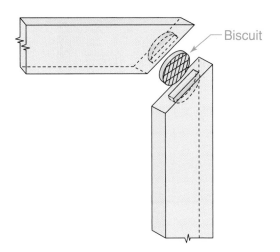

 Biscuit Joinery
Joint Strengthener Many joints, such as this miter joint, can be strengthened with a wood biscuit that is glued in place.

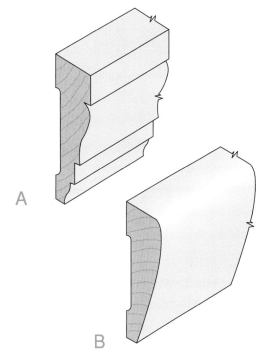

A

B

 Common Casing Profiles
Two Profiles Two common casings used for interior trim. **A.** Colonial. **B.** Ranch, or clamshell, casing.

Section 6.4) is used to cut a shallow groove in the ends of both pieces. A compressed biscuit is inserted into the groove and glued in place to reinforce the joint. When the pieces are brought together, the biscuit **absorbs** moisture from the glue and expands slightly, forming a tight joint. Biscuits are available in three standard sizes, as shown in **Table 26-1**.

Making a Miter Cut

Because two pieces of trim are often joined to form a 90° corner, the angle for most miter cuts is 45°. Use a miter saw to ensure accuracy when making these and other cuts (see Section 5.3).

If angles other than 90° or 45° are required, you must calculate them. To do this, divide 180 by the number of sides.

Table 26-1: Standard Sizes of Biscuit Joints	
Biscuit Size	**Dimension of Groove[a] (inches)**
#0	⅝ wide × 1¾ long
#10	¾ wide × 2⅛ long
#20	1 wide × 2½ long
[a] Lengths are approximate.	

Then subtract that answer from 90. The result will be the number of degrees for each miter cut. For example, to make cuts for a five-sided figure, you would make the following calculations:

$$180° \div 5 = 36°$$
$$90° - 36° = 54°$$

Door Casing
What is a reveal?

The most commonly used casings for interior doors vary in width from 2¼" to 3½". Thicknesses vary from ½" to ¾".

Installation

Casings are nailed to the door jamb and to the framing around it, allowing about a ³⁄₁₆" reveal on the face of the jamb, as shown in A Reveal on page 760. A **reveal** is a small offset between a piece of trim and the surface it is applied to. The small step this creates adds visual interest. It also

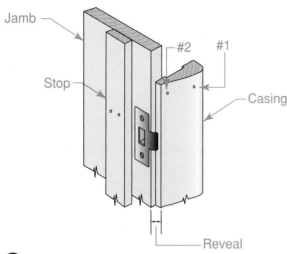

 A Reveal
Small Offset A reveal is visible along the inside edge of this door casing.

allows the trim carpenter to adjust the fit of the casing if the door is not perfectly square.

Nails are located in pairs and spaced about 16" apart around the opening. To nail into the framing, use either 6d or 7d finish nails, depending on the thickness of the casing, as shown in arrow 1. To fasten the thinner edge of the casing to the jamb, use 3d, 4d, or 5d finishing nails, as shown in arrow 2. With hardwood, the holes should be pre-drilled to prevent the wood from splitting. It is the trim carpenter's responsibility to countersink the nail heads. It is typically the painter's responsibility to fill them.

Door Rosettes

Miters are commonly used for joining molding at corners. However, molding can also be installed using *rosettes*, a type of plinth block. These add a decorative element to the room. They eliminate the need to miter the molding. They also conceal differences in thickness between baseboard and door casing.

Reading Check

Summarize Why are holes pre-drilled when using hardwood?

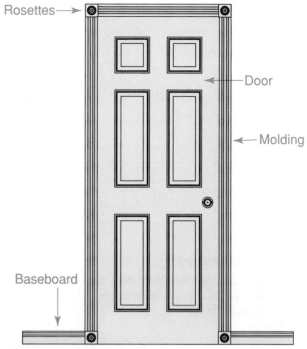

 Door Rosettes
Square Cut Molding that meets a door rosette can be cut square instead of being mitered.

Window Casing & Shutters

What is picture framing a window?

Casing for windows should be of the same pattern as that selected for the door. Windows may also require a stool and an apron. The *stool* is a horizontal member that laps the window sill and extends beyond the casing. An *apron* serves as a finish member below the stool.

Window trim is commonly applied in two different ways: with a stool and apron (A in Window Casing Methods illlustration) or with only casing, as shown by B in Window Casing Methods illlustration.

The Window Stool

The stool is normally the first piece of window trim to be installed. It is notched so that it fits between the jambs and butts against the lower sash. Refer to Window Stool Position, which is a section view.

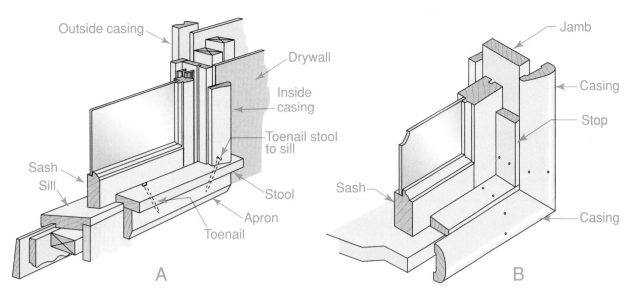

Window Casing Methods

Two Versions A. Window trim installed with a stool and apron. **B.** Window trim installed on four sides using casing.

The upper drawing shows the stool in place. The lower drawing shows it laid out and cut, ready for installation.

Note the three distances labeled A, B, and C. Distance A, the overall length of the stool, is equal to the distance between the outer edges of the side casings, plus the amount that each end of the stool extends beyond the casing's outer edges. Distance B is equal to the width of the finished opening.

Distance C is equal to the horizontal distance measured along the face of the jamb between its edge and the inside face of the lower sash. An allowance of about 1/32" should be deducted for clearance between the sash and the stool. A notch is then cut at each corner of the stool along the layout lines.

The stool is toenailed at the ends with 8d finish nails so that the casing at the sides will cover the nailheads. With hardwood,

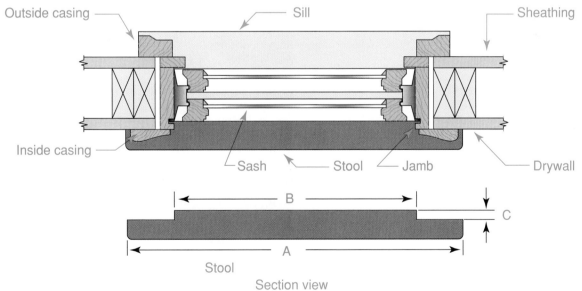

Section view

Window Stool Position

Under and Against Installation details for a window stool.

pre-drilling is required to prevent splitting. The stool should also be nailed at the center to the sill and to the apron when it is installed. Toenailing may be substituted for face-nailing to the sill.

The Casing

Apply the casing after installing the stool. Nail it as described for the door casing. Other types of windows, such as awning, hopper, or casement, are trimmed much like a double-hung window. Casings of the same type are used for all.

When just casing (and no stool or apron) is used to finish the bottom of a window, all four lengths are mitered (see page 761). This is called *picture framing* a window, because the four pieces form a continuous frame around it. The four pieces can be nailed in place one by one. An alternative is to lay the pieces face down on a clean, smooth surface and fasten them together from the back with corrugated fasteners. The assembled casing, much like a picture frame, can then be nailed as a unit to the window jambs and studs.

The Apron

Cut the apron to a length equal to the distance between the outer edges of the side casings. To avoid exposing endgrain at the ends of the apron, cut and nail a **return** in place. A return is a piece that continues the profile of trim or molding around the corner. Trim carpenters refer to this technique as returning the apron to the wall. Attach the apron to the rough sill with 8d finish nails, then glue the return into place at each end and hold it in place temporarily with painter's tape.

Interior Shutters

Movable interior shutters were popular in Western architecture from about 1700 to the early part of the nineteenth century. They were used in the great mansions of New Orleans and in many other fine homes of America. Shutters are once again popular. They are found most often in homes with traditional or country-style interiors.

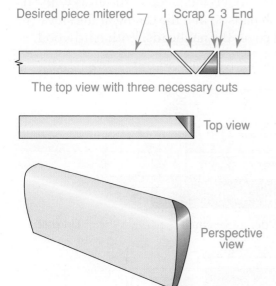

Desired piece mitered — 1 Scrap 2 3 End

The top view with three necessary cuts

Top view

Perspective view

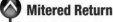

Mitered Return
Best Quality The ends of the apron should be mitered to continue the profile of the apron and conceal its end grain.

Window Shutters
Improved Privacy Louvered shutters may be used throughout a home instead of curtains.

To determine the size of the shutters to be installed in a window, measure the width of the opening between the side jambs. Measure its height from the top of the sill to the inside surface of the top jamb. Various methods can then be used to install the shutter.

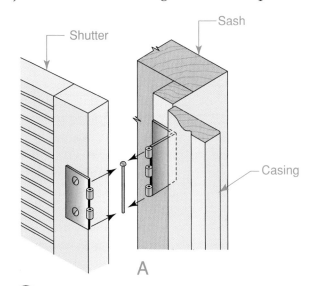

A

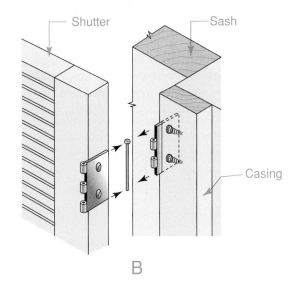

B

 Shutter Hinge Details
Two Mounting Methods If the edge of the window casing is thin, as in **A**, the shutter can be attached to a hinge strip secured to the window jamb. If the edge of the casing is thick, as in **B**, the hinge can be attached directly to it.

Section 26.2 Assessment

After You Read: Self-Check

1. What two basic cuts are used when installing door and window casing?
2. What is a reveal and why is it important?
3. To what length should the apron be cut?
4. How do you picture frame a window?

Academic Integration: Mathematics

5. **Miter Cuts** A carpenter is installing the trim around a window that is a regular hexagon. What is the angle of the miter cuts the carpenter should make?

   Math Concept A regular hexagon is a geometric figure with six sides. All the interior angles of a hexagon are congruent.

   Step 1: Divide 180° by the number of sides.

   Step 2: Subtract the result from 90°.

Go to **connectED.mcgraw-hill.com** to check your answers.

Baseboard, Ceiling, & Other Molding

Baseboard

What trades might need to know the type of baseboard planned for a house?

After the window and door casings are complete, the trim carpenter installs the other moldings in a room. These include flat moldings, such as baseboard or chair rail, and sprung moldings, such as crown molding. *Sprung moldings* are moldings that project out from the wall surface.

Baseboard, or base molding, is a board or molding used against the bottom of walls to cover their joint with the floor. It serves as a transition between the wall surface and the floor. It also covers the gaps that often occur at this location. It can be added after all the doors are trimmed and the cabinets are in place. It can be installed after the finish flooring and should be installed before any carpeting.

One-piece baseboard consists of a single piece of stock that varies in size from $\frac{7}{16}" \times 2\frac{1}{4}"$ to $\frac{1}{2}" \times 3\frac{1}{4}"$ or wider. It is the most common type of baseboard. A small molding called a *base shoe* is sometimes added to conceal the joint between the bottom of the baseboard and the floor. The shoe is nailed into the baseboard, not into the flooring. This prevents the shoe from being moved out of position as the flooring shrinks or expands.

When carpeting is to be installed, the baseboard is installed first, using temporary spacers to lift it slightly above the subfloor. A consultation with the carpet installer can determine how much clearance is needed. The edges of the carpet are then tucked beneath the baseboard. When wall-to-wall carpeting is used, the shoe is usually omitted, and in some cases, the entire baseboard is omitted.

Baseboard may have several parts. For example, two-piece baseboard consists of a

Builder's Tip

TRIMMING BASEBOARD To get a good fit, it is often necessary to cut a tiny amount off the end of a piece of baseboard. To do this, lower the miter saw until the blade's teeth are at table height. Hold the baseboard against the fence with your fingers well away from the blade. Then slide the molding under the blade guard until one end touches the teeth of the blade. DO NOT TURN THE SAW ON YET. Push the molding against the saw teeth slightly. This will nudge the blade slightly out of position. Now raise the saw blade, but without moving the molding. Turn on the saw and make the cut. A small fraction of wood will be removed. Test fit the cut. Repeat the process if necessary.

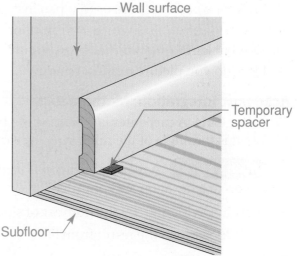

Wall surface

Temporary spacer

Subfloor

 One-Piece Baseboard
Planning Ahead Temporary spacers should be used to create a uniform space under baseboard in a room that will have wall-to-wall carpeting.

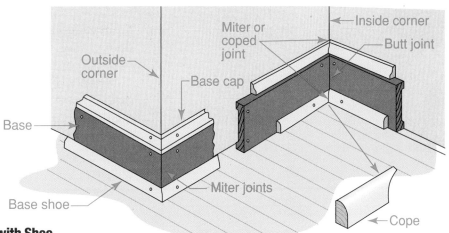

Two-Piece Baseboard with Shoe

Installation Details Baseboard molding installation details showing a simple two-piece baseboard with base shoe.

base topped with molding. When the wall covering is not straight and true, the base cap conforms more closely to the variations than a single wider baseboard would. Three-piece baseboard is shown on pages 752–753.

Square-edged baseboards should be installed with a butt joint at inside corners and a miter joint at outside corners. Profiled baseboards should also be mitered at outside corners, but they should be *coped* (shaped to fit each other) at inside corners. A coped joint looks similar to a mitered joint when complete, but it is a better joint to use in these locations. It forms a good joint even if walls are not perfectly square or plumb at the corners. Also, if the wood shrinks over time, a coped joint will not open up as visibly as a mitered joint will.

Baseboard cut to fit between walls should always be cut a little long. The stock can then be bowed slightly and sprung into place. This ensures a tight fit. When more than one length of baseboard is needed along a wall, the pieces are joined over a wall stud with a mitered lap joint. The angle of the miter is typically 45°. The baseboard is secured to each stud with two 8d finishing nails. The bottom nail should be close enough to the floorline to be covered by the base shoe molding.

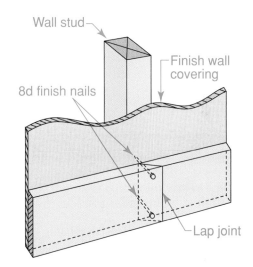

Baseboard Laps

For Long Walls Lengths of baseboard can be joined over a wall stud using a mitered lap joint.

Fitting a Joint

Though a length of baseboard can be measured precisely to fit into place on a wall, trim carpenters usually find it quicker and more **accurate** to mark a length of baseboard in place. The technique is to cut a board about an inch or two longer than necessary, then hold it in place for marking. This technique works particularly well when fitting baseboard against door casing. The trim carpenter saying is, "If you don't measure, you can't measure wrong." A similar technique is to

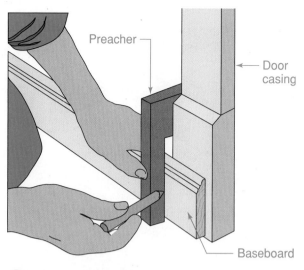

Preacher

Door casing

Baseboard

Marking the Baseboard

Jobsite Helper If casing is not perfectly plumb, a small piece of plywood scrap called a *preacher* can be used to transfer its angle to the baseboard.

hold a small piece of plywood scrap, sometimes called a *preacher*, against the edge of the door casing. This is very useful if casing is not perfectly plumb.

When walls are out of plumb, two lengths of square-edge baseboard that intersect at an inside corner will not fit together properly. One or both of the boards may have to be scribed and trimmed to fit. *Scribing* is a marking process that allows a piece of wood to be precisely fit against a surface that is irregular or not square. The process is shown in the Step-by-Step Application.

Mitering a Joint

Outside corners on baseboard are frequently mitered. The cut is made across the thickness of the material instead of across the width.

Step-by-Step Application

Scribing a Joint Variations of this technique can be used wherever one material must fit tightly against an irregular surface.

Step 1 Install the first length of baseboard.

Step 2 Set the second piece in position on the floor. Place the end to be joined against or near the face of the piece already installed.

Step 3 Using a compass, draw a line parallel to the face of the installed piece on the face of the piece to be joined. Be careful to hold the legs of the compass horizontally and at right angles to the baseboard being scribed. This will ensure a parallel line. If a compass is not available, the same results can be achieved by supporting a pencil flat on a scrap of wood and using them to draw the line.

Step 4 Cut the scribed piece along the line using a coping saw or a jigsaw fitted with a fine-tooth blade.

Step 5 Test the fit. Recut as needed, or use a woodfile or sandpaper to fine-tune the fit.

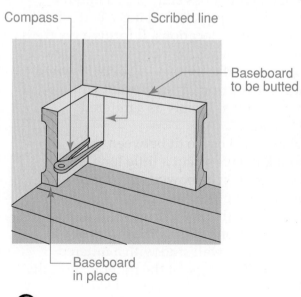

Compass

Scribed line

Baseboard to be butted

Baseboard in place

 Scribing a Joint

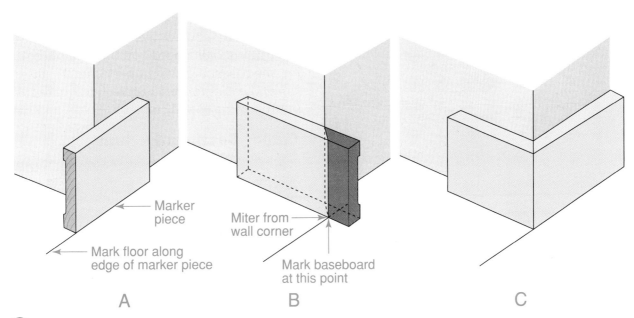

A

Marker piece

Mark floor along edge of marker piece

B

Miter from wall corner

Mark baseboard at this point

C

 Mitering Baseboard

Tight Fit This technique for locating the cut line on mitered baseboard ensures a good joint.

1. Set a piece of baseboard against the wall. Mark a layout line on the floor along the edge of the piece.

2. Repeat the process on the adjoining wall.

3. Hold the first piece to be mitered in place. Mark it where it intersects the layout line.

4. Set the miter saw to a 45° bevel angle and cut just outside the layout line.

5. Repeat Step 3 and cut the second piece at a 45° bevel angle.

6. Test fit the pieces and trim them as needed.

Coping a Joint

Inside corner joints between trim members are usually made by cutting the end of one member to fit against the face of the other. This shaping process is called **coping**. Coped joints are used in a variety of situations, but they are quite common when installing baseboard. This is because a coped joint will not open up after the baseboard is nailed in place. It is also less likely to show a gap if the baseboard shrinks after installation.

To prepare a coped joint, start by installing the first length of baseboard. This baseboard has a square cut on one end and is butted to the wall surface. Once this piece is nailed to the wall, the intersecting piece can be prepared. The end of this piece will be coped as shown on page 768.

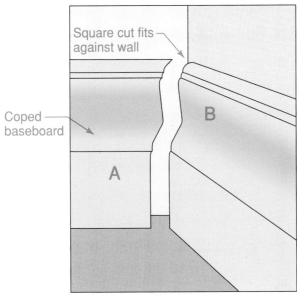

Square cut fits against wall

Coped baseboard

B

A

 Coped Baseboard Joint

Contoured Fit The end of a coped baseboard **(A)** fits against the face of the baseboard already installed **(B)**.

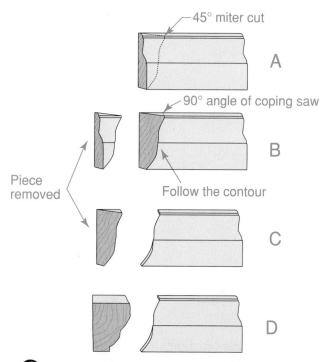

- 45° miter cut — A
- 90° angle of coping saw — B
- Piece removed
- Follow the contour
- C
- D

Coping a Joint
Careful Process To cope a joint, several steps are required using a miter saw and a coping saw.

1. Use a miter saw to miter the end at 45°.

2. Rest the coping saw blade against the edge of the miter cut. Hold the saw at 90° to the back of the molding and begin your cut. Then cut the molding along the inside edge left by the miter cut. As you cut, direct the saw slightly inward, away from the molding to be joined. This is known as back-cutting.

3. The end profile of the coped member should match the face of the intersecting molding. Fine-tune the fit as necessary for a tight joint.

Coping Strategy Baseboard is often coped at one end and butted into a wall or a door casing at the other end. The coped cut should be made first. There are two reasons for doing this.

1. It allows for recuts. A coped cut is more difficult than a square cut. A novice may have to cut it a second time to ensure a tight fit, and leaving the baseboard long allows for this.

2. Trimming is easier. When installing the coped baseboard, it is often necessary to trim a small amount off one end so that the baseboard will fit perfectly. It is faster and easier to trim a little off a square cut than off a coped cut.

Cutting Baseboard Returns

Ideally, the outside edge of the door casing will be thicker than the baseboard. This will prevent the end grain of the baseboard from showing. However, sometimes the baseboard is thicker than the casing. To provide a finished detail in this situation, the baseboard can be *returned* where it meets the casing. There are two methods for doing this: face mitering and edge mitering.

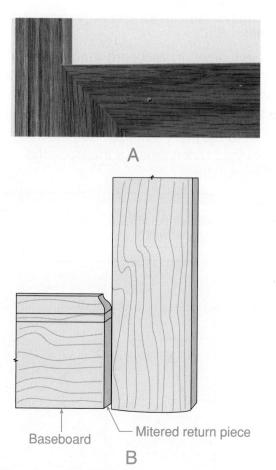

A

Baseboard — Mitered return piece

B

Mitered Returns
Two Methods A miter cut can be used where a thick baseboard meets a thinner door casing. **A.** Face mitering. **B.** Edge mitering.

The method shown in B in Mitered Returns is similar to the way the ends of a window apron are handled. Glue the small return piece into position and fasten it with brads or small finishing nails. Pre-drill the holes to avoid splitting the wood. When the face of the base shoe projects beyond the face of the door casing, the end of the base shoe can be returned in the same fashion. However, another way to handle this is to cut a reverse miter at a 45° angle. Note that the miter does not go completely through the thickness of the base shoe. Instead, a small stub should be left to fit against the casing for the sake of appearance. This method takes less time than a mitered return but exposes end grain.

Installing the Baseboard

Carefully plan the baseboard installation sequence before starting the job. Square cuts fit against wall surfaces. If the walls are unusually irregular, square cuts should be scribed to fit the wall surface (see page 766). Coped cuts fit against adjacent lengths of baseboard. Following are two methods for installing base molding, using square cuts and coped cuts.

Method 1

1. Cut and install a piece of molding to go along wall 1. It should have a square cut on each end.

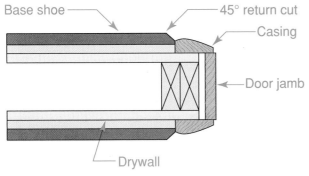

Base Shoe Detail
Partial Miter When the base shoe is thicker than the casing, a 45° return cut can be made on the base shoe.

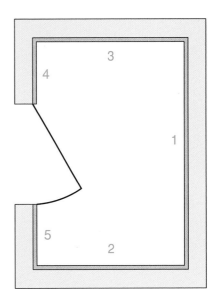

Baseboard Strategy
Sample Room A schematic view of a room with one door.

2. Cut and install the molding for wall 2. The end meeting wall 1 should be coped. The other end should be square.

3. Cut and install the molding for wall 3. The end meeting wall 1 should be coped. The other end should be square.

4. Cut and install the molding for wall 4. Cope one end to fit against wall 3. Cut the other end to fit against the door casing.

5. Cut and install the molding for wall 5. Cope the end that meets wall 2. Cut the other end to fit against the door casing.

6. Install the base shoe. The base shoe should be nailed into the baseboard itself, not into the finish floor.

JOB SAFETY

KNEE PROTECTION The work of installing baseboard can be made more comfortable with the use of kneepads. These protect a trim carpenter's knees from injury caused by prolonged contact with hard floors.

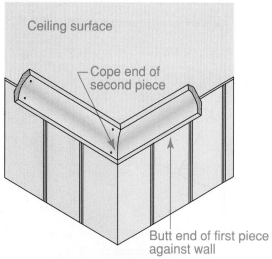

 Simple Ceiling Molding
Bridging Materials Installation of ceiling molding at an inside corner. The molding shown is a type of cove molding.

Method 2

If you are right-handed, you will find it easiest to work around the room in a counterclockwise direction.

1. Cut the first molding to fit along wall 5 between the door casing and the end wall. Make square cuts on each end.

2. Cope one end of the molding for wall 2 to fit against the molding on wall 5. Square cut the other end.

3. Cope one end of the molding for wall 1 to fit against the molding on wall 2. Square cut the other end.

4. Cope one end of the molding for wall 3 to fit against the molding for wall 1. Square cut the other end.

Builder's Tip

WALL PREPARATION Before installing large moldings, use a pencil to identify the location of ceiling joists and, if necessary, studs. This will make it easier for you to drive nails into these locations as the molding is being installed.

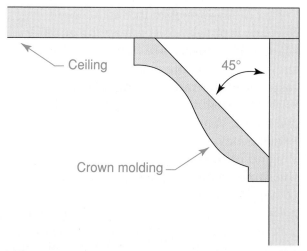

 Springing Angle
Wall Projection This crown molding has a springing angle of 45°.

5. Cope one end of the molding for wall 4 to fit against wall 3. Cut the other end to fit against the door casing.

Ceiling Molding

How is installing ceiling molding similar to installing baseboard?

Ceiling moldings are sometimes used at the junction of wall and ceiling for architectural effect. They are also used to cover any gaps between different materials on the wall and the ceiling.

Ceiling molding is cut and installed in similar fashion to baseboard, using a combination of square cuts and coped cuts. Coped cuts ensure tight joints even if the moisture content of the wood changes slightly. To secure ceiling molding, a finish nail should be driven through it and into the wall plates or studs behind the wall surface. For large moldings, a nail should be driven through the molding into each ceiling joist, if possible.

Crown Molding

Crown molding is a fairly large molding that usually includes both curved and angular surfaces. It calls for special cutting and installation techniques. This is because it is angled away from wall and ceiling surfaces, and its back is not in contact with either of them. The angle at which the molding projects away from the wall is called the **springing angle**. The springing angle is typically 38° or 45°. Moldings of this type are sometimes called *sprung* moldings.

Reading Check

Recall *How is ceiling molding similar to baseboard?*

Installing Backing It is a good idea to use a solid wood or plywood **backing** behind large moldings. Backing is a long strip of material that is nailed to the wall as support for large moldings. Nailing through molding and into backing is much easier than nailing into framing hidden behind plaster or drywall. This also provides more support for the molding. Pneumatic nailers are widely used to install crown molding because hand-nailing in awkward positions near the ceiling is difficult.

Some carpenters install backing made from 1× or 2× stock on page 772. Note the angled cuts on the edges of the backing. Another way to install backing is to attach a continuous ¾" thick plywood strip behind the molding. It should be positioned at the same springing

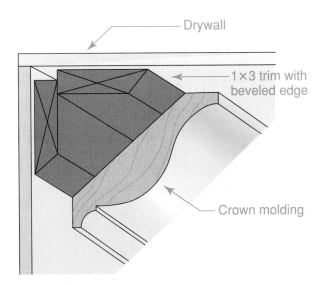

Backing for Crown Molding
Solid Support Solid wood or plywood backing should be nailed to the wall framing to support wide moldings.

Labels in figure: Drywall; 1×3 trim with beveled edge; Crown molding

Builder's Tip

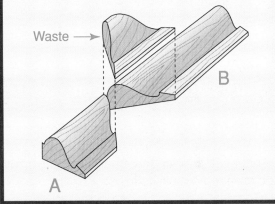

FIGURE CONTINUITY When molding and trim will be painted there is no need to worry about matching the figure of the wood where pieces intersect. However, if the wood will receive a stain or clear finish this becomes important. To install molding so that the figure appears to match on intersecting pieces, miter it as shown below. Once piece A and piece B are joined, the figure on the wood will look like it turns the corner.

Labels in figure: Waste; A; B

angle as the molding. When installing wide moldings against backing, the molding can be attached by nailing or by using trimhead screws. These screws have an unusually small head that can be countersunk below the surface of the molding.

Built-Up Crown Molding If a single piece of crown molding does not have the width desired for a wall, crown molding can be combined with other moldings, as shown on page 772. This approach is similar that used to create built-up baseboard (see page 765). Many assemblies are possible, but built-up crown usually requires some sort of backing or blocking. Often this is supplied by trim stock. The stock can be completely hidden, as on page 772, or portions of it can be exposed to form part of the visible profile of the crown. If built-up crown is specified for a room, large detail drawings will be found on the building plans. Trim carpenters may

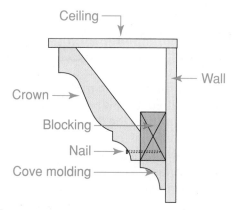

Ceiling

Wall

Crown

Blocking

Nail

Cove molding

Built-Up Crown

Combining Profiles Crown molding combined with cove molding.

even build a full-size mockup of the crown assembly as the house is being designed. This helps the client understand what it will look like, and helps the designer or estimator determine the cost of the work.

Cutting Crown Molding Crown molding can be mitered to fit an inside corner. However, such a joint may open up over time and is not recommended. It is better to use a combination of coped and square cuts similar to those used for baseboard. However, because crown molding is a sprung molding, the exact technique for cutting it differs from cutting baseboard and other types of flat moldings. There are two basic methods for cutting

---

Step-by-Step Application

Cutting Crown Molding Because of crown molding's shape, a compound-miter cut is required for a coped joint, instead of a simple miter cut as with baseboard. In a compound-miter cut, a miter and a bevel are cut simultaneously. Make a compound-miter cut on crown molding as follows:

Step 1 The springing angle of the molding determines the correct saw settings. Set a compound-miter saw for the correct miter and bevel angles as indicated in the table.

Step 2 Place the molding face up and flat on the saw table.

Step 3 After the molding is in position, make the compound-miter cut.

Step 4 Using the cut edge of the molding as a guide, cope the cut using a coping saw, as shown in the figure below. This is the technique shown on page 768.

Step 5 Test the fit against a scrap piece of crown molding. Fine-tune the fit as necessary with a file until it is tight.

Compound-Miter Saw Settings for Crown Molding*		
Type of Crown Molding	Miter Angle	Bevel (tilt of blade)
Cope on right end (Top edge of crown against saw fence)		
45°	35.3° (right)	30°
38°	31.6° (right)	33.9°
Cope on left end (Bottom edge of crown against saw fence)		
45°	35.3° (right)	30°
38°	31.6° (right)	33.9°
*Crown molding is flat on the saw table.		

crown molding. Both require a miter saw. One method calls for the molding to be positioned against the saw table and saw fence, as if these surfaces represented the ceiling and the wall. The second method, using a compound-miter saw, can be done without holding the crown upright. Instead, all cuts can be made with the molding flat on the saw table.

Other Uses of Molding

Where might built-in shelves be installed?

Moldings are used in many locations and for other purposes, such as for chair rails or shelving trim. An almost unlimited range of effects can be obtained by using or combining moldings and trim.

Chair Rail

Chair rail is a molding that runs horizontally across walls at 3' to 4' from the floor. It is often found in dining rooms, where it protects walls from damage caused by the backs of chairs. It may also serve as a transition between two different wall finishes. For example, a wall may be painted below the chair rail and wallpapered above. Chair rail can be installed in the same manner as baseboard. Inside corners may fit together in a coped joint but are sometimes mitered.

Reading Check

Explain *What is the purpose of a chair rail?*

Trimming a Clothes Closet

The baseboard in a closet is usually the same as the baseboard used in the adjoining room. Smaller moldings may be used to cover the front edge of closet shelving, especially if the shelf is made of plywood. Wood trim might also be used to support the closet shelf and clothes rod. Such trim, often a piece of 1×3 stock, may be continued around the inside of the closet to provide a solid base for attaching clothes hooks.

Photographs in the Carol M. Highsmith Archive, Library of Congress, Prints & Photographs Division, [LC-DIG-highsm-16118]

 Chair Rail
Well Trimmed Chair rail is visible on the wall to the right of the opening. Crown molding can be seen at the top of the wall bordering the ceiling.

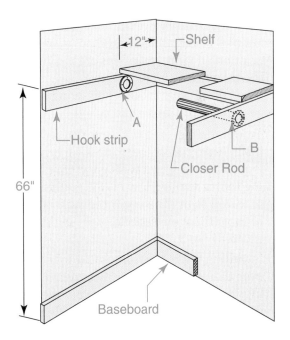

 Trimming a Closet
Function Over Form Installation details for **A** trimming a clothes closet and **B** are brackets.

Install closet trim as follows:

1. Cut the pieces of the hook strip to fit the closet. As you nail the hook strip into place, use a level to ensure correct position. Finish nails should be driven into the studs, not just into the drywall.

2. Measure 12" from the back wall, and center a closet rod bracket at this point on one side of the closet. These brackets support the closet rod. They are sometimes called *rosettes,* and may be made of plastic or wood. Screw one bracket into place but leave the other off for now.

3. Cut a closet rod to length. Place one end in the bracket attached in Step 2, and place the other bracket on the loose end of the rod. Slip the assembly into place, level the rod, and attach the second bracket to the hook strip.

4. Cut a shelf to length and set it on top of the hook strip. The shelf is not usually nailed. This allows it to be removed when the closet is painted.

If a closet is not needed for clothes, it can be put to other uses by adding shelving and table space, such as for a small home office. The doors can be closed when necessary to conceal office clutter.

Estimating and Planning

Molding and Trim

CERTIFICATION PREP

This estimating and planning exercise will prepare you for national competitive events with organizations such as SkillsUSA and the Home Builder's Institute.

Estimating Guidelines

Molding and trim are generally estimated by the lineal foot. However, they may sometimes be estimated based on board-foot calculations if large quantities are required.

The time needed to install molding depends on several factors. Following are some of the more important ones:

- Top-grade molding takes more time to install than paint-grade molding because joints must be fitted with great care. Molding that is applied low on the wall is easier to install than the same molding installed high on the wall, especially at heights greater than 8'.

- Installing wide crown molding takes more time than installing narrow crown molding. Narrow crown molding can often be nailed directly to the wall, but wide molding requires backing.

- Hardwood moldings are more difficult to install than softwood moldings because nail holes often must be pre-drilled.

- The style of trim also has a bearing on the installation time. For example, installing door and window casing with miter joints is more time consuming than installing trim that is butt jointed.

Trim carpenters sometimes base their rates on a per-lineal-foot figure. They may also charge on a per-window, per-door, or per-room basis.

In any case, measurements can be developed by studying the building plans. For example, a trim carpenter could review the floor plans to get a lineal-foot measurement for baseboard and ceiling molding. The carpenter could also check the interior elevation drawings or the window schedule to determine the lineal feet required for windows. Then he or she could review the finish schedule to find out exactly what types of woods and finishes have been specified.

Estimating on the Job

A living room has the following features: It measures 14' × 15'-6", has one 36" wide door and two 32" wide by 54" tall windows. How many lineal feet of baseboard and crown molding will be required? Add 10% to your figure to account for waste, and round the answer up to the nearest even number.

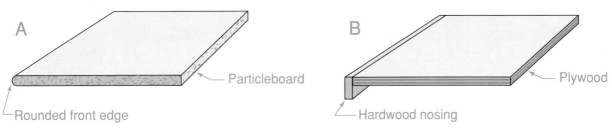

A — Particleboard
Rounded front edge

B — Plywood
Hardwood nosing

 Built-in Shelving
Adjustable Storage These two types of solid shelving are common in closets and built-in cabinets. **A.** Particleboard is inexpensive and installs quickly. **B.** Plywood requires a wood nosing to conceal the edge plies. This also stiffens the shelf.

Built-in Shelves

Closets, linen cabinets, and storage cabinets often require built-in shelves. These are adjustable shelves that are made to fit a specific space in the house. Built-in shelves are typically made of plywood, MDF, or particleboard. The front edge can simply be rounded over or it can be covered with a hardwood nosing. The shelves are typically supported by two or more wall standards. A **wall standard** is a perforated metal strip that can be screwed to a wall or to the inside of a cabinet. Metal shelf supports or small pegs fit into holes in the standard in order to support shelves at various heights. Wall standards are sometimes called *shelf track standards* or *adjustable shelf standards*.

Section 26.3 Assessment

After You Read: Self-Check

1. How should base shoe molding be nailed?
2. Name a technique carpenters use to mark a length of baseboard when fitting a joint.
3. When should baseboard be scribed to a wall?
4. What is crown molding?

Academic Integration: Mathematics

5. **Perimeter** A dining room measures 13' × 15', has one 42" × 84" door, and two 36" wide × 48" tall windows. How many lineal feet of baseboard, crown molding, and door and window casing will be required? Add 10% to your figures to account for waste. Round up to the nearest even number.

Math Concept The *perimeter* of a space or figure is the distance around the edge of that space or figure.

Step 1: Calculate the perimeters of the room, the door, and the windows.

Step 2: To calculate the crown molding needed, use the perimeter of the room plus 10%.

Step 3: To calculate the baseboard needed, subtract the width of the door from the perimeter of the room and add 10%.

Step 4: To calculate the door and window casing needed, add the perimeters of the door and windows, then subtract the width of the door before adding 10%.

Go to connectED.mcgraw-hill.com to check your answers.

Section 26.1
Chapter Summary

In wood construction, molding and trim are used both as decoration and for practical purposes, such as to protect walls. They are available in many patterns and shapes. Most are made of solid wood, but some are made from short pieces joined together or from synthetic materials. Cost varies depending on style and type of wood. Curved molding and trim can be made in various ways.

Section 26.2
Casing refers to all the trim around doors or windows. Square cuts and miter cuts are commonly used for joints. Window trim may consist of the casing alone or of the casing plus a stool and an apron. Shutters may also be added.

Section 26.3
Baseboard moldings may consist of a base, a small cap, and a shoe. Walls that are not plumb may make scribing a joint necessary in order to achieve a tight fit. Joints are coped when one member is trimmed to fit against the face of another. Cutting a return is done to create a finished look when one member is thicker than another. Molding may also be placed along the ceiling, used for a chair rail, or used to trim a closet.

Review Content Vocabulary and Academic Vocabulary

1. Use each of these content vocabulary and academic vocabulary words in a sentence or diagram.

Content Vocabulary

- molding (p. 751)
- trim (p. 751)
- casing (p. 758)
- side casing (p. 758)
- head casing (p. 758)
- reveal (p. 759)
- return (p. 762)

- baseboard (p. 764)
- coping (p. 767)
- crown molding (p. 770)
- springing angle (p. 771)
- backing (p. 771)
- wall standard (p. 775)

Academic Vocabulary

- reinforces (p. 751)
- absorbs (p. 759)
- accurate (p. 765)

Speak Like a Pro

Technical Terms

2. Work with a classmate to define the following terms used in the chapter: *S4S* (p. 751), *trim out* (p. 751), *mop board* (p. 751), *rosettes* (p. 760), *stool* (p. 760), *apron* (p. 760), *picture framing* (p. 762), *sprung moldings* (p. 764), *base shoe* (p. 764), *scribing* (p. 766), *preacher* (p. 766), *return* (p. 768), *chair rail* (p. 773).

Review Key Concepts

3. Describe two non-decorative uses for molding and trim.

4. Identify the two types of cuts made for window and door casings.

5. Describe three types of molding and trim.

6. List the steps involved in scribing molding.

7. Describe how to trim to an uneven surface.

8. Identify where coping joints are used.

Critical Thinking

9. Synthesize What problems may arise if carpeting were installed before baseboard?

Academic and Workplace Applications

STEM Mathematics

10. Identifying Operations in Word Problems Specifications for the remodeling of an apartment call for 300 lineal feet of window and door casing. The owners of the apartment are trying to decide if they want painted trim using paint grade casing at $0.77 per lineal foot, or stainable red oak trim at $1.26 per lineal foot. Calculate the difference between the total cost of each type of trim. Round up to the nearest $0.01.

Math Concept The word *difference* in a math story problem refers to the operation of subtraction.

Step 1: Calculate the cost of each type of trim by multiplying the price by the total number of lineal feet.

Step 2: Subtract the smaller number from the larger number to obtain the difference.

STEM Engineering

11. Creating a Schematic Locate a room within a structure you know where a chair railing could be installed. Write out a plan for installing the chair rail. Create a sketch of the room, including all measurements, materials, and procedures necessary to complete the installation.

21st Century Skills

12. Information Literary Skills Wood molding and trim details have a long history of use in houses. Research the following American architectural styles: Greek Revival and Craftsman Style. Identify the types of molding and trim used around windows and doors in these styles. In addition, compare and contrast the types of baseboard and ceiling trim used in each type of architecture. Record your findings in a one-page report. Include simple sketches of relevant types of trim.

Standardized TEST Practice

Multiple Choice

Directions Choose the phrase that best answers the following questions.

13. What are two species of wood that are characteristically smooth, close-grained, and free from pitch streaks?

 a. Sitka spruce and northern pine
 b. southern pine and yellow poplar
 c. northern pine and yellow poplar
 d. oak and northern pine

14. What is the name for the horizontal member that laps the window sill and extends beyond the casing?

 a. apron
 b. stool
 c. casing
 d. molding

15. Which type of cut ensures tight joints even if the moisture content of the wood changes slightly?

 a. coped cut
 b. miter cut
 c. crown cut
 d. all of the above

TEST-TAKING TIP

If the possible answers to a multiple-choice question have an "all of the above" option and you know that at least two of the choices are correct, select "all of the above."

*These questions will help you practice for national certification assessment.

CHAPTER 27

Cabinets & Countertops

Section 27.1
Planning for Cabinets

Section 27.2
Choosing & Installing Cabinets

Section 27.3
Countertops

Chapter Objectives

After completing this chapter, you will be able to:

- **Identify** the five basic kitchen layouts.
- **Describe** the difference between frameless and face-frame cabinet construction.
- **Explain** the process for installing a base cabinet.
- **Explain** the process for installing a wall cabinet.
- **Demonstrate** how to install postformed countertop.
- **Apply** plastic laminate to a surface.

Discuss the Photo
Cabinetry Cabinets make a house more livable by providing storage space as well as support for work surfaces. *Which rooms in your home have cabinets? How are these cabinets used?*

Writing Activity: Summarize Information
Find out more about kitchen and bathroom cabinets. Arrange an interview with a local cabinet maker, carpenter, or residential remodeling specialist. Prepare a list of questions you will ask in the interview. Make notes during the interview. Summarize your findings in a one-page document.

Chapter 27 Reading Guide

Before You Read Preview

Cabinets make a house more livable by providing storage as well as support for work surfaces. They are available in a wide variety of styles and in several basic configurations. Choose a content vocabulary or academic vocabulary word that is new to you. When you find it in the text, write down the definition.

Content Vocabulary

- wall cabinets
- base cabinets
- universal design
- work triangle
- stock cabinets
- semi-custom cabinets
- custom cabinets
- carcase
- face-frame cabinet
- frameless cabinet
- substrate

Academic Vocabulary

You will find these words in your reading and on your tests. Use the academic vocabulary glossary to look up their definitions if necessary.

- design
- increments
- bond

Graphic Organizer

As you read, use a chart like the one shown to organize information about the five basic layouts commonly used in kitchen design.

Type of Layout	Description
U-Shape	
L-Shape	
Parallel Wall	
Side Wall	
Island	

Go to connectED.mcgraw-hill.com to download this graphic organizer.

Planning for Cabinets

Planning for Kitchens & Baths

What is universal design?

The kitchen of a house usually contains more cabinetry than any other room. However, cabinets (sometimes called *casework*) are found also in bathrooms, laundry rooms, and family or recreation rooms.

Cabinetry is installed just before interior trim, or sometimes at the same time. In the past, most cabinets were built on site by finish carpenters. This is rarely done today. Instead, cabinets are either custom built in small cabinet shops or produced by regional or national manufacturers.

Because cabinets are an essential part of kitchens and baths, planning these rooms is largely a matter of planning and placing the cabinets in these rooms.

The Planning Process

The basic arrangement of cabinets in any room is shown on the building plans (see Chapter 2). However, these plans usually provide only the location of cabinets, appliances, and related plumbing. The choice of specific cabinets is made at a later date by the builder or the client. A professional kitchen or bathroom designer might review the plans at this stage and make recommendations. The designer usually develops computer-generated renderings showing

 Kitchen Cabinetry
Kitchen Planning The kitchen is the room that has the largest number of cabinets. This is part of the reason a kitchen is often the most expensive room to build in a house.

Robin Matthews/Ingram Publishing

exactly how the kitchen would look with a particular style or brand of cabinetry. Many home centers and cabinet suppliers can also do this. Cabinet planning software can also show the stock numbers of each cabinet on a floor plan. This improves ordering accuracy.

Once this phase of the planning is complete, the cabinets can be ordered. This should be done well before construction of the house is complete. Depending on the style and complexity of the cabinetry and how busy the manufacturer is, it can take as little as six weeks or as long as six months to receive cabinets once they have been ordered.

Planning Kitchens

Kitchen **design** concepts have changed a great deal over the years. At one time, kitchens were actually in small buildings separate from the main house. At other times, the kitchen area was arranged around a central chimney, which served both for cooking and as a heat source. Today's kitchen is often combined with the family room to create a center for everyday living or informal entertaining. Kitchens are more beautiful, functional, and efficient than ever before.

Kitchens generally feature two basic cabinet types. **Wall cabinets**, also called *upper cabinets*, hang on a wall. **Base cabinets**, often called *lower cabinets*, rest on the floor and support the countertops.

Universal Design For many years, the design of kitchens was based on the assumption that the kitchens would only be used by able-bodied adults. Cabinet heights and layouts were designed accordingly. However, kitchens are used by individuals of many different needs and abilities. Increasingly, kitchen designers are using the concept of universal design as the foundation for kitchen design. **Universal design** is design that aims at making a house usable and safe for the widest variety of people, including elderly adults, children, teens, and individuals with disabilities. Research indicates that by the year 2020, more than 20 percent of the population in the United States will be

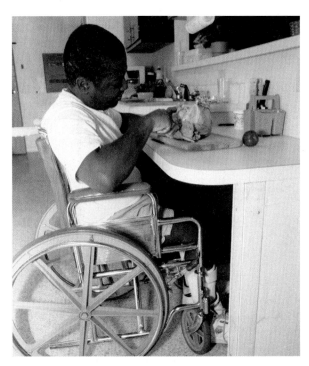

 Universal Design in a Kitchen
Design for All Universal design makes living spaces suited to people of varying physical abilities. *What kitchen features might be good for a person who uses a wheelchair?*

65 years of age or older. A properly designed kitchen, increases the ability of older adults to live independently. One characteristic of a universal design kitchen is that it features countertops set at various heights.

Kitchen Layouts

Five basic layouts are commonly used in kitchen design, as shown on page 782 (*REF* stands for refrigerator).

U Shape The U-shaped kitchen, with the sink at the bottom of the U and the range and refrigerator on opposite sides, is very efficient.

L Shape The L-shaped kitchen locates the sink and range on one leg of the L and the refrigerator on the other. Sometimes the dining space is located in the corner opposite the L.

Parallel Wall The parallel-wall kitchen is often found where there is limited space. The parallel-wall kitchen is sometimes called a *two-wall galley kitchen.* This type of

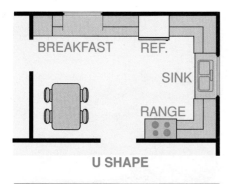

U SHAPE

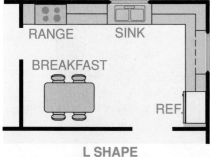

L SHAPE

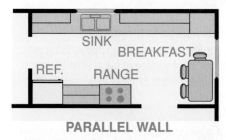

PARALLEL WALL

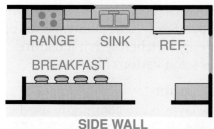

SIDE WALL

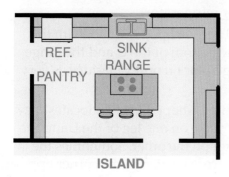

ISLAND

Common Kitchen Layouts
Most Popular These plans show the most popular kitchen layouts used in residential construction.

kitchen can be quite efficient with the proper arrangement of sink, range, and refrigerator.

Side Wall The side-wall kitchen is also called a *galley kitchen.* It is usually found in small apartments. The cabinets, sink, range, and refrigerator are all located on one wall. This type of kitchen usually has limited counter space.

Island The island kitchen features a cabinet "island" that is separate from the main cabinetry in the rest of the room. A range is usually placed in the island, along with storage for pans. This layout sometimes makes it difficult to provide a ventilating fan for the range. To solve this problem, some ranges have built-in downdraft fans that exhaust air outdoors through ducts in the floor.

Reading Check

Synthesize *Which type of kitchen layout would you recommend for a small apartment?*

Work Centers

Designers try to arrange a kitchen in a way that allows a meal to be prepared efficiently. One planning principle that leads to an efficient kitchen is called the *work triangle.* A **work triangle** represents the shortest walking distance between the refrigerator, the primary cooking surface, and the sink. The three sides of the triangle should add up to no more than 26'. A triangle with 15' to 22' is desirable, with 12' being the absolute minimum. No leg of the triangle should be shorter than 4' or longer than 9'.

Each point of the work triangle is associated with related cabinetry and countertop space that forms a work center. All equipment, storage space, and surface work areas for each activity should be located in the respective work centers. These work centers are:

Food Storage Center This is located around the refrigerator and specialized storage cabinets such as a pantry.

Cooking Center This is located around the primary cooking surface.

Cleanup Preparation Center Ideally, the cleanup/prep center (sink and dishwasher, — D.W. in The Work Triangle) should be located between the food storage center and the cooking center.

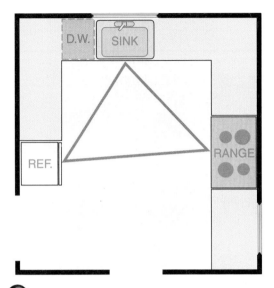

The Work Triangle
Efficient Arrangement In a work triangle, the sink or a major appliance is the focal point of each work center. The three sides of the triangle should add up to no more than 26'.

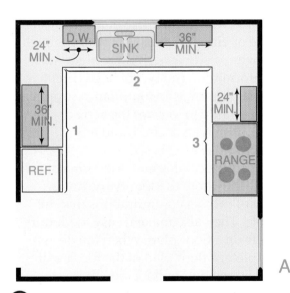

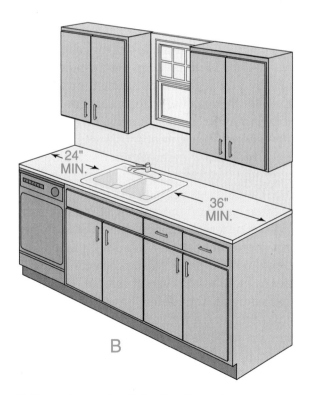

A   B

Work Centers
Minimums A. The three basic work centers: 1. Food storage. 2. Cleanup/preparation. 3. Cooking. The minimum counter space needed for each area is shown. **B.** A rendering of the cleanup/prep center is shown.

Kitchen Cabinet Dimensions

Ample storage in a kitchen is a necessity. Though the amount of cabinet space often relates to the size of the home, some basic guidelines are available. Even a small kitchen should have at least 158" of base cabinet frontage and 144" of wall cabinet frontage. This will ensure a sufficient amount of storage capacity.

It is essential to place the cabinets, countertops, and shelves at heights designed for efficiency, convenience, and comfort. Base and wall cabinets are usually installed at standard heights and depths, as shown on page 784. Clearances for wall cabinets over appliances and work centers must also be considered in the planning process.

Wall cabinets vary in height. Depending on the type of installation, they may be from 12" to 42" high. Wall cabinets are usually 12" deep and are often located beneath a soffit. A *soffit* is an area around the perimeter of a room that is lower than the rest of the ceiling. Kitchen soffits

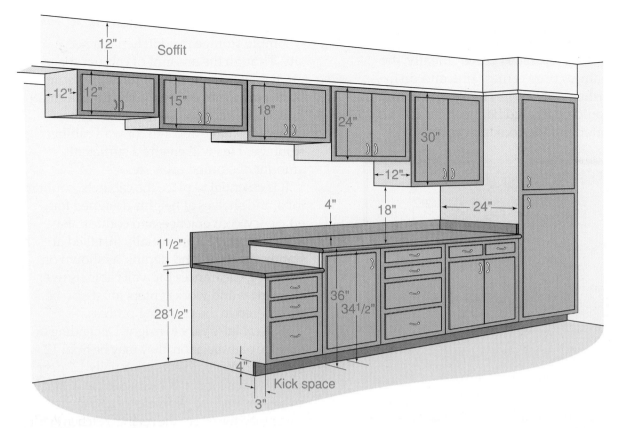

 Standard Dimensions
Cabinet Possibilities Basic dimensions for standard cabinets. Upper cabinets can come in a wide variety of heights.

are usually 12" below the rest of the ceiling. They may be 14" to 28" deep. Base cabinets are typically 34½" high, not including the countertop. They are usually 24" deep. Custom-built cabinets may have other dimensions.

When planning cabinetry, the designer typically consults manufacturer's catalogs to determine the dimensions they have available. The catalogs provide stock numbers that are often based on the cabinet's dimensions.

When ordering cabinets, be sure to use the correct product number listed with each illustration in the catalog. Most product numbers refer to the size and type of cabinet. The first letters indicate the cabinet type. For example, W would mean a wall cabinet. The first two numbers indicate the width in inches. The second pair of numbers, if any, indicates the cabinet height. For example, in Catalog Codes on page 785, the designation W-3012 would indicate a wall cabinet (W) that is 30" wide and 12" tall.

For single-door base cabinets, always indicate whether the door is to be hinged on the right or the left side. Provide the manufacturer with the size of any sink and the openings needed for built-ins such as an oven, dishwasher, and refrigerator. This information can be obtained directly from the plumbing and appliance suppliers. Finally, be sure to include the style of the cabinet, the finish desired, and any accessories.

The cabinet catalog codes are important because they help to identify the huge variety of cabinet configurations that are possible. They also make it easy to identify cabinets on a floor plan. When the design is complete, a floor plan of the room will be developed and will include these stock numbers. The plan will also determine the exact location of upper and lower cabinets. Drawings showing cabinet details are often done at a scale of ½" = 1'-0". They may also

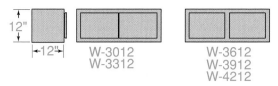

12" WALL CABINETS

12"
←12"→
W-3012
W-3312
W-3612
W-3912
W-4212

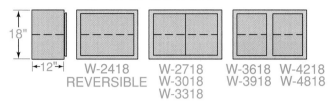

18" WALL CABINETS

18"
←12"→
W-2418
REVERSIBLE
W-2718
W-3018
W-3318
W-3618
W-3918
W-4218
W-4818

 Catalog Codes
Identification System A manufacturer's catalog shows kitchen cabinet stock numbers in a way that indicates their dimensions.

show special features of the kitchen, such as a valance over the kitchen window.

The standard method for showing wall and base cabinets on the same floor plan drawing is shown in the Cabinet Floor Plan. For example, look at the left side of the floor plan. Cabinet W3618 is a wall cabinet that is above cabinet SFRF36, a base cabinet. The base cabinet goes all the way to the wall.

Planning Bathrooms

Bathroom cabinets are sometimes referred to as *vanity cabinets.* Much less flexibility exists for placing cabinetry in bathrooms.

This is because the location of plumbing, drains, vents, and a tub or shower determines the cabinet layout. Also, bathrooms are much smaller than kitchens, and the need for storage is much less. Though large bathrooms may include upper cabinets, most feature only lower cabinets.

Bathroom Cabinet Dimensions Bathroom cabinets are planned and chosen in the same fashion as kitchen cabinets. The main difference between kitchen cabinets and bathroom cabinets is that bathroom cabinets tend to be smaller in size. Also, wall cabinets are not as common in bathrooms. Base cabinets are usually 30" high and 21" deep.

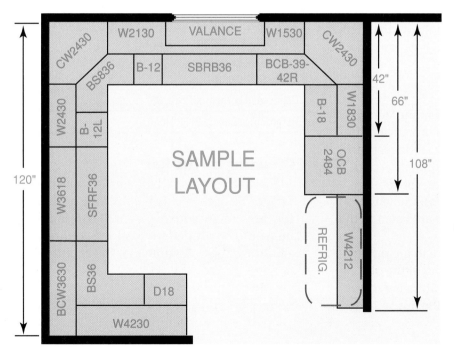

 Cabinet Floor Plan
Identity and Location
The cabinets for this kitchen layout are identified by the manufacturer's stock numbers.

When planning bathroom cabinets, it is important to note which cabinet will be the one that supports the sink. This is because the sink and the drain pipes beneath it take up a considerable amount of space in bathroom cabinets. Special cabinets for the location under the sink are typically fitted with false drawer fronts. This conceals the fact that there is no room under the sink for a drawer to slide into the cabinet. The drawer front is secured permanently in place.

Estimating and Planning

This estimating and planning exercise will prepare you for national competitive events with organizations such as SkillsUSA and the Home Builder's Institute.

Cabinetry

Materials

Approximate costs for wall and base cabinetry can be obtained from lineal-foot measurements taken from the plans. These represent the length of the cabinetry as measured at its front edge. However, these figures are used only for general planning purposes. A more accurate estimate must be made when the style and grade of cabinetry have been selected.

The precise costs for manufactured cabinets can be found on the manufacturer's current price lists. If higher-grade hardware will be used, make sure to include the additional cost. Site-built cabinets require a complete bill of materials with prices for each individual item, including hardware, glass, shelves, and any special trim.

Labor

The time required to install manufactured cabinets varies with the room layout and the cabinet type. An approximate labor cost can be determined by adding the times needed for installation and multiplying the total by the local hourly rate. The approximate times for installing various cabinets are listed in the table in the next column.

On remodeling jobs, be sure to include the cost of removing and disposing of the old cabinets. Sometimes the old cabinets can be reused in some other part of the house.

Approximate Installation Times for Factory-Built Cabinets	
Type of Cabinet	**Time (hours)**
Base cabinet containing one door and one drawer	½
Base cabinet containing two doors and two drawers	¾
Base corner cabinet	1
Broom closet	1
Drawer cabinet with four drawers	½
Oven cabinet	1¼
Sink cabinet	1½
Wall cabinet with two doors (refrigerator cabinet)	½
Wall cabinet with two doors (standard height)	½
China case corner unit with 36" front	2
Bathroom vanity up to 84" long	2

Estimating on the Job

Refer to the L-shaped kitchen layout shown on page 782. If the kitchen is 18' long and 12' wide, how many square feet of plastic laminate would be needed to cover the countertop excluding edges? Assume that the countertop is 2' wide.

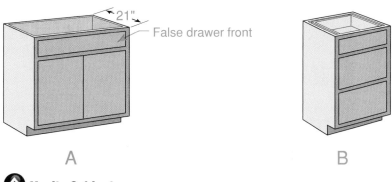

21"

False drawer front

A

B

Vanity Cabinet
Room for a Sink A. This base cabinet has an open top for placement of a sink. Compare this to the base cabinet shown in **B**, which has drawers.

Section 27.1 Assessment

After You Read: Self-Check

1. What information about cabinetry can be found on the building plans?
2. What is universal design?
3. Name the five basic kitchen layouts.
4. Why is there less flexibility for placing cabinetry in bathrooms?

Academic Integration: Mathematics

5. **Calculate Cabinet Size** An important part of a carpenter's job is being able to accurately calculate actual size from the scale of a floor plan. Assume two equal-sized cabinets are put together to make an island in a kitchen. If the scale used on the plans is ½ in. = 12 in., what are the actual dimensions of the island if a cabinet is shown as 1½ in. wide and 2 in. long on the floor plan?

   Math Concept To calculate actual width or length from scale dimensions, set up and solve a proportion. A proportion is an equation that sets two ratios equal to each other.

   Step 1: Let w equal the actual width of a cabinet. Think, "½ is to 12 as 1½ is to w."

   Step 2: Multiply each element of the proportion by 2 to eliminate fractions.

   Step 3: Solve for w, the actual width, using cross multiplication.

   Step 4: Repeat the procedure to find l, the actual length of a cabinet.

   Step 5: Calculate the dimensions of two cabinets placed side-by-side.

   Go to connectED.mcgraw-hill.com to check your answers.

Anatomy of a Cabinet

What techniques might be useful for making curved cabinets?

Manufactured cabinets may be stock cabinets, semi-custom cabinets, or custom cabinets. However, these are only approximate terms, because there are many overlaps between the categories. In general, the terms indicate how much input a client has in the final size and design of the cabinet.

- **Stock cabinets** are built in standard sizes and stored in a warehouse until ordered. They are the least expensive type of cabinet. The buyer has a modest number of choices regarding finishes and styles. Stock cabinets are available in width **increments** of 3". Where these increments do not quite fit the actual dimensions of a room, small wood pieces called *filler strips* are installed to make up the difference. Because they are built ahead of time, stock cabinets can be obtained fairly quickly.

- **Semi-custom cabinets** are built only after they are ordered for a specific kitchen. A buyer has more choices about style, finish, and hardware. The buyer also may work with a designer instead of picking cabinets from a catalog. Semi-custom cabinets are generally available only in width increments of 3".

- **Custom cabinets** can be built in any width or height to fit a kitchen exactly. Almost any size, style, finish, or hardware is possible. In this respect, manufactured custom cabinets are much like those made by local cabinetmakers. They are usually the most expensive type of manufactured cabinets and usually take the longest to arrive.

Cabinet Types

The **carcase** is an assembly of panels that forms a cabinet's basic shape. It is often made of plywood. It may also be made of particleboard or medium-density fiberboard (MDF) covered with wood veneer or plastic laminate. The carcase is assembled with nails, staples, glue, or a combination of these. It is sometimes reinforced with wood corner blocks or stretchers.

The two basic types of cabinet construction are *face-frame* (traditional) and *frameless* (often called *European-style*). Both types can be built from a variety of materials, using various types of joinery. Both types can be built in a factory as stock, semi-custom, or custom cabinets.

A variation of frameless cabinet construction is sometimes called the *32-mm system*. The number refers to a modular dimension used to locate the position of various cabinet features, including hinges. All cabinets based on the 32-mm system are frameless, but not all frameless cabinets use the 32-mm module. The advantage of this system is that it reduces the number of different dimensions needed when building cabinets.

Face-Frame Cabinets The face-frame cabinet is the traditional type of cabinet used in the United States. In the **face-frame cabinet**, the face frame fits around the front of the carcase and provides a mounting surface for hinges and drawer hardware. A face frame is usually made of ¾" thick hardwood. The joints of a face frame may be reinforced with dowels, biscuits, or screws. The cabinet rests directly on the subfloor. A hanger rail at the back is used to secure the cabinet to the wall. Base cabinets typically have one hanger rail. Wall cabinets have two hanger rails: one at the top of the carcase and one near the bottom.

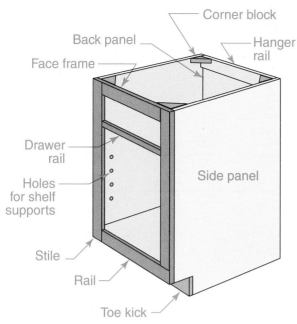

Face-Frame Cabinet
Traditional Parts of a typical face-frame cabinet.

Frameless Cabinets A **frameless cabinet** does not have a wood frame around the opening. Hinges are concealed and mounted on the side walls. This requires a special type

of hinge called a *cup hinge* (see page 792). Frameless cabinets do not rest directly on the subfloor. Instead, they are supported either on leveling feet or on a wood base frame. The base frame is often made of ¾" plywood. This is a good way to use narrow scraps that are often left after larger pieces are cut from a plywood sheet.

Cabinet Drawers

The drawers of any type of cabinet can be assembled from plywood, particleboard, MDF, or from a hardwood such as beech. The bottom panel of a drawer fits into a groove cut into the drawer sides, front, and back. The material and the methods used to assemble a drawer have a great impact on its strength and durability.

The drawer front is usually a different type of wood than the body of the drawer. Joinery details for drawers are shown on page 790. All of the methods shown are variations of a three-sided drawer box. In this method, the drawer box (back and two sides) is attached directly to the drawer front. In four-sided drawer box construction,

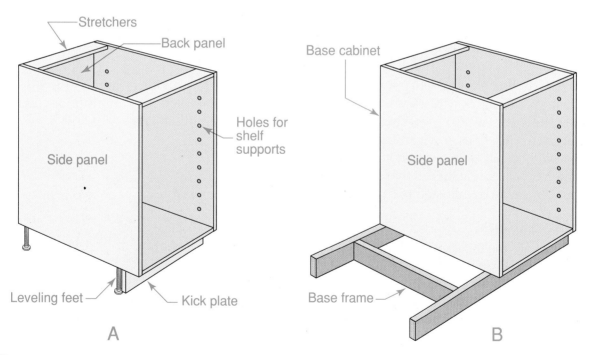

Frameless Cabinet
European-Style The base unit of a frameless cabinet rests on leveling feet (A) or on a wood frame (B).

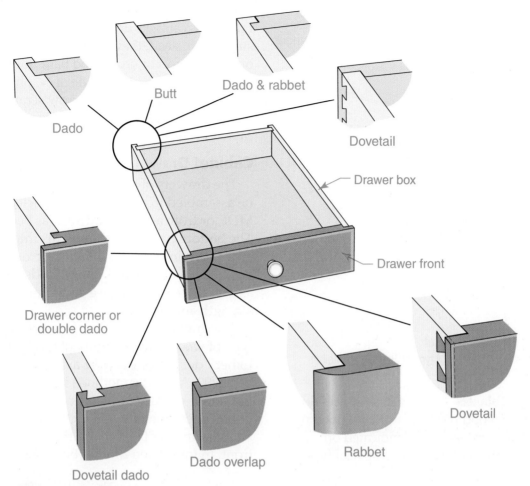

Dado

Butt

Dado & rabbet

Dovetail

Drawer box

Drawer front

Drawer corner or double dado

Dovetail dado

Dado overlap

Rabbet

Dovetail

Drawer Joinery
Many Options Common types of joinery used in three-sided drawer construction.

the drawer box has four sides (back, sub-front, two sides). The drawer front is attached to the sub-front with screws driven from inside the drawer. The disadvantage of four-sided construction is that it uses more materials. The advantage is that the drawer front can easily be adjusted to fit the cabinet.

Cabinet Doors

Doors are the most visible part of a cabinet. They can be made of solid wood, plywood, MDF, or particleboard. The type, style, and finish of the doors affect the cabinet's appearance. There are two basic types of doors:

Inset Doors *Inset doors*, also called *flush doors*, fit entirely within the door opening. A small gap is required between the door and the

face frame to provide clearance. Inset doors require tight tolerances and thus are more time-consuming to install.

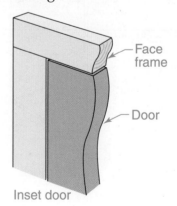

Face frame

Door

Inset door

Inset Door
Close Tolerances An inset door fits entirely within the opening formed by the face frame. The front surfaces are flush.

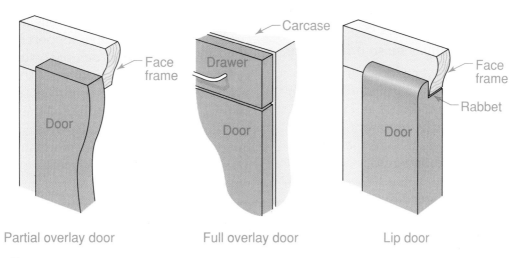

Partial overlay door   Full overlay door   Lip door

Overlay Doors

Easy to Fit The various types of overlay doors overlap the face frame or the carcase.

Overlay Doors *Overlay doors* fit over the edge of the carcase of a frameless cabinet. If the cabinet is a face-frame design, the doors fit over the face frame. Overlay doors may be constructed in several ways.

Door Panels Doors may have flat or raised panels. Both types can be used as inset doors or overlay doors. Both can be used on face-frame or frameless cabinets. Raised-panel doors have a more traditional look, but the added labor and material increase the price. Flat-panel doors are economical and easy to keep clean.

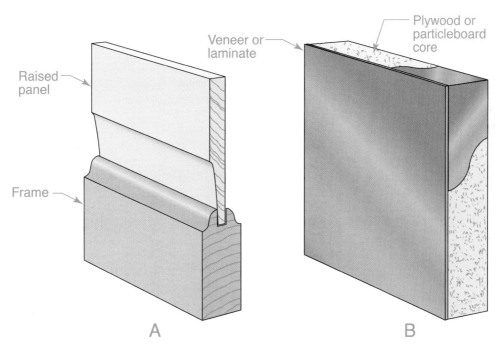

A   B

Door Panels

Two Options Details of panel door construction. **A.** A raised-panel door is built from solid wood. The panel is held in place by the frame. **B.** A flat-panel door is often covered on all sides and edges with plastic laminate.

Cabinet Hardware

Why must cup hinges be so adjustable?

To perform well over a long period of time, any cabinet must be fitted with quality hardware. Good door hinges and drawer guides are particularly important, but quality and cost vary widely. Some manufacturers offer several different grades of hardware for use with their cabinets. Others offer only one grade of hardware with each grade of cabinet. Makers of custom-built cabinets offer unlimited hardware choices. Most cabinets may be made more useful with the addition of accessories. These include spice racks, drawer organizers, and slide-out storage trays.

Drawer Guides

Drawers may be mounted on one or two guides, which are sometimes called *slides*. A single, center-mounted guide can be located beneath the drawer along its centerline. It can be used only on cabinetry that has a face frame because the front of the guide needs support. It is generally attached to the inner surface of the face frame. To keep the drawer from tipping to either side, small rollers or plastic guides can be attached to the face frame on the sides of the drawer opening. In manufactured cabinetry the center guide is usually made of metal.

Side-mounted guides are stronger than center guides because they support both sides of the drawer. They can be used on either face-frame or frameless cabinets. Full-extension guides are the most useful because they allow the drawer to be pulled all the way out for full access to the back. However, three-quarter extension slides are more common. The sliding mechanism may feature nylon wheels or ball bearings.

Reading Check

Summarize *Why are full-extension drawer guides considered to be the most useful?*

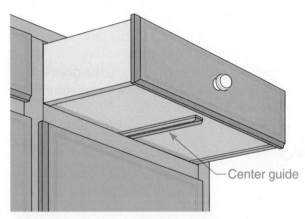

—Center guide

Center-Mounted Drawer Guide
Inexpensive A center-mounted drawer guide requires support by a face frame.

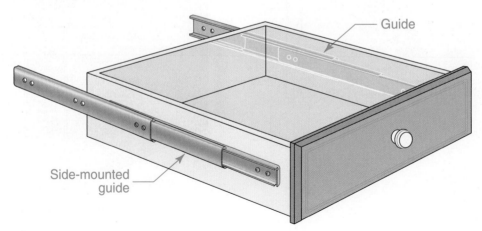

—Guide

Side-mounted guide

Side-Mounted Drawer Guides
Heavy-Duty Side-mounted guides are the best choice when drawers will be heavily loaded. This makes them well suited for kitchens.

Door Hinges

Door Hinges come in a variety of styles and designs. The three basic types are *barrel hinges, knife hinges,* and *cup hinges.*

Barrel Hinges are a very common type of hinge used on cabinetry in the United States. The hinge consists of two plates connected with a pin. One plate is screwed to the face frame and the other is screwed to the cabinet door. Some plates are L shaped and wrap around the edge of the face frame. Only the barrel of the hinge is visible on the outside of the cabinet.

Barrel Hinge
Common Style The large plate of this barrel hinge wraps around the cabinet face frame.

Knife Hinge
Door Closer A knife hinge is unobtrusive and holds a door in place against the face frame.

Knife Hinges feature two plates connected with a simple spring mechanism. When the cabinet door is closed, the knife hinge holds the door against the face frame without needing a separate latch. One plate is screwed to the face frame, while the other is screwed into a small slot cut into the door.

Cup Hinges, which are sometimes called *concealed hinges* or *European-style hinges,* were devised for use on frameless cabinetry. They have a mounting flange and a metal cup connected by a pivoting mechanism. The cup is inserted into a 35-mm diameter hole bored into the door. The mounting flange is screwed to the inside wall of the cabinet. There is no need for a face frame. Cup hinges have several advantages. They can be adjusted in several planes (up/down, side-to-side, in/out) simply by turning one or more screws on the hinge. In addition, they are quite strong and can be installed quickly. A cup hinge is not visible on the outside of the cabinet.

In the United States, the most common cabinet is one built with a face frame, partial overlay doors, and barrel hinges. In European homes, the most common cabinet is a frameless model with full overlay doors supported on cup-type hinges.

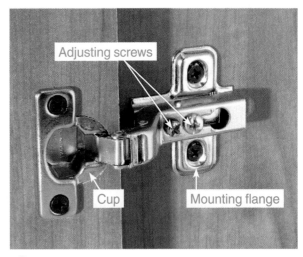

Adjusting screws

Cup

Mounting flange

Cup Hinge
Hidden Hinge Some models have a plastic cap that covers the hinge-adjusting screws.

Knobs and Pulls

Knobs and pulls for drawers and doors are sometimes installed by the cabinet manufacturer. However, this hardware is more commonly added after the cabinets are installed. This allows the client to have the greatest choice of style, color, and material.

The principles of universal design also apply to cabinet hardware. Many people find it difficult to grasp small round knobs when opening drawers and cabinet doors. Pulls are generally easier to use.

Knobs and pulls are held in place by one or two bolts inserted through the back of the drawer front or door. The installer must carefully measure the location of mounting holes. The holes for pulls may be 3", 3½", or 4" OC.

Installing Cabinets

What are recessed cabinets?

Factory-built cabinets may be installed in one of two ways. Some cabinetmakers prefer to install the base cabinets first and then the wall cabinets. Others prefer to install the wall cabinets first. This is sometimes more convenient because it allows the installer to stand close to the wall while working, rather than having to reach over or climb onto the base cabinets. With either approach, the end result must be cabinets that are plumb and level. The following methods describe how to install face frame cabinets and also apply to frameless cabinets.

Installing Base Cabinets

After the cabinets have been delivered, unpack them and check their dimensions against the plans and the original order. Store the cabinets near the room where they will be installed. Then begin the layout.

1. Use a straightedge and a laser level or standard level to locate the highest part of the floor in the area where the cabinets will be placed. This will be the starting point for the layout.

2. Measure up the wall from this high point a distance equal to the height of the base cabinet. This is usually 34½" from the floor (assuming a 1½" thick countertop). Draw a mark.

3. Use a standard level or laser level to locate additional layout marks at the same elevation around the perimeter of the installation. Connect the layout marks with a pencil line or chalk line.

4. Locate the position of studs in the area. Mark their centerlines.

5. Cabinet installation generally begins with a corner cabinet, if there is one. However, the installer may decide to begin with other cabinets depending on the layout of the room. In either case, align the top back edge of the first cabinet with the layout line. Use wood shims as necessary until the cabinet is plumb and level. If the floor and/or wall is uneven, the base of the cabinet may have to be scribed and material removed. For more on scribing, see page 767 in

⌃ **Cabinet Layout**
Mark the Height Measure upward from the high point of the floor and draw a layout mark.

Chapter 26 ("Scribing a Joint"). Set the compass to the amount needed to lower the cabinet to the layout line.

6. Screw the cabinet to the studs with 2½" or 3" wood screws. Run them through the mounting rail at the back of the cabinet. If studs cannot be reached, use hollow-wall anchors instead. Be sure to avoid any electrical wires or plumbing pipes in the wall.

7. Install the adjacent cabinets in the same way. As you install each cabinet, clamp it to the previous cabinet and fasten the two together permanently. Some carpenters run two screws through the edge of one face frame into the face frame next to it. They drill and countersink pilot holes first. Then they install slender trim-head screws that are just long enough to penetrate ¾" into the face frame. Other carpenters prefer to bolt the cases together. The bolt heads and nuts should be countersunk.

8. Where a run of cabinets must fit between walls, it is often necessary to insert a filler strip at one or both ends of the run. The filler strip will close up any gap between the cabinet and the wall. It must have the same color and finish as the cabinets. Scribe the filler strip to fit the gap. To secure the strip, screw it to the side of the last cabinet to be installed.

Frameless cabinets are often mounted on a separate continuous base. This eliminates joints in the base area and provides a stable platform for the cabinets. The base must be leveled and secured to the floor before the cabinets are installed.

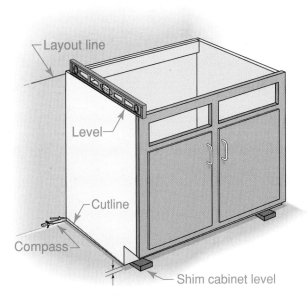

 Leveling the Cabinet
Shim or Scribe As necessary, add shims to make the cabinet level. If the floor is uneven, scribe the base as shown and cut along the scribed line using a jigsaw or circular saw.

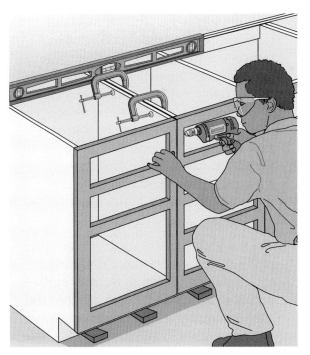

Connecting Cabinets
Level, Then Fasten Fasten the cabinets together with bolts or wood screws.

In North America, cabinets are permanently attached to the walls. In Europe, however, cabinets are often considered to be more like furniture. When a house is sold, the owner may remove the cabinetry and reinstall it in the new house.

Installing Wall Cabinets

A wall cabinet must be mounted securely so that it can bear heavy loads. It is attached to the wall with wood screws. Use at least #10 roundhead screws that are long enough to go through the ¾" back rail and the wall covering and extend at least 1" into the studs. A minimum of four screws should be used for each cabinet. Never rely on nails for hanging wall cabinets or on screws driven only into the wall covering. They will not hold securely. Also, do not use drywall screws, because the weight of the cabinet can cause them to shear off. Install wall cabinets as follows:

1. Measure up from the base cabinets (or down from the soffit) and draw a mark representing the bottom of the wall cabinet. Extend the layout line around the area where the wall cabinets will be installed. The most common distance between the countertop and the bottom of an upper cabinet is 18". Cabinets over a range require more clearance. Check local codes. Always refer to manufacturer's instructions when installing range hoods. Be sure to account for the thickness of the countertop when measuring from base cabinets.

2. Locate the positions of the studs and mark their centerlines.

3. Determine where the cabinet mounting rails will cross over the centerlines of the studs. Drill through the rails at these points using a drill bit that matches the shank diameter of the mounting screw. The screw should slip through this hole with little resistance.

4. Place a cabinet in position and brace it securely. Make certain the cabinet is tight

Builder's Tip

SUPPORTING CABINETS Commercial cabinet jacks can be used to support upper cabinets during installation. You can also make a sturdy support stand from plywood or framing lumber. Size the stand to rest on the base cabinets and support the wall cabinets at a standard height.

against the wall and the soffit. If there is no soffit, be sure the tops of the cabinets are level and aligned.

5. Drill pilot holes through the existing mounting rail holes and into the studs. Use at least two screws in the upper rail and two in the lower rail. If it is impossible to screw directly into studs,

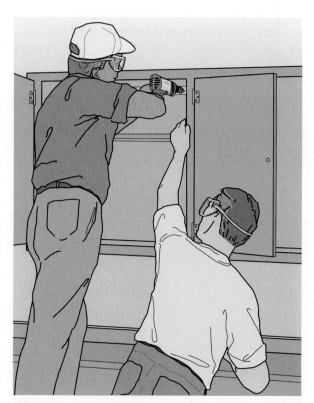

Drilling Pilot Holes
Teamwork One person should support the cabinet as another person drills the pilot holes and installs fasteners.

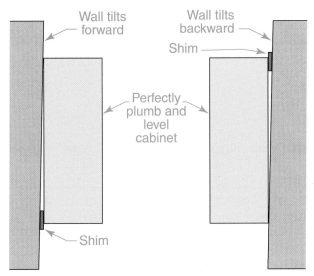

Wall tilts forward

Wall tilts backward

Shim

Perfectly plumb and level cabinet

Shim

 Shimming Wall Cabinets

Uneven Walls Scribe the cabinet to the wall, or place wood shims behind it to ensure that it remains plumb and level.

some carpenters install wood blocking in the wall. If the cabinet spans only one stud, use wood screws to fasten it to the stud and ³⁄₁₆" by 3½" toggle bolts to hold it against the wall surface.

6. Uneven walls sometimes make it difficult to obtain proper alignment and get a snug fit. If enough material is available on the back of the cabinet, the cabinet should be held in place, scribed, and cut to fit the irregular wall surface. Otherwise, it should be shimmed. Molding can be used to cover any gaps.

7. Screw the cabinet to the wall. As the installation progresses, attach cabinets to each other as in Step 7 under Installing Base Cabinets on page 795. Small cabinets can be screwed together before being lifted into place. Make certain that the joining faces of the stiles are flush and the tops and bottoms of the cabinets are aligned before drilling pilot holes.

Recessed Cabinets Small cabinets, called *recessed cabinets* or *medicine cabinets,* are

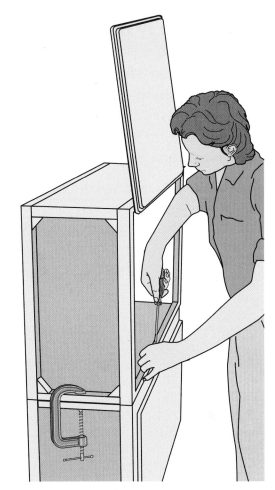

 Pre-Assembling Small Cabinets

Time Saver When two or more narrow wall cabinets are placed side by side, fasten them together on the floor and then mount them on the wall as one unit.

sometimes installed in bathrooms. These small, shallow cabinets are designed to fit in a wall cavity between studs. This makes them especially suitable for small bathrooms.

The carcase of a recessed cabinet is only 3½" to 4" deep. These cabinets typically need little in the way of support because they are attached directly to the adjacent studs, or to blocking. To install a recessed cabinet, drive screws through both sides of the carcase and into the framing or blocking. Screw heads can be concealed with plastic caps.

After You Read: Self-Check

1. List the three types of manufactured cabinets.
2. What types of materials are used for cabinet drawers?
3. What is a frameless cabinet?
4. Which type of screws should be used to attach a wall cabinet to the wall?

Academic Integration: English Language Arts

5. **Cup Hinges** Cup hinges were devised for use on frameless cabinetry. Find out more about cup hinges and the advantages they offer. Write a paragraph discussing these advantages.

 Go to connectED.mcgraw-hill.com to check your answers.

Section 27.3

Countertops

Types of Countertops

What features must a countertop surface have?

Countertops for cabinets are usually covered with plastic laminate, solid surfacing (acrylic- or polyester-based materials such as Corian, Gibraltar, or Avonite), solid stone, or ceramic tile. In modern house construction, these materials are generally installed by a subcontractor who specializes in these products. However, one type of plastic laminate countertop is often installed by carpenters after the cabinets have been installed. This is a manufactured product called a *postformed countertop*. The installation of postformed countertops will be described in this section.

Countertop Height While the limits for countertop heights range from 30" to 38", the standard height in a kitchen is 36". However, a kitchen designed according to universal design principles might contain countertops at several different heights to encourage use by anyone, including someone using a wheelchair.

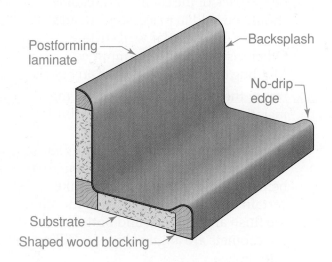

Postformed Countertop
Preassembled A postformed countertop includes a backsplash as well as a special front edge.

Preparation for Countertops

The preparation required prior to installing countertops varies according to the material. Sometimes the cabinet installer or general contractor will reinforce cabinets or provide a sub-base for the countertop material. In most cases, the countertop installer will handle this work.

Solid Stone Natural stone, such as granite, marble, limestone, and soapstone, is a popular countertop surface. The stone is cut to size, formed, and polished offsite, then delivered and installed by the fabricator. Solid stone is heavy. To ensure proper installation, cabinets must be securely anchored and the tops of the base cabinets must be perfectly flat. To make sure the stone will fit, the fabricator visits the site and makes a plywood template of the countertop area once the cabinets are in place.

Ceramic Tile Tile comes in a variety of shapes, sizes, and thicknesses, and is typically installed over cement backerboard. The backerboard must be supported by a plywood substrate. The substrate may be provided by the cabinet installers or by the tile contractor. The general contractor should coordinate their efforts.

Solid Surfacing Like solid stone, this synthetic material is usually applied directly to the tops of the cabinets. However, the cabinet installer may be asked to install wood cleats in areas to provide additional support for the material.

Plastic Laminate No particular preparation is needed prior to the installation of laminate countertops, but it is important that the tops of the cabinets are flat.

Plastic Laminate Countertops

Would universal design make postformed countertops easier or more difficult to install?

Plastic laminate, also called *high-pressure laminate,* is a durable and versatile countertop material. During manufacture, decorative surface papers soaked with melamine resins are pressed over kraft-paper core sheets soaked with phenolic resin. These

 Countertop
Bathroom Countertops
Bathroom countertops are often made of stone, tile, or plastic laminate.

papers are then bonded together at pressures of at least 1,000 pounds per square inch (psi) and at temperatures approaching 300°F (149°C). After bonding, the sheets are trimmed to size. Because its surface papers can be printed before assembly, plastic laminate comes in a large number of colors and patterns.

The backs of the sheets are sanded to improve the **bond** with the substrate. A **substrate** is a material that serves as a base for another material. A substrate is used because plastic laminate itself is quite brittle. Common substrate materials are plywood and particleboard at least ¾" thick with no defects or voids in the surface. In residential construction, the laminate is usually adhered to the substrate with contact cement.

The following three grades of laminates are used in residential construction:

General Purpose This is approximately ⅟₁₆" thick and is used for countertops.

Vertical Surface This is a somewhat thinner product designed for use on surfaces that will receive less wear and impact.

Postformed This is the thinnest grade and can be used on vertical or horizontal surfaces. It is thin enough to follow curves, as on a bullnose edge. This makes it suited for use on postformed countertop sections.

Reading Check

Recall *What is a substrate and what is it used for?*

Postformed Countertops

Postformed countertops come with both a backsplash and a countertop edge, and feature post-formed grade plastic laminate that has been attached to a substrate at the factory. Postformed countertops come ready to install. If the end of a countertop will be exposed, pre-made end pieces are available. Postformed countertops are generally available in 8', 10', and 12' lengths, and they come in a wide variety of patterns and colors.

Builder's Tip

SCRIBING COUNTERTOPS Postformed countertops must sometimes be trimmed to fit between two existing walls that are not perfectly square. To solve this problem, obtain a 2' × 3' piece of cardboard or hardboard. Use it to represent one end of the countertop. Lay it in place on the cabinets, scribe it to the wall, and cut it. Then transfer the pattern to the countertop. Repeat the procedure at the other end of the countertop. After the countertop is cut, it should fit precisely.

Plastic laminates may be sawed, routed, and drilled. Because laminate dulls tools more quickly than wood, cutting edges must be sharpened often. Dull tools may chip the laminate. Whenever possible, use carbide-tipped cutting tools when working on postformed countertops.

Installation To install postformed countertops, perform the following steps.

1. Trim the countertop on site to fit the space exactly. You can do this using files or a power tool such as a jigsaw, router, or circular saw. You may have to scribe the top edge of the countertop backsplash to fit variations in the wall.

2. Once you trim the countertop to fit, fasten it in place by driving wood screws through cleats on the underside of the cabinet and into the underside of the countertop. First, drill the correct size pilot hole. Take care not to drill entirely through the countertop. Use a wood screw of the correct length so that it does not pierce the laminate when the countertop is pulled down snug against the top of the cabinets.

3. After the top is secured to the cabinets, apply a small bead of caulking compound at the joint between the wall and the back

Table 27-1: Approximate Installation Times for Postformed Countertops

Countertop	Time (hours)
Postformed plastic laminate countertop	¼ (per lineal foot)
25" wide plastic laminate countertop with a 4" backsplash and self-edge	1 (per lineal foot)
Postformed plastic laminate mitered corner (L-shaped or U-shaped kitchen)	1
End cap on a postformed plastic laminate countertop	⅓

top edge of the backsplash. Some laminate manufacturers can supply caulk that is color-matched to the laminate.

4. Where sections of countertop meet, join them together with draw bolts and seal the joints with caulk.

Estimating Postformed countertops are sold by the lineal foot. The approximate installation times for postformed countertops are shown in **Table 27-1**.

Installing Laminate On Site

Though the application of plastic laminate to countertops is usually done off site, the contractor may encounter situations in which individual sheets of the material must be installed on a substrate on site. This might occur in remodeling projects, for example.

Plastic laminate sheet is sold by the square foot. Sheets are available in widths of 24", 30", 36", 48", and 60". The most common width is 24". The most common lengths range from 5' to 12'.

Applying the Adhesive Once the substrate has been attached to the top of the base cabinets, fill small holes and cracks with a spackling compound and then sand it flush with the surface. Vacuum the dust off all surfaces. For best results, all materials should be at room temperature (70°F [21°C] or more) before installation. The substrate should be clean, dry, and free of oil, grease, or wax.

1. Cut the laminate to approximate dimensions only. It will be trimmed to exact size after it has been adhered to a substrate.

2. Stir the adhesive thoroughly from the bottom of the can. Pour a small amount onto the back of the laminate. Spread the adhesive evenly, using a roller or a brush. A brush is recommended when applying adhesive to vertical surfaces or edges, or whenever the use of a roller is impractical. If you are using a brush, apply two coats to ensure proper coverage. Be sure to allow the adhesive adequate drying time between coats.

JOB SAFETY

USING CONTACT CEMENT When using contact cement, wear the appropriate safety equipment and follow all label precautions on the can. Low-VOC contact cement is not as hazardous as solvent-based contact cement but still requires care in handling and application. Make sure the work area is well ventilated. Use solvent-impervious gloves and a NIOSH-approved respirator if necessary.

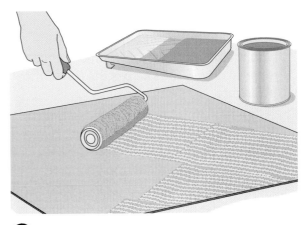

Spreading Adhesive
Even Coverage Spreading contact adhesive with a roller.

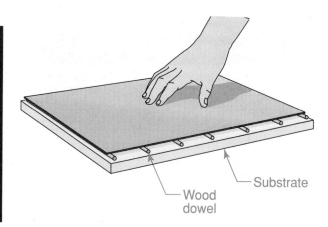
Mathematics: Calculation

Calculating Amount and Cost
Determine the amount and cost of materials for a kitchen island countertop that measures 37" wide and 47" long. The top has 1" edges around all 4 sides. The laminate costs $1.55 per square foot.

Starting Hint Find the area of the countertop, including its sides, in square feet. Then multiply times the cost per square foot.

3. Apply the adhesive to the substrate, and spread it with a brush or roller.

4. Let the adhesive dry according to instructions on the product label. Test the dryness of the adhesive by pressing a piece of paper lightly against it and pulling it away. If the adhesive sticks to the paper, more drying time is needed.

5. To make certain that you have applied enough adhesive, look across the surface into the light after the adhesive is completely dry. With most products, the surface will appear glossy. Spots that are dull after drying require additional adhesive.

Bonding and Trimming Laminate Install the laminate as soon as possible after the adhesive is dry.

1. Surfaces coated with contact cement will bond instantly when they touch. To prevent this from happening while you are positioning the laminate, place several dowel rods across the top of the substrate.

2. Align the laminate with the substrate so that an equal amount of laminate hangs over all edges. Use extreme care, because bonding is immediate upon contact.

3. Gently slip the center dowel rod out from beneath the plastic laminate, leaving the others in place. The two adhesive surfaces will come in contact with each other. Press to complete the bond.

 Positioning the Sheet
Getting Ready Use wood dowel rods to support the laminate while aligning it with the substrate. Be sure to use enough dowel rods to prevent sagging.

4. Remove the other dowel rods one at a time, working from the center toward the ends. This technique also helps to prevent air bubbles from being trapped beneath the laminate.

5. As you work, roll the surface outward from the center in all directions, using a wide, hard-rubber roller. If a roller is not available, use a block of soft wood with eased edges and corners. Place it at the center and work toward the edges, tapping sharply with a hammer. Tap or roll the entire surface to ensure a complete bond.

6. Use a router or laminate trimmer to remove the excess laminate that hangs over the edges of the substrate. A small

JOB SAFETY

TRIMMING LAMINATE When a router or laminate trimmer is used to trim plastic laminate, small sharp bits of the material are ejected from the tool at high speed. Wear suitable eye protection.

router can be fitted with a carbide trimming bit for this purpose. A type of router often used to remove excess laminate is called a *laminate trimmer*. It can be fitted with specialized attachments that make it a more versatile trimming tool than a standard router. Do not apply adhesive to other portions of the laminate until you have removed the debris from trimming.

 Laminate Trimmer
One-Handed Router A laminate trimmer is essentially a small router with a specialized base to guide the tool.

Section 27.3 Assessment

After You Read: Self-Check

1. What is the standard height of a countertop in a kitchen?
2. What is done to plastic laminate sheets to improve the bond with the substrate?
3. What makes a postformed countertop easy to install?
4. What technique is often used when installing a postformed countertop between existing walls that are not perfectly square?

Academic Integration: Mathematics

5. **Labor Cost** Write an equation you could use to find the cost of labor for installing a postformed plastic laminate countertop with one mitered corner and an end cap. The countertop measures 14 lineal feet (refer to Table 27-1 on page 801).

   Math Concept The cost for labor can be found by multiplying the rate at which the installer is paid by the time it takes to install the countertop, rounded to the nearest hour. The rate of pay can be represented using the letter r.

   Step 1: Look at Table 27-1. Calculate the time by adding three elements: the time per lineal feet multiplied by the number of lineal feet, the time required for a mitered corner, and the time required for an end cap.

   Step 2: Round the time to the nearest hour. Multiply by the rate.

   Go to connectED.mcgraw-hill.com to check your answers.

Chapter Summary

Section 27.1
The basic arrangement of cabinets for a room is shown on the building plans. Kitchen layouts include U shape, L shape, parallel wall, side wall, and island. Work centers include those for food preparation, cooking, and cleanup. Bathroom layouts are much less flexible because of plumbing requirements.

Section 27.2
Manufactured cabinets may be stock cabinets, semi-custom cabinets, or custom cabinets. The carcase is the cabinet's basic framework. The face frame provides a surface for mounting hinges and other hardware. Doors may be of the inset or overlay type. Cabinet hardware includes drawer guides, door hinges, knobs, and pulls. Cabinets must be plumb, level, and securely attached to wall studs. Several cabinets that run together must be attached to one another.

Section 27.3
Many countertops are made of plastic laminate glued to a substrate. The most common type of laminate countertop is a postformed countertop. However, conventional laminates can be purchased separately and installed on site.

Review Content Vocabulary and Academic Vocabulary

1. Use each of these content vocabulary and academic vocabulary words in a sentence or diagram.

Content Vocabulary
- wall cabinets (p. 781)
- base cabinets (p. 781)
- universal design (p. 781)
- work triangle (p. 782)
- stock cabinets (p. 788)
- semi-custom cabinets (p. 788)
- custom cabinets (p. 788)
- carcase (p. 788)
- face-frame cabinet (p. 788)
- frameless cabinet (p. 789)
- substrate (p. 800)

Academic Vocabulary
- design (p. 781)
- increments (p. 788)
- bond (p. 800)

Speak Like a Pro

Technical Terms

2. Work with a classmate to define the following terms used in the chapter: *casework* (p. 780), *two-wall galley kitchen* (p. 781), *galley kitchen* (p. 782), *vanity cabinets* (p. 785), *filler strips* (p. 788), *32-mm system* (p. 788), *inset doors* (p. 790), *flush doors* (p. 790), *overlay doors* (p. 791), *slides* (p. 792), *barrel hinges* (p. 793), *knife hinges* (p. 793), *cup hinges* (p. 793), *concealed hinges* (p. 793), *recessed cabinets* (p. 797), *postformed countertop* (p. 798), *laminate trimmer* (p. 803).

Review Key Concepts

3. Describe the five basic kitchen layouts.

4. Explain how frameless cabinet construction differs from face-frame construction.

5. Demonstrate how to install a base cabinet.

6. Demonstrate how to install a wall cabinet.

7. Practice how to install a postformed countertop.

8. Describe how plastic laminate is applied to a surface.

Critical Thinking

9. Discuss When applying adhesives, it is recommended that all materials be at room temperature. Discuss some potential difficulties that may arise if some materials are at room temperature and others are above or below room temperature.

Academic and Workplace Applications

STEM Mathematics

10. Converting Fractions and Estimating As a cabinet installer, you will be doing a number of installations for a homeowner. An important part of your job will be to estimate how long it will take you to complete the work. Figure out how long it will take you to install the following items: one sink cabinet, one base corner cabinet, one 84" bathroom vanity, and one oven cabinet. Refer to the table in the Estimating and Planning feature on page 786 to make your estimate. Then figure out how much you should charge the homeowner if your hourly rate is $25 per hour.

Math Concept To multiply a whole number by a fraction, you can convert the fraction to a decimal and multiply. Or, think of the whole number as having a denominator of 1. Multiply the numerators, then the denominators, then simplify.

Step 1: Add the total number of hours required to complete all installations.

Step 2: Multiply the total hours by the hourly rate.

STEM Engineering

11. Properties of Materials Imagine that you are building a woodworking shop at home in your garage or basement. When considering material for your countertops, you have many choices of material. Wood, granite or other stone, tile, laminate, even cement are all possibilities. Research the characteristics and attributes of your various choices and decide which would be best. Write one or two paragraphs explaining why you would choose one material in favor of others.

21st Century Skills

12. Communication Skills Assume you are remodeling a bathroom in a private home. The client has chosen base cabinets that measure 40" high and 32" deep. Through your experience as a cabinet installer, you know that these dimensions will not work in the bathroom. How would you convince the client to change his or her mind? Write a paragraph explaining why the cabinet choice is not appropriate. Tell which type of cabinet would be a better fit. Be convincing, but remember to explain your recommendations to the client in a way that will not upset or offend him or her.

Standardized TEST Practice

True/False

Directions Read each of the following statements carefully. Mark each statement as either true or false by filling in **T** or **F**.

Ⓣ Ⓕ **13.** The kitchen is the room that has the largest number of cabinets.

Ⓣ Ⓕ **14.** Custom cabinets are built in standard sizes and stored in a warehouse until ordered.

Ⓣ Ⓕ **15.** There is only one way to install factory-built cabinets.

TEST-TAKING TIP

Do not leave true/false questions unanswered if you do not know the answer. Remember that you have a 50 percent chance of getting the question right if you guess.

*These questions will help you practice for national certification assessment.

Wall Paneling

Section 28.1
Sheet Paneling

Section 28.2
Board Paneling

Chapter Objectives

After completing this chapter, you will be able to:

- **Identify** the three basic types of paneling.
- **Explain** the difference between full-height paneling and wainscoting.
- **Explain** the proper methods for storing and conditioning paneling.
- **Demonstrate** how to install sheet paneling.
- **Differentiate** between horizontal paneling and vertical paneling.
- **Estimate** the amount of sheet paneling required for a room.

Discuss the Photo
Paneling Raised paneling is made out of solid wood. *Why might raised paneling be simulated?*

Writing Activity: Identify Purpose
Before you read, you should decide on your purpose for writing. Is it to inform, persuade, inquire, or describe? Write a one-page e-mail or letter to a remodeling supply store asking them what types of wall paneling products they sell most frequently. State the purpose of your letter.

Jill Braaten/McGraw-Hill Education

Chapter 28 Reading Guide

Before You Read Preview

The texture and versatility of wood paneling make it popular as an indoor wall finish. Wood paneling can dramatically change the look of a room and can be installed over drywall, plaster, and masonry surfaces. Choose a content vocabulary or academic vocabulary word that is new to you. When you find it in the text, write down the definition.

Content Vocabulary

- full-height paneling
- wainscoting
- sticker
- box extender
- board paneling
- blind nailing

Academic Vocabulary

You will find these words in your reading and on your tests. Use the academic vocabulary glossary to look up their definitions if necessary.

- randomly
- contraction
- simulated

Graphic Organizer

As you read, use a chart like the one shown to organize information about content vocabulary words and their definitions, adding rows as needed.

Content Vocabulary	Definition
full-height paneling	paneling that runs from floor to ceiling

Go to connectED.mcgraw-hill.com to download this graphic organizer.

Sheet Paneling Basics

What trades might be affected by the installation of paneling?

Paneling made of engineered wood or solid wood offers an alternative to painted walls. It is typically applied over an existing wall surface such as drywall, so installation occurs near the end of home construction.

Paneling that runs from floor to ceiling is referred to as **full-height paneling**. Paneling that runs partway up the wall from the floor is called **wainscoting**. It is usually about 32" high.

Sheet paneling is the most common type of wall paneling. Most sheet-panel products are made of plywood, hardboard, or medium-density fiberboard (MDF). MDF paneling is sometimes referred to as *panelboard*. Sheet

 Sheet Paneling
Room Changer The installation of full-height wall paneling covers walls quickly and can have a dramatic effect on a room.

JOB SAFETY

MOVING SHEET PANELING Sheet paneling is unwieldy, and even thin sheets can be awkward to handle. Never move more than one sheet of paneling at a time unless you have help. When lifting panels, lift with your knees. To do this, lift while keeping your back straight. When carrying panels, avoid any twisting motion that may strain your lower back.

paneling is most commonly found in 4×8 sheets, but 4×3, 4×6, and 4×10 sheets are also available. Thicknesses include ⁵⁄₃₂", ³⁄₁₆", ¼", ³⁄₈", ½", ⅝", and ¾". The edges along the panel's length may be square or rabbeted. The edges along its width are square.

Sheet paneling is typically supplied prefinished. It is available in many textures and patterns, including saw-textured, relief-grain, embossed, and grooved. Plywood and medium-density fiberboard (MDF) paneling are available with a wide variety of surface veneers, including domestic and tropical hardwoods. Some paneling is finished so that it appears to be very old.

Suitable Locations

Sheet paneling made of plywood shares characteristics with other plywood products, including strength and stability (see Section 13.3). MDF panels can be damaged by moisture and high humidity. They should not be used in unheated rooms or in humid areas such as basements and bathrooms. Panels that are ⁵⁄₃₂" thick or less should always be installed over a backing

Plywood Paneling

An Alternative to Painted Walls Plywood paneling can be supplied with various surface veneers, from light to dark shades or appearing aged and distressed.

that will not burn easily, such as drywall. Thicker panels can be attached directly to studs if local building codes permit the practice. Panels 5/32" thick should never be placed over masonry. Other thicknesses can be installed over masonry according to the manufacturer's instructions.

Storage and Conditioning

Always store paneling indoors. Stack it on the floor, using stickers to separate the sheets. A **sticker** is a long, slender piece of scrap wood that separates layers of wood products and allows air circulation between them. Stacking panels in this way prevents them from warping. If you must store sheet paneling on edge, make sure that the panels rest on a long edge. Place stickers beneath the panels to raise them off the floor.

Paneling must be conditioned before installation. *Conditioning* means that the paneling has been in the room in which it will be used for a certain period of time. This allows the paneling to become accustomed to the temperature and humidity of the room. Condition sheet paneling for at least 48 hours.

Installing Sheet Paneling
What is mastic?

The methods for installing most sheet paneling products are similar. It is always wise, however, to consult the manufacturer's instructions first for any special precautions. The following instructions cover the installation of plywood paneling.

Local codes may allow the installation of thick paneling directly to studs. However, it is often recommended that any thickness of sheet paneling be applied over drywall. Drywall provides a fire-resistant base and solid support. Sheet paneling is secured to the wall with finishing nails, brads, or a combination of nails and panel mastic. *Mastic* is a thick adhesive that can be applied with a notched trowel or with a caulking gun.

Wall Preparation

Drywall must be taped and sanded before paneling is installed. Otherwise, irregularities in the surface will make installation difficult. Nail or screw dimples should be filled and the drywall should be primed.

Remodeling Paneling is often installed in remodeling projects. Remove all plates around wall switches and receptacles and save them for reuse. Although thin sheet paneling can be installed without removing existing trim, this method requires great precision. It is sometimes easier in the long run to remove window and door casings.

Builder's Tip

REMOVING MOLDING AND TRIM When removing trim that must be reused, pry it carefully away from the wall. Pull out any remaining nails through the backside of the trim, using nippers. This prevents the head of the nail from splintering the face of the trim as it exits. You can fill in the nail holes with wood putty.

Ingram Publishing

AVOIDING KICKBACK Sheet paneling is often cut using a table saw. However, thin panels are very flexible and may bend as they are being cut. This can cause the panel to pinch the saw blade, leading to a very dangerous condition called *kickback*. To prevent kickback, support thin paneling adequately across its width during the cutting operation. It must also be supported as it leaves the saw. For more on preventing kickback, see Section 5.2.

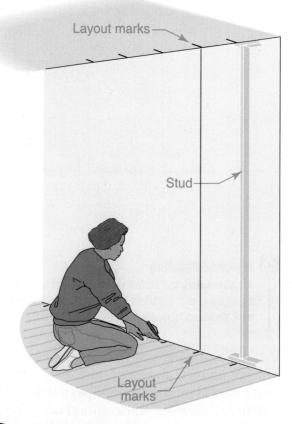

It is nearly always best to remove baseboard trim. Trim can be reinstalled unless it is damaged during removal.

New Construction Walls must be clean and flat. Sand or scrape down high spots. Locate the position of studs in the existing wall. They are generally on 16" centers. Lightly mark the stud center locations on the floor and ceiling to serve as a guide when nailing each panel into position.

Wall Preparation
Nailing Guides Mark the locations of wall studs at the ceiling and at the floor.

Reading Check

Explain *Why is it best to install paneling over a drywall surface?*

Planning the Installation

Panels are generally applied with the long dimension running vertically. Stand the panels on edge side by side around the room. Arrange them by natural color variations into a pleasing pattern.

For most rooms, it is practical to start paneling from one inside corner and then work around the room in only one direction. After you have established their order, number each panel on the back and set it aside.

If the panel has vertical grooves, their locations may appear to be **randomly** spaced. However, they will usually be spaced to align with standard 16" stud spacing. By nailing the panel through the grooves, the nail heads are hidden.

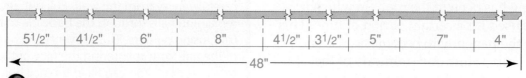

Groove Spacing
Measuring System When the dimensions are added together, the grooves fall on 16" and 24" centers. In this way, the grooves can be used to identify the location of studs.

However, for this spacing arrangement to work, the first panel in a wall must be set into place carefully. It must be plumb and the grooves must be aligned with the layout marks on the floor and ceiling.

Cutting Paneling

When cutting panels with a crosscut handsaw or table saw, place the face side (the "good" side) of the panel up. This reduces splintering. If you are using a circular saw or a saber saw, cut the panel with the face side down. The best blade to use in a circular saw or a table saw is a plywood-cutting blade (see Chapter 5).

Another way to reduce splintering is to place a strip of masking tape where you will make the cut. This is particularly helpful when crosscutting paneling. When making cutouts in a panel, such as for an electrical box, another way to reduce splintering is to score the cutting line with a utility knife. Then cut the panel on the waste side of the scored line.

Installing the First Panel

Positioning and installing the first panel determines the layout of all subsequent panels. The Step-by-Step Application on page 812 shows how to install the first panel.

Installing Additional Panels

Once the first panel is secure, position the second panel to the first. The panels should not be butted tightly together. To allow for expansion, a $\frac{1}{16}$" gap is recommended for hardboard and MDF panels. A smaller gap is recommended for plywood panels. If the first panel has been properly positioned, the edges of all the other full-width panels will also land on stud centers. This assumes that the stud spacing is uniform across the wall.

Making Cutouts

A hole must often be cut in a panel to accommodate an electrical box. It is possible to measure the distance from the floor to the box, but it is quicker and generally more accurate to use the following method.

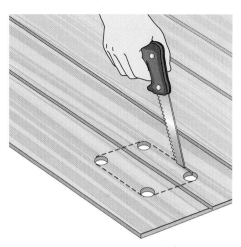

 Making Cutouts
Cut Between Holes Cutting an opening for an electrical outlet box.

1. Rub carpenter's chalk against the edges of the box. Then, hold the panel in place against it.

2. Strike the face of the panel sharply several times with the heel of your hand to transfer the box outline to the back of the panel.

3. Drill pilot holes in the corners of the marked outline. A plunge cut can also be made, eliminating the need for pilot holes (see Chapter 5).

4. Cut along the outline with a keyhole saw, or use a jigsaw fitted with a blade that reduces splintering. Special woodworking blades are available for this but many trim carpenters find that a blade intended for cutting metal works just as well.

Box Extenders Building codes require that the front of any electrical box should be flush with the surface of the wood paneling when the job is complete. This prevents combustible materials from being exposed to possible short circuits within the box. If the box is not flush, it should either be repositioned or fitted with a box extender. A **box extender** is a metal or plastic fitting that is screwed to the front of the outlet box, bringing it forward.

Positioning the First Panel Positioning and fastening are usually done together.

Step 1 Measure the height of the wall in several places. If the height is less than 8', subtract ½" from this dimension and cut the first 4×8 panel to length. This will provide ¼" at top and bottom for expansion and contraction. If the height is greater than 8', start with a 4×10 panel.

Step 2 Place the panel in position in a corner and butt it to the adjacent wall. The room corners may be irregular, particularly in an older house. Make sure the panel is perfectly plumb and its outer edge is directly over the centerline of a stud. If this edge does not fall directly on the stud, trim the other side of the panel so it will.

Step 3 Position the panel at the proper height by shimming it to allow for ¼" clearance at the bottom. When the panel is set perfectly plumb and at the correct height, check for gaps between the panel and the corner. If the panel fits tightly, proceed to Step 7. If there is an irregular gap between the panel and the corner, the panel must be scribed.

Step 4 Set a compass for an amount equal to the widest gap between the panel edge and the corner. Scribe a line on the panel. To scribe the panel edge, hold the compass tip against the wall and move the compass downward. The compass should be level. The tip will follow imperfections in the wall and the pencil will record them.

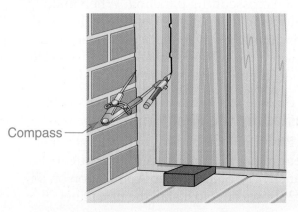

Compass

Step 5 Using a jigsaw or circular saw, cut the panel along this line so it will fit the corner.

Step 6 Set the panel back in place against the wall and again shim it to the correct height. Now it should fit the corner exactly.

Step 7 Nail the first panel to the wall. Fastening techniques will be discussed on the next page.

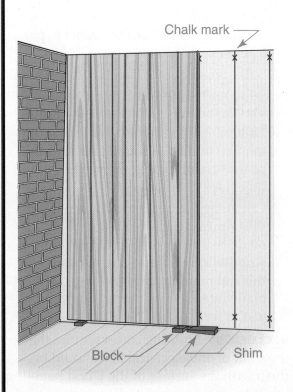

Chalk mark

Block

Shim

Fastening Panels

The nailing patterns for plywood paneling depend on the spacing of supports and the thickness of the panels (see **Table 28-1**).

1. Use colored ringshank nails that blend with the wood finish. This will eliminate the need for countersinking and puttying nail holes. Space the nails every 6" along panel edges. Space them every 12" along intermediate studs.

Builder's Tip

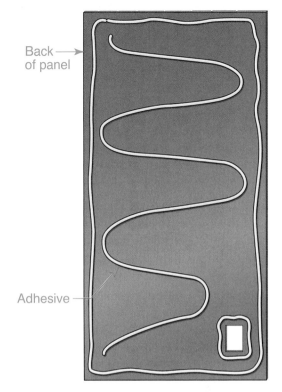

Back of panel

Adhesive

2. When nailing is complete, check that all nails are set properly and that the paneling is tight against the wall.

Prefinished paneling may also be nailed with standard finishing nails. If this technique is used, countersink the nails slightly below the surface of the paneling. Fill the holes using a putty stick that matches the color of the paneling.

Panel adhesive may be used instead of nails, though at least a few nails are usually necessary to hold a panel in place as the adhesive cures. Be sure to follow the adhesive manufacturer's instructions. Use only a latex, water-based adhesive with MDF paneling. Solvent-based adhesives may discolor finishes on this product.

1. After the panels have been properly cut and fitted, apply the adhesive to the wall surface or the back of the panel. Use a caulking gun to create a continuous

Adhesive Bonding

Nail Supplement Apply adhesive to the back of the panel. Keep it approximately 1" or more from the edges to avoid adhesive squeeze-out.

$\frac{3}{16}$" – $\frac{1}{4}$" wide bead around the perimeter of the panel and around any cutouts.

2. Apply additional adhesive in a zigzag pattern in the middle.

3. Position the panel and press it firmly against the adhesive.

4. Place three or four finishing nails across the top of the panel to hold it in place.

Table 28-1: Nailing Recommendations for Interior Plywood Paneling

Plywood Thickness (inches)	Maximum Support Spacing (inches)	Nail Size (Use casing or finishing nails)	Nail Spacing (inches)	
			Panel Edges	Intermediate
$\frac{1}{4}$	16[a]	4d	6	12
$\frac{5}{16}$	16[b]	6d	6	12
$\frac{3}{8}$, $\frac{11}{32}$, $\frac{1}{2}$, $\frac{15}{32}$	24	6d	6	12
$\frac{5}{8}$, $\frac{19}{32}$, $\frac{3}{4}$	24	8d	6	12

[a] Can be 20" if face grain of paneling is across supports.
[b] Can be 24" if face grain of paneling is across supports.

5. Place a padded block of wood against the panel and tap the block with a hammer or rubber mallet to achieve full-surface contact. Be sure to round over (ease) the corners and edges of the block so that they will not leave marks on the panel.

6. If necessary, use small finishing nails to hold the panel flat until the adhesive reaches full strength.

Seams and Joints

Seams and joints between panels can be handled in a variety of ways. Some panels are designed with interlocking edges so that the joint needs no special treatment. Grooved panels are designed so that there are ½" width grooves at the vertical edges. When two panels are butted together, the joint nearly disappears. If desired, seams can be covered with wood or plastic molding that is color-matched to the paneling.

Inside Corners If panels fit very well in a corner, a butt joint is acceptable. However, to ensure a good fit, the edge of one panel may have to be scribed to fit against the other. Back-cutting the edge of the scribed panel improves the fit even more. Another way to handle the corner is to cut the edges of the intersecting panels so that they will fit into a preformed plastic corner. The corner should be installed before either panel is nailed into place.

Outside Corners If the paneling is unfinished and relatively thick, outside corners can sometimes be mitered. The sharp edge of the corner must be eased, however, to prevent damage over time. Some carpenters use biscuit joinery to strengthen the joint and ensure that the mitered edges meet uniformly (see Section 6.4). If panels meet at 90°, the joint can also be covered with a wood corner guard. Another approach is to use a prefinished outside corner molding. Instead of mitering the paneling, the edges can be cut square.

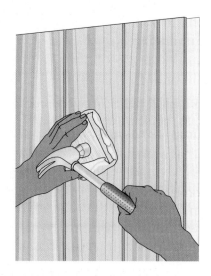

 Bedding the Panel
Finish Protection Tap the panel with a hammer and a block to press it firmly into place. Put a cloth under the block to protect the panel finish.

Estimating

Determining the number of panels needed for a room is based on the square footage of the walls. The method is explained in the following Estimating and Planning feature.

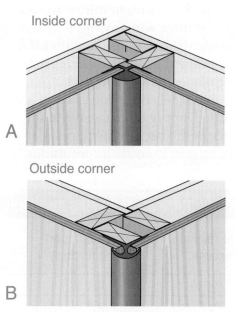

Inside corner

A

Outside corner

B

Prefinished Corner Molding
Inside and Outside Corner joints can be concealed with prefinished molding. The molding may have to be installed before the panels are placed.

Estimating and Planning

This estimating and planning exercise will prepare you for national competitive events with organizations such as SkillsUSA and the Home Builder's Institute.

Sheet Paneling

Estimate Perimeter

Step 1: Figure the perimeter of the room to be paneled by adding the lengths of the four walls together.

Step 2: Divide the perimeter by the width of the panel in inches.

$$18 + 18 + 15 + 15 = 66'$$

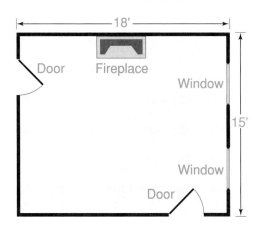

Step 3: Multiply this number by the waste allowance, which is usually 5 percent. Round up the result to the nearest whole number. This will be the number of panels you will need. In the floor plan that is shown here, 18 4×8 panels are required.

Step 4: If the ceiling height is more than 8', determine the additional height. For example, if the room shown in the first column has a 10' high ceiling, 2' of additional height are required. You would use 10' panels (if available) or cut four 2' pieces from an 8' panel. Since 18 panels are required to go around the room, 4½ (or, rounded up, 5) additional panels will be required (18 ÷ 4 = 4½), for a total of 23 panels.

Step 5: Deduct from the panel count for any large areas that will not be paneled, such as a fireplace. Do not deduct for windows and doors.

Estimating on the Job

Consult the floor plan in the first column. Figure the number of panels required if this room had a cathedral ceiling. Assume the following:

- The low walls are 8' in height.
- The ridge runs perpendicular to the fireplace wall.
- The roof over this room has a ¹²⁄₁₂ pitch (see Chapter 17 to review information about pitch).
- The paneling will be applied to all wall surfaces.
- There is no fireplace.

Section 28.1 Assessment

After You Read: Self-Check

1. Over what type of surface is paneling typically applied?
2. Why must drywall be taped and sanded before installing paneling?
3. Describe the process of scribing a panel for an inside corner.
4. How must paneling fit around electrical outlet boxes?

Academic Integration: English Language Arts

5. **Create a "How To"** Create a "how-to" guide for cutting out a shape in a panel to fit around an electrical outlet. List all the materials you will need. In addition, list the steps to follow in order. Make use of transitional words and phrases.

Go to connectED.mcgraw-hill.com to check your answers.

Board Paneling Basics

Where should paneling be conditioned?

Board paneling is a type of wall paneling made of solid wood. It comes in the form of individual boards that are applied to the wall one by one.

Only thoroughly seasoned wood should be used. The boards should not be too wide, or the gaps created by expansion and **contraction** will be excessive. A nominal 8" is the maximum width recommended in most parts of the United States. The boards are usually applied vertically, but they can be applied horizontally or diagonally.

 Installing Board Paneling

Board by Board The interlocking edges of board paneling provide a gap-free joint. *What is board paneling made of?*

Types and Sizes

Many kinds of wood are made into boards for paneling, including Douglas fir and various species of pine. Solid oak or cherry is also used. A rustic or informal look can be obtained with knotty pine or recycled barnwood. A more formal look can be achieved with a hardwood such as cherry, walnut, or mahogany. Hardwood ranges from ⅜" to ¾" thick and comes in lengths of 8' to 12' and longer. The edges of each board interlock with adjoining boards in either a lap joint or a tongue-and-groove (T&G) joint. Square-edged boards can also be used, but then the joints are often covered with molding. A T&G edge lessens appearance problems caused by expansion and contraction. It also makes it easy to align the panels.

Storage and Conditioning

Always store paneling indoors. Stack the wood flat on the floor using stickers to separate the boards. Condition solid-wood paneling for seven days, if possible. It should be conditioned by being stored in the room that it will be installed in.

REGIONAL CONCERNS

Wood Availability The woods used for board and raised paneling may vary based on local availability. In the Northwest, for example, Douglas fir and Western red cedar are widely available. In northern California, redwood is sometimes used, particularly in period architecture. In the southern United States, cypress is more available than elsewhere. In New England, oak and white pine are popular. Each wood is generally less expensive within its home region.

Ron Chapple Stock/Fotosearch RF/Glow Images

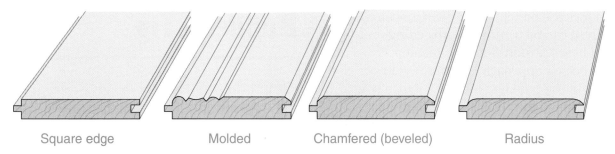

Square edge   Molded   Chamfered (beveled)   Radius

 Board Patterns

Standard Patterns These are some tongue-and-groove paneling patterns. Most retail lumberyards carry two or three patterns in stock. *Why is it important that paneling boards are not too wide?*

Installing Board Paneling

What is blind nailing?

Solid-wood paneling can be installed on a wall vertically, horizontally, or at an angle. The orientation of the wood can have a dramatic effect on the look of a room.

Installing Boards Vertically

When paneling is to be attached directly to studs (if allowed by code), adequate blocking must be placed between the studs to provide nailing support. The blocking should not be more than 24" OC.

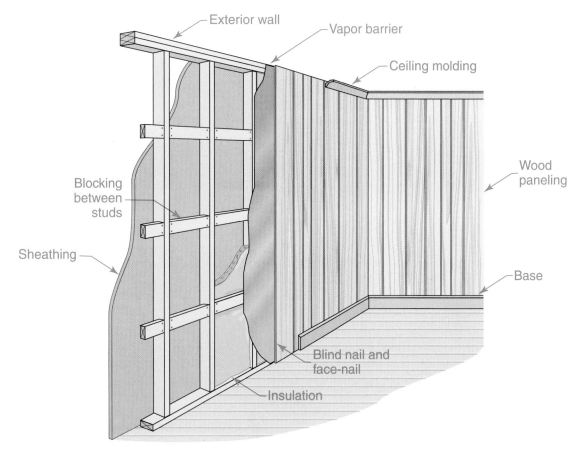

Exterior wall — Vapor barrier — Ceiling molding — Wood paneling — Base — Blind nail and face-nail — Insulation — Sheathing — Blocking between studs

 Blocking Details

Added Support Blocking provides a nailing surface for boards that are located between studs. This installation is on an exterior wall.

Once the paneling has been installed vertically, it can be trimmed with ceiling molding and baseboard in standard fashion.

Another approach is to butt the lower ends of the paneling into a 1×8 trim board. The ends of the vertical paneling will rest on the top edge of the 1×4 base. This detail is sometimes used when installing paneling over a drywall surface.

Positioning and Nailing

Once the horizontal blocking has been installed, the individual boards can be nailed in place. As with sheet paneling, the first board installed determines the layout of all following boards. If it is not plumb, this will make it difficult or impossible for any of the other boards to be plumb.

1. Starting at a corner, hold the first board in the corner and plumb it using a level. The grooved edge should be against the corner.

2. If the corner is not straight or plumb, you will have to scribe the board to fit. Undercut the edge about 5° to ensure a snug fit.

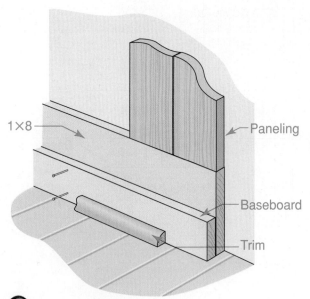
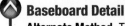

Baseboard Detail
Alternate Method This arrangement is possible when paneling is installed over drywall. The square-edged or molded baseboard is nailed to a furring strip.

Builder's Tip

MOISTURE CONTENT The moisture content of solid-wood paneling should be near the average it will reach in service—about 8 percent in most areas. However, in the dry southwestern United States, it should be about 6 percent. In the southern and coastal areas of the country, it should be about 11 percent. You can measure moisture content with a moisture meter.

3. Face nail the first board along the edge that is against the corner. Then blind nail the panel by driving 5d or 6d finishing nails at an angle through the tongue edge. In **blind nailing**, the nails are driven at an angle through the tongue of the board and into framing or furring

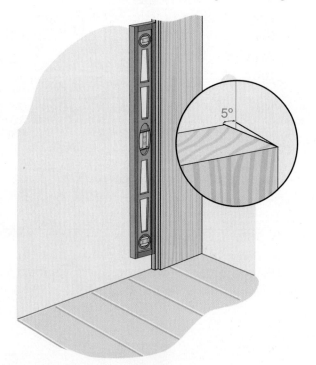

Setting the First Board
The First Board Is the Most Important The first board to be installed is scribed in a plumb position to the adjacent wall and undercut about 5° to provide a tight joint in the corner.

strips. When the next board is installed, it will conceal the nail heads.

4. Continue to install boards, checking them for plumb frequently. A slightly out-of-plumb board can be repositioned and the slight difference in spacing will not be noticeable. When nearing the opposite corner, leave the second-to-last board loose.

5. When installing the last board on a wall, scribe the edge that is to fit into the corner and undercut it at an angle of about 5°. Then lift the edge of the preceding board slightly, position the last board, and snap both boards into place. Face nail the edge of the last board. (If the last board does not easily snap into place, cut off the bottom lip of its grooved edge, then press the board into place.)

6. Continue around the room in this fashion. Cut holes for outlets and other features as you encounter them. Where boards meet at outside corners, miter them or conceal the joint with molding.

7. Add ceiling and baseboard trim to complete the job.

Installing Boards Horizontally

Horizontal board paneling, while not as common as vertical paneling, has some advantages. Blocking is not required. Instead, boards can span the distance between studs. Because longer boards can be used, it is less time-consuming to apply.

Measure from the ceiling to the floor in several places to make sure all measurements are equal. If they are not equal, make sure the narrower board will be at the bottom.

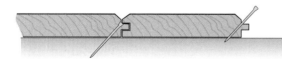

Blind Nailing
Invisible Nailing Blind nailing each paneling board conceals the nails.

Builder's Tip

RIGHTY OR LEFTY? The direction in which you move across a wall while installing vertical boards is a matter of personal preference. Right-handed people find it easier to start at the left side of a wall and work to the right. Left-handed people find the reverse to be easier.

1. Begin the paneling at the floor line, making certain that the first piece is level and the tongue edge is up.

2. Scribe the bottom edge to the floor to eliminate any gaps. The gaps can also be covered later with baseboard or trim.

3. Undercut the ends of each board about 5° to provide a tight joint at the inside corners of the wall.

4. Miter the boards at the outside corners or trim them with molding.

5. Blind nail each board, checking periodically to make certain the boards remain level.

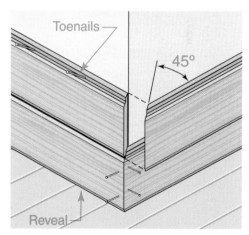

Toenails

45°

Reveal

Mitered Corners
Lots of Fitting Wood paneling applied horizontally can be mitered at outside corners. In this installation, instead of a baseboard, a reveal is shown between the bottom board and the finished floor.

6. If no molding is to be used at the ceiling, scribe and undercut the last panel edge at a 5° angle to ensure a snug fit.

Installing Boards at an Angle

Application of boards in the herringbone (chevron) pattern is quite demanding. The boards can be nailed to existing studs, although studs may not be in the right place to support the ends. Instead, you may have to apply vertical furring strips so the

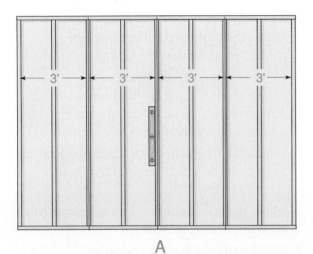

A

B

 Furring Strip Placement
Careful Planning To install paneling in a herringbone style, furring strips should be installed (**A**) where they will support the ends of each board. To provide a symmetrical installation (**B**), locate furring strips accurately.

space between studs is evenly divided. For example, if the wall is 12' long, you can place the strips 3' OC. Make sure that each furring strip is plumb.

1. Draw a plumb line at the center of every other furring strip. For a 12' wall, these lines should be 36" apart, or as close to that as possible.

2. Cut two pieces of paneling in the shape of a 45° triangle, with the tongue on the long edge.

3. Do not assume that the floor is level. Align the triangles with a vertical plumb line and to a level chalk line snapped across the wall at this point. The chalk line will serve to align this pair of starting triangles with other pairs. A molding strip will be applied later to cover the vertical joint.

4. Cut the next pair of boards at a 45° angle and fit them into place on the next centerline. Continue to work across the wall along the other centerlines, building toward the top. A miter saw is extremely useful for repetitive angle cutting such as this.

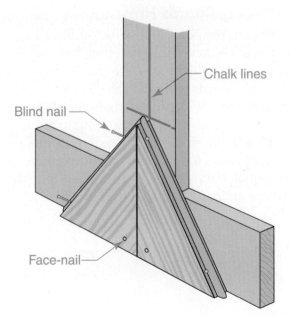

 Starting the Boards
Small Pieces Cut the starting triangles and align them with the chalk lines. Then nail the triangles into place.

5. Blind nail the boards to each furring strip. Use the play in the tongue-and-groove joint to keep the boards aligned along the vertical joint. Check their horizontal alignment frequently.

6. After you have installed all the pieces, apply ceiling and baseboard molding. Then apply molding over the vertical joints in the paneling. A cove or quarter-round may be used at the corners and a base shoe at the floor if necessary.

Raised Paneling and Wainscoting

Raised paneling is constructed much like raised-panel cabinet doors (see Chapter 27). The panels are made of solid wood, such as oak or cherry. Individual raised panels are held in place by a grid of stiles and rails

secured to the walls with nails or screws. Raised paneling is very expensive, so it is sometimes simulated with wood molding applied to sections of sheet paneling.

Board and sheet paneling is commonly applied from floor to ceiling, but a more formal look can be achieved by running it only on the lower portion of a wall. This is called wainscoting. A piece of wood molding runs along the top edge of wainscoting to conceal cut edges of the wood.

Estimating

It is possible to estimate the number of paneling boards necessary by calculating the square footage of the walls, but two other methods result in more accurate figures. These methods are explained in the Estimating and Planning feature on page 822.

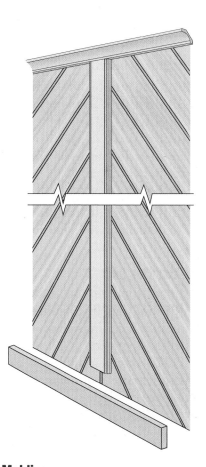

Wainscoting
Raised Paneling Floor to Ceiling Wood Raised paneling lends an expensive, rich look to a room, with its panels of solid wood extending to the ceiling.

Molding
Final Detail Apply a molding strip at the vertical joint of the paneling. The molding should extend from the baseboard to the ceiling molding.

Jill Braaten/McGraw-Hill Education

Board Paneling

Estimating Material

Board paneling may be estimated by the board foot or by the lineal foot.

Board-Foot Method

Step 1: Figure the wall area to be covered by multiplying the perimeter times the ceiling height. Suppose a room has a perimeter of 66' and a ceiling height of 8'. Its wall area is 528 sq. ft.

$$66 \times 8 = 528$$

Step 2: Subtract the area for windows, doors, and fireplaces. Assuming that those in our example total 112 sq. ft., a total of 416 sq. ft. is to be covered by wood paneling.

$$528 - 112 = 416$$

Step 3: Multiply the total area to be covered by the area factor shown in **Table 28-2**. For example, suppose we are using tongue-and-groove 1×8 paneling. Its area factor is 1.16. If we multiply that by the 416 sq. ft. of wall space, we need 483 board feet of paneling.

$$416 \times 1.16 = 482.56, \text{ or } 483$$

Step 4: Add an allowance for trim and waste. For straight paneling, add 5 percent; for herringbone, add at least 10 percent. In our example, $483 \times 0.05 = 24.15$, or 24. Then $483 + 24 = 507$. The total amount of paneling required is 507 board feet.

Step 5: Multiply this figure by the cost per board foot to determine the total cost of the paneling.

Lineal-Foot Method

Step 1: Measure the height and length of the wall to be paneled.

Step 2: Determine the *face width* of the paneling. This is the portion of the board that will be visible when the paneling is in place. On a tongue-and-groove board, for example, this dimension does not include the tongue. The face width can be measured on a board. It is also given by the manufacturer or supplier in product specifications. For example, 1×6 boards have a face width of 5¼".

Step 3: Assume that a wall 14' long and 8' high will be paneled vertically. Convert the wall length to inches, for a total of 168".

$$14' \times 12 = 168"$$

Step 4: Divide this total by the face width. The answer will be 32.

$$168" \div 5.25 = 32$$

This is the number of boards you will need.

Step 5: Multiply the number of boards by their length: $32 \times 8 = 256$. This is how many lineal feet of wood you will use. If the wall is 8'-6" high, however, you will have to order 10' boards and cut them down to size. You will still need 32 boards, but your lineal-foot total will be 320. For each board, the waste will be 18".

Step 6: Multiply the total lineal feet by the cost per foot to determine the total cost of the paneling.

Estimating on the Job

Using the board-foot method, estimate the amount of board paneling needed for a 15' × 20' room with 8' ceilings. Assume 126 sq. ft. will be used for doors and windows. The 1×6 panels will be placed tongue-and-groove in the herringbone pattern. How many board feet of paneling are needed?

Table 28-2: Estimating Coverage of Board Paneling

Paneling	Nominal Size	Width (inches)		Area Factor[a]
		Dress	Face	
Shiplap	1×6	$5\frac{7}{16}$	$4\frac{15}{16}$	1.22
	1×8	$7\frac{1}{8}$	$6\frac{5}{8}$	1.21
Tongue-and-groove	1×4	$3\frac{7}{16}$	$3\frac{3}{16}$	1.26
	1×6	$5\frac{7}{16}$	$5\frac{3}{16}$	1.16
	1×8	$7\frac{1}{8}$	$6\frac{7}{8}$	1.16
S4S	1×4	$3\frac{1}{2}$	$3\frac{1}{2}$	1.14
	1×6	$5\frac{1}{2}$	$5\frac{1}{2}$	1.09
	1×8	$7\frac{1}{4}$	$7\frac{1}{4}$	1.10

[a] Allowance for trim and waste should be added.

Note: For most installations, an allowance of 5 percent will be adequate for trim and waste. Sometimes, rather than add 5 percent, the area of the doors and windows is not subtracted but is used as a trim and waste allowance.

Section 28.2 Assessment

After You Read: Self-Check

1. What is the recommended maximum nominal dimension for solid-wood paneling?
2. When solid-wood paneling is applied vertically directly to studs, what extra preparation is necessary?
3. What is blind nailing?
4. Is blocking required for horizontal paneling? Why or why not?

Academic Integration: Mathematics

5. **Estimating Lineal Feet** Using Table 28-2, estimate the amount in lineal feet of 1×4 tongue-and-groove board paneling that would be required for wainscoting on two walls of a room. One wall is 12 ft. long. The other wall is 14 ft. 6 in. long. Assume that the wainscoting will be 36 in. high and that there are no doors or windows in these walls.

   Math Concept When solving measurement problems, convert quantities to the same unit of measure. When using a calculator, convert fractions and mixed numbers to decimals.

   Step 1: Determine the combined length of the two walls that comprise the job. Convert to inches.

   Step 2: Find the face width using Table 28-2. Convert to a decimal.

   Step 3: Divide the length in inches by the face width. This is the number of boards you will need.

   Step 4: Multiply the number of boards by their length in feet.

   Step 5: Add a 5 percent allowance for trim and waste.

   Go to connectED.mcgraw-hill.com to check your answers.

Review and Assessment

Chapter Summary

Sheet paneling is the most common type of wall paneling and usually comes in 4×8 sheets. Drywall makes the best backing for sheet paneling. Panels can be applied with nails or adhesives. Molding is used to cover joints and seams.

Board paneling is made of solid wood, and boards interlock with lap or tongue-and-groove joints. Raised paneling is made of solid wood. Paneling should be conditioned by placing it for at least 48 hours in the room in which it will be used. Board paneling may be installed vertically, horizontally, or at an angle. Blind nailing allows for nails placed through the tongue of one board to be hidden by the next. Edges are undercut to ensure a tight fit in corners.

Review Content Vocabulary and Academic Vocabulary

1. Use each of these content vocabulary and academic vocabulary words in a sentence or diagram.

Content Vocabulary

- full-height paneling (p. 808)
- wainscoting (p. 808)
- sticker (p. 809)
- box extender (p. 811)
- board paneling (p. 816)
- blind nailing (p. 818)

Academic Vocabulary

- randomly (p. 810)
- contraction (p. 816)
- simulated (p. 821)

Speak Like a Pro

Technical Terms

2. Work with a classmate to define the following terms used in the chapter: *sheet paneling* (p. 808), *conditioning* (p. 809), *mastic* (p. 809), *raised paneling* (p. 821), *face width* (p. 822).

Review Key Concepts

3. Name the three basic types of paneling.

4. Describe the ways full-height paneling and wainscoting are different.

5. Tell how paneling should be stored and conditioned.

6. Explain how sheet paneling is installed.

7. Describe horizontal and vertical paneling.

8. Create an estimate for the amount of sheet paneling required for your room.

Critical Thinking

9. Explain Would you expect to find wood paneling made of Douglas fir or Western red cedar in a home in New England? Why or why not?

Academic and Workplace Applications

STEM Mathematics

10. Perimeter and Cost A carpenter has been asked to submit a bid for remodeling a recreation room that is 16' × 28' with 8' high walls. This bid includes removing the old paneling and installing ¼" × 4' × 8' oak paneling at $10.19 per sheet of paneling plus 5.75% sales tax. How many sheets of paneling will be needed, and what is the cost of the paneling?

Math Concept To find the perimeter of a rectangle, multiply the length and the width by 2 and find the sum. To add a percentage of a number to itself, multiply the number by the percent plus 100.

Step 1: Find the perimeter of the room.

Step 2: To obtain the surface area, multiply the perimeter of the room by the height of the room.

Step 3: To figure 5% allowance for trim and waste, multiply the surface area by 105.

Step 4: Divide the surface area by the area of one sheet of paneling. Round up to the next whole sheet.

Step 5: Determine the cost of the sheets and add 5.75% for sales tax.

STEM Science

11. Condensation Condensation occurs when water vapor is changed into liquid form because it has been cooled. For example, during winter months, you might notice water droplets forming on windowpanes inside your home. Given this information, explain what might happen if you use paneling that has been stored in an unheated garage during winter without first conditioning it.

Starting Hint Condensation occurs when water vapor in the air contacts a surface whose temperature is lower than the dewpoint, the temperature at which water condenses out of the air.

21st Century Skills

12. Career Skills: Connecting School to Work List the connections between academic subjects you are taking and the work tasks in this chapter. For example, in what way would a carpenter benefit from studying geometry? How would a house remodeling construction manager benefit from studying a foreign language? Record your ideas in a bulleted summary or a one-page chart.

Standardized TEST Practice

True/False

Directions Read each of the following statements carefully. Mark each statement as either true or false by filling in **T** or **F**.

(T) (F) **13.** The moisture content of solid-wood paneling should be near the average it will reach in service—about 8 percent in most areas.

(T) (F) **14.** For most rooms, it is practical to start paneling from the center and then work around the room in both directions.

(T) (F) **15.** It is possible to estimate the number of paneling boards necessary by calculating the square footage of the walls.

TEST-TAKING TIP

If you are allowed to use a book during a test, make sure that the book you are planning on bringing is authorized before the day of the test.

*These questions will help you practice for national certification assessment.

Presenting a Professional Estimate

Your Project Assignment

An eco-conscious homeowner has asked for your bid to remodel a 200-square-foot kitchen. You will install new cabinetry, crown molding, and appliances in the new kitchen. You will create an estimate, then draw up and present your bid sheet to your potential client.

- **Predict** the total cost for the job.

- **Build It Green** **Research** and price environmentally friendly materials and appliances.

- **Calculate** cost, overhead, profit, mark-up, and gross profit margin.

- **Draw** up a bid sheet, attaching a spreadsheet showing your calculations.

- **Interview** a local member of the National Association of the Remodeling Industry.

- **Create** a three- to five-minute presentation for the homeowner that explains your bid.

Applied Skills

Some skills you might use include:

- **Research** the costs involved in a construction project.

- **Identify** certified green materials and appliances.

- **Calculate** costs based on quantity and type of materials.

- **Check** cost-estimate predictions.

The Math Behind the Project

The traditional math skills for this project are estimation, data analysis, and measurement. Remember these key concepts:

Estimation

Accurate estimates rely on accurate calculations. If you offer a bid that is too low, you may lose money.

If you offer a bid that is too high, you may not be offered the job. When estimating direct costs, make sure to include labor, materials, temporary power hook-ups, and insurance. Also include any necessary building permits.

Percentage

Profit, overhead, and mark-up are all calculated as percentages of cost. The typical overhead rate is 10%. Profits generally range from 5% to 20%, and mark-up runs from 5% to 50%. Mark-up is calculated on items that are purchased from a supplier and sold to the customer, such as a refrigerator or stove.

To calculate overhead, multiply your cost by your overhead rate. To calculate profit, add cost and overhead and multiply this total by the profit rate. Then figure your profit margin on the job by dividing the profit amount by the total amount of the job. Convert each percentage to a decimal before multiplying. For example, if the cost of the job is $15,000, your overhead rate is 10%, and your profit rate is 20%, use the following steps:

1. Calculate overhead.	$15,000 × .10 = $1,500
2. Add overhead to cost.	$1,500 + $15,000 = $16,500
3. Calculate profit on cost and overhead.	$16,500 × .20 = $3,300
4. Add profit to cost and overhead.	$3,300 + $16,500 = $19,800
5. Calculate profit margin.	$3,300 ÷ $19,800 = .167

Convert profit margin to a percentage by moving the decimal point two places to the left. In this case, your profit margin is 16.7%.

Project Steps

Step 1 Research

- Find a sample bid sheet for a construction or remodeling job.
- Consult the index of this book to find information on estimating costs for windows, cabinetry, drywall installation, painting, and flooring.
-  **Build It Green** Find and price green supplies and appliances, and products with certifications from programs such as Green Seal, the Forest Stewardship Council, and Energy Star.
- Predict the total cost of the job based on your research.

Step 2 Plan

- Choose the basic design for the new kitchen.
- Calculate the amount and price of materials, such as flooring, cabinetry, countertops, and paint.
- Choose the appliances you will use and your mark-up rate.

Step 3 Apply

- Itemize and calculate the costs for all materials you have selected.
- Calculate mark-up on the appliances.
- Estimate your direct costs for the job, including labor and materials.
- Calculate overhead, profit, and profit margin and create a spreadsheet showing all your calculations.
- Find a remodeling professional through the National Association of the Remodeling Industry. Ask him or her to review your calculations and offer suggestions and corrections based on local conditions.

- Create your bid sheet.
- Compare the final estimate on your bid sheet to your original prediction.

National Association of the Remodeling Industry

Mission: To advance and promote the remodeling industry's professionalism, product, and vital public purpose.

Step 4 Present

Prepare a presentation for the homeowner that shows and explains your bid sheet, demonstrates your research, and justifies your estimate.

PRESENTATION CHECKLIST

Did you remember to…

✓ Show your bid sheet and explain how to read it?

✓ Explain your choices for green materials and appliances?

✓ Show that your overhead, profit, and mark-up are fair?

✓ Demonstrate that your calculations are accurate?

✓ Show the thoroughness of your research and planning?

Step 5 Technical and Academic Evaluation

Assess yourself before and after your presentation.

1. Did you use an accepted format for your bid sheet?
2. Were labor, materials, and other costs itemized?
3. Were your costs based on solid research?
4. Did you select materials with environmental certifications?
5. Were your cost, overhead, and profit calculations accurate?

 Go to connectED.mcgraw-hill.com for a Hands-On Math Project rubric.

UNIT 7
Construction Specialties

In this Unit:

Hands-On Math Project Preview

Professional Green Painting

After completing this unit, choose paint for your room and calculate how many gallons you will need based on the room's size and the paint's spread rate. You will also estimate how long it will take to complete the job based on the number of coats needed and the forecast temperature and humidity.

Project Checklist

As you read the chapter in this unit, use this checklist to prepare for the unit project:

✓ **Measure** the square footage of the walls and ceiling of the room you will paint.

✓ **Identify** calculations, such as calculating spread rate or paint, used in determining materials required for a specific job.

✓ **Think** about how temperature and humidity can affect the work of different specialists.

➤ Go to connectED.mcgraw-hill.com for the Unit 7 Web Quest activity.

©PBNJ Productions/Blend Image

Construction Careers

Profile Painters apply paint, stain, varnish, and other finishes to buildings and other structures. Paperhangers cover walls and ceilings with decorative wall coverings made of paper, vinyl, or fabric.

Academic Skills and Abilities

- geometry
- mathematics
- blueprint reading
- interpersonal skills

Career Path

- on-the-job training
- apprenticeship programs
- trade and technical school courses

Explore the Photo

Finishing Touches Many construction specialties require specialized skills and equipment. *What safety issues do you see in this photo?*

CHAPTER 29

Steel Framing Basics

Section 29.1
Steel as a Building Material

Section 29.2
Steel Framing Tools

Section 29.3
Steel Framing Methods

Chapter Objectives

After completing this chapter, you will be able to:

- **Describe** the three types of steel frame construction.
- **Identify** tools used in steel framing.
- **Tell** the difference between welding and clinching.
- **Describe** how to lay out steel floor joists.
- **Explain** why steel framed structures use in-line framing.
- **Explain** how to set steel ceiling joists.

Discuss the Photo

Working with Steel Steel framing is used extensively in office buildings, but is relatively new in residential construction. *What might some advantages of steel framing be?*

Writing Activity: Experiment Results

Screw together two scrap pieces of steel framing stock with one or two screws of the same size and length. The flat sides of the pieces should be together and the open side should face out. Test the pullout capacity of the screws by trying to pry the two pieces apart. Try this same test again with screws of a different length and diameter. Use good safety practices. Report the results in a one-page summary.

Chapter 29 Reading Guide

Before You Read Preview

Steel frame construction is similar to wood frame construction. Choose a content vocabulary or academic vocabulary word that is new to you. When you find it in the text, write down the definition.

Content Vocabulary

- cold-formed steel
- performance method
- prescriptive method
- feathering
- pullout capacity
- welding
- clinching
- in-line framing
- axial load
- joist tracks
- clip angle
- roof rake

Academic Vocabulary

You will find these words in your reading and on your tests. Use the academic vocabulary glossary to look up their definitions if necessary.

- sequence
- remove

Graphic Organizer

As you read, use a chart like the one shown to organize content vocabulary words and their definitions, adding rows as needed.

Content Vocabulary	Definition
cold-formed steel	sheet steel that is bent and formed without using heat

Go to connectED.mcgraw-hill.com to download this graphic organizer.

Steel Framing

Where is steel framing common?

The steel that is used for residential steel framing is cold-formed steel. **Cold-formed steel** is sheet steel that is bent and formed without using heat. This type of framing system is sometimes referred to as light-gauge steel framing. Unlike wood building materials, steel is not damaged by insects and it is treated with a hot-dipped galvanized coating to resist rust and corrosion. Steel is 100 percent recyclable. New steel-framing materials contain at least 25 percent recycled content.

Sizes and Shapes

Steel framing is formed into shapes and sizes that are similar to what builders are accustomed to seeing in dimensional lumber (2×4, 2×6, 2×8, 2×10, 2×12, and so forth). Steel framing members are formed by passing the sheet metal through a sequence of rollers to form the bends that make the shape, such as the web, flanges, and lips of a stud or C-shape. Because this process is done without heat (also called *cold forming*), the studs and joists are stronger than the sheet steel.

When ordering steel framing materials, it is important to be aware of the variety and applications of the various shapes, encapsulated by the acronym **STUFL**. These letters stand for **S**tud, **T**rack, **U**-channel, **F**urring, and **L**-header.

A stud includes wall studs, joists, and rafters because they are all of the same shape. Track is the top and bottom *plates* of a steel wall or the rim of floors and rafters. U-channel can be used for bridging, blocking, and customized for cabinet backing. Furring channel is used as purlins, bridging, backing, and for subassembly sound separation.

L-headers are members that can be doubled and used as headers.

Cold-formed steel is specified by a universal designator system called out by web dimension, shape, flange dimension, and thickness. In addition, yield strength and anti-corrosion coating are required for each order. Web and flange sizes are expressed in $\frac{1}{100}$ths of an inch and thickness is expressed in $\frac{1}{1000}$ths of inch, or *mils*. Typical dimensions are shown in **Table 29-1**.

Common mil thickness for load-bearing walls are 43 and 33 mil, roof rafters are typically 33 mil, while floor joists often range between 97 and 54 mil. Steel framing also may be called *light gauge*, a term that has been used in the past to describe the thickness of the material. A comparison of "mil" and "gauge" thickness is shown below:

- 12 gauge (97 mil)
- 14 gauge (68 mil)
- 16 gauge (54 mil)
- 18 gauge (43 mil)
- 22 gauge (27 mil)
- 25 gauge (18 mil)
- 29 gauge (33 mil)

Steel Framing Design

Architects and engineers design steel frame houses using either the performance method or the prescriptive method.

Performance Method The **performance method** is a method of framing that depends upon established engineering principles and design-load specifications. Architects and engineers use these principles and specifications to calculate size and strength for individual steel-framing members.

The performance method is time consuming. Standard sizes of framing members are not always used, adding to costs and inefficiency.

Table 29-1: Typical Wood and Steel Dimensions				
Wood	Steel	Web Depth	Flange Size	Thickness (mils)
2×4	350S162-43	3½"	1⅝"	43
2×6	600S162-43	6"	1⅝"	43
2×8	800S162-43	8"	1⅝"	43
2×10	1000S162-43	10"	1⅝"	43

Prescriptive Method The **prescriptive method** is a method of framing that uses standardized tables that give specifications and other information. These tables, such as the one in **Table 29-1**, are created using regional or national design codes, regional design-load data, structural limitation data, and knowledge of engineering practices. Data for specfic elements such as earthquake, snow, and wind load are also determined by using tables for specific geographic regions.

The prescriptive method provides architects and engineers with:

- specifications for standard cold-formed steel members
- an identification system for labeling the members
- minimum corrosion protection requirements
- floor joist, ceiling joist, and roof rafter span tables
- wall stud specifications
- wall bracing requirements
- connection requirements
- construction details.

The prescriptive method has several advantages. It helps reduce the engineering costs for steel frame houses. Builders can pre-select stud, joist, and rafter sizes. Building inspectors can easily check and identify stud, joist, and rafter sizes by their labeling. Manufacturers can determine the framing members that need to be supplied for specific markets and geographic areas.

The National Association of Home Builders (NAHB) Research Center developed the prescriptive method for residential cold-formed steel framing in 1995. Since then it has been significantly expanded and is now recognized in the International Residential Code. The prescriptive method applicability limits are shown in **Table 29-2** on page 834.

Types of Steel Construction

Residential steel-framing construction is categorized into three types: stick-built, panelized, and pre-engineered.

Stick-Built Construction Stick-built steel framing is similar to wood framing. The names for the basic steel-framing members are the same as for wood members: stud, joist, and header.

JOB SAFETY

PERSONAL PROTECTION Protective clothing and safety devices must be worn when working with steel framing members. Work gloves help to prevent cuts and punctures, burns from steel exposed to heat or direct sunlight, and freeze burns from steel exposed to cold weather. Thin protective gloves are recommended. Thick gloves make precise movements and placement of materials more difficult. Ear protection, such as sound-reducing ear plugs, is required when noise levels are higher than normal conversational levels. The high-pitched noise caused by steel cutting saws can cause permanent hearing loss. Safety glasses prevent injuries caused by flying bits of metal from chop saws, circular saws, drills, and grinders.

Table 29-2: Prescriptive Method Applicability Limits

Attribute	Limitation
General	
Building dimension	Maximum width[a] is 36–40 feet Maximum length[b] is 60 feet
Number of stories	2 story with a basement
Design wind speed	150 mph
Wind exposure	Exposures C (open terrain) Exposures A/B (suburban/wooded)
Ground snow load	70 psf[b] maximum ground snow load
Seismic design category	A, B, and C
Floors	
Floor dead load	10 psf[b] maximum
Floor live load 　First floor 　Second floor (sleeping rooms)	40 psf[b] maximum 30 psf[b] maximum
Cantilever	24 inches maximum
Walls	
Wall dead load	10 psf[b] maximum
Load-bearing wall height	10 feet maximum
Roofs	
Roof dead load	12 psf[b] maximum total load (7 psf[b] maximum for roof covering only)
Roof/Snow live load	70 psf[b] maximum ground snow load (16 psf minimum roof live load)
Ceiling dead load	5 psf[b] maximum

[a]Building width is in the direction of horizontal framing members supported by the wall studs.
[b]Building length is in the direction perpendicular to floor joists, ceiling joists, or roof trusses.

Spacing of studs and joists for stick-built framing is set at the standard 16" or 24" OC intervals.

Cutting and assembly are performed on the job site. Instead of being nailed, the steel pieces are screwed, welded, or fastened together with pneumatic pins or rivets.

Panelized Construction Panelized construction is used to pre-build flat components such as walls and floors. These components are built to engineering specifications and tolerances on platform tables and jigs. Multiple components can be made using the same templates.

Components may be built at the job site or at an off-site production facility. The panelized components are set in place as units.

Wall panelization has several benefits. Straight walls can be consistently produced and are fairly easy to set in place. They can be completely constructed on a panel table. Assembly line methods allow faster construction and better quality control.

Pre-Engineered Construction In pre-engineered construction, individual steel studs, joists, headers, and roof members are used to assemble pre-built columns, beams, and rafter assemblies. Each assembly is

Stick-Built Steel Framing

Stick-Built Note the standard spacing.

Panelization Table

Pre-Built Panels can be built to exact specifications and tolerances.

designed and constructed for a specific purpose. This method closely resembles metal-frame commercial construction using structural columns.

Pre-engineered construction may produce a rigid or semi-rigid frame. Engineered columns may be spaced at 4' or 8' OC or greater in designs that call for large open areas.

Pre-engineered assemblies are designed using the American Iron and Steel Institute's Specification for the Design of Cold-Formed Steel.

Reading Check

Explain *How are steel pieces fastened together? How is this different from how wood pieces are fastened together?*

After You Read: Self-Check

1. What is cold-formed steel?
2. What two steel-framing design methods do architects and engineers use?
3. Panelized construction is used to pre-build which types of components?
4. What safety equipment must be worn when working with steel framing members?

Academic Integration: English Language Arts

5. **Record Information** Make a list of the special safety equipment needed for steel framing. Find out how much it costs and where you can buy it locally (if anywhere). Record your findings in a one-page chart.

Go to connectED.mcgraw-hill.com to check your answers.

Section 29.2 **Steel Framing Tools**

Safety and Tools

What safety issues relate to cutting steel framing?

Before beginning any framing job, the framers should create a list of all the tools that may be needed. With the proper tools in place, framing can proceed quickly and efficiently. The tools used in steel framing may look like their wood-framing counterparts. However, their use and operation may differ greatly. Always follow the instructions and safety precautions supplied with the tools you use. (Note: *A Builder's Field Guide to Cold-Formed Steel Framing,* published by the Steel Framing Alliance, includes a guide to the tools that builders may choose to use. It is a free download at their Web site.)

Power Tools

Most tools used in steel framing are electrically, hydraulically, or pneumatically powered. In some situations, a gasoline engine may power larger tools, such as plasma cutters.

Screw Gun An electric screw gun is specifically designed for attaching screws. It is not an electric drill, although its appearance is similar. Drills are not designed to apply screws to steel framing.

A screw gun has variable speeds. An industrial screw gun can operate at speeds as high as 4,000 rpm. This variable speed prevents damage from friction and overheating.

A screw gun will have a maximum rpm of 1,800 and have a clutch mechanism built into it that allows screws to be feathered. **Feathering** is the process of attaching a screw to the bit without stopping the screw gun. The screw spins only when pressure is applied to the bit and the tip of the screw.

Attachments can feather strips of collated screws automatically to the bit. Stand-up attachments are also available that allow the framer to remain standing while applying roof and floor sheathing.

The two types of screw guns that are used on a construction site are the framing screw gun and the drywall screw gun. A framing screw gun is designed to connect steel members, such as studs, to bottom and top tracks. A drywall screw gun is designed to attach plywood, OSB sheathing, or wallboard to steel members. It operates at faster

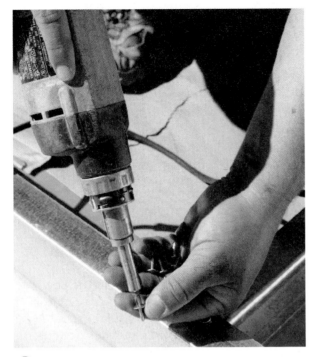

Feathering Screws
Attaching Screws Feathering screws with a power screw gun.

Drill Attachment
Remain Standing This drill attachment drives collated screws. It allows the worker to stand while drilling. After this worker positions the tool, she will straighten it to a perpendicular position. Screws should always be driven perpendicular to the floor sheathing.

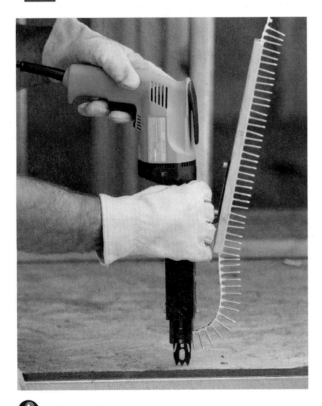

Collated Screws
Feathering Strips Using an attachment that feeds collated screws into the screw gun.

speeds than a framing screw gun. It also has a depth-sensitive nosepiece. The nosepiece prevents the bit from damaging the surface of the sheathing or wallboard while seating the screw.

Nailer Nailers are commonly used to attach plywood, oriented-strand board (OSB), and other sheathing materials to steel wall and roof members.

Nailers are sometimes called *pneumatic nail guns*. They use compressed air to fire the nails and pins into the sheathing materials.

Portable Shears Portable hydraulic shears are attached to a source of hydraulic power by high-pressure hoses. The hydraulic pressure powers a sharp blade that cuts through the steel. Shears produce a clean, straight cut.

Electric handheld shears are similar in operation to hydraulic shears. An electric motor powers the sharp blades. Electric shears can cut steel as thick as 68 mil (about $\frac{1}{16}$").

Chop Saw Chop saws resemble compound-miter saws used for cutting wood. The chop saw for cutting steel can be mounted on a table.

Chop saws require an abrasive (non-toothed) blade for cutting steel. The cut

JOB SAFETY

POWER CORDS When using electric power tools, be aware of the dangers of electrical shock. Check all power cords for exposed wiring. Check the cord's insulation for cuts and breaks. Be sure that a power cord is not wrapped around or passing over steel framing members. The sharp edges of the steel can snag and damage the insulation.

Drywall Screw Gun
Protecting Sheathing A drywall screw gun has a depth-sensitive nosepiece.

Nailer
Attaching Sheathing A nailer used to attach sheathing to steel studs.

Electric Shears
Cutting Steel Electric shears can cut sheet steel.

made is very rough. Sharp burrs remain on the edge. It may be necessary to grind these rough edges to **remove** the burrs.

Chop saw blades cut quickly and produce hot metal chips that can be a safety hazard. They are effective for making square cuts and for cutting bundled studs. However, they are very noisy and ear protection is required.

Plasma Cutter Plasma cutters produce a very hot arc of *plasma* between the tip of the cutter and the steel. Plasma is a heated gas that conducts electrical current. The plasma cuts very quickly and smoothly through steel stock like the flame of a welder's torch.

It may be necessary to grind the edges of a plasma cut. In harsh environments, the edge may require treatment with a metallic coating to resist corrosion.

Circular Saw Circular saws for cutting steel are similar to those used for wood. They must be equipped with a proper blade with carbide-tipped teeth. The carbide tips provide a very hard cutting surface. Abrasive blades are also available. Blades with

Plasma Cutter
Heated Gas Heated gas creates smooth cuts in steel.

 Circular Saw
Cutting Tool A circular saw used for cutting steel.

carbide-tipped teeth are expensive. Cost, safety, durability, and cutting ability are important considerations when choosing a blade.

Press Brake A press brake can create straight-line bends in steel. Builders and framers use a press brake to shape flat sheet steel for use as fascia material, ridge caps, and for other applications. Press brakes can bend flat stock as long as 10 feet.

 Press Brake
Taking Form A press brake forms sheet steel.

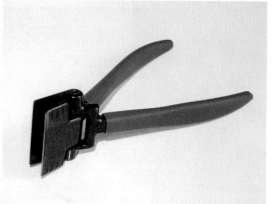

 Hand Seamers
Hands On These hand seamers make small bends.

Hand Tools

Manually operated tools used in steel framing include hand seamers, clamps, aviation snips, and hole punches.

Hand Seamer Hand seamers are used to make small bends in metal stock. They have a 3½" flat jaw and are often called *duck-billed pliers*. The tool is useful for bending steel webs or flanges. It is also useful when forming pieces of sheet steel around windowsills and door openings.

Clamps Clamps are used to temporarily hold steel members together while they are fastened. When layers of steel are screwed together, the first layer can "climb" the threads of the screw and pull away from the second layer. This is called *jacking*. Locking C-clamps prevent jacking by holding the layers firmly together. This allows the screw to penetrate both layers. Bar clamps are

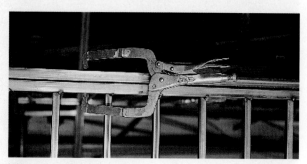

 Locking C-Clamps
Prevent Jacking These clamps temporarily hold members together.

(tl, bl, br)Tim Fuller; (tr)McGraw-Hill Education

 Aviation Snips
Right or Left A. Red-handled aviation snips are for right-handed people. **B.** Green-handled snips are for left-handed people. **C.** Yellow-handled snips should be used for straight cuts.

often used to hold headers in place after they are fitted into the top track.

Aviation Snips Aviation snips can cut steel as thick as 43 mil. They are especially useful for coping and making small cuts on such things as flanges. Snips are available in three different models to fit individual users.

Hole Punch Hole punches are used to create holes in the web of steel framing members that are as thick as 33 mil. Hole punches are

 Hole Punch
Packs a Punch A hole punch makes holes in steel members.

designed to fit around the flange of C-shaped members.

Holes made with punches can measure up to 1" in diameter, and larger holes also may be cut. However, the locations of holes must be in accordance with building codes. This prevents the studs from being weakened. Holes are often used to route pipe or wiring. In such cases, they must be lined with a grommet to prevent rough edges from causing damage or to isolate copper pipes from the steel framing.

Fastening Methods
What is pullout capacity?

Steel framing members can be attached with mechanical fasteners such as screws, nails, and pins. They can also be joined by welding or clinching.

More time is needed to join steel members than wood members. Selecting the correct fastening method for the task is important. If the wrong one is used, the connection may fail. The framer must know about the different methods that are available.

Steel Framing Screws

All steel framing screws have a point, a head, a drive type, threads, and plating. They are sized according to length and diameter. They come in many sizes, shapes, and head and thread styles.

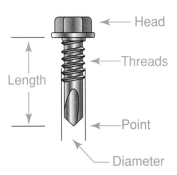

 Steel Framing Screw
Parts of a Screw Note the point, head, threads, and plating.

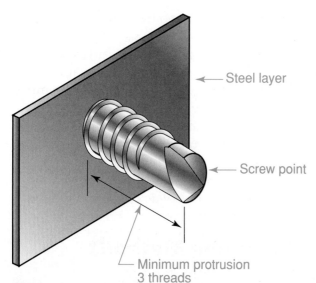

← Steel layer

← Screw point

Minimum protrusion
3 threads

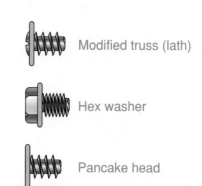

Modified truss (lath)

Hex washer

Pancake head

Steel Framing Screw Head Types
Thickness Matters Three types of heads.

Pullout Capacity
Holding Strong Screw penetration indicates pullout capacity.

Screws are rated on their pullout capacity. The **pullout capacity** of a screw is the screw's ability to resist pulling out of the connection. Pullout capacity is based on the number of threads penetrating and holding the connection.

Framing screws are used to fasten steel to steel. Sheathing screws attach exterior sheathing, such as plywood or OSB, to steel members. The tip of the screw penetrates both the sheathing and the steel. The head

and the threads hold the sheathing tightly against the steel.

Point Types Steel framing screws are self tapping. Self-tapping screws create their own holes. A pre-drilled hole is not needed.

Two types of self-tapping points are used: self-drilling and self-piercing. Self-drilling points have drill tips. The point must be as long as the steel is thick. If the point is too short, the top layer of steel will climb the threads of the screw. Self-piercing points are sharp and can pierce thin steel layers. They are used to attach plywood or wallboard to steel studs that are up to 33 mil thick.

Head Types The hex washer head is the most common style for steel-to-steel connections.

Self-drilling

Self-piercing

Screw Points
Self-Tapping Self-drilling and self-piercing points.

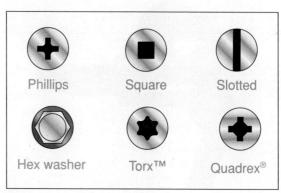

Phillips   Square   Slotted

Hex washer   Torx™   Quadrex®

Six Screw Head Designs
Common Heads Screw head design determines the type of driving tool used.

Builder's Tip

It provides the most positive drive connection. This means that the bit used to drive it fits securely and will not easily slip.

When sheathing and wallboard are applied over a screw head, the modified truss or pancake head styles are preferred. They have a very thin profile. A thin profile allows the sheathing to lie flat.

Drive Types The type of head determines the bit used to drive and turn the screw. The bit needs to fit securely and release quickly. An incorrect bit can become lodged in the screw head.

The most common drive types are Phillips and hex washer. Phillips-head bits are used with modified-truss head screws.

Reading Check

List In what ways are steel framing members attached and joined?

Using Drive Pins and Nails

Drive pins and nails are applied with a pneumatic nailer. Instead of being screwed into the layers, the pins and nails are fired with air pressure. They are used to attach sheathing materials to steel members.

Plywood and OSB sheathing for walls and roofs can be applied using 1" to 1½" pins. Sheathing is usually attached with screws along its edges and with pins or nails in its field. Sheathing attached to walls must be held firmly against the steel during installation. This is because the fasteners do not draw the sheathing against the steel as a screw does.

Welding and Clinching

Welding and clinching are used to attach steel to steel.

Welding is the process of melting the steel and adding filler metals to fuse the pieces at the point of attachment. Cold-formed steel that is 18 to 33 mil thick should never be welded together due to the likelihood of "burn-through" of the material, and caution should be exercised when welding 43 mil material. Welding is permanent.

Although the galvanized coating on the studs will continue to protect against corrosion for most types of cuts, welding or severe grinding may remove the layer of zinc. In these cases, the affected area should be re-coated with zinc-enriched paint.

Clinching is the process of joining two layers of steel with pressure. A powered clinching tool is used. A clinched joint takes more time and is also permanent.

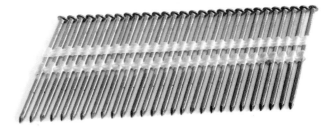

 Collated Drive Pins
Drive Pins Collated drive pins used with a pneumatic nailer.

After You Read: Self-Check

1. What is feathering?
2. What is special about the nosepiece of a drywall screw gun and why?
3. Why must sheathing be held tightly against the steel member before firing a pin or nail?
4. How is a weld formed?

Academic Integration: English Language Arts

5. **Compare Tools** At your local tool supplier, research the differences among the various electric drills, framing screw guns, and drywall screw guns. Research the difference between a metal chop saw and a compound-miter saw for wood. Summarize the different uses in a two-paragraph report.

 Go to **connectED.mcgraw-hill.com** to check your answers.

Section 29.3

Steel Framing Methods

Floors

Why is it important to brace floor joists?

As in wood construction, the frame is the supporting structure of a steel-frame house. It supports the weight of the house and defines its shape. The frame includes the side walls, end walls, floor and ceiling joists, and roof frame.

In-line framing is typically used in steel-frame construction. **In-line framing** is framing that aligns all vertical and horizontal load-bearing structural members. Because the members are aligned, all axial loads are transferred from the roof through the walls and floor joists to the foundation. The **axial load** is the load carried along the length of a structural member.

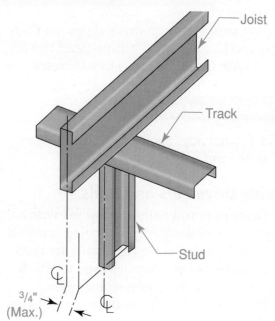

Joist

Track

Stud

3/4" (Max.)

 **In-Line Framing**
Frame Up Framing details for a joist track.

 Basement Foundation

Steel Joist A basement foundation with steel floor joists in place.

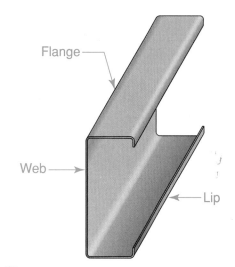

 C-Shaped Joist

Floor Framing Joists are used to frame floors.

The foundation acts as an anchor, as well as a support, for the frame of the house. The foundation may be a concrete slab-on-grade, poured concrete walls, or concrete block walls.

Steel joists, like wood joists, are used to frame floors. The joists are supported by the foundation and by posts and girders. Joists are attached to the foundation with C-shaped members called **joist tracks**. Each end of the joist is inserted into the track and screwed in place.

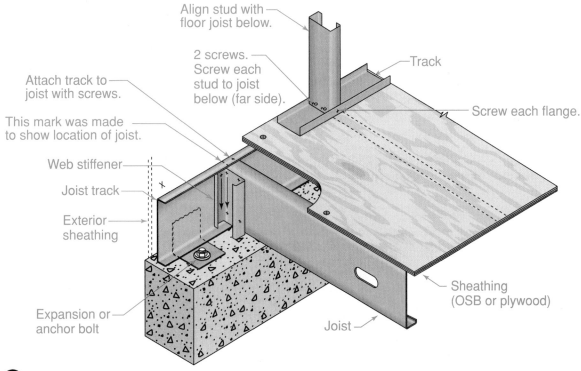

 Joists and Studs

Aligned This drawing shows how the steel joists and studs will be placed in relation to each other.

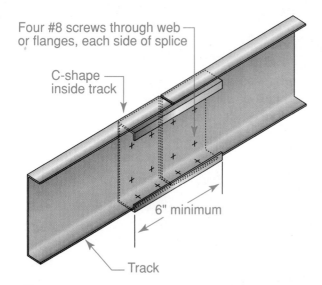

Four #8 screws through web
or flanges, each side of splice

C-shape
inside track

6" minimum

Track

Track Splice
Minimum Length A track splice showing the 3" lap of each section.

Joist tracks may also be called *rim tracks* because they are attached to the rim of a foundation (or a header).

Track splices are used whenever a single section is not long enough to extend the entire length of a foundation wall or header. The minimum length of a track splice is 6". This allows for 3" of lap on each section.

Laying Out Floor Joists

Floor joists may be laid out from one end wall to another or from one side wall to another. The floor joists should run in the same direction that the roof trusses or rafters will.

When beginning the joist layout, place joist tracks for both sides on one side of the structure. Temporarily clamp their webs together. This allows you to mark both tracks for layout at the same time.

Start the layout by marking for a joist at one end. Mark the location of the next joist on the joist track so it will be in line with the first roof member. It will be 24" or less from the end joist.

Continue from that mark along the length of the joist tracks, making a new mark every 24". Place an X on the side of each mark where the flange of that joist will be. The flanges must all be oriented in the same direction, or on layout. The open side of the joist should face away from the starting point.

Continuous-span joists span the entire floor opening. The Xs are all on the same side of the joist location marks.

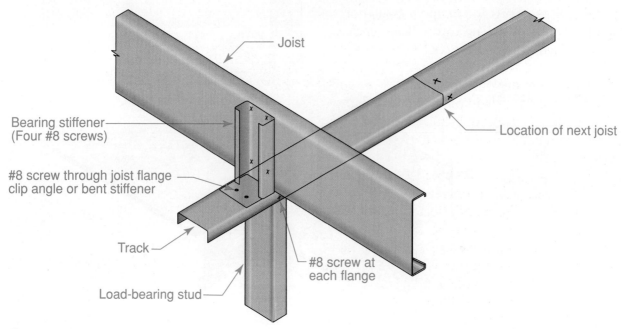

Joist

Bearing stiffener
(Four #8 screws)

#8 screw through joist flange
clip angle or bent stiffener

Track

Load-bearing stud

Location of next joist

#8 screw at
each flange

Continuous-Span Joist
Floor Support A continuous-span joist spans the entire floor.

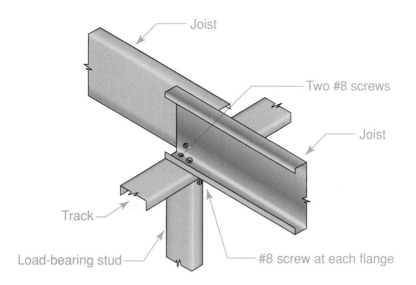

Joist

Two #8 screws

Joist

Track

Load-bearing stud

#8 screw at each flange

 Lapped Joist
Intermediate Support Lapped joist across a load-bearing stud.

A non-continuous joist, or *lapped joist,* is in two pieces. The pieces meet and overlap over an intermediate support. The Xs for the opposing track are on the side opposite the joist location marks.

Fastening Joist Tracks

A steel frame must be securely attached and anchored to the foundation. This is done with either embedded or epoxied anchor bolts. The bolts extend through a hole that is punched in the bottom of the joist track. Washers and nuts are tightened onto the bolts to hold the frame in place. Embedded anchor bolts are set in place before a concrete foundation is poured. When the concrete cures and hardens, it holds the bolt securely in place.

Epoxied anchor bolts are installed in cured concrete or in concrete block. A hole is drilled into the foundation and filled with epoxy. Epoxy is a type of adhesive. A threaded bolt is placed into the hole. When the epoxy hardens, it provides a strong bond that holds the bolt in place.

The joist tracks must then be fastened to the foundation. Place one track on each end wall or side wall. Stand the track on one flange with the web toward the outside of the foundation wall, as on page 848. Keep the web of the joist track aligned with the edge of the foundation wall by tacking the track to the wall. Place a clip angle over the anchor bolts at each anchor bolt location. A **clip angle** is a small piece of galvanized steel attached to a structural member to accept a

 Anchor Bolt
Anchored Embedded anchor bolt.

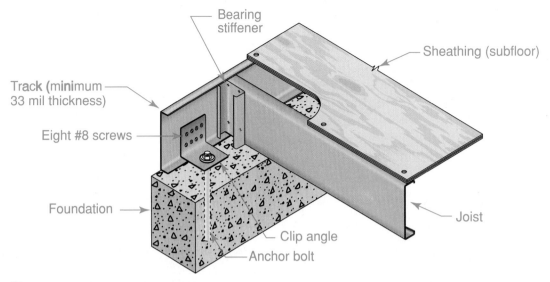

Bearing stiffener

Sheathing (subfloor)

Track (minimum 33 mil thickness)

Eight #8 screws

Foundation

Joist

Clip angle

Anchor bolt

 Clip Angle and Anchor Bolt
Firm Foundation Track attached to foundation with clip angle and anchor bolt. The web faces the outside of the foundation wall.

structural load. If the clip angle that is used is made on site (rather than a pre-engineered clip being used), it must be made with material that is the same thickness as the studs or joists it is being used to connect. Clamp the clip angles to the joist track to hold them in place. Attach them to the joist track with eight #8 screws as shown above. Place a washer and nut over the anchor bolt to secure the clip angle to the foundation.

Setting the Joists

The joists are set into the track after the track is secured. Turn the joist at an angle to fit between the flanges of the joist track. Twist the joist into position and set it inside the track. Keep the web sides of the joists oriented in the same direction. Position them on the same side of the mark on the track.

Use a triangular framing square to adjust the end of each joist so it is perpendicular to the track. Allow a ⅛" gap between the track and the end of the joist to prevent squeaks.

Screw the joist into position through its top and bottom flanges and the top and bottom track flanges. Use a minimum of one #8 screw at each point. Use screws with low-profile heads on the top flange to allow for the subflooring.

Floor Joists and Joist Tracks
On Track Fastening the floor joists in the joist tracks.

For lapped joists, lap the second joist toward the wall from which the layout was started. Otherwise, the distance between the starting point and the lapped joists will be greater than 24", which will cause problems when laying the subfloor. Check the layout at the intermediate supports and screw the lapped joist into the support. Set the remaining joists across the entire structure.

Bracing the Joists

Braces prevent joists from rolling or twisting in the tracks. The top flanges are braced with sheathing or subflooring. The bottom flanges are braced with gypsum board or a steel strap and blocking or bridging. Floor spans of 12' or less do not require bracing on bottom flanges.

Web Stiffeners Web stiffeners are added to joists to prevent them from bending under the weight of floor loads. A web stiffener is made from stud or track material. It should be the same thickness and depth as the floor joist.

Stiffeners must be added when the joists and joist tracks are installed. The stiffener is screwed to either side of the web. A stiffener is required under every load-bearing wall. Stiffeners are also required where non-continuous joists lap at an intermediate support.

Subflooring Be sure to check your local building code for the type of material to use for the subflooring and how it should be attached. When the joists are spaced 24" OC, $^{23}/_{32}$" tongue-and-groove APA-rated sheathing plywood is used. Generally, the sheets of plywood are attached to the floor joists with #8 screws, 6" OC on the edges, and 10" OC on the joists in between. The sheathing should be tight against the joists. Use *bugle-head screws* and a drywall screw gun with an adjustable depth setting.

Reading Check

Analyze *What may occur if floor joists are not braced properly?*

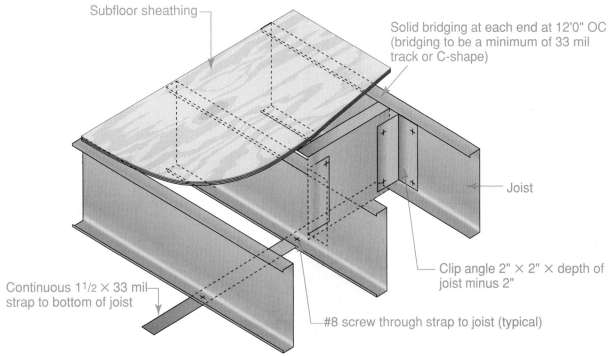

Subfloor sheathing

Solid bridging at each end at 12'0" OC (bridging to be a minimum of 33 mil track or C-shape)

Joist

Clip angle 2" × 2" × depth of joist minus 2"

Continuous 1¹/2 × 33 mil strap to bottom of joist

#8 screw through strap to joist (typical)

Joist Bracing
Roll Prevention Joist bracing using a steel strap.

Builder's Tip

Walls

What is in-line framing?

Construction and placement of the walls can begin after the floor joists, joist bracing, and subflooring are installed. Walls are either load bearing or nonload bearing (partitions).

Load-Bearing Walls

Load-bearing walls help support the weight of the house above them. Steel framed structures typically use in-line framing, meaning that all load-bearing studs must be aligned with the trusses, joists, or rafters above and below the wall. To properly carry the load, each stud must butt tightly inside its track. If the stud is not tight, the load will placed directly on the screws that attach the stud to the track. This could cause them to fail. Special instructions for panelized walls are included in the Step-by-Step Application on page 855.

Wall Length The maximum length of a wall depends mostly on the workforce available at the job site. The longer the wall, the larger the workforce needed to build and place it. Longer walls tend to twist if they are not placed properly, which may result in damage to the wall.

Walls are easier to frame in short sections and require a smaller workforce to put in place. Tracks for short sections are often spliced together to make longer walls. Care must be taken when splicing wall sections. The tracks must be kept aligned and straight. Two walls meet at an intersection.

Intersections Intersecting walls occur at the corners of the house where wall sections meet. Depending on the stud sizes being used—3½" or 5½"—the bottom track will be shorter than full length.

To form an intersecting wall, cut the top and bottom tracks to the correct length. The bottom track should be 7 or 11 inches shorter (two times 3½" or 5½").

Installed Stud
Inside Track A properly installed stud is butted firmly inside the track.

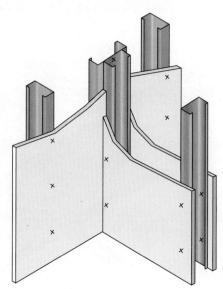

Intersecting Wall
Wall Sections Two walls meet at an intersection.

Braced Walls
Proper Support Temporary wall bracing.

Setting and Placement Load-bearing walls must be prepared, placed, spliced, and connected. When anchor bolts are used, measure their locations in the foundation. Make holes in the bottom track of the wall panels at these locations. Place temporary bracing material near the foundation so it is ready. Move the wall section into position. Tilt the wall up and position the bottom track over the foundation bolts.

Clamp the temporary bracing material to the wall studs in two or three locations. Install a brace every 8' to 12' along the wall. Secure each brace to a stud with a #10 hexhead screw before removing the clamps. Secure the bottom of the braces to a stake driven into the ground to hold them in place. Adjust the bracing to plumb the walls. Repeat this process until all load-bearing walls are in place.

Builder's Tip

THERMAL PERFORMANCE The thermal efficiency of a steel building is dependent upon several things. All joints must be tight to prevent air infiltration. Steel studs must be spaced properly, not clustered or grouped. Clustering forms cold spots. Insulation should fill the entire wall cavity, including the open sides of the C-shaped steel studs. Additional exterior foam sheathing should be used in colder climates.

JOB SAFETY

RAISING AND PLACING WALLS Raising and placing long wall sections require extra safety precautions. Be sure the on-site workforce is adequate to lift the length of wall being placed. Have temporary bracing and support beams readily available. Do not raise walls when strong winds could cause a lack of control. Securely brace the wall after it is in position.

Splicing and Connecting Sections Splicing wall sections must conform to accepted engineering practices. Center a 6" or larger piece of C-shaped material inside the two sections of track to be spliced. Screw the splice material through the flanges on both sides of the track with #8 screws. When each section is properly positioned, plumb it. Attach the walls to the frame at the corners with #10 screws. Leave the bracing and bottom track in any door openings until roof framing and permanent bracing are completed.

Attaching Sheathing After the wall is up and braced, Type II plywood or OSB sheathing can be attached. If openings in the wall are not extensive, the sheathing can act as bracing and protect the wall from racking and twisting. *Racking* occurs when the wall shifts and studs are forced out of plumb.

To be effective, the sheathing must cover the full height of the wall from top track to bottom track. It should be installed with the long dimension parallel to the stud framing. It should also be fastened tightly to the steel members with #8 self-piercing screws to draw the sheathing tight against the studs.

Interior Load-Bearing Walls Interior load-bearing walls increase the capacity of the house to resist shear forces, such as those

Bracing an Exterior Wall
Panel Bracing Sheathing can brace an exterior wall and prevent twisting and racking.

that occur during earthquakes. That is because they stiffen the structure. They may also help increase the load capacity of the floors above the main floor. Mark the top and bottom track for layout. Anchor the bottom track in place. Then secure studs in each end of the bottom track. Next, position the top track at the ends with intermediate studs. Install the remaining wall studs. After the wall is standing and properly positioned, install headers, X-bracing, or sheathing.

Nonload-Bearing Walls

Interior nonload-bearing walls, or *partitions,* do not support or carry the weight of the structure. They are built to enclose rooms, closets, and other spaces. In-line framing is not required for these walls. The

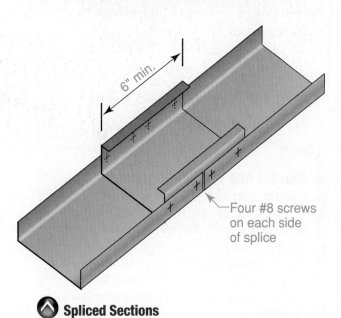

Four #8 screws on each side of splice

6" min.

Spliced Sections
Tied Together Splicing two wall sections together.

Jon Muzzarelli

structural members of nonload-bearing walls can be of a thinner gauge material than those of load-bearing walls. However, residential structures may need 33-mil (29-gauge) studs to prevent bending or damage during construction.

Nonload-bearing wall framing is similar to load-bearing wall framing. Generally, walls are stick built. However, they may also be panelized and then raised into position.

Cabinets and shelving may be attached directly to partition studs and to blocking materials in the walls with #8 2" self-drilling screws. Wall studs must be 33 mil (29 gauge) or thicker. Wood cabinet blocking is installed as shown in Cabinet Blocking. One end of the wood block is notched so it fits over the lip of the stud. When steel blocking is used, the flanges of the track material are notched.

Stick-Built Framing Stick-built framing of a nonload-bearing wall begins with attaching the bottom wall track to the floor. The top track of the wall is positioned using a stud and a level. (The location of the top track for sloped walls or a cathedral ceiling can be set with a plumb bob.) The top track is screwed to the ceiling joists, the bottom chord of a truss, or a second-story floor joist.

Where interior walls run parallel to the joists or trusses, pieces of track or stud material are placed every 24" as blocking material. Blocking material should be cut 2" longer than the distance between the trusses. One inch is then clipped from the flanges on each side of the blocking material. This allows the webs to overlap. The blocking is screwed on both ends with #8 self-drilling screws. The track is then marked for a stud spacing of 16" to 24".

The C-shaped studs must open toward the starting point of the layout, especially if they are 18-mil studs. The studs are twisted into the wall layout and the flanges on both sides of the tracks are secured with #6 or #8 self-piercing screws.

Rough Openings The rough openings in nonload-bearing walls do not need to be as strong as those in load-bearing walls. When framing rough openings, allow 1½" on each side of a door opening to install wood studs. Line all interior and closet door openings with wood studs to provide extra support at

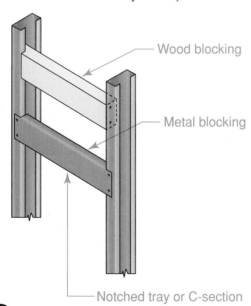

Cabinet Blocking
Blocking Wood or metal blocking can be used to support cabinets.

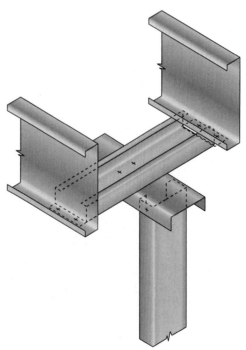

Metal Blocking
Track or Stud Blocking material between joists is used when the wall is parallel to the joists or trusses.

the hinge and strike plate. Place steel studs behind the wood studs with the flat side facing the opening.

Use wood or steel for the header of the opening. When using steel, cut the web back so the flanges provide an extra inch on each side. Install the cripple studs above the rough opening as necessary.

Reading Check

Recall *Why is in-line framing not required for interior nonload-bearing walls?*

Panelized Walls

As discussed in Section 29.1, panelized walls are pre-assembled. The location of the studs in the wall is governed by the roof and floor layout. It is important to accurately lay out the wall studs with reference to the roof and joist framing.

The following information applies only to panelized walls. Setting and placement are the same as for stick-built walls.

Wall Layout Panelized walls are built to engineering specifications and tolerances on templates such as platform tables and jigs.

Panelized Walls
Pre-Assembled Sections These panelized wall sections were built on a flat surface.

Builder's Tip

PUNCHOUTS Holes and punchouts are placed in steel studs to provide paths for utilities. When placing these paths, make sure both pre-punched and new holes line up. Place punchouts in the studs under windows if utilities will be run through them. Remember that the number of holes and punchouts will reduce the strength of a stud. Never punch a hole in the flange of a stud, joist, or track. Use grommets to cover the edges of holes and punchouts. The grommets prevent damage to wires. They also prevent the electrolytic corrosion of steel caused by contact with copper pipe or pipe hangers.

When laying out a panelized wall on a platform table, place the top and bottom tracks on the straight edge. Start with a stud at one end of the wall. Place a line on the flanges of the top and bottom tracks where the web of the stud will be located. Place an X on one side of this line to indicate the location of the stud flanges. Continue marking the locations of all studs every 16" or 24" on center.

Arrange and temporarily clamp wall members with all of the webs facing in the same

Stud Location
Locate Flanges Note the line and the X.

(l & r)Tim Fuller

direction. The studs must fit tightly against the straight edge at the end of the wall.

Rough Openings After the wall is laid out, the rough openings for doors and windows are marked. Locate the door and window locations on the architectural drawings. Check the size of the openings. Mark the center of each opening on the top and bottom tracks. Add 12" to the width of the window openings. Using a tape measure, center the dimensions over the marks on the track. Mark the location at each end of the tape to indicate the location of the king studs. Place an X on the side of the mark away from the window. The webs of the king studs will be on the rough-opening side.

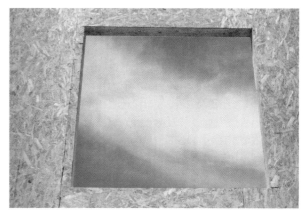

 Rough Openings
Rough Opening for a Window The measurements for a rough opening must be made carefullly to ensure proper fitting for the window.

Step-by-Step Application

Wall Stud Assembly After the layout is complete, assembly of the panelized wall can begin.

Step 1 Separate the top and bottom track members that were placed on the straight edge of the platform table.

Step 2 Install a stud at each end of the wall between the top and bottom tracks.

Step 3 Temporarily clamp the stud flanges to the track flanges at each end with locking C-clamps.

Step 4 Tap the tracks with a hammer to seat the top and bottom of the studs as tightly as possible.

Step 5 To prevent the studs from twisting, attach the flange of each stud to the flange of the track on each side of the wall using a #8 low-profile screw.

Step 6 Install the studs so all open sides face the same direction on parallel load-bearing walls.

Step 7 Align the punchouts in the studs to provide straight paths for plumbing and electrical runs.

Step 8 Install the king studs at the rough openings. Do not install studs at the markings between the king studs. These markings indicate the position of the cripple studs.

Step 9 Continue down the length of the wall until all studs are screwed into place. Do not remove the wall panel from the table until the header framing is complete.

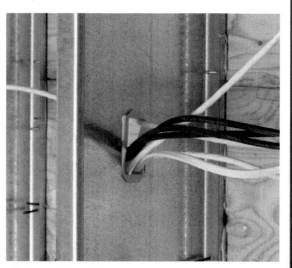

 Punchouts
Aligned Punchouts are aligned for straight runs of electrical wiring. Notice the grommets that cover rough edges.

Adding 12" of width simplifies header assembly. It allows for two trimmer (jack) studs, one on each side of the header, and a wood stud on each side of the opening.

Framers may vary the length of the headers. Two trimmer studs may not be required at every opening. However, standardizing header length helps to simplify cut lists.

For wall stud assembly, see the Step-by-Step Application on page 855.

Box Header Assembly A box header is a common header that is built from standard C-shaped steel framing members. A header jig is used to support the steel members and to keep them straight during assembly. A header jig can be built from C-shapes attached to a table and placed at the exact dimensions of the rough opening. The header can be built between these members.

To frame a box header, cut a section of wall track 2" longer than the header. Snip the flanges of the track back 1" at each end. Bend the web toward the flanges with a hand seamer. Make sure the bend is clean and straight.

Clamp the header track to the jig. Screw the web of the header track to the flanges of the C-shapes with two #8 screws spaced 24" OC.

Cut web stiffeners from 3½" stud material. Install the stiffeners in each end of the header with the flat side of the stiffener facing out. Attach four #8 screws through the web of the header pieces into each side of the stiffener flanges. Insulate the header before it is installed in the wall.

L-Headers The L-header consists of one or two angle pieces that fit over the top track and one leg extending down the side of the wall above window or door openings.

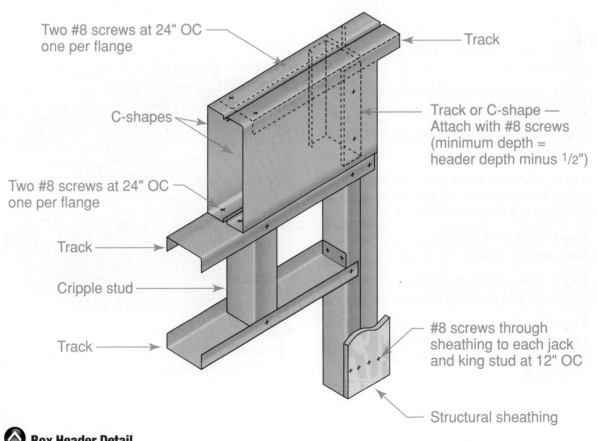

Two #8 screws at 24" OC one per flange

C-shapes

Two #8 screws at 24" OC one per flange

Track

Cripple stud

Track

Track

Track or C-shape — Attach with #8 screws (minimum depth = header depth minus 1/2")

#8 screws through sheathing to each jack and king stud at 12" OC

Structural sheathing

Box Header Detail
Box Header A box header is built from C-shapes.

Each angle is fastened to the top track above an opening with minimum #8 screws spaced at 12" OC. The "L" angle is placed on both sides of the wall opening to form a double angle L-shaped header (double L-header).

The L-header saves labor because unlike box headers, no special fabrication is required and the number of screws is reduced. The L-shape itself spans the opening for the header.

Bracing Before a panelized wall is removed from the platform table, it must be checked for squareness. Then it must be braced. Measure the panel diagonally from corner to corner. If these measurements are the same, the wall is square.

Lay extra bracing, studs, or truss material across the opening. X-bracing is the most effective at this stage. X-bracing consists of diagonal steel straps. The bracing is attached to the walls with gusset plates and screws so

it is permanent. Installation of the bracing straps must be inspected to ensure that the correct number of fasteners is used.

Reading Check

Explain How is the location of the studs in the wall determined?

Roofs

What precautions should be taken with framing installed in high wind areas?

Steel roof framing has several advantages over traditional wood-frame construction. With minimal support bracing it can provide more attic space. Fewer members are required. Complex roof designs cost less than when framed with wood.

Installing Ceiling Joists

The procedure for setting ceiling joists is similar to that for floor joists on page 858. Mark the layout for the top track. Start your layout at the same end for both sides. Next, measure and mark the layout over the headers. Also, mark any locations where there are no wall studs and the layout is not obvious.

Install the ceiling joists in the tracks one at a time. Move from where you began the layout to the end of the structure. Anchor the joists at the top of the track using two #10 screws. Install 2×4×33 mil C-shape

Bracing
Diagonal Support X-bracing using steel straps.

JOB SAFETY

CEILING JOISTS Once secured to the tracks with screws, the ceiling joists may be used temporarily as a work platform when installing the rafters. Before placing any weight on the joists, make sure that they are properly braced. Make sure all load-bearing walls below the ceiling joists are secured in place.

Tim Fuller

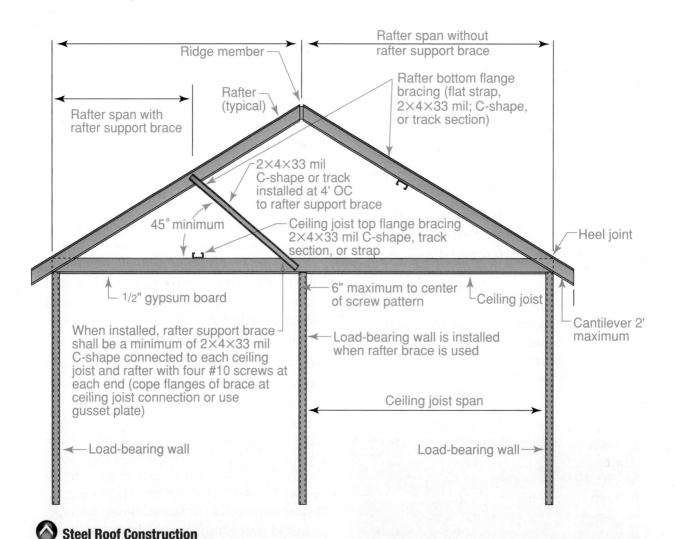

Rafter span without rafter support brace

Ridge member

Rafter (typical)

Rafter bottom flange bracing (flat strap, 2×4×33 mil; C-shape, or track section)

Rafter span with rafter support brace

2×4×33 mil C-shape or track installed at 4' OC to rafter support brace

45° minimum

Ceiling joist top flange bracing 2×4×33 mil C-shape, track section, or strap

Heel joint

1/2" gypsum board

6" maximum to center of screw pattern

Ceiling joist

When installed, rafter support brace shall be a minimum of 2×4×33 mil C-shape connected to each ceiling joist and rafter with four #10 screws at each end (cope flanges of brace at ceiling joist connection or use gusset plate)

Load-bearing wall is installed when rafter brace is used

Cantilever 2' maximum

Ceiling joist span

Load-bearing wall

Load-bearing wall

Steel Roof Construction
Roof Framing Steel roof construction with joists and rafters.

top-flange bracing on the joists. Blocking must be installed every 12' OC. The blocking keeps the joists from rolling in the tracks.

Preparing the Rafters

All rafters must be installed with the flat sides facing the same direction. The flat side of the rafter at the top track must be in contact with the flat side of the ceiling joist.

Cut the rafters to length. Cut the top end of the rafter to match the slope of the roof. The ridge plumb cut allows the rafter to lie flush against the ridge member. Next, use 2" × 2" clip angles to attach the rafters to the ridge member.

Rafter tails may be cut in advance or after roof framing is complete. However, the fascia material will remain straighter if the tail cuts are made after the roof is framed.

Setting Ridge Height

The ridge height can be determined by the same methods as for wood construction. See Chapter 17, "Basic Roof Framing," for complete instructions.

Common Rafter Method The common rafter method uses the length of the common rafters to set the ridge height. The length of the rafters must be accurate. The plumb cut at the top of the rafter must match the slope of the roof.

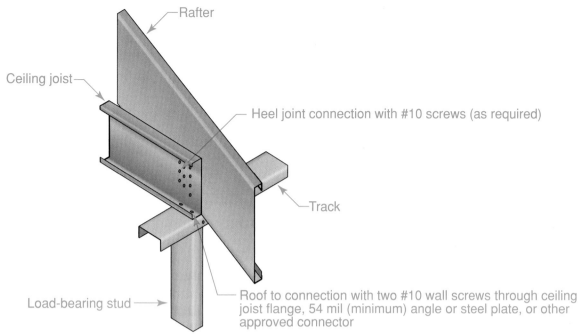

Rafter

Ceiling joist

Heel joint connection with #10 screws (as required)

Track

Load-bearing stud

Roof to connection with two #10 wall screws through ceiling joist flange, 54 mil (minimum) angle or steel plate, or other approved connector

 Rafter Connection
Flat to Flat The flat side of the rafter must contact the flat side of the joist.

Calculation Method With the calculation method the pitch of the roof and the rafter length are used to determine the ridge height. The steel rafters rest on the top outside edge of the top track.

A steel rafter must not be notched as is done with a wood rafter. Notching will reduce its strength. The lack of a notch must be considered when calculating ridge height.

Ridge member consists of C-shape inside a track section fastened with #10 screws at 24" OC through top and bottom flanges

#10 screws in each leg of clip angle

Clip angle

Rafter (typical)

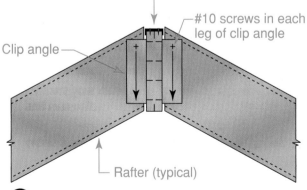 **Attach with Clip Angles**
At the Ridge Attaching the rafters to the ridge member with clip angles.

Framing with Trusses

Steel roof trusses can be manufactured off site or custom built on site.

Manufactured Trusses Manufactured steel trusses are made in different shapes and styles. One style is shown on page 860. Several factors are important when choosing trusses. They should be as cost effective as wood trusses. They must be durable enough to withstand normal shipping and handling. They must be able to support the weight of workers framing and finishing the roof. They must be light enough for work crews to lift, move, and place.

Site-Built Trusses Site-built trusses must be built according to approved engineering designs. They are framed with C-members having either mitered cuts or gusset plates at the connection points, as on page 860. When trusses are mitered, they may be assembled with the members in one plane. The thickness of the truss is the same as the thickness of a C-member. *Gusset plates* are made from pieces of track and have only one flange.

 Roof Trusses
No Gussets Roof trusses are built in many styles and shapes.

The top and intermediate chord members are positioned flat side up. Bottom chord members are positioned flat side down.

General Framing Details

The following details apply to framing with either rafters or trusses.

Roof Hold-Downs The roof framing is fastened to the top plate with screws. The number and type of screws that make this connection are determined by wind-load data. Two #10 screws are sufficient for 70 mph Exposure C wind loads or 90 mph Exposure A or B wind loads.

Uplift connectors or hold-down clips are required in higher wind conditions. Prescriptive-method tables and charts should be used to determine uplift load and connector requirements.

 Heel Gusset
Site Built Using a heel gusset to join the chords.

 Roof Hold-Downs
Connectors Clips and connectors must be solidly screwed to framing.

 Fascia
Plumb Fascia framing that is parallel to the side of the house.

Roof Fascia The roof fascia provides a finished look to the end of the rafter tails. The fascia can be installed so that it is parallel to the side of the house (perpendicular to the ground).

Roof fascia can also be installed so that it is perpendicular to the rafters.

Mathematics: Computation

Using a Construction Calculator
Determining the length and number of studs needed for a roof as well as finding the length or a hip or valley rafter is much easier when you use a construction calculator. Determine the length of a hip or valley rafter for a 9/12 roof that has a run of 16'-6".

Starting Hint Input the pitch, then the run.

 Alternate Fascia
Angled Fascia framing that is perpendicular to the rafters.

(t & b)Tim Fuller

Roof Rake The **roof rake** is that portion of the roof frame that extends beyond the walls on the gabled ends. The rake may overhang as much as 12". Rakes are formed using standard C-shaped members.

The rake supports the roof sheathing. Uplift loads determine the size of the rake. Deep rakes require that an engineer approve the design using uplift load data.

Using lookouts is another way of framing a rake. The lookouts are installed 2' OC from the gable end to the barge rafter.

 Roof Rake
Extended Rake A roof rake extends beyond the house walls.

 Lookouts
Extra Support Lookouts can be used to form a rake.

Builder's Tip

SOFFIT DEPTH The International Residential Code restricts the distance that light-gauge steel rafter tails can cantilever (extend unsupported) past the walls. The maximum length of a rafter tail is 24". This distance is measured horizontally from the face of wall to the lowest point of the rafter tail.

Enclosed Soffits A soffit is the underside of the roof overhang. Soffits cover the truss or rafter tails. Enclosed soffits are usually covered with aluminum, vinyl, or wood. The enclosure may be formed to match the pitch of the roof. The bottoms of other enclosed soffits may be horizontal and parallel to the ground. This surface often has ventilation openings, which allow air into the building.

 Enclosed Soffit

Horizontal Method An enclosed soffit with ventilation openings. *What is the purpose of the ventilation openings?*

Section 29.3 Assessment

After You Read: Self-Check

1. What is the difference between continuous-span and non-continuous joists?
2. At what intervals should temporary bracing be installed?
3. What are the advantages of steel roof framing over traditional wood-frame construction?
4. Name the two methods used to determine roof ridge height.

Academic Integration: Mathematics

5. **Fastening Sheathing** Sheathing plywood comes in 8' by 4' sheets and is used as subflooring. It is fastened parallel to the joists along its length. The joists are 24" OC. How many screws will you use if you fasten the plywood with screws that are 6" OC along the joists at the sides and 8" OC along the joists in the middle?

Math Concept Drawing a sketch of a problem situation is an effective problem solving strategy for two- and three-dimensional geometric problems.

Step 1: Draw a sketch of the project.

Step 2: Figure the number of screws needed for one side. Use the sketch to check your work. Multiply by 2, the number of sides.

Step 3: Use the same technique to figure the number of screws needed for the middle joists. Use the sketch to check your work.

Step 4: Add to find the total number of screws needed.

Go to connectED.mcgraw-hill.com to check your answers.

Tim Fuller

Chapter Summary

Steel frame houses are designed using the performance or the prescriptive method. Residential steel framing is either stick-built, panelized, or pre-engineered.

Most tools used in steel framing are electrically, hydraulically, or pneumatically powered. In some situations, a gasoline engine may power larger tools. A few hand tools, such as clamps, are also used. Mechanical fasteners, such as screws, nails, and pins, attach steel framing members. Framing screws attach steel to steel. Sheathing screws attach exterior sheathing to steel. Processes such as welding and clinching are also used.

Floor joists are laid out starting from the same end of the building as the roof members. The layout may be from one end wall to another or from one side wall to another. Joist flanges must all be oriented in the same direction. Load-bearing walls support the weight of the house above them. Each stud must butt tightly inside its track to properly carry the axial load. Interior nonload-bearing walls are not intended to carry axial loads and are typically used to enclose rooms. All rafters must be cut with the flat sides facing in the same direction. The flat side of the rafter at the top track must contact the flat side of the ceiling joist. Either the common rafter method or the calculation method can determine the height of the roof ridge.

Review Content Vocabulary and Academic Vocabulary

1. Use each of these content vocabulary and academic vocabulary words in a sentence or diagram.

Content Vocabulary

- cold-formed steel (p. 832)
- performance method (p. 832)
- prescriptive method (p. 833)
- feathering (p. 836)
- pullout capacity (p. 842)
- welding (p. 843)
- clinching (p. 843)
- in-line framing (p. 844)
- axial load (p. 844)
- joist tracks (p. 845)
- clip angle (p. 847)
- roof rake (p. 862)

Academic Vocabulary

- sequence (p. 832)
- remove (p. 839)

Speak Like a Pro

Technical Terms

2. Work with a classmate to define the following terms used in the chapter: *plates* (p. 832), *mils* (p. 832), *plasma* (p. 839), *framing screws* (p. 842), *rim tracks* (p. 846), *lapped joist* (p. 847), *bugle-head screws* (p. 849), *racking* (p. 852), *partitions* (p. 852), *gusset plates* (p. 859).

Review Key Concepts

3. State the three types of steel frame construction.

4. List the tools used in steel framing.

5. Differentiate between welding and clinching.

6. Demonstrate how to lay out steel floor joists.

7. Explain the importance of in-line framing.

8. Tell how to set steel ceiling joists.

Critical Thinking

9. Analyze What size wall would you expect a small workforce to be most efficient at constructing? Explain your answer.

Academic and Workplace Applications

STEM Mathematics

10. Estimating Labor Costs Installing a floor system typically takes 72 man hours per 1,000 sq. ft. of floor area. A supervisor, a carpenter, a helper, and a laborer work together to install the floor system in a storage building that is 54 ft × 32 ft. Estimate the total number of man hours needed to complete the job.

Math Concept A proportion is an equation involving two equivalent ratios.

Step 1: Calculate the square footage of the floor area in storage building.

Step 2: Let x represent the unknown quantity, the total man hours needed for the job. Set up a proportion with two equivalent ratios. One compares hours and the other compares square footage.

Step 3: Solve for the unknown quantity.

Step 4: Divide the total hours among the four workers. Round up to the next hour.

STEM Science

11. Shear Force Both steel and wood interior load-bearing walls increase the capacity of a house to resist shear forces. They may also help increase the load capacity of the floors above the ground floor. Shear force often affects houses during earthquakes. Use the Internet to learn more about shear forces. Write one or two sentences describing how shear force or stress differs from perpendicular stress.

21st Century Skills

12. Media Literacy Read through your local newspaper ads and trade industry magazines that advertise and sell framing tools. Select two ads. Try to determine what audience the ads are directed toward. State how the ad draws the reader's attention. What techniques are used to create desire or arouse interest? Write a short paragraph summarizing your discoveries.

Standardized TEST Practice

Multiple Choice

Directions Choose the word or phrase that best answers the following questions.

13. Which of the following is not a type of steel frame construction?

   a. stick-build
   b. prescriptive
   c. panelized
   d. pre-engineered

14. Aviation snips can cut steel up to which thickness?

   a. 55 mil
   b. 43 mil
   c. 66 mil
   d. 69 mil

15. Interior load-bearing walls increase the capacity of the house to do what?

   a. resist extreme weather
   b. resist shear forces
   c. maintain shape
   d. minimize fire damage

TEST-TAKING TIP

If you have time, review your notes shortly before you take the test.

*These questions will help you practice for national certification assessment.

Mechanicals

Section 30.1
The Plumbing System

Section 30.2
The Electrical System

Section 30.3
HVAC Systems

Chapter Objectives

After completing this chapter, you will be able to:

- **Describe** or sketch a simple plumbing system.
- **List** the common materials used by a plumber.
- **Recognize** the various types of piping used for water supply and DWV systems.
- **Describe** the basic elements of an electrical system.
- **Identify** the three basic kinds of circuits.
- **Explain** how split-system air conditioners work and identify basic parts.

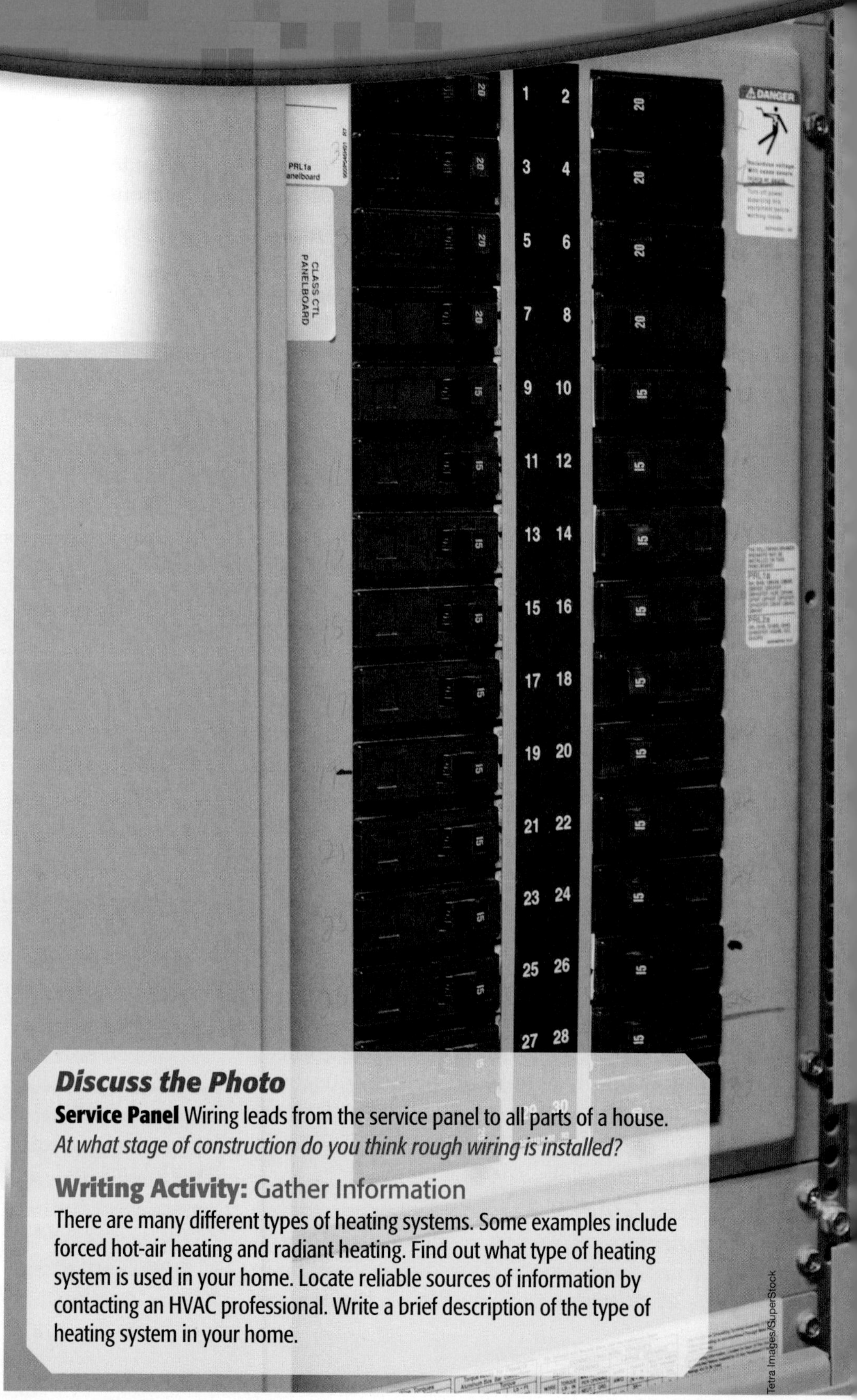

Discuss the Photo

Service Panel Wiring leads from the service panel to all parts of a house. *At what stage of construction do you think rough wiring is installed?*

Writing Activity: Gather Information

There are many different types of heating systems. Some examples include forced hot-air heating and radiant heating. Find out what type of heating system is used in your home. Locate reliable sources of information by contacting an HVAC professional. Write a brief description of the type of heating system in your home.

Tetra Images/SuperStock

Chapter 30 Reading Guide

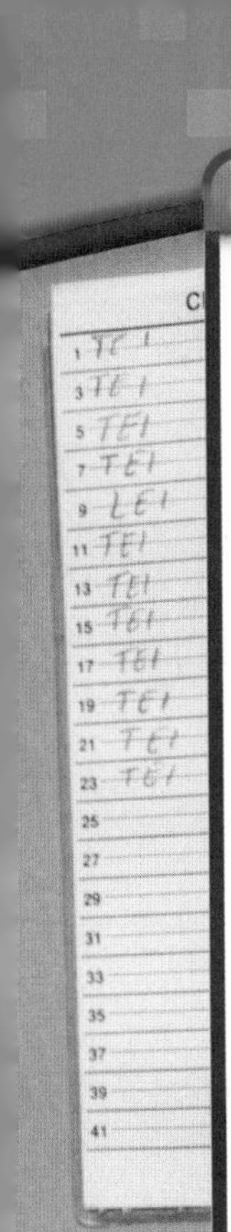

Before You Read Preview

The word *mechanical* will appear regularly throughout this chapter. Before reading the chapter, write down a definition of the word based on your knowledge. After you have finished reading the chapter carefully, rewrite your definition.

Content Vocabulary

- mechanicals
- fixture
- service main
- trap
- drain field
- circuit
- receptacle
- cell
- humidifier
- heat pump
- refrigerant
- infiltration
- air conditioning

Academic Vocabulary

You will find these words in your reading and on your tests. Use the academic vocabulary glossary to look up their definitions if necessary.

- rural
- circulates

Graphic Organizer

As you read, use a chart like the one shown to organize information about content vocabulary words and their definitions, adding rows as needed.

Content Vocabulary	Definition
fixture	any plumbing device that receives or drains water

Go to connectED.mcgraw-hill.com to download this graphic organizer.

The Plumbing System

Plumbing System Basics

What are some of the skills that a plumber must have?

The term **mechanicals** refers in general to plumbing, electrical, and heating/ventilating/air-conditioning (HVAC) systems. Mechanical systems should always installed by properly licensed subcontractors trained to do the work.

The various mechanical trades work at different times during the construction process. For example, plumbers are generally the first to arrive. They begin work during the rough framing stage to install rough drain lines. They return later to install finish plumbing. Because each trade requires multiple visits, their work must be carefully scheduled. The general contractor is responsible for coordinating the work of mechanical trades.

Each mechanical trade requires specialized tools and a great deal of knowledge. This chapter provides only a general introduction to mechanical systems.

The plumber installs the piping system for water and drainage, including all of the fixtures. In plumbing, a **fixture** is any device that receives or drains water. A bathtub is one example. The plumber must know the sizes of fixtures so that pipes will be in the correct location for each one.

Plumbers work with many different materials. Because of this, they must possess a wide variety of skills. These skills include woodworking, metalworking, welding, *brazing*, soldering, caulking, and pipe threading. Their hand tools include wrenches, reamers, drills, braces and bits, hammers, chisels, and saws. Plumbers also must be able to use power tools, such as portable drills and reciprocating saws. They also use oxyacetylene and propane torches for welding, brazing, and soldering.

As a house is being planned, the designer or architect determines the general location and type of fixtures. This information is shown on floor plans. However, the plumber determines the exact position of each fixture during installation. The plumber is also responsible for locating, sizing, and installing the pipes to serve those fixtures. Detailed specifications for installing plumbing systems are outlined in various publications, including the Uniform Plumbing Code (UPC), International Plumbing Code (IPC), and the International Residential Code (IRC). There may also be local codes that are suited to a particular town or region.

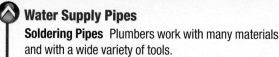

Water Supply Pipes
Soldering Pipes Plumbers work with many materials and with a wide variety of tools.

©Image Source

On the plans, individual fixtures are represented by standardized symbols. (A Plumbing Symbols table is in the **Ready Reference Appendix**.) The color, model number, and manufacturer of every fixture can be found on the plumbing fixture schedule.

Most plumbing installations must be checked by a plumbing inspector at two stages of construction: the rough-in stage and the finish stage. If the house has a concrete slab foundation, a plumbing inspection must also be performed prior to placing the concrete.

Understanding a Plumbing System

A plumbing system brings fresh water into the house and removes solid and liquid wastes. The two portions of the system are referred to as the supply side and the waste side. The waste side is also called the *drain/waste/vent (DWV) system*.

A typical plumbing system consists of three basic types of pipes:

- *Supply pipes* are small-diameter pipes that distribute hot and cold water to fixtures. The pipes are usually made of copper. Normal operating pressures inside the house range from 40 psi to 80 psi.

- *Waste pipes* are large-diameter pipes made of plastic or cast iron. They convey liquid and solid wastes away from the house under atmospheric pressure only.

- *Vent pipes* are large-diameter plastic pipes. They encourage drainage and remove gases by balancing atmospheric pressure in the waste pipes.

A **service main** is a pipe that brings water to the house. It is sometimes called a *water service pipe* and is connected at the street to the municipal water system. Water pressure inside the service main can be considerably higher than pressure inside the house but is typically less than 160 psi.

A water meter connected to the service main records the amount of water used. Waste flows out of the house by means of

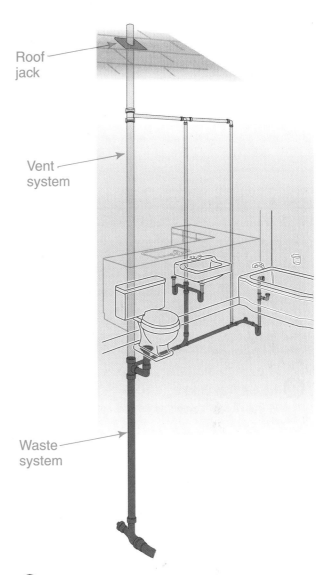

Roof jack

Vent system

Waste system

A Basic Waste System

Drain, Waste, Vent This schematic drawing shows the rough-in plumbing with the finished fixtures set in place. The plumbing is installed in a partition wall. A countertop sink is on one side, a lavatory sink is on the other side.

gravity to the public sewer or to a septic tank. Any horizontal lengths of waste pipe must be sloped so that they will drain properly.

An important component of the DWV system is a collection of traps. A **trap** is a curved section of drainpipe that is located beneath a fixture. It is sometimes called a *P-trap* due to its shape. A trap prevents gases in the waste pipes from entering the house

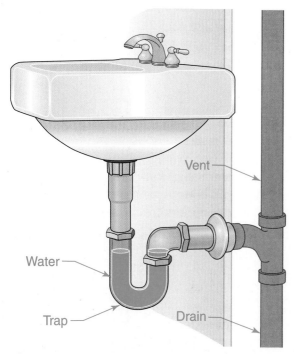

 How a Trap Works

P-Trap A trap prevents gases from entering the house.

but does not block drainage. A small amount of water in the bottom of each trap serves as a gas plug.

Wells and Septic Systems

Houses, cities, and most towns are connected to a municipal water system. Water travels to the house from a centralized source. However, houses in rural areas and some small towns are not usually connected to a municipal water system. Instead, water

Builder's Tip

VENT PIPES AT THE ROOF Vent pipes penetrate the roof and are sealed with flashing called a *roof jack,* as on page 869. Beginning with the 2006 IRC, vent pipes must extend at least 6" above the roofing. In regions where snow is common, the pipe must extend at least 6" above the anticipated level of snow accumulation.

is supplied by a water pump located near the house at the bottom of a narrow well. The well shaft is lined with pipe that is capped at the surface. Underground pipes lead from the pump to a tank located inside the house. Fresh water in the tank is kept under pressure and distributed by supply pipes as needed. In this system, each house in a neighborhood has an individual well.

Houses served by a well are not usually connected to a public sewer. Instead, wastes flow through the DWV system into a below-grade *septic system* on the property. The system collects solid waste in an underground tank and breaks it down with bacteria much like a municipal water treatment plant. The tank must be pumped out periodically to remove accumulated sludge. Liquid wastes in a septic system flow into a filtering area called a drain field. A **drain field** is a network of perforated pipes embedded in sand and gravel. The location of the drain field is determined primarily by soil conditions and grade. However, a well should always be located on the opposite or uphill side of the house from the drain field. It should be as far from the septic system as practical, but typically no less than 100 feet away.

Plumbing Costs

Pipes are the least expensive portion of a plumbing system. Fixtures and their related parts are much more costly. A deluxe faucet, for example, can easily cost ten times as much as a modest faucet. When preparing cost estimates for the house, the contractor generally looks closely at plumbing system costs. One way to reduce overall costs is to use fixtures of lower quality. Arranging the fixtures efficiently can also reduce installation costs.

Framing Requirements

Supply pipes are relatively easy for the plumber to position. Their diameter is small and they are pressurized. This means they can be run in a way that avoids obstacles. However, waste pipes are not as easy to position. The pipes are large and must slope for proper drainage. When framing

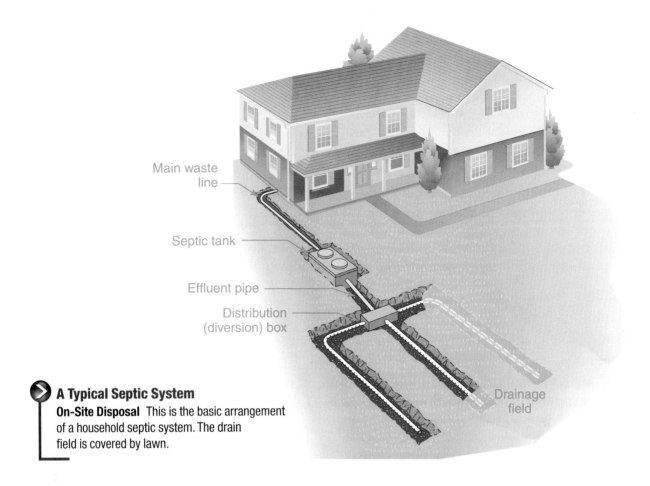

A Typical Septic System

On-Site Disposal This is the basic arrangement of a household septic system. The drain field is covered by lawn.

a house, carpenters should provide adequate space for drain and waste pipes. This is particularly important at bathtub and toilet locations.

Special framing is sometimes needed to support unusually heavy items such as large bathtubs. For details of this type of framing, see Section 16.4. Blocking may also be required to support heavy piping. This is usually considered the plumber's responsibility to install.

Cutting Floor Joists Plumbers often need to bore large holes through a series of joists in order to install waste pipes. However, building codes specify the limits for cutting holes and notches in joists and studs. The allowable size of holes and notches is based on the depth of the joist. This means they can

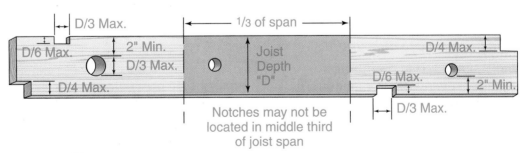

Holes and Notches in Wood Joists

Code Restrictions These are the maximum allowable sizes for notches and holes. Each hole or notch is sized based on the depth (D) of the joist.

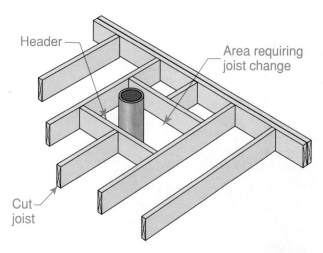

Header

Area requiring
joist change

Cut
joist

 Supporting a Cut Joist
Always Provide Support Header joists are used
to support joists that must be cut to make room for
plumbing drain lines.

be larger as the size of the joist increases. It
is generally best to avoid notching if at all
possible.

If greater than allowable notching is
unavoidable, joists must be reinforced by
nailing a 2× scab to each side of the altered
member, using 12d nails. A scab is a length
of wood used to reinforce another piece. In
extreme cases, an additional full-length joist,
called a *sister,* must be nailed to the notched
joist in order to maintain the strength of the
floor system.

Sometimes a joist must be cut through
completely. If this is necessary, the cut ends
must be supported by headers. Proper planning during framing can usually eliminate
the need to alter joists.

Basic Plumbing Materials

What is the purpose of soldering?

Pipes and tubing used in plumbing systems are made of several different materials
and are joined in different ways. Most new
supply systems have copper piping (sometimes called *copper tubing*), while most DWV
pipes are now made of plastic.

Supply Piping

Copper pipes are joined with copper
fittings that slide over the pipes. The fittings
are then soldered to create a leak-free joint.
Pipes come in 20' lengths. The main water
distribution lines in a typical house are ¾" or
1" in diameter. Branch distribution lines lead
to individual fixtures. They are typically ½"
in diameter. Fittings come in many shapes
and in each of the three common pipe diameters. Copper pipe also comes in three wall
thicknesses:

- *Type M* Thin wall. This is most common
in residential construction. These pipes
are labeled using red ink.

- *Type L* Medium wall. These are labeled
using blue ink.

- *Type K* Thick wall. These are used for
underground water lines and are labeled
using green ink.

The joints between copper pipes and
fittings are soldered (sweated) to seal the
pipes and fittings together. Solder was once
a combination of lead and tin. However, lead
in water supply systems is now considered a
health hazard. Lead-free solder has therefore
been required since 1988.

Reading Check

Explain *What may occur if soldering is not done
properly?*

Builder's Tip

COPPER PIPE DIMENSIONS The actual
outside diameter of a copper pipe is always
⅛" larger than the standard size designation. For example, ½" pipe has an actual
outside diameter of ⅝".

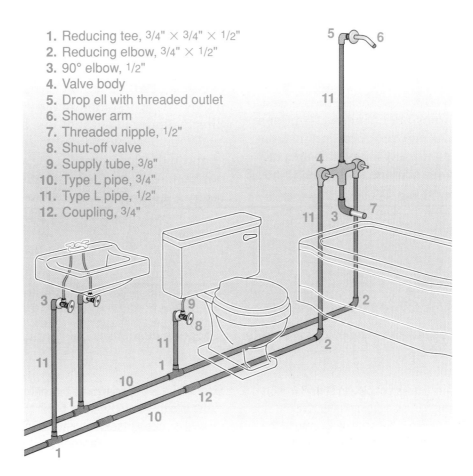

1. Reducing tee, 3/4" × 3/4" × 1/2"
2. Reducing elbow, 3/4" × 1/2"
3. 90° elbow, 1/2"
4. Valve body
5. Drop ell with threaded outlet
6. Shower arm
7. Threaded nipple, 1/2"
8. Shut-off valve
9. Supply tube, 3/8"
10. Type L pipe, 3/4"
11. Type L pipe, 1/2"
12. Coupling, 3/4"

Water Distribution Piping

Common Fittings These are the fittings most often used for the water supply piping in a bathroom.

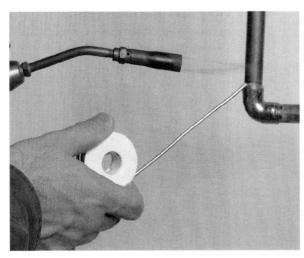

Lead-Free Solder

Water Supply Lead-free solder is a combination of tin, copper, and silver. Heat can be supplied by a propane torch.

Flexible copper tubing is much smaller in diameter than rigid copper piping. It comes in coils up to 100' long and is used to supply appliances that use water, such as dishwashers and refrigerator icemakers. It is connected by friction fittings instead of being soldered.

Various types of plastic pipe can now be used for water supply piping. Plastic supply pipe is lightweight, easy to handle, and resists fracture in freezing conditions. Connections between pipes and fittings are made with solvent welding or threaded fittings, or by using crimping rings. In the 2006 IRC, the following plastic supply piping was approved for use:

- chlorinated polyvinyl chloride (CPVC)
- cross-linked polyethylene (PEX)
- polyethylene (PE)
- polypropylene (PP)

Always check for local code approval before using plastic supply pipe. Some materials may be approved only in limited areas or for limited uses. For example, they may be approved for use inside the house as distribution piping but not allowed under a concrete slab. As new products and installation techniques are developed, this area of construction will continue to change.

Soldering Copper Piping The basic technique for soldering copper pipe is described in the Step-by-Step Application below. The technique must be repeated for each joint in a supply system. When installing copper supply piping, it is often necessary to solder overhead. Protect your eyes and skin from molten drips of solder. Avoid standing directly under the fittings you are working on. When possible, assemble and solder joints on the floor, then lift the assembly into position. After the entire supply system has been assembled, it should be pressurized and checked for leaks. Faulty joints must be resoldered.

Step-by-Step Application

Soldering Copper Pipe The process of soldering copper pipe joints is one of the essential skills of a plumber.

Step 1 Measure a length of pipe to fit the location, and cut it to length using a tubing cutter. Cut the pipe by rotating the tubing cutter repeatedly around the pipe. The cutter's wheel will gradually slice through the copper. Use the reamer on the back of the tubing cutter to smooth out any burrs left on the inside of the pipe. Test fit the pipe into the fitting.

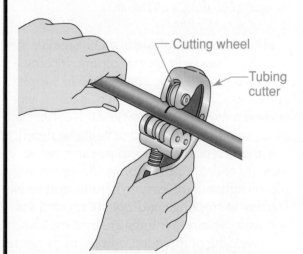

Cutting wheel

Tubing cutter

Step 2 The outside surface of the pipe and the inside surface of the fitting must be shiny-clean where they meet. Use fine emery cloth or a wire pipe-cleaning brush to clean the surfaces.

Step 3 Flux is a special paste that helps to draw solder into the joint. Use a flux brush to coat the cleaned copper surfaces with flux. Assemble the joint and twist the pipe back and forth briefly to spread the flux evenly.

Step 4 Use a torch to heat the joint area evenly. Touch a length of lead-free solder to the area frequently. When the solder starts to melt, quickly turn off the torch and hold the tip of the solder against the joint. As the solder melts, capillary action will draw it into the joint.

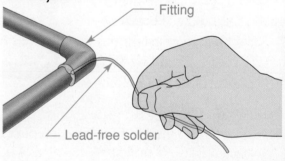

Fitting

Lead-free solder

Step 5 After the joint cools, wipe off surplus flux with a damp rag.

DWV Piping

Cast-iron drainpipe (sometimes called *soil pipe*) is used for waste systems in high-quality construction. It is harder and more expensive to install than plastic drainpipe but muffles the sound of water rushing through the pipes. One type of cast-iron pipe has a flared fitting, called a *bell*, at one end of the pipe. The other end of the pipe is called the *spigot end*. The spigot end fits into the bell end. The joints are sealed using various methods. Another type of cast-iron pipe is called *hubless cast iron*. Each length of pipe is joined to another with polyethylene clamping rings.

The black plastic pipe used in DWV systems is acrylonitrile-butadiene-styrene (ABS). It is inexpensive, lightweight, and easy to cut. It is joined using a solvent cement.

White plastic pipe is made from polyvinyl chloride (PVC). It has considerably lower thermal expansion characteristics than ABS. This makes it particularly suitable for long pipe runs. It is sometimes joined using a solvent-type primer followed by PVC cement. One-step primer-cements are also available.

Before connecting ABS or PVC pipe to a fitting, cut the pipe square with a fine-tooth saw. Plumbers use saws intended specifically for cutting plastic because they result in a clean, square cut. However, a hacksaw will also work. Be sure to remove any burrs from the end of the pipe after cutting it. Any leaks discovered in ABS or PVC fittings during testing generally require the fitting to be cut out and replaced.

Section 30.1 Assessment

After You Read: Self-Check

1. What does the abbreviation HVAC stand for?
2. At what stages must a plumbing inspector check a plumbing system while it is being installed?
3. What is a trap and what is its purpose?
4. How can joists be reinforced after notching?

Academic Integration: Mathematics

5. **Pipe Length** What is the length of piping needed to connect two offset pipes when the offset angles are both 45° and form an isosceles right triangle whose base and altitude are both 15"?

   Math Concept The length of the sides of a right triangle are related according to the Pythagorean Formula, $a^2 + b^2 = c^2$.

   Step 1: Plug the known values into the formula: $15^2 + 15^2 = c^2$

   Step 2: Solve for c, the unknown value: $\sqrt{15^2 + 15^2} = c$

Go to connectED.mcgraw-hill.com to check your answers.

The Electrical System

Electrical System Basics

How does electricity use affect the environment?

The general scheme of a home's wiring is shown in the wiring plan portion of the floor plans. The wiring plans use symbols to indicate the type of electrical devices to install at each location. (An Electrical Symbols table is in the **Ready Reference Appendix**.) However, the location of wires, as well as the exact placement of switches and receptacle outlets, is the electrician's responsibility.

The wiring plans should consider present and future needs of the homeowner. This is increasingly important due to the home use of computers and other electrical equipment once found only in office buildings. Specialized wiring is often added during this phase. This includes cable for television or Ethernet cable for high-speed Internet access. Additional wiring may serve phone, audio,

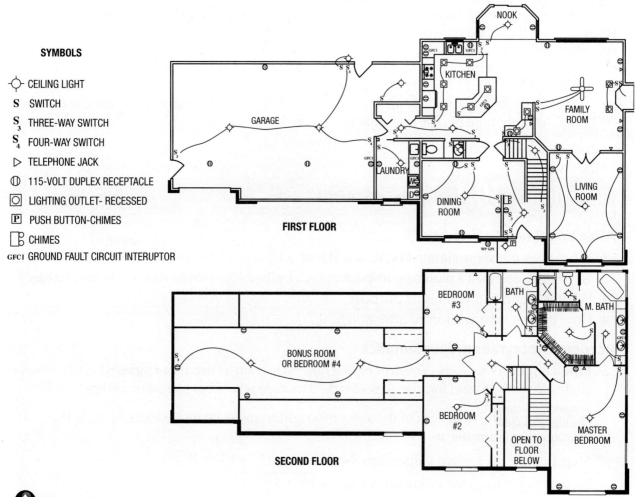

SYMBOLS

- $-\bigcirc-$ CEILING LIGHT
- **S** SWITCH
- **S₃** THREE-WAY SWITCH
- **S₄** FOUR-WAY SWITCH
- ▷ TELEPHONE JACK
- ◑ 115-VOLT DUPLEX RECEPTACLE
- ▣ LIGHTING OUTLET- RECESSED
- Ⓟ PUSH BUTTON-CHIMES
- ▯ CHIMES
- **GFCI** GROUND FAULT CIRCUIT INTERUPTOR

FIRST FLOOR

SECOND FLOOR

Wiring Plan

Power Arrangement The electrician uses an electrical plan as a guide when installing the various components of an electrical system. *Why is it important for all electricians to know how to read electrical symbols?*

Home Office

Technology at Home Many people now work partly or entirely at home. Wiring for a home office must allow for equipment such as computers, fax machines, scanners, and printers.

intercom, and other systems. Electricians install the general electrical system. Other contractors may be called on to install specialized wiring.

Understanding an Electrical System

All power comes into a building through the service entrance wires. These may be overhead wires or an underground cable called a *lateral*. New houses are generally supplied with 200-ampere service. As explained in Chapter 6, an *ampere* (amp) is a measure of electrical current. The service entrance wires run first to the watt-hour meter. This records how much electricity is used within the house. In newer homes, digital watt-hour meters are installed to make recording usage easier.

From the meter, the wires run to a master distribution panel, called the *service panel*. The panel is usually located in the basement or in a utility area. At the top of the service panel is a master switch. It is used to cut off all electricity in the house. This would be done in an emergency or when various parts of the wiring system are being worked on.

Smaller wires lead from the service panel to points throughout the house. The wires are organized into circuits. A **circuit** is a cable or group of cables that supplies electricity to a specific area or appliance. It can be connected or disconnected without affecting any other circuit. Each circuit is connected to an individual device called a *circuit breaker*. Circuit breakers are located inside the service panel below the master switch. A circuit breaker is like a fast-acting switch. It shuts off power in a circuit if it detects overloads that might lead to a fire. The circuit breaker can also be turned off manually if maintenance work must be performed on the circuit.

A house could easily have as many as 25 or more separate circuits. There are three basic kinds of circuits: appliance circuits, general-purpose circuits, and special-purpose circuits.

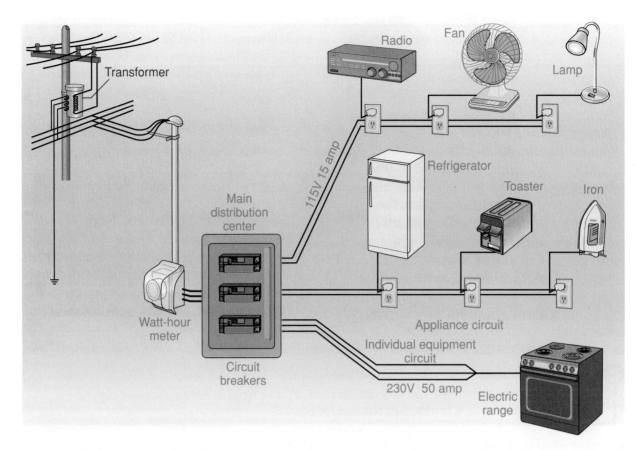

A Basic Electrical System

Distributing Power This drawing illustrates how electricity is distributed to the house and to the circuits inside.

- A small *appliance circuit* is wired with No. 12 wire and is connected to a 20-ampere circuit breaker. At least two small appliance circuits are needed in the kitchen. An appliance circuit might also be added in a basement to provide electricity for shop tools.

- A *general-purpose circuit* is wired with No. 14 wire and connected to a 15-ampere circuit breaker. It also may use No. 12 wire and a 20-amp circuit breaker. These circuits lead to lighting and to all receptacles.

- A *special-purpose circuit* supplies the needs of stoves, air conditioners, furnaces, and other appliances that use large amounts of electricity. It often serves a single appliance. This circuit uses thicker wire than other circuits and is connected to a 30-ampere or greater circuit breaker.

Electrical Materials & Systems

What are common types of wire?

Electrical materials include wires and outlet boxes.

Wires

Several different kinds of wire and wiring systems are allowed by electrical codes. Electrical wires are referred to as *conductors*. This distinguishes them from standard utility wire. Conductors are usually made of copper, although in some cases aluminum is allowed.

- *Nonmetallic sheathed cable wiring (NMC)* is the most common, the simplest to install, and the least expensive. The cable consists of two or three insulated copper conductors and one bare copper conductor within a thermoplastic covering. All wires within a given cable are always the same size.

- *Armored cable* is used in exposed locations where mechanical damage might be expected. It is also used where local codes do not allow NMC. This hollow cable, commonly called *BX,* has a flexible metal exterior. Individual insulated conductors are contained within the cable.

- *Rigid metal* or *plastic conduit* is used in exposed locations and sometimes underground. Like BX cable, it protects the conductors inside. Metal conduit can be bent around corners, using a special tool called a *conduit bender.* Plastic conduit is joined with solvent and plastic fittings.

Reading Check

Summarize *Why are electrical wires referred to as conductors?*

Outlet Boxes

Wiring leads from the service panel to various types and sizes of metal or plastic *outlet boxes.* Outlet boxes provide a convenient location for joining wire. They also prevent dust and debris from collecting on the connectors. They limit damage from short circuits and other wiring faults that could cause fires. They provide a solid mounting surface for switches, receptacles, and other devices.

Wiring enters a box through pre-scored holes called *knockouts.* Where wires enter, they must be held securely by cable clamps or by some other method. This prevents the wires from pulling away from receptacles. A **receptacle** has a combination of slots and grounding holes sized to accept the prongs of an electrical plug. Each box must be nailed securely to the framing. The electrician must position each box so that its front edge will be flush with the final wall covering.

An outlet box is required wherever wiring will be connected to a device, such as a switch, a ceiling light, or a receptacle. An outlet box is also required wherever lengths of wiring are spliced together. Wiring must never be spliced outside a box.

Wiring a House

Wiring is done in two stages: the rough-in stage and the finish stage. The rough-in wiring is done after the exterior of the house has been completed but before the insulation

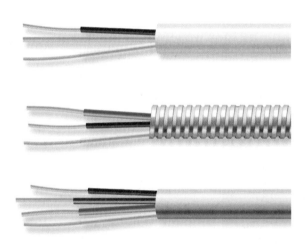

Common Wire Types
Basic Types Three kinds of wiring are used between the circuit breakers and the outlet (junction) boxes: Nonmetallic sheathed cable (top); armored cable (middle); and conduit (bottom).

Outlet Boxes
Metal and Plastic Outlet boxes are available in a wide variety of sizes. Electrical codes determine the number and size of wires that can enter a particular size of box.

has been installed. The electrician installs the service panel and breakers. Then cables are routed from the panel into outlet boxes throughout the house. This is sometimes referred to as *pulling the cables.*

Inside each outlet box, the outer sheathing of the cable is stripped off to expose individual conductors. The conductors are left exposed until the finish wiring stage.

After the rough-in wiring is completed and has been approved by a building inspector, insulation and wall finishes can be installed. After the interior of the house has been painted, the electrician returns to complete the finish wiring. All switches, receptacles, and lighting fixtures are connected at this stage, both inside and outside the house. The electrician connects conductors

Builder's Tip

 Service Panel
Distribution Center Circuit breakers are located in the main service panel. Never work on the panel unless you are absolutely certain that power is not being supplied to it.

Royalty-Free/Corbis/age fotostock

by holding their bare ends together and then twisting on a threaded cap called a *wire nut* or *wire connector*.

When the finish wiring is completed, the electrician tests it. Outlet covers must then be attached over all switches and receptacles. The inspector then returns to check and approve, or *final*, the installation.

Reading Check

Explain *When is rough-in wiring done?*

A

B

Wiring a Switch

Basic Steps **A.** During the rough-in stage, boxes are positioned and cable is run into them. The wires are left exposed until later. Once interior finishes are complete, the electrician will return to install and test the switch. **B.** In this photo, the drywall was left off to show how wires are secured to the stud.

(l)Jodi Jacobson/Getty Images; (r)Blend Images/Alamy Stock Photo

After You Read: Self-Check

1. What is an ampere?
2. What is the purpose of a circuit breaker?
3. What is a receptacle?
4. During the rough-in phase, what happens to cable after it is routed to the outlet boxes?

Academic Integration: Science

5. **Circuit Breaker** What size circuit breaker is needed for a dishwasher with a 115 volt circuit that carries a resistance of 8 ohms?

   Math Concept Voltage (E) is the force that causes electricity to flow through a conductor. We measure the resistance to that force in ohms (R). We measure the rate of electron flow through a conductor in amperes, or amps, (I). These three are related according to the formula $E = IR$, or $I = E \div R$.

   Step 1: Plug the known values into the formula. $I = 115 \div 8$

   Step 2: Solve for I, the unknown value.

   Go to connectED.mcgraw-hill.com to check your answers.

Section **30.3** HVAC Systems

Heating Systems

What are the different types of heating systems?

Without effective temperature control, houses in many areas would be either too hot or too cold. Because temperature control systems can be costly to operate, a great deal of research is under way to make them more energy efficient. This means getting more energy from less fuel. New HVAC systems are far more efficient than older systems. Thermostats that control the systems are also more efficient. They can be programmed to allow temperatures in a house to drop at night, or when the house is not occupied. The thermostat will automatically increase temperatures to a comfortable level when necessary.

There are many types of heating systems. Some are combined with air-conditioning and ventilation systems. Heating systems are categorized primarily by the way the heat is distributed, not by the fuel they use. Various fuels can supply the heating energy needed.

Reading Check

State *Why should homeowners maintain up-to-date HVAC systems in their home?*

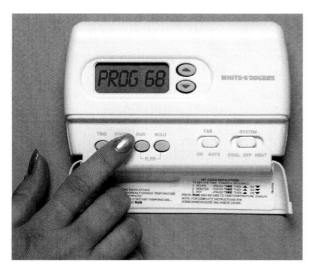

⌄ Electronic Thermostat

Programmable Electronic thermostats can be set to change temperatures on an hourly or daily basis. They are sometimes called *set-back thermostats*.

Forced Hot-Air Heating

A forced hot-air heating system consists of a furnace, ducts, and registers. The system is popular because it responds quickly to changes in outdoor temperatures. It can be used in many types of houses. The ducts and registers can also be used to distribute cool air created by a central air conditioner.

Fuel is burned inside the furnace. A blower **circulates** the warm air to the rooms through *supply ducts.* The ducts may be made of sheet metal, flexible insulated tubes, or rigid fiberglass insulation. *Supply registers* are located along the outside walls of the house. There is usually a *return-air register* in each room. It is usually located across the room from the supply registers. As air within the room cools, it sinks to the floor and flows into the return-air register. Return registers and ducts carry cooled room air back to the furnace. There, it is reheated and recirculated.

Heated air is filtered through replaceable or washable filters, sometimes called *media filters.* It is important for the homeowner to remove them for inspection on a regular basis during the heating season. Clogged filters dramatically reduce the effectiveness of the system and make it less efficient. As an alternative to media filters, electronic air cleaners can be installed in some heating systems. They are very effective at removing pollen, fine dust,

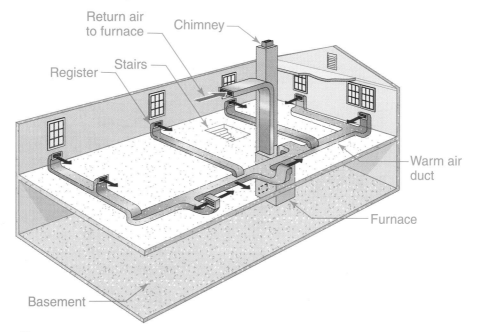

⌄ Forced Hot-Air System

Circulating Warmed Air Some forced hot-air systems have a cold-air return in each room (except the bathroom and the kitchen).

and other irritants that normally pass through standard filters. The part within an electronic air cleaner that actually removes contaminants is called a **cell**. Cells should be cleaned on a regular basis and can be reused.

A humidifier is another device that can be added to a hot-air system. A **humidifier** adds moisture to the air inside the house and counteracts the drying effects of hot air. Water is constantly fed to the humidifier by way of copper tubing. The tubing is connected to the main water supply.

In houses with a slab foundation, heating equipment and duct work are usually installed in the attic. However, cylindrical ducts may be installed in the slab as a *perimeter loop system*. If this is done, the ducts must be positioned within the foundation formwork before the concrete is placed.

Heat Pumps A **heat pump** is a device that can heat and cool a house. Heat pumps are energy-efficient because they do not burn fuel to make heat. Instead, they rely on electricity to pump refrigerant through a closed-loop system. **Refrigerant** is a material that absorbs heat as it becomes a gas and gives up heat as it becomes a liquid. The heat pump itself, is located outside the house. It contains a compressor and a fan. Pipes containing the refrigerant connect the heat pump to a unit inside the house. This unit contains evaporator coils and a blower. The blower sends heated or cooled air through the house using a standard duct system.

The ability of the system to heat and cool relies on the fact that the refrigerant absorbs heat or gives off heat as it changes back and forth from a liquid to a gas. In the summer,

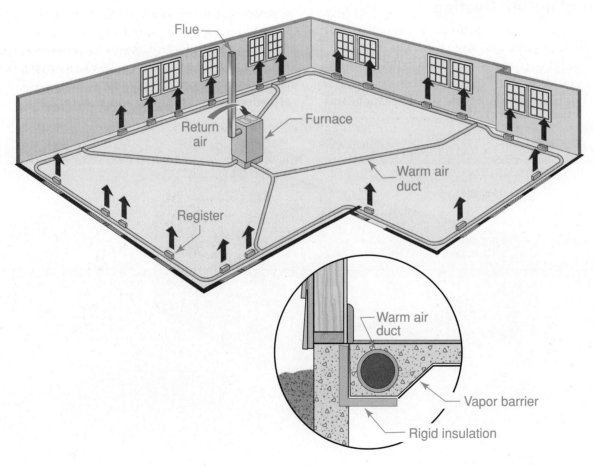

Perimeter Loop System

 Suitable for Slabs Perimeter loop hot-air systems are sometimes used in houses that are built on a concrete slab.

refrigerant is pumped through the system so that it absorbs excess heat from the house and releases it in the heat pump outdoors. In the winter, the process is reversed. The refrigerant is pumped through the system so that it extracts heat from outdoor air and releases it indoors. This is why heat pump systems are not suited to climates with cold winters. When the outdoor temperature drops below about 45° F, the system becomes less efficient.

Hydronic Heating

Hydronic heating systems, or hot-water systems, consist of a boiler, pipes, and room-heating units (convectors or radiators). Hot water generated in the boiler is pumped through copper pipes to the *convectors* or *radiators*. Heat then radiates into the room.

REGIONAL CONCERNS

Heating & Fuel Climate and the type of heating fuel used vary from region to region. These factors affect the type of heating system that is most common. In the northeastern United States, for example, oil is the most common heating fuel. In much of the Northwest, it is common to heat with electricity. This is because the climate is relatively mild and electricity is fairly inexpensive (compared to oil) due to plentiful hydroelectric resources. In still other parts of the country, natural gas is preferred. In regions with consistently clear skies, solar-assisted heating systems are a workable option.

⌃ Heat Pump
Double Duty A heat pump such as this one is part of a system that can heat and cool a house.

DonNichols/Getty Images

Boilers are made of steel or cast iron. When fuel is burned inside the boiler, this generates heat that is transferred to pipes containing water. A pump circulates this water to the room-heating units. As the water cools, it is returned to the boiler and heated again. Boilers can be designed to use various fuels, including electricity, coal, natural gas, or oil. Boilers designed for remote areas can even use wood as the basic fuel.

One problem with any system based on a boiler is that corrosion can shorten boiler life. Boilers should be inspected at the beginning of each heating season.

Convectors usually consist of tubes with fins. They are enclosed in a housing that has openings at the top and bottom. Hot water circulates through the tubes. The fins maximize the transfer of heat to the surrounding air. Convectors usually run along

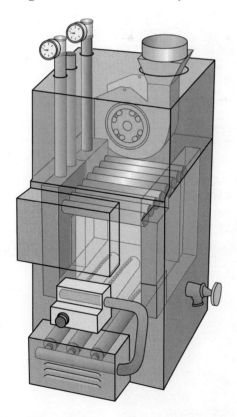

 Hot-Water Boiler
Cutaway View Boilers heat water and circulate it through pipes to heat the house. Heat is extracted from the water as it flows through convectors or radiators.

the baseboards and are often placed under windows. Low-profile convectors can be placed in locations that would otherwise not be suitable.

Radiant Heating

In *radiant heating* systems, heating coils, tubes, or cables are buried within ceilings, floors, or walls. No registers or ducts are required. This makes the system very quiet. Rather than heating air, as in a forced-air or hydronic system, a radiant system heats a material. This material then radiates the heat directly into the room. Many people find this type of heat very comfortable. There are two basic types of radiant systems: electric and hydronic.

Electric Systems Many types and designs of electric radiant heating systems are available. In one system, electric heating cable is laid back and forth across the ceiling surface. It is then covered with plaster or a second layer of drywall. As the cover material heats up, it radiates warmth to the room. Radiant panel units can also be placed directly on the finished surface of the ceiling. A thermostat located in each room generally controls heat levels.

Hydronic Systems In a radiant hot-water system, water heated in a boiler is circulated through continuous coils of polyethylene tubing. The tubing is embedded in a concrete floor. As the heated water circulates, it conducts its heat to the masonry. The floor then radiates heat to the room.

Preparing for Radiant Heat
Heat in the Slab In a radiant floor heating system, concrete is poured over a network of hot water distribution tubing. The red material is epoxy-coated rebar.

Heat Recovery Ventilation

Fresh air leaks into a house through cracks around windows, doors, and framing in a process called **infiltration**. This air must be heated in cold weather. However, heated air leaks out as easily as cold air leaks in. Builders reduce air infiltration by building "tight" houses. This means that there are few gaps in the house that can let in cold air. Of course, this also means that fresh air cannot get in. Moisture and indoor pollutants such as formaldehyde, tobacco fumes, and combustion byproducts can build to unhealthy levels in a tight house.

One solution is to install a device called a *heat recovery ventilator (HRV)*. Heat recovery ventilators are sometimes called air-to-air heat exchangers or energy recovery ventilators. A cutaway view of an HRV is shown on page 888.

An HRV removes the heat from stale indoor air before exhausting the air outdoors. That heat is transferred to fresh air drawn into the house. To accomplish this, a fan within the HRV pulls in fresh air from outdoors through a duct. A second fan removes stale air from inside the house through a separate duct. Both sets of ducts meet at the HRV. There, heat is transferred from one air stream to the other. Each air stream is kept separate. By using heat from the outgoing air to warm the incoming air, less energy is required to raise the temperature of the incoming air.

Cooling Systems

What factors affect how easily a house can be cooled?

In some areas, cooling a house is far more necessary than heating it. Energy efficiency is just as important when a house is being cooled as when it is being heated. Two types

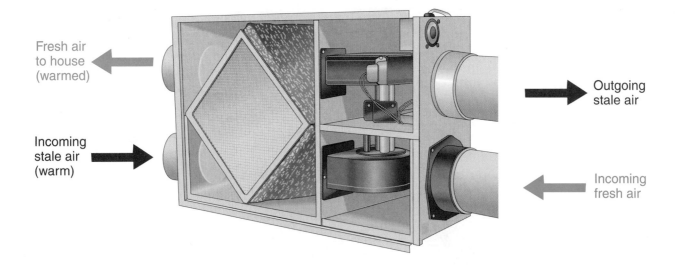

 Heat Recovery Ventilator
Energy Saver An HRV brings in fresh air and exhausts stale air. *What happens to the incoming fresh air?*

of systems—central air conditioning and whole-house ventilation—can be used to cool a house.

Central Air Conditioning

Air conditioning is a process of extracting heat from air and then releasing the heat outside the house. It relies on the same principles used by heat pumps. Small air-conditioning units can be placed in a window to cool a room. All the devices needed are contained in the unit. However, cooling an entire house calls for a central air-conditioning system. These units are sometimes called split systems because part is located outdoors and part is located indoors.

There are three basic elements in a split system:

- *Refrigerant Coils* These coils of copper tubing hold a liquid refrigerant. The condenser coil is located outside the house. The evaporator coil is located inside the house. The coils are connected by additional tubing to form a closed loop.

- *Air Handler* This unit contains the evaporator coil. It also contains a blower to move air through an insulated duct system.

- *Compressor* This unit contains the condenser coil. It includes a fan but does not include ductwork.

Here is how the system works:

1. The system is turned on when a thermostat indicates the temperature in a room has risen to a preset level.

2. The air handler draws in warmed house air through ducts and blows it over the evaporator coils. Refrigerant in the coils absorbs heat from the air. The cooled air is then distributed to the house.

3. As the refrigerant absorbs heat, it turns into a gas. The gas travels to the condenser coil outside the house, where it gives up its heat. The fan helps this process by circulating air over the condenser coil.

4. As the vapor cools, it condenses back to a liquid and returns to the evaporator coils. The process then repeats.

Whole-House Ventilation

If ventilation is desirable but a central air-conditioning system is not justified, a house can sometimes be cooled using a centralized

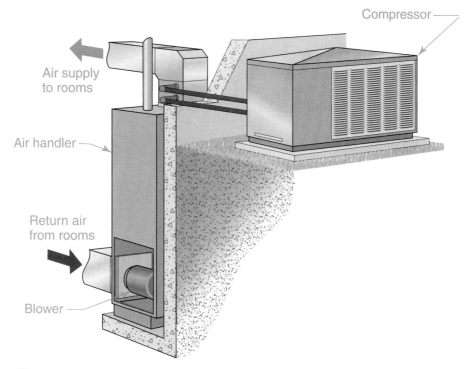

 A Central Air-Conditioning System
Same Ducts Central air conditioning can use the same ductwork as the central heating system.

fan. This powerful enclosed fan is mounted in the highest ceiling in the house, which is often above a stairwell. The fan draws relatively cool air into the house through open windows while exhausting hot air into the attic. Attic vents exhaust the heat outside. The system requires no ducts.

Some whole-house ventilating fans are designed to be mounted on top of ceiling joists. This eliminates the need to cut joists for installation. However, the unit should be mounted on rubber pads. This limits any noise and vibration that might otherwise be transmitted through the framing. Some fans have variable speed controls.

Section 30.3 Assessment

After You Read: Self-Check

1. Name four types of boiler fuel.
2. In a forced-air system, what happens after heated air is delivered to a room?
3. Describe the purpose of fins on a convector.
4. Describe the interaction of the air handler and the evaporator coils in a split-system air conditioner.

Academic Integration: Science

5. **From Liquid to Gas** Central air-conditioning systems, or split systems, are used in many homes. These systems use refrigerant coils that hold a liquid refrigerant. Refrigerant is a material that changes from a liquid to a gas. Explain how a refrigerant is converted to a gas.

Go to connectED.mcgraw-hill.com to check your answers.

Chapter Summary

Section 30.1
A basic plumbing system consists of a supply side and a DWV side. Supply pipes are pressurized, while DWV pipes are not. Traps are simple devices that prevent sewer gases from entering the house.

Section 30.2
An electrical system consists of wires, called conductors, which lead from circuit breakers in a service panel to individual outlet boxes. Each set of wires connected to a circuit breaker is called a circuit and leads to a particular portion of the house.

Section 30.3
Heating systems are classified according to how they distribute heat, rather than by what fuel they use to create it. Air-conditioning systems are sometimes incorporated with forced-air heating systems so that they can use the same ducts.

Review Content Vocabulary and Academic Vocabulary

1. Use each of these content vocabulary and academic vocabulary words in a sentence or diagram.

Content Vocabulary

- mechanicals (p. 868)
- fixture (p. 868)
- service main (p. 869)
- trap (p. 869)
- drain field (p. 870)
- circuit (p. 877)
- receptacle (p. 879)
- cell (p. 884)
- humidifier (p. 884)
- heat pump (p. 884)
- refrigerant (p. 884)
- infiltration (p. 887)
- air conditioning (p. 888)

Academic Vocabulary

- rural (p. 870)
- circulates (p. 883)

Speak Like a Pro

Technical Terms

2. Work with a classmate to define the following terms used in the chapter: *supply pipes* (p. 869), *waste pipes* (p. 869), *vent pipes* (p. 869), *septic system* (p. 870), *copper tubing* (p. 872), *service panel* (p. 877), *circuit breaker* (p. 877), *appliance circuit* (p. 878), *general-purpose circuit* (p. 878), *special-purpose circuit* (p. 878), *conductors* (p. 878), *outlet boxes* (p. 879), *knockouts* (p. 879), *pulling the cables* (p. 880), *wire nut* (p. 881), *wire connector* (p. 881), *final* (p. 881), *supply ducts* (p. 883), *supply registers* (p. 883), *return-air register* (p. 883), *media filters* (p. 883) *hydronic heating* (p. 885), *convectors* (p. 885), *radiators* (p. 885), *radiant heating* (p. 886).

Review Key Concepts

3. List the components of plumbing system.

4. Name the tools and materials used in basic plumbing.

5. Describe the various types of piping used for water supply and DWV systems.

6. List the components that make up an electrical system.

7. Describe the three basic kinds of circuits.

8. Identify the basic parts of a split-system air conditioner and describe how it works.

Critical Thinking

9. Discuss Explain the difference between supply pipes and waste pipes. Why should you not interchange supply and waste pipe materials?

Academic and Workplace Applications

STEM Engineering

10. Electrical Circuitry Could both of the following be run on the same 115 volt circuit?

- An oven that has 16.5 ohms of resistance and runs on 12.25 amps.
- A saw that runs on a 15 amp circuit and has 7 ohms of resistance.

Starting Hint Voltage (E), ohms (R), and amps (I) are related according to the formula $E = IR$.

Step 1: Use the formula to determine the voltage requirement of the oven and the saw.

Step 2: Compare the results to the voltage of the circuit.

STEM Science

11. Water and Aquifers Most houses in rural areas are not connected to a municipal water system. Instead, water is supplied by a water pump located near the bottom of a deep but narrow well. Wells often draw water from an underground aquifer. Use the Internet or your school library to find out more about aquifers. Write a few sentences describing what an aquifer is. Identify some types of rocks that make good aquifers.

21st Century Skills

12. Interpersonal Communication Assume that you are a journeyman HVAC professional in the eastern United States. Your firm has just hired an individual from Washington state. Your supervisor has asked you to work with the new employee for a few weeks. Each day you notice that the employee is having a hard time adjusting to the differences in procedure and installation of HVAC units. The individual continues to refer to the electricity-based HVAC systems he is used to installing in the Northwest. Write a short paragraph explaining how you would deal with this situation. Explain why different heating fuels are used in the East and in the Northwest. Describe the interpersonal tactics you could use to help the new employee. How can you remain professional and courteous while informing the employee of the differences?

Standardized TEST Practice

Short Response

Directions Write one or two sentences in response to the following questions.

13. How are waste materials broken down in a septic system?

14. What is the purpose of a watt-hour meter?

15. Why is it important to maintain clean filters in a forced hot-air heating system?

TEST-TAKING TIP

Use complete sentences when answering short-answer questions. A complete sentence contains both a subject and a predicate (verb phrase).

*These questions will help you practice for national certification assessment.

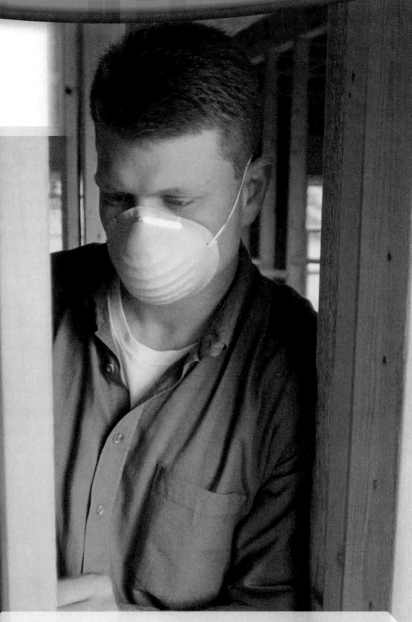

CHAPTER 31
Thermal & Acoustical Insulation

Section 31.1
Thermal Insulation

Section 31.2
Acoustical Insulation

Chapter Objectives

After completing this chapter, you will be able to:

- **Identify** several types and forms of insulation.

- **Describe** how an insulator's R-value determines its effectiveness as insulation.

- **Define** the main function of insulation.

- **Identify** the best uses for common types of insulating materials.

- **Explain** the importance of vapor retarders and ventilation.

- **Describe** several types of wall construction that reduce noise transmission.

Discuss the Photo

Insulation The individual in the photo is installing insulation. *What purposes can you think of for installing insulation?*

Writing Activity: Write a Letter

New methods are being designed to reduce heating and cooling costs. Write a one-page letter to a local builder asking about the newest techniques for insulating new homes. In your letter, introduce yourself and be clear about what information you are requesting. Proofread your letter to correct any errors.

Stockbyte / SuperStock

Before You Read Preview

Different regions require different types of insulation. Before reading this chapter, write down which U.S. regions you would expect to place a high priority on insulation. After you have read the chapter carefully, revisit your answers and correct them if necessary.

Content Vocabulary

- thermal insulation
- R-value
- building envelope
- batt

- condensation
- vapor retarder
- radiant heat
- emissivity

- acoustic insulation
- Sound Transmission Class (STC)
- Impact Noise Rating (INR)

Academic Vocabulary

You will find these words in your reading and on your tests. Use the academic vocabulary glossary to look up their definitions if necessary.

- byproduct
- expands
- flexible

Graphic Organizer

As you read, use a chart like the one shown to identify issues and their solutions.

Issues	Solutions
How do I determine the right type of insulation for a home in a given location?	Consider the climate, the R-value of the insulation, and how the house is constructed.

Go to **connectED.mcgraw-hill.com** to download this graphic organizer.

Insulation Basics

Why do U.S. building codes divide the country into zones?

Insulation is any material that slows the transmission of heat, sound, or electricity. Different uses require specific types of material. For example, the material that insulates electrical wires would not be suitable for insulating walls. Sound insulation is not necessarily effective at slowing heat loss.

This chapter will cover primarily thermal insulation. **Thermal insulation** slows the transmission of heat through walls, floors, and ceilings. Acoustical (sound) insulation is discussed in Section 31.2.

Properties of Thermal Insulation

Thermal insulation increases the comfort of home occupants. It also reduces the cost of utilities by heating or cooling climate.

Heating Climate Smaller, less expensive furnaces are required and less energy or fuel is needed to maintain a certain temperature level.

Cooling Climate A properly insulated house can be served by smaller, less expensive cooling equipment. In addition, less electricity is needed to maintain a house at comfortable temperature levels.

Most building materials and even the air space between studs have some insulating properties, as shown in **Table 31-1**. However, to meet current standards for energy efficiency, additional insulation is needed. The amount of thermal insulation required in a house varies greatly by region. Houses in mild climates need less insulation, while houses in severe climates need more.

Materials are rated according to their insulating abilities. The most common method is to rate materials according to R-value.

R-value is a measure of a material's ability to resist heat transmission. R-value varies according to a material's thickness, but it is cumulative. For example, one type of insulation might have a value of R-5 per inch. Two inches would have a total R-value of R-10. This is why R-value figures are often given per inch of material thickness.

When choosing the type and amount of insulation, climate is the primary factor to consider. The map on page 895 shows the lowest temperatures throughout the continental United States during an average winter. Such information is useful in figuring the amount of insulation needed for walls, ceilings, and floors. Generally, local codes specify the minimum amount of insulation required.

> **Reading Check**
>
> **Explain** What does R-value measure?

Table 31-1: Thermal Properties of Various Building Materials per Inch of Thickness	
Material	**Thermal Resistance (R)**
Wood	1.25
Air space[a]	0.97
Cinder block	0.28
Common brick	0.20
Face brick	0.11
Concrete (sand and gravel)	0.08
Stone (lime or sand)	0.08
Steel	0.0031
Aluminum	0.00070

[a] Thermal properties apply to air spaces ranging from ¾ inch to 4 inches in thickness.

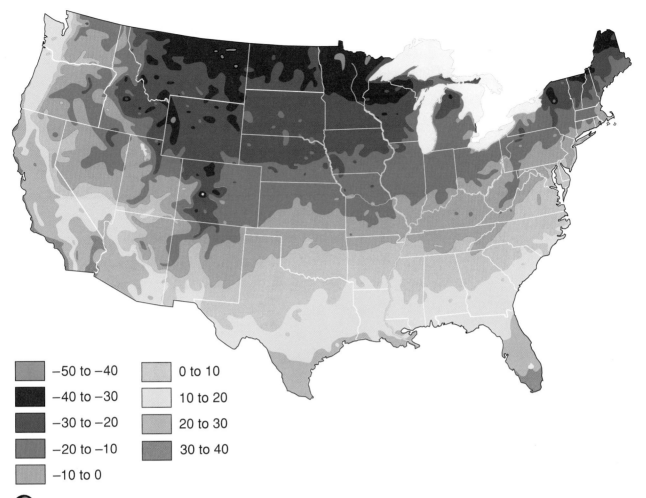

![-50 to -40]	-50 to -40
![-40 to -30]	-40 to -30
![-30 to -20]	-30 to -20
![-20 to -10]	-20 to -10
![-10 to 0]	-10 to 0
![0 to 10]	0 to 10
![10 to 20]	10 to 20
![20 to 30]	20 to 30
![30 to 40]	30 to 40

Lowest Temperatures
Coldest Regions This map of the continental United States indicates the lowest temperature (°F) occurring in each zone during an average winter.

The IRC identifies different climate zones as a way of identifying the amount of insulation various areas need. The 2006 IRC divides the United States into eight climate zones, as shown on page 896. The edges of the zones are defined by county lines. The map also identifies four climate types: Marine, Dry, Moist, and Warm-Humid. These regions influence the way insulation and related building components are installed. Compare Lowest Temperature map to Climate Zone Map on page 896 to see how the climate of an area relates to the zone map.

In each of the new zones, a minimum amount of insulation is required, as shown in **Table 31-2** on page 896. It is important to understand that the code identifies the *minimum* amount of insulation required in a house. A builder can increase insulation levels above the minimum amounts required. In fact, many builders participate in national, local, and state programs that certify above-code homes. These homes are especially energy-efficient.

Where to Insulate

To reduce heat loss during cold weather in most climates, all walls, ceilings, roofs, and floors that separate heated from unheated spaces should be insulated. This continuous layer of insulation is referred to as the **building envelope**. Everything inside the building envelope will be heated and/or cooled. Everything outside the envelope is exposed to outdoor temperatures.

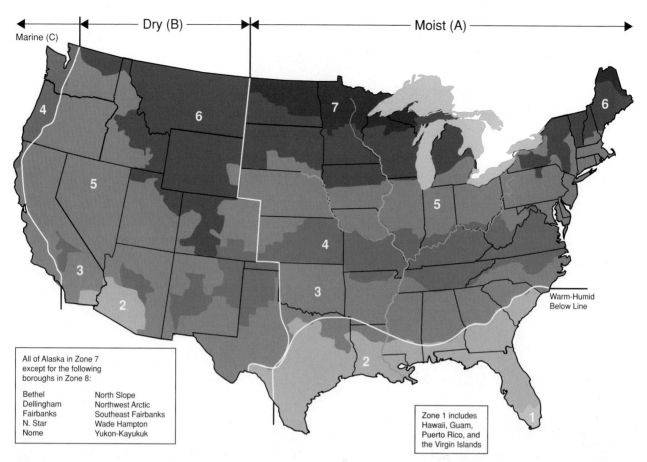

Dry (B) ———— Moist (A) ————

Marine (C)

4

6

7

6

5

5

3

4

3

2

2

1

Warm-Humid
Below Line

All of Alaska in Zone 7
except for the following
boroughs in Zone 8:

Bethel   North Slope
Dellingham   Northwest Arctic
Fairbanks   Southeast Fairbanks
N. Star   Wade Hampton
Nome   Yukon-Kayukuk

Zone 1 includes
Hawaii, Guam,
Puerto Rico, and
the Virgin Islands

Climate Zone Map

Eight Zones The 2009 IRC climate zone map. Each zone requires a certain minimum amount of thermal insulation.

Climate Zone	Ceiling R-Value	Wood Frame Wall R-Value	Floor R-Value	Basement[b] Wall R-Value	Slab[c] R-Value and Depth
1	30	13	13	0	0
2	30	13	13	0	0
3	30	13	19	0	0
4 except Marine	38	13	19	10/13	10, 2ft
5 and Marine 4	38	19 or 13 + 5[e]	30[d]	10/13	10, 2ft
6	49	19 or 13 + 5[e]	30[d]	10/13	10, 4ft
7 and 8	49	21	30[d]	10/13	10, 4ft

Table 31-2: Insulation and Fenestration Requirements by Component [a]

[a] R-values are minimums. R-19 insulation shall be permitted to be compressed into a 2×6 cavity.

[b] The first R-value applies to continuous insulation, the second to framing cavity insulation; either insulation meets the requirement.

[c] R-5 shall be added to the required slab edge R-values for heated slabs.

[d] Or insulation sufficient to fill the framing cavity. R-19 minimum.

[e] "13 + 5" means R-13 cavity insulation plus R-5 insulated sheathing. If structural sheathing covers 25% or less of the exterior, R-5 sheathing is not required where structural sheathing is used. If structural sheathing covers more than 25% of exterior, structural sheathing shall be supplemented with insulated sheathing of at least R-2.

Types of Insulation

What types of insulation would be hardest to seal against wall studs?

Insulation is manufactured in a variety of types. Each type has advantages for specific uses. The basic types are batts and loose-fill, rigid sheet, and spray-foam insulation. There are also hybrid insulation systems that use materials and installation techniques that do not fit neatly into any single category.

Batts and Blankets

The most common type of insulation used in houses comes in the form of a batt. A **batt** is any thick insulation material that comes in precut widths designed to fit between framing members. Extra-wide batts designed to cover large areas of an attic are sometimes

called *blankets*. Batts and blankets are packaged in compressed bundles or rolls. Each product must be cut to length on site. Batts and blankets come in different thicknesses and widths. Material intended for a standard 2×4 wall, for example, will be 15" wide and 3½" thick. This allows it to fit the wall cavity formed by studs spaced 16" OC. The cavities are actually 14½" wide, but batt insulation is slightly wider than this to make up for slight variations in stud position. The extra width also ensures a snug fit against the sides of the studs. Batts are also available for other OC spacings and wall thicknesses.

Materials The most common material for batts and blankets is fiberglass. Fiberglass insulation is made of natural ingredients such as sand and recycled glass. The materials are melted and formed into thin strands. When matted together, the strands are sometimes referred to as *glass wool*. Fiberglass insulation sometimes has a kraft paper facing on one side. Continuous tabs on the sides of the facing allow the batts to be stapled to studs or joists. The facing also serves as a vapor retarder, which will be discussed later in this section. Batts and blankets are also available without a kraft paper facing. *Unfaced insulation,* sometimes referred to as *friction-fit insulation,* is simply pressed into the wall cavities and not stapled. If this is done, 4-mil thick plastic film must be stapled over the walls to serve as a vapor retarder. As an alternative to fiberglass, batts can be made from recycled cotton, natural wool, and rock wool (a fibrous material made from molten stone and slag, a **byproduct** of steel making). Cotton and natural wool must be chemically treated to improve resistance to fire, mold, and rodents.

Batt Insulation
Common Choice Fiberglass insulation in blankets and batts is the most common type used in residential construction.

R-Value Fiberglass insulation has a value of about R-3 per inch. Low-density batts intended to insulate 2×4 walls, for example, are usually rated R-11; high-density batts are rated R-15. Low-density batts intended for 2×6 walls are rated R-19; high-density batts are rated R-21. Blankets intended for floors and ceilings may be rated as high as R-38. The R-value of cotton and natural wool insulation is similar to fiberglass. The R-value of rock wool is slightly higher than fiberglass. The higher the R-value, the better a material insulates.

Loose-Fill Insulation

Loose-fill insulation consists of materials that are usually supplied in bags or bales. Depending on where it will be installed, the material can be poured into place manually or blown into place using specialized pumping equipment. In an attic, blowing insulation over the joists covers a large area quickly. Loose-fill insulation is often used in attic floors where HVAC pipes and wiring make it difficult to install blanket insulation. It can also be pumped into the walls of older houses that were not insulated during construction. This involves drilling holes through the siding. In new construction, loose-fill can also be installed into open wall cavities. The technique calls for installing a special membrane over the studs. The loose fill can be blown through the membrane and into the wall cavities.

Materials and R-Value Materials used include shredded fiberglass, rock wool, and cellulose-based products. Cellulose-based products are made primarily from recycled paper, and must be treated with chemicals to make them insect and fire resistant. Because loose-fill materials can be applied in various thicknesses, their R-value is typically given per inch. Multiply the R-value per inch by the thickness of the installation to get an overall R-value. The approximate R-values for each material are:

- Fiberglass: R-2.5 per inch
- Rock wool: R-3.2 per inch
- Cellulose: R-3.5 per inch.

Rigid Insulation

Rigid insulation is manufactured in various sizes and thicknesses of solid panels. They are often 4 ft. by 8 ft. in size. Panels often have a reflective surface on one side. These thin panels pack more R-value into a smaller space than most other types of insulation. Rigid insulation is sometimes used as a nonstructural sheathing on walls during construction. It also serves as a substrate for other materials. Some types of rigid insulation are suitable for use below grade for the exterior surfaces of basement walls.

 Rigid Insulation
Insulating Layer Rigid insulation can be applied to the outside of a house prior to the installation of the siding. All seams should be sealed with aluminized tape.

Glow Images

Materials and R-Value Rigid insulation is made from one of the following materials:

Expanded Polystyrene (EPS) Polystyrene is formed into beads that puff up when exposed to steam. The beads are then molded into blocks of insulation and sliced into sheets. EPS is rated from R-3.6 to R-4.2 per inch.

Extruded Expanded Polystyrene (XEPS, sometimes called XPS) This material is similar to EPS but has greater compressive strength for use as foundation insulation below grade. It is rated at R-5 per inch.

Polyurethane and Polyisocyanurate These materials are often faced with foil to slow the loss of the blowing agent used in their manufacture. They are rated at R-5.6 per inch.

Spray-Foam Insulation

Spray-foam insulation is made from material that expands as it is installed. It is sprayed into open wall and ceiling cavities as a wet material. Once exposed to air, the material foams rapidly to fill the cavity. The amount of expansion can be a hundredfold or more. After the material expands, it typically solidifies within an hour. The exact amount of expansion is difficult to control, so excess material must be sliced off flush with the surface of the framing once the insulation hardens. The shavings are then collected and recycled. Because spray-foam insulation is applied as a liquid and then expands, it seals and fills small gaps and seals cavities much better than other types of insulation.

Materials and R-Value Spray foam is available in various formulations. Slow-curing types are designed to flow over and around obstructions before curing. Open-cell foam allows moisture to pass through the cured material. Closed-cell foam allows very little moisture to pass through the cured material. Materials used for foam include polyisocyanurate, polyurethane, and cementitious products made from minerals extracted from seawater. R-values range from approximately R-3.9 per inch for cementitious products to as much as R-8 for the other types.

Hybrid Insulation Systems

The importance of thermal insulation increases as the cost of electricity and heating fuels rises. Some insulation materials and methods combine aspects of the four basic insulation types noted above.

Wet-Spray Insulation Loose-fill materials such as cellulose or fiberglass can be mixed with water and sprayed into wall cavities. An adhesive is sometimes mixed in to help the material stick to surfaces. Once the moisture content of the insulation drops to less than 19 percent, it can be covered with drywall. Drying takes at least 24 hours.

Combined Insulation Some builders spray a 1"-thick layer of spray-foam insulation over wall cavities to fill gaps, then install fiberglass batt insulation in the cavities once the foam cures.

Spray-Foam Insulation
Superb Sealer Spray-foam insulation expands to fill gaps and seals the areas between framing members.

Controlling Moisture

Where might you see condensation on a hot, humid day?

Insulation is used to maintain comfortable temperatures within a house. However, a great deal of moisture vapor is generated inside a typical house. It comes from the breath of people as well as activities such as cooking, dishwashing, laundering, and bathing. If the insulation is not installed properly, moisture-vapor can become trapped inside the walls. This can eventually:

- Reduce the effectiveness of the insulation.
- Encourage decay in wood materials.
- Promote the formation of mold.
- Attract wood-destroying insects.

Most building materials allow water vapor to pass through them. In cold weather, the warm vapor from inside the house will eventually reach a part of the wall that is cold. This might be the inside surface of the wall sheathing, or it might even be somewhere inside the insulation itself. Cold air cannot hold as much moisture as warm air, so when warm air cools, it releases some of the moisture it carries. The moisture forms on surfaces as a liquid. The process by which a vapor turns into a liquid is called condensation. For example, on a cold day, water droplets often appear on the inside surface of windows. This occurs because warm, moist indoor air is cooled when it contacts the glass, and has to unload some of its moisture. This moisture is deposited on the cold surface. The liquid water formed by the process is also called condensation.

There are various ways to solve this problem. One way would be to eliminate all sources of moisture inside a house, but this is not realistic. A second method is to prevent moisture vapor from getting into the walls. Builders can do this by placing a barrier between the insulation and the moisture.

Vapor Retarders

A **vapor retarder** is a material that reduces the rate at which water vapor can move through a material. An older term for this is *vapor barrier.* However, the term is not quite accurate because it implies that moisture transfer is stopped completely. This is not true. Every material allows some moisture to pass through it.

Among the effective vapor retarder materials are asphalt laminated papers, aluminum foil, and plastic films. The kraft paper facing on batt insulation also acts as a vapor retarder.

The effectiveness of a vapor retarder is rated by its *perm value.* Perm value is a measure of water vapor transmission through a material. Low perm values indicate vapor retarders with high resistance to vapor transmission. A value of 0.50 perm is adequate. However, it is good practice to use barriers that have values less than 0.25 perm.

One common method for installing a vapor retarder is to staple wide rolls of plastic film over studs, plates, and window and door headers after insulating the walls. This is called *enveloping* and is used over insulation that does not already include a vapor retarder, such as unfaced batts. The plastic should fit tightly around outlet boxes and seams should be sealed. A ribbon of sealing compound around an outlet or switch box limits vapor transmission at this area.

It is extremely important to install a vapor retarder with great care. If the installer leaves gaps and untaped seams, moisture vapor will get into the wall cavities at these points.

Science: Energy & Matter

Conduction Most insulation is used to prevent the conduction of heat. Generally speaking, metals are considered to be good conductors, while wood products are poor conductors. Which type of material is a better insulator? Why?

Starting Hint Poor conductors make good insulators.

No vapor retarder resists all vapor. Some leakage into the wall can be expected. Therefore, the flow of vapor to the outside should not be slowed by materials of high resistance on the cold side of the barrier. For example, sheathing paper should be waterproof but not highly vapor resistant.

Attic Ventilation

A third approach to controlling moisture is to provide ventilation. This approach cannot be used in wall cavities, but it is very important in attics and roof framing. Ventilation removes water vapor but it also reduces ice-damming problems and helps to keep temperatures inside the house cooler in the summer.

Removing Water Vapor Water vapor in air leaking from heated sections of the house will condense when it comes in contact with cold surfaces in the attic. Even when vapor retarders have been installed, some vapor will work through spaces around pipes and other poorly sealed areas. Some vapor will also work through the vapor retarder itself. Although the amount of water vapor may be unimportant if it is evenly distributed, water vapor can cause damage if concentrated in cold spots. Moisture vapor can escape through wood shingle and wood shake roofing systems. However, it cannot escape through roofing systems topped with asphalt, fiberglass, metal and other impervious materials. This is why attic spaces and roof systems should be ventilated. Air flow carries away excess moisture vapor.

Preventing Ice Dams Another reason to ventilate an attic or roof is to reduce the formation of ice at the eaves. An attic that is poorly ventilated and poorly insulated tends to be warmer than outside air in the winter. The warmth melts snow on the roof. Water running down the roof freezes when it reaches the colder surfaces of the eaves, often forming into ice at the gutter. This ice dam may cause water to back up at the eaves and into the wall and ceiling cavities. With a well-insulated ceiling and enough

ventilation, attic temperatures are low. This greatly reduces the melting of snow on the roof.

Reducing Temperatures In hot weather, ventilation of attic and roof spaces allows hot air to escape. This lowers the attic temperature and helps the house to stay cooler.

Build It Green In regions where a house must be cooled for much of the year, thermal insulation helps the house to stay cool. But shade trees are also important in such climates. By shading walls and roof surfaces, trees can reduce a home's annual heating and cooling bill by 20 percent or more. Shading the west side of a house is particularly effective in reducing cooling needs. This is because the west side catches the afternoon sun.

It was once a common practice to install louvered ventilation openings in the end walls of gable roofs. However, air movement through this system varies and is often inefficient. A much more efficient approach is to provide openings in the soffit as well as a ridge vent at the top of the roof. As heated air rises in the attic, it is exhausted through the ridge vent. This process draws cooler air in through

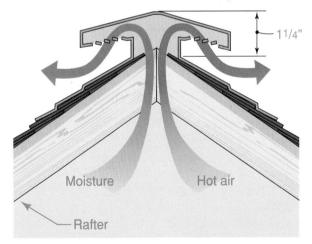

11/4"

Moisture   Hot air

Rafter

Ridge Venting
Effective Ventilation A continuous ridge vent can be capped with shingles to help blend the vent with the roof. *Why is ridge venting becoming more common?*

the soffit vents. Where a sloped ceiling is insulated, there should be a free opening of at least 1½" between the sheathing and the insulation to encourage air movement. (For more on ventilating roofs, see Chapter 19 and Chapter 22).

Crawl-Space Ventilation

Many houses are built on a crawl space foundation. Moisture rising through the soil can turn to vapor, which can move through the flooring system and into the house.

Soil Cover Crawl spaces must be protected from ground moisture by a vapor retarder, sometimes called a *soil cover*. The vapor retarder is often a 6-mil or 8-mil thick plastic film. Any joints must be lapped at least 6" and be sealed with appropriate tape or by some other method. Such protection helps to prevent decay of wood framing members. It also prevents floor insulation from becoming saturated with moisture.

Vents Just as with an attic, a crawl space should be vented to remove any moisture vapor that gets through the soil cover. The minimum net area of ventilation for a crawl space is 1 sq. ft. of venting for each 150 sq. ft.

of crawl-space area. *Net area* refers to the amount of air that actually gets through a screened vent. The net area rating is typically stamped on the vent itself. For a house with an area of 1,500 sq. ft., for example, this would be 10 sq. ft. This area should be distributed evenly between vents located around the crawl space. At least one vent must be located within 3 ft. of each corner of the foundation. Vents should be covered with a corrosion-resistant screen to keep insects out.

Unvented Crawl Spaces In some cases, there are advantages to eliminating vents in a crawl space. The 2006 IRC allows this type of construction but with some restrictions. Exposed earth must be covered with a continuous vapor retarder. The crawl space area must also be mechanically ventilated with a fan or supplied with conditioned air.

Radiant-Heat Barriers

What materials would make poor radiant heat barriers?

In the summer, outside surfaces exposed to direct sunlight may reach temperatures of 50°F (10°C) or more above shade temperatures. These surfaces tend to transfer this heat toward the inside of the house. Insulation in the attic slows the flow of heat, improving summer comfort and reducing the need for air conditioning. However, additional steps are sometimes taken to reduce radiant heat gain. **Radiant heat** travels in a straight line away from a hot surface and heats anything solid it meets. If radiant heat gain is reduced, the air-conditioning system does not have to work as hard to maintain a comfortable temperature in the house. Installing a radiant-heat barrier in the attic is one way to stop radiant heat gain.

A radiant-heat barrier is a thin, **flexible** sheet material with at least one reflective surface, usually of aluminum. Some radiant-heat barriers have a reflective coating on both sides. Installed properly, a radiant-heat barrier can reduce heat transfer into the attic

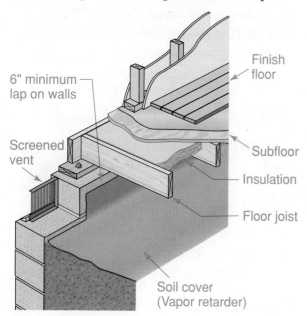

6" minimum lap on walls

Finish floor

Screened vent

Subfloor

Insulation

Floor joist

Soil cover (Vapor retarder)

Crawl-Space Vapor Retarder
Moisture Removal A ventilated crawl space that has a vapor barrier and soil cover.

 Radiant-Heat Barrier
Heat Blocker A radiant-heat barrier reduces heat gain in a house.

by about 95 percent. In climates where saving heating energy is the main concern, however, radiant-heat barriers are rarely cost-effective.

Reading Check

Explain *What is the relationship between radiant heat and the air-conditioning system in a home?*

How Radiant-Heat Barriers Work

All materials emit energy by thermal radiation. The amount emitted depends on the surface temperature and the material's emissivity. **Emissivity** is a measure of radiation that is expressed as a number between 0 and 1. The higher the emissivity, the greater the emitted radiation. Another important property is *reflectivity*. Reflectivity measures how much radiant heat is reflected by a material. It is expressed as a number between 0 and 1 or as a percentage between 0 and 100. A material with high reflectivity has low emissivity, and vice versa. To perform properly, radiant-barrier materials must have high reflectivity (usually 0.9 or 90 percent, or more), low emissivity (usually 0.1 or less), and face an open air space.

For example, on a sunny day, a roof absorbs solar energy. This heats the roof sheathing, which causes the underside of the sheathing and the roof framing to radiate heat downward toward the attic floor. Placing a radiant-heat barrier on the underside of the rafters reflects much of the heat back toward the roof. Thus the top surface of the attic insulation stays cooler and so do the rooms below.

Radiant-heat barriers may be installed in attics in two ways. One method is to attach the barrier to the underside of the rafter framing. Another method is to drape the barrier loosely over the rafters just before the roof sheathing is applied. In this method, the barrier should droop so that there is at least 1" of air space between it and the underside of the sheathing. The air space makes the barrier more effective. It also creates an air channel that allows the soffit and ridge vents to work more effectively. Do not install a radiant-heat barrier by spreading it over attic insulation. Dust accumulating on the barrier can reduce its effectiveness. Also, the barrier might trap moisture vapor rising through the ceiling.

ozgurcoskun/Getty Images

Installing Insulation

What factors can make insulation difficult to install in a wall?

Installation of insulation is a job in which good craftsmanship pays off in money saved for the homeowner. When installing any insulation, it must fill all the wall and ceiling cavities completely. There should be no gaps between the insulation and the framing or around other objects. As shown in Installation Gap, gaps form a ready passage through which heat and moisture vapor can escape.

Where to Insulate

The proper placement of insulation is important. The basic concept is to insulate all portions of the building envelope.

Walls All walls that separate living space from outdoor air must be insulated. In a one-and-a-half-story house, however, it is sometimes difficult to establish the building envelope. In such cases, second floor *knee walls* should be treated just as if they were exterior walls. Knee walls are half-height walls often found in attics.

Floor Systems In houses with unheated crawl spaces, insulation should be placed

Installation Gap
How Not to Install Insulation This shows a poor insulation job. *Why should gaps be avoided between insulation and framing?*

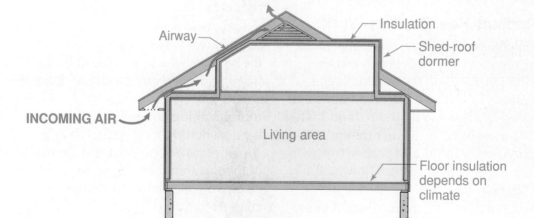

EXHAUST AIR

Airway

INCOMING AIR

Living area

Insulation

Shed-roof dormer

Floor insulation depends on climate

The Building Envelope
Surrounding Living Spaces Insulating a one-and-a-half-story house. In some climates, the floor system might also be insulated.

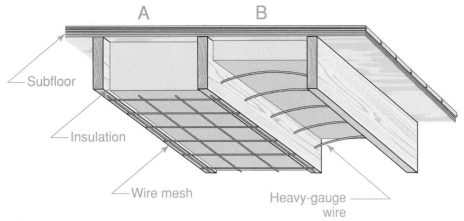

 Supporting Batt Insulation
Two Methods Methods of installing insulation between floor joists. **A.** Wire mesh stapled to the joists. **B.** Heavy-gauge wire pointed at each end and sprung into place.

between the floor joists. If batt insulation is used, it should be well supported by a galvanized wire mesh or a rigid board. The vapor barrier should be installed toward the subflooring. Press-fit or friction insulation fits tightly between joists and requires only a small amount of support to hold it in place.

Attics and Cathedral Ceilings Heat rises, so it is important to slow down heat loss during cold weather by insulating the attic. Where attic space is unheated and a stairway leads to the attic, insulation should be installed around the stairway as well as in the first-floor ceiling. The door or hatchway leading to the attic should be weatherstripped and insulated to prevent heat loss. Walls adjoining a garage or unheated porch should also be insulated. Vaulted ceilings above conditioned space must be insulated as much as possible. It is often necessary to provide an air space above the insulation. This would allow any moisture vapor to be exhausted. Ventilation is not necessary, however, if the ceiling cavities are insulated with spray-foam insulation. Check with local codes.

Foundation Walls and Slabs In Zones 4 and higher, these surfaces must be insulated if they form part of the building envelope. Slabs are typically insulated by a layer of rigid insulation that is installed just before the slab is placed.

Installing Batt Insulation

Blanket or batt insulation with a vapor retarder should be placed between framing members so that the tabs of the facing lap the edges of the studs as well as the top and bottom plates. This is generally preferred to stapling the tabs to the sides of the studs. A hand stapler or hammer tacker is commonly used to fasten the insulation and the barriers in place.

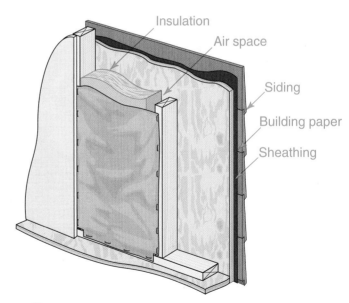

Installing Faced Batts
A Good Seal The insulation tabs should be stapled to the edge of each stud.

 Cutting Batts

Straight Cuts To cut insulation, place it on a piece of scrap plywood. Compress the material with a wood scrap and cut it with a utility knife.

Batt insulation can be cut with a utility knife. Batts and blankets should always be compressed while being cut. This ensures that cut edges will be smooth, improves cutting accuracy, and prevents too many fiberglass fibers from escaping into the air. Something as simple as a 2×6 board or some other wood scrap can be used to compress the insulation near the cutting area. It can also serve as a straightedge for guiding the utility knife.

When friction-fit batt insulation is installed, a plastic-film vapor retarder such as 4-mil polyethylene is commonly used to envelop the entire exposed wall and ceiling. It covers the openings as well as window and door headers and edge studs. This system is one of the best for blocking

vapor movement. After the drywall has been installed or plastering has been completed, the film can be trimmed around the window and door openings.

Blanket insulation is often used in ceilings. The vapor retarder should be placed against the back of the ceiling finish. Unfaced blankets may be layered to provide a suitable thickness. In this case, place one layer of insulation at 90° to the first. This also insulates the ceiling framing.

Insulation should be placed behind electrical outlet boxes and other utility connections in exposed walls to limit condensation. It is important to understand that some types of recessed light fixtures require a specific amount of clearance between the fixture and the insulation. This allows heat generated within the fixture to escape. A recessed fixture that overheats is a fire hazard.

Narrow gaps around doors and windows also require insulation. This is best done by spraying expanding foam sealant into the gaps. However, be sure that the foam does not push window and door jambs out of position. To minimize this problem, use low-expansion foam designed for this purpose.

Estimating and Planning

This estimating and planning exercise will prepare you for national competitive events with organizations such as SkillsUSA and the Home Builder's Institute.

Batt Insulation

Materials

To estimate the amount of insulation required, you must first figure the square footage of the wall or ceiling to be insulated. Refer to the house plan below.

Round off the outside dimensions of the heated portion of the home to a width of 28' and a length of 52'. The perimeter of the house is thus 160'.

$$(2 \times 28) + (2 \times 52) = 160$$

If the wall height is 8', the walls will have an area of 1,280 sq. ft.

$$8 \times 160 = 1{,}280$$

Subtract the area of the window and door openings, which equals about 150 sq. ft., from the total area.

$$1{,}280 - 150 = 1{,}130$$

The total wall area to be insulated is 1,130 sq. ft.

Using figures supplied by the insulation manufacturer, determine how many square feet each roll or bundle of insulation will cover. Divide the area to be covered by the coverage per roll or bundle. The answer will be the number of rolls or bundles required. Add approximately 5 percent to this figure to allow for waste.

Estimating on the Job

The ceiling of a home must also be insulated. Figure the area of the ceilings by multiplying the width times the length. Divide the number by the number of square feet in each roll. Add approximately 5 percent to this figure to allow for waste.

1. If one roll of insulation covers 100 square feet, how many rolls of insulation will be needed for the ceilings in this house?

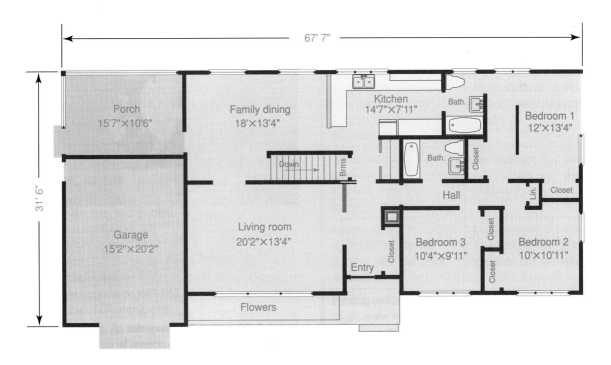

After You Read: Self-Check

1. What type of insulation is the most common?
2. What is the typical R-value for low-density batt insulation for 2×4 and 2×6 walls?
3. What is the building envelope?
4. Define *emissivity*.

Academic Integration: Science

5. **Condensation** You may have noticed the concentration of moisture on the inside of a window on a cold day. This moisture appears through the process of condensation. Write one or two sentences describing the process of condensation.

Go to **connectED.mcgraw-hill.com** to check your answers.

Section 31.2 Acoustical Insulation

Understanding Acoustics

How do windows affect sound transmission?

Acoustical insulation is insulation that slows the transmission of sound. Acoustical insulation has always been important in apartments, motels, and hotels. However, the use of household appliances, television, radio, and stereo systems has increased the noise levels in single-family homes. Today, sound insulation between the active areas (such as recreation rooms and home theaters) and sleeping areas is often desirable. Insulation against outdoor sounds is also important where houses are close together or where they are near highways. As a result, sound control has become an important part of house design and construction.

How Sound Travels

Sound is transmitted by waves. It travels readily through the air and also through some materials. A noise inside a house, such as music, a loud conversation, or a barking dog, creates sound waves. These radiate outward until they strike a wall, floor, or ceiling. The surface vibrates as a result of the pressure of the sound waves. When airborne sound strikes a conventional wall, the studs act as sound conductors unless they are separated in some way from the covering material.

The resistance of a building element, such as a wall, to the passage of airborne sound is described by its **Sound Transmission Class (STC)** number. Sound Transmission Class is a numerical rating that indicates the ability of a material or combination of materials to reduce sound transmission. The higher the STC number, the better a material is as a sound barrier.

Flanking Paths Faulty construction, such as poorly fitted doors, can allow sound to pass around a material without actually going through it. This type of sound transmission follows what is called a *flanking path*. Heating ducts, wiring chases, and plumbing runs can also allow sound to travel freely through the air within wall and ceiling assemblies. In fact, a hole as small as 1 square inch in a wall rated at STC 50 can reduce that wall's performance to STC 30. Plumbers, electricians, and others who regularly cut holes in framing should keep this fact in mind as they work.

Sound Absorption

To reduce sound levels in a house, sound-absorbing materials can be added. Sound-absorbing materials do not necessarily resist airborne sounds. However, they can reduce noise by preventing sound from being reflected back into a room. Perhaps the most commonly used sound-absorbing material is acoustical ceiling tile or panels. Numerous holes or fissures on the surface, or a combination of both, trap the sound. Acoustical tile and panels are most often used where they are not subject to too much mechanical damage, such as in the ceiling. Paint or other finishes that fill or cover the tiny holes or fissures greatly reduce their efficiency.

Sound Insulation Techniques

What types of wall construction might affect the work of other trades?

The addition of sound-absorbing materials is only one way to reduce sound levels in a house. The most effective methods usually involve the structure of a house. For this reason, they must be decided upon at the earliest stages of house design. Many of the materials will be difficult or impossible to add later.

Acoustical Walls

Thick walls of dense materials such as masonry can stop sound quite effectively. In a wood-frame house, however, an interior masonry wall results in increased costs. A less expensive system is to combine sound-deadening insulating board and gypsum board outer covering. This provides good sound control at only slight additional cost. A number of combinations, providing different STC ratings, are possible with this system.

Good STC ratings can be obtained in a wood-frame wall by using the combination of materials and techniques shown on page 910.

A double wall, which may consist of 2×6 or wider plate and staggered 2×4 studs, is sometimes constructed for sound control. However, the extra effort and planning

 Acoustical Ceiling Panels
Absorbing Sound A suspended ceiling system with acoustical panel inserts will absorb some sound.

Arnold & Brown

WALL DETAIL	DESCRIPTION	STC RATING
A	1/2" gypsum wallboard	32
	5/8" gypsum wallboard	37
B	3/8" gypsum lath (nailed) plus 1/2" gypsum plaster with whitecoat finish (each side)	39
C	8" concrete block	45
D	1/2" gypsum wallboard (each side)	45
E	1/2" sound-deadening board nailed 1/2" gypsum wallboard laminated (each side)	46
F	Resilient clips to 3/8" gypsum backer board 1/2" fiberboard laminated (each side)	52

Acoustical Performance of Walls

Quiet Options A standard wall is shown in **A**. The STC rating of a wall can be improved in various ways (**B–F**).

required for double-wall construction does not necessarily warrant its expense.

Acoustical Floors and Ceilings

Sound insulation between an upper floor and the ceiling of a lower story involves not only resistance to airborne sounds but also to impact noises. *Impact noise* results when an object strikes or slides along a wall or floor. Footsteps, dropped objects, and furniture being moved all cause impact noise. It may also be caused by the vibration of a dishwasher, food disposal, or other equipment. In all instances, the floor is set into vibration by the impact or contact, and sound is radiated from both sides of the floor.

The impact noise resistance of a floor system is described by its Impact Noise Rating. The **Impact Noise Rating (INR)** is a measure of the resistance of a floor system based on decibels (dB), a measure of sound

Builder's Tip

UNDERSTANDING DB The softest sounds humans can hear range from 0 to 1 decibel (dB). Except for thunder and erupting volcanoes, no sound that is found in nature exceeds 100 dB. Noise levels produced by saws, routers, and other tools and equipment range from 87 to 108 dB.

intensity. The higher the INR, the better the impact sound reduction. Another rating, the *Impact Insulation Class (IIC)*, is sometimes used instead because IIC figures are easier to determine. Like the INR, it measures the resistance to sound transmission due to impact. The higher the IIC number, the better the impact insulation.

Section 31.2 Assessment

After You Read: Self-Check

1. One wall assembly has an STC rating of 40, while another has an STC rating of 55. Which wall would be best at reducing sound transmission?
2. What is a flanking path?
3. Why is it important to limit the number and size of holes in a wall?
4. On what unit is the INR rating based?

Academic Integration: Science

5. **The Velocity of Sound** The velocity of sound is the product of wavelength and frequency (velocity = wavelength $\times$ frequency). In air, sound travels at a constant speed of 1,100 feet per second. If the value of the frequency is large, the value of the wavelength is small. For example, if the frequency is 1,100 hertz, then the wavelength would be 1 foot. What is the frequency of a sound wave whose wavelength is 0.5 feet?

Starting Hint: If a product is constant and one of two factors is doubled, the other factor is halved.

Step 1: Write an equation to represent the situation in the problem:
(Velocity = frequency $\times$ wavelength).

Step 2: Plug in the known values (velocity and wavelength).

Step 3: Solve for the unknown value (frequency). Express your answer in hertz.

Go to **connectED.mcgraw-hill.com** to check your answers.

Review and Assessment

Chapter Summary

Thermal insulation is available as flexible batts and blankets, rigid board, loose fill, and spray foam. Insulation helps to reduce heat gain as well as cooling loss. R-value is the measure of an insulator's effectiveness. A vapor retarder limits water vapor penetration.

Acoustic insulation slows the transmission of sound. The amount of sound transmitted through a house can be reduced with the proper wall and ceiling construction. The STC and INR ratings measure a material's resistance to sound transmission.

Review Content Vocabulary and Academic Vocabulary

1. Use each of these content vocabulary and academic vocabulary words in a sentence or diagram.

Content Vocabulary

- thermal insulation (p. 894)
- R-value (p. 894)
- building envelope (p. 895)
- batt (p. 897)
- condensation (p. 900)
- vapor retarder (p. 900)
- radiant heat (p. 902)
- emissivity (p. 903)
- acoustical insulation (p. 908)
- Sound Transmission Class (STC) (p. 908)
- Impact Noise Rating (INR) (p. 911)

Academic Vocabulary

- byproduct (p. 897)
- expands (p. 899)
- flexible (p. 902)

Speak Like a Pro

Technical Terms

2. Work with a classmate to define the following terms used in the chapter: *blankets* (p. 897), *glass wool* (p. 897), *unfaced insulation* (p. 897), *friction-fit insulation* (p. 897), *vapor barrier* (p. 900), *perm value* (p. 900), *enveloping* (p. 900), *soil cover* (p. 902), *net area* (p. 902), *reflectivity* (p. 903), *knee walls* (p. 904), *flanking path* (p. 909), *impact noise* (p. 911), *Impact Insulation Class (IIC)* (p. 911).

Review Key Concepts

3. Name the types and forms of insulation.

4. Explain how insulation effectiveness is determined by the R-value.

5. Discuss the primary purpose of insulation.

6. Describe the basic types of insulation.

7. Describe how vapor retarders and ventilation control moisture.

8. Explain the types of wall construction that will reduce noise transmission.

Critical Thinking

9. Explain How can tradesmen such as plumbers and electricians affect the STC rating of a wall?

Academic and Workplace Applications

10. Estimating Batt Insulation A rectangular house with outside dimensions of 26' × 44' is to be insulated. For the 8' exterior walls, 4" thick batts measuring 15" × 48" are to be used. Assume that the area of the doors and windows is 15 percent of the floor area. How many batts will be needed?

Math Concept Perimeter is the distance around a shape. The perimeter of a house is usually measured in feet. Area is the space covered by a shape. The area of exterior walls and floors is usually measured in square feet.

Step 1: Calculate the area of the exterior. Multiply the perimeter times the height. Subtract the area of the doors and windows.

Step 2: Calculate the area of one batt. Divide to find the number of batts required.

Step 3: Add a waste allowance of 5 percent.

11. Moisture and Ventilation Roofing plans play an important role in the ventilation of a house. Wood shingle and wood shake roofs do not resist vapor movement, but asphalt shingles are highly resistant to vapor movement. In this case, the most practical method of removing moisture is by ventilating the attic or roof. Discuss some of the reasons to ventilate a roof or attic in a two-paragraph summary. Explain some potential problems that may arise if a roof or attic is not properly ventilated.

12. Productivity and Accountability Your health has an impact on the work you do. Some illnesses are not preventable, but there are things you can do to prevent illness. Discuss some ways in which poor health can affect your productivity as an employee. Then name some things you can do to maintain excellent health and productivity on the job.

Standardized TEST Practice

Multiple Choice

Directions Choose the word or phrase that best completes the following statements.

13. A measure of a material's ability to resist heat transmission is called _____ .
   a. STC-value
   b. H-value
   c. R-value
   d. INR-value

14. The insulation commonly used in ceiling areas is called _____ .
   a. flexible insulation
   b. loose-fill insulation
   c. friction-fit insulation
   d. non-flexible insulation

15. _____ is a measure of water vapor transmission through a material.
   a. Vapor value
   b. Perm value
   c. R-value
   d. P-value

TEST-TAKING TIP

Do not just stop working during a test if you get stuck on a question. Take a 30-second break. If you are still stuck, move on to another question.

*These questions will help you practice for national certification assessment.

Wall & Ceiling Surfaces

CHAPTER 32

Section 32.1
Drywall

Section 32.2
Plaster

Section 32.3
Suspended & Acoustical Ceilings

Chapter Objectives

After completing this chapter, you will be able to:

- **Name** and describe the various types of drywall.
- **Identify** the different types of fire-code drywall.
- **Describe** a nail pop and explain the methods used to prevent it.
- **Explain** potential safety and health problems related to installing drywall and explain preventative measures.
- **Identify** the basic materials used in three-coat plaster work.
- **Construct** a suspended ceiling.

Discuss the Photo
Drywall Drywall panels are often handled by two or more individuals. *What might some of the reasons for this practice be?*

Writing Activity: Compare and Contrast
Team up with a classmate and examine the walls and ceilings in different parts of your school. See if you can tell which walls are plaster and which are drywall. Take notes about the texture, appearance, and feel of each type of wall. Highlight the similarities and differences between the drywall surface and the plaster surface.

Purestock/SuperStock

Chapter 32 Reading Guide

Before You Read Preview

Drywall and plaster are the two most common wall and ceiling finishing systems used in residential construction. Suspended ceilings are sometimes installed when access is needed. Choose a content vocabulary or academic vocabulary word that is new to you. When you find it in the text, write down the definition.

Content Vocabulary

- corner bead
- feathering
- veneer plaster
- lath
- ground
- suspended ceiling
- acoustical ceiling

Academic Vocabulary

You will find these words in your reading and on your tests. Use the academic vocabulary glossary to look up their definitions if necessary.

- stable
- enables

Graphic Organizer

As you read, use a chart like the one shown to organize information about types of walls and ceiling materials and their characteristics, adding rows as needed.

Type of Material	Characteristics
drywall	made of gypsum and covered with paper

Go to connectED.mcgraw-hill.com to download this graphic organizer.

Drywall Basics

Why are there different types of drywall?

After the mechanical systems have been roughed in and the house has been insulated, interior wall and ceiling materials can be applied. The choice of surface must be made as the house is being designed. However, drywall is the material most commonly used in residential construction. Plaster is another choice and, in some cases, the two can be used together. Wood paneling, another finishing choice, is discussed in Chapter 28.

Understanding Drywall

Drywall consists of sheets, or panels, made of a noncombustible gypsum core covered with paper. It is also known as *gypsum wallboard*, *gypsum board*, or by various trade names. Natural gypsum is a mineral rock. After it is mined, it is ground to a powder and baked. The resulting material is mixed with water and other ingredients. It is then sandwiched between the sheets of recycled paper to form drywall panels. A smooth-finish paper called *face paper* is applied to the front of the panel. A somewhat rough paper called *liner paper* is applied to the back.

Unlike plaster, the large sheets of drywall can be applied quickly. They do not require lengthy drying time before other work can progress. Strongest in the long dimension, drywall is dimensionally **stable** and inexpensive. Like plaster, it has fire-resistant properties. It can serve as a substrate for other finish materials, such as paint, wood paneling, or wallpaper. (A *substrate* is a material that serves as a base for another material.) On small projects drywall may be installed by general contractors or other trades, but in larger projects drywall is installed by specialized drywall contractors.

Types of Drywall

Drywall panels are available in many types, sizes, and thicknesses for a variety of conditions. The following types of drywall panels are common in residential construction.

Standard Drywall Drywall sheets are commonly 4' wide and 8' long. Panels are also available in lengths of 9', 10', 12', and 14'. Long panels speed construction but can be unwieldy to work with. Panels that are 4'-6" wide are sometimes applied horizontally

Installing Drywall
Specialized Skills Drywall contractors have the tools and equipment to install drywall quickly and efficiently.

Purestock/SuperStock

to reduce installation costs where walls are 8'-6" or 9' tall. The most common thickness for drywall is ½". However, ⅜" and ½" panels are used for covering old surfaces, for covering curved walls, and when layering to reduce sound transmission. Drywall is also available in thicknesses of ⅝" and ¾".

Face paper wraps around the long edges of the panel but does not cover the short edges. The edges along the length of a panel are tapered and, on some types, the ends are also tapered. Tapering allows the joints between panels to be filled and smoothed. Sheets with square-cut long and short edges are used as a substrate for paneling and other materials.

Fire-Code Drywall When certain additives are mixed with the gypsum, drywall becomes more fire resistant. The resulting product is generally referred to as *fire-code drywall*. It is important to understand that a fire-resistant product slows the passage of fire but does not completely stop it. Type-X fire-code drywall is ⅝" thick. It contains glass fibers that prevent it from crumbling in extreme heat. This drywall is required by building code on the outer surface of walls separating an attached garage from the house. This improves the fire resistance of the wall. It is also required on garage ceilings if the garage is beneath a habitable room. Type-C drywall is even more fire resistant than Type-X. It contains vermiculite and comes in thicknesses of ½" and ⅝". Type-C is sometimes used to provide extra fire resistance to ceilings. In commercial construction, multiple layers of fire-code drywall are sometimes installed to increase the fire-resistance of walls.

Moisture-Resistant Drywall In areas of high humidity, such as bathrooms, MR (moisture-resistant) drywall should be used. This material is often called *green board* for the color of its paper facing. The core is a water-resistant type of gypsum. The face and back paper are chemically treated to reduce moisture penetration. MR drywall comes in ½" and ⅝" thicknesses. Fire-code MR drywall

(Type X and Type C) is also available. Green board was once considered suitable as a backing for ceramic tile in shower stalls and in tub/shower surrounds. However, starting with the 2006 IRC, green board is no longer allowed in these locations, except on ceiling surfaces. Instead, cement, fiber-cement backer board, and glass-mat gypsum board may be used. Glass-mat gypsum board is sometimes called *paperless drywall* because the front and back surfaces are covered by a thin fiberglass material instead of paper.

Specialty Drywall Various types of drywall are available to solve specific building problems. Unlike standard drywall, which is readily available, specialty products may have to be ordered in advance.

Foil-Back Drywall This features a layer of aluminum foil on the back surface that serves as a vapor retarder.

Flexible Drywall This is typically installed as double-layer system to cover curved interior walls. The ¼ in. thick sheets are more flexible than standard ¼" drywall sheets.

Sound-Resistant Drywall This is a gypsum product that contains a layer of sound-absorbing polymers that significantly reduce sound transmission through the material.

Reading Check

Recall *What two types of paper are applied to the front and back of a drywall panel?*

Fasteners

Drywall panels must be securely fastened to wood or steel studs with special nails or corrosion-resistant screws. Drywall nails have thin, flat heads so that they can be slightly countersunk by the rounded nose of a drywall hammer. This will not damage the surface of the panel. Standard drywall nails have smooth shanks, but annular-ring (ring-shank) drywall nails offer better holding power. Drywall ½" thick requires a nail at least 1¼" long.

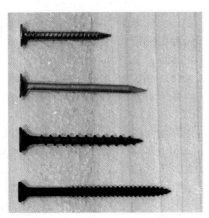

Annular ring nail

Drywall nail

Type-W screw

Type-S screw

 Drywall Fasteners
Nails and Screws Common types of drywall fasteners.

Drywall screws provide much better holding power than nails. They have a Phillips-type bugle head and an unusually sharp point. Type-W drywall screws are used for wood framing and must be long enough to penetrate wood framing at least ⅝" into wood framing. They have wide threads that drive quickly and grip the wood aggressively. Type-S drywall screws are used for steel framing and should be long enough to penetrate steel studs at least ⅜". They have narrow threads. Neither type of screw requires a pre-drilled pilot hole.

Joint Compound

Joint compound is a thick, paste-like material. It is used in combination with joint reinforcing tape made of perforated paper or self-adhesive fiberglass mesh to conceal the joints between panels. By itself, joint compound is used to fill nail dimples and is sometimes used for texturing the panel's surface. It can be purchased in ready-mixed or powder form.

There are many different options when it comes to choosing joint compound. Contractors may choose a particular joint compound based on various factors, including strength, sandability, and curing time. For example, some compounds are very strong but difficult to sand. Others are relatively weak but easy to sand. Various compounds might be used on the same job. For example, a

contractor may use a strong compound, often called a *setting compound,* to embed paper joint tape. Strength and crack-resistance are particularly important at this stage. However, the contractor might layer a different compound over the first one so that the joint will be easier to sand. The additional layers would be a type of compound called a *topping compound.* A topping compound shrinks very little and finishes smoothly. In other situations, a contractor might decide to use a single compound for bedding and topping. This type of compound is called an *all-purpose joint compound.* It is fairly strong and fairly sandable. All compounds, however, fall into one of two basic categories based on how they harden.

Drying-Type Compounds This type of compound cures as it looses moisture. Differences in house humidity and temperature can slow or speed up drying time. Drying-type compounds are available in ready-mixed (pre-mixed) and powdered forms.

Setting-Type Compounds This compound cures through a chemical process. It is less affected by humidity and temperature and cures more quickly than drying-type compounds. It can also be recoated even if it is not completely dry. Setting-type compounds are stronger than drying-type compounds so they are sometimes preferred when

Builder's Tip

MIXING JOINT COMPOUND Powdered joint compound stored in open bags can absorb moisture from the air. This will cause it to form lumps when it is mixed with water. Always use water at room temperature because cold water can also make it difficult to reach an even consistency. Do not pour water into dry compound. Instead, mix compound into the water, gradually adding more powder until the right consistency is reached.

embedding tape. However, they are more difficult to sand and they take more time to prepare. Setting-type compounds are available only in powdered form and must be mixed with water just before use.

Ready-mixed joint compound is easy to use because it does not have to be mixed. However, it is heavy to transport and will freeze if stored in a cold area. Frozen compound that is slowly thawed at room temperature will not be damaged. Repeated freeze/thaw cycles make the material more difficult to work with. Powdered compound must be mixed with clean water before use. However, it has a long shelf life and can be stored at any temperature.

Trim Accessories

A wide variety of metal and vinyl shapes can be used to cover and protect the edges of drywall sheets after they have been installed. At outside wall corners, one edge of the drywall overlaps the intersecting edge. Corner bead is then nailed, screwed, or crimped (using a special tool) over the entire length of the corner. **Corner bead** is a vinyl or galvanized metal strip that reinforces and protects the corner. It comes in lengths of 8' and 10'. Standard corner bead forms a square 90° corner. *Bullnose corner bead* forms a rounded 90° corner. Other trim can be used to finish or protect drywall edges near window and door jambs and where drywall meets another material. One example of this is J-trim.

Repairing Drywall

If drywall panels are not properly stored, installed, or finished, various problems may become evident. Problems are nearly always found along joints or directly above fasteners. However, mechanical damage can damage other areas of the panel. Often, problems will appear shortly after the drywall has been taped and finished. In some cases, they may not appear until much later. Following are some common problems, along with suggested repairs.

Improperly Fitted Panels Drywall panels must not be butted tightly against each other or against other materials. This can stress the panel and prevent it from fitting against the framing. Fasteners will then puncture the paper as they are driven. To repair the panel, remove it and cut it to fit properly. Always hold the panel tightly against framing while driving fasteners. Do not drive new fasteners through the old holes.

Damaged Face Paper The face paper of a panel can be damaged during storage or during installation. If the damage affects only the paper and little of the underlying panel, repair is simple. Small punctures should be filled with joint compound. If the paper is torn or loose, peel it back to solidly adhered paper and topcoat the area with joint compound.

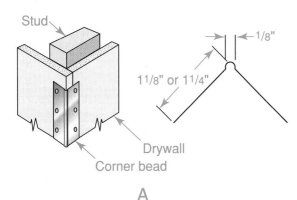

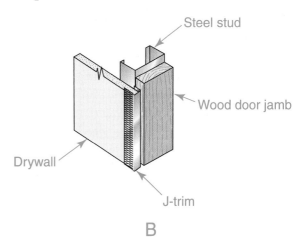

Drywall Trim
Edge Protection **A**. Corner bead protects the outside corners of a wall. Note how the drywall panels overlap. **B**. Products such as J-trim can be used to provide a finished edge where drywall meets another material.

Loose or Unseated Fasteners If a nail or screw has not been fully driven into the framing, it may become loose. In other cases, it may stick up above the surface of the drywall (called an *unseated* fastener). To repair the panel, loose nails should be removed, and a new nail or screw should be driven nearby. Unseated nails or screws can be driven to a proper depth as long as the drywall is tight against the framing.

Large Holes Panels are sometimes damaged after installation by subsequent construction activities. A hole that goes completely through a panel can be repaired in place. However, if the damage is extensive, it may be easier to remove the entire panel and replace it. The repair method depends on the extent of the damage. In general, the damaged area must be cut away so that a patch of new drywall can be fit into place. Cut the patch first, then use it as a template for cutting out the damaged area. The edges of the patch must be supported by drywall repair clips, wood furring strips, or by scraps of drywall. Fit the patch into place and screw it to the supports.

Blistered Tape Various problems can result in blistered joint tape, including using too much or too little compound, not pressing the tape into place firmly, or scraping the joint dry with taping knives. If the blistered area is small, slit the tape to open it up, fill the area with joint compound, and press the tape into the compound. If blistering affects an entire joint, remove the tape and loose joint compound. Then retape the joint.

Cracked Tape Over Flat Joint This is generally related to stresses in the panel caused by structural movement. It may also be due to the expansion and contraction of long walls. Once the causes of structural movement have been corrected, retape the joint.

Cracked Tape Over Inside Corner Joint This could have several causes: too much compound in the corner, corner tape damaged during installation, or structural movement between intersecting surfaces. Hairline cracks

can be filled with joint compound. If joint tape is slit, remove it and retape. If structural movement is suspected, remove fasteners closer than 6" to the angle and retape. This allows the joint to flex somewhat.

Cracking Over Corner Bead or Trim Several causes are possible: too rapid drying in hot weather, use of topping compound instead of taping compound, or application of unusually cold or wet compound. To repair the panel, remove cracked compound. Make sure corner bead is securely fastened, then reapply fresh compound.

Defective Joints Excess buildup of joint compound can create obvious lumps over the joints. Applying compound over previous layers that are not yet dry can result in concave surfaces. To repair the panel, sand down excess layers but do not scuff up the joint paper beneath. Then feather out the joint. Always let compound dry thoroughly before adding another layer. Concave joints must be filled with compound.

Nail Pops Repeated shrinkage and expansion of wood framing can cause nails to back out of the framing, resulting in a raised area called a *nail pop*. To repair, remove the nail and drive a screw nearby.

Bulge Around Fasteners If fasteners are driven too deeply, the face paper of the drywall will be punctured. Joint compound may then cause the edges of the paper to swell. If this happens, drive a screw near the damage, remove the damaged face paper, and repair the area using a setting-type compound.

Drywall Tools
What are the benefits of specialized lifting equipment?

Special installation and finishing tools are required for drywall work. In addition, specialized lifting and transport equipment should be considered for carrying panels from the delivery truck directly into the house. This speeds up construction. It also improves safety because the heavy, awkward

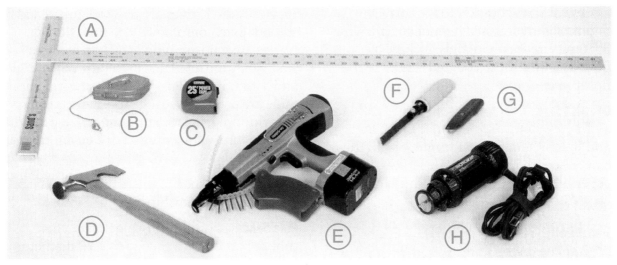

Installation Tools

Basic Tools Drywall installation tools. **A.** Layout square **B.** Chalk line **C.** Tape measure **D.** Drywall hammer **E.** Screw gun **F.** Drywall saw with replaceable blade **G.** Utility knife **H.** Drywall router *How can a drywall utility saw and a drywall router be used for a similar purpose?*

awkward sheets do not have to be carried by hand. It also reduces the chance that sheets will be damaged as they are carried. In some cases, truck-mounted drywall booms can be used to transfer sheets directly from the truck to various locations within the house.

Installation Tools

Installation tools are used to lay out, cut, and attach drywall panels to structural members.

Tape Measure, Chalk Line, and Drywall Square These tools are used to measure and mark drywall panels for cutting. The metal drywall square can also be used to guide a utility knife.

Utility Knife Drywall is easily cut with a utility knife. When very straight cuts are required, guide the utility knife with a drywall square.

Drywall Utility Saw Small cutouts for electrical boxes and other openings are made with a drywall utility saw, sometimes called a *jab saw.* It is similar to a keyhole saw but has a stiffer blade and larger, sharper teeth. Most saws have a sharp, stiff point that can pierce

drywall when starting a cut. Others have a replaceable blade.

Drywall Router Some workers prefer to use an electric tool for making holes in drywall panels. The routers bit can be plunged into the drywall. The bit can follow the contours of an electrical box.

Drywall Hammer The domed striking surface drives nails just below the surface of the drywall without tearing the face paper. The hatchet-type head can be used for cutting large holes.

Screw Gun A screw gun is similar to an electric drill but has a depth-sensitive nosepiece instead of an adjustable chuck. A Phillips bit fits into the head. The tool drives

JOB SAFETY

WATCH YOUR FINGERS! A utility knife is very sharp. Many workers are cut when a knife slips unexpectedly. When cutting drywall, keep your hands well away from the knife's path. Also, keep the knife in a leather sheath when not using it, rather than in a pocket.

a drywall screw quickly to the correct depth (slightly recessed) without overdriving it. Some screw gun models, include a magazine that feeds collated screws to the nosepiece.

Drywall Jack This tool is sometimes called a *panel hoist* or a *drywall lift*. It lifts panels and holds them against framing until they can be secured with nails or screws. It is especially useful when installing drywall on ceilings.

Finishing Tools

Some finishing tools designed to allow workers to reach awkward locations. Others are used to conceal joints between panels and fill dimples left by nails and screws.

Taping Knives Knives are used to spread and smooth joint compound. The thin, flexible blades are made of blued steel or stainless steel. They are available in depths of 2¼" and 3" and in various widths. Knives 6" or 8" wide are used for setting tape. Knives up to 20" wide are used for applying finish coats. Handles are made of wood or a cushioned material.

Corner Trowel The blade is angled at 103° and flexes to 90° for finishing inside corners.

Pole Sander This sanding block has a foam rubber pad that is attached to a ¾" diameter pole. It holds strips of sandpaper or sanding screen (an open grid coated with carbide grit). The wood pole **enables** the user to reach all parts of a wall or ceiling safely.

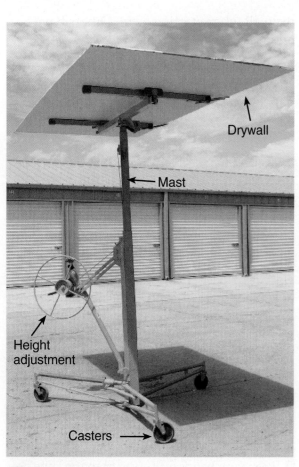

Drywall

Mast

Height adjustment

Casters

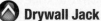
Drywall Jack
Back Saver A drywall jack can be used to hold panels against a ceiling as they are being fastened.

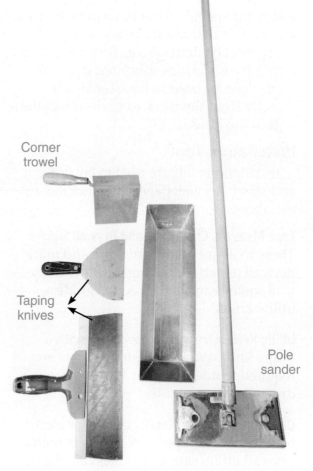

Corner trowel

Taping knives

Pole sander

Finishing Tools
Basic Tools Drywall finishing tools.

(l)Arnold & Brown; (r)McGraw-Hill Education

Dust Mask/Respirator A NIOSH-approved dust mask reduces exposure to sanding dust that can cause eye, nose, throat, or upper-respiratory irritation.

Safety Glasses or Goggles Properly fitted, these protect eyes from sanding dust.

Vacuum Sander This vacuum-assisted device is used to sand drywall and collect the dust.

Stilts Stilts allow the user to reach ceilings without repeatedly climbing a ladder or scaffold.

 Drywall Stilts
No Ladder Required Drywall stilts in use while installing drywall.

Installing Drywall

When would you install a double layer of drywall?

Drywall is generally attached directly to framing in a single layer. The maximum spacing of framing members for various thicknesses of drywall is shown in **Table 32-1** on page 924. However, double-layer installations are sometimes used to dampen sound transmission between rooms. In such cases, the first layer is fastened to the studs, and the second layer is attached to the first using drywall adhesive and a minimum number of fasteners. The basic techniques for installing drywall on wall and ceiling surfaces apply to other applications as well, including applying drywall to cabinet soffits and using it to conceal I-beams and posts. In each application, wood or metal framing blocking must be in place as a support system for the drywall. For more on blocking, see Chapter 16.

Installing Drywall Over Masonry

Drywall can also be applied to concrete or concrete block walls. With interior walls that are above grade, the drywall can be adhered directly to the masonry, usually by using setting-type joint compounds as the adhesive. Where masonry walls are below grade or form exterior walls, they must first be properly dampproofed. Then metal furring channels must be attached to the walls and drywall is screwed to the channels. This creates a continuous 1" minimum airspace behind the drywall that helps to prevent moisture problems.

Cutting Drywall

Drywall is easily cut to fit a wall. Most cuts can be made with a utility knife or drywall saw. Panels can be cut one by one off the top of a stack. However, it is generally more convenient to cut a panel when it is leaning against a wall surface, with its long edge on the floor. There are three basic types of cuts that are made in drywall.

Table 32-1: Nailing and Structural Support for Drywall

Thickness of Gypsum Board (inches)	Application	Orientation of Gypsum Board to Framing	Maximum Spacing of Framing Members (inches OC)	Maximum Spacing of Fasteners (inches)	
				Nails	Screws [a]
Application without Adhesive					
3/8	Ceiling [b]	Perpendicular	16	7	12
	Wall	Either direction	16	8	16
1/2	Ceiling	Either direction	16	7	12
	Ceiling [b]	Perpendicular	24	7	12
	Wall	Either direction	24	8	12
	Wall	Either direction	16	8	16
5/8	Ceiling	Either direction	16	7	12
	Ceiling [c]	Perpendicular	24	7	12
	Wall	Either direction	24	8	12
	Wall	Either direction	16	8	16
Application with Adhesive					
3/8	Ceiling [b]	Perpendicular	16	16	16
	Wall	Either direction	16	16	24
1/2 or 5/8	Ceiling	Either direction	16	16	16
	Ceiling [b]	Perpendicular	24	12	16
	Wall	Either direction	24	16	24

[a] Screws shall be Type-S or Type-W per ASTM C 1002 and shall be sufficiently long to penetrate wood framing not less than 5/8 inch and metal framing not less than 3/8 inch.

[b] 3/8-inch-thick single-ply gypsum board shall not be used on a ceiling where a water-based textured finish is to be applied, or where it will be required to support insulation above a ceiling. On ceiling applications to receive a water-based texture material, either hand or spray applied, the gypsum board shall be applied perpendicular to framing. When applying a water-based texture material, the minimum gypsum board thickness shall be increased from 3/8 inch to 1/2 inch for 16-inch OC framing, and from 1/2 inch to 5/8 inch for 24-inch OC framing or 1/2-inch sag-resistant gypsum ceiling board shall be used.

[c] Type-X gypsum board for garage ceilings beneath habitable rooms shall be installed perpendicular to the ceiling framing and shall be fastened at maximum 6 inches OC by minimum 1 7/8 inches 6d coated nails or equivalent drywall screws.

Cuts Across the Sheet This cut is the most common and the easiest cut to make. It is made from one edge to an adjacent or opposite edge using a sharp utility knife.

After marking the length on the edge of a sheet, score through the face paper using a utility knife guided by a drywall square. It is not necessary to cut entirely through the drywall. Be sure that your fingers are out of the way when scoring. Experienced installers anchor the bottom of the square with a foot to prevent it from slipping. When the entire cut line has been scored once, snap the drywall backwards along the score line and fold the two sections away from the scored line. To separate the pieces, slice through the backing paper, then pull the pieces apart.

Cuts Within the Sheet When an electrical box, heating duct, or other object will penetrate

A   B   C

 Cutting Drywall

Score, Snap, Separate Cutting drywall. **A.** Scoring with a utility knife. Note that the left hand is out of danger. **B.** Snapping the sheet along the scored line. **C.** Separating the pieces after cutting through the backing paper.

the drywall, a hole must be cut for it that is just slightly larger than the object. One way to do this is to measure the position of the object and then transfer its position and shape to a sheet of drywall. The hole can then be cut using a drywall saw. The point of the saw is simply pushed through the surface of the drywall until the saw's

teeth are able to cut. Contractors often prefer a quicker method when cutting holes for electrical boxes. With the drywall sheet held in place on the wall, they use a drywall router to cut along the outer perimeter of the box.

Cuts at the Edge of a Sheet When notching drywall to fit around obstructions such as windows, use a drywall saw to cut through the panel along one or more layout lines. Then score the intersecting line with a utility knife and snap the piece off.

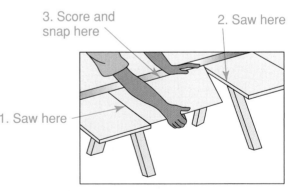

3. Score and snap here

2. Saw here

1. Saw here

 Cutting for Electrical Boxes

Quick and Accurate Holes for electrical boxes and other objects can be cut with a drywall router.

 Notching Drywall

Cuts Made from the Edge Make two saw cuts. Then score the drywall and snap it downward.

Fastening Drywall

Once the drywall has been cut to fit, it can be nailed or screwed to the wall. Always fasten drywall beginning at the center of the sheet, and then work toward the ends. This prevents the sheet from buckling. Hold the sheet tight against the framing as the fastener is driven, but never force drywall into position.

Drywall is generally applied to the ceiling first. The advantage of this method is that the edges of the ceiling drywall will be supported by drywall on the walls. Drywall ½" thick is common on walls and ceilings. However, some contractors prefer ⅝" thick drywall on ceilings because it is stiffer and better able to support the weight of ceiling insulation.

Vertical Application The edges of drywall should be supported by framing members. On walls, a standard drywall sheet can typically be positioned vertically without being trimmed to length. If the stud spacing is standard, most sheets will not have to be trimmed to width. Where sheets must be cut, however, align the sheet so that its edges fall on framing members. If necessary, install blocking to provide adequate support.

Horizontal Application The building code allows drywall to be installed with unsupported edges in certain cases. For example, if drywall sheets are installed with the long dimension perpendicular to the studs, no blocking is required for the horizontal joint between sheets. However, some contractors provide nailing blocks anyway. When sheets are installed horizontally, the upper sheet is generally installed first to provide the best fit at the ceiling/wall corner. Any modest gaps at the floor will be covered by baseboard. Horizontal application is generally preferred for the following reasons:

- The lineal footage of joints is reduced by up to 25 percent.
- Horizontal panels can more easily bridge studs that are not precisely aligned.

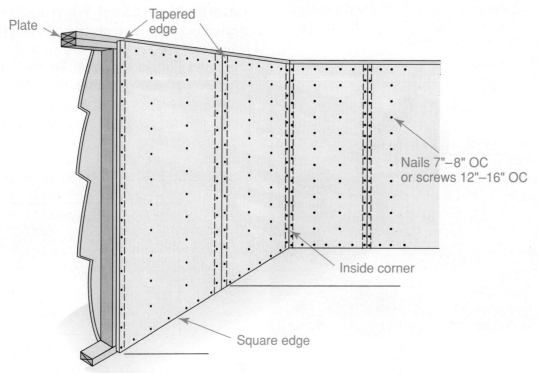

Plate

Tapered edge

Nails 7"–8" OC or screws 12"–16" OC

Inside corner

Square edge

Vertical Application
All Edges Supported Installation of full drywall sheets on walls.

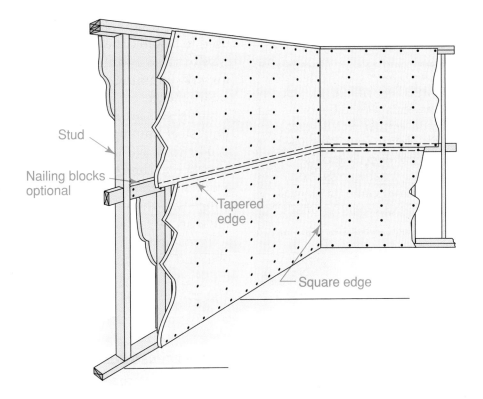

Horizontal Application
Optional Blocking Drywall can be installed horizontally on walls. Blocking behind the horizontal joint is optional.

- The strongest dimension of the panel runs across the studs.
- Horizontal joints are at a convenient height for finishing.

Fastener Spacing

Drywall can be installed using nails or screws. Nails are inexpensive and install quickly, but screws provide a much stronger connection. Where an even stronger connection is desired, a bead of construction adhesive can be run along framing members just before a sheet is installed. Using adhesive reduces the number of fasteners that must be installed. The recommended spacing for fasteners is shown in Table 32-1 on page 924.

Framing Flaws

Wall studs and ceiling joists must be in alignment to provide a smooth, even dry-wall surface. In the case of wood framing, bowed or twisted studs should not be used because the drywall will not seat properly. In addition, the framing lumber must have a low moisture content to prevent *nail pops*. These result if wood framing members dry out and shrink away from the nails.

Twisted Framing
What to Watch For Drywall must fit tightly against framing. **A**. A twisted stud increases the possibility of nail pops. **B**. A properly fastened connection. Note the slight dimple around the nail head.

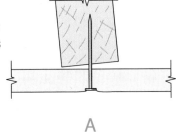
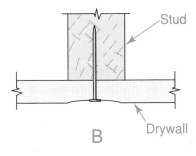

A

B

Builder's Tip

This can cause the nail head to pop above the drywall surface, causing a bump. Nail pops are greatly reduced if the moisture content of the framing is less than 19 percent when the drywall is applied. The use of screws nearly eliminates the problem.

Another type of framing problem that can interfere with drywall is misaligned blocking or bridging. This prevents the drywall from lying flat. This can damage the drywall as it is fastened. Before installing drywall, inspect the framing and fix any such problems.

In the case of houses framed with steel studs, many of the details for drywall support and placement are the same as those applied to wood framing. However, some factors are different. Only screws can be used to install drywall on steel framing. The type of screw is different than the one used for wood framing.

Another factor that differs is the order in which sheets are installed around a room. With steel framing, it should follow a particular plan. The sheets must be installed in a particular direction as the installer moves around the room. This direction depends on which way the stud flanges are facing. Plan the work so that drywall panel edges are screwed first to the open (unsupported) side of a stud. The edge of the next panel should then be screwed to the web side of the stud. This prevents the open side from deflecting as the screws are driven, which would result in uneven joints.

Trim Accessories

In order to prepare drywall for finishing, any trim accessories should be installed. Accessories are vinyl or metal products

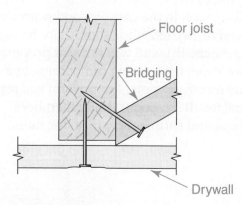

Misaligned Blocking
Always Check the Framing This bridging, which projects beyond the edge of the joists, prevents the back of the drywall from being brought into contact with the nailing surface. A puncture can occur.

Drywall and Steel Framing
Screw Gun This contractor is installing drywall over metal framing using a drywall screw gun.

Arnold & Brown

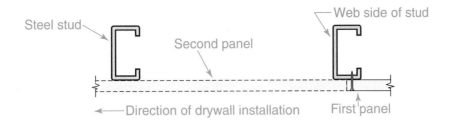

Steel stud   Second panel   Web side of stud

Direction of drywall installation   First panel

Order of Installation

Plan the Job Drywall should be fastened to steel studs by working around the room in one direction only.

that protect corners, decorate wall intersections, or conceal rough edges. The most common accessory is corner bead. This is a shaped length of vinyl or metal that is nailed or screwed over outside corners. Metal corner bead can also be attached by using a special tool that clinches the edges of the bead to hold it in place. Corner bead protects a corner and provides an even, smooth edge that makes the corners easier to finish.

Corner bead is often V-shaped so that it forms a 90° corner when it is installed. However, in some areas, it is common to use bead that has a rounded surface. This is sometimes called *bullnose corner bead*. Large-radius bullnose gives the corners a look similar to that of adobe walls. Smaller radius bullnose gives the walls the look of plastered walls. Segmented bead is another type of corner bead. It is used to finish off the edges of arches and similar curved surfaces.

Finishing Drywall

Why is good ventilation important when working with joint compounds?

After corner bead and other accessories have been installed, joints between the panels and at inside corners are filled in a multi-step process called *taping the joints*. In this process, layers of joint compound are applied over a single layer of perforated paper joint tape. Joint tape reinforces the joint. When the last layer of joint compound dries, it must be sanded smooth.

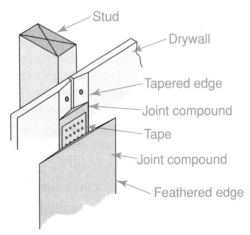

Stud

Drywall

Tapered edge

Joint compound

Tape

Joint compound

Feathered edge

Taping a Joint

Layers over Layers The tapered edge of the drywall is filled with joint compound and tape. Additional joint compound is then applied and feathered out to provide a smooth surface.

Mathematics: Calculation

Estimate Drywall Panel Calculate the number of 4×8 drywall panels necessary to drywall a room measuring 12' × 14' with 10' high ceilings.

Starting Hint First determine the square footage of the walls.

A method for taping joints, inside corners, and covering nail dimples using all-purpose joint compound and paper reinforcing tape is described in the Step-by-Step Application on the next page. At the same time that joints are taped, joint layers of joint compound are applied to accessories such as the corner bead.

Estimating and Planning

This estimating and planning exercise will prepare you for national competitive events with organizations such as SkillsUSA and the Home Builder's Institute.

Estimating Drywall Materials

Estimating and "Figuring Solid"

The amount of materials required for a room is based on the square footage of walls and ceilings to be covered. Contractors typically calculate the square footage and translate that into the number of panels. Each 4×8 panel represents 32 sq. ft, and each 4×10 panel represents 40 sq. ft. When calculating square footage, walls are often "figured solid." That means that the overall square footage of the wall is calculated without subtracting for door and window openings. This is done partly to simplify calculations and partly because the large cutouts required for doors and windows are generally considered waste. These pieces are often difficult to reuse elsewhere in the project. However, if a wall contains a large picture window, the estimator may decide to subtract its area from the overall square footage of the wall.

Sheets

To determine square footage, multiply room perimeter by room height. (Obtain this information from the plans.) Do not subtract door and window openings from the figure. This provides a small allowance for waste. For example, a 10' × 12' room with 9' high ceilings would contain 516 sq. ft.:

Walls: 12 + 12 + 12 + 12 = 44 ft.

$$44 \times 9 = 396 \text{ sq. ft.}$$

Ceiling: 10' × 12' = 120 sq. ft.

$$396 + 120 = 516 \text{ sq. ft.}$$

Each 4×8 drywall sheet covers 32 sq. ft., so the total number required for this room would be approximately 17 sheets:

$$516 \div 32 = 16.12$$

Fasteners

The quantity of nails is estimated by pounds per 1,000 sq. ft. of drywall. For example, ½" dry-wall requires 4.5 lbs. per 1,000 sq. ft. when applied to wood framing 16" OC. The room in our example would need about 2.5 lbs.:

$$516 \div 1,000 = 0.516$$
$$0.516 \times 4.5 = 2.32 \text{ lbs.}$$

Joint Treatment

To finish 1,000 sq. ft. of drywall, 370 lineal feet of joint tape and 138 lbs. of ready-mixed all-purpose joint compound or 83 lbs. of conventional drying-type powder are required.

To determine the quantities, divide the total square footage of wall and ceiling by 1,000. Then multiply this figure by the amounts per 1,000 sq. ft.:

$$516 \div 1,000 = 0.516$$
$$370 \times 0.516 = 190.9 \text{ l.f. of joint tape}$$
$$138 \times 0.5163 = 71.2 \text{ lbs. of all-purpose joint compound}$$

Finishing Drywall The following describes the use of all-purpose joint compound and paper reinforcing tape. Using other products would mean a change in the order of steps. Also, drywall contractors use specialized tools that make the work go faster.

TAPING FLAT JOINTS

Step 1 Use a 5" wide taping knife and firm pressure to force compound into the joint. Scoop the compound from a mud box.

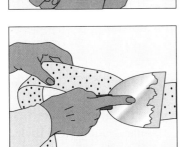

Step 2 Press joint tape into the fresh compound with a 6" or 8" taping knife until the excess compound is forced through the holes in the tape.

Step 3 Immediately cover the tape with additional compound, **feathering** (smoothing) the outer edges so there are no ridges.

Step 4 After the first layer of compound has dried, apply a second coat, using a wider knife.

Step 5 For best results, apply a third coat, feathering the edges beyond the second coat.

Step 6 After the final coat is completely dry, sand the joint smooth and even with the wall surface. Use a pole sander or vacuum sander whenever possible.

TAPING CORNERS AND ANGLES

Step 1 For an inside corner, apply joint compound along both sides of the corner.

Step 2 Cut joint tape to the length of the corner. Fold the tape lengthwise down the center and crease it to form a right angle.

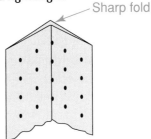

Sharp fold

Step 3 Where drywall meets at angles that are not 90°, angled trim accessories can be used instead of standard corner bead. Another method is to use metal-reinforced paper tape. This is installed much as standard tape, but a thin strip of metal adhered to the tape on both sides of the crease helps to keep it straight when it is applied over joints that are not 90°.

Step 4 Press the tape into the compound and follow Steps 2 through 5 for flat joints.

Step 5 For an outside corner, apply compound over the edges of the corner bead on both sides. Joint reinforcement tape is not necessary.

Step 6 Apply more layers as necessary, always feathering the edges.

Step 7 Sand the joint smooth. Be careful not to sand through the compound. This will damage the paper.

FILLING DIMPLES

Nails and screws must be driven slightly below the surface of the drywall. The resulting "dimples" must be filled with joint compound.

Step 1 To hide dimples, fill them with joint compound using a 6" knife. Apply additional layers as necessary after each earlier layer dries.

Step 2 Sand the dimple areas smooth when they are dry.

Table 32-2: Approximate Drying Time for Joint Compound

Temperature (°F)	Relative Humidity								
	0%	20%	40%	50%	60%	70%	80%	90%	98%
40°	28H	34H	44H	2D	2½D	3½D	4½D	9D	37D
60°	13H	16H	20H	24H	29H	38H	2½D	4½D	18D
80°	6H	8H	10H	12H	13½H	19½H	27H	49H	9D
100°	3H	4H	5H	6H	8H	10H	14H	26D	5D

Note: H = Hours, D = Days (24 hours).

Temperature and humidity have a direct effect on the drying time of drying-type joint compounds (setting-type compounds dry by chemical reaction). Each layer must be thoroughly dry before more coats are applied. In all cases, good ventilation speeds drying, and increased humidity slows drying. **Table 32-2** shows the approximate drying periods for joint compound under different temperature and humidity conditions.

Surface Textures

The surface of standard drywall is smooth once it had been taped, sanded and primed, but various products can be used to give it a decorative texture. For example, ceilings are often sprayed with a product containing fine, medium, or coarse polystyrene aggregrate. This results in a heavily textured surface creating what is sometimes called a *popcorn ceiling*. This finish masks minor surface defects and is usually left unpainted. Other finishes can be used on walls as well as ceilings. Some are created with joint compound applied with a flat trowel. This is sometimes referred to as skim-coating.

Section 32.1 Assessment

After You Read: Self-Check

1. Along which dimension is a drywall panel strongest?
2. *Type-X* and *Type-C* refer to which kind of drywall?
3. Which type of compound would be used when fast curing is important?
4. What type of drywall can be used where walls will receive very hard wear?

Academic Integration: Mathematics

5. **Estimating Drywall** Estimate the number and cost of 4 ft. by 8 ft. sheets of drywall at $5.55 each needed to cover the walls and ceiling of a 58 ft by 16 ft. rectangular recreation room with an 8 ft. high ceiling.

   Math Concept When solving a word problem, turn the wording into a numeric equation. You may need more than one equation.

Go to **connectED.mcgraw-hill.com** to check your answers.

Types of Plaster

How can the use of plaster slow the construction process?

Plaster is a material that can be applied to various substrates to form a dense, hard surface. It can also be given a wide variety of textures. Because it is applied as a wet material, it must dry before other work in the house can continue. This can be inconvenient when construction must be finished quickly. In fact, drywall was developed because of the need to provide a similar looking material that did not require as much drying time. However, many builders and homeowners feel the inconvenience of installing traditional plaster is worthwhile. That is primarily because its surface is harder and more durable than drywall.

Veneer Plaster

One variation of traditional plastering systems is called *veneer plaster.* **Veneer plaster**, also called thin-coat plaster, is a specially formulated gypsum plaster that is applied to a type of drywall called *gypsum base.* Gypsum base is sometimes called *blue board* because of the color of its face paper. It comes in 4' × 8' sheets. The moisture-absorbent face paper of gypsum base is treated to provide an excellent bond with the plaster. The plaster is applied by trowel either in one layer ¹⁄₁₆" to ³⁄₃₂" thick or in two layers totaling ⅛".

Traditional Plaster

Traditional plaster, sometimes referred to as lath-and-plaster, is a wall and ceiling material made from sand, lime or prepared plaster, and water. Plaster is applied by skilled contractors who specialize in its installation. Many of the tools used in plastering are similar to those used for drywall.

However, unlike veneer plaster, traditional plaster must be applied to a base material called **lath**. Lath is any base material for plaster that has qualities that encourage the plaster to stick to it.

In old houses, lath consisted of slender wood strips nailed perpendicular to framing members. Layers of plaster were then applied to the closely spaced lath and forced into gaps between the strips. As the plaster dried, it adhered to the back side of the lath. This provided a strong mechanical connection, sometimes called a *keyed connection.* Remodelers frequently find wood lath when they work on old houses, but it is no longer used for new construction. The most common types of lath used today are made of gypsum panels or expanded metal.

Materials & Techniques

How are drywall and plastered edges similar?

Before the development of drywall, plaster was the most common method for finishing interior wall and ceiling surfaces. Drywall is now the most common system, but plaster materials and techniques have continued to evolve.

Gypsum Lath

Gypsum lath has a core of gypsum surrounded by a multilayered paper face specifically designed for plaster. Gypsum lath comes in 16" by 48" panels with square edges. Gypsum lath is applied horizontally across the framing members. For stud or joist spacing of 16" OC, a ⅜" thickness is used. For 24" spacing, the lath should be ½" thick. Where joints do not fall on framing members, a formed metal clip can be used to support and align joints.

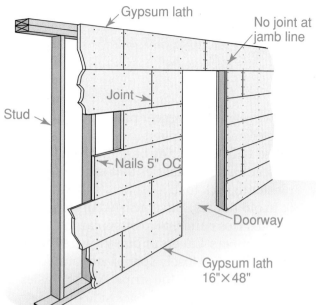

Gypsum lath

No joint at jamb line

Joint

Stud

Nails 5" OC

Doorway

Gypsum lath 16"×48"

 Details for Gypsum Lath
Staggered Joints Gypsum lath is nailed or screwed horizontally across studs. Note that the joints are staggered and that there is no joint at the jamb line in the doorway.

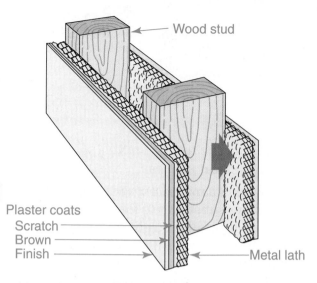

Wood stud

Plaster coats
Scratch
Brown
Finish

Metal lath

 Details for Metal Lath
Three-Coat Plaster Work A cross section of plaster on metal lath showing the buildup of the various coats. Notice how the plaster is keyed to the metal lath in the area indicated by the arrow.

Over wood studs, lath can be attached with either flat head 13-gauge gypsum-lathing nails 1⅛" long or 16-gauge galvanized staples. The staples have a flat ⁷⁄₁₆" wide crown and 1" divergent-point legs. Gypsum lath can be secured to metal studs by 1" long type-S screws.

Vertical joints should be made over the center of studs or joists. The nails should be spaced 5" OC, or four nails for the 16" height, and used at each stud or joist crossing. Joints over heads of openings should not occur at the jamb lines.

Metal Lath

Metal lath is made from sheet metal. The metal is slit and expanded during manufacture to form various patterns, such as flat ribs or a diamond mesh. Openings in the lath create gaps for the plaster to "grip."

Metal lath is usually 27" × 96" in size and galvanized to resist rust. It is usually installed on studs or joists spaced 16" OC.

Metal lath is often used around tub recesses and other bath and kitchen areas, even if the gypsum base is installed elsewhere. In such cases, Portland cement plaster is sometimes used instead of gypsum plaster. It provides a substrate more suitable for ceramic tile. When used in wet areas, metal lath must be backed with water-resistant sheathing paper.

Trim Accessories

Like drywall edges, the edges of a plastered wall can be covered and protected by trim accessories. These accessories also provide a ground. A **ground** is a material permanently or temporarily attached to a surface to be plastered. It provides a straight edge and helps the plasterer gauge the thickness of the plaster. It may be made of wood but is most often made of metal. Common accessories include metal corner bead, Cornerite, casing bead, and reinforcing lath.

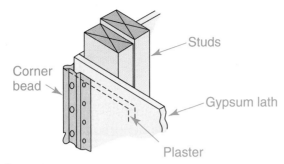

Studs

Corner bead

Gypsum lath

Plaster

Corner Bead
Corner Protector A corner bead is installed at outside corners to serve as a leveling edge when the plaster is applied. It also protects the corner.

- *Metal corner bead* is required at all outside corners. It has a solid galvanized metal edge and expanded flanges. Each flange is approximately 2⅞" wide.
- *Cornerite* is an angled length of metal lath that is used to strengthen interior corners.
- *Casing bead* is used around wall openings. It is also used where plaster meets other finishes. It can eliminate the need for wood trim around doors and windows.
- *Reinforcing lath* is a flat length of expanded lath 4" or 6" wide. It is used to reinforce areas that might crack, such as corners around doors and windows.

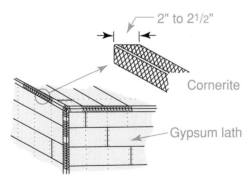

2" to 2½"

Cornerite

Gypsum lath

Cornerite
Reinforcer Cornerite is installed at inside corners for reinforcement and to limit plaster cracks.

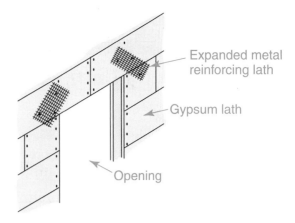

Expanded metal reinforcing lath

Gypsum lath

Opening

Reinforcing Lath
Crack Resister Expanded metal lath is used to help limit plaster cracks.

Installation

Plaster is applied in three layers over metal lath. This process is called three-coat plaster work. Three-coat plaster work is usually at least ¾" thick. It is applied in two or three layers over gypsum lath. The minimum thickness over ⅜" gypsum lath should be about ⅜".

The first plaster coat is called the *scratch coat*. After a slight set has occurred, the plaster is scratched to ensure a good bond with the second coat. The second coat is called the *brown coat*, or leveling coat. The plaster is brought to level during its application. The third coat is the *finish coat*. It provides the finished wall surface. Two-coat work over gypsum lath, sometimes called *double-up work*, combines the scratch and brown coats.

Plaster receives one of two basic finishes: the sand-float and the putty finish. For the sand-float finish, lime is mixed with sand, which produces a texture. The putty finish, made without sand, is smooth. It is common in kitchens and bathrooms where a gloss or enamel paint will be used.

Plastering should not be done in freezing weather without a source of constant, even heat. In normal construction, portable heating units are sometimes in place before plastering begins.

Estimating and Planning

This estimating and planning exercise will prepare you for national competitive events with organizations such as SkillsUSA and the Home Builder's Institute.

Gypsum Lath, Nails, and Labor

Materials and Labor

Gypsum lath is packaged in bundles of six 16" × 48" pieces. A standard lath bundle therefore contains approximately 32 sq. ft. of lath.

Step 1 See the floor plan below. To determine the number of bundles required, divide 32 into the total area to be covered. Suppose that the walls and ceiling of the room are to be finished with lath and plaster. Assume the ceiling is 8 ft. high. The total wall and ceiling area equals 798 sq. ft.:

Walls: 66 × 8 = 528
Ceiling: 15 × 18 = 270
528 + 270 = 798 sq. ft.

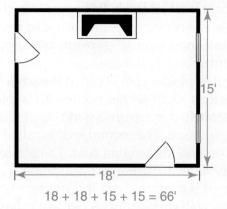

18 + 18 + 15 + 15 = 66'

Step 2 Divide the number of square feet to be covered by the number of square feet in a bundle of gypsum lath (32) for a total of 24.94, or 25 bundles of gypsum lath.

798 ÷ 32 = 24.94

Step 3 To estimate the amount of nails required for installing the gypsum lath, figure that 5 lbs. are needed for every 100 sq. ft. About 800 sq. ft. of lath are to be installed. Therefore, 8 × 5 = 40 lbs. of nails.

Step 4 A plasterer calculates the labor cost of a job by the number of square yards to be covered. Convert the square feet in the room to square yards by dividing by 9 (1 sq. yd. equals 9 sq. ft.). In our example, 798 ÷ 9 equals 88.66, or 89 sq. yd. This figure would then be multiplied by the plasterer's labor rate.

Estimating on the Job

Suppose a 9' by 12' room with 9 ft. ceilings must be plastered. How much gypsum lath and nails will be needed? If a plasterer charged $22 per square yard, how much would the labor cost?

Section 32.2 Assessment

After You Read: Self-Check

1. Name three types of lath used as a base for plaster.
2. What is a plaster ground?
3. What is Cornerite and what is it used for?
4. Name the two plaster finishes.

Academic Integration: English Language Arts

5. **Gypsum** One of the most common types of lath used today is gypsum. Research the history of the use of gypsum as a building material. Write a two-paragraph summary about your findings.

Go to **connectED.mcgraw-hill.com** to check your answers.

Suspended & Acoustical Ceilings

Suspended Ceilings

In what remodeling situations would suspended ceilings be especially useful?

Drywall and plaster are the most common materials used for residential ceilings. However, they prevent access to the areas above. If ducts and water pipes are routed between floor joists, a suspended ceiling is sometimes installed. A **suspended ceiling** consists of panels held in place by a metal or plastic grid. A suspended ceiling conveniently covers bare joists, exposed pipes, and wiring. Panels can be removed easily for access to valves, switches, and controls. A suspended ceiling may also be used to lower the ceiling level. This is sometimes done when walls are unusually high. Recessed fluorescent lighting can be installed at most locations in a suspended ceiling.

Suspended ceilings are widely used in commercial and light commercial construction. In houses, they are most common in finished basements. They are also common in commercial construction. In commercial construction, the exact layout of a suspended ceiling may be determined by a special drawing called a *reflected ceiling plan.* However, this drawing is rarely necessary in residential construction.

 Suspended Ceiling

Hanging Ceiling A suspended ceiling consists of flat or profiled panels loosely supported by a ceiling grid.

Installing a Suspended Ceiling Be sure to plan the layout before you begin. This can be done by sketching a basic grid plan using graph paper or by using online planning software. Always follow the ceiling manufacturer's instructions.

Step 1 Nail or screw the wall molding to the walls at the height desired for the ceiling. A laser level is the fastest and most accurate way to align wall molding, but a standard level and chalkline can also be used. At inside corners, cut the vertical leg of each molding to fit into the corner, but lap one lower leg over the other. Form outside corners by mitering the wall moldings together or by overlapping them.

Step 2 Determine where the main runners will go based on your grid plan sketch. Mark the location of runners and the hanger-wire screws that will support them by snapping a chalk line across the ceiling joists. Drive screws 4' OC (or as recommended by the ceiling manufacturer) into the bottom edge of joist at the chalked lines. The screw should be centered. Now loop a length of hanger wire through every screw and wrap it back around itself at least three times. Allow the wires to hang loose for now. They will be used in a later step to support the main beams.

Step 3 Check the ceiling layout for the location of the first main beam and the first row of cross tees. Stretch a string across the length of the room and another across the width to represent these locations. Keep these reference strings very tight. Check them with a framing square to ensure that they meet at a 90º angle. Adjust them as needed.

Step 4 Measure up ⅞" from the bottom of the wall molding under each joist that carries hanger nails. Drive a nail into the wall at each location. Stretch leveling strings across the room between opposing nails and tie them to the nails. Now use pliers to make a sharp 90º bend in every hanger wire where it intersects a leveling string. When all the wires have been bent, remove the leveling strings. (Do not remove the two reference strings.)

Step 5 Cut the beams one by one to fit into place, starting at the appropriate guide string. Slip the ends of the hanger wires into existing holes or slots in the beams. Twist the end of the wire around the vertical portion to secure it. Repeat the process with the remaining beams.

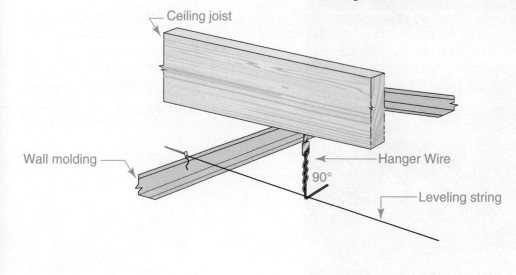

Step 6 Starting at the appropriate reference string, install the first row of cross tees between the main beams. Lock the ends of each cross tee into slots in the main beams. Install the remaining cross tees in the same way. Make sure they are spaced properly to accept the desired size of the ceiling panel.

Step 7 Measure each of the border ceiling panels individually. Cut each panel face up, using a sharp utility knife. After all the border panels are in place, slip the remaining ceiling panels into place. Take care when handling ceiling panels to avoid marring the surface. Handle the panels by the edges, keeping your fingers off the finished side as much as possible. Lightweight cotton gloves can be worn to prevent the panels from being smudged.

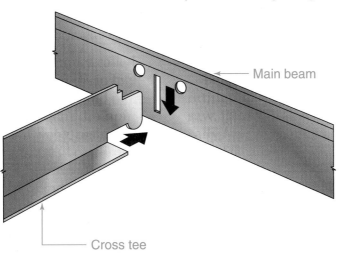

Main beam

Cross tee

Materials

Ceiling panels are made of plastic or mineral board, which is a lightweight material. It consists of mineral components, binders, and inert filler materials, such as recycled newsprint and mineral wool. Each panel is 2' × 2' or 2' × 4'. The grid system that supports these panels includes main beams (sometimes called *runners*), cross tees, and wall molding. Main beams are usually 12' long and are spaced 2' or 4' OC. Cross tees are installed at right angles to the main beams. Grid components are made of metal or vinyl, and come in various configurations.

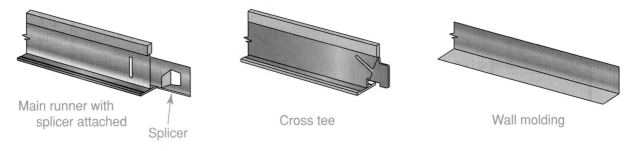

Main runner with splicer attached

Splicer

Cross tee

Wall molding

 Grid Components

Basic Parts The main runners support cross tees. Wall molding runs along the perimeter of the ceiling.

Installation

Before installing a suspended ceiling, establish the finished ceiling height. In general, the top edges of the grid system must be at least 3" below the bottom of the ceiling framing. This space is necessary for the insertion of the panels after the grid system is in place.

A suspended ceiling looks best if the panels on opposite sides of the room are the same width. To achieve this, you must accurately plan the layout of the system. Manufacturers provide planning instructions that include layout details. Follow these instructions carefully. Once layout has been planned, installation can begin.

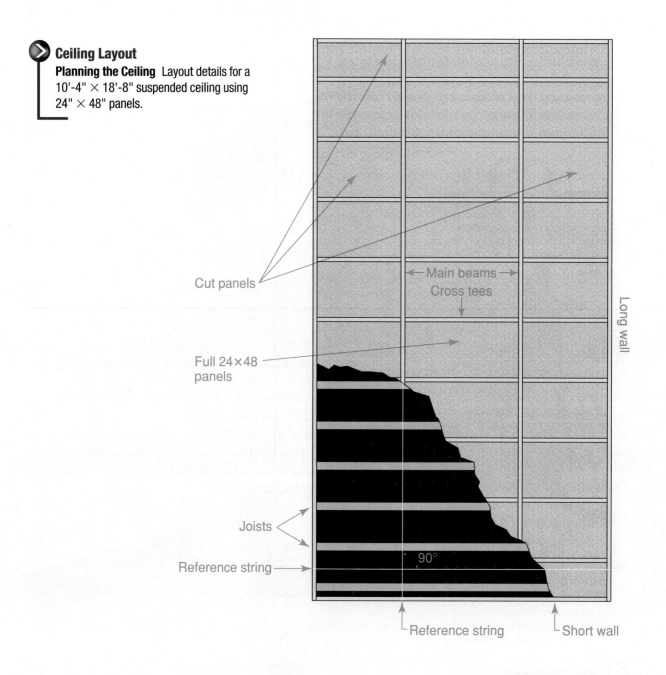

Ceiling Layout
Planning the Ceiling Layout details for a 10'-4" × 18'-8" suspended ceiling using 24" × 48" panels.

Cut panels

Main beams
Cross tees

Full 24×48 panels

Long wall

Joists

Reference string

90°

Reference string

Short wall

Acoustical Ceilings

Which installation method do you think would be faster?

In rooms without enough height to hang a suspended ceiling, an acoustical ceiling may be installed instead. An **acoustical ceiling** consists of panels glued directly to the ceiling surface or stapled to wood furring strips nailed to the ceiling joists.

The panels consist of 12" × 12" squares, or sometimes 12" × 24" rectangles. Each panel has a tongue on two edges, and a groove on two edges. This allows the panels to interlock. Panels are typically made of fiberboard. The surface may be embossed or textured. Some panels are made specifically to reduce sound levels in a room.

There are two methods for installing acoustical ceiling panels: the mastic method and the staple method.

Mastic Method

If the ceiling is flat and in good shape, each panel can be adhered directly to the ceiling surface. A dollop of mastic or construction adhesive placed near each corner and in the middle holds the panels in place.

Staple Method

Where a ceiling is uneven, or where there is no existing ceiling surface, panels must be stapled to wood furring strips. The strips are nailed to the joists and shimmed so that they are in the same plane.

Section 32.3 Assessment

After You Read: Self-Check

1. What are the advantages of suspended ceilings?
2. What materials are ceiling panels made of?
3. What supports a suspended ceiling grid?
4. Name the two methods for installing acoustical ceiling panels.

Academic Integration: Mathematics

5. **Determine Border Dimension** Determine the length of the top and bottom borders shown in this acoustical ceiling if you are using 4' × 2' tile.

   (**Math Concept**) To find the dimension of equal borders in an acoustical ceiling within a single axis, subtract the same dimension of the shape and divide the remainder by two.

   Step 1: Subtract the remaining amount of length (5'-2") from the total length to determine the total length of the borders (9'-2' − 4' = 5'-2" or 62").

   Step 2: Divide the remainder by 2 to determine the length of each border. Convert feet and inches into inches if this is helpful (5'-2" = 62").

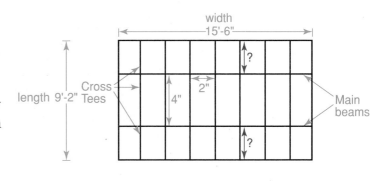

   Go to **connectED.mcgraw-hill.com** to check your answers.

Review and Assessment

Chapter Summary

Drywall is made of a gypsum core covered with special paper. It comes in three basic types: standard drywall, fire-code drywall, and moisture-resistant drywall. There are also specialty drywalls. Joint compound and tape are used to fill joints. Panels can be installed vertically or horizontally. Installation differs for wood framing and steel framing.

Plaster must dry before other work can progress. This can slow the construction schedule. Plaster is applied over gypsum or metal lath in two or more layers. It can be given a smooth or textured finish.

Suspended ceilings consist of panels held in place by a metal or plastic grid. They cover joists, pipes, and wiring. Suspended ceilings are also used to lower a ceiling or to provide sound insulation.

Review Content Vocabulary and Academic Vocabulary

1. Use each of these content vocabulary and academic vocabulary words in a sentence or diagram.

Content Vocabulary

- corner bead (p. 919)
- feathering (p. 931)
- veneer plaster (p. 933)
- lath (p. 933)
- ground (p. 934)
- suspended ceiling (p. 937)
- acoustical ceiling (p. 941)

Academic Vocabulary

- stable (p. 916)
- enables (p. 922)

Speak Like a Pro

Technical Terms

2. Work with a classmate to define the following terms used in the chapter: *face paper* (p. 916), *liner paper* (p. 916), *substrate* (p. 916), *joint compound* (p. 918), *setting compound* (p. 918), *topping compound* (p. 918), *nail pops* (p. 927), *popcorn ceiling* (p. 932), *metal corner bead* (p. 935), *Cornerite* (p. 935), *casing bead* (p. 935), *reinforcing lath* (p. 935), *scratch coat* (p. 935), *brown coat* (p. 935), *finish coat* (p. 935), *double-up work* (p. 935).

Review Key Concepts

3. Compare the different types of drywall and how they are used.

4. List the properties of fire-code drywall.

5. Explain how to repair a nail pop.

6. Describe the safety and health precautions that need to be taken when installing drywall.

7. List the basic elements of three-coat plaster work.

8. Demonstrate how to install a suspended ceiling.

Critical Thinking

9. Explain Which rooms in a house would moisture-resistant drywall be most appropriate?

Academic and Workplace Applications

STEM Mathematics

10. Estimating Cost To finish 1,000 sq. ft. of drywall, 370 lineal feet of joint tape and 138 pounds of ready-mixed all-purpose joint compound are needed. Tape comes in rolls of 250 ft. at $1.45, and joint compound comes in drums weighing 61.7 pounds each at $8.98. How much will tape and compound cost for the walls and ceiling of a rectangular cafeteria that is 65 ft by 45 ft with a wall height of 9 ft.?

Math Concept Solving some problems requires many steps. Take each step one at a time and write down your calculations as you go.

Step 1: Multiply the length times the width of the ceiling to find its area.

Step 2: Compute the perimeter of the room and multiply by the height of the wall to find the area of the walls.

Step 3: Add the two areas and divide by 1,000 sq. ft.

Step 4: Multiply the result from Step 3 by 370 ft., and then divide the result by 250 ft. Round up to the nearest whole number to find the number of rolls of tape needed.

Step 5: Multiply the result from Step 3 by 138 and divide by 61.7. Round up to find the number of drums of joint compound needed.

Step 6: Multiply to find the costs of each component and add to find the total cost of the materials.

21st Century Skills

11. Information Literacy Drywall panels must be securely fastened to wood or steel studs with special nails or screws. Discuss the properties and characteristics of drywall nails and screws that make them effective fasteners.

21st Century Skills

12. Collaboration Skills Each phase of the construction process has the potential to affect the processes that follow. Write two to three sentences explaining how the work of framing carpenters can affect the work of drywall installers.

Standardized TEST Practice

True/False

Directions Read each of the following statements carefully. Mark each statement as either true or false by filling in T or F.

(T) (F) **13.** In commercial construction, the exact layout of a suspended ceiling may be determined by a special drawing called a reflected ceiling plan.

(T) (F) **14.** Traditional plaster, sometimes referred to as lath-and-plaster, is a wall and ceiling material made from water and granite, coral or prepared plaster.

(T) (F) **15.** Drywall is never attached directly to framing.

TEST-TAKING TIP

If you are unsure about an answer, eliminate the choices you know are incorrect. Then make your best guess from the remaining answer choices.

*These questions will help you practice for national certification assessment.

CHAPTER 33

Exterior & Interior Paint

Section 33.1
Finishing Basics

Section 33.2
Exterior Painting

Section 33.3
Interior Painting

Chapter Objectives

After completing this chapter, you will be able to:

- **Describe** the differences between the two basic types of finishes.
- **List** the basic ingredients of paint.
- **Identify** the steps in painting a house exterior.
- **List** the steps in painting an interior.
- **Examine** problems with painted finishes.
- **Explain** how to paint windows and doors.

Discuss the Photo
Painting The painter in the photo is painting the exterior of a house. *What concerns might arise when attempting to protect exterior paint?*

Writing Activity: Create an Advertisement
Brushes and rollers are the primary interior painting tools. Write an advertisement that you feel would be effective in attracting buyers to a high-quality brush. Your ad should draw attention, arouse interest, create desire, and cause action.

©Kim Steele/Blend Images LLC

Chapter 33 Reading Guide

Before You Read Preview

Paints, coatings, and other finishes have been used for thousands of years to decorate and protect surfaces. Choose a content vocabulary or academic vocabulary word that is new to you. When you find it in the text, write down the definition.

Content Vocabulary

- summerwood
- sheen
- binder
- carrier
- solvent
- primer
- flagged bristles
- edging

Academic Vocabulary

You will find these words in your reading and on your tests. Use the academic vocabulary glossary to look up their definitions if necessary.

- synthetic
- compatibility

Graphic Organizer

As you read, use a chart like the one shown to organize information about content vocabulary words and their definitions.

Content Vocabulary	Definition
summerwood	the dense, dark-colored portion of the wood

Go to connectED.mcgraw-hill.com to download this graphic organizer.

Understanding Wood Finishes

What factors determine the ability of a wood to accept a finish?

Paints, coatings, and other finishes have been used for thousands of years to decorate and protect many materials. Finishes are particularly important for wood because they prolong its life and improve its appearance. For example, exterior surfaces of a home require finishes that will protect against weathering, sunlight, and moisture. Interior finishes must be durable and easy to clean. As a general rule, hardwoods are given a clear finish, while softwoods are given either an *opaque* (not clear) finish or a clear finish.

Wood is the most common building material. Therefore, this chapter is about paint and painting techniques for wood. The information generally applies to new construction. When repairing, removing, or recoating an existing finish, many additional surface preparation steps are required that are not covered here.

While all woods can be finished, some take finishes better than others. The ability of a wood to accept a finish is determined by four basic factors: species, grade, grain, and manufacture.

Species

In general, denser species of woods are less accepting of finish than less dense woods. Cedar and cypress hold paint best of all the woods used for siding and trim. Northern white pine, western white pine, and sugar pine are almost as good. Western yellow pine, white fir, and hemlock come next. Serious flaking of paint occurs soonest on southern yellow pine, Douglas fir, and western larch.

Grade

Top-quality grades of wood accept finishes better than lower-quality grades that contain defects such as knots and pitch pockets. The knots of yellow and white pines cause more trouble than the knots of such woods as cedar, hemlock, white fir, and larch.

Grain

Quartersawn boards hold paint much better than plain-sawn boards because the bands of summerwood are very narrow. **Summerwood** is the dense, dark-colored portion of the wood. Its cells have thick walls and small cavities. The more porous, light-colored springwood accepts finish more readily. Flat-grained boards hold paint better on the bark side than on the pith side.

Manufacture

The natural expansion and contraction of solid wood can reduce the durability of a finish. Engineered-wood products, however, are manufactured in ways that reduce this problem. From a finishing standpoint, engineered wood has some advantages over solid wood, particularly when used for exterior trim. It behaves predictably and its surface is uniform. It is dimensionally stable over a wide range of widths and thicknesses. It does not have the defects commonly found in solid lumber.

Some engineered-wood trims are primed at the factory. Others may have an unusually smooth finish that takes paint well. For more on this topic, see Chapter 13.

Reading Check

Recall *How do hardwood finishing options differ from softwood finishing options?*

Types of Finishes

Why are film-forming finishes the most popular of all interior finishes?

Many types of finishes can be used to protect and beautify wood. Finishing technology is constantly improving. New products are introduced to the market each year. It is therefore important to review manufacturer's recommendations for each type that you use. Pay particular attention to any instructions about health and safety.

Finishes can have many different characteristics. For example, a finish can be clear or opaque. It might be suited for exterior use or only for interior use. It can have different levels of sheen. **Sheen** is a description of how shiny a surface is. However, all finishes fall into two basic categories. *Film-forming finishes* coat the wood surface. *Penetrating finishes* soak into the wood.

Build It Green Finishes containing VOCs (volatile organic compounds) are discouraged in many parts of the country. VOCs contribute to air pollution. Regulations restricting their use have been enacted in some states, and other states are now taking steps to restrict the amount of VOCs that can be used in paints and other finishes. Low-VOC finishes are increasingly available and help to minimize the harmful health and environmental effects of VOCs.

Film-Forming Finishes

Many finishes protect wood by leaving a coating, or film, on the wood. The most common type of film-forming finish is paint surface. Some clear finishes, such as varnish and polyurethane, also fit into this category. The film protects the wood against moisture and seals in natural resins. Pigments may be added to protect wood from ultraviolet (UV) rays. They also add color.

Paint Any paint contains ingredients that make it suitable for a particular use. However, all paints contain the following:

Pigments *Pigments* are either finely ground natural minerals or synthetics. A pigment gives paint color and makes it opaque. A greater percentage of pigment increases opacity.

Binder A **binder** is a resin that holds particles of pigment together. The particles form a film after the liquid evaporates.

Carrier A **carrier** (sometimes called the *vehicle*) is a liquid that keeps the pigments and binders in suspension. A carrier also keeps pigments and binders evenly dispersed (spread out) during application.

Oil-Base Paint Paints that have oil-base binders suspended in a mineral spirit carrier are referred to as *oil-base paints,* or *oil paints.* There are two types of oil-base binders. *Vegetable oil binders* are chiefly linseed oil,

Protection From the Weather
Protection and Color Paint protects wood from the weather and also adds color and architectural interest to a house.

a yellowish oil pressed from flaxseed. *Alkyd binders* are **synthetic**. They are sometimes mixed with linseed oil.

Oil-base paints are less flexible than latex paints. This is an advantage where a tough, stable surface is required. However, oil-base paints tend to become brittle over time. This can cause the paint film to crack. Special solvents are needed to clean tools and equipment used with oil-base paints. A **solvent** is a material that dissolves another material. Solvents include mineral spirits (made from petroleum distillates) and turpentine (made from the resin of pine trees). Check the paint label for manufacturer's recommendations for suitable thinners or solvents.

Latex Paint *Latex paints* have latex-base binders suspended in water. There are two types of latex-base binders. *Acrylic latex* is a synthetic resin that is flexible and very durable. *Vinyl latex* is a synthetic resin that is somewhat less durable than acrylic latex.

Latex paints were first developed for interior use but are now readily available for exterior painting. In fact, research indicates that a good-quality acrylic latex outdoor house paint will generally outlast a good-quality oil-base outdoor house paint. In general, latex paints have the following characteristics:

Versatility Latex paints are easy to apply, even on slightly damp surfaces.

Flexibility The paint film expands and contracts slightly with wood movement. It is less likely to crack than an oil-base film.

Permeability Latex paints do not trap moisture within the wood. This makes it unlikely that water vapor will cause the paint to bubble.

Quick Drying This helps to speed construction.

Easy Cleanup Tools can be washed clean with water before the paint on them dries.

Exterior latex paints have one disadvantage, however. Siding woods such as redwood and cedar contain water-soluble extractives that can bleed through latex paint. This sometimes creates dark stains. To prevent this, paint all surfaces of the wood with an oil-base primer before installation. Then top coat it with acrylic latex paint.

Primer Most paint manufacturers make a primer, or undercoat, for use with their house paints. A **primer** is a paint that has a higher proportion of binder than standard paint. This enables it to hold particularly well to unpainted wood surfaces. Because a primer does not block UV radiation, it must be covered with two coats of standard paint.

Primers are available in oil-base or latex forms. They are typically white but may be tinted slightly for use under dark-colored paints. When painting metal, use a special rust-preventative primer.

Solid-Color Stain Solid-color stain is similar to a thin paint. It comes in latex and oil-base forms and is applied in almost the same way as paint. It is not as durable as paint, however. Solid-color stains are used mainly where they can be recoated frequently.

Penetrating Finishes

Unlike film-forming finishes, penetrating finishes actually soak into the wood. They fill the wood's surface pores. Some are clear, while others contain pigments. Penetrating finishes are very easy to apply. They allow the wood grain to show.

There are several types of penetrating finishes. Some are used primarily on exterior wood. These include oil-base semi-transparent stains and clear water-repellent finishes. (Latex semi-transparent stains are available. They are actually a type of film-forming finish and do not soak into the wood.) Some penetrating finishes are used primarily on interior wood, including furniture. These include Danish oil and tung oil.

Semi-transparent stains work very well on rough surfaces, such as plywood siding and some types of beveled siding. They can also be applied to weathered surfaces without much surface preparation. The pigment in semi-transparent stain protects the wood from UV damage. Penetrating finishes that do not contain pigment are not as effective in protecting wood from UV radiation. High-quality products also contain wood preservatives and water repellents.

Penetrating Finish
The Beauty of Wood A penetrating finish allows the grain of the wood to show. It can be protected with a coat of clear finish.

Section 33.1 Assessment

After You Read: Self-Check

1. What is the purpose of pigments?
2. What is a binder and how does it behave?
3. What is the basic difference in mixtures between oil-base paint and latex paint?
4. Why is it important to cover primer with standard paint?

Academic Integration: Mathematics

5. **Word Problems** Estimate the number of gallons and the cost of acrylic wall paint needed to paint all walls of a rectangular 42' × 58' basement that has walls 7' high. The paint costs $22.27 per gallon and has a spread rate of 250 sq. ft. per gallon.

     Math Concept Some word problems ask for more than one solution component. This problem asks you to find both the number of gallons and the cost of the paint.

    Step 1: Calculate the surface area of all walls, floor, and ceilings to be painted.

    Step 2: Divide the surface area by the spread rate. Round up to the nearest gallon.

    Step 3: Multiply the number of gallons needed by the per-gallon cost of the paint.

Go to **connectED.mcgraw-hill.com** to check your answers.

Exterior Painting

Preparing to Paint
How can exterior paint reduce cooling costs?

The most common type of exterior finish is paint. In part, this is because paint is available in many colors. Light-colored paints reflect heat away from the house. A white house, for example, can reflect almost 90 percent of the sun's rays. This can reduce interior temperatures and thus reduce cooling costs. See **Table 33-1** for differences in how colors reflect light.

Paint lasts longer than other exterior finishes. In general, it lasts seven to ten years before requiring recoating. A solid-color

Table 33-1: Light Reflectivity of Colors

Color	10%	20%	30%	40%	50%	60%	70%	80%	90%	100%
Black	■									
Light Brown	▓▓									
Apple Green	▓▓▓									
French Blue	▓▓▓▓									
Light Gray	▓▓▓▓▒									
Silver Gray	▓▓▓▓▒									
Coral	▓▓▓▓▓▓									
Sea Green	▓▓▓▓▓▓▒									
Cream	▓▓▓▓▓▓▒									
Light Buff	▓▓▓▓▓▓▒									
Pastel Green	▓▓▓▓▓▓▒									
Oyster White	▓▓▓▓▓▓▓▒									
Light Cream	▓▓▓▓▓▓▓▒									
Sunlight Yellow	▓▓▓▓▓▓▓▒									
Ivory	▓▓▓▓▓▓▓▓▒									
Light Orchid	▓▓▓▓▓▓▓▓▒									
White	▒▒▒▒▒▒▒▒▒									

stain generally lasts only three to seven years. For best appearance and maximum durability, three coats of exterior paint are best over bare wood. This means a primer followed by two finish coats of standard paint. To ensure compatibility between primer and finish coats, choose a primer and finish paint of the same brand and type.

Many manufacturers make several paints of differing quality and cost. However, the cost of the paint is a fairly small portion of the total cost of painting a home, and problems can be expensive to correct. It is therefore wise to use only top-quality products.

Supplies and Equipment

Equipment needed for painting exteriors includes the following basic items.

- A stepladder for lower areas and an extension ladder for the highest spots. Include attachments like a paint hook.

- Drop cloths to protect plants and walks from drips

- Caulking gun for sealing joints

- Hammer, nail set, putty, and putty knife

- Mixing pails

- Brushes

- Solvents for cleaning brushes and other equipment

- Cleaning cloths

- Rubber or latex gloves for use when using solvents

- Safety glasses or goggles for use with solvents or when preparing surfaces

- Roller, roller cover, and paint tray for painting large, flat surfaces

Exterior paint can also be sprayed on using a compressor, a paint pot, and a spray gun.

Reading Check

Recall *What color exterior paint would you recommend using to reduce cooling costs? Why?*

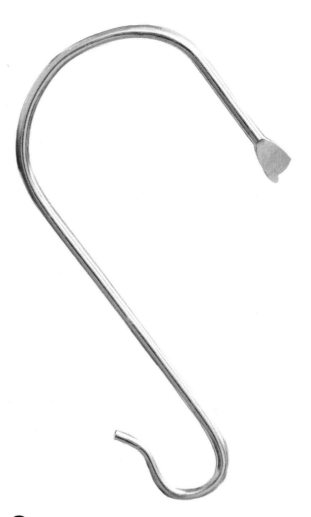

Using a Paint Hook
Comfort and Safety When brush painting from a ladder, use a paint hook to support the paint can. The hook hangs on a ladder rung.

Equipment for Spray Painting
Speedy and Consistent A rig for spray painting should include a pressure gauge on the compressor (at right) as well as on the paint pot (at left).

Builder's Tip

Planning the Job

Exterior construction should be complete before painting begins. Surfaces to be painted must be properly prepared. Usually, the following steps can be completed before the primer is applied.

Place drop cloths under the area you are about to paint. Cover nearby walks and shrubs. This will shorten cleanup time.

Nail heads may be left at the surface of the wood or sunk below it. Use a nail set and hammer to sink nail heads. Fill them with an exterior-grade wood putty. Seal any knots with primer to prevent stains from bleeding through the finish later.

JOB SAFETY

Use paintable caulk or high-quality sealant around door and window frames where necessary. Tightly sealed joints help weatherproof the house and prevent moisture damage. Joints between dissimilar materials, such as brick and wood, require extra attention.

Painting Techniques

When would it be useful to work from scaffolding?

Many painters prefer to apply primer and finish coats with a brush. This works paint into the wood surface and ensures that every surface and edge is coated. However, the large, flat areas of panel siding are sometimes painted using rollers.

Spray application is faster than brush painting but tends to deposit paint only on the very top surface of the wood. To ensure proper coverage on rough surfaces (such as plywood siding), *roll in* sprayed paint by going over the area with a dry roller. This works the paint into the uneven surface. Rolling in should be done a section at a time immediately after spraying.

Morning dew or water from a brief shower should be wiped off and at least an hour of warm sunshine should follow

 Sealing Joints
Blocking Water Always caulk around window and door moldings to prevent moisture from rotting the wood.

Arnold & Brown

before any painting is done. After a hard rain, several days may be needed before a surface is dry enough to paint. Always avoid painting when a surface will be heated by full sun.

Applying Paint

The outdoor temperature must stay above 40°F (4°C) for at least 24 hours after oil-base paints are applied. The temperature must stay above 50°F (10°C) for at least 24 hours after latex paints are applied. When using paintable water-repellent wood preservatives prior to painting, best results are obtained when the temperature is above 70°F (21°C).

Reading Check

Summarize *List one advantage spray application has over brush painting.*

Cleanup

Protect your tools by cleaning them immediately after use, especially brushes and rollers. After using oil-based paints, thoroughly work solvent into the brush bristles. Be sure to wear rubber gloves to protect your skin. Squeeze out as much paint and solvent as possible, then repeat the process of working it into the bristles. Repeat this operation until the paint disappears. Give the brushes a final rinse in clear solvent. Then wash them in soapy water, rinse thoroughly, and spin them dry.

To clean a roller cover, remove it from the roller frame and scrape off as much paint as possible. Then immerse the cover in a generous amount of the correct solvent. Work the solvent into the roller cover until it is clean. Then wash the cover in a mild detergent solution and rinse it in clear water. Disposable roller covers have cardboard cores and generally cannot be cleaned. Solvents used for cleanup of oil-base paints are flammable. Always use them in a well-ventilated area away from pilot lights and other flames.

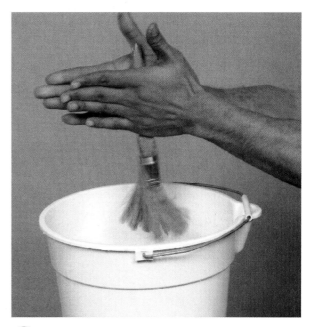

Spinning a Brush
Spin Clean Spin brushes to remove excess water or solvent. Center the brush in a bucket to contain the spray.

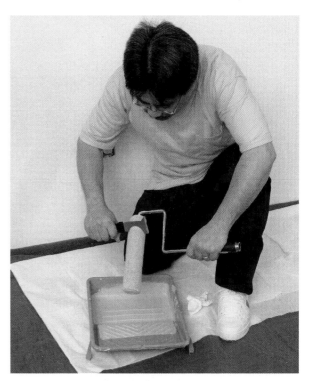

Cleaning a Roller Cover
5-in-1 Tool A 5-in-1 tool, sometimes called a painter's tool, can be used to scrape excess paint from a roller cover prior to cleaning.

(t & b)Arnold & Brown

Basic Painting Technique Primer and finish coats call for the same basic techniques.

Step 1 Even if the paint dealer has mixed the paint mechanically, mix it again just before and during painting. Stir the contents of the can from the bottom up. Then *box* the paint by pouring it from one can into a larger can and stir it again. This evens out any slight variations in color between cans.

Step 2 Load the brush by dipping it about two inches into the paint and tapping the excess off against the inside of the can. Repeat several times.

Step 3 Start painting at the top of the house and work down. This prevents drips and splatters from spoiling previously painted areas. Wearing safety glasses will protect your eyes from paint splatters.

Step 4 Apply the paint generously along siding joints, distributing it evenly. Do not bear down too hard. *Feather* the ends of your brush strokes. This helps avoid a distinct edge and ensures smoothness where one painted area meets another. Always paint with the grain.

Step 5 After painting the gable end of the house, start at a corner and work across. It makes no difference whether you work from the left or right. However, before you move or shorten the ladder, finish an area about four or five feet square.

Step 6 Paint windows with a narrow sash brush. Paint the mullions first, then the rails, and then the stiles. Paint the casing and trim last. Move the sash up and down before the paint dries to prevent sticking.

Step 7 For a panel door, first paint the molding and then the panels. Paint the rails next and finally the stiles.

Step 8 Paint shutters separately. Install them after the rest of the job has been completed.

JOB SAFETY

SPONTANEOUS COMBUSTION Any material that has been in contact with a solvent-based product must not be disposed of in the regular trash. This also applies to materials saturated with oil-base or alkyd-base paints and stains. Under the right conditions, rags, paper towels, newspaper, or steel wool can ignite (burst into flame) without being exposed to a heat source or a spark. The process is called *spontaneous combustion,* and it has caused many fires on construction sites. Contaminated materials must be stored in an airtight metal container. An alternative is to soak them thoroughly in water and leave them outdoors to air dry. Proper storage or disposal of contaminated materials will reduce the risk of spontaneous combustion.

To clean brushes used with latex paints, follow the same steps as on page 953, but use soapy water instead of solvent. Rinse the brushes with clear water. Allow them to dry thoroughly before storing.

To protect bristles as they dry, wrap brushes in heavy paper or a cardboard sheath and lay or hang them in a dry place. Some painters hang brushes to dry. Store roller covers on end so that their nap is not

flattened. Allow drop cloths to dry, if damp, before folding them for storage.

If paint has fallen on walkways, scrub it out with a suitable solvent and a stiff brush. Scrub off spatters from latex paint with soapy water before the paint dries.

Paint Problems

Why is mildew such a persistent problem for painters?

Problems caused by improper painting may not show up for months or even years. To avoid these problems, it is important to understand the ways in which paint fails. This section describes some common problems, their causes, and their solutions. Before applying any primer or house paint, always check the label. There you will find recommendations on surface preparation and compatibility with caulks, sealants, and primers.

Cracking and Alligatoring

If paint cracks it may have been applied in several heavy coats without sufficient drying time between coats. Also, the primer may not be compatible with the finish coat.

Protecting Bristles
Brush Sheath A loose-fitting cardboard cover protects bristles as they dry. Good-quality brushes are usually sold with such a cover.

Cracking and Alligatoring
Heavy Coats This can occur when bottom paint layers did not dry sufficiently.

This estimating and planning exercise will prepare you for national competitive events with organizations such as SkillsUSA and the Home Builder's Institute.

Exterior Painting

Paint Jobs

Estimating quantities and expenses for an exterior painting job calls for calculations of materials and labor.

Materials

Step 1 To estimate the amount of paint needed for the exterior of a house, first determine the number of square feet to be covered. Figure the siding area below the roofline by measuring the total distance around the house and multiplying this figure by the height. For the house shown below, the perimeter is 120'; 40' + 40' + 20' + 20' = 120'. Multiply this number by the height to determine the area: 120' × 12' = 1,440 sq. ft.

Step 2 For the gables, multiply the height of the gable at its highest point by half the width of the gable. Do this for each gable. In the example, 6' (gable height) × 10' (half the gable width) × 3 (number of gables) = 180 sq. ft.

Step 3 Add the area for gables to the area for siding below the roofline:

$$180 + 1{,}440 = 1{,}620 \text{ sq. ft.}$$

Step 4 Primer and topcoats typically cover different amounts per gallon. Always check the coverage recommendations on the product label. In general, though, divide the total number of square feet by 450 to find how many gallons of primer will be needed. Divide by 500 to find the number of gallons required for each finish coat:

$$1{,}620 \div 450 = 3.6 \text{ gal. of primer}$$
$$1{,}620 \div 500 = 3.24 \text{ gal. of paint}$$

Labor

To estimate labor for exterior painting, refer to **Table 33-2** to determine the number of hours required. Multiply this number by your local labor cost per hour to find the total cost.

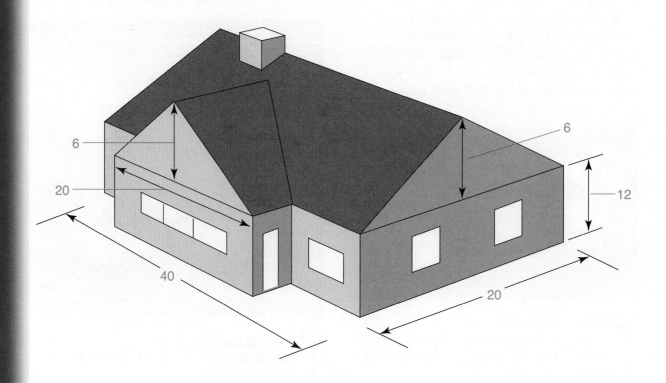

Table 33-2: Estimating Labor for Exterior Painting	
Preparation of siding and trim (sanding and puttying) Preparation of trim only (brick veneer or masonry construction)	175 sq. ft. per hr. 100 sq. ft. per hr.
Brushing windows and door frames Brushing wood siding	175 lineal ft. per hr. 175 sq. ft. per hr.
Brushing asbestos shingle siding Brushing wood shingle siding	75 sq. ft. per hr. 150 sq. ft. per hr.
Note: No allowance is included for preparatory work or for setting up scaffolding.	

To correct the problem:

1. Sand the cracked or alligatored surface smooth.

2. Apply one coat of primer and one top coat of house paint.

Reading Check

Explain *What can cause paint to crack?*

Localized Peeling

Peeling results when moisture trapped in siding is drawn from the wood by the sun's heat and pushes the paint from the surface. One cause is improper installation of a vapor barrier beneath the siding. To correct the problem:

1. Locate and eliminate sources of moisture. Is the area near a poorly ventilated bathroom or a kitchen? Is there seepage or leakage from eaves, roof, or plumbing?

2. Reduce future moisture by installing bathroom and kitchen exhaust fans.

3. Scrape off the old paint. Scrape down to the wood over the entire board or for a distance of 12" around the peeling area.

4. Sand the surface to fresh wood and spot prime with a recommended primer.

5. Apply a top coat of house paint.

Localized Peeling
Peeling Paint Moisture and heat cause peeling.

 Flaking
Flaking Paint Moisture contributes to flaking.

Flaking

Flaking is caused by the alternate swelling and shrinking of siding as the moisture behind it is absorbed and then evaporates. Brittle paint cracks under the strain and pulls away from the wood. To correct the problem:

1. Locate and eliminate sources of moisture. Is the affected area near a bathroom or kitchen? These areas generate a great deal of moisture vapor. Is there seepage or leakage from eaves, roof, or plumbing?

2. Scrape off the flaking paint to expose the wood for about 12" around the area.

3. Sand the surface to fresh wood and spot prime with a recommended primer.

4. Seal all seams, holes, and cracks against moisture, using suitable caulk.

5. Apply a top coat of house paint.

Mildew

Mildew is a microscopic fungus that thrives on many household surfaces, including painted siding. A warm, wet, or humid environment provides the best conditions for its growth. Although mildew is unattractive, it does not cause the wood to decay. However, if painted over, it will

 Mildew
Fungus Mildew thrives in humid environments.

grow through the new coat of paint. To correct the problem:

1. Gently scrub the entire surface with a solution of ⅓ cup of trisodium phosphate (TSP) or a comparable substitute, ½ cup of household bleach, and 4 quarts of warm water. Standard TSP is not available in some areas that limit phosphate-based detergents for environmental reasons.

2. Apply one coat of primer. Add mildew-resistant additives to a primer if the likelihood of mildew is high.

3. Apply one top coat of mildew-resistant latex house paint. This product contains a *biocide* that discourages mildew growth.

Extractive Staining

Staining is caused by moisture in redwood and cedar siding that dissolves extractives in the wood. Extractives are natural chemicals found in these woods and tend to be dark in color. The colored moisture seeps into the paint through breaks in the paint film. A stain forms when the water dries. To prevent the problem:

1. Locate and eliminate moisture sources before painting.

2. Back prime the siding boards before installation.

 Blistering
Blister in the Sun Moisture is pushed to the surface by the sun's heat, which pushes the paint away from the surface.

 Nail Head Staining
Rust Stains A nail head stain on siding. *What causes nail head stains?*

To correct the problem:

1. Wash stained surfaces with a mixture of 50 percent denatured alcohol and 50 percent clean water.
2. Allow the surface to dry for 48 hours. Then apply two coats of the house paint.

Blistering

Blistering, like peeling, is caused when moisture trapped in the siding is drawn from the wood by the sun's heat. This pushes paint from the surface. To correct the problem:

1. Locate and eliminate the sources of moisture.
2. Scrape off the old paint for a distance of about 12" around the blister condition.

3. Sand the surface to fresh wood and spot prime with primer.
4. Use caulk to seal all seams, holes, and cracks against moisture entry.
5. Apply a top coat of house paint.

Nail Head Staining

Nail head stains are caused when excessive moisture rusts uncoated or poorly coated steel nails used to install the siding. To correct the problem:

1. Sand the stained paint and remove the rust down to the bright metal of the nail head.
2. Countersink the nail head ⅛" below the surface of the siding. Immediately spot prime the nail head.
3. Fill primed, countersunk holes with exterior-grade putty. Apply two top coats of house paint.

Section 33.2 Assessment

After You Read: Self-Check

1. Why is it important to use good-quality exterior paint?
2. What does it mean to *box* paint and why is boxing paint important?
3. What type of paintbrush is used to paint the mullions of a window?
4. What type of environment is best when using solvents for cleanup of oil-based paints?

Academic Integration: English Language Arts

5. **Preventing Staining** Staining is caused by moisture in redwood and cedar siding that dissolves extractives in the wood. Write a set of step-by-step instructions that will help a first-time painter prevent the problem of staining. Use short, clear sentences and include effective transition words.

Go to **connectED.mcgraw-hill.com** to check your answers.

(l)©Ingram Publishing/SuperStock; (r)Ingram Publishing

Interior Painting

Preparing to Paint

In which rooms would you use a low-sheen paint?

Many different kinds of interior paints are available. The most popular are latex paints, which are easy to apply and dry quickly. Interior paints can make surfaces easy to clean and give them wear resistance. They seal surfaces against moisture and vapor penetration. They also add to the room's attractiveness.

Interior paints are available in various sheens. High-sheen paints are easier to clean, but low-sheen paints have a softer, less glaring appearance. High-sheen paints are used where cleanup is important, such as in kitchens and baths. Low-sheen paints are used in living rooms and bedrooms. Following is a list of paint types in order of their sheen, from greatest to least:

- enamel
- semi-gloss enamel
- pearl
- eggshell
- flat.

Supplies and Equipment

Brushes and rollers are the primary interior painting tools. Other items may be needed to prepare the surface, protect floors and furniture, mix the paint, and cleanup.

Good-quality brushes are expensive but worth the money. With a good brush, you get better results with less effort. A 3" or 4" wide brush is recommended for painting trim and for *cutting in* corners and edges. Cutting in means to brush paint carefully along a straight line, such as along the edge of trim. Brushes should be 5" to 7" long and have dense bristles with flagged, not square-cut, ends. **Flagged bristles** are slightly splayed at the tips.

Rollers are easier to use and faster than brushes for painting large flat areas. Paint is

Utility   Square sash   Wall   Angular sash

Types of Brushes

Most Common These brushes are the ones most often used for interior painting.

held in a tray instead of a can. It is mounted on a metal roller frame. Short-nap roller covers are suitable for most paints and surfaces. Lambswool covers are used for flat finishes on rough or imperfect surfaces. Mounting the roller on an extension pole has several advantages. The painter stands several feet away from the surface being painted so that he or she can better see areas that have not yet been covered properly. Loading the roller or painting low portions of the wall do not require bending over. A ladder is not needed to reach high portions of most walls.

Preparing Surfaces

Good preparation makes the painting job much easier and faster. On remodeling projects, many experienced painters spend more time preparing surfaces than actually painting. Be sure the room is dry, well ventilated, and at a comfortable temperature. Cover finished floors with drop cloths. Mask hardware such as doorknobs and hinges. Also mask wall switches and receptacles. If ceiling fixtures are in place, lower the canopy (the domed portion covering the electrical box) so that you can paint under it.

Start with a clean surface. On new construction, the walls do not have to be washed. A thorough dusting of the surfaces is usually enough. Fine cracks in walls or nail holes in wood trim should be filled with spackling compound or painter's putty.

When painting ceilings and walls, protect windows and other areas from being splattered with paint by *masking off* with tape. This is even more important when surfaces will be painted with spray equipment. Do not use standard masking tape for masking off. It will leave a sticky residue when removed and interfere with paint adhesion. Instead, use painter's masking tape. It is easily removed and will not mar glass or painted surfaces. To mask off large areas quickly, use masking rolls. This product combines painter's masking tape with a continuous length of either plastic or paper. However, many painters feel that it is better to rely on painting skill than on masking products. They feel that it is faster to carefully cut in than to mask off.

Painting a Room

Why is it important to paint the ceiling before the walls?

When surfaces have been prepared and the room is clean, painting can begin. Brushes are used to paint trim and to paint into corners. Rollers are generally used to paint all other surfaces. Sometimes paint may be applied to a room with a combination of spray equipment, rollers, and brushes.

Even if the paint has been mechanically shaken at the paint store, mix it well just before using. Stir rapidly, working pigment up from the bottom of the can. Professional painters buy paint in five-gallon pails. Mixing paddles driven by a heavy-duty, variable-speed electric drill can be used to stir it, as shown on page 962. When a great deal of paint must be mixed, a heavy-duty paint mixer can be clipped to a five-gallon paint bucket.

Roller Covers

Choosing the Nap Roller covers come with naps of various lengths, such as 1¼", 1", ¾", and ½". Longer naps are best for surfaces with a rough texture. Shorter naps are best for smooth surfaces.

Mixing Paint

Time Saver A paint mixing paddle and an electric drill can be used to mix paint. A detail of the paddle is shown on the left.

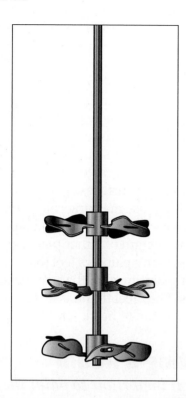

Always start with the ceiling, and then paint the walls. Complete the job by painting wood trim and doors. Clean up as follows:

- Wipe up spatters and spills immediately.
- Clean brushes, rollers, and other tools as soon as you finish using them.
- Wear rubber gloves when cleaning brushes and rollers. The gloves will protect your hands and make cleanup faster.

JOB SAFETY

AVOID FALLS Drop cloths are often used to protect floor surfaces when walls and ceilings are being painted. Plastic sheets can be used but can be slippery. For this reason, many painters prefer to use canvas tarps as drop cloths. Canvas is more durable and slip resistant than plastic.

Interior surfaces are normally under longer and closer observation than are exterior surfaces. Hence, the brushing on, smoothing out, and leveling off of the paint must be done with care. Most plaster and drywall surfaces are finished with two coats of flat paint over a single coat of primer or sealer. Primers and sealers reduce penetration of succeeding coats, so less paint will be needed for good coverage.

Ceilings

Generally ceilings are painted first. This prevents paint splatter from falling on finished walls. Scaffolding or stepladders may be needed if ceilings are unusually high. However, a paint roller mounted on an extension pole is often all that is needed. You will need a small brush for edging the ceiling. **Edging** is using a brush to paint along the corners between large flat surfaces, where a roller cannot reach.

Arnold & Brown

Roll paint in two- or three-foot strips across the shortest dimension of the ceiling. By doing this, you can paint the next strip before the last edge is dry. Overlapping a dry edge sometimes leaves a mark that shows later. Light strokes help to eliminate lap marks.

Walls

Cut in the edges of a wall by first painting a narrow strip around doors, windows, baseboards, and any other adjoining surfaces. Then edge the top of the wall adjacent to the ceiling. Finally, fill in the large areas with a roller. Finish one entire wall before beginning the next one.

When using a roller, pour paint into the deep portion of the tray. Work the paint into

Builder's Tip

WHAT TO PAINT FIRST There is no general agreement among painters about whether trim or walls should be painted first. Many painters feel that it is easier to cut in when painting trim. Others prefer to paint the trim first and then cut in when painting the walls. Novice painters should try both methods and see which method works better for them.

the roller by moving it back and forth in the tray until the paint is evenly distributed on the roller.

Edging
Outlining with Paint
A roller cannot reach into corners or get close to intersecting surfaces, so a brush is more effective.

Interior Painting

Two-Part Process

Calculating material requirements for interior painting is a two-part process. Wall and ceiling areas are calculated based on square footage. Trim is calculated based on lineal footage. Primer or sealer over smooth walls will cover 575 to 625 sq. ft. per gallon. The first and second coats of paint will cover 500 to 550 square feet. Trim may require a different paint, such as enamel.

Labor costs are based on hourly wage rates or figured as a portion of square footage rates.

Materials

To determine the amount of paint for a room, first calculate the total wall area.

Step 1 To find the area of one wall, multiply the length of the wall by the height of the wall.

If the entire room is to be painted, multiply the perimeter of the room by the height. Windows and doors are not usually subtracted from the total paint requirement unless they are unusually large or numerous.

As an example, see the bedroom in the lower right corner of the floor plan below. Assume that the ceiling height is 8' and that the entire room is to be painted. The end walls are each 10' long and the front wall and closet wall are each 14' long. The perimeter of the room is therefore 48 lineal feet:

$$10 + 10 + 14 + 14 = 48$$

Multiply this figure by the room height to obtain the total wall area:

$$48 \times 8 = 384 \text{ sq. ft.}$$

This room will require 1½ to 2 gallons of paint, depending on the coverage.

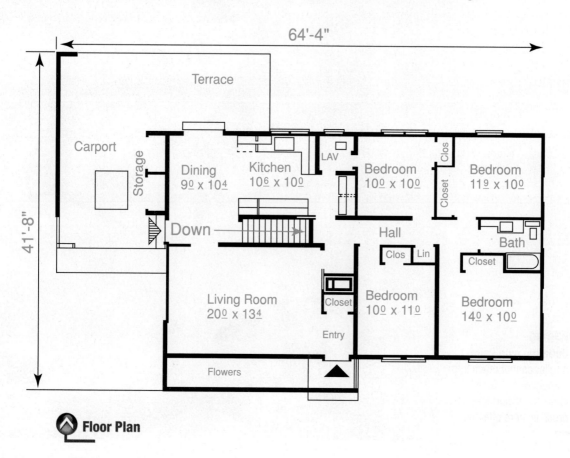

Floor Plan

Step 2 A window's trim and frame require ¼ pint of paint. The bedroom has three windows, for a total of ¾ pint. A door frame and door require ½ pint. The closet door is equal to two doors. A total of three doors would then require 1½ pints of paint. Doors and windows together require a total of 2¼ pints:

$$(¾ + 1½ = 2¼)$$

Like the walls, the trim will need both primer and finish coats. For our example, the trim will take one coat of primer (2¼ pints) and two coats of finish (4½ pints). Since paint is sold in cans no smaller than 1 quart, it will be necessary to buy 2 quarts of primer and 3 quarts of finish. However, this will provide enough extra for painting the baseboard, with some left over to allow for future touch-ups by the homeowner.

Step 3 To determine the amount of paint needed for the ceiling, calculate the area by multiplying the length of the room by its width. In the example, the bedroom ceiling area is 140 sq. ft.:

$$(10 \times 14 = 140)$$

Labor

Step 1 To estimate labor for interior painting, refer to **Table 33-3**. For example, the time needed to apply one coat of paint to one window is about ¾ of an hour. If there are ten windows and two coats of finish paint are to be applied, the total time will be 15 hours:

$$10 \times ¾ = \frac{30}{4}$$

$$\frac{30}{4} \times 2 = \frac{60}{4} \text{ or } 15$$

Step 2 Multiply this figure by the labor rate per hour to find the total labor cost.

Estimating on the Job

Calculate the amount of paint needed for a room that measures 9' × 12'. The ceilings of the room are 9' high, and it has one door and two windows. Include one coat of primer and two coats of finish paint for walls, ceiling, doors, and windows.

Table 33-3: Estimating Labor for Interior Painting	
Preparation of trim (including sanding and spackling) Molding (chair rails and other trim up to 6" wide)	115 lineal ft. per hr. 150 lineal ft. per hr.
Windows (including sash, trim, sills, and apron) Paneled door (including door and trim) Flush door (including door and trim)	Each coat ¾ hr. per window Each coat ¾ hr. per door Each coat ½ hr. per door

Finishing walls and ceiling:
   Brush
   Roller
	150 sq. ft. per hr.
	300 sq. ft. per hr.

Note: No allowance for preparatory work or for setting up scaffolding is included.	

Rolling a Wall
Basic Technique After the wall has been edged, roll paint on the wall in a W pattern. Note the use of an extension handle.

Next, start on one side of the wall and paint a **W** on the surface. Use slow, smooth strokes. Quick strokes and heavy uneven pressure may cause bubbles or spatters. When you have covered a few square feet, use parallel vertical strokes to spread the paint evenly.

Trim

Paint interior trim and woodwork using a 1½" sash brush for windows and a 2" brush for other parts of the trim. Complete one small area at a time, brush on the paint with back-and-forth strokes. Level the paint with even strokes in one direction. Work quickly but carefully. Never go back to touch up a spot that has started to dry, because this will mar the surface.

In general, trim is painted from the top down. For example, crown molding would be painted first and baseboards last. This

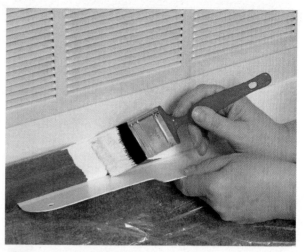

Using a Paint Guard
Floor Protector When painting baseboard, use a paint guard to protect the floor and to keep the brush from picking up dirt.

prevents finished work from being splattered by paint from above. A cardboard, metal, or plastic guard held flush against the bottom edge of the baseboard protects the floor. It will also prevent the brush from picking up dirt.

Painting windows calls for particularly careful work. Adjust a double-hung window so that you can first paint the lower part of the upper sash. Then raise the upper sash almost to the top to finish painting it. Paint the lower sash next. With the window open slightly at the top and bottom, it can be finished easily. Paint the recessed part of the window frame next, then the frame, and finally the windowsill.

When painting a door, paint the jambs and casing first. Then paint the edges of the door itself. Finally, paint the front and back face of the door. When painting the face of a raised-panel door, paint the panel molding first, starting at the top. Keep a clean cloth handy to wipe off any paint that gets on the area surrounding the panels. Then paint the remainder of the door.

Reading Check

Recall *In general, how should trim be painted?*

(l)Design Pics/Darren Greenwood; (r)Arnold & Brown

Cleanup

The cleanup of tools and equipment used for interior painting is no different than that for exterior painting (see Section 33.2). Remember that some paints and solvents (such as mineral spirits) are flammable.

- Use paints and solvents in a well-ventilated area.

- Store or dispose of rags and newspapers in the proper manner.

- Wipe up spatters and spills immediately.

- Clean brushes, rollers, and other tools as soon as you finish using them.

- Wear rubber gloves when cleaning brushes and rollers. The gloves will protect your hands and make cleanup faster.

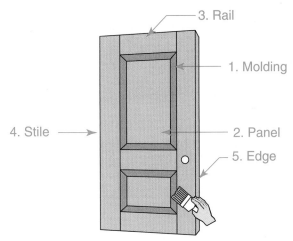

3. Rail
1. Molding
4. Stile
2. Panel
5. Edge

Painting a Door
Painting Strategy To paint a raised-panel door, follow the sequence shown here. In general, try to work from the center out, and paint rails before stiles.

Section 33.3 Assessment

After You Read: Self-Check

1. Which rooms in a home are low-sheen paints typically used in?
2. What is edging?
3. Describe two ways to prevent the brush from picking up dirt from the floor when painting baseboard.
4. Which part of a raised-panel door should be painted first and which part second?

Academic Integration: Mathematics

5. **Knowns and Variables** To paint the walls and ceiling of an average room in a house, it takes 45 minutes to move or cover furniture and assemble materials, 15 minutes of preparation per 100 sq. ft., 45 minutes of painting per 100 sq. ft., and 30 minutes to clean up. Write an algebraic equation that could be used to compute the total time it takes to paint any room.

   Math Concept An algebraic equation uses knowns and variables. Variables are quantities that may change. They are represented by symbols such as letters.

   Step 1: Convert time measurements to the same unit of measure.

   Step 2: Let T equal the total time in hours. Let A equal the area to be painted.

   Step 3: Write the equation using known and variable values. Simplify.

Go to **connectED.mcgraw-hill.com** to check your answers.

Review and Assessment

Chapter Summary

Film-forming finishes coat the wood surface. Penetrating finishes soak into the wood. In each type of finish are pigments, binders, and carriers that make that finish suitable for a particular use. Paint is the most common type of film-forming finish.

Surface preparation is important when painting wood. To get the best results, finishes are best applied by brush. On new wood, the first coat of paint should be primer, followed by two top coats of standard paint. Paint should not be applied if temperatures are not suitable. Proper application of paint avoids many problems that are difficult to correct later.

Interior painting procedure should minimize paint splatter and cleanup. Various tools, such as rollers, can be used to make the work go more quickly. These tools can improve the quality of the job as well and minimize strain on the painter.

Review Content Vocabulary and Academic Vocabulary

1. Use each of these content vocabulary and academic vocabulary words in a sentence or diagram.

Content Vocabulary

- summerwood (p. 946)
- sheen (p. 947)
- binder (p. 947)
- carrier (p. 947)
- solvent (p. 948)
- primer (p. 948)
- flagged bristles (p. 960)
- edging (p. 963)

Academic Vocabulary

- synthetic (p. 948)
- compatibility (p. 951)

Speak Like a Pro

Technical Terms

2. Work with a classmate to define the following terms used in the chapter: *opaque* (p. 946), *film-forming finishes* (p. 947), *penetrating finishes* (p. 947), *pigments* (p. 947), *vehicle* (p. 947), *oil-base paints* (p. 947), *vegetable oil binders* (p. 947), *alkyd binders* (p. 948), *latex paints* (p. 948), *roll in* (p. 952), *box* (p. 954), *feather* (p. 954), *spontaneous combustion* (p. 955), *biocide* (p. 958), *cutting in* (p. 960), *roller cover* (p. 961), *masking off* (p. 961).

Review Key Concepts

3. Identify the two basic types of finishes.

4. Name the basic ingredients of paint.

5. List the steps in painting a house exterior.

6. Identify the steps in painting an interior.

7. Identify problems with painted finishes.

8. Demonstrate how to paint windows and doors.

Critical Thinking

9. Discuss Why is it important to clean and store painting tools immediately after use?

Academic and Workplace Applications

STEM Mathematics

10. Buying Enough Paint Sharon estimated the amount of acrylic latex ceiling paint needed to paint the ceilings in her new house. The living room measures 13 ft. by 18 ft., the hallway measures 4 ft. by 18 ft., the dining room measures 13 ft. by 14 ft., 3 bedrooms measure 12 ft. by 12 ft., and the master bedroom measures 16 ft. by 20 ft. The paint has a spread rate of 400 sq. ft. per gallon. Sharon bought 3 gallons and 1 quart of paint. Will she have enough paint?

Math Concept When solving a word problem, translate the wording into smaller numeric equations using key words.

Step 1: Calculate the total area of the ceilings to be painted.

Step 2: Calculate the spread rate of 3 gallons and 1 quart of paint.

Step 3: Compare the total area of the ceiling to the spread rate of the paint.

STEM Science

11. Natural Paints Though many paints are now made with zero or low VOCs, they may still contain toxic chemicals such as formaldehyde. Natural paints can be made with bases of milk, clay, or plant oils. Research information about one type of natural paint. Summarize your findings in a one-page report.

21st Century Skills

12. Communication Skills As you know, each trade involved in the construction of a new home has the potential to affect the work of other tradespeople. Assume you were hired to paint the interior of a newly constructed home. In your preparation, you notice that the drywall surfaces in several rooms are uneven. Write a one-page letter to the homeowner explaining the problem. Explain to them the possible reasons why the drywall is uneven. Suggest which tradespeople they could contact to fix the deficient work.

Standardized TEST Practice

Short Response

Directions Write one or two sentences in response to the following questions.

13. Why are finishes important to wood?

14. How long should one wait before painting an exterior surface after a hard rain?

15. What size brush is recommended for painting trim and for cutting in corners and edges?

TEST-TAKING TIP

Before the day of the test, ask your teacher if the test will include multiple choice, true or false, and/or essay-style questions. Just knowing the type of questions will help you prepare for a test.

*These questions will help you practice for national certification assessment.

Finish Flooring

Section 34.1
Wood Flooring Basics

Section 34.2
Installing Hardwood Flooring

Section 34.3
Vinyl, Tile, & Carpet Flooring

Chapter Objectives

After completing this chapter, you will be able to:

- **List** the three most common forms of wood flooring.
- **Describe** the major kinds of wood used in flooring and how they are graded.
- **Explain** how to install wood strip and parquet flooring.
- **Estimate** the quantity of resilient flooring needed for a room.
- **Perform** the basic methods of installing ceramic tile and carpeting.

Discuss the Photo

Finish Flooring Installation of finish flooring is the last large construction operation in a house. *What properties do you think finish flooring should have?*

Writing Activity: Research and Summarize

Research jobs in the flooring industry in your community. Make a chart of the jobs, their average pay and working conditions, and the training required for each. Then summarize which job interests you most, and why.

Lev Dolgachov/Alamy Stock Photo

Before You Read Preview

Finish flooring is the topmost surface of a floor system and the last large construction operation in a house. Choose a content vocabulary or academic vocabulary word that is new to you. When you find it in the text, write down the definition.

Content Vocabulary

○ plank
○ parquet
○ acclimation
○ wear layer

○ mastic
○ sleeper
○ volatile organic compounds (VOCs)

○ underlayment
○ bisque
○ backerboard
○ grout

Academic Vocabulary

You will find these words in your reading and on your tests. Use the academic vocabulary glossary to look up their definitions if necessary.

■ uniform   ■ eliminate   ■ inclines

Graphic Organizer

As you read, use a cluster diagram like the one shown to organize information, adding circles as needed.

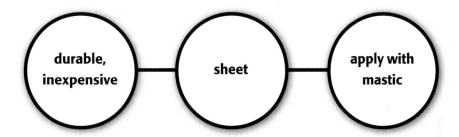

durable, inexpensive — sheet — apply with mastic

Go to **connectED.mcgraw-hill.com** to download this graphic organizer.

Wood Flooring Basics

Solid-Wood Flooring

What is finish flooring?

Finish flooring is the topmost surface of a floor system. It should be durable and easy to clean. It should resist wear and be comfortable and attractive as well. Finish flooring is installed after the plumbing, electrical wiring, and plastering are completed, but before the final interior trim work. It is the last large construction operation in a house.

Solid wood is the most common wood flooring material. Hardwoods are particularly popular because they resist wear. Strips or planks of solid wood are fastened to the subfloor with nails, staples, or adhesives to create different effects. The wood is most often applied in rows parallel to the long dimension of a room. You can also install wood flooring in other patterns. For example, you can lay it diagonally, with a mitered or stacked border. You can make an especially decorative border by using contrasting woods.

Solid-wood lengths of flooring generally have tongue-and-groove edges and ends that interlock with adjoining lengths. This flooring is most often installed by blind nailing it to the subfloor. Blind nailing is the process of driving fasteners at an angle through the edge of each board, so that each fastener is concealed by the next board.

Forms of Solid-Wood Flooring

Solid-wood flooring is available in three basic forms, or types.

Strips Strip flooring consists of narrow strips of wood, generally 3¼" wide or less. Strips are the most widely used type of solid-wood flooring. Many wood species are available in this form. You will learn how to install strips later in the chapter.

Planks A **plank** is any solid-wood board that is at least 3" wide. Plank flooring has been popular for centuries. Modern plank flooring comes in various widths. The edges of the planks may be beveled slightly to look like hand-hewn planks. You can simulate the wood pegs that fastened old plank floors by

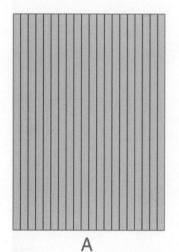
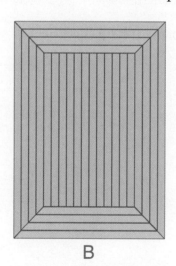
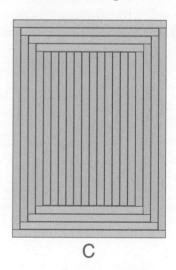

A   B   C

Wood Flooring Patterns

Creative Patterns Three wood flooring patterns: **A.** Standard installation **B.** Mitered border **C.** Stacked border

Plank Floor
Traditional Flooring Wood planks are fastened with screws, which are then covered with plugs.

Parquet Flooring
Decorative Options Parquet flooring comes in a variety of geometric patterns.

gluing wood plugs into shallow holes in the ends of the planks.

Parquet **Parquet** (par-KAY) is to any flooring assembled with small, precisely cut pieces of wood in a geometric pattern, such as squares, rectangles, and herringbone patterns. Parquet floors have been used since the fourteenth century.

Parquet flooring can be made from individual pieces, but prefabricated parquet tiles are more common and much easier to install. Parquet tiles are made by gluing individual pieces of solid wood to a plywood backing, then machining the edges of the tiles to form tongues or grooves. You can use parquet for a whole floor or as a decorative border around other forms of wood flooring.

Unfinished and Prefinished Flooring

A protective finish makes wood flooring durable and easy to clean. Unfinished flooring must be finished after installation, which is a time-consuming process. Many manufacturers now produce wood flooring that is prefinished at the factory. Prefinished floors do not come in as many colors and finishes as floors finished on site. However, prefinished flooring has a **uniform** finish and is ready for use immediately after installation. It also benefits from *curing* methods and finish types that are possible only in a factory.

Prefinished wood flooring is easier to install than unfinished wood flooring. However, estimating how much you will need requires greater care. Special trim such as thresholds, border strips around fireplace

hearths, and transition pieces must all be prefinished to match. Make sure to order these pieces with the rest of the flooring so that all pieces match in color and finish exactly.

Grades of Wood Flooring

Through trade associations, the major North American producers of solid-wood flooring have adopted grading rules for various species of wood. Every bundle of flooring identified by this grading system is guaranteed to meet certain standards of quality and uniformity.

Oak Oak is graded according to its appearance and how sawing methods reveal the grain during manufacturing. The three main sawing methods used on oak are:

Plain sawn The end grain of the board runs from 0° (parallel) to 45° to the board's face, or surface.

Quarter sawn The end grain runs between 60° and 90° (perpendicular) to the board's face.

Rift sawn The end grain runs between 30° and 60° to the board's face.

Grading for appearance of oak is based on the top, or best, face of a board. It does not consider the board's strength. Red oak and white oak have five appearance grades:

Clear This wood has the best appearance and the most uniform color. It is mostly *heartwood*. Limited small character marks are permitted.

Select The face may contain color variations typical of heartwood and sapwood. It can include slight milling imperfections, small tight knots, and a modest number of slightly open checks.

No. 1 Common Prominent variations of color are allowed, as well as broken knots less than ½" wide and other imperfections.

No. 2 Common A greater number and degree of natural and manufacturing imperfections are allowed. This grade is used for a utility floor or where character marks and contrasting appearance are acceptable.

Shorts Pieces 9" to 18" long are bundled together in either of two subgrades: No. 1 Common & Better and No. 2 Common. These pieces can be used to fill in rows to avoid cutting longer pieces.

Maple, Beech, and Birch Maple, beech, and birch are governed by almost identical grading rules. In order of quality, these grades are First, Second, Third, Second & Better, and Third & Better. Neither sapwood nor varying natural color is considered a defect.

Each of these woods is also available in a special grade selected for uniformity of color. For maple flooring, this is First Grade White. It is the finest grade of maple flooring. For beech and birch flooring, the special grade is First Grade Red.

Sizes of Strip Flooring

Oak strip flooring is commonly ½" or ¾" thick. Maple, beech, and birch flooring comes in thicknesses of ¾", $^{25}/_{32}$", and $^{33}/_{32}$". Widths for all species range from 1½" to 3¼". Flooring ⅜" thick is sometimes available by special order, but only in widths up to 2".

Strip flooring is packaged in bundles. A bundle contains strips of various lengths, positioned end to end. Lengths in a bundle vary, but average lengths are specified for each grade. For example, clear oak averages 60" long. Select oak averages 48" long, and No. 1 common oak averages 33" long.

Reading Check

Analyze *What imperfections cause oak flooring to be assigned a lower grade?*

Moisture Content and Acclimation

Moisture content greatly affects the performance of wood flooring. When wood absorbs moisture, it expands. When wood dries out, it contracts. Hardwood flooring is kiln-dried at the factory to reduce its moisture content but must be allowed to *acclimate* before installation. To permit **acclimation**, flooring should always be stored in the building in which it will be installed for at least four or five days before being laid. During acclimation, the moisture content of the wood rises or falls to match the moisture level of the building it is in. Proper acclimation prevents problems caused by wood movement after installation. One common problem is the appearance of open cracks between floor boards. **Table 34-1** shows how much a 2¼"-wide board will expand when installed in a building that has higher humidity than it does.

Humidity in the subfloor can also cause the underside of the flooring to expand. This causes boards to cup slightly. If the boards are then sanded flat, they will become crowned when the flooring later dries out. For more on moisture and wood behavior, see Chapter 16, "Wood as a Building Material."

CUPPED FLOORING

Subfloor

CROWNED FLOORING

 Cupped and Crowned Flooring
Moisture Damage Solid-wood flooring boards can become cupped or crowned, creating an uneven surface.

Storage and Handling of Wood Flooring Always protect wood flooring from the elements during storage and delivery. This helps to prevent excessive shrinkage or expansion, which could cause the floor to crack or buckle after it has been laid. Storage in a garage at the job site is not sufficient. While a garage protects the flooring from rain, it does not protect it from extremely dry or humid air. Take the following precautions as well:

- Never unload wood flooring when it is raining or snowing. Cover wood flooring with a tarp in foggy or damp conditions.
- Store wood flooring only in well-ventilated and weather-tight buildings.
- Store and install wood flooring in dry buildings only. Wait until the plaster and concrete work have given off most of their moisture.
- Heat the building to 70°F (21°C) before installing flooring.
- Never store wood flooring directly in contact with a concrete floor.

Table 34-1: Acclimation of a 2¼" Wood Flooring Strip	
Moisture Differential	**Width Increase (approximate)**
1%	$^1/_{128}$"
3%	$^1/_{64}$"
5%	$^1/_{32}$"
7%	$^3/_{64}$"
9%	$^1/_{16}$"
18%	$^1/_8$"

Reading Check

Summarize *How does moisture affect wood?*

Engineered-Wood Flooring

How are wood layers arranged?

Engineered-wood flooring is made of three, five, seven, or more layers of wood veneer or thin wood strips bonded together much like plywood. Unlike plywood, however, the top layer of engineered flooring is ⅛" to nearly ¼" thick. This top layer is called the **wear layer**.

Engineered-wood strips and planks look like solid wood flooring after the flooring has been installed. However, the individual pieces of flooring are thinner, usually less than ½" thick. The wood layers are arranged with their grains at right angles to each other, which makes the planks extremely stable. Engineered-wood flooring is less likely than solid-wood flooring to be affected by excess moisture. Engineered-wood flooring can also be sanded and refinished like solid-wood flooring. However, the flooring cannot be sanded as many times as solid wood flooring because the wear layer is relatively thin.

An Engineered-Wood Plank
Strength Through Layers Cross section of an engineered-wood flooring plank.

Installing Engineered-Wood Flooring

There are three basic installation methods for engineered-wood flooring. Always carefully follow the manufacturer's recommended installation methods to ensure coverage by any warranty on the product. It is especially important to use the exact type of adhesive called for in the installation instructions. Problems caused by incorrect adhesive usually do not show up until well after the flooring has been installed.

Nail-Down Method Blind nail or blind staple the planks or strips to the wood subfloor, just as you would do with solid wood flooring.

Floating Method Secure the lengths of flooring together with aliphatic resin (yellow) glue. Then float or lay the floor on a thin sheet of closed-cell foam padding.

Glue-Down Method Glue each length of flooring directly to the subfloor with mastic. **Mastic** is a thick, premixed adhesive that you spread with a notched trowel. You can sometimes use this technique to apply engineered-wood flooring directly to a concrete subfloor.

REGIONAL CONCERNS

Radiant Floor Heating This technique is popular in some parts of the United States. Engineered wood flooring and parquet flooring can be installed successfully over radiant systems. However, to improve results with solid wood flooring consider these guidelines. Use dimensionally stable woods such as American cherry, American walnut, mesquite, and teak, among others. Keep boards to 3" width or less. Use quartersawn stock if possible. Before installation, make sure the wood is at a moisture content suitable for the climate.

Joe Mallon

After You Read: Self-Check

1. What are the two main factors that determine the grade of an oak flooring strip?
2. Which woods are graded using the same system?
3. What happens if wood flooring absorbs moisture from the subfloor?
4. Compare and contrast solid wood flooring and engineered-wood flooring.

Academic Integration: Mathematics

5. **Estimate a Mitered Border** You are installing a red oak floor with a mitered border. The room is 9' × 12', and the mitered border is 1.5' wide. How many square feet is the total border area?

   Math Concept A mitered border is the area of the space between two concentric squares. First, calculate the area of the entire room. Then subtract 3' from both the width and the length to account for the border, and multiply these smaller numbers to discover the square footage inside the border. The difference between the larger and the smaller square footage represents the border area.

Go to **connectED.mcgraw-hill.com** to check your answers.

Installing Hardwood Strip Flooring

How many people usually install wood flooring?

In the past, hardwood flooring was usually installed by carpenters. Today, it is frequently installed by flooring subcontractors. Each type of wood flooring requires a different method of installation.

Hardwood strip flooring is the most common type of wood flooring. Most hardwood floors are created with strips of equal width. However, you can also create an attractive design by using strips of different widths, colors, or patterns. Most hardwood strip flooring today has tongue-and-groove edges. Each piece slides in snugly against the next one.

Fastening Techniques

Proper nailing of floorboards is essential for safety and to prevent squeaks. Tongue-and-groove flooring is blind nailed, except for the first one or two courses (rows). These courses are difficult to blind nail because they are close to the wall, so they are face-nailed instead. To blind nail, drive the nails through the edge of the board at an angle of 45° to 50° at the point where the tongue leaves the shoulder.

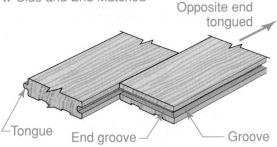

A. Side and End Matched

Opposite end tongued

Tongue   End groove   Groove

B. Side Matched

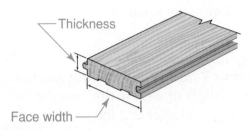

Thickness

Face width

Strip Flooring
Tongue and Groove Edges Two methods of installing strip flooring: **A.** Side and end matched. **B.** Side matched.

Using a Pneumatic Nailer
Precision Fastening A lip on the nailer's base plate enables it to be quickly positioned at a precise location. Note that the installer's hammer has two different faces.

Wood flooring was once blind nailed by hand. Today, most installers use a pneumatic power nailer to make sure that nails are placed consistently. The installer slides the nailer along the tongue of a board and strikes the tool's plunger with a mallet. This pushes the board tightly into place and drives a nail at an angle through the tongue of the board.

Types of Fasteners
The three most common types of flooring nails on the next page. Always follow the nailing recommendations of the flooring manufacturer. As a general rule, use 7d or 8d threaded (screw) nails or cut steel nails for flooring $^{25}\!/_{32}$" thick and 1½" or more wide. If you use steel wire flooring nails, they should be 8d and preferably cement coated.

Preparing the Wood Subfloor
Strip flooring is usually laid over a plywood subfloor. Before you install, examine the subfloor carefully and correct any defects. Drive down raised nails and scrape any bits of plaster, joint compound, or dried adhesives off the subfloor. Then sweep the area thoroughly.

Next, staple 15 lb. asphalt building paper to the subfloor to protect the flooring from moisture that might come from below and to help prevent squeaks. Overlap the seams by 2" to 4". Nail the flooring boards into joists rather than into the floor sheathing.

Planning the Installation
Make sure to lay strip flooring at right angles to the floor joists. Plan the work so that the flooring flows from one room to another, and be sure to consider closets and hallways.

If floor coverings in adjoining rooms are different, end the flooring of each room at the center of a doorway. What if different floorings meet where there is no door? Lay the wood floor through the opening to a point even with the wall line of the adjacent room.

 Flooring Nails

The Right Nail for the Job Common types of flooring nails: **A.** Barbed **B.** Screw **C.** Cut steel

Laying Strip Flooring

Many walls are not perfectly true (aligned), even in new construction. To ensure a straight course, stretch a string the length of the room between two nails placed 8" from a side wall. Line up the first courses at a uniform distance from the string rather than from the wall itself. Some installers snap a chalk line on the floor instead of measuring to a string.

After lining up the first courses, place a long piece of flooring with the grooved edge ½" to ⅝" from a side wall and the grooved end nearest an end wall. The space allows the flooring to expand without binding against the wall. You will hide the space later with baseboards or shoe molding.

Face-nail the board into place. Maintain the same distance from the guide string as you install each board in the first course. Drive one nail at each joist crossing, or every 10" to 12" if the joists run parallel to the flooring. Fit the groove end of each board over the tongue end of the previous board.

Depending on the width of the flooring, you may have to face-nail the second course as well. However, you can install subsequent courses with a power nailer or pneumatic nailer. Hold a board in place against the previous course and use the nailer to blind nail through the board's tongue. Slide the device along the board's tongue, nailing at each joist location. For best results, stand on

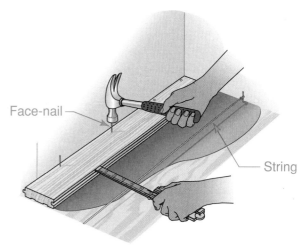

 Aligning the First Course

Align and Fasten To line up the first course of flooring, stretch a string or snap a chalk line about 8" from the wall. Then face-nail the first course.

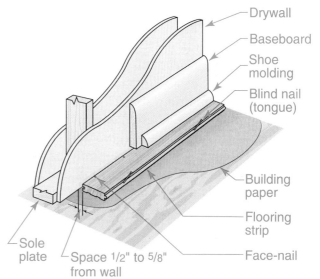

 Cross Section of Wall and Floor

Where Floor Meets Wall This cross section shows the first piece of strip flooring nailed in place. Note: With lath and plaster walls, sometimes the plaster ground (guide used when installing plaster) is kept about ⅞" above the subfloor. The edge of the first piece of strip flooring is set about even with the wall line. The flooring is then allowed to expand under the plaster ground. *Why is it important to leave a space of ½" to ⅝" between the grooved edge and the wall?*

Marking Flooring to Fit
Measuring Precisely Fitting a piece of flooring to the remaining space in a course.

Racking the Floor Take care to stagger the end joints of the flooring pieces so that joints are not grouped closely together. A joint should be 6" or more from a joint in an adjacent course. Arranging the strips in this way is called racking.

Two people usually work together to install a wood floor. One racks the pieces and cuts end boards to length. The other fits the racked boards together and nails them.

Reading Check

Analyze Why should you align flooring courses to a string and not a wall?

Installing Closing Boards

When you reach the opposite side of the room, you will find that there is no space between the wall and the flooring

the flooring strip with your toes in line with the outer edge, and strike the nailer from a stooped position.

If you cannot easily find a length of flooring to fit the remaining space in a course, cut one to approximate length. Position the piece in place with the groove end touching the wall and draw a line where it should be cut. Make sure to cut off the tongue end of the piece. You will need the groove end to join with the tongue end of the previous piece.

Builder's Tip

ALIGN LIKE A PRO Indicate the location of floor joists by marking the base of the walls before laying the building paper. Then snap chalk lines on the paper to indicate where the joists are. Use these chalk lines to guide you as you nail in the strip flooring.

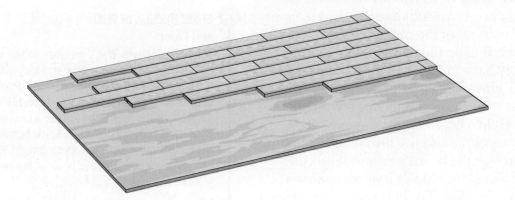

 Racking the Floor
Planning for Top Appearance It is important to lay out strip flooring so that the joints are staggered. An assistant usually works ahead of the installer to rack the floor, which allows the installer to concentrate on nailing.

to allow you to blind nail the last two or three courses. Place the last few courses in position. You may need to rip the last course to width on a table saw. Then face-nail the courses while pulling the flooring up tightly by exerting pressure against it with a prybar or a piece of scrap.

When prying the pieces with the crowbar or scrap length of flooring, put pieces of scrap stock against the wall to protect the wall surface. The scrap should span several wall studs to prevent damage to the wall. Remember to leave a space for expansion. If the last course is narrow, drill holes for the nails to prevent splitting.

If flooring continues into the adjacent room, fit floorboards around openings as needed. You can continue the same orientation of tongues and grooves unless you need to make a change for convenience. Just slip a spline, also called a slip tongue, into place to connect the grooves of two boards.

Reading Check

Explain Why should you face-nail the last courses in a room instead of blind nailing them?

Installing Wood Flooring over Concrete

Concrete gives off moisture as it cures, which can be harmful to wood flooring. To prevent problems, allow the slab to cure for at least two months. Then test it for excessive moisture. Tape a square foot of clear polyethylene to the slab. Seal the edges with duct tape and leave the plastic in place for twenty-four hours. Then check the underside of the plastic for signs of moisture. If there are no water droplets or moisture fog beneath the plastic, the slab is dry enough for the installation of a wood floor. Hardwood flooring can be installed over concrete slabs on grade or above grade. However, it should not be installed over concrete that is below grade, as on a basement floor.

You can use two methods to install strip flooring over concrete: flooring over sleepers and flooring over plywood. However, before you begin either method, must install a vapor barrier.

Vapor Barriers Even small amounts of moisture can harm wood flooring. Use asphalt felt building paper or polyethylene plastic as a vapor barrier between the concrete and the wood.

Asphalt Felt To use asphalt felt, sweep the slab clean. Then apply cut-back asphalt mastic to the slab with a notched trowel. Cut-back mastic has been thinned slightly with a solvent. About two hours after spreading the mastic, roll out strips of 15-lb. asphalt felt over the entire slab. Lap the edges 4". Spread a second layer of mastic over the felt. Then add a second layer of felt on top of that. Both layers of felt should run in the same direction, but the rows should be offset.

Polyethylene Plastic To use a *polyethylene* plastic vapor barrier, sweep the slab clean. Then spread cut-back mastic over the slab with a notched trowel. After the mastic has dried, spread 4-mil or 6-mil polyethylene over the slab. Use a weighted floor roller to press it into the mastic.

JOB SAFETY

CREATE SAFE LEVERAGE Never use chisels or screwdrivers to lever floorboards into place. These tools are not designed to take lateral loads and could snap suddenly under stress. Instead, use a prybar or similar device to pry with. Some flooring installers temporarily screw a wood block to the floor and use that to pry against.

Builder's Tip

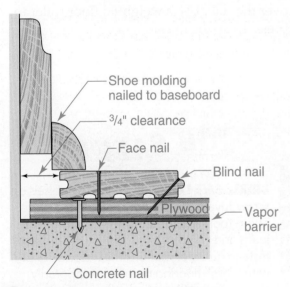

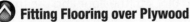

Fitting Flooring over Plywood
Using Plywood Plywood can also serve as a base for installing wood flooring over concrete. Note the use of concrete nails.

Installing Wood Flooring Over Sleepers and Plywood

Wood flooring should not be installed directly on concrete. Instead, use either a plywood or a sleeper base. A **sleeper** is a length of lumber that supports wood flooring over concrete. Choose sleepers that are preservative-treated. See the Step-by-Step Application on the next page for instructions on installing wood flooring over sleepers.

Installing Wood Flooring Over Plywood First prepare the surface with a vapor (moisture) barrier as described on page 981. Lay ¾" exterior grade plywood over the vapor barrier, staggering the end joints by 4". When all the plywood is in place, fasten it to the slab with concrete nails or powder-actuated fasteners. Use at least nine nails per panel. To ensure that the panels stay flat, nail them at the center first. Then work toward the edges. To allow for expansion, leave a gap of about ⅜" between each panel and a gap of ¾" at the walls.

Install the strip flooring as you would over a standard plywood subfloor. Nails should be slightly less than 1½" long.

Installing Parquet Flooring

What is the first step in installing parquet flooring?

Like hardwood strip flooring, parquet flooring can be installed unfinished or prefinished. Unlike hardwood strip flooring, parquet flooring does not require blind nailing. Instead, it is attached to the subfloor using an adhesive paste called mastic. In this respect, planning and installing a parquet floor is like planning and installing a vinyl or ceramic tile floor.

Read and follow the manufacturer's installation instructions, which are usually boxed with the flooring. Not following these instructions may void any warranty that applies to the product or its finish. Always check with the manufacturer if the conditions on the job site are not covered in the instructions. In general, however, you can apply parquet flooring to the same types of wood subfloors as hardwood strip flooring. Some parquet installers install ¼" luaun underlayment plywood over the existing wood subfloor. This provides an unusually smooth surface and increases adhesion. Some types of parquet flooring can be adhered directly to a concrete slab floor. The slab must be tested in several locations first to ensure that it is dry. Slabs less than 60 days old are typically too wet to install flooring.

Installing Parquet Tiles

The process for installing parquet tiles is similar to the process for installing a vinyl tile floor, which is discussed in Section 34.3.

First, measure the room and establish layout lines on the subfloor to guide the installation, as shown on page 989.

Attach individual tiles or sections of parquet flooring to the subfloor with mastic. Apply the adhesive to the subfloor and spread it with a notched trowel. Choose the size and shape of the notch based on the type of flooring and the brand of mastic. The configuration of the notch is designed to maximize coverage and ensure a uniform thickness of mastic.

Apply mastic to one small area of the floor at a time. This prevents the adhesive from "skinning over," or partially drying, which reduces its effectiveness. The temperature and humidity in the room affect how long the adhesive takes to skin over. The adhesive manufacturer will indicate open time on

---

Step-by-Step Application

Installing Strip Flooring over Sleepers Both sleepers and plywood can serve as a nailing base for installing wood flooring over concrete. Make sure to install a vapor barrier before you begin the installation.

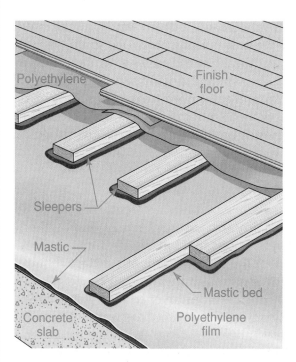

Polyethylene

Finish floor

Sleepers

Mastic

Concrete slab

Mastic bed

Polyethylene film

Step 1 Snap chalk lines 12" apart at right angles to the direction the flooring will run. Cover the lines with rivers of asphalt mastic about 4" wide.

Step 2 Embed 2"× 4" sleepers in the mastic. Lap the ends of the sleepers at least 4". Leave ¾" between the ends of sleepers and the walls. You do not need to nail the sleepers in place.

Step 3 After the mastic has cured, spread a 6-mil polyethylene vapor barrier over the sleepers. Lap all edges. This adds a layer of moisture protection to the system.

Step 4 Install the strip flooring at right angles to the sleepers. Adjoining courses of flooring should not have joints on the same sleeper. Where a flooring board runs over a lapped sleeper joint, nail the board into both sleepers. Provide at least ½" of clearance between flooring and the walls to allow for expansion.

the product packaging. "Open" time is the amount of time the adhesive can be used before it starts to skin over. Mastic that has skinned over should be scraped off the subfloor and discarded.

Lay tiles in the mastic immediately after spreading it. Push the tile into firm contact with the mastic and seat it further by tapping it with a rubber-faced mallet. As you work, clean off any mastic that squeezes up between the tiles using a soft cloth moistened in the recommended solvent.

After you cover an area of mastic with tiles, spread mastic over an adjacent area and lay another group of tiles. Cut tiles as needed to fit at the end of each row. Leave ½" of clearance between tiles and any wall or obstruction. To avoid disturbing recently set flooring, do not walk on it for at least twelve hours. Some installers recommend waiting for at least 36 hours.

Reading Check

Connect *What should you do if the special conditions at the job site are not covered in the manufacturer's instructions?*

Installing Parquet Tiles
Working with Mastic Tap each tile with a mallet to ensure a good connection with the mastic. These unfinished parquet tiles have shaped notches that help to maintain proper alignment.

Applying a Floor Finish
What are the various types of finishes?

After unfinished strip flooring or parquet flooring is in place, a finish must be applied. Finishing is a time-consuming process that stretches over several days.

Sanding

Unfinished hardwood flooring is smoothly surfaced by the manufacturer, but scratches and other marks caused by handling usually become visible after the floor has been laid. You may also see slight differences in the height of adjacent boards. This difference is called overwood. You must **eliminate** all of these imperfections. You can remove imperfections with a drum-type floor sander or a newer type of floor sander with random-orbit sanding pads. You may also need a smaller, handheld electric sander called an edger, which is designed to reach in corners and other small spaces. Finishing a new floor calls for several sandings, which are called cuts. Use coarse sandpaper for the first cut and progressively finer sandpaper for subsequent cuts.

When using a drum sander, make overlapping passes along the wood grain, not across the grain. Use the edger to sand close to the walls, where the drum sander cannot reach. You may also need to sand or scrape by hand if there are obstructions that interfere with the edger. After sanding, remove all dust with a vacuum.

Filling

Paste wood filler is recommended to fill the tiny surface crevices in oak and other hardwoods that have large pores. It gives the floor the perfectly smooth surface required for a lustrous appearance. Wood filler may be colorless or it may contain pigment to bring out the grain of the wood. Apply filler either before the last sanding cut or after staining. Let it dry for 24 hours before moving on to the next step.

Sanding a Hardwood Floor
Perfecting the Surface Although flooring is sanded at the mill, additional sanding with a drum sander or commercial random-orbit sander is required after the floor is laid. *Why should you sand along the wood grain, rather than across the wood grain?*

Staining

Stains are used to give the floor a color different from the color of the natural wood. Apply the first coat of stain or other finish on the same day as the last sanding. This prevents the grain from becoming raised, which roughens the surface slightly. Apply stain evenly, preferably with a high-quality brush that is 3" or 4" wide, and before applying other finishes.

Reading Check

Summarize *In what order should you perform the following finishing tasks: stain, sand, fill?*

Types of Finishes

In recent years, modern synthetic finishes have replaced many traditional finishes. For example, polyurethane has replaced varnish. Lacquer is another finish that is not used as much as it once was. An ideal finish is attractive and durable, is easy to apply and maintain, safe to apply, and can be retouched in worn spots without looking patchy.

However, no finish has all these characteristics. Some finishes are more durable than others. Some cannot be retouched easily. Others are so toxic prior to curing that they should be applied only by specialists. Choose a finish based on the characteristics that are most important for the situation.

In some areas, particularly large cities, environmental restrictions limit the use of floor finishes containing liquid solvents made from **volatile organic compounds (VOCs)**, a type of chemical that evaporates into the air. The purpose of a solvent is to dissolve various ingredients in the floor finish. When the floor finish is stirred, the ingredients are evenly distributed in the solvent. This ensures that the finish can be applied evenly. After the finish has been installed, the solvent evaporates. When this happens, the VOCs are released into the air. They react with sunlight and with other materials in the air to form ozone. Ozone is a major air pollutant. Many floor finishes have therefore been reformulated to reduce or eliminate VOCs.

Penetrating Finishes Penetrating finishes differ from other finishes in one important respect. Rather than forming a surface coating, the finish penetrates the wood fibers, becoming a part of the wood itself. It wears only as the wood wears and does not chip or scratch. It does not provide as shiny a surface as other finishes, but it can be easily retouched. Refinishing worn spots does not create a patched appearance. Penetrating finishes are available either clear or slightly tinted with color.

Applying a Penetrating Finish
Working Safely with Chemicals Penetrating finishes are tough and wear resistant and can be applied easily with long- or short-handled tools. Safe working practices are important. *What safety equipment is this worker wearing, and why?*

You can use a squeegee, a wool applicator, or a wide brush to apply penetrating finishes. Wipe off excess material with clean cloths or a rubber squeegee. For best results, buff the floor with No. 2 steel wool. You can also use penetrating finishes as a base for a surface finish such as varnish.

Urethane Finishes Several floor finishes are included in the general category of urethanes. These are durable finishes that cure to a hard film. They are fairly resistant to moisture.

Moisture-cured urethanes are the hardest and most moisture resistant. They offer a glossy look that resists abrasion. They cure by reacting with humidity in the air. The proper amount of humidity is critical, however. This makes these finishes difficult to apply, except by professional floor finishers.

Oil-modified urethanes, sometimes called polyurethanes, are easier to apply than moisture-cured urethanes. They provide a durable coating with a gloss, semigloss, or matte finish. They are widely available and frequently used.

Water-based urethanes are relatively new. The solid portions of the product are suspended primarily in water, rather than in a volatile *solvent*. Water-based urethanes are durable but require more coats to build up a thickness that compares to other urethanes.

Natural Finishes Varnishes were once widely used to finish floors, but they are seldom used now. They have been replaced by synthetic finishes that are easier to install and more durable.

Natural oil finishes can be used on a floor, though they require much time to fully cure. Oils are less durable than other finishes but are easy to repair. Linseed oil and tung oil are common types.

Shellac finishes result in floors of great beauty, but a shellac finish is difficult to maintain. Shellac water-spots easily and does not hold up against common household solvents.

Wax is usually applied over some other type of finish. It not only gives a lustrous sheen to a floor but also forms a film that protects the finish beneath. When wax becomes dirty, it is easily removed and replaced. However, wax can make a floor slippery and should be used with care.

Estimating and Planning

CERTIFICATION PREP

This estimating and planning exercise will prepare you for national competitive events with organizations such as SkillsUSA and the Home Builder's Institute.

Strip Flooring Materials and Labor

Materials

To estimate materials, first determine the number of square feet to be covered.

1. Multiply the length of the room by the width of the room. Figure any offsets or closets separately and add them to the total. For example, a room that is 10' × 12' has 120 sq. ft. of floor area (10 × 12 = 120). A closet that is 2' × 8' has 16 sq. ft. of floor area (2 × 8 = 16). The total area of room and closet is 136 sq. ft.

2. Use the table below to calculate the amount of strip flooring you will need. Select the size you will use in the first column, then read across for the amount. For example, if you select $^{25}/_{32}$' × 2¼' strip flooring, you will need 138.3 bd. ft. to cover 100 sq. ft. of floor area. The room in our example has 136 sq. ft. of floor space. Divide that number by 100:

$$136 \div 100 = 1.36$$

3. Multiply the result by 138.3:

$$1.36 \times 138.3 = 188.09 \text{ bd. ft.}$$

You now know that you will need 188.09 bd. ft. of strip flooring.

4. To figure the cost of the material, multiply the cost per board foot times the total number of board feet required.

5. The table shows that it will take 3 lbs. of nails to lay 100 sq. ft. of 2¼' strip flooring. Multiply 3 lbs. by 1.36:

$$1.36 \times 3 \text{ lbs.} = 4.08 \text{ lbs.}$$

6. Multiply the cost per lb. times the number of lbs. to find the cost of the nails.

Labor

1. The labor columns in the table show that a worker can lay 100 sq. ft. of $^{25}/_{32}$" × 2¼" strip flooring in 3 hours.

2. Multiply 3 hours by 1.36:

$$1.36 \times 3 \text{ hrs.} = 4.08 \text{ hrs.}$$

3. Multiply this number by the hourly rate to determine total labor cost.

Estimating on the Job

Estimate the quantity of $^{25}/_{32}$" × 2" strip flooring needed for a 9' × 13' bedroom and a 5' × 6' closet. Then estimate the hours of labor required to lay, sand, and finish the floor. Round your answers to the nearest tenth.

Estimating Strip Flooring Materials and Labor						
Strip Flooring	Material		Nails per 100 Sq. Ft. (Lbs.)	Labor Hours per 100 Sq. Ft.		
Size (inches)	Bd. Ft. per 100 Sq. Ft.	1,000 Bd. Ft. Wll Lay (Sq. Ft.)		Laying	Sanding*	Finishing*
$^{25}/_{32}$ × 1½	155.0	645.0	3.7	3.7	1.3	2.6
$^{25}/_{32}$ × 2	142.5	701.8	3.0	3.4	1.3	2.6
$^{25}/_{32}$ × 2¼	138.3	723.0	3.0	3.0	1.3	2.6
$^{25}/_{32}$ × 3¼	129.0	775.2	2.3	2.6	1.3	2.6
⅜ × 1½	138.3	723.0	3.7	3.7	1.3	2.6
⅜ × 2	130.0	769.0	3.0	3.4	1.3	2.6
½ × 1½	138.3	723.0	3.7	3.7	1.3	2.6
½ × 2	130.0	769.2	3.0	3.4	1.3	2.6

* Sanding and finishing times are averages.

After You Read: Self-Check

1. Which floor-nailing tool must be struck by a mallet?
2. Why is wood flooring laid on building paper?
3. What standard tools should be used to lever floorboards into place? Which should not be used?
4. What impact do VOCs have on the choice of a floor finish?

Academic Integration: Science

5. **Chemical Fumes and Vapors** Research the possible hazards involved in applying various floor finishes, including penetrating finishes and urethanes. What health-related problems might result from exposure to these materials? What personal protective equipment should you wear to work with each? Make a chart of your findings.

Go to **connectED.mcgraw-hill.com** to check your answers.

Section 34.3 Vinyl, Tile, & Carpet Flooring

Vinyl and Other Resilient Flooring

What are two types of resilient flooring?

Many other types of flooring are available besides wood, including vinyl, ceramic tile, and carpet. Vinyl is flexible and practical, ceramic is durable, colorful, and beautiful, and carpeting is comfortable and insulating.

Vinyl is the most common type of resilient flooring. Resilient flooring is flexible and less than ³⁄₁₆" thick. It is available in many forms, colors, patterns, and textures. Some types of resilient flooring, such as vinyl flooring, are made of synthetic materials. Others, such as linoleum, are made primarily of natural materials.

Vinyl flooring is often used in kitchens, bathrooms, and recreation rooms because it is durable, stain resistant, and fairly inexpensive. Vinyl flooring comes in sheet and tile form. Sheet vinyl comes in large rolls 12' wide. It is often laid without seams and is usually installed by specialists.

 Installing Sheet Vinyl
Resilient Flooring Sheet vinyl comes in 12"-wide rolls and is usually laid without seams.

Ernest Prim/Getty Images

Vinyl tiles are usually 9" or 12" square. This makes them easier for homeowners to install themselves. Sheet vinyl and vinyl tiles are usually applied to a wood subfloor using mastic. However, vinyl can be laid over almost any solid, dry surface, including concrete.

Installing a Sheet Vinyl Floor

Follow the manufacturer's specific recommendations when installing a vinyl floor. The following are general instructions.

First, cover the wood subfloor with plywood underlayment. **Underlayment** is a thin panel product whose surface is smoother than standard plywood or OSB subflooring. Underlayment prevents small flaws in the subfloor from showing through to the vinyl. It also provides firm, clean, and void-free support. Underlayment should be at least ¼" thick and have a sanded face.

Preparing the Surface Apply nails or staples 6" apart in the field, or center area of the underlayment panel, and 4" apart at the edges. The joints between underlayment panels should be staggered and butted together. Fill any gaps larger than ⅟₃₂" with latex patching compound and sand them flush.

To estimate the amount of sheet vinyl you will need, calculate the square footage of floor you need to cover. Add an extra amount to account for waste, seams, and pattern matching. The extra amount you will need depends on the type of flooring and the pattern. Consult the manufacturer or distributor for information.

Sweep the floor thoroughly to remove any dust and debris. Make sure that fasteners are flush with the surface of the underlayment. One way to check this quickly is to run a wide-blade putty knife over the floor. You will hear a metallic ring when the blade hits protruding fasteners.

Measuring and Cutting Measure the room and determine if the sheet vinyl will require seams. Because of the width of vinyl sheets, it is often possible to avoid a seam. If you will need a seam, snap a chalk line where the seam will fall. Then snap two parallel lines about 8" away on either side of the first line. These chalk lines identify the width of the seam area.

Unroll the flooring and use a utility knife to cut as many pieces as you will need to cover the floor. Each piece should be about 3" longer than the length of the room. Flooring that is cold is more difficult to work with than flooring that is at room temperature.

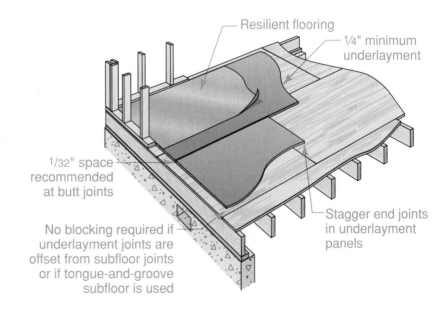

Vinyl and Underlayment
Preparing the Surface
Underlayment panels create a smooth surface for vinyl flooring.

Spreading Adhesive Spread mastic with a notched trowel over half of the area to be covered by a single strip. This ensures that the adhesive will be a uniform thickness. Remove excess mastic with the trowel. Keep mastic out of the seam area. Place the flooring in the wet mastic and roll it smooth with a weighted floor roller. One edge should overlap the center seam line by about 1". Lift the second half of the vinyl. Apply mastic to the underlayment beneath it, then lower the vinyl into the mastic and roll it smooth. Trim the ends of the vinyl to final length.

Repeat the procedure with the next strip of vinyl. Make sure that one edge overlaps the previous piece by several inches.

After the adjoining strips of vinyl flooring are in place, there will be a 16" wide "dry zone" beneath the overlapping edges. Cut through the overlapping edges with a utility knife. Guide the cut with a steel straight-edge. Lift each edge of the flooring, apply mastic to the underlayment, and roll the seam smooth.

Reading Check

Explain *Why should you use a notched trowel to spread mastic?*

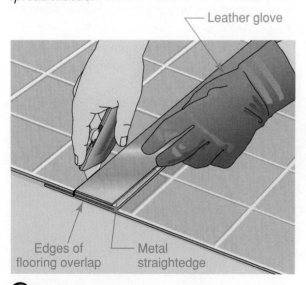

Leather glove

Edges of flooring overlap

Metal straightedge

Cutting Sheet Vinyl
Removing Overlap Double-cutting an overlapped area using a steel straightedge as guide.

Installing a Vinyl Tile Floor

Vinyl tile can be laid directly over wood flooring or over a plywood floor. It is important to lay out the tile correctly. After measuring the room, determine how many tiles will fit across its length and width. Then snap chalk lines to indicate one of the three basic layouts shown in Room Layout for Vinyl Tile on page 991. The layout you choose will govern the rest of the installation. Lay tiles in adhesive as the work progresses, cutting tiles to fit around obstructions such as pipes.

Ceramic Tile

How long has ceramic tile been used in construction?

Ceramic tile is often used where a highly durable, scratch-resistant flooring is desired, such as a kitchen. However, it can be used in any room of the house. It comes in a large variety of shapes, sizes, and colors. Tile has been used for over 6,000 years for roofs, paving, walls, and decoration. In new construction, tile is usually installed by tile contractors or tile setters.

Manufacturing Ceramic Tile

Custom lots of ceramic tile are still made by hand, but most tile is made in highly automated factories. It is usually made from a combination of pure clay or pure gypsum and other ingredients that extend the clay and control shrinkage. After the clay has been refined and mixed with water and these additives, it is shaped into a bisque. A **bisque** is a tile without the glaze. To form a bisque, the clay mixture is either extruded, dust-pressed, cut from a sheet, or formed by hand. Most commercial tile is made by the dust-press method. In this method, the ingredients are mixed with so little water that only high pressure can bond them together.

The bisque is then dried before being fired in a kiln at temperatures up to 2,200°F (1,204°C). A glaze (glassy finish) may be applied at this time. The glazed surface of a tile is waterproof after firing.

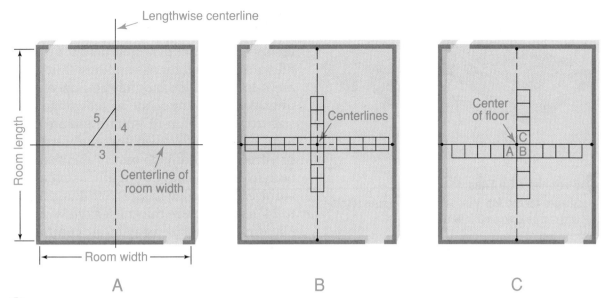

Lengthwise centerline

Room length

5
4
3

Centerline of room width

Room width

A

Centerlines

B

Center of floor

C
A B

C

 Room Layout for Vinyl Tile

Planning from Center Locate the center of the floor and establish centerlines as shown in **A**. Check the centerlines for square by using the 3-4-5 rule and adjust them as needed. Determine how many tiles will be required across the length and width of the room. If an odd number is required, establish a layout in **B**. If an even number is required, use **C** as your model. *Why should you plan the layout of tiles from the center of the room, rather than the edge?*

Characteristics of Ceramic Tile

Tile can be classified according to several characteristics. The three most important are permeability, placement, and use.

Permeability *Permeability* is the ability of a substance to allow water to pass through. Temperature and firing time determine the permeability of a tile. Highly permeable tiles are the least waterproof, because they absorb the most water. From most permeable to least permeable, the four types of tile are nonvitreous, semi-vitreous, vitreous, and impervious. (*Vitreous* means glasslike.)

Permeability is important because it determines the best use of the tile. Tiles that will be exposed to water in a bathroom, for example, should be less permeable. To test for permeability, turn a tile over and put a drop of water on the unglazed (back) portion. If the drop is absorbed immediately, the tile is highly permeable. If the drop remains on top of the tile, the tile is less permeable.

Placement Tile is often categorized by where it is used, either on walls or floors. Wall tile

is generally nonvitreous, with a relatively soft glaze that makes it unsuitable for foot traffic. It is usually about ¼" thick and 4" or 6" square.

Floor tile can be any kind of tile (from nonvitreous to impermeable and glazed or unglazed) that is strong enough to hold up in use on the floor. Floor tile can be used on walls, but wall tile should not be used for flooring.

 REGIONAL **CONCERNS**

Tile Use Varies by Region If you live in an area with a cold climate, you will see tile used mainly in bathrooms and kitchens. If you live in an area with a hot climate, you will see tile used throughout entire houses. Houses in hot climates are often built on a concrete slab foundation, which provides an excellent base for the installation of ceramic tile.

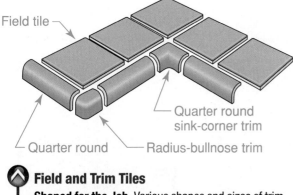

Field tile

Quarter round
sink-corner trim

Quarter round   Radius-bullnose trim

◆ Field and Trim Tiles
Shaped for the Job Various shapes and sizes of trim tiles can be used around field tiles.

Use Another way to categorize tile is by use either in the field or as trim. Trim tile is specially shaped to form a border and is often bullnosed in shape. It may also be *radiused* to form a curve or ridged to form a pattern.

Field tile is the tile contained within the borders formed by trim tile. Field tile is flat rather than shaped. It is glazed on the top surface only.

Reading Check

Connect What determines the permeability of a tile?

◆ Lugged Tile
Perfect Alignment The lugs on the edges of these tiles ensure the proper spacing between tiles. The lugs will later be covered with grout.

Types of Ceramic Tiles

Many different types of ceramic tiles are used in residential construction. Paver tiles are intended for use on floors. They are at least ½" thick and may be glazed or unglazed. Machine-made pavers range in size from 4" by 6" to 12" square and are up to ⅝" thick. They are usually semi-vitreous or vitreous. Handmade pavers are usually nonvitreous and have a slightly uneven surface. They range in size from 4" square to 24" square and in thickness from ½" to 2". Handmade, unglazed pavers are commonly known as Mexican tiles or Mediterranean tiles.

Quarry tile is generally unglazed, semi-vitreous or vitreous clay tile that ranges in thickness from ½" to ¾". It is excellent for use on floors because it is so dense. Mosaic tile is any tile 2" square or smaller. It is usually vitreous, and it ranges in thickness from ³⁄₃₂" to ¼". Lugged tile is tile that has spacing lugs built into its sides. When the tiles are placed edge-to-edge, the lugs automatically determine the proper spacing.

Substrates for Installing Ceramic Tile

The substrate for ceramic tile must be stiff enough to prevent flexing, which could crack tiles or joints. Tile can be adhered

Science: Chemical Reactions

Mixing Cement and Water Most grout is made from Portland cement, an unstable chemical compound that becomes stable when you mix it with water. You are making grout that calls for 14 ounces of water for every 5 pounds of dry grout. How many ounces of water would you need to mix with a 10-pound package of dry grout?

Starting Hint Solve for ? in the following equation:

$$\frac{14\ oz.}{5\ lbs.} \times 10\ lbs. = ?oz.$$

directly to plywood, but this is generally not recommended, particularly on floors. Tile can also be adhered directly to a drywall surface.

Cement-based sheets called **backerboard**, or cement board, provide an excellent base for tile, particularly on floors and in wet installations such as shower stalls. You can nail or screw backerboard to a subfloor or to a plywood countertop. It comes in sheets ½" or ⅝" thick, 32", 36", or 48" wide, and 4', 5', 6', or 8' long. Apply the sheets to a wood sub-floor with 1½" hot-dipped roofing nails or 1¼" corrosion-resistant screws. These screws, sometimes called cementboard screws, have small fins on the underside of the head that help to countersink it.

Installation Materials

You will need adhesives, grout, and waterproofing membranes to install ceramic tile properly.

Adhesives Ceramic tile is adhered to the substrate with dry-set mortar or mastic. Apply the adhesive with a trowel, then "combed out", or spread out, with a notched trowel so that it is distributed evenly. The even thickness helps to support the tile, preventing breakage caused by point loads.

Dry-set mortar, also called thin-set mortar, is a very effective adhesive. It is a mixture of Portland cement, sand, and additives that strengthen the bond. Dry-set mortar can be mixed with water, with a latex- or acrylic-modified liquid, or with *epoxy* resins, as shown in **Table 34-2** on the next page.

Mastic is an organic adhesive. It comes premixed in cans and is often preferred by nonprofessionals because it is easy to use.

Grout After attaching tiles to a substrate with adhesive, you will need to fill the spaces between the tiles with grout. **Grout** is a thin mortar used for filling spaces. It can be mixed with water, latex- or acrylic-modified liquids, or epoxy resin. It prevents moisture

Laying Ceramic Tile
Uniform Coverage To create an even surface for laying tile, spread adhesive on the surface with a notched trowel and then comb it out to produce uniform coverage.

Spreading Grout
Filling In Spread grout over the entire tiled surface with a grouting trowel. Pack it into the joints, then remove the excess grout with the trowel and a sponge.

(t & b)Arnold & Brown

Table 34-2: Adhesives for Ceramic Tile

Adhesive	Characteristics
Mastic	• Good grip strength (useful for setting wall tiles) • Lacks the strength or flexibility of mortar • Not heat resistant • Least expensive and easiest to use adhesive • Suitable for use on drywall • No mixing required
Dry-set mortar mixed with water	• Good bond strength • Good compressive strength • Not flexible • Heat resistant • Inexpensive • Easy to use
Dry-set mortar mixed with latex-modified or acrylic-modified liquid	• Excellent bond strength • Excellent compressive strength • Somewhat flexible • Heat resistant • Can be applied to most surfaces except steel • Resistant to frost damage
Dry-set mortar mixed with epoxy resin	• Very high bond strength • Somewhat flexible • Heat resistant • Expensive • Can be applied to almost any surface, including plastic laminate and steel • Can be hard to work with • Very high resistance to impact

latex- or acrylic-modified liquids, or epoxy resin. It prevents moisture and dirt from getting between the tiles. Grout comes in a wide array of colors.

Tile grout comes in two forms: plain and sanded. Plain grout is mixed with additives to make it smooth and creamy. It is generally used when the spaces between tiles are less than ⅟₁₆" wide. Sanded grout is simply plain grout to which sand has been added for strength. Sanded grout is used for joints wider than ⅟₁₆".

Waterproofing Membranes Properly installed, ceramic tile creates a durable and water-resistant surface. Water may still penetrate below the tile around bathtubs and in other areas exposed to large amounts of water. Prevent this by placing a water-proofing membrane beneath the substrate. This membrane can be any flexible, waterproof sheet material, such as tarpaper. Nail or staple it in place, sealing the edges with asphalt adhesive.

Other products, such as chlorinated polyethylene (CPE), provide even greater thick that comes in large rolls. Attach it directly to the substrate with dry-set mortar and use a roller to ensure a proper bond.

Reading Check

Explain What are the two functions of grout?

Tools for Ceramic Tile

Two basic types of tools are used to cut ceramic tile: snap cutters and tile saws. Tilesetters often rely on tile saws for most tile cutting.

Portable Snap Cutter This tool is used to cut tiles in a straight line. Place the tile on the bed of the tool (beneath the two rails) and against the guide. Then lift the long handle of the tool and pull it toward you. This draws a small scoring wheel across the tile's surface. You can then snap the tile apart at the scored line by putting modest pressure on each side of the line. One advantage of the snap cutter is that it can be operated without the need for water or electricity.

Tile Saw A type of tile saw called a wet saw is often used to cut tile, particularly when large quantities must be cut accurately. The tool includes a circular diamond-grit blade, a water pump, and a moisture-proof motor. The pump sprays a continuous stream of water on the blade during the cut. This lubricates and cools the blade. It also prevents the

JOB SAFETY

GUARD AGAINST DUST AND VAPORS Dry-set mortars and grout are very fine powders that must be mixed with a liquid. Wear a suitable dust mask when mixing them. Some mastics give off fumes that may be harmful as well, so always check the instructions on the can for any health warnings. Work in a well-ventilated area and remember that a standard dust mask offers no protection against vapors.

diamond abrasive from becoming clogged with clay particles. A wet saw creates a very smooth cut edge. It should be set up in an area where splashes of water will not harm nearby surfaces.

Nippers and Nibblers These simple hand tools look like pliers or small nail pullers, but they have straight, hardened cutting edges. They are used to cut curves in tile by nibbling, or eating, away at the tile edges. The resulting edge is rough, so it must be ground smooth or hidden behind some type of fitting or cap. Nibblers are often used to cut notches in tile to fit around shower pipes.

Knee Pads These pads help to protect the tile setter's knees from injury during long hours of setting floor tiles.

Trowels A tile setter uses a variety of flat and notched trowels with steel blades. Trowels with notched edges are used to spread adhesive. The size, shape, and depth of the notch determines the thickness of the adhesive layer.

Scoring Tool This hand tool has a carbide tooth mounted on a steel blank. It is used to score through the fiberglass reinforcement of cement board.

Methods of Installing Ceramic Tile

There are many methods for installing tile. Each job requires a different combination of tile, adhesive, grout, and setting methods. The proper methods and materials depend on several factors, including the stiffness of the floor system and whether or not the tile will be exposed to water. The two basic methods of installing tile are known as thick-bed and thin-bed. Thick-bed installation is a traditional method but it is time-consuming to install. Thin-bed installation goes more quickly, particularly on large areas of flooring.

Thick-Bed Installation In a thick-bed installation, you apply tiles over a mortar setting bed that is ¾" to 1¾" thick. Make sure to place a waterproofing membrane beneath

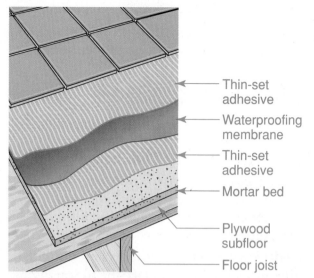

 Thick-Bed Installation
Five Layers A thick-bed installation of ceramic tile over a floor.

the setting bed. You should also add metal lath or wire mesh reinforcement. These products help to prevent the setting bed from cracking. In this way, they are much like reinforcing bar that is used in concrete slab construction. Apply the mortar in three layers: a scratch coat, a bed, and a bond coat.

After the setting bed has cured, you apply the tiles to it with dry-set adhesive. This method helps create accurate **inclines** and planes that can improve water drainage. It can also be used to level uneven substrates. The result is structurally strong and is not affected by prolonged contact with water.

Thin-Bed Installation In a thin-bed installation (also called thin-set installation), you adhere tiles directly to backerboard with adhesive. The adhesive may be mastic or dry-set mortar. Thin beds are less costly than thick beds, are relatively light, and are easier and quicker to install. However, the substrate must be very flat and very well prepared, and the surface cannot be easily sloped.

> **Reading Check**

Connect *What installation method would you use for a tile floor that slopes to a drain?*

Carpeting

What is one benefit of carpeting?

Carpeting is available in a wide variety of styles, colors, and construction types. Unlike hard surface flooring such as hardwood,

vinyl, and ceramic tile, carpeting offers a soft feel underfoot. This makes it particularly suited for use in living areas and bedrooms. Carpeting is slip-resistant and helps muffle sound.

Carpet Construction

The performance and appearance of carpeting depends on the fiber and on how the fiber is assembled on the carpet backing.

Types of Carpet Fibers The type of fiber is critical in determining how durable the carpet is and how easy it is to clean. Carpeting made of natural fibers such as wool is durable and naturally stain resistant. It is generally more expensive than other types of carpeting. *Synthetic* fiber carpeting is made from nylon, polyester, olefin, or acrylic. Nylon is the most durable and stain resistant carpet fiber.

Structure of Carpet Fibers Carpet fibers are twisted to help withstand crushing and matting. Carpets with dense, tightly packed fibers are more wear-resistant. Fibers can be sheared to create a soft, plush walking surface, or looped to increase the carpet's durability and its ability to conceal dirt.

Carpet Installation Basics

Wall-to-wall carpeting is installed by carpeting subcontractors, who often work exclusively for a particular carpet outlet or store. They have the specialized skills and tools to install carpeting efficiently.

The first step is to nail tackless strip (also called "tack strip") around the perimeter of the room. This is a narrow strip of thin plywood that contains many angled tacks. Always wear leather gloves when handling tackless. The installer nails the strip to the subfloor so that the points of the tacks are exposed and angled towards the wall. The tacks will grip the underside of the carpeting.

When the tackless strip is in place, the installer cuts carpet padding to fit just inside the tackless strip, then cuts and positions carpeting over the pad. The carpeting should be 4" to 6" larger than the room's dimensions. The excess material will be trimmed off later.

The installer places the carpet in position, hooks it on the tackless strip at one end or corner of the room, and stretches it toward the opposite end of the room using a power stretcher. This hand tool uses lever action to stretch the carpet and hook it onto the tackless strip. Once the carpet has been stretched in all directions, the installer trims the excess material and pushes the cut edges under the baseboard.

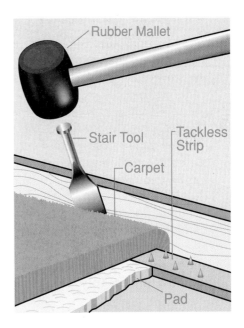

Tackless Strip for Installing Carpet
Wall to Wall Tackless stripping holds the stretched carpet in place. Trimmed carpet edges should be tucked under the baseboard.

Section 34.3 Assessment

After You Read: Self-Check

1. What are the advantages of vinyl flooring?
2. Name two types of synthetic flooring and two types of natural flooring discussed in this section.
3. Which type of adhesive is used to install resilient flooring?
4. When choosing tile, why is it important to know the tile's permeability?

Academic Integration: English Language Arts

5. **Ceramic Tile** Research the manufacture and use of ceramic tile in past centuries. What were the principal uses of tile in the past? Do those uses differ from the uses for ceramic tile today? Present your research in a one-page report.

 Go to **connectED.mcgraw-hill.com** to check your answers.

Chapter Summary

Section 34.1
Solid-wood flooring is available in several forms, including strips, planks, and parquet. The grade of the wood depends on the wood species, the manufacturing process, and the surface appearance. Allowing wood flooring to acclimate before installation prevents warping.

Section 34.2
Hardwood strip flooring is installed by blind nailing it into the floor joists. It is most often installed over a wood subfloor. It may also be installed over concrete as long as no moisture problems exist. Sanding a floor removes imperfections. After all sanding dust has been removed, the floor can be finished for color, shine, and durability.

Section 34.3
Resilient flooring is flexible and thin. It is attached to a smooth, solid underlayment using mastic. The most common types are sheet vinyl and vinyl tile. Ceramic tile can be classified by its water permeability, placement, and use. It can be applied to various substrates using different types of adhesives. Wall-to-wall carpet is installed by hooking it onto a tackless (plywood) strip.

Review Content Vocabulary and Academic Vocabulary

1. Use each of these content vocabulary and academic vocabulary words in a sentence or diagram.

Content Vocabulary

- plank (p. 972)
- parquet (p. 973)
- acclimation (p. 975)
- wear layer (p. 976)
- mastic (p. 976)
- sleeper (p. 982)
- volatile organic compounds (VOCs) (p. 985)
- underlayment (p. 989)
- bisque (p. 990)
- backerboard (p. 993)
- grout (p. 993)

Academic Vocabulary

- uniform (p. 973)
- eliminate (p. 984)
- inclines (p. 996)

Speak Like a Pro

Technical Terms

2. Work with a classmate to define the following terms used in the chapter: *curing* (p. 973), *heartwood* (p. 974), *polyethylene* (p. 981), *moisture-cured urethanes* (p. 986), *oil-modified urethanes* (p. 986), *water-based urethanes* (p. 986), *solvent* (p. 986), *permeability* (p. 991), *vitreons* (p. 991), *radiused* (p. 992), *epoxy* (p. 993), *synthetic* (p. 996).

Review Key Concepts

3. List the common forms of wood flooring.

4. Describe the major kinds of wood used in flooring and how they are graded.

5. Explain how to install wood strip flooring.

6. Explain how to estimate the quantity of resilient flooring needed for a room.

7. Describe the basic methods for installing ceramic tile and carpeting.

Critical Thinking

8. Explain A typical house contains several types of finish flooring, each with a different thickness. However, the top of the surfaces should all be at the same level. How might you achieve this?

Academic and Workplace Applications

STEM Mathematics

9. Front-End Estimation Your clients are trying to decide between bamboo and reclaimed wide-plank fir flooring for 950 square feet of their remodel. Bamboo costs $3.09 per square foot uninstalled, and reclaimed wide-plank fir costs $6.05 per square foot uninstalled. Calculate the cost of materials, then use front-end estimation to make a quick estimate of the difference.

Math Concept To make a quick estimate of the sum or difference between two numbers, you can use front-end estimation. Just add or subtract the digits of the two highest place values, and replace the other place values with zero. This will give you an estimate of the solution to a problem.

Step 1: Calculate the cost of both flooring options. Bamboo costs $2,934.50 and reclaimed fir costs $5,747.50.

Step 2: Front-estimate both numbers in the problem ($2,934.50 to $2,900 and $5,747.50 to $5,700).

Step 3: Now subtract using the new numbers.

21st Century Skills

10. Communication Skills Manufacturer's instructions are bundled with most flooring materials. Contact a flooring manufacturer to obtain an instruction booklet. Read all text and diagrams in the booklet. What installation method, tools, and materials are recommended? What is the proper process for preparing the subfloor? When should you lay a vapor barrier, and what kind?

Summarize in your own words in a one-page document.

STEM Science

11. Climate and Acclimation The proper moisture content of wood flooring changes by location. Determine the proper percentage for where you live. Consult an acclimation table, available from flooring manufacturers and flooring trade associations, then measure the moisture content of the flooring with a moisture content meter.

Standardized TEST Practice

Multiple Choice

Directions Choose the phrase that best completes the following statement.

12. It is important to install underlayment beneath sheet vinyl flooring because _____.

   a. it keeps the vinyl from peeling away from the subfloor

   b. it reduces the square footage of vinyl needed to cover the subfloor

   c. it blocks moisture from the subfloor

   d. it provides firm support and a smooth surface

13. Applying tiles over a mortar setting is called _____.

   a. thin-bed installation

   b. thick-bed installation

   c. baseboard installation

   d. mortar installation

14. The thickness of oak strip flooring is commonly _____.

   a. ½" or ¾"

   b. ⅜" or ⅝"

   c. ¼" or ½"

   d. 1½" or 3¼"

TEST-TAKING TIP

Read all of the questions and all of the answers and eliminate all statements that you know are incorrect before choosing.

*These questions will help you practice for national certification assessment.

CHAPTER
35

Decks & Porches

Section 35.1
Deck Materials

Section 35.2
Planning & Construction

Chapter Objectives

After completing this chapter, you will be able to:

- **Name** the basic types of materials used for decking.
- **List** the basic elements of a deck.
- **Describe** how to lay out piers.
- **Demonstrate** how to plumb a post.
- **Summarize** the proper way to handle and cut preservative-treated wood.
- **Describe** two methods for installing concrete porch steps.

Discuss the Photo

Decks Decks allow homeowners to enjoy views and the outdoors. *At which stage of construction are decks installed?*

Writing Activity: Career Profile

Decks and porches are popular remodeling projects. Contact a contractor who specializes in these types of projects. Ask him or her what skills, experience, and training are necessary in order to complete projects successfully. Summarize your findings in a two-paragraph career profile.

stevecoleimages/Getty Images

Chapter 35 Reading Guide

Before You Read Preview

Decks are constructed of several types of materials. Before reading the chapter, use the knowledge you have gained throughout the textbook to guess which materials are typically used for decks. After you have read the chapter, review your guess and revise it if necessary.

Content Vocabulary

- deck
- ACQ (Alkaline Copper Quaternary)
- heartwood
- sapwood
- galvanizing
- pier
- ledger
- stoop
- precast

Academic Vocabulary

You will find these words in your reading and on your tests. Use the academic vocabulary glossary to look up their definitions if necessary.

- exposed
- maintains

Graphic Organizer

As you read, use a chart like the one shown to organize main ideas and supporting details.

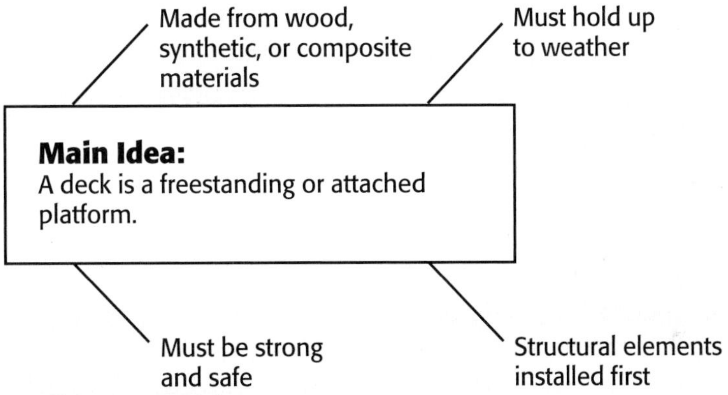

Made from wood, synthetic, or composite materials
Must hold up to weather
Main Idea:
A deck is a freestanding or attached platform.
Must be strong and safe
Structural elements installed first

Go to connectED.mcgraw-hill.com to download this graphic organizer.

Structural Materials

What are the basic types of decks?

Decks are often built by the carpenters who framed the house or by contractors who specialize in the construction of decks. A porch is another desirable feature. Depending on the house's architectural design, a porch may be built in various ways. It may resemble a deck with a roof over it, or it may be more enclosed.

A **deck** is a platform made from wood, synthetic, or composite materials fastened with nails, screws, bolts, and metal brackets. The primary requirements of a deck are weather resistance, strength, and a safe design. Because a deck is entirely exposed to the weather, joints must not trap water and the structure must resist wind uplift forces.

Deck construction techniques are straightforward. The structural elements are installed first. Deck boards, often called *decking*, are then attached to the structural elements. A railing ensures the safety of people using the deck, and steps or stairs allow access from the deck to the yard.

The decking itself is the most visible portion of a deck project. Many new types of decking materials have become available in recent years. These include tropical hardwoods as well as a wide variety of plastics and composites.

Types of Decks

There are two basic types of decks. A *freestanding deck,* or grade-level deck, is not attached to the house. It is low to the ground and does not require a foundation. It is usually on only one level. An *attached deck,* or elevated deck, has at least one side permanently connected to the structure of the house. It is partly supported by the house and partly by a network of concrete piers that extend below the frost line. This type of deck may have more than one level.

Posts, Beams, and Joists

Materials used for the structural elements beneath the decking are chosen primarily for their strength and durability. Appearance is usually a lesser concern. Preservative-treated wood is durable, inexpensive, and readily available. That is why it is generally used for structural members. It can also be used for decking itself. However, some treatment chemicals are more corrosive than others. This is a factor that must be considered when specifying metal fasteners, connectors, and other metal hardware that will be in contact with the wood.

Preservatives Prior to 2004, the most common preservative for lumber was CCA (chromated copper arsenate). However, studies indicated that the arsenic in CCA-treated wood could leach out and contaminate

An Attached Deck

Safety Is Important An attached deck is supported by posts and anchored to the house. A strong and well-designed railing makes a deck safe to use.

©iStockphoto.com/wests

soil and ground water. Beginning in 2004, CCA-treated wood was no longer allowed for use in residential construction.

Other non-arsenic preservatives have replaced CCA in the wood treatment process. One that is used extensively is **ACQ (Alkaline Copper Quaternary)**. Wood that is treated with ACQ does not contain arsenic. It is sometimes brownish in color. Many different softwoods can be treated with ACQ, but Southern yellow pine and Western hemlock-fir are common. Preservative-treated lumber is graded and stamped to indicate its suitability for various uses.

Three levels of preservative treatment can be applied. Treatment levels are specified by the number of pounds of treatment chemical used per cubic foot (pcf) of wood.

- *Above Ground* Wood in this category is treated at 0.25 pcf. This is the lightest treatment level. It is used for decking lumber and other wood that will be exposed to the weather but not to soil contact.

- *General Use* This is a utility grade that can be in moderate ground contact. The wood is treated at 0.40 pcf.

- *Ground Contact* The highest standard treatment level is 0.40 pcf. It is used for wood that will be in constant contact with the ground and where maximum durability is required.

Other treatment chemicals are available in addition to ACQ. They include CA (copper azole), micronized-copper (MCQ), and various trademarked formulations. Each treatment chemical imparts specific characteristics to the wood.

Softwood Decking

The most popular types of solid lumber decking are made from softwood lumber. Three different softwoods are commonly used.

Redwood This wood is highly resistant to decay and insect attack. However, it is fairly expensive and not readily available in all parts of the country.

Cedar Several species of cedar are resistant to decay and insect attack. Western red cedar is the species used most often for decking. It is medium-priced and readily available throughout the country.

Preservative-Treated Wood When preservatives are forced into softwoods such as Southern yellow pine, the wood becomes very resistant to decay and insect attack.

Sizes Softwood decking lumber is specified by its nominal size. The most commonly available sizes are 2×4 and 2×5 (actual thickness 1⅝"). Wider stock is more likely to cup as it weathers. In some areas, preservative-treated decking lumber is also available in an actual thickness of 1¼". Decking lumber is readily available in lengths of 8', 10', 12', and 14'. Softwood joist lumber is readily available in sizes from 2×5 through 2×12 and in lengths up to 14'.

Reading Check

Explain *What is the difference between a freestanding deck and an attached deck?*

JOB SAFETY

WORKING WITH WOOD PRESERVATIVES Any wood preservative contains chemicals that can be harmful to workers. Always wear work gloves when handling preservative-treated lumber and wear a dust mask when cutting it. To prevent chemicals in the sawdust from contaminating the soil, cut preservative-treated wood over a tarp. Dispose of the collected sawdust as directed by local regulations. Avoid sanding preservative-treated wood. Always wash your hands thoroughly with soap after working with it, particularly before eating.

Grades Grading policies for exterior lumber are not uniform. Redwood lumber, for example, is available in over 30 different grades.

Cedar lumber is graded with a different system and preservative-treated wood with yet another system. In general, however, exterior softwood lumber is graded according to certain characteristics.

Appearance This describes the size, type, and number of knots permitted in a board. Other surface flaws may be identified as well. Wood of the highest appearance grade is completely free from knots and is sometimes referred to as *clear*. This grade is often used for deck railings and skirt boards and sometimes for the decking itself.

Strength This describes the lumber's ability to support loads. Higher strength grades are important for joists and beams.

Moisture Content Deck lumber is often kiln-dried to a moisture content of either 19 or 15 percent. This reduces the tendency of the wood to shrink after installation. Some preservative-treated lumber is kiln-dried twice: once before and once after treatment. This grade is stamped KDAT (Kiln-Dried After Treatment).

Decay Resistance The most decay-resistant portion of a tree is called the heartwood. **Heartwood** is the portion of a tree nearest the core. It is dark in color. The least decay-resistant part of the tree is the sapwood. **Sapwood** is the outer growth layer. It is lighter in color than heartwood. When maximum decay resistance is required, grades containing larger proportions of heartwood should be used. Because heartwood and sapwood differ in color, decay resistance also has some bearing on appearance grading. The highest grades are the most uniform in color. For more on the decay resistance of woods, see Chapter 12.

Hardwood Decking

Many tropical hardwoods, such as mahogany, teak, and Ipe, are strong and highly resistant to decay and insect attack. The trees grow in Central and South America but are now readily available in North America.

Most hardwood decking ranges from ¾" to 1¼" thick. No special tools are required when cutting these woods. However, holes for nails and other fasteners must be predrilled because the wood is so dense. Also, some tropical hardwoods are not easy to finish because of their density.

Other Decking Materials

Many synthetic decking products are now available, and more are introduced each year. These products usually require little maintenance. However, they can be used only for decking, not for the structural portions of a deck. Check local codes before specifying synthetic decking. There are two basic types of synthetic decking products.

Plastic Decking These products are made entirely of plastic. Products are shaped into boardlike planks that are hollow or partially hollow. Chemical additives in the plastic improve its durability outdoors.

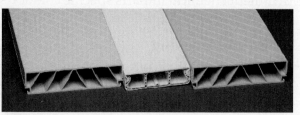

 Plastic Decking
Special Fastening Plastic decking consists of hollow extrusions. Each type uses a fastening system supplied by the manufacturer.

Arnold & Brown

This is important to help the material resist the harmful effects of UV radiation. Plastic cannot be nailed, so various concealed fastening systems are supplied by the manufacturers. These secure the decking to wood joists. Decking ranges from about 4" to 8" in width.

Composite Decking This material is a blend of recycled plastic and wood dust or fibers. It is denser and heavier than solid wood but usually not as stiff. Boards should be cut with carbide-tipped saw blades. Sizes include $\frac{5}{4} \times 5$, 2×4, 2×5, and 2×8 planks. Some composite boards are solid. Solid products can be attached to wood joists with nails or screws. However, some types of screws cause the surface of composite boards to deform slightly around the screw head. The deformation is sometimes called a *volcano*. Always follow the manufacturer's recommendations when choosing fasteners. Special composite lumber screws are available.

Build It Green Extensive cutting of tropical hardwoods can cause environmental damage. To discourage this, programs have been developed to certify that tropical hardwoods have been harvested responsibly.

Composite Decking
Solid or Hollow Composite decking is available in various dimensions, shapes, and colors. Skirt boards or special end caps are used to conceal the open ends of hollow composite decking.

Certification means that the wood comes only from well-managed forests or plantations that adopt sustainable and ecologically sound forestry practices.

Reading Check

Recall *Which grading characteristic should be of greatest concern when choosing wood for joists and beams?*

Hardware
What is the most common fastener used in deck construction?

The parts of a deck are fastened together with nails, screws, lag bolts, through bolts, and structural metal connectors. Structural connections should be made in a way that maintains the strength of the connections over time. Any joint that tends to trap moisture against the wood should be avoided.

Types of Fasteners
The most common fastener used to assemble a wood deck is the nail, but screws and bolts are also commonly used. See examples on page 1006. Connections between posts and beams or joists and rim joists are often made with metal connectors, such as brackets or joist hangers. These provide a stronger connection than nails or screws alone. They are often required by code in areas exposed to earthquakes or severe weather. For more on metal framing connectors, see Section 14.3.

Hidden Fasteners The traditional method for attaching decking to joists is face nailing. This method of attachment is easy and quick, but many people do not like the appearance of exposed nail heads. Another problem is that the nails create paths that allow water to soak into the wood. Manufacturers have developed a wider range of specialty fasteners that avoid these problems. They are generally called *hidden deck fasteners*, or sometimes *blind-nailing systems*.

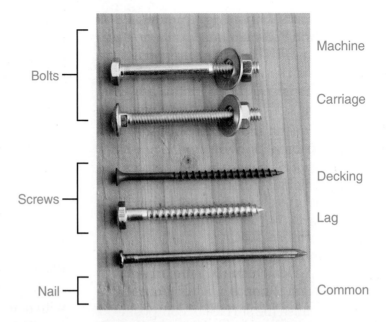

Bolts — Machine

Carriage

Screws — Decking

Lag

Nail — Common

Some hidden deck fasteners are installed as continuous metal strips that are screwed to the underside of the decking. However, most are attached to the top edge of the joists and to the edge of the deck boards.

Corrosion Resistance

Some wood treatment chemicals are highly corrosive to metal. In addition, the risk of environmental corrosion is high in many regions. For example, the salty air along the Pacific and Atlantic coasts can be very corrosive. To prevent deck fasteners and connectors from failing, always use corrosion-resistant products, especially when using preservative-treated wood.

Galvanized Steel The least expensive and most common type of corrosion resistance is provided by galvanizing. **Galvanizing** is the process of coating the steel with a protective layer of zinc. The thicker the coating, the better the protection. There are two methods for applying zinc to steel.

Electroplating This type of galvanizing coats the steel with a thin, smooth, and very uniform layer of zinc. It is available on all metal products, including metal connectors. Standard galvanized deck hardware is electroplated with 0.90 ounce of zinc for every square foot of metal surface area. This may be referred to as G90 Zinc. This level is weather resistant but

A

B

(t)McGraw-Hill Education; (bl & br)Arnold & Brown

Hidden Decking Fasteners
Strong But Invisible A hidden decking fastener can be made of plastic or metal. Some types (**A**) are fastened to the underside of decking. Others (**B**) are fastened to the edge of each deck board.

Builder's Tip

Science: Chemical Reactions

Salt and Corrosion Corrosion is an electrochemical process. Fill two clear glasses with water. Stir one tablespoon of salt into one glass. Label the glasses "unsalted" and "salted" with pieces of tape. Put a piece of aluminum foil in each glass and leave the glasses alone for one day. Record your results.

Starting Hint Be sure to not get any salt into the unsalted glass with your fingers.

not suitable for use with treated wood. G185 galvanizing provides 1.85 ounce of zinc per square foot of steel. It is the minimum recommended for use with treated lumber.

Hot-Dip Galvanizing Hot-dip galvanizing produces a slightly irregular layer of zinc that is thicker than an electroplated finish. It offers greater protection but is more expensive than electroplating. However, it is not used on through bolts because the thicker coating clogs their threads.

Stainless Steel Hardware made of stainless steel does not require galvanizing. Stainless steel is used where maximum corrosion resistance is necessary. It is ideal for use in coastal areas, particularly where salt spray is a factor. Stainless steel hardware is available in many forms, including through bolts and framing connectors. However, stainless steel is much more expensive than galvanized steel. Stainless steel structural connectors should always be fastened using stainless steel nails or screws.

Section 35.1 Assessment

After You Read: Self-Check

1. What supports an attached deck?
2. Name the three basic types of softwood lumber used for decking.
3. What preservative has been phased out of use for residential construction?
4. What is galvanizing?

Academic Integration: Science

5. **Sustainability and Certification** Certain programs have been developed to certify sustainable building materials. For example, one certification program might certify that only a certain percentage of trees in an area may have been harvested within a specific time period. What does it mean to say that a resource is *sustainable*? Brainstorm examples of how to increase the sustainability of wood as a building material.

Go to **connectED.mcgraw-hill.com** to check your answers.

Planning & Construction

Codes & Layout

Why are there various sources for span information?

Decking usually runs parallel to the house. However, this is primarily for installation convenience and appearance. It is not a requirement. Decking is supported by joists and usually runs perpendicular to them.

Building and Zoning Requirements

Decks are governed by local building codes as well as zoning restrictions. Building codes are concerned with such details as:

- The span of beams, joists, and decking
- The diameter and depth of foundation piers
- The design of railings and steps
- The deck's connection to the house.

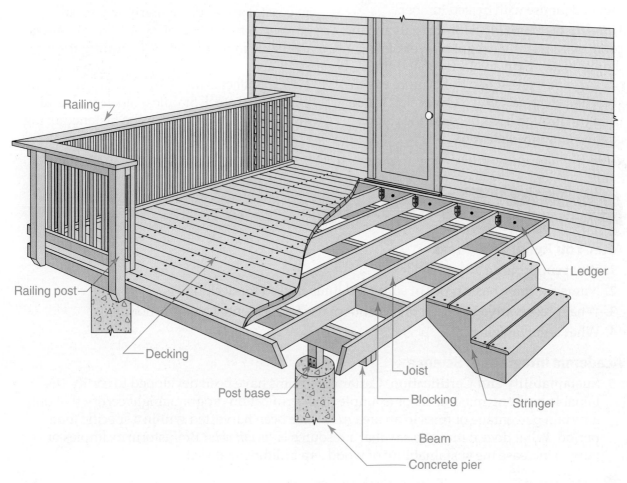

A Basic Deck

Outdoor Floor An attached deck is supported on at least one side by the house. Joists, beams, and concrete piers complete the support system.

Construction details for the deck should be included in the construction drawings for the house. The location of the deck would show up in a plan view. Assembly details would be included on a detail sheet. Check these details carefully, particularly in regions of high winds or frequent seismic activity. Deck construction in these areas calls for additional measures to prevent deck collapse.

Zoning ordinances affect deck location. These restrictions differ from community to community. However, they specify the minimum allowable distance between the deck and such features as streets, lot lines, septic systems, wells, and utility easements. They may also limit the height of a deck.

Deck Planning Unlike a house, a deck is typically planned from the top down. Once the overall shape and size of the deck has been designed, each deck material chosen determines the size and spacing of the supporting materials. For example, the type and dimension of deck boards determines their maximum span. Their span determines how closely spaced the joists must be. The maximum span of the joists determines where beams must be located, and their location determines where the posts and piers must be positioned. Depending on the materials chosen, span information may come from various sources, including the following:

- code books
- trade associations, such as the California Redwood Association or the Southern Forest Products Association
- manufacturers of specific decking products.

The process of building the deck proceeds from the bottom up. Once all the elements of the deck's structure have been double-checked to ensure that they meet or exceed building codes, construction begins with the piers.

Reading Check

Summarize *Why do most decks run parallel to the house?*

Locating Piers

Unlike the house, a deck does not require a continuous concrete foundation. Instead, it rests on piers. A **pier** is a concrete column that supports a concentrated load, such as a post. It is a type of foundation. Piers are typically cylindrical due to the method in which they are formed. However, the shape of the pier is less important than its ability to support loads. The bottom of a pier is generally wider than the rest of the pier in order to distribute loads to the soil. In this sense it serves as a footing and should be approximately 6" below the frostline. The frostline is determined by local climate and its depth can be found in local building codes.

The approximate location of piers is determined by the shape of the deck as well as the location of beams and posts. It is specified on building plans. However, it is up to the deck builder to determine the precise location of each pier based on measurements taken at the site.

An exact location for each pier can be determined by using the 3-4-5 method to ensure a right-angle layout in relation to the house (see Chapter 9). When the centerlines of the outermost piers have been determined, string lines can then be used to locate the centerlines of other piers as shown on page 1010.

Once the string lines are in place, the exact location of individual piers can be determined. This requires the use of a tape measure and a plumb bob as shown on page 1010. A small stake or marker is placed at the center point of the pier location to identify it for excavation.

Deck Construction
What factors might affect the height chosen for a pier?

When the basic layout of the deck is complete, string lines can be removed and excavation can begin. The tools used to excavate holes for piers will depend on the depth and diameter of the pier, as well as

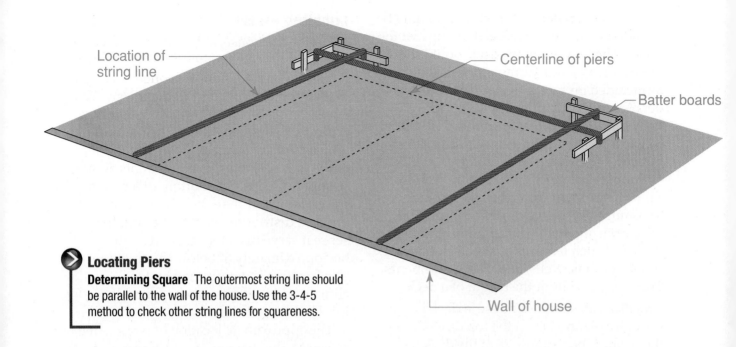

Location of string line

Centerline of piers

Batter boards

Wall of house

 Locating Piers
Determining Square The outermost string line should be parallel to the wall of the house. Use the 3-4-5 method to check other string lines for squareness.

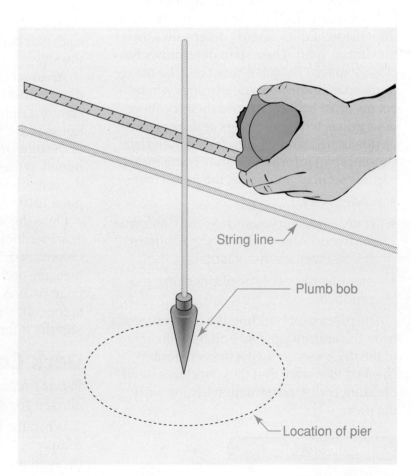

String line

Plumb bob

Location of pier

 Positioning Piers
Exact Position After string lines have been set up, use a tape measure and a plumb bob to locate the center point of individual piers.

how many are required. Hand-digging with shovels and post-hole diggers is sufficient where piers are shallow or only a few are necessary. Power equipment is generally preferred in other cases. Augers of various diameters are available.

Installing Piers

Forms for piers can be made of removeable steel or fiberglass sections. However, the most common method is to use inexpensive cylindrical single-use forms. These products are made of multiple layers of recycled paperboard laminated with adhesive. The interior surface is very smooth, and the outside surface has a moisture-resistant coating. It comes in various lengths and diameters that range from 6" to 60". Forms are typically 12' long but can easily be cut to shorter lengths.

Once a form has been cut to length and placed in an excavated hole, it must be supported by temporary support braces at the desired height. The braces also keep the form steady. Once the form has been plumbed, it can be backfilled just enough to hold it in position. Immediately after the concrete has been placed, a metal post anchor should be embedded in the top. Check local codes to see if reinforcing bar must also be inserted.

Some form manufacturers provide online volume calculators to help builders determine how much concrete to order for a given diameter and length of pier. Another way to calculate concrete volume for a cylindrical form is to consult **Table 35-1** on page 1012.

Posts

Posts should be made of solid lumber graded for structural use. Common dimensions are 4×4, 4×5, and 6×5. Most often posts are made from preservative-treated lumber, but redwood is common in some areas of the country. The bottom of a post must be secured by a metal post anchor embedded in a pier. Once a post has been plumbed, it can be secured to a beam in various ways. In some cases, it can be sandwiched between a pair of joists that serve as a beam. In this case, holes

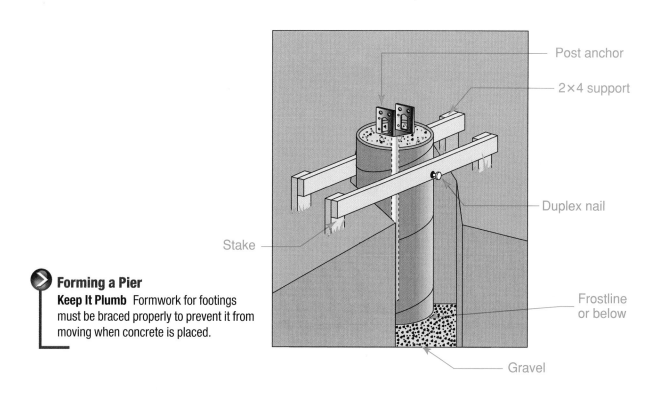

Post anchor

2×4 support

Duplex nail

Frostline or below

Gravel

Stake

> **Forming a Pier**
> **Keep It Plumb** Formwork for footings must be braced properly to prevent it from moving when concrete is placed.

Table 35-1: Estimating Concrete for Cylindrical Forms	
Form Diameter (inches)	Concrete Required per Lineal Foot (cubic yards)
6	.0073
8	.0129
10	.0202
12	.0291
14	.0396
16	.0617
18	.0654

Builder's Tip

STRIPPING FORMS Most portions of a single-use form are biodegradable and eventually disintegrate. Portions above grade may be stripped off after the concrete has cured for at least 24 hours, though most builders wait longer. However, do not wait longer than five days to strip above-grade portions of the forms.

for through bolts would be drilled through each joist and the top of the post. Another method to secure posts is to use a metal post connector.

Beams

Beams and girders may be solid wood or a built-up assembly of 2× lumber. Solid beams come in limited lengths. Nominal beam depths of 6" and 8" are common. Built-up beams are made of two or three layers of

2×8, 2×10, or other dimension lumber. The layers should be spiked or bolted together. Built-up beams are sometimes preferred because they are assembled in place. This makes them easier to position. Another advantage is that a built-up beam can be any length. One disadvantage of a built-up beam is that water can be trapped between the pieces of stock. This can cause rot. To eliminate this problem, two-layer beams are often assembled with an airspace between the pieces. This can be done by inserting

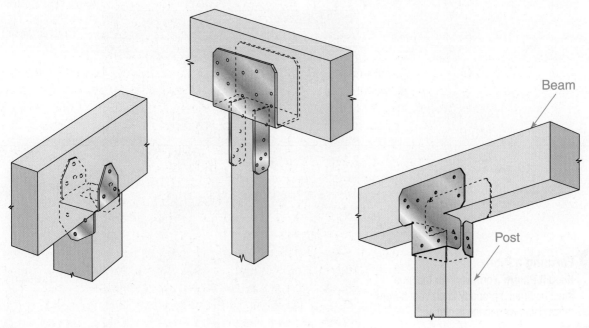

Beam

Post

 Post Connectors
Strong Connection Post caps come in various shapes. Always use the number and type of fasteners recommended by the post cap manufacturer.

Plumbing a Post An out-of-plumb post puts unnecessary stresses on the deck structure. Plumb posts with a 3' level or by using the following method:

Step 1 Tack a small wood block to the side of the post at its top. Then place a plumb line over the block so that it hangs alongside the post.

Step 2 At the bottom of the post, measure the distance from the post to the line. If the distance is not the same as the thickness of the block, the post is not plumb.

Step 3 Tilt the post as needed until the distance from the post to the line is exactly the same along the entire length of the line.

Step 4 Nail a temporary brace to the post to hold it in position. Repeat Steps 1, 2, and 3 on an adjacent face of the post. When this face is plumb, secure a second brace.

Step 5 Double-check the post on two adjacent faces. If they are both plumb, the post is plumb.

---

treated-wood spacers or stacked washers between the pieces as they are nailed or bolted together.

Whatever the type of beam, it should be straight and made of structural-grade wood. If the beam will be visible when the deck is complete, its appearance should also be considered. The cut ends of solid or built-up beams should be coated with a water repellent to increase their durability.

Reading Check

Recall How are the tools chosen for excavating holes for piers?

The Ledger

One of the most important but least understood parts of a deck's substructure is the ledger. The **ledger** is the length of lumber that connects the deck to the house, as shown on page 1008.

Proper installation of the ledger is critical. If not attached properly, the ledger can rip away from the house when stressed. This can cause the entire deck to collapse. Never secure a ledger to a house with nails alone. Nails are not strong enough to prevent the ledger from pulling away. Instead, use lag bolts or through bolts connected to studs, plates, or rim joists, as shown on page 1014. Do not rely on any connections made to the sheathing alone. Such connections do not provide sufficient strength. Metal flashing prevents water from rotting the siding, the sheathing, or the structural framing of the house.

A number of decks have failed in recent years due to inadequate ledger connections, and in some cases people on the deck were killed in the collapse. That is why a great deal of research is being devoted to this subject, including full-scale laboratory testing of ledger connections. Specific recommendations for ledger installation may become part of the International Residential Code in the near future.

To avoid problems with ledger connections, some builders avoid ledgers entirely. Instead of connecting the deck to the house, they install an extra row of concrete piers along the house wall to support the deck independently of the house. A slight gap between the house siding and the deck allows water to drain between them.

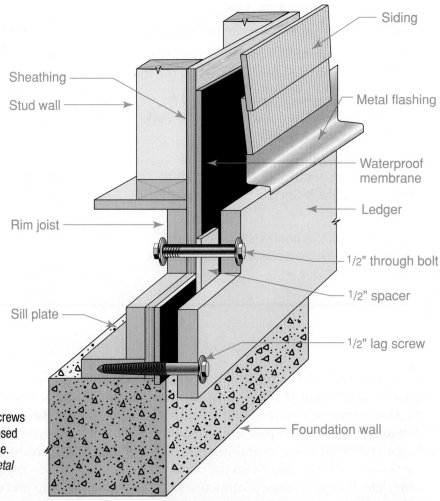

Siding

Sheathing

Stud wall

Metal flashing

Waterproof
membrane

Ledger

Rim joist

1/2" through bolt

1/2" spacer

1/2" lag screw

Sill plate

Foundation wall

Joists

Deck joists are laid out and installed much
like floor joists (see Chapter 15). They are
usually spaced 16" OC but may also be 12", 20",
or 24" OC. When synthetic decking is used,
always consult the manufacturer's instructions
for joist requirements. Some types of synthetic
decking are not as stiff as wood, and thus joists
may have to be closer together.

Joists are generally connected to other structural elements using metal joist hangers, brackets, or by toe-nailing joists to support beams.
Continuous solid blocking is often required
between joists that are more than 8" in depth.

Decking

Softwood decking with a nominal 2" thickness is often surface-nailed with 10d galvanized or stainless steel nails. Decking that is
1¼" thick is nailed with 8d nails. Use one nail

at every joist connection when installing 2×4
decking. Use two nails when installing 2×5
decking.

Galvanized or stainless steel screws make
a stronger connection than nails. Screws
made especially for attaching decking have a
slender shaft, a sharp self-drilling point, and
a fairly small head that sinks flush with the
surface of the decking. They should be long
enough to penetrate at least 1" into the joists.
When installing synthetic decking, follow
manufacturer's instructions carefully for
using nails or screws.

Spacing Gaps between deck boards ensure
that water will drain freely. Generally this
space is about ⅛". If hardwood, composite,
or kiln-dried decking is used, gaps must be
created as the boards are installed. However,
when installing preservative-treated lumber,

butt boards tightly during installation. As the boards shrink across their width, suitable gaps will eventually open up between them.

The thickness of a 12d or 16d nail can be used to gauge gaps between decking boards. Drive a nail through a small scrap of thin plywood to create a spacing jig. This prevents the nail from slipping through the gap as the board is positioned. Use several of these reusable devices to maintain a uniform gap thickness. Always follow the manufacturer's spacing recommendations when installing plastic or composite decking.

Stairs, Railings, & Porches

Why do stairs for decks require less finishing work than interior stairs?

All elevated decks require stairs and railings. Because these elements play a large role in the safety of a deck, local codes should be followed carefully.

Stairs

Most decks will require at least a few steps down to grade level. In the case of low decks such as the one on page 1016, the steps can be constructed simply. Instead of supporting the treads on stringers, as in standard stair construction, they are often supported by a box-like assembly made of framing lumber. This is called *platform stair construction.*

Elevated decks require steps as well as stair railings. The steps are supported by stringers made of preservative-treated lumber or a naturally decay-resistant wood such as redwood. These stringers are laid out just as those for interior stairs (see Chapter 25). However, the degree of finish work is not as great. This is because stairs to a deck are exposed to the weather.

Exterior stairs may be of the cut-stringer type or the cleat-stringer type. Both types are discussed in Chapter 25. Stringers can be cut on site, but pre-cut pressure treated stringers are also available. This ensures that treatment chemicals will protect the cut edges

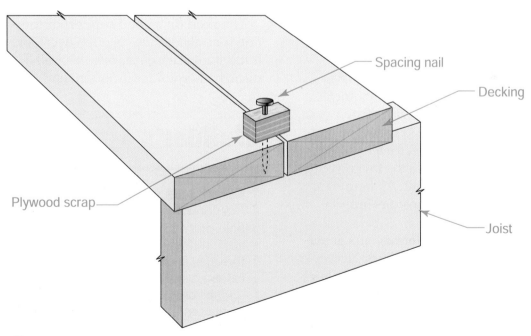

Spacing nail

Decking

Plywood scrap

Joist

Spacer for Decking
Inexpensive Spacer Use this site-made jig to ensure consistent spacing between deck boards.

 Platform Steps
Simple Construction A low-level deck sometimes incorporates steps that form a border around the deck. This deck features brick and columns.

of the stringer. Stringers are often attached to adjacent structures using metal framing connectors.

An exterior stair often does not have risers. This allows water to drain quickly and snow to be removed with relative ease. To aid drainage, each tread can slope up to two percent. For example, for a tread with a depth of 12½", the back of the tread would be ¼" higher than the front. All hardware used to assemble exterior stairs must be weather resistant.

The basic layout of exterior stairs generally follows that of interior stairs. However, building codes applicable to exterior stairs differ in some respects and should be checked. For example, there may be more flexibility in the dimensions of treads and risers when laying out porch steps or the approaches to low-level decks.

The need for a good support or foundation for outside steps is often overlooked. If the steps are located over backfill or disturbed ground, the bottom end of the

Builder's Tip

MAINTAINING CONNECTIONS Wood exterior stairs are often assembled using lag bolts or through bolts. As the wood weathers, it shrinks somewhat. This can cause threaded fasteners to loosen over time. Fasteners should be tightened securely during construction. They should then be tightened again later. Tightening such bolts should be a part of the homeowner's annual maintenance for an exterior stair.

stringers should rest on concrete piers or on a small concrete support slab. In any case, stair stringers should never rest directly on the ground because this will encourage rot. Such placement would also allow the stair to move up and down slightly during weather cycles of freezing and thawing.

Railings

The primary purpose of a railing is safety. It should be installed on both sides of a stairway, as well as around elevated decks. Local codes determine the minimum height of the railing and the spacing of balusters. However, height generally ranges from 32" to 38" above the decking.

A typical railing consists of a handrail, or cap, a series of support posts, and balusters. The railing posts provide strength for the system, so they must be solidly screwed or bolted to the deck structure. Though some builders notch the bottom of posts to fit over the top of the decking, this weakens the post and should generally be avoided.

Reading Check

Recall *When is platform stair construction used?*

Porches

A porch is a roofed structure that is attached to a house. It is often open on the sides or front. To aid drainage, the flooring is often installed like decking, with a gap between each board. If the porch flooring is made of tongue-and-groove boards, the surface must be sloped slightly away from the house to aid drainage. Be sure to use a rot-resistant wood.

A porch often serves as the main entry to a house. Construction often involves wall framing, roof framing, roofing, and concrete slab methods described elsewhere in this book.

Full-Length Entry Porch
Double Duty This porch fits the architectural style of the house. It serves as a protected main entry. It is also large enough to use as a sitting area.

Denise McCullough

Consider the following when building a porch:

- Porches supported on continuous foundation walls should have a clearance of at least 8" between the exterior finish grade and the nearest wood. Floor joists and beams should have a clearance of 18" or more from the bottom of the joists to the grade, unless preservative-treated lumber is used.

- Porch columns should be designed to avoid any details or joints that might trap water. Treated structural posts are often cased with untreated finish lumber for better appearance.

- It is important to protect the end grain of finished trim wood at joints, because this area absorbs water easily and is prone to rotting. The ends of porch flooring should be brushed, dipped, or soaked in a water-repellent preservative.

Concrete Steps and Stoops

Many porches, particularly those that serve as the main entrance, feature concrete steps and a stoop. A **stoop** is an enlarged landing at the top of the steps. Concrete is a durable, low-maintenance material that is ideal for this use. Horizontal concrete surfaces should be sloped to promote runoff of water.

Many builders prefer to install precast steps and stoops. **Precast** refers to any concrete object that is cast in a factory, cured under controlled conditions, and then delivered to the job site. They are lifted into place by a small crane mounted on the delivery truck. Precast units are hollow to reduce their weight. They are available in various sizes but are usually 48" wide. Precast units rest on footings or piers.

Porch steps may instead be cast in place. This work is done by masons or carpenters who build the formwork for risers and treads on site. After the concrete is placed

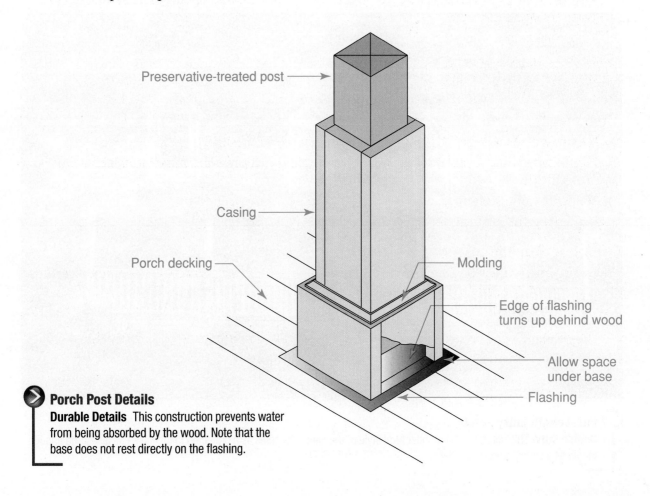

Preservative-treated post

Casing

Porch decking

Molding

Edge of flashing turns up behind wood

Allow space under base

Flashing

Porch Post Details
Durable Details This construction prevents water from being absorbed by the wood. Note that the base does not rest directly on the flashing.

and has partially cured, the forms are removed. Formwork may be made of lumber, but reusable forms called *edge forms* are typically made of metal. Wood formwork calls for a high degree of craftsmanship to ensure that the forms are strong and properly designed. For example, each tread and the stoop must be level side-to-side but

sloped forward slightly to encourage drainage. Risers are sometimes slanted inward toward the next lower tread. Like the nosing of a wood step (see Chapter 25), this provides clearance for using the step.

Precast Steps
Ready to Use Precast step units are hollow to reduce weight. However, they are still heavy enough to require placement by crane.

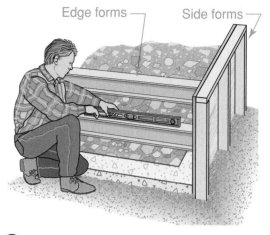

Forming Concrete Steps
Careful Craftsmanship The risers are leveled from side to side and positioned so the back of each tread is ¼" higher than the front.

Section 35.2 Assessment

After You Read: Self-Check

1. Name two advantages that built-up beams have as compared to solid beams.
2. What step should be taken with the cut ends of built-up or solid beams to increase their durability?
3. When installing a ledger, what factors should be kept in mind?
4. How are porches designed to allow for proper drainage?

Academic Integration: Mathematics

5. **Estimating Concrete** Use the formula for the volume of a cylinder to estimate the amount of concrete needed for 6 piers that are 12" in diameter and 4 ft. long. Round your answer up to the nearest ¼ cubic yard. Then show how to check your work using Table 35-1.

   Math Concept The volume of a cylinder is found by first finding the area of the circular opening, then multiplying by the length of the cylinder. $V = \pi r^2 h$.

   Step 1: Find the volume of one pier. Use 3.14 for π.

   Step 2: Multiply the result by 6 piers. Express your answer in cubic yards.

   Step 3: Round up to the nearest ¼ cubic yard. Check your work using Table 35-1.

   Go to **connectED.mcgraw-hill.com** to check your answers.

Review and Assessment

Chapter Summary

Section 35.1
In addition to traditional softwoods, decking materials include hardwoods, plastic decking, and composite decking. Decking materials should be weather and decay resistant. The hardware used to install decking must be corrosion resistant.

Section 35.2
Construction of a deck calls for several layers of structural support. All construction details must be installed to minimize the decay caused by trapped water. A ledger, the length of lumber that connects the deck to the house, is a very important element of an attached deck. It must be installed with great care. Porches are built with framing, roofing, and concrete-slab work that are similar to parts of a house.

Review Content Vocabulary and Academic Vocabulary

1. Use each of these content vocabulary and academic vocabulary words in a sentence or diagram.

Content Vocabulary

- deck (p. 1002)
- ACQ (Alkaline Copper Quaternary) (p. 1003)
- heartwood (p. 1004)
- sapwood (p. 1004)

- galvanizing (p. 1006)
- pier (p. 1009)
- ledger (p. 1013)
- stoop (p. 1018)
- precast (p. 1018)

Academic Vocabulary

- exposed (p. 1002)
- maintains (p. 1005)

Speak Like a Pro

Technical Terms

2. Work with a classmate to define the following terms used in the chapter: *decking* (p. 1002), *freestanding deck* (p. 1002), *attached deck* (p. 1002), *volcano* (p. 1005), *platform stair construction* (p. 1015), *edge forms* (p. 1019).

Review Key Concepts

3. Identify the basic types of materials used for decking.

4. Recognize the basic elements of a deck.

5. Demonstrate how to lay out piers.

6. Show the proper method to plumb a post.

7. Tell how to handle and cut preservative-treated wood safely.

8. Summarize the process of installing concrete porch steps.

Critical Thinking

9. Explain Why are porch steps sloped when they are cast in place?

Academic and Workplace Applications

STEM Mathematics

10. Planning a Deck How many support posts are needed for a 12 ft. × 15 ft. deck, attached to the house along its 12 ft. length, if the posts are laid 3 ft. OC?

Math Concept Use the problem solving strategy "Draw a Picture" when solving problems involving the use of space.

Step 1: Draw a rough plan of the deck. Place a post at each unattached corner.

Step 2: Divide the length of each unattached side by the desired spacing between the posts.

Step 3: Subtract 1 from each side because the corner post is already placed.

Step 4: Add the number of side posts and corner posts.

Step 5: Check your work by completing your drawing showing the location and spacing between the support posts.

STEM Science

11. Corrosion An ion is an atom or molecule which has lost or gained one or more electrons, making it positively or negatively charged. When a solution contains free ions, it is called an electrolyte. When two metals come into contact in the presence of an electrolyte, corrosion occurs. Corrosion of metal fasteners used to hold a deck together was less of a problem with CCA, the pressure treated lumber no longer used in residential construction because of its arsenic content, than with newer preserved lumber such as ACQ.

Research what happens to galvanized nails in exterior wood during the corrosion process. Describe your findings.

Starting Hint Look for information about rust and how it occurs.

21st Century Skills

12. Career Skills: Public Speaking As the newest member of a safety committee, your first assignment will be to give a five-minute talk to a group of carpenters about the dangers of working with wood preservatives. The main objective of your talk will be to highlight safety guidelines for working with wood preservatives. To prepare for your talk, write an outline on what you will say to the group of carpenters. Share your talk with the class.

Standardized TEST Practice

True/False

Directions Read each of the following statements carefully. Mark each statement as either true or false by filling in T or F.

Ⓣ Ⓕ **13.** Posts are rarely made from preservative-treated wood.

Ⓣ Ⓕ **14.** Materials used for the structural elements beneath the decking are chosen primarily for their strength and durability.

Ⓣ Ⓕ **15.** One problem with precast steps is that they are only available in one size.

TEST-TAKING TIP

Analyze multiple-choice questions carefully. Note key terms. Use your knowledge and anticipate what the answer should be. Find an answer choice that looks like the one you predict.

*These questions will help you practice for national certification assessment.

UNIT 7
Hands-On Math Project

Professional Green Painting

Your Project Assignment

You will spruce up a room at your home or school with new paint this weekend. You will choose a paint and estimate how long the job will take.

- **Measure** the square footage of the walls and ceiling of the room you will paint.

- *Build It Green* **Research** paints certified by Green Seal for quality and safety.

- **Select** a paint and calculate how many gallons you will need based on the room's size and the paint's spread rate.

- **Estimate** the time needed for the job based on the manufacturer's recommendations and the forecast temperature and humidity.

- **Create** a three- to five-minute presentation.

Applied Skills

Some skills you might use include:

- **Compare** the features and advantages of various brands of paint.

- **Explain** why temperature and humidity are important factors in planning a paint job.

- **Determine** the square footage of walls and ceilings minus doors, windows, and trim.

- **Calculate** the time required to complete a multi-step project.

- **Describe** the steps in making a time estimate.

The Math Behind the Project

The traditional math skills for this project are geometry and algebra. Remember these key concepts:

Square Footage

Estimating a paint job requires calculating the number of square feet to be painted. First calculate the square footage of the walls and ceiling. For walls,

multiply length by height. For the ceiling, multiply length by width. Then measure the square footage of the areas you will not paint or that will require a different paint, such as doors, windows, and trim. Subtract these areas from the total. To calculate how many gallons of paint you will need, divide the square footage to be painted by the paint's spread rate. For example, if you are painting 480 square feet and your chosen paint has a spread rate of 350 square feet per gallon, you will need 1.37 gallons. You will therefore need to buy two gallons.

Algebra

At 75°F and 50% percent relative humidity, flat latex paint usually dries to the touch in about one hour. Eggshell and satin latex paint usually dry in about two hours. High humidity and low temperature slow the drying process. Consult the weather forecast for the day you plan to paint. For this project, we will assume one added minute of drying time for every 1% of humidity above 50% and one added minute of drying time for every 1° of temperature below 75°F. You do not need any extra drying time if the temperature is above 75°F or the humidity is below 50%.

We can use algebra to express this relationship as an equation. Let m represent additional minutes of drying time, let t represent forecast temperature, and let h represent forecast humidity. The equation is: $m = (75 - t) + (h - 50)$. If the forecast indicates 52°F and 69% humidity, we would solve for m using the following steps:

1. Set up the equation with the stated values.	$m = (75 - 52) + (69 - 50)$
2. Complete the operations within parentheses.	$m = (33) + (19)$
3. Calculate the total.	$m = 52$

You will need to let the paint dry for 52 minutes (≈ 1 hour) beyond the standard drying time.

Green Seal

Mission: To safeguard the environment and transform the marketplace by promoting the manufacture, purchase, and use of environmental responsible products and services.

Project Steps

Step 1 Research

- Choose a room to paint. Create a sketch that represents the three dimensions of the room and indicates the position of windows and doors.
- Determine an appropriate color and finish (flat, eggshell, satin, semi-gloss, or gloss).
-  *Build It Green* Research interior latex paints certified by Green Seal. Compare at least three paint brands on color selection, spread rate, scrubbability, durability, toxicity, ease of application, and price. Choose one that meets your needs.
- Locate the forecast temperature and humidity for your planned painting day.

Step 2 Plan

- Select a paint brand, color, and finish.
- Determine how many coats of paint you will need to apply based on the paint and the surface.
- Measure the walls, ceiling, windows, and doors. Label your sketch with these measurements.
- Calculate the total square footage to be painted. Multiply the total by the number of coats required.
- Read the label to determine how long each coat of paint will need to dry.

Step 3 Apply

- Determine how many gallons of paint you will need by dividing the square footage by the spread rate of your chosen paint.
- Calculate the total price of the paint.
- Estimate how long it will take to apply a single coat.

- Multiply your time estimate by the number of coats required. Allow for the recommended drying time between each coat.
- Use the forecast temperature and humidity to calculate any necessary additional drying time.
- Create a total time estimate for the paint project.

Step 4 Present

Prepare a presentation combining your research and calculations using the checklist below.

PRESENTATION CHECKLIST
Did you remember to...
✓ Describe how you chose a paint brand, color, and finish?
✓ Determine the number of paint coats needed?
✓ Demonstrate how you calculated any extra drying time?
✓ Use a spreadsheet for your cost and time calculations?

Step 5 Evaluate

Assess yourself before and after your presentation.

1. Did you measure and calculate square footage?
2. Was your research thorough?
3. Was your paint choice suitable for the room?
4. Did you take temperature and humidity into account?
5. Was your presentation concise and easy to follow?

Go to **connectED.mcgraw-hill.com** for a Hands-On Math Project rubric.

Corbis/Superstock

Ready Reference Appendix

ARCHITECTURAL ABBREVIATIONS

This list identifies some of the most common abbreviations found on architectural drawings. No period is needed after an abbreviation unless it might be confused with a whole word. Note that some abbreviations, such as AC, can stand for more than one term. Some terms, such as *beam*, may have more than one acceptable abbreviation.

A

AB Anchor bolt
AC Air condition; alternating current
ADH Adhesive
AG Above grade; against the grain
AGGR Aggregate
AL Aluminum
ALLOW Allowance
ALT Alternate
AP Access panel
APPROX Approximate
ASPH Asphalt
AVG Average

B

B Bathroom; beam
BALC Balcony
BATT Batten
BD Board
BET Between
BF Board feet
BL Building line
BLDG Building
BLK Block
BLKG Blocking
BLR Boiler
BM Beam
BOT Bottom
BR Bedroom
BRG Bearing
BRK Brick
BS Both sides
BSMT Basement

C

CAB Cabinet
CAT Catalog
C CONC Cast concrete
CEM Cement
CER Ceramic
CHIM Chimney
CI Cast iron

CIR; CKT Circuit
CIR BKR Circuit breaker
CIRC Circumference
CKT; CIR Circuit
CL Centerline
CLG Ceiling
CLKG Caulking
CLO Closet
CLR Clear
CO Cleanout
COL Column
COMB Combination
COMP Component; composition
CONC Concentric, concrete
CONST Construction
CONT Continue
CONTR Contractor
CORR Corrugate
CS; X-SECT Cross section
CSG Casing
CSK Countersink
CTD Coated
CTR Center; counter
CW Cold water

D

D Dryer
DBL Double
DC Direct current
DEG Degree
DET Detail
DH Double-hung
DIAG Diagonal
DIM Dimension
DISP Disposal
DK Decking
DL Dead load
DMPR Damper
DN Down
DP Dampproofing
DR Dining room; door; drain
DS Downspout
DW Dishwasher

E

EA Each
ELEC Electric
ENAM Enamel
ENT Entrance
EQ Equal
EST Estimate
EXC Excavate
EXT Extension; exterior

F

FA Footing area
FAB Fabricate
FD Floor drain
FDN Foundation
FIN Finish
FIX Fixture
FL Flashing; floor
FL Flooring
FLUOR Fluorescent
FOS Face of studs
FPRF Fireproof
FR Frame
FS Full size
FTG Footing

G

G Gas; girder
GA Gauge
GALV Galvanize
GAR Garage
GB Glass block
GFCI Ground-fault circuit interrupter
GL Glass; grade line
GND Ground
GR Grade
GYP Gypsum

H

H Hall
HD Head
HDR Header
HDW Hardware
HOR Horizontal

ARCHITECTURAL ABBREVIATIONS, continued

HTR   Heater
HVAC   Heating, ventilating, air conditioning
HW   Hot water

I

I   I-beam; iron
IB   I-beam
ID   Inside diameter
INCL   Include

J

JST   Joist
JT   Joint

K

KIT   Kitchen
KD   Kiln-dried; knocked down
KO   Knockout
kW   Kilowatt

L

LAM   Laminate
LAU   Laundry
LAV   Lavatory
LBR   Lumber
L CL   Linen closet
LH   Left hand
LIN   Linear
LL   Live load
LOA   Length overall
LR   Living room
LT   Light
LTL   Lintel
LV   Louver

M

MATL   Material
MAX   Maximum
MECH   Mechanical
MEMB   Membrane
MET   Metal
MIN   Minimum
MIX   Mixture
MLDG   Molding
MN   Main
MOD   Model
MRTR   Mortar
MULT   Multiple

N

NAT   Natural
NO   Number
NOM   Nominal
NTS   Not to scale

O

OA   Overall
OC   On center
OD   Outside diameter
OPNG   Opening
OPP   Opposite
OR   Outside radius
OVHD   Overhead

P

PAR   Parallel
PC   Piece; pull chain
PERM   Permanent
PERP   Perpendicular
PL   Plaster; plate; property line
PLMB   Plumbing
PLYWD   Plywood
PNL   Panel
PRCST   Precast
PREFAB   Prefabricated
PRO   Property
PT   Part; pressure-treated
PTN   Partition

R

R   Radius, range; riser
RAD   Radiator
RD   Round
RECP   Receptacle
REF   Reference; refrigerator
REG   Register
REINF   Reinforce
REQD   Required
RET   Return
RF   Roof
RFG   Roofing
RH   Right hand
RM   Room
RO   Rough opening

S

SCH   Schedule
SDG   Siding
SECT   Section
SERV   Service
SEW   Sewer
SH   Sheet; shower
SHTHG   Sheathing
SIM   Similar

SP   Soil pipe
SPEC   Specification
SST   Stainless steel
ST   Stairs; steam; street
STD   Standard
STG   Storage
STK   Stock
STL   Steel
SUP   Supply
SUR   Surface
SYM   Symbol; symmetrical
SYS   System

T

T&G   Tar and gravel; tongue and groove
TC   Terra-cotta
TEMP   Temperature
TER   Terrazzo
THERMO   Thermostat
THRU   Through
TOL   Tolerance
TOT   Total
TR   Tread
TUB   Tubing
TYP   Typical

U

UNFIN   Unfinished

V

V   Vacuum, valve; volt
VAP PRF   Vapor-proof
VENT   Ventilate
VERT   Vertical
VP   Vent pipe
VS   Vent stack

W

W   Watt
W/   With
WC   Water closet
WD   Wood
WDW   Window
WH   Water heater; weep hole
WI   Wrought iron
WM   Washing machine
W/O   Without
WS   Weatherstripping

X, Y, Z

X-SECT; CS   Cross section

PLUMBING SYMBOLS

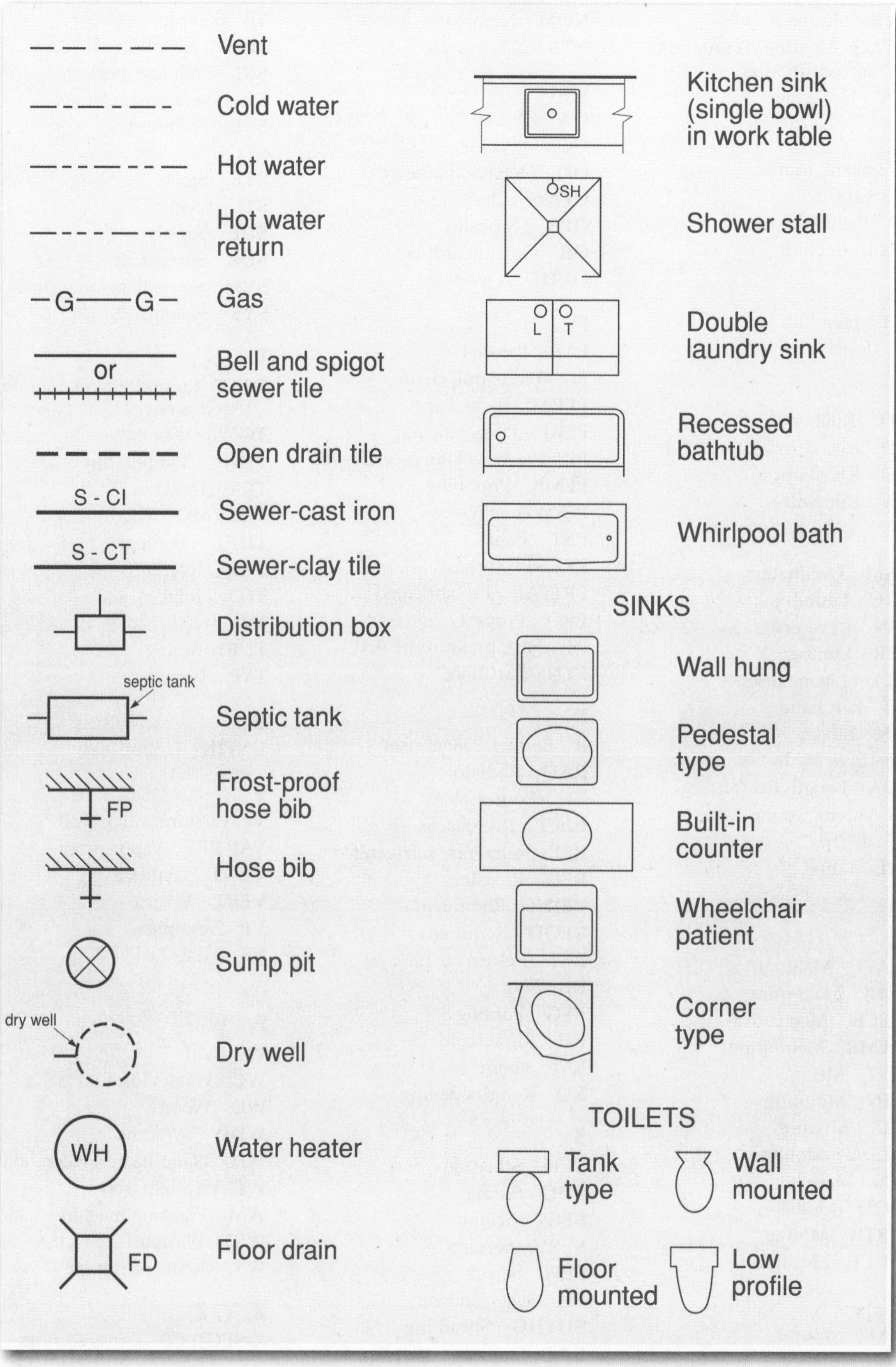

— — — — — Vent

— · — · — · — Cold water

— — — — — Hot water

— · · — · · — Hot water return

—G——G— Gas

or — Bell and spigot sewer tile

— — — — — Open drain tile

S - CI Sewer-cast iron

S - CT Sewer-clay tile

Distribution box

septic tank Septic tank

FP Frost-proof hose bib

Hose bib

Sump pit

dry well Dry well

WH Water heater

FD Floor drain

Kitchen sink (single bowl) in work table

SH Shower stall

L T Double laundry sink

Recessed bathtub

Whirlpool bath

SINKS

Wall hung

Pedestal type

Built-in counter

Wheelchair patient

Corner type

TOILETS

Tank type

Wall mounted

Floor mounted

Low profile

ELECTRICAL SYMBOLS

—○ Wall fixture outlet	⊖ Duplex convenience outlet	– – – – – Branch circuit; exposed
Ⓑ Blanked ceiling outlet	⊖₁,₃ Convenience outlet other than duplex 1=single, 3=triple, etc.	Home run to panel board; indicate number of circuits by number of arrows
—Ⓑ Blanked wall outlet	⊖wp Weatherproof convenience outlet	——— Feeders
Ⓓ Drop cord	⊖GR Grounded outlet	⊡ Push button
Ⓕ Ceiling fan outlet	⊜ Split wired outlet	▱ Buzzer
—Ⓕ Wall fan outlet	⊖R Range outlet	▱○ Bell
Ⓙ Ceiling junction box	⊖AC Air conditioner outlet	S⌐ Single-pole switch
—Ⓙ Wall junction box	⊖s Switch and convenience outlet	S₂ Double-pole switch
Ⓛ Ceiling lamp holder	⊖Ⓡ Radio and convenience outlet	S₃ Three-way switch
—Ⓛ Wall lamp holder		S₄ Four-way switch
Ⓛps Ceiling lamp holder with pull switch	◓ Special purpose outlet (design in specifications)	Sᴅ Automatic door switch
—Ⓛps Wall lamp holder with pull switch	⊙ Floor outlet	Sₚ Switch and pilot lamp
Ⓢ Ceiling pull switch	⊡ Floor single outlet	Sᴋ Key-operated switch
—Ⓢ Wall pull switch	⊟ Floor duplex outlet	Scв Circuit breaker
▭O Surface or drop individual fluorescent fixture	▬ Lighting panel	Swcв Weatherproof circuit breaker
▭OR Recessed individual fluorescent fixture	▨ Power panel	Sʀc Remote-control switch
▭O▭▭ Surface or drop continuous fluorescent fixture	——— Branch circuit; concealed in ceiling or wall	Swp Weatherproof switch
▭OR▭▭ Recessed continuous fluorescent fixture	– – – Branch circuit; concealed in floor	Sʟ Low-voltage switch
		Sᴛ Time switch

METRIC CONVERSION FACTORS

When you know:	You can find:	If you multiply by:
Length		
inches	millimeters	25.4
feet	centimeters	30.48
yards	meters	0.91
miles	kilometers	1.6
millimeters	inches	0.04
centimeters	inches	0.4
meters	yards	1.09
kilometers	miles	0.62
Area		
square inches	square centimeters	6.45
square feet	square meters	0.09
square yards	square meters	0.84
square miles	square kilometers	2.59
acres	hectares	0.4
square centimeters	square inches	0.16
square meters	square yards	1.2
square kilometers	square miles	0.4
hectares	acres	2.5
Mass		
ounces	grams	28.3
pounds	kilograms	0.45
short tons	metric tons	0.9
grams	ounces	0.04
kilograms	pounds	2.2
metric tons	short tons	1.1
Liquid Volume		
ounces	milliliters	30
pints	liters	0.47
quarts	liters	0.95
gallons	liters	3.8
milliliters	ounces	0.03
liters	pints	2.1
liters	quarts	1.06
liters	gallons	0.26
Temperature		
degrees Fahrenheit	degrees Celsius	0.6 (after subtracting 32)
degrees Celsius	degrees Fahrenheit	1.8 (then add 32)

CUSTOMARY/METRIC CONVERSIONS

Customary/ English (inches)	Metric (millimeters)
$\frac{1}{32}$	0.8
$\frac{1}{16}$	1.6
$\frac{1}{8}$	3.2
$\frac{3}{16}$	4.8
$\frac{1}{4}$	6.4
$\frac{5}{16}$	7.9
$\frac{3}{8}$	9.5
$\frac{7}{16}$	11.1
$\frac{1}{2}$	12.7
$\frac{9}{16}$	14.3
$\frac{5}{8}$	15.9
$\frac{11}{16}$	17.5
$\frac{3}{4}$	19.1
$\frac{13}{16}$	20.6
$\frac{7}{8}$	22.2
$\frac{15}{16}$	23.8
1	25.4
5	127.0
12	304.8
18	457.2
24	609.6
36	914.4
48	1219.2

RIGGING

Building a house calls for a many different types of materials and pieces of equipment to be delivered to the site. Many of these items are heavy or awkward. Moving them by hand, piece by piece, is slow and time consuming. Whenever possible, builders use equipment such as mobile cranes or lifts to move heavy or awkward loads. Various types of slings or cables will be used to connect the load to the crane. This makes the work go more quickly and reduces the physical effort required.

The process of lifting and moving heavy loads is generally referred to as rigging. This term is also used to refer to any use of ropes to lift or secure loads.

A load that is improperly rigged could shift or fall. This would present an extreme hazard to anyone in the area. For this reason, rigging should be done only by those trained in the specific techniques necessary.

When a load has been properly rigged, it is often necessary for someone on the ground to signal the crane operator to lift, lower, or move the load. A construction site is noisy, so these signals are often given by hand instead of voice. Some of the standard hand signals are noted here. However, only people who are properly trained in using all the basic hand signals should direct a crane. Whenever possible they should stay in one location so that the crane operator knows where to look for them.

HOIST (RAISE) With forearm vertical, forefinger pointing up, move hand in small horizontal circles.

LOWER Extend arm downward, forefinger pointing down, and move hand in small horizontal circles

MOVE SLOWLY Use one hand to give any motion signal and place other hand motionless above hand giving the motion signal (Hoist slowly is shown as example).

STOP Extend arm, palm down, hold position rigidly.

EMERGENCY STOP Extend arm, palm down, moving hand rapidly right and left.

STANDARD SIZES FOR FRAMING LUMBER, NOMINAL AND DRESSED

Product	Description	Nominal Size Thickness (inches)	Nominal Size Width (inches)	Dressed Dimensions Thicknesses and Widths Surfaced Dry inches	Surfaced Dry mm	Surfaced Unseasoned inches	Surfaced Unseasoned mm
Dimension	S4S	2	2	1 ½	38	1 $\frac{9}{16}$	40
		3	3	2 ½	64	2 $\frac{9}{16}$	65
		4	4	3 ½	89	3 $\frac{9}{16}$	90
			5	4 ½	114	4 $\frac{5}{8}$	117
			6	5 ½	140	5 $\frac{5}{8}$	143
			8	7 ¼	184	7½	191
			10	9 ¼	235	9 ½	241
			12	11 ¼	289	11 ½	292
			over 12	¾ off nominal	19 off nominal	½ off nominal	13 off nominal

Product	Description	Nominal Size		Thickness (unseasoned)		Width (unseasoned)	
Timbers	Rough or S4S (shipped unseasoned)	5 and larger		½" (13 mm) off nominal (S4S) [See 3.20 of WWPA Grading Rules for Rough.]			

Product	Description	Thickness	Width	Thickness (dry) inches	Thickness (dry) mm	Width (dry) inches	Width (dry) mm
Decking	2" (Single T&G)	2	5	1 ½	38	4	102
			6			5	127
			8			6 ¾	172
			10			8 ¾	222
			12			10 ¾	273
	3" and 4" (Double T&G)	3	6	2 ½	64	5 ¼	133
		4		3 ½	89		

Western Wood Products Association

Notes: Based on Western Lumber Grading Rules
Metric equivalents are provided for surfaced (actual) sizes.
Abbreviations: FOHC–Free of Heart Center
Rough Full Sawn–Unsurfaced lumber cut to full specified size

T&G–Tongued-and-grooved
S4S–Surfaced four sides

ESTIMATING MATERIAL FOR FOR PARTITION STUDS

Size of Studs	Spacing on Centers (inches)	Bd. Ft. per Sq. Ft. per Area	Lbs. Nails per 1000 Bd. Ft.
2×3	12	0.91	25
	16	0.83	
	24	0.76	
2×4	12	1.22	19
	16	1.12	
	24	1.02	
2×6	16	1.48	16
	24	1.22	

Note: With this table you can figure the board feet of lumber needed for partition studs. In addition to giving the amount of material for the partition construction, this table also gives consideration to the need for extra lumber for headers, trimmers, and other special framing.

ESTIMATING MATERIAL FOR FOR EXTERIOR WALL STUDS

Size of Studs	Spacing on Centers (inches)	Bd. Ft. per Sq. Ft. per Area	Lbs. Nails per 1000 Bd. Ft.
2×4	12	1.09	22
	16	1.05	
	20	0.98	
	24	0.94	
2×6	12	1.66	15
	16	1.51	
	20	1.44	
	24	1.38	

Note: This table tells you how you can figure the number of board feet and nails for exterior wall framing.

PARTITION STUDS NEEDED

Partition Length in Feet	Number of Studs Required
2	3
3	3
4	4
5	5
6	6
7	6
8	7
9	8
10	9
11	9
12	10
13	11
14	12
15	12
16	13
17	14
18	15
19	15
20	16

Notes: This table shows studs required for framing 16" OC with single top and bottom plates.
For dbl. plate, add per sq. ft.
For 2×8 studs, double above quantities.
For 2×6 studs, increase above quantities 50 percent.

GIRDER SPANS[a] AND HEADER SPANS[a] FOR INTERIOR BEARING WALLS
(MAXIMUM SPANS FOR DOUGLASS FIR-LARCH, HEM-FIR, SOUTHERN PINE AND SPRUCE-PINE-FIR[b] AND REQUIRED NUMBER OF JACK STUDS)

Headers And Girders Supporting	Size	Building Width[c] (feet)					
		20		28		36	
		Span	NJ[d]	Span	NJ[d]	Span	NJ[d]
One Floor Only	2-2×4	3-1	1	2-8	1	2-5	1
	2-2×6	4-6	1	3-11	1	3-6	1
	2-2×8	5-9	1	5-0	2	4-5	2
	2-2×10	7-0	2	6-1	2	5-5	2
	2-2×12	8-1	2	7-0	2	6-3	2
	3-2×8	7-2	1	6-3	1	5-7	2
	3-2×10	8-9	1	7-7	2	6-9	2
	3-2×12	10-2	2	8-10	2	7-10	2
	4-2×8	9-0	1	7-8	1	6-9	1
	4-2×10	10-1	1	8-9	1	7-10	2
	4-2×12	11-9	1	10-2	2	9-1	2
Two Floors	2-2×4	2-2	1	1-10	1	1-7	1
	2-2×6	3-2	2	2-9	2	2-5	2
	2-2×8	4-1	2	3-6	2	3-2	2
	2-2×10	4-11	2	4-3	2	3-10	3
	2-2×12	5-9	2	5-0	3	4-5	3
	3-2×8	5-1	2	4-5	2	3-11	2
	3-2×10	6-2	2	5-4	2	4-10	2
	3-2×12	7-2	2	6-3	2	5-7	3
	4-2×8	6-1	1	5-3	2	4-8	2
	4-2×10	7-2	2	6-2	2	5-6	2
	4-2×12	8-4	2	7-2	2	6-5	2

For SI: 1 inch = 25.4 mm, 1 foot = 308.4 mm.

[a] Spans are given in feet and inches.

[b] Tabulated values assume #2 grade lumber

[c] Building width is measured perpendicular to the ridge. For widths between those shown, spans are permitted to be interpolated.

[d] NJ—Number of jack studs required to support each end. Where the number of required jack studs equals one, the header is permitted to be supported by an approved framing anchor attached to the full-height wall stud and to the header.

PLYWOOD WALL SHEATHING APPLICATION DETAILS

Panel Identification Index	Minimum Thickness (inches)	Maximum Stud Exterior Covering Nailed to:		Nail Size (a)	Nail Spacing (inches)	
		Stud	Sheathing		Panel Edges (when over framing)	Intermediate (each stud)
12/0, 16/0, 20/0	5/16	16	16(b)	6d	6	12
16/0, 20/0, 24/0	3/8	24	16, 24(b)	6d	6	12
24/0, 32/16	1/2	24	24	6d	6	12

(a) Common smooth, annular, spiral thread, galvanized box or T-nails of the same diameter as common nails (0.113" dia. for 6d) may be used. Staples also permitted at reduced spacing.

(b) When sidings such as shingles are nailed only to the plywood sheathing, apply plywood with face grain across studs.

Notes: This is a general guide. Be sure to follow local building codes.

Look for these APA grade-trademarks on wall sheathing.

BOARD-FOOT MEASURES

Nominal Size (Inches)	Actual Length in Feet								
	8	10	12	14	16	18	20	22	24
1 × 4	2 2/3	3 1/3	4	4 2/3	5 1/3	6	6 2/3	7 1/3	8
1 × 6	4	5	6	7	8	9	10	11	12
1 × 8	5 1/3	6 2/3	8	9 1/3	10 2/3	12	13 1/3	14 2/3	16
1 × 10	6 2/3	8 1/3	10	11 2/3	13 1/3	15	16 2/3	18 1/3	20
1 1/4 × 4		4 1/6	5	5 5/6	6 2/3	7 1/2	8 1/3	9 1/6	10
1 1/4 × 6		6 1/4	7 1/2	8 1/3	10	11 1/4	12 1/2	13 3/4	15
2 × 4	5 1/3	6 2/3	8	9 1/3	10 1/3	12	13 1/3	14 2/3	16
2 × 6	8	10	12	14	16	18	20	22	24
2 × 8	10 2/3	13 1/3	16	18 2/3	21 1/3	24	26 2/3	29 1/3	32
2 × 10	13 1/3	16 2/3	20	23 1/3	26 2/3	30	33 1/3	36 2/3	40
4 × 4	10 2/3	13 1/3	16	18 2/3	21 1/3	24	26 2/3	29 1/3	32
4 × 6	16	20	24	28	32	36	40	44	48
4 × 8	21 1/3	26 2/3	32	37 1/3	42 2/3	48	53 1/3	58 2/3	64
4 × 10	26 2/3	33 1/3	40	46 2/3	53 1/3	60	66 2/3	73 1/3	80

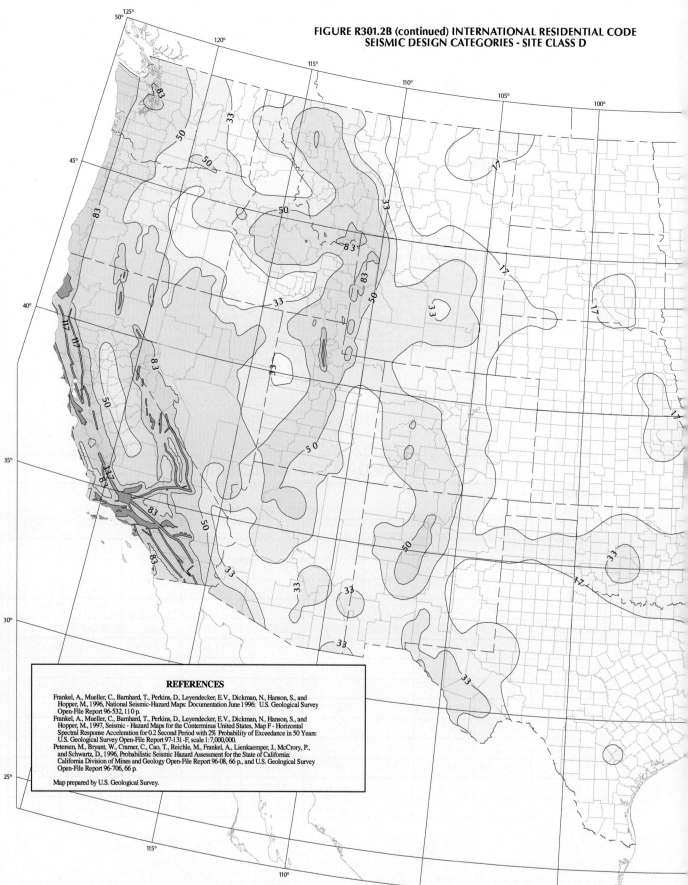

FIGURE R301.2B (continued) INTERNATIONAL RESIDENTIAL CODE
SEISMIC DESIGN CATEGORIES - SITE CLASS D

REFERENCES

Frankel, A., Mueller, C., Barnhard, T., Perkins, D., Leyendecker, E.V., Dickman, N., Hanson, S., and
    Hopper, M., 1996, National Seismic-Hazard Maps: Documentation June 1996: U.S. Geological Survey
    Open-File Report 96-532, 110 p.
Frankel, A., Mueller, C., Barnhard, T., Perkins, D., Leyendecker, E.V., Dickman, N., Hanson, S., and
    Hopper, M., 1997, Seismic - Hazard Maps for the Conterminus United States, Map F - Horizontal
    Spectral Response Acceleration for 0.2 Second Period with 2% Probability of Exceedance in 50 Years:
    U.S. Geological Survey Open-File Report 97-131-F, scale 1:7,000,000.
Petersen, M., Bryant, W., Cramer, C., Cao, T., Reichle, M., Frankel, A, Lienkaemper, J., McCrory, P.,
    and Schwartz, D., 1996, Probabilistic Seismic Hazard Assessment for the State of California:
    California Division of Mines and Geology Open-File Report 96-08, 66 p., and U.S. Geological Survey
    Open-File Report 96-706, 66 p.

Map prepared by U.S. Geological Survey.

This map shows areas where an earthquake is most likely to occur. Consult building codes to determm
which seismic design category is used locally. Construction in earthquake hazard zones will require ext
measures for strengthening a house, such as installing metal connectors (see Section 14.2).

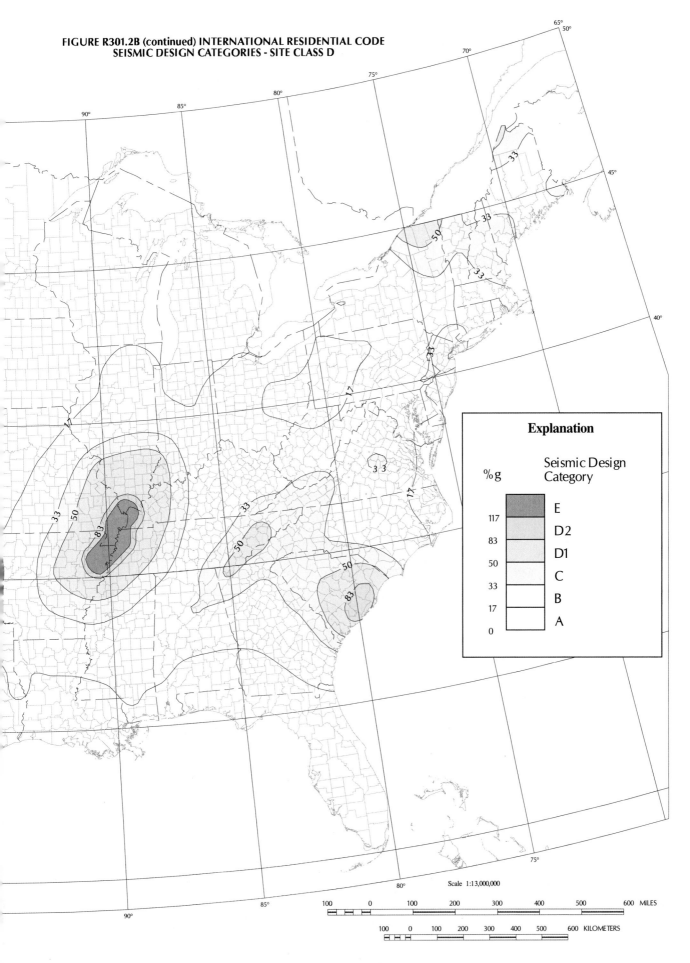

SPAN TABLE

This is an example of a span table.
You should refer to the span tables appropriate for your area.

TABLE R502.5(1) GIRDER SPANS[a] AND HEADER SPANS[a] FOR EXTERIOR BEARING WALLS
(Maximum header spans for douglas fir-larch, hem-fir, southern pine and spruce-pine-fir[b] and required number of jack studs)

HEADERS SUPPORTING	SIZE	GROUND SNOW LOAD (psf)[e]											
		30						50					
		Building width[c] (feet)											
		20		28		36		20		28		36	
		Span	NJ[d]	Span	NJ[d]	Span	NJ[d]	Span	NJ[d]	Span	NJ[d]	Span	NJ[d]
Roof and Ceiling	2-2×4	3-6	1	3-2	1	2-10	1	3-2	1	2-9	1	2-6	1
	2-2×6	5-5	1	4-8	1	4-2	1	4-8	1	4-1	1	3-8	2
	2-2×8	6-10	1	5-11	2	5-4	2	5-11	2	5-2	2	4-7	2
	2-2×10	8-5	2	7-3	2	6-6	2	7-3	2	6-3	2	5-7	2
	2-2×12	9-9	2	8-5	2	7-6	2	8-5	2	7-3	2	6-6	2
	3-2×8	8-4	1	7-5	1	6-8	1	7-5	1	6-5	2	5-9	2
	3-2×10	10-6	1	9-1	2	8-2	2	9-1	2	7-10	2	7-0	2
	3-2×12	12-2	2	10-7	2	9-5	2	10-7	2	9-2	2	8-2	2
	4-2×8	7-0	1	6-1	2	5-5	2	6-1	2	5-3	2	4-8	2
	4-2×10	11-8	1	10-6	1	9-5	2	10-6	1	9-1	2	8-2	2
Roof, ceiling and one center-bearing floor													
Roof, ceiling and one clear span floor													
Roof, ceiling and two center-bearing floors	3-2×12	8-5	2	7-4	2	6-7	2	8-2	2	7-2	2	6-5	3
	4-2×8	4-10	2	4-3	2	3-10	2	4-9	2	4-2	2	3-9	2
	4-2×10	8-4	2	7-4	2	6-7	2	8-2	2	7-2	2	6-5	2
	4-2×12	9-8	2	8-6	2	7-8	2	9-5	2	8-3	2	7-5	2

> ### READING SPAN TABLES
>
> A span is the horizontal distance between supports. In carpentry, various framing members are used across openings. Span tables enable a carpenter to determine what dimension of header, girder, joist, or rafter will be needed over a given opening or area. For example, this span table gives the dimensions of the girders and headers needed to span a given opening in an exterior bearing wall. Many span tables carry notes at the bottom. For a fuller understanding of the table, always refer to these notes. To use this table, you will need the following information:
>
> - The Ground Snow Load in your location. This may be found on-a map or chart-in the code book.
> - The Building Width. This will be given on the building floor plans.
> - The Span of the header. This will be given on the building floor plans.
>
> Refer to the highlighted cells in this table. The highlighted information identifies the-construction project as having:
>
> - A Ground Snow Load of 30 psf (pounds per square feet).
> - A Building Width of 28 feet.
>
> If a header must span 7'3" in this house, the table indicates that it should be made of two 2210s.
>
> Four other representative span tables follow this one. These tables are provided only as examples. On an actual construction project, refer to the span tables in your local code.

For SI: 1 inch = 25.4 mm, 1 pound per square foot = 0.0479 kN/m2.
a. Spans are given in feet and inches.   b. Tabulated values assume #2 grade lumber.   c. Building width is measured perpendicular to the ridge. For widths between those shown, spans are permitted to be interpolated.   d. NJ—Number of jack studs required to support each end. Where the number of required jack studs equals one, the headers are permitted to be supported by an approved framing anchor attached to the full-height wall stud and to the header.   e. Use 30 psf ground snow load for cases in which ground snow load is less than 30 psf and the roof live load is equal to or less than 20 psf.

SPAN TABLE

This is an example of a span table.
You should refer to the span tables appropriate for your area.

TABLE R502.3.1(2) FLOOR JOIST SPANS FOR COMMON LUMBER SPECIES
(Residential living areas, live load = 40 psf, L/Δ = 360)

JOIST SPACING (inches)	SPECIE AND GRADE	DEAD LOAD = 10 psf				DEAD LOAD = 20 psf			
		2×6	2×8	2×10	2×12	2×6	2×8	2×10	2×12
		Maximum floor joist spans							
		(feet - inches)	(feet - inches)	(feet - inches)	(feet - inches)	(feet - inches)	(feet - inches)	(feet - inches)	(feet - inches)
12	Douglas fir-larch SS	11-4	15-0	19-1	23-3	11-4	15-0	19-1	23-3
	Douglas fir-larch #1	10-11	14-5	18-5	22-0	10-11	14-2	17-4	20-1
	Douglas fir-larch #2	10-9	14-2	17-9	20-7	10-6	13-3	16-3	18-10
	Douglas fir-larch #3	8-8	11-0	13-5	15-7	7-11	10-0	12-3	14-3
	Hem-fir SS	10-9	14-2	18-0	21-11	10-9	14-2	18-0	21-11
	Hem-fir #1	10-6	13-10	17-8	21-6	10-6	13-10	16-11	19-7
	Hem-fir #2	10-0	13-2	16-10	20-4	10-0	13-1	16-0	18-6
	Hem-fir #3	8-8	11-0	13-5	15-7	7-11	10-0	12-3	14-3
	Southern pine SS	11-2	14-8	18-9	22-10	11-2	14-8	18-9	22-10
	Southern pine #1	10-11	14-5	18-5	22-5	10-11	14-5	18-5	22-5
	Southern pine #2	10-9	14-2	18-0	21-9	10-9	14-2	16-11	19-10
	Southern pine #3	9-4	11-11	14-0	16-8	8-6	10-10	12-10	15-3
	Spruce-pine-fir SS	10-6	13-10	17-8	21-6	10-6	13-10	17-8	21-6
	Spruce-pine-fir #1	10-3	13-6	17-3	20-7	10-3	13-3	16-3	18-10
	Spruce-pine-fir #2	10-3	13-6	17-3	20-7	10-3	13-3	16-3	18-10
	Spruce-pine-fir #3	8-8	11-0	13-5	15-7	7-11	10-0	12-3	14-3
16	Douglas fir-larch SS	10-4	13-7	17-4	21-1	10-4	13-7	17-4	21-0
	Douglas fir-larch #1	9-11	13-1	16-5	19-1	9-8	12-4	15-0	17-5
	Douglas fir-larch #2	9-9	12-7	15-5	17-10	9-1	11-6	14-1	16-3
	Douglas fir-larch #3	7-6	9-6	11-8	13-6	6-10	8-8	10-7	12-4
	Hem-fir SS	9-9	12-10	16-5	19-11	9-9	12-10	16-5	19-11
	Hem-fir #1	9-6	12-7	16-0	18-7	9-6	12-0	14-8	17-0
	Hem-fir #2	9-1	12-0	15-2	17-7	8-11	11-4	13-10	16-1
	Hem-fir #3	7-6	9-6	11-8	13-6	6-10	8-8	10-7	12-4
	Southern pine SS	10-2	13-4	17-0	20-9	10-2	13-4	17-0	20-9
	Southern pine #1	9-11	13-1	16-9	20-4	9-11	13-1	16-4	19-6
	Southern pine #2	9-9	12-10	16-1	18-10	9-6	12-4	14-8	17-2
	Southern pine #3	8-1	10-3	12-2	14-6	7-4	9-5	11-1	13-2
	Spruce-pine-fir SS	9-6	12-7	16-0	19-6	9-6	12-7	16-0	19-6
	Spruce-pine-fir #1	9-4	12-3	15-5	17-10	9-1	11-6	14-1	16-3
	Spruce-pine-fir #2	9-4	12-3	15-5	17-10	9-1	11-6	14-1	16-3
	Spruce-pine-fir #3	7-6	9-6	11-8	13-6	6-10	8-8	10-7	12-4
24	Douglas fir-larch SS	9-0	11-11	15-2	18-5	9-0	11-11	14-9	17-1
	Douglas fir-larch #1	8-8	11-0	13-5	15-7	7-11	10-0	12-3	14-3
	Douglas fir-larch #2	8-1	10-3	12-7	14-7	7-5	9-5	11-6	13-4
	Douglas fir-larch #3	6-2	7-9	9-6	11-0	5-7	7-1	8-8	10-1
	Hem-fir SS	8-6	11-3	14-4	17-5	8-6	11-3	14-4	16-10[a]
	Hem-fir #1	8-4	10-9	13-1	15-2	7-9	9-9	11-11	13-10
	Hem-fir #2	7-11	10-2	12-5	14-4	7-4	9-3	11-4	13-1
	Hem-fir #3	6-2	7-9	9-6	11-0	5-7	7-1	8-8	10-1
	Southern pine SS	8-10	11-8	14-11	18-1	8-10	11-8	14-11	18-1
	Southern pine #1	8-8	11-5	14-7	17-5	8-8	11-3	13-4	15-11
	Southern pine #2	8-6	11-0	13-1	15-5	7-9	10-0	12-0	14-0
	Southern pine #3	6-7	8-5	9-11	11-10	6-0	7-8	9-1	10-9
	Spruce-pine-fir SS	8-4	11-0	14-0	17-0	8-4	11-0	13-8	15-11
	Spruce-pine-fir #1	8-1	10-3	12-7	14-7	7-5	9-5	11-6	13-4
	Spruce-pine-fir #2	8-1	10-3	12-7	14-7	7-5	9-5	11-6	13-4
	Spruce-pine-fir #3	6-2	7-9	9-6	11-0	5-7	7-1	8-8	10-01

For SI: 1 inch = 25.4 mm, 1 foot = 308.4 mm, 1 pound per square foot = 0.0479 kN/m².

a. Check sources for availability of lumber in lengths greater than 20 feet.

b. End bearing length shall be increased to 2 inches.

Copyright 2000, International Code Council, Inc., Falls Church, Virginia. 2000 International Residential Code. Reprinted with permission of the author. All rights reserved.

SPAN TABLE

This is an example of a span table.
You should refer to the span tables appropriate for your area.

TABLE R802.4(1) CEILING JOIST SPANS FOR COMMON LUMBER SPECIES
(Uninhabitable attics without storage, live load = 10 psf, L/Δ = 240)

CEILING JOIST SPACING (inches)	SPECIE AND GRADE		DEAD LOAD = 5 psf			
			2×4	2×6	2×8	2×10
			Maximum ceiling joist spans			
			(feet - inches)	(feet - inches)	(feet - inches)	(feet - inches)
12	Douglas fir-larch	SS	13-2	20-8	Note[a]	Note[a]
	Douglas fir-larch	#1	12-8	19-11	Note[a]	Note[a]
	Douglas fir-larch	#2	12-5	19-6	25-8	Note[a]
	Douglas fir-larch	#3	10-10	15-10	20-1	24-6
	Hem-fir	SS	12-5	19-6	25-8	Note[a]
	Hem-fir	#1	12-2	19-1	25-2	Note[a]
	Hem-fir	#2	11-7	18-2	24-0	Note[a]
	Hem-fir	#3	10-10	15-10	20-1	24-6
	Southern pine	SS	12-11	20-3	Note[a]	Note[a]
	Southern pine	#1	12-8	19-11	Note[a]	Note[a]
	Southern pine	#2	12-5	19-6	25-8	Note[a]
	Southern pine	#3	11-6	17-0	21-8	25-7
	Spruce-pine-fir	SS	12-2	19-1	25-2	Note[a]
	Spruce-pine-fir	#1	11-10	18-8	24-7	Note[a]
	Spruce-pine-fir	#2	11-10	18-8	24-7	Note[a]
	Spruce-pine-fir	#3	10-10	15-10	20-1	24-6
16	Douglas fir-larch	SS	11-11	18-9	24-8	Note[a]
	Douglas fir-larch	#1	11-6	18-1	23-10	Note[a]
	Douglas fir-larch	#2	11-3	17-8	23-0	Note[a]
	Douglas fir-larch	#3	9-5	13-9	17-5	21-3
	Hem-fir	SS	11-3	17-8	23-4	Note[a]
	Hem-fir	#1	11-0	17-4	22-10	Note[a]
	Hem-fir	#2	10-6	16-6	21-9	Note[a]
	Hem-fir	#3	9-5	13-9	17-5	21-3
	Southern pine	SS	11-9	18-5	24-3	Note[a]
	Southern pine	#1	11-6	18-1	23-1	Note[a]
	Southern pine	#2	11-3	17-8	23-4	Note[a]
	Southern pine	#3	10-0	14-9	18-9	22-2
	Spruce-pine-fir	SS	11-0	17-4	22-10	Note[a]
	Spruce-pine-fir	#1	10-9	16-11	22-4	Note[a]
	Spruce-pine-fir	#2	10-9	16-11	22-4	Note[a]
	Spruce-pine-fir	#3	9-5	13-9	17-5	21-3
24	Douglas fir-larch	SS	10-5	16-4	21-7	Note[a]
	Douglas fir-larch	#1	10-0	15-9	20-1	24-6
	Douglas fir-larch	#2	9-10	14-10	18-9	22-11
	Douglas fir-larch	#3	7-8	11-2	14-2	17-4
	Hem-fir	SS	9-10	15-6	20-5	Note[a]
	Hem-fir	#1	9-8	15-2	19-7	23-11
	Hem-fir	#2	9-2	14-5	18-6	22-7
	Hem-fir	#3	7-8	11-2	14-2	17-4
	Southern pine	SS	10-3	16-1	21-2	Note[a]
	Southern pine	#1	10-0	15-9	20-10	Note[a]
	Southern pine	#2	9-10	15-6	20-1	23-11
	Southern pine	#3	8-2	12-0	15-4	18-1
	Spruce-pine-fir	SS	9-8	15-2	19-11	25-5
	Spruce-pine-fir	#1	9-5	14-9	18-9	22-11
	Spruce-pine-fir	#2	9-5	14-9	18-9	22-11
	Spruce-pine-fir	#3	7-8	11-2	14-2	17-4

For SI: 1 inch = 25.4 mm, 1 foot = 304.8 mm, 1 pound per square foot = 0.0479 kN/m².

[a] Check sources for availability of lumber in lengths greater than 20 feet.

SPAN TABLE

This is an example of a span table.
You should refer to the span tables appropriate for your area.

TABLE R802.5.1(2) RAFTER SPANS FOR COMMON LUMBER SPECIES
(Roof live load = 20 psf, ceiling attached to rafters, L/Δ = 240)

RAFTER SPACING (inches)	SPECIE AND GRADE		DEAD LOAD = 10 psf					DEAD LOAD = 20 psf				
			2×4	2×6	2×8	2×10	2×12	2×4	2×6	2×8	2×10	2×12
			Maximum rafter spans[a]									
			(feet-inches)	(feet-inches)	(feet-inches)	(feet-inches)	(feet-inches)	(feet-inches)	(feet-inches)	(feet-inches)	(feet-inches)	(feet-inches)
12	Douglas fir-larch	SS	10-5	16-4	21-7	Note[b]	Note[b]	10-5	16-4	21-7	Note[b]	Note[b]
	Douglas fir-larch	#1	10-0	15-9	20-10	Note[b]	Note[b]	10-0	15-4	19-5	23-9	Note[b]
	Douglas fir-larch	#2	9-10	15-6	20-5	25-8	Note[b]	9-10	14-4	18-2	22-3	25-9
	Douglas fir-larch	#3	8-7	12-6	15-10	19-5	22-6	7-5	10-10	13-9	16-9	19-6
	Hem-fir	SS	9-10	15-6	20-5	Note[b]	Note[b]	9-10	15-6	20-5	Note[b]	Note[b]
	Hem-fir	#1	9-8	15-2	19-11	25-5	Note[b]	9-8	14-11	18-11	23-2	Note[b]
	Hem-fir	#2	9-2	14-5	19-0	24-3	Note[b]	9-2	14-2	17-11	21-11	25-5
	Hem-fir	#3	8-7	12-6	15-10	19-5	22-6	7-5	10-10	13-9	16-9	19-6
	Southern pine	SS	10-3	16-1	21-2	Note[b]	Note[b]	10-3	16-1	21-2	Note[b]	Note[b]
	Southern pine	#1	10-0	15-9	20-10	Note[b]	Note[b]	10-0	15-9	20-10	25-10	Note[b]
	Southern pine	#2	9-10	15-6	20-5	Note[b]	Note[b]	9-10	15-1	19-5	23-2	Note[b]
	Southern pine	#3	9-1	13-6	17-2	20-3	24-1	7-11	11-8	14-10	17-6	20-11
	Spruce-pine-fir	SS	9-8	15-2	19-11	25-5	Note[b]	9-8	15-2	19-11	25-5	Note[b]
	Spruce-pine-fir	#1	9-5	14-9	19-6	24-10	Note[b]	9-5	14-4	18-2	22-3	25-9
	Spruce-pine-fir	#2	9-5	14-9	19-6	24-10	Note[b]	9-5	14-4	18-2	22-3	25-9
	Spruce-pine-fir	#3	8-7	12-6	15-10	19-5	22-6	7-5	10-10	13-9	16-9	19-6
16	Douglas fir-larch	SS	9-6	14-11	19-7	25-0	Note[b]	9-6	14-11	19-7	24-9	Note[b]
	Douglas fir-larch	#1	9-1	14-4	18-11	23-9	Note[b]	9-1	13-3	16-10	20-7	23-10
	Douglas fir-larch	#2	8-11	14-1	18-2	22-3	25-9	8-6	12-5	15-9	19-3	22-4
	Douglas fir-larch	#3	7-5	10-10	13-9	16-9	19-6	6-5	9-5	11-11	14-6	16-10
	Hem-fir	SS	8-11	14-1	18-6	23-8	Note[b]	8-11	14-1	18-6	23-8	Note[b]
	Hem-fir	#1	8-9	13-9	18-1	23-1	Note[b]	8-9	12-11	16-5	20-0	23-3
	Hem-fir	#2	8-4	13-1	17-3	21-11	25-5	8-4	12-3	15-6	18-11	22-0
	Hem-fir	#3	7-5	10-10	13-9	16-9	19-6	6-5	9-5	11-11	14-6	16-10
	Southern pine	SS	9-4	14-7	19-3	24-7	Note[b]	9-4	14-7	19-3	24-7	Note[b]
	Southern pine	#1	9-1	14-4	18-11	24-1	Note[b]	9-1	14-4	18-10	22-4	Note[b]
	Southern pine	#2	8-11	14-1	18-6	23-2	Note[b]	8-11	13-0	16-10	20-1	23-7
	Southern pine	#3	7-11	11-8	14-10	17-6	20-11	6-10	10-1	12-10	15-2	18-1
	Spruce-pine-fir	SS	8-9	13-9	18-1	23-1	Note[b]	8-9	13-9	18-1	23-0	Note[b]
	Spruce-pine-fir	#1	8-7	13-5	17-9	22-3	25-9	8-6	12-5	15-9	19-3	22-4
	Spruce-pine-fir	#2	8-7	13-5	17-9	22-3	25-9	8-6	12-5	15-9	19-3	22-4
	Spruce-pine-fir	#3	7-5	10-10	13-9	16-9	19-6	6-5	9-5	11-11	14-6	16-10
19.2	Douglas fir-larch	SS	8-11	14-0	18-5	23-7	Note[b]	8-11	14-0	18-5	22-7	Note[b]
	Douglas fir-larch	#1	8-7	13-6	17-9	21-8	25-2	8-4	12-2	15-4	18-9	21-9
	Douglas fir-larch	#2	8-5	13-1	16-7	20-3	23-6	7-9	11-4	14-4	17-7	20-4
	Douglas fir-larch	#3	6-9	9-11	12-7	15-4	17-9	5-10	8-7	10-10	13-3	15-5
	Hem-fir	SS	8-5	13-3	17-5	22-3	Note[b]	8-5	13-3	17-5	22-3	25-9
	Hem-fir	#1	8-3	12-11	17-1	21-1	24-6	8-1	11-10	15-0	18-4	21-3
	Hem-fir	#2	7-10	12-4	16-3	20-0	23-2	7-8	11-2	14-2	17-4	20-1
	Hem-fir	#3	6-9	9-11	12-7	15-4	17-9	5-10	8-7	10-10	13-3	15-5
	Southern pine	SS	8-9	13-9	18-1	23-1	Note[b]	8-9	13-9	18-1	23-1	Note[b]
	Southern pine	#1	8-7	13-6	17-9	22-8	Note[b]	8-7	13-6	17-2	20-5	24-4
	Southern pine	#2	8-5	13-3	17-5	21-2	24-10	8-4	11-11	15-4	18-4	21-6
	Southern pine	#3	7-3	10-8	13-7	16-0	19-1	6-3	9-3	11-9	13-10	16-6
	Spruce-pine-fir	SS	8-3	12-11	17-1	21-9	Note[b]	8-3	12-11	17-1	21-0	24-4
	Spruce-pine-fir	#1	8-1	12-8	16-7	20-3	23-6	7-9	11-4	14-4	17-7	20-4
	Spruce-pine-fir	#2	8-1	12-8	16-7	20-3	23-6	7-9	11-4	14-4	17-7	20-4
	Spruce-pine-fir	#3	6-9	9-11	12-7	15-4	17-9	5-10	8-7	10-10	13-3	15-5

SPAN TABLE

This is an example of a span table.
You should refer to the span tables appropriate for your area.

TABLE R802.5.1(2) — continued RAFTER SPANS FOR COMMON LUMBER SPECIES
(Roof live load = 20 psf, ceiling attached to rafters, L/Δ = 240)

RAFTER SPACING (inches)	SPECIE AND GRADE		DEAD LOAD = 10 psf					DEAD LOAD = 20 psf				
			2×4	2×6	2×8	2×10	2×12	2×4	2×6	2×8	2×10	2×12
			Maximum rafter spans[a]									
			(feet - inches)	(feet - inches)	(feet - inches)	(feet - inches)	(feet - inches)	(feet - inches)	(feet - inches)	(feet - inches)	(feet - inches)	(feet - inches)
24	Douglas fir-larch	SS	8-3	13-0	17-2	21-10	Note[b]	8-3	13-0	16-7	20-3	23-5
	Douglas fir-larch	#1	8-0	12-6	15-10	19-5	22-6	7-5	10-10	13-9	16-9	19-6
	Douglas fir-larch	#2	7-10	11-9	14-10	18-2	21-0	6-11	10-2	12-10	15-8	18-3
	Douglas fir-larch	#3	6-1	8-10	11-3	13-8	15-11	5-3	7-8	9-9	11-10	13-9
	Hem-fir	SS	7-10	12-3	16-2	20-8	25-1	7-10	12-3	16-2	19-10	23-0
	Hem-fir	#1	7-8	12-0	15-6	18-11	21-11	7-3	10-7	13-5	16-4	19-0
	Hem-fir	#2	7-3	11-5	14-8	17-10	20-9	6-10	10-0	12-8	15-6	17-11
	Hem-fir	#3	6-1	8-10	11-3	13-8	15-11	5-3	7-8	9-9	11-10	13-9
	Southern pine	SS	8-1	12-9	16-10	21-6	Note[b]	8-1	12-9	16-10	21-6	Note[b]
	Southern pine	#1	8-0	12-6	16-6	21-1	25-2	8-0	12-3	15-4	18-3	21-9
	Southern pine	#2	7-10	12-3	15-10	18-11	22-2	7-5	10-8	13-9	16-5	19-3
	Southern pine	#3	6-5	9-6	12-1	14-4	17-1	5-7	8-3	10-6	12-5	14-9
	Spruce-pine-fir	SS	7-8	12-0	15-10	20-2	24-7	7-8	12-0	15-4	18-9	21-9
	Spruce-pine-fir	#1	7-6	11-9	14-10	18-2	21-0	6-11	10-2	12-10	15-8	18-3
	Spruce-pine-fir	#2	7-6	11-9	14-10	18-2	21-0	6-11	10-2	12-10	15-8	18-3
	Spruce-pine-fir	#3	6-1	8-10	11-3	13-8	15-11	5-3	7-8	9-9	11-10	13-9

For SI: 1 inch = 25.4 mm, 1 foot = 304.8 mm, 1 pound per square foot = 0.0479 kN/m^2.

[a]. The tabulated rafter spans assume that ceiling joists are located at the bottom of the attic space or that some other method of resisting the-outward push of the rafters on the bearing walls, such as rafter ties, is provided at that location. When ceiling joists or rafter ties are located higher in the attic space, the rafter spans shall be multiplied by the factors given below.

H_c/H_R	Rafter Span Adjustment Factor
2/3 or greater	0.50
1/2	0.58
1/3	0.67
1/4	0.76
1/5	0.83
1/6	0.90
1/7.5 and less	1.00

where: H_c = Height of ceiling joists or rafter ties measured vertically above the top of the rafter support walls.

H_R = Height of roof ridge measured vertically above the top of the rafter support walls.

[b]. Check sources for availability of lumber in lengths greater than 20 feet.

Number and Operations

▶ *Understand numbers, ways of representing numbers, relationships among numbers, and number systems*

Fraction, Decimal, and Percent

A percent is a ratio that compares a number to 100. To write a percent as a fraction, drop the percent sign, and use the number as the numerator in a fraction with a denominator of 100. Simplify, if possiblet. For example, $76\% = \frac{76}{100}$, or $\frac{19}{25}$. To write a fraction as a percent, convert it to an equivalent fraction with a denominator of 100. For example, $\frac{3}{4} = \frac{75}{100}$, or 75%. A fraction can be expressed as a percent by first converting the fraction to a decimal (divide the numerator by the denominator) and then converting the decimal to a percent by moving the decimal point two places to the right.

Comparing Numbers on a Number Line

In order to compare and understand the relationship between real numbers in various forms, it is helpful to use a number line. The zero point on a number line is called the origin; the points to the left of the origin are negative, and those to the right are positive. The number line below shows how numbers in percent, decimal, fraction, and integer form can be compared.

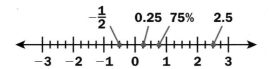

Percents Greater Than 100 and Less Than 1

Percents greater than 100% represent values greater than 1. For example, if the weight of an object is 250% of another, it is 2.5, or $2\frac{1}{2}$, times the weight.

Percents less than 1 represent values less than $\frac{1}{100}$. In other words, 0.1% is one tenth of

one percent, which can also be represented in decimal form as 0.001, or in fraction form as $\frac{1}{1,000}$. Similarly, 0.01% is one hundredth of one percent or 0.0001 or $\frac{1}{10,000}$.

Ratio, Rate, and Proportion

A ratio is a comparison of two numbers using division. If a basketball player makes 8 out of 10 free throws, the ratio is written as 8 to 10, 8:10, or $\frac{8}{10}$. Ratios are usually written in simplest form. In simplest form, the ratio "8 out of 10" is 4 to 5, 4:5, or $\frac{4}{5}$. A rate is a ratio of two measurements having different kinds of units—cups per gallon, or miles per hour, for example. When a rate is simplified so that it has a denominator of 1, it is called a unit rate. An example of a unit rate is 9 miles per hour. A proportion is an equation stating that two ratios are equal. $\frac{3}{18} = \frac{13}{78}$ is an example of a proportion. The cross products of a proportion are also equal. $\frac{3}{18} = \frac{13}{78}$ and $3 \times 78 = 18 \times 13$.

Representing Large and Small Numbers

In order to represent large and small numbers, it is important to understand the number system. Our number system is based on 10, and the value of each place is 10 times the value of the place to its right.

The value of a digit is the product of a digit and its place value. For instance, in the number 6,400, the 6 has a value of six thousands and the 4 has a value of four hundreds. A place value chart can help you read numbers. In the chart, each group of three digits is called a period. Commas separate the periods: the ones period, the thousands period, the millions period, and so on. Values to the right of the ones period are decimals. By understanding place value you can write very large numbers like 5 billion and more, and very small numbers that are less than 1.

Scientific Notation

When dealing with very large numbers like 1,500,000, or very small numbers like 0.000015, it is helpful to keep track of their

MATH APPENDIX

value by writing the numbers in scientific notation. Powers of 10 with positive exponents are used with a decimal between 1 and 10 to express large numbers. The exponent represents the number of places the decimal point is moved to the right. So, 528,000 is written in scientific notation as 5.28×10^5. Powers of 10 with negative exponents are used with a decimal between 1 and 10 to express small numbers. The exponent represents the number of places the decimal point is moved to the left. The number 0.00047 is expressed as 4.7×10^{-4}.

Factor, Multiple, and Prime Factorization

Two or more numbers that are multiplied to form a product are called factors. Divisibility rules can be used to determine whether 2, 3, 4, 5, 6, 8, 9, or 10 are factors of a given number. Multiples are the products of a given number and various integers.

For example, 8 is a multiple of 4 because $4 \times 2 = 8$. A prime number is a whole number that has exactly two factors: 1 and itself. A composite number is a whole number that has more than two factors. Zero and 1 are neither prime nor composite. A composite number can be expressed as the product of its prime factors. The prime factorization of 40 is $2 \times 2 \times 2 \times 5$, or $2^3 \times 5$. The numbers 2 and 5 are prime numbers.

Integers

A negative number is a number less than zero. Negative numbers like –8, positive numbers like +6, and zero are members of the set of integers. Integers can be represented as points on a number line. A set of integers can be written {..., –3, –2, –1, 0, 1, 2, 3, ...} where ... means "continues indefinitely."

Real, Rational, and Irrational Numbers

The real number system is made up of the sets of rational and irrational numbers. Rational numbers are numbers that can be written in the form a/b where a and b are integers and $b \neq 0$. Examples are 0.45, $\frac{1}{2}$, and $\sqrt{36}$. Irrational numbers are non-repeating, non-terminating decimals. Examples are $\sqrt{71}$, π, and 0.020020002....

Complex and Imaginary Numbers

A complex number is a mathematical expression with a real number element and an imaginary number element. Imaginary numbers are multiples of i, the "imaginary" square root of –1. Complex numbers are represented by $a + bi$, where a and b are real numbers and i represents the imaginary element. When a quadratic equation does not have a real number solution, the solution can be represented by a complex number. Like real numbers, complex numbers can be added, subtracted, multiplied, and divided.

Vectors and Matrices

A matrix is a set of numbers or elements arranged in rows and columns to form a rectangle. The number of rows is represented by m and the number of columns is represented by n. To describe the number of rows and columns in a matrix, list the number of rows first using the format $m \times n$. Matrix A below is a 3×3 matrix because it has 3 rows and 3 columns. To name an element of a matrix, the letter i is used to denote the row and j is used to denote the column, and the element is labeled in the form $a_{i,j}$. In matrix A below, $a_{3,2}$ is 4.

$$\text{Matrix A} = \begin{pmatrix} 1 & 3 & 5 \\ 0 & 6 & 8 \\ 3 & 4 & 5 \end{pmatrix}$$

A vector is a matrix with only one column or row of elements. A transposed column vector, or a column vector turned on its side, is a row vector. In the example below, row vector b' is the transpose of column vector b.

$$b = \begin{pmatrix} 1 \\ 2 \\ 3 \\ 4 \end{pmatrix}$$

$$b' = \begin{pmatrix} 1 & 2 & 3 & 4 \end{pmatrix}$$

▶ Understand meanings of operations and how they relate to one another

Properties of Addition and Multiplication

Properties are statements that are true for any numbers. For example, $3 + 8$ is the same as $8 + 3$ because each expression equals 11. This illustrates the Commutative Property of Addition. Likewise, $3 \times 8 = 8 \times 3$ illustrates the Commutative Property of Multiplication.

When evaluating expressions, it is often helpful to group or associate the numbers. The Associative Property says that the way in which numbers are grouped when added or multiplied does not change the sum or product. The following properties are also true:

- **Additive Identity Property:** When 0 is added to any number, the sum is the number.

- **Multiplicative Identity Property:** When any number is multiplied by 1, the product is the number.

- **Multiplicative Property of Zero:** When any number is multiplied by 0, the product is 0.

Rational Numbers

A number that can be written as a fraction is called a rational number. Terminating and repeating decimals are rational numbers because both can be written as fractions. Decimals that are neither terminating nor repeating are called irrational numbers because they cannot be written as fractions. Terminating decimals can be converted to fractions by placing the number (without the decimal point) in the numerator. Count the number of places to the right of the decimal point, and in the denominator, place a 1 followed by a number of zeros equal to the number of places that you counted. The fraction can then be reduced to its simplest form.

Writing a Fraction as a Decimal

Any fraction $\frac{a}{b}$, where $b \neq 0$, can be written as a decimal by dividing the numerator by the denominator. So, $\frac{a}{b} = a \div b$. If the division ends, or terminates, when the remainder is zero, the decimal is a terminating decimal. Not all fractions can be written as terminating decimals. Some have a repeating decimal. A bar indicates that the decimal repeats forever. For example, the fraction $\frac{4}{9}$ can be converted to a repeating decimal, $0.\overline{4}$

Adding and Subtracting Like Fractions

Fractions with the same denominator are called like fractions. To add like fractions, add the numerators and write the sum over the denominator. To add mixed numbers with like fractions, add the whole numbers and fractions separately, adding the numerators of the fractions, then simplifying if necessary. The rule for subtracting fractions with like denominators is similar to the rule for adding. The numerators can be subtracted and the difference written over the denominator. Mixed numbers are written as improper fractions before subtracting. These same rules apply to adding or subtracting like algebraic fractions.

Adding and Subtracting Unlike Fractions

Fractions with different denominators are called unlike fractions. The least common multiple of the denominators is used to rename the fractions with a common denominator. After a common denominator is found, the numerators can then be added or subtracted. To add mixed numbers with unlike fractions, rename the mixed numbers as improper fractions. Then find a common denominator, add the numerators, and simplify the answer.

Multiplying Rational Numbers

To multiply fractions, multiply the numerators and multiply the denominators. If the numerators and denominators have common factors, they can be simplified before multiplication. If the fractions have different signs, then the

MATH APPENDIX

product will be negative. Mixed numbers can be multiplied in the same manner, after first renaming them as improper fractions. A fraction that contains one or more variables in the numerator or denominator is called an algebraic fraction. Algebraic fractions may be multiplied using the same method described above.

Dividing Rational Numbers
To divide a number by a rational number (a fraction, for example), multiply the first number by the multiplicative inverse of the second. Two numbers whose product is 1 are called multiplicative inverses, or reciprocals. $\frac{7}{4} \times \frac{4}{7}$ = 1. When dividing by a mixed number, first rename it as an improper fraction, and then multiply by its multiplicative inverse. This process of multiplying by a number's reciprocal can also be used when dividing algebraic fractions.

Adding Integers
To add integers with the same sign, add their absolute values. The sum then takes the same sign as the addends. The equation $-5 + (-2) = -7$ is an example of adding two integers with the same sign. To add integers with different signs, subtract their absolute values. The sum takes the same sign as the addend with the greater absolute value.

Subtracting Integers
The rules for adding integers are extended to the subtraction of integers. To subtract an integer, add its additive inverse. For example, to find the difference $2 - 5$, add the additive inverse of 5 to 2: $2 + (-5) = -3$. The rule for subtracting integers can be used to solve real-world problems and to evaluate algebraic expressions.

Additive Inverse Property
Two numbers with the same absolute value but different signs are called opposites. For example, −4 and 4 are opposites. An integer and its opposite are also called additive inverses. The Additive Inverse Property says that the sum of any number and its additive inverse is zero. The Commutative, Associative,

and Identity Properties also apply to integers. These properties help when adding more than two integers.

Absolute Value
In mathematics, when two integers on a number line are on opposite sides of zero, and they are the same distance from zero, they have the same absolute value. The symbol for absolute value is two vertical bars on either side of the number. For example, $|-5| = 5$.

Multiplying Integers
Since multiplication is repeated addition, $3(-7)$ means that −7 is used as an addend 3 times. By the Commutative Property of Multiplication, $3(-7) = -7(3)$. The product of two integers with different signs is always negative. The product of two integers with the same sign is always positive.

Dividing Integers
The quotient of two integers can be found by dividing the numbers using their absolute values. The quotient of two integers with the same sign is positive, and the quotient of two integers with a different sign is negative. $-12 \div (-4) = 3$ and $12 \div (-4) = -3$. The division of integers is used in statistics to find the average, or mean, of a set of data. When finding the mean of a set of numbers, find the sum of the numbers, and then divide by the number in the set.

Adding and Multiplying Vectors and Matrices
In order to add two matrices together, they must have the same number of rows and columns. In matrix addition, the corresponding elements are added to each other. In other words $(a + b)_{ij} = a_{ij} + b_{ij}$. For example,

$$\begin{pmatrix} 1 & 2 \\ 2 & 1 \end{pmatrix} + \begin{pmatrix} 3 & 6 \\ 0 & 1 \end{pmatrix} = \begin{pmatrix} 1+3 & 2+6 \\ 2+0 & 1+1 \end{pmatrix} = \begin{pmatrix} 4 & 8 \\ 2 & 2 \end{pmatrix}$$

Matrix multiplication requires that the number of elements in each row in the first matrix is equal to the number of elements in each column in the second. The elements of the first row of the first matrix are multiplied by the corresponding elements of the first column of the second matrix and then added together to get the first element of the product matrix. To get the second element, the elements in the first row of the first matrix are multiplied by the corresponding elements in the second column of the second matrix then added, and so on, until every row of the first matrix is multiplied by every column of the second. See the example below.

$$\begin{pmatrix} 1 & 2 \\ 3 & 4 \end{pmatrix} \times \begin{pmatrix} 3 & 6 \\ 0 & 1 \end{pmatrix} = \begin{pmatrix} (1\times3)+(2\times0) & (1\times6)+(2\times1) \\ (3\times3)+(4\times0) & (3\times6)+(4\times1) \end{pmatrix} = \begin{pmatrix} 3 & 8 \\ 9 & 22 \end{pmatrix}$$

Vector addition and multiplication are performed in the same way, but there is only one column and one row.

Permutations and Combinations

Permutations and combinations are used to determine the number of possible outcomes in different situations. An arrangement, listing, or pattern in which order is important is called a permutation. The symbol P(6, 3) represents the number of permutations of 6 things taken 3 at a time. For P(6, 3), there are $6 \times 5 \times 4$ or 120 possible outcomes. An arrangement or listing where order is not important is called a combination. The symbol C(10, 5) represents the number of combinations of 10 things taken 5 at a time. For C(10, 5), there are $(10 \times 9 \times 8 \times 7 \times 6) \div (5 \times 4 \times 3 \times 2 \times 1)$ or 252 possible outcomes.

Powers and Exponents

An expression such as $3 \times 3 \times 3 \times 3$ can be written as a power. A power has two parts, a base and an exponent. $3 \times 3 \times 3 \times 3 = 3^4$. The base is the number that is multiplied (3). The exponent tells how many times the base is used as a factor (4 times). Numbers and variables can be written using exponents. For example, $8 \times 8 \times 8 \times m \times m \times m \times m \times m$ can be expressed $8^3 m^5$. Exponents also can be used with place value to express numbers in expanded form. Using this method, 1,462 can be written as $(1 \times 10^3) + (4 \times 10^2) + (6 \times 10^1) + (2 \times 10^0)$.

Squares and Square Roots

The square root of a number is one of two equal factors of a number. Every positive number has both a positive and a negative square root. For example, since $8 \times 8 = 64$, 8 is a square root of 64. Since $(-8) \times (-8) = 64$, -8 is also a square root of 64. The notation $\sqrt{}$ indicates the positive square root, $-\sqrt{}$ indicates the negative square root, and $\pm\sqrt{}$ indicates both square roots. For example, $\sqrt{81} = 9$, $-\sqrt{49} = -7$, and $\pm\sqrt{4} = \pm2$. The square root of a negative number is an imaginary number because any two factors of a negative number must have different signs, and are therefore not equivalent.

Logarithm

A logarithm is the inverse of exponentiation. The logarithm of a number x in base b is equal to the number n. Therefore, $b^n = x$ and $\log_b x = n$. For example, $\log_4(64) = 3$ because $4^3 = 64$. The most commonly used bases for logarithms are 10, the common logarithm; 2, the binary logarithm; and the constant e, the natural logarithm (also called $ln(x)$ instead of $\log_e(x)$). Below is a list of some of the rules of logarithms that are important to understand if you are going to use them.

$$\log_b(xy) = \log_b(x) + \log_b(y)$$
$$\log_b(x/y) = \log_b(x) - \log_b(y)$$
$$\log_b(1/x) = -\log_b(x)$$
$$\log_b(x)y = y\log_b(x)$$

MATH APPENDIX

▶ *Compute fluently and make reasonable estimates*

Estimation by Rounding

When rounding numbers, look at the digit to the right of the place to which you are rounding. If the digit is 5 or greater, round up. If it is less than 5, round down. For example, to round 65,137 to the nearest hundred, look at the number in the tens place. Since 3 is less than 5, round down to 65,100. To round the same number to the nearest ten thousandth, look at the number in the thousandths place. Since it is 5, round up to 70,000.

Finding Equivalent Ratios

Equivalent ratios have the same meaning. Just like finding equivalent fractions, to find an equivalent ratio, multiply or divide both sides by the same number. For example, you can multiply 7 by both sides of the ratio 6:8 to get 42:56. Instead, you can also divide both sides of the same ratio by 2 to get 3:4. Find the simplest form of a ratio by dividing to find equivalent ratios until you can't go any further without going into decimals. So, 160:240 in simplest form is 2:3. To write a ratio in the form *1:n*, divide both sides by the left-hand number. In other words, to change 8:20 to *1:n*, divide both sides by 8 to get 1:2.5.

Front-End Estimation

Front-end estimation can be used to quickly estimate sums and differences before adding or subtracting. To use this technique, add or subtract just the digits of the two highest place values, and replace the other place values with zero. This will give you an estimation of the solution of a problem. For example, 93,471 − 22,825 can be changed to 93,000 − 22,000 or 71,000. This estimate can be compared to your final answer to judge its correctness.

Judging Reasonableness

When solving an equation, it is important to check your work by considering how reasonable your answer is. For example, consider the equation $9\frac{3}{4} \times 4\frac{1}{3}$. Since $9\frac{3}{4}$ is between 9 and 10 and $4\frac{1}{3}$ is between 4 and 5, only values that are between 9×4 or 36 and 10×5 or 50 will be reasonable. You can also use front-end estimation, or you can round and estimate a reasonable answer. In the equation 73×25, you can round and solve to estimate a reasonable answer to be near 70×30 or 2,100.

Algebra

▶ *Understand patterns, relations, and functions*

Relation

A relation is a generalization comparing sets of ordered pairs for an equation or inequality such as $x = y + 1$ or $x > y$. The first element in each pair, the x values, form the domain. The second values in each pair, the y values, form the range.

Function

A function is a special relation in which each member of the domain is paired with exactly one member in the range. Functions may be represented using ordered pairs, tables, or graphs. One way to determine whether a relation is a function is to use the vertical line test. Using an object to represent a vertical line, move the object from left to right across the graph. If, for each value of x in the domain, the object passes through no more than one point on the graph, then the graph represents a function.

Linear and Nonlinear Functions

Linear functions have graphs that are straight lines. These graphs represent constant rates of change. In other words, the slope between any two pairs of points on the graph is the same. Nonlinear functions do not have constant rates of change. The slope changes along these graphs. Therefore, the graphs of nonlinear functions are *not* straight lines. Graphs of curves represent nonlinear functions. The equation for a linear function can be written in the form $y = mx + b$, where m represents the constant rate of change, or the slope. Therefore, you can determine whether a function is linear by looking at the equation. For example, the equation $y = \frac{3}{x}$ is nonlinear because x is in the denominator and the equation cannot be written in the form $y = mx + b$. A nonlinear function does not increase or decrease at a constant rate. You can check this by using a table and finding the increase or decrease in y for each regular increase in x. For example, if for each increase in x by 2, y does not increase or decrease the same amount each time, the function is nonlinear.

Linear Equations in Two Variables

In a linear equation with two variables, such as $y = x - 3$, the variables appear in separate terms and neither variable contains an exponent other than 1. The graphs of all linear equations are straight lines. All points on a line are solutions of the equation that is graphed.

Quadratic and Cubic Functions

A quadratic function is a polynomial equation of the second degree, generally expressed as $ax^2 + bx + c = 0$, where a, b, and c are real numbers and a is not equal to zero. Similarly, a cubic function is a polynomial equation of the third degree, usually expressed as $ax^3 + bx^2 + cx + d = 0$. Quadratic functions can be graphed using an equation or a table of values. For example, to graph $y = 3x^2 + 1$, substitute the values -1, -0.5, 0, 0.5, and 1 for x to yield the point coordinates $(-1, 4)$, $(-0.5, 1.75)$, $(0, 1)$, $(0.5, 1.75)$, and $(1, 4)$. Plot these points on a coordinate grid and connect the points in the form of a parabola. Cubic functions also can be graphed by making a table of values. The points of a cubic function from a curve. There is one point at which the curve changes from opening upward to opening downward, or vice versa, called the point of inflection.

Slope

Slope is the ratio of the rise, or vertical change, to the run, or horizontal change of a line: slope = rise/run. Slope (m) is the same for any two points on a straight line and can be found by using the coordinates of any two points on the line:

$$m = \frac{y_2 - y_1}{x_2 - x_1}, \text{ where } x_2 \neq x_1.$$

Asymptotes

An asymptote is a straight line that a curve approaches but never actually meets or crosses. Theoretically, the asymptote meets the curve at infinity. For example, in the function $f(x) = \frac{1}{x}$, two asymptotes are being approached: the line $y = 0$ and $x = 0$. See the graph of the function below.

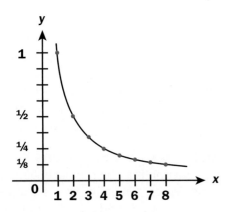

▶ Represent and analyze mathematical situations and structures using algebraic symbols

Variables and Expressions

Algebra is a language of symbols. A variable is a placeholder for a changing value. Any letter, such as x, can be used as a variable. Expressions such as $x + 2$ and $4x$ are algebraic expressions because they represent sums and/or products of variables and numbers. Usually, mathematicians avoid the use of i and e for variables because they have other mathematical meanings ($i = \sqrt{-1}$ and e is used with natural logarithms). To evaluate an algebraic expression, replace the variable or variables with known values, and then solve using order of operations. Translate verbal phrases into algebraic expressions by first defining a variable: Choose a variable and a quantity for the variable to represent. In this way, algebraic expressions can be used to represent real-world situations.

Constant and Coefficient

A constant is a fixed value unlike a variable, which can change. Constants are usually represented by numbers, but they can also be represented by symbols. For example, π is a symbolic representation of the value 3.1415.… A coefficient is a constant by which a variable or other object is multiplied. For example, in the expression $7x^2 + 5x + 9$, the coefficient of x^2 is 7 and the coefficient of x is 5. The number 9 is a constant and not a coefficient.

Monomial and Polynomial

A monomial is a number, a variable, or a product of numbers and/or variables such as 3×4. An algebraic expression that contains one or more monomials is called a polynomial. In a polynomial, there are no terms with variables in the denominator and no terms with variables under a radical sign. Polynomials can be classified by the number of terms contained in the expression. Therefore, a polynomial with two terms is called a binomial ($z^2 - 1$), and a polynomial with three terms is called a trinomial ($2y^3 + 4y^2 - y$). Polynomials also can be classified by their degrees. The degree of a monomial is the sum of the exponents of its variables. The degree of a nonzero constant such as 6 or 10 is 0. The constant 0 has no degree. For example, the monomial $4b^5c^2$ had a degree of 7. The degree of a polynomial is the same as that of the term with the greatest degree. For example, the polynomial $3x^4 - 2y^3 + 4y^2 - y$ has a degree of 4.

Equation

An equation is a mathematical sentence that states that two expressions are equal. The two expressions in an equation are always separated by an equal sign. When solving for a variable in an equation, you must perform the same operations on both sides of the equation in order for the mathematical sentence to remain true.

Solving Equations With Variables

To solve equations with variables on both sides, use the Addition or Subtraction Property of Equality to write an equivalent equation with the variables on the same side. For example, to solve $5x - 8 = 3x$, subtract $3x$ from each side to get $2x - 8 = 0$. Then add 8 to each side to get $2x = 8$. Finally, divide each side by 2 to find that $x = 4$.

MATH APPENDIX

Solving Equations With Grouping Symbols

Equations often contain grouping symbols such as parentheses or brackets. The first step in solving these equations is to use the Distributive Property to remove the grouping symbols. For example $5(x + 2) = 25$ can be changed to $5x + 10 = 25$, and then solved to find that $x = 3$.

Some equations have no solution. That is, there is no value of the variable that results in a true sentence. For such an equation, the solution set is called the null or empty set, and is represented by the symbol $\varnothing$ or {}. Other equations may have every number as the solution. An equation that is true for every value of the variable is called the identity.

Inequality

A mathematical sentence that contains the symbols < (less than), > (greater than), ≤ (less than or equal to), or ≥ (greater than or equal to) is called an inequality. For example, the statement that it is legal to drive 55 miles per hour or slower on a stretch of the highway can be shown by the sentence $s \leq 55$. Inequalities with variables are called open sentences. When a variable is replaced with a number, the inequality may be true or false.

Solving Inequalities

Solving an inequality means finding values for the variable that make the inequality true. Just as with equations, when you add or subtract the same number from each side of an inequality, the inequality remains true. For example, if you add 5 to each side of the inequality $3x < 6$, the resulting inequality $3x + 5 < 11$ is also true. Adding or subtracting the same number from each side of an inequality does not affect the inequality sign. When multiplying or dividing each side of an inequality by the same positive number, the inequality remains true. In such cases, the inequality symbol does not change. When multiplying or dividing each side of an inequality by a negative number, the inequality symbol must be reversed. For example, when dividing each side of the inequality $-4x \geq -8$ by -2, the inequality sign must be changed to ≤ for

the resulting inequality, $2x \leq 4$, to be true. Since the solutions to an inequality include all rational numbers satisfying it, inequalities have an infinite number of solutions.

Representing Inequalities on a Number Line

The solutions of inequalities can be graphed on a number line. For example, if the solution of an inequality is $x < 5$, start an arrow at 5 on the number line, and continue the arrow to the left to show all values less than 5 as the solution. Put an open circle at 5 to show that the point 5 is *not* included in the graph. Use a closed circle when graphing solutions that are greater than or equal to, or less than or equal to, a number.

Order of Operations

Solving a problem may involve using more than one operation. The answer can depend on the order in which you do the operations. To make sure that there is just one answer to a series of computations, mathematicians have agreed upon an order in which to do the operations. First simplify within the parentheses, and then evaluate any exponents. Then multiply and divide from left to right, and finally add and subtract from left to right.

Parametric Equations

Given an equation with more than one unknown, a statistician can draw conclusions about those unknown quantities through the use of parameters, independent variables that the statistician already knows something about. For example, you can find the velocity of an object if you make some assumptions about distance and time parameters.

Recursive Equations

In recursive equations, every value is determined by the previous value. You must first plug an initial value into the equation to get the first value, and then you can use the first value to determine the next one, and so on. For example, in order to determine what the population of pigeons will be in New York City in three years, you can use an equation with the birth, death, immigration, and emigration rates of the birds. Input the current population size into the equation to determine next year's population size, then repeat until you have calculated the value for which you are looking.

▶ Use mathematical models to represent and understand quantitative relationships

Solving Systems of Equations
Two or more equations together are called a system of equations. A system of equations can have one solution, no solution, or infinitely many solutions. One method for solving a system of equations is to graph the equations on the same coordinate plane. The coordinates of the point where the graphs intersect is the solution. In other words, the solution of a system is the ordered pair that is a solution of all equations. A more accurate way to solve a system of two equations is by using a method called substitution. Write both equations in terms of y. Replace y in the first equation with the right side of the second equation. Check the solution by graphing. You can solve a system of three equations using matrix algebra.

Graphing Inequalities
To graph an inequality, first graph the related equation, which is the boundary. All points in the shaded region are solutions of the inequality. If an inequality contains the symbol $\leq$ or $\geq$, then use a solid line to indicate that the boundary is included in the graph. If an inequality contains the symbol $<$ or $>$, then use a dashed line to indicate that the boundary is not included in the graph.

▶ Analyze change in various contexts

Rate of Change
A change in one quantity with respect to another quantity is called the rate of change. Rates of change can be described using slope:

$$\text{slope} = \frac{\text{change in } y}{\text{change in } x}$$

You can find rates of change from an equation, a table, or a graph. A special type of linear equation that describes rate of change is called a direct variation. The graph of a direct variation always passes through the origin and represents a proportional situation. In the equation $y = kx$, k is called the constant of variation. It is the slope, or rate of change. As x increases in value, y increases or decreases at a constant rate k, or y varies directly with x. Another way to say this is that y is directly proportional to x. The direct variation $y = kx$ also can be written as $k = \frac{y}{x}$. In this form, you can see that the ratio of y to x is the same for any corresponding values of y and x.

Slope-Intercept Form
Equations written as $y = mx + b$, where m is the slope and b is the y-intercept, are linear equations in slope-intercept form. For example, the graph of $y = 5x - 6$ is a line that has a slope of 5 and crosses the y-axis at $(0, -6)$. Sometimes you must first write an equation in slope-intercept form before finding the slope and y-intercept. For example, the equation $2x + 3y = 15$ can be expressed in slope-intercept form by subtracting $2x$ from each side and then dividing by 3: $y = -\frac{2}{3}x + 5$, revealing a slope of $-\frac{2}{3}$ and a y-intercept of 5. You can use the slope-intercept form of an equation to graph a line easily. Graph the y-intercept and use the slope to find another point on the line, then connect the two points with a line.

Geometry

▶ *Analyze characteristics and properties of two- and three-dimensional geometric shapes and develop mathematical arguments about geometric relationships*

Angles

Two rays that have the same endpoint form an angle. The common endpoint is called the vertex, and the two rays that make up the angle are called the sides of the angle. The most common unit of measure for angles is the degree. Protractors can be used to measure angles or to draw an angle of a given measure. Angles can be classified by their degree measure. Acute angles have measures less than 90° but greater than 0°. Obtuse angles have measures greater than 90° but less than 180°. Right angles have measures of 90°.

Triangles

A triangle is a figure formed by three line segments that intersect only at their endpoints. The sum of the measures of the angles of a triangle is 180°. Triangles can be classified by their angles. An acute triangle contains all acute angles. An obtuse triangle has one obtuse angle. A right triangle has one right angle. Triangles can also be classified by their sides. A scalene triangle has no congruent sides. An isosceles triangle has at least two congruent sides. In an equilateral triangle all sides are congruent.

Quadrilaterals

A quadrilateral is a closed figure with four sides and four vertices. The segments of a quadrilateral intersect only at their endpoints. Quadrilaterals can be separated into two triangles. Since the sum of the interior angles of all triangles totals 180°, the measures of the interior angles of a quadrilateral equal 360°. Quadrilaterals are classified according to their characteristics, and include trapezoids, parallelograms, rectangles, squares, and rhombuses.

Two-Dimensional Figures

A two-dimensional figure exists within a plane and has only the dimensions of length and width. Examples of two-dimensional figures include circles and polygons. Polygons are figures that have three or more angles, including triangles, quadrilaterals, pentagons, hexagons, and many more. The sum of the angles of any polygon totals at least 180° (triangle), and each additional side adds 180° to the measure of the first three angles. The sum of the angles of a quadrilateral, for example, is 360°. The sum of the angles of a pentagon is 540°.

Three-Dimensional Figures

A plane is a two-dimensional flat surface that extends in all directions. Intersecting planes can form the edges and vertices of three-dimensional figures or solids. A polyhedron is a solid with flat surfaces that are polygons. Polyhedrons are composed of faces, edges, and vertices and are differentiated by their shape and by their number of bases. Skew lines are lines that lie in different planes. They are neither intersecting nor parallel.

Congruence

Figures that have the same size and shape are congruent. The parts of congruent triangles that match are called corresponding parts. Congruence statements are used to identify corresponding parts of congruent triangles. When writing a congruence statement, the letters must be written so that corresponding vertices appear in the same order. Corresponding parts can be used to find the measures of angles and sides in a figure that is congruent to a figure with known measures.

Similarity

If two figures have the same shape but not the same size they are called similar figures. For example, the triangles below are similar, so angles A, B, and C have the same measurements as angles D, E, and F, respectively. However, segments AB, BC, and CA do not have the same measurements as segments DE,

EF, and *FD* , but the measures of the sides are proportional.

For example, $\dfrac{\overline{AB}}{\overline{DE}} = \dfrac{\overline{BC}}{\overline{EF}} = \dfrac{\overline{CA}}{\overline{FD}}$.

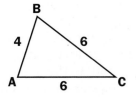

Solid figures are considered to be similar if they have the same shape and their corresponding linear measures are proportional. As with two-dimensional figures, they can be tested for similarity by comparing corresponding measures. If the compared ratios are proportional, then the figures are similar solids. Missing measures of similar solids can also be determined by using proportions.

The Pythagorean Theorem

The sides that are adjacent to a right angle are called legs. The side opposite the right angle is the hypotenuse.

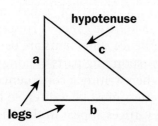

The Pythagorean Theorem describes the relationship between the lengths of the legs *a* and *b* and the hypotenuse *c*. It states that if a triangle is a right triangle, then the square of the length of the hypotenuse is equal to the sum of the squares of the lengths of the legs. In symbols, $c^2 = a^2 + b^2$.

Sine, Cosine, and Tangent Ratios

Trigonometry is the study of the properties of triangles. A trigonometric ratio is a ratio of the lengths of two sides of a right triangle. The most common trigonometric ratios are the sine, cosine, and tangent ratios. These ratios are abbreviated as *sin*, *cos*, and *tan*, respectively.

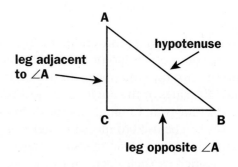

If $\angle A$ is an acute angle of a right triangle, then

$$\sin \angle A = \frac{\text{measure of leg opposite } \angle A}{\text{measure of hypotenuse}},$$

$$\cos \angle A = \frac{\text{measure of leg adjacent to } \angle A}{\text{measure of hypotenuse}}, \text{ and}$$

$$\tan \angle A = \frac{\text{measure of leg opposite } \angle A}{\text{measure of leg adjacent to } \angle A}.$$

MATH APPENDIX

▶ Specify locations and describe spatial relationships using coordinate geometry and other representational systems

Polygons

A polygon is a simple, closed figure formed by three or more line segments. The line segments meet only at their endpoints. The points of intersection are called vertices, and the line segments are called sides. Polygons are classified by the number if sides they have. The diagonals of a polygon divide the polygon into triangles. The number of triangles formed is two less than the number of sides. To find the sum of the measures of the interior angles of any polygon, multiply the number of triangles within the polygon by 180. That is, if n equals the number of sides, then $(n-2)$ 180 gives the sum of the measures of the polygon's interior angles.

Cartesian Coordinates

In the Cartesian coordinate system, the y-axis extends above and below the origin and the x-axis extends to the right and left of the origin, which is the point at which the x- and y-axes intersect. Numbers below and to the left of the origin are negative. A point graphed on the coordinate grid is said to have an x-coordinate and a y-coordinate. For example, the point $(1,-2)$ has as its x-coordinate the number 1, and has as its y-coordinate the number -2. This point is graphed by locating the position on the grid that is 1 unit to the right of the origin and 2 units below the origin.

The x-axis and the y-axis separate the coordinate plane into four regions, called quadrants. The axes and points located on the axes themselves are not located in any of the quadrants. The quadrants are labeled I to IV, starting in the upper right and proceeding counterclockwise. In quadrant I, both coordinates are positive. In quadrant II, the x-coordinate is negative and the y-coordinate is positive. In quadrant III, both coordinates are negative. In quadrant IV, the x-coordinate is positive and the y-coordinate is negative. A coordinate graph can be used to show algebraic relationships among numbers.

▶ Apply transformations and use symmetry to analyze mathematical situations

Similar Triangles and Indirect Measurement

Triangles that have the same shape but not necessarily the same dimensions are called similar triangles. Similar triangles have corresponding angles and corresponding sides. Arcs are used to show congruent angles. If two triangles are similar, then the corresponding angles have the same measure, and the corresponding sides are proportional. Therefore, to determine the measures of the sides of similar triangles when some measures are known, proportions can be used.

Transformations

A transformation is a movement of a geometric figure. There are several types of transformations. In a translation, also called a slide, a figure is slid from one position to another without turning it. Every point of the original figure is moved the same distance and in the same direction. In a reflection, also called a flip, a figure is flipped over a line to form a mirror image. Every point of the original figure has a corresponding point on the other side of the line of symmetry. In a rotation, also called a turn, a figure is turned around a fixed point. A figure may be rotated 90° clockwise, 90° counterclockwise, or 180°. A dilation transforms each line to a parallel line whose length is a fixed multiple of the length of the original line to create a similar figure that will be either larger or smaller.

▶ *Use visualizations, spatial reasoning, and geometric modeling to solve problems*

Two-Dimensional Representations of Three-Dimensional Objects

Three-dimensional objects can be represented in a two-dimensional drawing in order to more easily determine properties such as surface area and volume. When you look at the triangular prism below, you can see the orientation of its three dimensions, length, width, and height. Using the drawing and the formulas for surface area and volume, you can easily calculate these properties.

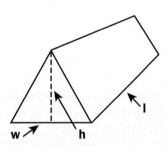

Another way to represent a three-dimensional object in a two-dimensional plane is by using a net, which is the unfolded representation. Imagine cutting the vertices of a box until it is flat then drawing an outline of it. That's a net. Most objects have more than one net, but any one can be measured to determine surface area. Below is a cube and one of its nets.

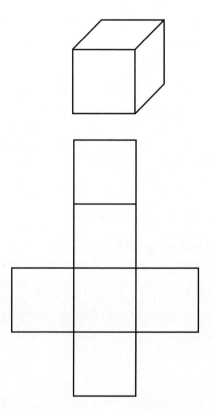

MATH APPENDIX

Measurement

▶ *Understand measurable attributes of objects and the units, systems, and processes of measurement*

Customary System
The customary system is the system of weights and measures used in the United States. The main units of weight are ounces, pounds (1 equal to 16 ounces), and tons (1 equal to 2,000 pounds). Length is typically measured in inches, feet (1 equal to 12 inches), yards (1 equal to 3 feet), and miles (1 equal to 5,280 feet), while area is measured in square feet and acres (1 equal to 43,560 square feet). Liquid is measured in cups, pints (1 equal to 2 cups), quarts (1 equal to 2 pints), and gallons (1 equal to 4 quarts). Finally, temperature is measured in degrees Fahrenheit.

Metric System
The metric system is a decimal system of weights and measurements in which the prefixes of the words for the units of measure indicate the relationships between the different measurements. In this system, the main units of weight, or mass, are grams and kilograms. Length is measured in millimeters, centimeters, meters, and kilometers, and the units of area are square millimeters, centimeters, meters, and kilometers. Liquid is typically measured in milliliters and liters, while temperature is in degrees Celsius.

Selecting Units of Measure
When measuring something, it is important to select the appropriate type and size of unit. For example, in the United States it would be appropriate when describing someone's height to use feet and inches. These units of height or length are good to use because they are in the customary system, and they are of appropriate size. In the customary system, use inches, feet, and miles for lengths and perimeters; square inches, feet, and miles for area and surface area; and cups, pints, quarts, gallons or cubic inches and feet (and less commonly miles) for volume. In the metric system use millimeters, centimeters, meters, and kilometers for lengths and perimeters; square units millimeters, centimeters, meters, and kilometers for area and surface area; and milliliters and liters for volume. Finally, always use degrees to measure angles.

▶ *Apply appropriate techniques, tools, and formulas to determine measurements*

Precision and Significant Digits
The precision of measurement is the exactness to which a measurement is made. Precision depends on the smallest unit of measure being used, or the precision unit. One way to record a measure is to estimate to the nearest precision unit. A more precise method is to include all of the digits that are actually measured, plus one estimated digit. The digits recorded, called significant digits, indicate the precision of the measurement. There are special rules for determining significant digits. If a number contains a decimal point, the number of significant digits is found by counting from left to right, starting with the first nonzero digit.
If the number does not contain a decimal point, the number of significant digits is found by counting the digits from left to right, starting with the first digit and ending with the last nonzero digit.

Surface Area
The amount of material needed to cover the surface of a figure is called the surface area. It can be calculated by finding the area of each face and adding them together. To find the surface area of a rectangular prism, for example, the formula $S = 2lw + 2lh + 2wh$ applies. A cylinder, on the other hand, may be unrolled to reveal two circles and a rectangle. Its surface area can be determined by finding the area of the two circles, $2\pi r^2$, and adding it to the

area of the rectangle, $2\pi rh$ (the length of the rectangle is the circumference of one of the circles), or $S = 2\pi r^2 + 2\pi rh$. The surface area of a pyramid is measured in a slightly different way because the sides of a pyramid are triangles that intersect at the vertex. These sides are called lateral faces and the height of each is called the slant height. The sum of their areas is the lateral area of a pyramid. The surface area of a square pyramid is the lateral area $\frac{1}{2}$ bh (area of a lateral face) times 4 (number of lateral faces), plus the area of the base. The surface area of a cone is the area of its circular base (πr^2) plus its lateral area (πrl, where l is the slant height).

Volume

Volume is the measure of space occupied by a solid region. To find the volume of a prism, the area of the base is multiplied by the measure of the height, $V = Bh$.

A solid containing several prisms can be broken down into its component prisms. Then the volume of each component can be found and the volumes added. The volume of a cylinder can be determined by finding the area of its circular base, πr^2, and then multiplying by the height of the cylinder. A pyramid has one-third the volume of a prism with the same base and height. To find the volume of a pyramid, multiply the area of the base by the pyramid's height, and then divide by 3. Simply stated, the formula for the volume of a pyramid is $V = \frac{1}{3}bh$. A cone is a three-dimensional figure with one circular base and a curved surface connecting the base and the vertex. The volume of a cone is one-third the volume of a cylinder with the same base area and height. Like a pyramid, the formula for the volume of a cone is $V = \frac{1}{3}$ bh. More specifically, the formula is $V = \frac{1}{3}\pi r^2 h$.

Upper and Lower Bounds

Upper and lower bounds have to do with the accuracy of a measurement. When a measurement is given, the degree of accuracy is also stated to tell you what the upper and lower bounds of the measurement are. The upper bound is the largest possible value that a measurement could have had before being rounded down, and the lower bound is the lowest possible value it could have had before being rounded up.

MATH APPENDIX

Data Analysis and Probability

▶ *Formulate questions that can be addressed with data and collect, organize, and display relevant data to answer them*

Histograms

A histogram displays numerical data that have been organized into equal intervals using bars that have the same width and no space between them. While a histogram does not give exact data points, its shape shows the distribution of the data. Histograms also can be used to compare data.

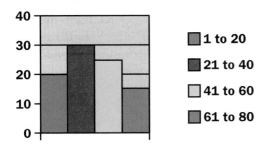

1 to 20
21 to 40
41 to 60
61 to 80

Box-and-Whisker Plot

A box-and-whisker plot displays the measures of central tendency and variation. A box is drawn around the quartile values, and whiskers extend from each quartile to the extreme data points. To make a box plot for a set of data, draw a number line that covers the range of data. Find the median, the extremes, and the upper and lower quartiles. Mark these points on the number line with bullets, then draw a box and the whiskers. The length of a whisker or box shows whether the values of the data in that part are concentrated or spread out.

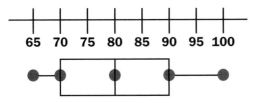

Scatter Plots

A scatter plot is a graph that shows the relationship between two sets of data. In a scatter plot, two sets of data are graphed as ordered pairs on a coordinate system. Two sets of data can have a positive correlation (as *x* increases, *y* increases), a negative correlation (as *x* increases, *y* decreases), or no correlation (no obvious pattern is shown). Scatter plots can be used to spot trends, draw conclusions, and make predictions about data.

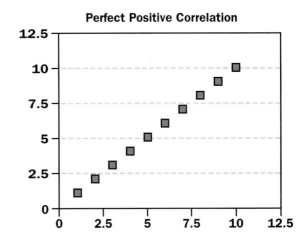

Perfect Positive Correlation

Randomization

The idea of randomization is a very important principle of statistics and the design of experiments. Data must be selected randomly to prevent bias from influencing the results. For example, you want to know the average income of people in your town but you can only use a sample of 100 individuals to make determinations about everyone. If you select 100 individuals who are all doctors, you will have a biased sample. However, if you chose a random sample of 100 people out of the phone book, you are much more likely to accurately represent average income in the town.

Statistics and Parameters

Statistics is a science that involves collecting, analyzing, and presenting data. The data can be collected in various ways—for example through a census or by making physical measurements. The data can then be analyzed by creating summary statistics, which have to do with the distribution of the data sample, including the mean, range, and standard error. They can also be illustrated in tables and graphs, like box-plots, scatter plots, and histograms. The presentation of the data typically involves describing the strength or validity of the data and what they show. For example, an analysis of ancestry of people in a city might tell you something about immigration patterns, unless the data set is very small or biased in some way, in which case it is not likely to be very accurate or useful.

Categorical and Measurement Data

When analyzing data, it is important to understand if the data is qualitative or quantitative. Categorical data is qualitative and measurement, or numerical, data is quantitative. Categorical data describes a quality of something and can be placed into different categories. For example, if you are analyzing the number of students in different grades in a school, each grade is a category. On the other hand, measurement data is continuous, like height, weight, or any other measurable variable. Measurement data can be converted into categorical data if you decide to group the data. Using height as an example, you can group the continuous data set into categories like under 5 feet, 5 feet to 5 feet 5 inches, over 5 feet five inches to 6 feet, and so on.

Univariate and Bivariate Data

In data analysis, a researcher can analyze one variable at a time or look at how multiple variables behave together. Univariate data involves only one variable, for example height in humans. You can measure the height in a population of people then plot the results in a histogram to look at how height is distributed in humans. To summarize univariate data, you can use statistics like the mean, mode, median, range, and standard deviation, which is a measure of variation. When looking at more than one variable at once, you use multivariate data. Bivariate data involves two variables. For exam-

ple, you can look at height and age in humans together by gathering information on both variables from individuals in a population. You can then plot both variables in a scatter plot, look at how the variables behave in relation to each other, and create an equation that represents the relationship, also called a regression. These equations could help answer questions such as, for example, does height increase with age in humans?

▶ Select and use appropriate statistical methods to analyze data

Measures of Central Tendency

When you have a list of numerical data, it is often helpful to use one or more numbers to represent the whole set. These numbers are called measures of central tendency. Three measures of central tendency are mean, median, and mode. The mean is the sum of the data divided by the number of items in the data set. The median is the middle number of the ordered data (or the mean of the two middle numbers). The mode is the number or numbers that occur most often. These measures of central tendency allow data to be analyzed and better understood.

Measures of Spread

In statistics, measures of spread or variation are used to describe how data are distributed. The range of a set of data is the difference between the greatest and the least values of the data set. The quartiles are the values that divide the data into four equal parts. The median of data separates the set in half. Similarly, the median of the lower half of a set of data is the lower quartile. The median of the upper half of a set of data is the upper quartile. The interquartile range is the difference between the upper quartile and the lower quartile.

Line of Best Fit

When real-life data are collected, the points graphed usually do not form a straight line, but they may approximate a linear relationship.

A line of best fit is a line that lies very close to most of the data points. It can be used to predict data. You also can use the equation of the best-fit line to make predictions.

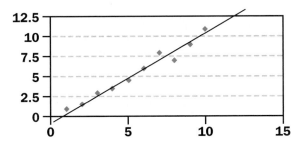

Stem and Leaf Plots

In a stem and leaf plot, numerical data are listed in ascending or descending order. The greatest place value of the data is used for the stems. The next greatest place value forms the leaves. For example, if the least number in a set of data is 8 and the greatest number is 95, draw a vertical line and write the stems from 0 to 9 to the left of the line. Write the leaves from to the right of the line, with the corresponding stem. Next, rearrange the leaves so they are ordered from least to greatest. Then include a key or explanation, such as 1|3 = 13. Notice that the stem-and-leaf plot below is like a histogram turned on its side.

```
0|8
1|3 6
2|5 6 9
3|0 2 7 8
4|0 1 4 7 9
5|1 4 5 8
6|1 3 7
7|5 8
8|2 6
9|5
```

Key: **1|3 = 13**

▶ Develop and evaluate inferences and predictions that are based on data

Sampling Distribution

The sampling distribution of a population is the distribution that would result if you could take an infinite number of samples from the population, average each, and then average the averages. The more normal the distribution of the population, that is, how closely the distribution follows a bell curve, the more likely the sampling distribution will also follow a normal distribution. Furthermore, the larger the sample, the more likely it will accurately represent the entire population. For instance, you are more likely to gain more representative results from a population of 1,000 with a sample of 100 than with a sample of 2.

Validity

In statistics, validity refers to acquiring results that accurately reflect that which is being measured. In other words, it is important when performing statistical analyses, to ensure that the data are valid in that the sample being analyzed represents the population to the best extent possible. Randomization of data and using appropriate sample sizes are two important aspects of making valid inferences about a population.

▶ Understand and apply basic concepts of probability

Complementary, Mutually Exclusive Events

To understand probability theory, it is important to know if two events are mutually exclusive, or complementary: the occurrence of one event automatically implies the non-occurrence of the other. That is, two complementary events cannot both occur. If you roll a pair of dice, the event of rolling 6 and rolling doubles have an outcome in common (3, 3), so they are not mutually exclusive. If you roll (3, 3), you also roll doubles. However, the events of rolling a 9 and rolling doubles are mutually exclusive because they have no outcomes in common. If you roll a 9, you will not also roll doubles.

Independent and Dependent Events

Determining the probability of a series of events requires that you know whether the events are independent or dependent. An independent event has no influence on the occurrence of subsequent events, whereas, a dependent event does influence subsequent events. The chances that a woman's first child will be a girl are $\frac{1}{2}$, and the chances that her second child will be a girl are also $\frac{1}{2}$ because the two events are independent of each other. However, if there are 7 red marbles in a bag of 15 marbles, the chances that the first marble you pick will be red are $\frac{7}{15}$ and if you indeed pick a red marble and remove it, you have reduced the chances of picking another red marble to $\frac{6}{14}$.

Sample Space

The sample space is the group of all possible outcomes for an event. For example, if you are tossing a single six-sided die, the sample space is {1, 2, 3, 4, 5, 6}. Similarly, you can determine the sample space for the possible outcomes of two events. If you are going to toss a coin twice, the sample space is {(heads, heads), (heads, tails), (tails, heads), (tails, tails)}.

Computing the Probability of a Compound Event

If two events are independent, the outcome of one event does not influence the outcome of the second. For example, if a bag contains 2 blue and 3 red marbles, then the probability of selecting a blue marble, replacing it, and then selecting a red marble is $P(A) \times P(B) = \frac{2}{5} \times \frac{3}{5}$ or $\frac{6}{25}$.

If two events are dependent, the outcome of one event affects the outcome of the second. For example, if a bag contains 2 blue and 3 red marbles, then the probability of selecting a blue and then a red marble without replacing the first marble is $P(A) \times P(B \text{ following } A) = \frac{2}{5} \times \frac{3}{4}$ or $\frac{3}{10}$. Two events that cannot happen at the same time are mutually exclusive. For example, when you roll two number cubes, you cannot roll a sum that is both 5 and even. So, $P(A \text{ or } B) = \frac{4}{36} + \frac{18}{36}$ or $\frac{11}{18}$.

Academic Vocabulary Glossary

A

absorbs v. soaks up or take up. *(p. 759)*

access n. ability to enter; to gain a way in to a space or area. *(p. 490)*

accommodate v. to make fit. *(p. 606)*

accurate adj. free from error, conforming to a standard. *(p. 765)*

adjacent adj. lying near or contiguous to something else; adjoining. *(p. 109)*

adjusted adj. arranged to meet a particular situation or requirement. *(p. 300)*

allocation n. something set apart or designated for a special purpose. *(p. 64)*

angles n. figures created by two lines that meet at a common endpoint, or vertex. *(p.108)*

approximate adj. nearly exact, but not perfect. *(p. 163)*

arc n. a curved line or object. *(p. 110)*

assemble v. to fit together or bring together. *(p. 340)*

B

benefit n. advantage. *(p. 481)*

bond n. an adhesive, cementing material. *(p. 800)*

byproduct n. something that is produced in addition to the original product, a side-effect. *(p. 897)*

C

chemicals n. substances produced by or used in a chemical process. *(p. 81)*

circulates v. flows, moves through from one point back to the same point. *(p. 883)*

compatibility adj. ability to exist together in harmony. *(p. 951)*

components n. parts of something. *(p. 86)*

consistent adj. regular; continuous. *(p. 618)*

contraction n. reduction, tightening, narrowing. *(p. 816)*

criteria n. a standard on which a decision may be made. *(p. 398)*

crucial adj. important or essential. *(p. 378)*

D

derived adj. drawn or deduced from a source.

design n. the arrangement of elements or details; purposeful planning. *(p. 781)*

dexterity n. skill and ease in using the hands. *(p. 600)*

diameter n. the width of a circular- or cylinder-like object. *(p. 128)*

dimensions n. measurement in length, width, and thickness. *(p. 457)*

distributed v. given out or delivered to the appropriate people or location. *(p. 482)*

E

ecosystems n. An ecological community that functions, along with its environment, as a unit. *(p. 318)*

eliminate v. get rid of; remove. *(p. 984)*

enables v. to provide with the means or opportunity; to make possible. *(p. 916)*

ensure v. to make certain or guarantee. *(p. 513)*

equilibrium n. a state of balance. *(p. 323)*

evaluate v. to judge or determine the value or significance of; to assess. *(p. 9)*

exceeds v. extends beyond. *(p. 34)*

expands v. to increase, open up, grow in volume or size. *(p. 899)*

exposed adj. being without a covering or shelter. *(p. 1002)*

F

features n. prominent or special parts or characteristics. *(p. 8)*

flexible adj. bendable, supple, elastic. *(p. 902)*

framework n. a structure designed to support or enclose something. *(p. 258)*

function v. to perform a specified task; work; operate. *(p. 384)*

H

horizontal adj. parallel to the level ground. *(p. 111)*

hypotenuse n. the side of a right-angled triangle that is opposite the right angle. *(p. 505)*

I

inclines n. graded or sloped surfaces. *(p. 996)*

incorporate v. to form an indistinguishable whole. *(p. 289)*

increments n. small increases in quantity or size. *(p. 788)*

indicate v. to show, make known, express. *(p. 678)*

initial adj. placed at the beginning; first, early, original. *(p. 694)*

injure v. to cause harm; to damage or hurt. *(p. 205)*

intermediate adj. being in the middle place, stage, or degree. *(p. 530)*

involves v. includes, entails, engages. *(p. 736)*

Academic Vocabulary Glossary

L

layer n. a thickness of material laid on or over a surface. *(p. 298)*

locate v. to find or show the place of. *(p. 237)*

M

maintains v. to keep in an existing state, to preserve. *(p. 1005)*

methods n. procedures or processes for attaining something. *(p. 236)*

minimum n. the least amount or value possible. *(p. 226)*

O

occurs v. to happen or come into existence. *(p. 298)*

offset v. to balance, complement, or compensate for. *(p. 406)*

overall adv. as a whole. *(p. 670)*

P

parallel adj. extending in the same direction; equidistant at all points. *(p. 741)*

perpendicular adj. being at a right angle to a certain line or plane. *(p. 403)*

photosynthesis n. The process in green plants where sunlight, carbon dioxide and water are converted into energy. This process often releases oxygen as a byproduct. *(p. 319)*

physical adj. of or affecting the body. *(p. 79)*

precaution n. an action taken in advance to protect against possible danger or injury. *(p. 631)*

precise adj. being exact; neither more or less. *(p. 155)*

primary adj. first in order. *(p. 432)*

process n. a procedure; a series of actions leading to some end. *(p. 736)*

prolonged v. lengthened in scope or range. *(p. 705)*

R

randomly adj. lacking a definite pattern. *(p. 812)*

ranges v. varies between specific limits. *(p. 162)*

ratio n. a comparison of two numbers. *(p. 205)*

reinforces v. strengthens with additional material or support. *(p. 751)*

remove v. to eliminate, push aside, or take away. *(p. 839)*

retain v. to hold secure or intact. *(p. 360)*

rural adj. the country, rustic. *(p. 870)*

S

scale n. the ratio between the size of the object as it is represented and the actual size of the object. *(p. 44)*

sequence n. a continuous or connected series. *(p. 832)*

series n. a number of objects arranged one after another. *(p. 222)*

significant adj. important, vital, imperative. *(p. 510)*

simulated adj. made to look real; an imitation. *(p. 821)*

species n. a biological classification of organisms having some common qualities. *(p. 318)*

specific adj. having a special application or reference. *(p. 12)*

specify v. to name or state clearly or in. *(p. 495)*

stable adj. steady, established, not changing. *(p. 922)*

succeeding adj. coming after in position. *(p. 638)*

sufficient adj. adequate for the purpose; enough; plenty. *(p. 576)*

suspended v. hanging free on all sides except at the point of support. *(p. 532)*

synthetic adj. formed through a chemical process instead of being of natural origin. *(p. 948)*

T

techniques n. a particular method of accomplishing a task. *(p. 222)*

technology n. a technological method, invention, process. *(p. 578)*

temporary adj. lasting for a short time. *(p. 356)*

traditional adj. customary, established. *(p. 661)*

transmission n. spread, diffusion, process of transmitting sound waves from one place to the next. *(p. 696)*

U

uniform adj. having an unvaried appearance of surface, pattern, or color. *(p. 973)*

V

versatile adj. having or capable of having many uses. *(p. 165)*

version n. a form or adaptation of the original. *(p. 552)*

vertical adj. being upright or plumb. *(p. 111)*

visible adj. capable of being seen. *(p. 240)*

visualization n. the formation of mental visual pictures. *(p. 6)*

Content Vocabulary Glossary

A

acclimation when wood reaches a moisture content equal to that inside the building where it is installed. *(p. 975)*

acoustical ceiling a ceiling that consists of panels glued directly to the ceiling surface or stapled to wood furring strips nailed to the ceiling joists. *(p. 941)*

acoustical insulation insulation that slows the transmission of sound. *(p. 908)*

ACQ (Alkaline Copper Quaternary) a chemical compound used to treat wood. *(p. 1003)*

addition a roof that intersects the main roof. *(p. 516)*

admixture an ingredient other than cement, aggregate, or water that is added to a concrete mix to change its physical or chemical characteristics. *(p. 221)*

air conditioning the process of extracting heat from air inside a house and releasing it outside. *(p. 888)*

amperage the strength of an electric current. *(p. 163)*

apprentice an inexperienced worker who learns a trade by working under the guidance of an expert worker. *(p. 6)*

architect's scale a tool that architects use when making scale drawings. *(p. 44)*

axial load the load carried along the length of a structural member. *(p. 844)*

B

backerboard cement-based sheets which provide a base for tile, particularly on floors and in wet installations; also called cement board. *(p. 993)*

backing a long strip of material that is nailed to a wall as support for large moldings. *(p. 771)*

backing the hip beveling the upper edge of the hip rafter so roof sheathing can be installed without hitting the corners of the hip rafter. *(p. 513)*

back-priming priming the back surface of siding on site. *(p. 663)*

balloon-frame construction framing in which the studs run from the sill attached to the foundation to the top plate of the second floor; also called balloon framing *(p. 370)*

balusters the slender vertical members that support the handrail on a stairway. *(p. 726)*

base cabinets cabinets that rest on the floor and support the countertops; also called lower cabinets. *(p. 781)*

baseboard a board or molding used against the bottom of walls to cover their joint with the floor; also called a base molding. *(p. 764)*

batt any thick insulation material that comes in pre-cut widths designed to fit between framing members. *(p. 897)*

batter board a board fastened horizontally to stakes placed to the outside of where the corners of a building will be located. *(p. 244)*

bearing capacity a measure of how well the soil can support the weight of a house. *(p. 247)*

bearing wall a wall that supports loads in addition to its own weight. *(p. 411)*

bed joint a horizontal mortar joint. *(p. 275)*

bench mark a basic starting point from which measurements in building layout can be made using a transit or level; also called a point of reference (POR). *(p. 239)*

bevel cut a type of miter cut that is made an an angle through the thickness of a board. *(p. 136)*

bid a signed proposal to do work and/or supply material for a specified price. *(p. 59)*

binder a resin that holds particles of pigment together. *(p. 947)*

biscuit a small, flat piece of compressed wood. Also called a plate. *(p. 184)*

bisque an unglazed ceramic tile. *(p. 990)*

blind nailing driving nails at an angle through the tongue of a board and into framing or furring strips. It allows subsequent boards to conceal the nail heads. *(p. 818)*

board foot a unit of measure that represents a piece of lumber having a flat surface area of 1 sq. ft. and a thickness of 1" nominal size. *(p. 63)*

board paneling a type of wall paneling that is made of wood and comes in the form of individual boards. *(p. 816)*

Boston ridge created when hip and ridge shingles overlap on a roof. *(p. 648)*

box extender a metal or plastic fitting that is screwed to the front of an outlet box, bringing it forward. *(p. 811)*

box sill a type of construction consisting of a sill plate that is anchored to the foundation wall, floor joists and rim joist, and subflooring. *(p. 402)*

brace a member used to stiffen or support a structure. *(p. 544)*

Content Vocabulary Glossary

bridging a method of bracing between joists to distribute loads, prevent the joists from twisting, and add stability and stiffness. *(p. 410)*

building code a standard set of regulations that govern the procedures and details of construction. *(p. 34)*

building envelope a continuous insulation layer that separates heated spaces from unheated spaces in walls, ceilings, roofs, and floors. *(p. 895)*

building permit a formal, printed authorization for the builder to begin construction. *(p. 35)*

bull float a wide, flat metal or wood pad that is pushed back and forth over concrete to make the surface even. *(p. 305)*

business plan gives specific information about the business, including vision, goals, strategies, and plan of action. *(p. 12)*

butt edge the exposed edge of a shingle. *(p. 626)*

C

camber a curve in a beam. *(p. 361)*

cambium a layer of living tissue that produces new wood, called sapwood, along its inner surface. *(p. 318)*

cantilever a supporting member that projects into space and is itself supported only at an end. *(p. 412)*

carcase an assembly of panels that forms a cabinet's basic shape. *(p. 788)*

career clusters groups of related occupations. *(p. 6)*

career pathway an area of concentration within a career cluster. *(p. 6)*

carrier a liquid that keeps pigments and binders in suspension, and keeps them evenly dispersed (spread out) during application. *(p. 947)*

casing the basic molding or trim around a window or door. *(p. 758)*

cell the part within an electronic air cleaner that removes contaminants. *(p. 884)*

certification a formal process that shows that an individual is qualified in a particular job task. *(p. 9)*

chair a small device that supports the wire fabric at a particular height as concrete is poured. *(p. 230)*

chamfer a beveled edge. *(p. 171)*

chord the top or bottom outer member of a roof truss. *(p. 492)*

circuit a cable or group of cables that supplies electricity to a specific area or appliance. *(p. 877)*

clinching the process of joining two layers of steel with pressure in steel-frame construction. *(p. 843)*

clip angle a small piece of galvanized steel attached to a structural member to accept a structural load. *(p. 847)*

cold joint a joint occurring where fresh concrete is poured on top of or next to concrete that has already begun to cure. *(p. 266)*

cold-formed steel sheet steel that is bent and formed without using heat. *(p. 832)*

collar tie a horizontal roof framing member that prevents opposing rafter pairs from spreading apart. *(p. 542)*

collated fasteners fasteners arranged into strips or rolls, with each fastener connected to the fasteners on either side. *(p. 191)*

collet the part of a router that holds a bit. *(p. 171)*

common difference the difference in length of gable-end studs based on a single figure that depends on the pitch of the roof. *(p. 546)*

competent person someone who has been trained to identify existing and predictable hazards on the job site. *(p. 205)*

composite panel product a panel product made from pieces of wood mixed with adhesive. *(p. 347)*

compound miter saw a saw in which the head of the saw pivots up and down and from side to side. *(p. 147)*

concrete a hard, strong building material that is made by mixing cement, fine aggregate (a granular material, such as sand), coarse aggregate (usually gravel or crushed stone), and water in the proper proportions. *(p. 218)*

concrete flatwork flat, horizontal areas of concrete placed either directly on the ground or over compacted gravel or sand. *(p. 294)*

condensation the process by which a vapor turns into a liquid. *(p. 900)*

conductor a material that electricity readily flows through. *(p. 85)*

coniferous tree a tree that produces seeds in cones and has needlelike or scalelike leaves. *(p. 320)*

consolidation a process that removes air pockets and forces the concrete into all parts of the forms. *(p. 227)*

control joint in a concrete slab, a joint that helps minimize random cracks; in a wall, a joint that controls movement caused by stress. *(p. 283)*

coping the process of shaping inside corner joints between trim members by cutting the end of one member to fit against the face of the other. *(p. 767)*

corbel a course of brick offset to extend past the course below it. *(p. 712)*

corner bead a vinyl or galvanized metal strip that reinforces and protects corners of drywall. *(p. 919)*

corner post an assembly of full-length studs at the corner of a building. *(p. 441)*

cornice the exterior trim of a structure consisting of a fascia, a soffit, and various types of molding. *(p. 552)*

countersink a bit with beveled cutting edges. It creates a funnel shape at the top of a drilled hole, allowing the head of a wood screw to be flush with the wood surface. *(p. 166)*

crazing the appearance of fine cracks in irregular patterns over the surface of concrete. *(p. 222)*

cripple stud a stud that does not extend all the way from the bottom plate to the top plate of the wall because of an opening such as framing for a window. Also called a cripple. *(p. 434)*

crosscut a cut made across the grain of a piece of lumber, and at a 90° angle to the edge. *(p. 128)*

crown the outermost curve of the bow in a joist. *(p. 406)*

crown molding a fairly large molding that usually includes both curved and angular surfaces. *(p. 770)*

custom cabinets manufactured cabinets built in any width or height to match a kitchen exactly. *(p. 788)*

cutterhead a solid metal cylinder on which three or four cutting knives are mounted on a jointer. *(p. 179)*

D

dead load the total weight of a building, including the frame and anything permanently attached. *(p. 383)*

deciduous tree a tree that sheds its leaves annually, during cold or very dry seasons. *(p. 320)*

design value a number assigned to how well a particular wood resists stresses. *(p. 379)*

differential leveling the process of determining differences in elevation between points that are remote from each other. *(p. 246)*

doghouse dormer a gable dormer with side walls that protrudes horizontally outward from a sloping roof and has its own gabled ends. *(p. 519)*

door frame the assembly around a door, attached to the wall framing and consisting of two side jambs and a head jamb. *(p. 597)*

dormer an upright projection on a roof, usually with a window. *(p. 518)*

draft the upward movement of air within a chimney. *(p. 709)*

drain field a filtering area within a septic system into which liquid wastes flow. *(p. 870)*

drainage plane a gap or series of gaps behind the siding that allows water to drain freely. *(p. 682)*

dropping the hip deepening the bird's mouth on a hip rafter to bring the top edge of the hip rafter in line with the upper ends of the jacks. *(p. 513)*

dry rot brown, crumbly decay in affected wood; this is a misnomer because wood must be damp for rotting to occur. *(p. 328)*

E

eaves the portions of a roof that project beyond the walls. *(p. 551)*

edging using a brush to paint into the corners between large flat surfaces where a roller cannot reach. *(p. 963)*

efflorescence a whitish crystalline deposit that sometimes appears on the surface of concrete or mortar. *(p. 222)*

electrical circuit when electricity flows from one point of origin and returns to that point of origin through a conductor. *(p. 85)*

elevation a side view that allows you to see the height and width of objects. *(p. 53)*

emissivity a measure of how much radiation something is emitting. *(p. 903)*

engineered lumber any manufactured product made of solid wood, wood veneer, wood pieces, or wood fibers in which the components have been bonded together with adhesives. *(p. 352)*

engineered panel any manufactured sheet product, including plywood, that is made of wood or wood pieces bonded with a natural or synthetic adhesive. *(p. 338)*

entrepreneur a person who creates and runs his or her own business. *(p. 11)*

ergonomics the science of designing and arranging things to suit the needs of the human body. *(p. 90)*

ethics your inner guidelines for telling right from wrong. *(p. 24)*

excavation a cut, cavity, trench, or depression made by removing earth. *(p. 82)*

exposure the amount of surface (as in a shingle or siding) exposed to the weather. *(p. 626)*

Content Vocabulary Glossary

F

face-frame cabinet a cabinet having a frame that fits around the front of a carcase, providing a mounting surface for hinges and drawer hardware. (p. 788)

fascia a board that is nailed to the ends of rafter tails. (p. 552)

feathering in taping drywall joints, the process of smoothing the outer edges of the joint compound so that there are no ridges (p. 931); in using a screw gun, the process of attaching a screw to the bit without stopping the gun. (p. 836)

feed rate the speed at which stock is pushed through the saw blade. (p. 142)

fiber-saturation point the point at which the cell walls of wood have absorbed all the water they can hold. (p. 321)

fines finely crushed or powdered materials. (p. 299)

finger joint a joint having a closely spaced series of wedge-shaped cuts made in the wood. (p. 363)

firebox the area of a fireplace where burning takes place. (p. 704)

first aid the initial help and care given to an injured person. (p. 83)

fixture any device, such as a bathtub, that receives or drains water. (p. 868)

flagged bristles bristles, as on a paintbrush, having slightly splayed tips. (p. 960)

flashing a piece of metal that protects a roof against water seepage. (p. 626)

flight a straight run of stairs. (p. 729)

floor plan a scale drawing showing the size and location of rooms on a given floor. (p. 37)

flue the passage inside the chimney through which the air, gases, and smoke rise. (p. 709)

footing a base that provides a surface that distributes weight over a wide area of soil. (p. 256)

frameless cabinet a cabinet that does not have a frame around the opening. Its hinges are concealed and mounted on the side walls. (p. 789)

framing connector a formed metal bracket that is installed at framing connections using nails. (p. 386)

free enterprise an economic system in which businesses or individuals may buy, sell, and set prices for goods and services. (p. 12)

frost depth the depth in any climate below which the soil does not freeze. (p. 294)

full-height paneling paneling that runs floor to ceiling. (p. 808)

G

gain a mortise (notch) cut into a door or jamb for a hinge. (p. 599)

galvanizing a process of coating steel with a protective layer of zinc. (p. 1006)

girder a large principal horizontal member used to support floor joists. (p. 396)

glazing the clear glass or plastic portions of a window. (p. 577)

glulam short for "glue laminated," as in "glue-laminated beam." (p. 360)

grade a general indication of the quality and strength of a piece of lumber. (p. 323)

grade stamp a permanent mark that identifies a board's species, quality, mill source, and a general indication of strength. (p. 324)

ground a material permanently or temporarily attached to a surface to be plastered. (p. 934)

grounding provides a path for electricity to flow safely from electrical equipment to the earth. (p. 85)

grout a thin mortar used for filling spaces, especially between ceramic tiles. (p. 944)

H

hand the direction in which a door swings. (p. 601)

head casing the horizontal piece at the top of a door or window. (p. 758)

head joint a vertical mortar joint. (p. 275)

header in flooring, a horizontal member that carries loads from other members and directs them around an opening (p. 411); in wall framing, a wood beam placed at the top of an opening. (p. 432)

headroom the clearance directly above a step. (p. 730)

hearth the floor of a firebox, plus the fireproof area in front of the fireplace. (p. 706)

heartwood the dark-colored portion of a tree nearest the core. (p. 1004)

heat pump a device that can heat and cool a house. (p. 884)

hip rafter a rafter forming a raised area, or hip, usually extending from the corner of the building diagonally from plate to ridge board. (p. 504)

hollow-core construction door construction consisting of a light framework of wood or corrugated cardboard faced with thin plywood or hardboard. *(p. 597)*

humidifier a device that adds moisture to the air inside a house and counteracts the drying effects of hot air. *(p. 884)*

hydration a chemical reaction that occurs when water combines with cement. *(p. 218)*

I

ice dam ice formed by melting snow freezing at the eave line on a roof. *(p. 632)*

Impact Noise Rating (INR) a measure of the resistance of a floor system based on decibels (dB). *(p. 911)*

independent slab a slab that is used in areas in which the ground freezes fairly deep during winter; also called ground-support slab. *(p. 295)*

indirect cost an overhead cost that relates to the organization and supervision of a project, which includes the cost of office supplies, construction tools and equipment, office payroll, and taxes. *(p. 64)*

infiltration the passage of fresh air into a building through cracks around windows, doors, and framing. *(p. 887)*

in-line framing framing that aligns all vertical and horizontal load-bearing structural members in both platform-frame construction and steel-frame construction. *(p. 373, p. 844)*

internal cut the technique of cutting a large hole in a material without starting at the edge. *(p. 150)*

interview a meeting between an employer and a job applicant. *(p. 22)*

J

jack rafter a shortened common rafter that may be framed to a hip rafter, a valley rafter, or both. *(p. 504)*

J-channel a J-shaped piece of plastic or metal that is used to support trim. *(p. 677)*

job application a form that asks questions about a job applicant's skills, work experience, education, and interests. *(p. 19)*

jointer a simple metal bar with a shaped end used to form and compact mortar joints; also a type of woodworking machine. *(p. 688)*

joist tracks C-shaped members by which joists are attached to the foundation; also called rim tracks. *(p. 845)*

K

kerf the width of a cut. *(p. 132)*

kick plate a short piece of framing lumber that is used to anchor the bottom of a stair. *(p. 744)*

kickback a reaction that occurs when a spinning blade encounters something that slows or stops it while the saw is under full power. *(p. 129)*

kiln an oven in which moisture, airflow, and temperature are carefully controlled. *(p. 321)*

kneeboard a board measuring placed on concrete to support the weight of the finisher. *(p. 307)*

L

ladder a structure made up of two long side pieces joined by multiple crosspieces on which you can step. *(p. 198)*

laminated-veneer lumber (LVL) an engineered wood product in which the basic element is wood veneer glued together. *(p. 353)*

lath any base material for plaster that has qualities that encourage the plaster to stick to it. *(p. 933)*

lead corner a partially constructed corner of brick. *(p. 699)*

ledger a length of lumber that connects a deck to the house. *(p. 1013)*

level a long wood, metal, or fiberglass instrument with several glass vials that measures the levelness or plumbness of a surface. *(p. 111)*

lifeline a rope intended to prevent a worker from falling. *(p. 209)*

lift a uniform and fairly shallow layer of fill. *(p. 298)*

line block a small L-shaped device that hooks over the edge of a brick and is held in place by the tension of the string. *(p. 701)*

lintel a horizontal support for masonry installed above an opening such as a window, door, or masonry fireplace. *(p. 699)*

live load weight that is not permanently attached to a building, such as furniture. *(p. 383)*

load a force, such as weight or wind, that creates stresses on a structure. *(p. 379)*

lockset an assembly of knobs, latch, and a locking mechanism for a door. *(p. 600)*

Content Vocabulary Glossary

lookout a horizontal member that extends from a rafter end to a nailer or the face of the wall sheathing; lookouts form a horizontal surface to which the soffit material is attached. *(p. 552)*

lumber pieces of wood having a uniform thickness and width, and sawn from a log. *(p. 321)*

M

makeup air air that replaces the air exhausted by a combustion appliance such as a fireplace. *(p. 705)*

mastic a thick, premixed adhesive spread with a notched trowel or with a caulking gun. *(p. 976)*

mechanicals a general term for plumbing, electrical, and HVAC systems. *(p. 868)*

medium-density fiberboard (MDF) a panel product made of compressed wood fibers mixed with urea-formaldehyde adhesive. *(p. 350)*

miter cut a cut made across the grain of a board at an angle other than 90°. *(p. 136)*

moisture content (MC) the amount of water wood contains. *(p. 321)*

molding narrow lengths of wood with a shaped profile used for decorative and framing purposes. *(p. 751)*

monolithic slab consists of a footing and floor slab that are formed into one pour; also called a unified slab, a thickened-edge slab, or a slab with a turned-down footing. *(p. 295)*

mortgage a long-term (15 to 30 years) loan that is secured by the property. *(p. 40)*

mullion strips vertical wood pieces separating windows of various styles and sizes combined to make up a larger unit. *(p. 587)*

muntin a short vertical or horizontal piece used to hold a pane of window glass. *(p. 578)*

musculoskeletal disorder (MSD) a disorder of the muscles, tendons, ligaments, joints, cartilage, or spinal discs which can be caused by lifting, fastening materials, and other tasks. *(p. 90)*

N

networking making use of personal connections to achieve career goals. *(p. 17)*

nominal dimension the width and thickness of rough-sawn lumber (not its length); also called nominal size. *(p. 325)*

O

Occupational Safety and Health Administration (OSHA) a federal agency which issues standards and rules for safe and healthful working conditions, tools, equipment, facilities, and processes. It conducts workplace inspections to ensure the standards are followed. *(p. 79)*

offcut a waste piece of cut wood. *(p. 141)*

on center (OC) the distance from the centerline of one structural member to the centerline of the next closest member. *(p. 382)*

open valley a type of roof valley in which shingles are not applied to the intersection of two roof surfaces, leaving the underlying flashing exposed along the length of the valley. *(p. 641)*

oriented-strand board (OSB) a wood product made from wood strands bonded with adhesive under heat and pressure. *(p. 349)*

overdig the term used to describe the additional excavation needed to provide clearance for work. *(p. 248)*

P

parging the process of spreading mortar or cement plaster over a block. *(p. 287)*

parquet any flooring assembled with small, precisely cut pieces of wood in a geometric pattern, such as squares, rectangles, and herringbone patterns. *(p. 973)*

passage door a door that swings open and closed on two or more leaf hinges mounted along one side. *(p. 596)*

performance method a method of framing that depends upon established engineering principles and design-load specifications to calculate size and strength for individual steel framing members. *(p. 832)*

pier a concrete column that supports a concentrated load, such as a post. *(p. 1009)*

pilot hole a hole drilled in wood to start and guide a screw. *(p. 166)*

pitch in roof framing, the ratio of total rise to span. *(p. 470)*

plan view a top view, or bird's eye view, which allows you to see the width, length, and location of objects as if you were standing on a platform high above them and looking down. *(p. 50)*

plank any solid-wood board that is at least 3" wide. *(p. 972)*

plate a horizontal framing member used to tie together interior and exterior wall framing. *(p. 432)*

platform-frame construction a construction technique in which each level of the house is constructed separately; also called platform framing. *(p. 370)*

pliers a hand tool with opposing jaws that are designed to hold things. *(p. 122)*

plies very thin, pliable sheets of wood that have been sawed, peeled, or sliced from a log and used in plywood. *(p. 338)*

plot plan the part of the house plans that shows the location of the building on the lot, along with related land elevations. *(p. 236)*

plywood a building material that consists of layers of wood veneer and sometimes other materials that have been glued together. *(p. 338)*

pneumatic tool a tool powered by compressed air. *(p. 187)*

porosity the measurement of the ability of water to flow through soil. *(p. 248)*

Portland cement a manufactured cement used in modern concrete. *(p. 219)*

post-and-beam framing a framing system in which fewer, but larger, pieces of wood are spaced farther apart than those used in conventional framing. *(p. 373)*

precast any concrete object that is cast in a factory, cured under controlled conditions, and then delivered to the job site. *(p. 1018)*

prescriptive method a method of framing that uses standardized tables that give specifications and other information relating to steel frame construction. *(p. 833)*

primer a paint that has a higher proportion of binder than standard paint. *(p. 948)*

pullout capacity the ability of a screw to resist pulling out of a connection. *(p. 841)*

pump jack a metal device with a foot pedal that a worker pumps to make it slide up and down on a wood or aluminum post. *(p. 208)*

purlin a horizontal structural member that supports roof loads and transfers them to structural supports. *(p. 544)*

Q

quantity takeoff a cost estimate in which every piece of material required to build the house is counted and priced. *(p. 60)*

R

rabbet a cut or groove along or near the edge of a piece of wood which allows another piece to fit into it to form a joint. *(p. 740)*

radiant heat heat that travels in a straight line away from a hot surface and heats anything solid it meets. *(p. 902)*

radon a colorless and odorless radioactive gas that travels through soil and can be extremely toxic to humans. *(p. 288)*

rafter an inclined framing member that supports the roof. *(p. 467)*

rails the vertical supports on a ladder to which the rungs or steps are attached. *(p. 198)*

rake the part of a gable roof that extends beyond the end walls. *(p. 558)*

receptacle a contact device with a combination of slots and grounding holes sized to accept the prongs of an electrical plug. *(p. 879)*

refractory cement a cement resistant to high temperatures. *(p. 705)*

refrigerant a material that absorbs heat as it becomes a gas and gives up heat as it becomes a liquid. *(p. 884)*

regulator a valve that controls the air pressure reaching a nailer or stapler. *(p. 189)*

repetitive stress injury (RSI) when a task done over and over causes minor irritation to nerves and tissues. *(p. 90)*

résumé a brief summary of a job applicant's contact information, education, skills, work experience, activities, and interests. *(p. 18)*

retempering the process of adding water to a batch of mortar that has become too stiff to work. *(p. 695)*

return a piece that continues the profile of trim or molding around a corner. *(p. 762)*

reveal a small offset between a piece of trim and the surface to which it is applied. *(p. 759)*

ridge a roof framing member placed at the intersection of two upward-sloping surfaces. *(p. 530)*

Content Vocabulary Glossary

ridge beam a horizontal roof framing member to which the tops of rafters are fastened. (p. 530)

ridge board a horizontal piece that connects the upper ends of rafters. (p. 468)

rip cut a cut made in the direction of the wood grain. (p. 142)

risers the spaces between treads enclosed by vertical boards. (p. 726)

roof rake that portion of a roof frame that extends beyond the walls on the gabled ends. (p. 862)

rough opening (RO) the space into which a door or window will fit. It allows room for the door or window and its frame. (p. 440)

rough sill a horizontal member placed at the bottom of a window opening to support the window. It connects the upper ends of the cripple studs. (p. 434)

R-value in insulation, a measure of a material's ability to resist heat transmission. (p. 894)

S

sapwood the outer growth layer of a tree. (p. 1004)

sash the part of a window that holds the glazing. (p. 577)

scaffold a raised platform used for working at a height. (p. 203)

scaffold planks the horizontal parts of a scaffold on which a worker stands. (p. 204)

schedule a list or chart that provides information too detailed to include in drawings or plans. (p. 56)

screed a long, straight length of metal or wood that is used to "strike off" (level) concrete. (p. 301)

seasoning the process of drying wood by air drying or kiln drying. (p. 321)

seat cut the cut in a rafter that the rafter is to rest on. (p. 508)

semi-custom cabinets manufactured cabinets built only after they are ordered for a specific kitchen. (p. 788)

service main a pipe that brings water to the house and is connected at the street to the municipal water system; also called a water service pipe. (p. 869)

shear wall a wall engineered to withstand unusual lateral (sideways) stresses. (p. 378)

sheathing rigid 4×8 or larger panels that are attached to the outside surface of the exterior wall framing. Sheathing adds great stiffness and strength to the walls. (p. 430)

sheen a description of how shiny a surface is. (p. 947)

side casing a vertical piece at the side of a door or window. (p. 758)

side lap the amount that adjacent roofing sheets overlap each other horizontally; also called end lap. (p. 626)

siding the exterior wall covering of a house. (p. 658)

site layout the process of marking the location of a building on the land itself. (p. 236)

skirtboard a finished board that is nailed to the wall before the stringers are installed on a stairway to protect the wall from damage. (p. 739)

sleeper a length of lumber that supports wood flooring over concrete. (p. 982)

slope in roof framing, a ratio of unit rise to unit run. (p. 470)

slump test a test to measure the consistency of concrete. (p. 226)

soffit on the interior, an area around the perimeter of a room that is lower than the rest of the ceiling; on the exterior, the underside of the eaves. (p. 460, p. 552)

solid-core construction a type of construction used for exterior doors in which strips of wood, particleboard, rigid foam, or other core material are covered with a thin outer material, such as wood veneer. (p. 596)

solvent a material that dissolves another material. (p. 948)

Sound Transmission Class (STC) a numerical rating that indicates the ability of a material or combination of materials to reduce sound transmission. (p. 908)

span in roof framing, the distance between the outer edges of the top plates. (p. 470)

span table a table listing the maximum spacing allowed between different sizes of joists or rafters. (p. 382)

specifications written notes that give instructions about materials and methods of work. (p. 57)

spline a thin strip of wood used to reinforce a joint. *(p. 375)*

spreader a device that holds a ladder open and prevents it from closing accidentally. *(p. 202)*

springing angle the angle at which molding projects away from a wall. *(p. 771)*

square a tool that is used primarily to measure or check angles; the amount of roofing required to cover 100 sq. ft. of rood surface. *(p. 108, p. 626)*

square the amount of roofing material required to cover 100 sq. ft. of roof surface. *(p. 628)*

stairway a series of steps along with all the related elements, including structural elements such as stringers and finish elements such as handrails and balusters. *(p. 726)*

stairwell the vertical shaft inside of which a stairway is built. *(p. 724)*

station mark the point over which a level for laying out a site is directly centered. *(p. 239)*

step a tread and a riser on a stairway. *(p. 726)*

stepladder a common type of folding ladder that has flattened steps instead of rungs. *(p. 199)*

sticker a long, slender piece of scrap wood that separates layers of wood products and allows air circulation between them. *(p. 809)*

stiles the vertical side members in raised-panel wood doors. *(p. 597)*

stock cabinets manufactured cabinets built in standard sizes and stored in a warehouse until ordered. *(p. 788)*

stock plan a standard house plan that can be adapted to fit many different lots. *(p. 37)*

stoop an enlarged landing at the top of steps. *(p. 1018)*

story pole in carpentry, a measuring device made on site to ensure a uniform layout all around the house *(p. 280)*; in masonry, a board used to gauge the top of the masonry for each course. *(p. 664)*

strike plate a metal plate inserted into an opening in the door jamb into which the latch slips. *(p. 596)*

stringer a long piece of 2× lumber that supports a stair. *(p. 726)*

structural insulated panel a rigid panel of 3½" thick expanded polystyrene (EPS) foam insulation between sheets of exterior plywood or oriented-strand board (OSB); also called a foam-core panel. *(p. 375)*

stud a vertical framing member. Conventional construction commonly uses 2×4 studs spaced 16" on center (OC). *(p. 432)*

subflooring engineered wood sheets or construction grade lumber that is used to construct a subfloor, which is a rough floor laid on floor joists as a base for the finish floor. *(p. 402)*

subgrade the earth below the foundation slab. *(p. 298)*

substrate a material that serves as a base for another material. *(p. 799)*

summerwood the dense, dark-colored portion of wood. *(p. 946)*

suspended ceiling a ceiling that consists of panels held in place by metal or plastic grid. *(p. 937)*

T

tail joist a floor joist interrupted by a header. *(p. 412)*

template a guide made from metal or thin wood. *(p. 172)*

temporary bracing bracing that prevents the walls from tipping as they are being erected, and holds walls in position after they have been plumbed and straightened. *(p. 450)*

theodolite a transit that reads horizontal and vertical angles electronically. *(p. 238)*

thermal insulation material that slows the transmission of heat through walls, floors, and ceilings. *(p. 894)*

top lap the portion of a shingle not exposed to the weather. *(p. 626)*

torque a twisting force that produces rotation. *(p. 130)*

total rise in roof framing, the vertical distance from the top of the top plate to the upper end of the measuring line. *(p. 470)*

total run in roof framing, one-half the span. *(p. 470)*

trap a curved section of drainpipe located beneath a fixture that prevents sewer gases in the waste pipes from entering the house but does not block drainage; also called P-trap. *(p. 826)*

treads the parts of the stair upon which you step. *(p. 725)*

Content Vocabulary Glossary

trend a general development or movement in a certain direction. *(p. 8)*

trestle a portable metal frame with rungs that is used to support scaffold planks at various heights. *(p. 207)*

trim a length of wood used for decorative and framing purposes with square edges that is surfaced on four sides (S4S). *(p. 751)*

trimmer joist a joist used to form the sides of a large opening. *(p. 412)*

trimmer stud a short beam that supports the header over a window or door opening to transfer structural loads from the header to the bottom plate. Also called a trimmer or jack stud. *(p. 434)*

U

undercourse a low-grade layer of side-wall shingles that will not be exposed to the weather. *(p. 670)*

underlayment in roofing, a material, such as roofing felt, applied to the roof sheathing before shingles are installed *(p. 627)*; in flooring, a thin panel product whose surface is smoother than standard subflooring. *(p. 989)*

unit dimension the overall size of a window, including casings. *(p. 586)*

unit rise the number of inches that a roof rises for every 12" of run (the unit run). *(p. 470)*

unit run in roof framing, a set length that is used to figure the slope of rafters; also called unit of run. *(p. 470)*

universal design a design concept aimed at making a house usable and safe for the widest variety of people, including older adults and those with disabilities. *(p. 781)*

V

valley rafter a sloping beam that forms a depression in the roof instead of a hip. *(p. 504)*

vapor retarder a material that reduces the rate at which water vapor can move through a material. *(p. 900).*

veneer match the arrangement of pieces of veneer to create different patterns and effects. *(p. 343)*

veneer plaster a specially formulated gypsum plaster that is applied to a type of drywall called gypsum base; also called thin-coat plaster. *(p. 933)*

volatile organic compound (VOC) a type of chemical that evaporates into the air. *(p. 985)*

W

wainscoting paneling that runs partway up the wall from the floor. *(p. 808)*

wales horizontal bracing members for a reusable form. *(p. 264)*

wall cabinets cabinets that hang on a wall; also called upper cabinets. *(p. 781)*

wall standard a perforated metal strip that can be screwed to a wall or to the inside of a cabinet to support shelves. *(p. 775)*

warp a general description of any variation from a flat surface. *(p. 325)*

wear layer the top layer of engineered flooring. *(p. 976)*

web the member between the chords on a roof truss. *(p. 492)*

weep hole in masonry, a hole that provides drainage near the bottom of a wall. *(p. 698)*

welding the process of melting steel and adding filler metals to fuse the pieces at the point of attachment. *(p. 843)*

wind chill a combination of temperature and wind speed. *(p. 98)*

winders radiating treads that can be used instead of a platform to turn a stair. *(p. 730)*

window schedule a portion of the building plans that contains descriptions of the windows, and the sizes for the glass, the sash, and sometimes the rough opening. *(p. 585)*

wood preservative a chemical that protects wood. *(p. 332)*

work ethic the belief that work has value. *(p. 15)*

work triangle a kitchen planning principle based on the shortest walking distance between the refrigerator, the primary cooking surface, and the sink. *(p. 782)*

wrench a hand tool designed for turning a fastener, such as a bolt or a nut. *(p. 120)*

A

acclimation/aclimatación condición que se produce cuando la madera alcanza un contenido de humedad igual al del edificio donde está instalada. *(p. 975)*

acoustical ceiling /cielo raso acústico cielo raso compuesto de paneles pegados directamente en la superficie del cielo raso o engrapados a listones de madera clavados en las viguetas del cielo raso. *(p. 941)*

addition/agregación techo que forma una intersección con el techo principal. *(p. 516)*

agregado ingrediente aparte del cemento, conglomerado o agua, que se agrega a la mazcla de concreto para modificar sus características físicas o químicas. *(p. 221)*

Alkaline Copper Quaternary (ACQ)/cuaternario de cobre alcalino (ACQ, por sus siglas en inglés) compuesto químico que se usa para tratar maderas. La madera tratada con ACQ no contiene arsénico. *(p. 1003)*

amperage/amperaje intensidad de una corriente eléctrica expresada en amperios. *(p. 163)*

apprentice/aprendiz empleado sin experiencia que aprende un oficio al trabajar bajo la tutela de un profesional experto. *(p. 6)*

architect's scale/escalímetro herramienta que los arquitectos usan cuando realizan dibujos a escala. *(p. 44)*

axial load/carga axial carga que se transporta por la longitud de un miembro estructural. *(p. 844)*

B

backerboard/tablero de soporte placa con base de cemento, que brinda un excelente soporte para losas, especialmente en pisos e instalaciones húmedas. También se denomina tablero de cemento. *(p. 993)*

backing/refuerzo tira larga de material que se clava a la pared como soporte para molduras grandes. *(p. 771)*

backing the hip/rellenar la limatesa biselar el borde superior del cabio de limatesa, esto permite que la cubierta del techo se instale sin tocar las esquinas del cabio de limatesa. *(p. 513)*

back-priming/imprimación posterior imprimar la superficie posterior del tablero en la obra, si el fabricante no imprimó el revestimiento en todas las superficies. *(p. 663)*

balloon-frame construction/construcción de armadura sin rigidez *(balloon-frame)* armazón en el cual las vigas van desde la solera instalada en la base a la placa superior del segundo piso. También se conoce como armazón provisional o *balloon framing.* *(p. 370)*

balusters/balaustres miembros verticales y finos que sostienen el pasamanos. *(p. 726)*

base cabinets/gabinetes base gabinetes que se ubican en el piso y sostienen las encimeras. También se denominan gabinetes inferiores. *(p. 781)*

baseboard/zócalo tabla o moldura que se usa contra la parte inferior de las paredes para cubrir su unión con el piso. Sirve como transición entre la pared y el piso. También cubre las imperfecciones que suele haber en esta ubicación. También se denomina moldura de base. *(p. 764)*

batt/bloque de fibra cualquier material aislante grueso que viene en anchos precortados, diseñados para ajustarse entre los miembros de un armazón. *(p. 897)*

batter board/tabla para marcar la línea de excavación tabla que se sujeta horizontalmente a las estacas colocadas por fuera de donde se ubicarán las esquinas del edificio. *(p. 244)*

bearing capacity/carga admisible medida de la capacidad del suelo para soportar el peso de una casa. *(p. 247)*

bearing wall/pared de soporte pared que sostiene cargas además de su propio peso. *(p. 411)*

bed joint/junta de asiento junta de mortero horizontal. *(p. 275)*

bench mark/cota de referencia punto de inicio básico desde el cual se pueden realizar las medidas en el diagrama de construcción, usando un teodolito o un nivel. También se denomina punto de referencia (POR, por sus siglas en inglés). *(p. 239)*

bevel cut/corte en bisel tipo de corte a inglete que se realiza en ángulo a través del grosor de una tabla. *(p. 136)*

binder/aglutinante resina que une las partículas de pigmento. *(p. 947)*

biscuit/biscuit galleta pieza pequeña y plana de madera comprimida. También se denomina placa. *(p. 184)*

bisque/bizcocho de porcelana pieza de loza sin el esmaltado. *(p. 990)*

blind nailing/clavadura invisible colocación de clavos en ángulo a través de la espiga de la tabla y al armazón o a listones. Permite que las tablas siguientes oculten las cabezas de los clavos. *(p. 818)*

board foot/pie tablar unidad de medida que representa una pieza de madera con un área de superficie plana de 1 pie² y un grosor de tamaño normal de 1". *(p. 63)*

Boston ridge/instalación en cumbrera se crea cuando la limatesa y las tejas de cumbrera se superponen. *(p. 648)*

box extender/extensor de caja accesorio de metal o plástico que se atornilla al frente de la caja de salida, llevándola hacia delante. *(p. 811)*

box sill/solera de caja tipo de solera que se compone de una placa de solera (también se denomina durmiente de apoyo o simplemente solera) que se ancla a la pared de cimentación, viguetas de piso y viguetas de amarre, y contrapisos. *(p. 402)*

brace/jabalcón miembro que se usa para reforzar o apoyar una estructura. *(p. 544)*

bridging/arriostramiento método de apuntalamiento entre vigas. Se realiza para distribuir cargas, evitar que las vigas se tuerzan y para mejorar la estabilidad y la resistencia. *(p. 410)*

building code/código de construcción conjunto estándar de normas que regulan los procedimientos y detalles de la construcción. *(p. 34)*

building envelope/cerramiento exterior capa aislante continua que separa los espacios calefaccionados de aquellos que no lo están en paredes, cielos rasos, techos y pisos. *(p. 895)*

building permit/permiso de construcción autorización formal e impresa para que el constructor dé inicio a la obra. *(p. 35)*

bull float/aplanadora de mango largo placa ancha y plana de metal o madera que se pasa hacia atrás y adelante sobre el concreto para nivelar la superficie. *(p. 305)*

business plan/plan de negocios ofrece información específica sobre el negocio. *(p. 12)*

butt edge/canto de cabeza borde expuesto de la teja. *(p. 626)*

C

camber/alabeo ligera curva ascendente en una viga de maderas laminadas encoladas. La viga se instala con la curva orientada hacia arriba. *(p. 361)*

cambium/cámbium capa de tejido vivo que produce madera nueva, denominada albura, por su superficie interna. *(p. 318)*

cantilever/ménsula miembro de apoyo que se proyecta en el espacio y que se sostiene solamente de un extremo. *(p. 412)*

carcase/carcasa conjunto de paneles que crea la forma básica de un gabinete. *(p. 788)*

carrier/soporte líquido que mantiene los pigmentos y aglutinantes en suspensión. También los mantiene dispersos uniformemente (diseminados) durante la aplicación. *(p. 947)*

casing/marco moldura básica alrededor de una ventana o una puerta. *(p. 758)*

certification/certificación proceso formal que demuestra que un individuo está capacitado para un trabajo en particular. *(p. 9)*

chair/silla pequeño dispositivo que sostiene la tela metálica a una altura en particular mientras se vierte el concreto. *(p. 230)*

chamfer/chaflán borde biselado. *(p. 171)*

chord/cordón miembro exterior superior o inferior de la viga de celosía. *(p. 492)*

circuit/circuito cable o grupo de cables que suministra electricidad a un área o electrodoméstico específico. *(p. 877)*

clinching/remachado proceso de unión de dos capas de metal con presión en una estructura de acero. Se usa una herramienta de remachado con motor. *(p. 843)*

clip angle/ángulo sujetador pieza pequeña de acero galvanizado instalada en un miembro estructural para aceptar una carga estructural. *(p. 847)*

cold joint/junta fría junta que se produce donde se vierte concreto fresco encima o a continuación de concreto que ya ha comenzado a curar. *(p. 266)*

cold-formed steel/acero conformado en frío acero laminado que se dobla y se forma sin usar calor. *(p. 832)*

collar tie/falso tirante miembro de un armazón horizontal que impide que los pares de cabios opuestos se separen. *(p. 542)*

collated fasteners/sujetadores secuenciales sujetadores dispuestos en cintas o rollos, con cada sujetador conectado a los sujetadores de cualquiera de los lados. *(p. 191)*

collet/mandril de pinzas parte de un enrutador que sujeta una broca. *(p. 171)*

common difference/diferencia común vigas de un hastial que tienen el mismo espaciado de centro a centro que las vigas de una pared estándar. Sin embargo, cada viga tiene una longitud diferente de las vigas de cualquiera de los lados. Las diferencias de longitud se basan en una sola cifra que depende del punto extremo del techo. Esta cifra se denomina diferencia común. *(p. 546)*

competent person/persona competente alguien que ha sido capacitado para identificar los riesgos existentes y predecibles del lugar de trabajo. *(p. 205)*

composite panel product/producto de paneles compuestos producto de paneles hechos de piezas de madera mezclada con adhesivo. *(p. 347)*

compound miter saw/sierra angular compuesta sierra cuya cabeza pivota hacia arriba y abajo y de un lado al otro. *(p. 147)*

concrete/concreto material de construcción duro y resistente que se hace al mezclar cemento, un conglomerado fino (un material granular, como arena), un conglomerado grueso (por lo general, gravilla o piedra triturada) y agua en las proporciones adecuadas. *(p. 218)*

concrete flatwork/pavimento de concreto consiste en áreas de concreto planas y horizontales, que suelen tener un grosor de 5" o menos. El pavimento se coloca directamente sobre el terreno o sobre gravilla o arena compactada. *(p. 299)*

condensation/condensación proceso por el cual el vapor se convierte en un líquido. *(p. 900)*

conductor/conductor material por el cual la electricidad fluye sin problemas. *(p. 85)*

coniferous tree/árbol conífero árbol que produce semillas en cono y que tiene hojas en forma de agujas o escamas. Las maderas blandas provienen de árboles coníferos. *(p. 320)*

consolidation/consolidación proceso que elimina las bolsas de aire y mete el concreto en todas las partes de los moldes. Hace que las partículas finas de la mezcla se desplacen hacia los moldes y que el conglomerado más pesado se aleje, lo que deja un acabado liso en las paredes cuando se retiran los moldes. *(p. 227)*

control joint/junta de control en un bloque de concreto, junta que ayuda a minimizar las grietas aleatorias. En una pared, junta que controla el movimiento provocado por la tensión. *(p. 283)*

coping/rematar proceso de dar forma a las juntas en ángulo internas entre los miembros de las molduras, al cortar el extremo de un miembro para que se ajuste contra la cara del otro. *(p. 767)*

corbel/cartela hilada de retallo de ladrillos que se extiende fuera de la hilada inferior. *(p. 712)*

corner bead/guardaesquina franja de vinilo o metal galvanizado que refuerza y protege la esquina. Viene en largos de 8' y 10'. Un guardaesquina estándar forma una esquina cuadrada de 90°. Un guardaesquina redondeado forma una esquina redondeada de 90°. *(p. 919)*

corner post/esquinal montaje de vigas completas en la esquina de un edificio. Normalmente, un esquinal está hecho de tres o más vigas que brindan mayor resistencia. *(p. 441)*

cornice/cornisa ribete exterior de una estructura que se compone de una imposta, un sofito y varios tipos de molduras. *(p. 552)*

countersink/avellanador broca con bordes cortantes biselados. Crea una forma de embudo en la parte superior de un agujero perforado. Esta forma de embudo permite que la cabeza de un tornillo para madera quede a ras con la superficie. *(p. 166)*

cripple stud/pie derecho cojo viga que no se extiende por completo desde la placa inferior a la placa superior de la pared, debido a una abertura como el marco de una ventana. También se denomina elemento corto. *(p. 434)*

crosscut/corte transversal corte realizado a través de la veta de un madero y en un ángulo de 90° con respecto al borde. *(p. 128)*

crown/vértice curva exterior del arco. Cualquier vigueta que tenga un ligero arco de costado se debe colocar siempre con el vértice en la parte superior. *(p. 406)*

crown molding/moldura de cornisa moldura bastante grande que suele incluir tanto superficies curvas como angulares. *(p. 770)*

cutterhead/cabezal portacuchillas cilindro de metal sólido en el cual se instalan tres o cuatro cuchillas. El cabezal portacuchillas se instala debajo de la base de la máquina. A medida que el cabezal gira, las cuchillas cortan pequeñas astillas de madera, dejando una superficie lisa. *(p. 179)*

D

dead load/carga permanente peso total del edificio. (p. 383)

deciduous tree/árbol caducifolio árbol que muda sus hojas anualmente, durante las estaciones frías o de sequía. Las maderas duras se extraen de árboles caducifolios de hojas anchas. (p. 320)

design value/valor de diseño número que se asigna a la capacidad de una madera en particular para resistir las tensiones. (p. 379)

differential leveling/nivelación diferencial proceso para determinar las diferencias de elevación de puntos remotos entre sí. (p. 246)

doghouse dormer/buhardilla tipo caseta buhardilla con techo a dos aguas y paredes laterales que sobresale de forma horizontal por un techo inclinado. (p. 519)

door frame/marco para puerta montaje alrededor de una puerta. Se instala en el armazón de la pared y se compone de dos jambas laterales y una jamba superior. (p. 597)

dormer/buhardilla proyección vertical de una ventana en un techo que da luz y ventilación a las habitaciones del segundo piso o al ático. (p. 518)

draft/tiro movimiento ascendente de aire al interior de la chimenea. El tiro atrae aire, lo que ayuda en la combustión y expulsa el humo y los gases tóxicos. (p. 709)

drain field/campo de drenaje área de filtrado de un sistema séptico hacia el cual fluyen los desechos líquidos. Un campo de drenaje contiene una red de tuberías perforadas enterradas en arena y gravilla. (p. 870)

drainage plane/plano de drenaje espacio o serie de espacios detrás del revestimiento que permite que el agua fluya libremente. (p. 682)

dropping the hip/bajar la limatesa profundizar la barbilla para que el borde superior del cabio de limatesa quede alineado con los extremos superiores de los enchufes. (p. 513)

dry rot/pudrición seca zona deteriorada de la madera, de color marrón y que se desmenuza. Este es un término inexacto, puesto que la madera debe estar húmeda para que suceda la putrefacción. (p. 328)

E

eaves/alero porciones de un techo que se proyectan más allá de las paredes. (p. 551)

edging/bordear usar una brocha para pintar en las esquinas entre superficies planas grandes, adonde un rodillo no puede llegar. (p. 963)

electrical circuit/circuito eléctrico cuando la electricidad se transmite desde un punto de origen y vuelve a ese mismo punto a través de un conductor. (p. 85)

elevation/alzado vista lateral que permite ver la altura y el ancho de los objetos. (p. 53)

emissivity/emisividad número que simboliza la cantidad de radiación que emite un objeto. (p. 903)

engineered lumber/madera procesada cualquier producto fabricado con madera sólida, chapa de madera, piezas de madera o fibras de madera, cuyos componentes se han pegado con adhesivos. (p. 352)

engineered panel/procesado cualquier producto fabricado con láminas, incluida la madera contrachapada, que está hecha de madera o piezas de madera pegadas con adhesivo natural o sintético. (p. 338)

entrepreneur/empresario persona que crea y dirige su propio negocio. (p. 11)

ergonomics/ergonomía ciencia del diseño y organización de las cosas para ajustarse a las necesidades del cuerpo humano. (p. 90)

ethics/ética conjunto de normas personales para distinguir lo correcto de lo incorrecto. (p. 24)

excavation/excavación corte, cavidad, zanja o depresión hecha al remover la tierra. (p. 82)

exposure/exposición cantidad de la superficie (como de una teja o un revestimiento) expuesto a la intemperie. (p. 626)

F

face-frame cabinet/gabinete de marco delantero gabinete que cuenta con un marco delantero que se ajusta alrededor de la abertura frontal de la carcasa, de esta forme ofrece una superficie para instalar las bisagras y los herrajes de las gavetas. (p. 788)

fascia/imposta tabla que está clavada en los extremos de los cabios. Protege el contrahilo de los cabios y sirve como superficie de montaje para los canales de desagüe. (p. 552)

feathering/biselado/carga ininterrumpida en el relleno de juntas de tablarroca (biselado), proceso de alisar los bordes externos del compuesto para juntas, de manera que no queden protuberancias. En el uso de una pistola para tornillos (carga ininterrumpida), proceso de instalar un tornillo en la punta sin detener la pistola. *(p. 836)*

feed rate/velocidad de avance velocidad a la que el material pasa por la hoja de la sierra. *(p. 142)*

fiber-saturation point/punto de saturación de la fibra punto en el cual las paredes celulares de la madera han absorbido toda el agua que pueden retener. *(p. 321)*

fines/finos materiales triturados o pulverizados finamente. *(p. 299)*

finger joint/empalme de cola de pescado ensambladura que tiene una serie de cortes con forma de cuña hechos en la madera y espaciados rigurosamente. Estos cortes crean un área de superficie amplia que mejora la unión del pegamento entre las dos piezas. *(p. 363)*

fixture/artefacto de plomería cualquier dispositivo, como una bañera, que recibe o drena agua. *(p. 868)*

flagged bristles/cerdas flexibles cerdas, como las de una brocha, con las puntas ligeramente separadas. *(p. 960)*

flight/tramo distancia recta de las escaleras. *(p. 729)*

floor plan/plano de la planta dibujo a escala que muestra el tamaño y ubicación de las habitaciones de una planta determinada. *(p. 37)*

flue/conducto de humos canal al interior de la chimenea por el cual suben el aire, los gases y el humo. Sus dimensiones, altura, forma y suavidad interior determinan la efectividad de una chimenea para crear un tiro suficiente. *(p. 709)*

footing/lecho de cimentación base que proporciona una superficie que distribuye el peso sobre un área amplia de suelo. Por lo general, los lechos de cimentación están hechos de concreto. *(p. 256)*

frameless cabinet/gabinete sin marco gabinete con marco alrededor de la abertura. Sus bisagras están ocultas e instaladas en las paredes laterales. *(p. 789)*

framing connector/conector de armazón soporte de metal formado que se instala con clavos en conexiones de armazón. Tales conectores constan de un soporte de metal formado. *(p. 386)*

free enterprise/libre empresa sistema económico en el cual las empresas o las personas pueden comprar, vender y fijar los precios de los bienes y servicios. *(p. 12)*

frost depth/profundidad de congelación profundidad en cualquier clima por debajo de la cual el suelo no se congela. *(p. 294)*

G

gain/caja entalladura (muesca) con una profundidad igual al grosor de una hoja. *(p. 599)*

galvanizing/galvanizar proceso de cubrir el acero con una capa protectora de zinc. *(p. 1006)*

girder/viga principal miembro horizontal grande que se usa para apoyar las viguetas del piso. *(p. 396)*

glazing/cristales porciones de vidrio o plástico transparente de una ventana. *(p. 577)*

glulam/glulam abreviatura en inglés para "laminado encolado" *(glue laminated)*, por ejemplo, viga laminada encolada. Cuando las capas de madera se unen con pegamento, su resistencia y dureza es mayor que la madera sólida de la misma dimensión. *(p. 360)*

grade/clasificación indicación general de la calidad y resistencia de una pieza de madera. *(p. 323)*

grade stamp/sello de clasificación marca permanente que identifica la especie, calidad, aserradero de origen y la resistencia general de una tabla. Los sellos de clasificación permiten a los inspectores de construcción saber si una casa se construye con madera de la calidad adecuada. *(p. 324)*

ground/base material instalado de manera permanente o temporal en una superficie que se va a enyesar. Proporciona un borde recto y ayuda al yesero a medir el grosor del yeso. Puede estar hecha de madera, pero es más frecuente que sea de metal. *(p. 934)*

grounding/conexión a tierra proporciona una ruta para que la electricidad pase de forma segura de la herramienta a la tierra. *(p. 85)*

grout/lechada mortero fino que se usa para rellenar espacios. *(p. 994)*

H

head casing/marco superior pieza horizontal en la parte superior de una puerta o ventana. *(p. 758)*

head joint/junta al tope junta de mortero vertical. *(p. 275)*

header/cabecero en la instalación de pisos, miembro horizontal que transporta las cargas de otros miembros y las dirige alrededor de una abertura. En el armazón de una pared, viga de madera que se coloca en la parte superior de una abertura. El cabecero soporta las cargas estructurales sobre la abertura y las transfiere al armazón en cada lado de la abertura. También se denomina dintel. *(p. 432)*

headroom/altura libre espacio que queda directamente sobre un peldaño. Se mide desde el borde externo del mamperlán al cielo raso directamente por encima. *(p. 730)*

heartwood/duramen porción de color oscuro de un árbol, cerca del núcleo. *(p. 1004)*

hip rafter/cabio de limatesa cabio que forma un área elevada, o limatesa, que normalmente se extiende desde la esquina del edificio en diagonal desde la placa al tablón de la cumbrera. *(p. 504)*

hollow-core construction/estructura con centro hueco estructura compuesta por un marco liviano de madera o cartón corrugado recubierto con madera contrachapada o madera prensada. *(p. 597)*

horizontal/horizontal superficie nivelada. *(p. 111)*

hydration/hidratación reacción química que se produce cuando el agua se combina con el cemento. *(p. 218)*

I

ice dam/reborde de hielo hielo que se forma al congelarse nieve derretida en la línea del alero. A medida que se derrite más nieve, el agua se acumula detrás del hielo y se filtra por debajo de las tejas. *(p. 632)*

Impact Noise Rating (INR)/calificación del ruido por impacto (INR, por sus siglas en inglés) medida de resistencia de un sistema de pisos basado en decibeles (dB), que es una medida de intensidad del sonido. *(p. 911)*

independent slab/losa independiente losa que se usa en áreas en las que el suelo se congela a bastante profundidad en el invierno. También se denomina losa de soporte del suelo. *(p. 295)*

indirect cost/costo indirecto costo general que se relaciona con la organización y supervisión de un proyecto, que incluye el costo de los suministros de oficina, herramientas y equipo de construcción, nómina de pagos e impuestos. *(p. 64)*

infiltration/infiltración paso de aire fresco a un edificio, a través de grietas alrededor de las ventanas, puertas y marcos. *(p. 887)*

in-line framing/armazón en línea armazón que alinea todos los miembros estructurales de soporte de carga verticales y horizontales en una estructura de madera y en una estructura de acero. *(p. 373)*

internal cut/corte interno técnica para hacer un agujero grande en un material sin comenzar en el borde. *(p. 150)*

interview/entrevista reunión entre un empleador y un postulante a un trabajo. *(p. 22)*

J

jack rafter/cabio corto cabio común recortado que puede estar unido a un cabio de limatesa, a un cabio de limahoya o ambos. Hay cabios cortos de limatesa y cabios cortos de limahoya. *(p. 504)*

J-channel/canal J pieza en forma de J de plástico o metal que se usa para sujetar molduras. *(p. 677)*

job application/solicitud de empleo formulario que hace preguntas acerca de las capacidades, experiencia laboral, educación e intereses de un postulante a un trabajo. *(p. 19)*

jointer/garlopa barra de metal simple con un extremo afilado. También se denomina cepillo de rejuntar. *(p. 688)*

joist tracks/pistas de viguetas miembros con forma de C mediante los cuales las viguetas se instalan en la base. También se denominan pistas de bordes. *(p. 845)*

K

kerf/entalla ancho de un corte. *(p. 132)*

kick plate/placa de protección pieza corta de madera de enmarcar que se usa para anclar la parte inferior de una escalera. *(p. 744)*

kickback/contragolpe reacción que se produce cuando una hoja giratoria encuentra algo que reduce su velocidad o la detiene mientras la sierra está a plena potencia. *(p. 129)*

kiln/horno de cochura horno en el cual la humedad, el flujo de aire y la temperatura se controlan rigurosamente. *(p. 321)*

kneeboards boards/tablas de apoyo tablas que miden aproximadamente 12" por 24", que se colocan sobre el concreto para soportar el peso del cementista. *(p. 307)*

L

ladder/escalera estructura hecha de dos piezas laterales largas, unidas por varios travesaños en los cuales se puede pisar. Las escaleras se usan para subir y bajar. Son fáciles de colocar y de retirar, pero se deben mover frecuentemente. *(p. 198)*

laminated-veneer lumber (LVL)/madera enchapada laminada (LVL, por sus siglas en inglés) producto de madera manufacturada cuyo elemento básico es chapa de madera encolada. *(p. 353)*

lath/listón cualquier material base para yeso que tiene cualidades que hacen que el yeso se le adhiera. *(p. 933)*

lead corner/esquina de referencia esquina de ladrillos parcialmente construida. *(p. 699)*

ledger/larguero longitud de la madera que conecta la plataforma de la casa. *(p. 1013)*

level/nivel herramienta con un instrumento largo de madera, metal o fibra de vidrio con varias ampollas de nivelación de vidrio, que mide el nivel o verticalidad de una superficie. *(p. 111)*

lifeline/cuerda de salvamento soga cuyo diseñada para evitar que un trabajador se caiga. Uno de los extremos de la cuerda de salvamento se sujeta en un punto seguro de la estructura y el otro extremo a un arnés que usa el trabajador. *(p. 209)*

lift/hormigonada capa de relleno uniforme y bastante baja. *(p. 298)*

line block/bloque de línea pequeño dispositivo en forma de L, hecho de madera o plástico. Se engancha sobre el borde de un ladrillo y se mantiene en su lugar por la tensión de la cuerda. Los bloques de línea opuestos se mueven de forma ascendente a medida que la pared gana altura. *(p. 701)*

lintel/dintel miembro de apoyo horizontal que se instala encima de una abertura como una ventana, una puerta o una chimenea de mampostería. *(p. 699)*

live load/carga móvil peso que no está conectado de forma permanente. *(p. 383)*

load/carga fuerza que crea tensiones en una estructura. El peso y el viento son diferentes tipos de carga. *(p. 379)*

lockset/cerradura conjunto de perillas, pasador y mecanismo de bloqueo. *(p. 600)*

lookout/mirador miembro horizontal que se extiende desde un extremo de un cabio a un listón para clavar o a la cara de la cubierta de la pared. Los miradores forman una superficie horizontal en la cual se puede instalar el material de sofito. *(p. 552)*

lumber/maderos piezas de madera que tienen un grosor y ancho uniforme y se obtienen al aserrar un tronco. *(p. 321)*

M

makeup air/aire de relleno aire que reemplaza el aire que expulsa un aparato de combustión. Los códigos locales exigen cada vez más que todos los tipos de chimeneas cuenten con aire de relleno. *(p. 705)*

mastic/mástique adhesivo grueso y premezclado que se aplica con una llana dentada o con una pistola de calafatear. *(p. 976)*

medium-density fiberboard (MDF)/tablero de fibra de densidad media (MDF, por sus siglas en inglés) producto de paneles hechos de fibras de madera comprimida mezcladas con adhesivo de úrea-formaldehído. *(p. 350)*

miter cut/corte angular corte realizado a través de la veta de una tabla y en un ángulo que no es de 90°. *(p. 136)*

moisture content (MC)/contenido de humedad (MC, por sus siglas en inglés) cantidad de agua que contiene la madera. Se expresa como un porcentaje de lo que la madera pesaría si estuviera completamente seca. *(p. 321)*

molding/moldura listones estrechos de madera con un perfil con formas. *(p. 751)*

monolithic slab/losa monolítica consta de un lecho de cimentación y una placa de piso que se forman en un colado. Se denomina losa unificada, losa de borde grueso o losa con lecho de cimentación invertido. *(p. 295)*

mortgage/hipoteca préstamo a largo plazo (15 a 30 años) con la propiedad por garantía. *(p. 40)*

mullion strip/listón de separación pieza de madera vertical que separa las ventanas de diversos estilos y tamaños combinados para formar una unidad mayor. *(p. 587)*

muntin/ montante central pieza corta vertical u horizontal, que se usa para sostener un panel de vidrio. *(p. 578)*

musculoskeletal disorder/trastorno musculoesquelético afección a los músculos, tendones, ligamentos, articulaciones, cartílago o discos vertebrales, que puede ser provocado por levantar, sujetar materiales y otras tareas. *(p. 90)*

N

networking/establecimiento de contactos hacer uso de todas las conexiones personales para lograr los objetivos profesionales. *(p. 17)*

nominal dimension/dimensión nominal ancho y grosor de madera aserrada áspera (no su largo). También se denomina tamaño nominal. *(p. 325)*

O

Occupational Safety and Health Administration (OSHA)/Administración de Seguridad y Salud Ocupacional (OSHA, por sus siglas en inglés) organismo federal que publica normas y reglamentos para que las condiciones laborales, las herramientas, los equipos, las instalaciones y procesos de trabajo sean seguros y saludables. Lleva a cabo inspecciones en los lugares de trabajo para constatar el cumplimiento de las normas. *(p. 79)*

offcut/recorte pieza de madera de desecho que puede introducirse entre la hoja y la guía de corte cuando se corta la madera con un medidor angular y una guía de corte. *(p. 141)*

on center (OC)/centro a centro (OC, por sus siglas en inglés) distancia desde la línea centra de un miembro estructural a la línea central del siguiente miembro más cercano. *(p. 382)*

open valley/limahoya abierto tipo de limahoya de techo en el que las tejas no se aplican en la intersección de las dos superficies del techo. Esto deja el tapajuntas subyacente expuesto a lo largo del limahoya. *(p. 641)*

oriented-strand board (OSB)/tablero de virutas orientadas (OSB, por sus siglas en inglés) producto de madera hecho de virutas de madera unidas con adhesivo bajo presión y calor. *(p. 349)*

overdig/sobreexcavación término que se usa para describir la excavación adicional necesaria para tener espacio para trabajar. *(p. 248)*

P

parging/revocar proceso de distribución de mortero o yeso sobre un bloque. *(p. 287)*

parquet/parquet cualquier piso armado con pequeñas piezas de madera cortadas con precisión, siguiendo un diseño geométrico como cuadrados, rectángulos y diseños en espiga. *(p. 973)*

passage door/puerta de paso puerta batiente que se abre y se cierra gracias a dos o más bisagras de hoja instaladas en uno de los lados. *(p. 596)*

performance method/método de rendimiento método de armado que depende de principios de ingeniería establecidos y especificaciones de carga prevista, para calcular el tamaño y la resistencia de los miembros individuales del armazón de acero. *(p. 832)*

pier/pilar columna de concreto que soporta una carga concentrada, como un poste. Es un tipo de cimiento. *(p. 1009)*

pilot hole/agujero guía agujero perforado en la madera para introducir y guiar un tornillo. *(p. 166)*

pitch/inclinación en el armazón de un techo, proporción de la altura total con respecto al vano. *(p. 470)*

plan view/vista en planta vista desde lo alto, o vista a vuelo de pájaro, que permite ver el ancho, largo y ubicación de los objetos como si se estuviera parado en una plataforma por encima de ellos y viéndolos hacia abajo. *(p. 50)*

plank/tablilla cualquier tabla de madera sólida que tenga al menos 3" de ancho. *(p. 972)*

plate/placa miembro horizontal del armazón que se usa para unir el armazón de la pared interior y exterior. El ancho de la placa corresponde al grosor de la pared. Cada pared tiene tres placas: una placa inferior y dos placas superiores. *(p. 432)*

platform-frame construction/construcción con sistema plataforma técnica de construcción en la que cada nivel de la casa se construye por separado. El piso es una plataforma que se construye de manera independiente a las paredes. También se denomina armazón de plataforma. *(p. 370)*

pliers/alicates herramienta manual con mordazas opuestas diseñadas para sujetar cosas. *(p. 122)*

plies/láminas pliegos de madera muy delgados y flexibles, que fueron serrados, desprendidos o cortados de un tronco y que se usan en la madera contrachapada. *(p. 338)*

plot plan/plano de terreno parte de los planos de la casa que muestra la ubicación de la edificación en el terreno, junto con las cotas de explanación relacionadas. *(p. 236)*

plywood/madera contrachapada material de construcción compuesto por capas de chapa de madera y, en ocasiones, por otros materiales que se unen con pegamento. *(p. 338)*

pneumatic tool/herramienta neumática herramienta que funciona con aire comprimido. *(p. 187)*

porosity/porosidad medición de la capacidad del agua de fluir a través del suelo. Si el suelo no tiene buen drenaje, es posible que se deba excavar material adicional y reemplazar por un mejor suelo. *(p. 248)*

Portland cement/cemento Portland cemento manufacturado que se usa en el concreto moderno. Su nombre proviene de la similitud con el color de la piedra Portland, una piedra caliza que se extrae en la Isla de Portland, fuera del cordón de Inglaterra. *(p. 219)*

post-and-beam framing/armazón de pie derecho y viga sistema de armazón en el cual menos piezas de madera, pero más largas, se ubican a mayor distancia entre sí que las que se usan en un armazón convencional. *(p. 373)*

precast/prefundido cualquier objeto que se funde en una fábrica, se cura en condiciones controladas y, a continuación, se envía al lugar de trabajo. *(p. 1018)*

prescriptive method/método prescriptivo método de armado que usa tablas estandarizadas que proporcionan especificaciones y otros datos relacionados con la estructura de acero. *(p. 833)*

primer/imprimación pintura que tiene una proporción de aglutinante mayor que la pintura estándar. *(p. 948)*

pullout capacity/capacidad de sujeción capacidad de resistencia de un tornillo en una conexión. La capacidad de sujeción se basa en el número de roscas que penetra y sostiene la conexión. *(p. 841)*

pump jack/caballete de bombeo dispositivo de metal con un pedal que un trabajador bombea para que se deslice de forma ascendente y descendente en un poste de madera o aluminio. Dos o más caballetes juntos pueden sostener tablones para que un trabajador los use como andamio. Los caballetes de bombeo se usan normalmente en las aceras de una casa, durante las operaciones de revestimiento o pintado. *(p. 208)*

purlin/correa miembro estructural horizontal que soporta las cargas del techo y las transfiere a los soportes estructurales. *(p. 544)*

Q

quantity takeoff/estimación de costos análisis de costos en el cual se contabiliza y se fija el precio de cada material necesario para construir la casa. *(p. 60)*

R

rabbet/rebajo corte o ranura cerca o por el borde de una pieza de madera. Permite que se encaje otra pieza para formar una junta. *(p. 740)*

radiant heat/calor radiante calor que se desplaza en línea recta desde una superficie caliente y que calienta cualquier cosa sólida que encuentre. *(p. 902)*

radon/radón gas radioactivo incoloro e inodoro que se desplaza a través del suelo. *(p. 288)*

rafter/cabio miembro inclinado del armazón que sostiene el techo. Los cabios cumplen la misma función en el techo que las viguetas del piso o las vigas de la pared. *(p. 467)*

rails/rieles soportes verticales en los que se instalan peldaños o travesaños. *(p. 198)*

rake/cornisa inclinada parte del techo a dos aguas que se extiende más allá de las paredes finales. Puede ser cerrada o extendida. *(p. 558)*

receptacle/tomacorriente dispositivo de contacto con una combinación de ranuras y agujeros de conexión a tierra para aceptar las puntas de un enchufe eléctrico. *(p. 879)*

refractory/cemento refractario cemento resistente a las altas temperaturas. *(p. 705)*

regulator/regulador válvula que controla la presión de aire que llega a una pistola de clavos o una engrapadora. *(p. 189)*

repetitive stress injury (RSI)/lesiones por esfuerzos repetitivos (RSI, por sus siglas en inglés) cuando una tarea se realiza una y otra vez, se produce una irritación menor en los nervios y los tejidos. *(p. 90)*

résumé/currículum vitae resumen breve de la información de contacto, educación, capacidades, experiencia laboral, actividades e intereses de un postulante a un trabajo. *(p. 18)*

retempering/reamasar proceso de agregar agua a una cantidad de mortero que está demasiado rígida para trabajar. *(p. 695)*

return/retorno pieza que continúa el perfil de moldura alrededor de la esquina. *(p. 762)*

reveal/telar pequeña compensación entre una pieza de moldura y la superficie en la cual está aplicada. Agrega un atractivo visual y permite que el carpintero ajuste el marco si la puerta no está perfectamente cuadrada. *(p. 759)*

ridge/cumbrera miembro del armazón del techo que se coloca en la intersección de dos superficies con inclinación ascendente. *(p. 530)*

ridge beam/viga de la cumbrera miembro horizontal del armazón al cual se sujetan las partes superiores de los cabios. Una viga de cumbrera está hecha de LVL, madera laminada encolada o nominal de 4". Los cabios se apoyan en la parte superior de la viga de cumbrera o se sujetan con soportes metálicos clavados por el costado. También se denomina cumbrera estructural. *(p. 530)*

ridge board/tablón de la cumbrera pieza horizontal que conecta los extremos superiores de los cabios. *(p. 468)*

rip cut/corte a lo largo corte hecho en la dirección de la veta de la madera. *(p. 142)*

riser/contrahuella en una escalera, tablero vertical que cierra los espacios entre los peldaños. *(p. 726)*

roof rake/moldura del techo parte del armazón del techo que se extiende más allá de las paredes en los extremos a dos aguas. *(p. 862)*

rough opening (RO)/abertura en bruto (RO, por sus siglas en inglés) espacio en el cual se colocará una puerta o una ventana. Deja espacio para la puerta o ventana y el marco. *(p. 440)*

rough sill/peana miembro horizontal que se coloca en la parte inferior de la abertura de una ventana para sostener la ventana. Conecta los extremos superiores de los pies derechos cojos. *(p. 434)*

R-value/valor R en aislamiento, medida de la capacidad de un material para resistir la transmisión de calor. *(p. 894)*

S

sapwood/albura capa de crecimiento exterior. Es de color más claro que el duramen. *(p. 1004)*

sash/ hoja móvil parte que sujeta los cristales. El marco es la porción fija del conjunto que recibe la hoja móvil. *(p. 577)*

scaffold/andamio plataforma elevada que se usa para trabajar en altura. *(p. 203)*

screed/maestra listón de metal o madera largo y recto que se usa para enrasar (nivelar) el concreto. *(p. 301)*

seasoning/curación proceso de secar la madera con aire o en un horno industrial. *(p. 321)*

service main/tubería principal tubería que suministra agua a la casa. Se conecta en la calle al sistema municipal de agua. También se denomina tubería de servicio de agua. *(p. 869)*

shear wall/pared de carga pared diseñada para soportar tensiones laterales inusuales. Las paredes de carga se suelen usar en áreas donde son comunes los terremotos y las tormentas intensas. Es posible que requieran fijaciones y/o pernos de anclaje especiales. *(p. 378)*

sheathing/cubierta paneles rígidos de 4×8 o más grandes, que se instalan en la superficie externa del armazón de la pared exterior. La cubierta agrega dureza y resistencia a las paredes. *(p. 430)*

sheen/resplandor descripción de cuán brillante es una superficie. *(p. 947)*

side casing/marco lateral pieza vertical al costado de una puerta o ventana. *(p. 758)*

side lap/recubrimiento lateral superposición horizontal que tienen entre sí las láminas adyacentes del techo. Esto se aplica principalmente al techo en rollo y al recubrimiento inferior. También se denomina recubrimiento longitudinal. *(p. 626)*

siding/revestimiento la cubierta de la pared exterior de una casa. Su objetivo es repeler el agua y proteger las partes estructurales de las paredes. *(p. 658)*

site layout/diagrama del emplazamiento proceso de marcar la ubicación de un edificio en el terreno mismo. Esto lo puede realizar un topógrafo o un constructor que conozca los métodos básicos de topografía. *(p. 236)*

skirtboard/rodapié tabla acabada que se clava a la pared antes de instalar los tirantes. Protege la pared de daños y proporciona un borde acabado contra la pared, lo que facilita pintar o empapelar las áreas adyacentes. *(p. 739)*

sleeper/durmiente largo de la madera que soporta el piso de madera sobre el concreto. *(p. 982)*

slope/inclinación en el armazón de un techo, proporción de la elevación unitaria con respecto a la longitud unitaria. *(p. 470)*

slump test/prueba de revenimiento prueba para medir la consistencia del concreto. *(p. 226)*

soffit/sofito en el interior, un área alrededor del perímetro de una habitación que es más baja que el resto del cielo raso. En el exterior, la parte inferior de los aleros. A veces está encerrado. Se puede dejar abierto, con los extremos de los cabios expuestos. *(p. 460, p. 552)*

solid-core construction/estructura con centro sólido tipo de estructura que se usa en las puertas de exterior, en la cual se usan listones de madera, paneles de partículas, espuma rígida u otro material central, que se cubren con un material exterior delgado, como chapa de madera. *(p. 596)*

solvent/solvente material que disuelve otro material. *(p. 948)*

Sound Transmission Class (STC)/clasificación por transmisión del sonido (STC, por sus siglas en inglés) calificación numérica que indica la capacidad de un material o una combinación de materiales para reducir la transmisión del sonido. *(p. 908)*

span/vano distancia entre los bordes externos de las placas superiores, medida en los ángulos rectos con respecto al tablón de la cumbrera. *(p. 470)*

span table/tabla de separación tabla con un listado del espaciado máximo permitido entre los diferentes tamaños de viguetas o cabios. Con las tablas de separación, un carpintero puede encontrar rápidamente el espaciado correcto para la especie, calificación y dimensiones de la madera que usa. *(p. 382)*

specifications/especificaciones notas escritas que pueden organizarse en forma de lista. *(p. 57)*

spline/cuña listón fino de madera que se usa para reforzar una junta. *(p. 375)*

spreader/separador dispositivo que mantiene abierta una escalera y evita que la misma se cierre por accidente. *(p. 202)*

springing angle/ángulo de alabeo ángulo en el cual la moldura se aleja de la pared. *(p. 771)*

square/escuadra herramienta que se usa principalmente para medir o verificar ángulos. *(p. 108, p. 626)*

square/escuadra cantidad de material que se requiere para cubrir 100 pies2 de superficie de techo. *(p. 628)*

stairway/escalera serie de peldaños junto con todos los elementos relacionados, incluidos elementos estructurales como tirantes y elementos de acabado como pasamanos y balaustres. *(p. 726)*

stairwell/cubo de la escalera hueco vertical dentro del cual se construye una escalera. *(p. 724)*

station mark/marca de estación punto sobre el cual se centra directamente el nivel. *(p. 239)*

step/peldaño escalón o contrahuella. *(p. 726)*

sticker/listón separador retazo de madera largo y delgado que separa las capas de los productos de madera, y permite que el aire circule entre ellos. *(p. 809)*

stiles/montantes miembros laterales verticales de puertas de madera de panel en relieve. *(p. 597)*

stock plan/plano de serie plano de la casa estándar que se puede adaptar a muchos terrenos diferentes. *(p. 37)*

stoop/galería rellano ampliado en la parte superior de los descansos. *(p. 1018)*

story pole/nivelador de carpintero en carpintería, dispositivo de medición hecho en la obra para garantizar un diseño uniforme en toda la casa. En albañilería, tablón con marcas a 8" de separación. Se puede usar para medir la parte superior de la mampostería de cada hilada. También se denomina poste de hilada. *(p. 664)*

strike plate/placa de impacto placa de metal insertada en una abertura de la jamba de la puerta, en la cual se desliza el pasador. *(p. 596)*

stringer/tirante partes de una escalera que soportan los peldaños. *(p. 726)*

structural insulated panel/panel aislado estructural panel rígido de aislante de espuma de poliestireno expandido (EPS, por sus siglas en inglés) de 3½" de grosor, entre las láminas de madera contrachapada exterior o de tablero de virutas orientadas (OSB, por sus siglas en inglés). También se denomina panel con centro de espuma. *(p. 375)*

stud/viga miembro vertical del armazón. Las estructuras convencionales usan normalmente vigas de 2×4, espaciadas a 16" centro a centro (OC). *(p. 432)*

subflooring/contrapiso láminas de madera procesadas o madera con clasificación de construcción que se usa para construir un contrapiso, que es un piso preliminar que se coloca sobre las viguetas del piso, como base para el piso acabado. *(p. 402)*

subgrade/subsuelo tierra debajo de la losa. El subsuelo debe estar compactado de manera uniforme para evitar cualquier asentamiento irregular de la placa de piso. *(p. 298)*

substrate/sustrato material que sirve como base para otro material. *(p. 799)*

summerwood/ madera de estío porción densa y de color oscuro de la madera. Sus células tienen paredes gruesas y cavidades pequeñas. *(p. 946)*

suspended ceiling/cielo raso suspendido cielo raso compuesto por paneles sostenidos por una rejilla de metal o plástico. *(p. 937)*

T

tail joist/vigueta de cola vigueta de piso interrumpida por un cabecero. *(p. 412)*

template/plantilla guía hecha de metal o madera delgada. *(p. 172)*

temporary bracing/refuerzo temporal refuerzo que tiene los siguientes objetivos: Evita que las paredes se inclinen a medida que se erigen. Mantiene las paredes en posición después de haberlas aplomado y enderezado. *(p. 450)*

theodolite/teodolito instrumento que lee electrónicamente ángulos horizontales y verticales. Muestra las medidas en una pantalla LCD (de cristal líquido). *(p. 238)*

top lap/recubrimiento superior parte de la teja que no está expuesta a la intemperie. *(p. 626)*

torque/torsión fuerza de giro que produce rotación. *(p. 130)*

total rise/elevación total distancia vertical desde la parte superior de la placa superior al extremo superior de a línea de medición. *(p. 470)*

total run/longitud total la mitad del vano (excepto cuando la inclinación del techo es irregular). *(p. 470)*

trap/trampa sección curva del drenaje que se ubica debajo de un artefacto de plomería. Impide que los gases del alcantarillado ingresen a la casa por las tuberías de desagüe, pero no bloquean el drenaje. También se conoce como sifón. *(p. 869)*

treads/peldaños piezas de la escalera sobre las cuales se pisa. *(p. 725)*

trend/tendencia desarrollo general o movimiento en una dirección determinada. *(p. 8)*

trestle/caballete armazón de portátil de metal con travesaños, que se usa para soportar tablones de andamio en diversas alturas. *(p. 207)*

trim/moldura pieza larga de madera con bordes cuadrados, cepillada por los cuatro lados (S4S). *(p. 751)*

trimmer joist/vigueta brochal vigueta que se usa para formar los costados de una abertura grande. *(p. 412)*

trimmer stud/viga brochal viga corta que soporta el cabecero sobre la abertura de una ventana o una puerta, para transferir las cargas estructurales desde el cabecero a la placa inferior. También se denomina brochal o viga de caballete. *(p. 434)*

U

undercourse/hilada interna capa de tejas de baja clasificación que no estará expuesta a la intemperie. *(p. 670)*

underlayment/recubrimiento inferior en techos, material, como el fieltro para techo, que se aplica a la cubierta del techo antes de instalar las tejas. En pisos, producto de panel delgado cuya superficie es más lisa que el contrapiso estándar. Impide que pequeñas imperfecciones del contrapiso se noten a través del revestimiento elástico, y brinda un soporte firme, limpio y sin vacíos. *(p. 989)*

unit dimension/dimensión de la unidad tamaño general de la ventana, incluido el marco. *(p. 586)*

unit rise/elevación unitaria número de pulgadas que se eleva un techo por cada 12" de longitud (la longitud unitaria). Cuando la elevación unitaria varía, la inclinación del techo cambia. *(p. 470)*

unit run/longitud unitaria longitud establecida que se usa para calcular la inclinación de los cabios. La longitud unitaria de un cabio que está en un ángulo de 90° con respecto a la cumbrera (un cabio común) siempre es de 12". La longitud unitaria de un cabio que está en un ángulo de 45° con respecto a la cumbrera es de 17". También se denomina unidad de longitud. *(p. 470)*

universal design/diseño universal concepto de diseño orientado a construir una casa que sea práctica y segura para la más amplia variedad de personas, incluidos adultos mayores y gente con discapacidades. *(p. 781)*

V

valley rafter/cabio de limahoya viga inclinada que forma una depresión en el techo en lugar de una limatesa. Como el cabio de limatesa, se extiende de forma diagonal desde la placa al tablón de la cumbrera. Un cabio de limatesa se usa solamente cuando se arma un techo a cuatro aguas, pero un cabio de limahoya se necesita en techos a dos y cuatro aguas, siempre que se intersectan los planos del techo. *(p. 504)*

vapor retarder/retardador de vapor elemento que reduce la velocidad a la que el vapor de agua se filtra por un material. *(p. 900)*

veneer match/combinación de enchapado disposición de piezas de enchapado para crear diferentes patrones y efectos. *(p. 343)*

veneer plaster/yeso para enchapado enlucido de yeso formulado especialmente que se aplica a un tipo de tablarroca denominada base de yeso. También se conoce como yeso de capa fina. *(p. 933)*

volatile organic compound (VOC)/compuesto orgánico volátil (VOC, por sus siglas en inglés) tipo de producto químico que se evapora en el aire. *(p. 985)*

W

wainscoting/material de revestimiento paneles que se colocan en la pared a cierta distancia del piso. *(p. 808)*

wales/vigas longitudinales miembros horizontales de refuerzo para una forma reutilizable. *(p. 264)*

wall cabinets/gabinetes de pared gabinetes que cuelgan de una pared. También se denominan gabinetes superiores. *(p. 781)*

wall standard/cremallera listón de metal perforado que se puede atornillar a una pared o al interior de un gabinete. *(p. 775)*

warp/alabeo descripción general de cualquier variación de una superficie plana. Incluye la inclinación, la torcedura y el abarquillamiento o cualquier combinación de estos. *(p. 325)*

wear layer/capa de desgaste capa superior del piso procesado. Puede tener un grosor de $1/8$" a aproximadamente $1/4$". *(p. 976)*

web/armazón miembro entre los cordones de una viga de celosía. *(p. 492)*

weep hole/agujero de alivio agujero de drenaje ubicado en la mampostería, formado normalmente al omitir el mortero en parte de una junta vertical. *(p. 698)*

welding/soldar proceso de fundir el acero y agregar metales de relleno para fusionar las piezas en el punto de conexión. *(p. 843)*

wind chill/efecto enfriador del viento combinación de la temperatura y la velocidad del viento que aumenta el efecto de enfriamiento. *(p. 95)*

winders/escalones de vuelta peldaños radiados que se pueden usar en lugar de una plataforma para girar una escalera. *(p. 730)*

window schedule/apéndice sobre las ventanas parte de los planos de construcción que contiene descripciones de las ventanas y los tamaños de los cristales, las hojas y en ocasiones de la abertura en bruto. *(p. 585)*

wood preservative/preservante para madera producto químico que protege la madera. *(p. 332)*

work ethic/ética del trabajo creencia de que el trabajo tiene valor. *(p. 15)*

work triangle/triángulo de trabajo el recorrido más corto entre el refrigerador, la superficie para cocinar principal y el fregadero. El tamaño del triángulo de trabajo es una medida de la eficiencia de la cocina. *(p. 782)*

wrench/llave herramienta manual diseñada para hacer girar un sujetador, como un perno o una tuerca. *(p. 120)*

Index

A

ABCs in first aid, 84
Abrasive wheels, 88
Abstract of title, 37, 67
Acceleration of gravity, 208
Acclimation, 975
Acoustical ceilings, 941
Acoustical insulation, 908–911
  acoustical floor and ceilings, 911
  acoustical performance of walls, 909–911
  ceiling panels, 909
  flanking paths, 909
  Impact Insulation Class (IIC), 911
  Impact Noise Rating (INR), 911
  sound absorption, 909
  Sound Transmission Class (STC) number, 908–909
ADA. *See* Americans with Disabilities Act
Addition roofs, 536–538
Advanced carpentry skills, 531
AFCIs. *See* Arc-fault interrupters
Air conditioning, 888–889
Air pollution, 947, 948
American Society for Testing and Materials (ASTM), 219
Americans with Disabilities Act (ADA), 34
Amperage, 163
Ampere (amp), 877
Angle cutting, 151
Annual rings of a tree, 319
APA. *See* Engineered Wood Association
Apprentices, 67, 8, 10
Arc-fault interrupters (AFCIs), 880
Architectural drawing, 41–45
architect's scale, 44, 45
  computer-aided drafting and design (CADD or CAD), 48
  dimensions, 46
  drawing elements, 45–48
  lines, 45–46
  measuring systems, 42–44
  notes, 48
  scale, 44–45
  symbols, 46–47
  types of drawings, 41
Architectural plans, 41, 49–58
  cutting planes, 51
  detail drawings, 49, 55
  electrical plans, 52

elevation views, 53
engineering drawings, 55–56
fire ratings, 57–58
floor plans, 51, 52
foundation plans, 50–51
framing plans, 51–52
general drawings, 49
isometric drawings, 55
landscaping plans, 53
mechanical plans, 52–53
plan views, 50
plot plans, 50
reflected ceiling plans, 51
renderings, 56
room-finish schedule, 57
schedules, 56–57
section views, 54–55
site plans, 50
specifications, 57–58
specific sections, 55
typical sections (TYPs), 55
Arsenic, 94
Articulated ladders, 198
Asbestos, 93
Assembling and disassembling tools, 117–123
  adjustable wrench, 120
  Allen wrench, 122
  bar clamp, 117
  box wrench, 121
  cat's paw, 119
  claw hammer, 118
  groove-joint pliers, 122
  hammer tacker, 120
  hand sledge, 118
  lineman's pliers, 123
  locking pliers, 121
  mallet, 118
  metal snips, 123
  nail set, 110
  needle-nose pliers, 122
  one-handed bar clamp, 117
  open-end wrench, 120
  pipe wrench, 121
  pliers, 122–123
  pry bar, 119
  rip hammer, 118
  ripping bar, 119
  screwdrivers, 119
  slip-joint pliers, 122
  socket wrench set, 121
  staplers, 120
  Warrington hammer, 118

wrenches, 120–122
ASTM. *See* American Society for Testing and Materials
Attic ventilation, 901

B

Backfilling, 68, 272–274
Backfill timing, 68
Back injuries, 91, 92
Balloon framing, 370, 371
Baseboards, 764–770
Basement floors, 201
Basement stairways, 745
Basic joinery, 758–759
Bathroom floor framing, 413–414, 415
Bathroom planning, 785–787
Batter boards, 244–245
Battery disposal, 164
Batts and blankets, 897–898
Bay windows, 456–457
Beam pockets, 271
Bearing walls, floor framing under, 411
Beetles, 331, 332
Belt cleaning, 178
Bench mark or point of reference (POR), 239
Bevel cuts, 136, 143, 155
Bids, 59
Bifold doors, 615–616
Bird's mouth layout, 481
Blade installation, 140
Blocking, 424
Blood-borne pathogens, 85
Blueprints, 41
Board feet, 63
  calculation of, 63
  in standard lumber, 418
Board paneling, 816–823
Box cornice, 552, 554–558
Box sill, 402–403, 404
Brackets, 206–207
Break lines, 46
Brick and masonry siding, 686–717
  bricklaying, 700
  building lead corners, 699–700
  cold weather, 702
  construction details, 696–699
  corner poles, 701
  cutting brick, 690
  estimating brick, 702–703
  flashing and drainage, 698
  laying brick to a line, 701

Index

Index

Index

Index

G

H

Index

678

Dangerous Current, 881

Drywall Dust, 929

Dust and Vapors in Grout Mix, 995

Dust Exposure, 683

Dust Masks, 350, 352

Dust Removal, 179

Electric Drills, 163

Excavation Escape Routes, 249

Excavation Falls and Cave-Ins, 274

Exposed Rebar Ends, 229

Extension Ladders, 952

Eyestrain, 48

Fiberglass, 906

First Aid Basics, 84

Handling Masonry Products, 691

I-Joists, 358, 359

Jigsaw and Reciprocating Saws, 157

Joiner Safety, 180

Kickback, 810

Knee Protection, 769

Ladder Safety, 200

Levering Floorboards into Place, 981

Lifting Glulams, 361

Lifting Ladder Framing, 559

Lifting Rafters, 473

Lifting with a Crane, 377

Local Codes and Earthquakes, 795

Miter Gauge Removal, 138

Miter-Saw Safety, 148

Mortar Precautions, 279

Moving Sheet Paneling, 808

Nailers and Staplers, 188

Nailing Safety, 483

National Institute for Occupational Safety and Health (NIOSH), 25

OSHA's Hazard Communication (Right-to-Know Law), 25, 93

Personal Protection When Working with Steel Framing, 835

Placing Concrete Safely, 300

Planer Safety, 181

Plate Joiner Safety, 185

Pneumatic Nailers, 410

Portable Electric Plane Safety, 182

Powder-Actuated Tools, 401

Power Cords, 838

Power Nailers and Staplers, 450

Power Sander Safety, 177

Power Saw Safety, 130

Power Tool Maintenance, 132

Power Tools, 86–87

Preservatives, 1003

Push Stick, 143

Radial-Arm Saw Safety, 152

Raising and Placing Walls, 453, 851

Respirator Use, 226

Roofing Hazards, 563

Router Safety, 174

Scaffolding Safety, 205

Scaffold Stability, 204

Sharp Edges, 634, 652

Sharper Is Safer, 116

Silica Dust, 277

Skin and Eye Protection, 264

Soldering Safely, 875

Spontaneous Combustion, 955

Stress Hazards, 48

Surfacing Tools, 179

Table Saw Safety, 141

Treated Wood, 332

Trimming Laminate, 802

Utility Knives, 921

Wet Saws, 996

Working on a Roof, 631, 713

Working With MDF, 351

Work Safely With Chemicals, 986

see also Hazardous materials; Personal safety and health; Safety of job site

Joiner safety, 180

Joint compound, 918

Joists, 486–489, 490

  decks and porches, 1014–1015

  framing, 402–414

  hangers, 388–389

  orientation, 850

  positioning, 388

Journey-level workers, 7

K

Kerf, 132

Kickback, 129, 134, 140, 345, 810

Kiln-dried lumber, 321

King-post truss, 493

Kitchen cabinet dimensions, 783–785

Kitchen layouts, 781–782

Kitchen planning, 781–785

Kneeboards, 308

Knee protection, 769

L

Labor unions, 26

Ladders, 198–203

  grades of, 199

  levelers and stabilizers, 203

  materials for, 199

  rails, 198

  safety, 199–201

  types of, 198–199

  using, 201–203

Ladder safety, 200

Lally columns, 397

Laminated-strand lumber (LSL), 364

Laminated-veneer lumber (LVL), 353–355, 433

Land elevations, 245–246

Landscaping, 69

Landscaping fabric, 263

Landscaping plans, 53

Large openings, framing, 411–412, 414

Laser instruments, 237

Latex paint, 948

Lath-and-plaster, 933

Layout on a concrete slab, 439

Lead, 94

Leader lines, 45

Ledgers, 1013–1014

Legal documents, 37

Leveling a transit head, 239

Leveling rods, 237, 240

Levels, 237–238, 238–242

Levering floorboards into place, 981

Licenses, 132

Lifelines, 109

Lifting glulams, 361

Lifting heavy objects, 92

Lifting ladder framing, 559

Lifting rafters, 473

Lifting with a crane, 377

Light reflectivity of colors, 950

Line-of-site observations, 237

Lines and dimensions, 45–46

Lines and grades, 243–251

Lintels and bond beams, 287–288

Load-bearing walls, 430

Loan providers, 40

Local career information, 20

Location of stairways, 724

Index

Index

Index